Invertebrate Zoology

Invertebrate Zoology

PAUL E. LUTZ
University of North Carolina at Greensboro

ADDISON-WESLEY PUBLISHING COMPANY

Reading, Massachusetts • Menlo Park, California
Don Mills, Ontario • Wokingham, England • Amsterdam
Sydney • Singapore • Tokyo • Mexico City
Bogotá • Santiago • San Juan

Library of Congress Cataloging in Publication Data

Lutz, Paul E.
Invertebrate zoology.

Includes bibliographies and index.
1. Invertebrates. I. Title.
QL362.L85 1985 592 85-4040
ISBN 0-201-16830-8

ABCDEFGHIJ-HA-898765

Preface

Almost 97 percent of all animal species and more than three out of four species of all living things—plant and animal combined—are invertebrates. Invertebrates possess an immense number of adaptations that have enabled them to be extraordinarily successful in countless ecological niches and in every conceivable habitat and ecosystem. The enormous invertebrate diversity is clearly and dramatically reflected in their numerical preponderance, innumerable adaptations, and ecological variability and opportunism. However, a fundamental problem presents itself: How does one deal with this overwhelming diversity in invertebrates in the preparation of a text of manageable size?

UNIFYING THEMES

I chose to present a comprehensive (but not encyclopedic) approach by basing this text on three basic, fundamental concepts: unity amid diversity, evolution, and ecology. Perhaps the major theme, an important and recurring thread of continuity woven throughout the entire fabric of this text, is that of "unity amid diversity." By that I mean that I earnestly want to provide the readers with an appreciation of the immense diversity within invertebrates and yet, at the same time, to imbue them with an equal sense of the great, important, unifying principles and recurring characteristics present within major groups. I stress the foremost unifying features of representatives of major taxonomic groups while, concurrently, reflecting the vast diversity present in both form and function.

Evolution is a second major concept in this text. One can gain a profound appreciation of evolutionary processes and phylogenetic relationships by studying invertebrates, for it is in them that a biologist can best see the ontogeny of all major animal systems and functions, classic examples of adaptive radiations, and a clear historical picture of the exploitation of the principal ecological habitats.

The third great organizing principle of this text, ecology, provides students with a meaningful framework on which they can develop a keener appreciation of the functional roles various animals play in the natural world. I stress the habitats of particular groups, their roles in those ecosystems, the morphological and physiological adaptations they possess that equip them to flourish in a particular niche and habitat, the life forms they represent, and other general dimensions of their ecology.

PEDAGOGY

Invertebrate Zoology, a one semester text, is *written for* students—it is not simply a book students use. Careful attention has been paid in the entire development of this book to reflect good pedagogy, to keep in mind the students' perspectives, and to entice the student intellectually into a progressively engrossing study of invertebrates. The content is up-to-date and authoritative and reflects the most recent discoveries and research. The presentation of this material reflects the pedagogic concerns; this book

begins each chapter with an overview.

follows the overview with a detailed section dealing with the overreaching principles, features, and characteristics common to the entire group. Careful attention has been given to the interrelationships of general morphological and physiological features as important adaptations to environmental conditions.

includes a somewhat briefer treatment of the principal diagnostic features of the main subordinate groups.

contains a concise narrative of the phylogeny of the principal groups covered.

concludes the coverage with a thumbnail taxonomic resume of the most important characteristics of each major group, usually down to ordinal level.

Several special features enhance the presentation; these include

boxes used as a mechanism for exploring interesting, provocative ideas.

copious illustrations with 313 out of 1165 figures being photographs, most of which are original.

a glossary with about 350 entries, plus a comprehensive index of approximately 4000 entries.

SCIENTIFIC FEATURES

Finally, the scientific presentation reflects current thinking; this book

stresses the basic divisions of invertebrates: protozoans and metazoans, radiates and bilaterates, protostomates and deuterostomates.

contains a discussion of all newly discovered groups including that for the Phylum Loricifera and the Subclass Remipedia.

employs the most recent taxonomic frameworks, especially those for protozoans, annelids, molluscs, and crustaceans.

gives primal evolutionary significance to *Trichoplax adhaerens* as the most primitive known metazoan.

discards the concept of the Kingdom Protista and redefines all protozoans as heterotrophic animals at the protistan level of complexity.

treats brachiopods and phoronids as deuterostomates; this decision is based on recent embryological information.

ACKNOWLEDGMENTS

One of the major factors in a book of this nature is its graphic materials. I would like to express my appreciation to those authors, publishers, and to many colleagues for their willingness to share their photographs and line drawings with me. Notable is the cooperation I have received from Ward's Natural Science, Inc., Rochester, N.Y., for allowing me to use a number of their excellent photographic materials.

I would like to thank Sandie B. Bateman, Eleanor L. Duggan, Pat M. Lutz, and Tara A. McCracken, who patiently took my ideas and crude sketches and turned them into superb line drawings. I would also like to express my deep appreciation to Peter Loewer, a professional illustrator, who converted the artists' line drawings into finished works of art. I am immensely grateful for the fine photography done by John S. Curtis, who was responsible for all the photographs in the book that are not otherwise credited. His enthusiasm for capturing photographically just the right perspective for each animal and his indefatigable patience in achieving that goal are remarkable. Finally, I wish to thank Dick Morton, Art Editor at Addison-Wesley, whose attention to details, patience with me, and interest in this project have been outstanding.

A number of colleagues and professionals have reviewed one or more chapters. Their comments, constructive criticisms, corrections, and suggestions have made this text much stronger. They are

Dr. James Barrett, Marquette University
Ruth Bauman, Writing Services
Dr. Steven A. Bloom, University of Florida
Dr. Phyllis C. Bradbury, North Carolina State University
Dr. Robert R. Bryden, Guilford College (Emeritus)
Dr. Paul R. Burton, University of Kansas
Dr. Kenneth H. Bynum, University of North Carolina at Chapel Hill
Dr. Michael Cipolla, Holy Cross College
Dr. Ronald V. Dimock, Jr., Wake Forest University
Dr. Norman R. Dollahon, Villanova University
Dr. Perry C. Holt, Virginia Polytechnic Institute and State University (Emeritus)
Dr. J.G. Humphreys, Indiana University of Pennsylvania
Dr. Charles E. Jenner, University of North Carolina at Chapel Hill
Dr. Raymond B. Manning, Smithsonian Institution
Dr. Edward McCrady, University of North Carolina at Greensboro
Dr. Elizabeth A. McMahan, University of North Carolina at Chapel Hill
Dr. Grover C. Miller, North Carolina State University
Dr. David H. Phillips, University of California–Davis
Dr. Earl Segal, California State University–Northridge
Dr. Robert H. Stavn, University of North Carolina at Greensboro
Dr. Mary K. Wicksten, Texas A&M University
Dr. Carl E. Wood, Texas A&M University

I would also like to thank the many persons at Addison-Wesley Publishing Company who individually and corporately contributed their time and many talents to make this project a success and Laura Skinger, Margaret Pinette, Hilary Pierce, and Emily Silverman in particular.

I am most grateful to the staff of the Jackson Library at the University of North Carolina at Greensboro whose acts of helpfulness and assistance seem to have no limits. Special thanks are extended to Patricia McCarron who typed the entire manuscript and helped to eliminate a great many errors. I would also like to thank my colleagues in the Department of Biology at the University of North Carolina at Greensboro, my family, and my friends for their continued support during this protracted project. Last, but certainly not least, I wish to thank all those students of mine who have been a major source of inspiration and motivation for me to undertake and complete this task. Their unseen, unknown, and subtle influences permeate this text.

I personally believe that one of the essential features of learning is that the student becomes infected with enthusiasm for a particular subject or course. An important and serious goal in the writing of this text was to interject some of my personal interests and enthusiasm about invertebrates. It is my fervent hope that this text will encourage, motivate, and stimulate students of invertebrate zoology.

Greensboro, N.C. *P.E.L.*

Contents

1

Introduction to the Invertebrates

OVERVIEW

It is a formidable and challenging task to introduce the reader properly to the study of invertebrates. The invertebrates are so preeminently diverse and heterogeneous, are adapted to every conceivable habitat, and constitute over 95% of all animal species. They are a source of wonder, awe, fascination, and exquisite beauty, and they are uncommonly important to the overall economy of nature.

Three cardinal biological principles are stressed throughout all the succeeding chapters of this book: evolution, environments, and unity amid diversity. The reader will find that these principles provide important themes for continuity.

Evolution, a compelling and pervasive concept that is basic to all of biology, can be seen in the invertebrates with striking clarity. Their phylogenetic relationships and evolutionary adaptations to particular environments will be stressed.

Marine environments are characterized by their salinity and by many constant physicochemical properties of seawater. Freshwater animals experience fluctuating temperatures, often encounter anaerobic conditions, and must compensate for living in a hyposmotic (relatively salt-free) medium. Terrestrial invertebrates must cope with climatic extremes, the ever-present possibility of dehydration, and the absence of a buoyant aqueous medium. Many invertebrates live as symbionts; they must survive in the internal or the immediate external environments of their hosts, and their feeding, integumental, and life-cycle adaptations characterize these associations.

Even though invertebrates are diverse, and one of their most important attributes is their many and different variations, some unifying concepts are generally applicable to all invertebrate creatures. These unifying principles involve the structures of invertebrates, the ways in which they maintain themselves physiologically, their basic activity systems, and their general methods of reproduction.

SOME INITIAL CONCEPTS

Welcome to the wonderful, incredible, fascinating world of invertebrates! You are about to embark on a study of some of the most enchanting, remarkable, and diverse living things known, and I certainly hope to generate in you some unbridled enthusiasm for this captivating group of beasts. Invertebrates are found everywhere—one simply cannot escape them—and they are much more important to our lives than can possibly be realized at first. So a study of the ubiquitous and indispensable invertebrates is a most desirable and productive endeavor. In fact, since they are so numerous and varied, a study of invertebrates is, in essence, a study of animal life.

It is, unfortunate that invertebrates are known by a name that has such a strong negative connotation: the term "invertebrate" means a nonvertebrate animal, or one lacking a vertebral column or backbone. There are over 1,100,000 described animal species, and since almost 1,070,000 species are invertebrates, the vertebrates represent only a paltry 3.4% of the total. Designating animals without backbones as "invertebrates" is about as nonsensical as referring to all animals as "non-mosses." This remarkable group of nonvertebrate animals deserves a better label than it has, but the term "invertebrate" is so ingrained in our culture, literature, and scientific knowledge that we are stuck with it. Nonetheless, I have registered my complaint about it.

There is so much that could be said in introducing both the invertebrates and the principal generalizations that apply to them. But rather than present an exhaustive list of even the most fundamental principles or concepts, I would like to highlight just three that the reader will soon see as major signposts for the rest of this text. First, the reader will get a strong sense of evolutionary change as we progress through the various groups. Second, it is extremely crucial to understand the environments or ecology of a particular group: where do these animals live, what adaptations enable them to be successful in their niches, and what roles do they play in their particular habitats and in the biosphere? Third, I want to stress both the enormous variations found in invertebrates and those unifying features that are possessed by most or all invertebrates. Perhaps the best description of this principle is "unity amid diversity." In the discussions that follow, these three threads of continuity—evolution, environments, and unity amid diversity—run throughout the fabric of the entire text. They are the hallmarks, the guideposts, the recurring themes that are to be the principal unifying concepts in the study of the engrossing, delightful invertebrates.

Evolution

The cornerstone of any biological study is the fundamental precept of organic evolution. Evolution, a tenet first espoused by Charles Darwin and his contemporary, Alfred Russel Wallace, is the foundation on which the entirety of biology or any study of biological systems is based. That organisms have evolved or changed during the long history of life on this planet is a comprehensive and cardinal principle for our study of invertebrates.

There are two central dimensions to the modern perspective of evolution. First is the idea that living things change with time, and second is the inference that these changes are directed by natural selection. When evolutionists talk about change over **time**, they are referring not to those changes that are experienced by an individual organism during its lifetime, but rather to those changes that occur in populations over many generations. There have been countless generations over millions of years for small, subtle changes to take place.

Precisely when or how the universe came into being is not known, but a generally held theory is that our universe is about 15 billion years old. Our solar system of the sun, the earth, and the other planets is thought to have been formed about five billion years ago from a cloud of cosmic gas and dust. During the first two billion or so years of its existence, our planet cooled and underwent monumental geomorphometric changes, lighter elements came to the surface, an atmosphere formed, and torrential rains eroded materials into future oceans. All these events slowly changed the cosmic mass into a terrestrial ball that was conducive to supporting life. Biologists believe that the first forms of life appeared at least three billion years ago. As soon as these primeval organisms were able to perpetuate themselves by some form of reproduction (probably asexual at first), life on earth was here to stay. With the advent of living things, the stage was set for these ancient creatures to evolve eventually into squids, earthworms, sea urchins, plants, and human beings.

The oldest-known fossils are procaryotes, such as bacteria, that have been found in chert from South Africa that has been dated as being about 3.1 billion years old. However, there is some recent fossil evidence that microorganisms flourished on Australian mud flats as long as 3.5 billion years ago (Groves et al., 1981). There are some traces of more advanced eucaryotic cells (those with a nucleus and other membrane-bounded organelles) that lived about 1.5 billion years ago. The fossil record becomes fairly abundant about 600 million years ago at the beginning of the Cambrian Era (Table 1.1). Fossils representing most animal phyla appear in Cambrian strata, so a great amount of evolution must have taken place in earlier periods. These very ancient rocks have undergone tremendous morphometric changes, which have tended to obliterate fossil remains. Also, several major crustal upheavals have brought a great many of the Proterozoic strata to the surface, where these rocks and any fossils they contained have eroded away. Therefore rocks formed during the Proterozoic Era provide us with very little fossil evidence about the early forms of life and almost nothing of the basic or primal evolutionary trends that were taking place then. It is as though the first three fourths of your favorite mystery novel was removed before you ever bought the book—you know how it ends, but you are left completely in the dark about how the characters developed and almost all the important details of the ontogeny of the plot.

The geological time scale, shown in Table 1.1, is divided into five eras; the last three have subordinate periods whose beginnings and endings are based both geologically, on types of rocks formed and major crustal events, and paleontologically, on the types of fossils present. It is most unfortunate that the oldest strata in which we would desperately like to find fossils of primitive and ancient invertebrates either are not productive or are no longer available.

So invertebrates have been around for well over two billion years, and during that time their evolution has followed many different pathways to produce the prodigious diversity of life around us. We think in terms of evolution occurring over millions of years, but even the concept of a million of anything is difficult—probably impossible—to grasp, especially if one tries to conceptualize about millions of time units. For example, do you know what point in history a million days ago was? If we use the year 1986 as a reference point, a million days ago was 751 B.C.! Can you imagine what was happening a million hours ago? Again using the year 1986 as a reference, 1 million hours ago was in the year 1872, and the United States was recovering from the Civil War! My point is that a million units of time, especially a million years, is extraordinarily difficult to comprehend, and invertebrates have been evolving for almost 3000 million years. That is an inconceivably long period of time and certainly long

TABLE 1.1. □ **THE GEOLOGICAL TIME SCALE AND MAJOR ANIMAL GROUPS PRESENT AS FOSSILS**

Era	*Period*	*10⁶ Years from Start of Period to Present*	*Animal Life*
Cenozoic	Quaternary	2.5	Appearance of *Homo sapiens*
	Tertiary	65	Most modern invertebrate orders present
Mesozoic	Cretaceous	136	Second great radiation of insects; new bryozoans, pterobranchs
	Jurassic	190	First mammals; many ammonites in the seas
	Triassic	225	First dinosaurs; most modern orders of insects; new crinoids and anthozoans
Paleozoic	Permian	280	End of trilobites and eurypterids; rise of modern insects
	Pennsylvanian	320	First great radiation of insects; first appearance of pulmonate snails
	Mississippian	345	Zenith of crinoids; first ophiuroids and winged insects
	Devonian	395	First amphibians, freshwater bivalves, spiders, isopods, decapods
	Silurian	430	First insects, scorpions, millipedes; zenith of brachiopods; first invasion of land by arthrpods
	Ordovician	500	First vertebrates, bryozoans; corals, trilobites abundant; many different kinds of molluscs (scaphopods, bivalves, gastropods), echinoderms (crinoids, homalozoans), brachiopods (articulates)
	Cambrian	600	Trilobites, brachiopods were dominant; most phyla were already established; ostracods, xiphosurans, radiolarians, foraminiferans, nautiloids, crinoids were present
Proterozoic		1600	Various protozoans, molluscs, worms, other marine invertebrates present; fossilized sponge spicules, worm burrows, jellyfishes (multicellular forms only during the last 1–2 $\times$ 10^6 years)
Archeozoic		3600	No recognizable fossils

enough for extensive evolution to have occurred.

The second principle on which modern evolutionary theory is based that of natural selection. **Natural selection**, sometimes defined as nonrandom reproduction, is the selective process that results as the environment acts on the phenotypic expression of individuals in a population so that gene pool changes take place with time. Capitalizing on the substantial variations present in a population, natural selection rejects, or selects against, certain variants and favors other variants. When this happens—and it happens all the time—the genome of the population changes or evolves subtly over many generations.

The entire study of natural selection has taken on far greater meaning in recent years with the rapid development of the important field of population genetics. Population genetics has shown that genetic equilibrium, a condition that would foster no changes in a population, simply never exists in nature because populations are usually large, because genes are constantly mutating, and because of natural selection. All three conditions are hallmarks of every population. Since genetic equilibrium in a natural population is impossible, then evolutionary change at the population level is inevitable. What results from natural selection is a complement of **adaptations** that enables a particular group of organisms to thrive in a given environment. Adaptations can take many forms—morphological, physiological, behavioral, and ecological—and always increase an organism's fitness. In short, a given group of organisms represents a series of adaptations resulting from natural selection over their long evolutionary history, and the adaptations make these organisms distinctive from all others. The most important adaptations, or the way in which organisms are fit, will be given prominence in this text.

Phylogeny

Intimately associated with the concept of evolution is that of **phylogeny**, the process of establishing the evolutionary history of a group of organisms. Phylogenetic studies are attempts to reconstruct ancestral lineages and to understand degrees of evolutionary relationships. Such activities are beset with great problems, since one has to judge so many independent characters and make some value judgments as to which are more important and which are less significant in showing evolutionary relationships. Phylogeny is heavily based on the fossil record which is very incomplete. This is compounded by the fact that usually only hard parts like shells, teeth, and bones are preserved, giving the paleontologist an incomplete picture of very ancient life forms.

Similar features in two different groups always have to be evaluated on the basis of whether they were inherited from a common ancestor and are **homologous** or whether they are simply functionally similar but have quite different evolutionary sources and are therefore **analogous**. Some of the most important phylogenetic information and insights can be gleaned by studying embryological development. Developmental similarities often are more meaningful phylogenetically than are similar or dissimilar features in adults. As will be evident in later chapters, embryological evidence is given primary consideration in determining phylogenetic affinities.

In any study of phylogeny it becomes necessary to use terms that need to be carefully defined so as not to be misunderstood. Primitive and advanced, generalized and specialized, and lower and higher are three pairs of terms that are frequently encountered. **Primitive** means older or more like the ancestral condition; primitive species are those whose particular adaptations are thought to have been possessed by their ancestors. **Advanced** means newer or less like the ancestral condition; advanced species are those who have adaptations not possessed by their ancestors. For example, trilobites were primitive arthropods, and honeybees are very advanced arthropods. **Generalized** is a concept that refers to an organism's being broadly adapted to a wide range of environments, and **specialized** refers to an organism's being adapted to a rather specialized condition. For example, house flies and garden spiders are generalists, but human liver flukes are specialists. Generalized characters are often more primitive, and specialized features are more likely to be advanced, but this correlation does not always hold true. **Lower** and **higher** refer to comparative positions or levels at which species or other taxa have evolved in relation to the ancestral stock. Lower and higher are often used synonymously with primitive and advanced, especially when used to compare species within groups. But, for example, a Portuguese man-of-war is an advanced cnidarian but is phylogenetically lower than a primitive insect like a cockroach. It should be clearly stated that these three pairs of terms are never meant to imply a value such as good or bad. Primitive, generalized, and lower neither mean nor imply inferiority or worse or bad, and advanced, specialized, and higher never mean superiority or better or more perfect or good. All are comparative terms without value and should be treated as such.

SPECIES AND THE CLASSIFICATIONAL HIERARCHY

Over the millennia, people have recognized that variations in living things do not form a continuum; rather there are interruptions or discontinuities in the biological spectrum so that discrete types or kinds of things exist. For example, green darning-needle dragonflies, as a group, are clearly different from white skimmer dragonflies, even though there are subtle variations between individuals within each group. These separate discrete entities are termed species. A **species**, one of the most basic units in biology, is defined as a group of very similar individuals wherein each individual is capable of interbreeding. Conversely, a species is reproductively isolated from all other such groups. Because individuals in a given species can interbreed or have some form of genetic recombination, they have a common gene pool; therefore such a group is the principal unit of evolutionary change. Often, a species is defined on the bases of morphological, functional, or other adaptive features, but the overriding criterion is whether a group has actual or potential gene flow. In subsequent chapters, numbers of species are given for major groups as a measure of their diversity; those numbers represent recognized, described, reproductively isolated discontinuities on the biological spectrum of variations.

With about 1.5 million species of plants and animals, biologists have had to develop a system by which species can be grouped in a logical and orderly manner. Any biological or taxonomic system should associate species together that are similar in structure and function. Even more important, this system should reflect evolutionary tendencies so that forms that are closely related phylogenetically would be closely related in the scheme. The classificational scheme of living things has a hierarchy of large categories, each containing subordinate groups at various levels as follows:

Kingdom
Phylum
Class
Order
Family
Genus
Species

Each category or level in this hierarchy is a collective unit of one or more groups from the next lower level. For example, a class is composed of one or more orders, each of which contains one or more families, and so on. All animals belong to the Kingdom Animalia, which in turn consists of more than three dozen phyla. The remainder of this text is organized to treat each major group of invertebrates in a sequence that reflects its individual phylogeny.

An interesting sidelight to this taxonomic system is that higher levels, that is, kingdom and phylum, are normally much less subjectively defined than are the intermediate levels. Most biologists now accept the 40 phyla mentioned on the following pages as distinctly different animal groups. As was mentioned above, the objective realities of a species and how that grouping of organisms is defined are generally accepted. Therefore while the highest category (phylum) and the lowest category (species) are quite objective in definition and generally recognized and accepted, the intermediate levels are correspondingly subjective, arbitrary, and open to debate. These intermediate-level taxa are based on inferences about genetic groupings rather than on absolute or proven genetic, evolutionary, or phylogenetic relationships. Therefore the student should not become unduly alarmed by several different taxonomic schemes proposed by individual investigators for a given phylum. Different schemes reflect the personal interpretations of relationships by specialists and are based on their own best judgments and insights. Just remember that any taxonomic scheme is a humanly contrived, arbitrary, and an open-to-question means of grouping organisms according to their evolutionary or phylogenetic affinities.

Environments of Invertebrates

Invertebrates are found in every natural habitat on this planet. They are especially common in the oceans, where all animal phyla originated and where their ancestors spent most of their early evolutionary history. From the seas, invertebrates have invaded fresh water, and others have been successful on land. Some species, distributed in most phyla, have adapted to still another environment—that of living on or in other organisms as symbionts. Since the invertebrates' adaptations to ecological conditions constitute a fundamental theme of this text, it will be most helpful for the reader to understand some of the important dimensions of each of the major environments at this point.

THE MARINE ENVIRONMENT

It is appropriate that we begin our study of environments with the marine habitat, since it is the cradle of evolution. Biologists believe that it was in some ancient sea that life arose, and life has continued to evolve in the sea for the last 3–3.5 billion years. Since freshwater and terrestrial environments were not invaded until relatively recently (ca. 500 million years ago), invertebrates evolved exclusively in the oceans for most of the history of life.

There are several essential, physical features that characterize the marine environment and that substantially influence the organisms found there. First, the marine environment is enormously large; about 71% of the earth's surface is covered by oceans. Because of the vastness of the oceans, it is with considerable ethnocentrism that we call our planet Earth; it properly should be called Ocean. Second, the oceans are quite deep and range from shallow coastal regions to the Mariana Trench in the western Pacific Ocean, where the water is 10.9 km (6.8 mi) deep! Surprisingly, life is found, though not uniformly, at all depths from coastal regions to the greatest depths. Third, seawater is in continuous circulation, since oceans are not separated completely by land masses. This central feature accounts for the often global distribution of a great number of marine organisms. Water is circulated by heat from the sun, the spin of the earth (the Coriolis force), upwellings, waves, and tides. Fourth, the oceans are relatively homeostatic in many of their essential properties, which are necessary for life.

The homeostatic nature of seawater has been perhaps the overriding feature both in providing an optimal environment in which life could originate and for its subsequent evolution. The salt content of seawater is, on the average, 35 parts per thousand (ppt) or 3.5%. Salinity varies slightly with latitudes and certain currents, and the seas have become progressively more saline during the long history of our planet. Salinity is an extremely crucial parameter. Invertebrates evolved in the salty isosmotic seas, which allowed ancient invertebrates to avoid the critical problems of water and salt balance that they would otherwise have needed to handle physiologically in rather complex ways. Even though water temperatures range from 0°C at the poles to about 35°C at the equator, seawater temperatures usually do not vary substantially in a given location during an annual cycle. Oxygen is generally abundant in seawater and is hardly ever a limiting factor. In summary, the oceans are characterized by rather constant conditions of salinity, temperature, and oxygen, all of which are ideal for invertebrates to live, thrive, and evolve.

Stratification and zonation. Lest one is misled into thinking that all conditions are uniform throughout the marine environment, there are three critical, interrelated variables that prohibit the uniform distribution of life in the seas. First, the penetration of sunlight into the water is certainly not uniform, since light is attenuated and finally extinguished as it traverses into deeper water. The rate at which light is diminished depends upon certain cosmic events such as the angle of the sun (a seasonal phenomenon) and cloud cover, a number of physical properties of water such as salinity and transparency, and the relative frequencies of small floating organisms. Therefore the seas are stratified into an upper lighted layer and a deeper, perpetually dark layer. It is immediately obvious to biology students that photosynthesis can take place only in the upper lighted stratum. It also follows that most other organisms will be restricted to that layer because universally they would depend directly or indirectly upon photosynthetic plants as a source of food. A second physical feature is water depth. Coastal areas around continents and islands are very shallow, whereas the water is much deeper in areas away from land. In fact, extending beneath the coastal zones of the seas for distances normally up to 100 km from the shore, the oceans are relatively shallow; by definition these **continental shelves** extend seaward until a water depth of 200 m is reached. At the seaward edge of the continental shelf, the sea floor drops away rapidly to the abyssal plain, which is often interrupted by tremendous underwater trenches and impressive mountain ranges. As depths increase, the external pressure increases substantially, so many organisms cannot survive in deeper layers. Water temperature, progressively colder with increasing depths, is another limiting factor. The third chief factor is the availability of nutrients especially nitrogen and phosphorus. Coastal areas are particularly productive because of the extensive amounts of these two elements that have been washed into the sea from the land. Away from land, however, the water is often nutrient poor, so very few plants or animals can live there even in the sunlit stratum. This means that most open-sea zones are virtual deserts owing to the relative unavailability of chemicals essential for life.

Because of these three paramount factors—light penetration, water depth, and availability of nutrients—life is not uniformly distributed in the seas. Rather, biotic diversity and biomass are great in lighted, shallow, nutrient-rich areas and are much less in dark, deeper, nutrient-poor areas. Using light penetration and water depth, we can subdivide the marine habitat into vertically layered strata and horizontally arranged zones (Fig. 1.1).

There are two principal horizontal zones: the littoral or neritic and the oceanic (Fig. 1.1). The **littoral zone**, or the zone of the continental shelf, is bounded on the seaward side by a point where the water depth is 200 m and on the land side by the intertidal zone. The **intertidal zone**, which is extremely interesting and productive, is the coastal area that is covered by high tides and exposed at low tides; successive high or low tides occur about every 12½ hours. A great many different invertebrates live in the intertidal zones, but many, many more inhabit the littoral zone. Some basically marine animals (for example, ghost crabs) spend most of their time on the shore above the tide lines in a zone called the **supralittoral zone** (Fig. 1.1). The **oceanic zone** includes all the marine areas away from shore where the water depth exceeds 200 m.

The principal vertical strata are, of course, the **euphotic** or lighted **stratum** and the **aphotic** or dark **stratum** (Fig. 1.1). The euphotic stratum extends downward to a point where only 1% of the total light remains, and this level is the **compensation point** below which no photosynthesis is possible. The compensation point varies greatly in depth owing to the physical

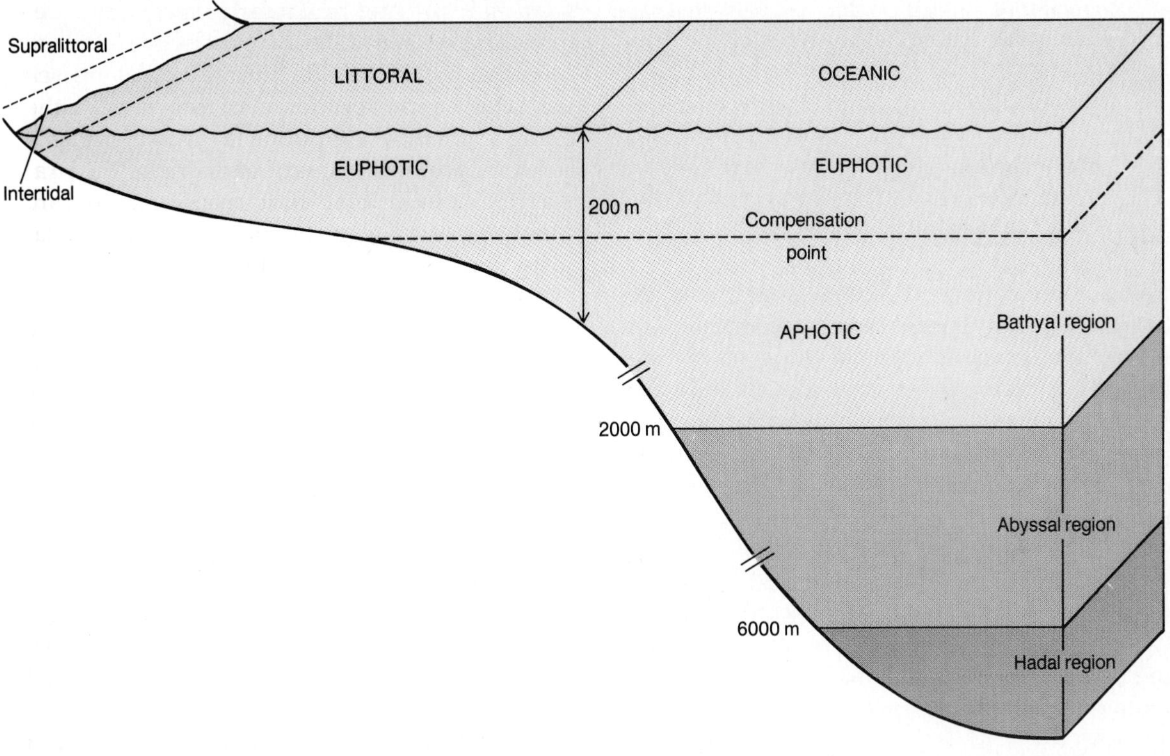

Figure 1.1 *A longitudinal section of coastal and oceanic regions of the marine environment, illustrating major strata and zones.*

conditions mentioned earlier that govern the amounts of light penetration. Below the euphotic stratum is the aphotic stratum, which is characterized by total and continual darkness. The oceanic zone down to a depth of 2000 m is the **bathyal region**, that between water depths of 2000 and 6000 m is the **abyssal region**, and the **hadal region** is where the water depth exceeds 6000 m (Fig. 1.1).

Life forms. Even the casual observer recognizes that some animals like fishes swim, others like jellyfishes float, and many others like clams live on the bottom. Ecologists have given formal names to groups of aquatic organisms that have the same **life forms** or **habits**. The principal ones are as follows.

Benthos—organisms that are attached to or resting on the bottom or that burrow into bottom sediments.

Nekton—organisms that are able to swim or navigate at will, even against a current; they often are referred to as being **pelagic**.

Neuston—organisms that are associated with and live on the surface film.

Periphyton (Aufwuchs)—organisms that live on surfaces of plants or other objects that project above the bottom such as rocks, shells, pilings, and jetties.

Plankton—organisms that float and whose movements are more or less dependent upon water currents. There are two types of plankton:

> **Phytoplankton**—photosynthetic or autotrophic unicellular plants.

Zooplankton—heterotrophic animals that are mostly small. There are two types of zooplankton:

Holoplankton—organisms that spend their entire life as plankters.

Meroplankton—animals that have one or more stages (usually larvae) as planktonic and other stages (adults) that are nonplanktonic.

Using combinations of zones, strata, and life forms, we can speak more specifically and meaningfully about various kinds of invertebrates. For example, referring to a creature as a littoral zooplankter (copepod), a bathyal benthic organism (a sea lily), and an oceanic nektonic creature (a squid) is far more descriptive than simply giving the names of the animals. For each major marine group, considerable care will be given to pointing out where representatives live (zones, strata) and what their life forms are. Only by using this information can one begin to understand the precise role of a given invertebrate in the ecology of a specific marine ecosystem.

Estuaries. There are many coastal bays, tidal marshes, sounds, and river mouths where fresh water from the land mixes with seawater; these special kinds of marine habitats are called estuaries. An **estuary** is normally extensive and often covers thousands of hectares, is partially surrounded by land masses, and is very shallow with water depths at high tide of no more than 1–3 m. An estuary is strongly affected by tides. At low tide most of the water flows into the sea, only to have the estuary refilled again in about 6¼ hours by the rising high tide levels plus its freshwater tributaries. The vast majority of estuarine organisms is marine, but all must have efficient osmoregulatory mechanisms to cope with widely fluctuating salinities. They also have important adaptations that permit them to be out of water about twice a day at low tides. Because of the availability of nutrients from the land and minerals from the sea, estuaries are extraordinarily productive both in biomass and in species diversity. Estuaries are essential to the economy of the seas' biota, being vast nursery grounds for many species of nektonic animals. Estuaries are one of the most exciting, productive, and interesting environments for invertebrates and invertebrate zoologists. Because of their importance, these environments must be protected at all costs.

The Freshwater Environment

Lakes, ponds, brooks, rivers, and swamps are quite familiar freshwater environments inhabited by a wide variety of invertebrates. Freshwater biologists recognize ponds, lakes, and swamps as standing-water or **lentic habitats** and brooks, creeks, and rivers as running-water or **lotic habitats**. One fundamental difference between freshwater and marine environments is in their salinities, a feature that has far-reaching consequences for organisms. Fresh waters have very small amounts of dissolved salts; therefore organisms living in fresh waters must be able to compensate physiologically for the continual osmotic gain of water from the hyposmotic environment. Efficient osmoregulating systems and organs are necessary prerequisites for animals to live in such an osmotically different environment from that of the oceans or their own internal environment.

Stratification, zonation, and life forms. The strata of lentic environments, based on the transparency of the water, are the upper lighted **euphotic stratum** and a lower dark **profundal stratum**; separating these two is the **compensation point** (Fig. 1.2a). The **littoral zone**, or the zone of rooted plants, is that area around the shore where the water depth does not exceed that of the compensation point. The **limnetic**

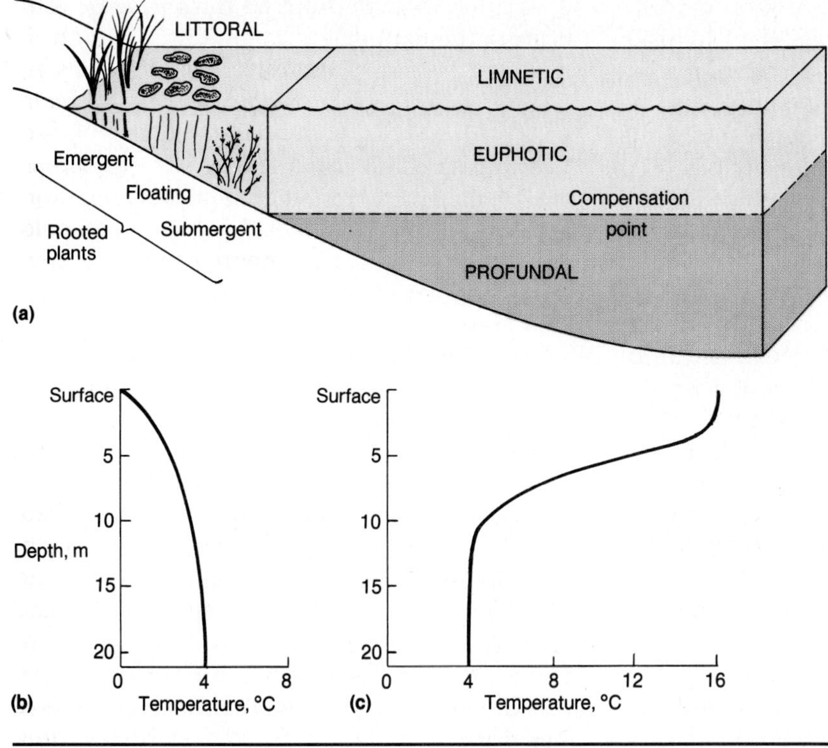

Figure 1.2 *Some dimensions of a freshwater lentic habitat: **(a)** a diagrammatic section through a portion of a lake illustrating major zones and strata; **(b)** a thermal profile of a temperate lake during winter; **(c)** a thermal profile of the same lake during summer.*

zone is the open-water area away from shore where rooted plants are not found (Fig. 1.2a).

The basic life forms present in the marine environment are also present in freshwater environments. One fundamental difference is that there are many fewer meroplankters in fresh water. Also, insects and rooted plants, almost totally absent in marine environments, abound in freshwater habitats.

Some important variables. A number of other obvious differences exist between freshwater and marine environments, but the three variable properties that characterize all freshwater habitats are their temperature, dissolved gases, and transparency.

One of the most critical and changing parameters of fresh water is its temperature. Because bodies of fresh water are relatively small and shallow, the annual range of fluctuations in water temperature is much greater than that of the seas. In temperate zones the water temperature in a given pond or small lake might vary from 0°C in the winter to as much as 40°C in the summer. Therefore freshwater organisms must be able to adapt to these thermal extremes over relatively short periods of time. Water has several unique thermal properties, the most important of which is that its greatest density is at 4°C. As water either cools or warms from 4°C, it expands and becomes lighter. For this reason a lake neither freezes solid nor freezes from the bottom up, but rather freezes first on the top, where the colder, lighter water is (Fig. 1.2b). In the summer a definite thermal stratification exists from top to bottom in temperate lakes; the warmer lighter water is at the surface, and the colder heavier water is near the bottom (Fig. 1.2c). As we shall see, thermal stratification has far-reaching consequences in these environments and for the organisms living there.

The concentration of dissolved gases, especially that of oxygen, is variable at least in lentic environments, mostly because of conditions imposed by thermal stratification. In the summer, because there is very little vertical movement due to thermally induced strata, oxygen is often depleted by the organisms in the lower strata and is reduced substantially in upper strata by the fact that warmer water holds less oxygen. An ice cover in the winter also serves as an effective barrier against oxygen dissolving into water. Therefore, many lentic habitats often have deeper strata that experience anaerobic conditions for extended periods, which are seriously limiting for most organisms. The concentrations of other gases, such as those generated by decomposition and including carbon dioxide, hydrogen sulfide, and ammonia, increase in the lower strata in the summer because of thermal stratification; these gases are also limiting to invertebrates. Lotic habitats, on the other hand, normally are not anaerobic because water movement enhances the dissolution of oxygen into the water.

Another vital but variable property of fresh water is its transparency. As in the marine environment, the penetration of light in lakes and ponds is severely limited by suspended particles. And since fresh waters are so vulnerable to the erosion of soils into them by precipitation and runoff, freshwater bodies tend to be turbid. Therefore the euphotic stratum is normally rather narrow. Turbidity may reduce the thickness of the lighted stratum to 2–4 cm, but in relatively clear, nonturbid water this layer often is 10–20 m thick. The reduced transparency is exacerbated by thermal stratification, which does not permit the settling of suspended particles to the bottom.

THE TERRESTRIAL ENVIRONMENT

The terrestrial environment is, of course, the most familiar environment. It is also one that presents some of the most severe problems to organisms, since land is the most variable and extreme in terms of several physical conditions. First, temperature variations are much more pronounced than in water; often the range of temperatures between winter and summer is 100–150°C. Many organisms have evolved daily or annual cycles as means of coping with severe temperatures. Second, water becomes one of the most critical limiting factors on land. All terrestrial organisms are constantly confronted with the potentiality or probability of dehydration, and a great number of different adaptations for water conservation have evolved. Third, unlike water, which is buoyant, air offers very little in the way of mechanical support; this means that successful land animals have had to deal with this exceedingly crucial problem. Thus many adaptations have evolved: skeletons, legs, wings, fluid-filled body cavities, and the like equip animals to live in this environment. Fourth, land masses are not continuous as are the seas, so there are formidable barriers to dispersal. On the other hand, terrestrial environments always have a remarkably constant concentration of oxygen (21%) and carbon dioxide (0.03%), so these gases are never limiting to terrestrial invertebrates.

The nature of the soil in terrestrial ecosystems is particularly important because it is the source of most of the essential mineral nutrients. Invertebrates have been quite successful in the soil, since that substratum mitigates against thermal extremes and water loss and provides considerable mechanical support. It is the soil and climatic conditions like temperature, precipitation, light, and gases that constitute the paramount dimensions of the terrestrial environment.

The insects (see Chapter 13) are the only group of invertebrates that have been able to adapt fully to a terrestrial environment. Only insects have the ability to fly and so have opened up an entirely new dimension of the land environment. Since insects are so very numerous and have such great species diversity (about 75% of all animal species are insects), one would have to conclude that the terrestrial environment is especially ideal for insects.

THE ENVIRONMENTS OF SYMBIONTS

A surprisingly large number of invertebrates species, distributed in most phyla, live symbiotically with other animals. **Symbiosis** means living together, and symbiotic relationships between two different kinds of animals range all the way from casual informal associations to obligatory and highly specialized dependencies. During their long evolutionary histories, many organisms began to inhabit a variety of new specialized environments and occupy new and unfilled ecological niches on or in other organisms; in so doing, they could avoid a great deal of competition and live in a secure habitat. Therefore symbiosis has become a common but diversified life-style for many invertebrates.

Many different types of symbiosis are recognized, including predation and competition, both of which have important ecological and evolutionary ramifications. To discuss environments of symbionts, we need to restrict the treatment here to commensalism, parasitism, and mutualism, in which two different organisms live in very close proximity and interact mutually in rather intimate ways. **Commensalism** is a relationship in which one organism, the commensal, benefits while the other, the host, is unaffected. **Parasitism**, a relationship to be discussed more fully later (see Chapter 6), is an association in which one animal, the parasite, derives its nourishment from another organism, the host, which is harmed to some degree by the parasite. **Mutualism** is an obligatory and mutually beneficial association of two different kinds of organisms.

In all three cases, one organism has to adapt to the physiology, ecology, and habits of the other. If their association is an external one, that is, on the surface of one animal, the adaptations of both organisms are not usually extreme. But if one invertebrate lives internal to the other, then there are extreme problems that must be dealt with successfully. On the one hand, many of the physiological requirements of the visitor, such as nutrition, gases, and waste removal, are provided by the host. But on the other hand, essential adaptations must be developed by the visitor in response to maintaining position, resisting digestion or the immune response, and adapting to the internal physiology of the host. Additionally, an internal symbiont must also be able to overcome the severe problems of reproduction when its eggs and larval stages are normally on the outside of the host and perhaps in an intermediate host. Therefore life-cycle problems are a most formidable obstacle for many symbionts. These barriers and the adaptations of each group will be mentioned prominently.

Unity amid Diversity

That invertebrates are extremely diverse has already been stated forcefully and is reflected in the fact that almost 97% of all animal species, and approximately 75% of all living species, are invertebrates. Even a casual observation of one's biotic surroundings will attest to the extensive invertebrate diversity. Reflect for a moment on the striking differences between a honey bee, an octopus, a sea anemone, an ameba, a spider, a leech, and a sea star. The fantastic varieties exemplified in these seven examples extend to almost every dimension of their biology, including their habitats, means of food gathering, responses to a given environmental stimulus, methods of gaseous exchange and waste removal, and patterns of reproduction, to mention only a few. To say that invertebrates are diverse is a gross understatement, and their heterogeneity is an important, universal theme that is stressed in this text.

But if various kinds of invertebrates are so very different, are there any unifying features that many or most have in common? Is there any single thing that generally applies to all invertebrates? Yes! Invertebrates share all the basic unifying features that living things generally possess; they all must be provided with foodstuffs, process these nutrients metabolically, eliminate wastes, and perpetuate themselves reproductively. Chemically, all animals are constructed of the same basic elements; morphologically, they are built along similar lines; ecologically, the same environmental factors must be dealt with; and metabolically, the same processes characterize them. Therefore there are some very meaningful unifying themes that are common to invertebrates. Below are some generalizations, some unifying concepts, some pervasive themes that characterize invertebrates. This list, not necessarily arranged in any order of importance, is not meant to be exhaustive, but it does enumerate the most important trends:

1. Most animals are mobile, but a great many are sessile (sedentary).
2. While sessile animals tend to have radial symmetry so as to confront the environment equally from all directions, most invertebrates, being mobile, generally have bilateral symmetry with a leading anterior (head) end meeting the environment first and a ventral (under) surface that often is next to the substratum or bottom.
3. There is a progression in complexity from the unicellular condition to intermediate animals composed of tissues to those constructed of organs and systems (most invertebrate species).
4. All animals are heterotrophic and therefore consume living things or once-living things. Of the several means of food gathering, the vast majority are predatory, and the next most common method is filter feeding.
5. All animals exchange respiratory gases (oxygen, CO_2) dissolved in an aqueous medium that surrounds either the entire organism or specialized gas exchange structures such as gills.
6. All animals have some means for circulating foodstuffs, gases, and other materials internally.
7. Means of maintaining proper water and mineral concentrations (osmoregulation) are present in all freshwater and terrestrial invertebrates.
8. Some means of waste elimination, often tied functionally to osmoregulators, are present. Small aquatic animals eliminate ammonia; larger aquatic and some terrestrial forms produce urea; and many terrestrial animals eliminate uric acid.
9. All invertebrates respond to environmental stimuli, and most have efficient sensory receptors and a nervous system that coordinates and controls their activities.
10. Most invertebrates have a protective or supportive skeleton either external or internal; many worms have a fluid-filled hydrostatic skeleton.
11. Almost all invertebrates reproduce sexually, though some reproduce exclusively by parthenogenesis. Animals with separate sexes (dioecious forms) are the rule, although a great many are hermaphroditic (monoecious).
12. Many different invertebrates possess the capacity to reproduce themselves asexually by a variety of means.

Try to keep these unifying principles in mind as you progress through this text. It will be the path of least resistance to get lost sometimes in the maze of invertebrate diversity, but always try to remember this cardinal principle about invertebrates: unity amid diversity.

GENERAL WORKS FOR ADDITIONAL READINGS

Adiyodi, K.G. and R.G. Adiyodi, eds. 1983. *Reproductive Biology of Invertebrates*. Vol. I: *Oogenesis, Oviposition, and Oosorption*, 752 pp. Vol. II: *Spermatogenesis and Sperm Formation*, 648 pp. New York: John Wiley & Sons,

Autrium, I.H. 1981. *Comparative Physiology and Evolution of Vision in Invertebrates*. Part B: *Invertebrate Visual Centers and Behavior*, 632 pp. New York: Springer.

Boardman, R.S., A.H. Cheetham, and W.A. Oliver, Jr., eds. 1973. *Animal Colonies: Development and Function through Time*, 586 pp. Stroudsburg, Pa: Dowden, Hutchinson, & Ross, Inc.

Brusca, R.C. 1980. *Common Intertidal Invertebrates of the Gulf of California*, 2nd ed., 493 pp. Tucson: University of Arizona Press.

Cheng, T.C. 1973. *General Parasitology*, 917 pp. New York: Academic Press.

Chia, F-S. and M.E. Rice, eds. 1977. *Settlement and Metamorphosis of Marine Invertebrate Larvae*, 285 pp. New York: Elsevier.

Cloud, P. and M.F. Glaessner. 1982. *The Ediacarian Period and System: Metazoa Inherit the Earth. Science 217*:783–792.

Corning, W.C., J.A. Dyal, and A.O.D. Willows. 1973. *Invertebrate Learning*. Vol. 1: *Protozoans through Annelids*, 290 pp. Vol. 2: *Arthropods and Gastropod Mollusks*, 274 pp. New York: Plenum Press.

Crawford, C.S. 1981. *Biology of Desert Invertebrates*, 314 pp. New York: Springer.

Edmonson, W.T., ed. 1963. *Ward and Whipple's Freshwater Biology*, 2nd ed., 1202 pp. New York: John Wiley & Sons.

George, J.D. and J.J. George. 1979. *Marine Life: An Illustrated Encyclopedia of Invertebrates in the Sea*, 268 pp. New York: John Wiley & Sons.

Gosner, K.L. 1971. *Guide to Identification of Marine and Estuarine Invertebrates, Cape Hatteras to the Bay of Fundy*, 620 pp. New York: John Wiley & Sons.

Gould, S.J. 1982. Darwinism and the expansion of evolutionary theory. *Science 216*:380–387.

Groves, D.I., J.S.R. Dunlop, and R. Buick. 1981. An early habitat of life. *Sci. Amer. 245*:64–73.

Hammen, C.S. 1980. *Marine Invertebrates: Comparative Physiology*, 123 pp. Hanover, N.H.: University Press of New England.

Hanson, E.D. 1981. *Understanding Evolution*, 533 pp. New York: Oxford University Press.

Hart, C.W., Jr., ed. 1980. *Pollution Ecology of Freshwater Invertebrates*, 376 pp. New York: Academic Press.

Hyman, L.H. *The Invertebrates*. Vol. I: *Protozoa through Ctenophora*, 696 pp., 1940. Vol. II: *Platyhelminthes and Rhynchocoela*, 531 pp., 1951. Vol. III: *Acanthocephala, Aschelminthes, and Entoprocta*, 554 pp., 1951. Vol. IV: *Echinodermata*, 746 pp. 1955. Vol. V: *Smaller Coelomate Groups*, 767 pp., 1959. Vol. VI: *Mollusca: I*, 769 pp., 1967. New York: McGraw-Hill Book Co.

Kerfoot, W.C., ed. 1980. *Evolution and Ecology of Zooplankton Communities*, 794 pp. Special Volume 3, American Society of Limnology and Oceanography. Hanover, N.H.: University Press of New England.

Kozloff, E.N. 1974. *Keys to the Marine Invertebrates of Puget Sound, the San Juan Archipelago, and Adjacent Regions*, 213 pp. Seattle: University of Washington Press.

Levinton, J.S. 1982. *Marine Ecology*, 508 pp. Englewood Cliffs, N.J.: Prentice-Hall, Inc.

Mann, K.H. 1982. *Ecology of Coastal Waters: A Systems Approach*, 322 pp. Berkeley: University of California Press.

McDonald, A.G. 1975. *Physiological Aspects of Deep Sea Biology*, 434 pp. New York: Cambridge University Press.

Mill, P.J., ed. 1976. *Structure and Function of Proprioceptors in the Invertebrates*, 673 pp. New York: Halsted Press, John Wiley & Sons.

Miner, R.W. 1950. *Field Book of Seashore Life*, 864 pp. New York: G.P. Putnam's Sons.

Morris, R.H., D.P. Abbott, and E.C. Haderlie. 1980. *Intertidal Invertebrates of California*, 666 pp., 200 plates. Stanford, Calif.: Stanford University Press.

Murray, B.G., Jr. 1979. *Population Dynamics: Alternative Models*, 203 pp. New York: Academic Press.

Newell, R.C., ed. 1976. *Adaptation to Environment: Essays on the Physiology of Marine Animals*, 510 pp. Boston: Butterworths.

Newell, R.C. 1979. *Biology of Intertidal Animals*, 3rd ed., 757 pp. Faversham, Kent: Marine Ecological Surveys, Ltd.

Nybakken, J.W. 1982. *Marine Ecology, an Ecological Approach*, 446 pp. New York: Harper & Row.

Odom, E.P. 1983. *Basic Ecology*, 603 pp. Philadelphia: CBS College Publishing.

Parker, S.P., ed. 1982. *Synopsis and Classification of Living Organisms*. Vol. 1, pp. 389–1166; Vol. 2, pp. 1–830. New York: McGraw-Hill Book Co.

Pennak, R.W. 1978. *Fresh-water Invertebrates of the United States*, 2nd ed., 768 pp. New York: John Wiley & Sons.

Price, R.W. 1980. *Evolutionary Biology of Parasites*, 237 pp. No. 15, Monographs in Population Biology. Princeton, N.J.: Princeton University Press.

Scientific American Book. 1978. *Evolution*, 135 pp. San Francisco: W.H. Freeman and Co.

Smith, R.I. and J.T. Carlton, eds. 1975. *Light's Manual: Intertidal Invertebrates of the Central California Coast*, 3rd ed., 684 pp. Berkeley: University of California Press.

Smith, W.L. and M.H. Chanley, eds. 1972. *Culture of Marine Invertebrate Animals*, 318 pp. New York: Plenum Press.

Stancyk, S.E. 1979. *Reproductive Ecology of Marine Invertebrates*, 266 pp. Columbia, S.C.: University of South Carolina Press.

Stansfield, W.D. 1977. *The Science of Evolution*. 588 pp. New York: Macmillan Publishing Co.

Symposium on Invertebrate Behavior. 1972. *Amer. Zool.* 12:385–594.

Tasch, P. 1980. *Paleobiology of the Invertebrates: Data Retrieval from the Fossil Record*, 975 pp. New York: John Wiley & Sons.

Vernberg, F.J. and W.B. Vernberg. 1981. *Functional Adaptations of Marine Organisms*, 347 pp. New York: Academic Press.

Vernberg, W.B. and F.J. Vernberg. 1972. *Environmental Physiology of Marine Animals*, 338 pp. New York: Springer.

Wiley, E.O. 1981. *The Theory and Practice of Phylogenetic Systematics*, 440 pp. New York: Wiley-Interscience.

2

The Protozoans

OVERVIEW

Protozoans are the simplest of all eucaryotic organisms, and our discussion of animal groups appropriately begins with them. Protozoans are found in freshwater, marine, and damp terrestrial habitats, and some live symbiotically with other animals. The single most important protozoan feature is their small size, which has resulted in their having the highest surface-area-to-volume ratio among all animals. This characteristic alone has dictated that most of the details of their functional morphology are, in reality, adaptations of very small organisms to a watery environment.

Protozoans are single celled. Instead of organs, a protozoan has functionally equivalent subcellular organelles contained within the single cell membrane that carry out various organismic functions. Most protozoans are motile; important organelles of locomotion include flagella, pseudopodia, and cilia. In fact, a primary feature that distinguishes between the major protozoan phyla is the nature of their locomotive organelles. Protozoans ingest organic molecules, which are digested within food vacuoles. Many species possess contractile vacuoles, which are osmoregulatory organelles. Fission, a form of asexual reproduction, occurs in all protozoans, and methods of genetic recombination are known for many forms. Most freshwater and parasitic species form cysts in response to unfavorable environmental conditions.

Of the seven protozoan phyla, the Sarcomastigophora and Ciliophora are the largest in terms of numbers of species and contain the most familiar protozoans. The Phylum Sarcomastigophora includes protozoans that have monomorphic nuclei and that locomote with flagella, pseudopodia, or both; it includes the flagellates, amebae, radiolarians, and foraminifera. The Phylum Ciliophora, which includes the most complex of all protozoans, contains individuals that have dimorphic nuclei, cilia for locomotion, and a complex pellicle and infraciliature. Representatives of four other phyla—Apicomplexa, Microspora, Ascetospora, and Myxospora—are all parasitic in various metazoans, and each taxon contains an intermediate number of species. One other phylum,

Labyrinthomorpha, consists of only a very few species.

As a group, protozoans are very ancient—undoubtedly the oldest of all animal groups—but even the most fundamental details of protozoan phylogeny are lost in antiquity. Thus there is little direct evidence about ancient protozoans or early evolutionary pathways. Very likely, the first protozoans were primitive flagellates that had evolved from simply constructed procaryotes. On the basis of the almost universal occurrence of flagella in metazoan sperm and the great adaptive plasticity of mastigophorans, we can infer that primitive flagellates were most likely the stock from which arose other protozoan groups and, indeed, all animals. Specific phylogenetic affinities of the seven modern-day protozoan phyla are also obscure; even some fundamental questions remain regarding the evolutionary relationships of groups within several phyla. It is probable that protozoans had multiple origins from ancestral flagellates.

PROTOZOANS VS. PROTISTS

Our survey of the animal groups begins with the protozoans because they are the least complex of all animals. Protozoans are small, animallike, single-celled organisms whose phylogenetic relationships with other unicellular organisms have been debated over the years without any effective resolution of the issues, at least to the satisfaction of some biologists. Within the past several decades, protozoans and various other relatively simple organisms have generally been included in a heterogeneous grouping, the Kingdom Protista. This kingdom, as characterized by its proponents, includes the entirely or primarily unicellular eucaryotic forms, that is, those with a nuclear membrane and well-defined, membrane-bounded, cytoplasmic parts. But there has not been general agreement as to what groups should be designated as protists; along with protozoans, various schemes have included bacteria, unicellular and multicellular algae, and even fungi. The problems have been narrowed substantially because bacteria and blue-green algae are now considered to constitute the Kingdom Monera, and fungi are treated as a separate kingdom. Now this assemblage of protistans would include protozoans, many green algae, diatoms, dinoflagellates, euglenophytes, and slime molds (Fig. 2.1a). Such a lumping of disparate groups, however, appears to be rather artificial, is apparently one mostly of convenience, and is not supported by overwhelming phylogenetic evidence. It is important to keep in mind that the ultimate goal of any taxonomic scheme is to reflect as accurately as possible evolutionary affinities.

In a classificational scheme advocated by Leedale (1974) the concept of the Kingdom Protista is abandoned. Instead, four kingdoms are proposed: Monera, Plantae, Fungi, and Animalia. While discarding the notion of a separate protistan kingdom, this scheme reflects the fundamental importance of recognizing the protistan (single-celled) level of complexity. Leedale's system assigns each of the various protistan taxa to one of the three major kingdoms to which it seems most closely related. Thus all autotrophic protists are assumed to be plants, the slime molds are treated as fungi, and protozoans are considered as belonging to the Kingdom Animalia. This means that each of the plant, fungal, and animal kingdoms would contain various unicellular organisms that are at the protistan level of organization (Fig. 2.1b). Leedale's four-kingdom approach has a great deal of merit; it places the heretofore protistan taxa with the multicellular groups with which they appear to have the closest evolutionary relationships. Admittedly, there are also some good conventional reasons for retaining the five-kingdom approach in which protists are recognized as a kingdom. Yet taking all these reasons into

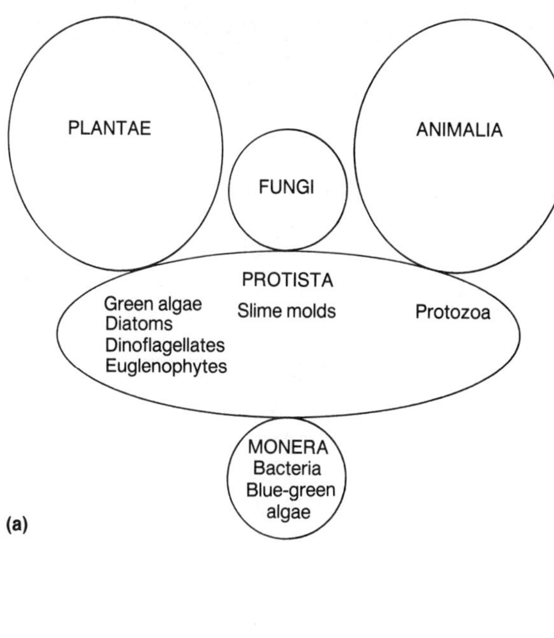

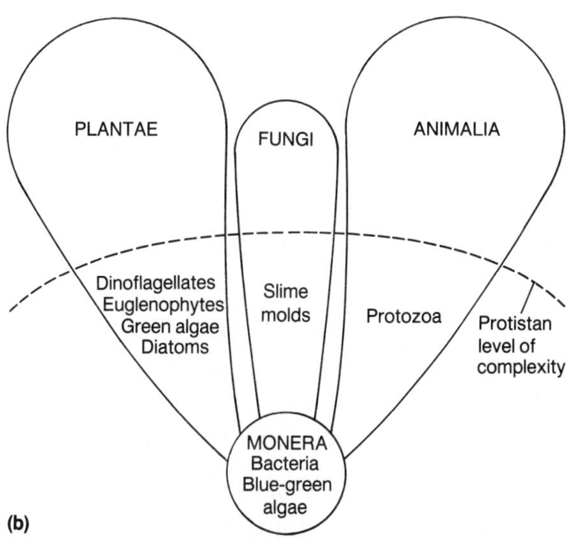

Figure 2.1 *Two different classificational plans:* ***(a)*** *the five-kingdom scheme in which a number of groups of small unicellular forms are treated as belonging to the Kingdom Protista;* ***(b)*** *the four-kingdom scheme of Leedale in which the same unicellular forms are considered to be at the protistan level of complexity but are placed in one of three major kingdoms.*

consideration, we shall break with tradition and follow the taxonomic outline of Leedale, since it appears that such a plan more accurately reflects phylogenetic affinities, especially those of the protozoans. Therefore protozoans will be considered as animals at the protistan level of complexity (Fig. 2.1b).

All representatives of the seven protozoan phyla, being exclusively heterotrophic, have been considered almost universally as animal-like organisms. Some biologists might not wish to call them animals, but no one would ever refer to them as plants. But problems lie within the Phylum Sarcomastigophora, since this taxon traditionally has included many autotrophic flagellates, including euglenophytes, diatoms, dinoflagellates, many chlorophytes, and even slime molds. There is a growing tendency among biologists to consider all autotrophs as plants. This concept would remove all the green flagellates from this phylum and treat them as plants at the protistan level of complexity, and we shall follow this plan (see Box 2.1 for a further discussion of this vexing problem). Similarly, the slime molds should be considered as fungi at the protistan level of complexity. Therefore protozoans, including sarcomastigophorans, are to be defined as being **heterotrophic**—that is taking in preformed organic molecules, bits of organic material, or entire organisms as nourishment—and as organisms that do not form spores with cellulose in the spore wall as do the slime molds.

Historically the term "protozoa" has had a dual meaning. First, protozoa (meaning *first animals*), is an appropriate term, since ancient protozoans were undoubtedly the first forms of animal life on earth, from which all other later animals have been derived. Even though the protozoan paleontological record is very sketchy, especially during their early history, owing to the absence of hard parts to fossilize, recent discoveries of protozoan fossils more than 1.5 billion years old are compelling evi-

BOX 2.1 □ WHERE DO WE PLACE THE AUTOTROPHIC FLAGELLATES TAXONOMICALLY?

It is very likely that ancestral green flagellates gave rise both to higher plants and animals. These plantlike flagellates, sometimes considered as protozoans and included in the Phylum Sarcomastigophora as the Class Phytomastigophorea, apparently are the stem evolutionary group for both plants and animals; this alone is justifiable reason for them to be considered as belonging to both kingdoms. Indeed, heated debates have ensued for decades as to whether they are plants, animals, protozoans, or protists. Understandably, botanists have important reasons for considering such a fundamental group as plants, and many zoologists have different, perhaps equally valid reasons for considering them as animals. As a result, representatives of phytomastigophoreans are often included in both botanical and zoological courses. The two pivotal characteristics are that they are autotrophic and have flagella. Autotrophism is a distinctly plant feature, and flagella are found in representatives of both kingdoms but are more characteristic of animals. Since it would appear that their autotrophism is perhaps a more fundamental character than their being flagellated, we have arbitrarily considered them to be plants. But there is even some evidence that the heterotrophic flagellates may, in fact, be more closely related to autotrophic flagellates than they are to other heterotrophic protozoans. Additionally, some green forms like *Euglena* (an euglenophyte) are autotrophic in light; when placed in continuous darkness, they lose their chloroplasts and become saprozoic! Furthermore, *Euglena* and others can survive saprozoically for indefinite periods under aphotic conditions as long as essential sugars and amino acids are supplied to their culture medium. Examples like this clearly point out that the distinctions between plants and animals or between autotrophism and heterotrophism are not always clear-cut or absolute. An important lesson to be learned from this example is: Not all organisms fit neatly and conveniently into any one scheme concocted by biologists!

dence of the antiquity of this group of animals. Second, Protozoa has, for decades, been the name of the phylum encompassing all animal-like protists. But recently it has become increasingly clear that each of the seven protozoan taxa probably had separate origins, and that each is sufficiently different from the others to warrant elevating all of these groups to phylum status. Therefore the term protozoa is used today as a collective noun but without the taxonomic status it once held. The most important diagnostic features of the protozoan phyla are set forth in Table 2.1.

GENERAL CHARACTERISTICS

By far the single most-important feature that both characterizes and has a dominant influence over the morphology of protozoans is their small size. In all animals big and small a great many functions, including absorption of nutrients, gaseous exchange, and excretion, take place across surfaces. One of the most important features in the evolution of higher animals has been that as animals became larger, the surface-area-to-volume ratio became progressively less, and natural selection has decidedly favored those organisms that could compensate most efficiently for the relative loss of vital surface areas. Many significant adaptations have evolved in larger animals to increase certain surfaces so as to maintain a high degree of functional efficiency. A good example is the tremendous surface area that is present in most gas exchange organs such as gills or lungs. The most important adaptations regarding surfaces will be treated later when we consider more advanced animals.

The fact that protozoans have the highest surface-area-to-volume ratio among all animals has been highly significant in influencing almost every aspect of their morphology, physiology, and ecology. Therefore it is essential that we explore those adaptive features most affected by the miniature size of protozoans, including body organization, several important functional dimensions, and some aspects of their ecology. Since the vast majority are from 1 μm to 1 mm in length, almost all are microscopic; but representatives of a few species can be seen with the unaided eye, and one exceptional fossil foraminifer (sarcomastigophoran) was almost 18 cm in diameter! Being small has minimized the problems associated with surfaces, but being small also has inherent draw-

TABLE 2.1. □ **THE MOST IMPORTANT CHARACTERISTICS OF THE SEVEN PROTOZOAN PHYLA***

Phylum	*Common Name(s)*	*Method of Locomotion*	*Method of Asexual Reproduction*	*Other Distinguishing Features*
Sarcomastigophora	Flagellates, amebae, foraminifera, radiolarians	Flagella or pseudopodia or both	Binary fission	Monomorphic nuclei; many are symbiotic; many secrete tests
Labyrinthomorpha	None	None, but some by gliding	Fission	Symbiotic or parasitic on algae; mostly marine
Apicomplexa	Gregarines, coccidians	None, but body flexion in some	Multiple fission	Have apical complex; all are parasitic; many do not form spores
Microspora	None	None	Multiple fission	Unicellular spores with extrusion apparatus; no mitochondria; all parasitic
Ascetospora	None	None	Multiple fission	Multicellular spores with 1 or more sporoplasms; no polar capsule; all parasitic
Myxozoa	None	None	Multiple fission	Multicellular spores; 1–3 valves on spores; all parasitic
Ciliophora	Ciliates	Cilia	Transverse binary fission	Highly complex; dimorphic nuclei; genetic recombination by conjugation

* Classificational scheme of Levine et al. (1980).

backs that will be evident in the following discussions. Someone once wrote that protozoans are prisoners of their own unicellularity. Wait until you have completed this chapter or even the entire book to agree or disagree with this statement.

Body Organization

Protozoans have been termed both unicellular and acellular, a distinction that is mostly semantic. **Unicellular** means that an organism is composed of only one cell, whereas **acellular** conveys the idea that the body is not comparted into cells as are the bodies of higher animals. The individual protozoan appears to be basically similar to a metazoan cell in both structure and function, and the usual subcellular parts present in a typical animal cell are also found in all protozoans (Fig. 2.2). Since protozoans are both acellular as entire organisms and unicellular as individual cells, both terms are technically correct; nonetheless, we shall use the more widely accepted term, unicellular or single-celled, to describe the protozoan condition.

In higher animals, specific organs or specialized surfaces have evolved for achieving a

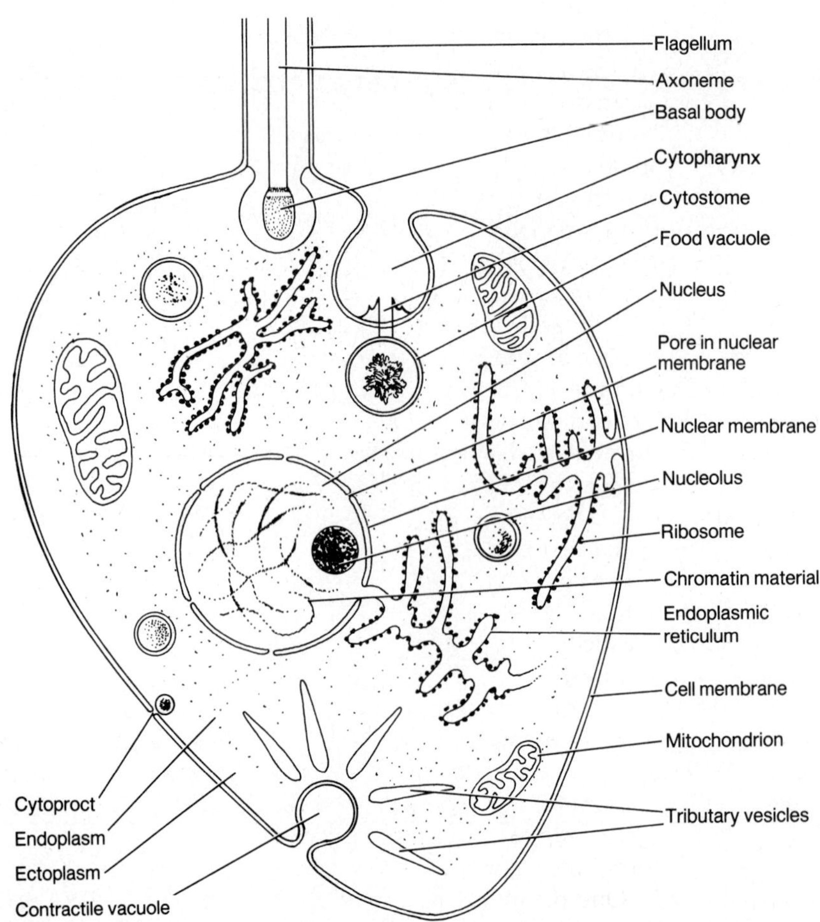

Figure 2.2 *A stylized and generalized protozoan with a flagellum illustrating basic protozoan morphology.*

variety of organismic functions. But because protozoans are small, are unicellular, and have a relatively large surface area, specialized surfaces are unnecessary, and all organismic functions are embodied within the one cell. Therefore protozoans have been forced to rely on specialized subcellular structures called **organelles** to carry out all organismic functions, including those associated with locomotion, food capture, osmoregulation, and reproduction (Fig. 2.2). High degrees of structural differentiation and functional specialization are reflected in the organelles. Because of them, protozoans exhibit considerable adaptive complexity. In recent years, electron microscopy and biochemical techniques have been responsible for a dramatic increase in protozoan literature dealing especially with the ultrastructure and function of organelles. Intensive research on all phases of protozoology in the past several decades has clearly demonstrated that what were once considered to be "simple" organisms are, in reality, remarkably complex. We shall give considerable attention to the ultrastructure and specific functions of certain organelles as important adaptations of protozoans.

In the unicellular condition, that part of the body which is in contact with the environment is the **cell membrane** or **plasma membrane** (Fig. 2.2). This membrane is of preeminent importance to a protozoan: through it pass all substances entering or leaving the organism. The membrane is a delicate, living, bilaminar structure composed of phospholipids and proteins. The protein molecules are thought to be located in an irregular mosaic pattern within the regularly arranged lipid layers. Proteins provide strength and elasticity to the membrane, and the degree of permeability is due principally to the phospholipids. Both characteristics are particularly necessary for a surface membrane to be functionally important in homeostatic maintenance. The membrane, a dynamic ever-changing structure, is intimately involved with osmosis, diffusion, and active transport of materials across it. One of the fascinating properties of a cell membrane is that it is differentially or selectively permeable, and further, its permeability can be altered quickly by the organism to reflect changes in the metabolic state of the cell. In some protozoans the plasma membrane plus an inner complex of membrane-lined alveoli form a thickened outer layer, the **pellicle**, which provides form and, in some, rigidity. Exterior to the cell membrane there may be a nonliving, secreted covering such as a lorica or test. A **lorica** is a protective sheath and tends to be a chamber in which the animal can move around freely; a **test** is a shell or armored case that fits snugly around the animal. Arenaceous species, cementing environmental particles together in a secreted matrix to form a protective envelope, include some sarcodinans and ciliophorans.

Internally, the cytoplasm of a protozoan can usually be divided into two distinct regions: an outer ectoplasm and an inner endoplasm. The clear **ectoplasm**, usually in a gelatinous or gel state, lies just beneath the cell membrane, and its physical properties often determine the protozoan's shape and form (Fig. 2.2). In many sarcodinans, for example, the ectoplasm permits the formation of ever-changing pseudopodia. In ciliates, on the other hand, the more rigid ectoplasm, along with the pellicle, confers a definite unchanging shape. Usually devoid of organelles and granules, the ectoplasm is often associated with the locomotive organelles to be described later. The opaque, granular sol **endoplasm** is more fluidlike, and most organelles are embedded within it (Fig. 2.2). As will be pointed out later when sarcodinan locomotion is described, ectoplasm and endoplasm often represent reversible colloidal states of the cytoplasm.

All protozoans have at least one **nucleus**, and many have two or more nuclei. Each nucleus is separated from the cytoplasm by a double-layered **nuclear membrane**, which is perforated at places. The resulting pores allow continuity between the nucleus and cytoplasm (Fig. 2.2). The nuclear and plasma membranes are continuous within the cytoplasm as the **endoplasmic reticulum**, and the ultrastructure of all three is essentially identical (Fig. 2.2). As in all eucaryotic cells, inside the nuclear membrane is contained **chromatin material**, which is present as discrete **chromosomes** at the time of cell division. The chromatin material and chromosomes are constructed chiefly of the well-known hereditary chemical, deoxyribonucleic acid (DNA). Characteristically, each nucleus contains one to several small bodies, the nucleoli. Each **nucleolus**, associated with one or more chromosomes, is rich in ribonucleic acid (RNA) and is involved with the formation of ribosomal RNA.

Several Important Protozoan Functions

Because of their small size and large surface area, protozoans have no separate organelles for gaseous exchange or excretion, since these are obviated by diffusion of materials across the cell membrane. Several important functions are especially dependent upon the organism's size. We shall explore the details of four such activities to see generally how a small protozoan functions.

FEEDING

All animals including protozoans are heterotrophic, but correlated with their diminutive size the adaptations for food intake in protozoans are quite different from those in higher animals. Many protozoans, especially many free-living flagellates and most parasitic forms, are able to absorb molecules in solution, and they are termed **saprozoic**. In saprozoic feeding, organic molecules enter the cell by diffusion or active transport. Often, large organic molecules in solution are taken in by a process called **pinocytosis** (= cell drinking). In pinocytosis the cell membrane invaginates, surrounds a small amount of solution, and pinches it off internally to form a pinocytic **food vesicle.**

Most other protozoans are able to ingest whole organisms such as bacteria, algae, or other protozoans or organic fragments such as particulate debris. These forms are termed **holozoic**. But a holozoic protozoan has problems in transporting particulate food to the interior of the cell simply because the cell membrane is not permeable to large particles. Therefore holozoic forms must ingest food without its actually passing through the cell membrane. Typically, the protozoan encloses the food substance or particle in a membrane-bounded, invaginated vesicle or **food vacuole** pinched off from the cell membrane. There are two methods by which food vacuoles are formed. Sarcodinans have cytoplasmic extensions of the body, the pseudopodia, which typically surround and engulf food and imprison it in a food vacuole (see Fig. 2.9c later in the chapter). A second method, present in ciliophorans and some flagellates, is to concentrate food particles on the surface and take them in *en masse* at some particular point on the pellicle, usually near the anterior end (Fig. 2.3a). These protozoans often have a definite opening, the **cytostome**, and a **gullet** or **cytopharynx** where bits of food are concentrated and formed into a food vacuole. The process by which the cell engulfs particulate food, termed **phagocytosis** (= cell eating) or **phagotrophy**, results in a food vacuole (Fig. 2.3a). Following ingestion, living prey is usually killed within an hour and often within a few seconds. The mechanism by which prey is killed is not fully known but undoubtedly is related to oxygen depletion of the prey and to digestive enzymes released into the food vacuole. Some protozoans may be exclusively pinocytic and others phagocytic, but the vast majority are probably both. Often, the soluble contents of a phagocytic vacuole are removed by minute pinocytic vesicles formed at its periphery.

Regardless of the manner by which food vacuoles are formed, digestion proceeds rapidly. **Lysosomes**, organelles surrounded by a single membrane, contain a collection of enzymes that are inserted into food vacuoles. The

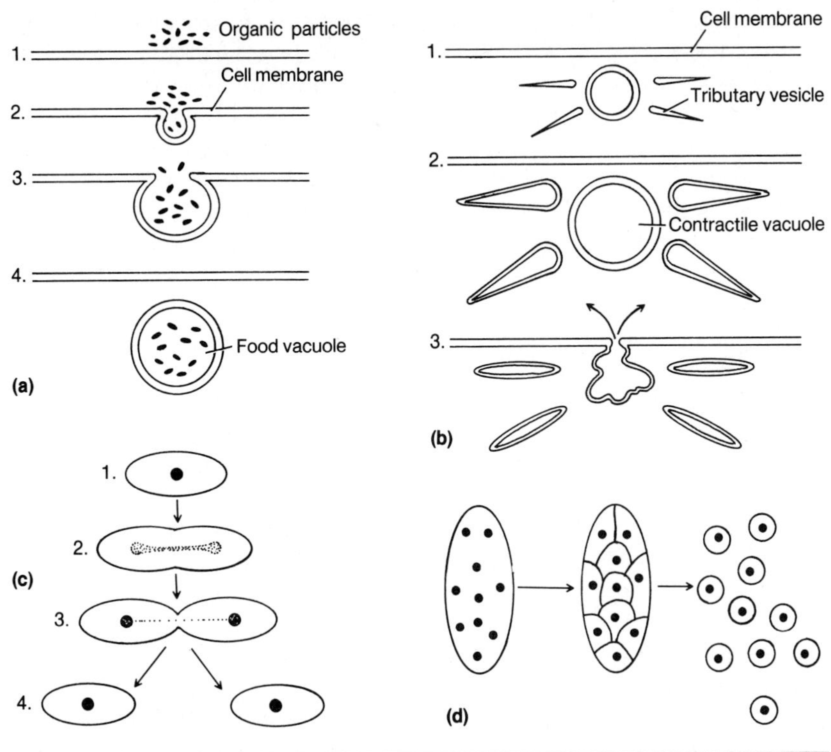

Figure 2.3 *Sequential stages of four protozoan activities:* ***(a)*** *the formation of a food vacuole or vesicle;* ***(b)*** *diastole and systole of a contractile vacuole;* ***(c)*** *binary fission;* ***(d)*** *schizogony.*

enzymes are responsible for the hydrolytic breakdown of food particles into simple soluble molecules such as amino acids and monosaccharides. Proteases and carbohydrases are known to be present in most, and lipases are probably present, although this has not been demonstrated. Shortly after a food vacuole is formed, its contents become acid, pH values of 1.5–5.0 being common; but as digestion proceeds, the contents become slightly alkaline. As simple molecules are absorbed through the vacuolar membrane, the vacuole shrinks until all that remains is an undigestible residue. The residue and vacuolar bag are then ejected to the outside through a temporary rupture of the plasma membrane. In those forms with an organized pellicle (ciliophorans) a definite opening, the **cytoproct**, is present through which the depleted food vacuole is voided. In all others, residual food vacuoles can be voided at any point on the cell membrane.

Osmoregulation

Because of the relatively large surface area, protozoans are especially vulnerable to subtle osmotic changes in their environment. Since their protoplasm is isotonic with seawater, marine forms normally do not have osmotic problems. But for protozoans to be able to live successfully in fresh water, some mechanism to cope with a hypotonic medium had to evolve. In such environments the animal is generally hypertonic to its surroundings and is therefore confronted with the problem of a continual influx of water from the environment—a fact exacerbated by the high surface-area-to-volume ratio. Such protozoans have adapted by developing an osmoregulator organelle, the **contractile vacuole**, an important maintenance pump that aids in the regular elimination of excess water (Fig. 2.3b). Each membrane-bounded vacuole contains a peripheral region supplied with numerous vesicles and mitochondria, the latter presumably present to supply the necessary energy for vacuolar elimination. The vacuole swells slowly (diastole) by the accumulation of fluids from the cytoplasm. In some the central space is fed either by vesicles or by radiating canals or tributaries. When filled, the vacuole collapses suddenly (systole), and the fluid is released to the outside through a pore in the cell membrane (Fig. 2.3b).

The sole function of the contractile vacuole is osmoregulation, but sometimes nitrogenous wastes (urea and ammonia) are incidentally eliminated also. A contractile vacuole is present in all freshwater sarcodinans and flagellates, in all ciliophorans both freshwater and marine, and also in a few parasitic protozoans. Sarcodinans have one or two contractile vacuoles, which have an indefinite position in the cell and thus no fixed pore. Flagellates have a single contractile vacuole that has a constant position, and ciliophorans have one to several, which are in definite constant locations and with pores in fixed positions.

Locomotion

Since all protozoans are aquatic or live in moist habitats, one of their most important adaptations is related to their motility. Several different solutions have evolved to the problem of moving a miniature organism through a watery medium or over a surface. In fact, a fundamental distinction between several protozoan phyla is the type of locomotion and the organelles employed in this activity. Two similar kinds of organelles, **flagella** and **cilia**, are hairlike structures extending from just beneath the plasma membrane to the exterior (Fig. 2.2). These organelles are adapted for swimming. As they beat, the small flagellates and ciliophorans are moved through the water by a sculling, rowing, or propellerlike motion. Sarcodinans have temporary flowing **pseudopodia**, which enable them to creep over a surface rather than swim. Protozoans belonging to four phyla, all of which are parasitic, are nonmotile as adults. Locomotion has been made unnecessary by the formation of cysts and by other aspects of their complex life cycles.

Reproduction

Protozoans reproduce both asexually and sexually. But because of their size and unicellularity, their patterns of reproduction are substantially different from those in other animals. Asexual reproduction is a phenomenon in all protozoans and is the sole form of multiplication in many species. As a protozoan grows, its surface-area-to-volume ratio decreases; and in ways still obscure, the increasing volume serves as a stimulus for division. Division of one individual into two or more progeny cells is called **fission**. In some small protozoans, fission potentially can take place every three or four hours, so their reproductive potential is enormous. Fission is always preceded or accompanied by nuclear division, usually involving mitosis. Nuclear fission without mitosis (amitosis) occurs commonly in one of two types of nuclei in ciliophorans. If fission results in the formation of two new cells, it is called **binary fission** (Fig. 2.3c). In some parasitic forms, **multiple fission** or **schizogony** follows repeated nuclear divisions and results in the production of multiple progeny cells (Fig. 2.3d). Certain ciliophorans (suctorians) engage in budding, a form of fission in which one or more small individuals are pinched off from a parent cell.

With fission many organelles must be replicated in one or both progeny. In some protozoans most parental organelles are degraded at fission, and both progeny cells regenerate all missing parts. But in most forms a given organelle is passed to one progeny cell, and the other cell regenerates it. Regeneration, a highly variable phenomenon, is totally dependent on the presence of some nuclear material, since an enucleate portion is usually not viable and will degenerate shortly.

Because protozoans are single celled, patterns of sexual reproduction reflect this unusual organismic condition. For many protozoans, no sexual process is known. Many other protozoans do not form gametes typical of sexual reproduction, but rather they have other methods of mixing genomes. It is important to remember that the essence of gametic fusion is, in reality, genetic recombination. Ciliophorans engage in a process of genetic recombination called **conjugation** in which nuclear materials are exchanged between two organisms. Some other species form male and female gametes in the same organism, which fuse in a phenomenon called **autogamy**. Both conjugation and autogamy are nonmultiplicative directly but are rejuvenating phenomena. Still other protozoans form gametes, so meiosis and fertilization must be events at some point in their life cycles. Most adults are diploid, and meiosis occurs in gametogenesis; but adult apicomplexans and some flagellates are haploid and have zygotic meiosis. If male and female gametes are of equal sizes, they are called **isogametes**, but they are **anisogametes** if unequal in size and motility and suggestive of eggs and sperm of higher animals. Haploid gametes fuse in a process of **fertilization** or **syngamy** to form a diploid zygote. The life cycles of the parasitic forms are especially complex and often involve two hosts.

Protozoan Ecology

Because of their small size and susceptibility to desiccation, protozoans are restricted to the oceans and fresh water or wherever moisture is present as in damp places such as moss beds, decomposing organic material, or the soil. Sometimes protozoans are present in prodigious numbers, but the free-living forms never create a nuisance. As adults, they can survive temperatures down to 0°C, and even though some can tolerate a temperature of up to 56°C, the thermal maxima for most are in the range of 35–40°C. Protozoans are mostly aerobic, but some forms live in anaerobic conditions such as in sewage or decomposing organic material.

A very important adaptation for survival of most protozoans is that they can form cysts, usually in response to drying, heat, cold, lack of food, or chemicals. To encyst, the organism produces a thick, double-walled envelope around itself. It can remain viable in this state for years. Once favorable environmental conditions are restored, rapid excystment normally follows. Cyst formation is of great adaptive significance for several reasons. First, it is effective protection against adverse environmental conditions. Second, it is a convenient form for dispersal, since protozoan cysts can be transported great distances by wind and water currents and in mud on the feet or bodies of many animals. In fact, most protozoans are cosmopolitan in distribution because of the ease with which cysts are transported throughout the biosphere. Finally, cyst formation is often a stage in the life cycle in which cytoplasmic reorganization and nuclear division may take place.

It is not generally realized that protozoans are abundant in moist soils to a depth of 15–18 cm, and a few may be found as deep as 0.5 m. Several hundred soil species, mostly sarcomastigophorans and ciliophorans, are known to thrive in the aquatic film between soil particles. Species diversity and numbers of individuals are correlated with types of soils, and both are directly proportional to numbers of bacteria (food organisms) present.

A rather large number of protozoans live in association with other organisms. These protozoan symbiotic relationships range all the way from commensalism to parasitism, external to internal, and obligatory to facultative. Representatives of four entire protozoan phyla are endoparasitic in animals, a majority of those of another class (Zoomastigophorea) are endosymbionts of animals, and certain species of the other two phyla are commensals, mutualists, or parasites. However, it should be noted that in protozoan symbioses, like those of other groups, it is exceedingly difficult or impossible to determine the full extent of benefit or harm to either or both symbionts. Symbiotic associations should be considered only as definable points on a continuum of degrees of association.

PROTOZOAN DIVERSITY

Protozoans are a heterogeneous and diverse group, and various schemes have been used in the past to group them taxonomically. The classificational scheme employed here is that of Levine et al. (1980), but with some important omissions. The phytoflagellates and slime molds have been omitted here on the basis of criteria previously discussed. A complete resume of the principal protozoan phyla appears at the end of this chapter.

PHYLUM SARCOMASTIGOPHORA

The sarcomastigophorans are characterized by most having monomorphic nuclei, locomotion by flagella or pseudopodia or both, and, in the sexually reproducing forms, gametes of the same size. To this large phylum belong about 10,000 modern-day species of flagellates and sarcodinans (amebae, radiolarians, foraminifera). Because this grouping includes three subphyla, perhaps the best way is to proceed with the characteristics of each group separately.

Subphylum Mastigophora

All mastigophorans bear one or more whiplike flagella as their principal locomotor organelles. It is from this important locomotive organelle that are derived both the subphylum name (mastigo = whip; phora = bearer) and their common name, flagellates.

FLAGELLA

The characteristic feature of mastigophorans that enables the organism to move through a watery medium is the whiplike undulations of their hairlike flagella. A **flagellum** is a thin, threadlike, cylindrical process about 10–50 μm in length and about 0.25 μm in diameter. Each flagellum, anchored within the ectoplasm by a **basal body**, extends outward from the cell membrane, and its undulations move the animal through the water. There are typically two flagella present in each mastigophoran, but many have only one, and others have several to many flagella. Flagella may be of equal or unequal length, and frequently one or more lead and the others follow.

Electron microscopy has revealed many utterly fascinating details about the ultrastructure of a flagellum. A cilium has an identical ultrastructure, and the discussion below will apply equally well to flagella or cilia. The flagellar ultrastructure consists of an **axoneme** or cylinder surrounded by a **sheath**, the latter being continuous with the plasma membrane (Figs. 2.2, 2.4a, b). The axoneme consists of a matrix in which are found fine longitudinal **microtubules** or **fibrils** in a precise pattern. Nine doubled microtubules are arranged symmetrically around two centrally located single tubules, forming a highly characteristic 9 + 2 pattern (Fig. 2.4 a-c). All eleven tubules apparently are straight and run along the length of the flagellum without twisting or spiraling. One member (subfibril A) of each pair of doublets has two rows of laterally directed arms pointing toward subfibril B of an adjacent pair. Microtubules are composed of a globular protein called **tubulin** which is similar in construction to actin, a protein found in muscles. Sometimes the central fibrils are longer than the doublets, and in other

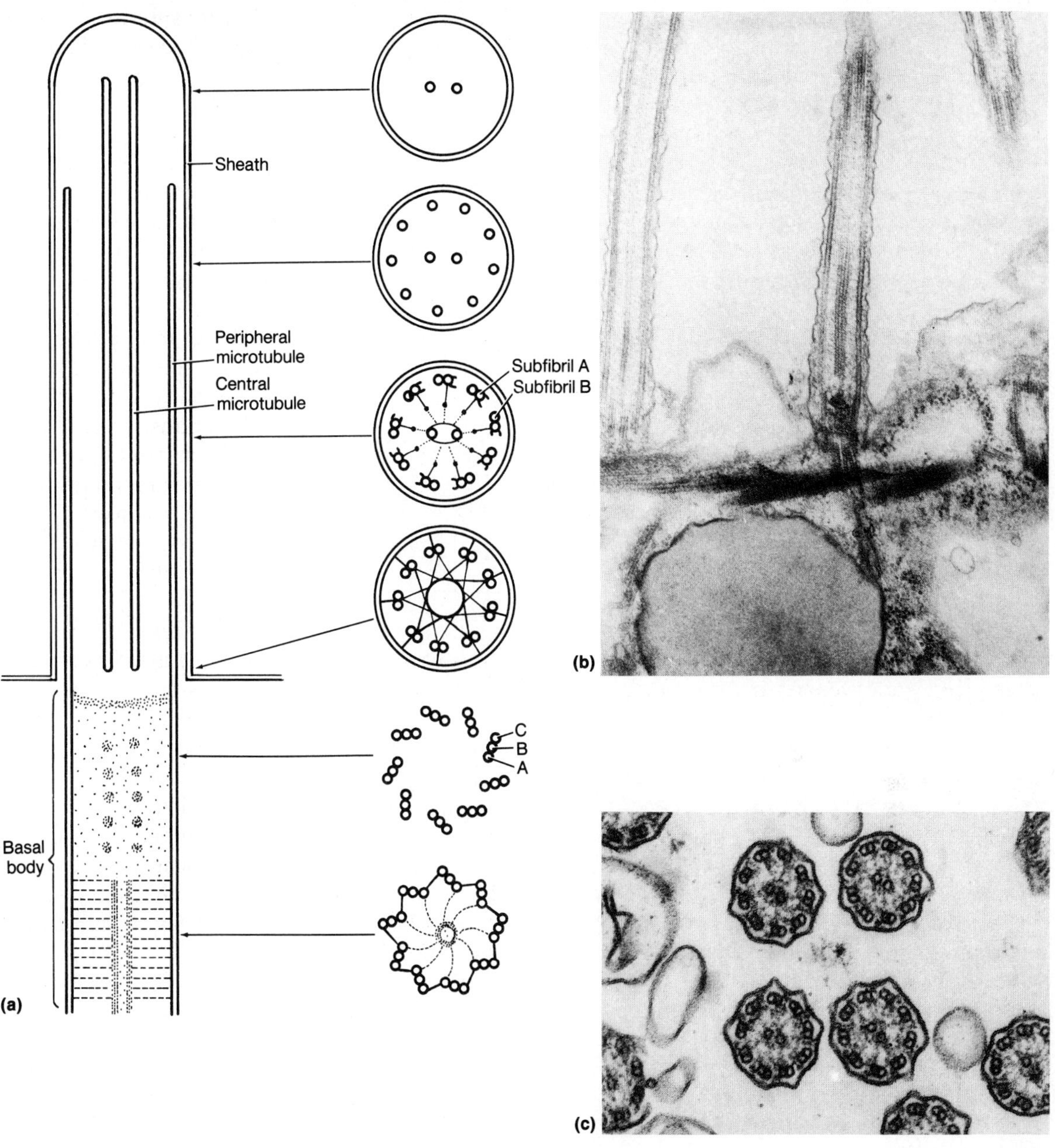

Figure 2.4 *The ultrastructure of flagella and cilia: **(a)** stylized diagrams of the features seen in a longitudinal and several cross-sectional views of a typical flagellum and basal body (from K. G. Grell,* Protozoology, *Springer-Verlag, New York, 1973); **(b)** an electron micrograph of a cilium seen in longitudinal section (courtesy of B. Grimes **(c)** an electron micrograph of several cilia seen in cross-section (courtesy of J. L. Carson).*

cases the peripheral doublets become single near the apex. Either type results in a flagellum normally narrowing toward its distal extremity (Fig. 2.4a,b).

The ultrastructure of flagella and cilia has been both a fascination and an enigma for protozoologists ever since their explication by the use of electron microscopy. The 9 + 2 arrangement is repeated faithfully with few exceptions in all flagella and cilia. In fact, the almost universal 9 + 2 condition in the flagella of sperm is thought to reflect the condition of the ancient flagellate ancestors from which all animals (and plants as well) evolved. But fundamental questions persist. Why the 9 + 2 arrangement rather than some other combination? Why are the peripheral microtubules doubled and the central ones single? What is the significance of arms on the lateral doublets? Why are there ultrastructural differences between a flagellum and its basal body? These and other questions remain unanswered.

Flagellar functions have been studied intensely in recent years. At the ultrastructural level it would appear that movement is achieved not by the shortening of microtubules as once was thought, but by the tubules sliding past each other in a way similar to the process of muscle contraction. It is hypothesized that there are cross bridges between fibrils and that these interfibrillar bridges act as hooks or levers that enable one microtubule to pull an adjacent microtubule by it, thus producing a sliding motion. As in all cellular events, the energy necessary for the sliding of the microtubules is based on the breakdown of adenosine triphosphate (ATP) to release energy, a mechanism thought to be involved in a manner similar to that in muscle contraction. The remarkable chemical similarities of the functional properties of tubulin and actin may be only coincidental, but they also suggest that the contractile machinery of muscles may have originated as an elaboration of the flagellar microtubules of ancient flagellates.

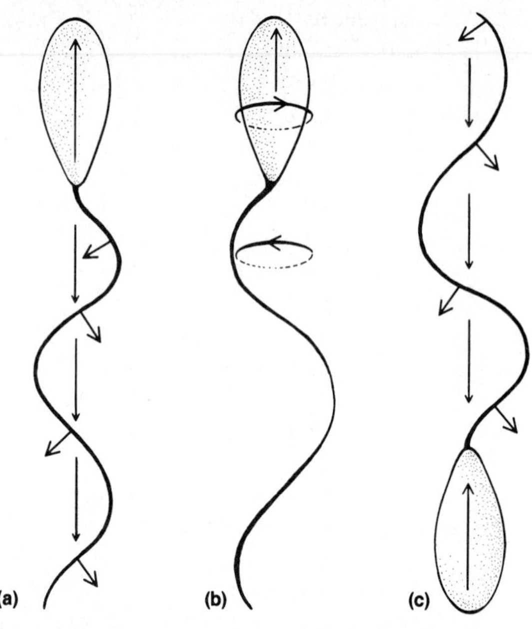

Figure 2.5 *Patterns of flagellar locomotion:* ***(a)*** *the posterior flagellum beats in one plane, and waves pass from proximal to distal;* ***(b)*** *the posterior flagellum beats in a helical pattern, and waves pass from proximal to distal;* ***(c)*** *the anterior flagellum beats in one plane, and waves pass from distal to proximal.*

At the organellar level, flagella function in a variety of ways such as pushing, pulling, steering, or even anchoring. Regardless of the action, flagellar movement of the animal follows essentially the same plan. A flagellum moves in an undulating manner either in a flat plane or in a helical or cylindrical pattern. Undulations most commonly proceed from base to tip, in which case the organism is pushed or propelled through the water by the posterior propellerlike flagellum (Fig. 2.5a, b). But in some the undulations proceed from the tip to base and thus pull the animal forward (Fig. 2.5c). The flagellum of most mastigophorans may have several undulations passing along its length at one time so that movement of the organism is continuous. The movement of the flagellum produces a resultant force on the water along the long axis of the flagellum and in the direction of the flagellar wave (Fig. 2.5).

The basal body, often called the **kinetosome** or **blepharoplast**, is continuous with the flagellum, and both have somewhat similar ultrastructures. In the basal body the central fibrils are absent, as are the lateral arms of subfibril A; a third subfibril (C) is present on each of the nine peripheral tubules; and the peripheral tubules are twisted so that subfibril A is nearest the center (Fig. 2.4a). A system of exceedingly thin fibrils connects each triplet microtubule to a central region. In every case studied thoroughly the basal body is surrounded by one or more mitochondria. This arrangement suggests that the basal body is actively involved in flagellar movement and that considerable energy, supplied by the mitochondria, is required. Interestingly, basal bodies and centrioles (organelles in animal cells that organize the spindle fibers during nuclear division) have the same detailed ultrastructure, suggestive of similar ancestral functions. In fact, the basal bodies of a few mastigophorans serve in forming the poles during mitosis in lieu of centrioles. A fibrillar "rootlet" structure extends inward from the basal body often to the nucleus, centriole, or other regions of the cytoplasm.

Other flagellate features

Flagellates are for the most part uninucleate; but if more than one nucleus is present, they are all **monomorphic**, i.e., all of the same type and shape. All free-living and many parasitic forms reproduce asexually by longitudinal binary fission, but schizogony occurs in some parasitic flagellates. Sexual reproduction is known for some species. Life-cycle patterns are quite varied and complex especially in parasitic forms.

Flagellate diversity

The Subphylum Mastigophora contains about 5,500 known species, all in the Class Zoomastigophorea. Zoomastigophoreans are an extremely important group, since they appear to be the most primitive of all protozoans. Ancestral zooflagellates were probably the stock from which arose modern-day protozoans, including modern zoomastigophoreans. They are quite diverse in shape, life cycles, details of reproduction, and habitats in which they live. Free-living zoomastigophoreans are found in both fresh and salt water, but the vast majority are symbiotic, and many are parasites. Discussions will cover the major characteristics of only certain taxa that are important in the phylogeny of metazoans, are relatively abundant, are human parasites, or are interesting symbionts.

Choanoflagellates are of particular interest, since they may have been ancestral to sponges; more will be said about this probable phylogenetic relationship in Chapter 4. Choanoflagellates are unusual in that each has a cylindrical **collar** surrounding the single flagellum (Fig. 2.6a). Electron microscopy has shown the collar to be composed of numerous parallel, rodlike **microvilli** arranged much like the vertical slats in a picket fence. The undulations of the flagellum bring suspended food particles, chiefly bacteria, over the collar, where the particles, filtered from the water by adhering to the collar, are transferred to a food vacuole in the cytoplasm. Choanoflagellates live in freshwater or marine habitats as solitary or, more often, colonial forms, some of which are stalked (Fig. 2.6a). Some planktonic marine forms secrete a rather complex lorica constructed of precisely arranged, siliceous rods around themselves. Some colonial forms like *Proterospongia* secrete a gelatinous mass in which are embedded the cells, each with its collar and flagellum at the periphery.

Kinetoplastids comprise an order of both free-living and parasitic forms, and some of the parasites are of particular concern to human beings. Representatives of this order have a special, DNA-rich organelle, the **kinetoplast**, which is larger than the basal body but often fused with it. Like basal bodies, the kinetoplast is always associated with at least one mitochondrion, and the entire complex undoubtedly functions in flagellar movement in some manner. Many species are polymorphic, having several distinct morphological stages in their life cycle. Some kinetoplastids have an undulating membrane constructed in part by a flagellum. The best-known genus, *Trypanosoma*, contains hundreds of species, all of which are parasitic in the guts of insects and blood of vertebrates

Figure 2.6 *Types of zoomastigophoreans:* ***(a)*** *a colony of* Codosiga, *a choanoflagellid;* ***(b)*** *a photomicrograph of* Trypanosoma, *a kinetoplastid, in a blood smear (courtesy of D. F. J. Hilton);* ***(c)*** Giardia, *a diplomonad;* ***(d)*** Trichomonas vaginalis, *a trichomonad found in the vagina of some women;* ***(e)*** *a photomicrograph of* Trichonympha, *a hypermastigid (courtesy of K. G. Grell,* Protozoology, *Springer-Verlag, New York, 1973, after Cleveland).*

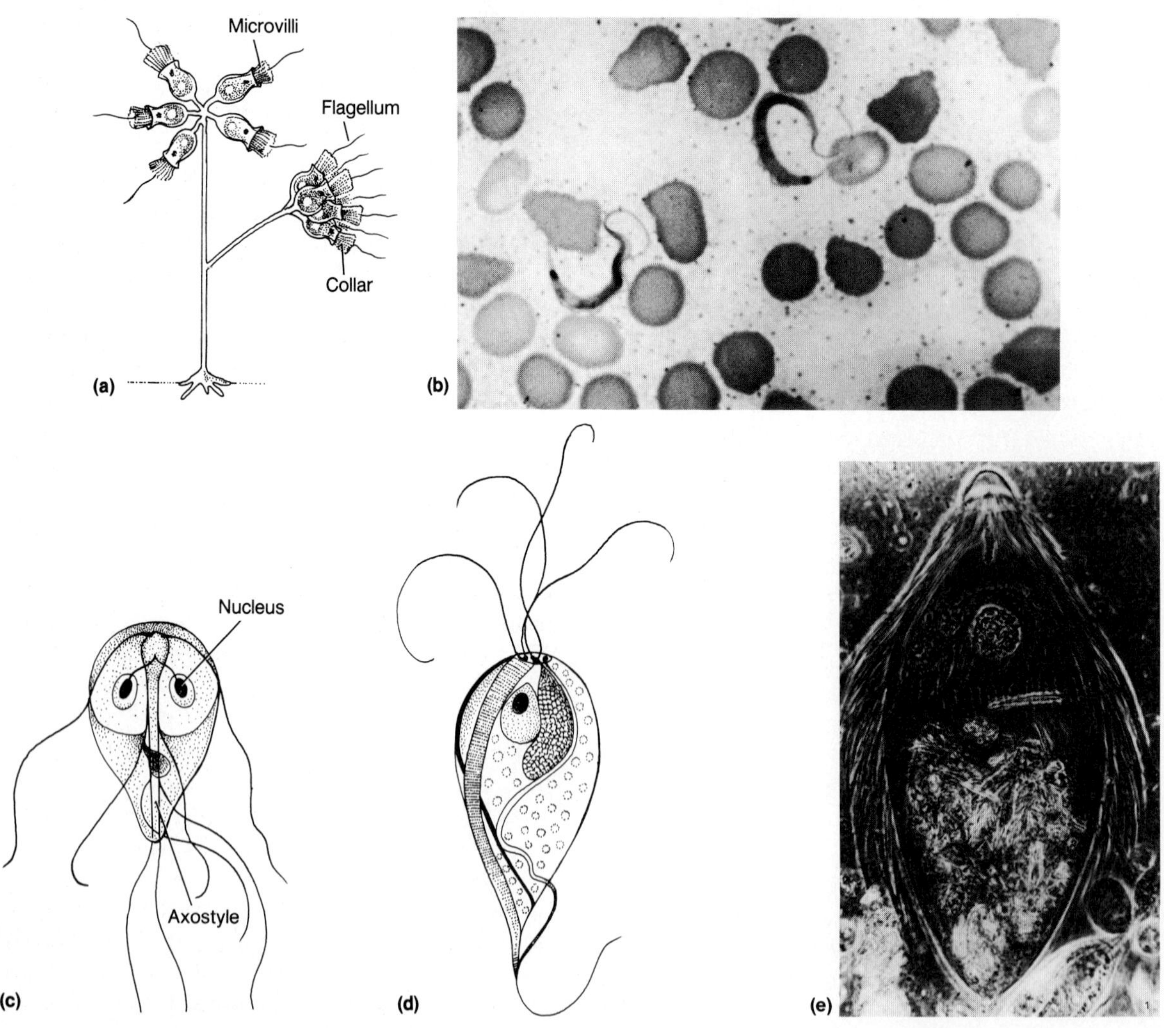

(Fig. 2.6b). These organisms are elongate and flattened and have a single flagellum whose basal portion forms the outer margin of an undulating membrane. All that remains of a second flagellum is its basal body. As the insect host bites, it transmits the trypanosomids to the blood of vertebrates, including human beings. In the plasma of the vertebrate host the organism grows, reproduces asexually, and is transmitted back to the insect at its next feeding. In human beings, African sleeping sickness (confined to Central Africa) and American trypanosomiasis, called Chagas' disease, are conditions caused by the presence of *Trypanosoma* in human blood. Symptoms include fever, lethargy, and insensitivity to pain, and the infestation may result in death. The insect vector for African sleeping sickness is the tsetse fly (*Glossina*); the vector for American trypanosomiasis is the large, blood-sucking, kissing bug (*Panstrongylus*). *Leishmania*, another kinetoplastid, causes visceral leishmaniasis or kala-azar, an often fatal disease present in areas around the Mediterranean Sea. Its insect host is the sandfly (*Phlebotomus*).

The best-known diplomonad is *Giardia*, an intestinal parasite found in a variety of vertebrates (Fig. 2.6c). The drop-shaped body has a blunt anterior end and tapers toward the opposite end. Present are two nuclei, two axostyles, and four pairs of flagella. An **axostyle** is a subcellular system of microtubules that may be in the form of a ribbon or, more commonly, an open cylinder. The axostylar microtubules often surround the nucleus and are associated with the basal body of one or more flagella. The particular function of this organelle is not fully known but is probably associated somehow with flagellar movement. A common species of *Giardia* inhabits the small intestine of both children and adults and causes recurrent enteritis, diarrhea, and stools with copious mucus. A few species of diplomonads, including some of the genus *Hexamita*, are free living in stagnant pools.

Trichomonads, mostly parasitic or symbiotic forms found in the alimentary or genital tracts of vertebrates, are characterized by the presence of an axostyle. *Trichomonas*, the most familiar genus, contains forms with four anterior flagella and a fifth that forms an undulating membrane (Fig. 2.6d). Different species of *Trichomonas* parasitize persons in the mouth, colon, vagina of females, and urinary tract of males. While none of the trichomonads is dangerous or fatal, their presence often irritates mucous membranes and may result in considerable pain.

Hypermastigids are all complex multiflagellate species living symbiotically in the alimentary canal of wood-eating insects, including termites, woodroaches, and cockroaches. Numbers of hypermastigids present in a single host are often prodigious. Each is relatively large and oval, and each possesses a single nucleus, several axostyles, and several centrioles. Each flagellum is individually attached to a basal body, and flagella numbering in the hundreds or even thousands are present either in longitudinal or spiral rows or in circles girdling the cell (Fig. 2.6e). The symbiotic relationship between insect and flagellates is a fantastic example of **mutualism**, an obligatory relationship from which both organisms derive distinct benefits. The insect ingests wood particles, the flagellates digest the wood, and both the insect and flagellates absorb the relatively simple, soluble sugars that are the end products of cellulose digestion. Without the hypermastigids the wood-eating insects would starve and die even when supplied with wood because they lack the enzymes (cellulases) necessary to digest cellulose. The insect, on the other hand, provides an optimal, homeostatic, moist environment without which the hypermastigids would perish. So intimate is this insect–flagellate symbiosis that the life cycles of the hypermastigids are incredibly attuned to the hormonal sequence associated with molting in the insect! In most insects the flagellates are lost at each molt, and the newly molted (as well as the newly hatched) insects obtain a new flagellate fauna by feeding on feces containing the flagellates or their cysts or by licking the rectal area of another insect. The symbiotic relationship between insects and hypermastigids is one of the most remarkable examples known of obligatory mutualism.

Subphylum Opalinata

The opalinids are an interesting group, but historically they have created perplexing taxonomic problems. Opalinids live symbiotically in the rectum of frogs, toads, and salamanders and occasionally in other ectothermic vertebrates. They are flattened and large for flagellates; some are up to 0.8 mm long and 0.3 mm wide. All surfaces of the body are densely covered with short flagella (cilia). The flagella sweep posteriorly in coordinated waves to move the animal forward. Since no cytostome is present, all are saprozoic and feed by pinocytosis. Large numbers of monomorphic nuclei are present in the most common genus, *Opalina* (Fig. 2.7), but other genera may have as few as two nuclei. Opalinids reproduce asexually by binary fission in which the resulting progeny may be of the same or very unequal sizes. Opalinids also reproduce sexually; this sexual reproduction is thought to be synchronized to the reproductive hormonal cycle of the amphibian. A diploid opalinid produces a large number of spherical forms, which soon encyst and are eliminated with the fecal material of the host into the water (Fig. 2.7). If eaten by a tadpole, the cyst hatches and, by a series of longitudinal fissions and meiosis, produces uninucleate haploid anisogametes. Fertilization ensues, and the zygote

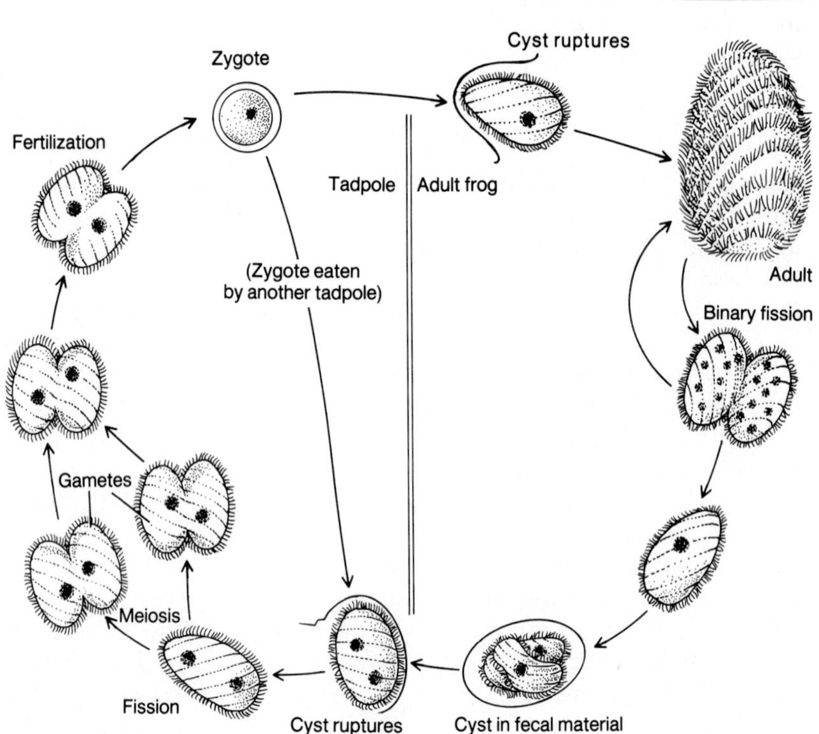

Figure 2.7 *An adult* Opalina *and its life cycle.*

undergoes nuclear fission to produce a multinucleate adult. In some species the zygotes are released in the fecal material of the tadpole, are eaten by another tadpole, and reproduce to increase the number of future adults (Fig. 2.7). Opalinids were for many years classified with the ciliophorans on the basis of the numerous locomotive organelles present. But because they possess monomorphic nuclei, they are now included as sarcomastigophorans; because of their own unique features, they are treated as a separate subphylum.

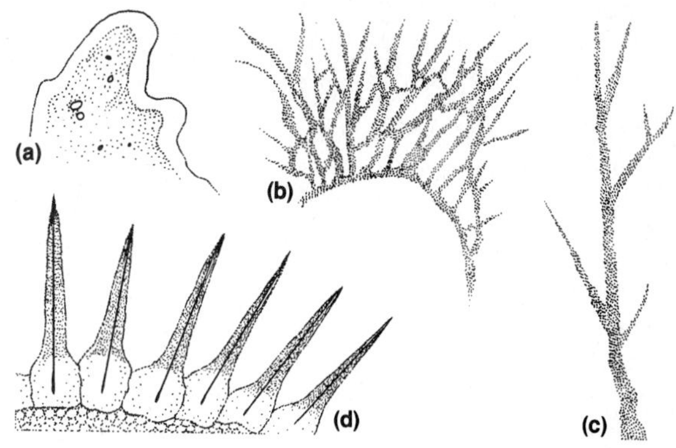

Figure 2.8 *Types of pseudopodia:* ***(a)*** *a lobopodium;* ***(b)*** *reticulopodia;* ***(c)*** *filopodia;* ***(d)*** *axopodia.*

Subphylum Sarcodina

The sarcodinans (= "flesh-colored") all possess organelles called pseudopodia that have significantly affected almost all aspects of the lifestyles of sarcodinans. To this group belong the sun animalcules, foraminifera, radiolarians, and the familiar amebae. Some 4000 living species are recognized. Representatives are common in fresh water, the oceans, and damp places, and some are parasitic.

Pseudopodia

The most characteristic feature of sarcodinans, and the one that generally sets them apart from all other protozoans, is the universal presence of pseudopodia. **Pseudopodia** are extensions of the cytoplasm that are used as organelles of both locomotion and food capture. Recently, the major research interests in this phylum have centered around the pseudopodia and the details of their functional morphology. Pseudopodia are diverse in shape, form, and function. For example, pseudopodia of amebae are broad, flowing extensions that enable the animal to creep or flow over a surface; amebae are thus asymmetrical, and a given pseudopodium is but an ephemeral organelle. In others like the spherically symmetrical sun animalcules and radiolarians, the pseudopodia are much thinner, have skeletal supports, and are more permanent than those of amebae. In foraminifera the thin pseudopodia form a network of anastomosing strands. It will be manifestly evident to the reader that pseudopodial formation is not a simple phenomenon and that no single explanation can account for all forms of movement. Undoubtedly, cytoplasmic microtubules contract in various ways to alter the shape of the different pseudopodial types. In fact, several different contractile properties of the cytoplasm may be involved in ways not yet clear. The functional ultrastructure of pseudopodia will continue to be the object of considerable research.

There are four main types of sarcodinan pseudopodia: lobopodia, reticulopodia, filopodia, and axopodia (Fig. 2.8). A discussion of the functional morphology of the four types will serve as a rather comprehensive overview of these definitive sarcodinan organelles.

Lobopodia. Under an ordinary light microscope, many sarcodinans such as amebae readily form one or more pseudopodia, and the granular cytoplasm can be seen conspicuously flowing into the new pseudopodia. These organelles are broad, lobose, and tubular and are called lobopodia. They are utilized in both locomotion and food capture (Figs. 2.8a; 2.9a).

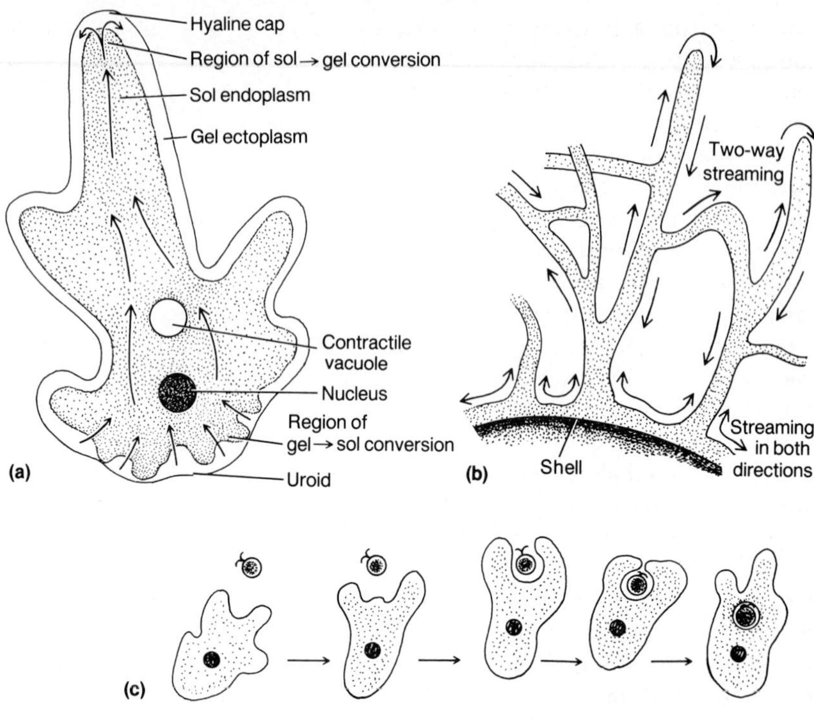

Figure 2.9 *Several pseudopodial functions:* ***(a)*** *the formation of a lobopodium in* Amoeba *to illustrate the fountain-flow hypothesis;* ***(b)*** *reticulopodia of a shelled sarcodinan illustrating constant, two-way, cytoplasmic streaming distally and streaming in both directions proximally;* ***(c)*** *sequential stages of phagotrophy in an ameba.*

Lobopodial formation has been most intensively studied in sarcodinans like *Amoeba* (Figs. 2.9a; 2.10a). In these forms the absence of a restrictive pellicle is important, but the interconversion of ectoplasmic gel to endoplasmic sol is of greater significance. At the tip of a newly forming lobopodium, the firmer ectoplasm (gel) is converted to a more liquid state (sol), and internal pressures cause the more centrally located endoplasm to flow into the extending pseudopodium. This typically forms a **hyaline cap** at the pseudopodial tip (Fig. 2.9a), where the more fluid endoplasm is turned toward the periphery of the pseudopodium in a manner analogous to water in a fountain as it is turned away from its central vertical source. As the fluid endoplasm is moved peripherally at the hyaline cap, it is converted into ectoplasmic gel. The conversion of sol endoplasm to gel ectoplasm results in the formation of a gelatinous ectoplasmic sleeve around the pseudopodium through which more endoplasm flows. The surface of a lobopodium has been shown to have adhesive properties that enable it to adhere to a substratum. This added traction makes locomotion a bit easier. Near the posterior end or **uroid** there is a corresponding zone of conversion from ectoplasmic gel back to endoplasmic sol (Fig. 2.9a). The net result of all these cytoplasmic events is to allow the sarcodinan to move forward by lobopodia. This is known as **ameboid locomotion**. Naked (nonshelled) sarcodinans with lobopodia can form one or more pseudopodia at any point on the cell surface. However, the shelled (testate) species usually possess a large shell aperture through which the lobopodia are extended.

The fountain-flow theory of lobopodial formation is generally accepted today. But what are the mechanisms behind the sol–gel conversion and the extension of endoplasm into a pseudopodium? The force necessary for the fountain-flow has for years been thought to be generated in the uroid by contraction of the posterior ectoplasm, pushing the sol endoplasm forward. Recent careful studies have shown that proteins in the ectoplasm are in a contracted state, whereas the same proteins in the endoplasm are relaxed (noncontracted). In the conversion of anterior endoplasm to ectoplasm it is now thought that the protein chains contract at the hyaline cap and thus pull the endoplasm anteriorly. In other words, according to the front contraction hypothesis, the animal is literally pulled along by the sol-to-gel conversion at the anterior end.

Most, if not all, sarcodinans with lobopodia also use them for food capture as they feed on bacteria, diatoms, flagellates, ciliophorans, or even small metazoans like nematodes or rotifers. As a sarcodinan approaches a prey organism, a lobopodium is formed on either side of the prey in a double flanking motion (Fig. 2.9c). Once the prey is surrounded on three sides, the pseudopodia are extended toward each other and fuse to engulf completely the prey into a food vacuole by phagotrophy. In the process of phagocytosis the food vacuoles of some amebae contain a considerable amount of water along with the prey. But in other amebae the plasma membrane is in intimate contact with the prey, and very little water is engulfed.

Reticulopodia. These pseudopodia, characteristic of foraminifera, are in the form of fine strands, the smallest of which is 1 μm or less (Figs. 2.8b, 2.9b). Composed solely of ectoplasm, the strands, even though they move independently of each other, have anastomosing connections that form a pseudopodial reticulum. Ectoplasmic granules, probably mitochondria, can be observed flowing in oppositely moving currents on either side of a reticulopodium (Fig. 2.9b). The reticulopodia, whose surfaces are adhesive, function as a food net, and small adhering prey can also be seen moving in opposite directions on either side of the pseudopodium. Obviously, the fountain-flow theory does not apply to reticulopodia. Numbers of longitudinal microtubules are found in these pseudopodia, which presumably have contractile properties. It has been proposed recently that adjacent cytoplasmic, proteinaceous microtubules move by sliding or shearing past each other, but confirmation of this hypothesis is yet to come. All foraminifera also have a heavy shell, and the contractile forces of the reticulopodia pull both the animal and its relatively heavy shell across a surface. Future studies on the mechanisms of reticulopodial contraction and cytoplasmic streaming should be quite fruitful.

Filopodia. Filopodia are present in a few small amebae and some radiolarians. Since they resemble reticulopodia, they may very well be shown to be reticulopodia. Filopodia are transparent, have only a few granules, and are composed exclusively of ectoplasm (Fig. 2.8c). Apparently, they do not anastomose as reticulopodia do, but they undoubtedly function in ways identical to reticulopodia.

Axopodia. These pseudopodia are thin, unbranched, and semipermanent. The distinguishing feature of axopodia is that they are strengthened or supported by an internal central microtubular complex that is skeletal and fibrous (Fig. 2.8d). In cross section the ultrastructure of the central complex consists of a double spiral of microtubules arranged in a precise arrangement and interconnected by cross-links. The microtubules are not present in the 9 + 2 pattern characteristic of flagella and cilia, and so axopodia should not be thought of as being similar to these other organelles. Covering the complex is granular cytoplasm, consisting mostly or entirely of ectoplasm. By following the ectoplasmic granules under the microscope a two-way cytoplasmic streaming pattern, similar to that of reticulopodia, can be observed along each axopodium. The axial fibers are all attached to the single nucleus (some), to several nuclei (others), or to a central body, the **centroplast** (most). The significance of this arrangement is not fully understood, but the ultrastructure of the centroplast is very much like that of centrioles and basal bodies of flagellates, and centroplasts may even be involved in mitotic spindle formation. Axopodia, present in heliozoeans and radiolarians, most of which are spherically symmetrical, radiate out in all directions from the central body. Axopodia can be extended or contracted, and both phenomena are thought to be due to the addition or disassembly of axial microtubules. Sarcodinans with axopodia are mostly carnivorous. Small protozoan prey adhere to the sticky axopodial surface and are engulfed in a food vacuole when the axopodia retract toward the body. Like foraminifera, radiolarians also have a shell, and the contractile properties of the axopodia pull the animal along over a surface.

Other sarcodinan features

Sarcodinans have the fewest organelles of any protozoan group—an indication that they are also primitive. Many are uninucleate; others are multinucleate, but their nuclei are monomorphic except that some foraminifera have nuclear dimorphism. It is not surprising that one contractile vacuole, mobile in the cytoplasm, is present in all freshwater forms. Asexual reproduction is usually by binary fission, although multiple fission is common in multinucleate sarcodinans. Sexual reproduction occurs in some groups and is particularly complicated in foraminifera and radiolarians.

Superclass Rhizopoda

This sarcodinan taxon includes those individuals that are asymmetrical and employ the pseudopodia both for locomotion and feeding. It includes the amebae, the most characteristic and best known of all sarcodinans, which typically have lobopodia; the foraminifera, which possess reticulopodia; and some small amebae that have filopodia. Since rhizopodans lack a pellicle, they usually lack a fixed shape.

The naked amebae (Subclass Gymnamoebia), of which *Amoeba* is best known, range from 3 μm to 3–4 mm in size (Fig. 2.10a, b). Most are free living and are found in fresh water, but some are marine, and others live in moist soils. The larger forms are carnivores and feed on ciliates and flagellates; herbivores feed on diatoms or other small algae.

Parasitic amebae are characteristically smaller than their free-living relatives and generally range from 5 to 50 μm. Perhaps the most important amebae parasitic in human beings belong to the genus *Entamoeba*. *Entamoeba histolytica*, responsible for amebic dysentery, is the most pathogenic species. It invades the intestinal mucosa and may be carried by the blood to other organs, principally the liver; this species often causes severe lesions. It has been estimated that 100 million people suffer from amebiasis caused by this ameba! *Entamoeba coli*, found in the intestine of many mammals, and *E. gingivalis*, a widely distributed species in tartar and debris around the teeth, are nonpathogenic in human beings.

The testate amebae (Subclass Testacealobosia) have a one-chambered **test** or shell, which is formed in one of two ways; in many it is composed of secreted organic or siliceous material, while in other species it consists of environmental particles such as sand grains or diatom shells (frustules) embedded in a hardened matrix (Fig. 2.10c). The ameba is attached to the interior of the test by means of fine protoplasmic strands. The shell always bears an aperture through which are extended the lobopodia. The contractile properties of the pseudopodia are used to drag the organism and shell over the substratum.

The foraminifera or forams (Class Granuloreticulosea), basically a marine group, are often present in astounding numbers. Aside from their having reticulopodia, forams are characterized by having a calcareous shell of one to many chambers (Fig. 2.10d,e). The shell or test is typically constructed of calcium carbonate crystals or calcite embedded within a hardened, secreted, organic matrix. In some families, arenaceous materials are incorporated into the shell. In a few foram families, shells are constructed of arragonite or silica. The walls of the shell contain numerous **pores** or **foramina**, the feature on which their name is based. Uniloc-

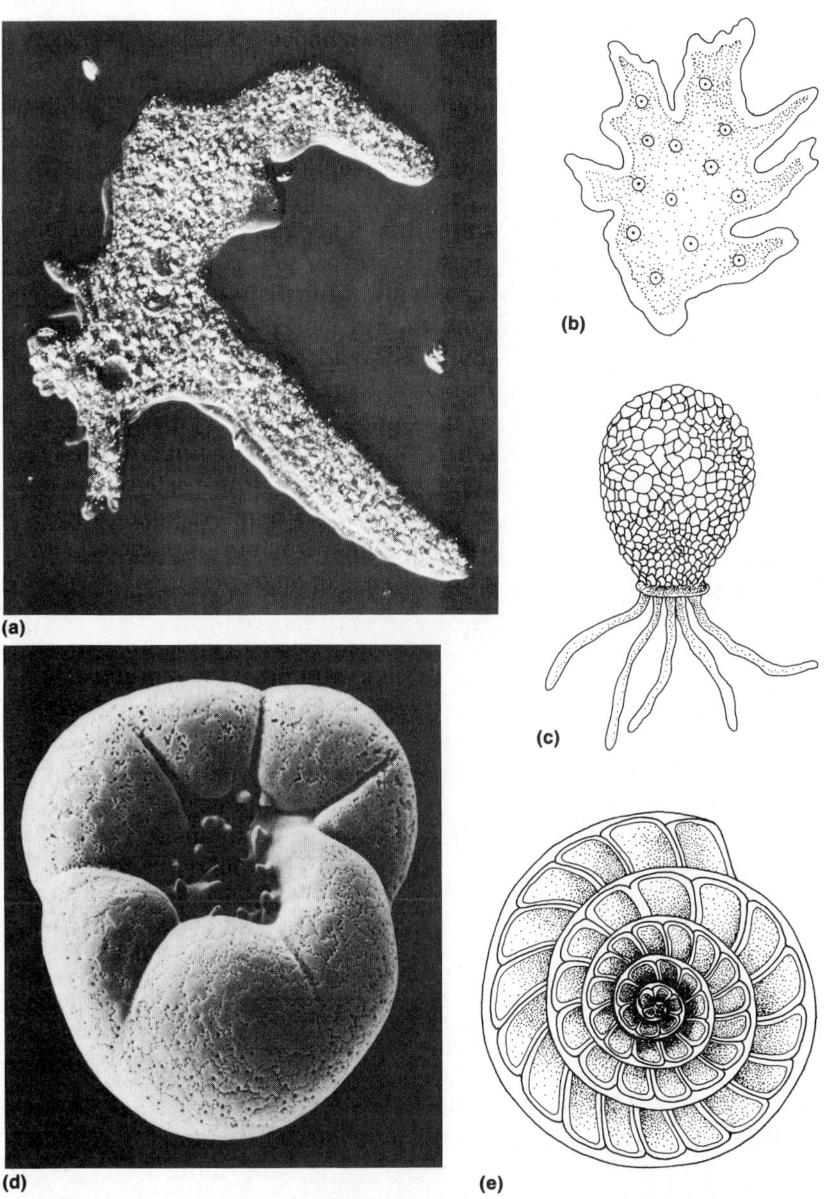

Figure 2.10 *Types of rhizopodans:* ***(a)*** *a photomicrograph of* Amoeba, *a gymnamoebian (courtesy of Carolina Biological Supply Co.);* ***(b)*** Pelomyxa, *a gymnamoebian;* ***(c)*** Difflugia, *a testacealobosian;* ***(d)*** *a photomicrograph of* Rotaliella, *a foraminifer, seen in ventral view (courtesy of K. G. Grell,* Protozoology, *Springer-Verlag, New York, 1973;* ***(e)*** Nummulites, *a foraminifer, with shell cut in vertical section.*

ular species have a shell of a single chamber, but many multilocular species are known. Multilocular species begin their existence in a single chamber or **proloculum**, which is equipped with a large opening; as the protoplasm increases in size, some flows outward through the aperture and secretes another chamber. This process is repeated to form a series of chambers, the newer ones being progressively larger. In some forams, each later chamber is secreted concentrically around the earlier chambers; in other species the chambers are in a straight line or are spirally or asymmetrically arranged (Fig. 2.10d, e). In multilocular forms the single-celled animal occupies all chambers and is continuous through the aperture and pores of each chamber.

In many foraminiferal species a thin layer of cytoplasm covers most or all of the external surface of the shell. Reticulopodia may originate from within this thin outer layer of cytoplasm, or they may be extended through the aperture of the most recent chamber. Some foram species contain symbiotic intracellular zooxanthellae

BOX 2.2 □ REPRODUCTION IN THE FORAMINIFERA

Foraminifera are unique among protozoans and, indeed, all animals. In all forms studied, there is a true alternation of generations between haploid and diploid stages. In addition to their fundamental differences in chromosomal numbers, the two stages are different in size. The diploid stage, also called the **agamont**, is smaller; because of its size, it is said to be **microspheric**. The haploid stage or **gamont** is the larger of the two and is termed the **megalospheric** generation (Fig. 2.11). Shells of both agamonts and gamonts also reflect differences in size manifested primarily in the sizes of the proloculum. The diploid microspheric agamont undergoes meiosis and multiple divisions (schizogony) to produce the larger megalospheric gamonts. By mitosis, each gamont produces a large number (several hundred) of haploid gametes, which are released into the surrounding seawater. Gametes, each equipped with two flagella of unequal length, are often of equal size for a given species (isogametes). Gametic fusion, occurring in the surrounding water, results in the formation of a diploid agamont or zygote (Fig. 2.11).

A number of variations on this basic plan are known. Some species are autogamous in that gametes of the same gamont unite in fertilization. In other species, two or more gamonts form an aggregate, and gametic fusion takes place in the common space between the aggregated gamonts. Some forams even have nonflagellated, ameboid gametes. Some species exhibit subtle differences in the gamontic stage suggestive of sexual dimorphism.

The fascinating reproductive patterns of forams are important features of a remarkable and interesting group of protozoans.

Figure 2.11 *The life cycle of the foraminifer,* Myxotheca, *showing the production of dimorphic individuals (adapted from K. G. Grell,* Protozoology, *Springer-Verlag, New York, 1973).*

(dinoflagellates), which, in a fantastic circadian rhythm, are moved from inside the shell (at night) to the distal reaches of the reticulopodia (during daylight). Part of the nutritional needs of a foraminifer comes from the photosynthetic products of these dinoflagellates. Reproduction in forams is unique for several reasons: there is an alternation of generations between haploid and diploid stages; the diploid stage (agamont) reproduces asexually, whereas the haploid stage (gamont) reproduces sexually; and the two stages (agamont and gamont) have distinctly different-sized shells (see Box 2.2 and Fig. 2.11 for further details).

Some species are planktonic and usually have delicate, spined shells, but most foraminiferal species are benthic and usually have heavier, nonspined tests. About one-half of all protozoan species are forams, but the vast majority of these are extinct and known only as fossils. Extensive oozes on the oceanic floors are composed of innumerable empty shells of foraminifera, principally those of *Globigerina*. Foram oozes are usually not found in oceanic

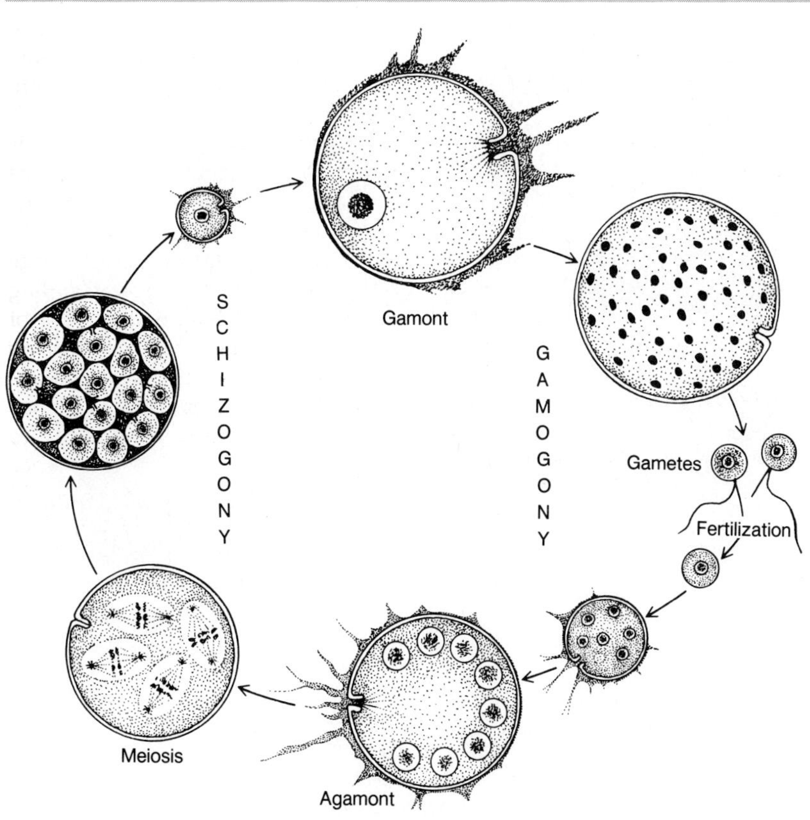

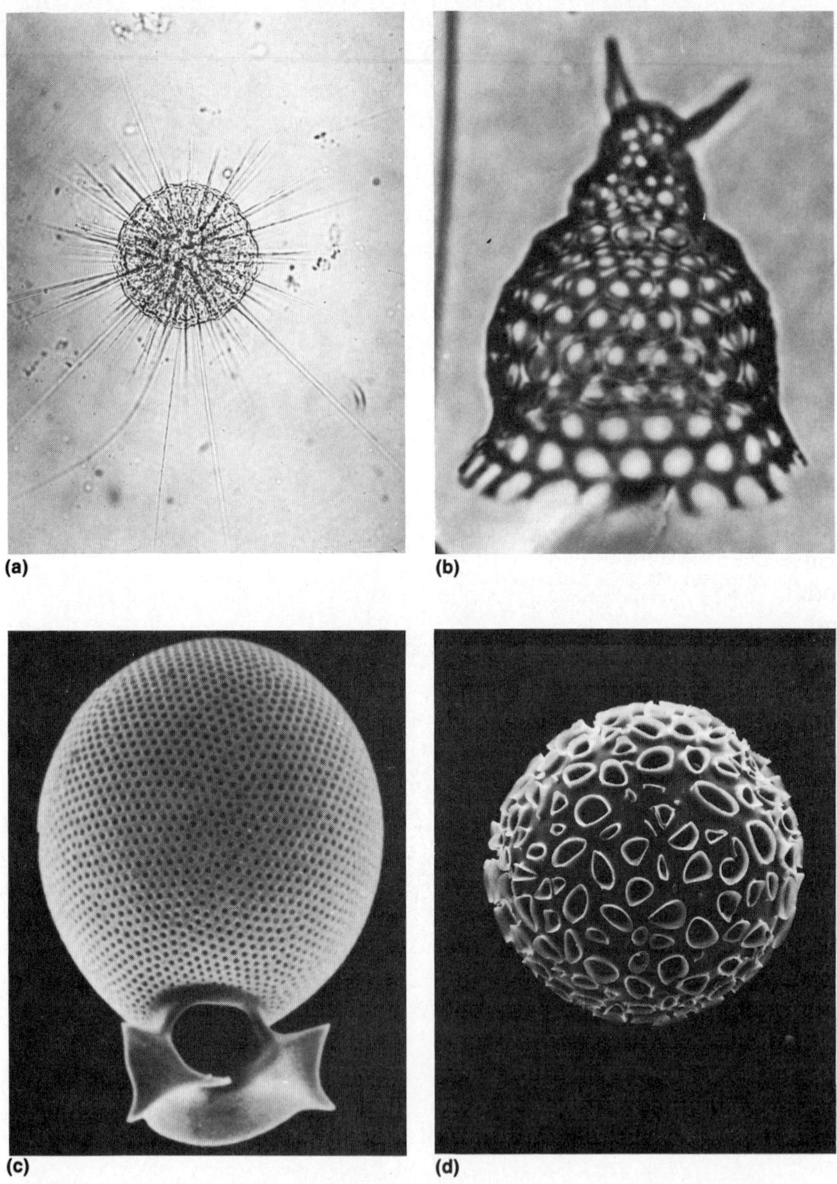

Figure 2.12 *Types of actinopodans:* ***(a)*** *photomicrograph of* Actinosphaerium, *a heliozoean (courtesy of Carolina Biological Supply Co.);* ***(b)*** *photomicrograph of the shell of* Theocalyptra bicornis, *a radiolarian;* ***(c)*** *scanning electron micrograph of the shell of* Protocystis murrayi, *a radiolarian (courtesy of K. Takahashi in O. R. Anderson,* Radiolaria, *Springer-Verlag, New York, 1983).* ***(d)*** *scanning electronmicrograph of the shell of* Siphonosphaera cyathina, *a radiolarian (parts b, d courtesy of O. R. Anderson,* Radiolaria, *Springer-Verlag, New York, 1983).*

waters at depths exceeding 4000 m; at greater depths there is an increased concentration of carbon dioxide, which in water forms carbonic acid that dissolves the calcareous tests. Fossil forams are of immense economic importance because their shells are good indicators of oil-bearing strata, and they are widely sought by stratigraphers and other geologists prospecting for oil. The tests of cold-water species are smaller and have fewer pores than warmer-water forms. Thus they are important to micro-paleontologists in determining the climates in past geological periods.

Superclass Actinopoda

Actinopodans are basically spherically symmetrical sarcodinans with numerous radiating axopodia. This superclass includes the heliozoeans and radiolarians.

The common name of heliozoeans, sun animalcules, comes from their spherical body with thin axopodia radiating in all directions (Fig. 2.12a). The axopodia contain granular cytoplasm, and the axial fibers arise from a centroplast rather than from the nuclear membrane. The outer cortical ectoplasm contains large numbers of vacuoles, which give it a "frothy" appearance, and one or several eccentrically located nuclei are situated in the medullary endoplasm. Many heliozoeans have a test composed of siliceous spicules or environmental particles embedded within a gelatinous matrix. Numerous shell pores permit the extension and contraction of axopodia. Heliozoeans are mostly freshwater forms and are planktonic or benthic. These sarcodinans are holozoic, and small prey such as other protozoans and rotifers adhere to the sticky axopodial surface and are ingested.

The radiolarians are entirely marine; some are benthic, but most are pelagic and planktonic. Their most striking feature is their intricately constructed test composed of silica spicules. The spicules in a radiolarian test are arranged in a wide variety of patterns and shapes such as spheres, lattices, spines, needles, and delicate plates that collectively form an absolutely exquisite skeleton (Fig,. 2.12 b–d). The diversity of shell patterns is enormous, and the shells are among the most beautiful of all natural structures. In some the shells are symmetric, but in others the spicules are tangentially arranged or fused to form irregular tests. The body of a radiolarian consists of an inner medullary region and an outer cortical region separated by a **central capsule**. The central capsule is perforated so that the cytoplasm is granular and contains vacuoles and one or more nuclei. Outside the central capsule, the outer layer of dense cytoplasm is surrounded by a cortical, highly vacuolated **calymma**. The skeleton or shell is internal because it is contained inside the calymma. The calymma contains, in addition to the siliceous skeleton, numerous vacuoles, and, in some forms, symbiotic zooxanthellae, which provide some nutrition for the radiolarian. The vacuoles of the calymma are thought to contribute in varying degrees to the buoyancy employed in vertical movements. The medullary cytoplasm is concerned with reproduction, while the calymma is concerned mostly with feeding. The axopodia (filopodia in some) originate in a central capsule of the calymma and extend outward, where they are involved in food capture and aid in flotation. Radiolarians are common in the upper strata of the oceans, and their shells are exceedingly common on the sea floor, where they form extensive radiolarian oozes. Chert is a type of rock composed of innumerable radiolarian skeletons compacted together.

The Class Acantharea contains organisms that resemble radiolarians and indeed are commonly referred to also as radiolarians. Like radiolarians, they are all marine and mostly pelagic, but they differ fundamentally from radiolarians in that their tests are composed of strontium sulfate spicules.

PHYLUM LABYRINTHOMORPHA

This new phylum was created for the taxonomic scheme used for protozoans. Individuals in this phylum have a feeding stage in which an ectoplasmic network is present composed of spherical, nonameboid cells. Some genera have ameboid cells that glide within the network. They have a unique cell surface organelle associated with the ectoplasmic network. Labyrinthomorphans are saprozoic and parasitic on algae in marine and estuarine ecosystems. Genera of this phylum include *Labyrinthula* and *Thraustochytrium*; as a group, they are not well known.

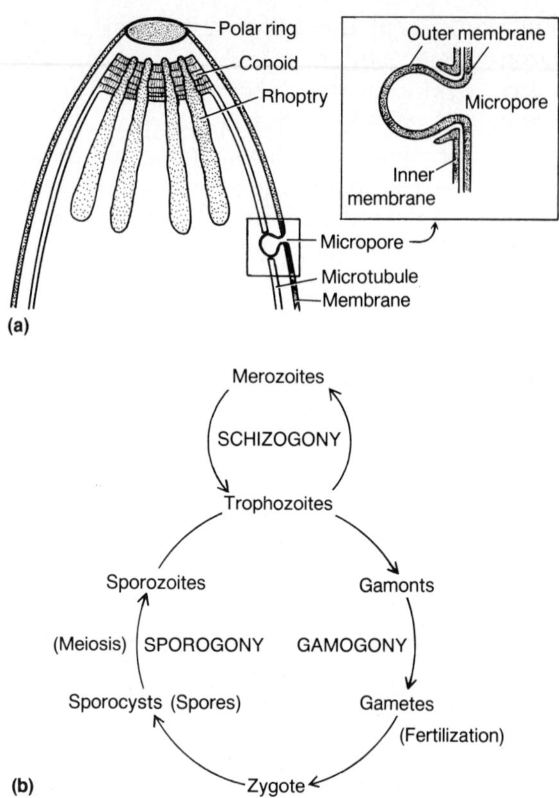

*Figure 2.13 Some sporozoean features: **(a)** a diagram of the apical end showing the polar organelle and a micropore; **(b)** a word diagram of a typical life cycle.*

PHYLUM APICOMPLEXA

This phylum is the largest of four phyla that were, at one time, considered to be the Phylum Sporozoa. The apicomplexans are endoparasites in many different invertebrates and vertebrates. Some live intracellularly, others are found in body cavities, and a few are blood parasites. The virulence with which they affect the host ranges from slight and characterized by an almost commensalistic relationship in which the host is basically unaffected to severely impairing and often killing the host. **Sporozoites** are elongate cellular stages that typically infect new cells or new hosts in the life cycle. The **apical complex**, a characteristic feature of at least one stage in the life cycle and usually the sporozoite, lies at one end and consists of a complex of rings, threads, bands, and sacs (Fig. 2.13a). The function of this organelle is not known, but it may be involved in the penetration of the host's cell. The universal presence of this apical complex is the basis for the phylum name. Apicomplexans also have **micropores**, which are small, short, dense cylinders located just beneath the outer cell membrane and may be sites for the extrusion of cellular materials or the ingestion of nutrients (Fig. 2.13a).

The basic distinction between the various subordinate apicomplexan taxa is found in details of their life cycles. There are some other subtle morphological distinctions, but these are secondary in importance to life cycles. Apicomplexans are subdivided into two classes, a preponderance of species belonging to the Class Sporozoea.

The apicomplexans can best be understood by their generalized life cycle. A typical life cycle is divisible into three distinct phases: **schizogony** or asexual reproduction following infection of the host, **gamogony** or gametogenesis, and **sporogony** or a meiotic (reduction division) multiplicative phase that forms infective sporozoites often within spores (Fig. 2.13b). The apicomplexan enters the host tissues as a **sporozoite**; once inside, the sporozoite enters a period of growth and is called a **trophozoite**. Trophozoites often undergo schizogony to produce many daughter cells or **merozoites**, which become trophozoites in the same or different host cells, or they may grow extracellularly. Eventually, some trophozoites undergo gamogony by transforming into **gamonts** and later in turn into **gametes**. Two gametes fuse, and the resulting **zygote** transforms into a **sporocyst**, which undergoes sporogony and then mitosis to produce a large number of infective sporozoites, each typically within the sporocyst.

There are three subordinate groups of sporozoeans: the gregarines, coccidians, and piroplasmians. We shall examine several life cycles for both illustrative and comparative purposes.

Monocystis (Subclass Gregarinia) is a common parasite in the seminal vesicles of earthworms. Earthworms ingest sporocysts containing sporozoites; in the intestine the sporozoites are released and migrate to the seminal vesicles. Each sporozoite enters one cell in a mass of developing sperm; as the trophozoite begins to grow, the host cell, unable to contain it, ruptures to free the trophozoite into the fluid-filled cavities of the seminal vesicle (Fig. 2.14). Pairs of large mature trophozoites associate together to form a **gamontocyst** in which each trophozoite is called a gamont, and the two gamonts secrete a common wall around themselves. Each gamont undergoes repeated nuclear divisions followed by cytoplasmic divisions to form

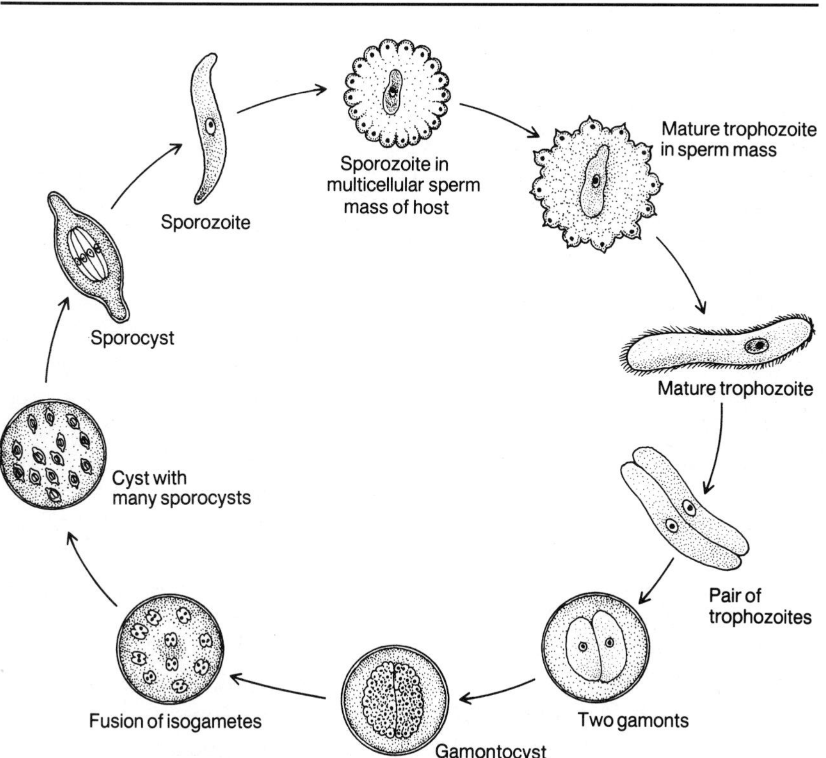

Figure 2.14 *The life cycle of* Monocystis, *a gregarine parasitic in earthworms (adapted with permission from T. C. Cheng,* General Parasitology, *Academic Press, New York, 1973.*

a number of uninucleate ameboid isogametes that fuse in pairs still within the gamontocyst. Each resulting zygote secretes a wall around itself and becomes a sporocyst. A diploid sporocyst undergoes sporogony (meiosis, then mitosis) to produce eight haploid sporozoites within each sporocyst. Presumably, sporocysts are released with the fecal material of the earthworm into the soil, where another earthworm ingest them. It is of interest to note that no schizogonic multiplication occurs in the life cycle of *Monocystis*.

Eimeria (Subclass Coccidia) is a large genus of sporozoeans whose individuals invade the epithelial cells lining the alimentary canal of a number of domesticated vertebrates. *Eimeria tenella,* representative of the coccidians, is found in chickens and causes a condition called coccidiosis, which often is fatal to the host. Sporocysts (often called oocysts) are ingested by the chicken, and the enclosed sporozoites escape through a minute pore in the sporocyst wall and into the intestinal mucosa (Fig. 2.15). Blood macrophages of the host engulf the sporozoites

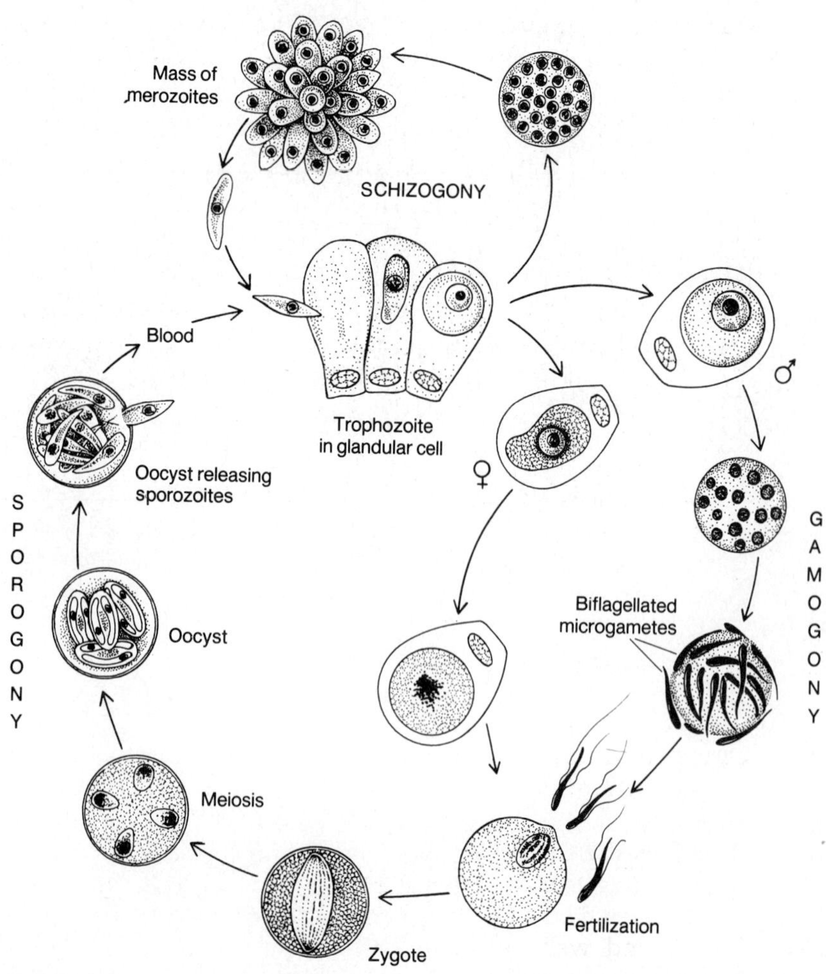

Figure 2.15 *The life cycle of* Eimeria tenella, *a coccidian parasitic in chickens (adapted from K. G. Grell,* Protozoology, *Springer-Verlag, New York, 1973).*

and transport them to certain glandular tissues of the intestine (glands of Lieberkühn), where each sporozoite invades a cell and becomes a trophozoite. Schizogony produces up to 900 small daughter cells or merozoites in each host cell. Merozoites escape from the host cell, invade other cells, become trophozoites, and repeat the schizogonic cycle. Eventually, some trophozoites develop into either male or female gamonts. The female gamont is slightly larger and develops into a single macrogamete, but the smaller male gamonts give rise to numerous biflagellated microgametes. Microgametes are released, find their way to macrogametes within a host cell, and fertilize them. Each resulting zygote secretes a cyst wall around itself and becomes an **oocyst**. Oocysts, voided in the feces of the host, mature outside the host. Each oocyst contains a single diploid cell that will undergo meiosis and mitosis to produce eight sporozoites, all within the oocyst. Ingestion of the oocyst by another host completes the life cycle.

Plasmodium (Subclass Coccidia) is a genus whose representatives are parasitic in various vertebrates including lizards, monkeys, and various amphibians, birds, rodents, and human beings. There are four species, *P. vivax*, *P. malariae*, *P. ovale*, and *P. falciparum*, which cause specific types of malaria in human beings, *P. vivax* being the most common species. Malaria is one of the most serious and debilitating parasitic conditions known; it is often fatal. In the 1940s there were an estimated 350 million cases of malaria, which resulted in approximately 3 million deaths annually. It was thus the Number 1 human disease. The incidence of malaria has decreased somewhat in the last three decades owing to extensive programs of mosquito abatement, better water drainage in swamps and marshes, and improvement of dwellings by adding screens to windows and doors. Even though this disease is now generally limited to Central and South America, Africa, and southern Asia, these are still the most populous areas of the world, and well over one-half of the world's population is potentially susceptible to the ravages of malaria.

Sporozoites are injected into the blood during the bite of a mosquito vector almost invariably belonging to the genus *Anopheles*. Since only female mosquitoes bite, they alone serve as vectors. The sporozoites are carried immediately by the blood plasma to the liver, where they infect hepatic cells and begin to grow (Fig. 2.16). After 5–15 days, each sporozoite undergoes schizogony there (exoerythrocytic) to produce numerous merozoites, which escape the host cell and invade erythrocytes. Once inside an erythrocyte, the single merozoite undergoes erythrocytic schizogony to form 6–24 new merozoites. The erythrocyte ruptures and releases the merozoites, and each merozoite enters another erythrocyte. This erythrocytic schizogonic cycle normally requires 48 hours to complete, but in *P. malariae* the schizogonic periodicity is 72 hours. A remarkable synchrony is achieved so that large numbers of erythrocytes burst and release their merozoites simultaneously at a periodicity of two (or three) days. When the erythrocytes burst, products of growth are released into circulation. Some of these materials are toxic and cause a characteristic regular fever that occurs predictably and with remarkable accuracy every 48 (or 72) hours. Typically, the onset of fever is in late afternoon on every second (or third) day. The rise in fever, accompanied by shivering, peaks at a temperature of about 40°C (104°F), then drops to subnormal levels as the patient perspires profusely.

Concomitant with the asexual or schizogonic cycle, some merozoites invade cells and do not multiply but become rounded gamonts as some differentiate to form **microgamonts** and the rest form **macrogamonts** (Fig. 2.16). The gamonts remain dormant within the erythrocytes until they are ingested by an anopheline mosquito. Once inside the alimentary canal of the mosquito, the microgamonts undergo a process called **exflagellation** during which each undergoes three nuclear divisions to produce eight flagellar projections each with a single nucleus.

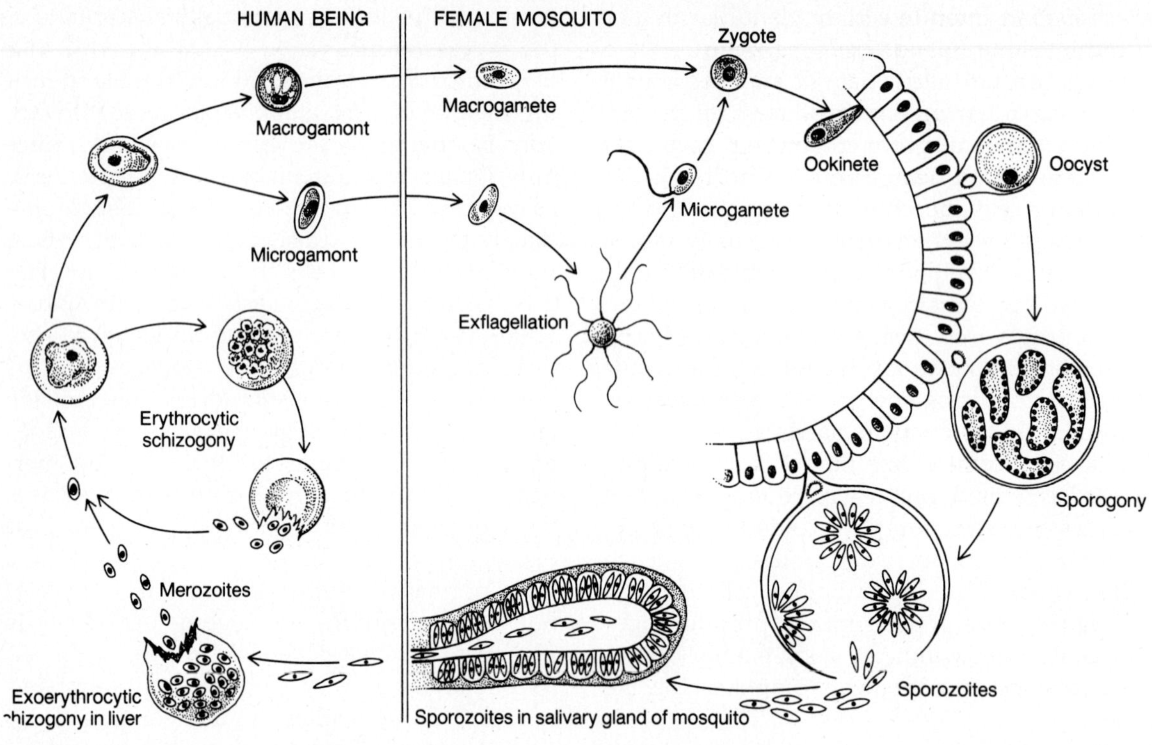

Figure 2.16 *The life cycle of* Plasmodium, *a hemosporian that causes malaria.*

They break away from each other, and each transforms into a **microgamete**. The macrogamont is unchanged and is called a **macrogamete**. Fertilization occurs in the insect's stomach; the zygote becomes elongate and active and is called an **ookinete**. The ookinete penetrates the stomach lining, secretes a cyst wall, and is termed an **oocyst**. The oocyst then undergoes sporogony to produce large numbers of sporozoites sometimes numbering over 10,000 per oocyst. As the oocyst ruptures and frees the sporozoites into the body cavity, they migrate to the salivary gland of the mosquito, where they remain until injected into another human host at the next feeding of the mosquito (Fig. 2.16).

The pathogenesis of malaria varies widely depending upon a number of characteristics of the host (age, sex, race, diet, etc.) and upon the species of *Plasmodium* involved. Three species (*P. vivax*, *P. malariae*, and *P. ovale*) are not usually fatal, but they can persist for years and cause debilitating recurrent fever. *Plasmodium falciparum*, on the other hand, induces a highly dangerous, often fatal condition. The presence of these coccidians in erythrocytes causes the red blood cells to stick together; in small capillaries the adhered erythrocytic mass often blocks circulation to tissues or causes the capillaries to rupture. Blockage to a large area of the brain usually results in death, and reduced blood supply to portions of other organs such as liver, lungs, or pancreas is almost as serious and is physically crippling.

Babesia (Subclass Piroplasmia) represents small parasites found in the erythrocytes of vertebrates; their intermediate hosts are ticks, and they are the causative organism for red-water or Texas fever in cattle (Fig. 2.17b). They differ from many other sporozoeans in that they do not form spores and do not produce pigments from metabolized hemoglobin. Introduced into a cow from the bite of a tick, sporozoites penetrate erythrocytes, where each forms 2–4 merozoites by schizogony. Eruption from the erythrocytes causes fever and the excretion of blood hemoglobin—hence the name, red-water fever. After another tick ingests a meal of infected blood, certain sporozoites transform into isogametes, fuse in pairs, and form zygotes. Each zygote forms a cyst wall within which are ultimately formed minute infective sporozoites, which are introduced into another cow as the tick feeds. A sexual phase like the one described above has not been established, however, for a majority of the piroplasmians. Since various species affect other mammals (dogs, horses, rodents, cats, and hogs), they are of considerable veterinary importance.

Figure 2.17 *Features of some small, miscellaneous, parasitic protozoans:* ***(a)*** *a sporozoite of* Toxoplasma, *a coccidian;* ***(b)*** *two sporozoites of* Babesia, *a piroplasmian, in a cow erythrocyte;* ***(c)*** *a microsporan spore;* ***(d)*** *three spores from various myxozoans (adapted from M. A. Sleigh,* The Biology of the Protozoa, *Contemporary Biology Series, Edward Arnold Publishers Ltd, London, 1973).*

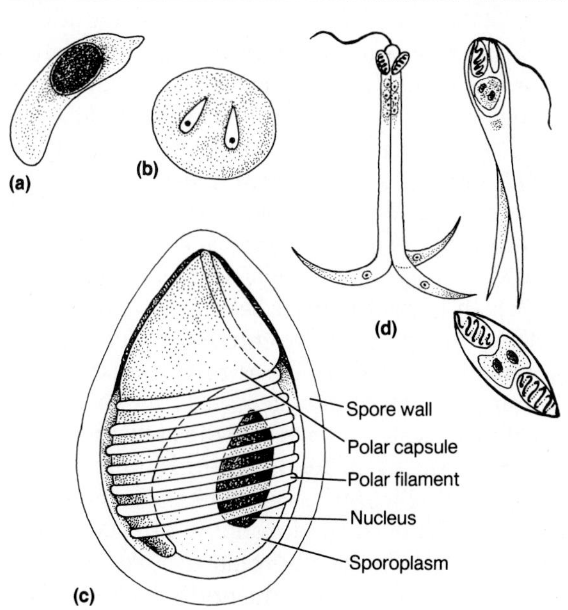

PHYLUM MICROSPORA

This phylum is represented by a small number of species, all of which are intracellular parasites mostly in invertebrates and fishes. Microsporans have a unicellular spore with a single sporoplasm. They also usually possess a single filament uncoiled within the polar capsule but extending backward from the capsule and coiling around the inside of the spore wall. The **polar capsule** and **polar filament** can be discharged and are used for penetration into the host cell and for anchorage (Fig. 2.17c).

Microsporans are especially common as parasites of insects and sometimes cause epidemic diseases. Some species infect honey bees and silkworms and are of considerable economic importance. *Nosema* is a microsporan that causes a destructive disease in honey bees. Spores are ingested by the insect, the polar filament is everted, and the sporoplasm enters an epithelial cell of the gut. Within the cell the parasite grows and undergoes multiplicative divisions so that the cell is destroyed. New spores infect other cells or pass out with the feces to infect new hosts.

PHYLUM ASCETOSPORA

This is a very small phylum with only a few poorly known species. These parasites infect invertebrates and some lower vertebrates. Ascetosporans form spores with one or more sporoplasms, but the spores lack a polar capsule and a polar filament. Two typical genera are *Urosporidium* and *Haplosporidium*.

PHYLUM MYXOZOA

Myxozoans are all extracellular parasites in the body cavities and tissues of some invertebrates and fishes. They are often found in the gall bladder, urinary bladder, liver, muscle, or cartilage of the host. Some myxozoan species are economically important, since they parasitize commercially important fishes. The parasite often reduces growth and ruins the flesh of the piscine host. The spores of myxozoans are unique in that each is a complex multicellular structure consisting of two or more valves, one or more sporoplasms, and several polar filaments coiled within the polar capsules (Fig. 2.17d).

Some myxozoan species like *Myxobolus* cause a "boil" disease in which large numbers of spores escape from ruptured surface cysts. Spores are ingested by the fish, the polar filaments evert, and the sporoplasms emerge and pass through the gut wall into body spaces. Either by migrating or by being carried by the blood, the sporoplasm reaches muscles and connective tissue of the body wall. After repeated nuclear divisions, a syncytial mass is formed, surrounded by a cyst wall formed from host tissues. Sporogony then follows; the spores mature and are released by rupture of the cyst.

PHYLUM CILIOPHORA

The ciliophorans (= cilia bearers) are the most highly organized and specialized of all protozoans, and it is in this group where the complexity of external shapes and the array of internal ultrastructures reach their zeniths. A majority of the 7500 species of ciliates, as they are commonly referred to, are free living and are common in most aquatic habitats. Ciliates are distinguished from all other protozoans by a number of adaptations, the most important of which are the presence of cilia, a complex pellicle and infraciliature, nuclear dimorphism, and their unique modes of reproduction.

Cilia

Certainly one of the most distinguishing features of ciliophorans is their cilia. However, one group, the suctorians, have cilia only in their early development and are aciliate as adults. Therefore ciliophorans all possess cilia at some stage in their life cycle. **Cilia** are short, slender, hairlike organelles that function basically in movement of the animal through the water. Cilia are quite similar structurally to flagella with the characteristic 9 + 2 pattern of microtubules, and some protozoologists even consider cilia as a specialized type of flagella. At the organellar level, however, distinctions between the two can usually be recognized, since cilia are typically shorter and more numerous than flagella. But there are exceptions to this generalization; thus their differences are not always sharp. Cilia normally range from 2 to 10 μm in length, and characteristically 200–1000 cilia are present in each ciliate, though some have up to 12,000 cilia and others have very few. Typically, cilia are arranged in longitudinal or spiral rows called **kineties**. In many ciliates, adjacent cilia combine together to form membranelles or cirri; in these ciliary complexes, cilia are usually functionally (and sometimes structurally) linked so as to operate as a single unit. A **membranelle** is composed of one or more kineties of closely positioned cilia that appear contiguous as a thin delicate membrane (Fig. 2.18c). In a great many ciliates, many membranelles are arranged into a rather prominent longitudinal row that is the **membranellar band**. Commonly, membranelles and membranellar bands are located adjacent to the cytostome and are used mostly in food gathering. The details of the oral ciliation and membranelles are important criteria in the taxonomy of ciliates (see Fig. 2.26 later in the chapter). A **cirrus** is a tuft of a few to more than a hundred cilia in the

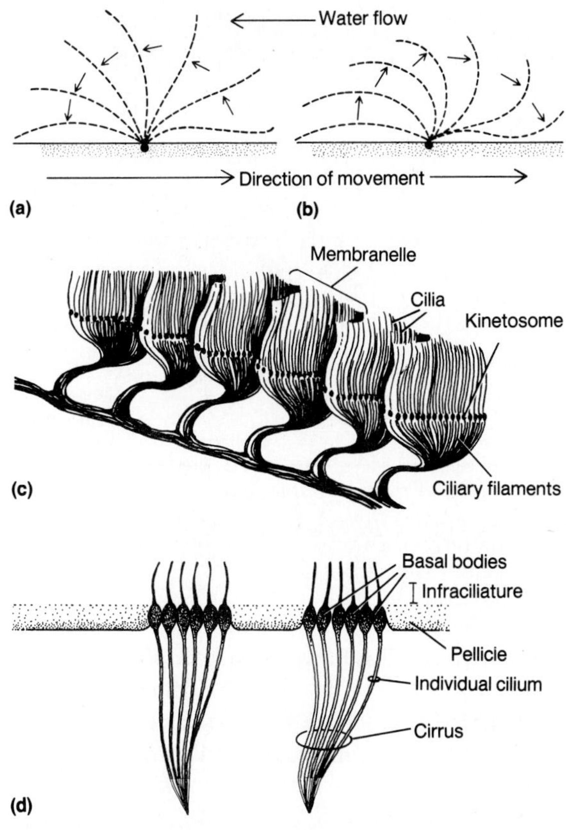

Figure 2.18 *Functional morphology of cilia:* ***(a)*** *stages in the effective stroke of a cilium;* ***(b)*** *stages in the recovery stroke of the same cilium;* ***(c)*** *a portion of a membranellar band with several individual membranelles;* ***(d)*** *a ventral cirrus.*

form of a thick, conical, compound organelle in which the constituent cilia, rather than bending metachronally, bend together as a single unit. Basically used in locomotion, cirri can function as limbs for walking or even jumping (Fig. 2.18d; also see Fig. 2.25c–e). Often, cirri are found on the ventral surface and are used as strong powerful levers against the substratum.

Cilia are organelles adapted for moving the fluid medium at right angles to the surface of the ciliary attachment. Of course, free-living forms cause the body to move in relation to the fluid medium, but in attached forms, cilia move water over the organism. Each cilium, acting as an oar, moves mostly in a single plane also at a right angle to its attachment surface. In its **effective stroke**, the relatively rigid cilium bends near its base and beats in a direction opposite that of the animal's path (Fig. 2.18a). At the completion of the effective stroke, the cilium then undergoes a **recovery stroke** in which it is held close to the animal's surface and is returned to its starting position (Fig. 2.18b). Recovery is effected by regions of bending being propagated up the ciliary shaft toward the tip. The effectiveness of the ciliary beat is dependent upon a number of factors, including the ciliary length, velocity of the beat, duration of the stroke, and viscosity of the medium. The duration of the recovery stroke in most ciliates is 2–6 times longer than that of the effective stroke.

A most important dimension of ciliary function is that cilia move in what appears to be a coordinated smooth rhythm. But rather than some intracellular nervous coordination as once was thought, the cilia are hydrodynamically linked, which results in their having a coordinated beat. This is best seen in membranelles where the ciliary bases are very close together so that there is a tight, viscous–mechanical coupling (= hydrodynamic linkage), which makes possible a synchronous beat. As a cilium in a kinety beats, its beat interferes with the water layers surrounding an adjacent cilium; this interference serves as a stimulus and causes it to beat, and so on down the line in a wave. This type of ciliary beating in which each later cilium is slightly out of phase with a former cilium is termed **metachronal beat**. If the bases of cilia in a kinety are farther apart and do not form a membranelle as do most somatic cilia, they still beat in coordinated fashion if there is sufficient hydrodynamic coupling between them. Smooth gliding locomotion by ciliary action is possible by the coordinated metachronal rhythm. Typically, the cilia in a kinety are slightly out of phase with those in the adjacent kinety; thus waves appear to pass over the surface metachronally in a direction opposite that of the effective stroke. Ciliary action is reversible and often is employed in avoidance reactions where, after encountering an obstacle, a ciliate backs up for a distance, then moves forward again at a slightly different angle, and repeats this action until the barrier is cleared. The reversal of ciliary beat usually takes place over the entire surface simultaneously—a truly remarkable feat when one considers the absence of any neural mechanism to achieve or coordinate such an action. Ciliary reversal is dependent upon complex biochemical and biophysical changes in the membrane, and calcium ions have been shown to be intimately involved.

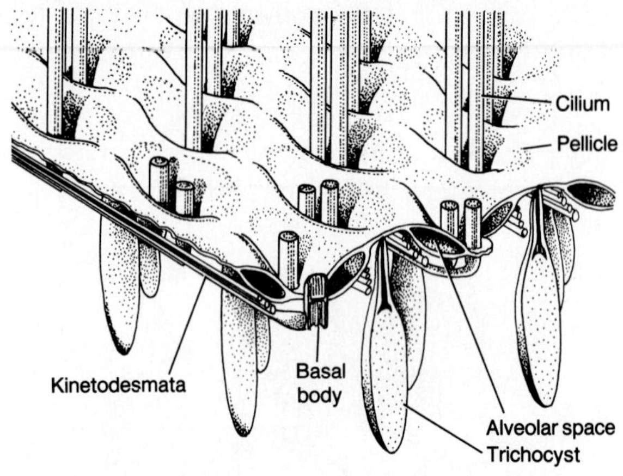

Figure 2.19 *A diagram of the pellicle and infraciliature of ciliates (adapted from A. Jurant and G. C. Selman,* The Anatomy of Paramecium Aurelia, *Macmillan, London and Basingstoke, 1969).*

Pellicle and Infraciliature

All ciliates have a complex series of surface membranes that collectively constitute the **pellicle**. The pellicle apparently affords the ciliate with considerable protection and support while, at the same time, permitting some flexibility to facilitate feeding. The pellicle of many ciliates, including that of *Paramecium*, consists of several membranes, the outer one of which is the plasmalemma and is continuous over each cilium. The surface of *Paramecium* is sculptured into series of regularly arranged, raised polygons. At the center of each polygon there is a shallow depression, in the middle of which arise one or two cilia. At the point where a polygon abuts another, the plasmalemma forms a ridge that produces the lattice or polygonal pattern. Just beneath each of the elevations is an **alveolus**, bounded laterally by an outer alveolar membrane and medially by one or two inner alveolar membranes (Fig. 2.19). Where the alveolar membranes join, a thickened membrane is formed surrounding the basal body of the cilium. Associated with the basal body are one or more cytoplasmic parasomal sacs whose function is not known. A basic feature of the outer regions of a ciliophoran is a complex arrangement of fibrils or microtubules called the **kinetodesmata**. These fibrils run in tracts parallel to the kineties and join basal bodies or join basal bodies to the pellicle (Fig. 2.19). The entire subpellicular complex or network, consisting of basal bodies, alveoli, and kinetodesmata, is called the **infraciliature**. The infraciliature is probably involved with anchoring the cilia or their basal bodies and is perhaps concerned with the ciliary beat, but it does not function as a neuromotor system. The infraciliature is a remarkable adaptation that is found universally in ciliates. Since it does not vary substantially in the various groups, it is a very basic, primal, ciliate feature.

Associated in the same outer layer of ectoplasm as the infraciliature, but functionally not a part of it, are various extrusive organelles, the most familiar of which are trichocysts. **Trichocysts,** characteristic of many ciliates, are rodlike or oval organelles arranged at right angles to the pellicular surface (Fig. 2.19). They may be either uniformily distributed over the surface or restricted to certain regions; some are toxic and others nontoxic. When appropriately stimulated, the trichocysts are discharged explosively through a preformed pellicular pore and into the water. As they discharge, they evert, increase about 10 times in length, and appear as very fine, often barbed threads with striated shafts (Fig. 2.20). Their discharge can be triggered by mechanical, electrical, or chemical stimulations usually in a few milliseconds. Trichocysts, derived from cytoplasmic vesicles, are anchored in the pellicle in part by the fibrils of the infraciliature. They can be quickly regenerated, but they are not reused after being discharged. Trichocysts are employed by the ciliate in defense, in food capture, and for achieving temporary anchorage while feeding.

Nuclear Dimorphism

With only a very few exceptions, all ciliates contain two very different kinds of nuclei, a condition that is often referred to as nuclear dimorphism. The presence of two nuclei in a given cell, with a pronounced division of functions, is an adaptation unique to ciliates. One type of nucleus, the **macronucleus,** is large and normally polyploid. Rapid replication of its chromosomes often provides it with 20–40 sets of chromosomes or genomes, and some forms may have as many as 13,000 genomes. The macronuclear chromatin material is often confined to numerous small, dense bodies, and this organelle contains nucleoli and large amounts of RNA. The macronucleus is the somatic or vegetative nucleus and regulates most somatic functions like locomotion, feeding, and osmoregulation. One of the its chief functions is in the production of RNA, which directs protein synthesis in the ribosomes. The other nuclear type, the **micronucleus**, is small, compact, and normally diploid, contains large amounts of histones and DNA, undergoes mitosis and meiosis, and usually lacks RNA and nucleoli. It functions in DNA synthesis, which is of paramount importance in reproduction. Therefore the micronucleus is the reproductive nucleus, and its genes are the ones passed along to the progeny cells. Some variations, however, occur in this nuclear dualism. Some ciliates may have many micronuclei, and a few have more than one macronucleus. In some the macronucleus is normally diploid, and some viable strains are known in which organisms lack a micronucleus completely.

Figure 2.20 *A scanning electron micrograph of discharged trichocysts each with a barb at the distal end of a long striated shaft (courtesy of M. A. Jacus and C. E. Hall,* Biol. Bull. *91: 141, 1946).*

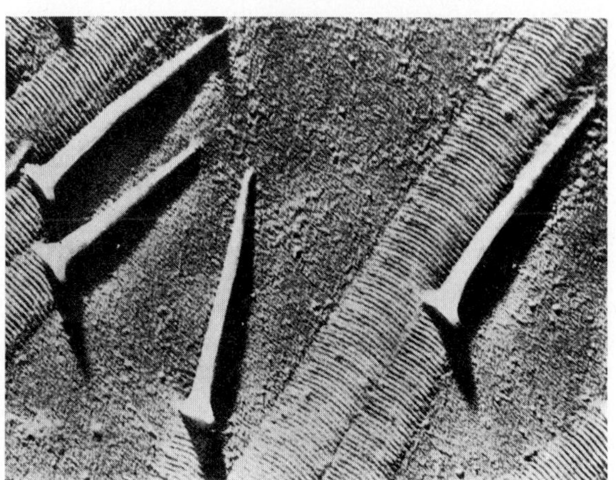

Patterns of Reproduction

Asexual reproduction in ciliophorans is by transverse binary fission, a distinctive feature of ciliates. In transverse fission the plane of division is perpendicular to the long axis of the body and to the longitudinal kineties. In ciliate fission the micronucleus divides mitotically, while the macronucleus divides amitotically.

Almost all ciliates undergo a peculiarly characteristic rejuvenative process called **conjugation** in which both conjugating ciliates swap micronuclear material. Two morphologically identical and sexually compatible ciliates randomly come in contact and adhere to each other

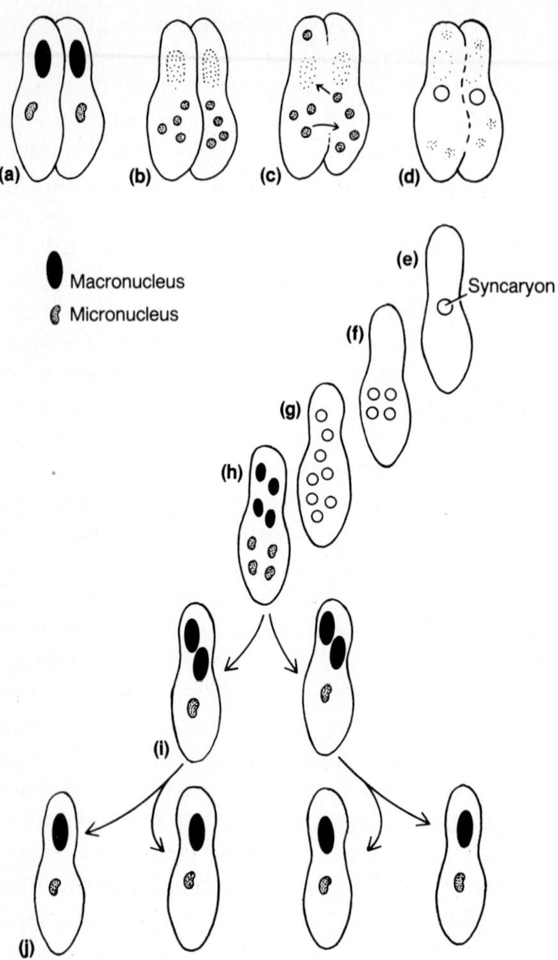

Figure 2.21 *Reproduction in* Paramecium caudatum *(a–d, conjugation; f–g, mitotic nuclear division; h–j, binary fission):* ***(a)*** *two conjugants adhere;* ***(b)*** *the micronucleus in each conjugant divides meiotically;* ***(c)*** *migratory nucleus from each conjugant crosses cytoplasmic bridge into other conjugant;* ***(d)*** *fusion of micronuclei to form a syncaryon in both conjugants;* ***(e)*** *an exconjugant;* ***(f)*** *the syncaryon divides twice to form four nuclei;* ***(g)*** *each of the four nuclei in (f) divides again;* ***(h)*** *four nuclei form macronuclei, one is retained as the micronucleus, the other three disintegrate;* ***(i)*** *daughter cells after the first binary fission;* ***(j)*** *daughter cells after the second binary fission restores the proper nuclear condition.*

at the oral region; adhesion is often augmented by a substance produced by the oral cilia (Fig. 2.21a). At these contact points the pellicles of both individuals dissolve, and a cytoplasmic bridge is formed. The conjugants adhere together for several hours while an absolutely remarkable process takes place. Let's follow the sequence of events in the conjugative process in *Paramecium caudatum*, a common familiar ciliate.

Paramecium caudatum has one macronucleus and one micronucleus. After the conjugants are adhered, the diploid micronucleus in each conjugant undergoes meiosis to produce four haploid micronuclei of which three disintegrate (Fig. 2.21b). The one remaining micronucleus divides mitotically to produce two equal-sized **pronuclei**, one of which remains in position and is called the **stationary** (female) **nucleus**, and the other is termed the **migratory** (male) **nucleus**. The migratory nucleus of each conjugant moves to the oral region, across the cytoplasmic bridge, and to the stationary nucleus of the other conjugant (Fig. 2.21c). The forces of movement of the migratory pronuclei are unclear but may be due to subtle influences of the plasma membranes. At this point the migratory and stationary nuclei fuse to form a **syncaryon** or zygote nucleus (Fig. 2.21d). During this time the macronucleus becomes pycnotic (disintegrates) and plays no role in conjugation.

Shortly after syncaryotic formation, the animals separate, and each is then called an **exconjugant** (Fig. 2.21e). In *P. caudatum* the syncaryon of an exconjugant divides three times mitotically to produce eight diploid nuclei of which four are destined to form the new macronuclei. Three other nuclei disintegrate, and their constituent compounds are resorbed by the cytoplasm (Fig. 2.21f–h). The lone remaining nucleus becomes the micronucleus in each exconjugant. Conjugation is usually followed shortly by two binary fissions in which the four macronuclei are sorted into separate cells (Fig. 2.21i, j).

The patterns by which both macronucleus and micronucleus are restored in the exconjugants vary in different ciliate species. The syncaryon may divide only once to produce two nuclei, one of which becomes the macronucleus and the other the micronucleus; but in others, two syncaryotic divisions produce four nuclei, which, in various combinations, become both kinds of nuclei. Some forms like *Vorticella* are dioecious in that one of the conjugants is smaller and contributes only a migratory nucleus, while the other larger individual forms only a stationary nucleus. The enucleate cell either dies or is absorbed by the other cell.

Some ciliates engage in **autogamy**, which is similar to the nuclear divisions and disintegrations described above, but the ciliates do not conjugate. Rather, following meiosis, the two haploid nuclei fuse with each other to form an autogamous syncaryon. This process of self-fertilization obviously does not promote genetic variability, but it does serve to enhance the vitality of the organism.

It has been known for years that only certain individuals of a given species of *Paramecium* can conjugate with another individual. Much research has elucidated that species of *Paramecium* are subdivided into complex varieties or syngens and mating types. Members of one **syngen** (= same gene pool) do not normally conjugate with members of another syngen; each syngen is therefore a genetically isolated species. Within each syngen there are two to several **mating types** so conjugation can only take place between individuals of different mating types but within the same syngen. Mating-type specificity is due primarily to characteristics of the cell membrane, but certain conjugative combinations are lethal, a feature suggesting that nuclear compatibility is also important. Some species other than *Paramecium* also have specific mating types and syngens. It is important to note here that while conjugation involves only the micronucleus, control of the mating type resides in the macronucleus.

What are the adaptive significances of conjugation? First, conjugation demands that there be cross-fertilization between mating types, and it is a method of genetic recombination that results in greater variability—the raw material on which natural selection operates. Although no new individuals result directly from conjugation, it is the one mechanism ciliates have of ensuring genetic variability. Second, conjugation is a rejuvenating phenomenon, since numerous binary fissions follow conjugation. Clones of many ciliates, wherein no conjugation can take place because of the absence of another mating type, undergo senescence and finally reach a state of genetic death. Conjugation therefore becomes an absolutely essential process for the future success of the individual ciliate in leaving genetically variable progeny.

Ciliophoran Ecology

Ciliates are found in all types of aquatic and damp habitats, including damp soils and even forest litter. They have wide geographical and climatic distributions due to both their adaptability and the fact that many form resistant cysts. Perhaps the single most important ecological factor controlling their distribution is the availability of food. Most ciliates are holozoic and feed on bacteria, other protozoans, algae, rotifers, and other minute invertebrates. Typically, ciliates are solitary and free living, but many are sessile or sedentary, and some are colonial. Only a few are parasitic, and only one genus, *Balantidium*, is parasitic in human beings. Most are relatively large; some may be up to 3 or 4 mm long. Some construct tests or loricae from arenaceous particles cemented together.

Class Kinetofragminophorea

Representatives of this class all have patches or fragments of short cilia called **kinetofragments** located in the general area of the cytostome, a

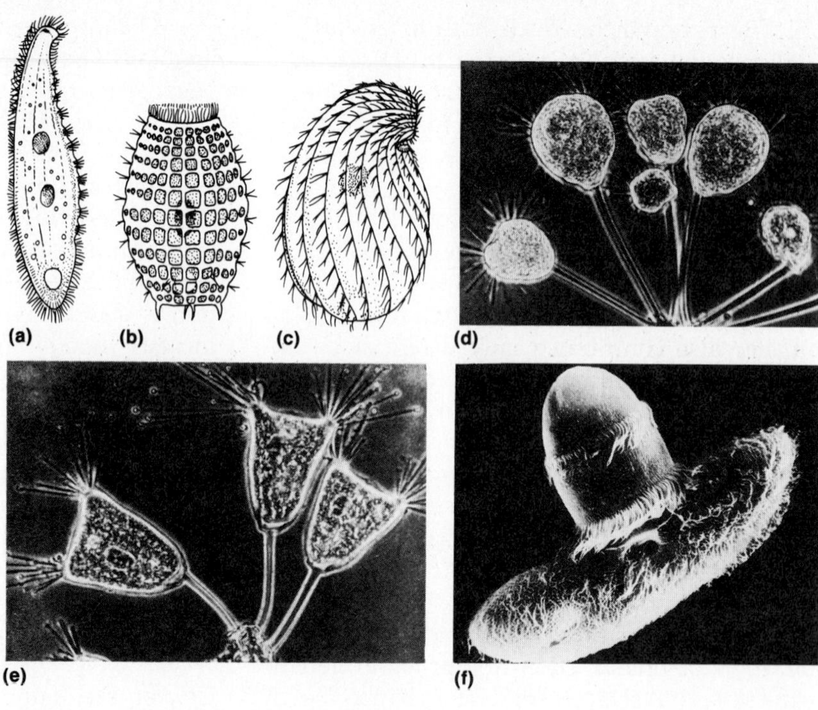

Figure 2.22 *Types of kinetofragminophoreans:* ***(a)*** Loxophyllum, *a gymnostomatian;* ***(b)*** Coleps, *a gymnostomatian;* ***(c)*** Colpoda, *a vestibuliferan;* ***(d)*** *a photomicrograph of* Tokophyra, *a suctorian;* ***(e)*** *a photomicrograph of* Acineta, *a suctorian;* ***(f)*** *a scanning electron micrograph of* Didinium, *a gymnostomatian, feeding on* Paramecium, *an oligohymenophorean (parts a–c from M. A. Sleigh,* The Biology of the Protozoa, *Contemporary Biology Series, Edward Arnold Publishers Ltd., London, 1973; parts d–e from K. G. Grell,* Protozoology, *Springer-Verlag, New York, 1973; part f from H. Wessenberg and G. A. Antipa,* Protozoology *17: 250, 1970, used with permission).*

Figure 2.23 *Role of the haptocysts in feeding in suctorians:* ***(a)*** *a haptocyst;* ***(b)*** *position of haptocysts on a tentacle;* ***(c)*** *haptocysts attach prey to tentacular tip;* ***(d)*** *prey contents sucked into the suctorian tentacle (parts a-d adapted from Rudzinska, Bardele, and Grell [1967], in M. A. Sleigh,* The Biology of the Protozoa, *Contemporary Biology Series, Edward Arnold Publishers Ltd., London, 1973).*

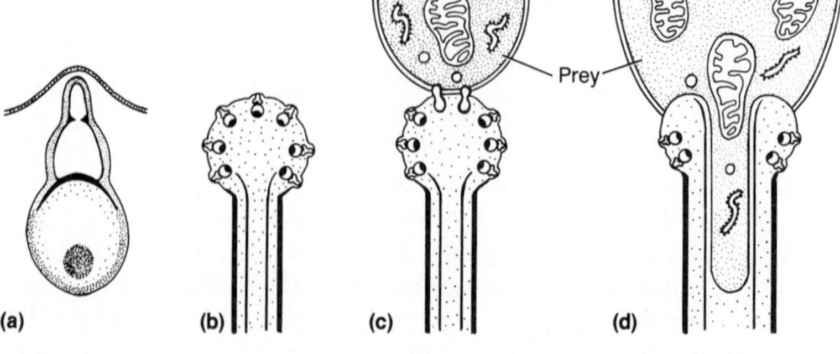

feature from which the class name is derived. Their scattered oral cilia may represent the very primitive stages of development of the buccal apparatus of higher ciliates. To this class belong the most primitive of ciliates, which do not possess distinct buccal ciliature and compound ciliature organelles (membranes, cirri) characteristic of higher ciliates (Fig. 2.22).

The suctorians deserve special mention for they are a puzzling and enigmatic group of kinetofragminophoreans whose phylogenetic affinities with other ciliates have been debated over the years. The suctorians are noteworthy, since as adults they lack cilia altogether, even though cilia are present in an immature stage; their infraciliature even persists in adults. They are normally sessile as adults and attached by means of a stalk (Fig. 2.22d, e). The body is equipped with several to many adhesive tentacles, which are supported internally by microtubules and equipped with special anchoring organelles, the **haptocysts** (Fig. 2.23). Their prey usually consists of ciliates that are caught by their tentacles, and the haptocysts discharge into the prey to hold it against their tentacles. The tentacles are used as sucking organelles through which nutrients are drawn to form vacuoles (Fig. 2.23c, d). Suctorians engage in a modified form of fission called budding by which ciliated larvae are formed. The presence of cilia and dimorphic nuclei clearly qualify them as ciliates, but their hierarchial position within the phylum is still contested by some protozoologists.

Class Oligohymenophorea

Representatives of this class are intermediate in levels of organizational complexity between the primitive kinetofragminophoreans and the advanced polymenophoreans. These ciliates possess a true buccal cavity surrounded by a few compound buccal organelles distinct from the uniformly arranged, somatic ciliature (oligohymenophora = bears few membranelles) (Fig. 2.24). To this large class belong the most familiar of all ciliates, including *Tetrahymena* and *Paramecium* (Fig. 2.24b, c). Extensive investigations of these two species and of other oligohymenophoreans have elucidated much important information on a wide spectrum of biological processes.

This taxon also includes the Subclass Peritrichia, which contains perhaps the largest number of ciliate species. Most peritrichians are sessile and live as ectocommensals on plants and animals, to which they attach by a **stalk**. The peritrichian body resembles an inverted bell with an expanded oral end uppermost. This disklike anterior end bears several well-developed oral membranelles and cilia arranged in rows that spiral counterclockwise when viewed from the anterior end. The cytostome lies at the basal end of a shallow-to-deep buccal cavity, the **infundibulum** (Fig. 2.24d, e; also see Fig. 2.26g). The cilia and membranelles sweep mostly bacteria into the infundibulum and into the cytostome. Perhaps the best-known and best-studied peritrichian is *Vorticella*, a common, solitary, attached form (Fig. 2.24d). *Vorticella* has endoplasmic myonemes in its stalk that enable the stalk to contract rapidly. Almost all events in living organisms require energy, which is derived from the breakdown of ATP; interestingly, the contraction of the stalk myonemes in *Vorticella* appears to be excited by calcium ions and does not involve the release of energy from ATP.

Class Polymenophorea

The polymenophoreans or spirotrichians represent the most advanced ciliates. They are all characterized by the presence of a conspicuous adoral (along one side of the cytostome) region of many membranelles (polymenophora = bearer of many membranelles) (Fig. 2.25a–d). The numerous adoral membranelles spiral

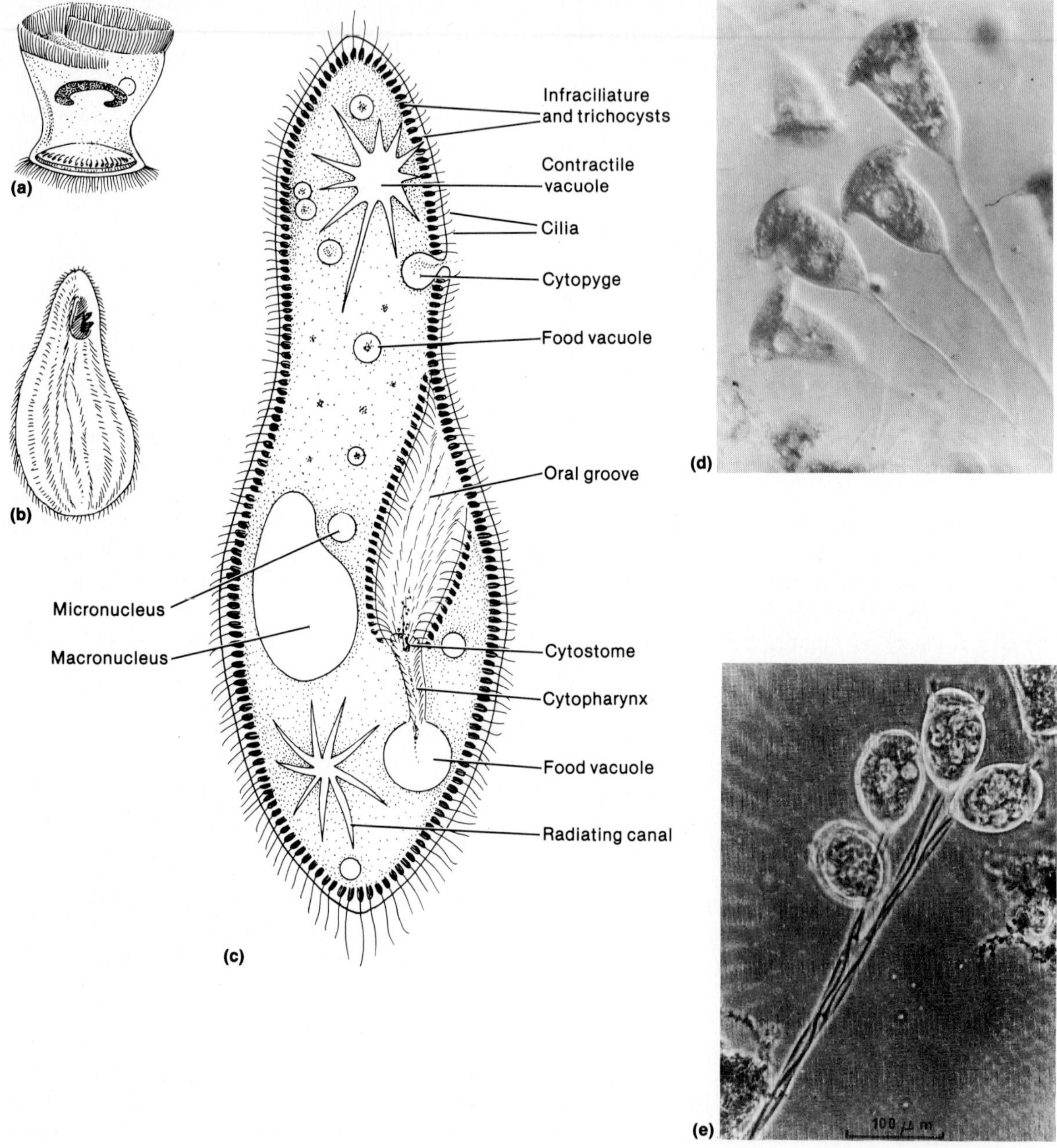

Figure 2.24 *Types of oligohymenophoreans:* ***(a)*** Trichodina, *a peritrich;* ***(b)*** Tetrahymena, *a hymenostomatian;* ***(c)*** Paramecium, *a hymenostomatian;* ***(d)*** *a photomicrograph of* Vorticella, *a peritrich (courtesy of Ward's Natural Science, Inc., Rochester, N. Y.);* ***(e)*** *a photomicrograph of* Carchesium, *a peritrich (parts a and e from M. A. Sleigh,* The Biology of Protozoa, *Contemporary Biology Series, Edward Arnold Publishers, Ltd., London, 1973).*

clockwise down to the cytostome. Some species have uniform somatic ciliation, while others possess unique locomotive cirri. Polymenophoreans usually have a large body size and are mostly free-living swimmers.

Figure 2.25 *Types of polymenophoreans:* ***(a)*** Spirostomum, *a heterotrichid;* ***(b)*** *a sessile* Stentor, *a heterotrichid;* ***(c)*** Urostyla, *a hypotrichid;* ***(d)*** Euplotes, *a hypotrichid;* ***(e)*** Stylonychia, *a hypotrichid, lateral view (parts a-d from M. A. Sleigh,* The Biology of Protozoa, *Contemporary Biology Series, Edward Arnold Publishers Ltd., London, 1973).*

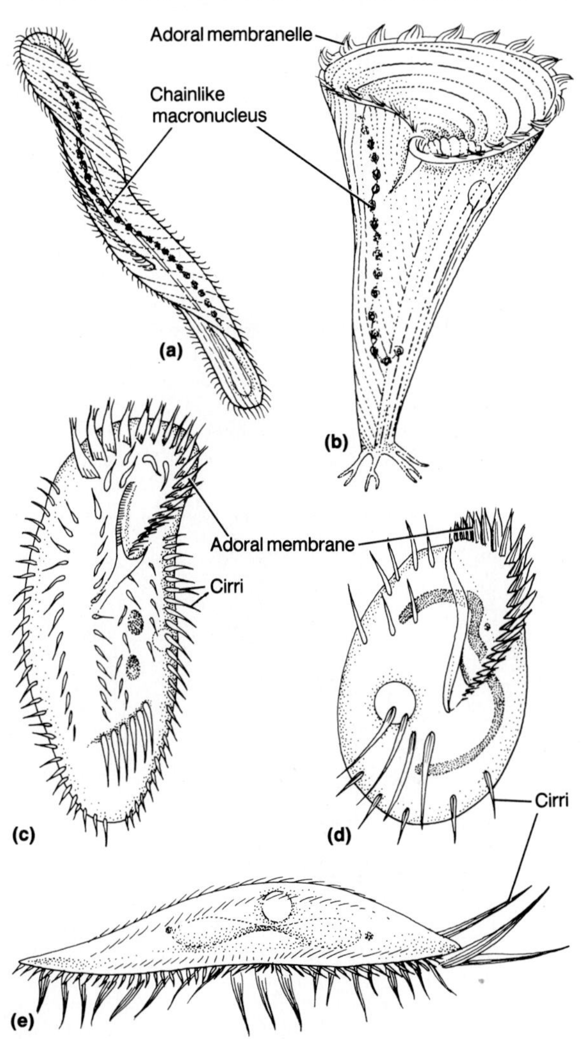

Representatives of the Order Heterotrichida, the least specialized of the polymenophoreans, all have uniform simple ciliation over the body in addition to the oral and adoral membranelles. This order contains many commensals found on or in invertebrates and vertebrates. Individuals of some species are quite large and visible to the unaided eye. *Stentor* is a large (1–2 mm) trumpet-shaped heterotrich that has a prominent row of oral membranelles leading to the cytostome (Fig. 2.25b). It is characterized by having a macronucleus in the form of a series of beads, a spiral cytopharynx, and, in some species, dispersed cytoplasmic pigments that give them a bluish hue. *Stentor* also has endoplasmic myonemes that enable it to contract or retract abruptly. *Blepharisma* contains intracellular pigments that give its individuals a pink color. *Spirostomum* is cylindrical and elongated and may be up to 3.5 mm long (Fig. 2.25a). *Nyctotherus* is commonly found as an endocommensal in the alimentary tract of frogs and certain insects like cockroaches.

The Order Hypotrichida includes the most complex of all ciliates. All are dorsoventrally flattened with few or no dorsal cilia. The ventral cilia are formed into complex **cirri** that are arranged in rows, patches, or scattered patterns (Fig. 2.25c–e). The cirri are unique in that they act as levers for the hypotrich to jump suddenly, and often spontaneously, backward through the water, possibly as an avoiding reaction. The cytostome is apically situated at the anterior end and surrounded by well-developed membranelles. Commonly encountered genera include *Stylonychia*, which is ovoid and has three posterior cirri (Fig. 2.25e); *Kerona*, an ectocommensal on hydras; and *Aspidisca*, with small oral membranelles and anterior and posterior clumps of cirri.

Evolution of Oral Organelles in Ciliophorans

Details of the oral or buccal region of ciliates are of great interest to ciliatologists, since the arrangement of buccal organelles is thought to be of great adaptive significance among the ciliate taxa. The position of the mouth and the arrangement of buccal ciliature and membranelles are of considerable phylogenetic importance, and they are the principal bases on which much of ciliate systematics and phylogeny are based.

Unlike the situation of somatic cilia, oral cilia differ in several important ways: their infraciliatural fibrils often extend deep into the endoplasm; additional rows of microfibrils are found near the cytostome; complex membranelles are often present; and, in some, the cytostome may be recessed at the inner end of a buccal cavity. The probable stages in the evolution of the buccal cavity are of particular interest, since this depression, an important adaptation in higher ciliates, functions as the external center for the aggregation of food particles prior to ingestion. A more efficient food-gathering depression would obviously be a decided asset, and evolution has undoubtedly favored those ciliates with the best-organized buccal organelles. An interesting dimension is that amid the rather extensive diversity found in buccal structures, there is considerable consistency within many subordinate groups.

In the simplest ciliates, representing the Class Kinetofragminophorea, the mouth is always apical or subapical (Fig. 2.26a–d). The complexity of oral structures ranges from forms with no compound cilia to others in which the cytostome lies in a depression, the **vestibule**, into which extend kineties of cilia; however,

Figure 2.26 *Probable phylogenetic stages in the evolution of oral organelles in ciliates:* ***(a)*** *cytostome simple and apical (gymnostomatians);* ***(b)*** *apical cytostome at base of vestibule (vestibuliferians);* ***(c)*** *cytostome at base of deeper subapical vestibule (vestibuliferians);* ***(d)*** *ventral cytostome within shallow atrium (hypostomatians);* ***(e)*** *shallow buccal cavity often with prebuccal oral groove (hymenostomatians);* ***(f)*** *deeper buccal cavity with prebuccal atrium (hymenostomatians);* ***(g)*** *specialized infundibulum (peritrichians);* ***(h)*** *conspicuous peristomial field and prominent buccal cavity or peristome (spirotrichians) (parts a–h adapted with permission from J. O. Corliss,* The Ciliated Protozoa, *2nd ed., Pergamon Press Ltd., Oxford, 1979).*

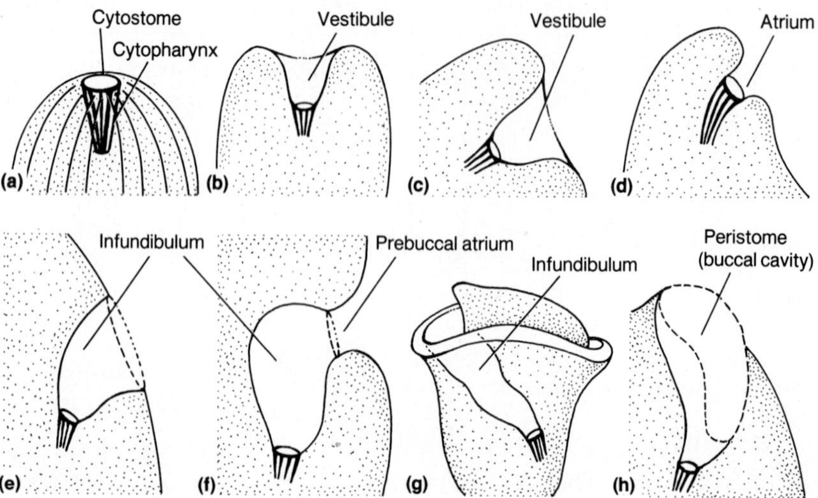

these vestibular cilia are never formed into membranelles.

The Class Oligohymenophorea, an intermediate group with respect to oral ciliation, includes the ciliates that have oral cilia distinctively different from the somatic cilia. The mouth typically is ventral in position and located inside a buccal cavity or **infundibulum** (Fig. 2.26e–g). The infundibulum, which ranges from shallow to deep, is lined by several membranelles, which are usually inconspicuous.

Representatives of the Class Polymenophorea, often referred to as spirotrichians, are the most advanced of all ciliates. Polymenophoreans are unique among ciliates in having a prominent buccal cavity or **peristome** (Fig. 2.26h) and a conspicuous adoral zone of buccal membranelles, reduced somatic cilia, and the presence of cirri. The oral structures are relatively constant among the various polymenophorean taxa.

PHYLOGENY OF THE PROTOZOANS

It is generally believed that the first forms of life were procaryotes lacking a nuclear membrane, endoplasmic reticulum, and other membrane-bounded cytoplasmic organelles. From these relatively simple procaryotic beginnings, eucaryotes arose whose individual cells were much more highly organized. These early primitive eucaryotes stand at the level of evolutionary bifurcation of plants and animals. At this pivotal point in the evolution of all higher organisms stand the flagellates, and from them were derived all other animals and plants (Fig. 2.27). There is little paleontological evidence to support this concept, since by the time the shells or tests of radiolarians and foraminifera appear in the fossil record, each was a well-developed and distinct group. Therefore consideration of protozoan phylogeny has had to depend mostly upon morphological, physiological, and developmental observations of present-day forms. The only logical protozoan group to be considered as the most primitive is clearly the mastigophorans. This fundamental assumption is based primarily on the almost universal appearance of a flagellated stage at some point in the life cycle of many plants and all animals (sperm in metazoans) plus the almost ubiquitous occurrence of the 9 + 2 pattern of microtubules. These two flagellar details are exceedingly strong evidence supporting this theory. Flagellates are also remarkably plastic from both a functional and a structural standpoint, are easily adaptable to various types of nutrition (autotrophic, saprobic, holozoic), have various forms of locomotion (flagella, pseudopodia), and are quite variable in the habitats in which they live. Their adaptability is a second strong piece of evidence supporting their antiquity and their primal role as the ancestral stock from which all higher organisms arose. It is now generally held that all metazoans arose from a colonial flagellate ancestor, a concept that will be developed more fully in Chapter 3. Even though some biologists believe that metazoans were derived from some type of multinucleate ciliate, the preponderance of opinion today supports the flagellate origin of all other animals.

It is now generally considered that the most primitive flagellates were simple autotrophic phytomastigophoreans (green algae). From these green ancestral flagellates arose the seven protozoan phyla independently of each other (Fig. 2.27). In fact, there is some evidence that certain groups within the Phylum Sarcomastigophora may have arisen independently themselves. The "sporozoan" phyla (Apicomplexa, Microspora, Ascetospora, Myxospora) also appear to be relatively closely related on the basis of their parasitic habits, their lack of mobility, and the presence of spores in the life cycles of most forms. Nonetheless, the protozoans as a group must definitely be considered to be polyphyletic (Fig. 2.27).

In a survey text such as this it is impossible to explore in detail the phylogeny of each protozoan phylum. In brief, in most of the protozoan phyla, none of the principal subordinate

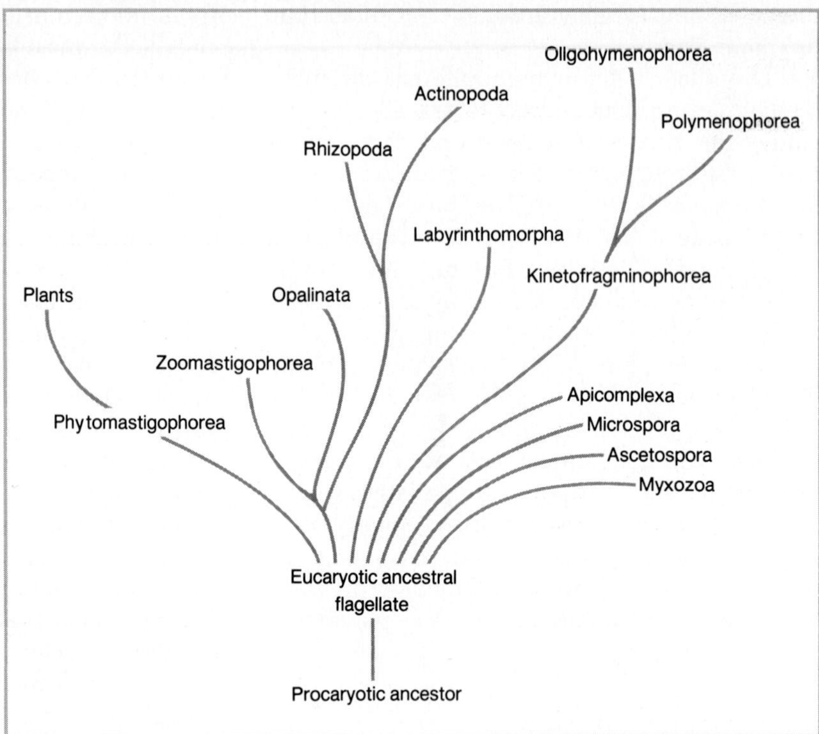

Figure 2.27 *Probable phylogenetic relationships of the protozoans.*

groups is decidedly more advanced than the others. In all cases, each subordinate group, representing a separate evolutionary line, contains some primitive forms as well as more advanced species. In the Phylum Ciliophora, the kinetofragminophoreans are clearly primitive, and from them arose the oligohymenophoreans, which are intermediate in complexity, and the highly advanced polymenophoreans (Fig. 2.27).

SYSTEMATIC RESUME

The Principal Protozoan Taxa

PHYLUM SARCOMASTIGOPHORA

Subphylum Mastigophora—locomotion by one or more flagella; reproduce asexually by binary fission; sexual reproduction known in certain groups.

Class Zoomastigophorea—animallike flagellates; one to many flagella; chloroplasts absent; some are freeliving, but most are symbiotic or parasitic.

Order Choanoflagellida—free-living, sessile, aquatic forms; have one flagellum surrounded by delicate cytoplasmic collar; some are colonial, others with loricae, includes *Codosiga* (Fig. 2.6a) and *Proterospongia*.

Order Kinetoplastida—free living or parasitic; small, with one or two flagella; body is flexible but not ameboid; many species are polymorphic; parasitic species include *Leishmania, Trypanosoma* (Fig. 2.6b), and *Leptomonas*.

Order Proteromonadida—one or two pairs of flagella of unequal length; one mitochondrion curved around nucleus; forms cysts; all parasitic; includes *Karatomorpha* and *Proteromonas*.

Order Retortamonadida—parasitic in alimentary canal of vertebrates; bear two or four flagella, one turned posteriorly and associated with cytostomal area; the better-known genera include *Chilomastix* and *Retortamonas*.

Order Diplomonadida—mostly parasitic; blunt anterior end and tapered posteriorly; two nuclei and four flagella present; some genera are *Giardia* (Fig. 2.6c) and *Hexamita*.

Order Oxymonadida—symbionts mostly in arthropods; bear 4–12 flagella; have well-developed axostyles, usually one nucleus; representative genera are *Oxymonas* and *Pyrsonympha*.

Order Trichomonadida—mostly symbiotic or parasitic in birds, cows, and human beings; 4–6 flagella, one usually forms the outer margins of an undulating membrane; best-known genera are *Trichomonas* (Fig. 2.6d), *Tritrichomonas*, and *Histomonas*.

Order Hypermastigida—all symbiotic in alimentary canal of wood-eating insects; many flagella often arranged in longitudinal rows; uninucleate; includes *Trichonympha* (Fig. 2.6e) and *Barbulanympha*.

Subphylum Opalinata—symbiotic in alimentary canal of amphibians and sometimes fishes; body covered with longitudinal rows of short flagella; two or more monomorphic nuclei; reproduce asexually by binary fission and sexually by flagellated gametes; *Opalina* (Fig. 2.7) is the best-known genus.

Subphylum Sarcodina—locomote by pseudopodia or protoplasmic flow; flagella sometimes present; body either naked or with external or internal test; most are free living.

Superclass Rhizopoda—have lobopodia, reticulopodia, or filipodia; most are naked, but some are testate.

Class Lobosea—most have lobopodia; usually uninucleate.

Subclass Gymnamoebia—without a test.

Order Amoebida—naked amebae; mostly free living in fresh water, damp places, or soil; some parasitic; *Amoeba* (Figs. 2.9a, 2.10a), *Paramoeba*, and *Entamoeba* are common.

Order Schizopyrenida—cylindrically shaped with hemispheric bulges; temporary flagellate stages in most forms; *Tetramitus* is typical.

Order Pelobiontida—thick cylindrically shaped; usually multinucleate; lacking mitochondria; no flagellate stages known; a common form is *Pelomyxa* (Fig. 2.10b).

Subclass Testacealobosia—body enclosed by test or complex membrane outside of plasma membrane.

Order Arcellinida—testate amebae; shell either secreted or of cemented environmental particles; mostly in fresh water or moist areas; *Arcella* and *Difflugia* (Fig. 2.10c) are well-known genera.

Order Trichosida—test of fibrous or calcareous materials with multiple apertures; all marine; *Trichosphaerium* is typical.

Class Acarpomyxea—small plasmodium or branching uninucleate and forming reticulum; lack a test; some form cysts; live in fresh water and soils; *Rhizamoeba* and *Stereomyxa* are typical.

Class Filosea—hyaline filiform pseudopodia; no flagellate stages known; pseudopodia often branched but not reticulate; most have a test; *Euglypha* is rather common.

Class Granuloreticulosea—mostly marine; benthic or planktonic; have reticulopodia; have calcareous shells of one to many chambers, walls of chambers with numerous pores; shells form extensive oozes and are important indicators of oil-bearing strata.

Order Foraminiferida—the foraminifera; characteristics same as those of the class; *Globigerina*, *Allogromia*, and *Discorbis* are frequently encountered.

Class Xenophyophorea—multinucleate body enclosed in branched-tube system; many barite crystals in cytoplasm; marine; *Psammetta* is representative.

Superclass Actinopoda—body spherical with

thin radiating axopodia; mostly planktonic; some forms with flagellated stages.

Class Acantharea—often called radiolarians; all marine and mostly planktonic; test composed of spicules of strontium sulfate; *Acanthometr* is typical.

Class Polycystinea—radiolarians; siliceous skeleton present in most; cytoplasm divided by central capsule into inner portion and outer calymma; marine; planktonic; *Collosphaera* and *Lithelius* are representative genera.

Class Phaeodarea—radiolarians; skeleton of silica and organic matter; thick capsular membrane divides cytoplasm into inner portion and outer calymma; marine; planktonic; *Aulacantha* and *Circoporus* are typical genera.

Class Heliozoea—sun animalcules; mostly in fresh water; lack a central capsule; many lack a skeleton, but others have a test of siliceous spicules; *Actinophrys* and *Actinosphaerium* (Fig. 2.12a) are commonly encountered.

PHYLUM LABYRINTHOMORPHA

PHYLUM APICOMPLEXA

Class Perkinsea—polar capsular tubules form incomplete cone; sporozoites flagellated; no sexual reproduction; *Perkinsus* is the only genus.

Class Sporozoea—when present, cone is complete; reproduction in most by both sexual and asexual means; infective stages or sporozoites are either naked or contained within a spore capsule.

Subclass Gregarinia—parasitic in invertebrates; trophozoites grow in extracellular body spaces; *Monocystis* (Fig. 2.14), *Gregarina*, and *Enterocystis* are common genera.

Subclass Coccidia—parasitic in both invertebrates and vertebrates; trophozoites small and intracellular in intestinal epithelium; typically alternate between sexual and asexual reproduction; spores usually highly resistant; *Eimeria* (Fig. 2.15), *Toxoplasma* (Fig. 2.17a), *Haemoproteus*, and the very important *Plasmodium* (Fig. 2.16) are well-known genera.

Subclass Piroplasmia—small parasites in erythrocytes of many domestic animals; intermediate hosts are ticks; do not form spores; sexual reproduction has not been established for many species; best-known genera are *Babesia* (Fig. 2.17b) and *Theileria*.

PHYLUM MICROSPORA

Class Rudimicrosporea—spores with rudimentary or simple extrusion apparatus; hyperparasites of gregarines in annelids; *Amphiacantha* is a typical genus.

Class Microsporea—spores with complex extrusion apparatus; most have polar filaments; parasitic in a wide range of invertebrate groups, especially insects; includes *Nosema* and *Glugea*.

PHYLUM ASCETOSPORA

Class Stellatosporea—spores with one or more sporoplasms; parasitic in molluscs and other invertebrates; includes *Marteilia* and *Urosporidium*.

Class Paramyxea—bicellular spores without oriface; *Paramyxa* is the type genus.

PHYLUM MYXOZOA

Class Myxosporea—parasitic and occur in body cavities and tissues of lower vertebrates and some invertebrates; have complex spores, each containing one or more sporoplasms and several polar filaments within polar capsules; includes *Ceratomyxa* and *Wardia*.

Class Actinosporea—spores with three valves and three polar capsules; includes *Triactinomyxon* and *Synactinomyxon*.

PHYLUM CILIOPHORA

Class Kinetofragminophorea—the more primitive ciliates; most have uniform ciliation, but some lack cilia as adults; have kinetofragments at oral end of body.

Subclass Gymnostomatia—no organized ciliature immediately adjacent to apically situated cytostome; usually contain trichocysts; five orders are recognized; *Bryophyllum, Holophyra*, and *Didinium* (Fig. 2.22f) are representative.

Subclass Vestibuliferia—have a vestibulum lined with distinctive cilia leading to cytostome; have primitive cytopharyngeal apparatus; contains three orders; representative genera include *Sonderia, Isotricha,* and *Colpoda* (Fig. 2.22c).

Subclass Hypostomatia—ventrally located cytostome surrounded by special cilia; have an advanced, complex, cytopharyngeal apparatus; holotrichous ciliation; this taxon contains six orders; common genera are *Nassula, Trochilia,* and *Chromidina.*

Subclass Suctoria—an enigmatic group; adults lack cilia, have feeding tentacles; all are sessile and usually attached by a stalk; have haptocysts; representative genera include *Acineta* (Fig. 2.22e) and *Podophyra.*

Class Oligohymenophorea—possess well-developed oral membranelles; cytostome located at base of buccal cavity.

Subclass Hymenostomatia—buccal organelles usually inconspicuous; uniform somatic ciliation; many are freshwater forms.

Order Hymenostomatida—all have several oral membranelles and well-defined buccal cavity; cytostome usually in anterior half of body; the best-known genera include *Tetrahymena* (Fig. 2.24b), *Paramecium* (Figs. 2.21, 2.22f, 2.24c), and *Frontonia.*

Order Scuticociliatida—small body with uniform ciliation; buccal cavity often shallow; mostly marine; many are endosymbionts in invertebrates; includes *Uronema* and *Pleuronema.*

Order Astomatida—large, elongate body; uniform ciliation; cytostome totally absent in all astomes; many with anterior elaborate holdfast organelle; includes *Anoplophrya* and *Mesnilella.*

Subclass Peritrichia—sessile or sedentary; body often bell-shaped and with an attaching stalk; conspicuous buccal ciliature spiraling counterclockwise toward oral region; most have prominent infundibulum.

Order Peritrichida—characteristics same as subclass; *Vorticella* (Fig. 2.24d), *Epistylis,* and *Trichodina* (Fig. 2.24a) are common genera.

Class Polymenophorea—most advanced ciliates, conspicuous oral ciliature and membranelles spiraling clockwise; little or no somatic ciliation, and some with prominent cirri; mostly free-swimming forms.

Subclass Spirotrichia—same as those of the class.

Order Heterotrichida—body usually large and often contractile; cilia present in even, dense rows; both free-living and symbiotic forms; familiar genera include *Spirostomum* (Fig. 2.25a), *Stentor* (Fig. 2.25b), and *Condylostoma.*

Order Odontostomatida—body small and laterally compressed; somatic cilia reduced; several oral membranelles present; most are freshwater forms; includes *Epalxella* and *Mylestoma.*

Order Oligotrichida—body elongate, reduced somatic ciliation; oral membranelles conspicuous; *Strobilidium* and *Epiplocylis* are often encountered.

Order Hypotrichida—body ovate and dorsoventrally flattened; conspicuous ventral cirri used in locomotion; prominent oral membranelles; widely distributed; many common genera including *Holosticha, Euplotes* (Fig. 2.25d), *Kerona,* and *Opisthotricha.*

ADDITIONAL READINGS

Anderson, O.R. 1983. *Radiolaria,* 328 pp. New York: Springer.

Bonner, J.T. 1983. Chemical signals of social amoebae. *Sci. Amer. 248*(4):114–120.

Chen, T., ed. *Research in Protozoology.* Vol. 1: 404 pp., 1967; Vol. 2: 382 pp., 1967; Vol. 3: 718 pp., 1969; Vol. 4: 391 pp., 1972. New York: Pergamon Press.

Cheng, T.C. 1973. *General Parasitology,* pp. 137–305. New York: Academic Press.

Corliss, J.O. 1979. *The Ciliated Protozoa,* 2nd ed., 430 pp. New York: Pergamon Press.

DeLaca, T.E. 1982. Use of dissolved amino acids by the foraminifer *Notodendrodes antarctikos. Amer. Zool.* 22:663–690.

Farmer, J.N. 1980. *The Protozoa: Introduction to Protozoology*, 693 pp. St. Louis, Mo.: C.V. Mosby.

Giese, A.C. 1973. *Blepharisma: The Biology of a Light-Sensitive Protozoan*, 351 pp. Stanford, Calif.: Stanford University Press.

Grell, K.G. 1973. *Protozoology*, 541 pp. New York: Springer.

Hammond, D.M. and P.L. Long, eds. 1973. *The Coccidia*, 458 pp. Baltimore: University Park Press.

Haynes, J.R. 1981. *Foraminifera*, 433 pp. New York: John Wiley & Sons.

Hedley, R.H. and C.G. Adams, eds. *Foraminifera*. Vol. I: 264 pp., 1974; Vol. II: 244 pp., 1976. New York: Academic Press.

Hyman, L.H. 1940. *The Invertebrates*. Vol. I: *Protozoa through Ctenophora*, pp. 44–232. New York: McGraw-Hill Book Co.

Jahn, T.L., E.C. Bovee, and F.F. Jahn. 1979. *How to Know the Protozoa*, 2nd ed., 279 pp. Dubuque, Iowa: W.C. Brown.

Jeon, K.W. 1973. *The Biology of Amoeba*, 580 pp. New York: Academic Press.

Leedale, G.F. 1974. How many are the kingdoms of organisms? *Taxon* 23:261–270.

Levandowsky, M. and S.H. Hutner, eds. *Biochemistry and Physiology of Protozoa*. Vol. 1: 420 pp., 1979; Vol. 2: 457 pp., 1979; Vol. 3: 384 pp., 1980; Vol. 4: 584 pp., 1981. New York: Academic Press.

Levine, N.D., J.O. Corliss, F.E.G. Cox, G. Deroux, J. Grain, B.M. Honigberg, G.F. Leedale, A.R. Loeblich III, J. Lom, D. Lynn, E.G. Merinfeld, F.C. Page, G. Poljansky, V. Sprague, J. Vavra, and F.G. Wallace. 1980. A newly revised classification of the Protozoa. *J. Protozool.* 27(1):37–58.

Long, P.L. ed. 1982. *The Biology of the Coccidia*, 502 pp. Baltimore: University Park Press.

Maramorosch, K. 1981. Spiroplasmas: Agents of animal and plant diseases. *Bioscience 31*: 374–380.

Nanney, D.L. 1980. *Experimental Ciliatology*, 292 pp. New York: John Wiley & Sons.

Pennak, R.W. 1978. *Fresh-water Invertebrates of the United States*, 2nd ed., pp. 19–79. New York: John Wiley & Sons.

Poag, C.W. 1981. *Ecologic Atlas of Benthic Foraminifera of the Gulf of Mexico*, 192 pp. Stroudsburg, Pa.: Hutchinson Ross Publishing Co.

Ristic, M. and J.P. Kreier, eds. 1980. *Babesiosis*, 589 pp. New York: Academic Press.

Sleigh, M.A. 1973. *The Biology of Protozoa*, 303 pp. London: Edward Arnold.

3

Introduction to the Metazoans

OVERVIEW

If we exclude the protozoans, all animals are metazoans and are characterized as being multicellular. The advent of multicellularity opened up a tremendous range of evolutionary potentialities. It has enabled modern-day metazoans to adapt morphologically and physiologically in an immense number of ways, thrive in every ecosystem, and fill innumerable ecological niches.

Even though metazoan origins are lost in antiquity, several theories have been proposed to explain the evolutionary stages. The most plausible theory holds that metazoans arose from a colonial flagellate. As the ancient primitive colony became composed of more cells with increasing cellular specialization and mutual dependence upon other cells, a trend was established that ultimately led to multicellular organisms. The stages in the ontogeny of multicellularity were probably similar to those in the embryology of modern metazoans. The earliest metazoans were most likely represented by a planula that was simple, ovoid, multicellular, and flagellated. The rediscovery in 1969 of a remarkably simple, planuloidlike creature, *Trichoplax adhaerens*, gave considerable support to the planuloid ancestor theory.

Early metazoan evolution resulted in three principal lines. One pathway led to sponges, a second to the radiates, and representatives of the third line were characterized by having bilateral symmetry. A fundamental dichotomy in the bilaterates led to the development of two major evolutionary lines: the Protostomia and the Deuterostomia. Protostomates, including molluscs, annelids, arthropods, and a number of smaller phyla, are characterized by having the developmental fate of each of the earliest embryonic cells determined (i.e., losing their potentiality for developing into a complete embryo), the arrangement of cells in the young embryo (spiral cleavage), and the mouth opening appearing before the anal opening. On the other hand, deuterostomates, or the echinoderms and chordates, have embryos in which cells are determined somewhat later in development, embryonic cells are produced in a characteristic fashion (radial cleavage), and the anal opening develops before the appearance of the mouth.

As embryonic development proceeds, the three germ layers (ectoderm, mesoderm, endoderm) are determined and formed. Two very important features found in many metazoans, the coelom or body cavity and metamerism or segmentation, are intimately associated with the mesoderm. The presence or absence of metamerism, types and methods of formation of the body cavity, embryonic patterns, and types of symmetry are the principal criteria for grouping various metazoan phyla.

CHARACTERISTICS OF METAZOANS

Excluding the protozoans, all animals, both invertebrate and vertebrate, are metazoans. Several diagnostic characteristics apply generally to them, the most important of which is that they are all multicellular. In fact, the most important event in the evolution of metazoans was the development of the multicellular condition. In metazoans there is an enormous range in degrees of cellular differentiation or specialization from that in very simple organisms with a minimum of only two different kinds of somatic cells to that in complex organisms constructed of many different types of cells in intricate groupings such as tissues and organs.

The advent of multicellularity made possible two sets of related and exceedingly important metazoan dimensions. First, it effectively removed the strict functional and geometrical limitations on size necessarily imposed by the unicellular condition. As a cell becomes progressively larger, a limit is reached on the volume of cytoplasm its nucleus can effectively control. Further, as size increases, the ratio of surface area to volume decreases to some critical level below which the cell cannot function properly because of its diminished relative surface area. Second, multicellularity made possible entirely new structural and functional patterns that simply were not possible in protozoans. In the unicellular condition, each cell obviously is almost totally independent of all other cells; even in some colonial protozoans a given cell retains for the most part considerable independence from all other cells. But some other ancient protozoans exhibited a rather high degree of cellular differentiation and mutual dependence between the cells—a most important adaptation that was to portend the development of a simple multicellular animal not substantially advanced over a differentiated colonial flagellate. In fact, there is a continuum of increasing complexity from the unicellular to the multicellular condition, and certain contemporary colonial flagellates and green algae and the earliest metazoans are located in intermediate positions on this continuum.

In the evolution of the multicellular condition a degree of cellular differentiation gradually developed in which certain cells became more specialized for a given function and, in turn, lost some other functions. Thus even in simple multicellular metazoans there is a division of labor among the somatic cells in which each cell is partially dependent on all other cells. All metazoans have at least two different kinds of somatic cells in contrast to all colonial flagellates and green algae, which have only a single somatic cell type. Further cell specializations in metazoans have resulted in the development of tissues (groups of like cells), organs (groups of different tissues), and systems (composed of several different organs), which have allowed for an almost infinite number of architectural patterns. With the advent of specialized tissues and organs such as those associated with gaseous exchange and circulation, even larger body sizes were possible. Moreover, larger, more-specialized organisms were able to adapt to a myriad of ecological niches and habitats possible in new environments that were never available to protozoans. In short, the ontogeny of multicellularity was the first, and certainly the most important, of several paramount evolutionary developments. The advent of multicellularity made possible almost limitless potentialities for morphological and functional adaptations.

Even though some metazoans are small, most are relatively large and possess a number of important external and internal adaptations that equip them for survival. In addition to their multicellularity, metazoans are also all heterotrophic, mostly mobile, generally produce gametes in multicellular gonads, and develop from embryos. The diversity of their functional morphology is almost limitless, and they are found in every habitat and ecosystem and have a great measure of adaptive plasticity.

THE ORIGINS OF METAZOANS

From what preexisting group did metazoans evolve? What were the important and fundamental stages in metazoan evolution? These and other vexing and immensely important phylogenetic questions have been the subject of extensive speculation by biologists over the years. Yet there is no concrete evidence, either as fossils or in any other form, about the metazoan ancestors or the stages in the evolution of multicellularity. Thus we are forced to rely solely on indirect evidence, inferences, and speculations about their origins. A number of theories or ideas about metazoan ancestry and phylogeny have been advanced, and two of the best known will be described briefly.

The Colonial Flagellate Theory

The **Colonial Flagellate Theory** holds that metazoans evolved from colonial mastigophorans. Adherents of this theory look to several lines of flagellates in which colonialism became progressively more pronounced and intimate until a type of colonial organism was evolved that, in effect, was multicellular. Strong supporting evidence comes from the fact that flagellates and most all metazoans have flagella and only one type of nucleus. An excellent model is to be found in the autotrophic volvocines, including *Volvox,* all of which form spherical hollow colonies. In the volvocines there is a series of colonial organisms that contains progressively larger numbers of individuals, beginning with a single cell (*Chlamydomonas*), including those with few to tens of cells (*Gonium*, 4–16 individuals; *Pandorina*, 16 cells, *Eudorina*, 32 cells; *Pleodorina*, 32–128 individuals), and ranging up to hundreds of cells (*Volvox*) (Fig. 3.1).

At each higher step in this series, cellular differentiations become more evident, and the

Figure 3.1 *Possible stages in the evolution of colonial flagellates from which metazoans ultimately arose; shown is the volvocine series of phytoflagellates illustrating a progressive increase in numbers of individuals per colony and in cellular specializations:* ***(a)*** Chlamydomonas; ***(b)*** Gonium; ***(c)*** Pandorina; ***(d)*** Pleodorina; ***(e)*** Volvox; ***(f)*** *an enlarged diagrammatic section of* Volvox *illustrating somatic and reproductive cells.*

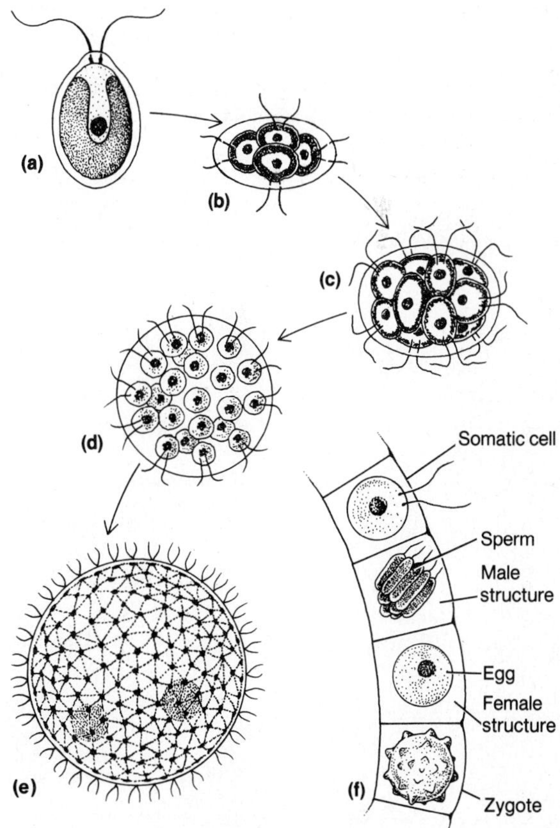

constituent cells become progressively more dependent on each other. The organisms in this series also reflect an increasing division of labor, especially between vegetative and reproductive cells; certain specialized reproductive cells in the interior of *Volvox* even produce two different kinds of gametes (eggs and sperm) (Fig. 3.1f). You may at this point be wondering why *Volvox* is not a multicellular organism rather than a colonial green algae. The answer is that it has only one type of somatic cells—not the two needed to qualify it as being multicellular. Some biologists believe that the flagellate ancestors of metazoans were indeed these freshwater volvocine phytoflagellates. But most phylogenists hold that the flagellate ancestors were now-extinct zooflagellates of some sort that had a colonial organizational pattern similar to that found in the volvocines.

The colonial flagellate theory is a modification of an earlier theory proposed in 1874 by the famous German biologist Ernst Haeckel. Haeckel's theory, known as the **Theory of Recapitulation** and one that prevailed for many decades, was the first serious attempt to hypothesize about metazoan origins (Fig. 3.2a). Haeckel held that ontogeny recapitulates phylogeny, that is, the embryonic development (ontogeny) repeats (recapitulates) the evolutionary pathways (phylogeny). His theory was based on embryological evidence in which the succession of embryonic stages of a given animal in theory represents its evolutionary stages. Thus the zygote represents the unicellular protozoan

Figure 3.2 *Possible phylogenetic stages in the early evolution of metazoans:* ***(a)*** *Haeckel's Recapitulation Theory;* ***(b)*** *the Syncytial Ciliate Theory;* ***(c)*** Trichoplax adhaerens; ***(d)*** *two stages in the formation of a temporary digestive chamber by* Trichoplax; ***(e)*** *the Planuloid Theory.*

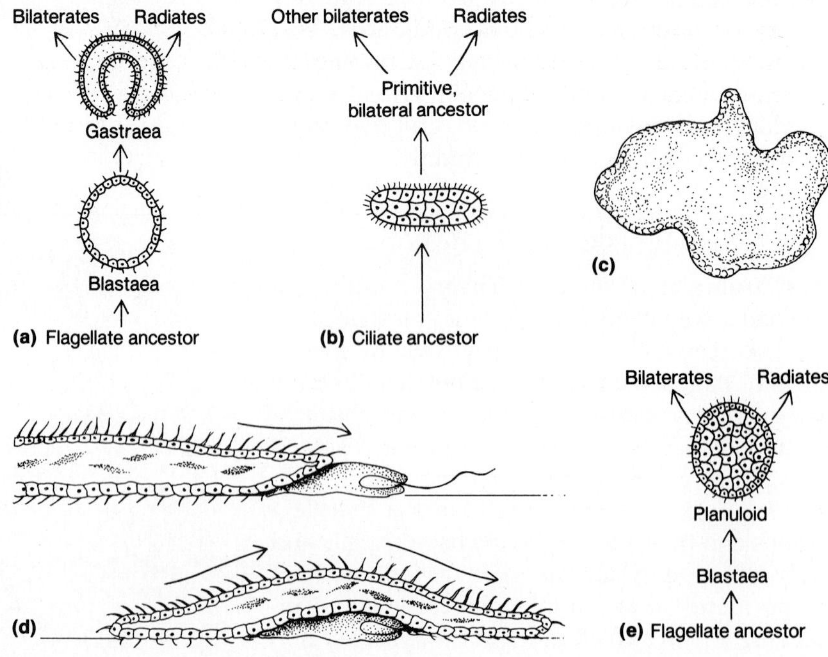

stage, and the 2-, 4-, 8-, and 16-cell stages resemble very ancient metazoan ancestors. Haeckel proposed a hypothetical ancestral metazoan, the **blastaea**, which is represented in the embryonic development of animals as the blastula. His blastaea was a hollow-sphered, colonial flagellate similar to *Volvox* but without the chloroplasts. Haeckel believed that radiates arose from this hypothetical ancestral blastaea by a process of invagination to form a **gastraea** similar to the embryonic gastrula stage. All other metazoans therefore arose from the gastraea by developing bilateral symmetry, and organisms with radial symmetry were secondarily derived from those with bilateral symmetry.

However, some of the points in Haeckel's theory have been invalidated on several legitimate bases. First, most modern-day zoologists do not hold that the embryonic stages of higher animals represent the adult ancestors as Haeckel believed, but rather that embryonic stages of a given animal resemble the embryonic stages of ancestors. Stated differently, we hold that ontogeny recapitulates ontogeny, but with modifications. Second, gastrulation in radiates, a very ancient and primitive group of metazoans, does not take place by invagination; rather, the endoderm is formed by inwandering cells. Third, most radiates form a solid blastula rather than a hollow one. In summary, most zoologists today have rejected Haeckel's overall theory on the basis of some of the specific objections already mentioned, even though the basic principle of multicellularity arising from colonial flagellates is generally supported.

The colonial flagellate theory supports Haeckel's theory up to the formation of blastulalike colonies. However, the blastaea is believed to have been solid rather than hollow as Haeckel theorized. Haeckel's idea of a subsequent gastraea stage is now abandoned; in its place a **planuloid** stage is proposed (Fig. 3.2e). The planuloid stage is theorized as being multicellular, ovoid, and radially symmetrical. Its outer cells were flagellated and functioned in locomotion. This stage, having no definite mouth, fed saprozoically, and the cells of the interior functioned both in reproduction and digestion. From this simple ancestor the lower metazoans are thought to have arisen. In fact, the basic larval stage of radiates is termed the **planula**, and it reflects the chief characteristics thought to have been present in the planuloid ancestor.

Support for the planuloid ancestor today is found in a fairly old theory and in the rediscovery in 1969 of a simple metazoan. In 1884, Otto Butschli proposed that the most primitive metazoan might have been a two-layered, flattened creature creeping over the marine bottom. Butschli called this the **plakula**, and his theory became known as the **Plakula Theory**. The upper surface of the plakula would be mainly concerned with protection and perhaps locomotion. Its undersurface would presumably come in contact with minute food organisms or particles. In fact, an important adaptation of the plakula would be for the organism to elevate part of its body to form a temporary digestive chamber, which would concentrate food particles and digestive enzymes, thus making digestion more efficient. Butschli proposed that in this manner the temporary digestive chamber would become the cavity of the gastrula—temporary at first and later a permanent chamber. Butschli also hypothesized that the plakula was bilaterally symmetrical and that radial symmetry would have been a secondary adaptation.

Butschli undoubtedly based much of this theory on a minute living marine organism discovered one year earlier and named *Trichoplax adhaerens*. *Trichoplax* is a two-layered benthic marine creature that in fact does form temporary ventral digestive chambers (Fig. 3.2c, d). *Trichoplax* and Butschli's plakula were remarkably similar except that *Trichoplax* has no constant shape, whereas the plakula was supposed to possess bilateral symmetry. *Trichoplax* was forgotten shortly after its initial discovery, since the zoological community assumed that it was a larval stage of a radiate. In 1969, K.G. Grell rediscovered *Trichoplax* and found that, incredibly, it is not a larval stage but a very primitive adult! *Trichoplax* is unlike anything else known, and Grell has placed it in its own phylum, Placozoa (see Chapter 4). It probably represents the most primitive living metazoan, and it has many characteristics of the hypothetical planuloid mentioned earlier.

The colonial flagellate theory, which holds that metazoans arose from a series of colonial flagellates that subsequently evolved into a blastaea and later into a planuloid ancestor, is the most widely adopted theory today (Fig. 3.2e). Such a theory would account for the ubiquitous occurrence of flagellated sperm in all metazoans, for the flagellated cells that are common in the bodies of most lower metazoans, and for the presence of only one kind of nucleus in all metazoans.

The Syncytial Ciliate Theory

A second theory, the **Syncytial Ciliate Theory**, is supported by a minority of zoologists. This theory holds that the protozoan ancestor from which the planuloid ancestor arose was a syncytium, that is, it was multinucleate without any internal cell membranes (Fig. 3.2b). Certain modern-day ciliates are indeed syncytial, and it is hypothesized that similar syncytial ciliates were the stock from which metazoans evolved. This theory holds the cell membranes were produced between the nuclei so that a multinucleate unicellular organism would become a multicellular wormlike creature with bilateral symmetry. An extension of this theory is that the most primitive metazoans were acoelomates, and from them arose the radiates rather than the reverse.

Several basic difficulties arise from this theory. First, it assumes that bilateral symmetry is primitive and that radial symmetry was derived from it, but most zoologists, accepting the colonial flagellate theory, believe the reverse, that is, that radial symmetry is the more primitive of the two. Second, it cannot account for the almost universal presence of flagellated cells in metazoans. Third, since all ciliates have two kinds of nuclei, the syncytial theory does not explain the evolutionary loss of the second nuclear type. Finally, true eggs and sperm have evolved in the flagellates but not in the ciliates from which, as the syncytial theory would maintain, the metazoans arose.

It is certainly within the realm of possibility that the development of multicellularity did not happen only once but may have occurred many different times along several paths. We shall return in several later chapters to the issues of **monophylogeny**, or the evolution of an entire group from one ancestral type, versus **polyphylogeny**, or having several different ancestral types of a large taxon.

SOME BASIC EVOLUTIONARY TRENDS IN METAZOANS

Primitive metazoans were probably present as early as one billion years ago, and it is very probable that planuloid ancestors like *Trichoplax* diverged early in the evolutionary history to form several different major paths leading to contemporary phyla. One line resulted in the sponges (Porifera), but there is a growing feeling among many invertebrate zoologists that poriferans evolved independently from choanoflagellates. A second line, whose representatives had evolved cylindrical (radially symmetrical) bodies with a tissue-grade construction, were basically sedentary and became the radiates (Cnidaria, Ctenophora). Ancestors of the third great line evolved bodies in which one side was a mirror image of the other (bilaterally symmetrical) and developed an active life style with an anterior head; these became known as the bilaterates. These ancient benthic marine creatures ultimately gave rise to all higher animals.

TABLE 3.1. □ **A SUMMARY OF THE BASIC FEATURES OF THE PROTOSTOMIA AND DEUTEROSTOMIA**

Feature	*Deuterostomia*	*Protostomia*
Major representative phyla	Echinodermata, Chordata	Platyhelminthes, Nematoda, Mollusca, Annelida, Arthropoda,
Cleavage patterns	Radial, indeterminate	Spiral, determinate
Development	Regulative	Mosaic
First embryonic opening	Anus	Mouth
Origin of mesoderm	Endoderm	Endoderm and ectoderm
Method of mesoderm formation	Hollow, lateral, enterocoelic outpocketings	Solid, lateral, cylindrical mass of cells derived from 4d cell
Type of coelom	Enterocoelom	Schizocoelom

Early in the history of bilaterates, a major evolutionary bifurcation took place that in effect divided all bilaterates into two very large groupings: the Protostomia and the Deuterostomia. The **Protostomia** includes the flatworms, annelids, arthropods, molluscs, and almost 20 smaller phyla. The **Deuterostomia** is represented by the echinoderms, chordates, and four smaller phyla. Great phylogenetic significance is attributed to the differences between these two basic metazoan groups. A number of characteristics divide bilaterates into either protostomates or deuterostomates; they are summarized in Table 3.1. Since these diagnostic characteristics basically concern embryonic development, a fuller enumeration of these developmental features follows.

EMBRYONIC DEVELOPMENT OF METAZOANS

Patterns of embryonic development in all bilaterates fall into two rather distinctive groups that are the bases for recognizing the Protostomia and Deuterostomia. One can only speculate about the mechanisms behind the ontogeny of two different embryological patterns; and even though the mechanisms perhaps will never be known, invertebrate zoologists believe the two developmental lines to be of cardinal importance in understanding metazoan phylogeny. To be sure, almost every metazoan phylum has evolved certain developmental characteristics that are unique to that group. But many developmental features are present in most or all phyla within one of the two large groupings, and these characteristics are seen as being of primal importance to our understanding of metazoan phylogeny. The following discussions will point out the fundamental differences and similarities in embryonic development of metazoans.

Cleavage

In almost all organisms and certainly in all metazoans, fertilization is characterized by the union of a sperm and an egg, resulting in a fertilized egg or **zygote**. Shortly after its formation, the zygote undergoes a series of cell divisions collectively called **cleavage** that continues until the blastula stage of 30–100 cells is formed. Cleavage differs in one respect from ordinary cell divisions because in cleavage, no cellular growth ensues between consecutive divisions, and therefore the resulting cells or **blastomeres** become progressively smaller as cleavage progresses.

In the early embryonic development of most deuterostomates the fate of each blastomere is not determined or fixed until after several cleavages have taken place. This means that each cell in the early stages of development retains the potentiality of producing a complete embryo if separated from other blastomeres. Such a pattern of delayed determination, termed **indeterminate cleavage**, is characteristic of deuterostomates. Embryos with indeterminate cleavage are said to undergo **regulative development**, that is, the embryo and its individual cells are capable of adjusting or regulating to missing portions of cells or cytoplasm. In many other animals, however, the fate of embryonic cells is determined in the very first cleavage. Thus a blastomere loses its potentiality to develop into a complete embryo at the first cleavage, and the resulting two cells do not have equivalent developmental potentialities. This means that following the first cleavage, each of the two blastomeres will develop into only a predetermined fixed half of the embryo as each has lost its potentiality of developing into the other half of the embryo. Such cleavage, termed **determinate cleavage**, is characteristic of many protostomate taxa. Embryos with determinate cleavage are said to undergo **mosaic development**, that is, the embryo is a mosaic of cells, each with a fixed, determined, prospective fate. But the differences between indeterminate and determinate cleavage are not absolute; in animals in which cleavage is indeterminate, a developmental stage is soon reached at which blastomeres are produced with different developmental potentialities.

Yolk distribution

The amount of yolk or nutritive material deposited within the cytoplasm of the egg often alters the patterns of cleavage substantially. Owing mostly to varying amounts of yolk present, eggs of invertebrates vary widely in size from microscopic to almost 2 cm in diameter. Many metazoan eggs contain only a minimal amount of yolk, which is evenly distributed; they are called **oligolecithal**. Some eggs contain a relatively large amount of yolk concentrated at one end of the egg (vegetal pole); they are termed **telolecithal**. In **centrolecithal** eggs, found in many other metazoans, the nonyolky cytoplasm is restricted to the center and outer cortex of the egg.

The amount and distribution of yolk in an egg greatly affect the pattern of the cleavage furrows. Yolk tends to retard cleavage by serving as a barrier to the normal assortment of chromosomes, the formation of mitotic spindles, and cytokinesis. In oligolecithal eggs, cleavage furrows pass through the entire egg; such total cleavage patterns are termed **holoblastic**. In telolecithal eggs, cleavage furrows pass more readily through the less yolky animal pole but are retarded by the concentration of yolk at the vegetal pole, so this cleavage is incomplete and is said to be **meroblastic**. In centrolecithal eggs, nuclear divisions take place in the central cytoplasm after which the nuclei migrate to the periphery, where they are individually surrounded by cell membranes; cleavage in centrolecithal eggs is called **superficial**.

Radial and spiral cleavage

Cleavage is also substantially different in the protostomates and deuterostomates. In radial cleavage, a pattern found in deuterostomates, the mitotic spindles and cleavage planes are either parallel with or at right angles to the basic axis extending through the animal and vegetal poles (Fig. 3.3). In radial cleavage the first two cleavages are both perpendicular to the animal–vegetal axis and at right angles to each other. The third cleavage is generally a horizontal one. Subsequent cleavage results in the production of blastomeres in strata or layers in which a

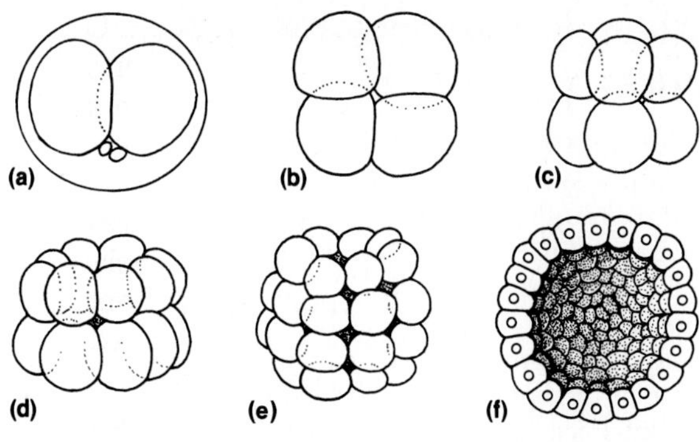

Figure 3.3 *Radial cleavage in a deuterostomate: **(a)** two-cell stage; **(b)** four-cell stage; **(c)** eight-cell stage; **(d)** 16-cell stage; **(e)** 32-cell stage or morula; **(f)** blastula shown in vertical section.*

blastomere is always situated directly above or below another blastomere (Fig. 3.3c–e). Such cleavage is generally indeterminate.

In **spiral cleavage**, characteristic of most protostomate phyla, a very different pattern is seen in which the mitotic spindles lie at an oblique angle to the basic animal–vegetal axis (Fig. 3.4). Thus cleavage produces cells that, rather than lying immediately above another blastomere, are situated over an underlying cleavage furrow. Thus spiral cleavage produces tiers of cells that come off in a spiral fashion. Since spiral cleavage is also determinate and embryos have mosaic development, it is possible to follow the fate of individual blastomeres in a wide variety of protostomates (see Box 3.1). In such a study of cell lineage, similar patterns are seen in quite different invertebrates, and the fate of a given blastomere is usually identical in all animals with spiral cleavage.

Blastula and Gastrula

Regardless of whether cleavage is radial or spiral, the mass of 30–100 cells is termed the **blastula** (Figs. 3.3f; 3.4f). Depending upon the amount of yolk present, the blastula is either

Figure 3.4 *Spiral cleavage in a protostomate with the cell progeny from each of the four original blastomeres shown in different patterns: **(a)** four-cell stage; **(b)** eight-cell stage seen from the animal pole; **(c)** eight-cell stage, lateral view; **(d)** 16-cell stage seen from the animal pole; **(e)** 32-40 cell stage, lateral view; **(f)** blastula seen from the vegetal pole showing the four macromeres and the "4d" cell.*

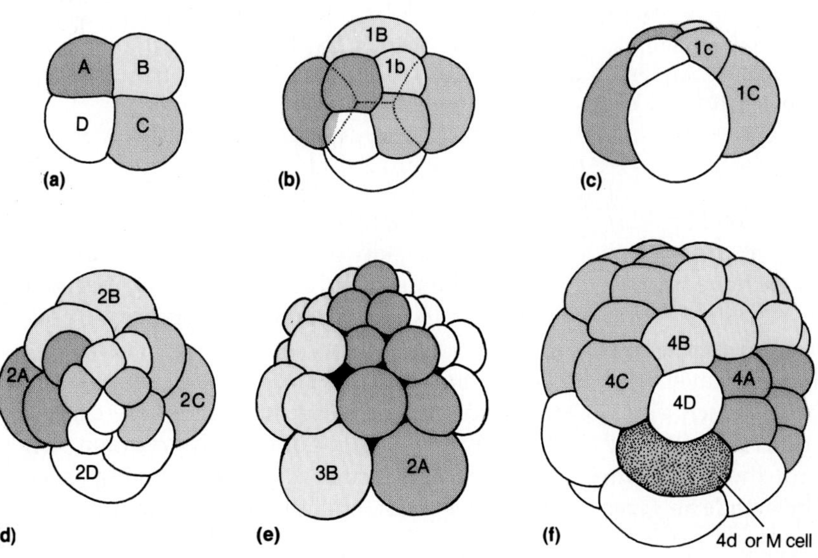

BOX 3.1 □ SOME DETAILS OF EARLY SPIRAL CLEAVAGE

Invertebrate embryologists have devised a system of designating and identifying individual blastomeres undergoing spiral cleavage and mosaic development. The first two cleavage planes pass along the animal–vegetal axis and produce four equally sized blastomeres designated A, B, C, and D (Fig. 3.4a). The third cleavage plane, horizontal or at right angles to the animal–vegetal axis, is shifted toward the animal pole so that the upper set of blastomeres is much smaller than the lower four (Fig. 3.4b, c). The smaller blastomeres or **micromeres** are at the animal pole, and the larger ones, termed **macromeres**, are at the vegetal pole. But since the spindle axes and cleavage planes are oriented obliquely to the polar axis, each of the four micromeres is situated, not above its corresponding macromere, but to one side and above one of the cleavage furrows. Macromeres are designated by a capital letter, and micromeres are given appropriate lowercase letter designations. The first micromeres are denoted in 1a, 1b, 1c, and 1d, corresponding to that cell of the original four from which they were derived, and the macromeres are now called 1A, 1B, 1C, and 1D (Fig. 3.4b, c).

The quartet of macromeres now cleaves to produce a second quartet of micromeres designated 2a, 2b, 2c, and 2d and a quartet of macromeres termed 2A, 2B, 2C, and 2D. If the initial quartet (1a, 1b, 1c, 1d) is given off clockwise from the macromeres, then the second quartet (2a, 2b, 2c, 2d) is produced in a counterclockwise fashion (Fig. 3.4d–f). The four macromeres undergo a horizontal cleavage for a third and fourth time to produce a third (3a, 3b, 3c, 3d) and a fourth (4a, 4b, 4c, 4d) quartet of micromeres. These four horizontal cleavages produce a five-tiered mass of blastomeres or **blastula** with 1a, 1b, 1c, and 1d at the animal pole and 4A, 4B, 4C, and 4D at the vegetal pole (Fig. 3.4e, f). The first-formed micromeres now begin to cleave, and their daughter cells are denoted by an exponent, i. e., $1b^1$ or $1b^2$, the exponent "1" being nearest the animal pole. When $1b^2$, for example, divides, another exponent is added to form $1b^{21}$ and $1b^{22}$. Subsequent divisions result in the addition of more exponents ($1b^{211}$ or $1b^{212}$, etc.).

In spiral cleavage the first three quartets of micromeres give rise to all the ectoderm and the ectomesoderm. The four macromeres and 4a, 4b, and 4c form the endoderm. The remaining micromere, 4d, gives rise to several cells that contribute to endoderm formation and all the endomesoderm (Fig. 3.4f). In gastrulation the most recent micromeres (4a, 4b, 4c, 4d) and the macromeres are located inside the gastrula.

hollow (coeloblastula) or solid (stereoblastula). By the blastula stage the cells that are to give rise to the future three germ layers have become determined.

Subsequent development produces a **gastrula** that gives rise to the three germ layers and the **archenteron** or primitive gut, which opens to the outside by the **blastopore**. The three germ layers are, from outside inward, the ectoderm, mesoderm, and endoderm. The **ectoderm** gives rise to the outermost portions of the body, including the skin or integument, nervous system, and mouth and anal canals. The **mesoderm** gives rise to most structures, including muscles, heart and circulatory organs, excretory and reproductive organs, and many others. The **endoderm** produces the linings of the digestive tract and related glands. Gastrulation takes place by ectodermal cells overgrowing the endodermal mass (epiboly), by invagination, or by a combination of these two processes as in most metazoans. In protostomates the blastopore persists and becomes the mouth. In deuterostomates the blastopore becomes the anus, and the mouth appears later in development some distance from the blastopore. It is from this highly significant difference in the ultimate fate of the blastopore that the nomenclature for these two large groups is derived (protostomates = first mouth; deuterostomates = secondary mouth).

Mesoderm

Once gastrulation has occurred, the basic morphogenetic patterns of the future animal are being determined, one of the most important of which is the formation of the mesoderm. Mesoderm may be formed either from ectoderm or endoderm or, in most animals, from both ectoderm and endoderm. The mesoderm of radiates (relatively simple, radially symmetrical animals) and some of that in protostomates is derived from inwandering ectodermal cells. However, most of the mesoderm in protostomates and all of it in deuterostomates is derived from endoderm.

Origin of the Mesoderm and Formation of the Coelom

In all higher metazoans there is developed a body cavity situated between the body wall and the digestive tract. This body cavity develops as a space within the mesoderm and is termed a **coelom**. The methods of both mesoderm and coelom formation are fundamentally different in protostomates and deuterostomates. In protostomates the endomesoderm arises as a solid mass of cells derived from a single blastomere, the **4d** or **mesendoblast cell**, located near the blastopore (Figs. 3.4f). The 4d cell produces two primordial mesodermal cells, one being located on either side of the blastopore. Each primordial mesodermal cell proliferates to form a cylindrical solid mass of mesodermal cells extending anteriorly (Fig. 3.5d). Later, a split develops inside these two mesodermal masses, and these cavities enlarge and coalesce to form a coelom surrounded on all sides by mesoderm (Fig. 3.5d–h). Such a method of coelom formation is called **schizocoely**; and all coelomate protostomates are referred to as **schizocoelomates**.

In deuterostomates the mesoderm arises as hollow lateral outpocketings from the archenteron in a process called **enterocoelic pouching** (Fig. 3.5b, c). Later, the pouches separate from the archenteron, and the cavities surrounded by the mesoderm fuse to become the coelom (Fig. 3.5g, h). Thus all deuterostomates, being coelomate, are called **enterocoelomates**.

Metamerism and Tagmatization

Metamerism refers to the phenomenon in which the bodies of many animals are composed of a series of **segments** or **metameres** arranged linearly along the anteroposterior axis. Primitively, each metamere is similar in both construction and function to all other metameres; but in more advanced metameric animals the segments may be substantially different. Metamerism has evolved independently in at least two different groups: the vertebrates, i.e., the noninvertebrate chordates, and the annelids and arthropods. Metamerism is fundamentally a property of the mesoderm. In embryonic development of all protostomates the mesoderm is derived from mesendoblasts located on either side of the blastopore. As new mesoderm is developed in metameric protostomates, it is done so in paired blocks, **metameres**, or **somites**, with one member of each pair on either side of the gut. Therefore in protostomates, new mesodermal metameres are added at the posterior end just anterior to the blastopore. The oldest segments are therefore anterior, and the progressively younger segments are located nearer to the posterior end.

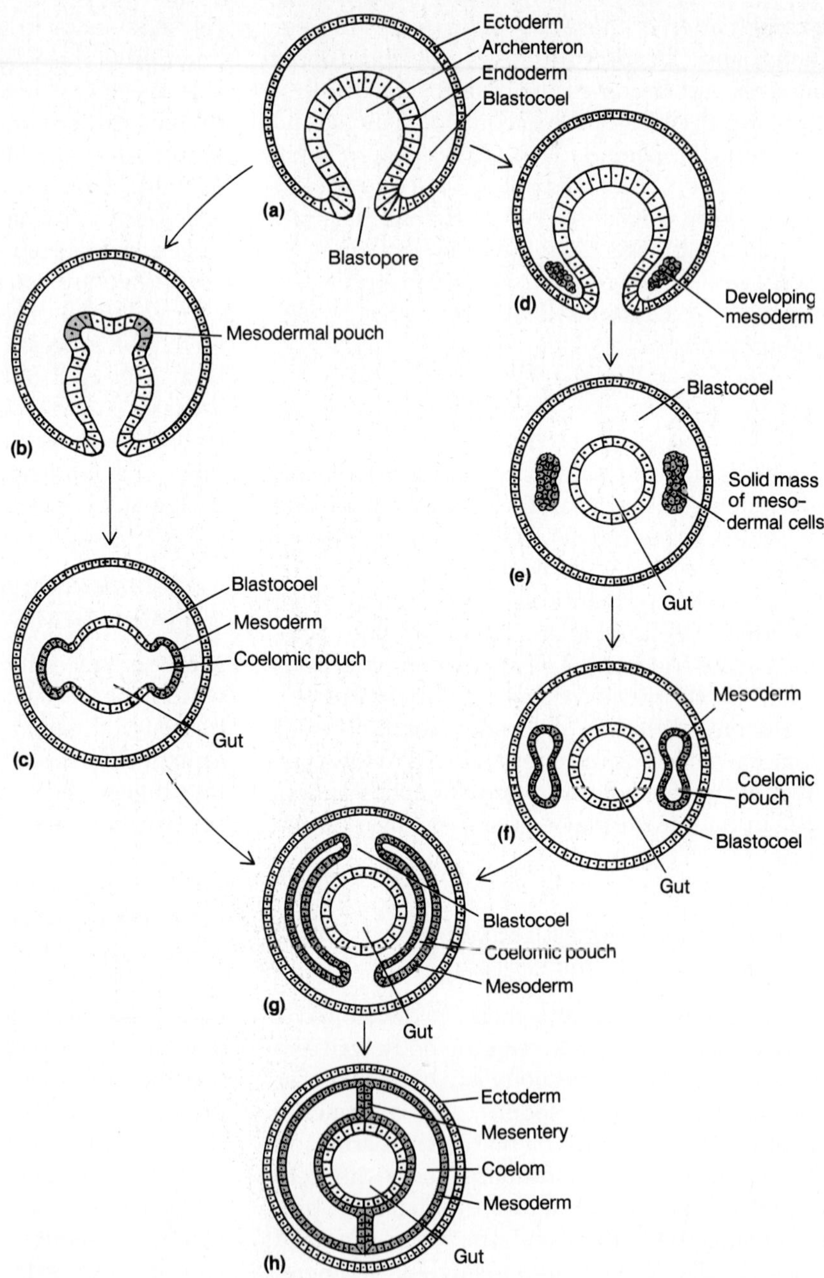

Figure 3.5 *Two methods of forming the coelom:* ***(a)*** *gastrula stage;* ***(b, c)*** *stages in enterocoelic pouching;* ***(d–f)*** *stages in schizocoely;* ***(g, h)*** *later stages of coelom formation in both enterocoelomates and schizocoelomates (parts a, b, and d are frontal sections; all others are cross sections.)*

In the metameric deuterostomates there is developed an anteroposterior series of paired enterocoelic pouches arising on either side of the archenteron, which gives rise to the metameres.

In a segmented animal, all definitive adult structures that are mesodermally derived are also segmented and include muscles, blood vessels, and excretory organs. Metamerism proceeds outward and eventually involves ectodermal derivatives such as the skin, brain, nerves, and nerve cord. Metamerism thus not only becomes an internal property but is very evident externally as exemplified by annelids and arthropods. Only the endoderm and its derivatives, i.e., the lining of the digestive tract, are not included in metamerism. Metamerism often is best seen in the embryonic condition; its presence frequently is partially masked in adults.

Although details of the evolution of metamerism are lost forever, there can be little doubt as to the significance of metamerism in locomotion. Segmentation permitted the musculature in each metamere to operate independently from that in adjacent metameres, and such localized muscular actions made possible the development of local specific changes in the body. Internally, the septa comparting any two adjacent segments effectively divide the coelom into a series of repeating, fluid-filled chambers. These watery chambers collectively form a hydrostatic skeleton that supports and gives form to the animal. Muscles of an individual segment contract and alter the shape of that hydrostatic unit so that local changes in shape can be effected. Localized muscles acting on individual, metamerically arranged, fluid-filled body cavities are of immense importance to locomotion, and metamerism was a most significant adaptation in a large number of metazoans. We shall explore further the functional significance of metamerism in Chapter 9.

Tagmatization is a pattern in metameric animals in which several to many segments are united in various functional groups. Each group of metameres or **tagma** is structurally and physiologically differentiated from other tagmata. For example, in an insect there are three tagmata: head, thorax, and abdomen, each consisting of three or more metameres. Each tagma is specialized for a particular function such as feeding, locomotion, or reproduction.

ORGANIZATION OF METAZOANS

Amid the enormous diversity of metazoans, it might appear that few if any unifying features are present or indeed possible. We have already explored several very important developmental features that unite large groups of metazoans. There are also some immensely important characteristics, both morphological and functional, that are present in all or most metazoans. In fact, zoologists normally group together various invertebrate phyla on the basis of the presence or absence of several fundamental features; excluding those already mentioned, the most important ones are discussed briefly below.

Symmetry

One of the most important and fundamental animal features pertains to their symmetry, that is, the arrangement of body parts relative to each other. Two basic metazoan patterns, radial and bilateral, are found. **Radial symmetry** is a type of symmetry in which the body is basically cylindrical and the parts are arranged radially around a single central axis running through both ends or surfaces (Fig. 3.6a). In a radially symmetrical metazoan there is no right or left side because each peripheral arc is identical to every other arc. In other words, an imaginary plane running through any two opposite radii divides a radially symmetrical organism into two similar halves (Fig. 3.6a). Such a symmetry is particularly useful in a sessile existence, since

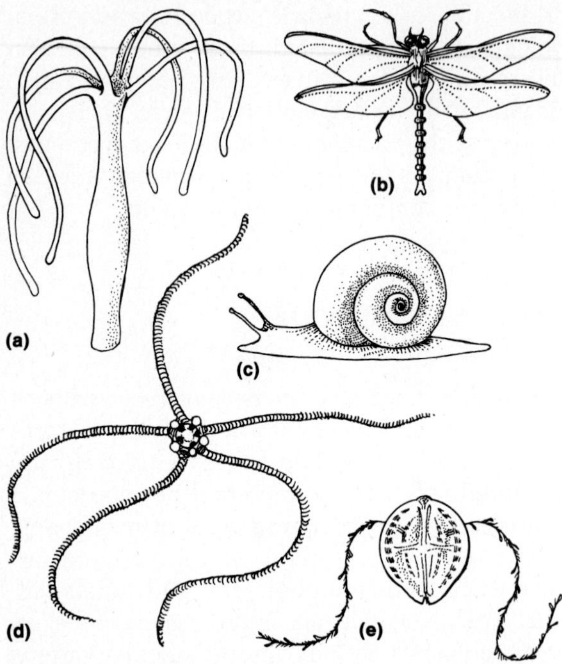

Figure 3.6 *Types of symmetry found in metazoans:* ***(a)*** *radial (hydra);* ***(b)*** *bilateral (dragonfly);* ***(c)*** *secondarily asymmetrical (snail);* ***(d)*** *secondarily radial (brittle star);* ***(e)*** *biradial (comb jelly) (parts a-d from J. W. Kimball,* Biology, *© 1983, Addison-Wesley, Reading, Massachusetts, pp. 813, 826, 828, adapted with permission).*

it enables an organism to process sensory information or gather food arriving from any point on its periphery. It is generally held that radial symmetry is primary rather than being secondarily derived from bilateral symmetry. Radial symmetry is found in some of the most primitive metazoans, and the cnidarians and ctenophores are grouped under the designation of **Radiates** because of their symmetry (see Chapter 5).

Bilateral symmetry, on the other hand, refers to the deployment of structures on either side of an imaginary longitudinal plane bisecting an animal (Fig. 3.6b–d). Thus in a bilaterally symmetrical animal there is an anterior end, which differs from the posterior end, and a right and left side, each of which is a mirror image of the other. Animals in all higher metazoan phyla (exclusive of the radiates) are characterized as being bilaterally symmetrical and are called the **Bilaterates**. In some forms like ctenophores and corals, their symmetry is basically radial; however, certain paired structures are present that modify this basic radial symmetry into a type of symmetry termed **biradial symmetry** (Fig. 3.6e).

Polarity

Polarity refers to a gradient of materials, activities, or functions along an axis. A basic overall gradient usually exists in all organisms along their major imaginary axis or plane. In radially symmetrical animals the oral end of the central axis is the point of highest activity in feeding and digestion, sensory reception and neural activity, and defense. In bilaterally symmetrical animals the head end is by far the more active functionally. Polarity is especially evident in metazoan embryology, in which the rates of development and metabolic activity proceed differentially along one or more axes. Many invertebrates can regenerate lost parts or entire body regions, and regeneration always involves the establishment of gradients or polarity along which the regenerative process is organized and proceeds.

Body Cavities

Metazoans are subdivided on the basis of whether or not they have a body cavity and, if present, what type of cavity it is. Several of the more primitive phyla, the most important of which includes the flatworms, have a solid mass of mesodermal cells and mesodermally derived muscles positioned between the outer epithelium and the gut lining (Fig. 3.7a). This means

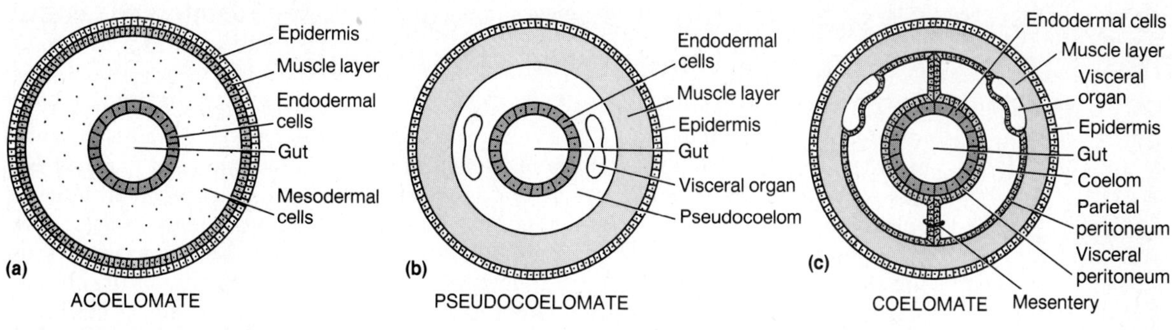

Figure 3.7 *Principal metazoan body plans based on the presence or absence of a body cavity:* ***(a)*** *the acoelomate condition;* ***(b)*** *the pseudocoelomate condition;* ***(c)*** *the coelomate condition.*

that they lack a body cavity, and they are termed **acoelomates**. The bodies of all higher metazoans contain a body cavity between the body wall and the digestive tract. Representatives of a group of seven phyla, often referred to as the aschelminth phyla and including nematodes and rotifers, have a body cavity without peritoneal (mesodermal) linings called a **pseudocoelom**; they are said to be the **pseudocoelomates** (Fig. 3.7b). The name pseudocoelom should connote the idea that such a cavity is not a true coelom rather than inferring that it is a false cavity. The pseudocoelom in most aschelminths represents the persistent blastocoel or cavity of the blastula; but in some the pseudocoelom is formed in several other ways. Regardless of its method of formation, a pseudocoelom is a body cavity lacking peritoneal linings, bounded on the outside by the body wall tissues and on the inside by the tissues of the digestive tract (Fig. 3.7b). This means that the internal organs lie free within the pseudocoelom.

All other metazoans are **coelomate**, that is, they have a body cavity or coelom that develops within the mesoderm and that is always bounded on either side by mesodermally derived peritoneal layers (Fig. 3.7c). That peritoneum on the outside of the coelom is the **parietal peritoneum**, and that on the inside covering the digestive tract is the **visceral peritoneum**. Often, the two peritoneal layers fuse dorsally or ventrally to form a **mesentery**, which suspends or holds in place the viscera so that they are not free within the coelom. It will be instructive to remember that a coelom may be formed as a split within the mesodermal mass (schizocoelom) or by outpocketings from the developing gut (enterocoelom).

The various spaces and body cavities within metazoans are usually filled with water and are a major component in what are known as **hydrostatic skeletons**. Hydrostatic skeletons are the subject of Box 3.2. Discussions of specific hydrostatic skeletons will appear in a number of places in succeeding chapters.

MAJOR PHYLOGENETIC RELATIONSHIPS AMONG THE METAZOANS

It should prove instructive to conclude this chapter, and to serve as the preface to the remainder of the text, by briefly reviewing the major steps in the evolution of metazoans. After the multicellular condition was developed, two

BOX 3.2 □ INVERTEBRATE HYDROSTATIC SKELETONS

Invertebrates uniformly lack a bony endoskeleton like that of vertebrates to which muscles can attach. Representatives of some invertebrate taxa have hard parts as exoskeletons (arthropods, molluscs) or endoskeletal plates (echinoderms) that serve as points of attachment for muscles that facilitate movement and locomotion. In most other invertebrate metazoans the bodies are soft and have no hard skeletal structures for somatic muscles. Therefore one might logically conclude, solely because of the absence of a hard skeleton, that these numerous metazoans would have extremely inefficient locomotive patterns.

However, a remarkably efficient skeletal system is present in most invertebrates that is called a **hydrostatic skeleton**. This type of skeleton is very common, extremely simple, and very effective. A hydrostatic skeleton is an internal, fluid-filled, closed space surrounded by layers of muscles. The fluid in the internal space is basically water, a uniformly available material that has three important properties: its incompressibility, its capacity for transmitting pressure changes equally in all directions, and the ease with which it can be deformed (its low viscosity). In most groups, circular and longitudinal muscles surrounding the fluid-filled space are arranged at right angles to each other, and one type has an antagonistic action to that of the other. Contraction of one set of muscles exerts pressure in the fluid, which is then transmitted in all directions to the remainder of the body. Contraction of the circular muscles forces the fluid toward either end and thus produces an elongated thin space and an extended animal. Contraction of the longitudinal muscles pulls the ends inward to force the fluid laterally so that the animal shortens and thickens in diameter. By alternate contractions of these muscle layers, invertebrates can creep, crawl, swim, and extend parts of the body in a surprisingly efficient means.

The invertebrate hydrostatic skeleton can take many forms and shapes such as a gastrovascular cavity (radiates, acoelomates); a rhynchocoel (nemerteans); a pseudocoelom (aschelminths); a coelom (annelids, many other groups); a hemocoel (molluscs); or a specialized water–vascular system (echinoderms). Invertebrates have capitalized on the principal of hydrodynamics to achieve various forms of movement and locomotion for various functions. Their hydrostatic skeleton is a prime example of adapting major body functions to this simple but efficient principle of hydrodynamics.

basic evolutionary lines were established: animals in one line evolved radial symmetry and gave rise to the radiates, while metazoans in the other pathway evolved bilateral symmetry and produced all the other metazoans. In the Bilateria, two basic fundamental lines arose, the Protostomia and the Deuterostomia. Some of their differences have already been mentioned. More specific characters include the presence or absence of metamerism, a lophophore (a specialized, tentacle-bearing, feeding organ), and the type (if any) of a body cavity. If one compares Table 3.2 and Fig. 3.8, the phylogenetic relationships will become much clearer. We shall return to phylogeny at the conclusion of each of the subsequent chapters.

TABLE 3.2. □ **OUTLINE OF THE METAZOAN PHYLA (Major* Phyla in Capital Letters)**

	Phylum
I. Poorly defined or no tissues; no organs	Placozoa PORIFERA Mesozoa
II. Well-defined tissues, organs; digestive systems usually present	
A. Radially symmetrical	CNIDARIA Ctenophora
B. Bilaterally symmetrical	
1. Protostomate	
a. Acoelomate	PLATYHELMINTHES Nemertea Gnathostomulida
b. Pseudocoelomate	Gastrotricha Rotifera Kinorhyncha NEMATODA Nematomorpha Acanthocephala Loricifera†
c. Coelomate	
i. With a lophophore	Bryozoa Entoprocta
ii. Without a lophophore	
(a) Nonmetameric	MOLLUSCA Priapulida Tardigrada Pentastomida
(b) Metameric (and related phyla)	ANNELIDA Pogonophora Sipunculida Echiura Onychophora ARTHROPODA
2. Deuterostomate	
a. Nonmetameric	
i. With a lophophore	Phoronida Brachiopoda
ii. Without a lophophore	ECHINODERMATA Chaetognatha Hemichordata
b. Mostly metameric	CHORDATA

* Major is defined as having 5000 or more species.

† A preliminary placement that will probably change with additional information.

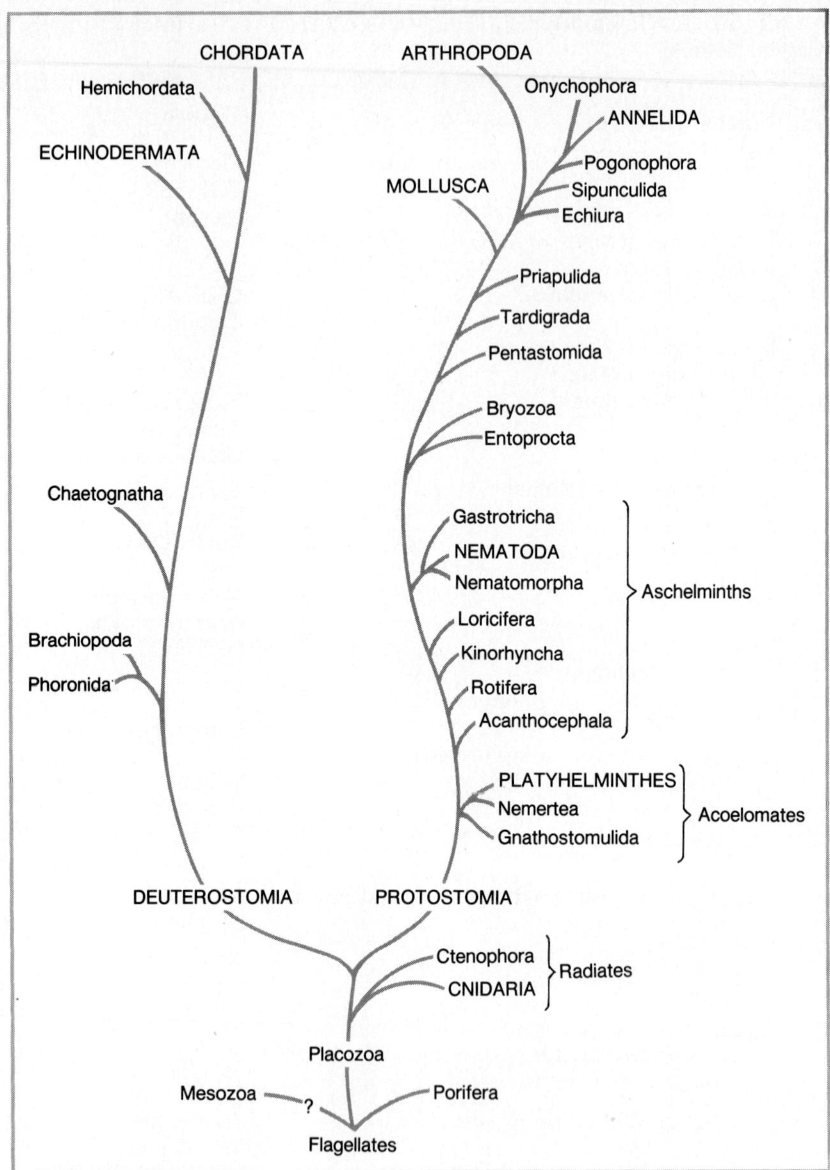

Figure 3.8 One interpretation of the phylogenetic relationships among the 33 metazoan phyla.

ADDITIONAL READINGS

Berrill, N.J. and G. Karp. 1976. *Development*, pp. 131–181. New York: McGraw-Hill Book Co.

Brusca, G.J. 1975. *General Patterns of Invertebrate Development*, 129 pp. Eureka, Calif.: Mad River Press.

Gilbert, L.I., ed. 1981. *Metamorphosis*, 2nd ed., 598 pp. New York: Plenum Publishing Co.

Grell, K.G. 1971. *Trichoplax adhaerens* F.E. Schulze und die Entstehung der Metazoen. *Naturwissenschaftliche Rundschau* 24:160–161.

Hadzi, J. 1963. *The Evolution of the Metazoa*, 466 pp. New York: Macmillan Co.

Hanson, E.D. 1977. *The Origin and Early Evolution of Animals*, 670 pp., Middletown, Conn.: Wesleyan University Press.

House, M.R., ed. 1979. *The Origin of Major Invertebrate Groups*, 494 pp. The Systematic Association Special Volume 12. New York: Academic Press.

Kristensen, R.M. 1983. Loricifera, a new phylum with Aschelminthes characters from the meiobenthos. *Zeit. Zool. Syst. Evolut.-forsch.* 21:163–180.

4

The Poriferans and Other Simple Metazoans

OVERVIEW

There are some uniquely simple and primitive metazoans comprising three phyla—Placozoa, Porifera, and Mesozoa—and they are distinguished from all other metazoans by the lack of tissues or organs.

Trichoplax adhaerens, the only species in the Phylum Placozoa, is a small, irregularly shaped marine organism. Because it is probably the simplest, most primitive of all metazoans, it has great phylogenetic significance.

Poriferans (sponges) are mostly small, sedentary, aquatic, and mostly marine animals whose bodies contain numerous small apertures. They are unique among metazoans because each cell functions mostly independently and because their functional morphology is built on creating an internal water current. The body wall consists of outer and inner cell layers and a middle mesohyl. The heterogeneous mesohyl contains several different cell types and a skeleton composed of organic fibers or microscopic inorganic spicules. Most sponges contain both fibers and siliceous spicules. Each cell of the inner layer, a choanocyte, bears a cytoplasmic collar surrounding a flagellum. The uncoordinated, nonsynchronized beat of all the choanocytic flagella drives the internal water current.

There are three levels of sponge architecture reflected basically in the complexity of the water channels. The asconoid body is a simple sac surrounding a large central cavity lined by choanocytes. The body wall of syconoid sponges is folded extensively, choanocytes line these radially arranged water channels, the central cavity is reduced, and the mesohyl is more extensive. Leuconoid sponges have the organizational pattern found in most sponges with numerous ovate chambers lined with choanocytes. The water channels and mesohyl are the most complex in leuconoid construction.

In the absence of tissues or organs, all organismic functions take place at the cellular level. Food particles are captured and ingested by phagocytic amebocytes of the mesohyl and by the screening effect of the choanocytic collars. Gametogenesis, fertilization, and early embryonic development typically occur in the mesohyl. A free-swimming, flagellated larva is produced that soon settles, attaches, and develops into a new sponge. Asexual reproduction is achieved by budding, fragmentation, and gemmules (special reproductive bodies); sponges have great powers of regeneration. The four classes are distinguished from each other principally on the types of skeletal elements present.

The Phylum Mesozoa consists of a small number of species of simple wormlike creatures parasitic in marine invertebrates. Each is simply constructed of a fixed number of outer cells enclosing one to several sex cells. Mesozoans have rather complicated life cycles including a free-swimming larval stage.

The phylogeny of these three groups is difficult, especially that of the enigmatic *Trichoplax* and the mesozoans. *Trichoplax* may indeed represent the ancestral form from which most metazoans evolved, and it probably is a significant evolutionary link between colonial flagellates and primitive radiates. The mesozoans may represent extremely degenerate flatworms, but most zoologists consider them to be an aberrant primitive group. They possibly evolved independently from other metazoans by having been derived from a ciliate ancestry. Sponges evolved either from choanoflagellates, a line different from that of most other metazoans, or very early from the metazoan line; in either case they represent an evolutionary dead end.

INTRODUCTION

We begin our study of multicellular animals by considering the most primitive, simplest metazoans, which are placed in three phyla: Placozoa, Porifera, and Mesozoa. The Phylum Porifera (sponges) contains by far the largest number of species and the most familiar organisms of the three phyla. The Phylum Placozoa consists of only one species, and there are fewer than 50 mesozoan species. The phylogenetic affinities of the three groups are not clear, but they probably are not closely related evolutionarily. The common feature of animals in all three phyla, however, is that they are very simple, primitively constructed metazoans.

PHYLUM PLACOZOA

The Phylum Placozoa is represented by only one known species, *Trichoplax adhaerens.* Its great phylogenetic significance as probably the most primitive living metazoan was discussed in Chapter 3. Following its initial discovery in 1883, it was dismissed as a larval stage of a cnidarian. Rediscovered in 1969, it was found to be an adult of a very simple, primitive metazoan.

Trichoplax is a very small, asymmetrical, marine organism. Its body is flattened, and its irregular shape can be altered like that of an ameba (Fig. 4.1a). The body consists of two layers of flagellated epithelial cells surrounding an inner sheet of loose stellate **fiber** or **mesenchymal cells** (Fig. 4.1b). The thin dorsal epithelium is constructed of **cover cells,** each bearing a flagellum and containing numerous vacuoles with lipid materials. The lower epithelium is much thicker and consists of flagellated **cylinder cells** and nonflagellated **gland cells** (Fig. 4.1b). As *Trichoplax* creeps over a surface, it overruns organic particles or very small organisms, which are digested and absorbed by the lower epithelium. *Trichoplax* can elevate its body and thus creates a concavity on its lower surface. In this shallow depression, food and digestive enzymes are concentrated so as to make digestion and absorption more efficient (see Fig. 3.2d). The inner fiber cells probably function in some manner in locomotion. *Trichoplax* reproduces asexually by both fission and budding when population densities are low. When conditions are crowded, sexual reproduction takes place; eggs

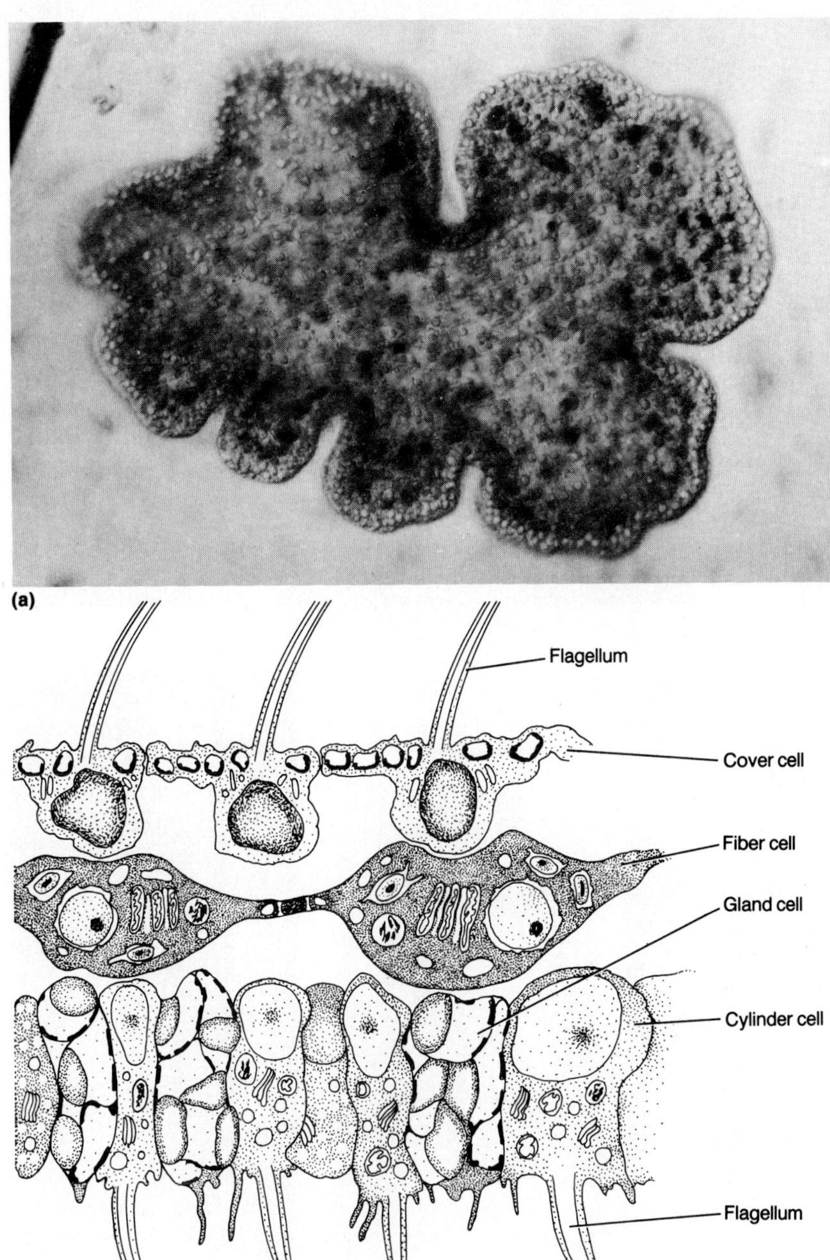

Figure 4.1 Trichoplax adhaerens, *the only known placozoan:* ***(a)*** *a photomicrograph, dorsal view (courtesy of K. G. Grell);* ***(b)*** *a diagrammatic section through a portion of the body.*

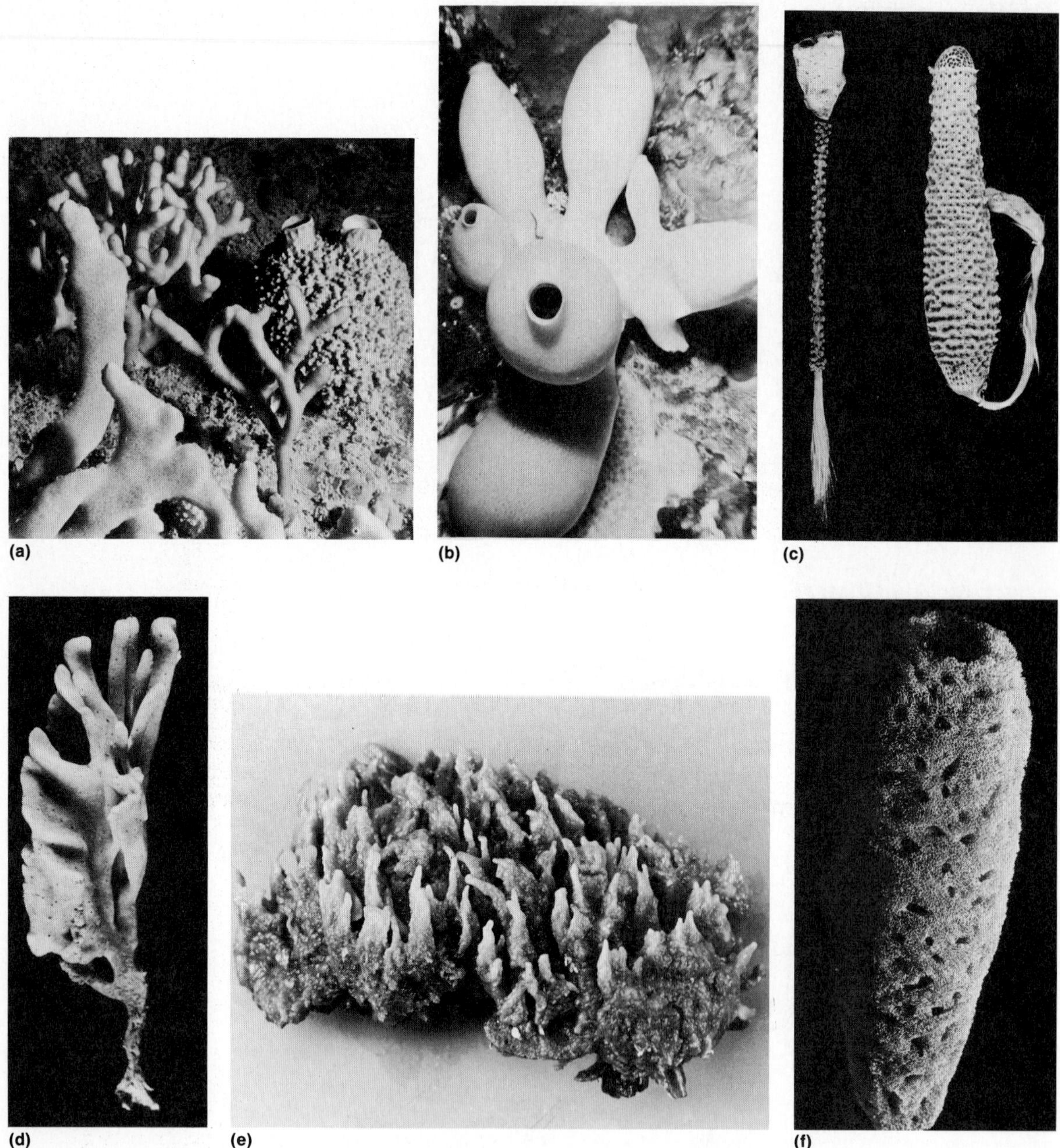

Figure 4.2 *A photographic collage of types of sponges:* ***(a)*** *a typical sponge community found on rocky reefs;* ***(b)*** Leucettusa, *a calcarean sponge;* ***(c)*** Hyalonema *(left) and* Euplectella *(right), hexactinellid sponges (courtesy of C. F. Lytle);* ***(d)*** *a branching marine demospongian;* ***(e)*** *a* Hymeniacidon, *a marine demospongian;* ***(f)*** *a vase-shaped marine demospongian; (parts a, b from P. R. Berquist,* Sponges, *Berkeley, University of California Press, 1978.)*

have been identified in the internal layer, but other details of sexual reproduction are poorly known.

Trichoplax is remarkably simple. It undoubtedly represents the simplest, most primitive living metazoan known.

PHYLUM PORIFERA

The Phylum Porifera is a moderately sized group of about 5000 species of asymmetrical aquatic metazoans, the sponges. Sponges were long considered to be plants. It was not until the middle of the 19th century that the animal nature of sponges was finally determined. All adult sponges are permanently sessile, although most forms have a motile larval stage. The vast majority are marine, but there are a number of freshwater species found in clean lakes, ponds, and streams. While representatives of one class (Hexactinellida) live at considerable depths in the oceans, marine sponges are generally found in shallow coastal waters. They are common on shells, rocks, corals, or almost any submerged object, and some live attached in soft sand or mud bottoms. Sponges vary greatly in size from 1 cm to a meter or more in diameter and up to 2 m in height. While some sponges are drab or have inconspicuous coloration, the majority are brightly colored. Many contain carotenoid pigments that give them a red, orange, or yellow color, and some are green because of the presence of symbiotic algae living within the sponge. Some diversity in poriferans is illustrated in Fig. 4.2, and the four classes are characterized in Table 4.1.

There are two fundamental distinguishing features that characterize sponges. First, a given sponge cell exhibits a very high degree of independence from other constituent cells. In fact, sponges represent a level of organization not substantially advanced over a complex colonial protozoan, and individual sponge cells are often considered to maintain an almost protozoan independence. As a result of this cellular autonomy, no true tissues or organs are present as in most other metazoans. Second, sponge architecture is preeminently concerned with maintaining a continuous water current through the animal. Sponges are adapted to move rather high volumes of water at low pressures through a system of internal water channels, and almost all of their functional morphology is related in some way to this unidirectional flow. As we shall see later, the

TABLE 4.1. □ **BRIEF CHARACTERIZATIONS OF THE CLASSES OF SPONGES**

Class	*Level of Organization*	*Nature of Skeleton*	*Other Important Features*
Calcarea	Ascon, sycon, leucon	Calcareous spicules	Small; spicules usually separate
Hexactinellida	Leucon	Hexactine siliceous spicules	Called glass sponges because spicules are fused to form an elaborate, glassy, skeletal latticework; no pinacoderm; mostly deep-water forms
Demospongiae	Leucon	Siliceous spicules, spongin, both, or neither	Includes most sponges and all freshwater forms; spicules usually not hexactines
Sclerospongiae	Leucon	Siliceous spicules, spongin fibers, and veneer of $CaCO_3$	Found in crevices in coral reefs

water current subserves all organismic functions of the poriferan including feeding, circulation, removal of wastes, gaseous exchange, and gametic transfer. In short, sponges are rather independent aggregations of cells organized more or less solely to generate a water current through the body, and these two important features dictate most of the facets of their biology.

What is this important sponge feature that is responsible for driving the current of water internally? **Choanocytes,** a type of metazoan cell unique to sponges, are these strategic and vital cells, since the combined action of their flagella is the hydrodynamic force. Choanocytes are irregularly or ovately shaped, and each bears an apical, centrally located flagellum (Fig. 4.3). The flagellum is situated toward the excurrent water channels, and the spiral flagellar motion progresses from base to tip, thus pulling water over the choanocyte. There is neither coordination of flagellar beat nor synchrony of the choanocytic action, but the net effect of all the choanocytes is a unidirectional water flow. Surrounding the flagellum is a circlet of 20–40 **microvilli** (sometimes referred to as pseudopodia or tentacles) that constitute a cytoplasmic collar, a feature that gives choanocytes their common name, collar cells (Fig. 4.3). The microvilli, connected by a fine mucous reticulum, are used in feeding, an important function to be described later. Choanocytes are very similar in construction to choanoflagellates (compare Figs. 2.6a and 4.3), a fact that strongly suggests that sponges may have evolved from these zooflagellates.

Figure 4.3 *A scanning electron micrograph of several choanocytes of* Clathrina *(courtesy of L. De Vos and N. Boury-Esnault).*

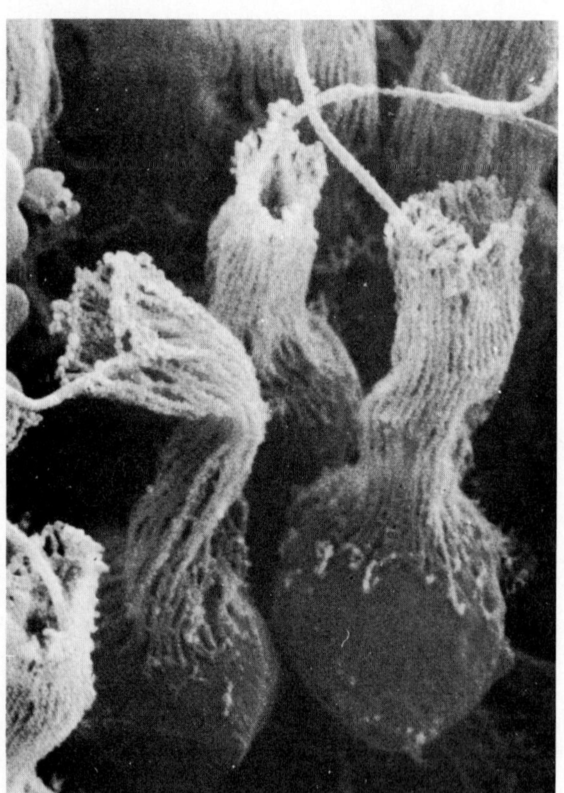

Sponge Structure and Levels of Organization

Even though the internal morphology of sponges is variable, the diversity of their architecture can be reduced to three levels of complexity: ascon, sycon, and leucon. The fundamental distinction between the three levels of organization resides in differences in their functional morphology associated with internal water currents.

The **asconoid** pattern, the simplest and most primitive of the three organizational patterns, is found only in adults of two primitive genera of the Class Calcarea (*Leucosolenia* and *Clathrina*). These sponges are always small and usually live in clusters or colonies. The body is a simple tubular unit or sac enclosing a central cavity, the **spongocoel** or **atrium,** which opens to the outside by way of a relatively large opening, the **osculum** (Fig. 4.4a). The body wall consists of an outer layer of flattened cells called

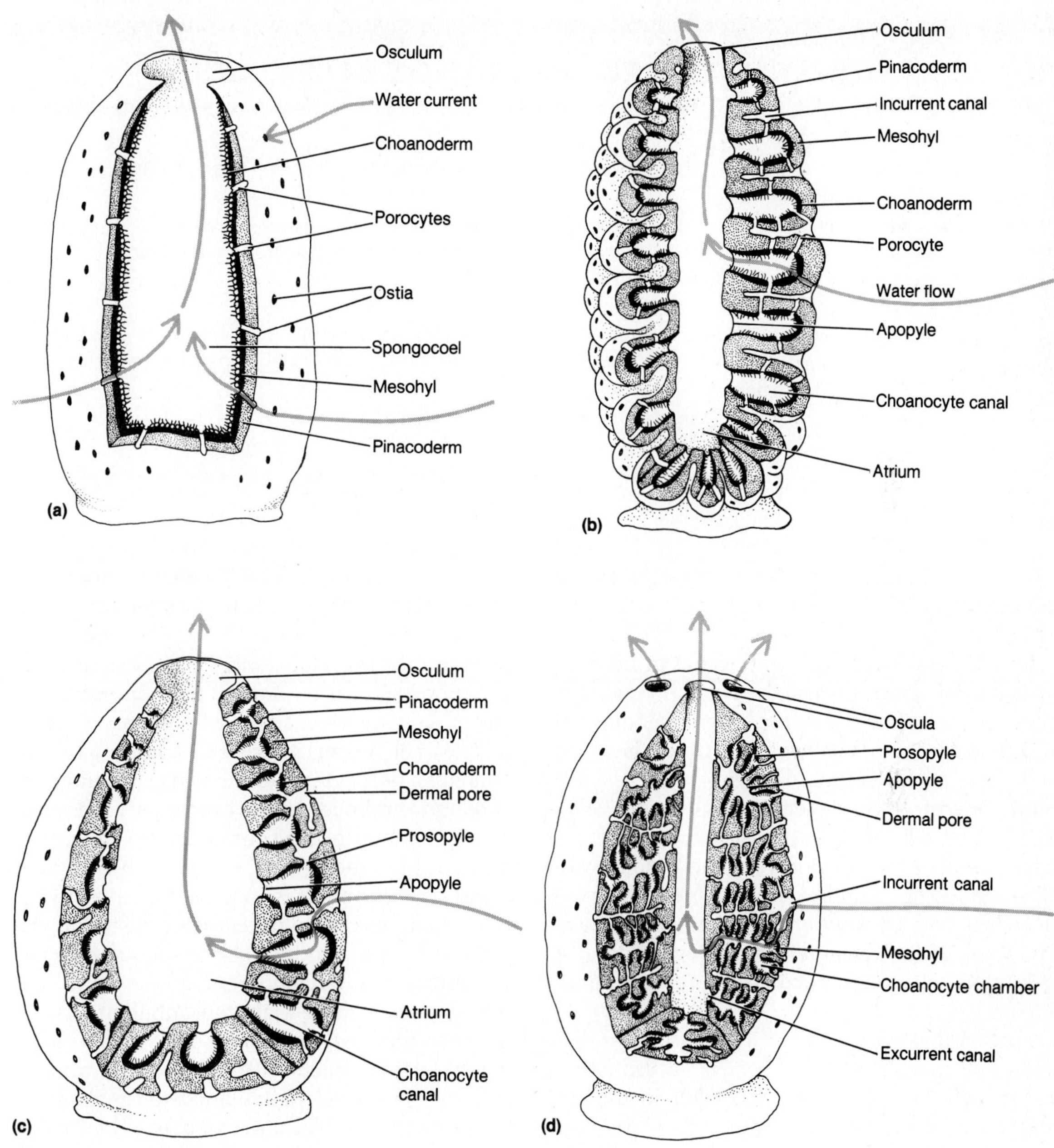

Figure 4.4 *Organizational patterns present in sponges:* ***(a)*** *asconoid;* ***(b)*** *simple syconoid;* ***(c)*** *more complex syconoid;* ***(d)*** *leuconoid.*

BOX 4.1 □ PATHWAYS OF WATER CIRCULATION IN SPONGES

A. Asconoid
Exterior → pore canal of porocyte → spongocoel → osculum → exterior

B. Syconoid
1. *Simple*
Exterior → pore canal of porocyte → choanocyte canal → apopyle → spongocoel → osculum → exterior
2. *Complex*
Exterior → superficial ostium → inhalent lacuna → porocyte or prosopyle → choanocyte canal → apopyle → reduced spongocoel → osculum → exterior

C. Leuconoid
Exterior → superficial pore → incurrent canal → prosopyle → choanocyte chamber → apopyle → excurrent canal → osculum → exterior

pinacocytes and collectively referred to as the **pinacoderm. Porocytes,** an abundant cell type, are scattered through the pinacoderm. Each is a tubular cell enclosing a **pore canal** or **ostium,** which is an incurrent water opening. It is from this feature that the phylum name was derived (Porifera = hole bearer). The inner layer of cells, i.e., that lining the spongocoel, is the **choanoderm** or the layer of choanocytes (Fig. 4.4a). The collective flagella of all choanocytes drive a flow of water in a very simple pathway through the sponge (Box 4.1). Between the pinacoderm and choanoderm is a heterogeneous layer called the **mesohyl** consisting of cells, a gelatinous intercellular matrix, and skeletal materials.

From the asconoid plan, folding of both the pinacoderm and choanoderm and a reduction in the size of the spongocoel result in the **sycon** level of organization found in a number of genera of the Class Calcarea. Syconoid organization represents an increase in complexity in the pathways of the internal current and enhancement of the water flow efficiency (Box 4.1). The choanoderm is folded outward to line a series of radially arranged **water** or **choanocyte canals** (Fig. 4.4b, c). The pinacoderm is also folded outward to invest these projections. This folding increases substantially the extent of the choanoderm, since the greater numbers of choanocytes present markedly enhance the efficiency of water flow. In simple syconoid sponges, porocytes permit water to enter the choanoderm-lined water canals as each water canal opens into the central atrium by way of an opening, the **apopyle** (Fig. 4.4b). In more advanced syconoid sponges the mesohyl is thickened, and the outer ends of the radial projections fuse to produce a smooth outer surface or cortex. The development of a cortical area necessitates a more elaborate inhalent system. In these sponges, superficial ostia open into inhalent lacunae lined with pinacoderm, and the lacunae open into the choanocyte canals either by porocytes or by intercellular pores called **prosopyles** (Fig. 4.4c). Since choanocytes are restricted to the water canals, the reduced spongocoel is lined with pinacoderm, and the atrium opens to the outside by way of a single osculum.

Leuconoid construction, found in most of the Class Calcarea and in all other sponges, results from further elaboration of the choanoderm and mesohyl and an increase in the complexity of the water canals. In leuconoid sponges the choanoderm is subdivided into numerous discrete spherical or ovate chambers called **choanocyte** or **flagellated chambers** (Fig. 4.4d). Both a marked increase in numbers of choanocytes and restriction of choanocytes to the chambers greatly increase the efficiency of water flow. Each chamber is supplied with one or more incurrent canals whose openings into the chamber are prosopyles, but a single apopyle leads into an exhalent channel (Fig. 4.4d). Exhalent channels coalesce to form larger channels that merge toward several oscula. (See Box 4.1 for water pathways.) In leuconoid sponges the volume of mesohyl is considerably larger and contains a greater diversity of cells and skeletal elements.

It should be kept in mind that the three levels of sponge construction refer to structural complexity rather than to taxonomic groups. Each of these three terms conveys a concept of the distribution of the choanoderm, folding of the pinacoderm, and complexity of the mesohyl, all of which are related to intricacies of water flow. In some sponges there is a developmental as well as a proposed evolutionary progression beginning with ascon, progressing to the syconoid level, and culminating with the leuconoid construction. In most other sponges, however, there is no phylogenetic or developmental sequence present or implied, since leuconoid construction has been evolved independently in several different lines of sponges.

SKELETON

All sponges possess a unique skeleton that is an integral feature of their functional morphology. Of what value is a skeleton in such simply constructed metazoans as sponges? A skeleton of some sort is absolutely necessary to prevent deformation of the sponge body in the creation and maintenance of an internal water flow. A skeleton is also essential for support of the sponge cells growing upward from the substratum. The nature and type of skeletal elements present in a sponge are the principal morphological features distinguishing various taxa. Subdivided into organic and mineral forms, the organic skeletal elements are collagen and spongin, and the mineral elements are minute spicules.

The fundamental component of the skeleton of most sponges is a fibrillar collagenous framework comprising much of the mesohyl. This intercellular proteinaceous matrix is secreted by certain cells, the **collencytes.** Collagen fibers give strength to and support for the extracellular mesohylar matrix. A specialized form of collagen called **spongin** is unique to sponges and is found in those of the Class Demospongiae; spongin is secreted by cells termed **spongocytes.** Usually present as spongin fibers, this skeletal element forms a complex, supportive, strengthening network (Fig. 4.5a).

An inorganic skeleton composed of spicules is present in most sponges. **Spicules** are microscopic, often discrete, mineral structures secreted by specialized cells, the **sclerocytes.** Members of the Class Calcarea have only calcareous spicules, usually in the form of calcite ($CaCO_3$), but the spicules in all other sponges are siliceous, usually as silicate (SiO_2). Most sponges in the Class Demospongiae, and therefore the vast majority of all sponges, have both siliceous spicules and spongin fibers. In most sponges, except those in the Class Calcarea, in which there is no spicule differentiation based on size, smaller spicules are termed **microscleres,** and the larger supportive spicules are **megascleres.** Spicules are quite diverse in their size, shape, number of prongs, and in methods of formation (Fig 4.5b, c). An elaborate system

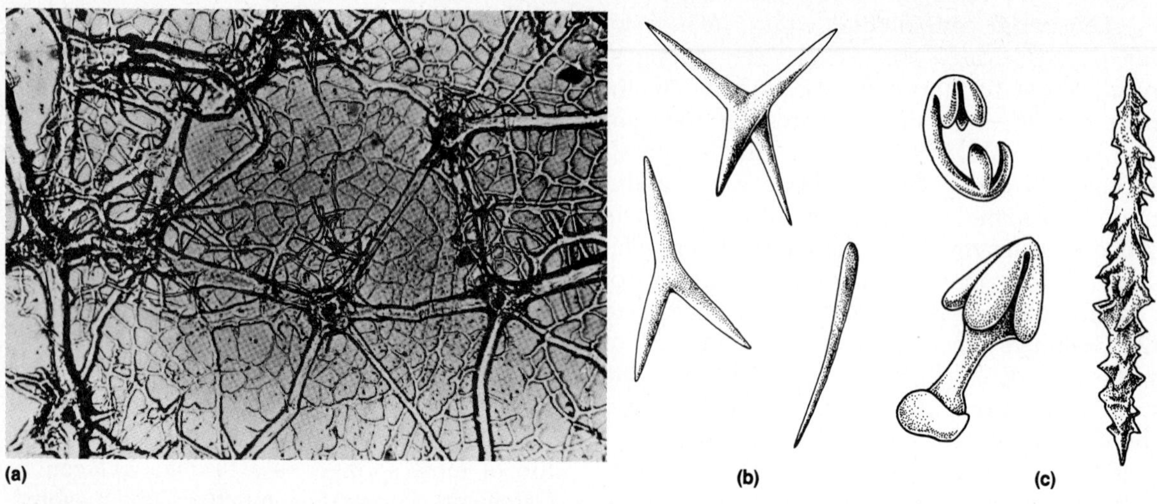

Figure 4.5 *Sponge skeletal elements:* ***(a)*** *a photomicrograph of the superficial dermal fiber skeleton of a demospongian (from P. R. Bergquist,* Sponges, *Berkeley, University of California Press, 1978);* ***(b)*** *several calcareous spicules;* ***(c)*** *various siliceous spicules from demospongians.*

of spicule nomenclature has been developed that is widely used in taxonomic identifications. Since this system is both very tedious (to many, boring) and beyond the scope of this book, only a few basic terms will be introduced. The number of axes is reflected in terms with the suffix "-axon," and the number of rays or prongs present is indicated by a number with the suffix "-actine." For example, perhaps the most common spicule type is a monaxonal diactine, a spicule that has a pointed terminus at either end of a single axis. Hexactinellid sponges have hexactinal or six-rayed spicules, and calcareous sponges have triaxonal triactines (Fig. 4.5b).

Spicules are formed either intracellularly within a sclerocyte or extracellularly by several sclerocytes. Sclerocytes have the remarkable capability of accumulating calcium or silicon to be used in spicule formation. This is especially unusual in the case of siliceous spicules; the concentration of silicon is so very low in seawater (about 3 ppm), and yet these sclerocytes manage to concentrate and accumulate considerable quantities of silicon and form intricately constructed spicules. In siliceous spicules the silicate is secreted around an organic filament, but no axial template is present in calcareous spicules. Box 4.2 and Fig. 4.6 provide more details on the formation of calcareous spicules by the cooperation of several sclerocytes.

Major Cell Types

Since the constituent cells of sponges are relatively independent of each other, it will be most instructive to understand the functional morphology of each principal cell type. The following discussions will be based primarily on cell types of sponges of the Class Demospongiae, since most sponges belong to this taxon and more information is available about them than about other types.

BOX 4.2 □ THE COOPERATIVE SECRETION OF A CALCAREOUS SPICULE BY SCLEROCYTES

Since sponges have no nervous system and the constituent cells operate more or less independently of each other, the cooperation of sclerocytes to produce a very precise spicule is nothing short of amazing. Let's examine closely the formation of a calcareous triaxon spicule to illustrate cellular cooperation in sponges.

Three sclerocytes aggregate together, the nucleus of each divides (Fig. 4.6a, b), and a thin ray or sliver of calcite is secreted between each nuclear pair (Fig. 4.6c). The binucleate cells then undergo cytokinesis in such a way that three of the aggregate cells are central and three are outside the central cluster. The inner sclerocyte of each pair then moves outward progressively, lengthening the prong; they are known as **founder cells,** since they add to the ray's length. The other three cells, known as **thickener cells,** remain for a time at the junction of the forming rays (Fig. 4.6d). Later, each thickener cell follows a founder cell, adding calcium carbonate to the ray. Eventually, both cells leave the ray and disintegrate.

There is no ready explanation as to how cooperation is achieved between independent sclerocytes to form a spicule that is very precise and predictable in its geometry. What is the genetic source of information that dictates the precision and regularity of this process? Is it in the genome of the sclerocytes, or does it represent an example of organismic coordination whose underlying control is yet to be discovered? Further research may elucidate some of these vexing questions.

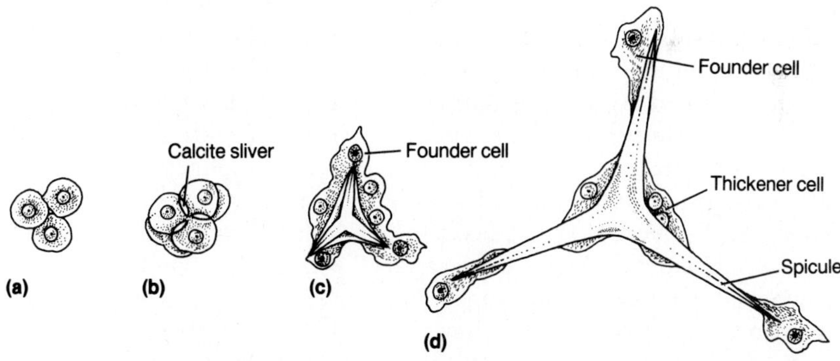

Figure 4.6 *Secretion of a calcareous triaxonal spicule:* ***(a)*** *three sclerocytes aggregate;* ***(b)*** *sclerocytes divide and begin to form calcite slivers;* ***(c)*** *calcite slivers are united centrally, and founder cells lengthen the spicule rays;* ***(d)*** *later stages of spicule formation with founder cells having completed the ray lengths; thickener cells remain in the axes of the spicule and later migrate along rays adding calcite (parts b–d from L. H. Hyman,* The Invertebrates. *Vol. I,* Protozoa through Ctenophora, *New York, McGraw-Hill, 1940, adapted with permission).*

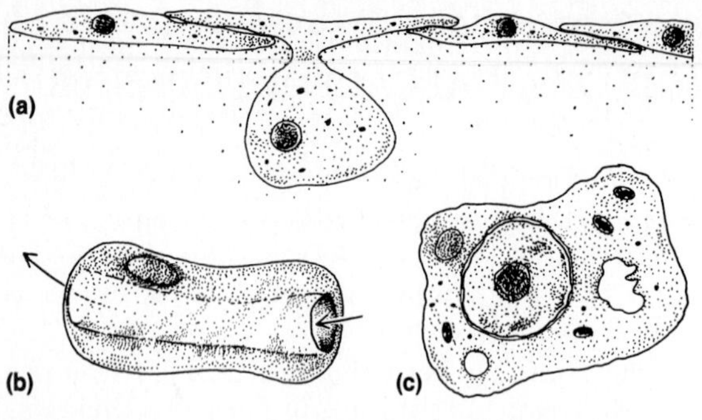

Figure 4.7 *Principal cell types found in sponges:* ***(a)*** *pinacocytes, with fusiform cells alternating with globular "T" cells;* ***(b)*** *a porocyte;* ***(c)*** *an archeocyte.*

Surface cells

Several different cell types are present on various surfaces of a sponge. **Pinacocytes** occur on the outer surfaces of sponges and line all exhalent and inhalent canals. Each cell is flattened, and its margin often joins or overlaps those of adjacent pinacocytes (Fig. 4.7a). The pinacoderm is fundamentally different from an epithelial layer of higher metazoans because it lacks a basement membrane. Pinacocytes function in protection of the underlying mesohyl, in phagocytosis, and in attachment of the sponge to a substratum. **Porocytes** serve as components of the inhalent system of asconoid and simple syconoid sponges. Derived from pinacocytes, each is a tubular cell surrounding a pore canal through which water enters the sponge (Fig. 4.7b). Since porocytes are contractile, they can regulate the diameter of the pore canal and thus control the inhalent water flow.

Choanocytes, a most important cell type, have already been described (see Fig. 4.3). In asconoid sponges, choanocytes line the spongocoel, but in syconoid and leuconoid forms they are found only in the choanocytic canals or chambers. Like the pinacoderm but unlike epithelial layers in other metazoans, the choanoderm lacks a basement membrane, and the choanocytes rest directly on the mesohyl.

Cells in the mesohyl

The mesohyl is a continuously changing, dynamic system for two reasons. First, most cells found in the mesohyl are freely motile. Second, pinacocytes and choanocytes often dedifferentiate, become motile, and move into the mesohyl. Therefore the mesohyl is a kinetic system and always in a state of flux.

The cells responsible for the formation of collagenous and spongin fibers and calcareous and siliceous spicules have already been mentioned. **Collencytes** are very similar to pinacocytes and are probably derived from them. **Spongocytes** are distinguished from collencytes by the fact that their secreted fibers, spongin, are slightly different from collagenous fibers. **Sclerocytes** are motile, spicule-secreting cells within the mesohyl (Fig. 4.6). Ameboid cells are more properly known as **archeocytes.** Indispensible as sponge cells, archeocytes are large motile cells that function in two important ways (Fig. 4.7c). First, all other cell types are derived from archeocytes; thus they are essential for growth and development. Second, archeocytes are active in phagocytosis and, along with choanocytes, are principal means of food acquisition. **Myocytes,** elongate or fusiform shaped and contractile in nature, tend to be grouped concentrically around oscula and water canals; by contraction they can regulate the diameter of these passageways and thus effect some control over water flow. Several other cell types are recognized basically by the nature of materials enclosed in intracellular vacuoles, and all are derived from archeocytes.

The Physiology of Sponges

Even though sponges lack cellular cooperation, several vital organismic functions are achieved by the activities of the collective cells. Given the fact that the most important sponge activity is undoubtedly the internal water flow, all other organismic functions are directly affected by it.

In leuconoid sponges the choanocytes are grouped into choanocytic chambers of 20–50 choanocytes each (Fig. 4.8). Their flagella are situated toward the apopyle, and the flagellar action pulls water into the chamber by way of prosopyles and out through the somewhat larger apopyles. Generated by the choanocytes, the water flow is fairly constant under normal conditions. But under adverse environmental conditions or by experimental stimuli, sponges immobilize the choanocytic flagella in several different ways, all of which retard water flow. In many sponges, each choanocytic chamber contains a specialized ameboid **central cell** (Fig. 4.8b). Under appropriate stimulation this cell moves from its central position to one near the apopyle. In so doing, the choanocyte flagella become trapped in the cell's pseudopodia and are immobilized. The immobilization of flagella and the partial obstruction of the apopyle effectively reduce current flow (Fig. 4.8b). Sponges can impede the internal current by closing the porocytic or superficial ostia, closing the oscula by contraction of the surrounding myocytes, and compressing or constricting the flagellated chambers and water channels. Moreover, in a few studies performed on several different sponges the volume and velocity of water flow were shown to be dependent on what appear to be endogenous factors. These studies do not imply the presence of an integrative system such as a nervous system, but they do suggest that sponges are capable of low-level coordinated activity.

Figure 4.8 *Choanocytic chambers:* ***(a)*** *a scanning electron micrograph of the inside of a choanocytic chamber (courtesy of L. De Vos and N. Boury-Esnault);* ***(b)*** *influence of a central cell on the operation of the chamber by inactivating the choanocytic flagella and by partially obstructing the apopyle (from P. R. Bergquist,* Sponges, *Berkeley, University of California Press, 1978).*

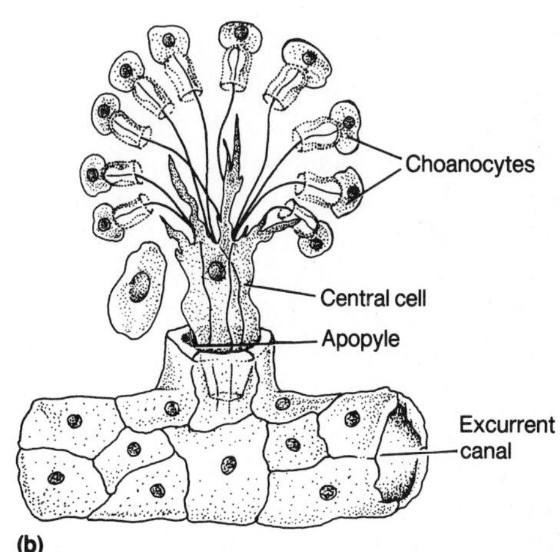

The water current is the vehicle by which most organismic functions are achieved. Archeocytes containing undigestible wastes within their vacuoles are voided in the exhalent canals. Excretory wastes, principally ammonia, diffuse into the exhalent water currents and are eliminated. Osmoregulation is not a problem in marine forms, but most cells of freshwater species contain osmoregulating contractile vacuoles. Gaseous exchange occurs by diffusion between the internal water canals and the adjoining cells made possible by the relatively large internal surfaces.

A few other features should be mentioned at this point. There is no trace whatsoever of any form of nervous system. No adult sponge is capable of any locomotion, even though the larvae of most are motile. No organs of any sort are present, and there are no means for storing food reserves.

FEEDING

Sponges feed on extremely fine particulate materials in the incurrent water. The sizes of the particles ingested are dependent first and foremost on the apertures of the ostia, which are maximally about 50 μm. (0.05 mm). Particles in the 5- to 50-μm range are trapped by the ostia, where they are phagocytized by archeocytes located near these inhalent pores. Archeocytes are therefore the initial capture system and function both in feeding and in preventing canal occlusion. There is very little selectivity of particles; detritus and small organisms such as dinoflagellates, bacteria, and other plankton are all phagocytized indiscriminately. The absence of particle selectivity means that some ingested material is nonnutritive and must be eliminated later.

The principal particle capture takes place at or on the choanocytic collars, since the vast majority of all nutrients enter the sponge via the choanocytes. As water flows into the flagellated chambers, minute detrital particles and many bacteria (0.1–2.0 μm in size) are screened out between the microvilli and on the reticular net between the microvilli. The particulate material is carried to the base of the collar, where endocytosis takes place at the choanocyte surface. The ingested food, now within a vacuole, is quickly passed to the base of the choanocyte and transferred to an archeocyte within the mesohyl. In both initial and principal particle capture systems, digestion is always intracellular within the mesohylar archeocytes. After digestion and assimilation have been completed, the archeocytes with their residual wastes are discharged into the exhalent system or directly to the exterior.

REPRODUCTION

All sponges have the capabilities of reproducing sexually; asexual methods are also common but are not ubiquitous. Generally, sponges are monoecious and produce eggs and sperm at different times, but some dioecious species are known. In forms studied carefully, choanocytes have been shown to transform into spermatogonial cells, often with every choanocyte within a flagellated chamber being involved. After losing their flagella and collars, choanocytes enter the mesohyl and become a mass of spermatogonial cells surrounded by a wall of follicle cells to form a **spermatic cyst.** Meiosis, often preceded by mitotic divisions, results in the production of numerous flagellated **sperm,** each of which is 10–15 μm in length. The origin of oogonial cells is apparently less specific. Studies have shown that in the Class Calcarea, oogonial cells are derived from transformed choanocytes, but oogonia in most Demospongiae arise from archeocytes. In either case an oogonium plus several to many special **nurse cells** are surrounded by a layer of follicle cells, all within the mesohyl. After meiosis the resulting **oocyte** engulfs the adjacent nurse cells. In excess of 1000 nurse cells can be accumulated within a single oocyte (Fig. 4.9a). Once inside the oocyte,

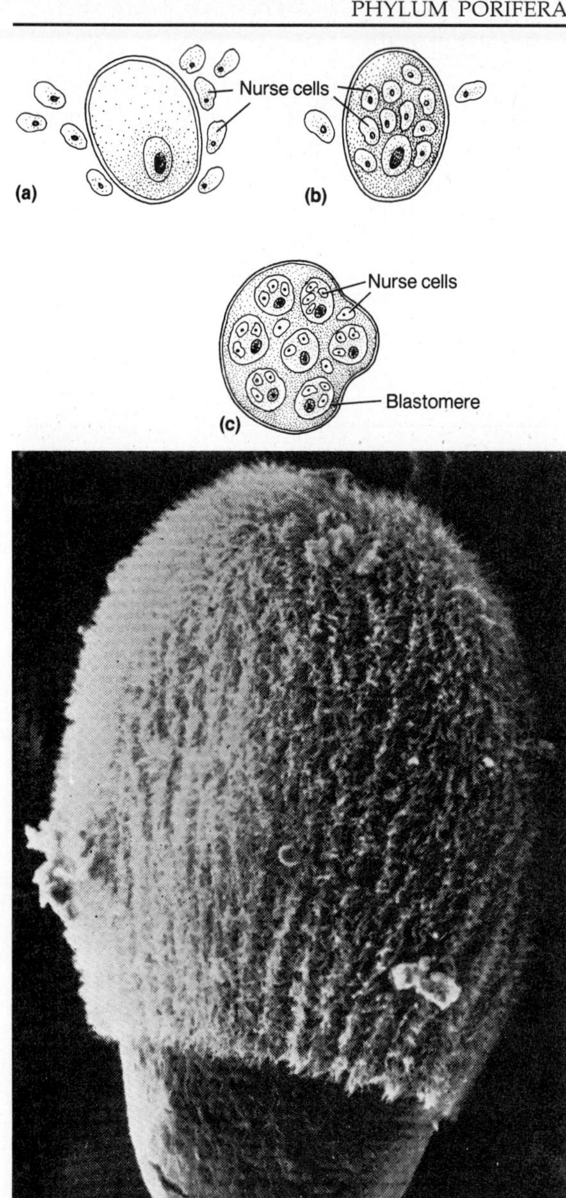

Figure 4.9 *Some stages in sexual reproduction in sponges: **(a)** nurse cells aggregate around the oocyte; **(b)** nurse cells incorporated into the cytoplasm of the zygote; **(c)** nurse cells incorporated in the blastomeres; **(d)** a scanning electron micrograph of a free-swimming parenchymella larva (from P. R. Bergquist,* Sponges, *Berkeley, University of California Press, 1978).*

nurse cells apparently retain their integrity until fertilization has taken place.

Sperm are variously released into the exhalent current and are carried to the exterior. Probably most sponges release their sperm slowly and inconspicuously. But in a few species, sperm are released suddenly in a large milky cloud; divers have reported "smoking sponges," which undoubtedly are accounts of such sudden sperm release. Sperm are carried to the interior of another sponge in its inhalent current. It is thought that when a sperm reaches a flagellated chamber, a choanocyte traps the sperm, engulfs it in a vacuole, and becomes a **carrier cell.** The carrier cell migrates through the mesohyl to the oocyte and transfers the vacuolated sperm to the oocyte. After oocytic maturation is complete, fusion of male and female nuclei result in a fertilized egg.

Early development takes place in the mesohyl of the parent sponge and is difficult to observe. In oviparous Demospongiae, fertilized eggs are released into the sea, where subsequent development takes place. Early cleavage is complete but may be regular or irregular. As blastomeres are formed, they are filled with engulfed nurse cells, the nurse cells are gradually fragmented, and their nutritive materials are slowly incorporated into the blastomeric cytoplasm (Fig. 4.9b, c). Two types of cells are recognized in sponge cleavage: smaller **micromeres** migrate to the periphery of the embryo and form a flagellated outer layer, and the larger **macromeres** move centrally and become archeocytes and other adult cells. Development proceeds rapidly to form a larval stage.

Most sponges produce a solid **parenchymella** larva, which is covered with flagellated cells (except for the posterior region) surrounding a central core of presumptive adult cells (Fig. 4.9d). A few sponges produce an **amphiblastula** larva, a hollow sphere in which one hemisphere contains flagellated cells and the other hemisphere is composed of macromeres. Both larval types are histologically very simple. The larva breaks out of the mesohyl (aided by enzymes?) and exits from the parent sponge via the exhalent current. Larval motility, which lasts for several days, is the principal dispersal mechanism.

As the larva begins its settlement process, it creeps along the bottom for a time, followed by attachment by its anterior end to a substratum. After attachment begins, a process takes place that is unique to the parenchymella larva and analogous to gastrulation in other metazoans. In brief, a reversal of layers occurs in which the inner cells move outward and the outer flagellated cells move to the interior. Some of the inner cells stream outward to broaden the point of attachment while other interior cells move to the periphery and become pinacocytes. The outer flagellated cells lose their flagella, move to the interior, develop collars and new flagella, and become choanocytes. Archeocytes differentiate into the various other cells as development ensues. In sponges with an amphiblastula larva, settlement and attachment proceed as in other sponges. Then cellular reorganization begins in which the larger macromeres overgrow by epiboly the smaller flagellated micromeres. The macromeres give rise to the pinacoderm, the micromeres produce the choanoderm, and both layers contribute to the production of archeocytes.

Asexual reproduction takes several forms in sponges. Most species reproduce asexually by the formation of buds or fragments. In all such cases the bud or fragment must include a minimum number of essential cells, mainly archeocytes. Some sponges, particularly most freshwater forms and certain marine species, reproduce asexually by forming gemmules. **Gemmules** are spherical masses of archeocytes surrounded by a thick coat of spongin and spicules and range in diameter from 0.1 to 1.0 mm (Fig. 4.10). These bodies are produced in the autumn and are capable of withstanding winter conditions. In freshwater sponges, in which gemmule formation has been studied most thoroughly, masses of archeocytes aggregate in the mesohyl and begin to divide actively. Other cells, called **nurse cells,** actively migrate to the archeocyte aggregate, where they are phagocytized by these ameboid cells. The archeocyte mass is then surrounded by a layer of ameboid spongocytes, which secretes a three-layered shell or envelope of spongin. Microscleres, secreted intracellularly in sclerocytes, are transported to the gemmule case and incorporated within it. At one end the gemmule case remains thin and devoid of spicules and becomes the **micropyle** (Fig 4.10). Because the gemmule case is composed of spongin and spicules, it is very resistant to most environmental conditions such as freezing and desiccation. In spring, after some initial development of the archeocytes near the micropyle, the gemmule hatches. As the sponge primordium emerges through the micropyle, the first cells to emerge spread out over the gemmule case and form pinacocytes and collencytes. The next archeocytes to emerge differentiate into most cell types including choanocytes. The last archeocytes to exit from the gemmule enter the developing sponge without having differentiated; these constitute a reservoir of totipotent cells for future use. The method of gemmule formation in those marine

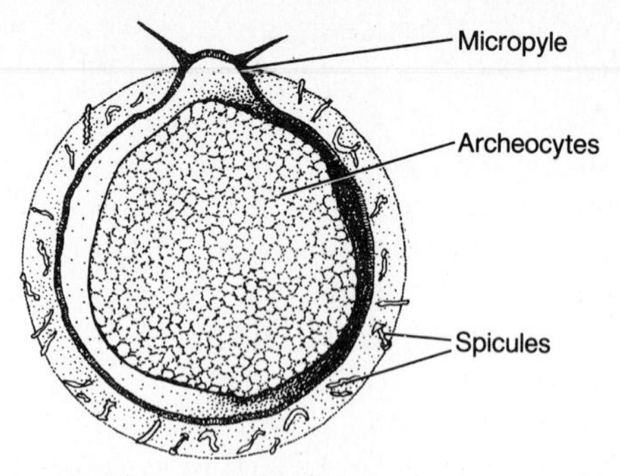

Figure 4.10 *A diagrammatic section through a gemmule of a freshwater sponge (from R. W. Pennak,* Freshwater Invertebrates of the United States, *2nd ed., New York, John Wiley & Sons, © 1978, reprinted by permission).*

sponges that produce them does not differ substantially from that just described for freshwater sponges. Gemmule production apparently is not triggered by the alteration of seasonal conditions, as once was thought. A current hypothesis is that gemmules are formed as a result of internal factors related to sponge growth and the availability of food.

Sponges have great powers of regeneration, a phenomenon that is very much related to asexual reproduction and the independence of sponge cells. Some sponges naturally will constrict regions or branches, and the detached fragments will attach and develop into new sponges. Small pieces of cultivated sponges are frequently "seeded" onto suitable substrata by commercial sponge fishermen as a means of propagating desirable sponges. Sponge cells, separated from one another by experimental means and placed in seawater, will aggregate and form several new sponges. This remarkable phenomenon was first demonstrated over 80 years ago by H. V. Wilson, who forced living sponges through a very fine silk mesh and into a seawater aquarium. Not only is it remarkable that the cells aggregate within several days, it is absolutely astounding that they ever find each other to begin with!

The Diversity of Sponges

The four classes of sponges are distinguished from each other principally on the nature of the skeletal elements. A brief discussion of each class follows.

Class Calcarea

Sponges of the Class Calcarea are known as calcareous sponges because they all have spicules composed of calcium carbonate. The spicules, mostly triactines or tetractines with some monaxons, are usually discrete, but some may be in a fused mass. Spicules are not differentiated into microscleres and megascleres. All three levels of construction—i.e., asconoid, syconoid, and leuconoid—are encountered in calcareous sponges. Both sperm and eggs are derived from choanocytes in calcareous sponges. They all have internal fertilization, and the larval stage, usually an amphiblastula, is retained by the parent until it is mature.

Most calcareous sponges are small, are comparatively simple in construction, and often live in clusters (Fig. 4.2b). They are exclusively marine, are more common to shallow coastal waters, and apparently are restricted to firm substrates. Most of the approximately 150 species are drab colored.

Class Hexactinellida

The hexactinellids are better known as glass sponges, owing to the nature and arrangement of their spicules. The spicules are always siliceous and are always six-rayed or hexactines, a feature on which the class name is based. Both microscleres and megascleres are always present. The skeleton of some species is absolutely exquisite; the spicules are arranged in intricate long fibers appearing very much like spun glass, hence their common name (Fig. 4.2c). The body is usually tubular or basket shaped and almost radially symmetrical. They range from 10 to 30 cm in height and are mostly pale colored. The approximately 100 hexactinellid species are exclusively marine and are characteristically found in seawater ranging in depth from 500 to 1500 m. Many are anchored in soft benthic material by means of long siliceous fibers at the lower end.

Hexactinellids are basically different in organization from all other sponges, a fact that strongly hints of a separate phylogenetic history for them. The tubular body surrounds a rather large central spongocoel. A single terminal osculum is often covered by a sieve plate formed from fibers constructed of fused spicules. All hexactinellids have choanocytes situated in numerous thimble-shaped, flagellated chambers in the body wall. For this reason they are more closely constructed along the leuconoid plan. But their structures are so different from those of other sponges that hexactinellids perhaps should not be defined as leuconoid; their morphology is simply unique among sponges. There is no pinacoderm in glass sponges; therefore cellular material is extremely sparse on the outer surface. Ostia are simple holes in a thin, filmlike, dermal membrane covering the skeleton, a feature that eliminates the possibility of ostial control over the incurrent water flow. The mesohyl is unusual in that the noncellular matrix is missing. The mesohyl is represented by a subdermal trabecular network formed by the union of long pseudopodia of the collencytes and archeocytes. Much of the biology of glass sponges is poorly known.

Perhaps the best known hexactinellid is *Euplectella,* whose common name is Venus' flower basket; its skeleton is the one usually appearing in museums illustrating hexactinellids (Fig. 4.2c). An interesting commensal relationship exists between individuals of this genus and certain shrimp *(Spongicola).* A very young male and female shrimp enter the sponge, grow, and become permanently trapped within the sponge. They spend their entire adult life imprisoned in the spongocoel and feeding on planktonic organisms brought to them by the sponge's water current.

Class Demospongiae

The Demospongiae includes those sponges having a siliceous skeleton that is usually supplemented or replaced by spongin. About 4700 species (95% of all sponges) belong to this class, making it the dominant and most familiar. These sponges are basically marine and are distributed from shallow coastal areas to great depths. The freshwater sponges are restricted to unpolluted water with low turbidity. Freshwater sponges are small, but many marine forms attain considerable size. Many different growth patterns are found in these sponges; upright branching, low and encrusting, irregular mounds, urn or vase-shaped, and leaflike patterns are frequently encountered (Fig. 4.2a, d-f). The diverse growth patterns are adaptations to the external water currents, nature of the substratum, and limitations in space.

The general characteristics of the phylum discussed earlier apply to sponges in this class and will not be repeated here, but the skeletal elements deserve some special mention. The megascleres are either monaxons or tetraxons, triaxons are generally absent, and the microscleres are diverse. Spongin fibers are also present in most sponges as a matrix in which the siliceous spicules are embedded. In other sponges, spongin is present as the only skeletal element in the absence of spicules. Representatives of two families have neither spicules nor spongin fibers.

Freshwater sponges, comprising two families, are typically encrusting and matlike on many submerged objects. Because of their drab coloration and small size, they frequently are not noticed in collections of benthic materials. About 25 species are found in the United States. They are most common in water less than 2 m deep and generally do not thrive in very rapidly flowing streams. Many other metazoans, including crustaceans, insects, mites, annelids, and nematodes, find freshwater sponges to be a favorable habitat.

Bath sponges possess only spongin fibers; their total lack of spicules obviously makes them highly desirable as bath sponges. After the sponges are collected, the tissues are allowed to decay, the dead cells are washed away, and the sponges are dried to produce a beautiful, serviceable, bath sponge.

Boring sponges are important ecological agents because they contribute substantially to the decomposition of shells and coral. Tunnels in the shells are created by dissolution of the calcareous material and by individual archeocytes that remove microscopic shell chips and void them in the exhalent current.

Class Sclerospongiae

The Class Sclerospongiae includes a small number of species of leuconoid sponges found mostly in crevices of coral reefs; for this reason they are called coralline sponges. These sponges have an internal mass of siliceous spicules and spongin fibers inside of a thin veneer of an external, calcareous, skeletal mass. The living components are entirely comparable to those of the Demospongiae. The numerous oscula are slightly raised from the sponge's surface. They are not a well-known group because of their cryptic habitats.

PHYLUM MESOZOA

The approximately 50 species of mesozoans are very simple multicellular animals, and all are endoparasitic in a variety of marine invertebrates. Mesozoans are wormlike and small and range from 0.1 to 9.0 mm in length. These animals are simply constructed, and for this reason most zoologists consider them to be very primitive metazoans. But because they are exclusively parasitic and have rather complicated life cycles comparable to those of trematodes (Phylum Platyhelminthes), other zoologists believe that they are extremely degenerate flatworms. Future research may resolve these differences in interpretation of the phylogenetic affinities of this enigmatic group, but we shall consider their primitiveness to be primary rather than secondary.

Mesozoans are normally divided into two classes: Rhombozoa and Orthonectida. Since details of morphology and life histories vary in the two classes, we shall consider the groups separately.

Class Rhombozoa

Rhombozoan mesozoans live within the nephridial cavities of squids, octopods, and cuttlefishes (Phylum Mollusca, Class Cephalopoda). The minute, elongated, cylindrical body is about 0.5–9 mm in length and is constructed of a very small number (20–30, depending upon the species) of ciliated cells, the **somatoderm,** surrounding one to more than 100 inner generative or reproductive cells, the **axoblasts** (Fig. 4.11a). The anterior somatodermal cells are used in attaching the parasite to the host's kidney tissues.

When the cephalopod host is still a juvenile, the axoblasts develop asexually into vermiform embryos that becomes **nematogens,** which emerge and attach to the renal cavity of the host. When the host becomes sexually mature, the nematogens become **rhombogens.** The axoblasts of rhombogens develop asexually into infusiform embryos that develop only into rhombogens.

Probably stimulated by crowded conditions within the host, the axoblasts develop asexually into a third form, an **infusorigen** (Fig. 4.11b). Infusorigens are peculiar organisms in that they have no somatic cells, are hermaphroditic, and meiotically produce both male and female gametes derived from the axoblasts. Apparently, there is no cross-fertilization, since eggs and sperm derived from the axoblast fuse in fertilization, and each resulting zygote develops into an **infusoriform larva.** Each ciliated larva escapes from the host via the host's urine and becomes a free-swimming planktonic form. Its fate in seawater is not known, but presumably it seeks out and enters another host by some unknown means. Once the infusoriform larva has infected another cephalopod host, its axoblast develops into a nematogen to complete the life cycle.

Class Orthonectida

Members of this class live in a variety of invertebrates including brittle stars, polychaetes, flatworms, nemerteans, and bivalves. The parasites live unattached in the gonadal tissues or body cavities of the host. Orthonectids are all small and seldom reach more than 0.3 mm in

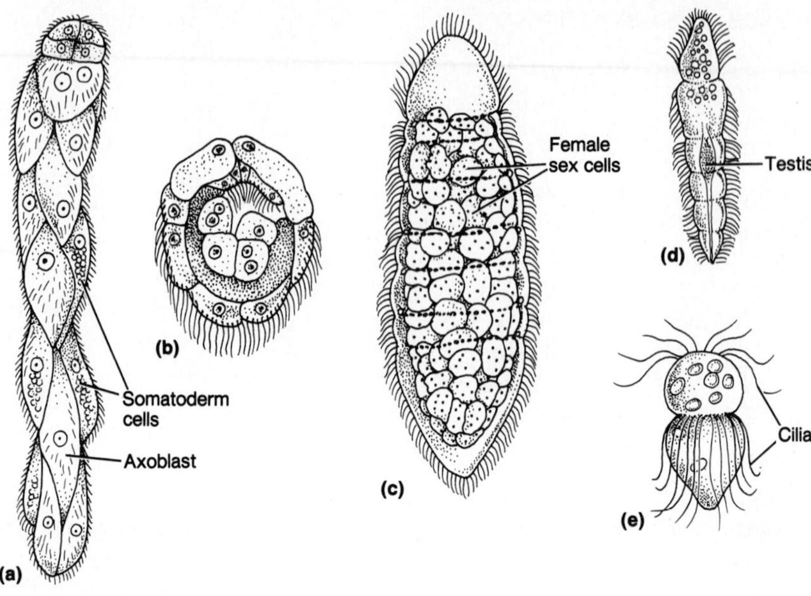

Figure 4.11 *Mesozoans: **(a)** a rhombogen adult; **(b)** an infusorigen of a rhombogen; **(c)** a female orthonectid; **(d)** a male orthonectid; **(e)** a ciliated larva of an orthonectid (parts a [after Whitman], b [after Nouvel], c–e [after Atkins] in L. H. Hyman,* The Invertebrates. *Vol. I,* Protozoa through Ctenophora, *New York, McGraw-Hill, 1940, adapted with permission).*

length. Orthonectids that live in the host's gonadal tissues often parasitically castrate their host.

The generations alternate in that asexual multiplication in plasmodia gives rise to free-swimming sexual individuals. The **plasmodium** is a multinucleate ameboid stage that feeds on host tissues. Asexual cell divisions within the plasmodium give rise to morulalike masses that often escape the plasmodium and spread the infection to other parts of the host. In some species the plasmodium is hermaphroditic; in other species the plasmodia are either female or male. Each morula then differentiates into a free-swimming form composed of a layer of ciliated surface cells, the **somatoderm,** and an inner mass of several hundred **sex cells,** which give rise to gametes. The bodies of both males and females elongate, but females are often several times longer than males (Fig. 4.11c, d). After a male and a female establish contact, sperm are transferred to the eggs, fertilization ensues, and cleavage produces a ciliated larva (Fig. 4.11e). The larvae escape from the female and locate a suitable invertebrate host. Once inside the host, the inner morula masses of the larva scatter and form separate plasmodia to complete the life cycle.

PHYLOGENY OF SPONGES AND OTHER SIMPLE METAZOANS

The ancestral origins of the Phylum Placozoa, Phylum Porifera, and Phylum Mesozoa are still being debated, and no definitive answers are readily available. It is generally agreed that all three arose from a protozoan ancestor, but each may have had a separate lineage, which would indicate that they are not closely related.

The Phylum Placozoa, and specifically *Trichoplax,* remains a singular enigma. On one hand, it may be only an interesting and phylogenetically unimportant creature. On the other hand, *Trichoplax* probably occupies an extraordinarily important position in the phylogeny of all other metazoans. Very likely, it evolved from a colonial zooflagellate and became a two-layered, asymmetrical, creeping organism. It, or a simple metazoan like it, may in fact have given rise to the radially symmetrical planuloid ancestor from which all higher metazoans arose (Fig. 4.12).

Sponges are an ancient group, and the important events in their evolution occurred in the Proterozoic Era, a period from which the fossil record is very sketchy (see Table 1.1). There are two basic theories as to the origin of sponges. One theory holds that poriferans evolved from a colonial choanoflagellate like *Proterospongia* (see Chapter 2). The presence of distinctive cytoplasmic collars in both choanoflagellates and sponge choanocytes is a strong uniting link. The second theory holds that sponges diverged very early from the same ancestral stock from which all higher metazoans arose. Such a simple, free-swimming, colonial, flagellated ancestor then evolved a sessile habit concomitant with the development of internal water channels. The flagellated exterior cells migrated to the interior and became choanocytes, a process that is recapitulated in the embryology of most sponges. Both theories have some merit, and at this time one cannot rule out either one. Regardless of their origins, sponges were not the ancestral form for any other metazoans. Undoubtedly, they represent an evolutionary blind end (Fig. 4.12).

Within the phylum the Class Calcarea contains the simplest sponges. From them sponges of the Class Demospongiae arose to produce the vast majority of sponges (Fig. 4.12). The Class Sclerospongiae is a somewhat specialized group but very similar to the Demospongiae. The Class Hexactinellida, however, is a fundamentally different group, and it is clear that they arose early from the early poriferan stem and have evolved some features unique to them (Fig. 4.12).

Whether the simplicity of mesozoans is primary or secondary remains a baffling puzzle to invertebrate phylogenists. That they possibly represent extremely degenerate flatworms has already been mentioned. But there is a growing theory that they were derived from ciliate protozoans by perhaps a multinucleate syncytial ciliate becoming a multicellular metazoan (Fig. 4.12). Many zoologists now feel that mesozoans developed their multicellularity from ciliates, independently from that in all other metazoans. This would mean that the metazoans are at least diphyletic—a most remarkable possibility!

Figure 4.12 *Possible phylogenies of the Phylum Placozoa, Phylum Porifera, and Phylum Mesozoa showing probable evolutionary relationships to the protozoans and to higher metazoans.*

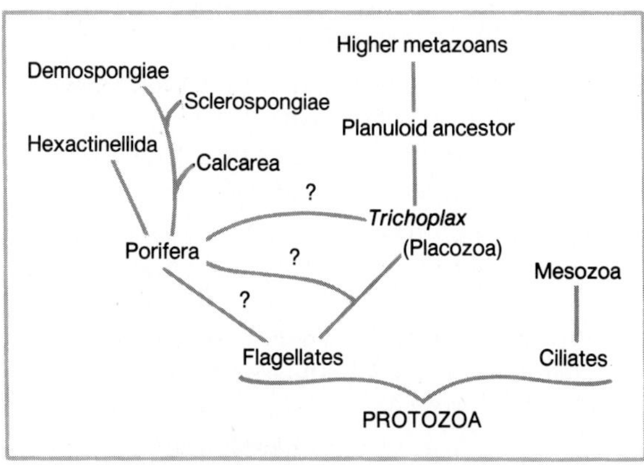

SYSTEMATIC RESUME

The Principal Taxa of the Phylum Porifera and Other Simple Metazoans

PHYLUM PLACOZOA

Only one known species, *Trichoplax adhaerens* (Fig. 4.1).

PHYLUM PORIFERA

Class Calcarea—exclusively marine; skeleton either a fused mass or of discrete spicules of $CaCO_3$; microscleres absent; most produce amphiblastula larvae.

Subclass Calcinia—choanocytic nuclei are basal; flagellum arises independently of choanocyte nucleus; includes *Clathrina* and *Leucettusa.* (Fig. 4.2b)

Subclass Calcaronia—choanocytic nuclei are apical; flagellum arises directly from choanocyte nucleus; *Leucosolenia* and *Sycon* are common genera.

Subclass Pharetronidia—calcareous skeleton in various forms; usually have peculiar tuning fork triaxons; includes *Petrobiona* and *Murrayona.*

Class Hexactinellida—glass sponges; exclusively marine and mostly deepwater forms with siliceous skeleton composed largely of hexactine spicules; megascleres and microscleres always present.

Subclass Amphidiscophora—megascleres never fused, often elaborately shaped; never fixed to hard substrata, but are anchored by basal spicules; *Hyalonema* (Fig. 4.2c) is representative.

Subclass Hexasterophora—normally grow attached to hard substrata but may be anchored by basal spicules; megascleres commonly fused into rigid framework; includes *Euplectella* (Fig. 4.2c) and *Farrea.*

Class Demospongiae—marine and freshwater forms; siliceous spicules with 1–4 rays; all are leuconoid; spongin fibers commonly present; some have lost all skeletal components; contains about 95% of all sponge species.

Subclass Homoscleromorpha—microscleres not always distinguished from megascleres; spicules all small; larvae are amphiblastulae; *Oscarella* and *Plakina* are representative genera.

Subclass Tetractinomorpha—oviparous; most larvae are parenchymellae; megascleres are tetractines or monactines; includes *Stelletta, Geodia,* and *Cliona.*

Subclass Ceractinomorpha—mostly viviparous; larvae are parenchymellae; megascleres are always monaxonoid; spongin is almost universally present; *Halichondria, Haliclona, Spongilla* (freshwater form), and *Hippospongia* (bath sponge) are common genera.

Class Sclerospongiae—produce complex skeleton of siliceous spicules and spongin fibers inside of a calcium carbonate mass; cells similar to those in Demospongiae; exclusively marine; *Ceratoporella* and *Astrosclera* are representative.

PHYLUM MESOZOA

Class Rhombozoa—parasitic in nephridia of cephalopods; monoecious; very simply constructed; *Pseudocyema* is a common genus.

Class Orthonectida—parasitic in a variety of worms, echinoderms, bivalves, and other invertebrates; dioecious; structurally more complicated than rhombozoans; *Rhopalura* is representative.

ADDITIONAL READINGS

Bayer, F.M. and H.B. Owre. 1968. *The Free-Living Lower Invertebrates,* pp. 1–24. New York: Macmillan Co.

Bergquist, P.R. 1978. *Sponges,* 256 pp. Los Angeles: University of California Press.

Cheng, T.C. 1973. *General Parasitology,* pp. 309–319. New York: Academic Press.

Frost, T.M., G.S. DeNagy, and J.J. Gilbert. 1982. Population dynamics and standing biomass of the freshwater sponge *Spongilla lacustris. Ecology 63:*1203–1210.

Fry, W.G., ed. 1970. *The Biology of the Porifera,* 512 pp. New York: Academic Press.

Grell, K.G. 1971. *Trichoplax adhaerens* F.E. Schulze und die Entstehung der Metazoen. *Naturwissenschaftliche Rundschau 24:*160–161.

Harrison, F.W. and R.R. Cowden, eds. 1976. *Aspects of Sponge Biology,* 347 pp. New York: Academic Press.

Hyman, L.H. 1940. *The Invertebrates.* Vol. I: *Protozoa through Ctenophora,* pp. 233–247, 284–364. New York: McGraw-Hill Book Co.

Kaestner, A. 1967. *Invertebrate Zoology.* Vol. I, pp. 20–42. New York: John Wiley & Sons.

Lapan, E.A. and H. Morowitz. 1972. The Mesozoa. *Sci. Amer. 227*(6)(December):94–101.

McConnaughey, B.H. 1967. The Mesozoa. *In* M. Florkin and B.J. Scheer, eds. *Chemical Zoology,* Vol. II, pp. 557–570. New York: Academic Press.

Pennak, R.W. 1978. *Fresh-water Invertebrates of the United States,* 2nd ed., pp. 80–98. New York: John Wiley & Sons.

5

The Radiates

OVERVIEW

The two radiate phyla—Cnidaria and Ctenophora—are notably successful, even though individual animals are simply constructed from sheets of cells (tissues). As their name implies, radiates are characterized by having radial symmetry, which, coupled with the fact that they are basically sedentary, has either directly or indirectly influenced all other facets of their functional morphology. All radiates are constructed on a plan that has an internal gastrovascular cavity with but a single opening—the mouth—which in turn is usually surrounded by several to many tentacles used in the capture and manipulation of food. The body wall surrounding the gastrovascular cavity consists of an outer epidermis, and inner gastrodermis, and a middle mesoglea. The mesoglea is variable in composition, ranging from a thin acellular layer to an extensive gelatinous mass replete with numerous cells, fibers, spicules, and other inclusions.

The cnidarians, having by far the greater species diversity of the two phyla, are characterized by the presence of stinging organelles or nematocysts contained within specialized cells, the cnidocytes. Nematocysts are discharged explosively in attachment, defense, or food capture, in which their poisons kill or immobilize prey. No other single feature characterizes every cnidarian as does the universal occurrence of cnidocytes. Another important cnidarian feature, although not a constant character, is polymorphism, a phenomenon in which two or more distinctly different forms are present in an animal's life cycle. In cnidarians the two principal morphs are a cylindrical-shaped, sedentary polyp and a disc-shaped, motile jellyfish or medusa. The four classes of cnidarians—Hydrozoa (hydroids), Cubozoa and Scyphozoa (jellyfishes), and Anthozoa (sea anemones, corals)—are sharply differentiated in details of their dimorphism and in the complexities of the gastrovascular cavity and mesoglea.

The ctenophores (comb jellies) are a rather small group of marine and mostly planktonic radiates. All ctenophores possess eight ciliated comb rows utilized in locomotion, adhesive collocytes used in food capture and located on a pair of nonoral tentacles, and an apical organ that functions in coordination of locomotion and in equilibrium.

The radiates are of particular phylogenetic interest, since they are ancient animals and probably are reasonably closely related to the ancestral stock from which all other metazoans arose. Radiates are also immensely important to a better understanding of metazoan evolution and the many metazoan adaptations, for in radiates were developed some primal features that were to be of unusual importance in the evolution of all higher animals. Radiates probably arose from a ciliated wormlike creature that had in turn been derived from a colonial flagellate. It is generally thought today that the earliest cnidarians were polyps. The almost universal occurrence of a polyp in all cnidarians is perhaps the strongest piece of evidence supporting the theory that the polyp is really an elaboration of the larval stage, and the medusa is therefore the ancestral adult. But strongly differing opinions can be found among some zoologists. Members of each of the classes of cnidarians have modified the basic plan, and each class is unique in its own individual ways. The close relationship between cnidarians and ctenophores is not seriously questioned and is based on comparable symmetry and body wall construction, the presence of a similar nervous system and gastrovascular cavity, and the occurrence of cnidocytes in one species of ctenophore.

GENERAL CHARACTERISTICS

The radiates are represented by two phyla: the Phylum Cnidaria, which includes the hydras, jellyfishes, and corals, and the Phylum Ctenophora or comb jellies. The radiates, especially the cnidarians, are a rather large group of exclusively aquatic and almost entirely marine animals. Remarkably, they are relatively successful in spite of their being simply constructed. But while they are quite simple and primitive, radiates have evolved some incipient adaptive features that were to be of immense importance in the subsequent evolution of all higher animals.

Radiates are very ancient animals, and they undoubtedly preceded higher metazoans by millions of years. For this reason, radiates are in a pivotal position in the evolutionary pathways among animals, for they are relatively closely related to the ancestral stock from which all other metazoans were eventually derived. Unfortunately, the fossil record of their early history is very scant; if these exceedingly important evolutionary stages were better documented, it would answer many of the elusive phylogenetic questions about early metazoans. Because of their antiquity, primitiveness, relative simplicity, and primal contributions to the evolution of animals, coupled with a number of unique features that they possess, radiates are of particular interest to a large number of biologists.

Several distinctive and important adaptations characterize radiates and have had a major impact on their modes of existence. Their symmetry has influenced almost all facets of their functional morphology, but it has had a particular influence on their basic body construction and feeding patterns. Each of these—symmetry, body construction, and feeding—will be discussed below in some detail.

Radial Symmetry

The most distinguishing and diagnostic feature of members of both radiate phyla is that they are radially symmetrical, a characteristic from which they derive their common name. As described in Chapter 3, a radially symmetrical organism has parts arranged radially around a central axis such that a plane running through any two opposite radii divides the organism into two similar and equal halves. This means that most anatomical parts, present in multiples ranging from four to several hundred, are arranged radially. Some of the more advanced radiates have bilaterally arranged parts amid the primitive radial plan, a condition termed **biradial symmetry.** Some radiates even have internal parts deployed in a bilaterally symmetrical pattern.

There is a general correlation between the type of symmetry present in an organism and its life habits. Bilateral symmetry is usually, but not invariably, associated with an active existence in which the animal moves freely from place to place. Radial symmetry, on the other hand, is associated with a sedentary or sessile existence. If an animal moves slowly or spends most of its existence attached to an immovable object, there are obvious advantages in being equally receptive and responsive to environmental conditions on all sides. But while this arrangement has its advantages, radial symmetry also restricts the development of very specialized structures that, if present, would have to be arranged repetitively and radially. The development of such specialized structures and their required maintenance would not be energetically efficient. Radiates are primarily a sedentary group; while some are capable of some movement, locomotion is slow and methodical. Because of radial symmetry and the obvious metabolic and maintenance costs associated with the ontogeny of specialized organs, radiates are constructed very simply from layers of cells.

A Simple Body Plan

The bodies of radiates vary widely in shape, but their body plan reflects their basic radial symmetry. In its simplest form the body is cylindrical, and one end contains the **mouth** and is termed the **oral end** while the other end is the **aboral end.** Thus an imaginary axis, running centrally through the cylinder and passing through both ends, is an oral–aboral one. In radiates that are ovate or disc shaped, the mouth is centrally (axially) located on the oral surface, the oral–aboral axis is still very evident, and parts are still arranged radially around the axis.

All higher animals have a highly complex organization with many different organs specialized for the various organismic functions. Radiates, however, are constructed on a much simpler body plan in which organs, for the most part, are absent. Instead, their body is constructed of sheets of cells or tissues in which the constituent cells often function in several different but important ways. Radiates are often said to be at the **tissue grade of construction,** a concept that means that the body is simply constructed of tissues. A simple radiate body is, in effect, a hollow sac with the mouth leading into a single internal cavity surrounded by the body wall. The body wall is built of two epithelial layers: an outer **epidermis** and an inner **gastrodermis.** Between these two cellular layers is a variably constructed **mesoglea.** In primitive radiates the mesoglea is nonliving and acellular and may contain only a few wandering cells called **amebocytes.** In more advanced forms, it is far more extensive and often forms the bulk of the body wall. The mesoglea in the more complex radiates is invaded by numerous amebocytes that often produce fibers, spicules, and an extensive, nonliving, gelatinous mass. So numerous are these cells in many radiates that they form a third cellular layer between the epidermis and gastrodermis. However, mesogeal cells do not constitute a true mesoderm as is present in all higher animals. A true mesoderm always persists as a permanent cellular layer and is derived, for the most part from endodermal tissues. Nevertheless, the nature of the mesoglea is of considerable importance to the more highly specialized radiates.

The saclike body wall surrounds a central gastrodermally lined **gastrovascular cavity.** Two important dimensions of a gastrovascular cavity are that it is a blind-end cavity and that it has a single opening, the **mouth,** leading into it from the outside. The gastrovascular cavity may be a simple undivided space, divided internally by

TABLE 5.1. □ **THE MOST IMPORTANT CHARACTERISTICS OF THE MAJOR RADIATE TAXA**

Taxon	*Common Names*	*Nature of Gastrovascular Cavity*	*Nature of Dimorphism*	*Other Important Features*
Phylum Cnidaria				
Class Hydrozoa	Hydroids, hydras	Simple, saclike, not partitioned	Most have both polyp and medusa	Gastrodermis without gametes and cnidocytes; acellular mesoglea; velum in medusa
Class Cubozoa	Sea wasps	Manubrium, stomach, gastric pouches	Medusa is large and dominant	All marine; medusa elongate in oral–aboral axis; stings of some are very virulent
Class Scyphozoa	Jellyfishes	Stomach, gastric pouches, numerous radial canals	Large, prominent medusa; small polyp	All marine; have marginal sense organs; gastrodermis with gametes and cnidocytes; mesoglea is cellular, fibrous
Class Anthozoa	Sea anemones, corals	Radially divided by septa	Exclusively polypoid	All marine; gastrodermis with cnidocytes and gametes
Phylum Ctenophora	Comb jellies, sea walnuts	Stomach, extensive canal system	Absent	All marine; eight comb rows; collocytes; apical organ; tentacles not oral

septa, or present as a system of canals or sacs. As its name implies, the gastrovascular cavity functions in both digestion and circulation; both functions will be more fully discussed in the next section.

Feeding

That radiates are both sedentary and radially symmetrical has had far-reaching implications for most of their activities. However, no single function has perhaps been so influenced by these two features as has feeding. All radiates are carnivores, but they lack the ability to pursue prey. Even the mobile radiates are very slow movers and usually do not move purposefully toward food organisms. A few radiates feed on particles suspended in water around them, but by far the majority are trappers in that they lie in wait for potential prey to come in close proximity to them.

Feeding in radiates has been augmented by three important adaptations: tentacles, specialized food-capturing cells, and cilia. **Tentacles,** present in many different metazoan taxa, are fingerlike extensions of the body wall usually near the mouth or at the anterior end. In most radiates the tentacles are arranged radially around the mouth. This arrangement is an important adaptation, for in sessile forms like radiates, food organisms approaching from any point on the periphery can be captured and utilized. All radiates have special stinging or adhesive cells used in food capture; **cnidocytes** in cnidarians and **collocytes** in ctenophores are effective adaptations for food acquisition. Present in various parts of the radiate body, these food-capturing cells are far more numerous on the tentacles and around the mouth. **Cilia** are commonly found on the tentacles and oral end. In some radiates, cilia create a continuous current of water over the tentacles and oral disc and, in so doing, bring plankton to the organism. In most other forms the stunned or adhered prey is carried to the mouth and into the gastrovascular cavity by the tentacles themselves or by tentacular cilia.

Prey organisms often enter the gastrovascular cavity already killed or in an immobilized condition, but in many radiates, prey is ingested live. Once inside the gastrovascular cavity of most cnidarians, ingested organisms are immobilized by cnidocytes within the gastrodermis. Digestion begins immediately as gastrodermal glandular cells secrete enzymes onto the food. Within the gastrovascular cavity occur both the preliminary stages of digestion and the circulation of particles of partially digested food, enabled in part by the flagella of the gastrodermal cells. Certain gastrodermal cells ingest the particles of partially digested food by phagocytosis, and final digestion is intracellular within these gastrodermal cells. Since there are no food-storing organs in radiates, regular feeding is absolutely essential for life. Undigested materials are voided to the outside through the mouth.

PHYLUM CNIDARIA

The Phylum Cnidaria, formerly called the Phylum Coelenterata, is a moderately sized phylum of almost 10,000 species that includes the hydras and other hydroids, jellyfishes, corals, and sea anemones. This taxon is decidedly marine, and most cnidarians are found in shallow, coastal, marine habitats. Even though some cnidarians are large, most forms are small, and a great many species are colonial. The classes of cnidarians are characterized in Table 5.1.

Since the vast majority of radiates are cnidarians, all characteristics mentioned above apply specifically to cnidarians. In addition to these radiate features there are two important characteristics that have had an immense impact on cnidarian existence. These are the presence of cnidocytes and the phenomenon of dimorphism.

Cnidocytes and Nematocysts

One of the most important and distinguishing cnidarian adaptations involves specialized cells called **cnidocytes,** each of which produces and contains a stinging structure or organelle called a **nematocyst.** The cnidocyte and its nematocyst are used for food capture, defense, and assisting in attachment. Since cnidocytes are found in all cnidarians but in no other animal except one species of ctenophore, they are diagnostic for the phylum. In fact, the phylum name is based on the universal presence of cnidocytes (cnidaria = nettle-bearers). Cnidocytes were formerly termed cnidoblasts, but this latter term should be dropped, since the suffix implies a developing cell. Derived from unspecialized cells, cnidocytes are present throughout the epidermis but especially that of the tentacles and around the mouth and also in the gastrodermis of cubozoans, scyphozoans, and anthozoans. Each cnidocyte is an ovoid cell with a basal nucleus and a **lid,** located at the epithelial surface, that covers the **nematocyst capsule,** which in turn contains the coiled, tubular, threadlike **nematocyst.** At the base of a typical nematocyst is a **butt** with large barbs inside; the butt tapers into the **tubule,** which in turn may bear minute barbs (Fig. 5.1). The tubule is pleated and has the appearance of a screw because of three left-handed spiral folds in its wall. In hydrozoans, each cnidocyte is equipped with a distal, short, stiff bristle, the **cnidocil** (Fig. 5.1a), which is probably a modified flagellum, since the ultrastructure of both is similar. The cnidocil is exposed to the exterior and functions as a trigger in nematocyst discharge. In all other cnidarians the function of the cnidocil is assumed by a cilium or a tuft of cilia.

At the time of discharge, an extremely high osmotic pressure is developed within the capsule, butt, and thread; and four important events are initiated almost simultaneously. First, as the osmotic pressure within the capsule reaches some critical level, the lid of the cnidocyte is forced open, and the butt and thread begin to turn inside out. Second, as the nematocyst everts, it lengthens considerably (up to several times its undischarged length), owing to the unfolding of the pleats. Third, a discharging nematocyst revolves because of the spirally arranged folds. Finally, the minute barbs in an undischarged tubule point forward, and as the tubule everts, the barbs are suddenly flicked backward (Fig. 5.1b).

What would a discharging, oncoming nematocyst look like to a potential prey? Try to imagine a double-walled tube everting toward you at a high velocity with the advancing edge of the everting tube rotating at an enormous speed. At each revolution, new barbs are exposed at the everting end, each barb flicking backward like the blade of a penknife (Fig. 5.1b). What a beautiful, but awesome, mechanism!

A small amount of poison, produced by the cnidocyte, travels the length of the discharged nematocyst and into the prey. Several different poisons have been isolated, and all appear to be proteins and function as paralyzing neurotoxins. If enough nematocysts are involved, these toxins are fatal to most very small animals, both invertebrate and vertebrate; in larger vertebrates the toxins, while usually not fatal, are quite effective as defensive chemicals. Stings of the Portuguese man-of-war and of many scyphozoans can inflict on human beings painful welts, severe burning sensations, and general irritations. Several species of small cubozoans may sting persons so severely as to cause death.

Nematocysts range in length from 5 μm to about 1.5 mm. Twenty-seven nematocyst types have been described in which differences are based on numbers of butt spines and tubular barbs, shape, and whether a nematocyst has an open or closed end. Most cnidarians contain more than one type of nematocyst. Functionally, nematocysts can be grouped into three categories. A **penetrant** or **stenotele** is one in which the tube is open at the tip, which typi-

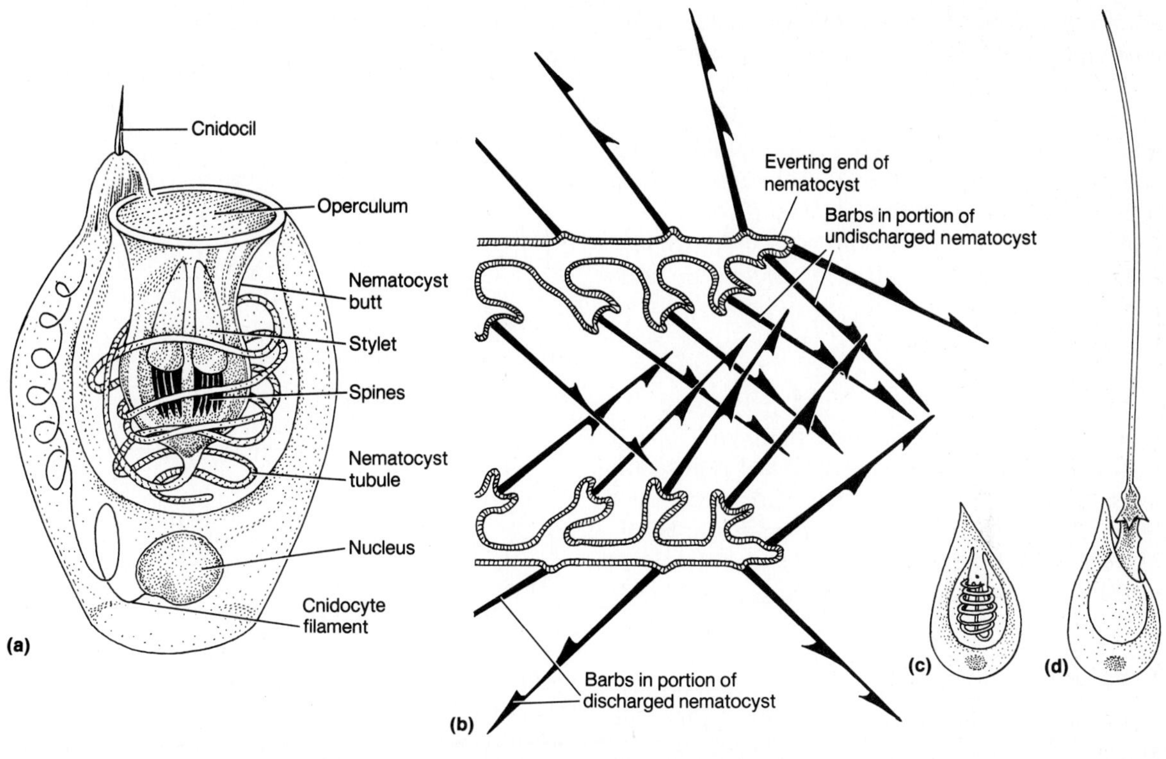

Figure 5.1 *Cnidarian cnidocytes and nematocysts:* ***(a)*** *a cnidocyte with an undischarged penetrant or stenotele;* ***(b)*** *a diagrammatic view of the everting end of a nematocyst (adapted with permission of Macmillan from* A Life of Invertebrates, *by W. D. Russell-Hunter. © 1979 by W. D. Russell-Hunter);* ***(c)*** *a cnidocyte with an undischarged nematocyst;* ***(d)*** *a discharged nematocyst (parts c, d from J. W. Kimball,* Biology, *© 1983, Addison-Wesley, Reading, Massachusetts, p. 809, Fig. 39.2. Reprinted with permission).*

cally is equipped with barbs, and whose poison induces paralysis as it penetrates the prey. A **glutinant** or **isorhiza** has an open sticky tubule and is employed by many cnidarians to anchor themselves to substrata. A **volvent** or **desmoneme** has a closed tubule, is without barbs, and, when discharged, coils around and entangles a prey.

Once discharged, a nematocyst and its cnidocyte are not reused, but new cnidocytes are quickly formed by cellular division of unspecialized cells. Cnidocytes are produced mainly near the bases of the tentacles, around the mouth, or in other mitotic locations. Many studies have shown that cnidocytes often are transported some distance from their production site to the point where they are eventually sited for use.

Many investigations have been conducted on the mechanisms of stimulation for nematocyst discharge. Some nematocysts can be discharged by mechanical stimuli, and others discharge in response to chemicals. Nematocysts are generally considered to be independent effectors; this means that each cnidocyte is stimulated directly by an appropriate environmental stimulus and not by the organism itself. However, recent studies suggest that the animal itself is able to modulate nematocyst discharge. One study showed that a local mechanical stimulation may promote general nematocyst discharge. In another study, cnidarians fed to repletion discharged significantly fewer nematocysts. However, there is little evidence to date as to whether such control is nervous or nonnervous.

Dimorphism

Dimorphism is a phenomenon in which two distinctly different morphological types are present, often at different phases in an animal's life cycle. Cnidarians are good examples of dimorphism, since in many taxa, two basic morphological types, polyp and medusa, are found; both are radially symmetrical, but each differs from the other in many important respects.

Prior to a discussion of polyp and medusa, a few general comments need to be made. Unlike the ubiquitous occurrence of cnidocytes in all cnidarians, dimorphism is not a constant character, and the degree of development of polyp and medusa varies widely between species or in the four classes. In most hydrozoans, both stages are present and are about equally balanced, even though some species are exclusively polypoid and others medusoid. In cubozoans and scyphozoans the medusa is strongly emphasized, and the polyp is suppressed. In anthozoans the polyp is the only morph, since there are no medusae.

There has been considerable debate as to which is the more primitive morph. Some workers hold that the polyp is the primitive adult form and that the medusa was developed later as a means of dispersal. Probably most cnidarian workers, however, view the polyp as an elaboration of the larval stage and the medusa as the primitive adult. This interpretation is strongly contested by some biologists, and future studies may help to resolve these phylogenetic and developmental questions.

Polyp

The **polyp** is cylindrical, tube shaped, and typically sessile. The body consists of an aboral basal disc, a column, and an oral disc (Fig. 5.2). The **basal** or **pedal disc** is variously adapted to attach the polyp to a substratum and often has an adhesive surface, a pointed end, or rootlike stolons. In colonial species the basal disc is usually absent. The **column** is cylindrical and makes up most of the body (Fig. 5.2). The **oral disc** bears the mouth and several to many tentacles. Often situated at the apex of a conical elevation or **hypostome,** the mouth ranges from oval shaped to being strongly elongate and elliptical. The mesoglea is thin and noncellular in hydrozoans, but it contains many cells in anthozoans. Most polyps live with the basal portion attached permanently to a substratum and the oral end extended upward in the water. Some polyps, however, are able to move in a limited way by sliding, creeping movements of the pedal disc, or even by swimming or somersaulting.

Figure 5.2 A diagrammatical view of a cnidarian polyp seen in an oral-aboral section.

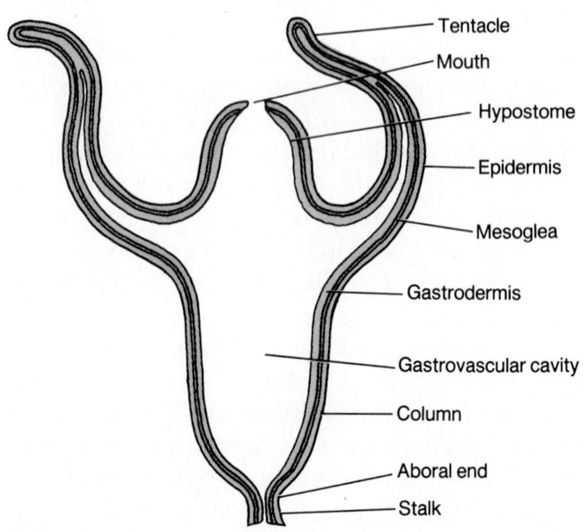

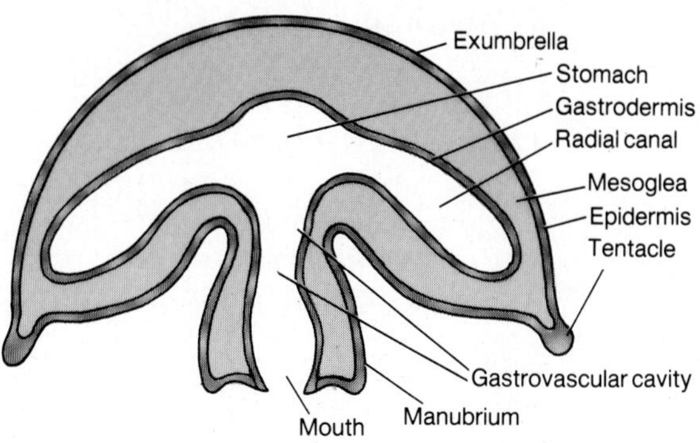

Figure 5.3 *A diagrammatic view of a cnidarian medusa seen in an oral-aboral section.*

MEDUSA

The **medusa** is bell shaped or umbrella shaped and built basically on the same plan as the polyp. The upper, aboral, convex surface is the **exumbrella,** while the lower, oral, concave surface is the **subumbrella** (Fig. 5.3). The mouth is at the lower or distal end of a centrally located subumbrellar **manubrium.** In scyphozoans, four or more **arms,** modified extensions of the manubrium, are employed in feeding. Rather than being located around the mouth, the small numerous tentacles are on the margins of the umbrella or bell (Fig. 5.3). Medusae generally have tetramerous symmetry in that most structures are in fours or multiples of fours. The medusoid mesoglea contains cells and fibers and is extensive, firm, resilient, and gelatinous (from which is derived the term *jellyfish*). The gastrovascular cavity is represented by a central **stomach** and, in many species, four **gastric pouches,** but because of the extensive mesoglea, the cavity usually extends to the umbrellar edges only as radiating **canals** (Fig. 5.3).

Medusae, being limited in their swimming abilities, are mostly at the mercy of water currents and are therefore planktonic, even though many can swim to some degree. Bell muscles contract and expel water out of the subumbrellar space. This causes the medusa to be forced in an aboral direction by jet propulsion. Then the muscles relax, the umbrella returns to its original shape because of the mesogeal elasticity, and the process is repeated. Medusae can move vertically or horizontally, but most are propelled at angles ranging from 15° to 50° away from the vertical plane. After each jet propulsion the animal sinks slowly, only to spurt again. The net effect is to achieve some horizontal movement.

Cellular Morphology and Physiology

The body wall of a cnidarian is composed of two epithelial layers and the mesoglea. It will be useful to have some knowledge of the types of cells found in each layer of the body wall, since their functions are those of the entire organism. A brief discussion follows on the major cell types found in each tissue. Tissues and cell types have been studied most extensively in *Hydra;* therefore the discussion below is based disproportionately on *Hydra.* However, this treatment generally applies to all cnidarians and equally well to polyp and medusa.

EPIDERMIS

The **epidermis** or outer epithelial layer is a single layer of cells derived embryologically from ectoderm. Its constituent cells vary widely in shape, structure, and function. Five cells types are distinguished: epitheliomuscular, interstitial, glandular, and nerve cells and cnidocytes. **Epitheliomuscular** cells, the most common epidermal cell, are columnar with a slightly expanded distal end (Fig. 5.4). At the basal end there are several extensions that are usually oriented parallel to the oral–aboral axis. Each extension contains a bundle of myonemes or contractile fibers, and while these are not true muscles, the collective contraction of many myonemes shortens the animal along its oral–aboral axis. **Interstitial cells** are small, rounded, and undifferentiated; one school of thought

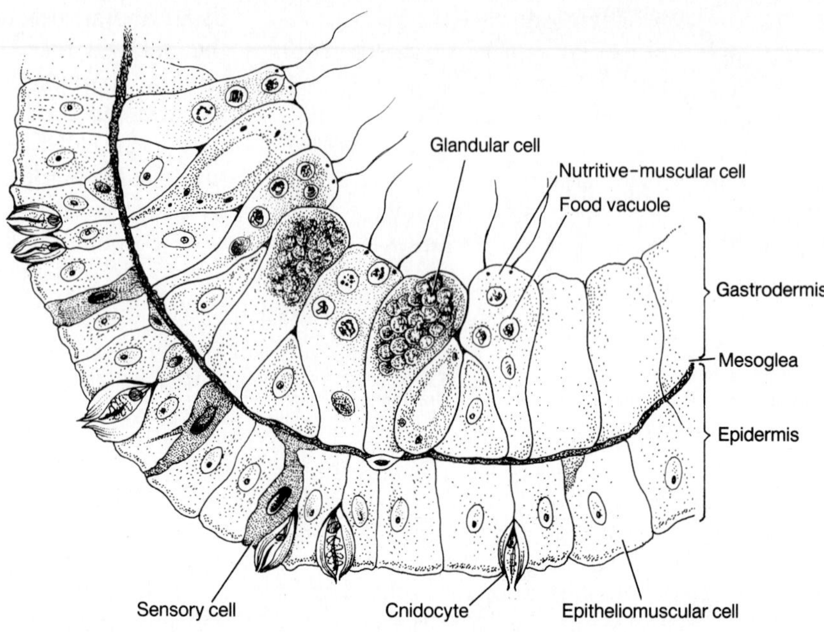

Figure 5.4 *A cross-sectional view of a portion of the column of* Hydra *to illustrate the principal cell types.*

maintains that interstitial cells have the potentiality of producing all other types of cells, including gametes. **Glandular cells** may have contractile basal extensions like those of epitheliomuscular cells, secrete mucus, and are common on tentacles (Fig. 5.4). In polyps, glandular cells are also concentrated at the basal disc, and their mucus aids in attachment to a substratum. **Nerve cells** are represented by both sensory cells and true neurons. **Sensory cells** are especially abundant on tentacles and sometimes are present wholly within epitheliomuscular cells. Distally, a sensory cell ends in a sensory bristle or tuft; proximally, there are several neural processes. **Neurons,** occurring between the epidermis and mesoglea, are elongate branching cells that are oriented parallel to the oral–aboral axis. Usually, neurons are present as a **nerve net** or plexus that tends to be concentrated around the mouth. Functions of the nerve net system will be more fully described later in a section on coordination. The very important **cnidocytes** were discussed earlier.

Gastrodermis

The gastrodermis, derived embryologically from endoderm, is histologically similar to the epidermis. **Nutritive–muscular cells,** the most common type and similar in many respects to the epitheliomuscular cells of the epidermis, are columnar, biflagellated, and have delicate contractile extensions that run mostly in a circular fashion (Fig. 5.4). Contractions of the myonemes of these cells elongate the animal along its oral–aboral axis. These cells, along with the epitheliomuscular cells and the gastrovascular cavity, represent one of the simplest examples of a hydrostatic skeleton. In addition, these cells are principally responsible for circulation within the gastrovascular cavity and for intracellular digestion. **Glandular cells** are wedge shaped with the tapered end pointing toward the mesoglea (Fig. 5.4). They produce several digestive enzymes, and some cells, especially those around the mouth, produce mucus. **Nerve cells,** including some sensory cells, and **interstitial cells** also are present in the gastrodermis. **Cnidocytes** are absent in the gastrodermis of hydrozoans but are present in certain specialized gastrodermal areas in cubozoans, scyphozoans, and anthozoans.

MESOGLEA

Some of the details of the mesoglea have already been mentioned. It is entirely acellular and very thin in hydrozoan polyps, and it reaches its greatest development in cubozoan and scyphozoan medusae and in anthozoans (Figs. 5.2, 5.3, 5.4). It may contain fibers, calcium carbonate spicules, and wandering stellate amebocytes; in medusae it is composed of up to 95% water. The mesoglea of medusae serves as a gelatinous skeleton providing support and firmness to the animal.

General Physiology

Cnidarians lack definitive organs associated with excretion, ventilation, or circulation. But they do engage in several important functions described briefly below and have tissues or organs specialized for these functions.

FEEDING

As was mentioned earlier, cnidarians are carnivores but have limited powers of movement. Therefore food in the form of small prey must come to them. Essential to the feeding process in all cnidarians are the thin flexible **tentacles.** Most cnidarian tentacles are unbranched, slender, and termed **filiform** (Figs. 5.2, 5.3); but some are short, bear a small cap at the distal end, and are called **capitate.** Tentacles may be hollow, having an internal space that is continuous with the gastrovascular cavity, while others are solid and filled with a core of gastrodermal tissue. Numbers of tentacles often vary between morphs of an individual and between individuals of the same species. Tentacles may be present in one or several whorls around the mouth. Adapted to capture food and manipulate it into the mouth, tentacles are liberally supplied with sensory cells as well as cnidocytes.

As a small organism contacts the tentacles, it is quickly immobilized, entangled, and often killed by the nematocysts. The tentacles, assisted by ciliary action, then manipulate the prey to the mouth. The reflexive feeding activity of tentacles bending toward an opening mouth has been shown to be triggered by the presence of reduced glutathione (a tripeptide) released by the injured prey. As the prey enters the gastrovascular cavity through the mouth, glandular cells in the gastrodermis secrete enzymes, mostly proteases, which initiate digestion. Later, most of the gastrodermal cells, but chiefly the nutritive–muscular cells, form pseudopodia and engulf the partially digested materials. The final phases of digestion occur intracellularly within food vacuoles. Apparently, lipids and carbohydrates can be digested only intracellularly. The end products of digestion then diffuse to all other cells. It is important to note here that cnidarians are the most primitive group (and the first group encountered in this text) to have an internal cavity devoted principally to digestion.

In some polyps, symbiotic dinoflagellates are found, usually within the cells of the gastrodermis. These symbionts are mentioned here because of their important contribution to the nutritional needs of the cnidarian. *Chlorohydra*, the green freshwater hydra, gets its distinctive color by the presence of zoochlorellae within gastrodermal cells. The reef corals contain symbiotic zooxanthellae within their body walls, and these algae are exceedingly important to the overall functioning of a coral reef. The more important dimensions of the coral–zooxanthellae symbiosis will be treated later in a section on coral reefs.

Coordination

Many fascinating details of the nervous system and neural transmission in cnidarians have been elucidated in the past two decades by neurophysiologists throughout the world. Space does not permit even a partial listing of the discoveries made on cnidarian nervous systems. The interested student is encouraged to utilize some of the references listed at the end of this chapter. Box 5.1, appearing later in this chapter, explores some of the activities and movement of sea anemones. A few details follow about the functional morphology of the nervous system and coordination in cnidarians generally.

Cnidarians are the most primitive animals to possess true nervous cells or tissues. Neurons are arranged in a plexus or as a nerve net just beneath the epidermis (Fig. 5.5a). Neurons are always dispersed, and cnidarians never have anything like a central nervous system. Cnidarian neurons are bipolar or multipolar in that they have two to many branches (Fig. 5.5b). Synapses occur between adjacent neurons as in other metazoans so that there is no cellular contact between adjacent neurons. Interestingly, in most cnidarians, nerve impulses can be transmitted across a synapse in either direction, and neurons also have a two-way conducting potential. Neuron fibers lack sheathing such as myelin or glial cells as in higher animals.

Stimulation of a given cnidarian neuron follows the all-or-none principle, which means that if a stimulus is strong enough to excite the neuron, the nervous impulse is carried maximally. Degrees of neural stimulation are achieved by varying the strength of the stimulus and thereby stimulating greater or lesser numbers of neurons. Some nerve fibers function as sensory cells, others as motor fibers that end in the muscular processes of epitheliomuscular cells, and still others as interneurons. Some cnidarians have a double nerve net, one functioning as a slow-conducting system of multipolar neurons and the other, composed of bipolar cells, conducting impulses rapidly as a through-system. Up to four nerve net systems have even been discovered in certain hydrozoans.

Studies of cnidarian nervous systems during the last two decades have shown that what was formerly considered to be a "simple" sys-

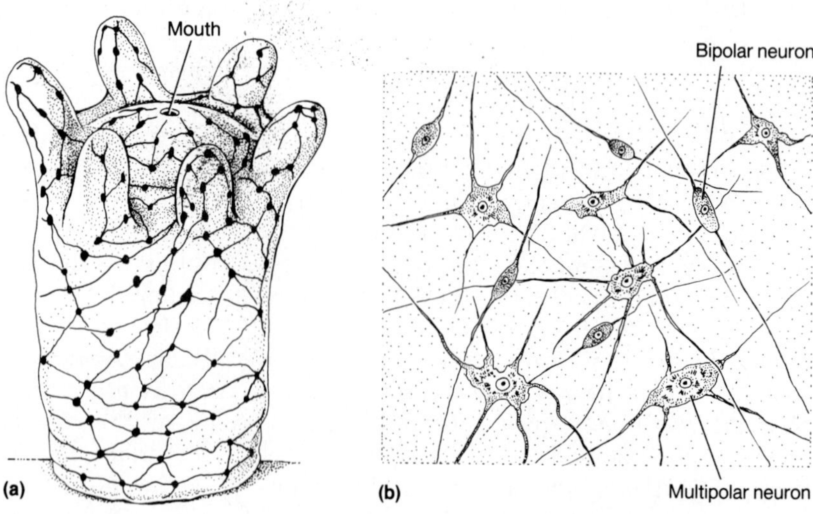

Figure 5.5 *Cnidarian nerve cells: **(a)** a nerve net in a hydra; **(b)** hydra neurons (part a from Hadji, part b after the Hertwig's in L. H. Hyman,* The Invertebrates. *Vol. I,* Protozoa through Ctenophora, *New York, McGraw-Hill, 1940, adapted with permission).*

tem is in reality a rather elaborate and complex mechanism.

Reproduction

Cnidarians reproduce both asexually and sexually. Asexual reproduction is common in the entire phylum and may be ubiquitous. Medusae are nearly always formed by budding of the polyps, but occasionally they are formed by buds from other medusae. Polyps are formed by fission or by budding from other polyps or from stolons connecting polyps; they are never formed by buds from medusae. Complex polypoid colonies are common in many hydrozoans and anthozoans as a result of repeated budding in which the polyps remain linked to each other by strands or sheets of tissues. Often, polyps within such a colony demonstrate cooperation or even a division of labor.

Most cnidarians are dioecious and reproduce sexually during the warmer periods of the year. Gametes arise from interstitial cells in either the epidermis or the gastrodermis. Following formation the gametes may aggregate together to form so-called **gonads.** Gametic masses escape either to the outside surface (hydrozoans) or into the gastrovascular cavity (cubozoans, scyphozoans, and anthozoans). Except for all anthozoans and a few hydrozoans (hydras plus a minor group), gametes are produced only in the medusa. Anthozoans, being exclusively polypoid, produce gametes on the edges of septal filaments in the gastrovascular cavity. Fertilization takes place in situ in females of some cnidarians or externally in the water. The fertilized egg divides holoblastically to form a hollow ciliated blastula. Subsequently, gastrulation produces a two-layered embryo, and the newly formed endodermal cells wander into the blastocoel and form a solid internal mass. This embryo develops into a ciliated free-swimming larva, the **planula.** After a few hours or days the planula settles onto a substratum and attaches by the larval anterior end. At about the same time an endodermal split forms the gastrovascular cavity, the mouth and tentacles form at the larval posterior end, and the larva develops into a polyp.

Associated with their ability to reproduce asexually, cnidarians, especially polyps, have great powers of regeneration. Most polyps exhibit a polarity gradient along the oral–aboral axis; this means that in a polyp cut transversely into several pieces the end of each piece originally nearest the oral end will develop a mouth and tentacles.

Cnidarian Diversity

There are four distinctively different classes of cnidarians, and a given species can easily be assigned to its appropriate class on the basis of clear characteristic features (see Table 5.1).

Class Hydrozoa

This class contains over 3000 species of mostly small, inconspicuous, encrusting, marine organisms. The few freshwater cnidarians are all hydrozoans; in the United States there are 14 species of hydras, one species of "freshwater jellyfish," a single species of a colonial polyp, and a rare primitive species found in brackish waters. Many hydrozoans are rather common as branching colonies on rocks, shells, and other submerged objects. Hydrozoans are found mostly in coastal waters, and their diversity generally increases nearer the tropics. Most individuals range from 0.5 to 3 mm, and colonies usually are no more than 2–10 cm in length, but one gigantic, solitary, deep-sea species, *Branchiocerianthus,* may approach 2.5 m in length! Hydrozoans are, for the most part, both drab colored and small; these features explain why hydrozoans are often overlooked by amateurs. Synopses of the several hydrozoan orders appear at the conclusion of this chapter.

General characteristics. Hydrozoans are constructed in the least complicated pattern of all cnidarians, and their simplicity is manifested in a number of ways. The hydrozoan gastrovascular cavity is always a simple space and not divided or comparted by septa. No cnidocytes are ever found within the gastrodermis. With the exception of tentacles no organs are present in hydrozoans. The mesoglea is always acellular, although wandering amebocytes may be found there. While a few species are nonpolymorphic, hydrozoans as a group ideally illustrate dimorphism, since polyp and medusa are about equal in size and importance.

Polyp. The polyp appears in the life cycle of most hydrozoans. In some, like the hydras, the polyp is solitary, and no colonies are developed. Most species, however, are colonial, and a colony may be composed of thousands of polyps connected by a network of living, tubular, horizontal **stolons.** Rootlike, anastomosing stolons, sometimes called **hydrorhizae,** anchor the colony into the substratum. In these colonies the hollow living tube in both polyps and stolons, consisting of epidermis, mesoglea, and gastrodermis surrounding the inner gastrovascular cavity, is collectively called the **coenosarc.** The coenosarc is continuous between polyps and stolons, and it is often difficult to determine where one individual ends and another begins. Typically, a polyp consists of a simple or branching stem or stalk, the **hydrocaulus,** and the terminal polyp proper, the **hydranth** (Fig. 5.6). The hydranth is usually vaselike and bears an elevated hypostome with an apical mouth and a circle of tentacles. In solitary forms there is an attached base or an adhesive pedal disk.

Polymorphism is evident even among polyps in all colonial hydroids in that two or more morphologically and functionally different polyps are always present. The most common type of hydranth, described above, is the feeding polyp or **gastrozooid** (Fig. 5.6). Gastrozooids furnish the collective nutritional needs for the entire colony. Circulation of partially digested food throughout the colony is through the common gastrovascular cavity and is augmented by rhythmical contractile waves in the coenosarc. A second type of polyp, a **gonozooid, gonophore,** or **blastostyle,** is a reproductive individual (Fig. 5.6). These nonfeeding, nontentacled buds may arise from any point on the colony. Gonozooids either are transformed asexually into medusae or develop medusa buds. It should be noted again that most hydrozoans are dioe-

Figure 5.6 *Examples of hydrozoan hydranths with and without a theca:* ***(a)*** *a thecate hydrozoan* (Campanularia); ***(b)*** *an athecate hydrozoan* (Pennaria).

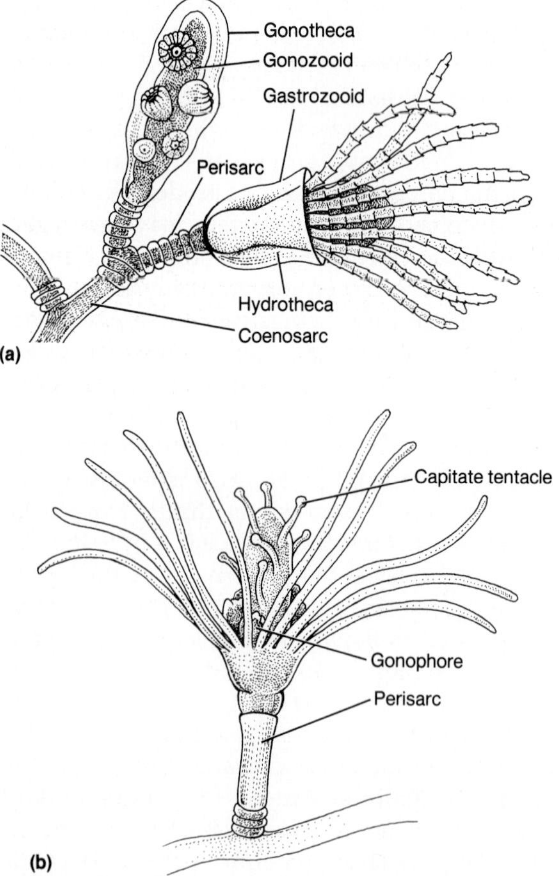

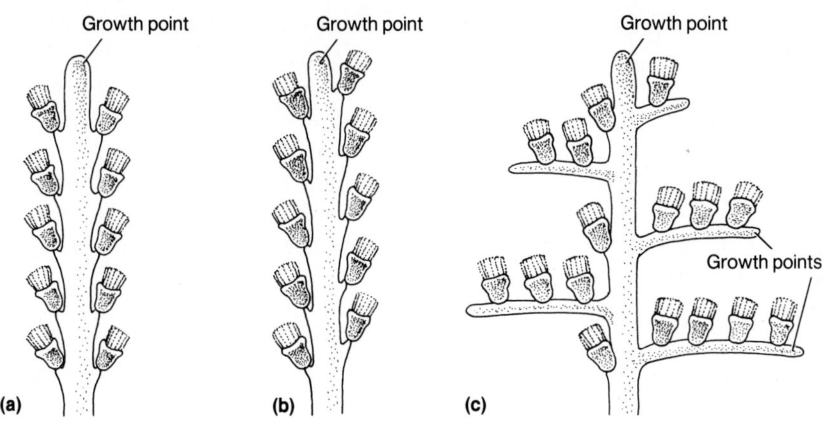

Figure 5.7 *Growth patterns in colonial hydrozoans:* ***(a)*** *monopodial growth with a terminal growth point* (Sertularia); ***(b)*** *sympodial growth with a terminal growth point* (Halecium); ***(c)*** *monopodial growth with many terminal growth points* (Plumularia).

cious; therefore a colony would normally bear either male or female medusa buds but not both. **Defensive polyps,** a third morph present in some, are usually located near gastrozooids.

Most hydrozoan colonies, especially those of the Order Hydroida, have a thin, hard, nonliving, chitinous **perisarc** covering much of the coenosarc (Fig. 5.6). The perisarc, secreted by the epidermis, may cover only hydrocauli and stolons, but in many hydrozoans the perisarc surrounds the hydranth and is termed the **hydrotheca.** Hydranths with a hydrotheca are called **thecate** (Fig. 5.6a), and those lacking a hydrotheca are termed **athecate** (Fig. 5.6b). In thecate hydranths the hydrotheca may be permanently open so as to expose the hydranth, or it may be equipped with a lid that closes when the polyp retracts.

New polyps arise as vertical or horizontal buds from existing polyps, hydrocauli, or stolons. In colonies with **monopodial growth** the hydrocaulus of the original polyp continues to grow and elongate; secondary polyps arise as lateral buds from the hydrocaulus, and they in turn often produce tertiary polyps (Fig. 5.7a, c). Other colonial species exhibit **sympodial growth** in which primary and secondary polyps do not continue to elongate, but rather the oldest polyps are at the base or in the main axis, and the youngest polyps are located apically (Fig. 5.7b).

Medusa. Hydrozoan medusae, or **hydromedusae,** are small individuals ranging in diameter from 5 to 70 mm. Most are free swimming, but some remain permanently attached to the gonozooids (polyps) that produce them. Markedly tetramerous, they are bell shaped or dome shaped (Fig. 5.8a, b), and the umbrellar margin bears four (or a multiple thereof) cnidocyte-equipped tentacles. At the umbrellar margin there is a unique inward projection or shelf, the **velum,** which reduces the opening of the subumbrellar space. The velum is a constant character for all hydromedusae and is absent in medusae of cubozoans and scyphozoans. The umbrellar margin also contains numerous sense organs of two basic types: ocelli and statocysts. Each **ocellus,** situated on the outer portions of the tentacular base, consists of some pigment and photoreceptor cells, usually in a shallow

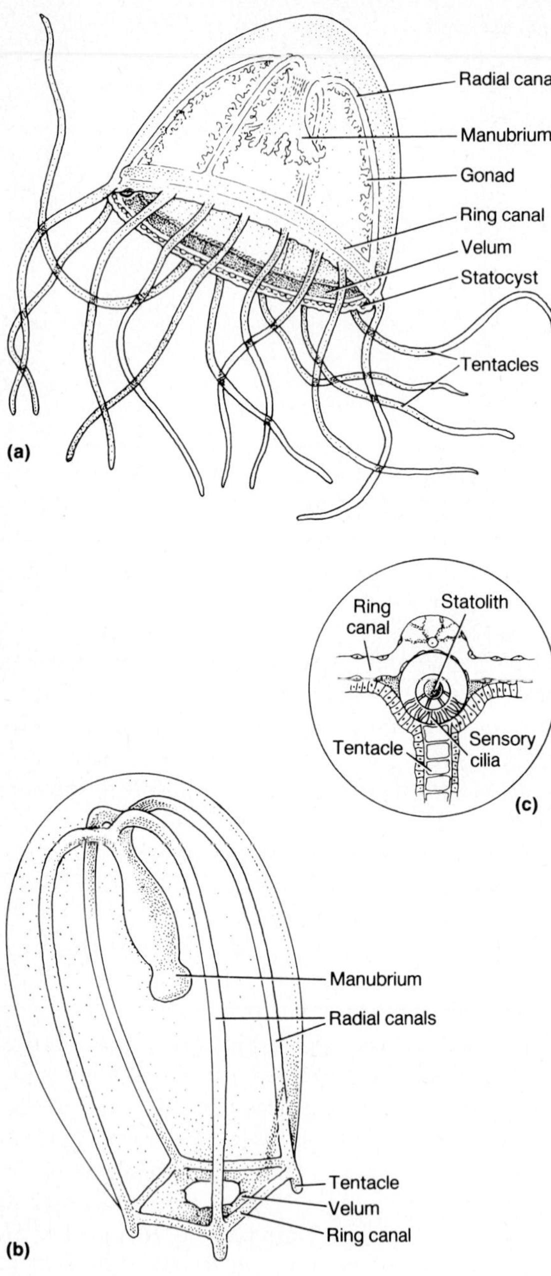

*Figure 5.8 The hydrozoan medusa: **(a)*** Gonionemus, *lateral view;* ***(b)*** Pennaria, *lateral view;* ***(c)*** *statocyst of* Obelia *(from the Hertwigs in L. H. Hyman,* The Invertebrates. *Vol. I,* Protozoa Through Ctenophora, *New York, McGraw-Hill, 1940, adapted with permission).*

depression, and provides the medusa with limited information about light intensities. **Statocysts** are organs of equilibrium and, like ocelli, are usually located at the tentacular bases. Each is a pit or closed vesicle containing one to several small calcareous **statoliths** surrounded by cells whose sensory bristles project into the interior (Fig. 5.8c). Displacement of the statolith stimulates the appropriate sensory bristles, and the coordinating effect of the nerve net allows the medusa to right itself. At the center of the subumbrella is a quadrangular tubelike **manubrium** at whose lower end is the mouth (Fig. 5.8a, b). The manubrium, sometimes supplied with lobes, frills, or arms, is also armed with cnidocytes. Like polyps, medusae capture and paralyze small zooplankters with which they come in contact.

Internally, the gastrovascular cavity is more complicated than that in polyps. Located at the base of the manubrium is a central **stomach** or **gastric pouch** from which arise four **radial canals** arranged 90° apart from each other (Fig. 5.8a, b). Each radial canal joins the **ring canal** at the umbrellar margin, and at the junction point of the two canals is a swollen **bulb.** The manubrium, stomach, and all canals are lined with gastrodermis, and their spaces constitute the gastrovascular cavity. The mesoglea is rather extensive, and although acellular, it contains fibers produced by both epithelial layers.

Muscular fibers associated with locomotion are more highly developed than those of the polyp. Composed chiefly of extensions of epitheliomuscular cells, muscles are especially developed around the umbrellar margin and subumbrellar surface, and they are arranged in both a longitudinal and a circular pattern. Contraction of the muscle fibers forces water out of the subumbrellar space. Rhythmic muscular pulsations propel the medusa upward in an aboral direction aided by the fact that the opening of this space is reduced substantially by the velum. Located in the umbrellar margin are two nerve rings; one coordinates muscular pulsations, and the other is associated with the statocysts.

All hydromedusae reproduce sexually, and most are dioecious. Epidermally produced gametes aggregate in the epidermis of the subumbrella or manubrium (Fig. 5.8a). Gametes are always released to the outside and never into the gastrovascular cavity.

Development and life cycles. Fertilization is often external in the water, but female medusae of many species retain the eggs, and therefore fertilization is in situ. Cleavage patterns are typical of those in all other cnidarians, and a ciliated **planula** develops. After a short, free-living existence it attaches and develops into a hydranth.

Most hydrozoans have both polyp and medusa represented in the life cycle, but life-cycle details vary among hydrozoans. A few species, like *Hydra,* are exclusively polypoid and never form medusae. Others, like *Liriope,* have no polypoid stage. In the freshwater jellyfish, *Craspedacusta,* the medusa is dominant, and the polyp is minute and primitive. Some species produce free-living hydromedusae, but in others the medusae remain attached to the gonophore. In still others, like the ostrich plume hydroids *(Aglaophenia),* the medusa morph has degenerated, apparently as an adaptation to high surf, so that only gonadal tissues remain on the gonophore. Four examples of species belonging to the Order Hydroida have been chosen to illustrate some of the diversity in morphology and life history details.

Hydra. *Hydra* is one of the most familiar of all laboratory animals, since it is often studied in introductory courses. The 14 species of hydras belong to three exclusively freshwater genera. They are rather common in littoral areas, where they can be found among vegetation or amid benthic debris. They all are solitary polyps without a perisarc and range in length from 2 to 25 mm. The body is cylindrical, and around the mouth are five or six hollow, extensible tentacles (Fig. 5.9). The body wall surrounds the

Figure 5.9 *Photomicrographs of* Hydra: ***(a)*** *an adult with three buds;* ***(b)*** *a female with ovaries;* ***(c)*** *a male with spermaries.*

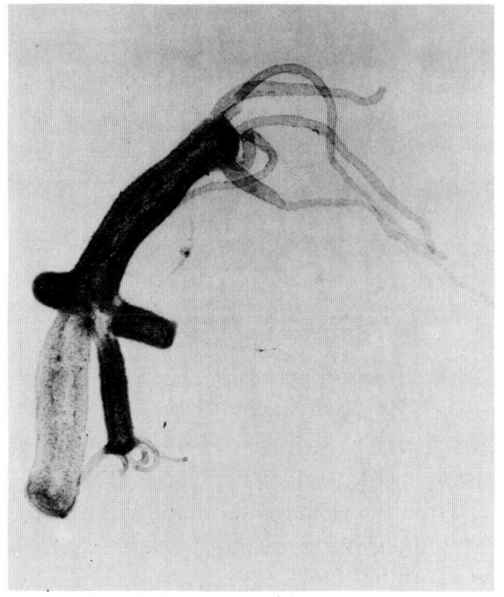

(a)

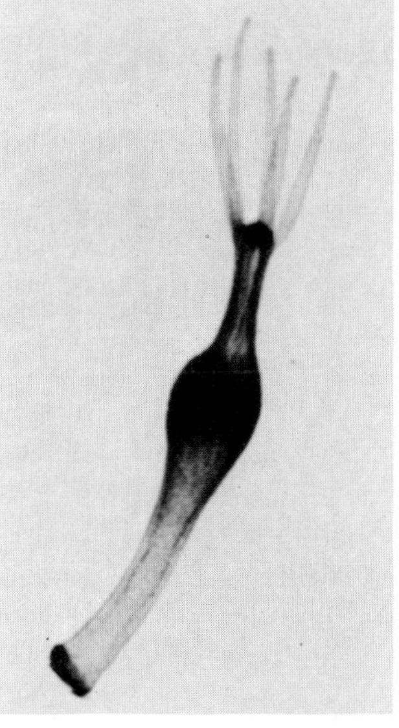

(b)

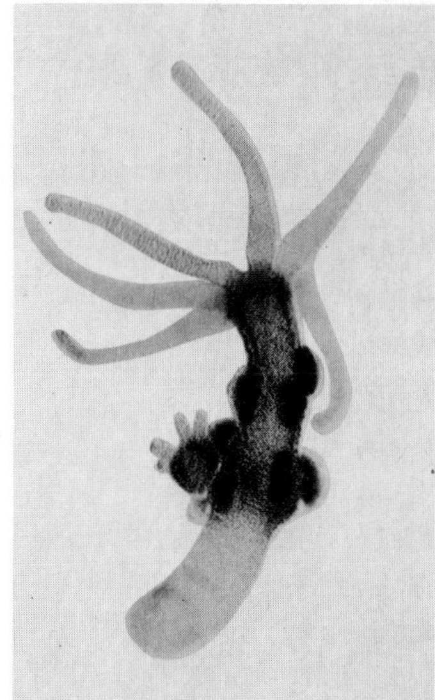

(c)

large, undivided gastrovascular cavity and consists of the typical tissues (epidermis and gastrodermis) and the acellular mesoglea. Several different types of nematocysts are present, and the cnidocytes are aggregated as **batteries** on the tentacles and around the mouth. A pedal disc enables the animal to attach temporarily to a surface. Locomotion is achieved by the gliding of the pedal disc over the surface or by the animal somersaulting oral end over aboral end.

Hydra reproduces commonly by budding during the spring and summer months, and one or two buds frequently appear on the side of the column (Fig. 5.9a). Later, tentacles develop, the base pinches off from the parent, and a new separate polyp is produced. Sexual reproduction is generally restricted to autumnal months and begins as clumps of presumptive gametes aggregate to form an ovary or testis (Fig. 5.9b, c). Most species are dioecious; interestingly, males usually outnumber females. In monoecious forms the testes are distal, and the ovaries are proximal on the column. The single egg is retained within the ovary, where it is fertilized. The developing embryo is surrounded by a chitinous coat, which enables it to endure adverse winter conditions. After a dormant period and the return of favorable environmental conditions the coat ruptures, and a juvenile miniature hydra emerges.

Tubularia. The several species of *Tubularia* are marine colonial hydroids whose hydranths have long hydrocauli. The individuals may be unbranched or arise from branches of the stolons. Each athecate hydranth bears two whorls of solid tentacles: a basal circle of relatively large elongate tentacles and an oral circle of shorter tentacles (Fig. 5.10). The polyp produces lateral budlike gonophores, which are found in grapelike clusters between the two tentacular whorls (Fig. 5.10). Each gonophore develops into a single medusa that remains attached to the parent polyp. The zygote develops into a planula and, later, into a second larval stage, the **actinula,** before leaving the female medusa. The actinula contains the rudiments of a mouth, tentacles, and gastrovascular cavity. The actinula settles to a substratum, transforms into a hydranth, and subsequently develops into a new colony.

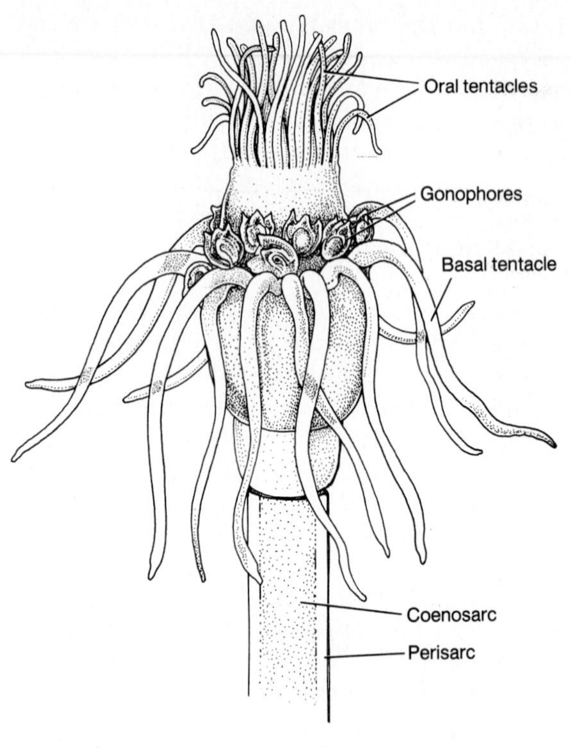

Figure 5.10 *A* Tubularia *hydranth with gonophores.*

Obelia. This genus contains many species of colonial hydroids found throughout marine habitats; they are particularly common in shallow coastal waters, where individuals are found attached to shells, rocks, and other submerged objects. Colonies range in size from 2 to 15 cm in height. Each hydranth, arising from a lateral hydrocaulus branch, contains about 20 solid tentacles in a single whorl around the mouth (Fig. 5.11). The coenosarc is extensive and continuous between individuals and in the hydrorhizal network. A tough perisarc surrounds all parts of the colony, including the hydranths.

Reproductive individuals are called gonangia. A **gonangium** consists of a central blastostyle that gives rise to many medusa buds (Fig. 5.11). The blastostyle is surrounded by an extension of the perisarc, the **gonotheca.** Young hydromedusae about 1 mm in diameter break away from the blastostyle, exit from the gonangium via a distal **gonopore** in the gonotheca, and mature.

Mature hydromedusae are free swimming, are about 5 mm in diameter, have numerous marginal tentacles, and have a small rudimentary velum. Male and female medusae shed gametes into the water, where fertilization takes place. Each zygote develops into a planula, which attaches to a substratum and develops into a polyp from which arise stolons, hydrocauli, and other polyps (Fig. 5.11).

Craspedacusta sowerbyi. This species is unusual in that it is the only freshwater jellyfish found in North America. It is found in many diverse habitats, including farm ponds, small lakes, and water-filled rock quarries. Medusae, the dominant morph, range in diameter from 5 to 22 mm. Their umbrellar margins bear numerous (50 to several hundred) tentacles of various lengths arranged in several series; short tentacles are used in feeding, and long tentacles

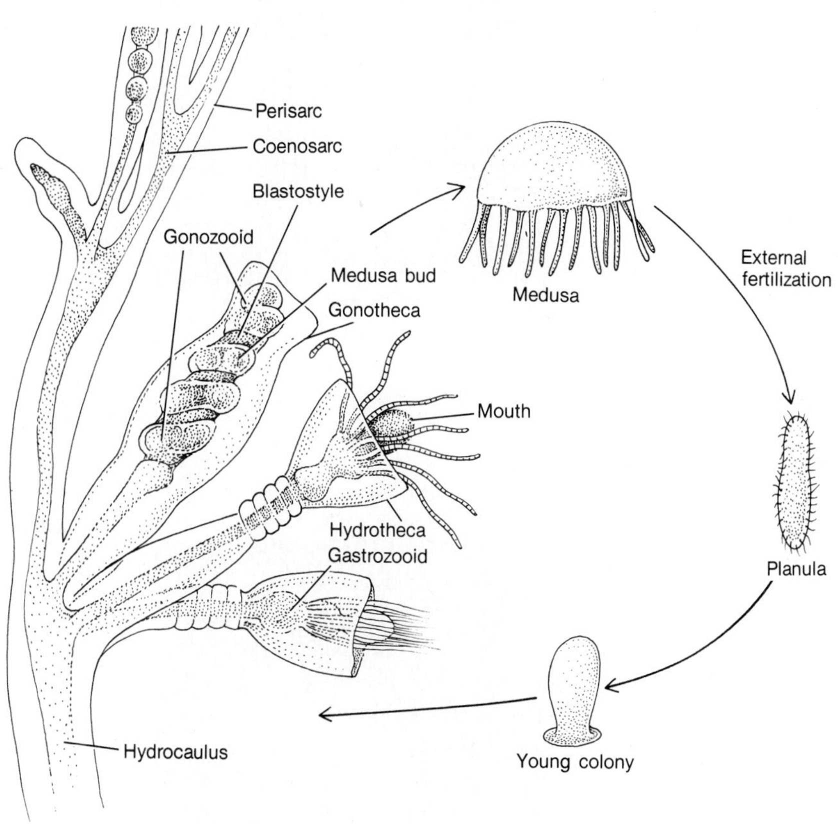

Figure 5.11 *A portion of a colony of* Obelia *and a diagramatic representation of the life cycle.*

probably are employed as stabilizers in swimming (Fig. 5.12). The umbrellar margin bears many statocysts. The velum is thick, and cnidocytes predominate near the mouth and on the umbrellar margin and tentacles.

Four masses of gametes are found on the subumbrellar surface. Following external fertilization a zygote develops into a small simple colony of several individuals, all of which lack tentacles (Fig. 5.12). The colony creeps over the bottom, feeding on minute invertebrates. New polyps, produced asexually by budding of the colony, may remain attached or may constrict off from the parent. Nonciliated, planulalike buds called **frustules** may be formed, which constrict off, creep about for a time, and develop into new polyps (Fig. 5.12). The colony may also produce medusa buds, which break away from the parent; these young medusae grow and develop rapidly into mature medusae.

The appearance of medusae of this species is sporadic, unpredictable, and often dramatic, and densities of medusae may be in excess of 50 per liter of water! Some ponds have an abundant population for a week or more for each of

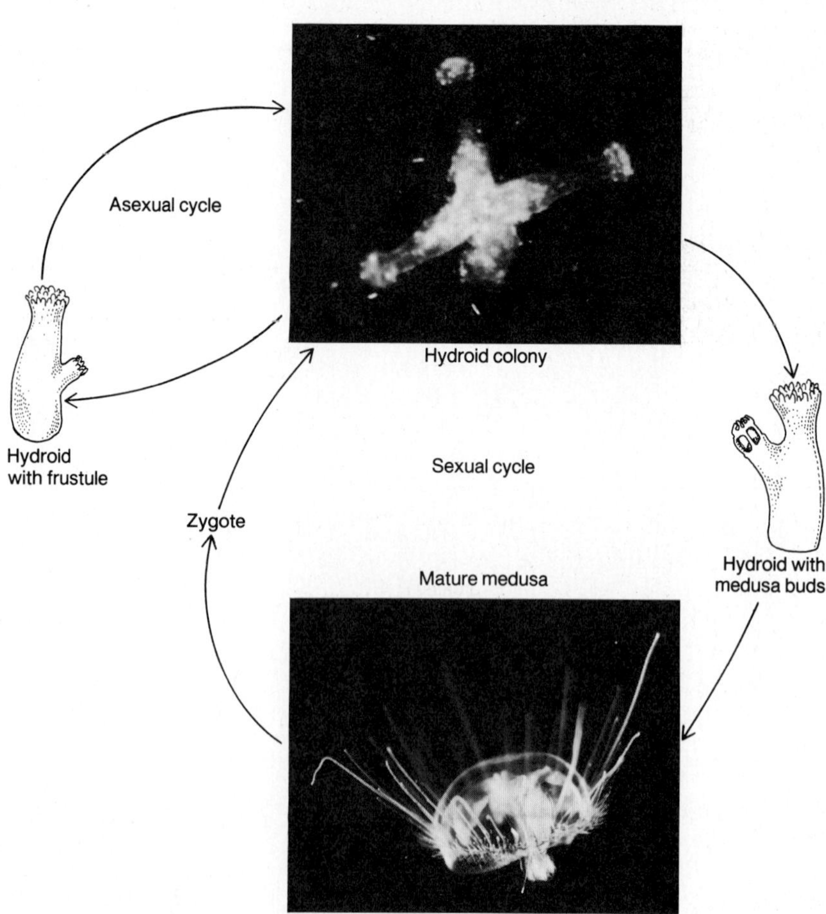

Figure 5.12 *Photomicrographs of the hydroid colony and a mature medusa of* Craspedacusta sowerbyi, *and a diagram of the life cycle (photographs courtesy of C. F. Lytle).*

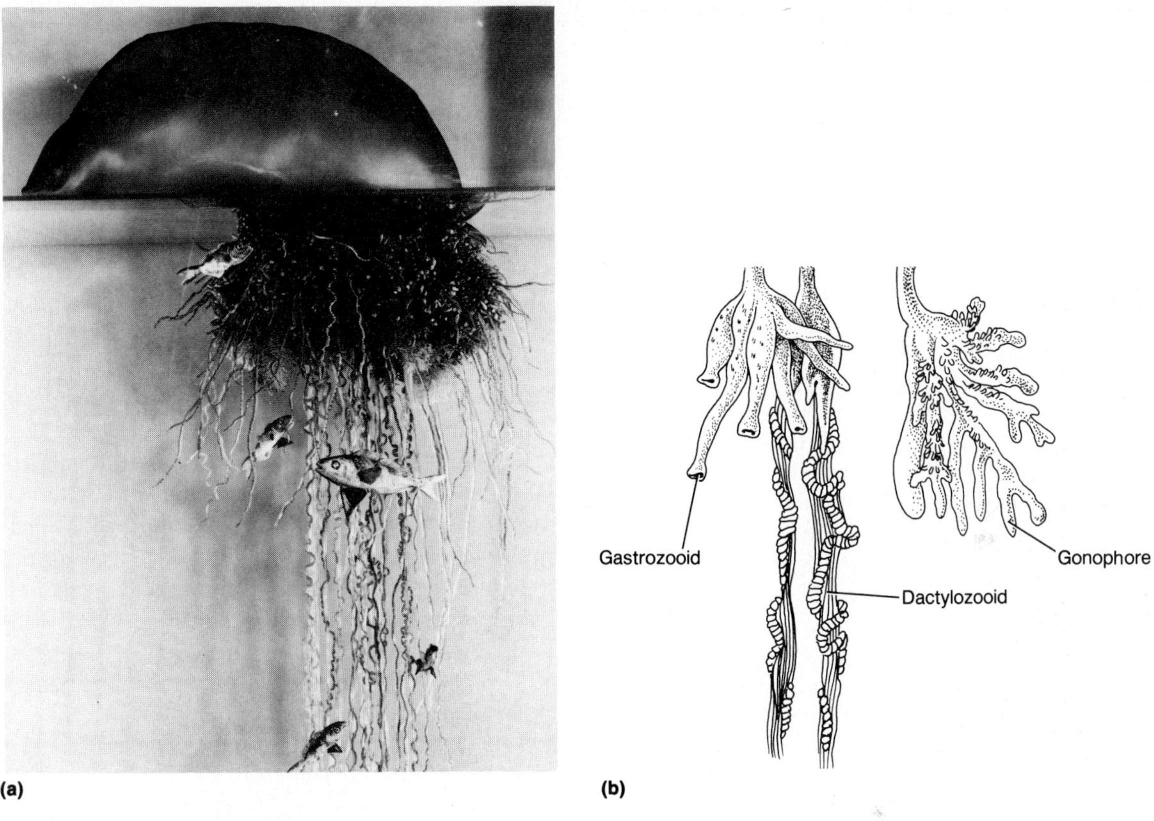

Figure 5.13 Physalia, *a representative siphonophore:* ***(a)*** *a photograph of a glass model of an entire organism (Neg. No. 321082 [Photo: Boltin] courtesy Department Library Services, American Museum of Natural History);* ***(b)*** *several different individuals (from Delsman in L. H. Hyman,* The Invertebrates. *Vol. I,* Protozoa through Ctenophora, *New York, McGraw-Hill, 1940, adapted with permission).*

a number of consecutive years followed by several years when they are absent. Medusae have been reported in bodies of water once and never found there again. One lake may be totally devoid of this species, while a nearby lake may have a large population.

The Siphonophores or oceanic hydrozoans. Members of the Order Siphonophora exist as large pelagic or floating colonies of polyps and medusae. They are often found in the open sea far removed from land and thus are called oceanic hydrozoans. The siphonophores include *Nectalia, Stephalia,* and the well-known Portuguese man-of-war, *Physalia* (Fig. 5.13). Siphonophores represent the highest degree of cnidarian polymorphism, and they are remarkably complex organisms because of the many different morphs that are present together as a colony. Siphonophores are often considered, perhaps rightfully so, as "superorganisms" because of pronounced division of labor exemplified by the various individuals. Because of their extreme polymorphism and colonial nature, their delicate coloration and exquisite beauty, and their potential for causing severe stings, they have attracted the attention of a number of zoologists.

The individuals in a colony are attached to a central column or disc of coenosarc, which is continuous between disc and all individuals. Up to eight distinctive morphs may be present; four are basically polypoid, and four are modified medusae. The polypoid morphs are:

Gastrozooids—feeding individuals that have a mouth; some lack tentacles, and others have a single, basal, contractile tentacle (Fig. 5.13b).

Dactylozooids—protective individuals usually lacking a mouth and having a single basal tentacle heavily armed with cnidocytes (Fig. 5.13).

Gonozooids—reproductive individuals homologous to the blastostyle or gonophore of colonial hydroids; they produce medusa buds.

Pneumatophore or **gas sac**—usually a single morph present in some as a sail or float into which gas is secreted. Once thought to be a medusa umbrella, it is now considered to be an inverted polyp (Fig. 5.13a).

The medusoid morphs are:

Nectophores—swimming individuals that lack manubrium and tentacles but have an umbrella, which contains strong muscles used to propel the entire colony.

Phyllozooids—often referred to as bracts, these leaflike medusae are equipped with cnidocytes, and they surround and protect the colony.

Gonophores—these gonad-bearing individuals are often borne on the gonozooids. They do not develop into free-swimming medusae, but rather they are retained as attached buds that produce gametes (Fig. 5.13b).

Oleocytes—oil floats that are similar to pneumatophores and are present in some siphonophores.

Certainly the best known siphonophores belong to the genus *Physalia;* a brief description is included here, since they are so familiar (Fig. 5.13). The pinkish blue pneumatophore or float is blown by the wind in tropical or subtropical areas. The colony lacks nectophores and must rely totally on wind or wave action for horizontal movement. The gas-filled float contains nitrogen gas or carbon dioxide that is secreted into the pneumatophore by special cells in the oval coenosarcal disc located beneath the float and common to the entire colony. Numerous gastrozooids, dactylozooids, and gonophores are present. Each dactylozooid bears an enormously extensible tentacle with many cnidocytes (Fig. 5.13). A 12-cm-long *Physalia,* for example, can have dactylozooid tentacles 9 m long. Further, after immobilizing a prey, the tentacles can shorten to about 10 cm, a remarkable 90-fold contraction! Prey is brought near to the mouths of the gastrozooids, the margins of each mouth spread out over the prey, and initial digestion begins. Later, bits are sucked into the gastrovascular cavity of separate gastrozooids and digested. *Physalia* nematocysts are unusually potent and are dangerous to human beings and can even be fatal.

Class Cubozoa

Representatives of this small class are probably evolutionarily intermediate between the Hydrozoa and Scyphozoa. Cubozoan medusae are rather elongate in the oral–aboral axis, and some are even up to 25 cm tall. Their medusae are usually colorless, strongly pelagic, and square in cross section. The short manubrial canal opens into a central stomach, from which arise four gastric pockets extending toward the edge of the medusa. At each corner of the medusa is a single tentacle or a group of hollow tentacles (Fig. 5.14). Situated between each of the four tentacles is a **marginal sense organ** that opens into the underlying gastric pouch by way of a gastrodermal canal. The sense organ consists of a statolith plus one to several ocelli. Medusae reproduce asexually, and the resulting larvae develop into polyps, though the polyp stage for only one species is known.

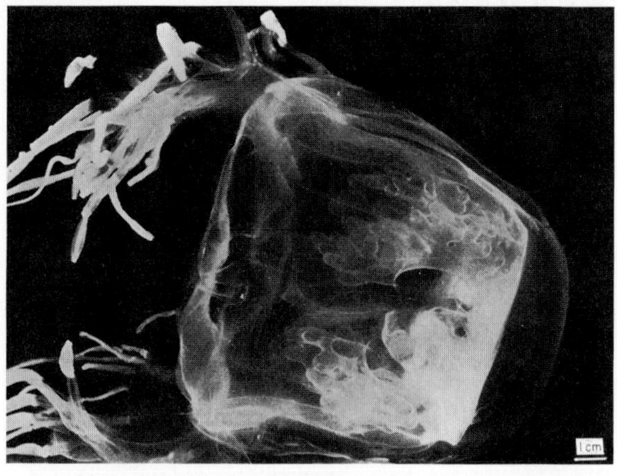

Figure 5.14 Chironex fleckeri, *a cubozoan, and one of the virulent sea wasps (J. H. Barnes in W. J. Rees [ed.],* The Cnidaria and Their Evolution, *Symposium #16, Zoological Society of London, 1966, reproduced with permission from the Zoological Society of London).*

Cubozoans are present in all tropical and subtropical seas. The nematocysts of these cnidarians contain a powerful poison, a feature that is the basis of the common name of some: sea wasps. Certain species of *Chiropsalmus* and *Chironex* are especially virulent; 50 human deaths attributed to their stings have been reported in Australia alone.

Class Scyphozoa

The scyphozoans are exclusively marine cnidarians in which the medusa is the dominant morph and the polyp is present as a minute larva. Scyphozoans commonly are called jellyfishes because of the large, conspicuous, medusoid stage. Scyphozoan medusae range in diameter from 2 to 40 cm, but the giant scyphozoan, *Cyanea capillata,* may have a bell diameter exceeding 2 m. Typically, medusae are free swimming, but in one order they are sessile and attached by an exumbrellar stalk. Jellyfishes are found in all seas, and while some are deep-sea or oceanic forms, most frequent coastal areas. There are fewer than 200 species, but often they seem to be a much more prominent group because they are sometimes present in enormous numbers, often are a nuisance, and are sometimes dangerous to bathers. Because of their large size and their nematocysts, they should be treated with caution by swimmers. In fact, one can be "stung" by stepping on or handling dead jellyfishes washed up on the beach; incredibly, the cnidocytes apparently remain functional for several days after the death of the organism. Lesions produced by jellyfish stings are usually painful and slow to heal. Colors range from delicately tinted to deep tones of pink, blue, purple, orange, brown, and milky. Brief resumes of the three orders of scyphozoans appear at the end of this chapter.

General characteristics. Scyphozoans are generally larger and constructed on a more complex level than hydrozoans. Most of the distinguishing scyphozoan features refer to the medusoid stage, and these adaptive characters will be highlighted in the discussion below about medusae.

Medusa. Medusae of scyphozoans, or **scyphomedusae,** are fundamentally similar to hydromedusae but differ in a number of details. The umbrella is large and varies in shape from a shallow saucer to a deep dome (Fig. 5.15). Irrespective of the bell shape, scyphomedusae have pronounced tetramerous symmetry. The umbrellar margin contains numerous tentacles and marginal sense organs. Tentacles may be either solid or hollow, and they are present in a definite number ranging from four to many. Tentacles are generally short but may be quite long in certain species and are supplied with numerous cnidocytes. The **marginal sense organs** are present as 4, 8, or 16 structures (Fig. 5.16a). Each is composed of a pair of **lappets** surrounding a sensory structure, the **rhopalium;** the lappets give the margin a scalloped appearance. Each rhopalium is a concentration of neurons and is often club shaped or tongue shaped. On each rhopalium is borne a statocyst,

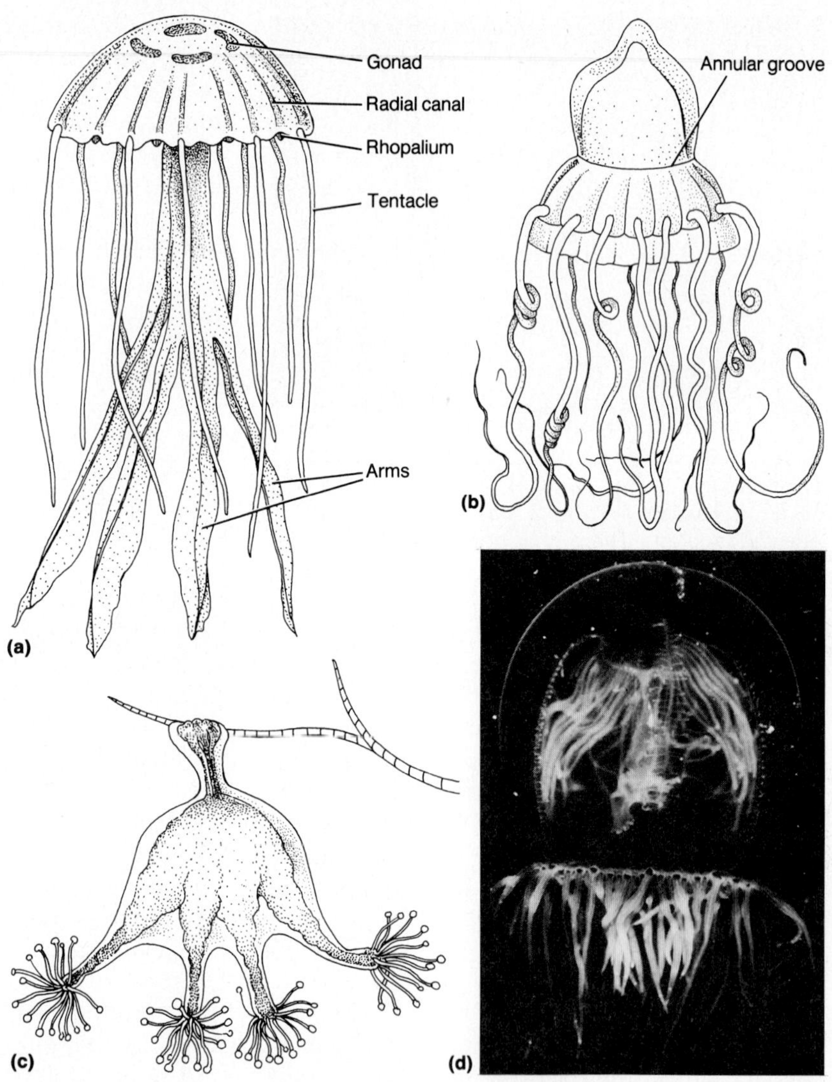

*Figure 5.15 Types of scyphozoan medusae: **(a)*** Pelagia, *a pelagic form; **(b)*** Periphylla; ***(c)*** Haliclystus, *a sessile form; **(d)** a photograph of* Polyorchis *(courtesy of Ward's Natural Science, Inc., Rochester, N.Y.).*

two sensory pits, and often an ocellus. The **statocyst** is located at the tip of the rhopalium, contains a concretion or statolith, and provides postural information to the medusa. The **sensory pits** lie at the rhopalial base, one being on the exumbrellar surface and the other on the subumbrellar surface (Fig. 5.16b,c). These pits contain concentrations of sensory cells that also are found generally over the umbrellar surface. When present, the **ocellus** is usually a simple pigmented area. Scyphomedusae are positively phototactic to dim light, but on bright sunny days they sink deeper in the water. A velum, characteristic of hydromedusae, is absent; but an analogous structure, the **velarium,** is present in some scyphomedusae. Representatives of the Order Stauromedusae are sessile and attached to an alga or some other substratum by an ex-

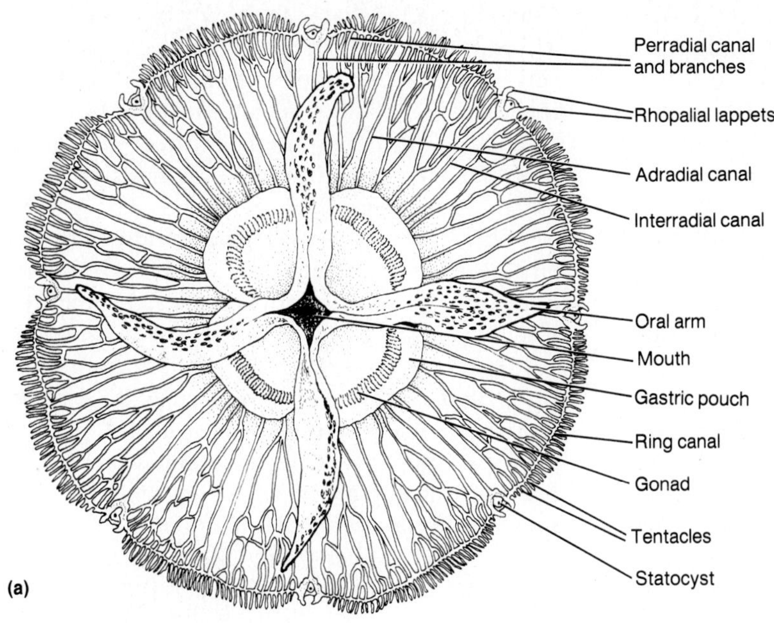

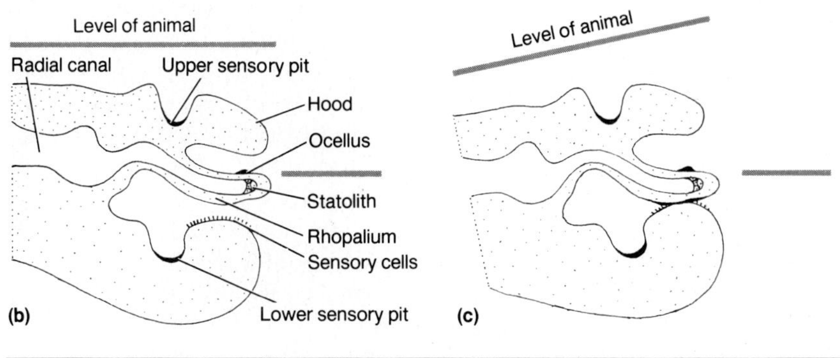

Figure 5.16 Structures of the scyphomedusa of Aurelia: ***(a)*** *an oral view;* ***(b)*** *a section through a rhopalium;* ***(c)*** *same section in an animal oriented out of the horizontal plane.*

umbrellar **stalk.** The stalk is a modified polyp, and the medusa is the distal, flared, funnellike end (Fig. 5.15c).

A prominent quadrangular **manubrium** hangs centrally from the subumbrellar surface. Bearing the mouth at the lower end, the manubrium often has four or eight frilly **manubrial** or **oral arms** (Figs. 5.15a, 5.16a). These arms are flexible and ciliated and bear cnidocytes. In representatives of the Order Rhizostomeae the oral arms are fused, and the primary mouth is closed and therefore nonfunctional. In place of the mouth, many minute suctorial openings or mouths lead into the gastrovascular cavity.

Scyphomedusae feed on a wide variety of animals, including other medusae and ctenophores, medium-sized fishes and crustaceans, and most other moderate-sized invertebrates.

Others, like *Aurelia,* are suspension feeders. Minute planktonic organisms are trapped in mucus on the subumbrellar surface and tentacles as the medusa slowly sinks. After sinking for a time, umbrellar contractions force the medusa aborally, and the sinking is resumed. The mucus-bound prey is moved by ciliary action to the umbrellar margins, transferred to grooves on the oral arms, and moved into the mouth.

The gastrovascular cavity is extensive in scyphomedusae and consists of the manubrial canal, stomach, gastric pouches (in some), and many canals, all of which are lined with gastrodermis. The manubrial canal opens into a centrally located **stomach** (Fig. 5.16a). In primitive scyphomedusae there are four **gastric pouches** that extend outward toward the margin. Each pouch is separated from its neighbor on either side by a **septum** composed mostly of mesoglea. The septa are perforate, and these openings facilitate circulation within the gastrovascular cavity. The inner margins of the septa facing the stomach contain a number of cnidocyte-bearing threads called **gastric filaments.** Their nematocysts are used to kill or stun prey still alive when ingested. Gastric filaments also contain concentrations of enzyme-secreting and absorptive cells.

In adult medusae of more advanced forms, the prominent gastric pouches extending to the bell margins and septa have been substantially reduced. In their place is an extensive canal system necessitated by the thick mesoglea. Three different types of canals, based on their positions, are found in higher scyphomedusae. Four **perradial canals** extend from the central stomach to the bell margins. These canals correspond in position to the oral arms and thus divide the umbrella into quadrants. Between each perradial canal there is an **interradial canal.** Between each interradial and perradial canal is an **adradial canal** (Fig. 5.16a). Therefore four perradial, four interradial, and eight adradial canals are usually present. Some or all of the canals may be branched near the umbrellar margin where they join a **ring canal** (Fig. 5.16a). Generally, after water is drawn into the stomach, it is circulated in the adradial canals to the ring canal. Water returns from the margin via interradial and perradial canals to the stomach. Circulation of materials within these gastrovascular canals is made possible by the flagella of gastrodermal cells.

The mesoglea is thick, firm, and gelatinous and contains some muscle fibers used in swimming and many wandering cells that constitute a cellular layer. The scyphomedusoid muscular system is restricted mainly to the epidermis; extensions of the epitheliomuscular cells, especially in the subumbrella and tentacles, are the principal muscles. A circular band of **coronal muscle,** running around the subumbrellar margin, is especially effective in swimming or pulsating. Tentacular muscles are mostly longitudinal and are used to contract the tentacles during feeding.

The nervous system is a nerve net typical of that in all cnidarians. In some orders there are two nerve nets: one controls the umbrellar pulsations, while the second is more diffuse and coordinates local reactions in food ingestion. In most scyphomedusae, however, pulsation control is centered in the mass of neurons within each rhopalium, and thus the rhopalia, located perradially and interradially, are important sensory–nervous structures.

Scyphozoans are generally dioecious, and gametes are derived from interstitial cells of the gastrodermis. Likewise, the gonads are situated in the gastrodermis lining the gastric pouches or stomach. In those scyphomedusae with septate gastrovascular cavities a gonad is located on each side of each septum, making a total of eight. In nonseptate medusae there are four horseshoe-shaped gonads located on the outer margins of the stomach.

In summary, the principal distinguishing features of scyphomedusae reflect their relatively complex structures. The medusa has a complex gastrovascular cavity represented typically by a central stomach and radially arranged gastric pouches and numerous canals. The mesoglea is greatly enlarged over that of hydromedusae and contains many amebocytes and fibers. The gastrodermis contains cnidocytes. The lack of a velum and the presence of rhopalia are also diagnostic.

Polyp. The scyphozoan polyp is very small, inconspicuous, trumpet shaped, and fastened aborally by an adhesive disc to a substratum. Often they even live "upside down" whereby the aboral end is attached to the undersurface of some object and the oral end hangs downward in the water. This morph is more properly referred to as a larva or **scyphistoma** (Fig. 5.17a). Like the medusa, this stage has tetramerous symmetry reflected in the tentacles, gastrovascular cavity, septa, peristomial pits, and manubrium. A scyphistoma resembles a miniature hydra with four tentacles located in the perradii, but some species develop tentacles in multiples of four. A short, four-sided **hypostome** is present bearing the **mouth,** and four shallow sensory **peristomial pits** are located

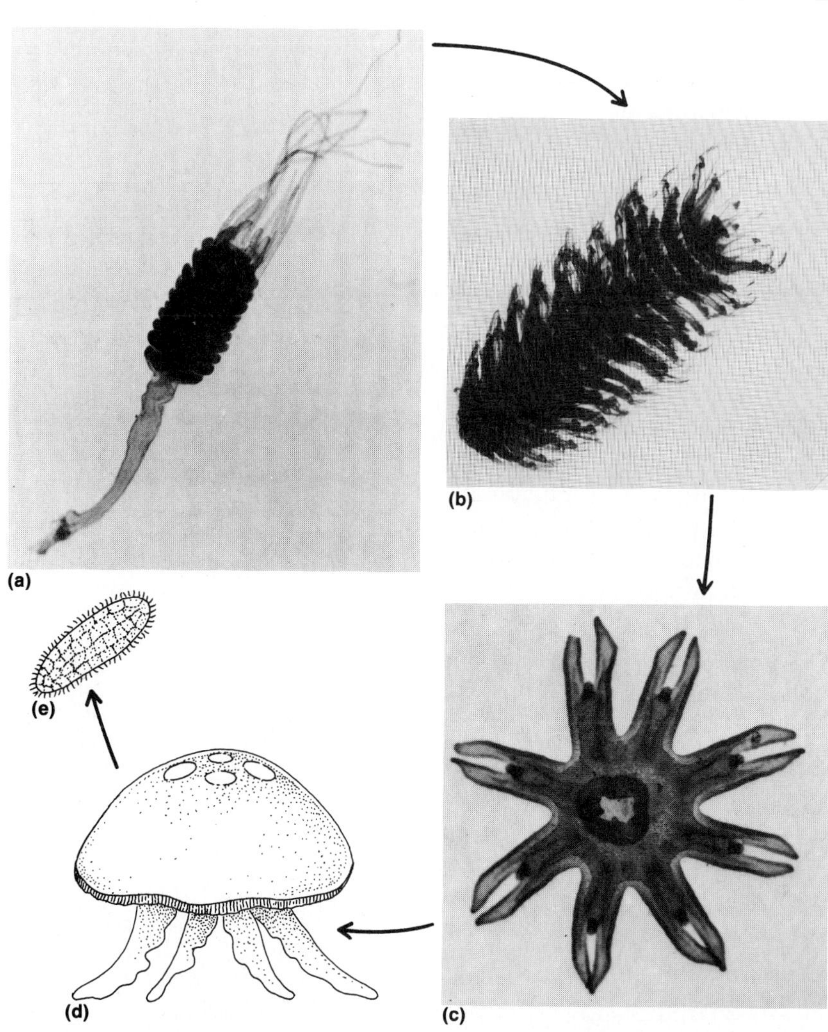

Figure 5.17 *The life cycle of a typical scyphozoan:* ***(a)*** *photomicrograph of a scyphistoma beginning to strobilate;* ***(b)*** *photomicrograph of a strobila;* ***(c)*** *photomicrograph of an ephyra;* ***(d)*** *medusa;* ***(e)*** *planula.*

around the mouth with one in each quadrant. The gastrovascular cavity is divided internally by four **septa.** Scyphistomae feed on minute plankton and produce more scyphistomae asexually by budding. Scyphistomae also produce young medusae by transverse fission, a phenomenon described below as a part of their life cycle.

Scyphozoan life history. When mature, the gonads rupture and free their gametes into the scyphomedusoid gastrovascular cavity. Gametes exit through the mouth into the water, where fertilization normally ensues. In some semaeostomes like *Aurelia,* unfertilized eggs are retained in pits on the oral arms of the female, where fertilization and early embryonic development take place. Each zygote cleaves to form a hollow blastula and then invaginates to form a typical ciliated **planula** (Fig. 5.17e). After a brief period of free-swimming existence, the planula settles onto a substratum, attaches by its anterior end, and develops into the **scyphistoma** (Fig. 5.17a). At certain periods of the year the scyphistoma typically undergoes repeated transverse fissions or **strobilation** to produce asexually a number of small immature medusae called **ephyrae.** A scyphistoma in the process of strobilation is called a **strobila** (Fig. 5.17b). Thus a strobila is composed of a basal portion and ephyrae arranged one on top of the other like a stack of plates in which the oldest ephyrae are at the oral end and the youngest are more basal. Ephyrae break away from the strobila one at a time and become part of the zooplankton. After strobilation the strobila may become a scyphistoma once more and strobilate again the following year. Some scyphistomae have been known to live for four or more years.

Each ephyra is a microscopic gelatinous creature about 1 mm in diameter (Fig. 5.17c). Most adult structures are present but incompletely developed. The margins are deeply incised to form eight lobes. Each lobe is bifurcated distally to form a pair of lappets surrounding a sensory rhopalium (Fig. 5.17c). A short quadrangular manubrium bearing the mouth is present. Food for the ephyra, mostly planktonic crustaceans, is caught on the lappets and transferred to the mouth. The intake of food energy enables the ephyra to expand the margins between the lobes. Manubrial arms and gastrovascular canals develop, and the medusa becomes sexually mature in about two months (Fig. 5.17d).

Some scyphozoans have modified their life cycle to various degrees. In some oceanic forms, like *Pelagia,* the planula transforms directly into an ephyra. In several other species the scyphistoma is retained in cysts within the female. A deep-sea form, *Stygiomedusa,* retains the scyphistoma within the gastric pouch, where it is nourished by special maternal tissues—a case, incredibly, of a viviparous cnidarian!

CLASS ANTHOZOA

Anthozoans, including sea anemones as well as hard and soft corals, are exclusively marine organisms inhabiting coastal waters worldwide; they are especially abundant in tropical regions. More than 6500 modern species have been described as well as a similar number of fossil species. Anthozoans are totally polypoid; no trace of a medusa is ever developed. Individual polyps range in diameter from 1 mm to about 1 m. They may exist as solitary polyps but frequently occur as complex colonies (Fig. 5.18).

Figure 5.18 *A photographic collage of various types of anthozoans:* ***(a)*** *the skeleton of* Tubipora, *the organ-pipe coral (alcyonarian);* ***(b)*** *an entire colony of* Ptilosarcus, *a sea pen (alcyonarian) (courtesy of B. Best);* ***(c)*** Acropora, *the staghorn coral (zoantharian);* ***(d)*** Dichosenia, *the star coral (zoantharian) with a Christmas tree worm (annelid);* ***(e)*** Diploria, *the brain coral (zoantharian);* ***(f)*** *the skeleton of* Fungia, *an ahermatypic, solitary, cup coral (zoantharian);* ***(g)*** Ceriantheomorphe (=Cerianthus), *a burrowing, tube-building sea anemone (zoantharian); individual on left side has been removed from its tube while the right-hand individual is still within its tube (parts c–e courtesy of J. M. Resing).* ▶

(a)
(b)
(c)
(d)
(e)
(f)
(g)

Colorations run the gamut from drab to brilliant shades of red, blue, orange, green, and cream, and they often have colors in absolutely exquisite combinations and shades.

Anthozoans are of particular interest to many biologists for several important reasons. First, anthozoans, especially the hard corals, are of considerable ecological importance because they are primarily responsible for the formation of distinct, huge, unique ecosystems—the coral reefs—about which more will be said later. Second, they are a very successful group, owing mostly to their many adaptations to a sessile existence. The functional morphology of anthozoan polyps can best be understood under four general categories: external features, pharynx, gastrovascular cavity, and body wall construction. There follows a brief discussion of these four groups of adaptations.

Externally, anthozoans exhibit classical radial symmetry. The body is typically polypoid with a basal disc, cylindrical column, and oral disc. The **basal disc** is present in solitary forms, where it forms a permanent or semipermanent attachment with a substratum. In many colonial species, one polyp is continuous with its neighbors by means of basal stolons, which obviate the basal disc. The **column** may be elongate and slender, or it may be short and squatty. The flattened **oral disc** bears the central slitlike **mouth** surrounded by a number of hollow ciliated tentacles. Radial muscles in the oral disc are instrumental in opening the mouth when the animal is feeding. The ovate or elliptical mouth is the only external hint of the biradial or bilateral symmetry to be found internally.

One of the most distinguishing features of all anthozoans is the presence of an invaginated, flattened, epidermal area immediately inside the mouth. This invagination, the **pharynx,** usually extends internally into the gastrovascular cavity more than one-half of the column length (Fig 5.19c, 5.20c). The pharynx is lined with ectodermally derived epidermis and is considered to be a stomodeum. All anthozoans have a pharynx, but no other cnidarians have anything comparable to it. Beginning at the mouth and extending the length of the pharynx are one or two ciliated grooves or gutters, the **siphonoglyphs,** present in most anthozoans (Figs. 5.19c, 5.20c, d). Siphonoglyph cilia direct food and a continuous flow of water from the outside through the mouth and pharynx and into the gastrovascular cavity.

Internally, anthozoan symmetry tends strongly toward bilateral symmetry. There is some recent embryological evidence to support a theory that anthozoan bilateral symmetry is primary rather than secondarily acquired. Nonetheless, we shall continue to consider anthozoans, like all cnidarians, to be radially symmetrical but with a clear tendency toward a secondary bilateral symmetry.

The gastrovascular cavity reaches a new level of complexity in the anthozoans, for it is in this group that this cavity is divided by a series of longitudinal or oral–aboral **septa** (Figs. 5.19, 5.20). The radially arranged septa are composed mostly of mesoglea, but a layer of gastrodermis lies on either side. Each septum extends from the oral to basal discs and projects medially from the body wall out into the gastrovascular cavity. Some or all may extend to the pharynx and fuse with it and thus are also appropriately called **mesenteries.** Those septa extending to the pharynx are called **complete** or **primary septa,** and those septa that do not reach centrally to the pharynx are called **incomplete septa.** Based on the degree to which incomplete septa extend into the gastrovascular cavity, the longest incomplete septa are **secondary septa,** the next longest are **tertiary septa,** and the shortest are **quaternary septa.** The net effect of these septa is to increase to a substantial degree the surface area of the gastrovascular cavity. Below the pharynx, all septa project into the gastrovascular cavity and divide it imperfectly into radially arranged gastrovascular compartments. The free edges of the septa below the pharynx are usually thickened and form sinuous, coiled **septal filaments.** These filaments are equipped with many cnidocytes, cells that

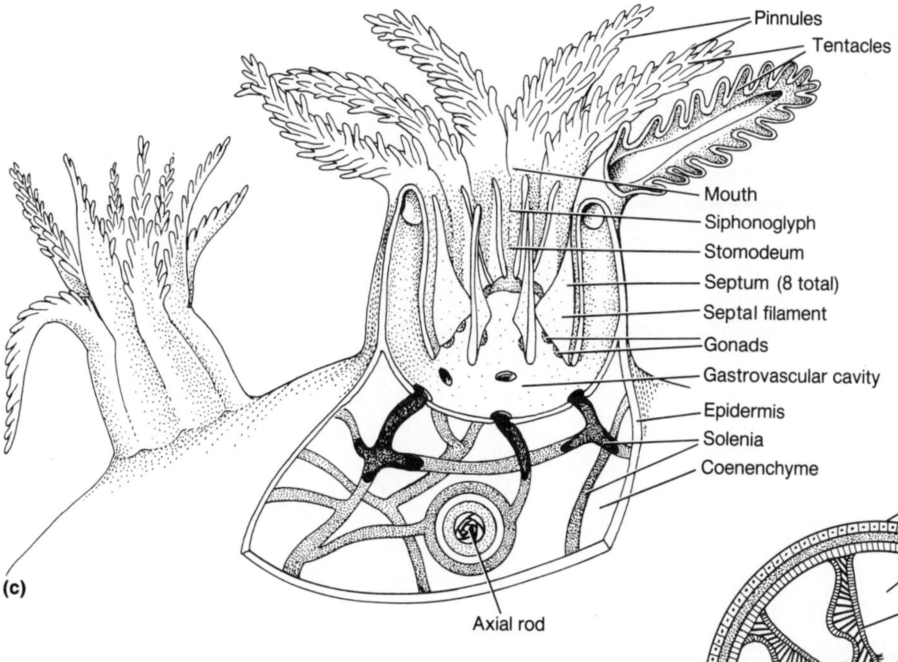

Pinnules
Tentacles
Mouth
Siphonoglyph
Stomodeum
Septum (8 total)
Septal filament
Gonads
Gastrovascular cavity
Epidermis
Solenia
Coenenchyme
Axial rod
(c)

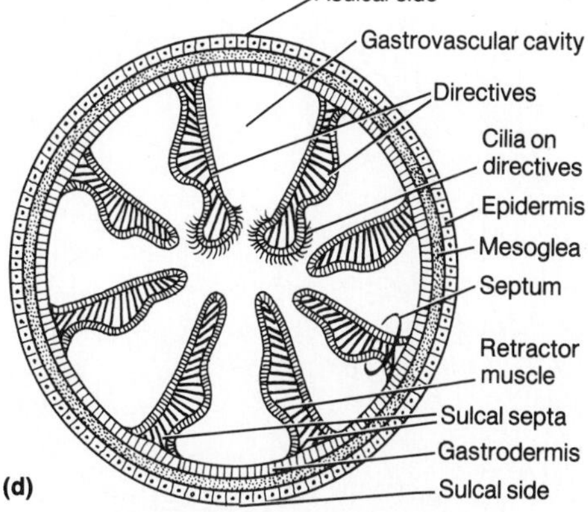

Asulcal side
Gastrovascular cavity
Directives
Cilia on directives
Epidermis
Mesoglea
Septum
Retractor muscle
Sulcal septa
Gastrodermis
Sulcal side
(d)

Figure 5.19 *Alcyonarians: **(a)** a photograph of a portion of a colony of a sea fan (courtesy of J. M. Resing); **(b)** a photograph of a portion of a colony of* Leptogorgia *showing several individual polyps extended; **(c)** a diagrammatic cross-sectional view of a portion of a colony of the Order Gorgonacea—polyps are shown in an oral-aboral plane; **(d)** a diagrammatic cross-sectional view of a gorgonian polyp below the pharynx.*

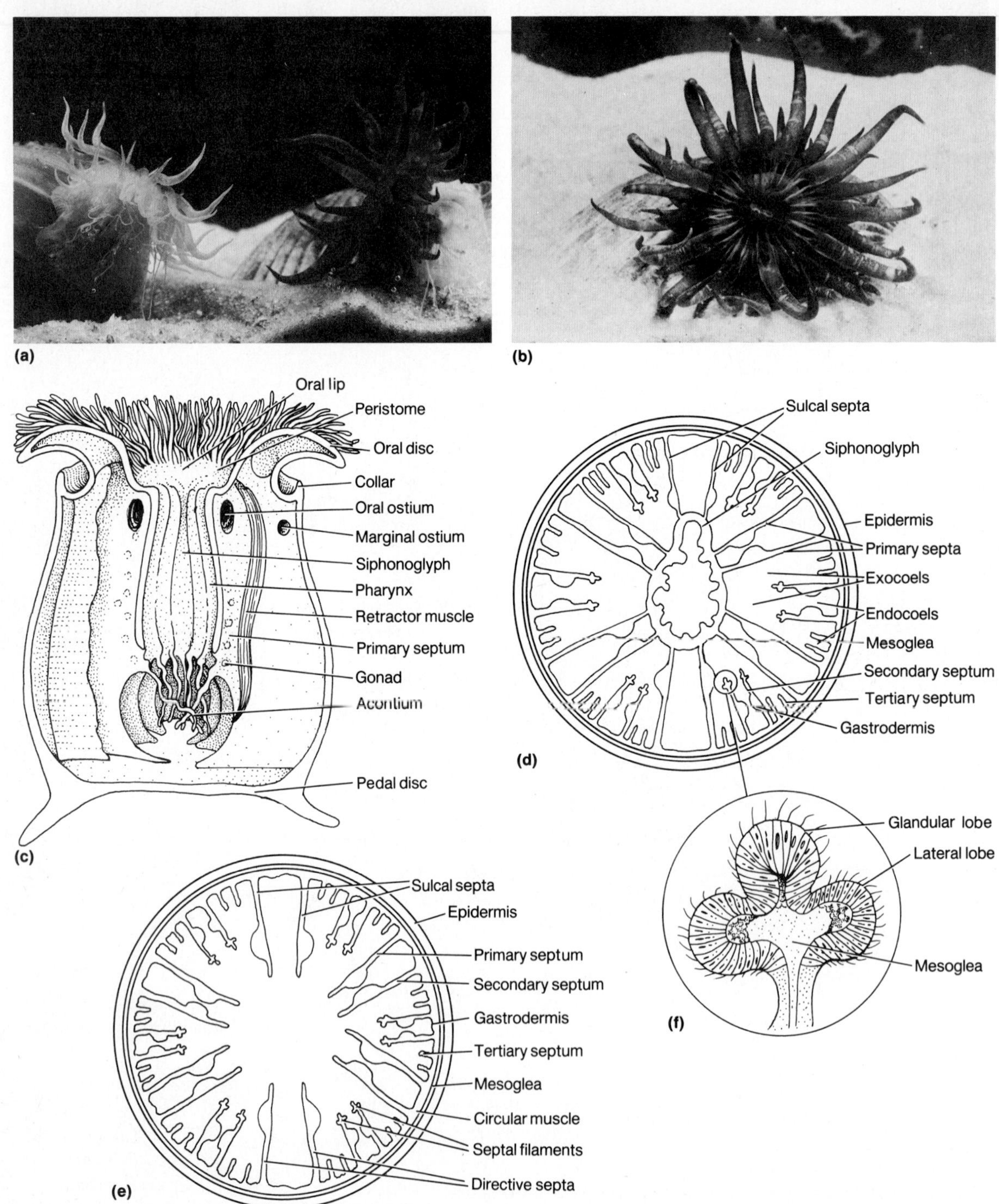

(a)
(b)
Oral lip
Peristome
Oral disc
Collar
Oral ostium
Marginal ostium
Siphonoglyph
Pharynx
Retractor muscle
Primary septum
Gonad
Acontium
Pedal disc
(c)
Sulcal septa
Siphonoglyph
Epidermis
Primary septa
Exocoels
Endocoels
Mesoglea
Secondary septum
Tertiary septum
Gastrodermis
(d)
Glandular lobe
Lateral lobe
Mesoglea
(f)
Sulcal septa
Epidermis
Primary septum
Secondary septum
Gastrodermis
Tertiary septum
Mesoglea
Circular muscle
Septal filaments
Directive septa
(e)

◀ ***Figure 5.20*** *Sea anemones: **(a)** a photograph of two different anemones; **(b)** a photograph of an anemone seen in oral view to illustrate the oral disc, mouth, and tentacles; **(c)** an oral-aboral section through the column of* Metridium *(from L. H. Hyman,* The Invertebrates. *Vol. I,* Protozoa Through Ctenophora, *New York, McGraw-Hill, 1940, adapted with permission); **(d)** a cross sectional view of the column of* Metridium *at the level of the pharynx; **(e)** a cross-sectional view of the column of* Metridium *below the pharynx. **(f)** a cross section of a septal filament.*

secrete digestive enzymes, and gamete-producing interstitial cells. Each septum is equipped with a longitudinal **retractor muscle** on only one side. The combined contractions of the longitudinal septal muscles retract the oral disc.

The anthozoan body wall is considerably more complex than that of hydrozoan polyps. The mesoglea is thickened and is a true cellular layer containing stellate cells, fibers, and a thick matrix. The body wall also contains some rather distinct muscles composed mainly of longitudinal epidermal fibers and circular gastrodermal fibers. Additionally, the longitudinal retractor muscles, composed of gastrodermal fibers and present in each septum, are well developed and are effective in retraction of the oral end and tentacles. In many anthozoans an internal or external skeletal system is present. If internal, as in the soft corals, the skeleton is composed of sclerites or spicules secreted by collenchymal cells located in the mesoglea. External skeletons like those of the hard corals, are calcareous and secreted by epidermal cells. The nervous system is composed of two plexuses: a subepidermal and a gastrodermal net. Sense organs are absent, and scattered sensory cells are present in both epithelial layers but are concentrated on the tentacles. The presence of fairly well developed muscles and a nervous system makes possible rather complex behavioral patterns, especially in sea anemones (see Box 5.1 later in this chapter).

The class is subdivided into two subclasses: Alcyonaria or Octocorallia and Zoantharia or Hexacorallia. The distinctions between the two are based chiefly on symmetry, nature of the tentacles, numbers of siphonoglyphs, type of skeleton, and arrangement of septa. A discussion of both subclasses follows, and some anthozoan diversity is illustrated in Fig. 5.18.

Subclass Alcyonaria (Octocorallia). Alcyonarians or soft corals often have descriptive common names like horny, soft, whip, and pipe corals. About 3000 living species are known. While some species are found in temperate regions, many more live in tropical seas. Alcyonarians are all colonial and possess octamerous symmetry reflected in numbers of tentacles and septa. While individual polyps are small, colonies may attain a length of 1 m or more.

Each polyp is seated in a thick mesogleal matrix called **coenenchyme** and covered by the epidermis (Fig. 5.19c). It is usually firm or fleshy because of its hornlike material or the calcareous spicules deposited in it. In some the calcareous spicules are fused, but in others the spicules remain separate from each other. The coenenchyme constitutes an internal skeleton characteristic of all alcyonarians. The beautiful shades of yellow, orange, red, and blue of many octocorallians result from the colors of the internal skeleton. Running through the coenenchyme are gastrodermally lined tubes or **solenia,** which connect the gastrovascular cavities of all the polyps in the colony. New polyps arise asexually as buds or sprouts from the solenia. In the Order Gorgonacea, individual colonies are built on a central axial rod composed of a proteinaceous substance called **gorgonin** secreted collectively by the polyps (Fig. 5.19c).

The shape and form of alcyonarian colonies are usually distinctive, and many common names are descriptive of the overall colony shape. For example, sea fans have a growth form in a flat plane with many latticelike cross branches. Sea feathers are more highly branched. Sea pansies are composed of a primary leaflike polyp and many secondary polyps. Sea pens have a primary polyp elongated into a long stemlike base penetrating the soft benthic material (Fig. 5.18b). *Tubipora,* the organ-pipe coral, has an endoskeleton built as a series of parallel tubes with internal partitions (Fig. 5.18a). Often, growth forms are oriented according to the direction of the prevailing water currents.

An individual alcyonarian polyp is vase-shaped, and on the oral disc are borne eight ciliated tentacles. They are termed pinnate because they bear many small side branches like a feather (Fig. 5.19a–c). Tentacular cnidocytes and cilia aid in food capture. The epidermis covering the coenenchyme is continuous on the outside of the polyp, and the mesoglea is quite thin in the polyp wall. The mouth opens into the pharynx bearing a single **siphonoglyph,** a distinguishing feature of octocorallians. The side bearing the siphonoglyph is termed the **sulcal** side, and the opposite side the **asulcal** side (Fig. 5.19d). Both terms are derived from the fact that the siphonoglyph was formerly called the sulcus.

Internally, the gastrovascular cavity is divided by eight complete septa (Fig. 5.19c, d). The two septa attached to the pharynx at the siphonoglyph, or on the sulcal side, are the **sulcal septa.** The two septa on the asulcal side attach similarly to the pharynx and are called **asulcal septa** or **directives** (Fig. 5.19d). The two asulcal septa are different from the others in that their free septal filaments contain numerous long cilia. These septal ends direct a current of water out of the gastrovascular cavity and upward through the pharynx. The septal filaments, exclusive of the asulcal ones, are supplied with cnidocytes, enzyme-secreting and absorptive cells, and gonads. The septal retractor muscles have an interesting arrangement in that they are always on the sulcal side of each septum including those of the sulcal septa, which face each other (Fig. 5.19d). This arrangement is probably of functional significance when the polyp retracts and ingests food down this side of the pharynx. The arrangement of the single siphonoglyph, the two sulcal and the two asulcal septa, positions of the retractor muscles, and the nature of the asulcal septal filaments confer a definite internal biradial symmetry on the polyp.

Subclass Zoantharia (Hexacorallia). This large taxon includes the sea anemones and hard corals. They are either solitary or live in colonies, and many have a protective exoskeleton. While most are sessile and attached, some can move to a limited degree. At least 3500 species, or a majority of all anthozoans, are zoantharians. Many forms are stunningly beautiful with delicate subtle shades of red, blue, green, and violet; the beauty exemplified by many zoantharians is rarely matched in any other group.

Two important features that distinguish zoantharians from other anthozoans are found in their hexamerous symmetry and paired septa. Hexamerous symmetry is particularly evident in the arrangement of tentacles and in the septa, since both types of structures are often present in some multiple of six. In some species these organs are present in multiples of some number other than six, but never as eight. The simple, unbranched (nonpinnate), hollow tentacles borne by the oral disc may be several hundred or more in number. Tentacles may be present in a single marginal whorl or in several circles on the oral disc. The elongated mouth opens into the pharynx, which extends well into the gastrovascular cavity. The pharynx may bear a siphonoglyph at both ends, but some forms have only one siphonoglyph, and in others, such as corals, it may be absent.

The gastrovascular septa may all be complete, or some may be incomplete. All septa in zoantharians are **coupled** so that they are arranged in a bilateral or biradial manner. In most hexacorallians they are also **paired,** that is, they occur in twos close together (Fig. 5.20d, e). That gastrovascular cavity between members of a septal pair is the **endocoel,** and the space between pairs is the **exocoel.** A pair of complete septa, the directives, is present at either end of the pharynx regardless of the presence or absence of siphonoglyphs. The longitudinal retractor muscles of both pairs of directives lie on that septal side facing the exocoel. In all other pairs, including the other complete septa and all incomplete septa, the retractor muscles are on the endocoelic side (Fig. 5.20d). The inner free ends of all septa are modified as **septal filaments.**

Zoantharians are divided into two large orders representing the sea anemones and corals plus several other orders, each of which contains a small number of species. A discussion follows on the two principal groups.

Sea anemones. Sea anemones, belonging to the Order Actiniaria, are widely distributed throughout the world. Almost invariably, they are found in shallow water below the low tide level and attached to rocks, shells, and other submerged objects. Some anemones are commensal on crabs or other cnidarians, some burrow in the benthic material, and others (ceriantipatharians) secrete mucous tubes. A great many symbionts live in association with sea anemones. Certain species of crabs attach commensal sea anemones to their dorsal surfaces or claws as a means of defense or protection. Certain shrimp and even small fishes often are found living symbiotically on the external surface or among the tentacles. Some burrow in the benthic material, and others live in mucous tubes. A skeleton, characteristic of the corals, is absent in all sea anemones. They are always solitary and are considerably larger than hydrozoan polyps. They are called anemones because of the flowerlike appearance of the numerous tentacles (Fig. 5.20a–c). They range in diameter from 0.5 cm to more than 1 m but more commonly range from 1 to 4 cm. Often, sea anemones are brightly or even spectacularly colored, and greens, reds, oranges, blues, and whites predominate.

The flattened pedal disc is used for permanent or mostly permanent attachment. The column is thick and cylindrical and often contains several circular grooves or folds including the **collar,** a fold just beneath the oral disc (Fig. 5.20c). When an anemone contracts, the oral disc is pulled down and inward, and the upper portions of the column, especially the collar, are retracted toward the center to cover and protect the oral disc. The oral disc bears the mouth and from six to several hundred tentacles, usually in some multiple of six, which are arranged in single or multiple whorls or in radiating rows.

The mouth opens into a long pharynx at either end of which is located a ciliated siphonoglyph, though some sea anemones have only one siphonoglyph. The shape of the mouth and pharynx and the presence of one or two siphonoglyphs confer on actiniarians a biradial symmetry (Fig. 5.20b–e).

As in all zoantharians, septa are coupled and paired, and six (or a multiple of six) pairs constitute a **cycle.** Thus in anemones there is a cycle of six pairs of primary or **complete septa,** a cycle of six pairs of **secondary septa,** and a cycle of 12 pairs of **tertiary septa** (Fig. 5.20d, e); some have another cycle of 24 pairs of **quaternary septa.** However, many exceptions and variations on this numerical pattern are found due mainly to regenerative anomalies described below. Septa are often perforate near the oral disc, and the **ostia** facilitate water circulation between endocoels and exocoels.

The septal filaments, trilobed in cross section, contain enzyme-secreting and absorptive cells and cnidocytes (Fig. 5.20f). The middle septal lobe is often continuous as a long cnidocyte-containing thread or **acontium.** Acontia usually project into the gastrovascular cavity, and they are sometimes protracted through the mouth to aid in food procurement. Most anemones are predaceous and feed on a variety of small animals. Some, however, are suspension feeders and trap minute organisms using tentacular cilia.

The muscles present are specialized for a broad range of movements. Special circular muscles, located below the oral disc and forming a sphincter muscle, aid in pulling the collar over the retracted oral disc. Some anemones actually use muscular action in effecting limited locomotion, some are able to glide over the surface by using the pedal disc or by walking on tentacles, and a few anemones even swim by a lashing movement of the tentacles. Still others float oral end down by means of a gas bubble within the pedal disc. (See Box 5.1 for additional details of sea anemone behavior.)

While all anemones reproduce sexually, many reproduce asexually as well. Fission is perhaps the most common asexual means, some dividing longitudinally and others transversely. Fragmentation of the pedal disc produces several to many pieces, each of which regenerates into another separate polyp. Since numbers of tentacles and septa are not always regenerated with a high degree of fidelity, this results in some wide variations in numerical symmetry of septa and tentacles found in actiniarians.

Anemones are either monoecious or dioecious. Gametes, originating from gastrodermal cells, form longitudinal gonadal bands on the septal filaments. Fertilization may be internal in the gastrovascular cavity or external in the seawater. Early development produces a free-swimming planula, which subsequently settles, develops tentacles and septa, and matures into an anemone.

Hard corals. The hard, true, or stony corals belong to the Order Scleractinia and are, for the most part, colonial and found in warmer seas. All corals produce a calcareous exoskeleton

BOX 5.1 □ SOME ASPECTS OF SEA ANEMONE BEHAVIOR

One's first impression of sea anemones is that they are seemingly simple in construction, especially their muscles and nerves, and that even a modicum of coordination, let alone any behavioral patterns, would be unexpected. Yet in spite of their comparatively uncomplicated body, anemones do exhibit rather complex patterns of behavior in both reflexive responses and spontaneous activities. A suitable mechanical or electrical stimulation on any part of an anemone's surface will elicit a rapid contraction of the entire body accompanied by the expulsion of most of the water from the gastrovascular cavity. Contrasted with this fast retraction of tentacles and contraction of the column, reextension is achieved very slowly because there are few or no extensor muscles. Extension of the column and protraction of the tentacles are achieved by the relaxation of the retractor muscles, the elasticity of the mesoglea, and a slow buildup of hydrostatic pressure within the gastrovascular cavity as seawater is permitted to enter via the mouth. Once the anemone is fully extended and the gastrovascular cavity filled with water, the closed pharynx acts as a valve to maintain the slight positive water pressure in the cavity. This fact is important, for if anemones make differential movements that do not involve total contraction, they do so around this aqueous mass in the gastrovascular cavity, which serves as a hydrostatic skeleton. This means that the hydrostatic skeleton has a fixed volume when the pharynx is closed. In this situation, contraction of the longitudinal muscles on one side stretches those of the opposite side, and contraction of all longi-

around the polyps. There are two types of corals: hermatypic and ahermatypic. **Hermatypic corals** are found in shallow tropical seas, produce massive exoskeletons, and are characterized by individual polyps containing symbiotic algae within the gastrodermal cells. **Ahermatypic corals** usually do not form huge exoskeletons, and individual polyps do not harbor symbiotic algae. Ahermatypic corals are widely distributed even in colder waters of higher latitudes and may be found in fairly deep water. Generally, hermatypic corals form reefs, and most ahermatypic corals do not, even though some ahermatypic forms like *Oculina* do form reefs in deep water. Individual polyps of both types are quite small and range from 1 to 10 mm in diameter.

Structurally, scleractinian polyps are very similar to sea anemones except that they are much smaller. Because of the calcareous exoskeleton, the pedal disc is absent, and the column is reduced somewhat. The oral disc protrudes upward from the exoskeleton and bears a number of tentacles in one or more cycles of six each. In the center of the oral disc is a slightly elevated **peristome** surrounding the mouth.

The mouth opens into the pharynx, which always lacks siphonoglyphs (Fig. 5.21c). Paired septa are present basically in an hexamerous plan, but numerous variations are found. Typically, each of the six pairs of complete septa alternates with a pair of incomplete septa, and tertiary septa may also be present. Coral septa are nonperforate, and all bear highly convoluted acontia, which can be protruded through the mouth in feeding. For some unexplained reason, each septal filament contains only one

---

tudinal muscles results in a shorter fatter column with stretched circular muscles.

Using electrical stimulations, one can quickly determine the threshold for reflexive excitation and see that varying the frequency and intensity of the stimulus will evoke graded responses. Slight stimulation causes movement of a tentacle, and stronger stimuli involve progressively more tentacles and the oral disc. As the strength of the stimulus increases, at some point it will evoke a feeding reaction in which the anemone attempts to ingest the electrode. Finally, a rather strong stimulus causes both general body contraction and expulsion of most of the water from the gastrovascular cavity.

Spontaneous, much slower activity can be observed in anemones and includes tentacle waving, column elongation and contraction, peristaltic movements in defecation, and actual movement along the substratum. Some anemones can actually swim by rhythmical contractions of the column or by lashings of the tentacles! The ability to move across a surface or to swim demands a rather sophisticated nervous system and requires pacemakers to integrate and modulate muscular actions. But anemones engage in both reflexive actions and spontaneous behaviors without any traces of a central nervous system so characteristic of most higher animals. Thus the behavior and coordination in sea anemones are of particular interest, since rather complicated patterns occur in comparatively simple animals. No wonder that neurophysiologists also find sea anemones so interesting.

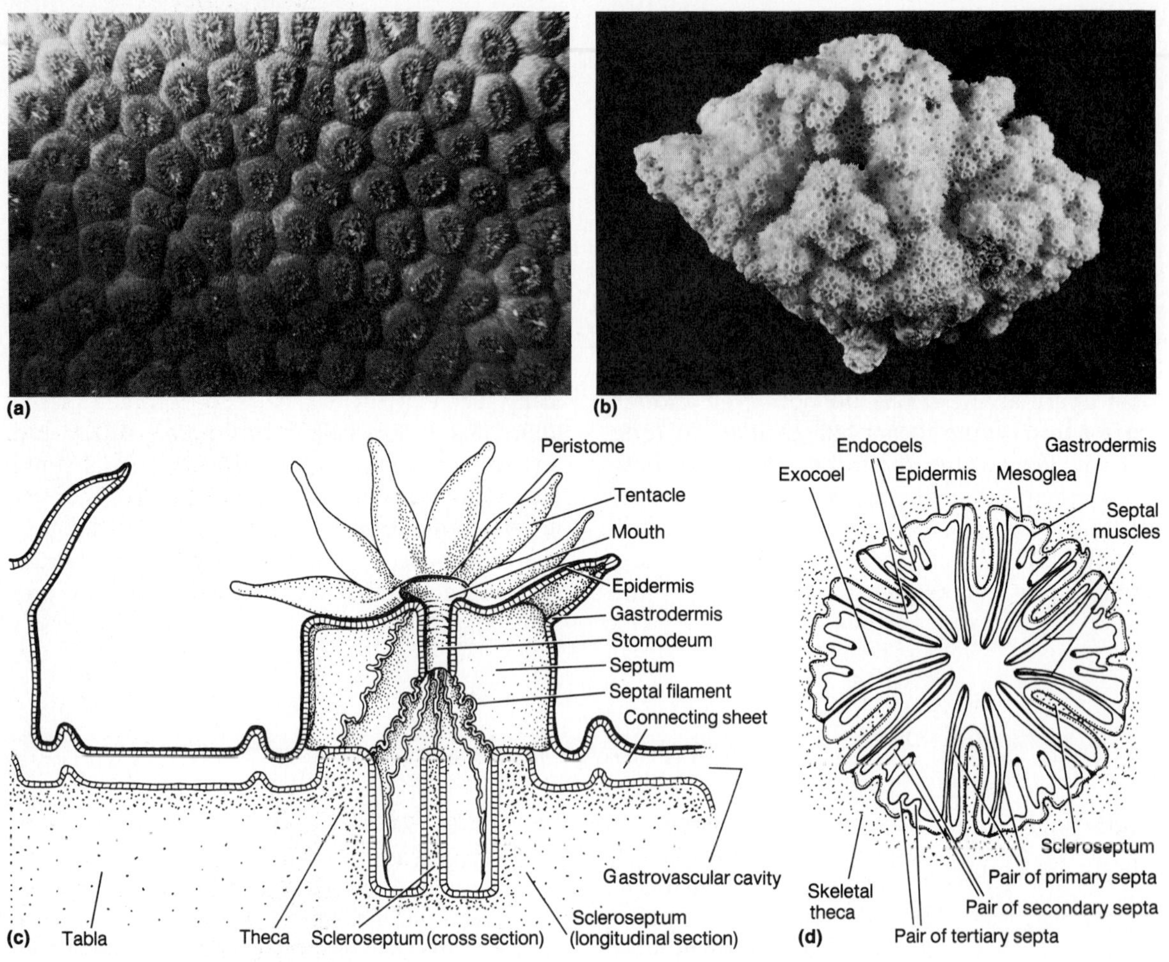

Figure 5.21 *Corals: **(a)** a photograph of* Montastrea, *a hermatypic reef coral (courtesy of J. M. Resing); **(b)** a photograph of a portion of the calcareous skeleton of a hermatypic coral (dried); **(c)** an oral-aboral section of a polyp seated in its thecal pocket; **(d)** a diagrammatic cross-sectional view of a portion of a polyp taken from below the pharynx.*

longitudinal lobe rather than three as in actiniarians.

The most obvious feature of corals is the extensive hard, nonliving exoskeleton (Fig. 5.21). As the coral planula settles, almost immediately it begins to secrete the skeleton. Because of the way in which the skeleton is secreted, each young polyp is fixed permanently within a skeletal cup. The exoskeleton, composed of calcium carbonate ($CaCO_3$) crystals, is secreted by the epidermis of the polyp base and lower part of the column. First secreted as a basal plate or **tabla,** the undersurface of the polyp develops radial invaginations or folds that secrete skeletal septa called **sclerosepta** between each septal pair; the sclerosepta collectively are called the **calyx** (Fig. 5.21c, d). At the same time a wall is built up by the column epidermis to form a cup or **theca.** The sclerosepta project upward into the endocoelic spaces of the gastrovascular cavity surrounded internally by the body wall (Fig. 5.21c). These calcareous sclerosepta are built up from the bottom, and the polyp is seated on top of them. Formation of new sclerosepta is usually in an alternating pattern between living septal cycles. In life, when a polyp expands, it is mostly clear of skeletal material, but when it contracts, it is appressed closely onto and between the sclerosepta, and the gastrovascular cavity is, in effect, squeezed out of existence.

Polyps of all colonial corals are interconnected, not by aboral solenia as in octocorallians, but by connecting sheets of tissue arising laterally from the column (Fig. 5.21c). This horizontal sheet represents a fold of the body wall and contains body wall tissues on the upper and lower surfaces with the gastrovascular cavity between them. The lower epidermis continues to secrete the skeleton beneath it. The colony grows because all polyps and the connecting sheets secrete more skeleton continuously beneath them, and thus the living portions are displaced slowly upward or outward.

Shapes of colonies are due both to the growth patterns and to the arrangement of polyps within the colony (Fig. 5.18c–f). Colony shapes vary from flat to upright and from rounded to dendritic. In *Astrangia,* the common eyed coral, the numerous thecae, each of which contains a polyp, are widely separated. In brain corals like *Diploria* (Fig. 5.18e), polyps are in widely separated rows, but polyps within a row are situated so close together that their thecae run together and fuse, forming the convoluted pattern of ridges and grooves. *Orbicella,* one of the common reef-builders, produces a massive irregular mass of coral.

Sexual reproduction is the same essentially as in sea anemones. Asexual reproduction is primarily by budding as new buds arise from oral discs or bases of existing polyps. In brain corals, for example, new buds arise from the oral disc, but they never separate from the parent polyp, and what results is one large, common, oral disc containing many mouths.

Solitary corals are called cup corals and may attain a diameter of up to 25 cm, but not all are totally sessile. In a few species, including *Fungia* (Fig. 5.18f), individuals are known to be able to slide themselves unaided by water currents over the substratum.

Coral reefs. Some of the most interesting and aesthetically beautiful ecosystems are the huge reefs formed by reef-forming or hermatypic corals. Normally, they require a minimum temperature of 20–22°C for survival; thus they are limited to such areas as the Indian and south Pacific oceans and the West Indies south of Bermuda. Reefs are restricted to shallow water where the depth normally does not exceed 30–50 m because of a most interesting symbiotic relationship described below. Reef corals are absent from areas in which the water is turbid because of erosional runoff from land masses. There are several hundred species of hermatypic corals. As many as 200 species are found on the Great Barrier Reef alone.

Four different types of reefs are found: barrier reefs, fringing reefs, atolls, and platform reefs. **Barrier reefs** are always associated with a landmass such as a continent or island and typically run parallel to the coastline. They are generally separated from shore by many miles of deep water. A familiar example is the Great Barrier Reef, a huge reef extending about 2000 km along the northern east coast of Australia. **Fringing reefs** are similar to barrier reefs except that a narrow strip of sea, usually with a depth of no more than 50–60 m, occurs between the reef and land. Fringing reefs are common in the seas around southern Florida and in the south Pacific. **Atolls** are usually oceanic and distantly removed from continental land masses. They consist of low reefs that may rise up to 10 m above sea level and enclose a central lagoon. Atolls most likely began as fringing reefs around ancient volcanic cones that have since eroded and sunk beneath the sea. **Platform reefs** are flat coral tables without a lagoon. Often present along continental shelves, they may be associated with atolls or lie between fringing or barrier reefs and land masses.

All hermatypic corals contain symbiotic algae called zooxanthellae, and all ahermatypic corals lack them. **Zooxanthellae** are the palmate or resting stages of dinoflagellates, the most common of which is *Symbiodinium microadriaticum*. Dinoflagellates also live symbiotically in certain other cnidarians, bivalves, and protozoans. However, in reef-forming corals the dinoflagellates are present within vacuoles in gastrodermal cells.

The interrelated roles of corals and zooxanthellae have been the subject of considerable research in the last 50 years. It is now clear that the zooxanthellae are very active photosynthetically, and they produce much of the nutritional needs of the entire reef. Even though hermatypic coral polyps are carnivores as are all cnidarians, the reef community is, on balance, autotrophic! This means that the rate of autotrophic production exceeds that of heterotrophic consumption.

Recent studies have elucidated more clearly the functional roles of both zooxanthellae and polyps in this symbiosis, and four important parameters deserve mention. First, it has been shown that the zooxanthellae are responsible to a great extent for the very high rates of $CaCO_3$ deposition in the exoskeleton. Second, zooxanthellae aid in the removal of metabolic wastes of the corals such as carbon dioxide, nitrogen, and phosphorus by using them as raw materials for their own growth. Third, much of the carbohydrates and the oxygen produced by the zooxanthellae in photosynthesis is used by the corals; in fact, recent studies have shown that direct transfer of carbohydrates occurs from the zooxanthellae to the gastrodermal cells of the coral. Fourth, the dependency by the corals on the zooxanthellae varies between coral species, but most corals either do not survive or are greatly diminished in their vitality when the zooxanthellae are experimentally removed. The dependency that polyps have on the zooxanthellae is the fundamental reason reefs are found only in shallow water where adequate solar radiation is available to the algae. Many coral polyps are also able to utilize organic materials that are produced by the algae and that are in dissolved or colloidal states in the surrounding water. In summary, the high rates of organic productivity of the reef communities are dependent upon the efficient exchange of energy and of recycling of nutrients between the zooxanthellae and polyps.

It has been found recently that the exoskeletons of hermatypic corals contain calcareous algae. These algae live just beneath the sheet of living tissue covering the nonliving skeleton, and thus they live outside the zoantharian colony. They add considerable amounts of $CaCO_3$ within the interstices of the reef as well as contributing substantially to the overall autotrophic nature of the reef. The gastrodermal zooxanthellae and the calcareous algae in the exoskeleton are the primary source of the enormous rate and volume of productivity of a coral reef. Coral reefs are not only enormously successful ecosystems, they are truly remarkable and fascinating assemblages of colonial anthozoans and symbiotic algae!

PHYLUM CTENOPHORA

The ctenophores, commonly known as comb jellies or sea walnuts, comprise a phylum of almost 100 species of marine radiates. Most are planktonic or pelagic, but a few representatives creep along the bottom. Ctenophores may be especially abundant at times in coastal waters. The body is spherical or ovoid and normally ranges in length from 1 to 6 cm exclusive of tentacles. Glassy transparent with subtle shades of orange, red, or purple, ctenophores are exquisitely irridescent by day, and most are luminescent at night. Once a person has seen a living ctenophore, it will be hard to convince them that anything else can be as delicate or interesting.

Brief characterizations of the major taxa appear at the conclusion of this chapter, and some ctenophoran diversity is shown in Figures 5.22 and 5.23.

External Anatomy

Even though there is no polymorphism in ctenophores, adults look superficially somewhat like cnidarian medusae. The globular biradial body is built around a central axis running through the mouth at the oral end and a sensory apical organ at the aboral end. The **mouth** is always sessile and never located on a manubrium as in cnidarians. Extending meridionally from the aboral pole nearly to the mouth are eight ciliated bands (Fig. 5.22a, b). Each band consists of a series of numerous, short, transverse plates, each of which contains long fused cilia (Fig. 5.22e). These plates are called **combs** or **ctenes,** and all combs in each of the eight rows form a **comb row.** It is from this feature that representatives of this phylum derive their name (cten = comb; phora = bearer). The comb rows are the locomotive force in ctenophoran movement. Cilia in the comb rows have an effective coordinated stroke toward the aboral end; thus the animal is driven forward with the oral end leading. The beat of cilia on the comb rows is reversible when the animal swims backward. The center for this fascinating ciliary coordination is the apical organ.

Located on each side near the aboral pole of most of these animals is a long **tentacle** (Fig. 5.22a). Unlike those of cnidarians, ctenophore tentacles are neither associated with the oral disc nor attached to the surface of the animal. Rather, they are attached at the bottom of a long, ciliated, epidermal invagination or **canal.** Each tentacle, emerging from the body at an angle of about 45°, is very flexible and can be contracted within its tentacular canal or extended for some distance. Each is solid and consists of an epidermal layer surrounding a core of mesoglea. Ctenophore tentacles contain numerous small branches or **pinnules** that are covered with special adhesive cells called **collocytes** (Fig. 5.22d). Each collocyte consists of a rounded head located at the distal end of a **straight filament** around which is coiled a contractile **spiral filament.** The straight filament is attached permanently to the tentacular mesoglea. On the head of the collocyte is produced a sticky substance on which small zooplankters are trapped as food. Ctenophores are carnivores, and prey are mostly small crustacean larvae or adults. Tentacles with attached prey are brought to the mouth, and food is wiped into the oral opening. A major fact about ctenophores is that individuals of one species, *Euchlora rubra,* possess cnidocytes in place of collocytes. The presence of cnidocytes in even one ctenophore species adds immensely to the theory that ctenophores evolved from a cnidarian ancestor. The pair of tentacles plus the eight comb rows confer upon the ctenophore body a definite biradial symmetry.

Internal Features

The body wall is similar in construction to that of cnidarians. The outer epidermis is either a syncytium or epithelium of cuboidal or columnar cells. Beneath the epidermis is a thick gelatinous **mesoglea** or **ectomesoderm.** The mesoglea contains amebocytes and muscle cells or fibers arranged in a branching anastomosing network. Interior to the mesoglea is the internal body cavity or **gastrovascular cavity** lined with **gastrodermis,** many of whose cells are flagellated.

The mouth opens into a long tubular **pharynx,** which is also a **stomodeum,** since it is lined with epidermis (Fig. 5.22a, b). Extracellular digestion is initiated in the pharynx, and the partially digested food is passed to a centrally

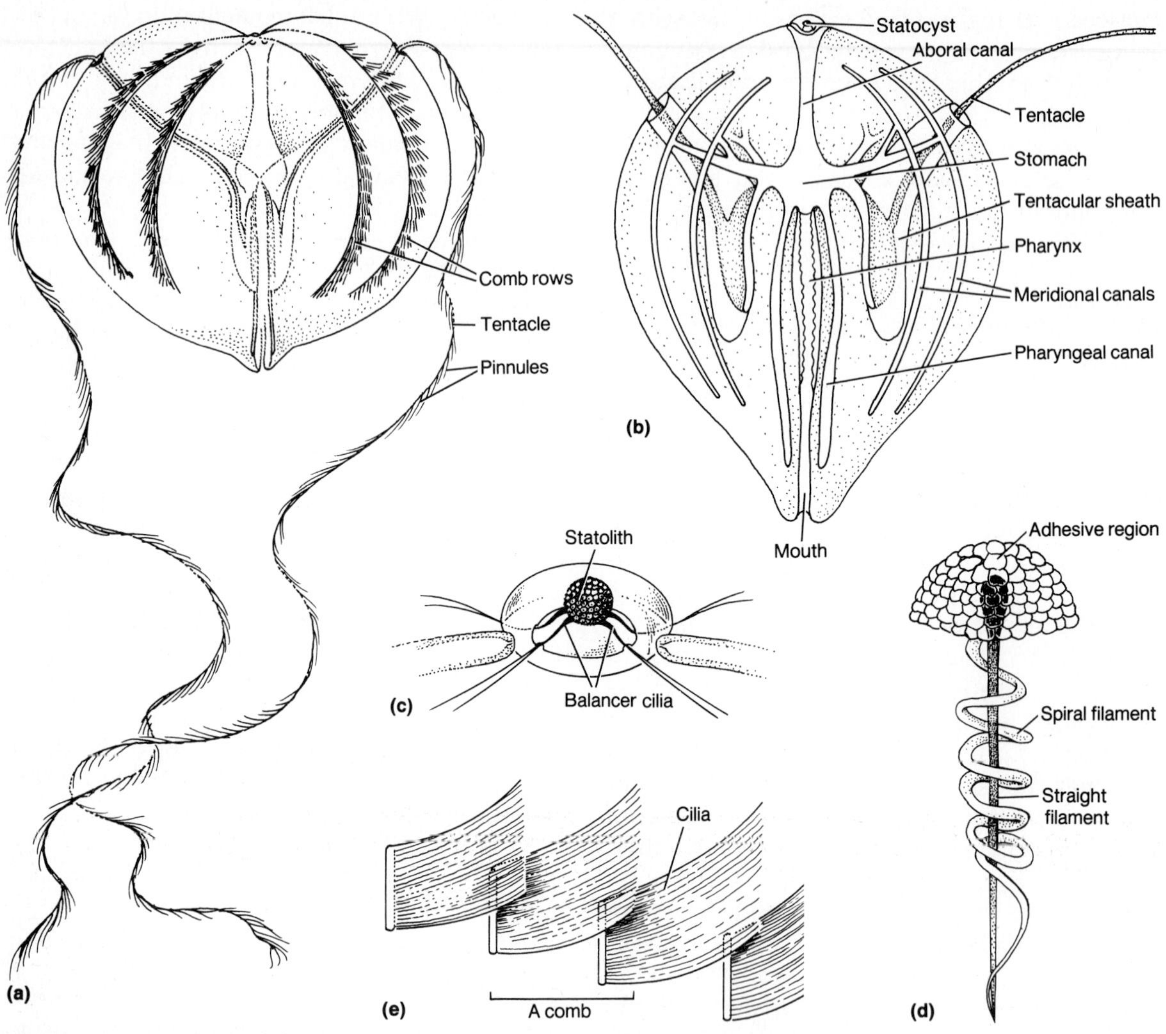

Figure 5.22 Structural features of ctenophores: ***(a)*** Pleurobrachia, *with tentacles extended;* ***(b)*** *a diagrammatic oral-aboral section through a ctenophore;* ***(c)*** *the aboral sense organ;* ***(d)*** *a single collocyte;* ***(e)*** *a portion of a comb row showing four ctenes (parts a–c [b, c from Chun] in L. H. Hyman,* The Invertebrates. *Vol. I,* Protozoa through Ctenophora, *New York, McGraw-Hill, 1940, adapted with permission).*

located, gastrodermally lined **stomach,** where extracellular digestion continues. Arising from the stomach is an extensive, biradially arranged canal system. Initially beginning as a transverse canal on each side of the stomach, each biradial canal branches into a number of smaller canals that run beneath and parallel to the comb rows, to the apical organ and tentacles, and into the general mesoglea (Fig. 5.22b). Intracellular digestion occurs in gastrodermal cells lining the stomach and canals.

The nervous system of ctenophores consists of a neural network located beneath the epidermis and is particularly well-developed beneath the comb rows. The **apical organ,** the only sense organ in ctenophores, provides postural information that has an important effect on the neuromuscular and comb row mechanisms (Fig. 5.22c). It contains a **statolith** balanced on four tufts of cilia called **balancers.** As the animal is tilted or rolled, the statolith is displaced accordingly, and the balancer cilia transfer nervous stimuli to ciliary tracts beneath the apical organ. Functionally like axons of motor neurons, these ciliary tracts transmit the information to the eight comb rows, the rate of beating is altered, and the animal rights itself. These ciliary tracts comprise one of four conducting systems. A second system is a subepidermal nerve net that appears to inhibit the comb row system but may stimulate certain muscle fibers. Another system involves a slow intercellular transmission between muscle cells, and a fourth is a gastrodermal system for propagating luminescent waves.

Ctenophores are monoecious, and both testes and ovaries develop as bands within each of the eight canals located beneath the comb rows. Mature gametes are shed into the gastrovascular cavity and through the mouth to the outside. Fertilization is external; but in a few species the eggs are retained by the female, and fertilization takes place in situ. Cleavage is complete, determinate, and mosaic. Following gastrulation there is developed in most forms a free-swimming larva, the **cydippid.** Possessing some adult features, this stage acquires additional features and becomes sexually mature. One genus is unique in that its representatives have a planula similar to that of cnidarians.

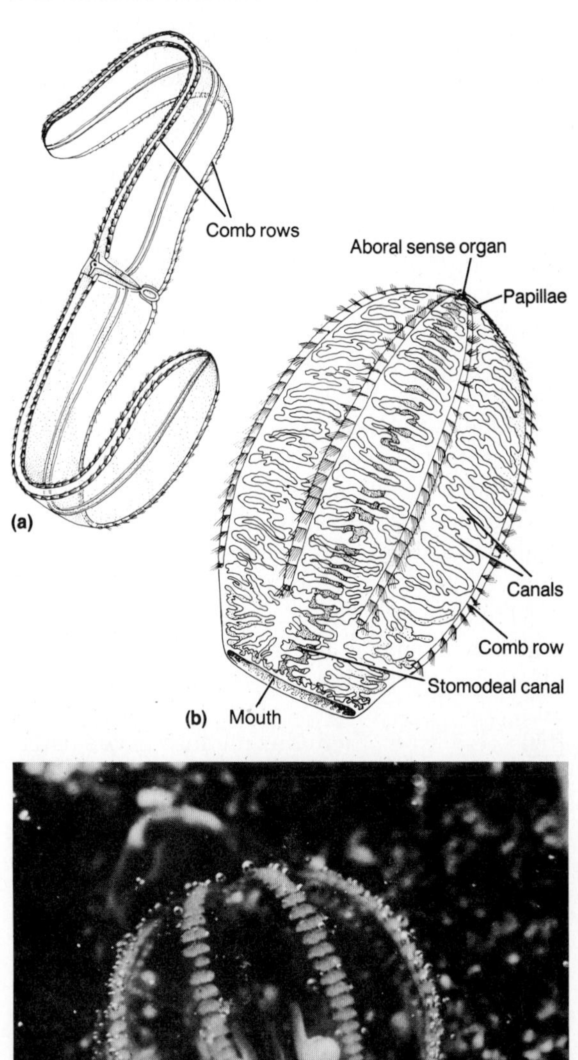

Figure 5.23 *Types of ctenophores:* ***(a)*** Velamen *(after Mayer in L. H. Hyman,* The Invertebrates. *Vol. I,* Protozoa Through Ctenophora, *New York, McGraw-Hill, 1940, adapted with permission);* ***(b)*** Beroë; ***(c)*** *a photograph of* Pleurobrachia *(courtesy of Ward's Natural Science, Inc., Rochester, N.Y.).*

PHYLOGENY OF THE RADIATES

The radiates occupy a unique position among animals for two principal reasons. First, they are nearest of all metazoans to the ancestral stock that gave rise to the bilaterates. Second, cnidarians developed incipient features that were destined to be of fundamental importance in the subsequent evolutionary pathways, including the organization of tissues, a mouth, definite symmetry, an internal gastrovascular cavity, muscles, tentacles, and other organs.

As was pointed out in Chapter 3, the origins of the Metazoa can only be speculated about, since there are no known fossils of those most important developmental stages. Thus there are no indisputable forms of evidence to describe the ancestral stock from which radiates arose. The stock may well have been a marine, ciliated, free-swimming, tetramerously symmetrical, planulalike organism containing cnidocytes. It very probably was similar to the actinula of some hydrozoans. Most certainly, it was not a sponge or a mesozoan. It probably represented a later evolutionary stage that began millions of years earlier with colonial flagellates.

But whether the ancestor was basically a polyp or a medusa has been intensely debated in recent years. One school of thought maintains that the ancient cnidarian was a polyp. If this were true, then the medusa arose later, probably as a convenient means of dispersal. Other specialists hold that the planulalike ancestor gave rise to medusae. This line of argument would maintain that the polyp represents a secondarily acquired elaboration of the larval stage. Today, most cnidarian experts support the theory that the ancestral adult was the medusa. The larval condition of the scyphistoma and the almost universal presence of the polyp in all life cycles are strong supporting evidence. In anthozoans the medusa was lost, and the larval polyp was developed extensively.

The ancestral cnidarian probably was adaptively very plastic and contained the potentialities to develop medusa or polyp morphs, or both. Cnidarian evolution produced three different pathways that resulted in the three modern classes (Fig. 5.24). Anthozoans capitalized on developing the polyp and never evolved anything resembling a medusa. Scyphozoans and cubozoans capitalized on the pelagic dispersal form, the medusa, while at the same time retaining the larval scyphistoma. Hydrozoans modified their inherent plasticity to develop both polyp and medusa as separate but equally important morphs. A good case can be made for the ancestral or stem cnidarian to be a hydrozoan trachyline that has both polyp and medusa morphs. Further debate, research, and thought may resolve the thorny problem of cnidarian phylogeny, but for now each of the several lines of argument has merit.

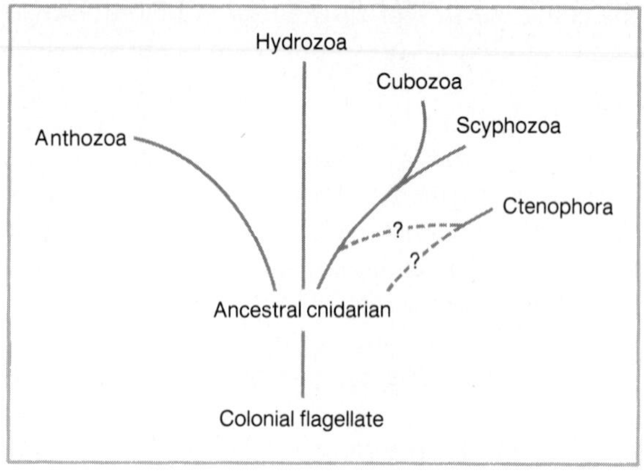

Figure 5.24 *Probable phylogenetic relationships among the radiates.*

The close phylogenetic relationship of ctenophores to cnidarians is not seriously questioned today. The radial or biradial symmetry, body wall construction, gastrovascular cavity, nerve net nervous system, lack of organs, and presence of cnidocytes in one ctenophore species are all considered to be strong evidence linking the two phyla. There is considerable uncertainty, however, as to the present cnidarian group from which ctenophores were derived. Their general morphology and functional adaptations suggest that they are most closely allied to scyphozoans (Fig. 5.24). Perhaps they, like the four cnidarian classes, arose independently from the ancestral stock and utilized their inherent plasticity to evolve features unique to them such as the comb rows and the apical organ.

SYSTEMATIC RESUME

The Major Taxa of Radiates

PHYLUM CNIDARIA

Class Hydrozoa—mostly marine; polyps solitary or colonial; simply constructed; gastrovascular cavity not partitioned; acellular mesoglea; medusa with velum.

Order Hydroida—includes most hydrozoans; colonial or solitary; polyp generation is well developed, almost always sessile, usually colonial; medusae are free swimming or attached; almost all reproduce sexually; *Hydra* (Fig. 5.9). *Obelia* (Fig. 5.11), *Tubularia* (Fig. 5.10), and *Gonionemus* (Fig. 5.8a) are familiar genera.

Order Milleporina—all marine, colonial; form massive erect or encrusting calcareous skeleton; called fire coral because of powerful polyp stings; medusae are small, free swimming; *Millepora* is the only extant genus.

Order Stylasterina—all marine, colonial; called hydrocorals; usually form a colorful erect or encrusting calcareous skeleton; medusae all sessile; *Stylaster* is a typical genus.

Order Trachylina—all marine; medusa present but with no polyp; velum is present; actinula larva develops directly into pelagic adults; about 40 species, including *Cunina* and *Pegantha*.

Order Actinulida—all marine; minute, solitary polyps with no medusoid stage; polyp resembles actinula larva; interstitial in marine sandflats; about ten species belonging to two genera: *Halammohydra* and *Otohydra*.

Order Siphonophora—all marine, pelagic, swimming, or floating colonies; polymorphism is greatly developed with up to four polyp and four medusa morphs present in colonies; many have a conspicuous float; *Stephalia* and the Portuguese man-of-war (*Physalia* [Fig. 5.13]).

Order Chondrophora—all marine, colonial; pelagic colonies lack stem but have float and polymorphic polyps; includes *Velella* and *Chondrophora*.

Class Cubozoa—medusae are tall, pelagic, and square in cross section; mostly tropical; stings are particularly virulent and may kill human beings; includes *Carybdea* and the sea wasps (*Chironex* [Fig. 5.14], *Chiropsalmus*).

Class Scyphozoa—the jellyfishes; all marine; medusa dominant, mostly free swimming, with marginal sense organs and extensive mesoglea; cnidocytes in gastrodermis.

Order Stauromedusae—sessile, mostly in shallow colder waters; trumpet-shaped medusa is attached to substratum by aboral stalk, which is the polyp; umbrella has eight short lobes, each tipped with 20–200 tentacles; most species attach to algae with oral end down; *Haliclystus* (Fig. 5.15c) and *Lucernaria* are two genera.

Order Semaeostomeae—includes the familiar jellyfishes; found in coastal regions; umbrella flat with scalloped margins bearing sense organs, rhopalia, and tentacles; four elongate oral arms; gastrovascular cavity with numerous radiating canals but usually with neither ring canal nor septa; polyp is small, square in cross section; *Aurelia* (Fig. 5.16a), *Chrysaora*, and *Cyanea* are commonly encountered.

Order Rhizostomeae—many-mouthed jellyfishes; found in tropical or subtropical coastal seas; umbrellar margin lacks tentacles; the eight branched oral arms are fused, thus obliterating the primary mouth; numerous suctorial secondary mouths communicate via canals with the gastrovascular cavity; *Cassiopeia* and *Rhizostoma* are perhaps best known.

Class Anthozoa—corals; exclusively marine and polypoid; many are colonial; have extensive mesoglea, stomodeum, gastrovascular septa, biradial and bilateral symmetry.

Subclass Alcyonaria (Octocorallia)—soft, leathery, or horny corals; all colonial; have internal skeleton, one siphonoglyph, and octamerous symmetry with eight pinnate tentacles and eight complete septa.

Order Protoalcyonaria—deep-water, solitary, monomorphic forms that reproduce exclusively by sexual means; *Hartea* and *Taiaroa* are typical.

Order Stolonifera—polyps arise singly from matlike stolons; no coenenchyme; skeleton of calcareous tubes or spicules; *Tubipora* (Fig.

5.18a) and *Clavularia* are two common genera.

Order Telestacea—lateral polyps on simple or branched stems; skeleton of tubes of fused spicules; *Telesto* is a representative genus.

Order Helioporacea—blue corals found on Indo-Pacific coral reefs; blue color of extensive calcareous skeleton due to iron salts in skeleton; *Heliopora* is the best-known genus.

Order Alcyonacea—fleshy or soft corals; polyps mostly concealed in a thick gelatinous coenenchyme; skeleton of separate calcareous spicules; *Alcyonium, Gersemia,* and *Xenia* are genera frequently encountered.

Order Gorgonacea—horny corals; axial filament of gorgonin; colonies are mostly upright, plantlike; thin coenchyme; spicules are often present; more common in tropical areas; includes red coral *(Corallium),* sea whips (*Leptogorgia* [Fig. 5.19b]), and sea fans *(Gorgonia).*

Order Pennatulacea—sea pens, sea pansies; all colonial and unbranched; one elongate, axial, primary polyp that forms a proximal peduncle and distal rachis; rachis bears secondary dimorphic polyps including autozooids and siphonozooids; includes sea pansies *(Renilla), Pennatula,* and *Chunella.*

Subclass Zoantharia (Hexacorallia)—sea anemones, hard corals; simple (nonpinnate) tentacles; hexamerous symmetry reflected in tentacles and septa; septa typically paired.

Order Zoanthinaria—without a skeleton or pedal disc; one siphonoglyph; solitary or, if colonial, polyps connected by solenia; some are epizoic; includes *Epizoanthus* and *Palythoa.*

Order Actiniaria—sea anemones; all solitary and mostly sessile; no exoskeleton; most have two siphonoglyphs; both complete and incomplete septa present; *Metridium* (Fig. 5.20c–e), *Actinia, Adamsia,* and *Calliactis* are common genera.

Order Scleractinia (Madreporaria)—stony or hard corals; mostly colonial, and polyps collectively secrete a massive exoskeleton; siphonoglyphs are absent; hexamerous sclerosepta; *Astrangia, Oculina, Mancina, Fungia* (Fig. 5.18f), and *Diploria* (brain coral [Fig. 5.18e]) are encountered frequently.

Order Ceriantipatharia—resemble sea anemones; burrowers; body is elongate, cylindrical, and housed in mucous tube with oral end protruding when feeding; one siphonoglyph; many single, complete septa; *Ceriantheomorphe* (= *Cerianthus*) (Fig. 5.18g) is the best-known genus.

Order Corallimorpharia—single or colonial polyps resembling anemones; no exoskeleton or siphonoglyphs; radially arranged tentacles; *Corallimorphus* and *Corynactis* are distributed widely.

PHYLUM CTENOPHORA

Order Cydippida—representatives are most typical and least modified; globular or egg shaped; the two tentacles are retractile and have lateral branches covered with collocytes; familiar examples are *Pleurobrachia* (Figs. 5.22a, 5.23c), *Mertensia,* and *Hormiphora.*

Order Lobata—body compressed in tentacular plane forming two large oral lobes on either side of that plane; tentacles reduced and without sheaths; *Mnemiopsis* and *Bolinopsis* are common on the Atlantic coast.

Order Cestida—elongate, ribbon shaped, and compressed in tentacular plane; tentacles reduced and near the mouth but with sheaths; some comb rows reduced; *Cestum* (Venus' girdle) and *Velamen* (Fig. 5.23a) are the only two genera.

Order Platyctenida—aberrant ctenophores; creep along bottom or are sedentary; body flattened in oral–aboral axis; tentacles with sheaths present; comb rows are reduced or absent; *Coeloplana* is perhaps the best-known genus.

Order Beroida—complete absence of tentacles or sheaths; enlarged pharynx; body compressed in tentacular plane; *Beroë* is the principal genus (Fig. 5.23b).

ADDITIONAL READINGS

Burnett, A.L., ed. 1973. *Biology of* Hydra, 453 pp. New York: Academic Press.

Deas, W. and S. Somm. 1976. *Corals of the Great Barrier Reef,* 127 pp. Sydney, Australia: Ure Smith.

Faulkner, D. and R. Chesher. 1979. *Living Corals,* 308 pp. New York: Clarkson N. Potter.

Gierer, A. 1974. Hydra as a model for the development of biological form. *Sci. Amer.* December 1974:44–54.

Goreau, T.F., N.I. Goreau, and T.J. Goreau. 1979. Corals and coral reefs. *Sci. Amer.* August 1979:124–136.

Hyman, L.H. 1940. *The Invertebrates: Vol. I, Protozoa through Ctenophora.* pp. 365–695. New York: McGraw-Hill Book Co.

Kaestner, A. 1967. *Invertebrate Zoology.* Vol. I, pp. 43–153. New York: John Wiley & Sons.

Lenhoff, H.M. 1983. Hydra *Research Methods,* 453 pp. New York: Plenum Press.

Lenhoff, H.M., L. Muscatine, and L.V. Davis. 1971. *Experimental Coelenterate Biology,* 271 pp. Honolulu: University of Hawaii Press.

Mackie, G.O., ed. 1976. *Coelenterate Ecology and Behavior,* 738 pp. New York: Plenum Press.

Odum, E.P. 1971. *Fundamentals of Ecology,* 3rd ed., pp. 344–349. Philadelphia: W.B. Saunders Co.

Pennak, R.W. 1978. *Fresh-water Invertebrates of the United States,* 2nd ed., pp. 99–113. New York: John Wiley & Sons.

Rees, W.J., ed. 1966. *The Cnidaria and Their Evolution,* 418 pp. Symposium of the Zoological Society of London, No. 16. New York: Academic Press.

Robinson, R.A. and C. Teichert, eds. 1981. *Treatise on Invertebrate Paleontology.* Part F: *Coelenterata,* Supplement 1, 2 vols., 762 pp. Geological Society of America. Lawrence, Kan.: University of Kansas.

Rosen, B.R. 1982. Darwin, coral reefs, and global geology. *Bioscience 32:*519–525.

Schlicter, D. 1982. Nutritional strategies of cnidarians: The absorption, translocation and utilization of dissolved nutrients by *Heteroxenia fuscescens. Amer. Zool. 22:*659–669.

Shih, C.T. 1977. *A Guide to the Jellyfish of Canadian Atlantic Waters,* 90 pp. Ottawa: National Museum of Canada.

Smith, F.G.W. 1971. *A Handbook of the Common Atlantic Reef and Shallow-water Corals,* 160 pp. Coral Gables, Fla.: University of Miami Press.

Symposium on the Developmental Biology of the Cnidaria. 1974. *Amer. Zool. 14:*440–866.

6

The Acoelomates

OVERVIEW

The acoelomates, represented by the phyla Platyhelminthes, Nemertea, and Gnathostomulida, are important phylogenetically, being transitional between radiates and more complex bilaterates. Three very important characteristics appeared initially in this group: bilateral symmetry, a true mesoderm that gives rise to muscles and many other organs, and a central nervous system with a brain and one or more nerve cords. Since the mesodermal mass completely fills the area between the outer epidermis and digestive tract, these animals lack a body cavity, hence their name, acoelomates.

The Phylum Platyhelminthes is a large group of dorsoventrally compressed animals, the flatworms. Their digestive system, functionally similar to the gastrovascular cavity of radiates, is a blind-end tract with the mouth as the single opening. Platyhelminths possess an osmoregulatory system of flame bulbs that functions mostly in water elimination. Generally, flatworms are monoecious and have great reproductive powers; many also reproduce asexually.

Members of the Class Turbellaria, the most primitive of all platyhelminths, are mostly free living and swim or creep by using cilia and muscles. They are effective predators that ensnare prey in mucus secreted by numerous glands. Representatives of the classes Trematoda (flukes) and Cestoda (tapeworms) are exclusively parasitic, and most adults are endoparasites in a number of vertebrates. Most flukes have two suckers, which they use to maintain their position within a host, and they have complicated life cycles usually requiring two or more hosts. Prodigious numbers of eggs are released to the outside via the host's feces. Eggs washed into water develop into miracidia or larvae that infect snails. Each miracidium transforms into a sporocyst, which asexually produces rediae, each of which in turn asexually generates a number of cercariae. Cercariae encyst as metacercariae; and after being ingested by the definitive host, they transform into adults. Cestodes are very specialized parasites that have completely lost the mouth and digestive tract; food materials are absorbed across the tegument. The body of most cestodes, which may be up to 15 m long, is composed of a holdfast scolex, a neck, and a strobila subdivided into hundreds of proglottids, each of which contains one or two sets of both reproductive systems. Since cestode larvae do not usually reproduce asexually and mostly are found in only one intermediate host, cestode life cycles are not as complicated as those of trematodes.

The Phylum Nemertea is a group of predominately marine, elongate, burrowing worms, each of which possesses a remarkable muscular proboscis that can be everted rapidly to capture prey. The proboscis may be equipped with a stylet used to impale prey, but some nemerteans wrap the distal end of the proboscis around the victim and pull it to the mouth. Similar to turbellarians, nemerteans have two other features not encountered in flatworms: a complete digestive system with a posterior anus and a circulatory system with contractile vessels and blood. The Phylum Gnathostomulida is a small, recently discovered group of minute marine worms living anaerobically in mud and sand. Their most important features are a pair of lateral oral jaws, a blind-end gut, and the absence of circulatory and excretory systems.

Acoelomates arose from a planuloid ancestor that evolved bilateral symmetry. The stem group for all acoelomates is the turbellarians, which are simply constructed yet possess many rudimentary features that were to be of profound importance in the evolution of higher animals. From something like a modern turbellarian, all other acoelomate taxa evolved through adaptive radiation.

GENERAL CHARACTERISTICS OF THE ACOELOMATES

The acoelomates are relatively simple, wormlike metazoans comprising three phyla: Platyhelminthes (flatworms), Nemertea (proboscis worms), and Gnathostomulida (a small, poorly known group). Acoelomates occupy a pivotal phylogenetic position; they represent the most primitive of the protostomial groups, and all other protostomates are generally believed to have evolved from an acoelomate ancestor. Although primitive and simply constructed, acoelomates evolved three extremely important adaptations that were to be of immense significance in the evolution of all higher metazoans: the acquisition of bilateral symmetry, the development of mesoderm, and the ontogeny of a central nervous system. All three adaptations first appeared in the acoelomates, undoubtedly evolved concurrently, and are related structurally and functionally to one another in very important ways.

Bilateral Symmetry

Acoelomates are the first animals in which bilateral symmetry appeared as the primary form of symmetry. As discussed in Chapter 3, a bilaterally symmetrical organism has an imaginary longitudinal plane that divides the animal into right and left halves, which are mirror images of each other. By its nature, bilateral symmetry means that an animal has an anterior and a posterior end and a dorsal and ventral surface (Fig. 6.1a). Acoelomates are worms, and the bodies of both larvae and adults are elongated in the anteroposterior axis. Bilaterality is directly associated with mobility: the anterior end functions as the leading end, while the ventral surface, usually applied to the substratum, is directly involved in locomotion. While they lack any form of lateral appendages, they are still able to move by several different means—a most important advance over the generally sessile radiates. A majority of all bilaterally symmetrical organisms are predatory, and most seek out food aggressively. Therefore bilateral symmetry is intimately associated with an active, mobile habit in which the animal's locomotion is constantly or frequently employed in

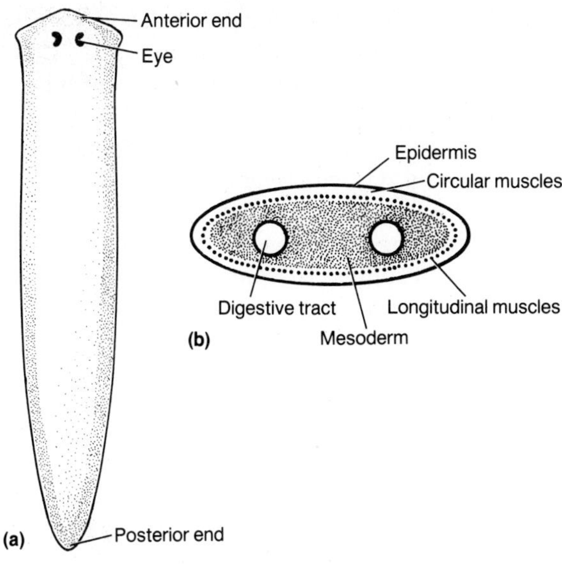

Figure 6.1 *Diagrammatic representations of the features of a generalized acoelomate:* ***(a)*** *dorsal view;* ***(b)*** *cross-sectional view.*

the search for food or a mate. Since acoelomates are the most primitive metazoans with bilateral symmetry, they represent a crucially important phylogenetic link between the radially symmetrical, planuloid ancestors and all other bilaterates.

Mesoderm

Acoelomates are the first metazoans to have evolved a true **mesoderm,** an important layer of cells that develops embryologically between the ectoderm and endoderm (see Chapter 3). The permanent cellular nature of the acoelomate mesoderm represents a most significant advance over the mesoglea of radiates or the mesohyl of poriferans, since in both latter cases either the middle layer is noncellular or the cells do not constitute a permanent tissue. The bulk of the mesoderm of acoelomates, indeed that of all bilaterates, is derived initially from endoderm. In acoelomates it is a persistent mass of parenchymous cells called the **mesenchyme** that completely fills what was once the embryonic blastocoel. Since the mesenchyme is a solid mass of parenchymous cells between the outer and intestinal epithelia, it is not possible to have a cavity located between the intestine and body wall as in all other bilaterates (Fig. 6.1b); therefore they are called **acoelomates** (= without a body cavity).

The advent of the mesoderm was of immense significance, since it opened up fantastic, new, and different architectural possibilities. The mesoderm relieved the ectoderm and endoderm from certain of their former functions. In fact, the mesoderm has become the principal germ layer in terms of numbers of different things derived from it, including muscles, excretory and reproductive systems, and, in higher animals, metamerism and a circulatory system (most metazoans) and bones (vertebrates).

The primal importance of muscles, all of which are mesodermally derived, cannot be overstated, since they are absolutely indispensable to bilaterates for movement and locomotion. In acoelomates the principal muscles are a layer of circular muscles outside of a band of longitudinal muscles (Fig. 6.1b); additionally, diagonal and dorsoventral muscles are usually present. Contractions of the circular muscles lengthen the animal, while those of the longitudinal muscles shorten the worm. Since both circular and longitudinal muscle layers lie outside of the mesenchyme, these layers can exert significant pressures on this deformable cellular mass. Therefore there exist important and intimate relationships between the mesoderm and its muscles, bilateral symmetry, and locomotion.

As locomotive, sensory, nervous, and muscular systems developed, it became necessary to amplify and elaborate the capabilities of feeding, gas exchange, osmoregulation, and internal transport. Such metabolic processes can be carried out efficiently on an organismic level only by the aggregation of specialized tissues to form **organs.** Therefore in a direct way the appearances of locomotion and bilateral symmetry were inherently associated with the ontogeny of organs and organ systems to perform specific metabolic functions. Acoelomates are the first metazoan group to have internal parts specialized as organs, and they and all higher metazoans are said to be at the **organ grade of construction.** Further, most organs in acoelomates are mesodermally derived. Acoelomates have developed a mesodermal osmoregulatory system of protonephridia, a foreshadow of very important excretory organs in higher animals. A prolific reproductive system, so critical especially to the parasitic flatworms, is of mesodermal origin. Even though there are no gas exchange organs in acoelomates, a mesodermal circulatory system first appears in the proboscis worms.

Yet the mesenchymal mass in which most organs are embedded is a mixed blessing, since these cells do interfere with internal transport. However, acoelomates are small and many are flattened, so interior cells are not too distantly removed from the surface. Even though the body shape in acoelomates obviates the necessity of an internal transport cavity, nonetheless the lack of a body cavity has imposed definite limits on the size and shape of acoelomates.

Central Nervous System

The development of a central nervous system was of exceptional importance in the evolution of all bilaterates. It is thought that certain nerve cells in the diffuse nervous system of the radially symmetrical ancestor aggregated near the oral surface to form a nerve plexus. Arising from the plexus were several meridional (oral–aboral) strands that became longitudinal nerve cords. With the advent of bilateral symmetry there was a concomitant shift forward of the nerve plexus to the anterior end. As more neurons were aggregated into the nerve plexus, this neural mass increased in complexity and become a **ganglion,** which made possible one central site for both interpretation of sensory information and coordination of motor responses. In fact, it is absolutely essential for a successful mobile animal to have such a single coordinating, controlling center. Further elaboration of this ganglion and the development of greater numbers of ganglionic neurons evolved into a **brain,** an organ of inestimable significance to higher animals.

Bilateral symmetry produced an anterior end that contains a concentration of sense organs and an aggregate of nerve cells to interpret sensory information. There are two strategically important advantages of this arrangement. First, in a crawling or swimming animal it is of great adaptive significance to have sensory organs located in the leading or anterior end, since that is the first part of the body to encounter new environments or conditions. Stated differently, it is better for an animal to sense where it is going than where it has been. Second, it is important to have the sense organs in close proximity to the brain because in such an arrangement the connecting neurons are short, thus minimizing the time required to react to environmental stimuli. The concentration of sensory, nervous, and other components at the anterior end is a phenomenon known as **cephalization,** a process that culminates in the formation of a **head** in most metazoans. Cephalization is already apparent in the acoelomates, since there is a concentration of nervous, sensory, and adhesive structures as well as the mouth at or near the anterior end. The development of a central nervous system is clearly associated with that of bilateral symmetry, mobility, and mesoderm. All were critical factors in determining the course of metazoan evolution, and all are of fundamental importance to acoelomates.

PHYLUM PLATYHELMINTHES

The Phylum Platyhelminthes, including the planarians, flukes, and tapeworms, is a diverse and phylogenetically important group. Some representatives, especially the planarians, are the darlings of almost everyone who has ever observed them. Platyhelminths are best known as flatworms (platy = flat; helminth = worm) because their bodies are dorsoventrally flattened and generally elongated. Of the approximately 15,000 species, about 80% are internal parasites in many different metazoan hosts. Since some flatworms are parasitic in human beings, this group is medically important. Most of the free-living platyhelminths are marine, some are common in fresh water, and a few species are found in moist terrestrial habitats. The length of most free-living flatworms is less than 2 cm, while that of some parasitic species exceeds 15 m! Most flatworms are brown, gray, black, or white, but the free-living forms are often brightly colored. The phylum is divided into three classes, two of which are composed exclusively of obligatory parasites. All three classes are characterized in Table 6.1.

Being vermiform and bilaterally symmetrical, flatworms exhibit some degree of cephalization, since the brain, most sensory organs, tentacles (when present), and often the mouth are located in the anterior end. Prominent on the external surface of most flatworms and toward the anterior end are organs of attachment or adhesion, which are most evident and better developed in parasitic forms. These organs—glandular, muscular, or in the form of hooks or spines—are effective holdfast structures or are employed in locomotion and food capture (Table 6.1).

Platyhelminths classically demonstrate the three basic features that first appeared in acoelomates and that are extremely important to them: bilateral symmetry, mesoderm, and the central nervous system. However, within the phylum the worms in each class are significantly different from those in the other two classes, and many distinguishing features can best be introduced as each class is discussed.

TABLE 6.1. □ **CHARACTERISTICS OF THE THREE CLASSES OF THE PHYLUM PLATYHELMINTHES**

Class	*Common Name*	*Habit*	*Adhesive Structures in Adults*	*Special Features*
Turbellaria	Turbellarians, planarians	Mostly freeliving	Frontal gland, other mucous and adhesive glands	Epidermal rhabdites, mostly predaceous
Trematoda	Flukes	Exclusively parasitic	Typically an anterior and a ventral sucker	Most have complicated life cycles with larvae parasitic in one or two intermediate hosts and adults in vertebrates.
Cestoda	Tapeworms	Exclusively parasitic	Scolex bearing suckers, hooks, or flaps	Most have a very elongated body or strobila composed of repetitive units called proglottids

Then are there any diagnostic or characteristic features of the phylum as a whole? Yes, indeed, there are three principal, common systems that most flatworms possess and that are of particular importance to the phylum. They are the digestive, osmoregulative, and reproductive systems, and each will be discussed as a unifying feature of the phylum.

Digestive System

The typical flatworm digestive space is a **gastrovascular cavity** that is morphologically quite different from but functionally similar to that of radiates (Fig. 6.2a). It is an internal cavity in which both digestion and circulation of food materials take place. This cavity, often referred to as the **intestine** or simply the **gut,** is a blind-end space with a single opening, the **mouth.** The form and shape of the intestine vary widely, but in a majority of flatworms the digestive cavity is dendritically shaped with numerous small tributaries arising from one to three principal branches (Fig. 6.2a). Secondary and tertiary diverticula serve only to increase the surface area of the gut, and there are no regional differen-

*Figure 6.2 Platyhelminth features: **(a)** a photomicrograph of* Dugesia *illustrating its digestive system; **(b)** flame bulbs of the protonephridial system; **(c)** a cutaway diagram of a flame bulb illustrating the ultrastructure of its wall (from G. Kummel,* Zeit. f. Zellforschung *37:174, 1962); **(d)** a diagram of the male reproductive system; **(e)** a diagram of the female reproductive system of a turbellarian.*

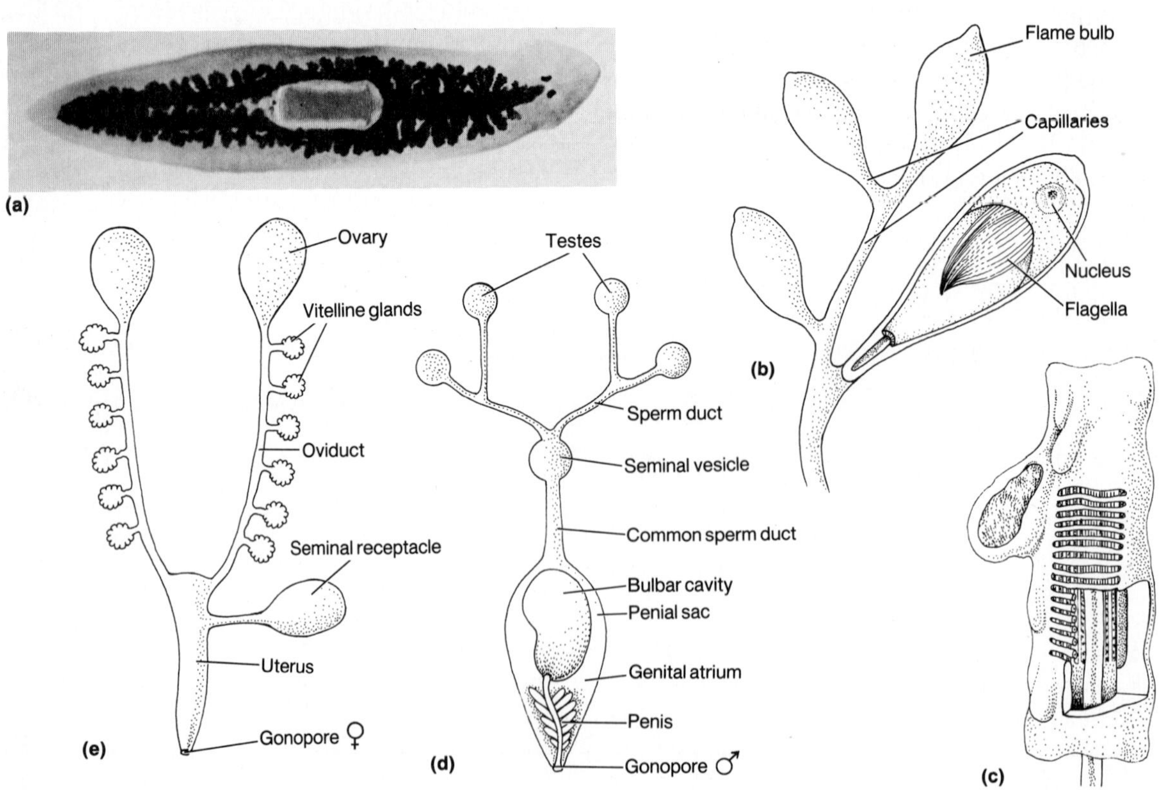

tiations of the intestine. The digestive tract is lined with a single layer of endodermally derived cells of which some are phagocytic and others secretory. Among some of the primitive turbellarians the intestinal cells are ciliated, and the cilia apparently aid in gut circulation. In one group of turbellarians (Order Acoela) there is no gastrovascular cavity at all, a condition that is considered to be primitive. In all tapeworms (Class Cestoda) the mouth and intestine are totally absent; their ancestral stock, without question, had a gastrovascular cavity, but it was secondarily lost in the evolution of their parasitic habit.

The mouth is situated either at the anterior end or on the ventral surface. Food, in the form of small invertebrates for the free-living flatworms and host cells and fluids for the flukes, is sucked into the intestine by peristaltic action generated by muscles. In some forms, initial digestion proceeds outside the animal, and partially digested bits are then drawn into the intestine. Initial digestion is always extracellular and facilitated by digestive enzymes produced by the intestinal secretory cells. Later, bits of food and large molecules are engulfed by phagocytic cells of the intestinal wall, and final digestion is intracellular within the phagocytes. Undigestible particles are voided to the exterior via the mouth opening.

Osmoregulative System

Excretory organs in most invertebrate taxa are called **nephridia,** and they eliminate nitrogenous wastes dissolved in body fluids. Even though the stages in the evolution of nephridia are lost in antiquity, it is clear that as metazoans grew larger, it became physically impossible to eliminate wastes by diffusion to the exterior alone. Therefore those organisms that evolved some means of waste elimination certainly had a positive selection value. The ontogeny of excretory organs freed metazoans from the limitations of a small size and the constraints imposed by the elimination of waste products by diffusion, and their development allowed animals to evolve larger body sizes. Since nitrogenous materials can be eliminated most expeditiously if they are dissolved in water, the developing nephridia assumed the additional role of osmoregulation or the elimination of excess water. Osmoregulation is a particularly important problem in freshwater organisms, since body fluids are hypertonic to the aquatic medium. Therefore for these animals to be successful they had to evolve some mechanisms to cope with the continual influx of water from the environment. Most likely, nephridia evolved initially as osmoregulatory organs, but clearly both osmoregulatory and excretory functions could be handled efficiently by the nephridia. This dual function is readily apparent in all higher metazoans.

The platyhelminths are a most interesting group, being the first phylum to have developed a system of primitive nephridia. Platyhelminth nephridia have closed inner ends, open directly to the environment, possess a current-producing mechanism, and are called **protonephridia** (Fig. 6.2b, c). Since protonephridia are more highly developed in freshwater turbellarians, in which osmotic problems would be the most severe, the protonephridia of flatworms function primarily in osmoregulation. Conversely, protonephridia are not so well developed in most marine turbellarians and in adult flukes and tapeworms. One interesting note is that protonephridia are completely absent in the most primitive turbellarians (Order Acoela), which are also marine.

The protonephridial system consists of several longitudinal **protonephridial canals,** each of which has a number of smaller tributary canals or ducts. Each duct in turn has secondary branches that ultimately end in microscopic capillaries, each of which is surrounded by a **protonephridium,** a blind-end, cup-shaped cell containing a tuft of several to many flagella (Fig. 6.2b). The flagella of each flame bulb extend for a very short distance in the lumen of the capillary. The beating of the collective flagella is reminiscent of a candle flame in a weak draft, and it is for this feature that they were given the name **flame bulbs.** The protonephridia seem either to secrete substances actively into the capillary or to allow fluids and materials to pass into the lumen. Recent electron micrographs have shown the wall of each flame bulb to be composed of a very fine meshwork, which apparently acts as an ultrafilter of the hypotonic body fluids (Fig. 6.2c). The action of each flagellar tuft is probably twofold: to create a negative pressure within the lumen to pull water into the capillary and to drive by hydrostatic pressure this watery solution through the capillary and down the protonephridial canal. Even though what is mostly lost by the flame bulb system is water and the system is primarily one of osmoregulation, the protonephridia probably function in excretion by eliminating some nitrogenous materials such as ammonia in solution in water. The protonephridial canals carrying water and other materials open to the exterior via openings, the **nephridiopores.** Typically, there is a single nephridiopore on either side of the animal toward the posterior end, but some platyhelminths have four or more pairs.

Reproductive Systems

Platyhelminths reproduce both asexually and sexually. Because of their high rate of fecundity, most flatworms, especially the parasitic forms, have enormous reproductive potentials. In fact, it often appears that most of the worms' energies are involved in their reproductive strategies. Asexual reproduction is rather common and typically involves fragmentation, transverse fission, or even budding. Flukes (Class Trematoda) have a curious phenomenon called **polyembryony** in which new larval stages are produced asexually in existing larvae.

Both reproductive systems are surprisingly complex in these comparatively primitive worms. The vast majority of flatworms are monoecious (hermaphroditic), since they contain both male and female reproductive systems; further, they are simultaneous hermaphrodites, since both ova and sperm are produced by an animal at the same time. In both systems, certain free mesodermal cells migrate to a number of locations where gonads are being formed, and these cells undergo gametogenesis to form eggs and sperm. Below are descriptions of the generalized male and female reproductive systems, but there are many variations on this basic plan, the most important of which will be included in the description of each class.

The male system consists of two to many small **testes** each of which produces many sperm (Fig. 6.2d). Interestingly, the flagellar tails of most flatworm sperm have an axoneme of 9 + 0 or 9 + 1 microtubules rather than the 9 + 2 pattern characteristic of the cilia and flagella present in almost every other animal. Additionally, sperm of most flatworms are biflagellated. The significance of these unique sperm features is not known. Two or several **sperm ducts (vasa efferentia)** convey sperm to a single or a pair of **seminal (spermiducal) vesicles,** where they are stored temporarily until copulation. A **common sperm duct (vas deferens)** transports sperm into the copulatory device (Fig. 6.2d). If the seminal vesicles are paired, a short duct from each then fuses with the other to form the common sperm duct.

The copulatory device, a mechanism that transfers sperm from one worm to another, is a means of ensuring cross-fertilization, which in turn promotes genetic variability. Mutual insemination is the rule among flatworms, although self-fertilization occurs in rare cases. In copulation the ventral surfaces of the two worms are applied so that the opening of the male system, the **male gonopore,** of each worm is in close proximity to the **female gonopore** of the sexual partner. Often, a single gonopore is the opening to both systems; obviously, in these worms the common gonopore of each copulating worm is closely applied to that of the other. The typical transfer organ is a penis or a cirrus, a muscular organ that at rest is contained within a cavity, the **genital atrium,** just inside the male gonopore (Fig. 6.2d). In animals with a **penis** this copulating organ is protruded but not everted through the male gonopore and into the female gonopore. A **cirrus,** a structure that is functionally similar to a penis, is both protrusible and eversible. Sperm are deposited within the female reproductive system, where they are stored until used in fertilization.

The female system typically has one or two pairs of ovaries (Fig. 6.2e). Unlike the situation in all other metazoans, where yolk materials are incorporated into the ovum's cytoplasm, most platyhelminths are uniquely different, since their **eggs** or **ova** are almost completely devoid of yolk. Instead, yolk materials are contained within **vitelline** or **yolk cells,** which are in reality modified ova. Therefore there are two types of ovarian organs: a germinarium and a vitellarium. The **germinarium** (used synonymously with ovary) produces the haploid ova, and the **vitellarium,** or **vitelline** or **yolk glands,** produces the yolk cells. The vitelline glands are numerous, small, and located along both sides of the animal. A group of yolk cells is attached to each ovum and is encapsulated with the ovum prior to its being laid. The condition in flatworms, in which yolk cells are adhered to the ova is said to be **ectolecithal,** whereas the ova of all other metazoans are **endolecithal** (yolk within cytoplasm of ovum).

Ova are discharged from the ovary into a short **oviduct,** and in most trematodes and cestodes the oviducts terminate in a bulbous chamber, the **ootype** or **bursa.** The ootype is, in reality, a staging area for the ova, since glands and the ducts from several different organs or structures discharge their materials into this chamber. Yolk cells often are released into a pair of **common vitelline ducts,** which transport them to the ootype. During copulation, sperm from the other worm are stored in a **seminal receptacle** until they are used in fertilization; a short duct runs from this sperm reservoir to the ootype, where fertilization takes place (Fig. 6.2e). After fertilization, certain materials secreted by the vitelline glands harden around each egg to form a **shell.** The secretions of other small unicellular glands, often called **Mehlis' gland,** are thought to lubricate the eggs during their passage to the outside. Cleavage of the egg is typically spiral and determinate, characteristic features of many protostomates. Development is direct in some platyhelminths; but in all trematodes and cestodes, one or several larval stages are produced.

Class Turbellaria

Turbellarians represent the archetype platyhelminth in both form and function, and they are of considerable interest to zoologists because of their simplicity. All free-living members of the phylum are turbellarians, but within this class of about 3100 species a small number of species is ectocommensal or parasitic. The vast majority are aquatic and are, for the most part, marine. Some marine species are pelagic (found in the open sea), but most live near shore amid rocks, shells, and vegetation, over whose surfaces they swim or creep. Many minute species live in the mud or sand and form an important element in these interstitial communities (Fig. 6.3a, b). Freshwater turbellarians are common in brooks, springs, ponds, and lakes, and some species live exclusively in cave streams. Two species, *Dugesia tigrina* and *D. dorotocephala,* belong to a family of freshwater flatworms called planarians (Fig. 6.2a). Both species have been used extensively for decades as laboratory animals, and they are the best-known representatives of this class. A few species of *Bipalium* are found in very moist terrestrial habitats, including greenhouses (Fig. 6.3e). Upon seeing one's first *Bipalium,* one can hardly believe that this

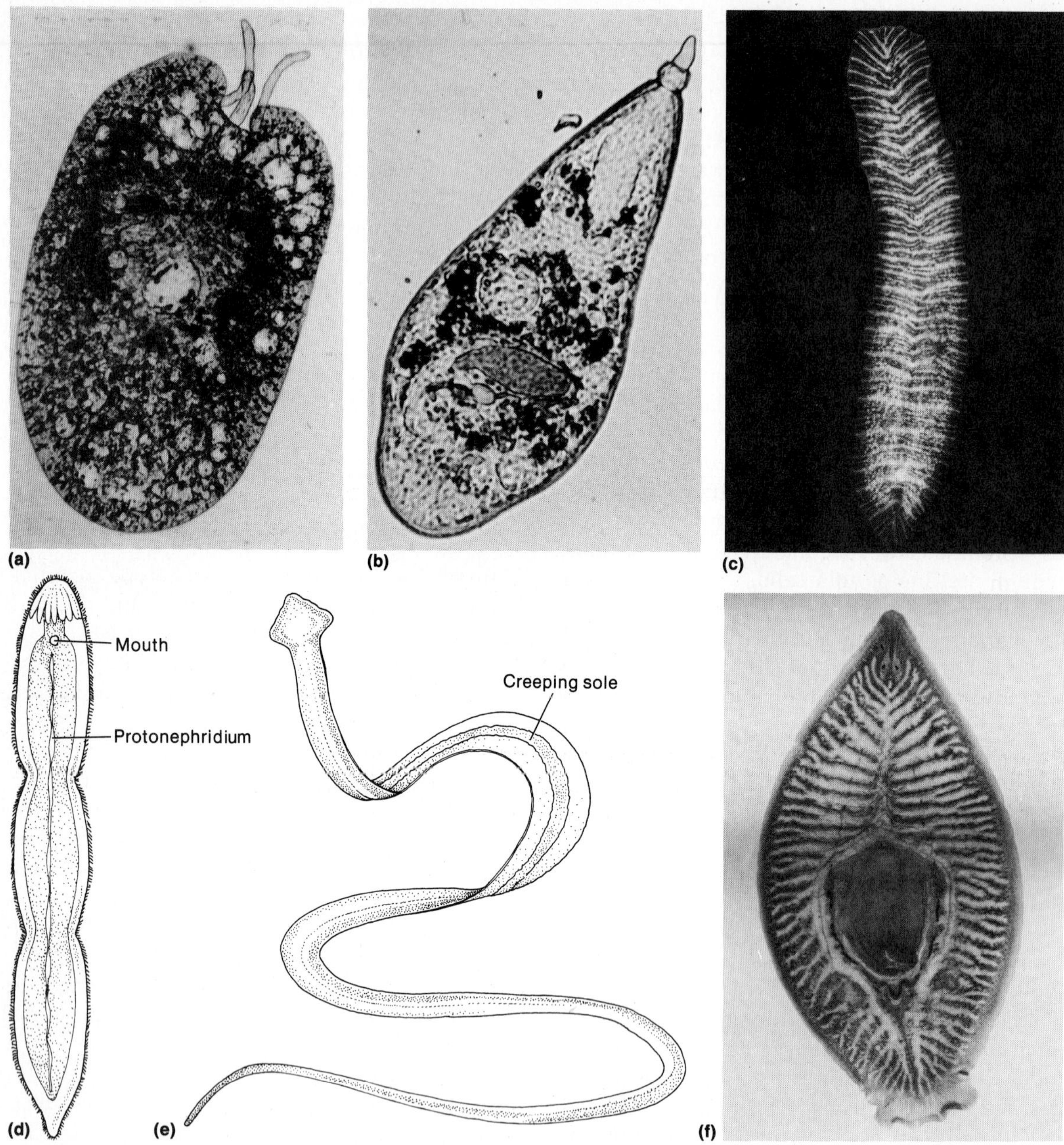

Figure 6.3 *Types of turbellarians:* ***(a)*** *a photomicrograph of* Polychoerus, *an acoel;* ***(b)*** *a photomicrograph of a neorhabdocoel;* ***(c)*** *a photomicrograph of* Stylochus, *a polycladid;* ***(d)*** Stenostomum, *a catenulid with three zooids;* ***(e)*** Bipalium kewense, *a terrestrial tricladid;* ***(f)*** *a photomicrograph of* Bdelloura, *a tricladid that is a commensal on the gills of horseshoe crabs (parts a, b courtesy of R. M. Rieger).*

animal of up to 60 cm in length can actually be a relative of the much smaller planarians.

Most turbellarians range in length from 0.5 to 5 cm. Many are broadly ovate or leaf shaped, while others are more elongate. Their colors range from white or cream to black or brown. Some species, especially those in tropical marine habitats, are brilliant orange, red, yellow, or blue, sometimes because of recently ingested food; others owe their green color to the presence of symbiotic algae (zoochlorellae) living in mesenchymal cells.

Even though turbellarians are quite similar in external features, they demonstrate considerable internal diversity, which is the basis for a number of subordinate taxa. But there still exists considerable uncertainty in this class's taxonomy based primarily on the evolutionary interpretations of internal structures. Even though a more complete listing of the orders of turbellarians appears at the conclusion of this chapter, it will be helpful at this point to mention two groups (sometimes referred to as subclasses) and the most important orders in each group. The two groups are differentiated primarily on their level of organization, especially that of the female reproductive system. The **archeophoran level** represents the more primitive level of organization and includes the acoels, macrostomids, and polycladids. To the **neoophoran level** belong those turbellarians with a more advanced level of organization such as the neorhabdocoels, proseriates, and tricladids (planarians). We shall refer to these groups at various points in the discussions below.

General body features

The outermost layer of the turbellarian body is a ciliated epithelium composed of flat, cuboidal, or columnar cells (Fig. 6.4). This layer may be a syncytium in more primitive groups. Cilia characteristically cover the entire surface, although in some species they are restricted mostly to the ventral surface, and in a few others, cilia are absent. These external cilia are used primarily for locomotion either in water or in slime produced by special glands. Just inside the epithelium is a noncellular basement membrane that often serves as an attachment point for muscles. Beneath the epidermis are three, delicately thin layers: an outer **circular muscle layer,** an inner **longitudinal muscle layer,** and a **diagonal muscle layer** between them (Fig. 6.4). All are responsible for various bodily movements.

Located in the epidermis are curious, rod-shaped bodies, the **rhabdites,** characteristic of most turbellarians (Fig. 6.4). Set mostly at right angles to the surface and occurring either singly or in bundles, the rhabdites are secreted by glandular cells. Their exact function is unknown, but they can be discharged to the outside and form a gelatinous, and perhaps a defensive, sheath around the animal. Some investigators have suggested a possible evolutionary relationship between the rhabdites and nematocysts of cnidarians, even though they are not the same structurally and are only vaguely similar functionally. It is possible that both structures evolved from a common source in animals ancestral to both phyla, but conclusive evidence has yet to be produced.

Turbellarians possess a number of epidermal, mostly unicellular glands, which are often sunk into subepidermal regions, and each is connected to the surface by a neck (Fig. 6.4). Secretions of these glands are used by the animal in adhesion, locomotion, and feeding. Apparently, there are two types of glands based on the products they secrete: one type secretes slime or mucus used in locomotion as well as to entangle prey, and a second type secretes adhesive materials. A typical characteristic feature of turbellarians is a **frontal gland,** an aggregation of unicellular glands situated at the extreme anterior end whose function is to secrete slime used to entangle prey. Turbellarians may also have aggregations of glandular cells

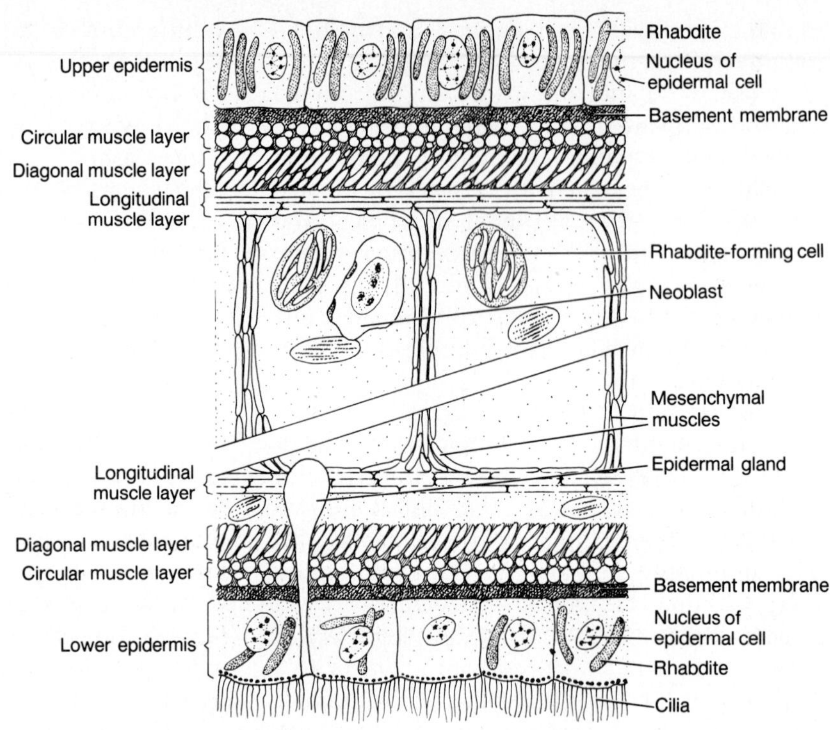

Figure 6.4 *A diagrammatic longitudinal representation of the upper and lower surfaces of a turbellarian illustrating the principal layers.*

on the lateral margins, at the posterior end, or along the ventral surface. A few species may have specialized adhesive structures such as plates, suckers, specialized cilia, or a grasping proboscis.

Internal to the muscles is the **mesenchyme,** an area of loosely packed, parenchymous cells with numerous, small, ameboid, undifferentiated cells called **neoblasts** (Fig. 6.4). Neoblasts of turbellarians are very similar in structure and function to poriferan archeocytes, but their similarities are probably due only to the fact that they are both undifferentiated cells. Neoblasts are primarily responsible for initiating regeneration of lost parts, but they also function in reproduction and internal transport. Embedded within the mesenchyme are the reproductive and osmoregulatory organs and discernible dorsoventral muscle strands (Figs. 6.4, 6.7).

A remarkable flatworm adaptation is found in a few species, characteristically those of *Microstomum,* in which cnidocytes are located in the epidermal layer. *Microstomum* feeds on *Hydra* and yet, incredibly, is able to ingest the hydrozoans without digesting the cnidocytes or discharging their nematocysts. The cnidocytes are engulfed by neoblasts and transported to the epidermis, where they are employed by the worm in defense or food capture! Apparently, the nematocysts are mechanically discharged when accidentally touched by another animal, but no neural control is effected by the turbellarian over them.

Locomotion

Different locomotive patterns are found in turbellarians, are basically related to the size of the animal, and usually involve the action of both cilia and muscles. Minute forms swim for the most part, and then are able to do so by the coordinated action of epidermal cilia, which are generally uniformally distributed. Most larger turbellarians do not swim but rather creep or glide over surfaces. In these forms there is a strong tendency to restrict the ciliation to the ventral surface. While cilia are still important in locomotion in larger turbellarians, muscles become the dominant propulsive force. Muscular movement is achieved by the rhythmical contraction of circular muscles alternating with that of the longitudinal muscles. Transverse waves of contraction are generated near the anterior end, and these pass posteriorly along the length of the body. Contractions of the circular muscles raise and elongate a given section of the body, while those of the longitudinal muscles lower and shorten a similar section. As a given section of the ventral surface is raised, it is set down again slightly ahead of its previous point of contract, and in effect, that region advances by a small step. The passage of a given wave over the length of the animal moves the entire animal forward for a short distance; concurrent waves make possible a smooth gliding motion. Terrestrial turbellarians secrete mucus at the anterior end, and muscular waves and cilia propel the animal along in the slime trail.

FEEDING AND DIGESTION

While a few turbellarians acquire food from their symbiotic autotrophic algae and a few others are herbivorous, the vast majority are carnivorous. Turbellarians are active and efficient predators because of their mobility and rather efficient sense organs. They are voracious feeders on many different invertebrates including rotifers, nematodes, bryozoans, tunicates, other turbellarians, and small crustaceans and annelids. Larger turbellarians can actually capture and ingest earthworms, snails, and insect larvae. They have several different methods for capturing live prey, but basically the worm uses its own body plus slime from the frontal gland and wraps its body around that of the prey; it then entangles its victim with mucus from frontal and ventral glands. Adhesive materials are also used by the flatworm to attach itself to a surface, since by having the body anchored, capture and ingestion of the prey can be made easier. A quiet, deadly struggle ensues with the worm eventually ingesting its prey whole by means of a muscular, often protrusible **pharynx.**

The ventral **mouth** is located typically about midbody at the site of the embryonic blastopore. The pharynx, connecting mouth and intestine, presents a wide range of complexities in construction and function. Three types of pharynges are found in this class: simple, plicate, and bulbous. A **simple** pharynx is a short length of invaginated ciliated epidermis without special muscles (Fig. 6.5a), this type is found in some acoels and in certain other primitive forms in several orders. A **plicate** pharynx is a cylindrical fold projecting into a deep spacious **pharyngeal cavity** (Fig. 6.5b, 6.7). The plicate pharynx can be thrust through the mouth and extended for some distance as a feeding **proboscis** (Fig. 6.5c). When protruded, the pharynx may point forward or backward or it may hang downward. Food is sucked into the intestine by a force created by peristaltic muscular waves within the pharynx. The plicate pharynx is found in polycladids, tricladids, and some individuals in other orders. A **bulbous pharynx** also has thickened walls due to the presence of special muscles. Distally, it opens into a more shallow pharyngeal cavity, this pharynx can be everted and protruded (Fig. 6.5d, e). The bulbous pharynx is thought to have been derived from the plicate type and is found mostly in neorhabdocoels. A few species may have two or more pharynges associated with one or more mouths.

The form of the intestine is an important and widely used taxonomic feature. In fact, the

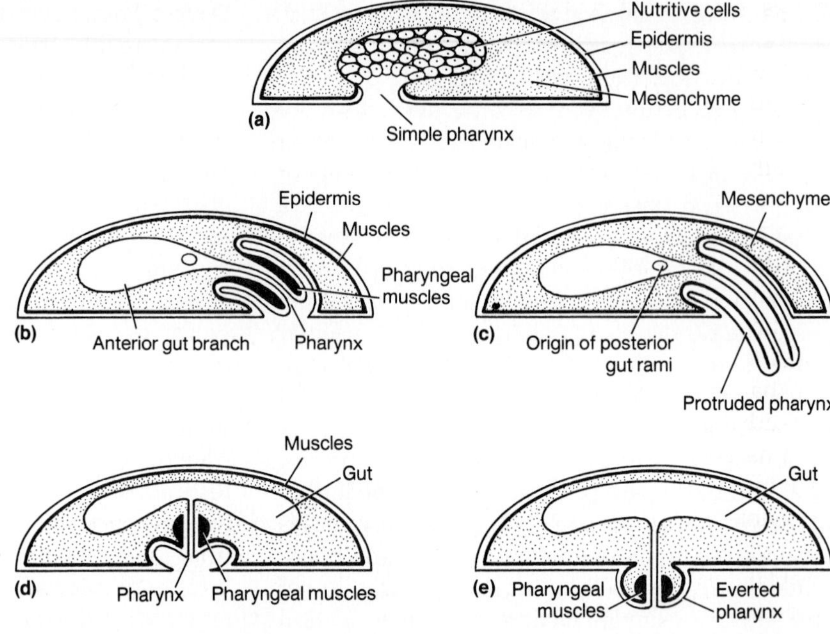

Figure 6.5 *Some variations in the nature of the turbellarian pharynx:* ***(a)*** *simple (some acoels);* ***(b)*** *tubular plicate (tricladids);* ***(c)*** *tubular pharynx protruded;* ***(d)*** *bulbous (neorhabdocoels);* ***(e)*** *bulbous pharynx everted (parts b–e adapted from J. B. Jennings,* Biol. Bull. *112:66, 72, 1960).*

Figure 6.6 *Variations in the form and shape of the intestine in turbellarians:* ***(a)*** *with no intestine but with nutritive cells (acoels);* ***(b)*** *a simple straight intestine (neorhabdocoels);* ***(c)*** *an intestine with three main trunks (tricladids);* ***(d)*** *an intestine with many small trunks (polycladids).*

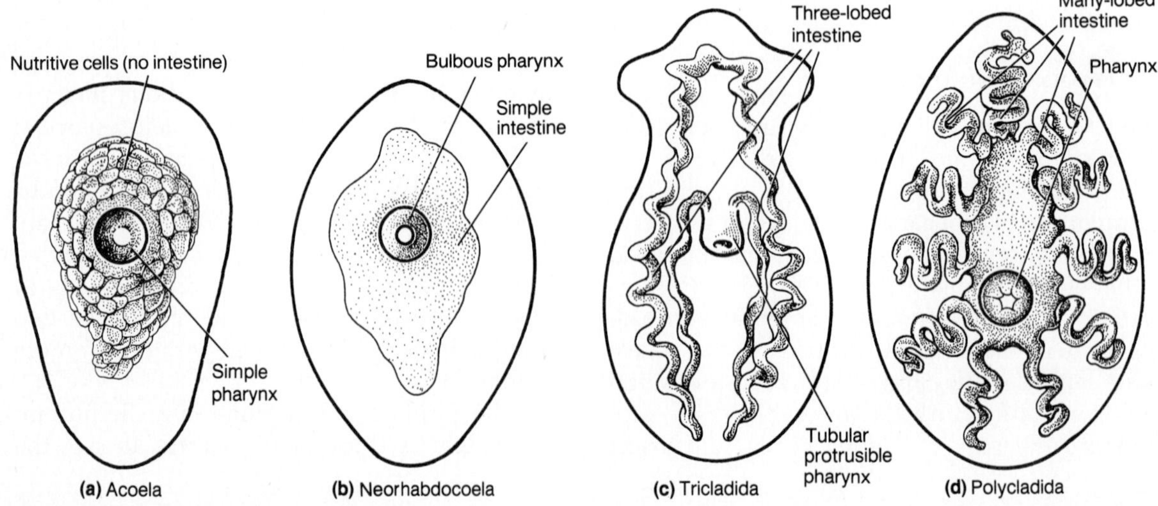

names of some of the orders characterize the nature of the intestine. The acoels, as their name would imply, have no intestine at all, and the endodermal nutritive cells form a solid syncytial mass (Figs. 6.5a, 6.6a). Some acoels can partially protract the endodermal mass through the mouth to engulf small food particles. Representatives of several orders, including the Neorhabdocoela, have a simple, straight intestine with no lateral branches (Fig. 6.6b). Lateral diverticula, present in larger turbellarians, increase the surface area for absorption and serve to enhance internal transport of food. The polycladids have many intestinal branches, hence their name (Fig. 6.6d). The tricladids have an intestine with an anterior trunk and two lateroposterior trunks, and all three bear numerous small diverticula (Figs. 6.6c, 6.7).

Many turbellarians have interesting feeding and nutritional habits. One interesting species of *Stylochus,* a polycladid, lives inside oyster shells, where it feeds on bits of oyster tissue, and another species of *Stylochus* preys on barnacles (Fig. 6.3c). *Bdelloura,* a tricladid, is an ectocommensal on the gills of horsehoe crabs (Fig. 6.3f). Several species of *Convoluta,* an acoel, contain zooxanthellae intracellularly in the mesenchymal cells, and they are reported to digest some of its symbionts as food. Certain neorhabdocoels are ectocommensal in the mantle cavity of molluscs, on the gills of crustaceans, and on turtles. Other neorhabdocoels are endoparasitic in molluscs and echinoderms. The ectocommensals share the food of their larger symbionts, and parasitic forms subsist on host fluids and tissues.

Nervous system, sense organs, and behavior

The nervous system of turbellarians illustrates various transitional stages between the ancestral, radially symmetrical system and a more advanced bilateral plan. A few primitive acoels have a nerve net system located just beneath the epidermis, but in the remaining turbellarians the nerve complex is located beneath the muscular layers. In other acoels the submuscular network consists of up to five pairs of

Figure 6.7 *A cross-sectional view of a planarian through the level of the pharynx to illustrate the arrangement of principal layers.*

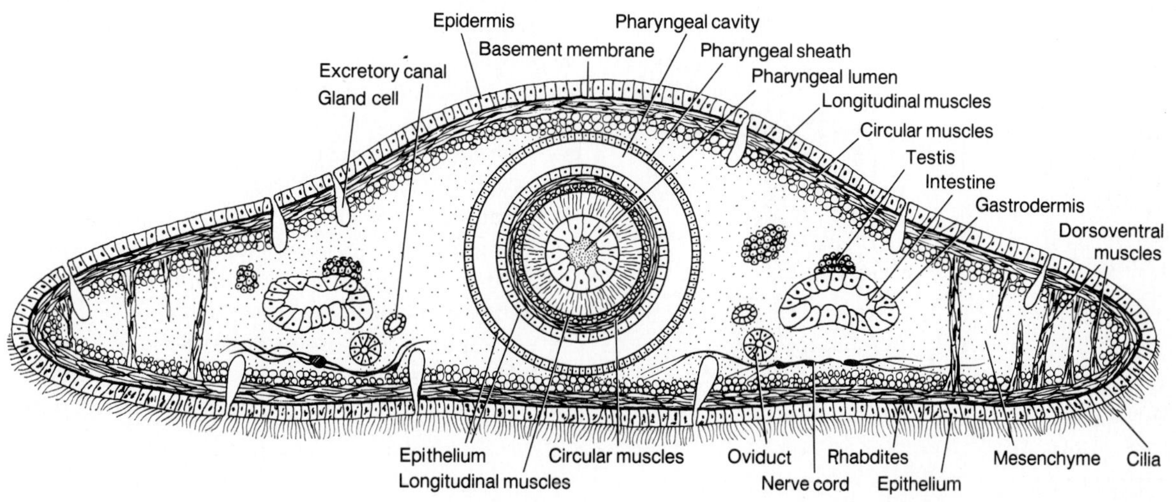

longitudinal cords: dorsal, dorsolateral, marginal (lateral), ventrolateral, and ventral, but one or more pairs are often missing. The cords are interconnected by ring commissures along the length of the body, the anterior one of which has assumed a dominant role in coordination and is called the **cerebral ganglion.** In most other turbellarians there is a reduction in numbers of nerve cords and an increase in the role of the ventral pair for nervous transmission as exemplified by polycladids, most neorhabdocoels, and many tricladids (Fig. 6.8a). Land planarians have developed a unique nerve plate containing the ventral nerve trunk and connecting the ventral margins of the nerve plexus. The **cerebral ganglion** or **brain** is characteristically a bilobed structure associated most closely with the ventral trunks.

Figure 6.8 *Turbellarian nervous system and eyes:* ***(a)*** *nervous system of a planarian (from L. H. Hyman,* The Invertebrates. *Vol. II,* Platyhelminthes and Rhynchocoela, *New York, McGraw-Hill, 1951, adapted with permission);* ***(b)*** *a simple, direct, pigment-spot eye;* ***(c)*** *an inverse, pigment-cup eye.*

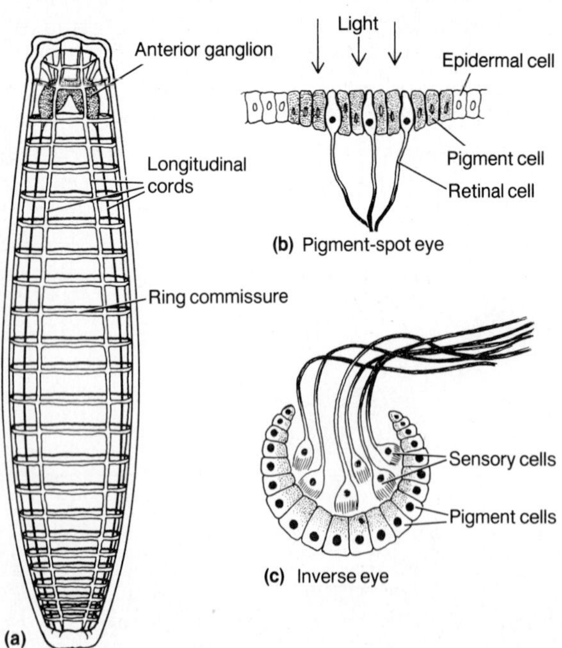

A characteristic feature of most protostomates is that the nerve cords are ventral in position, primitively doubled, and always solid. They are solid because they represent simple longitudinal strands of neurons present in ancestral forms. They are doubled because two longitudinal cords persisted from the meridional ancestral arrangement. The cords in protostomates are always ventral in position, since they evolved in close association with the ventral surface, which in turn, is directly involved with locomotion. Therefore a closer proximity between the nerve cords and the locomotive surface (and therefore shorter neurons) would make for a more intimate degree of control. In contrast, vertebrates all have a dorsal nerve cord that is never doubled and always hollow. It is hollow because of the longitudinal infolding manner in which it develops. The vertebrate nerve cord is not intimately associated with the ventral surface, since that surface has lost its basic direct association with locomotion. The dorsal position in vertebrates is thought to be more closely associated with the larger brain and more elaborate sense organs and not so much with locomotion per se.

Behavioral scientists have shown that turbellarians can modify their behavior on the basis of previous experiences (i.e., conditioning). To a limited degree, such learning has been achieved experimentally with a number of forms, including *Dugesia.* Planarians have been taught to avoid conditions they normally would not avoid and to respond positively to stimuli to which they normally would react negatively. Extirpation of the brain prevents learning and, in trained animals, voids the learned experience, thus implicating the brain as the primary locus for associative learning.

Compared to those in radiates, sense organs in turbellarians represent a substantial increase in complexity that is intimately associated with their mobility and predaceous life-style. Specialized receptors for light, touch, chemicals, currents, and gravity are concentrated mostly at the anterior end. **Photoreceptors** or pigment-cup **eyes** occur in most turbellarians usually as one pair, sometimes two or three pairs, or, as in land forms, many pairs. The eyes are light-detecting structures and do not form images. A few acoels have a simple pigment or eye spot that undoubtedly represents the primitive condition from which more advanced eyes evolved (Fig. 6.8b). More specialized eyes are present as either **inverse** eyes, in which the retinal cells are turned away from the light (Fig. 6.8c), or **direct** eyes, in which the retinal cells are oriented outward toward the light source. The eye is usually recessed into the mesenchyme; and although a lens is absent, the overlying epidermis is sometimes unpigmented and may serve as a cornea. Most turbellarians are negatively phototactic and, if provided with choices, will seek out the darkest location.

Chemoreceptors are important in locating and recognizing food. Such receptors are usually located in the head region, sometimes in the frontal gland, and most often concentrated in ciliated pits. Chemoreceptor cells bear both sensory and ordinary cilia, and the latter circulate water over the sensory cilia; this continuous monitoring system helps in providing directional responses. **Tactile cells,** also equipped with sensory cilia, are located throughout the epidermis but are concentrated near the anterior end and along the lateral body margins. The ventral surface is positively thigmotactic, and contact of this surface with the substratum is important so that, when turned over, the worm immediately rights itself. **Rheoreceptors** (current detectors) are found in some and perhaps have a fairly uniform occurrence in all turbellarians. Some primitive species have a single median **statocyst** located at or in the brain that provides the worm with information about its orientation with respect to gravity.

Reproduction

While a very few species are dioecious, all other turbellarians are monoecious. With the exception of the acoels, which have no definite gonads, the testes and ovaries are discrete organs embedded within the mesenchyme. A great number of variations are found in both male and female systems. The basic plan for both systems has already been presented, and only the most significant variations will be noted.

The copulatory mechanism for most turbellarians is a penis that may be equipped with minute barbs, hooks, or a single stylet, which is useful in sperm transfer. In other forms the penial stylets are used to inject sperm hypodermically into the mesenchyme of the partner, and sperm then migrate to the ovary. A few turbellarians are equipped with many penes or stylets, and some of these are used, incredibly, for both reproduction and defense! The female reproductive system is well developed except in the Order Acoela, in which most organs are much reduced and even absent. The female system is much more simple in the archeophoran level but is rather complicated in those worms at the neoophoran level of complexity. Turbellarians at the archeophoran level have no yolk glands, and their ova are endolecithal. Those turbellarians at the neoophoran level of organization have both vitellaria and ovaries, and their eggs are ectolecithal.

Eggs are usually laid singly, but some planarians like *Dugesia* produce several eggs in each of many stalked capsules or cocoons (Fig. 6.9a). In the Order Acoela, in which oviducts and even gonopores are absent, eggs escape through the mouth or by rupture of the body wall. Egg development is initiated almost at once after the egg has been laid. Endolecithal eggs undergo holoblastic, unequal cleavage that is both spiral and determinate. The developmental patterns of ectolecithal eggs, however, are quite different from those of endolecithal eggs. Ectolecithal cleavage is determinate but not spiral as blastomeres divide to form an intermingled mass of embryonic and yolk cells (Fig. 6.9c). Certain peripheral blastomeres form the epidermis, and inner groups of blastomeres coalesce to form several embryonic masses that, in turn, give rise to groups of primordial organs. In some a rudimentary pharynx is formed early through which yolk cells are ingested; but in others the yolk material is slowly digested extracellularly by adjacent blastomeres. Most polycladids develop into a free-swimming, ciliated stage called **Muller's larva** (Fig. 6.9b). This short-lived, nonfeeding stage soon settles to the bottom and develops into an adult. A few polycladids have a larval stage, **Gotte's larva,** which is similar to Muller's larva, and a curious, wormlike larval stage has recently been reported from a freshwater catenulid. But all other turbellarians lack a larval stage, and development proceeds within the capsule with the young worm emerging in 2–20 days.

Figure 6.9 *Embryonic stages in turbellarians:* ***(a)*** *a stalked egg capsule of* Dugesia; ***(b)*** *a dorsal view of Muller's larva (polycladid) (from Kato in L. H. Hyman,* The Invertebrates. *Vol. II,* Platyhelminthes and Rhynchocoela, *New York, McGraw-Hill, 1951, adapted with permission);* ***(c)*** *early embryonic development of an ectolecithal egg.*

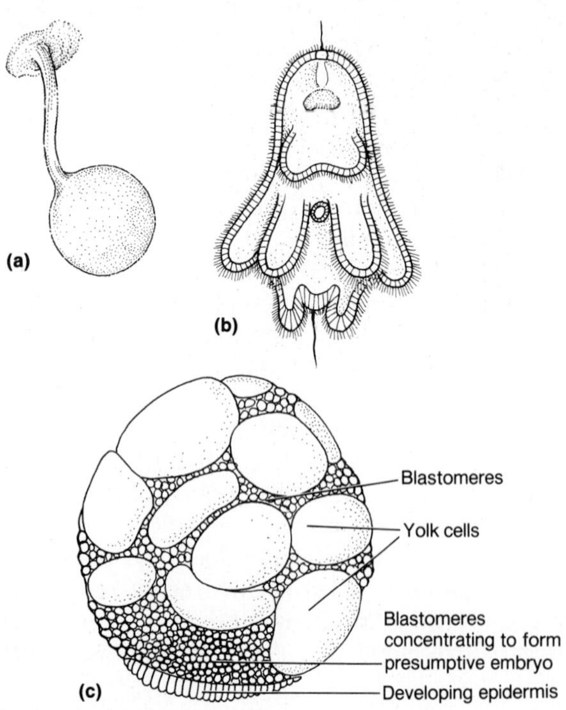

Many freshwater planarians produce both summer and winter eggs; the type of egg is influenced significantly by photoperiodic and thermal conditions. Summer eggs have a thin capsule and develop without delay. Winter eggs have a thicker capsule, which protects them during a prolonged, dormant, hibernal period. Parthenogenesis occurs in some turbellarians, and in several species, males have not been found; this lack would indicate permanent parthenogenesis.

Most turbellarians reproduce asexually, employing fission, fragmentation, or both. Transverse fission is perhaps the more common asexual process. Some species even reproduce exclusively by fission and are termed **fissiparous.** Often, representatives of the orders Catenulida and Macrostomida undergo successive transverse fissions to produce a chain of attached individuals or zooids (Fig. 6.3d). After reaching a certain level of maturity a zooid will detach itself from the chain and become an independent organism. Other freshwater turbellarians like *Dugesia* also undergo transverse fission but do not form chains of zooids. Some freshwater and land planarians may fragment into multiple pieces, each of which becomes a future adult. Turbellarians have great powers of regeneration, some details of which are briefly set forth in Box 6.1.

Class Trematoda

Trematodes are commonly known as flukes, a word that describes their wide, flat shape. Adult trematodes are all obligate parasites on or in a wide variety of vertebrates, and a few forms

BOX 6.1 □ THE FASCINATING PROPERTY OF REGENERATION IN TURBELLARIANS

That turbellarians possess the ability to regenerate lost parts is a well-known phenomenon. Even small fragments can regenerate an entire worm. Cultured pieces of *Dugesia dorotocephala* as small as 0.08 mm^3, representing an estimated 10,000 cells, can regenerate a complete worm. Regeneration principally involves neoblasts, dedifferentiated cells, and an anteroposterior physiological gradient or axis of polarity. Neoblasts quickly accumulate at the wound site along with dedifferentiated adult cells and form a mass of actively dividing, nonspecialized cells called a **blastema,** which will give rise to the missing parts. What the blastema produces depends upon the longitudinal gradient. The head end (or anterior end of a fragment) has a higher metabolic rate than that of more posterior regions, and the relative difference between the two establishes the gradient or polarity. One region, for example, the head, suppresses the regeneration of the same region elsewhere in the regenerating worm.

Many ingenious experiments have been performed to elucidate more clearly the details of turbellarian regeneration. Some of the more dramatic studies have dealt with grafting, studying the effects of tissue extracts on developing blastemas, inducing regeneration of multiple parts such as heads or pharynges, and reversing the axial polarity by the use of selected chemicals. But even the simplest of experiments will almost inevitably produce dramatic results. All the student will need are some turbellarians, pond water, a glass slide, and a sharp razor blade. Take a nonfeeding planarian, put it in a drop of pond water on a chilled slide, and make several longitudinal, transverse, oblique, or partial cuts with the razor blade; use your imagination to cut the animal into various pieces. Then transfer all pieces to clean pond water, do not feed the regenerating worms, and observe at intervals over the next several weeks. Since regeneration in turbellarians is fascinating, intriguing, and so easily studied in the laboratory, every zoology student should become familiar with it.

are parasitic in molluscs. Most adults are endoparasitic and are found in the gut, liver, and blood vessels of their vertebrate or final host, but a small number of species are ectoparasitic. All trematode species produce one or more larval stages that are also obligate parasites in a diversity of invertebrates. Most trematodes have unusually complicated life cycles with larvae typically living in one or more intermediate hosts. If there is a free-living phase, it is of extremely short duration. Adults of a number of trematodes are parasitic in human beings or domesticated animals and thus are of great medical and veterinary importance. The word *trematode* means to form a hole, presumably referring to holes or depressions made in the host by these helminths.

The body shape, similar to that of most turbellarians, is elongate or ovate and flattened, and most are from 0.1 to 4 cm in length, though one species found in the ocean sunfish has an unbelievable length of up to 7 m.! Estimates of trematode species vary widely among helminthologists (specialists in parasitic worms), but most experts recognize about 8000–8500 species. Flukes are generally nonpigmented and appear

BOX 6.2 □ THE NATURE OF PARASITISM

Parasitism is an intimate obligatory relationship between two different organisms in which the **parasite,** usually the smaller of the two, is metabolically dependent upon the **host.** The relationship may be permanent for the life of either the parasite or the host as in the cases of flukes or tapeworms, while in other situations the relationship is very temporary as exemplified by feeding mosquitoes and leeches. If the parasite is prevented from making contact with the host, it usually cannot survive for long because the parasite is dependent upon the host for its metabolic requirements (i.e., nutrition), and thus the relationship is obligatory for the parasite. Often, the presence of the parasite or its metabolic products in the host stimulates the host to produce antibodies, and the parasite responds by producing its own chemicals to inactivate the antibodies. Thus an immunological response by the host against the parasite is characteristic, especially of endoparasitism.

Parasites are usually quite successful, and their success is made possible by a variety of adaptations. It is adaptively significant that many parasites are small or are extremely flattened so as not to impede seriously the functions of the host. Parasites must be able to maintain their position on or in a host and therefore have evolved hooks, suckers, and other holdfast devices. Some have a muscular pharynx with which they can hold tightly onto the intestinal mucosa of the host. During their evolutionary history, adults of many parasites have lost many organs and structures because they were no longer selective adaptations to the parasite. Thus, for example, sense organs are reduced or lost altogether, locomotion in adults is almost completely lost, and tapeworms have no digestive system at all.

Parasitic physiological adaptations involve three basic areas: combating the immunological response of the host, resisting digestion for those that live in the host's intestine, and utilizing metabolic substances from the host. Immunological reactions against the parasite are quite varied and involve a variety of host mechanisms; these are reviewed by Cheng (1973) and will not be discussed here. Parasites, in turn, have evolved a number of mostly biochemical events that take place at the parasite's surface which prevent or inactivate both the host's immune response and digestive enzymes. Both

pale or cream colored. Since the Trematoda is the first large, multicellular group we have encountered in this text whose members are all parasitic, it will be most instructive to review some of the more important dimensions of host–parasite relationships. A brief discourse on parasitism appears in Box 6.2.

EXTERNAL FEATURES

The surface of adults is a nonciliated, syncytial layer called the **tegument.** The tegument was long thought to be a nonliving cuticle, but electron microscopy has demonstrated that it is neither an inert cuticle nor a typical epidermis but an extremely complicated living layer composed of two zones (Fig. 6.10). The **outer zone** is a cytoplasmic syncytium; located within it are various cytoplasmic organelles, including mitochondria, endoplasmic reticulum, vacuoles, and Golgi apparatus. The outermost surface of the outer zone is the **unit** or **plasma membrane,** which is thrown into numerous microscopic folds, the **microvilli,** which greatly increase the absorptive surface of the tegument (Fig. 6.10). In the **inner zone** are found circular and longi-

large and small parasites are able to exchange materials with their host through their general body surfaces. Others obtain nutritional substances by ingesting small bits of the host.

Some of the most striking parasitic adaptations involve their life cycles and reproduction. Most platyhelminth parasites exist in more than one host either as adults or larvae; sometimes as many as five sequential hosts are necessary to complete the life cycle. Therefore there are numerous adaptations associated with an extraordinarily complicated and perilous life cycle. Some of the more important adaptations include (1) mechanisms for locating and penetrating hosts, (2) existing as larvae in one or more hosts and adapting to the diverse physiology of these different hosts, (3) withstanding the immune response of intermediate hosts, and (4) infesting intermediate hosts that are in turn normally eaten by subsequent hosts. To compensate for the extremely hazardous life cycle in which at best only a few progeny from a given parasite are successful and survive, parasites possess enormous reproductive potentials. For example, *Ascaris*, a common roundworm of the human intestine, can produce up to 200,000 eggs per day and may live for years in the host! Other parasites compound their fecundity rate by having larvae that reproduce asexually (polyembryony) to engender one or several parthenogenetic generations. Thus the reproductive rate for many parasites is astronomical, and their biotic potential is staggering.

There are two important advantages in having various stages in a secondary host. First, within the secondary or intermediate host, the genome of the parasite is usually amplified many times over by the parthenogenetic production of other stages. Second, by having stages in other hosts, the primary host as a resource or habitat is not used up nearly as fast as it would be if all stages had to be completed in a single host. Thus adult and larval parasites have different hosts and different resources, and therefore their complex life cycle is a very important adaptation.

Why, then, you may ask, aren't we up to our ears in parasites? The answer, very simply, is that the odds are incredibly small that any single helminth egg will survive the rigors of the life cycle and eventually become an adult.

tudinal muscles and cellular units, the **cytons,** embedded within the mesenchyme and containing the usual cytoplasmic organelles. The line of demarcation between outer and inner zones is a **basement membrane** (or basal lamina), and distinct cytoplasmic bridges connect the outer syncytium with the inner cytons (Fig. 6.10). Excluding those materials ingested by the worm, all other metabolic substances exchanged between trematode and host pass through the tegument. The resistance of the worm to digestion (if it is an intestinal form) and to the immune reaction by the host resides in the remarkable cellular and biochemical properties of the tegument. Details of tegumental physiology are beyond the scope of this text, but it is a very active area of investigation, and future research will undoubtedly add considerably to our knowledge of this important layer.

The **mouth** of a trematode is always at the anterior end and is usually surrounded by an attachment organ, the **oral sucker** (Fig. 6.11). In addition to the oral sucker, trematodes typically have a **ventral** or **posterior sucker** or some other adhesive structure. A sucker, composed basically of subtegumental muscles, is a protrusible

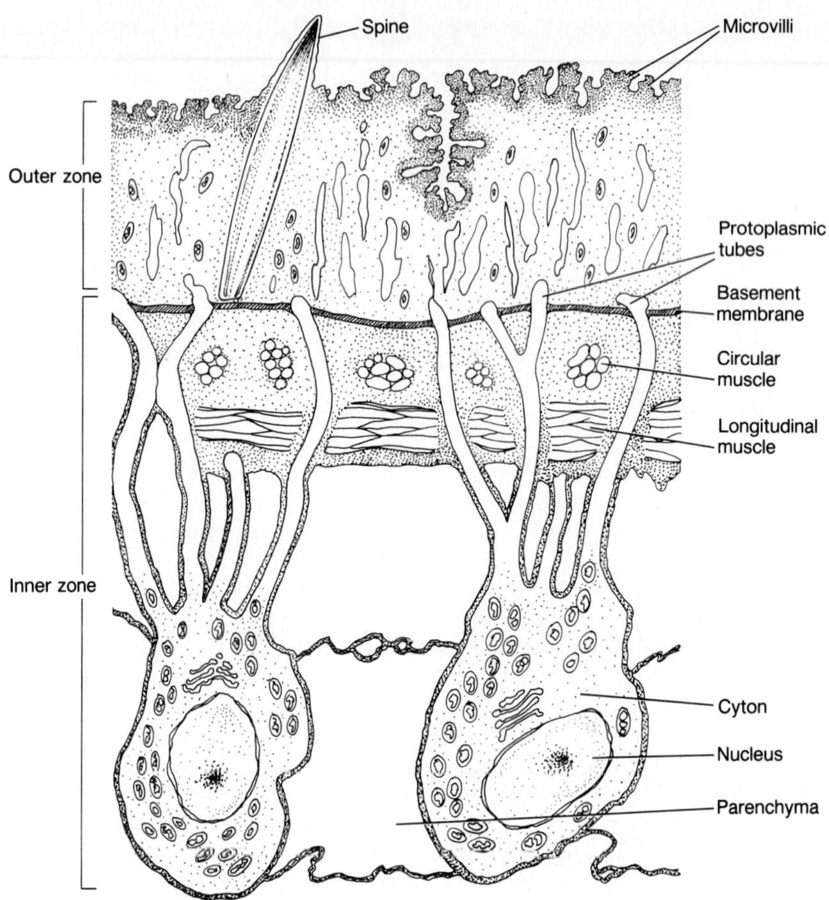

Figure 6.10 *A diagrammatic representation of the tegument of a trematode (adapted from L. T. Threadgold,* Quart. Journ. Microscop. Sci. *104:506, 1963, The Company of Biologists Ltd.).*

cup that, when applied to a host tissue, actually encloses and surrounds it to make an effective strong vacuum. Other attaching devices such as hooks and clamps are found in some trematodes.

INTERNAL FEATURES

The mesenchyme of trematodes is essentially like that of turbellarians. Muscles are better developed around the suckers, where the musculature is crucially important for sucker function. Within the mesenchyme of a few trematodes is found a branched system of fluid-filled canals that apparently is concerned with internal transport.

An anterior mouth, muscular pharynx, and short esophagus convey food to the intestine. The **pharynx** acts as a pump enabling the fluke to ingest host cells, mucus, fluids, and blood; tissues, injured by the attachment organs, are also commonly ingested. The **intestine** usually consists of two branches or ceca that extend posteriorly along the sides of the worm (Fig. 6.11). The ceca, either simple or bearing small lateral diverticula, are blind-end tubes. A very few flukes have a complete alimentary tract with a separate posterior **anus** that constitutes an aberrant situation among flatworms.

Details of the protonephridial system are used widely as taxonomic features. Usually, there is a pair of longitudinal collecting ducts that opens to the outside by one or two **nephridiopores** (Fig. 6.11). If a single nephridiopore is present, it is usually at the posterior end; but if a pair is present, each is often located on the lateral margins toward the anterior end. The protonephridial system is not as prominent or well developed in adult trematodes as in freshwater turbellarians.

Figure 6.11 *A stylized and generalized digenetic trematode.*

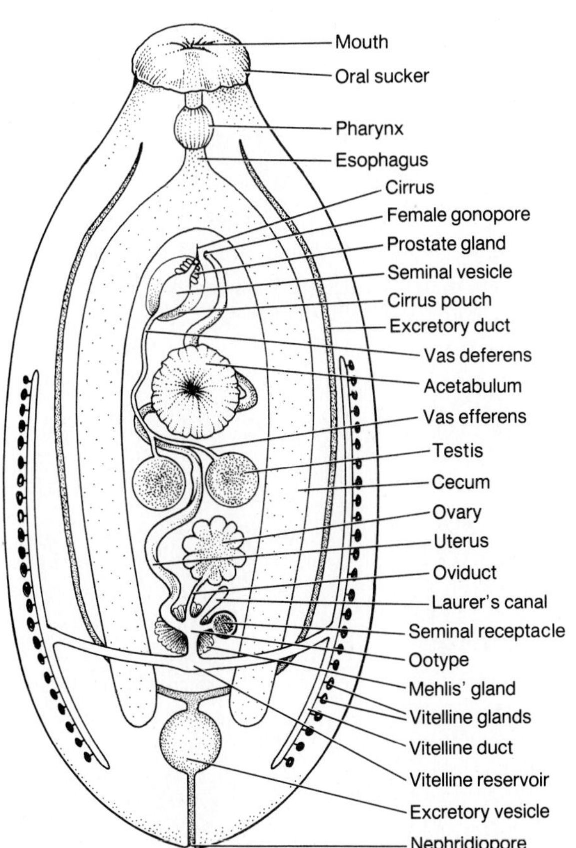

The nervous system is essentially like that of turbellarians. From the **brain** or **cerebral ganglia** arise three pairs of longitudinal nerve cords. The ventral pair tends to be the most highly developed, the dorsal and lateral pairs may be missing, and numerous ring commissures connect the nerve cords. The adhesive organs are richly innervated. Sense organs are poorly developed or are generally absent, but some of the adult ectoparasites possess one or two pairs of simple eyes.

Typically, trematodes are monoecious, although the schistosomes (family of blood flukes) are dioecious. Most trematodes have a pair of **testes,** but others may have one or many testes. The copulatory device, either a cirrus or a penis, has associated with it **prostate glands** whose secretions aid in sperm vitality. The copulatory organ is protruded through a **common genital atrium,** which opens to the exterior by a midventral **gonopore** usually located in the anterior half of the fluke (Fig. 6.11).

The female system generally consists of one **ovary,** and the yolkless ectolecithal ova travel through a short oviduct to the **ootype,** where fertilization occurs, yolk cells are added to each ovum, and a shell is formed. In some trematodes a copulatory pore separate from the gonopore is present, and a short duct, **Laurer's canal,** leads from the gonopore to the **seminal receptacle** (Fig. 6.11). The ootype is continuous as the **uterus** which is typically a long convoluted tubule that runs anteriorly and empties into the common genital atrium and to the exterior via the gonopore (Fig. 6.11). Thousands of eggs can sometimes be found in the distended uterus, which then fills most of the body. In some the distal part of the uterus is a muscular **metraterm** that is used to expel eggs.

Subclass Aspidogastrea

This is a small group of flukes that shows similarities to the other two orders. They are primarily endoparasites in molluscs, but they are

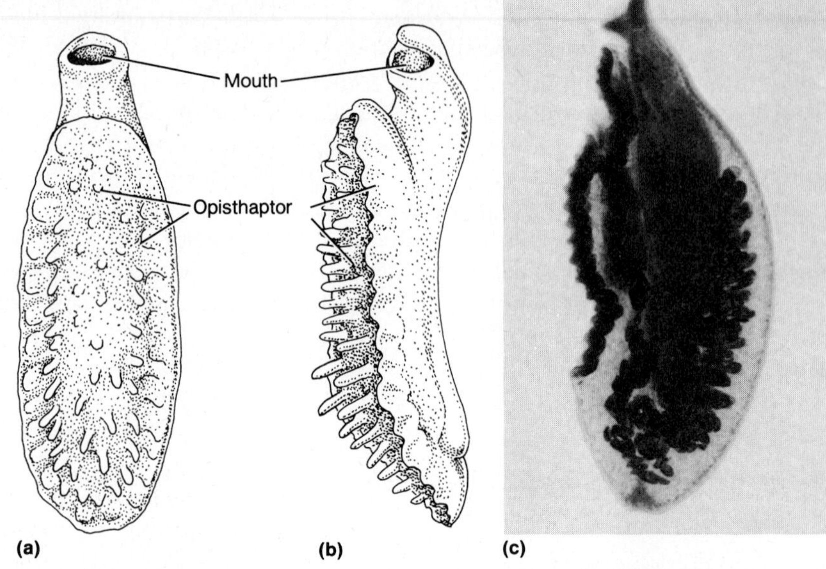

*Figure 6.12 Some representative aspidogastrean trematodes: **(a)**Lophotaspis, ventral view; **(b)** Lophotaspis, lateral view; **(c)** a photomicrograph of another aspidogastrean. (parts a, b from Ward and Hopkins in L. H. Hyman,* The Invertebrates. *Vol. II,* Platyhelminthes and Rhynchocoela, *1951, McGraw-Hill Book Co., adapted with permission).*

*Figure 6.13 Some monogeneid trematodes: **(a)*** Gyrodactylus; ***(b)*** Sphyranura *(from L. H. Hyman,* The Invertebrates. *Vol. II,* Platyhelminthes and Rhynchocoela, *New York, McGraw-Hill, 1951, adapted with permission).*

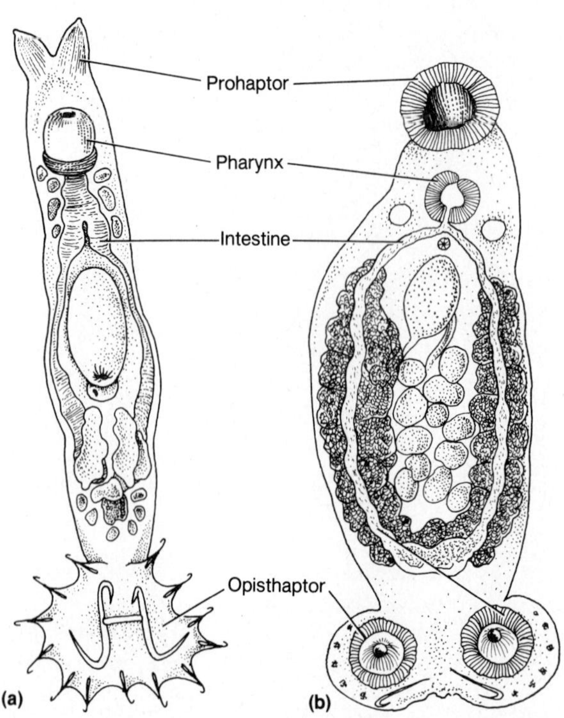

occasionally found in other invertebrates and in vertebrates, including fishes and turtles. All aspidogastreans are characterized by the presence of a large oval holdfast, the **opisthaptor,** that covers the entire ventral surface (Fig. 6.12). The opisthaptor, an extremely strong attachment organ, is often subdivided by ridges or septa. The oral sucker, characteristic of most other trematodes, is generally absent, although the arrangement of muscles around the mouth is suggestive of a sucker. The life cycle of aspidogastreans may involve only one host, a mollusc, or two hosts; in the latter case the final host is usually a vertebrate that becomes infected by ingesting a mollusc containing immature helminths.

Subclass Monogenea

Monogenetic trematodes are so named because there is only one host and thus only one form in the life cycle. These flukes are generally ectoparasitic on the gills, scales, or fins of fishes or on other ectothermous vertebrates, but they may occasionally be found on aquatic mammals, crustaceans, and molluscs. A few species have migrated through body orifices and are endoparasitic in the bladder, ureters, or buccal cavity, and one species inhabits the coelom of rays. Most adult monogeneans are less than 3 cm long and possess one or two pairs of eye spots.

A distinguishing feature of the monogeneans is the presence of a ventral, posterior, adhesive structure, the **opisthaptor,** which differs from that in the aspidobothrids (Fig. 6.13). In the monogenetic trematodes it is a complex structure separated by septa and may be equipped with sclerotized hooks and bars. In addition to the opisthaptor there is usually an anterior sucker or **prohaptor** located near the mouth (Fig. 6.13). A true oral sucker surrounding the mouth is only rarely present.

Most Monogenea possess a **vagina** into which the penis or cirrus is inserted via the vaginal pore during copulation. Eggs, discharged through the worm's gonopore, are often attached to the host by long filaments. Egg development is completed in several days, and each egg hatches into a ciliated, free-living larva, the **oncomiracidium.** Larvae seek out new hosts, which they must find within a day if they are to survive, attach themselves to the host by means of the larval opisthaptor, and gradually develop into adults.

Polystoma integerrimum, while not typical of monogeneans, will serve as an example of this order. This rather common parasite is found in the urinary bladder of various frogs and toads (Fig. 6.14). This species demonstrates a remarkable degree of synchrony in the life cycles of both host and parasite. It is possible that this fascinating synchronization is modulated by hormonal influences of the host. Worm eggs are released when the amphibian returns to the water to breed and lay its own eggs. The gyrodactylid larvae attach to the gills of tadpoles; and as the tadpole metamorphoses, the larvae migrate through the digestive tract of the amphibian and into the host's bladder (Fig. 6.14). Sexual maturity of the helminth requires about three years in the host.

Subclass Digenea

The vast majority of all trematodes (ca. 7500 species) belongs to this group, and the number of known species continues to increase as new forms are identified. At least two different forms, adult and one or more larval stages, are developed, a characteristic from which the name of the subclass was derived. Digenetic trematodes are characterized by having at least two different hosts required for the worm to complete its life cycle, and thus these worms possess by far the most complex life cycles of any platyhelminth. As adults, they are all endoparasites in the blood, digestive tract, ducts of auxiliary digestive glands, and other visceral organs in a wide variety of vertebrates that serve as definitive or final hosts. One or more

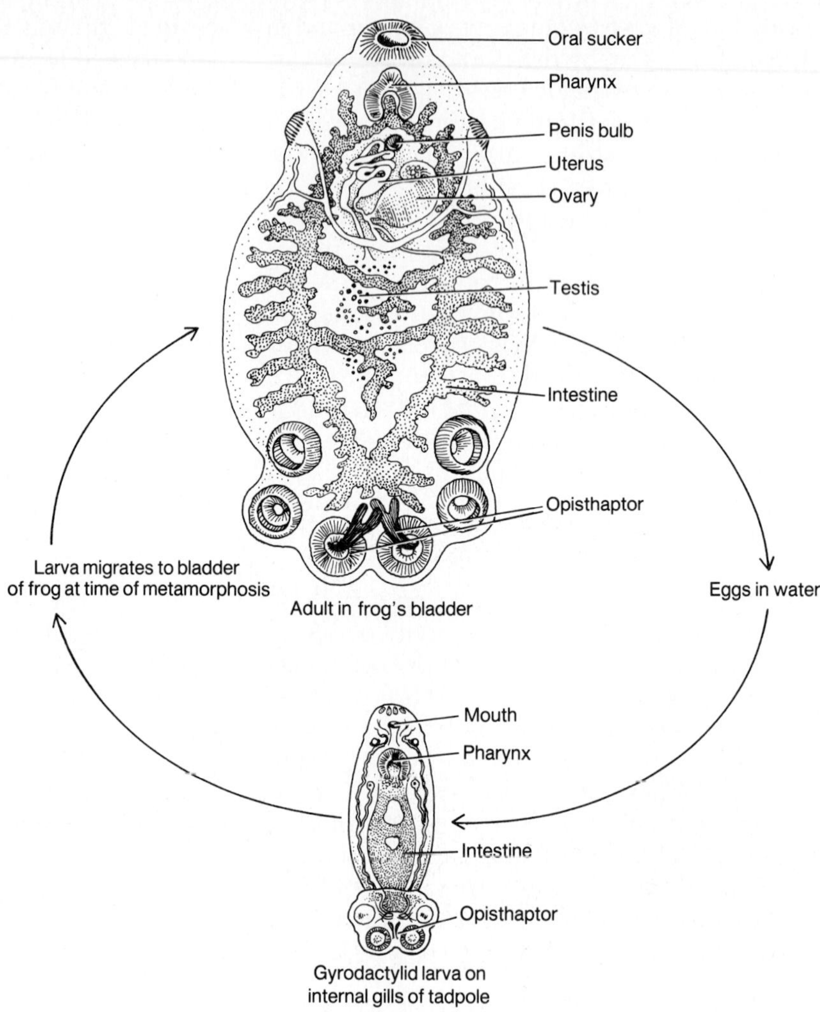

Figure 6.14 *The life cycle of the monogeneid trematode,* Polystoma integerrimum *(adult and larva adapted from J. B. Williams,* J. Helminthology, *Vol. 34 (1960), p. 153).*

intermediate hosts, almost always invertebrates, harbor the several larval stages. Considerable diversity exists in adult digenetic trematodes, some of which are illustrated in Figure 6.15. Two of the best-known flukes are *Fasciola hepatica,* a trematode found in the bile ducts of many domestic animals (Fig. 6.16), and *Opisthorchis sinensis,* the oriental liver fluke, which may also inhabit the bile ducts of human beings, other mammals, and birds (Fig. 6.15b).

An adult digenetic trematode possesses two holdfast suckers: an **anterior sucker,** which is called the **oral sucker** if it surrounds the mouth, and a highly muscular ventral or posterior sucker called the **acetabulum** (Figs. 6.11, 6.15, 6.16). In some, only the anterior sucker is present, and no opisthaptor is ever present in the digeneids.

Life cycles. Perhaps the most characteristic feature of the digenetic trematodes is their complex life cycles. The following discussion is about the life cycle of *Fasciola hepatica,* a rather common trematode present in the liver or bile ducts of many domestic animals, including sheep. Even though many exceptions and modifications occur in the life cycles of various trematodes, the life cycle of *F. hepatica* will serve as a model or basic plan.

Eggs, produced in prodigious numbers, are released into the host's digestive tract and are expelled to the outside with the host's feces. Typically, each shelled egg is equipped with a lidlike structure, the **operculum,** at one end that facilitates hatching (Fig. 6.16b). Eggs can survive for a year or longer with no subsequent development. Only if the egg finally reaches water does development proceed, and it hatches into a free-swimming, nonfeeding larval stage, the **miracidium** (Fig. 6.16c). The miracidium is covered with ciliated epidermal plates and swims actively in search of a mollusc, usually a snail, which it must enter within several hours or the larva will perish. Miracidia of

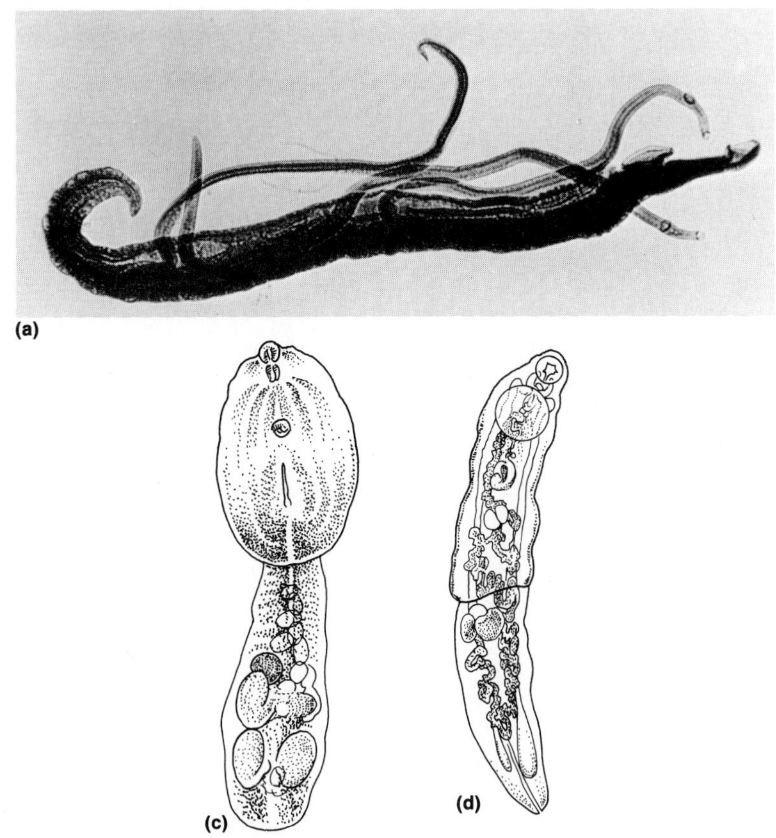

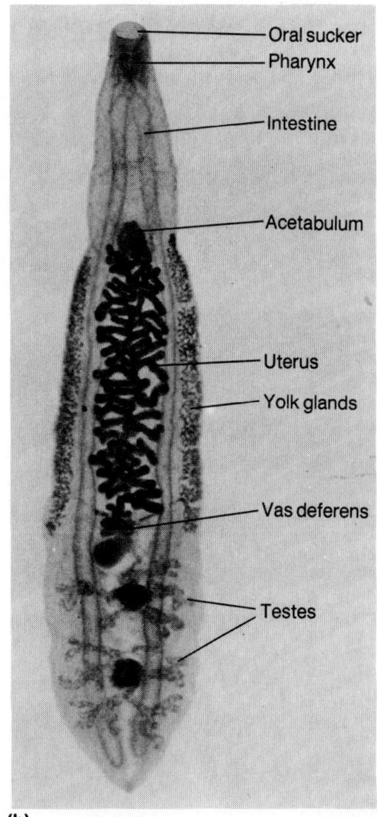

Figure 6.15 *Types of digenetic trematodes:* ***(a)*** *a scanning electron micrograph of* Schistosoma mansoni *in the blood of human beings—smaller female is copulating with larger male (courtesy of Ward's Natural Science, Inc., Rochester, N.Y.);* ***(b)*** *a photomicrograph of* Opisthorchis sinensis *in liver ducts of sheep;* ***(c)*** Diplostomum *from the intestine of gulls;* ***(d)*** Hemiurus *from the intestine of fishes and amphibians (part c [from Dubois], d [from Looss] in L. H. Hyman,* The Invertebrates. *Vol. II,* Platyhelminthes and Rhynchocoela, *New York, McGraw-Hill, 1951, adapted with permission).*

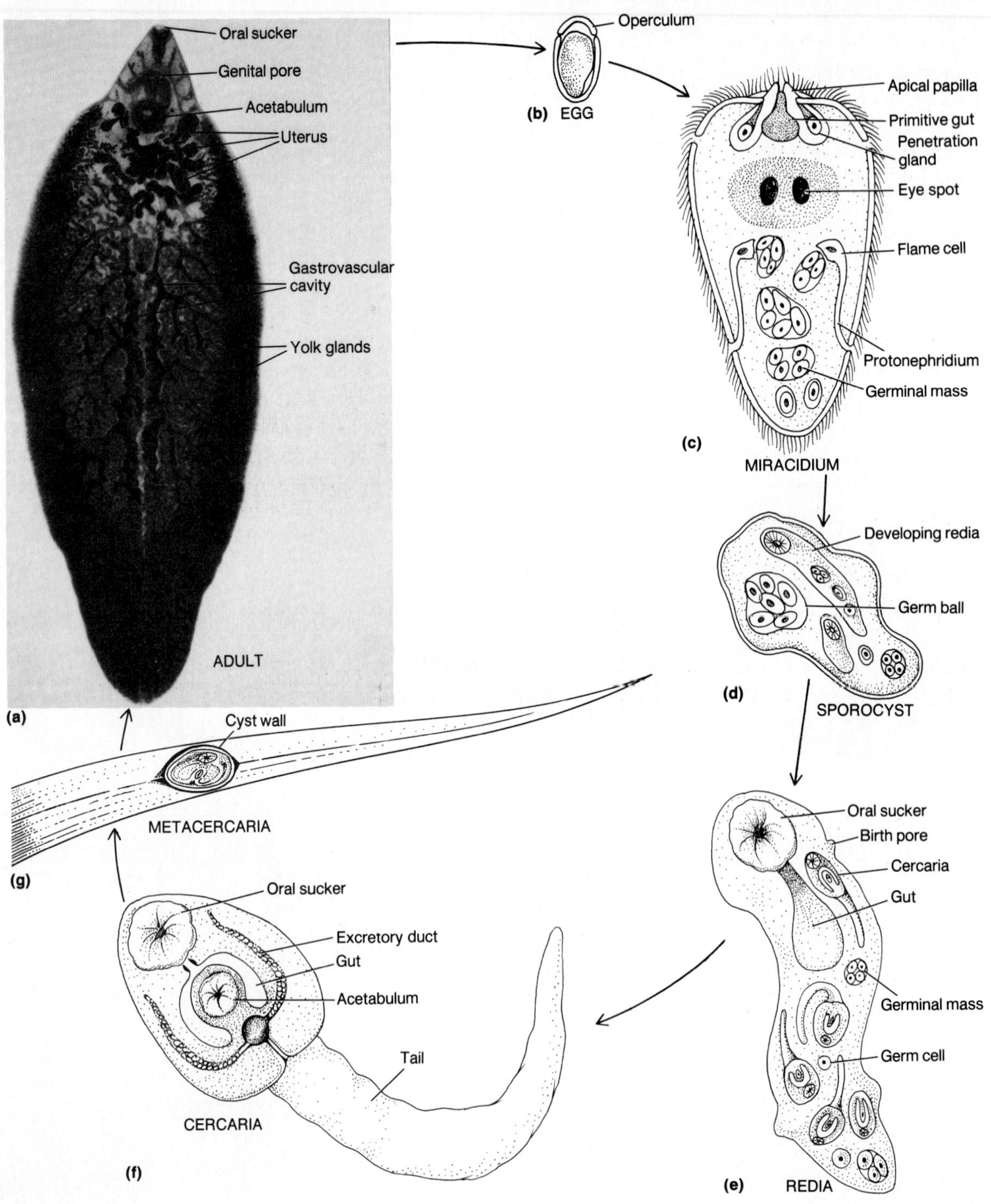
Oral sucker
Genital pore
Acetabulum
Uterus
Gastrovascular cavity
Yolk glands
ADULT
(a)
Operculum
(b) EGG
Apical papilla
Primitive gut
Penetration gland
Eye spot
Flame cell
Protonephridium
Germinal mass
(c)
MIRACIDIUM
Developing redia
Germ ball
(d)
SPOROCYST
Cyst wall
METACERCARIA
(g)
Oral sucker
Birth pore
Cercaria
Gut
Germinal mass
Germ cell
(e) REDIA
Oral sucker
Excretory duct
Gut
Acetabulum
Tail
CERCARIA
(f)

◀ ***Figure 6.16*** *The life cycle of* Fasciola hepatica, *a fluke commonly found in the liver and bile ducts of herbivorous mammals:* ***(a)*** *a photomicrograph of an adult;* ***(b)*** *an egg that exits to the outside with the host's feces;* ***(c)*** *a miracidium, developing only if the egg is washed into water; miracidium* ***(c)****, sporocyst* ***(d)****, redia* ***(e)****, and cercaria* ***(f)*** *all occur in fresh-water snails;* ***(f)*** *the cercaria exits from the snail and uses its muscular tail to move through the water to emergent vegetation at the shore;* ***(g)*** *a metacercaria encysted on vegetation growing at water's edge; herbivore ingests metacercaria on plant, metacercaria excysts, and develops into an adult.*

some trematodes exhibit great host specificity in that only one species of snail can serve as the first intermediate host. In other trematode species, however, this specificity is not so precise, and several to many snail species can serve as hosts. The miracidium is equipped with an anterior **apical papilla** containing an **apical gland** (Fig. 6.16c). Secretions from this gland plus those of a **penetration gland** lying on either side of the apical gland dissolve the host's tissues in penetration. The miracidium also bears a pair of eye spots, a relatively large bilobed brain, and one to three pairs of flame bulbs. Upon contracting the first compatible host, the miracidium penetrates the integument of the mollusc and sheds its ciliated epidermis. The naked larva or **sporocyst** migrates to the snail's digestive gland or, occasionally, to other organs such as the heart cavity, gonad, or gill chamber.

The sporocyst is elongate, ovate, or even branched and has a hollow brood chamber (Fig. 6.16d). Sporocysts lack digestive, sensory, and reproductive structures, but they are equipped with flame bulbs. In the brood chamber a few to several dozen germinal cells are present, each of which divides mitotically to form a **germ ball** (Fig. 6.16d). This is diploid parthenogenesis or **polyembryony.** Germ balls continue to increase in size, and each develops into a second larval stage, the **redia.** Some sporocysts may even give rise parthenogenetically to second or third generations of sporocysts.

The rediae emerge from the sporocyst brood chamber either via a **birth pore** when one is present or by rupture of the sporocyst body. Rediae are also normally found in the digestive gland but have been noted in gills and in several other host organs. Each redia is elongate and bears a pair of projections, the **ambulatory buds,** near either end (Fig. 6.16e). A mouth, pharynx, and blind-end cecum are present along with a brood chamber and usually a birth pore. Within the brood chamber, up to a dozen or more germ balls are parthenogenetically produced, each of which develops into either the next larval stage, the **cercaria,** or into another redial (daughter) generation. The production of cercariae is a classic example of polyembryony and further amplifies the biotic potential.

The cercariae escape from the redia either through the birth pore or by rupture of the redial body and then from the molluscan host into the water, where they become free swimming. Characteristic of the cercarial stage is a muscular posterior **tail,** which is its principal locomotor organ (Fig. 6.16f). The cercarial alimentary system consists of an anterior mouth surrounded by an oral sucker, pharynx, esophagus, and bifurcated intestine (Fig. 6.16f). Generally present are an acetabulum, a pair of eye spots, penetration glands located at the anterior end, and genital primordia situated near the acetabulum. Also present is a rather complicated protonephridial system with two lateral ducts, a common posterior excretory vesicle, and a nephridiopore in the tail. Numerous subtegumental **cystogenous glands** produce the encysting material as the cercaria encysts.

A cercaria penetrates a suitable second intermediate host, loses its tail, and encysts. In *Fasciola* the cercariae encyst on aquatic vegetation. However, in most digenetic trematodes, encystment occurs in a second intermediate host, which is most likely an arthropod. But the cercariae of some species encyst in other invertebrates, even in vertebrates, or in the slime trail produced by their terrestrial snail host. In still other cases, cercariae encyst in the mantle cavity or within the tissues of its molluscan host. The encysted cercaria, called a **metacercaria,** metamorphoses into an immature adult (Fig. 6.16g). Following encystment, cercarial structures, including the penetration and cystogenous glands and eye spots all disappear, and adult structures not already present soon develop. The final step in the life cycle is for the vegetation containing the encysted metacercariae, or the second intermediate host, to be ingested by the definitive host. In the digestive tract of the final host the metacercaria excysts, migrates to its definitive location, and gradually matures into an adult.

The life cycle of *Fasciola hepatica* appears in Figure 6.16 as that of a typical intestinal trematode. But not all trematodes live in the digestive tract or ducts of auxiliary glands. Many trematode species live as blood flukes in a variety of vertebrates. Undoubtedly the best known and the most serious parasites belong to the Family Schistosomatidae. Three human-infecting species belong to the genus *Schistosoma* and cause a condition called **schistosomiasis.** More than 200 million cases of human schistosomiasis exist today. It ranks only behind malaria as the world's most serious parasitic disease in terms of the degree of host debilitation and in numbers of deaths caused by this small helminth. The accumulation of eggs interferes with the flow of blood, and the eggs cause infection and destroy tissues in such vital organs as kidneys, liver, and lungs. The victim is severely weakened and debilitated and eventually dies if not treated.

Dioecious adults usually live in the inferior mesenteric vein of the definitive host. Each male is equipped with a ventral longitudinal groove into which fits a longer, more slender female (Fig. 6.15a). Following copulation, eggs are released into the blood, carried to the large intestine, penetrate the intestinal wall, and are expelled with the host's feces. Schistosomes lack a high degree of host specificity, so a number of pulmonate snails may serve as hosts for the miracidia. Since the redial stage is absent, sporocysts give rise directly to cercariae. Upon coming in contact with the definitive host the cercariae penetrate, lose their tails, and are termed **schistosomules.** The schistosomules are carried first to the lungs, then to the liver, and finally to the mesenteric veins, where the worms mature sexually.

Another interesting feature of schistosomes is that other members of this family normally infest birds and other mammals. Frequently, the cercariae of these species penetrate the skin of human beings, resulting in dermatitis or "swimmer's itch." People are not suitable definitive hosts for these other schistosomes; but even though the cercariae will perish in human skin, they can cause an itchy rash, edema, and a general or even severe discomfort. Such cercariae can be found in both freshwater and marine habitats. One of the most common schistosome species is a parasite of gulls; larvae live in the common mud snail, *Ilyanassa obsoleta,* and their cercariae are often present in large numbers in estuarine waters at certain times of the year.

Class Cestoda

The cestodes are the most highly specialized of all flatworms. All of the approximately 3500 species are endoparasites and usually reside in the digestive tract of many different vertebrates. Their color is often white with shades of yellow or gray. They are commonly known as tapeworms because they are narrow, thin, and extremely long. Some are very small and range from 1 mm to 10 cm, but some tapeworms are, unbelievably, up to 15 m in length. Many cestodes living in human beings and domesticated animals are regularly in excess of 10 m long, and it is certainly not a very comforting thought to imagine a helminth that long living parasitically in one's intestine!

Two of the most distinguishing cestode features are that they completely lack a mouth and digestive tract and that they have enormous powers of reproduction. How can worms of that length sustain themselves without any vestige of a digestive system? It is very simple—they absorb all their metabolic requirements through the body surface, the tegument. Almost all cestodes are monoecious, and they leave very little doubt that their primary function is reproduction.

There are two subclasses of tapeworms; since they differ substantially, a discussion of each taxon will best serve the treatment of the cestodes.

SUBCLASS CESTODARIA

Representatives of this small group of about 15 species are all endoparasitic in the intestine and coelom of some primitive fishes and occasionally in reptiles. Some may attain a length of up to 35 cm. They possess some trematode features in that only one set of both reproductive systems is present in each animal, some bear suckers similar to those in digenetic trematodes, and their bodies are not subdivided into units or proglottids as in other cestodes (Fig. 6.17a). Yet the complete absence of a digestive system, larval stages similar to those of cestodes, and the presence of mesenchymal muscles all suggest strong phylogenetic affinities with other cestodes.

Details of their life cycles are poorly known. Eggs hatch into a ciliated larval stage, the **lycophore** (Fig. 6.17b), characterized usually by having ten anterior hooks. If an intermediate host is necessary, it is an amphipod as in the life cycle of *Amphilina foliacia*, one of the few known life cycles. Several stages follow within the amphipod; and when the crustacean is eaten by a sturgeon, the larva burrows into the coelom and matures. In several other species, no intermediate host is necessary; the lycophores simply infect other vertebrate hosts.

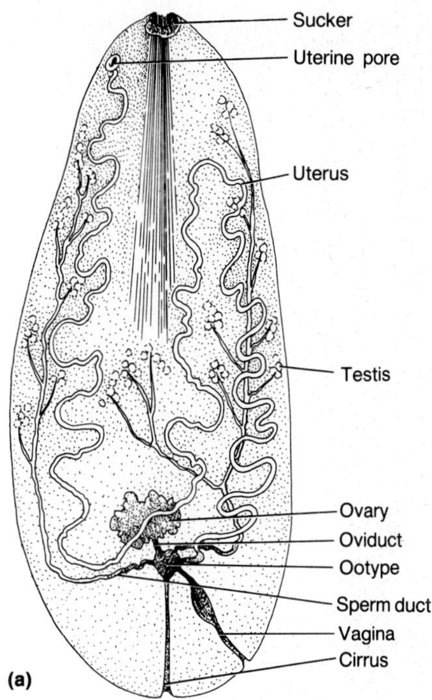

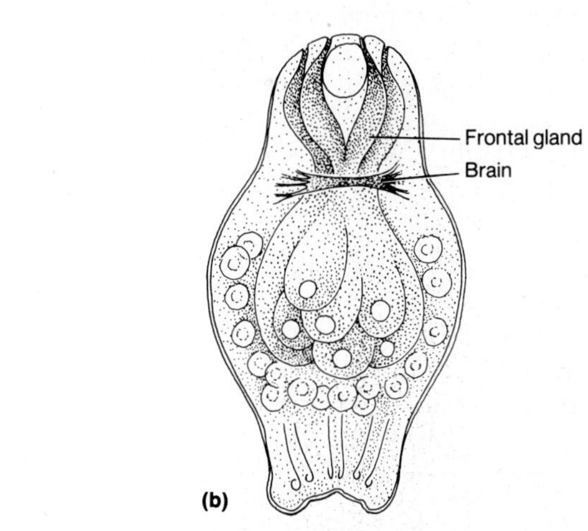

Figure 6.17 *Two principal stages in the life cycle of the cestodarian,* Amphilina: ***(a)*** *a diagram of an adult highlighting both reproductive systems;* ***(b)*** *a lycophore (from Janicki in L. H. Hyman,* The Invertebrates. *Vol. II,* Platyhelminthes and Rhynchocoela, *New York, McGraw-Hill, 1951, adapted with permission).*

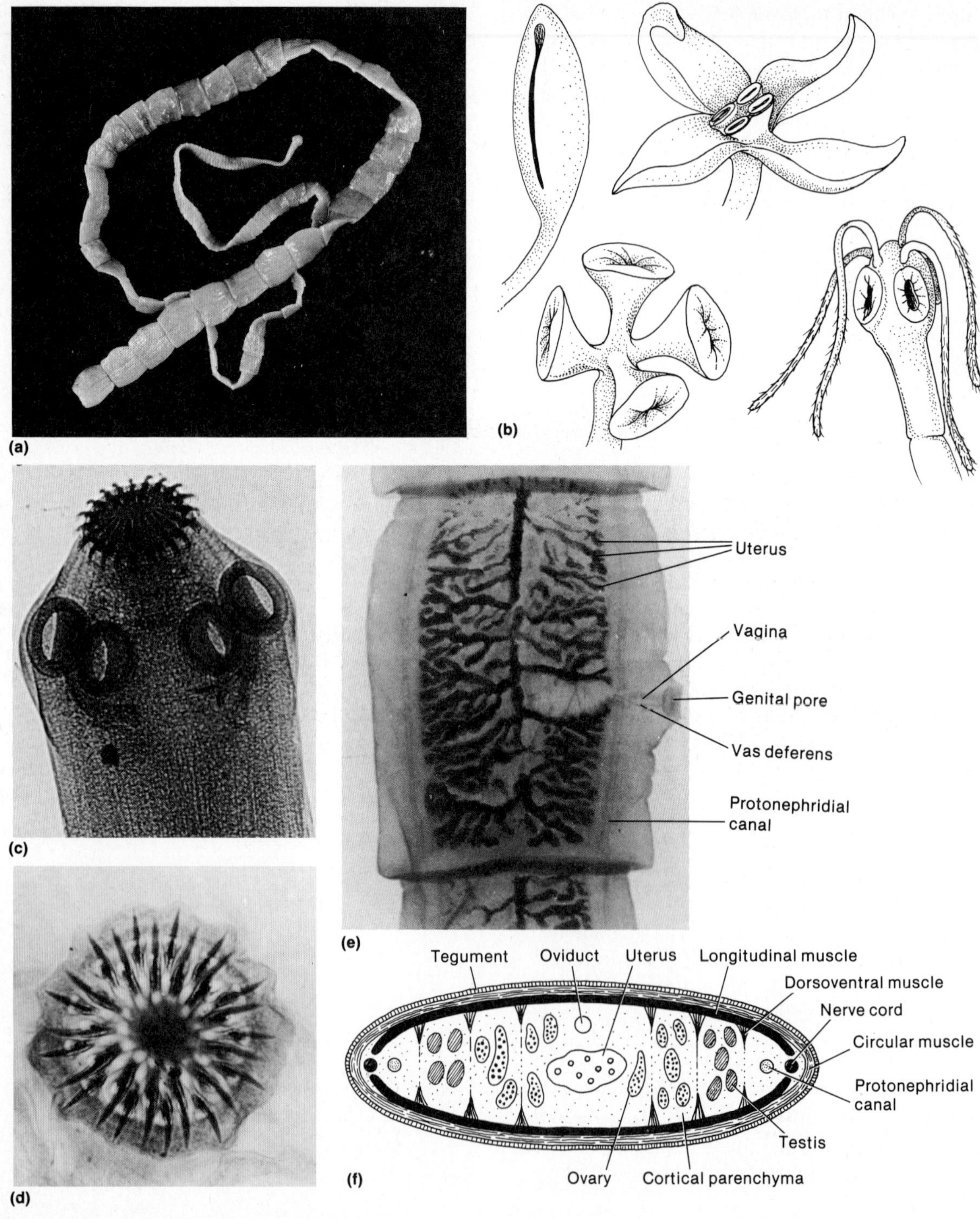
(a)
(b)
(c)
(d)
(e)
Uterus
Vagina
Genital pore
Vas deferens
Protonephridial canal
(f)
Tegument
Oviduct
Uterus
Longitudinal muscle
Dorsoventral muscle
Nerve cord
Circular muscle
Protonephridial canal
Testis
Ovary
Cortical parenchyma

SUBCLASS EUCESTODA

Almost all cestodes, about 3500 species, belong to this subclass. They are known as the true tapeworms. A large and medically important group, they represent the ultimate degree of specialization of any parasitic animal group. The body of a eucestode is divided into a scolex, neck, and strobila (Fig. 6.18a). The **scolex,** the anteriormost part of the worm, represents the attachment mechanism. One or several holdfast devices may be found on the scolex of a given species and include four suckers or **acetabula,** four flaplike outgrowths called **bothridia,** slitlike depressions called **bothria,** and ringed **hooks** borne on an elevation, the **rostellum** (Fig. 6.18b–d). Details of the holdfast structures are used extensively in cestode taxonomy. Burrowing deep into the intestinal mucosa of the host, the scolex frequently causes localized tissue destruction. Behind the scolex is a poorly differentiated region, the **neck,** usually narrower than the scolex and strobila. A short distance from the scolex is a region of the neck that is continuously differentiating into new proglottids (Fig. 6.18a).

The **strobila,** the main portion of the body, is composed of a chain of body units, each termed a **proglottid,** which may number from a few to several thousand (Fig. 6.18a). Since proglottids arise from transverse constrictions in the neck, younger proglottids are continuously being produced at the anterior end, and the older, more mature ones are displaced further away from the scolex and toward the posterior end. The youngest, most anterior proglottids are **immature** in that the reproductive systems are not yet functional. Older, more posterior, **mature** proglottids are fully formed sexually (Fig. 6.18e), while the oldest, **gravid** proglottids are completely filled with eggs. Each proglottid usually contains both male and female systems, and sometimes two complete sets are found in each. In a given proglottid of most eucestodes the male system develops earlier than the female system, producing a protandrous situation. Proglottids are termed segments by a few biologists, and thus tapeworms might be said to be metameric like arthropods and annelids. In these last two groups, however, new metameres are added at the posterior end, whereas in eucestodes they are added at the anterior end. Since metamerism was defined in Chapter 3 as new segments being added at the posterior end, tapeworms should not be considered to be segmented.

◀ *Figure 6.18 Morphology of eucestodes: **(a)** a photograph of an entire worm* (Taenia); ***(b)** four different types of scolices; **(c)** a photomicrograph of the scolex of* Taenia; ***(d)** a photomicrograph of the rostellum of* Taenia; ***(e)** a photomicrograph of a mature proglottid of* Taenia; ***(f)** a cross-sectional view of a mature proglottid (parts c, d courtesy of Ward's Natural Science, Inc., Rochester, N.Y.).*

The outermost layer of eucestodes is a **tegument** that is both morphologically and physiologically like that of trematodes (Figs. 6.10, 6.19). In eucestodes there are a greater number of **microvilli** or microtriches, which increase the absorptive surface and compensate for the lack of a mouth and digestive tract. Also common in the cestode tegument are pinocytic vesicles that take up fluids from the outside. Every component needed for growth and metabolism including amino acids, vitamins, sugars, gases, and water is absorbed by the tapeworm across its tegument. The tapeworm, in reality, parasitizes the fluids within the host's digestive tract rather than the host's tissues per se.

Beneath the tegument and usual platyhelminth muscles is the mesenchyme, which in cestodes contains thin strata of longitudinal, circular, and dorsoventral muscles. Specialized tegumental muscles equip the holdfast structures on the scolex. The osmoregulatory system is a typical protonephridial system. Two or four longitudinal collecting canals run the length of the strobila (Fig. 6.18e, f). In eucestodes that

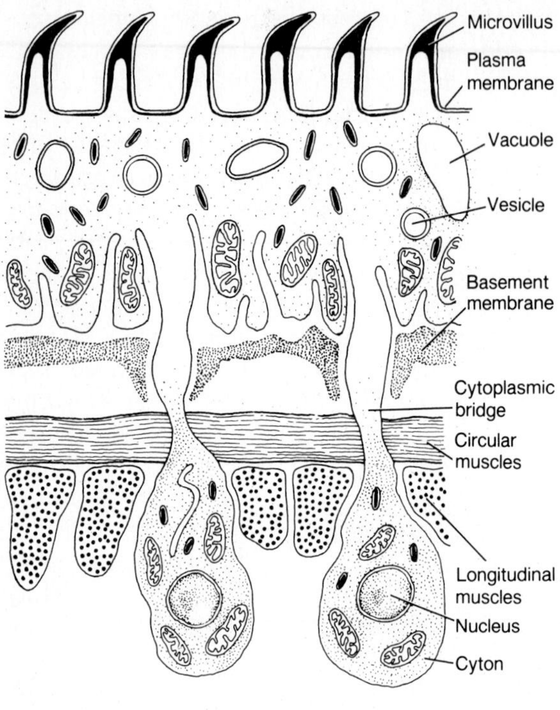

Figure 6.19 *A diagrammatic representation of the cestode tegument (after Beguin [1966] from* The Physiology of Cestodes *by J. D. Smyth. W. H. Freeman and Company., Copyright © 1969).*

are shedding proglottids the longitudinal canals open to the exterior at the exposed posterior ends. The nervous system consists of a bilobed brain or cerebral ganglia in the scolex and two large nerve trunks running the length of the strobila (Fig. 6.18f); ring commissures connect the two trunks in each proglottid. Special nerves run to the holdfast organs. Recently, a cestode neurosecretory system has been discovered that may be involved with synchronizing strobilation.

Members of one genus *(Dioecocestus)* are dioecious, but all other eucestodes are monoecious. The reproductive systems in each proglottid are similar to those found in the digenetic trematodes and will not be repeated here. In eucestodes, self-fertilization within a single proglottid occurs only rarely. Somewhat more common is swapping of sperm between different proglottids in the same strobila, a process called **selfing.** Self-fertilization and selfing are deleterious processes resulting in increased abnormalities in the offspring, and selfing has been shown to produce nonviable progeny after several generations. Certainly the best situation in terms of producing healthy offspring is when cross-fertilization takes place between two different worms, which results in genetically variable individuals.

Sperm transfer is achieved by the cirrus of one proglottid being everted through the genital pore and into the vagina of the other proglottid. The uterus transports the fertilized eggs to the genital atrium, where they are voided, or the uterus may serve to store the eggs temporarily. The posteriormost gravid proglottids break off from the strobila and rupture either within the host's intestine or after they are carried intact to the outside. The reproductive potential of tapeworms is enormous; a beef tapeworm, *Taeniarhynchus saginatus,* living in a human being can produce up to ten proglottids daily, each proglottid containing up to 80,000 eggs!

Life cycles. Life cycles of eucestodes are not as complicated as those of trematodes because the larvae of eucestodes do not usually reproduce asexually. Eucestodes normally require at least one and sometimes two intermediate hosts. Egg development, proceeding usually within the helminth's uterus, results in a larva with three pairs of hooks, called an **oncosphere** or **hexacanth embryo.** The life cycles of eucestodes, deviating significantly from each other, fall into four distinct patterns arbitrarily numbered I, II, III, and IV. Each type, distinguished both by the nature of the larval stages involved and by its overall pattern, will be discussed using a particular species as an example.

Dibothriocephalus latus, the broad fish tapeworm, exemplifies Type I (Fig. 6.20). Adults are found in terrestrial and marine fish-eaters like bears and seals. Eggs escape from the host and develop in water in 8–12 days; each hatches into a free-swimming, ciliated larva, the **coracidium** (Fig. 6.20c). The coracidium must be ingested by a specific copepod (crustacean) within a day or the coracidium dies. Following ingestion, the coracidium loses its ciliation and becomes a naked oncosphere, bores through the intestine, and comes to rest in the coleom of the first intermediate host. There it metamorphoses into another larva, the **procercoid,** which possesses six hooks contained in a caudal protuberance, the **cercomer** (Fig. 6.20e). When the first intermediate host is eaten by a fish, the procercoid bores through the intestine and finally becomes located in muscle tissues of the second intermediate host. In the fish's musculature it transforms into a **pleocercoid,** which bears an inverted scolex, or it may be encysted if it is found

Figure 6.20 *The life cycle of* Dibothriocephalus latus: ***(a)*** *several mature proglottids;* ***(b)*** *an egg;* ***(c)*** *a ciliated coracidium;* ***(d)*** *a naked oncosphere in a copepod;* ***(e)*** *a procercoid in a copepod;* ***(f)*** *a pleocercoid in the muscle of a fish.*

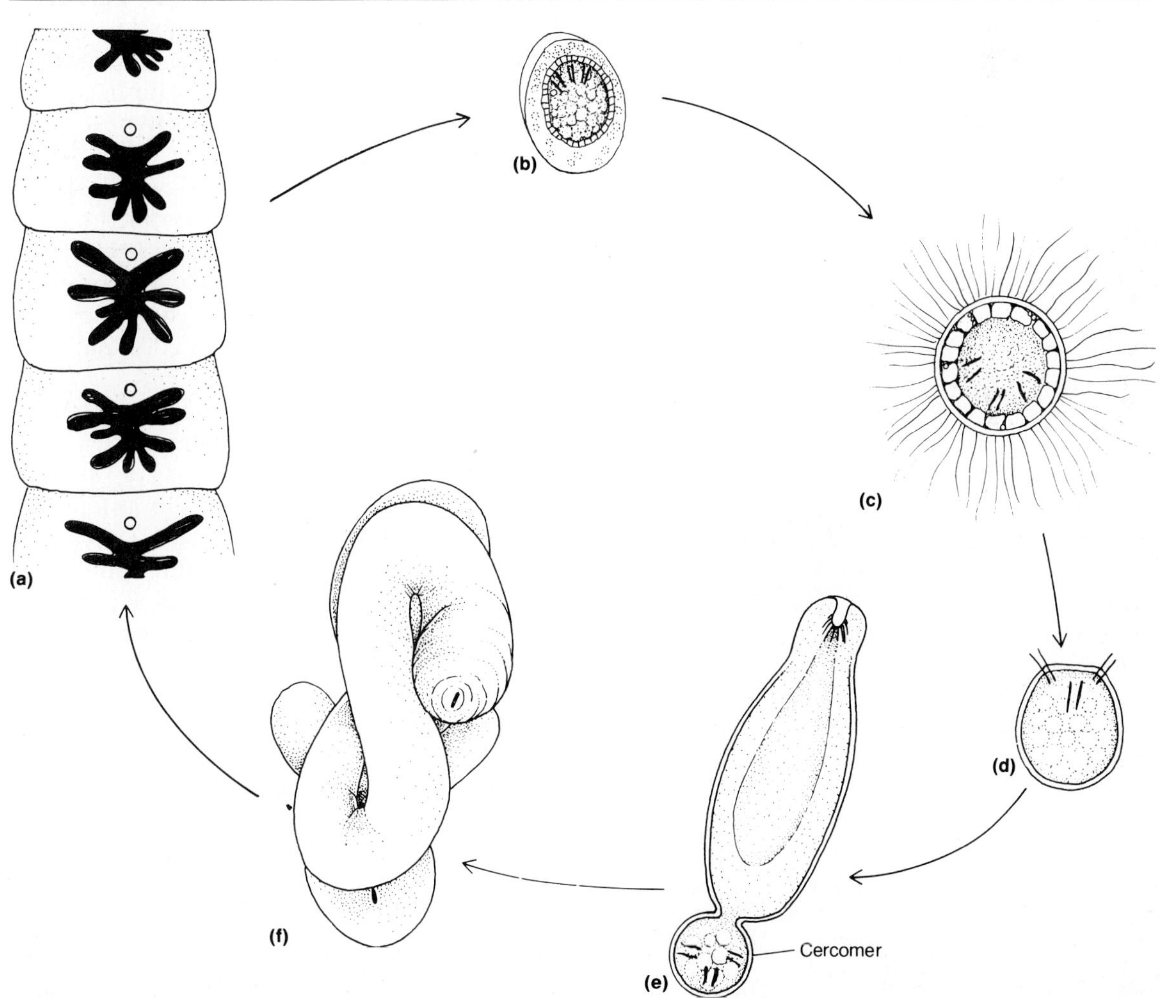

in visceral organs (Fig. 6.20f). When the fish is ingested by the final host, the scolex of the eucestode everts, attaches to the intestinal mucosa of the definitive host, and develops into an adult.

Hymenolepis nana, a parasite of human beings and rodents, is an example of Type II. This species has a one-host life cycle. Eggs containing oncospheres are ingested directly by another suitable host, each egg hatches, and the released oncosphere burrows into the intestinal mucosa of the host and develops into a larva, the **cysticercoid** (Fig. 6.21a). Fully developed cysticercoids migrate back to the intestinal lumen and develop into adults. A closely related species, *H. diminuta,* has a similar life cycle except that several different insects eat the eggs and serve as intermediate hosts in which the eggs develop into cysticercoid larvae. The intermediate host must obviously then be ingested by the definitive host for the life cycle to be completed.

Figure 6.21 *Immature stages of several different cestodes: **(a)** a cysticercoid of* Hymenolepis diminuta; ***(b)** a cysticercoid of* Dipylidium caninum; ***(c)** a photomicrograph of a cysticercus of* Taeniarhynchus.

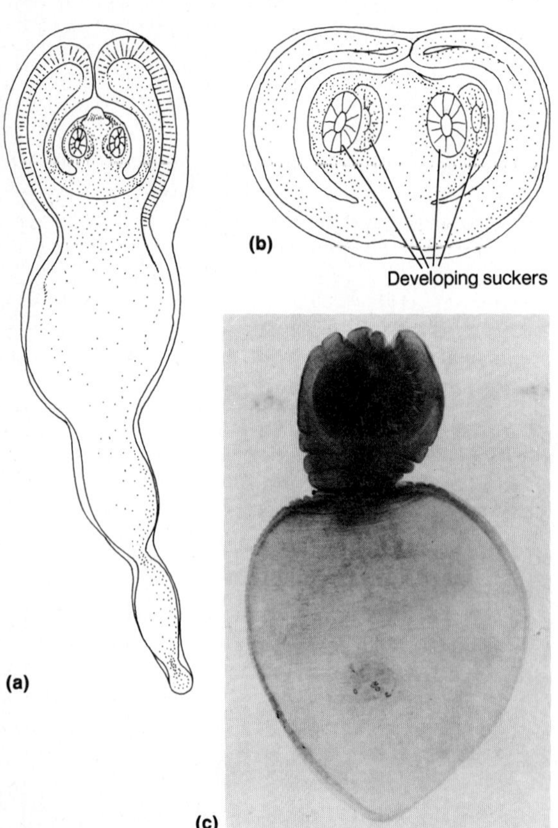

A modification of Type II life cycle is found in *Taeniarhynchus saginatus,* perhaps the most common of the larger human tapeworms, which is commonly called the beef tapeworm because cows serve as the intermediate host. Eggs within intact gravid proglottids are released by the host, and either entire proglottids or eggs are ingested by a cow. The hexacanth embryo emerges from the egg, burrows through the intestinal wall, is carried by the blood or lymph to muscles, and there develops into a **cysticerus** (Fig. 6.21c). It is often called a **bladderworm** because of the bladderlike cyst containing an inverted scolex. When improperly cooked, infected beef is eaten, the scolex evaginates, attaches to the human intestinal wall, and matures into an adult.

Dipylidium caninum, an example of Type III, is found in dogs, cats, and human beings. Eggs are ingested by flea larvae or by lice, and oncospheres hatch in the insect's intestine and burrow through the intestinal wall. When the flea metamorphoses into an adult, the oncospheres develop into **cysticercoids** (Fig. 6.21b). Ingestion of the flea, or when the pet has just nipped an infested flea and licks a child's mouth, completes the life cycle.

The hydatid worm, *Echinococcus granulosus,* illustrates the Type IV life cycle (Fig. 6.22). Adults primarily parasitize dogs and other canines, but larvae are occasionally found in people. The epidemiology of this species in people often goes undetected until major surgery is required to remove the very large cysts. Often, its presence results in death, so this worm

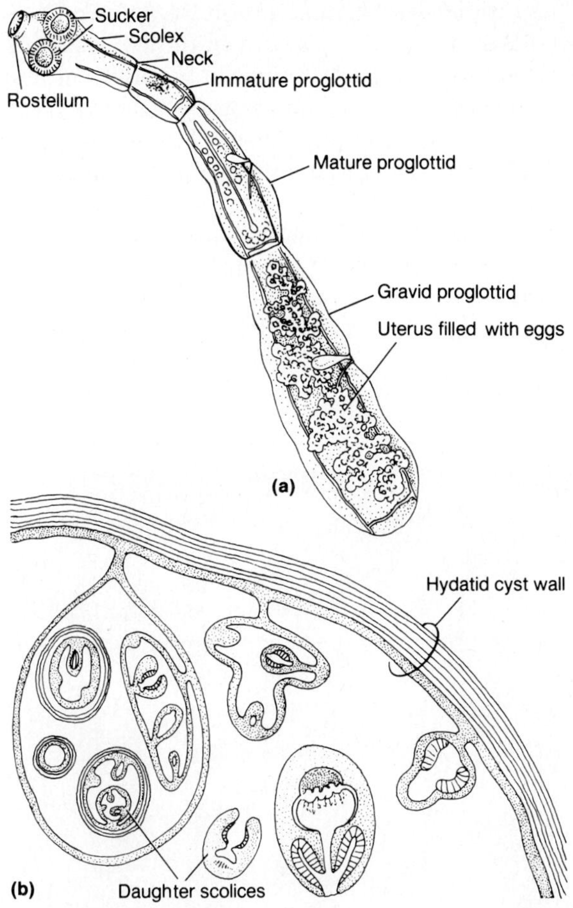

Figure 6.22 *The principal stages in the life cycle of* Echinococcus granulosus: ***(a)*** *an adult (from Southwell in L. H. Hyman,* The Invertebrates. *Vol. II,* Platyhelminthes and Rhynchocoela, *New York, McGraw-Hill, 1951, adapted with permission);* ***(b)*** *a diagram of a portion of a hydatid cyst in an intermediate host showing daughter and granddaughter scolices cut in various planes.*

should definitely be avoided. Adults live in the intestine of the final host, and eggs escape from the canine and enter an intermediate host, which can be a wide variety of mammals through ingested food or water. In the small intestine the hatched oncospheres burrow through the intestinal wall and migrate via the circulatory system to the liver, lungs, brain, and other organs. Each oncosphere develops into a **hydatid cyst** that, at this point, is about 1 cm in diameter. The walls of the cyst invaginate to produce **daughter (secondary) cysts** (Fig. 6.22b); in some daughter cysts, **granddaughter (tertiary) cysts** are formed. Each individual cyst contains a central cavity, the **brood chamber,** in which asexually produced, minute, inverted scolices project into the brood chamber. Thus what is formed is a large, growing, fluid-filled hydatid cyst often as large as an orange and containing up to three generations of budding scolices numbering in the hundreds or even thousands. Older hydatid cysts in human beings have been reported to be up to 30 cm in diameter! One can easily see that a cyst this size in any one of the vital organs would most certainly be lethal. The life cycle is completed when the intermediate host or the host's carcass is eaten by a canine. In the intestine of the definitive host the hydatid cyst is fragmented, and each scolex everts and develops into a new worm. Since ingestion of the hydatid cysts is essential for the life cycle to continue, human infestations are normally a blind end for the parasite.

PHYLUM NEMERTEA

The Phylum Nemertea is a group of about 650 species of mostly marine worms that burrow in sandy or muddy bottoms, live in cracks in reefs, or inhabit bivalve beds. They are often called ribbon or nemertine (= thread) worms, since they are greatly elongated and sometimes extraordinarily thin; lengths of up to 30 m have been reported in a worm having a diameter of only 9 mm! Most nemertines are less than 1 cm long and in cross section are flattened or cylindrical. They are also called proboscis worms, since they all possess a remarkable protrusible proboscis primarily used both in food capture and in defense. Several species of *Prostoma* are found in fresh water, and there is a tropical terrestrial genus found in damp habitats, but all the remaining nemerteans are benthic organisms and are found in littoral zones in temperate seas. While some are pale, other nemertines may be brightly colored with shades of red, orange, brown, yellow, or green.

Nemertines possess many flatworm features including mesenchyme between the outer epidermis and that of the digestive tract, creeping locomotion, a ciliated epidermis, a protonephridial system, and a primitive bilateral nervous system. (Fig. 6.23). Even though there are close phylogenetic affinities between them and turbellarians, nemertines possess characteristics substantially different from those of platyhelminths that more than qualify them to be placed in a separate phylum.

Two classes of nemertean worms are generally recognized. The principal feature separating the two classes is in the nature of the proboscis. The Class Anopla contains nemer-

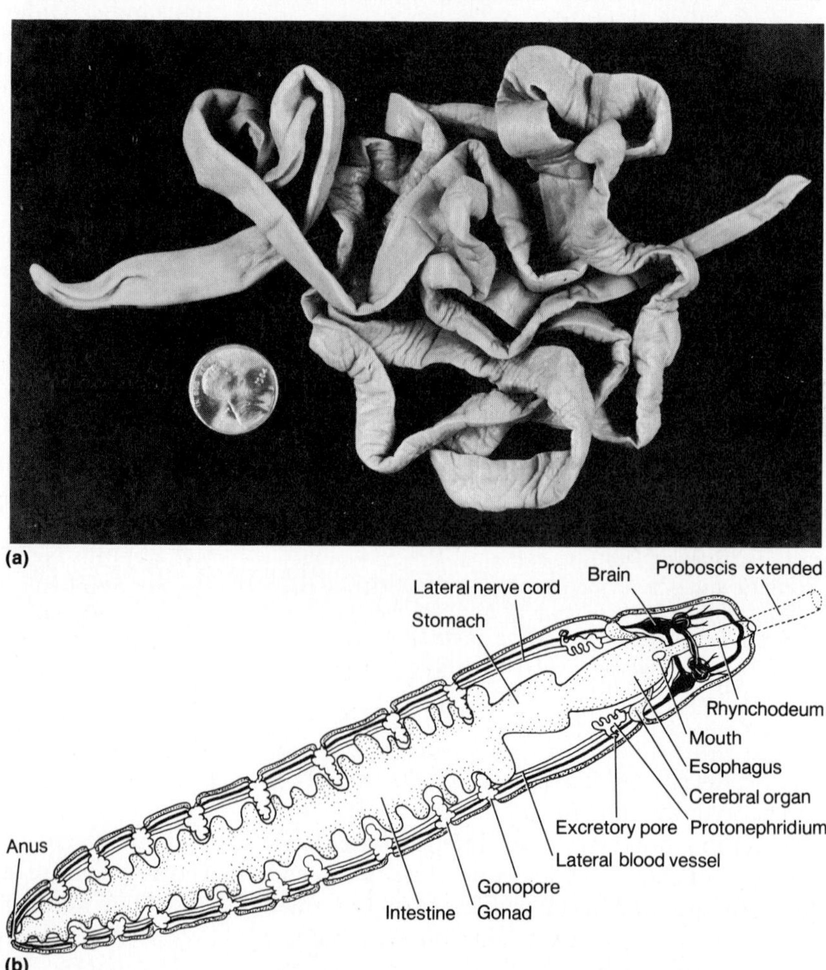

Figure 6.23 *The basic features of nemerteans:* ***(a)*** *a photograph of* Cerebratulus ***(b)*** *a diagrammatic dorsal view illustrating the visceral systems.*

tines whose proboscis lacks a stylet, and the mouth is beneath or posterior to the brain. The Class Enopla includes those nemerteans whose proboscis bears a stylet and whose mouth is situated anterior to the brain. A brief account of the principal nemertean orders appears at the end of this chapter.

The anterior end of a nemertine, with little cephalization, sometimes bears lobes or flaps. The body wall is reminiscent of that in turbellarians but with some added complications. The outer ciliated **epidermis** rests on a stratum of connective tissue, the **dermis** (Fig. 6.24d). Many unicellular mucous glands are found in the epidermis. Most nemertines have a cluster of **cephalic glands** that produce mucus and open near the anterior end; they may be homologous with the frontal gland of turbellarians. Beneath the dermis are found an outer **circular** and an inner **longitudinal muscle layer** (Fig. 6.24d). In one order (Heteronemertea) there is a middle circular layer located between two longitudinal layers. The mesenchyme occupies all the area between the intestine and body wall. Body muscles increase at the expense of mesenchyme; the latter may be much reduced or absent in those worms whose body muscles are highly developed. Locomotion is essentially like that in turbellarians as ciliary action, occurring in a mucous sheet, and the well-developed body muscles make possible many movements including undulations that enable some to swim.

Feeding and Digestion

The most diagnostic feature of these worms is their **proboscis.** Unlike that found in other invertebrate groups, the nemertine proboscis is situated dorsal to the alimentary canal and typically is not directly associated with the digestive tract. Its external opening, the **proboscis pore,** is anterior to the mouth, is slightly subterminal, and opens into a short canal, the **rhynchodeum,** which extends posteriorly to about the level of the brain (Fig. 6.24a). The walls of the rhynchodeum develop from an ectodermal invagination and are lined with epidermis. The epidermis and the rhynchodeal wall sometimes contain rhabdites like those in turbellarian epidermal layers—further evidence of their probable common ancestry.

The rhynchodeum is continuous posteriorly as the lumen of the proboscis (Fig. 6.24a). The proboscis is a long, often coiled tube lying in a completely enclosed internal cavity, the **rhynchocoel** (Fig. 6.24a). This cavity is of mesodermal origin and could be considered anatomically as a coelom, but in no way does it function as the coelom does in higher animals. A blind-end tube with glandular cells lining its lumen, the proboscis, often longer than the worm itself (Fig. 6.24b), is equipped in nemertines of the Class Enopla with a **stylet** or **barb** located in the proboscis wall. The stylet is located at a point about two-thirds of the way back from the proboscis pore and is associated with a swelling called the **diaphragm.** Extension of the proboscis is usually achieved hydrostatically when muscle contraction in the rhynchocoel wall exerts pressure on the rhynchocoel fluid, which in turn everts the proboscis; often, it is literally shot out at great speed. The armed harpoonlike proboscis stabs its prey, sometimes repeatedly. The stylet may also be equipped with glands producing toxic substances used to immobilize the prey. A **retractor muscle** runs from the posterior end of the rhynchocoel to the inner end of the proboscis and withdraws the protracted proboscis (Fig. 6.24b). The proboscis is never completely everted because of the retractor muscle. In nemertines with a stylet, only about two-thirds of the proboscis is extended so that when it is fully extended, the stylet is at the distal end. Prey organisms are brought toward the mouth as the proboscis is retracted. In the Class Anopla the proboscis coils around the prey, and glandular secretions from cells lining the proboscis lumen, now on the outside of the everted proboscis, assist in holding the prey.

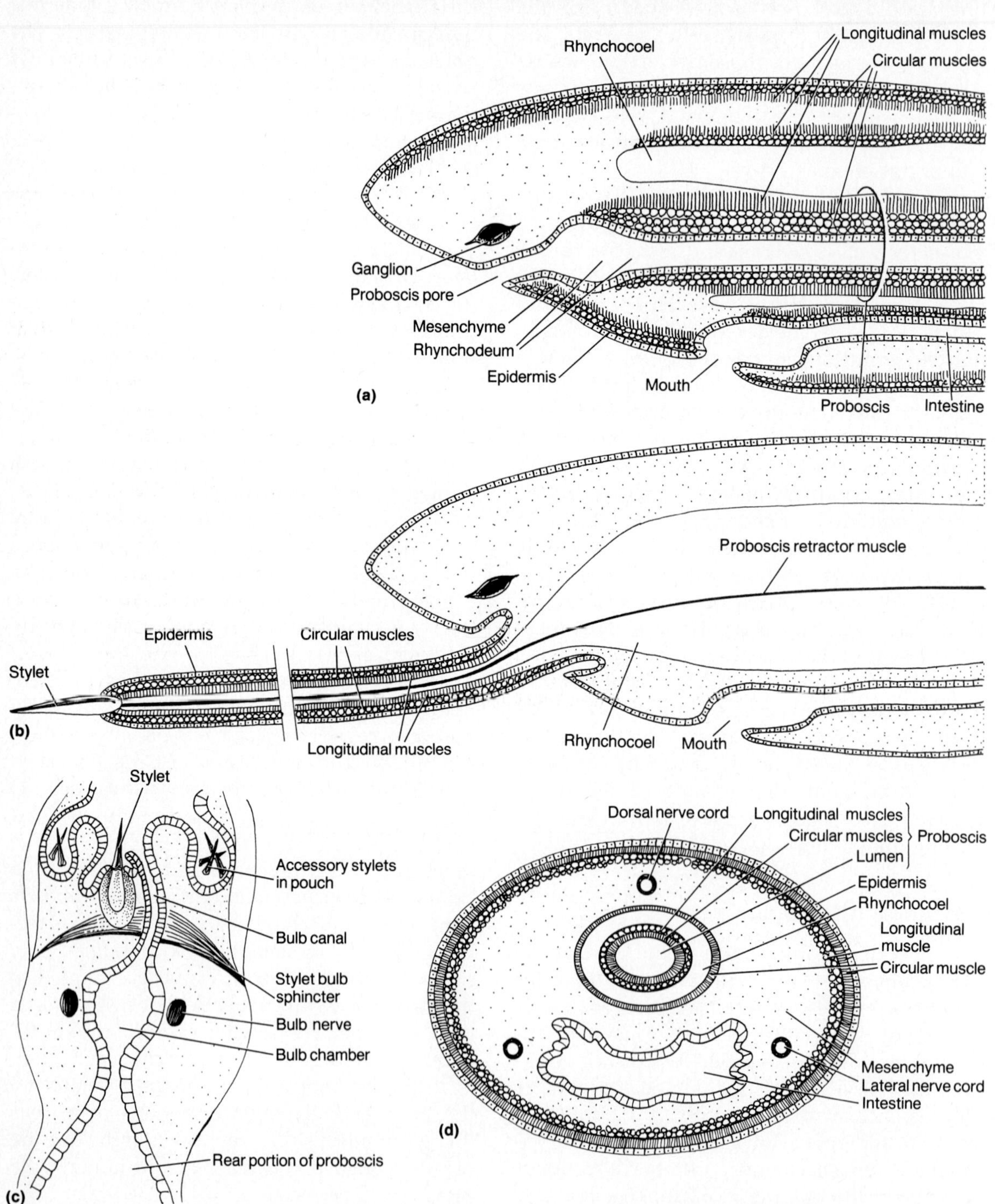
Rhynchocoel
Longitudinal muscles
Circular muscles
Ganglion
Proboscis pore
Mesenchyme
Rhynchodeum
Epidermis
Mouth
Proboscis
Intestine
(a)
Proboscis retractor muscle
Epidermis
Circular muscles
Stylet
(b)
Longitudinal muscles
Rhynchocoel
Mouth
Stylet
Accessory stylets in pouch
Bulb canal
Stylet bulb sphincter
Bulb nerve
Bulb chamber
Rear portion of proboscis
(c)
Dorsal nerve cord
Longitudinal muscles
Circular muscles
Lumen
Proboscis
Epidermis
Rhynchocoel
Longitudinal muscle
Circular muscle
Mesenchyme
Lateral nerve cord
Intestine
(d)

◀ *Figure 6.24 Details of the nemertean proboscis: **(a)** a longitudinal section through the anterior end of* Prostoma; ***(b)** a diagram of the proboscis extended; **(c)** the bulbous portion of the proboscis of* Prostoma; ***(d)** a cross-sectional view of* Tubulanus.

Nemertines are exclusively carnivorous and usually feed on annelids. They will eat anything reasonably small and usually moving, but some feed on dead invertebrates. Food is ingested whole, and digestion begins extracellularly in the gut but is completed intracellularly in phagocytic cells lining the intestinal lumen. Like turbellarians, nemertines can withstand long periods of starvation to the extent that the body size may be reduced to 5% of its original size.

The digestive system in nemertines is significantly more advanced than that in flatworms because it has an **anal opening** at the posterior end (Fig. 6.23b). Thus for the first time in animals we have explored, ingestion is completely separated spatially from egestion by a tubular gut in which digestion and absorption can occur sequentially from anterior to posterior. The ciliated digestive tract consists of an anteroventral mouth, buccal cavity, esophagus, glandular stomach, and long intestine terminating in the anus. In some nemertean worms the esophagus almost incidentally opens into the rhynchodeum, and there is no separate mouth. In many nemertines the intestine bears lateral diverticula or one to several large ceca.

Circulation and Excretion

Nemertines are unique among the acoelomates in that they have a well-defined, closed circulatory system complete with blood. Basically, the system consists of three longitudinal trunks: a **lateral vessel** on either side of the digestive tract and a **middorsal vessel** (absent in the Palaeonemertea) (Fig. 6.23b). These vessels, connected by cephalic and anal lacunae (lined mesodermal spaces), are contractile, but blood flows in no definite pattern as its direction is influenced by muscle contractions in the walls of both vessels and the body. Blood is usually colorless and contains amebocytes and nucleated corpuscles, which may contain one of several different pigments including hemoglobin.

One pair of protonephridial tubules is present, one tubule running posteriorly on either side of the foregut and ending in a **nephridiopore** some distance from the anterior end. Flame bulbs, usually restricted to the anterior part of the worm, are closely associated with the lateral blood vessels (Fig. 6.23b). In some the flame bulbs push in and indent the blood vessel wall, and in others the blood vessel wall has disappeared so that the bulbs are bathed directly with blood. This intimate association between the flame bulbs and blood indicates that the protonephridia probably function in both excretion and osmoregulation. Some nemerteans have a nephridial gland composed of a mass of protonephridial capillaries that have lost their flame bulbs. Some worms have flame bulbs associated with the digestive tract and with most other body tissues. In other nemertines the flame bulbs tend to open separately by individual nephridiopores, which sometimes number in the thousands.

Nervous System and Sense Organs

The nervous system is somewhat similar to that in turbellarians but with some notable exceptions. The four-lobed **brain** consists of a dorsal and ventral (lateral) ganglion on either side of the rhynchodeum (Fig. 6.23b). Dorsal and ventral commissures connect dorsal and ventral ganglia so that the rhynchodeum is surrounded by a nerve ring. A number of longitudinal nerve trunks are present, suggestive of the almost radial arrangement of these cords in some primitive turbellarians. The most prominent trunks are a pair of lateral cords originating from the ventral ganglia and a single, middorsal cord arising from the dorsal commissure, and from each arise many minute branches.

Sense organs include tactile cells, statocysts, inverse eyes, and chemoreceptors, all similar to those in turbellarians and generally situated toward the anterior end. Most nemertines bear a pair of **cerebral organs,** and each is composed of an invaginated epidermal tube or canal whose inner end, containing glandular and nerve cells, is in close proximity to one of the dorsal ganglia. The cephalic organs are richly innervated and are probably chemotactile, although they may have an endocrine function.

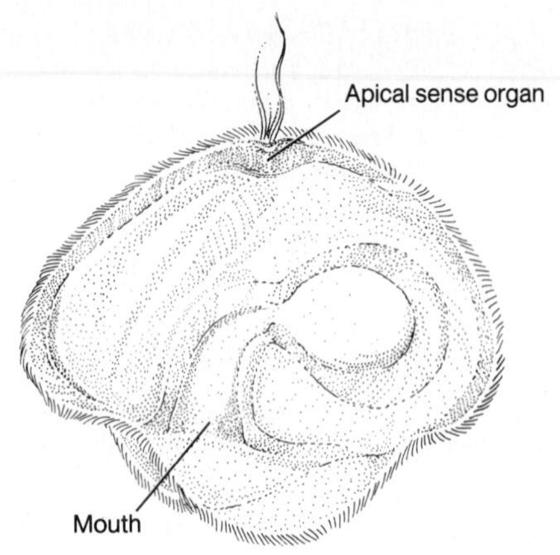

Figure 6.25 The pilidium larva of a nemertean (from L. H. Hyman, The Invertebrates. *Vol. II,* Platyhelminthes and Rhynchocoela, *New York, McGraw-Hill, 1951, adapted with permission).*

Reproduction and Development

Nemertines typically are dioecious, but the nonmarine forms are usually monoecious and often are self-fertilizing. Gametes originate from mesenchymal cells that aggregate to form one to many saclike gonads situated on either side of the intestine and alternating with the intestinal diverticula (Fig. 6.23b). When mature each gonad produces a short duct to the outside, and gametes are squeezed out by body contractions. Fertilization occurs externally either with or without contact between a male and female. Eggs are often spawned in strings of gelatinous material produced by epidermal glands. *Prostoma* is unique among nemertines in that it is viviparous.

Cleavage in nemerteans is spiral and determinate, and development is usually direct. In some heteronemerteans there develops a free-swimming, planktonic, ciliated, feeding larva, the **pilidium,** which later metamorphoses into a mature worm (Fig. 6.25).

Like turbellarians, nemertines possess great regenerative abilities. Multipotent mesenchymal cells quickly aggregate at a wound, form a blastema, and subsequently differentiate into any missing organ. The primary organization center is associated with the nerve cords, which apparently activate the regenerative cells and control their movements and development.

PHYLUM GNATHOSTOMULIDA

This recently discovered group of about 100 species of minute worms has a worldwide distribution, but they are especially abundant on the Atlantic coast of North America. Characteristically found in anaerobic mud and sand interstices in intertidal regions, they are easily overlooked because of their small size.

They are generally less than 1 mm in length, elongate, cylindrical, and mostly transparent. There is an anterior **head,** a constricted **neck,** and a **trunk** that terminates in a tapered **tail** (Fig. 6.26a). The body wall is ciliated, each epidermal cell having a single cilium whose beat is reversible. Scattered mesenchymal cells and delicate longitudinal and circular muscle strands are located between the epidermis and gut. Locomotion is by ciliary action, and body muscles produce wormlike movements.

A ventral **mouth** at the level of the neck is equipped with a pair of lateral **jaws** which give this phylum its name (Fig. 6.26b). Food consists of interstitial organisms, chiefly bacteria and fungi, that are scrapped off sand grains and

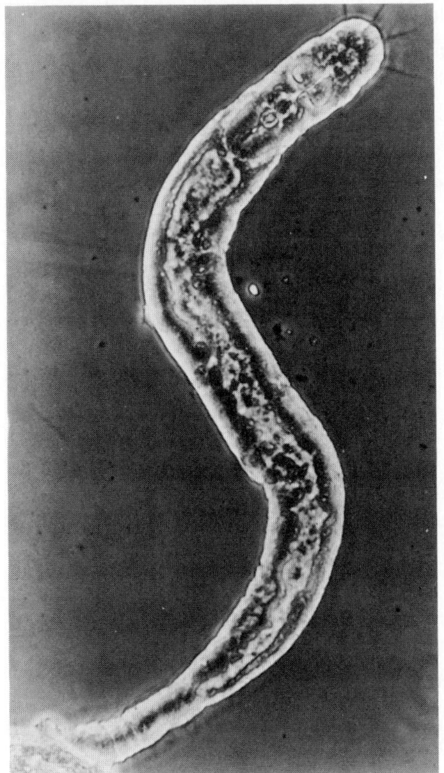

(a)

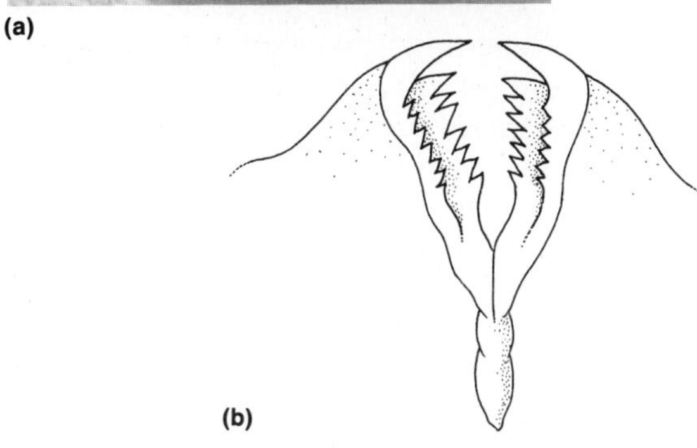

(b)

Figure 6.26 *Gnathostomulid structures:* ***(a)*** *a photomicrograph of an adult (courtesy of R. M. Rieger);* ***(b)*** *the paired lateral jaws.*

passed into the blind-end digestive tract. There is an anterior supraesophageal **brain** but no circulatory or protonephridial systems.

Most gnathostomulids are monoecious, although only one system may be functional in an organism. The female system contains a single **ovary** and a **bursa** for sperm reception and storage. The male system contains a pair of **testes** and a **copulatory organ.** Fertilization is followed by the release of a single large egg at each oviposition, usually by rupture of the body wall. Cleavage is spiral, and development is direct.

Gnathostomulids possess many features common to flatworms, but there are other features that are distinctly nonflatworm, including a 9 + 2 pattern of microtubules in flagella and cilia, reduced mesenchyme, and the fact that ciliary beat is reversible. Other features tend to link them to aschelminths such as rotifers and gastrotrichs. They are tentatively considered here as acoelomates, but their eventual phylogenetic affinities may be altered.

PHYLOGENY OF THE ACOELOMATES

The acoelomates are of particular evolutionary importance for two reasons. First, they are quite primitive and thus appear to be similar to the ancestral stock from which all higher metazoans evolved. Second, acoelomates were the first metazoans to develop a number of incipient but fundamental characteristics that were to be of primal importance in the development and evolution of all higher metazoans. The acoelomates are an extremely interesting group phylogenetically because they are the most important transitional group between the simpler protozoans and radiates and the more complex organisms like annelids, arthropods, and molluscs.

There is little doubt among most zoologists that the turbellarians are the most primitive of all acoelomates. Further, turbellarians at the archeophoran level of complexity (including the orders Acoela, Macrostomida, and Catenulida) are considered to be the simplest and most primitive of all turbellarians. But there are sharp differences of opinion both as to the nature and features of the ancestral turbellarian and as to which of the modern-day groups is most primitive. Let's digress for a moment to recall some details of the probable ancestral turbellarian. In Chapter 3 the idea was discussed that the ancestral metazoan stock is thought to have been a planuloid stage. This primitive stage is thought to have been multicellular, ovoid, ciliated, and radially symmetrical. One of the very early steps in the evolution of acoelomates (and indeed that of all bilaterates) was the development of bilateral symmetry. The gradual ontogeny of such a plan of symmetry had to be directly associated with those of locomotion, mesoderm and muscles, central nervous system, and sense organs. But since the fossil record of these stages is almost completely lacking, we can only advance a conjecture about the various phylogenetic steps.

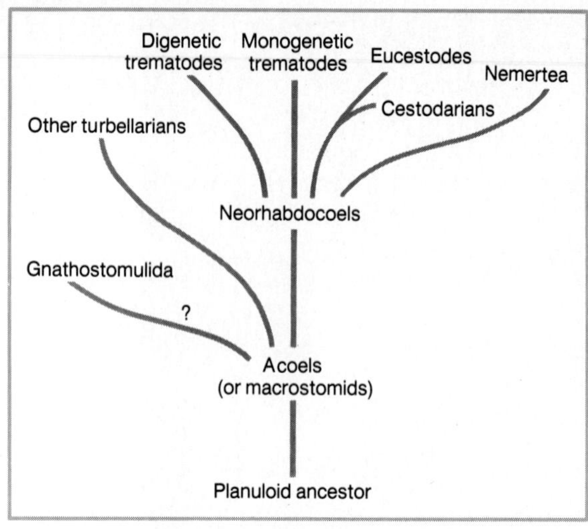

Figure 6.27 *Probable phylogenetic pathways in the acoelomates.*

Equally speculative are the characteristics of the very early turbellarians, and two different hypotheses have been advanced to answer this important and unresolved problem. One theory holds that the Macrostomida are probably the closest to the ancestral stock. Such an ancestor would possess a simple, ciliated, saclike intestine without diverticula; a simple pharynx; a radially arranged system of longitudinal nerve cords; and a protonephridial system. The other hypothesis holds that the Acoela are the most primitive of all acoelomates. An acoel ancestor would lack a gut and would have a solid core of nutritive endodermal cells, no protonephridia, and no ducts for the female reproductive system. Perhaps the weight of evidence supports the acoel ancestor theory, but the issue is still very much an open one.

From this ancient, primitive, turbellarian ancestor were evolved all other acoelomates (Fig. 6.27). The trematodes and cestodes probably arose independently from a turbellarian ancestor with each evolving parasitic habits. Many zoologists hold that a neorhabdocoel may have been the most likely precursor to both parasitic groups. In this turbellarian order there is a taxon (Dalyellioida) that contains many commensals and parasites associated with a variety of marine invertebrates. It is now suggested that the neorhabdocoels gave rise to four lines of evolution that culminated in the following groups: digenetic trematodes, monogenetic trematodes, eucestodes, and nemertean worms (Fig. 6.27). It is generally believed that the cestodarians represent an offshoot from the evolutionary line that led to eucestodes.

The dalyellioid neorhabdocoels are chiefly commensalistic or parasitic in molluscs and echinoderms. It is speculated that as vertebrates evolved from the same phylogenetic stem as that of echinoderms, vertebrates became the definitive hosts, and molluscs were retained as intermediate hosts. Alternatively, the primitive host for digenetic trematodes may have been a mollusc and only later were mollusc-eating vertebrates infested. Vertebrates then became the definitive hosts, and molluscs were retained as intermediate hosts. In the evolution of cestodes this primitive life-cycle tie with molluscs was severed, and tapeworms exist in vertebrates with other nonmolluscan animals serving as intermediate hosts.

The nemertean worms probably had a neorhabdocoel ancestor but have evolved into the modern group of worms that are substantially more advanced than turbellarians (Fig. 6.27). The gnathostomulids represent a phylogenetic enigma. While it is likely that they also evolved from a primitive turbellarian, more information will be necessary to determine more precisely their phylogeny.

SYSTEMATIC RESUME

The Major Acoelomate Taxa

PHYLUM PLATYHELMINTHES

Class Turbellaria—mostly free living and aquatic; external surface usually ciliated; predaceous; possess rhabdites, protrusible proboscis, frontal gland, and many mucous glands.

Archeophoran Level—a primitive level of organization; female gonads not divided into germinaria and vitellaria; yolk glands absent; eggs endolecithal; cleavage spiral.

Order Acoela—intestine is absent but with a mouth and a simple pharynx (some); lack protonephridia, distinct gonads, and oviducts; all are small, marine; *Convoluta, Nemertoderma,* and *Polychoerus* (Fig. 6.3a) are representative genera.

Order Catenulida—simple pharynx and unbranched ciliated intestine; animals usually sexually immature, reproduction usually asexual; gonads unpaired; female gonopore absent; two pairs of nerve cords; mostly freshwater; *Stenostomum* (Fig. 6.3d) and *Catenula* are common genera.

Order Macrostomida—simple pharynx and unbranched ciliated intestine; one pair of nerve cords; paired gonads with separate gonopores; marine and freshwater; *Macrostomum* and *Microstomum* illustrate this order.

Order Polycladida—have plicate pharynx, intestine with many diverticula, separate gonopores, many eyes, and two nerve cords; all are marine, relatively large, and many are brilliantly colored; *Planocera, Stylochus* (Fig. 6.3c), *Notoplana,* and *Thysanozoon* are often encountered.

Neoophoran Level—more advanced level of organization; female gonad composed of a germinarium and yolk glands; eggs ectolecithal; first cleavage stages spiral, but subsequent development not typically spiral.

Order Prolecithophora—pharynx is plicate or bulbose; intestine lacks diverticula; have a common gonopore; marine and freshwater; a typical genus is *Plagiostomum.*

Order Lecithoepitheliata—have a simple straight intestine and four pairs of longitudinal nerve cords; penis is a stylet; marine and freshwater; *Prorhynchus* is representative.

Order Neorhabdocoela—have bulbous pharynx, saclike intestine with no diverticula, and one pair of nerve cords; marine and freshwater with commensal and parasitic species; *Mesostoma, Gyratrix,* and *Dalyellia* are frequently found.

Order Proseriata—possess plicate pharynx and intestine with lateral diverticula; yolk glands paired and arranged serially among yolk duct; mostly marine; *Archotoplana* and *Tabaota* are typical.

Order Tricladida—have a three-branched intestine; includes the large marine, freshwater, and terrestrial forms; some of the more common genera are *Bdelloura* (Fig. 6.3f), *Dugesia* (Fig. 6.2a), and *Bipalium* (Fig. 6.3e).

Class Trematoda—flukes; all are parasitic; several holdfast devices present; have complicated life cycles.

Subclass Aspidogastrea—mostly endoparasitic in molluscs; possess large opisthaptor; most lack oral sucker; representative genera include *Aspidogaster* and *Cotylaspis.*

Subclass Monogenea—mostly ectoparasitic on ectothermous vertebrates; one life-cycle form in only one host; bear opisthaptor and prohaptor; *Discocotyle* and the endoparasite *Polystoma* (Fig. 6.14) are common examples.

Subclass Digenea—adults endoparasitic in vertebrates; at least two different life-cycle forms in two or more hosts; have oral sucker and acetabulum.

Order Strigeidida—miracidia with two pairs of flame cells; metacercariae found in aquatic animals; adult body divided into a fore- and hind-body; includes *Strigea, Diplostomulum,* the very important *Schistosoma* (Fig. 6.15a), and *Spirorchis.*

Order Azygiida—miracidia with one pair of flame cells; adults usually monoecious, parasitic in intestine of fishes; typical genera are *Hemiurus* (Fig. 6.15d), *Isoparorchis,* and *Azygia.*

Order Echinostomida—miracidia with one pair of flame cells; adults have a collar of spines around oral sucker; *Echinostoma, Notocotylus,* and *Fasciola* (a familiar liver fluke [Fig. 6.16]) are representative.

Order Plagiorchiida—miracidia with one pair of flame cells; adults are mostly intestinal parasites; cercariae with thin straight tails; *Plagiorchis, Cephalogonimus,* and *Prosthogonimus* are rather common.

Order Opisthorchiida—miracidia with one pair of flame cells; in adults, testes posterior to ovary; includes the familiar *Opisthorchis* (Fig. 6.15b), *Heterophyes,* and *Centrocestus.*

Class Cestoda—tapeworms; all endoparasitic in vertebrates; no mouth or digestive tract; great reproductive potentials.

Subclass Cestodaria—body not subdivided into proglottids; larva in crustaceans, adult in fishes; *Amphilina* (Fig. 6.17) is representative.

Subclass Eucestoda—true tapeworms; body divided into scolex, neck, strobila; strobila composed of many proglottids; each proglottid contains both reproductive systems; adults in digestive tracts of vertebrates.

Order Proteocephalidea—adults in freshwater fishes, amphibians, reptiles; procercoid in crustaceans, pleocercoid in other animals; *Proteocephalus* is representative.

Order Tetraphyllidea—adults in elasmobranch fishes; procercoid in crustaceans, pleocercoid in fishes; scolex has four bothridia; a common genus is *Phyllobothrium.*

Order Lecanicephalidea—adults in elasmobranch fishes; larvae in marine molluscs; scolex without bothria and bothridia; *Tetragonocephalum* is an example.

Order Pseudophyllidea—adults in fishes and mammals; procercoid in crustaceans; pleocercoid in vertebrates; scolex with bothria; *Dibothriocephalus* (Fig. 6.20) and *Spirometra* are well known.

Order Trypanorhyncha—adults in elasmobranch fishes; coracidia in crustaceans; procercoids in fishes; scolex with four spiny proboscides; *Grillotia* is representative.

Order Cyclophyllidea—largest and best-known order; adults in mammals or other vertebrates; larvae in many different animals; scolex with four acetabula and some with rostellum; well-known genera are *Dipylidium, Echinococcus* (Fig. 6.22), *Taeniarhynchus, Taenia* (Fig. 6.18a), and *Hymenolepis.*

Order Aporidea—adults in birds; larvae are not known for the two genera; scolex large and prominent; strobila not externally divided; *Gastrotaenia* is one of only two genera.

Order Diphyllidea—adults in elasmobranch fishes; scolex with two bothridia and small apical organ; *Echinobothrium* is representative.

Order Caryophyllidea—adults in fishes and annelids; strobila without proglottids; only one set of reproductive organs; *Archigetes* is well known.

Order Spathebothriidea—adults in marine fishes; procercoids in crustaceans and fishes; strobila lacks external divisions; includes *Schizocotyle.*

PHYLUM NEMERTEA

Class Anopla—proboscis without stylet; mouth is posterior to the brain.

Order Palaeonemertea—lateral nerves lie outside of muscles; eyes, cerebral organs, dorsal blood vessel are absent; some of the more common genera are *Tubulanus, Cephalothrix,* and *Carinina.*

Order Heteronemertea—circular muscle layer between two longitudinal layers; cerebral organs, eyes, dorsal blood vessel, intestinal diverticula are present; *Cerebratulus*

(Fig. 6.23a), *Lineus*, and *Micrura* are representative.

Class Enopla—proboscis almost always with stylet; mouth is anterior to brain and often opens into rhynchodeum.

Order Hoplonemertea—all have proboscidian stylet; intestine with lateral diverticula; dorsal blood vessel present; *Geonemertes*, *Amphiporus*, *Prostoma*, and *Tetrastemma* are common.

Order Bdellonemertea—no proboscis stylet, intestinal diverticula, eyes, or cerebral organs; all are commensal; one genus, *Malacobdella*.

PHYLUM GNATHOSTOMULIDA

ADDITIONAL READINGS

Aria, H.P., ed. 1980. *Biology of the Tapeworm* Hymenolepis diminuta, 733 pp. New York: Academic Press.

Arme, C. and P. Pappas, eds. 1983. *Biology of the Eucestoda*. Vol. 1, 390 pp., Vol. 2, 386 pp. New York: Academic Press.

Ball, I.R. and T.B. Reynoldson. 1981. *British Planarians: Synopsis of the British Fauna 19*, 141 pp. New York: Cambridge University Press.

Bayer, F.M. and H.B. Owre. 1968. *The Free-living Lower Invertebrates*, pp. 144–204. New York: Macmillan Co.

Best, J.B. 1963. Protopsychology. *Sci. Amer. 208*:54–62.

Cheng, T.C. 1973. *General Parasitology*, pp. 3–119, 323–541. New York: Academic Press.

Crowe, J.H., L.M. Crowe, P. Roe, and D. Wickham. 1982. Uptake of DOM by nemertean worms: Association of worms with arthrodial membranes. *Amer. Zool. 22*:671–682.

Erasmus, D.A. 1972. *The Biology of Trematoda*, 312 pp. New York: Crane, Russak, and Co.

Gibson, R. 1972. *Nemerteans*, 224 pp. London: Hutchinson University Library.

Gibson, R. 1982. *British Nemerteans: Keys and Notes for the Identification of the Species*, 212 pp. New York: Cambridge University Press.

Hyman, L. H. *The Invertebrates*. Vol. II: *Platyhelminthes and Rhynchocoela*, 531 pp., 1951. Vol. V: *Smaller Coelomate Groups*, pp. 731–739, 1959. New York: McGraw-Hill Book Co.

Kaestner, A. 1967. *Invertebrate Zoology*. Vol. I, pp. 154–200, 206–216. New York: Interscience.

Karp, G. and N.J. Berrill. 1981. *Development*, 2nd ed., pp. 653–663. New York: McGraw-Hill Book Co.

Kennedy, C.R., ed. 1976. *Ecological Aspects of Parasitology*. 462 pp. Amsterdam: North Holland.

MacGinitie, G.E. and N. MacGinitie. 1968. *Natural History of Marine Animals*, 2nd ed., 498 pp. New York: McGraw-Hill Book Co.

Pennak, R.W. 1978. *Fresh-water Invertebrates of the United States*, 2nd ed., pp. 114–146. New York: Ronald Press.

Riser, N.W. and M.P. Morse, eds. 1974. *Biology of the Turbellaria*, 530 pp. New York: McGraw-Hill Book Co.

Russell-Hunter, W.D. 1968. *A Biology of Lower Invertebrates*, pp. 66–86. New York: Macmillan Co.

Schockaert, E.R. and I.R. Ball, eds. 1980. *The Biology of the Turbellaria*, 302 pp. The Hague, Netherlands: Junk.

Smyth, J.D. 1969. *The Physiology of Cestodes*, 254 pp. San Francisco: W.H. Freeman and Co.

Von Brand, T. 1979. *Biochemistry and Physiology of Endoparasites*, 417 pp. New York: Elsevier/North-Holland Biomedical Press.

Wardle, R.A., J.A. McLeod, and S. Radinovsky. 1974. *Advances in the Zoology of Tapeworms, 1950–1970*, 259 pp. Minneapolis: University of Minnesota Press.

7

The Aschelminths (Pseudocoelomates)

OVERVIEW

There are seven aschelminth phyla: Gastrotricha, Rotifera, Kinorhyncha, Nematoda, Nematomorpha, Acanthocephala, and Loricifera. Most have had a separate evolutionary history; thus this grouping is one mostly of convenience. The principal unifying aschelminth feature is a pseudocoelom—a perivisceral, highly variable space that is not homologous in all taxa. Other generally occurring aschelminth features include a thin-walled complete digestive tract, a muscular pharynx, protonephridia, constant cell numbers, an outer cuticle, and adhesive glands.

Gastrotrichs are microscopic and aquatic, and each is composed of a head lobe, neck, trunk, and furca. Numerous adhesive tubes with adhesive glands are prominent. They lack a distinct pseudocoelom. Since males are rare, female parthenogenesis is common.

Rotifers, found mostly in fresh water, are planktonic or sessile; the sedentary forms often secrete an outer protective tube. The head bears a unique, highly specialized, ciliated corona employed in both locomotion and food capture. The trunk is usually saclike, and the foot is primarily an adhesive organ. A characteristic feature is a feeding and masticating mastax. Males are either small or unknown for most taxa. Females develop parthenogenetically or from fertilized eggs, and males develop from haploid unfertilized eggs.

Kinorhynchs are minute wormlike creatures living in littoral marine areas. The body is composed of 13 zonites embellished with cuticular scales, plates, and spines. The head or zonite I bears an oral cone, the neck is zonite II, and the trunk is composed of 11 zonites. Kinorhynchs produce a nonzonited larva.

Nematodes live in all aquatic and terrestrial habitats; many are parasitic, and some are of considerable medical importance. Amid their enormous ecological diversity, they are morphologically very similar, since all are slender, elongate, and circular in cross section. They possess a highly modifiable cuticle that is molted periodically, a capacious pseudocoelom, and a pumping pharynx used in feeding. Excretion and osmoregulation are achieved by unique glandular or tubular systems. Sexual dimorphism is pronounced.

Adult nematomorphans are threadlike and free living in fresh water, but they lack a functional alimentary canal. The small larvae, which are obligate parasites in arthropods, exit from their host and into the water to mature.

Acanthocephalans are all alimentary parasites in vertebrates. Characteristic of adults is an anterior retractile proboscis, a well-developed pseudocoelom, and the absence of a digestive system. Arthropods serve as intermediate hosts for the several parasitic larval stages.

The Phylum Loricifera was discovered and named only in the summer of 1983, and the phylogenetic position of these creatures is quite tenuous at the present time. Loriciferans are microscopic interstitial animals (meiofauna) that have a spiny head and thorax that is retractable into a lorica surrounding the abdomen; the anterior mouth is surrounded by eight oral stylets. Loriciferans are believed to share some characteristics with priapulids, kinorhynchs, and nematomorphans.

The phylogenies of the heterogeneous aschelminths are difficult to determine; at best we are forced to rely heavily on morphological and functional attributes of living forms. Most groups undoubtedly had separate origins, probably from an acoelid turbellarian that was ciliated, marine, lacking a cuticle, and monoecious. Of the seven phyla the gastrotrichs apparently are the central or most primitive group. Nematodes probably are most closely allied to gastrotrichs, and, in turn, nematomorphans are related to nematodes. The phylogenetic relationships of the rotifers, kinorhynchs, acanthocephalans, and loriciferans are even more obscure.

GENERAL ASCHELMINTH CHARACTERISTICS

The aschelminths are represented by seven phyla: Gastrotricha, Rotifera, Kinorhyncha, Nematoda, Nematomorpha, Acanthocephala, and Loricifera. In the not too distant past, these taxa were treated as subordinate groups within the Phylum Aschelminthes or Nemathelminthes. Recently, however, a general consensus has developed that each is different enough from the others to justify elevating all seven to phylum status. All are characterized in Table 7.1. The nontaxonomic but convenient term **aschelminth** will be used here when referring to the entire group.

The reason for grouping rather disparate taxa together under a common term usually is that they have either a common or a close phylogeny, but that is not exactly the case with aschelminths. As we shall see, some of these groups do not have close ancestral affiliations with the others; the aschelminths are clearly a heterogeneous group with polyphyletic origins. Then are there morphological, functional, or ecological similarities that might justify our lumping them together? There are, in fact, several general features that apply to most aschelminths but that are not invariable or universal in their occurrence. The most important of these characteristics is the nature of the body cavity.

Pseudocoelom

As described in Chapter 3, all higher metazoans have a body cavity situated between the digestive tract and the body wall. But noteworthy is the fact that aschelminths are the first group we have encountered that possess a distinct body cavity. In most animals this space is a coelom, but in aschelminths the body cavity lacks peritoneal linings and mesenteries, so visceral organs are free within the cavity. Such a cavity has been termed a **pseudocoelom** (Fig. 7.1; also see Fig. 3.7b), and aschelminths are commonly referred to as **pseudocoelomates.** Certainly the presence of a pseudocoelom would be a strong unifying feature for all aschelminths but again that is not exactly the case. The pseudocoelom in aschelminths is neither homologous in all groups nor of constant form. In most it repre-

TABLE 7.1. ☐ **CHARACTERIZATIONS OF THE ASCHELMINTH PHYLA**

Phylum	*Common Name*	*Habitat*	*Other Distinguishing Features*
Gastrotricha	Gastrotrichs	Aquatic	Many have adhesive tubes; distinct pseudocoelom is absent; have ventral locomotor cilia, epithelial cuticle
Rotifera	Rotifers, wheel animals	Mostly fresh water	Have ciliated corona and masticating mastax
Kinorhyncha	—	Marine, interstitial	Body of 13 zonites, each with recurved spines; mouth on extensible oral cone
Nematoda	Roundworms	Found everywhere in soils and aquatic habitats; many are parasitic	Possess a complex cuticle, longitudinal body wall muscles, and unique osmoregulatory–excretory systems
Nematomorpha	Horsehair or gordian worms	Adults free living in fresh water; larvae parasitic in arthropods	Adults are elongate, thin, and have a nonfunctional digestive system
Acanthocephala	Spiny-headed worms	All endoparasitic in gut of vertebrates as adults	Adults with spiny, holdfast proboscis; larvae parasitic in arthropods; adults lack a digestive tract
Loricifera	—	Marine interstices	Adults with a spiny head and thorax retractable into a lorica surrounding the abdomen; eight oral stylets

sents the persistent embryonic blastocoel, but in others it is formed in several different ways. Extremes in its variable form are illustrated on one hand by the nematodes, in which it is a large cavity, and on the other by the gastrotrichs, in which the pseudocoelom is absent except for some small intercellular spaces. Another unfortunate fact is that the pseudocoelom must be defined in negative terms, as a body cavity that does not meet the criteria of a true coelom. Nonetheless, the term pseudocoelom will be used here to refer to the noncoelomic, often ill-defined body cavity of aschelminths. The pseudocoelom, often fluid-filled or containing a gelatinous material, serves as a cavity for circulation and as an internal hydrostatic skeleton that functions in locomotion. Both types of body cavities, pseudocoelom and coelom, function primitively in locomotion in the same manner. Both are fluid-filled spaces surrounded by body wall muscles. Alternating contractions on either side of the body exert pressures on the pseudocoelomic fluid to produce the whiplike undulations that prove to be not very efficient for locomotion in these pseudocoelomate worms. The pseudocoelom also, indirectly, aids in digestion. Since the walls of digestive organs lack muscles, peristalsis is impossible. Instead, food is moved through the digestive tract by hydraulic pressures exerted by the pseudocoelomic fluid and generated initially by body wall muscles.

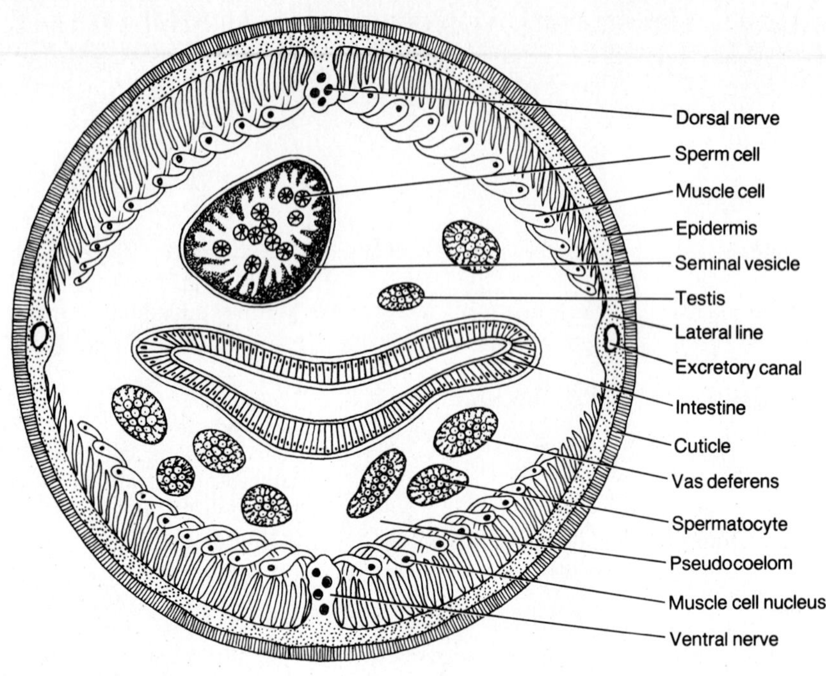

Figure 7.1 *A cross-sectional view of a nematode to illustrate the tissue layers and the pseudocoelom of aschelminths. (Note the absence of peritoneal layers or mesenteries.)*

Other Aschelminth Features

Most aschelminths have a complete, tubular digestive tract extending from the anterior **mouth** to the posterior **anus.** Such a tube-within-a-tube body plan, first encountered in the nemertean worms and characteristic of almost all higher bilaterates, enables mechanical breakdown of food, digestion, absorption, and fecal formation to proceed sequentially from anterior to posterior. The walls of the alimentary tract are usually thin, consist of only a single layer of endodermally derived cells, and mostly lack muscles. Several regional alimentary modifications are present, including the almost universal presence of a specialized muscular **pharynx** adapted for feeding.

An unusual characteristic of many aschelminths is the phenomenon of **eutely.** The eutelic condition is when the cell or nuclear number is constant within a species. (Box 7.1 provides a few more details of this interesting phenomenon.)

An osmoregulatory system of **protonephridia** is found in most aschelminths. There are typically two protonephridial canals, each with a nephridiopore. As in the acoelomates, the protonephridial system, being basically concerned with water balance, is best developed in freshwater aschelminths, in which osmotic problems are the greatest. No separate blood vascular or gas exchange systems are present. The nervous system is relatively simple with a brain located dorsal to the anterior alimentary canal and with one or two ventral nerve cords. Sensory organs are also simple. The vast majority of aschelminths are dioecious, and both reproductive systems are relatively uncomplicated. Cleavage is determinate, and the life cycle is usually simple, though it may be complex in the case of parasitic forms.

For the most part, aschelminths are microscopic, although some grow to a length of a meter or more. They are commonly wormlike, bilaterally symmetrical, unsegmented and cylindrical in cross section. There is some cephalization with the anterior end bearing the mouth, sensory organs, and brain. Cilia are generally absent from external surfaces, and a thin, tough, external cuticle is generally present. The **cuticle,** a scleroprotein, often bears spines, scales, or other ornamentations. Beneath the cuticle is a cellular or syncytial **epidermis** that secretes the cuticle. The epidermis often contains **adhesive glands** whose secretions aid in attachments. Beneath the epidermis are ill-defined muscle layers or bands (Fig. 7.1).

Aschelminths are basically a freshwater assemblage, even though many nematodes are terrestrial and other forms live on land in damp places. Only a limited number are found in marine habitats. Nematomorphans, acanthocephalans, and many nematodes are parasitic. The remainder are free living and solitary, but some rotifers are colonial.

One may rightfully conclude that no unifying features are present in all pseudocoelomates. Most aschelminths are similar in many respects, but in the final analysis and in the absence of close phylogenetic kinships, they are grouped together primarily for the sake of convenience.

BOX 7.1 □ THE PHENOMENON OF EUTELY AMONG ASCHELMINTHS

A very unusual property of many aschelminths is that they are eutelic. **Eutely** is a phenomenon in which numbers of cells, or nuclei in syncytia, are constant both for the entire animal and for a given organ for all the members of that species. For example, the number of somatic cells in adults of the nematode *Caenorhabditis elegans* is about 800, and the number of cells in the pharynx of every worm belonging to this species is precisely 80. However, variations between species may exist. The phenomenon of eutely is perhaps unique to aschelminths and is particularly characteristic of rotifers, gastrotrichs, kinorhynchs, and many nematodes. It is a provocative phenomenon and one that evokes the question, "I wonder why?" The answer is perhaps a structural rather than a functional one; small organisms have fewer numbers of cells, and the numerical range of cellular variability is very narrow in relation to that in larger, higher metazoans.

The fact that many aschelminths are eutelic makes them ideal models for the study of aging and gerontology for several reasons: (1) there is an exact number of cells present, and the lineage of each is known; (2) the animal is devoid of the capacity to repair cells; and (3) no cellular systems are being continually renewed. Cellular longevity appears to be a simple, measurable parameter, and the onset of senescence can be studied easily in eutelic animals. Some preliminary characteristics of aging have already been observed and include a decrease in cellular motility, increase in specific gravity, accumulation of an "age pigment," and progressive disorganization of nerve and muscle cells and their mitochondria. One of these days a lowly simple nematode or rotifer may provide science with many more secrets about the ubiquitous process of aging.

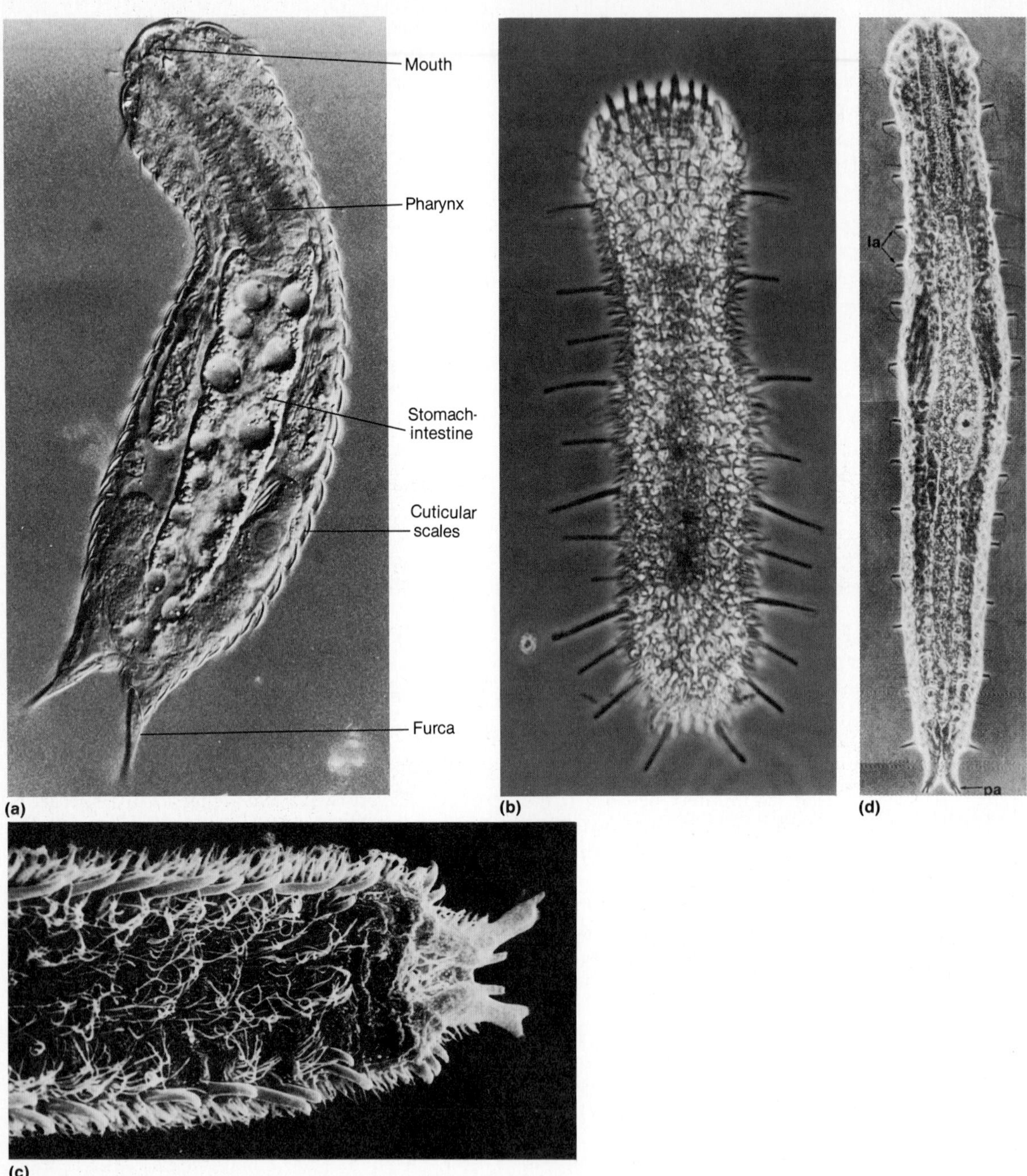

Figure 7.2 *Gastrotrich features shown by photographs:* ***(a)*** *an optical section (Nomarski) of* Lepidodermella squammata *(courtesy of M. J. Weiss, unpublished);* ***(b)*** Tetranchyroderma; ***(c)*** *a scanning electron micrograph of the posterior end of* Tetranchyroderma *to illustrate lateral and posterior adhesive tubes;* ***(d)*** Turbanella *(la = lateral adhesive tubules; pa = posterior adhesive tubules) (parts b–d courtesy of R. M. Rieger).*

PHYLUM GASTROTRICHA

The approximately 450 species of gastrotrichs are microscopic, range from 0.01 to 3 mm in length, and are delightful little creatures to observe. They are found in freshwater and marine habitats amid vegetation or in interstitial or detrital spaces. Gastrotrichs are all free living and feed on organic particles and small organisms such as bacteria, protozoans, and algae. Though they are generally colorless and transparent, gastrotrichs often appear the color of recently ingested food. There are two orders of gastrotrichs, Macrodasyida and Chaetonotida, which are differentiated basically on the presence or absence of head lobes, protonephridia, and pharyngeal pores.

The body is typically elongate with a convex dorsal surface and a flattened ventral surface. Most have a rounded **head lobe,** a slightly constricted **neck,** a relatively large **trunk,** and a forked posterior **furca** (Fig. 7.2a). Others lack the head lobe or have pointed or rounded posterior ends.

A thin external **cuticle** is present as abutting or overlapping scales or plates, as bristles, or as a smooth surface. Cilia, restricted to the ventral surfaces of the head and trunk, may be uniform or aggregated in tufts, longitudinal bands, or transverse rows and are responsible for the animal's smooth gliding locomotion. Specialized sensory cilia are found on the head, usually in tufts or as bristles.

Present in many gastrotrichs are cylindrical, cuticular **adhesive tubes** situated laterally or at the posterior end (Fig. 7.2b–d). Each movable tube, controlled by small muscles, contains epidermal **adhesive glands** that produce a sticky secretion. Some gastrotrichs move in a leechlike fashion, using the secretion from the adhesive glands to provide temporary attachments. Numbers of adhesive tubes range from one or two at the posterior end (chaetonotids) to more than 100 on each lateral surface (macrodasyids).

The body wall consists of the cuticle, epidermis, and thin layers of longitudinal and circular muscles. There is no pseudocoelom as such, but some small, slitlike, intercellular spaces devoid of any cellular materials are discernible.

The digestive system consists of a terminal or subterminal mouth, buccal cavity in some, pharynx, stomach-intestine, and anus. The **buccal cavity,** when present, is lined with ridges or teeth and is located between the mouth and pharynx. The muscular **pharynx** is elongate and tubular and may contain one or more bulbous enlargements. Food particles or organisms are sucked into the mouth primarily by the pumping action of the pharynx, but some gastrotrichs feed by sweeping particles into the mouth by ciliary currents. Macrodasyids have paired **pharyngeal pores** opening to the outside that presumably provide an exit for excess water ingested with the food. The midgut or **stomach-intestine** is a simple straight tube (Fig. 7.2a). Glandular epithelial cells lining the midgut secrete digestive enzymes, and both digestion and absorption take place in the stomach-intestine. The anus is located at the posterior end of the trunk.

Protonephridia are present in all chaetonotids, and nephridiopores are found on the ventral trunk; these osmoregulatory organs are absent in the macrodasyids. Gastrotrich protonephridia are morphologically different from those in the acoelomates in having the terminal cell composed of cytoplasmic rods and possessing a single flagellum. The **brain** is represented by a ganglion on either side of the pharynx connected by a dorsal commissure. From each ganglion a lateroventral nerve cord extends posteriorly the length of the trunk. Sensory structures include head cilia and bristles, ciliated lateral organs in some, tentacles, and even eye spots in a few forms.

Originally, gastrotrichs were probably hermaphroditic, but the male system has degenerated in all chaetonotids except for a very few genera. Therefore these gastrotrichs are exclusively female, and eggs develop parthenogenetically. The female system consists of one or two posterolateral **ovaries,** and ova are transported via oviducts to the female gonopores located near or in common with the anus. When present, **testes** lie in the anterior trunk, and sperm are conveyed via sperm ducts to the male gonopore on the ventral trunk. While some gastrotrichs have direct sperm transfer, individuals of most species flex the posterior body in such a manner that sperm or spermatophores are transferred to a posterior **caudal organ** that serves as a copulatory structure.

Each female (or hermaphrodite) produces from one to six eggs singly. Eggs are of two types: direct eggs hatch without delay in several days, and dormant eggs are attached to a submerged surface and are able to withstand adverse environmental conditions. With the return of more favorable conditions, development of dormant eggs proceeds rapidly. Both photoperiod and temperature are instrumental in determining which type of egg a female will produce. No larval stage is developed, and the newly hatched gastrotrichs attain sexual maturity within several days.

PHYLUM ROTIFERA

Rotifers are small animals that are abundant in most freshwater habitats. They are utterly fascinating creatures, and one can spend an engrossing several hours with a microscope simply watching them move and feed. Some are marine, and representatives of a few rotifer species live in aqueous films on mosses and marsh plants. There are about 1700 species, most of which are solitary and freemoving, but some are sessile and others are colonial. Commonly, they live on the surfaces of submerged vegetation or those of other animals; many others are planktonic, and a few are even parasitic. Rotifers are about the same size as larger ciliates and gastrotrichs, and though they range in length from 0.04 to 3 mm, most do not exceed 0.5 mm. Like gastrotrichs, rotifers are basically transparent but may appear the color of recently ingested food. The principal two classes of rotifers, Bdelloidea and Monogononta, are basically differentiated by the numbers of ovaries present.

External Features

The external surface of a rotifer is covered by a thin, epidermally produced cuticle, which is often thickened to form a **lorica.** In some rotifers the lorica may be poorly developed, thin, and restricted to the trunk; but in others it is thick, rigid, and well developed and covers the trunk, foot, and part of the head. Numerous variations in the nature of the lorica are found between these two extremes.

The typical rotifer body is elongate and cylindrical and consists of three regions: head, trunk, and foot (Fig. 7.3c). The **head** contains the corona, mouth, anterior digestive organs, most sense organs, and brain. One of the most characteristic rotiferan organs is the unique ciliated **corona.** Primitively, the corona consists of a large, ventral, ciliated area, the **buccal field.** From the buccal field a band or tract of cilia, called the **circumapical band,** extends around the circumference of the anterior end to form a ciliated crown (Fig. 7.3c). The ciliated band surrounds a nonciliated area, the **apical field.** Such a primitive corona is found in some of the creeping browsing monogononts (Order Ploima). From this basic form, the corona has become adapted in a number of ways. Most modifications have involved cilia of the buccal field and circumapical band; often, cilia have been altered to form cirri, bristles, or membranes or are lost altogether (Figs. 7.3; 7.4a, b). In other rotifers, ciliary tufts are borne by lateral projections called auricles. In some monogononts (Order Collothecaceae) the buccal field is modified into a funnellike **infundibulum** whose upper edge is folded and lobate. In sessile forms, the corona is often small and may be scalloped.

In the bdelloids the corona is composed of two prominent lobes called **trochal discs,** each bearing anterior and posterior rows of cilia

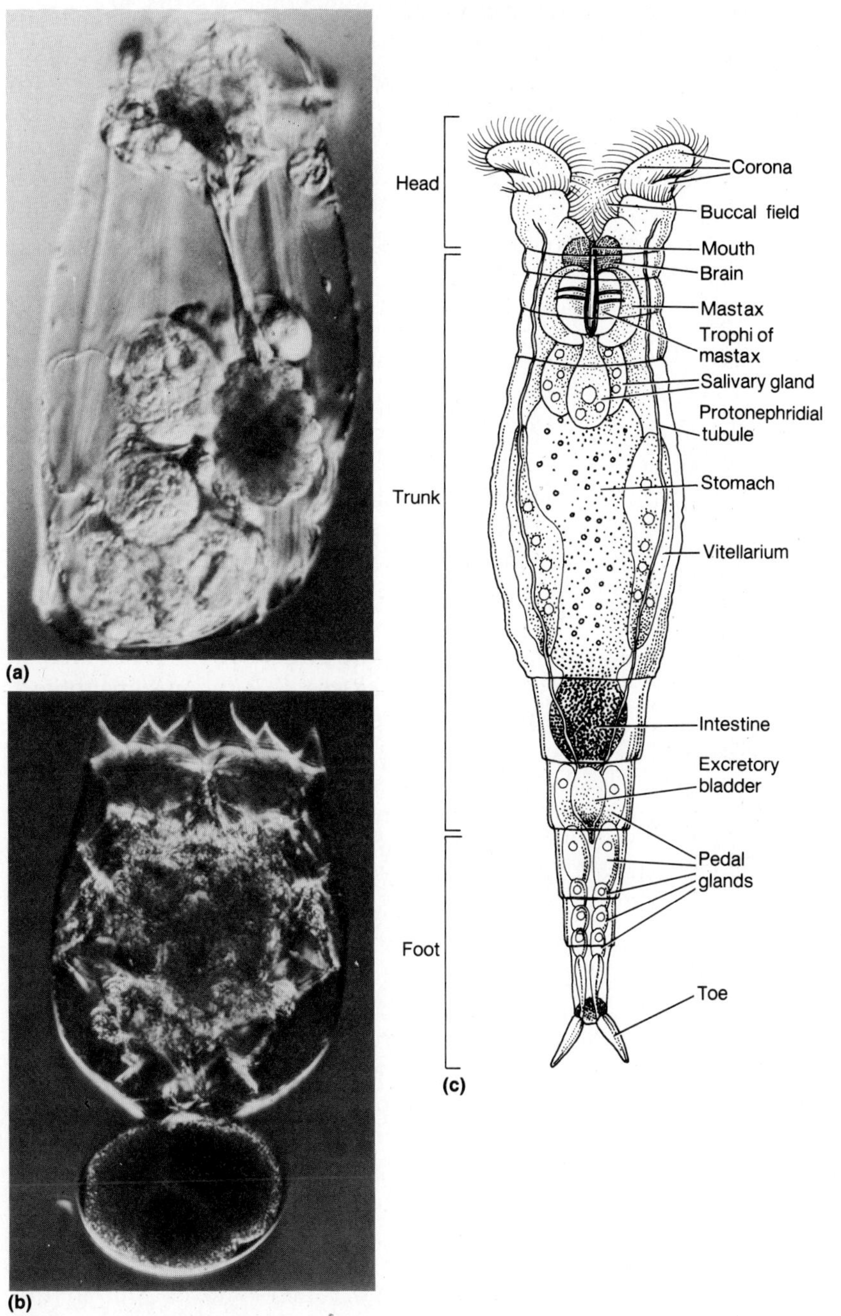

Figure 7.3 *Rotifer features:* ***(a)*** *a photomicrograph of* Asplanchna *(courtesy of J. J. Gilbert);* ***(b)*** *a photomicrograph of a marine rotifer (from Carolina Biological Supply Co.);* ***(c)*** Philodina, *a bdelloid (from Hickernell in L. H. Hyman,* The Invertebrates. *Vol. III,* Acanthocephala, Aschelminthes, and Entoprocta, *New York, McGraw-Hill, 1951, adapted with permission).*

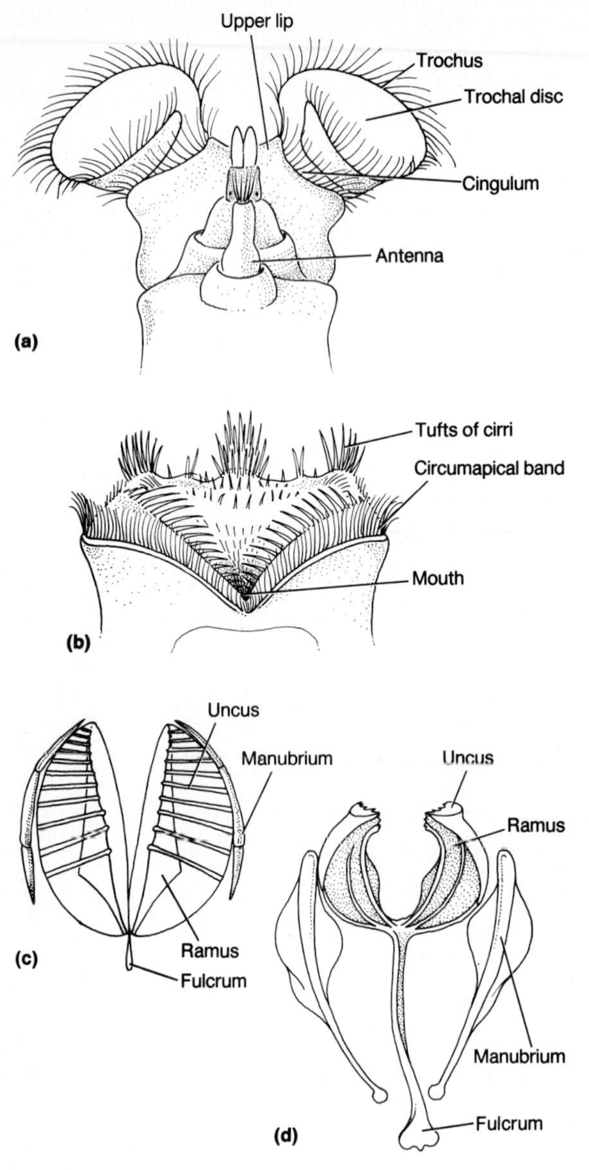

Figure 7.4 *Some variations in the corona and mastax of rotifers: **(a)** the corona of* Macrotrachela; ***(b)** the corona of* Epiphanes; ***(c)** a mastax with ramate trophi; **(d)** a mastax with malate trophi (parts b, c from Beauchamp in L. H. Hyman,* The Invertebrates. *Vol. III* Acanthocephala, Aschelminthes, and Entoprocta, *New York, McGraw-Hill, 1951, adapted with permission).*

(Figs. 7.3c, 7.4a). The anterior row or **trochus** is at the edge of the trochal discs, and the strokes of the trochal cilia on either disc sweep toward the mouth like two wheels spinning toward each other, a feature from which rotifers derive their common name, wheel animals. The posterior row or **cingulum** is located at the base of each trochal disc and extends ventrally beneath the mouth (Fig. 7.4a). In bdelloids the corona can be retracted to reveal an anterior suckerlike **rostrum.**

The head contains the mouth, which is terminal, subterminal, ventral, or in the center of the corona. The dorsal head surface may also bear a variable number of pigmented areas or eyes that are sensitive to light intensities, one or two short antennae equipped with tactile and chemical receptors, and two sensory **ciliated pits** situated on or near the corona (Fig. 7.4a, b).

The **trunk,** the largest part of a rotifer, is elongate or saclike (Fig. 7.3) and in some rotifers bears movable papillae and spines, lateral antennae, and sometimes anterior antennae. The anus occurs dorsally on the posterior trunk. The posterior narrow portion, called the **foot,** is often set off sharply from the trunk. The terminal portion of the foot usually bears one to several **toes** or **spurs** (Fig. 7.3c). At the bases of the toes are many **pedal glands** whose ducts open on the toes or foot. Adhesive secretions of the pedal glands aid in temporary attachment of the foot to a surface. In most planktonic forms, the foot and toes are reduced or absent. Often, the cuticle of the foot is ringed with joints that enable sections to be telescoped, and thus the entire foot can be retracted.

Locomotion and Tube Building

Often, planktonic rotifers move through the water in a spiral twisting path by the beating of the coronal cilia, but a few planktonic forms are able to move abruptly because of the sudden beating of long cilia. Freely moving nonplanktonic forms creep over surfaces using both coronal cilia and the toes. Bdelloids often move leechlike over a surface in a process called **looping.** With the toes adhered temporarily to a surface the head and trunk are extended anteriorly. With the corona retracted the rostrum adheres to the surface. The toes release their attachment, the body is flexed dorsally, the foot and toes are brought forward ventrally and anteriorly toward the rostrum, and the toes attach, thus completing one looping cycle. When the animal is swimming, the corona must be extended to allow the coronal cilia to beat, and the toes are used mainly as a rudder in creeping or swimming.

Many rotifers are sessile, and most of these produce a tube. Tube materials, secreted by the general body surface, are gelatinous and usually colorless. But to the gelatinous secretions many rotifers add debris from the environment including fecal pellets, minute sand particles, and organic detritus. Some forms like *Floscularia* form cylindrical pellets, which are positioned by movements of the head. Sometimes, tube particles are arranged in transverse rings or annuli.

Feeding and the Digestive System

Most rotifers feed on suspended organic particles or small planktonic organisms. The coronal cilia are the principal means for creating a current of water that brings food particles to the anterior end, and from there, particles are swept to the mouth. Other rotifers are carnivores and capture food by trapping it in the buccal field or elsewhere on the corona.

Like the corona, the unique mastax is characteristic of rotifers. The **mastax,** a modified posterior part of the pharynx, is a muscular organ in which food is ground or macerated. The inner walls of the mastax contain several large masticatory pieces called **trophi.** The trophi, composed of a cuticularized mucopolysaccharide, vary in morphological details, and these variations are used widely in taxonomic keys. Typically, there are seven interconnecting trophi: three pairs of lateral parts named **rami, unci,** and **manubria,** and a median unpaired **fulcrum** (Fig. 7.4c, d). In monogononts the mastax lacks rami and is termed nonramate. The trophi are moved by muscles to make the mastax an effective masticating structure. A number of mastax types are recognized that are based on the degree of specialization for biting, cutting, holding, and crushing food. In many predatory forms and some herbivorous rotifers the mastax or a pair of rami are protruded through the mouth and used to grasp, hold, or puncture food organisms (Fig. 7.4d). In others the mastax is used to suck food into the mouth. Secretions from a number of small salivary glands in the walls of the mastax facilitate the mechanical breakdown of food.

From the mastax, food passes through a short, often ciliated **esophagus** to the ciliated **stomach.** Opening into the anterior portions of the large bulbous stomach is a pair of **gastric glands** that secretes digestive enzymes. Digestion and absorption occur in the stomach. A short, tubular, often ciliated **intestine** extends posteriorly, where it becomes a nonciliated **cloaca,** which also receives water from the protonephridia and eggs from the ovaries. The cloaca opens to the outside as a dorsal **cloacal aperture** or **anus** at the base of the foot. In some sucking forms and in most males the anal opening is absent.

Other Visceral Systems

The pseudocoelom, filled with fluid and numerous interconnecting, ameboid cells, is a space in which all visceral organs lie. Osmoregulation is achieved by a protonephridial system located in the pseudocoelom. Each protonephridium is equipped with 4–50 flame bulbs and empties into a small median **bladder,** which in turn opens into the cloaca, which may function as a temporary water reservoir in the absence of a bladder (Fig. 7.3c). Discharging several times per minute, the bladder or cloaca eliminates water and probably metabolic wastes such as ammonia.

The nervous system is composed of a bilobed ganglionic **brain** located on the dorsal surface of the mastax and two principal lateral nerves extending posteriorly. Additionally, the **retrocerebral organ** lies dorsal to the brain and consists of a pair of subcerebral glands located behind the brain and connected to a median, pigmented, retrocerebral sac. The sac opens onto the apical field by way of two minute ducts. The specific functions of the retrocerebral organ are still unclear, but most likely this organ is sensory in nature.

Reproduction and Development

Rotifers are dioecious, and males are always smaller than females. Parthenogenesis is typical of most rotifers, and males appear only sporadically; in the bdelloids, incredibly, no males are known! Most rotifers (monogononts) have a single ovary and an attached syncytial **vitellarium** (Fig. 7.3c). Often, both ovary and vitellarium fuse to form a **germovitellarium.** The vitellarium produces yolk, which is incorporated into each singly produced egg. After fertilization, each egg travels through a short oviduct to the cloaca and out its aperture (or gonopore in those without an anus). The female system in the Bdelloidea is similar except that it is Y-shaped with paired germovitellaria and oviducts. In males the mouth, cloaca, and other digestive organs are usually degenerate or absent altogether. Sperm, produced by a single ovate testis, travel through a vas deferens to the male gonopore. Typically, male rotifers have an eversible **penis,** but in some the vas deferens or gonopore wall can be everted as a copulatory organ. Sperm are commonly injected hypodermically by the male into the pseudocoelom of the female.

In monogononts, two different types of females are produced that are indistinguishable morphologically but distinct functionally. One type, called **amictic females** and reproducing exclusively by parthenogenesis, forms **amictic eggs** (Fig. 7.5). Amictic or summer eggs are thin shelled, are produced without meiosis and thus are diploid, cannot be fertilized, and develop without delay into diploid amictic females. A second type, called **mictic females,** produces **mictic eggs,** which are thin shelled and haploid (Fig. 7.5). If a mictic egg is not fertilized, it develops parthenogenetically into a male. If fertilized, mictic eggs are provided with a thick heavy shell and become **dormant** or **winter eggs** capable of resisting severe environmental conditions. Dormant eggs require a latent period of a month or more before developing, and they always hatch into amictic females (Fig. 7.5). Most females lay either amictic or mictic eggs

Figure 7.5 *A word diagram of the life cycle of a typical rotifer.*

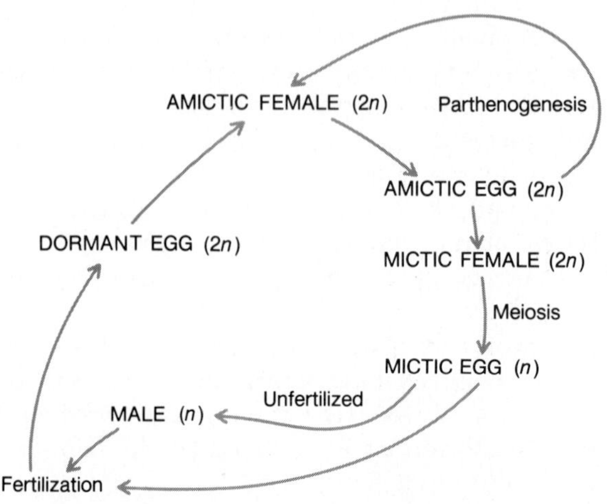

but not both. Apparently, during oocyte development the physiological condition of the female determines whether her eggs will develop into amictic or mictic females. Normally, one or two mictic and 20–40 amictic generations are produced each year, since the adults of many species do not survive the winter. Mictic females appear only at certain crucial periods as in the autumn, when environmental conditions such as photoperiod and temperature are changing rapidly or when population densities are high.

Females produce from 3 to 50 eggs but normally fewer than 10. The females of some species are ovoviviparous and retain the young within the oviduct or pseudocoelom; young are released through the cloacal aperture or by rupture of the trunk wall. Amictic eggs often adhere to the posterior end of the female, but mictic eggs may float or sink to the bottom, where they remain until hatching in the spring. Rotifer eggs undergo a modified pattern of determinate cleavage. Cell divisions occur rapidly to form all adult organs, but then mitosis ceases, and eutely is therefore pronounced in rotifers. No larval stage is developed, although sessile rotifers have a hatching stage that is free swimming for a short time prior to settling.

Rotifers exhibit cyclic patterns of population sizes and morphological shapes. Because generation times are short, many rotifer species exhibit regular predictable periodicities of abundance followed by periods of relative scarcity. Many planktonic species, however, have cycles of abundance that are extremely variable among and between species, from year to year, and between populations in nearby lakes or ponds. In many rotifer species there is a progression of morphological changes during one annual cycle. This phenomenon, called **cyclomorphosis,** prevalent in certain freshwater crustaceans (branchiopods), is poorly understood but is undoubtedly related to seasonal environmental changes, subtle changes in reproductive patterns, predation by other animals, and short but numerous generations.

PHYLUM KINORHYNCHA

Kinorhynchs are minute elongate aschelminths found in marine littoral zones burrowing in the mud, sand, and debris. There are fewer than 100 known species, and they range mostly from 0.4 to 1 mm in length. The body is transparent but often contains hues of yellow or brown.

The wormlike body is composed of 13 (14 in one species) superficial "segments" called **zonites** (Fig. 7.6a). Kinorhynchs are not segmented as metamerism was defined in Chapter 3. The external body surface, entirely devoid of cilia, is covered by a thin cuticle. The body of a kinorhynch is divided into three regions: head, neck, and trunk. The **head,** represented by zonite I, bears the mouth, an oral cone, and spines. The mouth is situated at the tip of the extensible **oral cone,** which is armed with a circlet of **oral spines.** The **neck** is represented by the second zonite; and at the junction of head and neck are several circlets of spines, called **scalids,** and plates, the **placids** (Fig. 7.6a). The head can be retracted into the neck. In some kinorhynchs, both head and neck can be retracted into the anterior trunk. The fact that the oral cone, head, and neck are retractile is the source of their name (kino = movable; rhynch = beak or snout). The **trunk** consists of 11 zonites and terminates with the anus. Each trunk zonite bears a middorsal spine and a pair of lateral spines (Fig. 7.6a). The scalids and zonital spines are recurved and movable, and their hollow cuticle is filled internally with epidermal tissues. The cuticle of each zonite, composed of a single dorsal and two or three ventral plates, is especially thin and flexible between plates and between zonites so as to facilitate movement. Several ventral or lateral adhesive tubes are borne by trunk zonites. Kinorhynchs do not swim but employ the spines as anchors to burrow into or creep through benthic material.

The body wall consists of the cuticle, epidermis, and zonite-arranged muscles. The epidermis is thickened as a middorsal and two

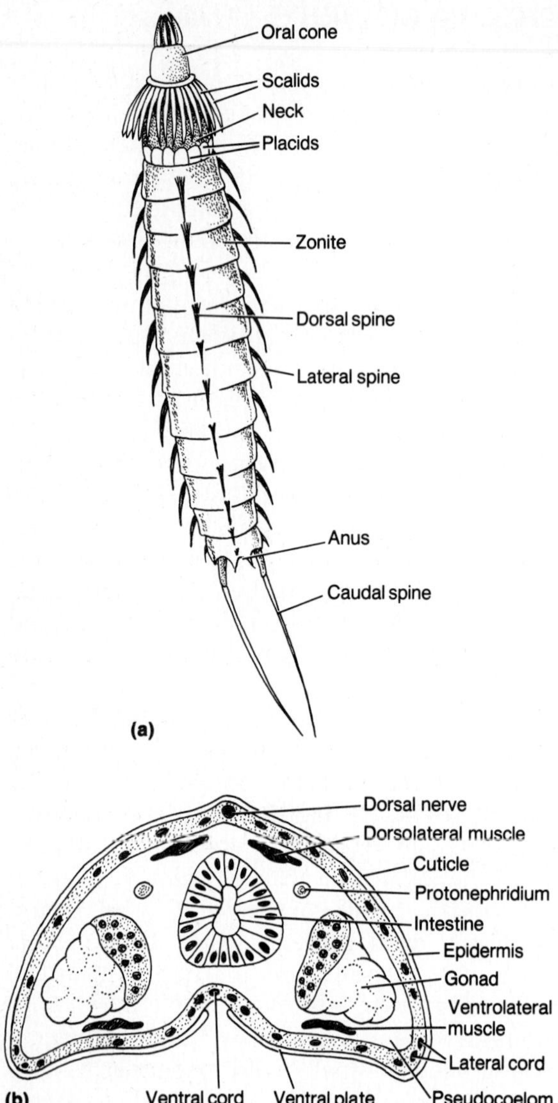

Figure 7.6 *Kinorhynch features:* ***(a)*** *an adult;* ***(b)*** *a cross-sectional view through a middle zonite.*

lateral **cords.** Two pairs of longitudinal muscle bands are present (Fig. 7.6b). The pseudocoelom is rather large and contains ameboid cells.

The digestive system is composed of a foregut, midgut, and hindgut, and both foregut and hindgut are lined with a thin cuticle. The **foregut** consists of the mouth, the buccal cavity with the oral cone, a muscular pharynx, and a short tubular esophagus. The **midgut,** a long, noncuticularized stomach-intestine, is the site of digestion and absorption. The short **hindgut** terminates in the posterior anus. Associated with the alimentary canal are two pairs of esophageal salivary glands and a pair of pancreatic glands opening at the junction of esophagus and stomach-intestine. Most kinorhynchs feed on diatoms, other algae, and organic particles.

A pair of protonephridia is located in the pseudocoelom in zonite X. Each protonephridium has a single flame bulb with but one flagellum, and both nephridiopores are located on the dorsum of zonite XI. The nervous system consists of a circumpharyngeal ring or brain and a ventral nerve cord with a ganglion in each zonite. Kinorhynchs are dioecious, and the saclike paired gonads are in midbody. Gametes are conveyed by gonoducts to gonopores situated in the posteriormost zonite. The male gonopore is surrounded by two or three spines, presumably used in copulation. Details of copulation and early development are poorly known, but the young hatch into larvae whose bodies are not divided into zonites. In successive molting of the cuticle, zonites and adult morphology are attained. Kinorhynchs may live for up to a year or more.

PHYLUM NEMATODA

Nematodes or roundworms are usually small worms that are morphologically quite similar but extraordinarily versatile in the habitats in which they live. They generally range in length from microscopic (250 μm) up to several millimeters. Parasitic and marine nematodes tend to be longer than freshwater or soil worms. Individuals of several species are reported to be in excess of 1 m in length, and representatives of one species, found in the placenta of sperm whales, are, incredibly, up to 9 m long! Many nematodes are parasites and live in many different plants and animals. But many others are free living in marine, freshwater, and soil habitats. Some ecological aspects of nematodes in soil ecosystems are explored in Box 7.2.

Nematodes live almost everywhere, surprisingly even in such severe habitats as deserts, deep-sea bottoms, polar seas, and the soil

BOX 7.2 □ SOME DIMENSIONS OF THE ECOLOGY OF SOIL NEMATODES

Nematodes are often incredibly abundant in decomposing organisms and in aquatic habitats, but they are particularly numerous and diverse in soils. Nematodes are plentiful and ubiquitous in all soils, ranging from those of deserts to those of arctic regions. What roles do they play in soil ecosystems? What are their positions in food chains in these complex ecosystems?

A great many soil nematodes are parasitic on the roots of various plants, and these worms often do considerable damage to garden, truck farm, woody, and ornamental plants. Nematodes alone are known to inhibit to a significant degree plant vitality, root growth, and the overall process of primary production. Some nematodes penetrate the root system, become immobile, and feed entirely endophytically. Often their presence induces the plant to form root knots. Other nematodes are ectoparasitic and remain more active on the root surfaces. Of course, many nematodes occupy intermediate positions between these two extremes. In any case, nematodes as a group are a principal, subterranean, primary consumer, and rather large amounts of plant energy are converted by nematodes into helminth reproductive energy and, more specifically, into egg production. Their fantastic biotic potential and fecundity are generated completely at the expense of the plants on whose tissues they feed.

However, a majority of all soil nematode species are not parasitic but are free living in the soil. They locomote between the soil particles and are important in the microbiology of the soil ecosystem. A great many nematodes, especially the rhabditid forms, are voracious feeders on bacteria and fungi, and it is thought that these bacteriophagous and fungiphagous nematodes act as important regulators of decomposing populations. Those nematodes that feed on bacteria, fungi, and the roots of plants are in turn attacked by predaceous nematodes, a phenomenon that makes them effective in biological control and indirectly influential on the rate of primary production. More recently, nematologists have discovered that certain nematodes are themselves intimately important in the entire process of decomposition. Apparently, a large number of species are omnivorous or saprozoic and thereby decompose litter, organic debris and detritus, and dead organisms. The abundant soil nematodes are in turn fed upon by soil arthropods (largely collembolans and mites), tardigrades, and earthworms.

Given the great logistical problems of trying to quantify the precise functions of nematodes in natural soil ecosystems, it is still safe to say that they are more important than was thought a decade ago. Nematodes are very essential in energy flow and nutrient cycling in these ecosystems, and future studies may indicate that these "lowly worms" are preeminently important in the proper functioning of most soil ecosystems.

of the tundra. Often, they are present in prodigious numbers; for instance, millions of individuals may be found in 1 m^3 of littoral mud, one spadeful of garden soil may contain a million nematodes, and up to 100,000 individuals can be routinely present in a single apple, pear, or sugar beet decomposing on the ground. Estimates on numbers of species range from 10,000 to 30,000 described species, and some nematologists estimate that the total number of species may be as much as 20 times higher! Two classes of nematodes are generally recognized: Adenophorea and Secernentea. The adenophoreans have a poorly developed excretory system and lack external olfactory organs (phasmids). Secernenteans possess both an excretory system and phasmids.

External Features

Representatives of this phylum are built on a surprisingly uniform morphological plan and are functionally similar. In general, marine nematodes are more primitive than terrestrial or freshwater forms, and parasitic forms possess a number of adaptations in keeping with their host-dependent existence.

A typical nematode body is slender, elongate, cylindrical, and tapered at both ends (Fig. 7.7). It has long been dogma that nematodes lack cilia and flagella completely, but recent studies on certain sense organs (amphids) indicate the presence of modified nonmotile cilia. There is little cephalization, and there are no discernible appendages and no external regions such as a head or trunk. Even though nematodes are bilaterally symmetrical, structures around the mouth are radially or biradially arranged, suggestive of the symmetry of sessile animals. Speculation is that ancestral nematodes were indeed sessile, attached by the posterior end, and that the anterior end protruded upward into the water.

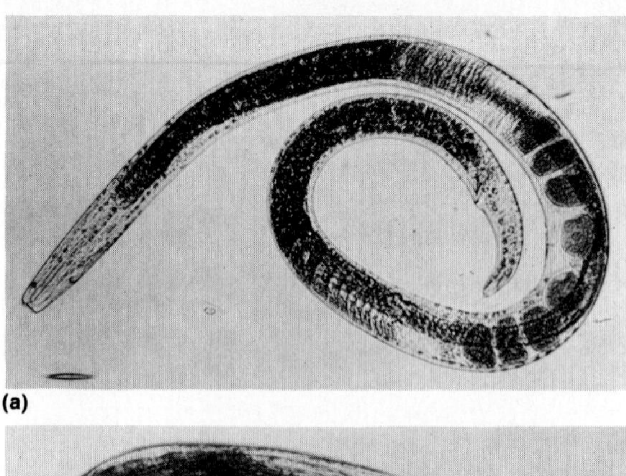

(a)

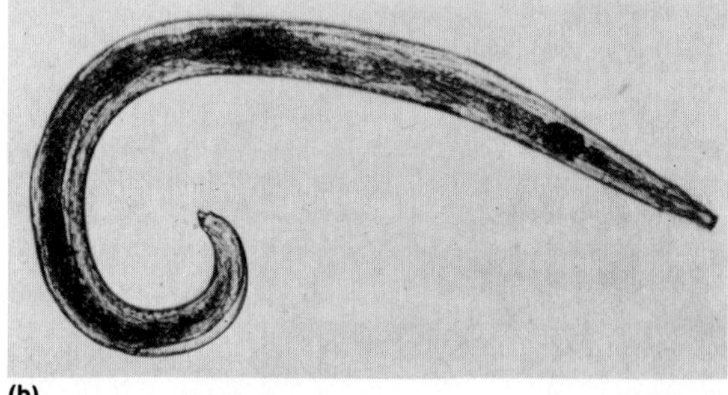

(b)

Figure 7.7 *Photographs of two different nematodes:* ***(a)*** *a marine species;* ***(b)*** *a male* Enterobius vermicularis, *a parasitic nematode known as the pinworm (parts a, b courtesy of Ward's Natural Science, Inc., Rochester, N.Y.).*

BODY WALL

Much of the success of nematodes is attributable to an outer **cuticle** of considerable adaptive plasticity. Not only does the cuticle cover the external body surfaces, it also lines the foregut and hindgut, the outer parts of the female reproductive system, and the sense organs. The cuticle is often smooth without ornamentations; but spines, bristles, papillae, warts, and ridges are commonly present and are of taxonomic significance. Chemically, it is composed chiefly of scleroproteins but with no chitin or keratin. The cuticle is subdivided into three primary layers: cortical, matrix, and basal layers (Fig. 7.8c). The outer **cortical layer** is composed of several layers, including a thin outer lipid layer in some. Beneath the cortex is the **matrix** or **medial layer** composed of albuminlike proteins and containing many minute branching canals. The innermost, striated, collagenous **basal layer** is constructed of three fibrous layers and a **basal lamella** (Fig. 7.8c). The cuticle functions in a number of ways, including maintaining internal turgor pressure, providing for mechanical protection, and aiding in resisting digestion in par-

Figure 7.8 *Body-wall and internal features of nematodes:* ***(a)*** *the internal features of a female* Rhabditis; ***(b)*** *the internal features of a male* Rhabditis; ***(c)*** *a diagram of a section of the cuticle;* ***(d)*** *a cross-sectional view through the pharyngeal region of* Ascaris *(parts a, b adapted from Chitwood and Chitwood in R.W. Pennak,* Freshwater Invertebrates of the United States, *2nd ed., 1978, New York, John Wiley, © John Wiley & Sons, used by permission).*

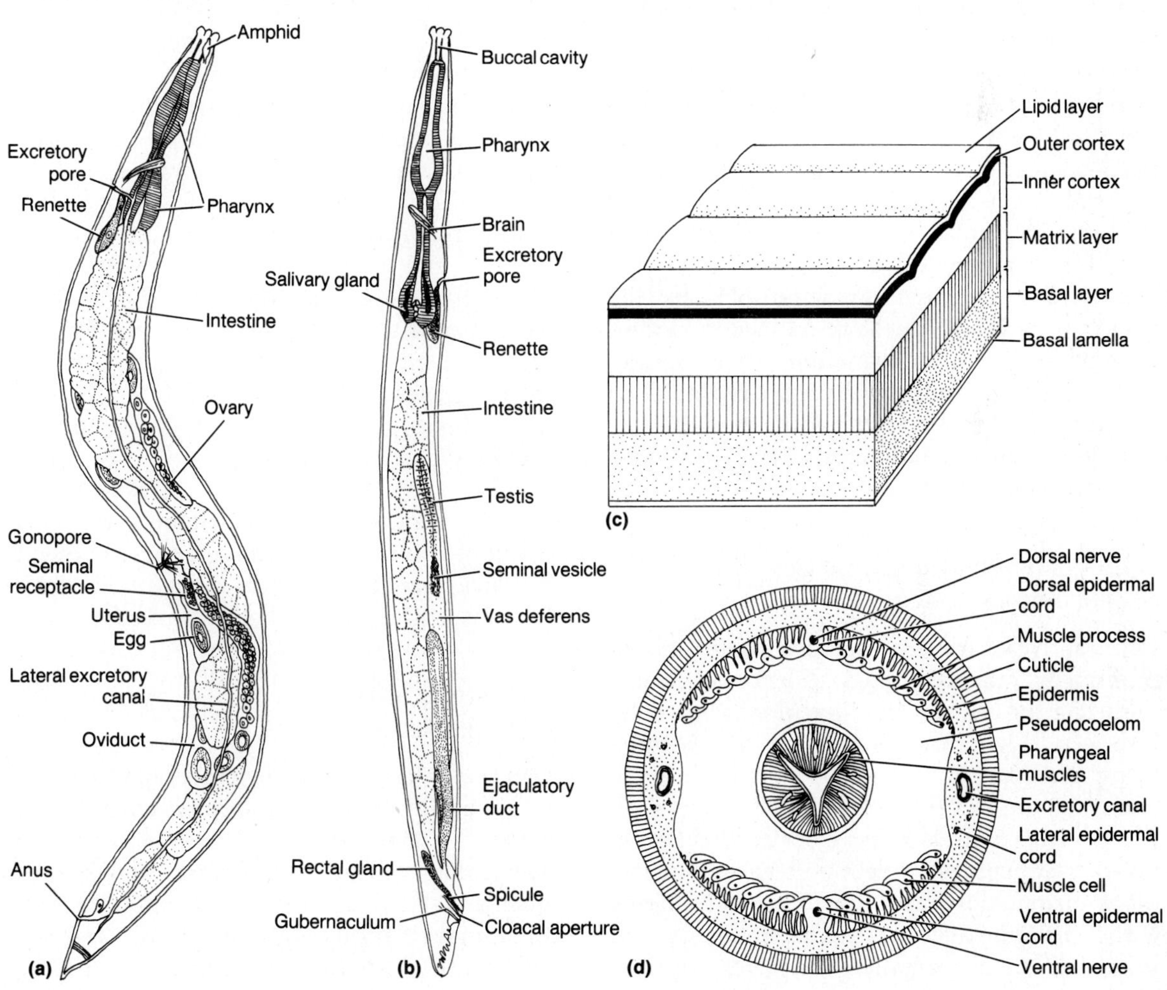

asitic forms. The cuticle is molted four times during growth of juvenile nematodes, a phenomenon that is controlled by neuroendocrine secretions as in arthropods. The old cuticle becomes detached from the underlying epidermis (apolysis) by a molting fluid, a new cuticle is secreted beneath the old, and the old cuticle is shed (ecdysis), usually in patches or fragments or, sometimes, entire.

Beneath the cuticle is the **epidermis.** This layer projects internally into the pseudocoelom as four **longitudinal cords** situated in the middorsal, midventral, and both lateral lines (Fig. 7.8d). A peculiar situation is that the nuclei of all epidermal cells are situated in these longitudinal cords. Medial to the epidermis is a relatively thick muscular layer of longitudinal fibers only. Minute fibers attached spindle-shaped muscle cell to the cuticle. Another peculiar arrangement is that all muscle cells have long, slender processes that extend to either a dorsal or a ventral nerve cord for innervation (Fig. 7.8d). The longitudinal muscles are the principal means for locomotion in nematodes. Locomotion is inevitably within a watery medium, and even terrestrial forms move in an aqueous film. Contraction of the longitudinal muscles results in undulatory waves that pass from anterior to posterior.

Specialized structures of the body surface

In primitive marine nematodes there are six prominent, radially arranged **lips** or lobes surrounding the mouth. In other forms, however, fusion or loss of lips has resulted in only three being present; one is always dorsal, and the other two are ventrolateral. In many others, all lips have disappeared. Some species bear a circular row of **spines** or **teeth** on or near the lips. Many nematodes have cuticular **head shields** that occur immediately posterior to the lips and protect them. Sensory structures include amphids, phasmids, and ocelli. **Amphids,** anterior depressions in the cuticle, are chemoreceptors containing sensory processes of recently identified, modified, nonmotile cilia. Paired unicellular **phasmids** are situated posteriorly near the anus and function as olfactory organs. Paired anterior **ocelli,** whose specific functions are unknown, are present in many aquatic nematodes. Other cuticular structures include **alae,** which are longitudinal folds or ridges found on either end or laterally; **caudal glands** whose secretions aid in temporary attachments; and various papillae and setae.

Internal Features

The nematode pseudocoelom is a spacious, fluid-filled cavity (Fig. 7.8d). The fluid is of fundamental importance to nematodes because it bathes all internal organs and forms an internal hydrostatic skeleton. The pseudocoelomic fluid and internal organs are always under pressure, owing to the tonicity of the body wall musculature. Therefore nematodes are round worms because of the outward force generated by the body wall muscles contracting against the hydrostatic pseudocoelomic fluid. The hydrodynamic forces are of great significance in locomotion, digestion, and excretion. Present in the pseudocoelom are one to three pairs of giant oval or branched **pseudocoelomocytes.** Their functions are unclear, although some evidence indicates that they may be phagocytic, excretory, absorptive, or synthetic in function.

Feeding and the digestive system

Nematodes feed on a wide variety of foods, and almost all forms of heterotrophic nutrition can be found. Carnivores feed on almost any small invertebrate, including rotifers, tardigrades, small annelids, and other nematodes. Herbivores ingest various algae and fungi, and parasites suck tissue fluids from their hosts. Many are omnivores or saprobes that consume bits of decomposing organisms. Feeding in nematodes is almost invariably aided by the pharynx serving as a pump. Some prey is swallowed whole; but in others, bits are broken away by the lips or teeth and ingested. Many predatory forms have a **stylet** or spear in the anterior alimentary canal with which they pierce the prey. The stylet may also serve as a tube through which the worm sucks up plant or animal juices. Even though nematodes as a group appear to be catholic in their diets, a given species usually feeds on only one type of food.

The alimentary canal consists of foregut, midgut, and hindgut, but there are few specialized digestive regions. The **foregut** is subdivided into the anterior mouth, a buccal cavity, and the pharynx (Fig. 7.8a, b). The **mouth,** often surrounded by lips or teeth, opens into a **buccal cavity** whose cuticle is strengthened by ridges, rods, and plates that aid in maintaining the shape of this chamber, but in some parasitic forms this cavity is reduced or absent. The buccal cavity, often armed with a stylet and teeth, is continuous posteriorly as the pharynx (or esophagus). The **pharynx** is an elongate, usually syncytial tube whose lumen is always triradiate in cross section as one angle is always midventral and the other two are dorsolateral in position (Fig. 7.8d). Beneath the cuticle of the pharyngeal lumen is a thick muscular layer. Numerous radial muscles run from its cylindrical outer wall to the cuticle of the body wall. Rhythmical contractions of the radial muscles enlarge the lumen and create an effective suction. Relaxation of the radial muscles, along with turgor pressure of the pseudocoelom, closes the pharyngeal lumen. Rates of the pharyngeal pump pulsations range from one to eight per second. Walls of the pharynx commonly contain three bulbous expansions whose ducts open into the pharyngeal lumen, and they apparently make possible a complex pumping action.

The **midgut** consists of a long tubular intestine comprising most of the length of the digestive system (Fig. 7.8a, b). A pharyngeointestinal valve regulates the one-way passage of food into the intestine, where both digestion and absorption take place. The intestine opens into a short tubular **hindgut** or **rectum,** which in males is called a **cloaca.** (A cloaca is defined as a posterior organ common to two or more visceral systems, one of which is always the digestive system.) Several unicellular rectal glands of unknown function open into the anterior rectum or cloaca of many parasitic forms. The **anus** (**cloacal aperture** in males) is terminal or subterminal (Fig. 7.8a). Hydrostatic pressures within the pseudocoelom are responsible both for the passage of food through the alimentary canal and for defecation.

Other visceral systems

Nematodes possess an excretory system that is different from that of any other group of animals, since protonephridial flame bulbs are missing altogether. Excretion and osmoregulation are achieved in nematodes in one of two unique systems: glandular and tubular. In the **glandular system,** found in most aquatic nematodes and some parasitic species, one or two ventral gland cells, called **renettes,** are positioned in the anterior body lateral to the pharynx (Fig. 7.9a). Each absorbs wastes from the pseudocoelomic fluid and empties to the outside through an **excretory pore** at the level of the brain (Figs. 7.8a, b, 7.9a). In more advanced nematodes a **tubular system** develops from the renette system (Fig. 7.9b). The two renette tubules unite to form an H, the crossbar of which is a transverse canal joining the lateral renette tubules. Sometimes the anterior horns of the lateral tubules are lost, leaving a system shaped like an inverted U. The intracellular canals are located in the lateral longitudinal cords (Figs. 7.8d, 7.9b). The system opens to the outside via a common midventral excretory pore. Some nematodes eliminate ammonia, urea, and peptides dissolved in water. But this system is not

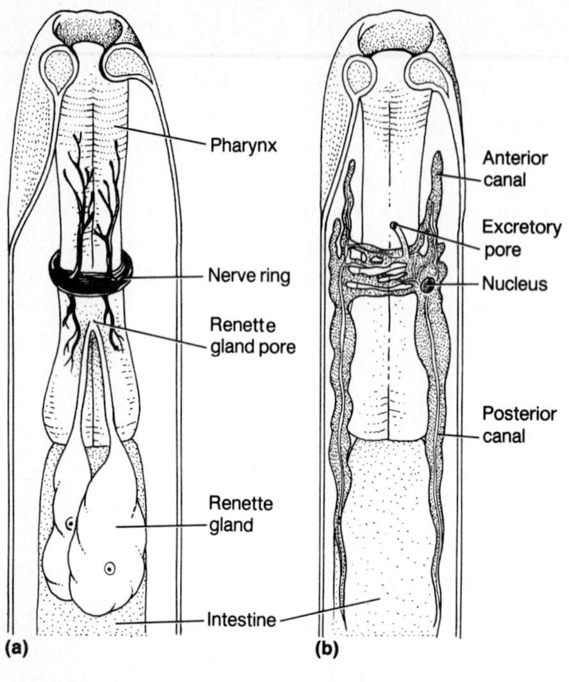

Figure 7.9 *Two types of nematode excretory systems: **(a)** glandular* (Rhabditis); ***(b)** tubular* (Ascaris).

a universal excretory or osmoregulatory mechanism; many species lack any form of a structural excretory system, and their metabolic wastes are voided through the cuticle and anus.

The nervous system is essentially like that in other aschelminths with a circumpharyngeal **brain** bearing four ganglia (Fig. 7.8a, b). Six nerves extend anteriorly and innervate structures around the mouth. A midventral nerve, a middorsal nerve, and paired lateral nerves extend posteriorly in the respective epidermal cords (Fig. 7.8d). In some nematodes a series of commissures interconnect these longitudinal nerves. Neuroendocrine functions have been investigated only superficially, but evidence is accumulating that hormones are involved in growth and development, molting, cuticle formation, and metamorphosis.

Reproduction and Development

Nematodes are basically dioecious, although some are monoecious and often protandric. A few nematodes reproduce parthenogenetically, and a few monoecious forms are capable of self-fertilization. External sexual dimorphism is always pronounced; males are smaller, have curved posterior ends, and bear bursae and other accessory reproductive structures. The gonads are usually long and coiled and lie free within the pseudocoelom.

In males the **testis,** usually unpaired, is found in midbody and is highly convoluted (Fig. 7.10b). Sperm arise from germinal cells located at the anterior free end or along the length of the testis. Nematode sperm are remarkably unique in that they are nonflagellated; sperm of some are totally nonmotile, and those of others can move ameboidlike only in a limited way. The **vas deferens** or sperm duct, a simple continuation of the testis, extends posteriorly and is expanded near its posterior terminus to form a **seminal vesicle.** The seminal vesicle communicates with the **cloaca** by way of an **ejaculatory duct.** Prostate glands secrete seminal fluids into the ejaculatory duct. Males are commonly armed with a bursa containing one or two copulatory, cuticular **spicules.** The spicules, slightly curved, pointed, nonhollow, and equipped with minute special muscles, aid in sperm transfer by spreading the female gonopore. When not in use, they are housed in small cloacal sacs. Often, an accessory cuticular structure, the **gubernaculum,** is present that functions as a guide for the spicules (Fig. 7.8b).

The female system usually consists of a pair of convoluted tubular **ovaries,** which, oddly enough, extend in opposite directions—one anterior and the other posterior. Each ovary is continuous as an **oviduct** whose proximal end is swollen to form a **seminal receptacle** (Fig. 7.10a). Each oviduct becomes a tubular **uterus,** and the two uteri unite to form a **vagina,** which in turn opens to the outside through a female **gonopore** located midventrally, usually in the middle third of the body (Fig. 7.10a).

Copulation involves the tail region of a male coiled around the gonoporic region of a female. Some nematodes have their anterior ends together or in opposite directions; in others the male is at a right angle to the female. Spicules are inserted into the female gonopore, and after sperm are introduced into the vagina the two worms separate. Sperm migrate through the vagina and uteri to the seminal receptacles, where fertilization occurs.

Each fertilized egg, ensheathed in several membranes, is moved to the gonopore by hydrostatic forces of the pseudocoelomic fluid and released to the outside. Some female nematodes produce several dozen eggs; others, especially

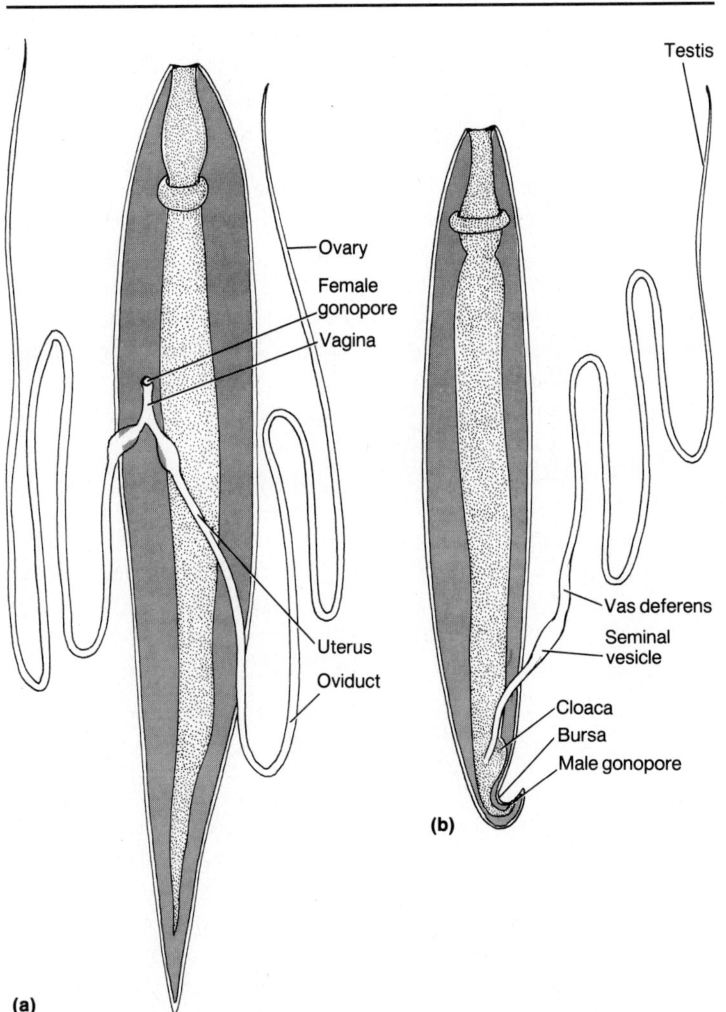

Figure 7.10 *Diagrammatic representations of the reproductive systems of* Ascaris *(sizes of reproductive systems are exaggerated): **(a)** female; **(b)** male.*

parasitic forms, may lay as many as 200,000 daily. Determinate cleavage follows an asymmetrical, nonspiral pattern. Development usually proceeds without delay, and the eggs hatch promptly. However, eggs of most parasitic forms can remain viable for years, and subsequent development depends upon ingestion by a suitable host.

Development and hatching are influenced by external factors such as temperature and moisture. Hatching produces a juvenile nematode or **larva** replete with most adult structures. The first one or two cuticular molts sometimes take place within the egg shell before hatching.

Nematodes as Human Parasites

Some of the more important aspects of parasitism in general were discussed in Chapter 6 and will not be repeated here. Representatives of many nematodes are parasitic in plants and animals, and some are important, widespread human parasites. Parasitic nematodes are very common in human beings; probably more than 95% of everyone who reads this text has had a nematode infection! Further, probably one out of every two readers at this very moment is a host to adult nematodes! It may be of some comfort to understand that most parasitic nematodes go unnoticed and cause little harm to their human host.

Nematodes possess a number of adaptations for their parasitic habit, including high fecundity rates, elaborations of the life cycles to ensure transmission from one host to another, an enzyme-resistant cuticle, dormant eggs, and encysted larvae. The life cycles of nematodes generally are not as complicated as those of trematodes or cestodes, since they typically involve only one host or, at most, two. Nematodes also lack the several morphologically distinct larval stages seen in trematodes. Discussions of the adaptations and life cycles of four of the most important human parasites follow.

ASCARIS LUMBRICOIDES

There are an estimated 700 million infections worldwide of this nematode. Adult females are from 20 to 40 cm in length, and adult males are 15–31 cm long. Adults live in the small intestine of human beings, where they feed on intestinal contents and blood sucked from punctures in the intestinal wall. Eggs, produced in prodigious numbers, exit from the host in its feces. A first-stage larva, developing rapidly in the shelled egg, molts and matures into a second-stage larva; eggs must contain second-stage larvae to be infective.

Infective eggs must be swallowed, and larvae hatch in the intestine. Normally, infective eggs can be ingested only if proper sanitary or hygienic practices are not followed, and a simple washing of the hands after forming a stool is remarkably effective. Larvae actively burrow through the intestinal wall and enter the circulatory system, where they are carried sequentially to the liver, right side of the heart, and lungs. Larvae bore through the lung's walls and enter lung alveoli. They are carried up the bronchioles, bronchi, trachea, and into the pharynx; are swallowed; pass through the esophagus and stomach; and enter the small intestine. During this tortuous migratory path, each larva molts two more times, and the immature worm attains sexual maturity in the intestine.

NECATOR AMERICANUS

The New World or American hookworm is distributed widely from the southern United States to central South America. Females measure 9–11 mm in length, while males are 7–9 mm long. Adults, living within the small intestine, hold onto the intestinal wall by means of a circlet of oral hooks or teeth (Fig. 7.11a) as they feed on

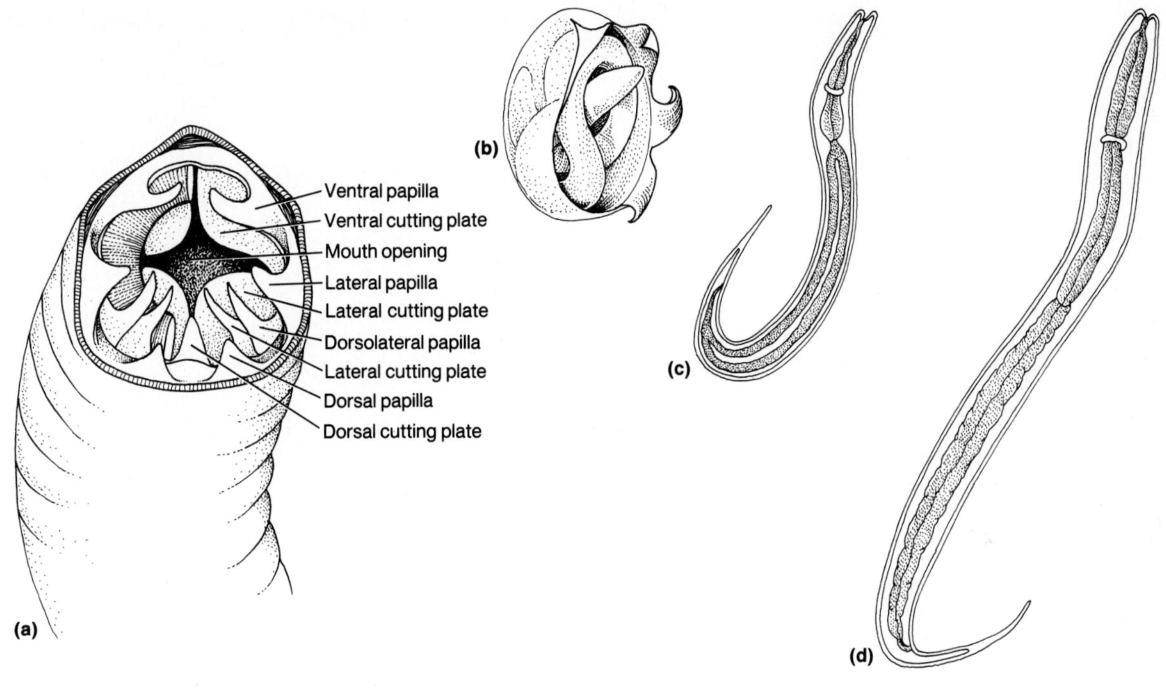

Figure 7.11 Necator americanus, *the hookworm: **(a)** the anterior end of an adult; **(b)** an embryonated egg; **(c)** a rhabditiform larva; **(d)** a filariform larva.*

blood and tissue fluids. Individual females may produce as many as 10,000 fertilized eggs daily! Embryonated eggs exit in the feces of the host, and development continues only in well-oxygenated, moist soil (Fig. 7.11b). Hatching takes place within 24 hours, producing a minute larva, a **rhabditiform** (Fig. 7.11c). It molts twice to become a second larval type, the **filariform,** which is free living, feeds on soil bacteria, and is the infective stage (Fig. 7.11d).

Infection is achieved when the filariform penetrates the skin, usually on the bottom of the foot or between the toes. Defecation out-of-doors and then, later, a prospective host's walking through the immediate area barefooted are two almost essential prerequisites for the life cycle to continue. The larva burrows rapidly through the skin and enters the circulatory system. From this point onward the passage of the larva is identical to that of *Ascaris lumbricoides.* Upon reaching the small intestine the last two molts take place, and the worm attains sexual maturity.

WUCHERERIA BANCROFTI

This small filarial worm causes millions of cases of elephantiasis in the Orient, Australia, the Near East, Mediterranean Africa, and Central and South America. The worms live in the human lymphatic ducts and glands. The condition of **elephantiasis** is caused by numerous worms blocking lymph vessels. Lymph accumulates distal to the blockage, resulting in grotesque-appearing, severe edema to an extremity such

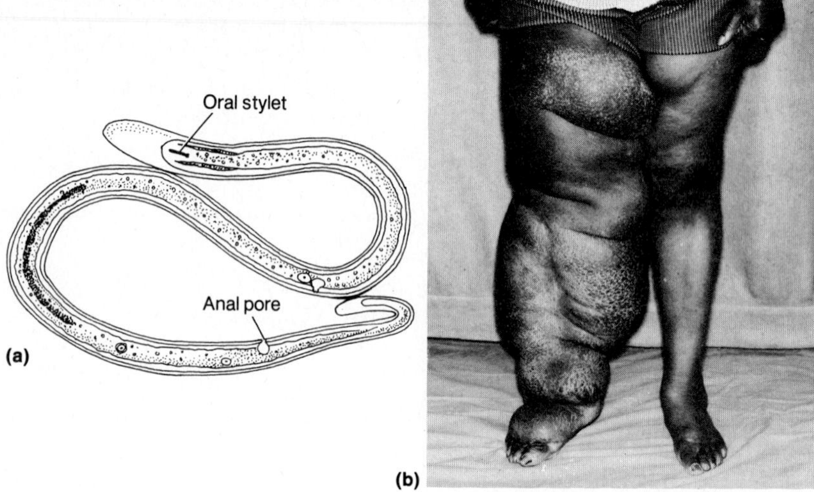

Figure 7.12 Wuchereria bancrofti, *a nematode that causes elephantiasis:* ***(a)*** *an adult worm (from Fulleborn in L.H. Hyman,* The Invertebrates. *Vol. III,* Acanthocephala, Aschelminthes, and Entoprocta, *New York, McGraw-Hill, 1951, adapted with permission);* ***(b)*** *a photograph of an example of elephantiasis (courtesy of the Armed Forces Institute of Pathology [AFIP], Negative No. 76-18002).*

as arm, leg, or scrotum (Fig. 7.12b). Worms are 200–300 μm in length, males being smaller than females.

After copulation, rapid embryonic development ensues, and the young larvae, called **microfilariae,** are released into the blood stream of the human host (Fig. 7.12a). With precise regularity, microfilariae migrate at night to the peripheral circulation in the skin. The underlying causes of this fantastic example of nocturnal periodicity within the human body are not known. The life cycle can continue only if females of certain species of mosquitoes bite the human host and ingest microfilariae, which are, strategically, in the peripheral circulation of that host at night when mosquitoes are feeding. Upon reaching the mosquito's midgut the worms penetrate the gut wall and enter thoracic muscles. At this point the worm undergoes a metamorphosis and molts twice. Third-stage larvae migrate to the mosquito's proboscis and are introduced into another human host when the mosquito feeds again. The final two molts take place as larvae enter the human lymphatic system.

TRICHINELLA SPIRALIS

This species is by far the most serious of the four examples of nematode parasites. Adult females, measuring about 3–3.5 mm in length, and males, about one-half the size of females,

Figure 7.13 *A photomicrograph of larvae of* Trichinella spiralis *encysted in striated muscle (courtesy of Ward's Natural Science, Inc., Rochester, N.Y.).*

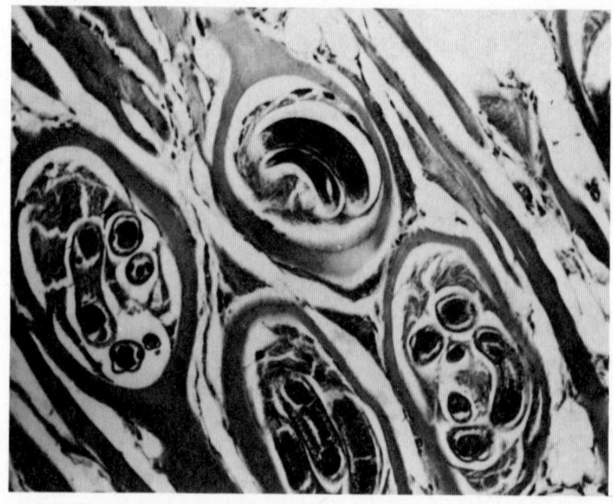

are situated in the small intestine of human beings. Additionally, they are present in many mammals, the most important of which are pigs. Adults are embedded in the intestinal mucosa, where they mate. Females are ovoviviparous, and a single female may give birth to as many as 1500 larvae; adults die shortly after reproduction. Minute, fourth-stage larvae bore through the intestinal wall, enter the circulatory system, and are carried by blood to muscles of the same host such as those of the appendages and trunk wall. Each larva bores only into striated muscles, where it soon becomes encapsulated within a calcified cyst (Fig. 7.13). Encysted larvae remain viable for as long as seven to eight years. Infective meat must be ingested by another host for the life cycle to continue. Once ingested, larvae excyst in the small intestine and mature into adults.

Trichinosis is a serious, sometimes fatal condition that is usually accompanied by muscular pains, stiffness, and some muscular atrophy. Most cases of human trichinosis occur from consuming improperly cooked pork. In the mid-1940s an estimated 28 million persons had this disease, but the incidence of trichinosis is considerably less today, owing to increased awareness of the condition and better education about the necessity of cooking pork thoroughly. Pigs may be infected by eating rats or garbage containing improperly cooked pork. Rats usually become infected by cannibalism.

PHYLUM NEMATOMORPHA

Adult nematomorphans or horsehair worms are free living in aquatic habitats. However, the larvae are obligatory parasites in arthropods. There are about 250 species; and since all but a few belong to the Class Gordioidea, the discussion of nematomorphans that follows is based on the gordioideans.

Adult nematomorphans are long, threadlike, cylindrical worms (Fig. 7.14a). Most range from 0.5 to 1 m in length and are no more than 3 mm in diameter. All are dioecious, and the smaller males have posterior ends that are curved ventrally as in nematodes. Adults are found writhing in shallow puddles, marshes, streams, and ponds. Because they live in entan-

Figure 7.14 *Nematomorphans:* ***(a)*** *a photograph of two adult worms;* ***(b)*** *the larva of a typical gordian.*

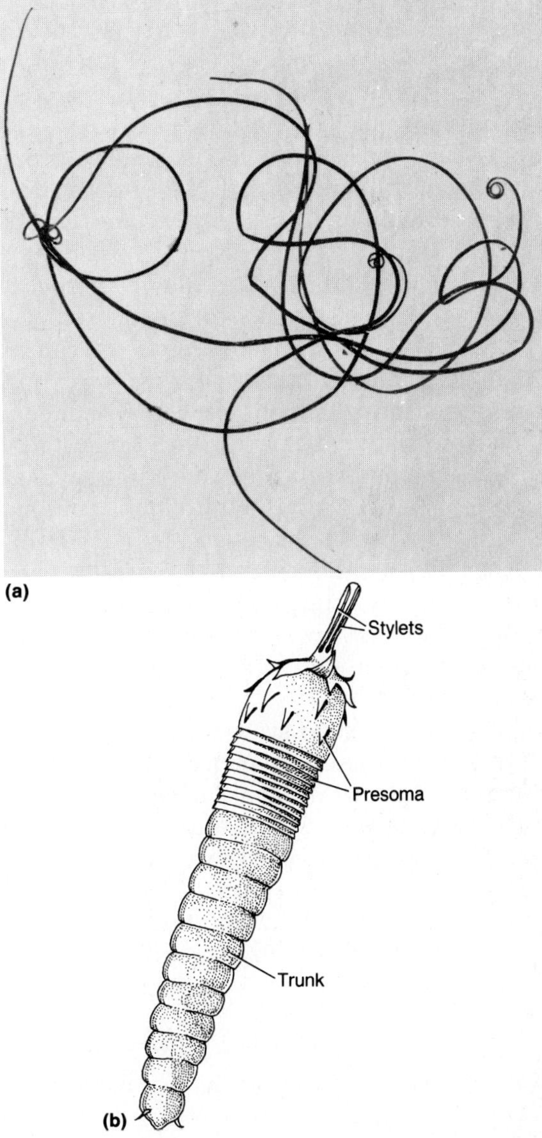

gled masses, they are called gordian worms from the Greek mythological Gordian knot. They may be yellow, gray, tan, dark brown, or black. Both ends of the body are blunt and rounded rather than tapered as in nematodes. The unpigmented anterior end or **calotte,** only slightly set off from the rest of the body, bears a terminal or subterminal mouth. The posterior end in many species bears two or three **lobes** surrounding the cloacal aperture.

The body wall consists of an outer cuticle, an epidermis, and a muscle layer. The cuticle is laminated, fibrous, and often rough, owing to the presence of sensory plates called **areoles** and minute papillae, bristles, and hairs. The epidermis is thickened in the midventral line to form a **ventral cord.** Beneath the epidermis is a layer of longitudinal muscles that are unusual in that the striated contractile fibers are located outside of the cytoplasmic portion. Internal to the muscle layer is a pseudocoelom packed with mesenchymal cells so that little space is evident.

Internally, a nonfunctional alimentary canal is present, and adults do not feed. The mouth opens into a pharynx filled with cells. Posterior to the pharynx is a hollow intestine that may function in excretion. The intestine is continuous posteriorly as a cloaca, which opens to the outside by a cloacal aperture. No distinct circulatory, gas exchange, or excretory organs are present. The nervous system consists of a circumenteric brain within the calotte and a midventral nerve cord.

The reproductive system of both sexes is essentially the same. The paired ovaries contain numerous lateral diverticula where egg maturation takes place. Eggs from each ovary travel through a separate oviduct to a bulbous chamber, the **antrum,** representing a part of the cloaca. Arising from the antrum is a slender seminal receptacle. In males, each of the two elongate, cylindrical testes communicates with the cloaca via a sperm duct.

Copulation usually takes place in the summer. The male coils its posterior end around that of the female, and sperm are transferred through the cloacal apertures of both worms. Sperm actively migrate to the seminal receptacle, and fertilization occurs in the antrum. Fertilized eggs are extruded through the cloaca of the female in long gelatinous strings, each containing up to several million eggs. Obviously, the reproductive potential of a single female is enormous. Males die after copulation, and females die after egg laying.

After incubating for 15–80 days, depending upon the temperature, the eggs hatch into small (250 μm) cylindrical larvae, each of which is divided into a large anterior **presoma** and a **trunk** (Fig. 7.14b). The presoma has three long stylets and three circlets of hooks. There are two methods by which a larva can enter the body of a suitable arthropod host such as a cricket, grasshopper, beetle, various aquatic insects, or even a centipede or millipede. Larvae of some species utilize the stylets and hooks to penetrate the host's body and enter the hemocoel. In most species, however, the larva encysts on vegetation or at the water's edge, where it is later ingested, excysts, bores through the wall of the alimentary canal, and enters the hemocoel. During a period of one to several months, each larva grows slowly and metamorphoses into a juvenile worm, involving the gradual loss of stylets, hooks, and other larval structures and the acquisition of adult structures such as the brain and reproductive organs. When mature, the juvenile molts, breaks through the body wall, and enters the water. Emergence from the host is apparently synchronized to the host's being wetted or to the arthropod's coming to the water to drink. The newly emerged nematomorphans are ready to copulate almost immediately. In the past, nematomorphans commonly appeared in watering troughs, since those were a convenient and available source of water. Nematomorphans are also known as horsehair worms because they were once thought to have developed from the detached hairs of horses who came to drink from the trough and because of their long threadlike bodies.

PHYLUM ACANTHOCEPHALA

Adult acanthocephalans, or spiny-headed worms, are all parasitic in the alimentary canal of a variety of vertebrates but mostly mammals, in which numbers of worms per host can be as high as 1500. They are small and usually do not exceed 40 mm in length, but one common form (*Macracanthorhynchus hirudinaceus*) can be up to 70 cm long. Females are always larger than males, and there are slightly more than 500 species.

The body of an adult acanthocephalan is elongate, cylindrical, tapered at both ends, and divided into two major regions: presoma and trunk. The anterior **presoma** is composed of the proboscis and neck. A distinguishing feature of acanthocephalans is the anterior retractile **proboscis,** which is armed with rows of spines and hooks (Fig. 7.15). The size, shape, number, and arrangements of these proboscidian spines are widely used criteria in species identification. The spiny proboscis, the source of their common name, is embedded within the intestinal wall of the host, where it is an effective holdfast organ. When not being used, the proboscis is retracted into an invagination, the **proboscis receptacle.** The ill-defined **neck** is immediately posterior to the proboscis. On either side of the neck the body wall is invaginated into the body cavity as two sacs, the **lemnisci,** which are not open to the exterior but are fluid filled. They function as a part of the hydrostatic mechanism responsible for proboscidian extension. The **trunk,** often equipped with minute spines, is separated from the smaller presoma by a fold in the body wall.

The body wall consists of a thin cuticle, an epidermis (or hypodermis), and two muscle layers. The epidermis contains a system of branching anastomosing canals, which are termed the **lacunar system.** The function of these canals is unclear, but they may be responsible for circulation or storage of nutrients. Beneath the outer circular and the inner longitudinal muscle layer is a well-developed pseudocoelom, which extends into the presoma.

The digestive system is completely absent

Figure 7.15 *Scanning electron micrographs of* Moniliformis, *an acanthocephalan:* ***(a)*** *an adult;* ***(b)*** *the proboscis or holdfast (parts a, b courtesy of D. R. Nelson).*

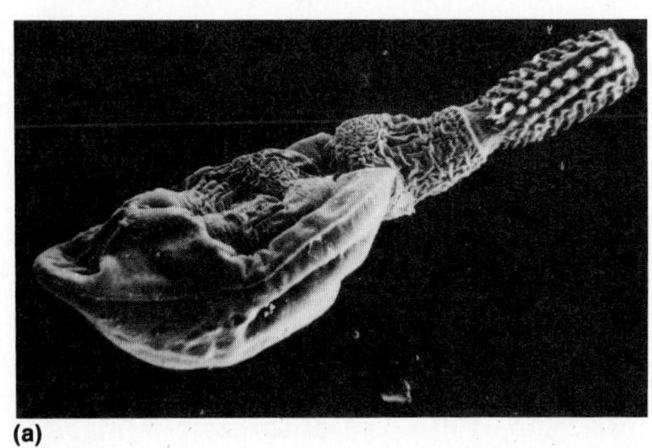

(a)

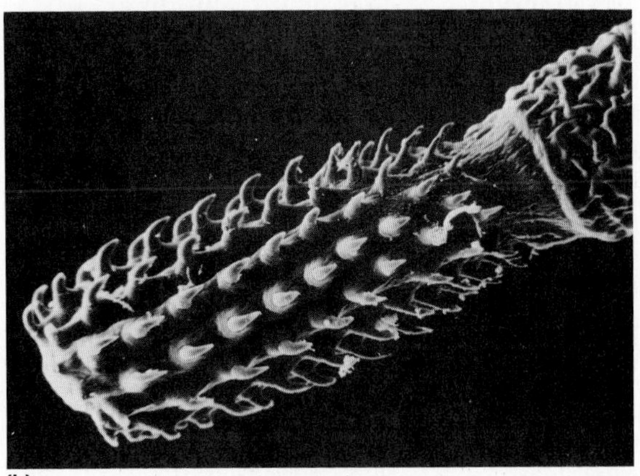

(b)

in all acanthocephalans. Likewise, the excretory system is missing in all except representatives of the Order Archiacanthocephala, in which a system of protonephridia is present. In all others, wastes simply diffuse through the body wall to the outside, perhaps through the lacunar system. The nervous system consists of an anterior ganglionic brain within the presoma and a number of nerves extending posteriorly in the trunk.

All acanthocephalans are dioecious, and the gonads are suspended within one or two **ligament sacs.** The transparent sac is attached anteriorly to the posterior end of the proboscis and posteriorly to the genital sheath (males) or uterine bell (females). In males the testes are suspended in tandem in the ligament sac. Arising from each testis is a sperm duct or vas deferens extending posteriorly that unites with its opposite to form a common sperm duct. As the common duct extends posteriorly, small cement glands open into it, and their secretion aids in copulation. The sperm duct, which may include a seminal vesicle, terminates in the bell-shaped, eversible cup, the **bursa.** In females, one or two ovaries are present only in the larvae; in adults they are fragmented into numerous **ovarian balls.** Ova, released from the ovarian balls into the ligament sac, may fill it completely. Incredibly, the ligament sac of a gravid *Macracanthorhynchus hirudinaceus* may contain up to 10 million embryonated eggs at any one time!

In copulation the bursa of the male is everted to surround the posterior end of the female, and sperm are inserted into the female gonopore. Sperm migrate anteriorly through the vagina and uterus to the uterine bell, where fertilization occurs. Only mature fertilized eggs are allowed to pass outward through the uterine bell to the female gonopore and into the intestinal lumen of the host.

Eggs escape in the fecal material of the host. As embryonic development proceeds, each egg contains a partially developed stage, the **acanthor.** Even in severe environmental conditions, eggs remain viable for months. For the life cycle of all acanthocephalans to continue, these embryonated eggs must be ingested by an invertebrate, usually an arthropod. Within an intermediate host, each egg hatches to release the acanthor, which has a spiny body and an anterior **rostellum** with hooks. The acanthor penetrates the gut wall of the intermediate host and enters the hemocoel. There the acanthor begins to transform into an immature adult; it is called an **acanthella** and, later, a **juvenile.** The juvenile either migrates to various organs or remains in the hemocoel, becomes surrounded by a sheath, and is called a **cystacanth.** Development within the invertebrate host requires from 1.5 to 3 months. When the invertebrate host is ingested by the appropriate definitive host, the cystacanth loses its sheath, attaches, and develops to sexual maturity. In a few species a second intermediate host is required.

PHYLUM LORICIFERA

This is the most recent phylum to be discovered; its first members were identified and named as recently as 1983. Loriciferans have been found in the interstitial spaces of shelly marine gravels in coastal areas of France, the Azores Islands, Denmark, Greenland, Florida, and North Carolina. They, along with other meiofauna such as kinorhynchs, gnathostomulids, and some priapulids, are adapted for existence in the interstices of many different marine, benthic habitats. Only one species of loriciferan, *Nanaloricus mysticus,* has been described to date, but specimens of several other species have been collected and will be described and named shortly.

Loriciferans are small (225–235 μm), bilaterally symmetrical, aschelminthlike animals. They have a spiny head or introvert, a neck or thorax, and an abdomen surrounded by a lorica (Fig. 7.16a, b). The **introvert** bears eight rigid **oral stylets** that are free only at their distal ends; they surround the terminal mouth. The introvert also bears nine rows of short appendages.

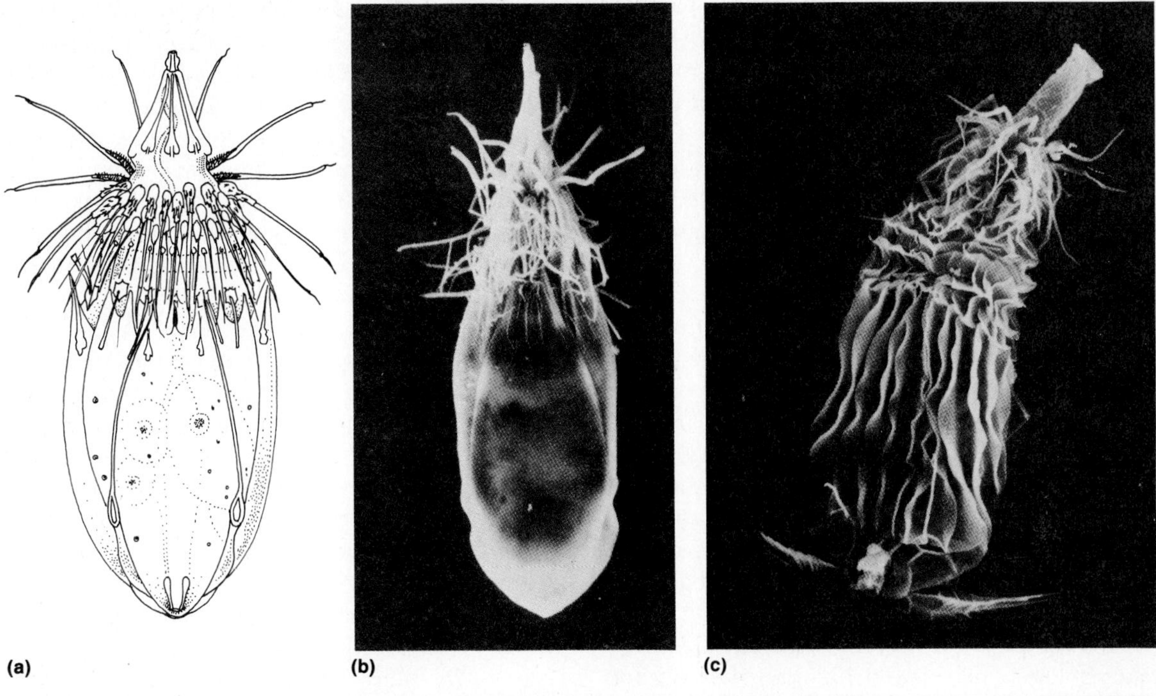

Figure 7.16 Nanaloricus mysticus, *a loriciferan and a representative of the most recently discovered phylum:* ***(a)*** *an adult female;* ***(b)*** *a scanning electron micrograph of an adult female, ventral view (parts a, b from R.M. Kristensen,* Zeitschr. f. zool. Systematik u. Evolutionsforschung 21:*167, 1983);* ***(c)*** *a scanning electron micrograph of a Higgins-larva (courtesy of R.M. Kristensen, Smithsonian Institution).*

The **thorax** is not sharply demarcated from the introvert. The introvert and neck can be retracted into the **abdomen,** which is surrounded by a lorica. The **lorica** consists of six cuticular plates with short, anteriorly directed spines (Fig. 7.16a). The lorical cuticle, which is periodically molted, may be homologous to that of kinorhynchs and even priapulids. The phylum name is based on the presence of a lorica (lorifera = to bear a lorica).

The digestive system is composed of a buccal tube in the introvert, a pharynx, esophagus, a straight midgut, short rectum, and a posterior anal cone (Fig. 7.16a). A large dorsal brain fills nearly half of the introvert, and the various introvert spines are individually innervated, suggestive that they are sensory in function. There are also several ganglia in the abdomen. There are very large internal retractor muscles extending to the introvert. The body cavity is presumed not to be a true coelom but rather a pseudocoelom. The loriciferans are dioecious, and the gonads are paired, but nothing is known at present about reproduction or early development.

There is produced a distinctively shaped **Higgins-larva** ranging in length from 120 to 185 μm, each having an introvert, thorax, and abdomen (Fig. 7.16c). The introvert may not be retracted into the lorica, but a prominent elongate **mouth cone** can be retracted into the introvert. The larval lorica is not as cuticularized as that of the adults. There are two prominent **caudal appendages** or toes that articulate with the abdomen by a ball-and-socket joint (Fig. 7.16c). The movable toes are employed as the locomotor organs in climbing or in swimming. Most adult systems and organs are present in the larva. The larva molts several times to become a **postlarval stage** before reaching adulthood.

The phylogenetic affinities of the loriciferans to other known groups are at best speculative at this point. The loriciferan cuticle is perhaps homologous to that in kinorhynchs and priapulids, details of the mouth cone are like those of kinorhynchs, details of the buccal canal and associated structures are similar to those of at least one genus of tardigrades, and larvae of loriciferans and those of certain marine nematomorphans have similar anterior digestive structures. The loriciferans may be the most closely related to the kinorhynchs, but their precise phylogenetic relationships to other similar phyla are obscure at this time.

PHYLOGENY OF THE ASCHELMINTHS

Among groups as diverse as are the phyla of aschelminths, phylogenetic affinities are at best obscure and confusing. Since there is meager paleontological evidence, we are forced to rely more on the structural and functional attributes of living forms. Often, morphological and physiological features are adaptations to specific ecological conditions, specialized habitats, or unique modes of life. Therefore evolutionary trends and ancestral origins are especially difficult to determine in such groups.

There are basically two theories about the phylogeny of the seven taxa of aschelminths. One argument is that the taxa are related to each other, since the presence of a pseudocoelom, cuticle, muscular pharynx, and adhesive glands are generally common to all groups. A contrasting theory holds that the aschelminth groups are not related to each other at all. The absence of any feature unique to all aschelminths strongly suggests independent evolution for each taxon. Moreover, the similarities observed among living aschelminths may simply be the result of convergent evolution as the animals in each group became adapted to similar environments and to the restrictions of their small size. It would appear that both theories have elements that, when combined, explain their relative positions more adequately than either of the theories alone. All groups are probably distantly related to each other, as evidenced by the few features that all share. True, convergent evolution may have produced some apparent nonphylogenetic similarities, but each taxon probably arose from a common ancestral stock and diverged early in their evolutionary history.

It is likely that the ancestral stock from which aschelminths arose was an acoelomate (Fig. 7.17). More specifically, a primitive, ciliated, acoelid turbellarian is the most likely source of the origin of aschelminths and, indeed, of the rest of the bilaterates. This would mean that the first aschelminths were ciliated, acoelomate, and marine; lacked a cuticle; and probably were monoecious.

Figure 7.17 *Possible phylogenetic relationships among the aschelminths.*

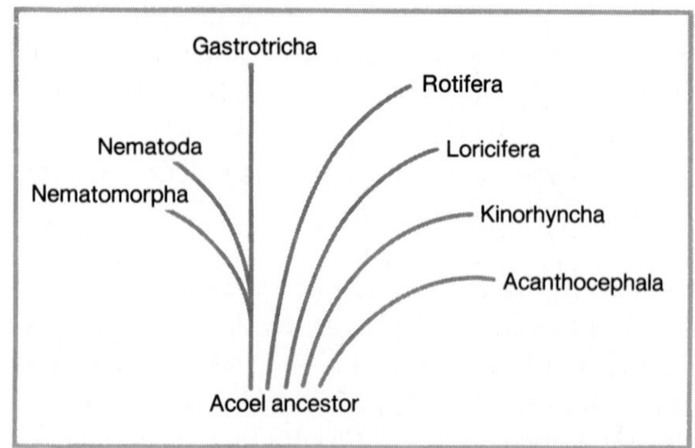

Of the seven aschelminth taxa the most primitive or central group is the Gastrotricha, since their ventral ciliation, hermaphroditism, small size, and being aquatic are primitive features. The aschelminth group apparently most closely allied to gastrotrichs is the Nematoda; similarities between these two groups include cuticular plates and pharyngeal cells with radially arranged myofibrils. Nematomorphans are probably more closely allied to nematodes than to any other group by virtue of representatives of both groups being wormlike, dioecious, and sexually dimorphic and having a relatively thick cuticle (Fig. 7.17).

Relationships among the remaining aschelminths (rotifers, kinorhynchs, loriciferans, and acanthocephalans) are even more obscure. Rotifers, like nematodes, are highly successful, and rotifers and gastrotrichs superficially appear to be similar, but rotifers are not closely related to either group. The evolutionary positions and phylogenetic affinities of the Kinorhyncha and Loricifera and especially the Acanthocephala are problematic. Future studies may show close affinities between the loriciferans and kinorhynchs, but that definite association simply cannot be confirmed at this time (Fig. 7.17). It is hoped that the grouping of aschelminths will be substantiated by phylogeny rather than simply being an association for convenience.

SYSTEMATIC RESUME

The Principal Aschelminth Taxa

PHYLUM GASTROTRICHA—aquatic; have a small body of head lobe, neck, trunk, and furca; have cuticular adhesive tubes.

Order Macrodasyida—marine or brackish; head lobe present or absent; lack protonephridia; possess pharyngeal pores and numerous lateral adhesive tubes; simultaneous or sequential hermaphrodites; commonly encountered genera include *Macrodasys* and *Urodasys*.

Order Chaetonotida—mostly freshwater forms; head lobe with tufts of cilia; lack pharyngeal pores and most have no male reproductive system; possess a pair of protonephridia and one or two pairs of posterior adhesive tubes; common genera are *Lepidodermella* (Fig. 7.2a) and *Chaetonotus*.

PHYLUM ROTIFERA—aquatic; body elongate with a ciliated corona, trunk, and foot; have a masticating mastax.

Class Seisonidea—elongate body; reduced corona; males well developed; marine; only one genus, *Seison*, which is ectosymbiotic on the crustacean, *Nebalia*.

Class Bdelloidea—mostly freshwater; two ovaries with vitellaria; no males known, and reproduction is exclusively by parthenogenesis; anterior and posterior ends highly retractile within telescopic body; two of the better-known genera are *Philodina* (Fig. 7.3c) and *Rotaria*.

Class Monogononta—only one ovary present; have a nonramate mastax; secreted tube and lorica are present or absent; males, known for some species, are small and degenerate.

Order Flosculariaceae—some free swimming and many sessile forms; when present, foot has no toes; corona is not very large; most have a tube or lorica; includes *Conochilus* and *Floscularia*.

Order Ploima—most are free swimming; corona not very large; have a posterior foot and two toes; most rotifers belong to this order; includes such species as *Asplanchna* (Fig. 7.3a), *Keratella*, and *Epiphanes* (Fig. 7.4b).

Order Collothecaceae—mostly solitary and sessile; have a very large corona; mouth at lower end of infundibulum; familiar genera are *Collotheca* and *Atrochus*.

PHYLUM KINORHYNCHA—marine, littoral; small; body with 13 zonites; have adhesive tubes.

Class Cyclorhagida—closing mechanism of retracted head consists of placids of zonite II or paired lateral plates on zonite III; *Echinoderes* is frequently encountered.

Class Homalorhagida—when head is retracted, opening is guarded by a dorsal and three ventral plates on zonite III; *Pycnophyes* is representative.

PHYLUM NEMATODA—roundworms; live in aquatic and soil habitats; many are parasitic; body slender, elongate, covered by a thin cuticle.

Class Adenophorea—excretory system is poorly developed or absent; phasmids absent; amphids modified externally.

Order Enoplida—predatory, primitive, free living, mostly marine; have many cephalic sense organs and a well-developed buccal cavity; includes *Enoplus* and *Cryptonchus*.

Order Dorylaimida—contains many plant parasites; lips usually well developed; includes *Longidorus, Dorella,* and *Mydonomus*.

Order Trichocephalida—pharynx with weak muscles but with one or two rows of large glands; mouth without lips; mostly parasitic in digestive tract of vertebrates; includes whipworms *(Trichuris)* and trichina worms (*Trichinella spiralis* [Fig. 7.13]).

Class Secernentea—possess excretory system and phasmids; amphids are simple; many are free living in soils; many are parasites.

Subclass Rhabditia—free living in soil; saprobic; lack an oral stylet; larvae and adults often have a complex, three-part esophagus.

Order Rhabditida—0–6 lips present; females with one or two ovaries; some are free living in soil, others are endoparasites; includes vinegar eel *(Turbatrix aceti)* and *Rhabditis* (Fig. 7.8a, b).

Order Strongylida—parasites in vertebrate guts; no lips, but oral teeth present; males have expanded caudal bursa supported by rays; includes the hookworms (*Necator americanus* [Fig. 7.11], *Ancylostoma duodenale*) and *Syngamus*.

Order Ascaridida—large; all parasitic in digestive tract of vertebrates; have three large lips; no buccal cavity; includes *Ascaris* (Fig. 7.10), *Toxocara,* and the pinworm (*Enterobius vermicularis* [Fig. 7.7b]).

Subclass Spiruria—parasites in vertebrates; many have complicated life cycles involving two hosts; includes elephantiasis worm (*Wuchereria bancrofti* [Fig. 7.12a]), African eye worm *(Loa loa), Thelazia,* and *Spirura*.

Subclass Diplogasteria—cuticle with longitudinal striae; muscular esophagus divided into two distinct parts; many representatives are of economic importance, since they are pests on agricultural plants; includes *Diplogaster, Tylenchus,* and *Criconema*.

PHYLUM NEMATOMORPHA—adults free living in fresh water, larvae parasitic in arthropods; adults lack digestive tract and certain other systems.

Class Gordioidea—includes most nematomorphans; larvae parasitic in terrestrial or freshwater arthropods; pseudocoelom filled with mesenchymal cells; paired gonads; only a ventral epidermal cord present; common genera include *Gordius* and *Paragordius*.

Class Nectonematoidea—includes only one genus, *Nectonema,* the adults of which are pelagic and marine; larvae parasitic in crustaceans; have an extensive hollow pseudocoelom and both a dorsal and ventral epidermal cords; unpaired gonads.

PHYLUM ACANTHOCEPHALA—adults are parasitic in digestive tract of vertebrates; have a spiny, anterior, retractile, holdfast proboscis.

Order Archiacanthocephala—adults parasitic in terrestrial hosts like birds and mammals; larvae parasitic in roaches and grubs; proboscidian spines are concentrically arranged; have protonephridia; females with two ligament sacs; includes the common *Macracanthorhynchus* in mammals and *Moniliformis* (Fig. 7.15) in rodents.

Order Palaeacanthocephala—adults parasitic mostly in fishes, a few aquatic birds, some mammals, and reptiles; larvae parasitic in crustaceans; proboscidian hooks in alternating radial rows; no protonephridia; one ligament sac; common genera are *Illiosentis* and *Leptorhynchoides*.

Order Eoacanthocephala—adults parasitic in fishes and reptiles; larvae parasitic in crustaceans; proboscidian hooks radially arranged; no protonephridia; females with two ligament sacs; includes *Neoechinorhynchus* in freshwater fishes and turtles and *Pallisentis* in fishes.

PHYLUM LORICIFERA—interstitial meiofauna in gravely marine beds; body of introvert, thorax, and abdomen; abdomen is surrounded by a cuticularized lorica of six plates; only one species, *Nanoloricus mysticus* (Fig. 7.16), described to date.

ADDITIONAL READINGS

Bird, A.F. 1971. *The Structure of Nematodes,* 318 pp. New York: Academic Press.

Cheng, T.C. 1973. *General Parasitology,* pp. 545–692. New York: Academic Press.

Chitwood, B.G. and M.B. Chitwood. 1974. *Introduction to Nematology,* rev. ed., 323 pp. Baltimore: University Park Press.

Croll, N.A., ed. 1976. *The Organization of Nematodes,* 424 pp. New York: Academic Press.

Croll, N.A. and B.E. Matthews. 1977. *Biology of Nematodes,* 193 pp. New York: John Wiley & Sons.

Crompton, D.W.T. 1970. *An Ecological Approach to Acanthocephalan Physiology,* 121 pp. Cambridge, England: Cambridge University Press.

Dumont, H.J. and J. Green, eds. 1980. *Rotatoria,* 264 pp. The Hague, Netherlands: Junk.

Freckman, D.W., ed. 1982. *Nematodes in Soil Ecosystems,* 206 pp. Austin: University of Texas Press.

Gilbert, J.J. 1980. Developmental polymorphism in the rotifer *Asplanchna sieboldi. Amer. Sci. 68:*636–646.

Hyman, L.H. *The Invertebrates.* Vol. III: *Acanthocephala, Aschelminthes, and Entoprocta,* pp. 1–520, 1951. New York: McGraw-Hill Book Co.

Kaestner, A. 1967. *Invertebrate Zoology.* Vol. I, pp. 217–263. New York: John Wiley & Sons.

Kristensen, R.M. 1983. Loricifera, a new phylum with Aschelminthes characters from the meiobenthos. *Zeitsch. Zool. System. Evolut.-forsch. 21:*163–180.

Lee, D.L. and H.J. Atkinson. 1977. *Physiology of Nematodes,* 2nd ed, 199 pp. New York: Columbia University Press.

Maggenti, A. 1981. *General Parasitology,* 372 pp. New York: Springer.

Nicholas, W.L. 1975. *The Biology of Free-Living Nematodes,* 209 pp. Oxford, England: Clarendon Press.

Pennak, R.W. 1978. *Fresh-water Invertebrates of the United States,* 2nd ed., pp. 147–238. New York: John Wiley & Sons.

Poinar, G.O., Jr. 1983. *The Natural History of Nematodes,* 324 pp. Englewood Cliffs, N.J.: Prentice-Hall.

Zuckerman, B.M., ed. 1980. *Nematodes as Biological Models.* Vol. I: *Behavioral and Developmental Models,* 304 pp., Vol. II: *Aging and Other Model Systems,* 295 pp. New York: Academic Press.

8

The Molluscs

OVERVIEW

Members of the Phylum Mollusca, including snails, clams, octopods, and many more, are truly remarkable animals because of their extensive diversity in both form and function. The more than 100,000 species are found in many habitats and occupy innumerable ecological niches. Even though most are marine, rather large numbers are found in fresh water, and slugs and many snails are terrestrial.

Amid this enormous diversity in external shapes, niches, and varied functions, all molluscs have a basic, unique body plan. The soft body is composed of three parts: mantle, head–foot, and visceral mass. The mantle both protects the enclosed visceral mass and secretes the protective, external, calcareous shell. Situated between the mantle and visceral mass is the mantle cavity, which contains the ctenidia (gills) and is variously adapted for many different functions. The muscular, highly variable head–foot is involved in feeding and locomotion. The visceral mass contains the internal organs, including an open circulatory system with an extensive hemocoel, a reduced coelom, paired kidneys, and regional ganglionic "brains." Two diagnostic digestive features are a feeding radula in the head and a crystalline style in the stomach.

Four molluscan classes contain few species, are marine, and are relatively unfamiliar. Aplacophorans are primitive, small, wormlike creatures lacking many molluscan features. Monoplacophorans, long thought to be extinct and represented today by *Neopilina* and *Vema*, possess a one-piece shell, broad foot, radula, style, and repetitive arrangement of many visceral organs. Polyplacophorans (chitons) are found on rocky coasts and have a broad creeping foot, eight overlapping shell plates, a reduced head, but no style. Scaphopods bear a tubular shell open at both ends and possess a burrowing foot.

Gastropods (snails) are extremely diverse and remarkably successful. Two fundamentally important events occur in their larval (veliger) stage: torsion and spiraling. All gastropods undergo torsion, a 180° counterclockwise rotation of the visceral mass so that the arrangement of some internal organs is twisted.

Most gastropods also undergo spiraling, a process in which the elongating visceral mass is spiraled into an asymmetrical coil. Gastropods also generally have a single spiraled shell, broad creeping foot, and radula. Most gastropods are prosobranchs, a very large group whose representatives are characterized by being mostly marine and strongly torted and having a pedal operculum, an elongated incurrent siphon, and internal fertilization. Opisthobranchs are all marine, have an elongated body, are mostly shell-less, exhibit no obvious coiling, are monoecious, and are able to swim. Pulmonates are freshwater or terrestrial snails and slugs that exhibit degrees of detorsion; all have a mantle cavity as a "lung" for gaseous exchange.

Bivalves (clams), which are laterally flattened to facilitate their burrowing habit made possible by a strongly compressed, muscular foot, lack a head and radula but have a shell composed of two equally sized valves. Most bivalves are lamellibranchs and have large, reflected, ciliated ctenidia used in filter feeding. A diverse group, bivalves are found buried in soft bottoms, attached to hard surfaces, or boring in other substrata. The protobranchs and septibranchs are small groups of atypical bivalves.

Cephalopods (squids, octopods), once numerous but now declining numerically, are mostly shell-less and use the muscular mantle in locomotion. They are efficient swimmers and have excellent sense organs and strong jaws; the foot is modified as both a swimming funnel and a ring of tentacles. Nautiloids have an external, coiled, bilateral shell; two pairs of ctenidia, auricles, and kidneys; and many tentacles. Decapods (squids) and octopods have one pair of ctenidia, auricles, and kidneys; decapods are torpedo shaped and have eight arms plus two tentacles, whereas octopods have eight arms and a globose body.

Because they are such an ancient group, most details of the evolutionary affinities of molluscs are obscured in history. Probably, the ancestral stock from which molluscs arose was a primitive turbellarian that gradually evolved the uniquely molluscan plan. Aplacophorans are thought to be very primitive and nearer to the ancestral stock than any other group. Polyplacophorans represent an early evolutionary offshoot that resulted in no other molluscan group. Monoplacophorans appear to be a more recent stem group; from them developed two principal evolutionary lines: one evolved into gastropods and cephalopods, and the other line produced the bivalves and scaphopods. Adaptive radiation was particularly dramatic in the gastropods, bivalves, and cephalopods, resulting in the enormous molluscan diversity.

INTRODUCTION TO MOLLUSCS

The Phylum Mollusca is an unusually large and extremely important phylum, and it contains many familiar representatives. To this eminently successful group belong clams and oysters, snails and slugs, squids and octopods, and many, many more. During their evolutionary history, molluscs have undergone extensive adaptive radiation, which has resulted in an incredible number of different kinds of animals.

It will be helpful to the reader at this point to begin to understand the taxonomic framework in which the many and varied molluscs are grouped. Taking into account both similarities and differences, molluscs have been grouped taxonomically into seven classes, which are characterized in Table 8.1. Almost four of every five mollusc species belong to the Class Gastropoda (snails, slugs), and the next largest taxon is the Class Bivalvia (clams, oysters); more than 98% of all mollusc species belong to these two groups. Each of two other classes, Polyplacophora and Cephalopoda, con-

TABLE 8.1. □ **CHARACTERISTICS OF THE CLASSES OF MOLLUSCS**

Class	*Common Name(s)*	*Nature of the Shell*	*Nature of the Foot*	*Other Distinguishing Features*
Aplacophora	Solenogasters	Absent	Absent	No head, mantle, or nephridia; many have longitudinal ventral groove
Monoplacophora	—	Conical, 1 piece	Flat, creeping	Multiple gills, retractor muscles, nephridia, gonads
Polyplacophora	Chitons	8 valves or plates	Flat, creeping	Thick mantle, no style
Gastropoda	Snails, slugs	1 piece, often coiled	Flat, creeping	All undergo torsion, some detorsion; most are coiled and asymmetrical
Bivalvia	Clams, mussels, oysters	2 hinged valves	Bladelike, burrowing	No head or radula; most are filter-feeders using large complex ctenidia
Scaphopoda	Tusk shells	Tubular, 1 piece	Burrowing	Head with tentacles or captacula
Cephalopoda	Squids, octopods, nautiluses	Absent, reduced, or spiral	Forms a ventral funnel, tentacles	Arms with adhesive suckers; have well-developed eyes, jaws, radula

tains only between 500 and 1000 species, and the remaining three classes are represented by even fewer species.

This phylum and its representatives are exceptional because of their numerical preponderance, familiarity, ecological importance, and immense evolutionary plasticity. In terms of numbers of species, molluscs rank second only to the arthropods. There are about 110,000 recognized modern species plus thousands of others known only as fossils. Often, numbers of individuals of a given species are quite large and occasionally staggering; for example, as many as 23,000 individuals of the common mud snail, *Ilyanassa*, have been found on 1 m^2 of littoral mud along the New England coast!

Molluscs are perhaps the best known and most familiar invertebrates because many are large, conspicuous, and common. Equally important in their familiarity is that the exquisitely colored and delicately sculptured shells of bivalves and gastropods appeal to professional and hobby collectors and even to casual beachcombers throughout the world. There is hardly a home in North America that does not contain decorative molluscan shells. They are also well-known paleontologically, since their hard mineral shells have been abundantly preserved in the fossil record, a fact that has richly contributed to our understanding of their evolutionary history. Most molluscs have lived in shallow aquatic habitats where their shells could be easily and quickly covered by sediments, thus preventing disintegration; quick burial of a hard part such as a shell is usually essential for fossilization. At least 45,000 extinct species have been discovered, owing primarily to the fossilization of their shells. The molluscan paleontological record is rather complete back into the Cambrian period, and molluscs undoubtedly thrived in the seas of the Proterozoic Era (see Table 1.1, Chapter 1).

Ecologically, molluscs are very important in many ecosystems. Basically an aquatic group, the vast majority is marine, but many species of gastropods and bivalves are found in fresh water, and many gastropods have been able to colonize terrestrial habitats. Limpets and chitons cling to rocks and other substrata, snails crawl or dig or swim, bivalves burrow or dig or bore, and cephalopods torpedo through the water or lurk in crevices. In terms of both numbers of species and numbers of individuals, molluscs are often the dominant organisms in many aquatic habitats, and they usually have a significant impact on these ecosystems. Exploiting a great measure of evolutionary plasticity, molluscs are adapted to fill a large number of ecological niches, and thus they bear the adaptive stamp of the environment in a far more obvious way than do other animals that move about actively. Perhaps the one feature that characterizes molluscs is their adaptability to many diverse ecological conditions, and their evolutionary opportunism has resulted in many different forms.

If one considers, for example, the extreme variations between a snail, slug, clam, and squid, one might arrive at an initial conclusion that molluscs have very little in common with each other. Yet amid this extensive morphological and physiological heterogeneity, molluscs are constructed on a remarkably consistent plan, and there are two methods by which we can better understand their common features. First, we can hypothesize about an ancestral mollusc that possessed all these features in a primitive, unspecialized way. Even though this ancestral form has not been found in the fossil record, such an archetypical mollusc may have existed, and careful analysis of existing molluscan taxa makes its existence at least plausible. The hypothetical ancestor represents a basic plan of the phylum, and it can be considered to be a generalized mollusc. The most important features of this archetypical mollusc appear in Box 8.1 and are illustrated in Figure 8.1.

A second, perhaps even more fruitful approach is to explore the basic plan on which modern-day molluscs are constructed and to emphasize those features common to most or all living molluscs. Clearly, there is no standard or common molluscan shape, since their vast diversity in form has already been stated. But a surprisingly uniform plan is present throughout the group, and the following description is one of this basic molluscan plan, a combination of unified form and function unique to molluscs.

THE MOLLUSCAN BODY PLAN

The molluscan body is invariably soft, owing to the lack of any sort of internal skeleton, and is usually contained within a hard protective shell; the phylum name is derived from this feature (mollis = soft). Upon closer examination we find that the body is divided into three distinct regions: the **head–foot**, a part that is basically concerned with locomotion and sensory reception; the **visceral mass** containing the organs of digestion, circulation, excretion, and reproduction; and the **mantle** that surrounds the visceral mass and secretes the shell (Fig. 8.1).

Before proceeding with a description of each of the body parts, it is important to understand a fundamental feature common to all three regions. Almost all external surfaces are covered with ciliated epithelium, which often contains mucous glands, and many epithelial layers also are ciliated. Molluscs use the cilia and mucus to carry out many of their basic functions. **Mucus**, secreted by epithelial mucous glands often in copious amounts, is utilized by the mollusc in food capture and feeding, locomotion, moistening of body surfaces, and cleansing surfaces or areas. **Cilia** are common both externally and internally and perform a variety of extremely important functions including the creation of water currents essential for ventilation of gas exchange surfaces, continuous monitoring of environmental conditions, and, in some, feeding. Cilia are frequently used in locomotion, for surface cleansing, and to move materials through the alimentary canal, kidneys, and gonoducts. Both cilia and mucus are of unusual importance in a great many of the functions of molluscs.

Particularly important is that molluscs have ciliated epithelial "**sorting surfaces**" or "**grading areas**," which function in sizing and diverting variously sized particles to different places. Grading areas are located in such places as within the alimentary canal and on accessory feeding organs, ctenidia (gills), and inner mantle surface. In its simplest form, the epithelium of sorting areas is thrown into a series of parallel ridges and grooves (Fig. 8.2a). Ciliary waves on the crests of the ridges sweep across or at right angles to the ridges, whereas cilia in the grooves beat along the grooves. Thus larger inert particles such as sand grains drop into the grooves and are usually rejected, while finer food particles are carried at right angles across the ridges to other areas for processing. In more complex sorting areas, separation of particles in up to four or five size categories is achieved by the creation of differently directed currents within subdivisions of the grooves (Fig. 8.2b). These important grading areas will be mentioned later in connection with their several different functions.

First, the mantle and its associated features will be explored, followed by a treatment of the head–foot complex and then a discussion of internal features of the visceral mass. It should be kept in mind that all of the following morphological and physiological adaptations are integral parts of the basic molluscan plan.

Mantle

Perhaps the most distinguishing feature of all molluscs is the presence of a characteristic **mantle** or **pallium** and the features associated with it. As a lateral outgrowth of the dorsal aspects of the visceral mass, the mantle surrounds and enfolds the visceral mass and some of the foot like a poncho or skirt so that its symmetry is almost radial. Thin, fleshy, and muscular, the pallium functions basically in protection of the animal and in the formation of the shell. The most important region of the mantle from a functional standpoint is its distal portion or margin. The mantle margin bears three lobes: an outer **secretory lobe**, which produces the shell; the middle **sensory lobe**, which often bears small tentacles and, sometimes, eyes; and the inner **muscular lobe**, which is capable of being extended into water current channels or in cross fusion with the same mantle part on the opposite side of the animal (Fig. 8.3).

Because of the mantle, a semiinternal space is formed between the mantle and visceral mass called the **mantle** or **pallial cavity**. This space, with its remarkable modifications in both form and function, is as much a diagnostic molluscan feature as the mantle itself. Its principal function is as a space in which gaseous exchange takes place, since it houses the **ctenidia** or gills. Additionally, the mantle cavity functions as a feeding chamber in bivalves, as a brooding chamber in many different forms, and in locomotion in cephalopods and a few bivalves. The tremendous molluscan diversity both in overall shape and form and in their shells is due in large measure to the adaptive plasticity of this remarkable organ and its cavity. We shall now explore two vitally important features associated with the mantle and mantle cavity: the shell and the ctenidia.

Shell

A hard, external, calcareous **shell** is the most obvious feature of molluscs and one of their most characteristic structures. Varying widely in form and shape throughout the phylum, the shell reaches its greatest development in snails and bivalves. It may be of one piece (gastropods and others), two equally sized valves (bivalves), or eight transverse valves (polyplacophorans);

BOX 8.1 □ THE HYPOTHETICAL ANCESTRAL MOLLUSC (HAM)

Scientists disagree on HAM, but the following is a likely scenario. The ancestral mollusc inhabited shallow inshore waters of the Proterozoic oceans and was most abundant on rocks or other hard surfaces in temperate or tropical habitats. HAM was small, slow moving, bilaterally symmetrical, and oval in shape when viewed from above or below. It possessed a dorsally arched body and a small head (Fig. 8.1). Ventrally, there was a flat, ovoid, muscular foot used for creeping. The foot's sole was equipped with mucous glands and cilia; secretions from the former served to lubricate the substratum, and the beating of the latter aided in locomotion. The dorsal body extended upward into a low visceral mass covered on all sides by the mantle, which served both to protect the underlying visceral organs and to secrete a protective outer covering. The outer epidermis of the mantle secreted calcium carbonate spicules, which, in time, became numerous and contiguous and formed a primitive calcareous exoskeleton. Alternatively, the mantle may have produced a rudimentary shell of little more than a tough horny cuticle that later became reinforced with $CaCO_3$. From this primitive outer covering evolved a true, protective, calcareous shell.

The small head probably bore short sensory tentacles and a pair of simple eyes. Inside the mouth a buccal cavity contained a radula sac housing the filelike tongue or radular apparatus (Fig. 8.1). Operated by a complex of small muscles, the radula loosened bits of encrusting algae on which HAM fed. Food particles were trapped in mucus to facilitate feeding. Several pairs of lateral pedal retractor muscles extended from the mantle and shell into the foot. When they contracted, the visceral mass and shell were pulled down tightly over the foot and against the substratum, thus providing a modicum of protection.

Between the mantle and the visceral mass and foot was the mantle cavity, which was shallow anteriorly but became progressively deeper toward the posterior end, where one to several pairs of bipectinate ctenidia or gills were located (Fig. 8.1). The posterior mantle cavity received wastes from the excretory organs and from digestion. Cilia on the ctenidial filaments created definite pathways for water circulation in the mantle cavity, thus facilitating both ctenidial aeration and waste removal.

From the mouth the mucus–food cord was pulled through a short esophagus and into a large stomach by the rotating crystalline style; the stomach contained several specialized areas for processing food. A pair of large digestive glands was responsible primarily for secreting digestive enzymes and for absorption. A long coiled intestine terminated posteriorly in the anus (Fig. 8.1).

The "open" circulatory system was represented by a middorsal heart consisting of a pair of posterolateral auricles and a median ventricle (Fig. 8.1). Oxygenated blood, brought from the ctenidia to the auricles by efferent blood vessels, then entered the muscular ventricle, which pumped it anteriorly through a single aorta. The aorta branched into smaller arteries that finally terminated in the ill-defined hemocoel. Blood was collected from the body by afferent blood vessels and returned to the ctenidial circulation. The small reduced coelom in the upper middorsal region surrounded the heart and a small portion of the intestine as the latter passed in close proximity to the

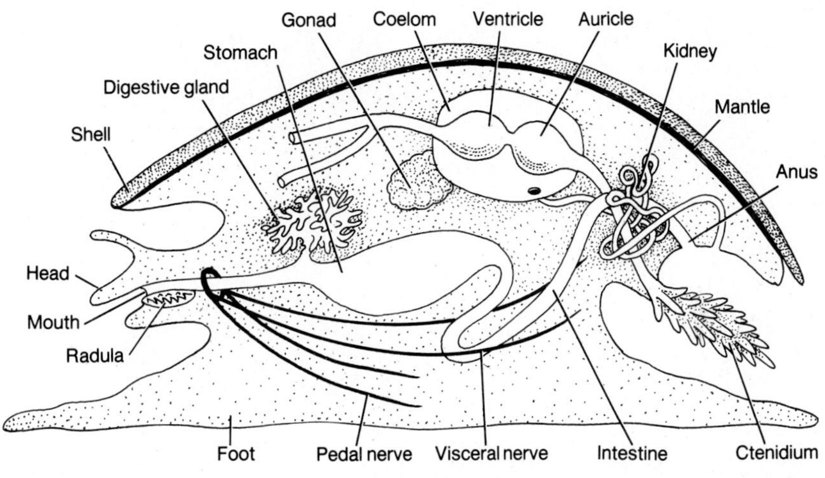

Figure 8.1 *A sagittal section of the hypothetical ancestral mollusc (HAM).*

heart (Fig. 8.1). Nitrogenous wastes, secreted into the coelom, were excreted to the outside via a pair of coelomoducts.

The simple nervous system was probably devoid of ganglia and thus consisted of dispersed neurons. A central ring of nerves encircled the esophagus, and two pairs of longitudinal nerves arose from this ring; one pair innervated the mantle and viscera, and the other pair supplied the muscles of the foot (Fig. 8.1). Sense organs were also simple and consisted of eyes on the head, osphradia near the ctenidia in the incurrent water pathway, and statocysts in the foot.

Mature eggs or sperm were shed into the coelom from a pair of laterally situated gonads, transported to the mantle cavity by coelomoducts, and expelled in the exhalent current. Fertilization took place in the surrounding seawater. The embryology of HAM logically followed the pattern of the primitive modern-day forms. Following fertilization the embryo developed quickly into a small, free-swimming trochophore larva, and a later veliger stage may have developed. After completing larval development in perhaps several weeks the natal animal sank to the bottom, metamorphosed, and began its benthic existence.

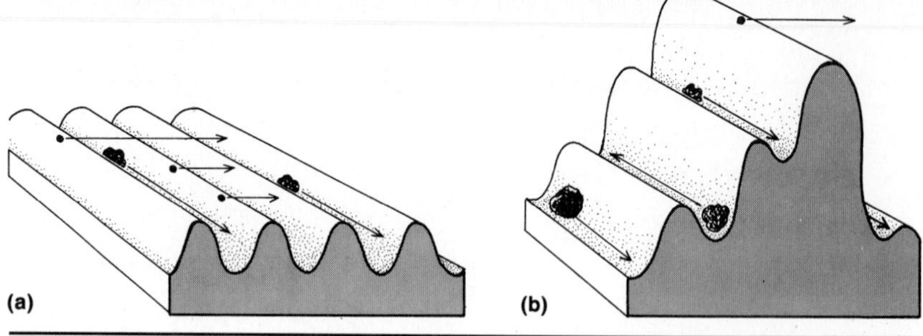

Figure 8.2 Diagrammatic representations of molluscan ciliated sorting areas: ***(a)*** a simple area on which lighter particles are carried across the ridges and the larger ones are swept along in the grooves; ***(b)*** a complex area in which particles of various sizes are sorted in different directions.

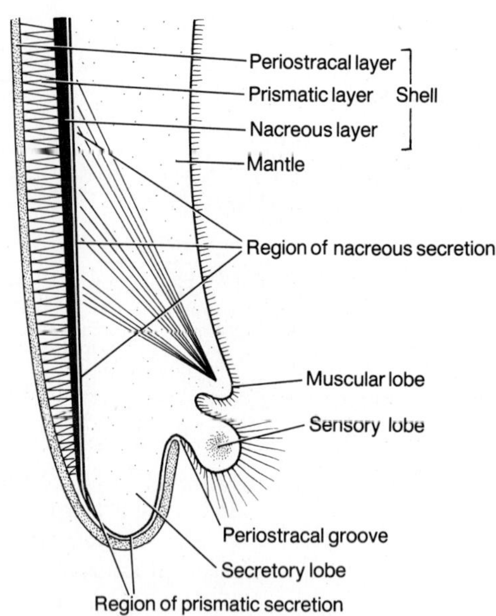

Figure 8.3 The margin of the molluscan shell and mantle.

it is reduced or absent in others (cephalopods and some gastropods) (Table 8.1, Fig. 8.4). Regardless of the shape or form of the shell, it is important to remember that it is always underlain by the fleshy mantle.

The shell, first appearing in the larval stage (veliger), is a product of the mantle, which continues to add new materials to it by accretion during the entire lifetime of the mollusc. Since the shell is secreted initially by the free, distal, mantle margins, new materials are laid down in concentric layers on the existing shell margin. Most molluscs do not produce new shell layers continuously but rather in spurts during warmer months. When shell layers are deposited in thick straight layers, the shell is heavy and porcelainlike, but when layers are thin and wavy, then the shell is glossy, pearly, and often iridescent. Shell deposition is also subtly affected by tidal, diurnal, and other seasonal events and rhythms.

The shell, constructed of calcium carbonate crystals embedded within a protein meshwork, is composed of three layers: an outer periostracum, a middle prismatic layer, and an inner nacreous layer. The **periostracum**, the first layer to be produced, is composed chiefly of a horn-like protein, **conchiolin**, which is fairly resistant to dissolution by water. The periostracum is secreted in a characteristic groove between the secretory and sensory mantle lobes (Fig. 8.3). The **prismatic layer**, thickest of the three layers, is composed of densely packed, polygonal prisms or columns of calcium carbonate either as calcite or aragonite. Prisms are deposited at right angles to the the periostracum and nacreous layer with thin membranes of conchiolin produced between them. The prismatic layer is secreted by a group of columnar cells on the outer surface of the secretory lobe (Fig. 8.3). The inner **nacreous layer**, usually laid down in sheets or laminae, is produced during the entire lifetime of the mollusc by scattered secretory cells in the outer pallial epithelium, with the result that the entire shell increases in thickness (Fig. 8.3).

Figure 8.4 *Sagittal diagrams of a representative of each of the five principal classes of molluscs to illustrate the extent of the shell (shown in bold lines and shaded pattern) and the foot:* ***(a)*** *a polyplacophoran (chiton);* ***(b)*** *a gastropod (snail);* ***(c)*** *a bivalve (clam);* ***(d)*** *a scaphopod (tusk shell);* ***(e)*** *a cephalopod (squid).*

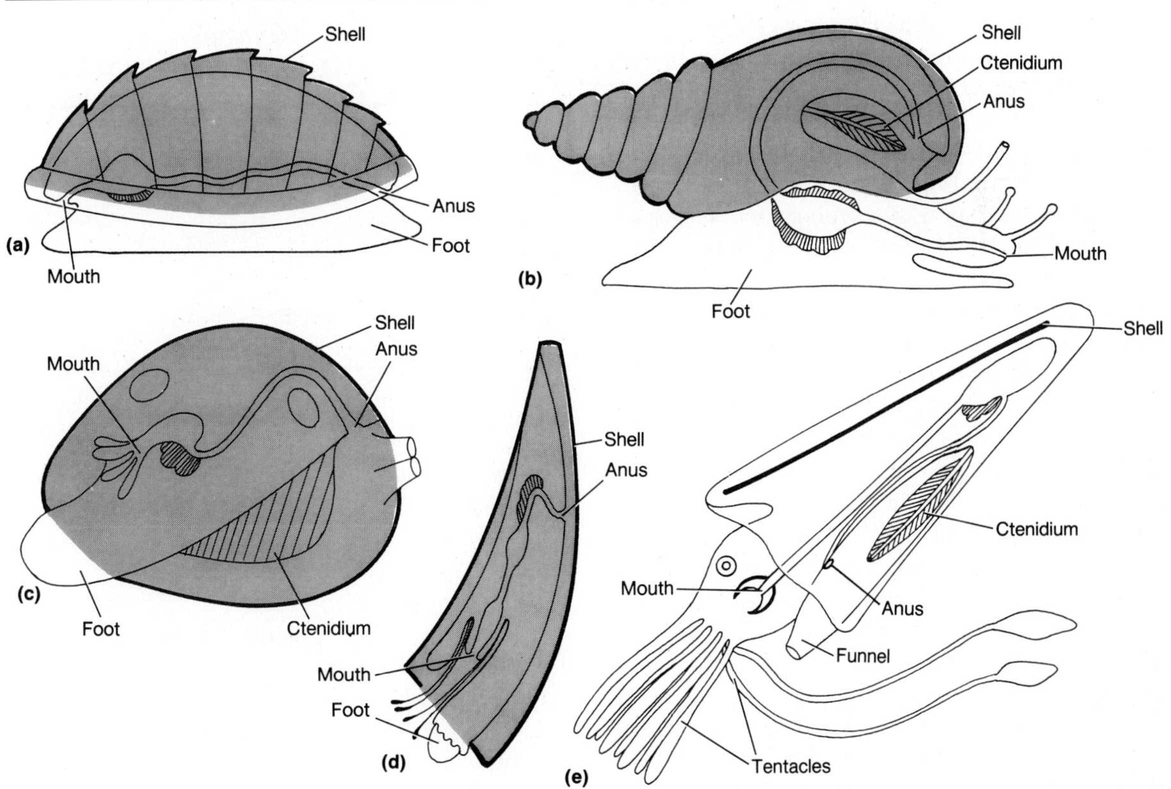

CTENIDIA

Molluscan gills, structurally and functionally different from those in other animals, are called **ctenidia**. Primitively, ctenidia are paired structures positioned in the posterior mantle cavity. Each ctenidium is constructed of a flattened axis bearing triangular platelike **filaments** in an alternating fashion on either side of the axis (Fig. 8.5a). The axis contains a **nerve**, a dorsal **afferent blood vessel** bringing deoxygenated blood from the body to the ctenidium, and a ventral **efferent blood vessel** carrying oxygenated blood to the heart. Blood enters each ctenidial filament from the axis, diffuses from dorsal to ventral through the filament, and is gathered into the efferent blood vessel (Fig. 8.5b).

All surfaces of ctenidial filaments are covered by ciliated epithelium, and the collective action of the cilia creates an essential flow of water through the mantle cavity and over the filaments. **Frontal cilia** cover the ventral or frontal ctenidial margins (Fig. 8.5). Along the surfaces or faces of the filaments are the **lateral cilia**, whose action generates a flow of water over the ctenidia. Water flow is always maintained in a ventral-to-dorsal direction (i.e., frontal-to-abfrontal) as it transverses across the filaments (Fig. 8.5a). Since blood flow in the filaments is always in a dorsal-to-ventral direction, this means that blood and water are flowing in opposite directions; such a countercurrent flow greatly enhances gaseous exchange in the ctenidia. Frontal and **abfrontal cilia** are concerned both with cleansing sediment particles from the ctenidium and with moving them toward the axis and ultimately into the exhalent current or sweeping food particles toward the mouth. A tiny, internal, chitinous **rod** is located near the frontal or leading edge and just beneath the frontal cilia and provides support for each ctenidial filament (Fig. 8.5b). These skeletal rods enable the filaments to remain separate from each other in a current of water, an important feature because it maximizes the very large ctenidial surface areas for gaseous exchange.

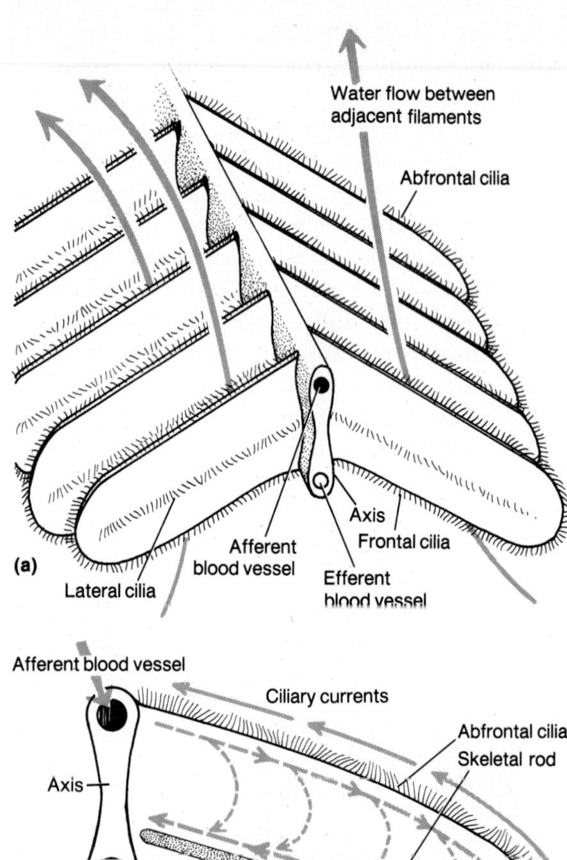

Figure 8.5 *Diagrams of molluscan ctenidia: **(a)** a portion of a ctenidium showing cilia, arrangement of ctenidial filaments, and direction of water flow; **(b)** a single ctenidial filament showing pathways of internal blood flow (broken arrows) and external, cleansing, ciliary currents (solid arrows).*

Because of their position in the mantle cavity and membranes that suspend the axis, the ctenidia are like a curtain that divides the cavity into two portions—a ventral space for the inhalent water current and a dorsal part for the exhalent current. The anal, excretory, and reproductive openings are located strategically in the exhalent channel, an arrangement that minimizes the chances of visceral products fouling the ctenidia as they would if the openings were located in the inhalent portion. Functioning to monitor the chemical composition of incoming water and the amounts of sediment it contains, one or more patches of sensory epithelia, the **osphradia**, are found on the mantle surface in the inhalent channel. Above the ctenidia in the excurrent channel near the anus are the **hyperbranchial glands** (often misnamed as the hypobranchial glands) that secrete mucus into which are accumulated sediments and fecal materials, thus facilitating their elimination from the mantle cavity.

Most molluscs have modified this basic pattern and function of ctenidia in some manner. In most gastropods, only one ctenidium remains. In other gastropods, ctenidia are absent, and the mantle cavity functions like a "lung"—an adaptation for terrestrial life. In bivalves the ctenidia have become greatly enlarged and adapted for food gathering.

Head–Foot

The head–foot is a muscular, bilaterally symmetrical complex whose muscles are involved principally in feeding, locomotion, and retraction into the shell. The **foot** is always ventral in position, and its massive muscles aid in locomotion, burrowing, or in attachment. The foot has undergone rather dramatic modifications in the various classes (Table 8.1, Fig. 8.4). In primitive forms the foot is broadly ovate and flat, and its ventral surface or **sole** contains mucous glands and is heavily ciliated. Mucus lubricates the substratum, and cilia assist in locomotion especially in small forms. In bivalves the foot has become laterally compressed and modified into an effective burrowing organ. The molluscan foot reaches its greatest modification in some cephalopods, where it forms a series of grasping tentacles and a muscular funnel used in locomotion (Fig. 8.4e). The foot can be protruded and its shape altered both by intrinsic foot muscles and by permitting blood to enter its extensive blood sinuses. But the foot is withdrawn by the contractions of one or more pairs of powerful retractor muscles originating in the visceral mass or mantle–shell complex and inserting into it.

While cephalopods and gastropods have a well-developed head, cephalization in the other molluscs is not very pronounced. The anterior end typically bears the terminal **mouth** and one or more pairs of **tentacles**, and at the base or tip of each tentacle may be an **eye**. Surfaces of the anterior end and tentacles are liberally supplied with tactile and chemical receptor cells. Like many other parts of the molluscan body, the **head** is a highly variable organ (Fig. 8.4). In some snails and slugs it is a fairly well-defined anterior region; in others, like chitons and scaphopods, the head is ill defined and little more than the anterior portion of the foot; and in bivalves it is absent altogether. But in cephalopods, cephalization is very pronounced with a highly developed head replete with image-forming eyes, a complex brain, and several masticatory structures. In these highly advanced molluscs, head and arms are intimately bound together as a single morphological and functional unit.

The head is, of course, associated with food procurement; the mouth opens into an ectodermally lined foregut. It is within the foregut that a uniquely molluscan feature, the radular apparatus, is housed. A discussion of this diagnostic organ will be found below in the section on feeding.

Visceral Mass

The visceral mass, situated beneath the protective mantle and shell, contains most of the internal organs, including those concerned with digestion, circulation, excretion, nervous transmission, and reproduction. We shall now look at the broad pattern of each of these as integral parts of the basic molluscan plan.

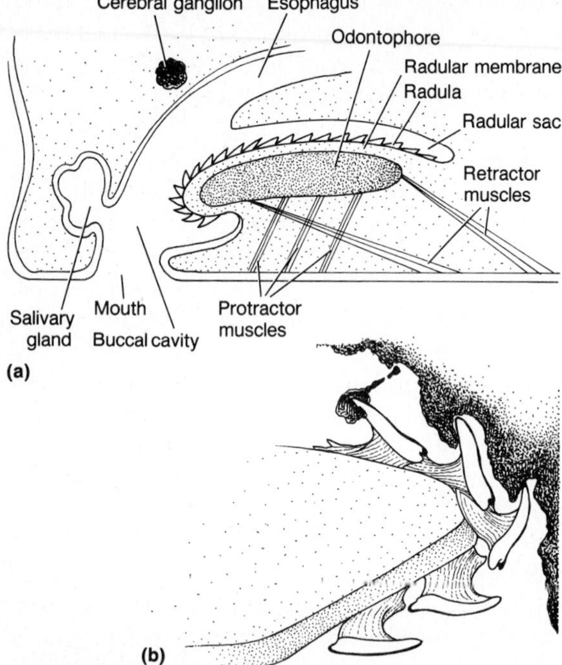

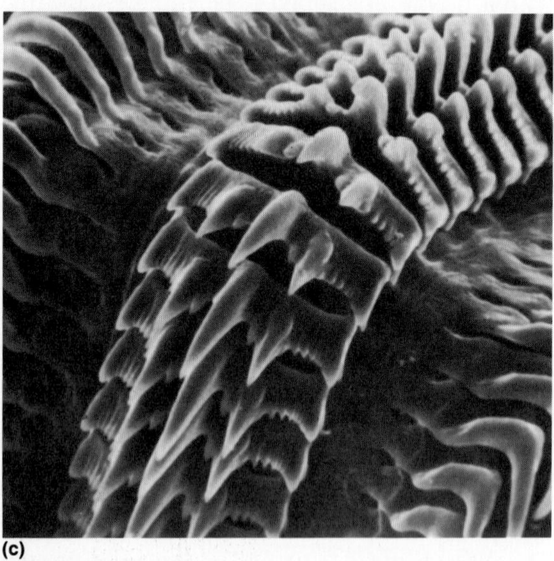

Figure 8.6 *The molluscan radula:* ***(a)*** *a sagittal section showing radula, radular sac, and the anterior digestive tract;* ***(b)*** *the cutting action of the radula (from* The Shell Makers—Introducing Mollusks, *A. Solem, 1974, © by John Wiley & Sons, Inc., used by permission);* ***(c)*** *a scanning electron micrograph of a portion of a radula of* Urosalpinx *(courtesy of M.R. Carriker).*

Feeding and Digestion

The nutritional habits of molluscs are some of their most diverse features, almost every form of heterotrophic nutrition being present. Herbivores feed on rooted plants, periphyton, and phytoplankton; food for carnivores includes annelids, nemerteans, other molluscs, echinoderms, crustaceans, and fishes. We shall explore some of the more specific food-gathering mechanisms as subordinate groups are discussed.

The mouth of most molluscs opens into the **foregut** or **stomodeum.** This anterior portion contains the very important radula apparatus. The **radula apparatus** is an essential food-gathering device in a majority of molluscs and consists of the radula, odontophore, and buccal muscles (Fig. 8.6a). The straplike, somewhat flexible, chitinous **radula** bears numerous transverse rows of minute chitinous **teeth** that are recurved, i.e., point backward (Fig. 8.6a, b). The radula, supported beneath by a cartilagenous **odontophore**, is normally protruded through the mouth when the animal feeds, and it rasps or tears away bits of food for ingestion. A complex of small protractor and retractor **buccal muscles**, attached to the odontophore and radula, moves the entire apparatus as a unit, though the radula can also be slid to some extent over the odontophore. The rhythmical to-and-fro movements, commonly averaging about 40 strokes per minute, tear food away and bring bits into the foregut (Fig. 8.6b). Mucous secretions from a pair of **salivary glands** trap the food particles and thereby aid in the feeding process. Teeth and the radular membrane at the forward end are constantly being worn away or broken, and new teeth and membrane, added posteriorly at the radular base by special secretory cells, progressively move forward. When not in use, the apparatus is housed in a **radular sac**, a ventral diverticulum of the buccal cavity.

All transverse rows of teeth on the radular ribbon are exactly alike in animals of the same species. Shapes of teeth within a given row are variable, and the tooth shapes, numbers of teeth per row, and dental formulae are widely used as taxonomic characters (Fig. 8.6c). Generally, primitive forms have the most teeth per row, and the more advanced forms have fewer teeth, but numbers of teeth are also correlated with food preferences and feeding habits. Cephalopods and some gastropods have, in addition to the radula, a pair of **jaws** for seizing and tearing prey. Bivalves, on the other hand, lack a radula.

A short, tubular, ciliated **esophagus** opens into the **stomach**, which in its simplest form is cone shaped, the enlarged portion being anterior and tapering posteriorly to the intestine. The morphology and physiology of the stomach depend in large measure upon feeding patterns. Many molluscs, such as most bivalves, polyplacophorans, and many gastropods, feed continuously either by filter feeding or by radula grazing, the archetypic method found in the more primitive molluscs. More specialized carnivores, like most gastropods and the cephalopods, feed intermittently followed by periods of digestion.

Two principal adaptations have evolved for processing a slow but continuous stream of finely divided particles. First, the entire alimentary canal is ciliated, and within the stomach and digestive gland are found extensive sorting areas. In general, larger particles are sorted toward the anterior stomach, where they are either broken down enzymatically or rejected into the intestine. Some of the finer particles at any one time enter diverticula of the digestive glands for further sorting. Probably no fine material can enter the hindgut without being recirculated in the stomach and passing over sorting areas at least twice.

A second, extremely important adaptation in continuously feeding molluscs is the presence of a unique crystalline style. The **crystalline style** is a hardened, hyaline, mucoprotein rod containing several digestive enzymes. In some primitive forms, the style is secreted by cells lining the lumen of the anterior intestine, but in most forms with a style it is secreted by special cells lining a **style sac**, a ventral diverticulum of the stomach (Fig. 8.7a). The free end of the style protrudes into the stomach, and the stomach wall opposite the style is lined with chitin to prevent damage by the protruding style. Cilia in the style sac (Fig. 8.7b) cause the style to rotate continuously at a rate of about one revolution every 2–12 seconds.

Figure 8.7 *The molluscan crystalline style:* ***(a)*** *a diagrammatic sagittal section of the stomach showing the style, mucus–food cord, and sorting areas;* ***(b)*** *a cross-sectional view of the style sac and style taken at plane A–A′.*

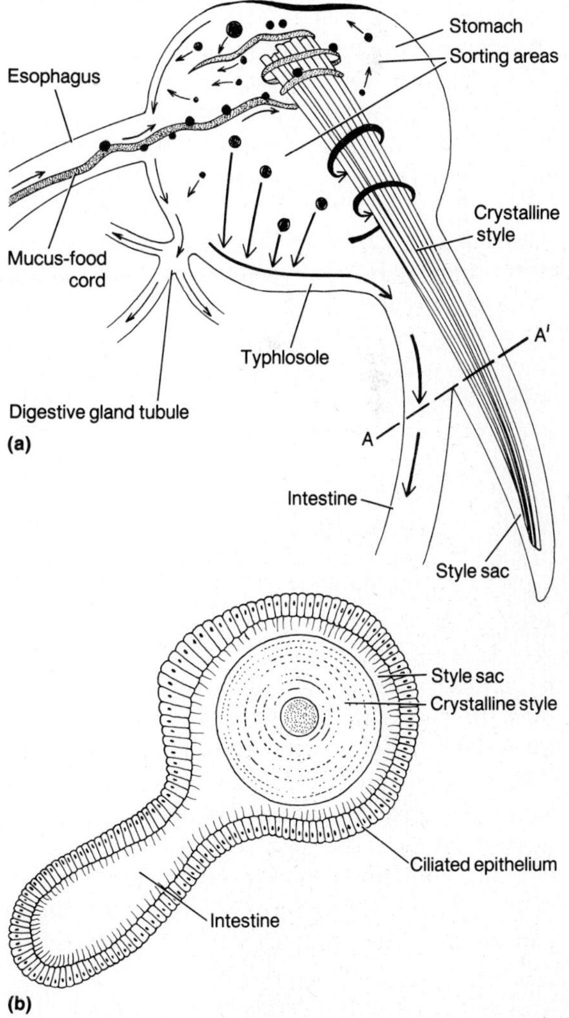

A rotating style is an important feature for procuring food and for triturating stomach contents. Initially, minute particles are enmeshed in the foregut in the mucous secretions of the salivary glands. As the mucus–food complex enters the alimentary canal, it takes the shape of a continuous cord or rope whose inner end is wrapped around the style (Fig. 8.7a). As the style is rotated like a windlass, the mucus–food rope is wound up and pulled into the stomach from the esophagus. Stomach fluids, the pH of which is usually lower than that of the mucous cord or style, reduce the viscosity of the mucus and free the food particles it contains. The stomach acidity also causes the slow progressive disintegration of the style at its free end, and in disintegrating, the style releases digestive enzymes, principally amylases, stored within it. Minute particles are then swept to the sorting areas for grading (Fig. 8.7a). In carnivorous molluscs the style is absent, and food is macerated by the jaws and radula prior to entering the stomach, where extensive enzymatic breakdown of prey tissues takes place.

Minute food particles are carried by cilia from the sorting areas into ducts of the two digestive glands. Each **digestive gland**, consisting of a number of ciliated, blind-end tubules, is a large organ that functions in both digestion and absorption. Some enzymatic digestion continues in the digestive gland canals, but the site of final digestion is in phagocytic cells lining the tubules. In many forms, there is a ciliated ventral tract, the **typhlosole**, along which particles are carried past the openings to the digestive glands and into the intestine (Fig. 8.7a). Heavier, larger particles, initially ingested and freed from the mucous cord by stomach acids, are carried ventrally to the typhlosole and then into the intestine. The long coiled **intestine** opens as a middorsal **anus** situated in the posterior mantle cavity and in the exhalent water canal (Fig. 8.8). Fouling is minimized by the materials being usually in the form of compact pellets, and viscous secretions from the hyperbranchial glands coat the pellets to retard their fragmenting before they are expelled to the outside.

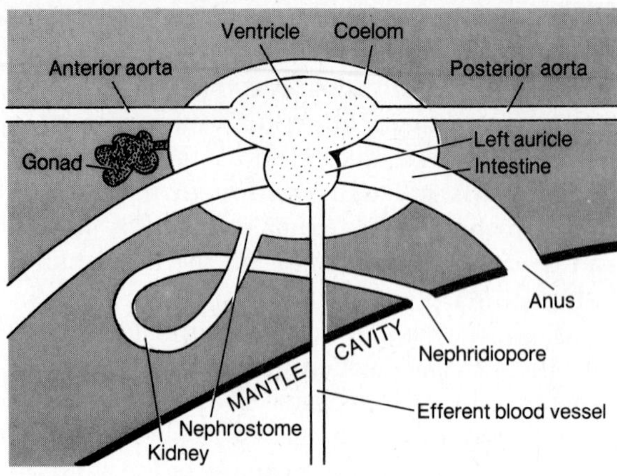

Figure 8.8 *A diagrammatic view from the left side of the molluscan coelom showing the various visceral organs associated with it.*

Coelom and excretion

The molluscan coelom develops embryologically from a cleft within the mesodermal mass and therefore is a **schizocoelom**. Because of the extensive hemocoel associated with the circulatory system, the **coelom** is much reduced and persists only as a fluid-filled chamber around the heart and a portion of the intestine, making it both a pericardial and perivisceral coelom (Fig. 8.8). Both waste products and gametes are often shed into the coelom, from which they are transported to the mantle cavity.

The excretory system typically consists of a pair of elongate tubular **coelomoducts** or **kidneys**. Each coelomoduct, opening internally into the coelom by a **nephrostome**, may contain regional differentiations such as secretory and absorptive areas and a bladder, and each opens into the posterior mantle cavity by a **nephridiopore** located laterally to the anus (Fig. 8.8). Waste products are either secreted into the coelomic fluid by special glandular cells in the pericardial lining or added as a filtrate from the heart wall. Wastes may also be secreted directly into the renal organ from the blood, and selective reabsorption takes place along the length of each coelomoduct. Ammonia is excreted in most aquatic forms and uric acid in terrestrial species. Because of osmotic problems, considerable water is also voided via the kidneys in freshwater species.

Sometimes molluscan coelomoducts are erroneously referred to as nephridia or metanephridia. Coelomoducts originate in the mesoderm at the coelom and grow outward to the body wall, whereas nephridia have ectodermal origins and grow inward to the coelom. A fuller treatment of both types of organs appears in Chapter 9. Even though molluscan kidneys are not nephridia, some terminology of nephridia is used, such as nephrostome and nephridiopore. Since protonephridia are developed in the larvae of some molluscs, this strongly suggests a phylogenetic link between primitive platyhelminths and molluscs.

CIRCULATION AND BLOOD

The circulatory system typically consists of a heart, vessels, hemocoel, and blood. The dorsal **heart**, situated in the pericardial coelom, is composed primitively of two auricles and a single ventricle (Fig. 8.8). Each thin-walled **auricle** is lateral to the ventricle and receives blood from a ctenidium via an efferent blood vessel. The middorsal muscular **ventricle** pumps blood anteriorly and posteriorly into one or two aortae (Fig. 8.8). Auriculoventricular and aortic valves prevent backflow. Each aorta branches into a number of smaller vessels that conveys blood into an extensive spongy system of ill-defined sinuses and lacunae collectively termed the **hemocoel**, where tissues are bathed directly. Such a system in which blood is not contained within a system of arteries, capillaries, and veins is called an **open circulatory system**. Blood eventually drains from the body toward the ctenidia, enters a ctenidial axis by an **afferent blood vessel**, diffuses through the filaments, gathers in an **efferent blood vessel**, and is returned to an auricle (Figs. 8.5, 8.8).

Throughout the phylum there is a high correlation between numbers of auricles and ctenidia present. If there is a single pair of ctenidia, then there is usually a single pair of auricles; but if there are two pairs of ctenidia, then there are two pairs of auricles. Even though many gastropods have lost both ctenidia, however, one auricle always remains.

Molluscan blood is basically colorless and contains a number of formed elements including amebocytes and corpuscles. In the blood of most molluscs is a copper-containing respiratory pigment, **hemocyanin**, dissolved in the plasma, but a few molluscs have hemoglobin instead of hemocyanin. Hemocyanin is pale blue when oxygenated but colorless when deoxygenated. Because of the low oxygen-carrying capacity of hemocyanin and the open vascular system, the molluscan circulatory efficiency is low. But in the highly mobile cephalopods the circulatory system is essentially a closed one, thus enabling these molluscs to live a far more active life.

Molluscan blood and the hemocoel have another extremely important function, since they are used as a hydraulic mechanism. Most molluscan parts such as head, foot, tentacles, and siphons are extended indirectly by muscles forcing blood into specific hemocoelic spaces. There are two obvious disadvantages to such a system. First, such extensions are slower than muscular contractions, since they require blood both to be transferred from one part to another and to generate sufficient pressures to effect extensions. Retraction of extended parts is much faster because it is achieved by direct muscle contractions. Second, there are restrictions in having a finite blood volume, which limits the number of body parts that can be extended at any one time. Molluscs have overcome both problems to some degree by being able to seal off or compart the hemocoel of various organs and then using local intrinsic muscles to enhance the hydraulic effect. The volume of fluid available for hydrodynamic actions can be supplemented in a few species by temporarily taking up seawater. Locomotion, feeding, and sperm transfer are among the more important functions augmented by the hydrostatic nature of the hemocoel and blood.

NERVOUS SYSTEM

Molluscs exhibit a wide range of arrangements of nerve cords and ganglia. In primitive molluscs like polyplacophorans the neurons are diffuse, and no ganglia are present even as a brain. Cephalopods, on the other hand, have an extraordinarily complex nervous system rivaling that of many vertebrates. Most molluscs, however, have a nervous system intermediate in complexity between these two extremes. The nervous system of primitive gastropods will serve as the prototype for the phylum. Generally, there is a nerve ring around the esophagus containing the **cerebral ganglia** associated with the radular apparatus and head (Fig. 8.9). Extending posteriorly from the nerve ring are two pairs of nerve cords:

1. the ventral **pedal cords**, which innervate the foot, each cord often bearing a **pedal ganglion** that is the center for coordination of foot movements, and
2. the dorsal **visceral cords**, each one having along its length a **pleural**, a **parietal**, and a **visceral ganglion** controlling the mantle, ctenidia, and visceral mass, respectively (Fig. 8.9).

Short interganglionic connectives join each cerebral and pedal ganglion to their opposite members and unite pedal and pleural ganglia. In effect, the molluscan nervous system has a number of regional "brains" connected by thin neural connectives. With some notable exceptions like cephalopods, certain gastropods (opisthobranchs), and a few bivalves, molluscs lack an efficient central nervous system and fast reflexes, so most molluscs are slow moving and sluggish (no pun intended). A discussion of sense organs will be included with the treatment of each class, since their distributions are variable.

Figure 8.9 *A diagram of the molluscan nervous system seen from the left side showing the principal cords and ganglia (ganglia on left side shown in open circles and those on the right side by solid circles).*

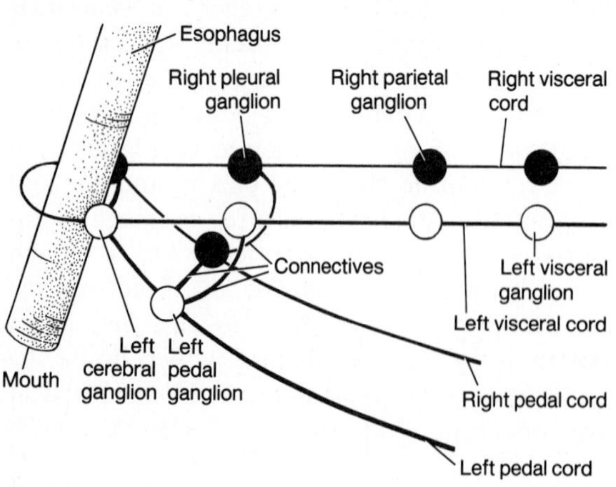

REPRODUCTION AND DEVELOPMENT

Basically, molluscs are dioecious even though a very large number of gastropods are monoecious. One or two dorsal **gonads** lie on either side of the coelom. When mature, the gametes of many molluscs are simply released into the coelom by rupture of the gonads (Fig. 8.8). From the coelom, gametes are transported to the posterior mantle cavity via the **coelomoducts** or kidneys, which also function as gonoducts. Many other molluscs have separate gonoducts, and gametes usually enter the gonoducts directly without passing through the coelom proper. Primitively, fertilization is external in seawater, but in some species it takes place

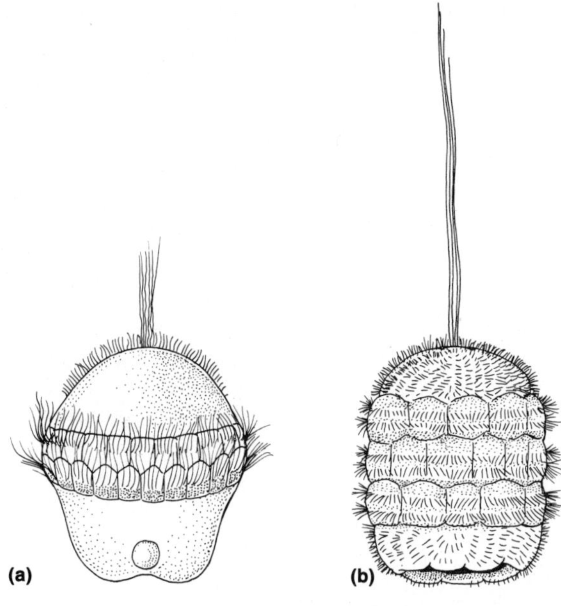

Figure 8.10 *Two examples of molluscan trochophore larvae: **(a)** a gastropod* (Patella) *(after Patten in L.H. Hyman,* The Invertebrates, *Vol. VI:* Mollusca I, *New York, McGraw-Hill, 1967, adapted with permission); **(b)** a bivalve* (Yoldia).

within the mantle cavity of the female. In most gastropods and all cephalopods, sperm are transferred to the female by a copulatory organ, and fertilization is internal. Many hermaphroditic gastropods practice cross-fertilization by mutual insemination.

Cleavage is spiral, determinate, and total. Early embryonic development may take place in the genital tract or mantle cavity among the female's ctenidia. Some species lay large yolky eggs in masses, and some are even viviparous. Representatives of some taxa develop a free-swimming planktonic stage, the **trochophore**. It bears a characteristic preoral ring of locomotive cilia or **prototroch** and an apical sensory plate with an **apical tuft** of cilia at the dorsal apex (Fig. 8.10). A trochophore stage is found in other marine invertebrates including annelids; a discussion of this stage and its phylogenetic importance appears in Chapter 9. Typically, the trochophore is succeeded in molluscs by a second larval stage, the veliger. A **veliger** is characterized by a large, ciliated, swimming organ or **velum**, derived from the trochophoric prototroch and consisting of two, large, semicircular lobes (Fig. 8.11). It is in this stage that foot, eyes, tentacles, mantle, and shell are initially developed. The veliger is planktonic and usually feeds by straining food particles from

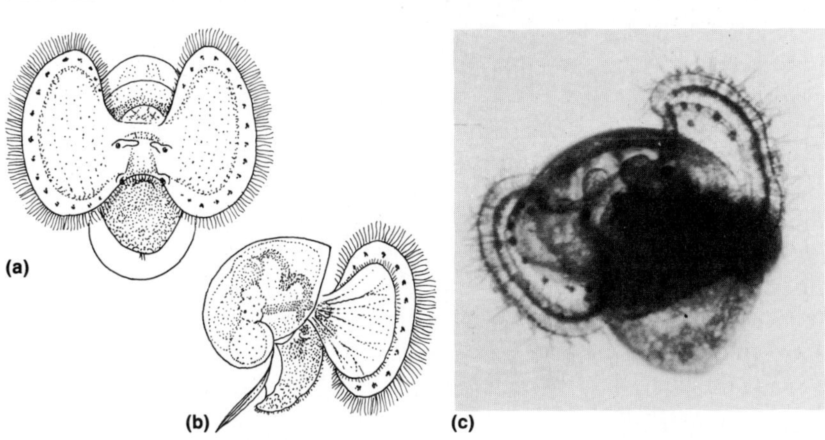

Figure 8.11 *A gastropod veliger representative of the advanced larval stages of marine molluscs: **(a)** frontal view; **(b)** lateral view; **(c)** a photomicrograph of a gastropod veliger (courtesy of A.B. McCrary). (parts a, b after Werner in L.H. Hyman,* The Invertebrates, *Vol. VI:* Mollusca I, *1967, McGraw-Hill Book Co., adapted with permission.)*

water brought over the long cilia of the velum. After a variable duration as a veliger, the larva settles onto a substratum and shortly thereafter metamorphoses; metamorphosis involves loss of the velum and acquisition of adult features.

CLASS APLACOPHORA

The aplacophorans, also called the solenogasters, are apparently very primitive molluscs, and their class name is derived from the fact that they lack a shell. They are benthic, wormlike, marine animals that live at depths of 50–9000 m, but they frequently burrow in muddy bottoms or are found creeping over sedentary organisms such as cnidarians and sponges. Most of the approximately 250 species are quite small and are seldom seen. Generally, they range from 1 to 40 mm in length, but some are up to 30 cm long.

Aplacophorans lack many of the features characteristic of molluscs in that a distinct head, mantle, discrete foot, shell, and kidneys are all absent (Fig. 8.12). Several layers of calcareous spicules are embedded within the thickened integument, and the buccal cavity contains a radula. A pair of coelomoducts and the anus open into a small posterior mantle cavity that may contain a pair of ctenidia or, in place of ctenidia, epithelial folds that function as gills.

Members of the Subclass Caudofoveata have a mantle containing overlapping scales (Fig. 8.12a). The mouth has an oral **pedal shield** associated with it. The posterior mantle cavity contains a single pair of bipectinate ctenidia. The caudofoveatans are dioecious and have external fertilization. Members of the Subclass Ventroplicida live on cnidarians or sponges and have a ventral, median, longitudinal groove called the **pedal groove** (Fig. 8.12b). It is from this groove that they get their common name (solen = channel; gaster = stomach). The pedal groove is paralleled laterally by **pedal folds**, which some specialists equate with the foot of other molluscs. They are all monoecious, and fertilization is in the sea.

Solenogasters are an evolutionary puzzle. In the past, most biologists considered them to be a rather aberrant and degenerate group. Some recent interpretations, however, see aplacophorans as extremely primitive molluscs. In fact, they are probably closer to the ancestral molluscan stock than is any other group. Their

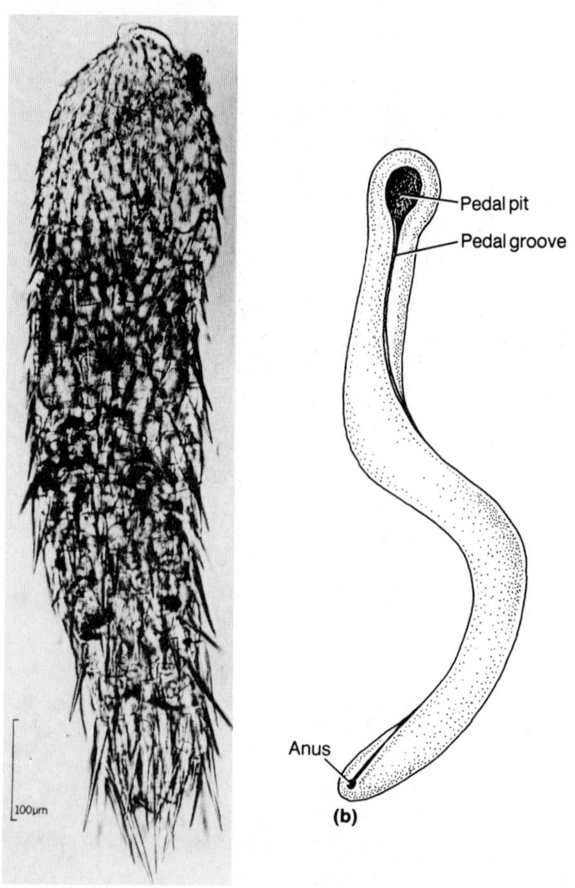

Figure 8.12 *Aplacophorans: **(a)** photograph of* Falcidens, *a caudofoveatan (from R.M. Rieger and W. Sterrer,* Zeitschr. f. Zool. Systematik u. Evolutionsforschung 13:259, 1975); ***(b)*** Proneomenia, *a ventroplicid.*

fossil record is very scant owing to the absence of a hard shell, and additional studies are needed to determine more precisely the phylogeny and evolutionary affinities of the Aplacophora.

CLASS MONOPLACOPHORA

Until the middle of this century, monoplacophorans were known only as fossils, and the group was thought to have become extinct near the close of the Devonian period about 350 million years ago. In 1952 the Danish ship *Galathea*, dredging in the Pacific Ocean west of Costa Rica, brought up from a depth of almost 3600 m —incredibly—ten living specimens that were identified as monoplacophorans and named *Neopilina galatheae*. Since this remarkable discovery, specimens belonging to four other species of *Neopilina* and three species of *Vema* have been collected from various deep-water marine localities.

Specimens of *Neopilina* are about 3 cm in length, bilaterally symmetrical, and elliptical when viewed from above or below. Monoplacophorans (= bearing one plate) possess a single dome-shaped shell whose apex, pointing anteriorly, gives these animals a superficial resemblance to limpets (Fig. 8.13b). On the undersurface there is a broad, flat, creeping foot into which are inserted eight pairs of **pedal retractor muscles**. A reduced head, lacking eyes or tentacles, is situated anterior to the foot. There is a median preoral fold, the **velum**, which is continuous on either side of the mouth as a wide ciliated **palp** (Fig. 8.13a). The shallow mantle cavity contains five or six pairs of single-branched gills, which are not true ctenidia but may be ctenidial precursors. The gill ciliation creates a posteriorly directed current over the gills.

The digestive system includes a radula, style, and a highly coiled intestine. The circulatory system consists of two pairs of auricles and a pair of ventricles; an aorta arises from each ventricle, and the aortas fuse shortly into a single anterior aorta. The coelom is a pair of dorsal compartments surrounding the heart. Six pairs of kidneys lie along the sides of the body

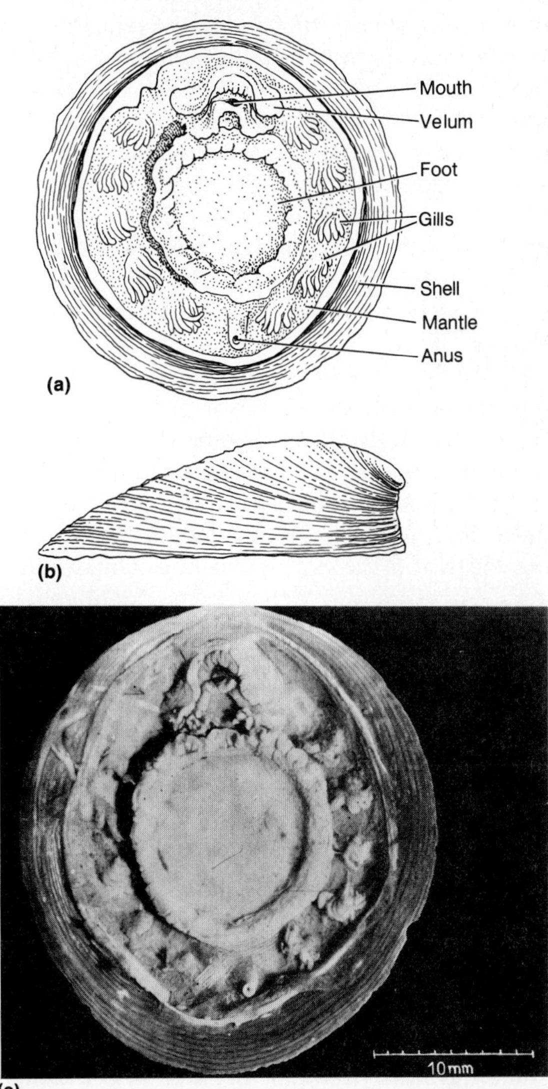

Figure 8.13 Neopilina, *a living monoplacophoran:* ***(a)*** *ventral view;* ***(b)*** *lateral view of the shell;* ***(c)*** *photograph, ventral view. (parts a-c from H. Lemche and K.G. Wingstrand,* Galathea Reports 3: *Plate I, 1959).*

of which the posterior five pairs open into the coelomic compartments and transport excretory products from the coelom to the pallial cavity. Monoplacophorans are dioecious and possess two pairs of medial gonads, each of which has a short gonoduct that opens into a kidney, which conveys gametes to the outside. Fertilization is external or perhaps takes place in the pallial groove of the female. The primitive nervous system is similar to that of the ancestral mollusc.

Neopilina and *Vema* are unique among molluscs because of the presence of multiple pairs of gills, kidneys, heart chambers, gonads, and retractor muscles. It might appear initially that these organs are segmentally arranged, and indeed, if *Neopilina* were segmented, the prevailing ideas of molluscan phylogeny would be altered drastically. Metamerism, defined in Chapter 3 and more fully discussed in Chapter 9, is a property of the mesoderm and affects almost all mesodermal derivatives. Little is known of the embryology of *Neopilina*, but there is no evidence that any mesodermally derived structures are metamerically arranged. In addition, no other mollusc is segmented either as adults or as larvae. Rather, the multiple organs are perhaps best considered to reflect serial repetition and not metamerism. *Neopilina* and *Vema* are considered today as interesting and provocative relics of an ancient, aberrant, monoplacophoran stock. They would seem to be neither the most primitive ancestral mollusc nor a "missing link" between molluscs and another invertebrate group such as annelids. However, additional evidence, obtained especially from developmental patterns, may clarify the phylogeny of monoplacophorans and the provocative members of *Neopilina* and *Vema*.

CLASS POLYPLACOPHORA

Polyplacophorans or chitons, representing about 600 species, are all marine and are found mostly in rocky intertidal zones. Chitons are often inconspicuous, and their drab colorations of gray, brown, yellow, green, or red enable them to blend in with the colors of stones or shells on which they live, but some Pacific Ocean forms are brightly colored. Most chitons are 2–12 cm long, although representatives of the Pacific *Cryptochiton* may attain a length of up to 30 cm.

External Features

Polyplacophorans are dorsoventrally flattened and elliptical when viewed from above. There is a shell of eight dorsal, transverse, overlapping plates or **valves**, a feature from which they derive their name (polyplacophoran = bearing many plates) (Figs. 8.4a, 8.14a, d). Unlike the typical molluscan shell, the valves are composed of four or five layers: the periostracum, tegmentum, mesostracum, articulamentum, and hypostracum. The anterior margin of each plate except for the anterior one extends below the posterior margin of the plate immediately anterior to it. Since valves are movable against each other, this arrangement permits the animal to curl up somewhat with the protective plates on the outside (Fig. 8.14b). The diverse sculptured patterns and designs of the valves are formed by minute nerve canals running from sensory cells called aesthetes or shell eyes. **Aesthetes**, unique to chitons, are light-sensitive organs that, in some species, may be relatively complex ocelli (Fig. 8.14c). The lateral edges of the calcareous plates are embedded in the **girdle**, the thickened mantle covering the dorsal and lateral aspects of the body (Fig. 8.14a, d). The girdle is thick, tough, heavy, and often equipped with bristles, spines, or calcareous spicules of various lengths. Valve patterns, color, and texture and ornamentations of the girdle are useful taxonomic characters.

The head–foot is ventral and for the most part consists of a broadly oval, creeping foot similar to that of the ancestral mollusc (Fig. 8.14b, d). If disturbed, the chiton pulls the girdle down against the substratum, the mantle

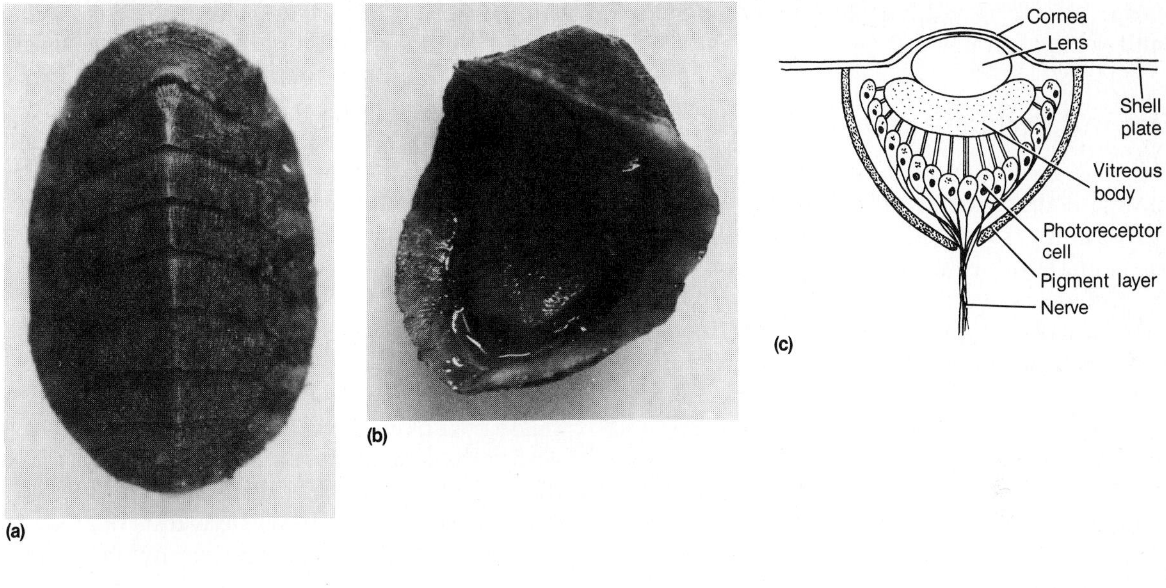

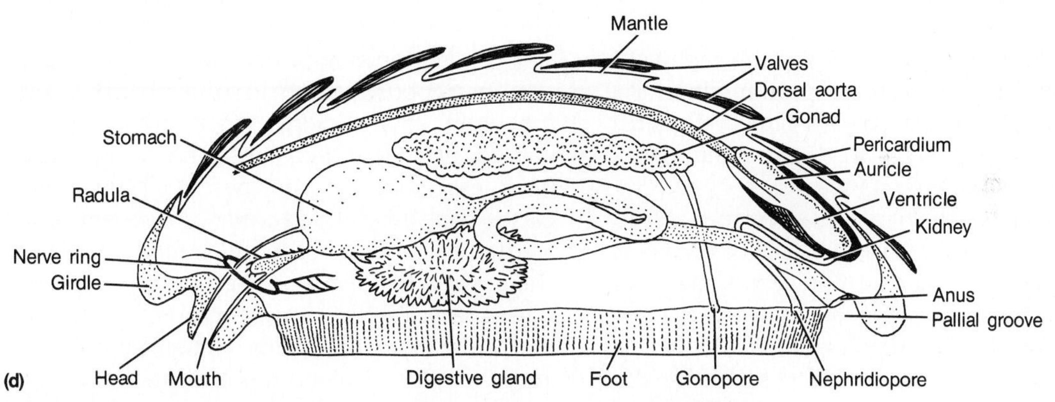

Figure 8.14 *Polyplacophoran features:* ***(a)*** *photograph of* Chaetopleura, *dorsal view;* ***(b)*** *photograph of* Chaetopleura, *ventral view (note how living animal curls up when dislodged from its attachment);* ***(c)*** *a complex aesthete;* ***(d)*** *longitudinal section through a typical chiton (from T.I. Storer and R.L. Usinger,* General Zoology, *4th ed., New York, McGraw-Hill, 1965, adapted with permission).*

and plates are arched dorsally, and the foot is retracted; this effectively forms a strong suction that enables the animal to adhere tenaciously to the substratum. The reduced head lacks both eyes and tentacles.

Lying between the foot and mantle is the narrow, shallow, mantle cavity or pallial groove completely encircling the foot. Attached to the roof of the mantle cavity are the ctenidia, whose numbers, ranging from 4 to 88 pairs, vary between species and even between individuals within a species. Ctenidia may be distributed uniformly along the length of the pallial groove on either side, or they may be clustered in the posterior portions. Water is drawn into the anterior mantle cavity, passes over the ctenidia, and is expelled posteriorly.

Internal Features

Most chitons are microphagous and feed on algae and other small organisms found on the substratum. The mouth opens into a buccal cavity containing a radula in which each radular row contains precisely 17 teeth. Mucus from a pair of salivary glands lubricates the food and aids in transporting it through the alimentary canal. Since a style is absent, the mucus–food complex is swept through the esophagus and into the stomach by ciliary action. Food is digested in the stomach, digestive glands, and anterior portions of the intestine (Fig. 8.14d). The intestine is separated into anterior and posterior portions by a sphincter valve that functions both in slowing down the passage of food within the intestine and in forming fecal pellets.

The circulatory system is quite similar to that described for the archetypical mollusc. The heart, located beneath the last two shell plates, pumps blood anteriorly through a single aorta (Fig. 8.14d). Nitrogenous wastes, secreted into the relatively large pericardial cavity, enter a coelomoduct located on either side of the heart. Each kidney is U-shaped and forms many diverticula; its nephridiopore is located in the posterior portion of the mantle cavity.

The nervous system, reflecting the sedentary life-style and reduced head, is decentralized and rather diffuse. Nerves are distributed in nerve cords, and ganglia are absent. The circumesophageal ring gives off nerve trunks to the principal body regions. Sensory structures are located around the mouth and in the mantle cavity and valves.

Polyplacophorans are dioecious, the single gonad is located anterior to the heart, and a pair of gonoducts runs posteriorly from the gonad to the gonopore located in the mantle cavity near the nephridiopore (Fig. 8.14d). Gametes are carried from the mantle cavity into the surrounding seawater by exhalent water currents, and gametic union takes place in the sea, even though in some species, fertilization occurs in the mantle cavity of the female. The young embryos develop quickly into planktonic trochophores. Chitons do not have a typical veliger stage, but rather, the trochophore metamorphoses directly into an immature chiton, which becomes benthic and matures sexually.

CLASS GASTROPODA

The Class Gastropoda, the largest of the mollusc classes with almost 85,000 living species, includes the snails, slugs, limpets, abalones, nudibranchs, whelks, and many others. This fantastic group exhibits the greatest degree of adaptive radiation found in the phylum. Snails are a very ancient group; they appeared in the fossil record in the early Cambrian period about 550 million years ago. About 15,000 fossil species of gastropods have been discovered to date.

Gastropods are unbelievably diverse in size, functional morphology, shell patterns, and many ecological dimensions. They range in size from less than 1 mm to over 50 cm in length. Gastropods vary widely in food preferences and feeding methods, modes of locomotion, gas exchange adaptations, and reproductive patterns. They are also well known because of the shell that most bear. Exhibiting almost infinite varieties of shapes, colors, designs, and sizes, gastropod shells are prized by collectors worldwide because of their beauty and innumerable patterns. Basically a marine group, many gastropods have adapted to freshwater habitats, and others have become terrestrial. Their morphological versatility and functional adaptability have enabled them to occupy widely different habitats and ecological niches, and these combinations have made them one of the most successful invertebrate groups.

Gastropods are economically important in several ways. Certain gastropods, such as abalones and conchs, are eaten by people in certain parts of the United States, and edible snails known as escargots are considered to be a delicacy. Some gastropods are pests as in the cases of oyster drills and the many slugs found around homes and gardens. Snails are also important in that they serve as intermediate hosts for many parasitic helminths (see Chapter 6).

Gastropods are subdivided into three well-defined subclasses—Prosobranchia, Opisthobranchia, and Pulmonata—that are characterized in Table 8.2. The Subclass Prosobranchia, by far the largest and most diverse of the three, is subdivided into three orders, whose most important features also appear in Table 8.2. Even though each taxon will be discussed more fully later in this chapter, it will be helpful to the reader to begin to differentiate among the various groups at this point.

Coiling, Torsion, and Spiraling

Gastropod morphology, body organization, and physiology, both externally and internally, have been dramatically modified by the phenomena of coiling, torsion, and spiraling. Hardly any gastropod structure or function remains unaffected by these three exceedingly important developmental processes. **Coiling**, perhaps the first of these three processes to evolve, refers to the arrangement of the body into a series of coils all in a single plane; coiling is analogous to arranging a length of rope on the floor such that each later coil is outside of the others. **Torsion**, on the other hand, is a unique process by which the visceral mass is rotated in relation to the head–foot. **Spiraling** describes the process by which the body of a great many gastropods is coiled, but successively later spirals do not lie in the same plane as previous spirals. Torsion and spiraling are responsible for the high degree of asymmetry that characterizes most gastropods. All gastropods are torted, and most, but not all, are spiraled.

Before torsion the larval gastropod body is bilaterally symmetrical, and the deepest portions of the mantle cavity with anus, urogenital openings, and ctenidia are at the posterior end (Fig. 8.15a). Torsion causes the visceral mass and mantle–shell to be twisted 180° in a counterclockwise direction when viewed from above (Fig. 8.15b–d). Torsion takes place behind the head in the neck region and basically involves neck tissues and those structures that run through it. Thus most of the visceral mass is

Figure 8.15 *Stages of torsion in gastropods: (a) pretorsion; (b) early torsion; (c) late torsion; (d) posttorsion.*

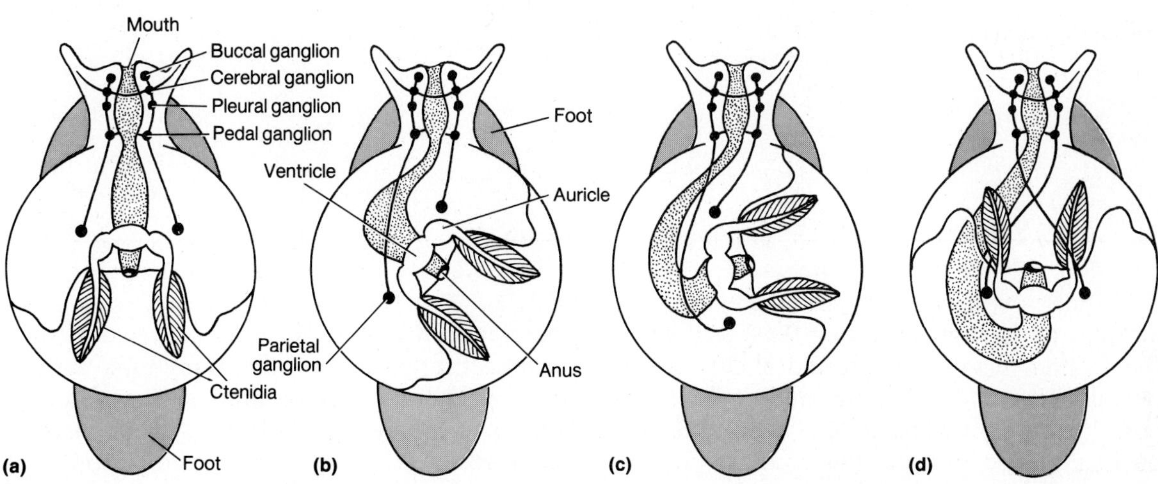

TABLE 8.2. □ **THE MOST IMPORTANT CHARACTERISTICS OF THE THREE SUBCLASSES OF GASTROPODS AND OF THE ORDERS OF PROSOBRANCHS**

Taxon	*Habitat*	*Body Shape*	*Gas Exchange Organs*	*Other Features*
Subclass Prosobranchia	Mostly marine	Torted, spiraled	1 or 2 ctenidia	Have operculum on foot, long incurrent siphon, conical shell, 1 pair of tentacles; are dioecious
Order Archeogastropoda	Mostly marine	Torted, spiraled	1 or 2 bipectinate ctenidia	Gametes exit via coelomoducts of functional kidneys
Order Mesogastropoda	Mostly marine	Torted, spiraled	1 monopectinate ctenidium	Largest order; functional separation of gonadal and renal ducts
Order Neogastropoda	Marine	Torted, spiraled	1 monopectinate ctenidium	Carnivorous; few teeth per radular row; long extensible proboscis; separate renal and genital ducts
Subclass Opisthobranchia	Marine	Detorted, non-spiraled	Cerata, secondary gills, general body surface	Lack operculum, style, ctenidia, mantle cavity, and shell (most); have 2 pairs of tentacles; are monoecious
Subclass Pulmonata	Mostly terrestrial	Spiraled, somewhat detorted	Pulmonary cavity or "lung"	Have pneumostome on right side, shell (most); lack style, operculum, and ctenidia; are monoecious

affected by torsion, including the positions of the mantle, mantle cavity, and shell and the overall shape of the digestive and nervous systems. The primitive bilateral symmetry is seriously modified because the anus, nephridiopores, position of the future ctenidia, and mantle cavity are brought from their former posterior location to an anterior position and dorsal to the head (Fig. 8.15d). Those visceral mass structures positioned on the pretorsional left side are now located on the posttorsional right side and vice versa. Hereafter, references to right or left organs will refer to the positions of adult structures, that is, their posttorsional locations. Internally, the digestive tract is twisted into a U-shape, and the torted nervous system describes a figure eight (Fig. 8.15d). The shell position immediately outside the mantle is rotated 180°, but the nascent shell is unaffected structurally.

Torsion can be a very rapid process requiring a few minutes, but in most forms it is a growth process that takes place over a period of a week or more. Some gastropods undergo torsion in two steps; an initial twisting of 90° brings the mantle cavity to the right side, and a subsequent 90° twist displaces the mantle cavity to its anterior position. The rotational process is effected initially by asymmetrical growth of muscles that run from the head–foot to the mantle–shell. The one muscle that is principally involved with torsion is the pretorsional right retractor muscle. Some of its fibers, however, originate on the left side, sweep across the visceral mass, and insert on the shell on the right. As it contracts, the visceral mass, mantle, and shell are drawn from right to left. Torsion always and inevitably proceeds as a counterclockwise event regardless of the direction of spiraling or coiling of the future adult. Torsion is ubiquitous in all living gastropods and therefore is obviously a very basic and primal gastropod feature.

What possible advantages could be gained by an ancient gastropod in undergoing such a revolutionary process as torsion? Some theories hold that torsion was of adaptive significance to the larval stage. A torted body might be advantageous to the pelagic larva in escaping predators, since the head could now be retracted quickly into the mantle cavity and beneath the visceral mass. Other ideas suggest that torsion is of special significance to the benthic adult by

1. bringing the center of gravity forward thus increasing stability,
2. bringing the mantle cavity forward, heading it into the current of undisturbed water, and thus facilitating ctenidial aeration, and
3. bringing the osphradia of the mantle cavity forward nearer to the sensory structures of the head.

None of these postulates, however, completely and satisfactorily explains the universality of torsion among gastropods.

Torsion has resulted in some rather severe complications especially to adult snails, and many of the gastropodan modifications represent adjustments to these difficulties. The most significant problem is the tendency for fecal and excretory materials to foul the ctenidia or for these products to be dumped onto the head. Gastropods have adapted in many ways to minimize the sanitation problem; these will be mentioned in the discussion of the mantle cavity.

In some gastropods like the opisthobranchs and pulmonates, torsion occurs in the veliger in the usual fashion, but this is followed subsequently by detorsion. **Detorsion**, a twisting process proceeding clockwise, results in the return of the mantle cavity, ctenidia, and anus toward the right side or to the posterior end. Detorsion is usually accompanied by a reduction or loss of the shell, mantle, mantle cavity, and visceral mass. Because of these morphological alterations, the bodies of detorted gastropods often become elongate as exemplified by the opisthobranchs and the familiar slugs. Degrees of detorsion, varying from slight to complete, impart a variety of new morphological patterns to many species.

Spiraling and coiling result because one side or the front of the visceral mass grows upward at a faster rate than that of the opposite side. This unequal growth causes the visceral mass to topple away from the faster-growing side and over onto the side of slower growth (Fig. 8.16a). In order to maintain a reasonably balanced yet unequally growing visceral mass the mass begins to spiral or coil with each loop wrapped

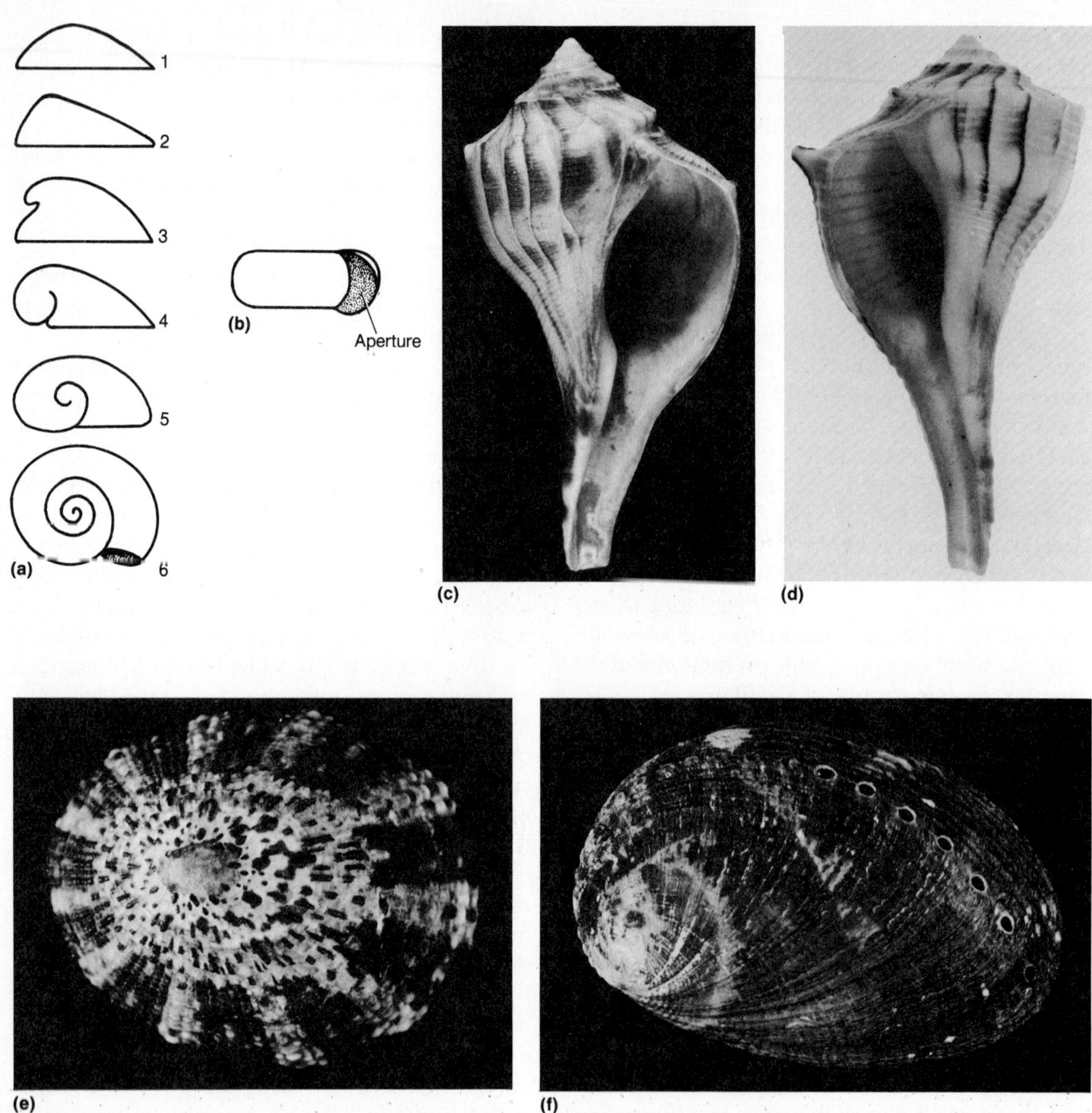

Figure 8.16 *The effects of coiling and spiraling on the shape of the gastropod shell:* ***(a)*** *sequential stages in the coiling of a planospiral shell;* ***(b)*** *the same planospiral shell, frontal view;* ***(c)*** *photograph of the dextral shell of* Busycon carina, *which is helically spiraled;* ***(d)*** *photograph of the sinistral shell of* Busycon contrarium, *which is helically spiraled;* ***(e)*** *photograph of the shell of a limpet, which is neither spiraled nor coiled;* ***(f)*** *photograph of the shell of* Haliotus, *an abalone, which represents a single, asymmetrical body whorl.*

around the preceding loop (Fig. 8.16a). By coiling or spiraling, the visceral mass and its protective mantle become more compact and provide better balance and stability. Coiling results in a **planospiral** coil or one that coils in a single plane (Fig. 8.16a, b). Spiraling, on the other hand, results in a **helical** spire in which one spiral is outside of or beneath the previous coil (Fig. 8.16c, d). The changing positions of the visceral mass and shell have resulted in dramatic morphological alterations in the mantle cavity and internal systems. The bulging visceral mass whorl has also resulted in the reduction or complete loss of certain internal organs on the inner side of the coil. Since most gastropods are spiraled to the right, the right-hand members of the paired auricles, kidneys, and gonads are reduced or lost altogether.

When viewed from the shell's apex, the direction of spiraling is clockwise in **dextral** snails and counterclockwise in **sinistral** forms (Fig. 8.16c, d). The dextral spiral is by far the more common, and only a small number of species is sinistrally spiraled; but in a few species, both dextral and sinistral individuals are found. The direction of spiraling is genetically determined and is foreshadowed even in cleavage, in which the spiraling pattern is evident in very young embryos.

External Features

Shell

The most obvious external feature of a majority of gastropods is the asymmetrical, one-piece shell, which is an effective portable retreat into which the animal can withdraw completely (Fig. 8.4b). From an evolutionary standpoint, as the visceral mass began to elongate, the shell design also had to be altered to accommodate the coiled visceral mass and to make the shell easier to transport. What resulted was a primitive shell that was bilaterally symmetrical with each later, progressively larger coil lying outside the earlier ones. Such a two-dimensional pattern is a **planospiral** shell (Fig. 8.16a, b). The main disadvantage of a planospiral shell is that as it gets larger, it is not very compact and becomes progressively more unwieldy to transport. Most gastropods circumvented this problem by evolving an asymmetrical spiraled pattern, which resulted in a more compact and easily managed helical spire and shell (Fig. 8.16c, d). In a **helical** spire the spirals are positioned around a central axis, later larger spirals being beneath and partially outside of earlier spirals. What results from a helically spiraled pattern is a shell that is cone shaped or turbinate. Adding to the problems, a helical shell cannot be carried in the same position as a planospiral shell because it would be mechanically unbalanced. To compensate for this and to readjust the location of the shell toward a more balanced position, the dextrally spiraled snails have shifted the shell position toward the left side with the shell's apex pointing upward, somewhat posteriorly, and to the right. However, this relocation of the shell's position had other repercussions; because of the shell's being skewed to the right side in dextrally spiraled snails, the mantle cavity was restricted to the left side, and only the left-hand members of certain paired organs were retained.

A typical gastropod shell begins as a **protoconch**, which is formed initially in the veliger stage as a simple cap. The protoconch is represented in the adult shell as the small apical portion, which is usually smooth or bears ornamentations different from the rest of the shell. The typical shell is a helical cone in which larger, younger whorls lie nearer to the shell **aperture**, an opening through which the head and foot protrude (Fig. 8.16b–d). The underlying mantle is directly responsible for secreting new shell materials all around the aperture. Gastropod shells can increase in size only by the addition of new shell materials secreted by the mantle at the aperture as a progressively larger, spiraled, calcareous tube. The whorls, each of which is joined to the whorl just inside it by a suture, are spiraled around a central axis or **columella**. In three families of gastropods, collectively the vermetids, the shells grow to be somewhat spiraled but nonsutured in an irregular serpentine fashion (see Fig. 8.25g).

Most opisthobranchs and the slugs possess a shell only as larvae or juveniles, and in these forms it is almost invariably lost or substantially reduced, a phenomenon correlated with detorsion. In still other forms (limpets, abalones, slipper shells, etc.) the shell loses its typically spiraled appearance and is represented by a single, body-sized whorl. Some single-whorled shells may retain the asymmetry (abalones [Fig. 8.16f]), or they may have a secondarily acquired bilateral symmetry (limpets [Fig. 8.16e]).

HEAD–FOOT

The **head** of a gastropod is well developed in comparison with that of most other molluscs and is especially so in the more active species. The head, pliable neck, and radula are actively used in feeding. One or two pairs of **tentacles** are borne on the head. If **eyes** are present, they are located either at the tentacular bases or at the tips. The tentacular surfaces contain numerous receptors for tactile and chemical stimuli.

The **foot** is a broadly flattened, elongated, heavily muscularized creeping organ (Fig. 8.4b). Muscle strands run in all directions, making possible a variety of movements. Locomotion, especially that of snails living on hard substrata, is achieved by short, rhythmical, transverse, muscular waves similar to the locomotive pattern of the larger turbellarians. Waves may pass in the direction of movement (direct waves) or in a direction opposite that of movement (retrograde waves) (Fig. 8.17b). At the crest of a pedal wave, that portion of the sole is lifted, extended forward somewhat, then returned to the surface, thus making a short step (Fig. 8.17a). Some snails have muscular waves extending across the width of the foot, but others have two waves moving alternately along either half of the foot. For small snails, mobility is enhanced by ciliary action in mucus secreted by pedal

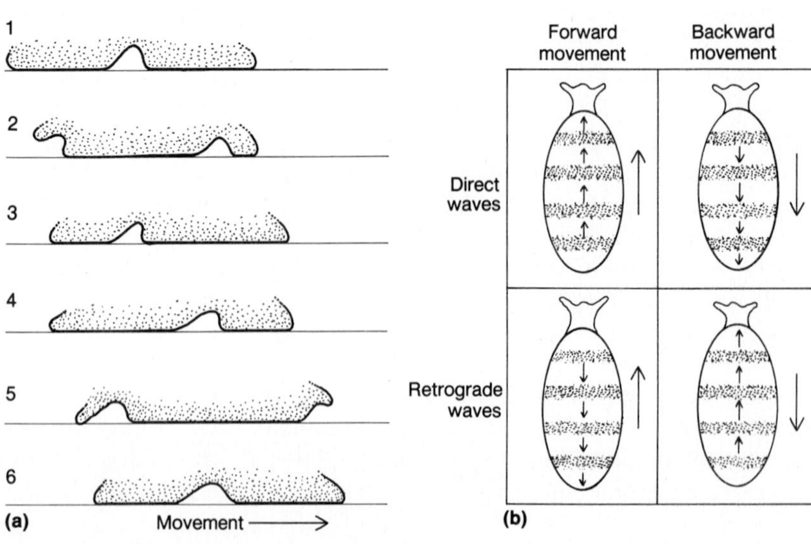

Figure 8.17 *Muscular waves in the foot of a gastropod during locomotion: **(a)** diagrams of sequential stages showing direct short steps in the foot; **(b)** direct and retrograde waves effecting forward and backward movements.*

glands. Some gastropods use the foot for burrowing or digging, and the foot of some pelagic prosobranchs and opisthobranchs is modified as an effective swimming organ. The foot of most prosobranchs and a few other gastropods bears a large **operculum**, which protects the retracted animal from predation or desiccation. Foot muscles are also important in maintaining posture, a feat made somewhat more difficult by the weight and position of the acentric visceral mass and shell.

Pedal extension is achieved by increased blood pressure within the pedal hemocoel. The posttorsional right retractor muscle is delayed in development until after torsion, and it then develops into the single, strong **columella muscle** that extends from the foot to encircle the columella. This powerful muscle is solely responsible for retracting the entire animal into the shell. Intrinsic pedal muscles, contracting in such a way as to produce a strong suction effect, enable many gastropods to hang tenaciously onto a surface.

Mantle cavity and ctenidia

The mantle cavity, substantially influenced by torsion, spiraling, and coiling, has undergone extensive modifications from the ancestral molluscan pattern. Torsion brings the mantle cavity forward along with the ctenidia and openings of the anus and urogenital ducts. But in so doing, there is a definite risk of contaminating the ctenidia and the head region with fecal materials. Therefore many gastropodan adaptations are designed to minimize fouling, especially of the ctenidial surfaces, by the products of egestion, excretion, and reproduction. Representatives of each subclass have approached these major problems of water circulation and gaseous exchange in fundamentally different ways, which we shall now explore.

The most primitive gastropods belong to the Subclass Prosobranchia. In this very large group are found basically three levels of complexity in the mantle cavity, ctenidia, and water circulation pathways. The least complicated and most primitive arrangement is found in members of the Order Archeogastropoda including abalones, keyhole limpets, and others. These snails have two ctenidia and two auricles; they are thus similar to the generalized mollusc or the hypothetical ancestral mollusc except that because of torsion, the mantle cavity and associated structures are in front of the visceral mass. The ctenidia are primitive in that they are **bipectinate**, that is, they have filaments on both sides of the ctenidial axis (Fig. 8.18a). Water enters the mantle cavity from either side, passes between the ctenidial filaments of both ctenidia because of the action of filamental cilia, and then travels upward in the mantle cavity and out medially through an opening in the shell by way of a cleft, a series of perforations (abalones), or an apical hole (keyhole limpets) (Fig. 8.18a, d). In all these cases the anal and urogenital openings are located medially in the exhalent current near to the point where the water current exits from the mantle cavity, an arrangement that effectively minimizes ctenidial fouling.

A second major adaptation involved the loss of the right ctenidium and auricle due to torsion; the left bipectinate ctenidium remained unaffected. The loss of the right ctenidium made possible more efficient pathways of water through the mantle cavity. Incurrent water is now brought in at the left side, passes over the ctenidium, and exits on the right side, thus making an oblique transverse path through the mantle cavity (Fig. 8.18b). The anal and urogenital openings are in the right side of the mantle cavity in the exhalent current. This type of arrangement is found in the more advanced archeogastropods like the top shells (*Trochus*) and certain littoral snails like *Nerita*. Even with the obliquely directed water current passing over the single bipectinate ctenidium there is still a tendency in this arrangement for sediments and other particulate materials to accumulate on the

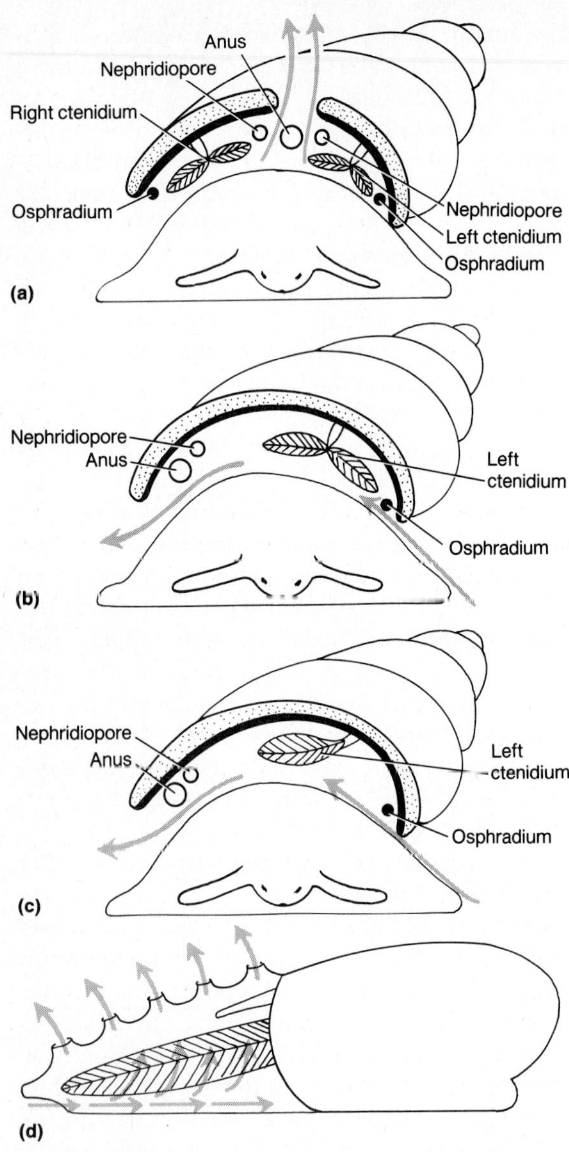

Figure 8.18 *Major alterations of the mantle cavity, ctenidia, and direction of water currents in prosobranch gastropods:* ***(a)*** *a diagrammatic, cross-sectional view of an unspecialized prosobranch (primitive archeogastropod);* ***(b)*** *a diagrammatic, cross-sectional view of a prosobranch (advanced archeogastropod) with one bipectinate ctenidium;* ***(c)*** *a diagrammatic, cross-sectional view of an advanced prosobranch (mesogastropod or neograstropod) with one monopectinate ctenidium;* ***(d)*** *a longitudinal section of an abalone (primitive archeogastropod) showing water pathways.*

ctenidium and in the mantle cavity owing to membranes suspending the ctenidium from the roof of the mantle cavity. Thus these snails are restricted ecologically to hard substrata so as to minimize the clogging of the mantle cavity by sediments.

A third major water circulation advance in prosobranchs involved a substantial modification of the ctenidium. This ctenidium became **monopectinate**, i.e., having filaments on only one side (Fig. 8.18c). The ctenidial axis is attached lengthwise along the mantle wall on the side where the filaments have been lost. Filaments of the other side, projecting into the mantle cavity, are basically unchanged in structure or function from the primitive bipectinate condition. Such an arrangement of ctenidial filaments greatly enhances the hydrodynamic efficiency of water circulation in the mantle cavity (Fig. 8.18c). With the ctenidium attached directly to the mantle roof and thus lacking a suspensory membrane, sediment clogging was no longer a serious problem, and these advanced snails could then colonize muddy bottoms or areas where other gastropods could not live.

The monopectinate animals, representing the orders Mesogastropoda and Neogastropoda and thus comprising the vast majority of all gastropods, illustrate various other adaptations as solutions to the problems of sediments clogging the mantle cavity. In higher mesogastropods and all neogastropods the frontal part of the shell aperture is drawn out into a short or long spout, the **anterior canal**. Through this runs a tubular but incomplete fold of the left mantle margin, the **inhalent siphon**. The opening of the siphon is held above the substratum or projected well ahead of the snail so as to have available for the animal clean inhalent water. The osphradium has been greatly enlarged and elaborated to form an organ of chemoreception that can quickly detect water with sediments or the presence of prey organisms some distance away. In many gastropods the rectum and ducts of the genital and excretory organs have elongated so that they open well toward the right-hand side of the mantle cavity in the exhalent current.

The opisthobranchs and pulmonates were both derived from a prosobranch stock that primitively had only the left ctenidium. The opisthobranchs, having undergone detorsion, have a mantle cavity that is moved back along the right side or to its posterior position; in so doing, the ctenidium and mantle cavity are reduced and eventually lost. Because of the reduction and loss of the ctenidium, opisthobranchs rely more on cutaneous exchange of gases. Some primitive forms have secondarily acquired pallial gills, and more advanced opisthobranchs bear tufts of anal gills or dorsal cerata.

In pulmonates the ctenidium has also disappeared completely, and the mantle cavity forms a vastly different sort of gas exchange organ, the "**lung**" or **pulmonary cavity**, from which this group derives its name. The inner walls of the mantle, especially those of the roof, have become heavily vascularized for gaseous exchange. In fact, the walls are thrown into ridges and grooves that may treble the gas exchange surface. The pulmonary cavity is typically air filled, but in some aquatic forms it is water filled. The mantle margin is fused to the neck except at one place or opening, the **pneumostome**. The pneumostome, usually situated on the right side near the anus, nephridiopore, and gonopore, may be a permanent opening, but in most species it is normally closed except when it is opened rhythmically to admit air or water for ventilation. In shelled pulmonates the volume of the pulmonary cavity is variable depending upon the degree of retraction of the animal into its shell. Ventilating muscular contractions within the floor of the pulmonary cavity are known for some species. Mucous secretions in the lining of the pulmonary sac keep it moist for gaseous exchange. The pulmonary cavity is potentially an important locus for water loss; and because of problems of desiccation, terrestrial forms are nocturnal or live in humid or damp areas. The closing of the pneumostome also serves to reduce evaporation. Additionally, all pulmonates are capable of cutaneous exchange. One family has lost the pulmonary cavity completely; this group exchanges gases solely through the body surfaces.

Internal Features

Feeding and the Digestive System

Primitive gastropods (archeogastropods) are very much like the ancient molluscs in that they are microphagous grazers on encrusting organisms living on hard substrata such as rocks. These gastropods have a radula with a great many teeth, at least 12 per radular row, which makes this organ an effective scraping and rasping organ. Food typically consists of encrusting algae, sponges, and other small organisms. Some of these animals also have a crystalline style and typically feed more or less continuously by the rotating style pulling in the mucus–food cord.

From this basic primitive habit, gastropods have become extraordinarily diverse in the types of food preferred and in methods of feeding. Among gastropods are found carnivores, parasites, herbivores, scavengers, filter- and detritus-feeders, stalkers, and hunters. This adaptive radiation in feeding habits among gastropods shows two important trends. First, there has been a decided shift from microphagous to macrophagous feeding. Second, there is a general tendency to change from being grazing herbivores to being predatory carnivores.

One of the most important trends has been the evolution of a carnivorous habit among gastropods. Carnivores, feeding on a wide variety of invertebrates and even some vertebrates like fishes, have developed a number of adaptations for locating and procuring food. Higher carnivorous gastropods have keener sensory organs for locating prey, including an elongated incurrent siphon and an enlarged osphradium (most prosobranchs), image-forming eyes (some opisthobranchs), and more sensitive chemoreceptors on tentacles. Some of the opisthobranchs are active swimmers, and they aggressively seek out prey in their locomotion. There is a reduction in numbers of teeth per radula row, and the fewer teeth have become sharper and specialized for cutting into prey or their shells. Carnivorous gastropods have lost the style and become intermittent feeders. The development of a muscular **proboscis**, a pliable organ that can be inserted into small crevices or between shells to feed, is an extremely important adaptation for many gastropods (Fig.

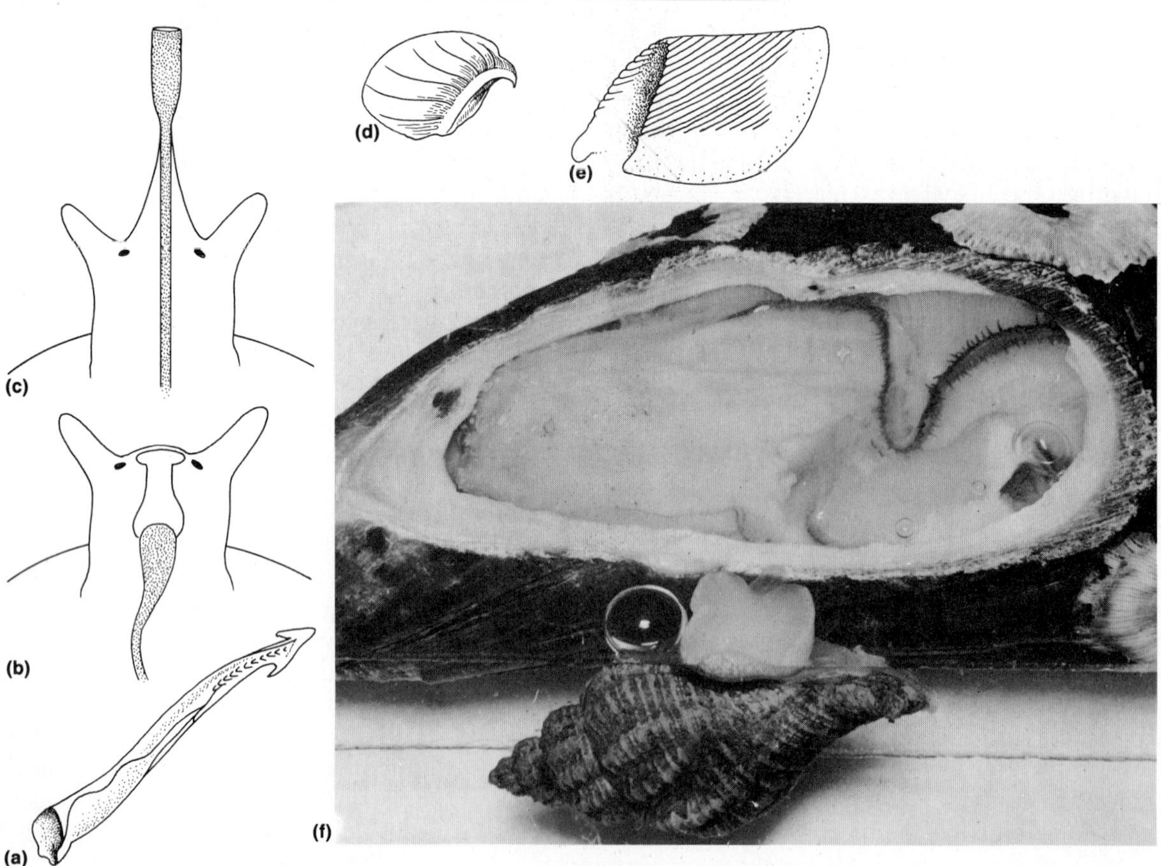

Figure 8.19 *Some structural feeding adaptations in gastropods:* ***(a)*** *the single radular tooth of* Conus*;* ***(b)*** *a retracted proboscis;* ***(c)*** *an extended proboscis;* ***(d)*** *a jaw;* ***(e)*** *paired jaws in contact dorsally;* ***(f)*** *photograph of the oyster drill,* Urosalpinx*, extending its proboscis between a glass slide and a mussel* (Mytilus) *cut in sagittal section; note the drill's proboscis extended almost halfway into the mussel's tissues (part a [after Shaw]; parts d, e [after Urban] in L.H. Hyman,* The Invertebrates, *Vol. VI:* Mollusca I*, 1967, McGraw-Hill Book Co., adapted with permission; part f courtesy of M.R. Carriker).*

8.19f). The proboscis, containing mouth and radula, is extended hydrostatically by hemocoelic fluids; but when not in use, it is retracted by paired muscles into a **proboscis cavity** (Fig. 8.19b, c). Many carnivores also have paired **jaws** with which they cut and shred the prey (Fig. 8.19d, e).

Several special cases of very interesting feeding patterns need to be noted here. Many different prosobranchs are adapted for drilling holes in the shells of prey. Perhaps the best known is the American oyster drill, *Urosalpinx*, a serious economic pest in the shellfish industry (Fig. 8.19f). On the anterior of the foot is a gland whose acid secretions soften the oyster shell, and the radula rasps away a thin layer of demineralized shell. The secretory-rasping cycle is repeated numerous times until the shell has a hole large enough to accommodate the gastropod's proboscis, which is then inserted into the oyster, and the radula tears away bivalve tissues (Fig. 8.19f).

Another group of tropical and subtropical carnivores belong to the genus *Conus*. These snails are remarkable in that the radula is modified into a single longitudinal row of long grooved teeth (Fig. 8.19a). At any one time a single tooth is used as a harpoon, spear, or dart with which the snail impales its prey. A neurotoxin, secreted by a poison gland whose duct opens near the tooth, is injected into and quickly immobilizes the prey, usually polychaete worms, fishes, or other gastropods. Some South Pacific cone species are potentially dangerous or even fatal to persons who might handle them carelessly or step on the gastropod buried in sand.

Some carnivorous prosobranchs, slowly developing a more sluggish habit, eventually became parasitic on worms, echinoderms, other molluscs, and crustaceans. Externally, ectoparasites are not very specialized for their parasitic life; their chief adaptations include the development of a piercing proboscis and a sucking muscular pharynx used to suck up bits of the host. Endoparasites are typically more specialized and severely modified for their parasitic existence including reduction of the shell, elongated body, and smaller size. **Jaws** are commonly present in parasitic gastropods and assist in feeding.

Herbivorous gastropods include some prosobranchs and opisthobranchs and many pulmonates. Marine forms feed on algae, and freshwater and terrestrial snails feed on tender or decaying vegetation. Jaws are also often present in herbivores and are used to tear off bits of vegetation (Fig. 8.19d, e). Some gastropods are suspension- or filter-feeders; they obtain their food by straining water, thus trapping particles and microorganisms on enlarged ctenidial filaments, mucous threads, mantle lobes, or mucus-covered parapodia.

Once ingested, food passes through a short **esophagus** into the **stomach** (Fig. 8.20b). Present in the stomach of continuously feeding, microphagous gastropods is a **crystalline style** that aids in feeding, triturates stomach contents, and releases digestive enzymes like amylases. Extensive grading or sorting areas are found in the stomach. In herbivores and microphagous feeders, extracellular digestion is mostly restricted to the tubules of the digestive gland. In macrophagous carnivores the style is absent, and strong proteolytic enzymes, produced by the digestive gland, are liberated into the stomach, where considerable extracellular digestion takes place, but final digestion is intracellular. Undigested materials leave the stomach and digestive gland along the **typhlosole**, enter the **intestine**, and are egested as fecal pellets.

Coelom, Excretion, and Circulation

The coelom is typically molluscan and primitively is involved to some degree with excretion. A pair of **kidneys** is present in the archeogastropods, but in all others the right kidney has been lost. The remaining kidney, located in the anterior visceral mass because of torsion, has

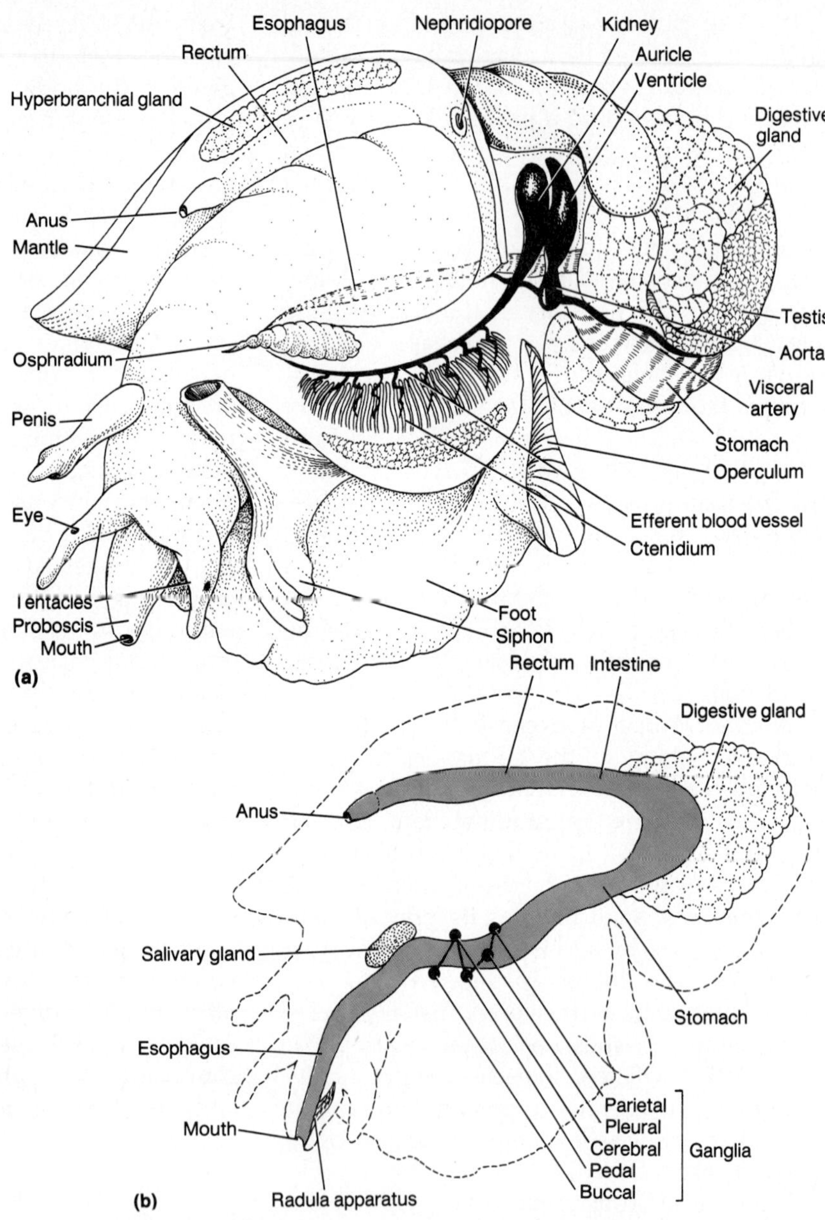

Figure 8.20 The internal features of Busycon, *a prosobranch gastropod:* ***(a)*** *a partially dissected male with shell and left mantle removed (from F.A. Brown, Jr., ed.,* Selected Invertebrate Types, *New York, 1950, © John Wiley & Sons, adapted with permission);* ***(b)*** *the digestive system and ganglia.*

been modified into a saclike organ whose walls contain numerous folds that increase its internal surface areas. Primitively, the inner or proximal end connects to the coelom via a nephrostome. In most other species, however, the connection between coelom and kidneys is a small **renopericardial canal**. The outer or distal portion of the kidney is the **ureter**, which opens into the right side of the mantle cavity through a nephridiopore (Fig. 8.20a).

Functionally, cells in the kidney wall secrete water and wastes into the kidney lumen.

Aquatic species excrete ammonia; freshwater forms eliminate copious amounts of water owing to osmotic gradients; and land pulmonates, confronted with problems of desiccation, eliminate uric acid, thus conserving precious supplies of water.

The **heart** is located in the anterior visceral mass because of torsion. Two **auricles** are present in those forms with two ctenidia (most archeogastropods); but in most others, only the left auricle remains (Fig. 8.20a). The muscular **ventricle**, now posterior to the auricle because of torsion, gives rise either to a posterior aorta or to a posterior and an anterior aorta supplying the visceral mass and head–foot, respectively. Blood in the hemocoel usually circulates near the kidney for waste removal before entering the ctenidium or "lung." Most gastropods possess hemocyanin as an oxygen-transporting pigment, though some have hemoglobin, and others lack a blood pigment.

NERVOUS SYSTEM AND SENSE ORGANS

The nervous system of gastropods is similar to that in primitive molluscs except that it has been dramatically complicated by torsion. To best understand the nervous system in adults, a review of the larval, pretorsional plan will be instructive (Fig. 8.21a). Three pairs of ganglia lie near the esophagus: **cerebral ganglia** lie in the circumesophageal nerve ring and innervate the head, **pedal ganglia** are situated below the esophagus in the anterior midline of the foot and innervate it, and **pleural ganglia** are positioned lateral to and above the esophagus and supply nerves to the mantle (Fig. 8.21a). Neural connectives, linking these three ganglia, form a triangle on either side of the esophagus. In addition, small, paired, anterior **buccal ganglia**, innervating the radula and nearby structures, are joined by connectives with the cerebral ganglia (Fig. 8.21a). A **visceral nerve** originates in each cerebral ganglion and runs posteriorly. Along the length of each visceral nerve are located three ganglia: a pleural ganglion already mentioned, a **parietal ganglion** that innervates the ctenidium and osphradium, and a terminal **visceral ganglion** that controls the functions of the visceral organs (Fig. 8.21a).

Torsion produces a twisting of the visceral cords between the pleural and parietal ganglia so that the posterior portions of the nervous system describe a figure eight (Fig. 8.21b, c).

Figure 8.21 *Diagrams of the nervous system of gastropods:* ***(a)*** *pretorsional plan seen from the left side;* ***(b)*** *posttorsional plan seen from the left side;* ***(c)*** *posttorsional plan seen from the dorsal side. (All ganglia originally located on left side are shown in open circles and those on the right by solid circles.)*

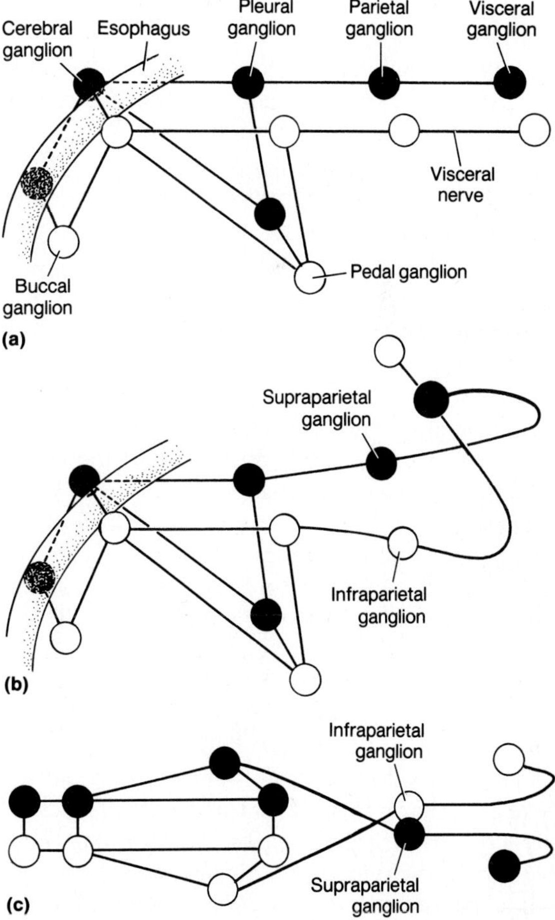

The right parietal ganglion becomes the left or **supraparietal ganglion** because of its new position, and the left parietal ganglion becomes the right or **infraparietal ganglion**. Torsion brings the visceral ganglia anteriorly and may result in some shortening of the visceral cords. The positions of those ganglia located around the esophagus remain essentially unchanged by torsion (Fig. 8.21b, c).

Because of torsion, the nervous system in most prosobranchs is twisted and asymmetrical. In more advanced gastropods, however, there are tendencies for the nervous system to become more bilaterally symmetrical again. These tendencies are due to

1. detorsion, which returns the parietal and visceral ganglia back toward their original positions;
2. shortening of the visceral nerves, which pulls the posterior ganglia back into a position of bilateral symmetry; and
3. concentration and fusion of ganglia.

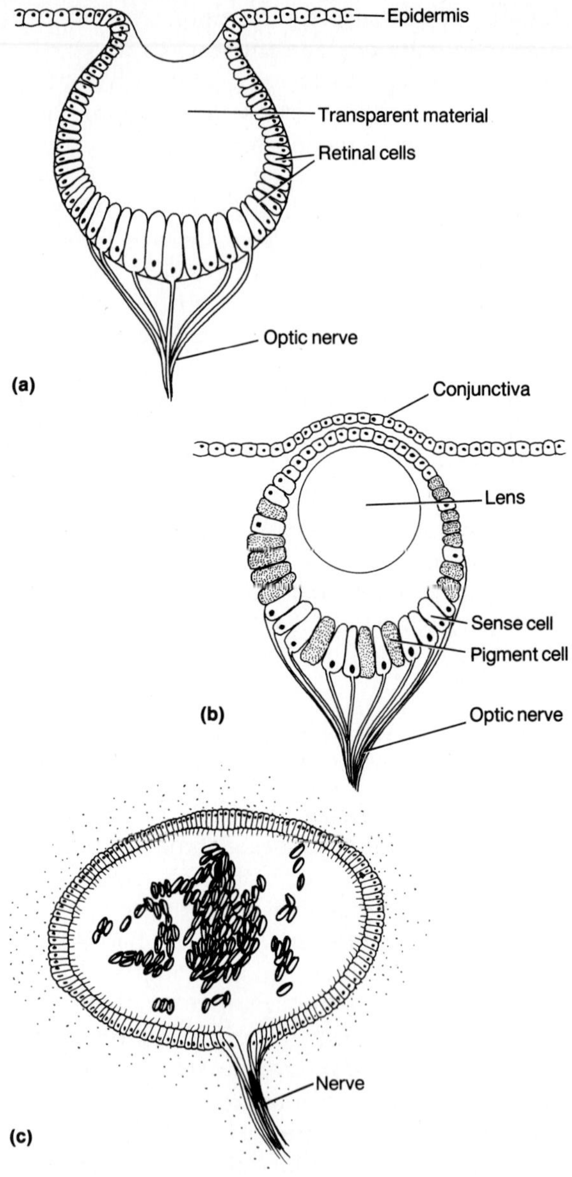

Figure 8.22 *Gastropod sense organs:* ***(a)*** *a simple eye;* ***(b)*** *a complex eye;* ***(c)*** *a statocyst with many statoliths.*

Sense organs include eyes, tentacles, statocysts, and osphradia plus numerous chemoreceptor and tactile receptor cells. Most gastropods have one pair of **eyes** with an eye located at the base of each tentacle, at the tips of tentacles, or on special optical stalks. An eye may be simple with only a cuplike invagination lined with light-sensitive cells, but some of the more advanced gastropods have complex eyes with cornea (conjunctiva), lens, and retina (Fig. 8.22a, b). Gastropodan eyes function generally only in detection of light intensities; but in some mesogastropods (heteropods), eyes provide the animal with a high degree of visual acuity. Sensory **tentacles** are present on the head either as one pair (prosobranchs) (Fig. 8.20a) or two pairs (opisthobranchs and most pulmonates). Some forms have additional tentacles on the mantle margins. **Statocysts**, absent in some snails, are present near the pedal or cerebral ganglia (Fig. 8.22c). One or two **osphradia** lie in the incurrent water passage, and numbers of osphradia are correlated with numbers of ctenidia. In the carnivorous gastropods the single remaining osphradium has become an efficient center for chemoreception used in locating prey.

Reproduction and development

A majority of gastropods are dioecious. In all gastropods the **gonad** is an unpaired organ lying in the dorsal visceral mass near the right kidney (Fig. 8.20a). Since gametes of most gastropods first enter a short duct that conveys them from the gonad to the kidney, the **gonoduct** consists of the short gonadal duct and distal portions of the right kidney. In archeogastropods, gametes are released into the mantle cavity and discharged into the water, where fertilization takes place.

In higher gastropods, part of the right kidney degenerates, and this organ no longer has any excretory function but is used solely for gametic transport. Because of the addition of a distal **pallial duct** formed from the mantle, the gonoduct becomes considerably lengthened and bears regional specializations. Concomitantly with the gonoductal elaborations, means have been developed for sperm transfer, internal fertilization, and production of large eggs with considerable yolk.

In most male prosobranchs (mesogastropods and neogastropods) the reproductive system consists of the testis, vas deferens, prostate gland, and penis (Fig. 8.23b). The **vas deferens**, a highly coiled tubule in which sperm are stored temporarily, usually opens at the tip of the **penis**, an elongated tubular extension of the body wall just behind the right cephalic tentacle (Fig. 8.20a). Seminal secretions of the **prostate gland** contribute to the viability of sperm.

In female prosobranchs the reproductive system is composed of the ovary, oviduct, accessory glands, and seminal receptacle (Fig. 8.23a). The **oviduct**, like the vas deferens, consists of renal and pallial portions. Secretions of a jelly gland surround the eggs with a jellylike material; and in snails who lay capsulated eggs, albuminoid materials from the albumin glands are added before the outer capsule materials from a capsule gland are applied. The **seminal receptacle** is a saclike diverticulum that receives and stores sperm after copulation until they are used for fertilization (Fig. 8.23a). As ova are liberated from the gonad, they are fertilized prior to receiving accessory coats or membranes.

All opisthobranchs and pulmonates are monoecious; some are protandric (i.e., males first, then functional females), but most are simultaneous hermaphrodites. Both eggs and sperm are produced by the single **ovotestis**. Most higher gastropods engage in cross-fertilization by mutual copulation, although in some snails, one member acts as a male and the other as a female. A period of courtship prior to copulation is practiced by some pulmonates (Fig. 8.24a). The reproductive tracts are rather complex and vary widely in details between species. Much of the genital tract complexity results from there being essentially three separate ductal systems: one for outgoing sperm, one for incoming sperm via copulation, and one for fertilized eggs to be laid (Fig. 8.23c). Sperm, produced by the ovotestis, travel down the common hermaphroditic duct and then into the vas deferens running the length of the penis. Incoming sperm are introduced by the penis of the other gastropod through a common or separate genital opening into a vagina, and the sperm then enter the seminal receptacle, where they are stored until used for fertilization (Fig. 8.23c). Eggs, produced by the ovotestis, travel through the common hermaphroditic duct to a fertilization chamber, where gametic union takes place and the various accessory coats and membranes are added. These eggs exit through the oviduct and vagina and, in some, a common genital duct.

Protandrous species like many limpets and slipper shells initially produce sperm and only later produce eggs. One of the most interesting protandric groups is the slipper shells, which belong to the genus *Crepidula*. Young specimens are always males, and the sex of older adults depends upon the association with other individuals. Most individuals of each of the several species of *Crepidula* live in stacks, the individual

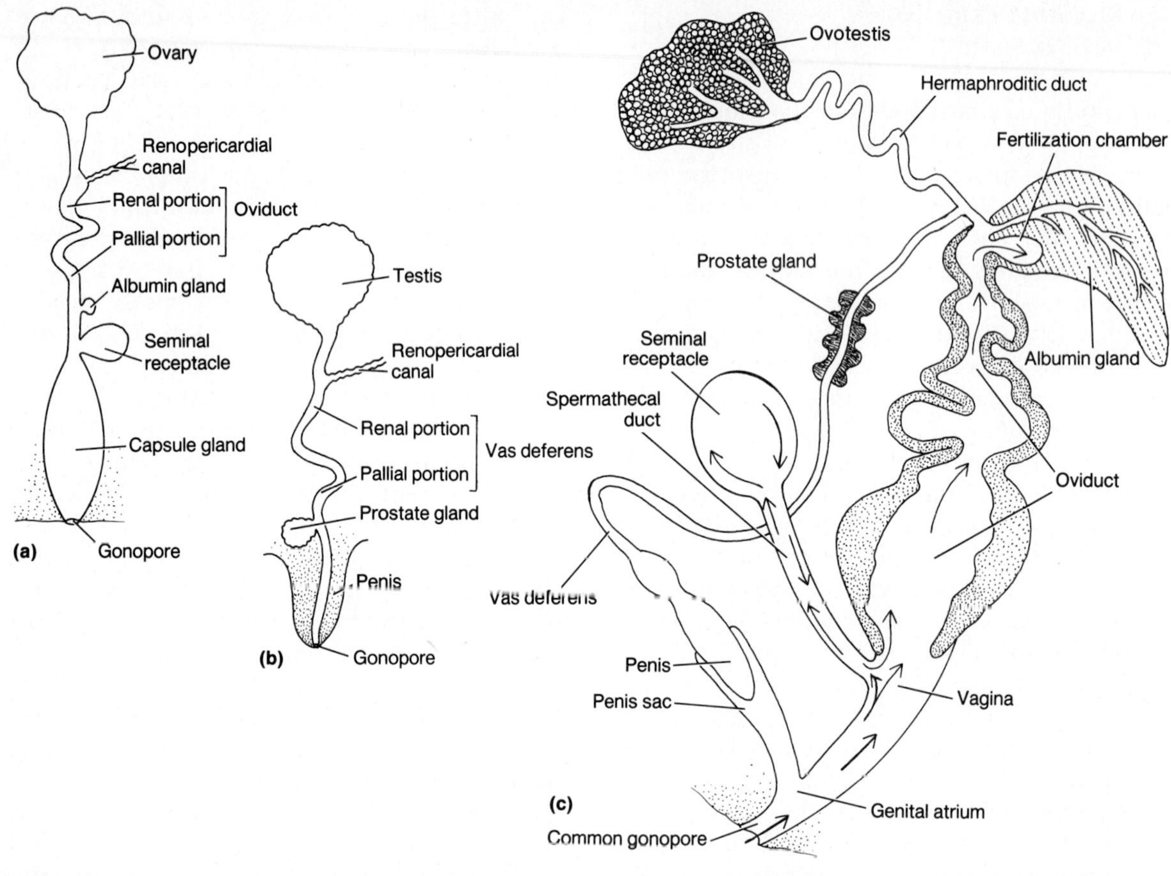

Figure 8.23 *The reproductive systems of gastropods:* ***(a)*** *a female;* ***(b)*** *a male;* ***(c)*** *a hermaphroditic individual with the path of the introduced sperm shown by arrows.*

living at the top being the youngest and always a male (Fig. 8.24b). Females are near the bottom of the stack, and individuals in the middle are hermaphrodites. As new males are added to the stack, the previous males become hermaphrodites, then females, then die at the bottom of the stack (Fig. 8.24b). The protandric sequence depends upon complicated interactions of both environmental and physiological factors. For example, isolated males develop into small females, and males kept in the presence of females remain males. But females never develop into males again. One of the more common species of slipper shells has the very descriptive name, *Crepidula fornicata*.

In all gastropods that have internal fertilization, eggs are laid in egg mass shapes characteristic of the species; these shapes include strings, ribbons, gelatinous or albuminoid masses, and in cases composed of leathery or gelatinous materials or even sand grains (Fig. 8.24c–g). Terrestrial gastropods produce large yolky eggs laid separately in damp places as under logs and in leaf mold. Some land snails are viviparous, and their eggs are brooded internally. A trochophore stage is absent except in primitive prosobranchs that have external fertilization.

Characteristic of marine gastropods is the **veliger** stage, during which they undergo torsion (Fig. 8.11). As metamorphosis proceeds, the young gastropod becomes progressively better adapted for its benthic existence, and sexual maturity is often attained in 6–24 months. Some prosobranchs and almost all opisthobranchs and pulmonates have no separate larval stage; juvenile gastropods hatch directly from the eggs.

Subclass Prosobranchia

Prosobranchs are immensely varied in both morphological and physiological adaptations that have enabled them to be a most successful group with perhaps 60,000 species having been

Figure 8.24 *Reproductive adaptations found in gastropods:* ***(a)*** *two courting slugs* (Arion) *(pulmonates) (from L.H. Hyman,* The Invertebrates, *Vol. VI:* Mollusca I, *New York, McGraw-Hill, 1967, used with permission);* ***(b)*** Crepidula, *a protandric prosobranch (from W.R. Coe,* J Exper. Zool. 72:457, 1936, used by permission of Alan R. Liss, Inc.); ***(c)*** *a portion of the egg mass of* Aplysia *(opisthobranch);* ***(d)*** *several egg masses of Conus (prosobranch);* ***(e)*** *the egg mass of* Polinices *(prosobranch);* ***(f)*** *an egg mass of* Physa *(pulmonate);* ***(g)*** *a photograph of the egg case of* Busycon *(prosobranch) (courtesy of C. F. Lytle).*

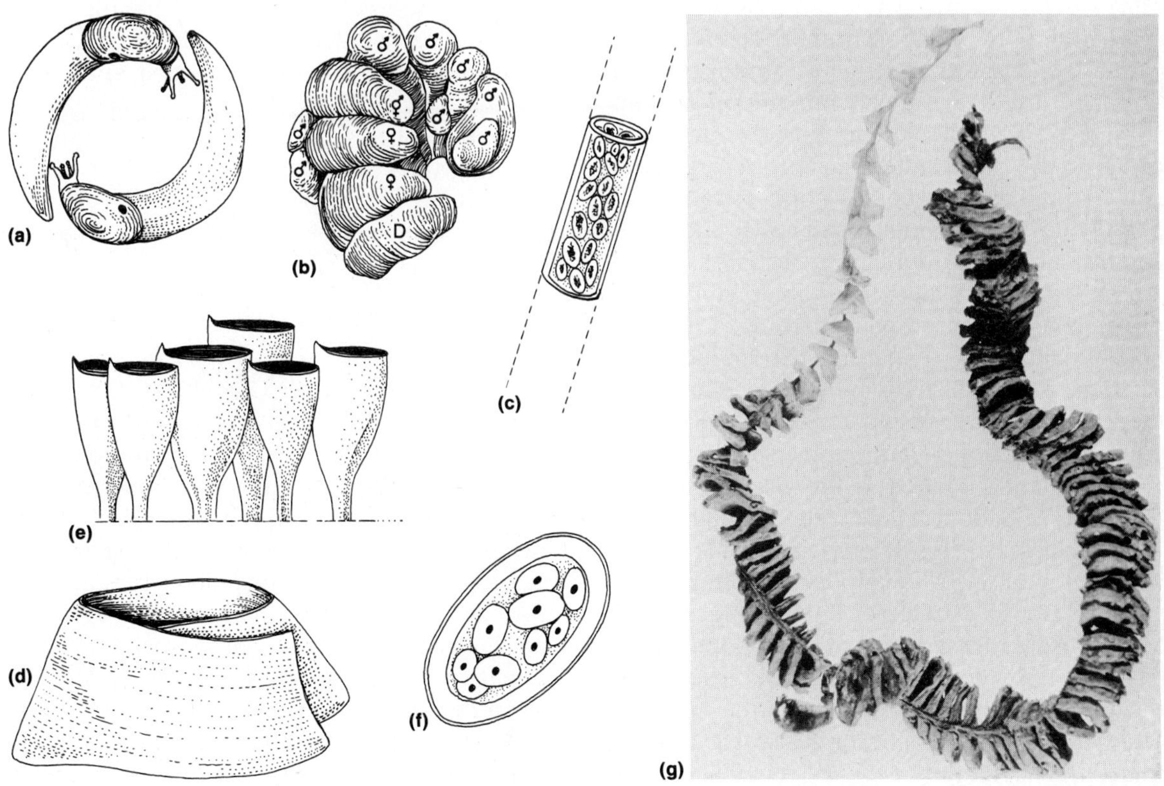

described to date. Most prosobranchs are marine animals, but some are found in brackish, freshwater, or even terrestrial habitats; some are even pelagic. They are especially common in the benthos in littoral zones near the tide marks. Probably most prosobranchs are carnivores, but many are herbivores, and a few are filter- or deposit-feeders or parasites.

Among gastropods, torsion is the most evident in adult prosobranchs. A definite head is present bearing one pair of tentacles and a pair of eyes located at the base of or on the tentacles. A shell is always present, and although many variations are found, it is usually of the conical helical shape. Shells of prosobranchs are especially varied in colors, patterns, spines, ridges, and other forms of sculpturing (Fig. 8.25).

The typical gastropodan foot is present, and on its dorsal posterior surface is borne a rounded or oval **operculum** constructed of hornlike conchiolin. When the foot and head are withdrawn, that portion bearing the operculum is the last to be retracted, and it fills the shell aperture like a door. Thus the operculum is an important adaptation, since it effectively protects the retracted animal from predators and also reduces the dangers of desiccation especially for those animals living in intertidal zones. The operculum, however, is absent in limpets, abalones, most heteropods, and those prosobranchs whose shell aperture is a long, narrow slit (e.g., cowries, olives, and cones). The larval stage of most nonoperculate prosobranchs does have an operculum, so its presence in this group is fairly universal.

The mantle is open to the exterior around the neck region to facilitate water circulation. In most species the left side of the mantle is drawn out to form an incurrent siphon, which may be an open trough or a closed tubule (Fig. 8.20a). Most prosobranchs have a single monopectinate ctenidium. In some prosobranchs, especially the more primitive ones, the mantle surfaces often are more important for gaseous exchange than is the ctenidium.

The digestive system, typically U-shaped because of torsion, contains a radula and a style in the continuously feeding forms. Many prosobranchs, especially the carnivores, have an extensible proboscis employed in feeding (Fig. 8.19f). Often the lateral walls of the buccal cavity are stiffened by a pair of hardened serrate plates or spines, the jaws. The basic prosobranch nervous system is torted into a figure eight, although there are some tendencies toward a bilaterally arranged pattern because of fusion and condensation of ganglia. Usually, only the left renal organ persists, although some lower forms have two.

Prosobranchs are, in general, dioecious, although most have no sexually dimorphic shell patterns. Most of the hermaphrodites function either as males or as females at any one time. In some lower forms, fertilization is in sea water and a trochophore develops, but in most prosobranchs, internal fertilization is the rule. Most males have a penis located on the right side of the head. Some females brood their eggs until hatching, but most eggs are laid in capsules or egg masses. The early development of prosobranchs produces a characteristic veliger.

Three large orders of prosobranchs are recognized. Each will be characterized briefly below.

ORDER ARCHEOGASTROPODA

Archeogastropods are an ancient and primitive marine group. In many archeogastropods the mantle cavity and related systems are bilateral with paired auricles, kidneys, osphradia, and bipectinate ctenidia. Some other members of this order possess only a single ctenidium and effect an oblique water current through the mantle cavity. Typically, these snails are microphagous continuous feeders. Internally, archeogastropods possess a radula with many radular teeth per row, a style, ciliary sorting surfaces in the stomach, and paired digestive glands. A copulatory organ is usually absent, fertilization is external, and some primitive forms have a free-swimming trochophore larva. Some familiar archeogastropods are keyhole limpets (*Fissurella*) and certain other limpets (*Patella*), abalones (*Haliotis* [Fig. 8.16f]), top snails (*Trochus* [Fig. 8.25a]), certain freshwater genera like *Theodoxus*, and a rather common operculate land snail, *Helicina*.

ORDER MESOGASTROPODA

This is the largest prosobranch order and contains many familiar snails such as conchs and periwinkles. Though it is predominately a marine group, there are a few representatives in fresh water and even on land. All mesogastropods have only the left ctenidium (monopectinate), auricle, and osphradium present. Almost every form of heterotrophic nutrition and means of procuring food is found in these gastropods, and a few species are even endoparasitic in echinoderms. Feeding is accomplished by a radula with only seven teeth per row and, in carnivores, paired jaws. The heteropods are active swimmers and have well-developed eyes on telescopic stalks. A penis is usually present for copulation. Some of the more common mesogastropods include conchs (*Strombus*), cowries (8.25d), heteropods (Fig. 8.25n), moon snails (*Polinices* [Fig. 8.25l]), the intertidal *Littorina*, slipper shells (*Crepidula* [Fig. 8.24b]), and others known commonly by the shape or color of their shell (tuns, violets, helmets, carriers, turrets, etc.).

ORDER NEOGASTROPODA

Neogastropods are almost all marine benthic carnivores and are aggressive feeders. They possess a number of features in keeping with their predatory habit including a long mantle siphon, a radula with only three or fewer teeth in each transverse row, a long extensible proboscis, a rather simple saclike stomach, and an osphradium as a chemoreceptor for "smelling" prey. They lack completely a style and primitive ciliary sorting areas in the stomach. Neogastropods have a prominent penis, and eggs are generally laid in tough capsules. Typical and familiar neogastropod genera include the whelks (*Busycon* [Fig. 8.25f], *Buccinum*), oyster and other bivalve drills (*Urosalpinx* [Fig. 8.19f]), the tropical *Conus* (Fig. 8.25i), many mud snails (*Ilyanassa, Nassarius*), the murexes (*Murex*), volutes, and olive (Fig. 8.25h), auger (Fig. 8.25j), and harp (Fig. 8.25k) shells.

Subclass Opisthobranchia

Opisthobranchs are represented by about 3000 species; thus this group is by far the smallest of the three subclasses. Nonetheless, they are highly variable, and many different structural and functional adaptations have enabled them to fill diverse niches (Fig. 8.26). They are almost always marine, but a few species are found in brackish water. Found generally in littoral areas, most burrow in or creep over the bottom material, but others glide, swim, or are planktonic. They range in length from 1 mm to 40 cm, but most are no longer than 10 cm. Opisthobranchs are often dramatically beautiful with striking shades of almost every color, which are often present in exquisite combinations. Resumes of the various orders of opisthobranchs are listed at the conclusion of this chapter.

Primitive opisthobranchs have an asymmetrical body because of torsion and a typically coiled shell. Representatives of one genus even have an operculum. But the evolutionary pathways of most opisthobranchs have taken them away from the torted, coiled body and heavy shell of prosobranchs to an elongate, slender body. These trends, correlated with detorsion, reduction or loss of shell and mantle cavity, and loss of the primitive ctenidium, have resulted in bilateral symmetry in adults.

The head, usually clearly delimited, bears the mouth, tentacles, and eyes. Characteristic

Figure 8.25 *Collage of photographs illustrating diversity in prosobranch shells (a–k) and living prosobranchs (l, m) and a drawing of a heteropod (n):* ***(a)*** Trochus, *a top shell (archeogastropod);* ***(b)*** Diodora, *a keyhole limpet (archeogastropod);* ***(c)*** Phalium, *the scotch bonnet (mesogastropod);* ***(d)*** Cypraea, *a cowrie, ventral view to show slitlike aperture (mesogastropod);* ***(e)*** Lambis, *the spider shell (mesogastro-*

*pod); **(f)*** Busycon, *a whelk (neogastropod); **(g)*** Vermicularia, *the lady's curl (mesogastropod); **(h)*** Oliva, *an olive shell (neogastropod); **(i)*** Conus, *the textile cone (neogastropod); **(j)*** Terebra, *an auger shell (neogastropod); **(k)*** Harpa, *a harp shell (neogastropod); **(l)*** Polinices, *the moon snail (mesogastropod); **(m)*** Fasciolaria, *the banded tulip (neogastropod); **(n)** a heteropod (mesogastropod).*

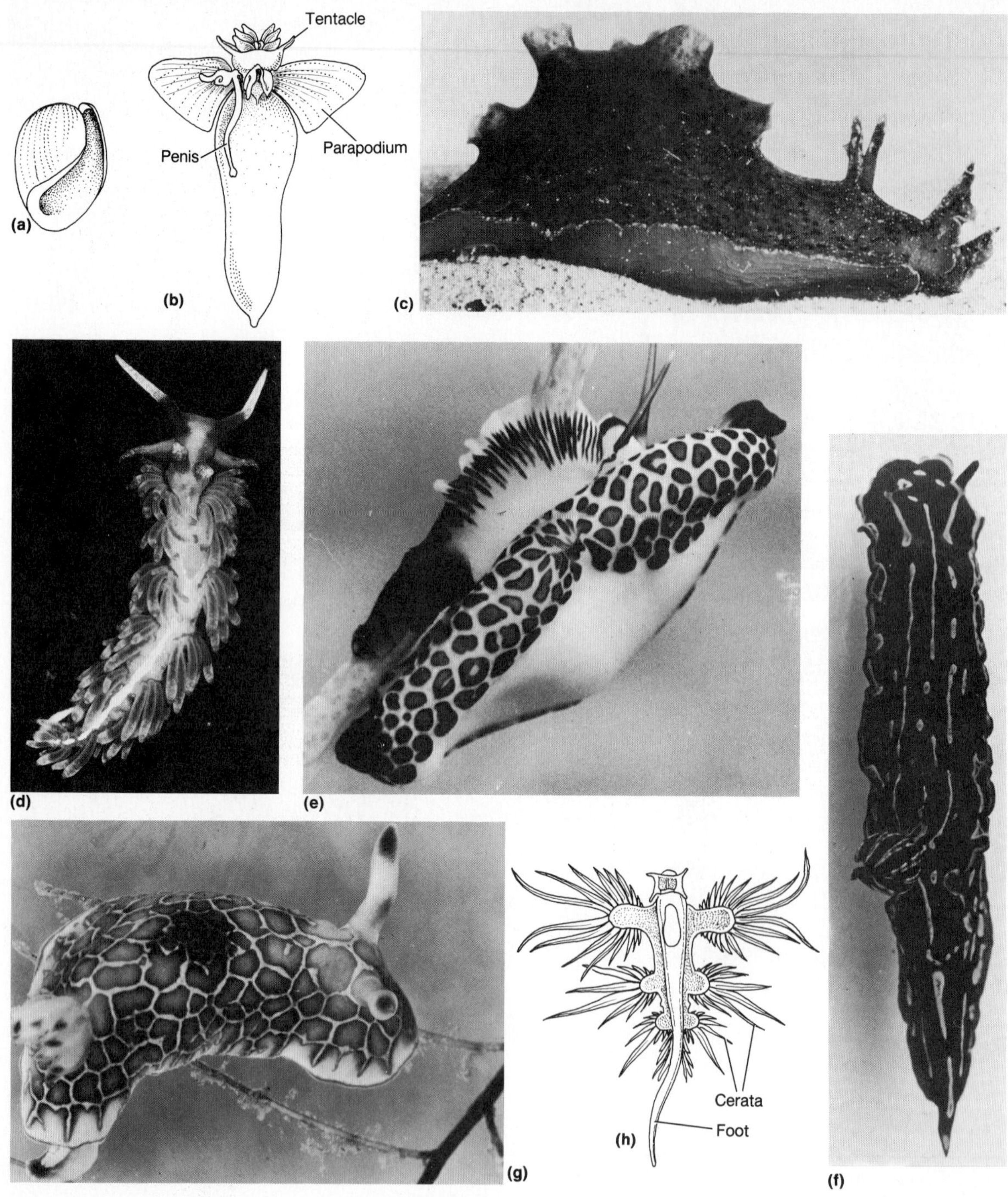

Tentacle
Penis
Parapodium
(a)
(b)
(c)
(d)
(e)
(f)
(g)
Cerata
Foot
(h)

of opisthobranchs is that most have two pairs of tentacles, which are of two types: **cephalic tentacles** situated just behind the oral region and the more posterior **rhinophores**, whose bases are often surrounded by a collarlike fold (Fig. 8.26d). A pair of eyes is found on the anterior head. The foot is broad, elongate, and of the creeping type, but there is a tendency for it to be reduced in some forms. Rapid creeping is achieved by finely controlled movements of the intrinsic foot muscles. Many opisthobranchs are pelagic, and swimming is achieved in several different ways; some employ undulations of the foot or lateral undulations of the entire body, but the vast majority of the swimmers use lateral extensions of the foot called **parapodia** (Fig. 8.26b, c). Often, the parapodia are moved vigorously up and down together like wings to achieve movement, and the pteropods even swim with the ventral side up. In other opisthobranchs the parapodia are slender and oarlike and are used in a rowing or sculling motion. In *Aplysia* the expanded parapodia have large waves of contraction passing posteriorly to achieve a graceful swimming action.

The shells of primitive opisthobranchs are often well developed and colorful. The shell is external in some, has become partially or wholly internalized in others, and is lost completely in all advanced opisthobranchs. A single ctenidium is located in the mantle cavity of primitive forms, and in others it is exposed as a lateral organ. In most species, however, the ctenidium is lost completely. Gaseous exchange in those without a ctenidium is achieved by several means, the most common of which is through the general body surface. Other opisthobranchs like the nudibranchs bear secondarily acquired gills usually in a ring or tuft near the anus (Fig. 8.26f, g). Other nudibranchs and sacoglossans possess dorsal and lateral fingerlike projections, the **cerata**, which also function in gaseous exchange (Fig. 8.26d, h). Cerata are usually brilliantly colored, and color patterns are often associated with the presence of subepidermal glands that secrete distasteful or toxic substances used for defensive purposes.

◀ ***Figure 8.26*** *Collage of drawings (a, b, h) and photographs (c–g) of opisthobranchs:* ***(a)*** *the shell of* Bulla, *a bubble shell;* ***(b)*** Clione, *a pteropod (ventral view);* ***(c)*** Tethys, *a sea hare;* ***(d)*** *a nudibranch, probably* Eolidia; ***(e)*** Cyphoma, *a nudibranch;* ***(f)*** Hypselodoris, *a nudibranch;* ***(g)*** Chromodoris, *a nudibranch;* ***(h)*** Glaucus, *a nudibranch. (part a [from Vayssiere], b [from Boas], h [from Pruvot-Fol] in L. H. Hyman,* The Invertebrates, *Vol. VI:* Mollusca I, *New York, McGraw-Hill, 1967, used with permission; part c courtesy of Ward's Natural Science, Inc., Rochester, N.Y.; parts e–g courtesy of C. E. Jenner).*

Some opisthobranchs are herbivorous, others are filter-feeders, and a few are parasites, but most species are carnivorous. Prey organisms include cnidarians, sponges, other opisthobranchs, barnacles, polychaetes, bivalves, and the eggs of many invertebrates and fishes. A radula is almost always present, and well-developed jaws are often found, but a crystalline style is always absent. Diverticula of the digestive glands extend into the cerata of opisthobranchs that possess them. Some nudibranchs feed on cnidarians in such a manner as not to discharge the nematocysts. Freed from other cnidarian cells in the gut of the opisthobranchs, the cnidocytes are transported to the ceratal tips, where they may become a useful defensive mechanism for the nudibranch.

Opisthobranchs are mostly simultaneous hermaphrodites with mutual insemination occurring between both members of a copulating pair. All have a large ovotestis, and most have an extensible penis for copulation. Some species engage in prenuptial courtship. Eggs are laid in a variety of gelatinous strings, ribbons, or sacs. Many opisthobranchs do not produce free-swimming veligers, but a stage similar to this is passed within the egg shell.

BOX 8.2 □ SOME PROBLEMS ASSOCIATED WITH GASTROPODS LEAVING THE SEA

The only eminently successful land animals are the arthropods and vertebrates, both of which possess jointed limbs for terrestrial locomotion and efficient water-conserving mechanisms. Molluscs are fundamentally adapted to an aquatic life and lack locomotive appendages and a water-conserving, impermeable skin. Gastropods are, for the most part, marine, even though many live in fresh water. But many gastropods, especially the pulmonates, have been rather successful on land and rank only behind arthropods and vertebrates in numbers of terrestrial species. Still other gastropods have returned secondarily to fresh water. This ecological diversity is indeed remarkable, especially if we remember how very much adapted gastropods are to the marine habitat.

Many different gastropod forms have left the sea, including a surprising number of prosobranchs; one entire group, the pulmonates, evolved in fresh water. In leaving the sea, gastropods had to utilize intermediate habitats such as estuaries or intertidal zones for generations before entering fresh water. Rarely did a gastropod leave the sea and become terrestrial without first becoming a freshwater form. Therefore the principal evolutionary trend for a number of diverse snails was first to adapt to freshwater habitats via estuaries. Apparently, this was relatively easy, judging by the large number of prosobranchs that were successful in this new environment. Freshwater snails had to evolve more efficient kidneys as osmoregulators to eliminate excessive amounts of water. They also had to produce eggs or egg masses that would be osmotically compatible with the developing embryos.

The advent from fresh water to a land existence was a major leap attained for the most part only by the pulmonates, which had evolved from prosobranchs. Land molluscs either had to live in very moist places where the ctenidium could still function (the terrestrial prosobranchs) or had to develop the mantle cavity as a pulmonary cavity or

Subclass Pulmonata

Pulmonates are mostly freshwater or terrestrial gastropods and include most land and freshwater snails, some limpets, and all slugs, but some have returned secondarily to an aquatic existence. A very successful group, the approximately 20,000 species are distributed widely in tropical, subtropical, and temperate regions. They range from 1 mm to nearly 30 cm, but most pulmonate snails are about 1–2 cm in shell height, and most slugs are around 1–4 cm long. Freshwater pulmonates are benthic or live on various submerged surfaces. Their ability to thrive in many different habitats has resulted in considerable external diversity. A brief treatment of some of the interesting solutions to problems faced by pulmonates as they became terrestrial or secondarily aquatic appears in Box 8.2. As a rule, pulmonates are herbivorous, and their diets are as variable as the habitats in which they live, but the carnivores feed upon earthworms, other snails, turbellarians, and various small invertebrates. Pulmonates are of special interest because they serve as intermediate hosts for many medically important parasites such as trematodes, cestodes, and strongyloid nematodes.

Many pulmonates possess a typical coiled shell into which the snail can withdraw. Most

"lung," a feat accomplished successfully by the pulmonates. Additionally, a terrestrial gastropod also had to have internal fertilization, storage capacities for sperm, and mechanisms for secreting desiccation-resistant egg capsules. Terrestrial snails were to be forever limited on land by the availability of calcium in the environment for shells, by widely fluctuating ambient temperatures, and above all by the lack of moisture. Most snails conserve water by being nocturnal; living in cool, damp, shady places; and being particularly sensitive in their daily cycles by responding to diurnal changes in temperature and moisture. For these reasons the land gastropods are, as a group, more active at night, when temperatures are lower and moisture is higher, and they also excrete nitrogenous wastes in the form of uric acid to conserve water even more. All the water-conserving adaptations have added to the success of terrestrial gastropods.

A remarkable turn of events has involved certain terrestrial pulmonates that have returned secondarily to fresh water. Now the "lung" must function as a water-filled chamber, and vascularizations of the lung wall have to extract oxygen from water in the lung. This is not difficult when one remembers that in air-breathing snails, the "lung" is lined with a thin film of water, and thus atmospheric oxygen must first be dissolved in water before absorption by the tissues can take place. However, these freshwater pulmonates lack the structurally specialized renal organs one would expect. But clearly the successful pulmonates in fresh water were derived from fully terrestrial ancestors. At least one prosobranch genus, *Smaragdia*, and a surprising number of pulmonate genera have returned secondarily to life in the seas.

In conclusion, the physiological stresses of such a transition clearly point up the evolutionary plasticity of this remarkable group of molluscs.

shelled pulmonates have a dextral shell, but sinestral forms are known (Fig. 8.27a, c–g); both planospiral and helical shapes are found. Amounts of calcium in food or otherwise taken up by the snail can result in shells of various thicknesses. In slugs the shell either is reduced or concealed by mantle tissues or is absent completely. Pulmonates, like all gastropods, are torted, but they exhibit various degrees of detorsion. The head, usually not well demarcated from the body, bears the mouth, one or two pairs of tentacles, and usually one pair of eyes located either at the bases or tips of the short tentacles. The foot has a typical creeping sole with cilia and mucous glands. A ctenidium is absent, although some have an accessory gas exchange organ called a **pseudobranch**. In primitive marine pulmonates a gill is present in the mantle cavity that may be homologous to the ctenidium of prosobranchs from which pulmonates evolved. But certainly the distinguishing features for this group as a whole is the modification of the mantle and mantle cavity into a "**lung**" or **pulmonary cavity**. It is within this chamber that gaseous exchange takes place between air or water in the "lung" and the vascularized walls of the mantle cavity. A **pneumostome** is the opening of the pulmonary cavity situated on the right side near the anus and nephridiopore.

(a)
(b)
(c)
(d)
(e)
(f)
(g)
(h)

◀ *Figure 8.27 Collage of photographs of living animals (a, b), shells (c–f), and drawings (g, h) of pulmonates:* ***(a)*** Lymnaea, *a common pond snail (basommatophoran);* ***(b)*** Limax, *a common garden slug (stylommatophoran);* ***(c)*** Liguus, *the banded Florida tree snail (stylommatophoran);* ***(d)*** Goniobasis, *a freshwater snail (basommatophoran);* ***(e)*** Helisoma, *a freshwater snail (basommatophoran);* ***(f)*** Cepaea, *a European garden snail (stylommatophoran);* ***(g)*** Physa, *a sinistrally spiraled pouch snail (basommatophoran);* ***(h)*** *a limpet (basommatophoran).*

Feeding is aided by a radula and, in some, by one or two jaws. A typical style is absent, but in some primitive forms, a mucous mass may itself be rotated as a windlass and thus aid in winding in the mucus–food cord. The nervous system reflects detorsion, and the ganglia are concentrated into a circumesophageal ring. Sense organs include paired cephalic eyes and pedal statocysts. The circulatory system is characterized by the presence of one auricle and ventricle and a capillary system in some organs. Blood flows from arterial capillaries into hemocoelic channels before entering the venous system. Hemocyanin is the most common respiratory pigment, and hemoglobin is present in the Planorbidae, a family of freshwater snails. Only the left coelomoduct remains as the single kidney.

All pulmonates are monoecious and usually produce both types of gametes simultaneously. In temperate regions, breeding is normally restricted to the warmer months. The single ovotestis lies amid the digestive gland, and sperm and eggs pass through separate, often complex ducts to the gonopores. A penis is usually present for sperm transfer, and some can self-fertilize but do not self-copulate. Some pulmonates, especially most terrestrial snails and the slugs, engage in a precopulatory courtship (Fig. 8.24a).

Freshwater pulmonates lay eggs in gelatinous capsules attached to submerged objects, but terrestrial pulmonates lay large yolky eggs one at a time at night in damp places such as under logs, bark, or leaves, or in damp soil depressions where numerous eggs may be deposited as a single clutch. A veliger stage occurs in marine pulmonates, but no free larval stages are developed in freshwater or terrestrial pulmonates, even though corresponding stages are passed within the egg capsule.

CLASS BIVALVIA

The bivalves, representing about 20,000 species and including such common and well-known animals as clams, scallops, mussels, and oysters, comprise the second largest of the molluscan classes. Like gastropods, bivalves are basically marine, although some species have adapted to brackish or freshwater habitats; there are no terrestrial bivalves. Even though certain species are found at oceanic depths of up to 5000 m, most inhabit littoral zones, where they are typically found burrowing in the soft benthic material. They range in size from about 2 mm to *Tridacna*, the giant clams, whose shell may be 1.3 m in length and weigh more than 1000 kg.!

Bivalves first appeared in the Cambrian seas. During the Ordovician and Silurian periods they underwent a great adaptive radiation and became widely distributed. Bivalves reached their zenith about 350 million years ago; since that time, numbers of species have slowly declined. About 20,000 fossil species have been discovered to date. While not as numerically diverse as gastropods, bivalves present a wide range of shell variations, as can be seen from Figure 8.28. Bivalves have been able to fill a rather large number of ecological niches, thus enhancing their variability.

While an earlier classification scheme divided the bivalves into three subclasses, a new taxonomic scheme has been widely adopted

(a)
(b)
(c)
(d)
(e)
(f)
(g)
(i)
(j)
(h)
(k)

◀ *Figure 8.28 Collage of a drawing (a) and photographs (b–k) of marine bivalve shells (a–c, e, g–j) and living animals (d, f, k):* ***(a)*** Solemya, *a protobranch (cryptodontan);* ***(b)*** Atrina, *the pen shell (pteriomorphan);* ***(c)*** Modiolus, *a mussel (pteriomorphan);* ***(d)*** *an* Argopecten, *the Atlantic bay scallop with valves open (pteriomorphan);* ***(e)*** Mytilus, *a blue mussel (pteriomorphan);* ***(f)*** Lima, *a file shell with elongate sensory tentacles (pteriomorphan) (courtesy of C.E. Jenner);* ***(g)*** Crytopleura, *an angel wing (heterodontan);* ***(h)*** Mercenaria, *the venus clam or northern quahog (heterodontan);* ***(i)*** Ensis, *a razor or jackknife clam (heterodontan);* ***(j)*** Cerastoderma, *a cockle (heterodontan);* ***(k)*** *several* Donax, *the coquina (heterodontan).*

that consists of six subclasses, which are characterized in Table 8.3. But for purposes of understanding the functional morphology of bivalves, it will be the best approach to understand the three basic, nontaxonomic groups: protobranchs, lamellibranchs, and septibranchs. The **protobranchs** are primitive bivalves with simple protobranch ctenidia; most have paired feeding proboscides, which are the principal feeding organs. Most bivalves are **lamellibranchs**, and they use their enlarged, complex ctenidia for food procurement. **Septibranchs** are carnivorous bivalves that, in place of ctenidia, have muscular septa used in food capture. It would also appear to be in the best interests of the reader to have a rather extensive

TABLE 8.3. ☐ **BRIEF CHARACTERIZATIONS OF THE FIVE SUBCLASSES OF BIVALVES**

Subclass	*Type of Ctenidia*	*Method of Feeding*	*Hinge Teeth Present or Absent*	*Other Features*
PROTOBRANCHS				
Palaeotaxodonta	Protobranch	Large labial palps and proboscides	Present	Primitive, small, marine
Cryptodonta	Protobranch	Labial palps	Absent	Primitive, small, marine; no proboscides
LAMELLIBRANCHS				
Pteriomorpha	Filibranch	Filter feeding	Present	Attached by byssus or cementation
Palaeoheterodonta	Pseudolamellibranch	Filter feeding	Present	Mostly freshwater
Heterodonta	Eulamellibranch	Filter feeding	Present	Mostly burrowing; valves with no nacreous layer; siphons present; includes most bivalves
SEPTIBRANCHS				
Anomalodesmata	Absent	Muscular septum	Reduced or absent	Unequally sized valves; carnivores; mantle margins are mostly fused

knowledge of the lamellibranchs first, since about 98% of all bivalve species belong to this group. A short treatment of the primitive protobranchs and aberrant septibranchs will conclude the section on bivalves.

Lamellibranchs

Most bivalves are found burrowing in a variety of soft substrata, but others bore into very hard materials like wood or rocks, still others live attached to hard surfaces, and a few are unattached. Nonetheless, most bivalve adaptations, especially their external features, reflect their burrowing habit.

SHELL AND MANTLE

A characteristic feature of bivalves is that they have a shell consisting of two, similar, laterally compressed, convex **valves** that are mirror images of each other. Even though bivalve shells are not as varied as those of gastropods, they are still remarkably diverse in shape, size, color, and in the ornamentations they bear (Fig. 8.28). The two valves do not develop independently; rather, a single mantle rudiment produces a first shell or **protoshell** in the veliger stage. The dome-shaped protoshell becomes folded, and each side develops into a valve that is separate but still attached to the other. A dorsal **mantle isthmus** secretes a periostracal strip containing tanned proteins and very little calcium carbonate. This band becomes the dorsal elastic **hinge ligament**, a uniquely bivalve feature, which attaches the two valves structurally (Fig. 8.29). The elasticity of the ligament is important, since it is the principal means for opening the valves. When the valves are closed by adductor muscles described below, the outer layers of the hinge are stretched, while its inner parts are compressed (Fig. 8.29b). The outer tensile and inner compressional forces tend to open the valves; thus the hinge ligament is an antagonist to the adductor muscles. Near the hinge line, most bivalves have hinge teeth, those of one valve fitting into sockets in the other to prevent front-to-rear slipping (Fig. 8.29a). The first or oldest portion of each valve persists in the adult as a dorsoanterior protuberance, the **umbo**, and new shell materials are deposited in concentric growth lines around the umbo.

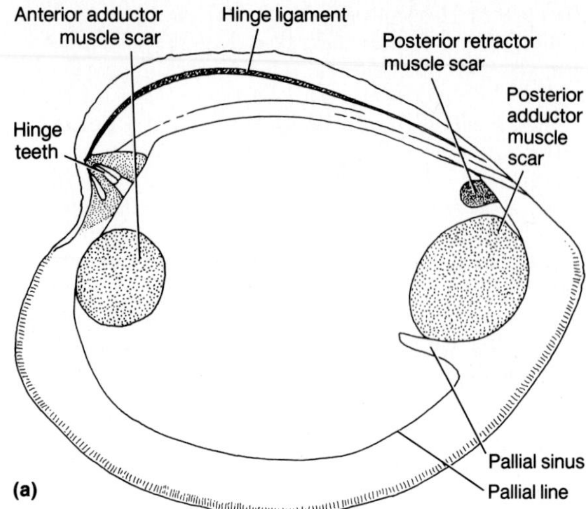

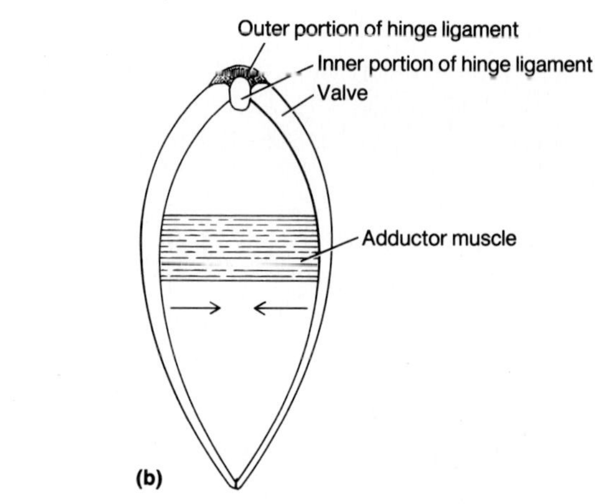

Figure 8.29 *Shell features of a typical bivalve:* ***(a)*** *an inner view of the right valve of* Mercenaria; ***(b)*** *a diagrammatic, cross-sectional view illustrating the hinge ligament and adductor muscle.*

Powerful **anterior** and **posterior adductor muscles** extend transversely or horizontally between the valves and close them (Fig. 8.29b). The attachments of both muscles leave scars on either valve as the **anterior** and **posterior adductor muscle scars** (Fig. 8.29a). Generally, both adductor muscles are about equal in size, but in certain specialized forms like scallops, the anterior adductor muscle is reduced or absent, and the posterior adductor muscle becomes more centrally located and is the sole source of adduction. Bivalve adductor muscles engage in two different sorts of contractions: first, most bivalves are able to close the valves rapidly; and second, the animals are able to remain closed for extended periods of time. To achieve both functions, adductor muscles normally contain two distinctly different types of muscle fibers—quick and catch. Quick muscle fibers contract rapidly but tire easily, and catch muscle fibers contract relatively slowly but can remain contracted without tiring for extended periods of time with a minimal expenditure of energy.

The secretory lobe of the large mantle margin is chiefly involved with shell production. A short distance from the shell margin, the mantle is attached to the shell by muscle fibers, which keep most foreign particles from coming to rest between the two (Fig. 8.3). But sometimes a small particle such as a sand grain lodges between mantle and shell and becomes a nucleus around which nacreous material is secreted. If the nacreous-covered particle remains free, it often forms an exquisite **pearl**, but if it adheres to the shell and is embedded in the nacreous layer, it is not of commercial value. Any shelled mollusc is theoretically capable of producing a pearl, but most pearls are secreted by bivalves, and those of superior quality are produced by several species of *Pinctada*, the pearl oyster. Cultured pearls are produced by first manually implanting a minute particle between the mantle and shell and then "transplanting" the year-old pearl to another oyster for three or more additional years of nacreous deposition. Pearls are among the most beautiful and exquisite of the precious gems, and a given pearl's commercial value is based on its luster, color, and degree of perfection.

In most lamellibranchs the posterior margins of the mantle have fused in such a way as to form distinct openings, the ventral or **inhalent** and the dorsal or **exhalent channels**. If the mantle tissues are elongated into tubes around these channels, they are called **siphons**. Siphons may be short or, when fully extended, may be longer than the body itself (Figs. 8.30, 8.34a, b). In some the siphons may be partially or completely fused to form a single tube with two separate channels. In burrowing lamellibranchs the siphons are often extended above the water–bottom interface and serve as water viaducts between the buried animal and the water above. In many burrowing bivalves the mantle margins on either side have fused completely except for the pedal aperture and the siphons.

FOOT AND BURROWING ADAPTATIONS

The ventral, muscular, laterally compressed **foot** is an effective and marvelously adapted burrowing organ (Figs. 8.4c, 8.30). Foot movements and alterations in its shape are achieved by intrinsic and extrinsic muscles and by hydrostatic means. The foot is extended or protracted by three different processes usually acting in concert with each other. First, two small **pedal protractor muscles** run from each valve into the foot so that their contractions pull the foot ventrally. Second, the foot is extended hydrostatically as the animal allows additional blood to enter the pedal hemocoel. Once in the foot, blood is sealed off by valves, and intrinsic muscles force the blood ventrally and further into the pedal hemocoel, thus extending the foot. Third, the foot is extended hydraulically by water in the mantle cavity so that when the

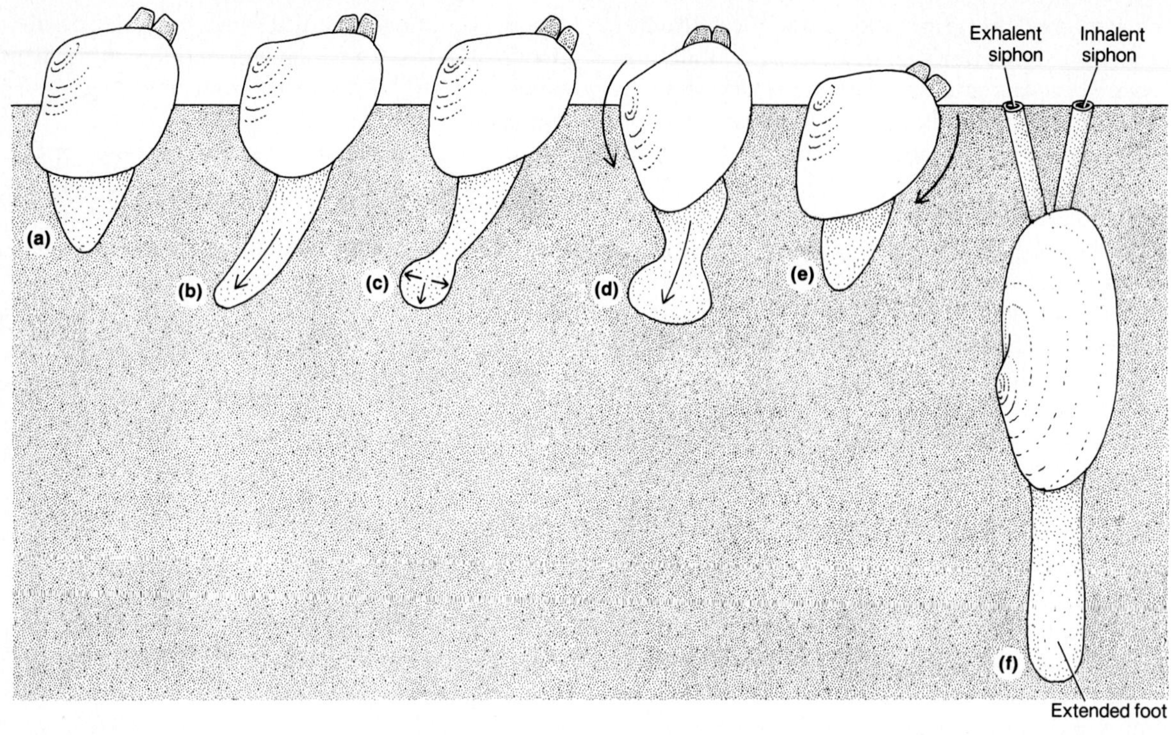

Figure 8.30 *Burrowing in bivalves:* ***(a)*** *a penetration anchor formed with foot probing downward;* ***(b)*** *the foot fully extended and siphons closed;* ***(c)*** *the adductor muscles contract forcing water from the mantle cavity; foot forms a pedal anchor;* ***(d)*** *the anterior pedal retractor muscle contracts rotating the animal forward;* ***(e)*** *the posterior pedal retractor muscle contracts rotating the animal backward;* ***(f)*** *a bivalve buried in the intertidal sand.*

valves are closed, the trapped water exerts external pressures on the foot, resulting in its extension. Foot retraction is accomplished chiefly by the contraction of one or two pairs of strong **pedal retractor muscles** and aided by a release of hydrostatic pressures in the hemocoel and mantle cavity.

Burrowing has become an essential feature in the life of most bivalves, and most of their adaptations are related to this life-style. The process of burrowing has been studied most extensively in bivalves living in soft benthic materials (Fig. 8.30f). Initially, a **penetration anchor** is created by the hinge ligament pulling the valves slightly agape and thereby anchoring them in the substratum (Fig. 8.30a). Contractions of the pedal protractor muscles and intrinsic foot muscles fully extend the foot into the bottom (Fig. 8.30b). The tip of the foot then dilates because of an influx of blood in the hemocoel to form a **pedal anchor** (Fig. 8.30c). Then the valves are suddenly closed by the adductor muscles, which frees the valves from the substratum, forces water out of the mantle cavity around the foot in powerful strong jets to loosen the sand around the valves, and forces more blood into the foot (Fig. 8.30c). With the valves closed, contraction of the pedal retractor muscles pulls the entire animal down toward the foot, and the cycle is repeated. In some bivalves the anterior retractor muscles contract slightly earlier than the posterior muscles, producing a rocking motion that facilitates burrowing even more (Fig. 8.30d, e). Many clams are able to burrow rapidly, especially in soft bottoms. Razor clams (*Ensis*), completing one burrowing cycle in less than one second, can burrow in soft sand to a depth of more than 50 cm in 10 seconds! To return to the surface, most forms back out by pushing against the pedal anchor, but others even turn around and burrow upward. Cockles (*Cerastoderma*) are unusual in that the foot is used both for burrowing and leaping. When the cockle is disturbed, the foot is extended rapidly against the substratum, thus catapulting the animal away from danger.

Once the burrowing habit became fully evolved, bivalves underwent adaptive radiation to colonize many different types of substrata. Perhaps the most primitive burrowing habit involves shallow burrowing like that in the cockles (*Cerastoderma*), hard-shell clams (*Mercenaria*), coquinas (*Donax*), and freshwater mussels (*Anodonta*). In these bivalves the valves are scarcely below the soft surface, the mantle margins are only slightly fused, the siphons are rather short, and the foot rather large.

Representatives of a rather large number of bivalves are deep burrowers and have a number of adaptations for living well below the water. Many deep burrowers are active, while others are more sedentary; in either case they are characterized by possessing very elongate siphons to provide contact with the water above (Fig. 8.30f). The siphons themselves may even develop sensory organs at their distal tips, and most deep burrowers have strong, siphon retractor muscles. Often, the shells are long, thin, flattened—in a word, streamlined—for burrowing. In bivalves like *Tellina*, which is an active burrower, the siphons function like vacuum hoses sweeping material from the sea bottom. Other bivalves like *Mya* are sedentary deep burrowers, and their burrows are often lined with mucus to reduce the influx of sediment into the mantle cavity. The Pacific geoduck, *Panopea*, burrows to a depth exceeding a meter, and its extremely elongated, muscular siphons of up to a meter in length cannot possibly be retracted into the mantle cavity.

The burrowing habit has been extended so that a large number of bivalves bore into relatively hard substrata such as peat or wood using the serrated anterior ends of the valves. Effective strokes for boring are achieved in a variety of ways and usually involve a combination of adductor and pedal muscles to force the shell effectively against the surface. Some produce an acid secretion that softens the substrata. Perhaps the best known of these boring bivalves is the shipworm, *Teredo*, which bores through wood to produce a network of tunnels. The siphons of a shipworm are greatly elongated and wormlike, and its two small valves are located at the anterior end and on either side of the small visceral mass (Fig. 8.31b). The valves

Figure 8.31 *Wood-boring bivalves;* ***(a)*** Xylophaga, *a deepsea wood eater, lateral view;* ***(b)*** Teredo, *the shipworm.*

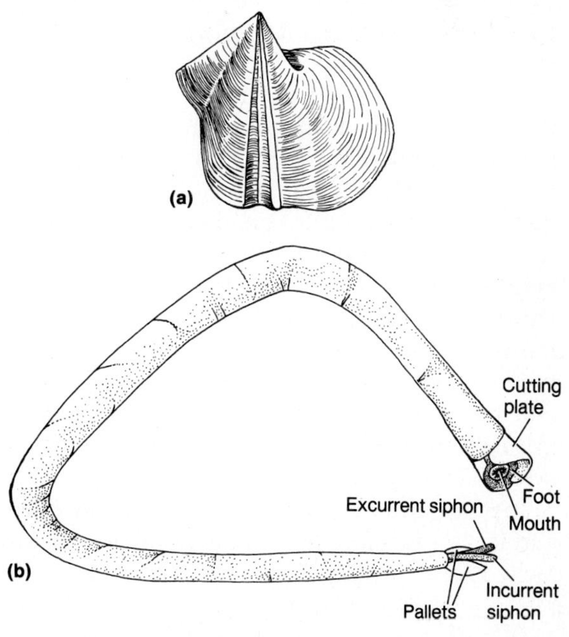

are moved by muscles in such a way as to tunnel into wood, and the sawdust created by their excavation is ingested and digested by the bivalves. *Teredo*, along with other forms like *Xylophaga* (Fig. 8.31a), can completely destroy most untreated wood structures submerged in saline water within several months or a year.

The boring habit has been carried to the ultimate in those bivalves that are capable of boring into rocks, coral, or concrete pilings. In the piddocks or true rock borers, the anterior edges of both valves are quite sharp and bear small abrasive teeth. By alternately opening and closing the valves with powerful adductor muscles, piddocks slowly abrade away small layers of rock. The modified foot is protruded anteriorly between the cutting valves and grips the inner end of the burrow like a sucker to enhance the action of the valves. Of course, the progress is slow for the piddocks, but considering their task and the nature of the substratum, they are relatively efficient borers.

Most pteriomorphs such as mussels and oysters are able to live on hard surfaces such as sea walls, jetties, coral, and rocks. The mechanism by which most are attached permanently to the surface is the **byssus** (Fig. 8.32). At the base of the foot of these bivalves are **byssal glands**, which secrete a proteinaceous fluid, conchiolin, that flows down a ventral groove to the tip of the foot and onto the surface. This material, hardening quickly by a tanning process and by being in contact with seawater, produces a hard, tough **byssal thread**. The animal then changes its position slightly and secretes another thread. Many threads, collectively constituting the byssus, attach the animal firmly to the substratum (Fig. 8.32). The best known of the byssus-bearing bivalves are called mussels, and common examples are the horse mussel (*Modiolus* [Fig. 8.28c]), common mussel (*Mytilus* [Figs. 8.28e, 8.32a]), and the pen shell (*Pinna*).

A number of other bivalves live on hard surfaces in which one valve is permanently attached to a surface. Attachment is achieved in

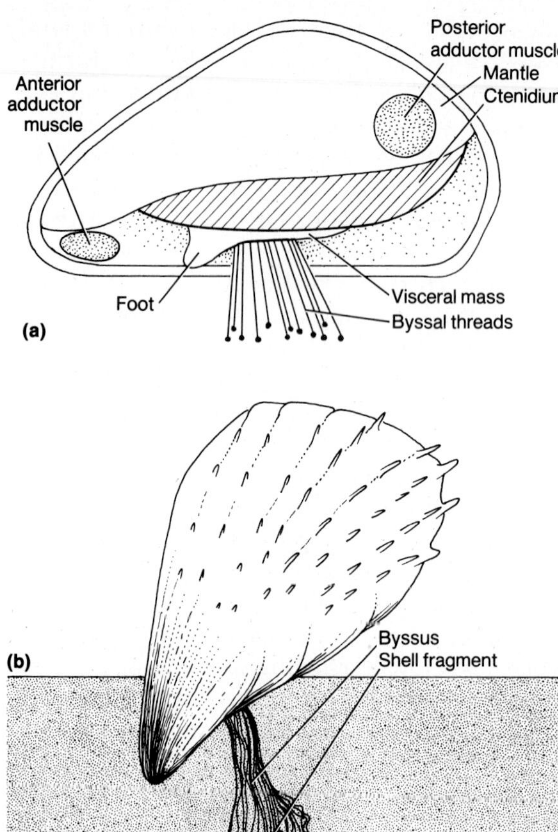

Figure 8.32 *The bivalve byssus:* ***(a)*** *diagrammatic view of* Mytilus *with left valve removed to show the byssus;* ***(b)*** *the byssus of* Atrina, *a pen shell.*

the larval stage by the formation of a byssus followed by permanent cementation; two of the best-known species are the American (*Crassostrea virginica*) and European (*Ostrea edulis*) oysters. In most oysters the valves are of unequal size, and the bottom one (left) is usually the larger. The giant Pacific clams (*Tridacna*) also are sedentary with the dorsal hinge side attached by a byssus, but each individual is held in place mostly by the weight of the animal and shell. All sedentary bivalves, whether attached by a byssus or by one valve, have undergone some modifications in keeping with their sessile habit. There is a tendency for the foot and siphons to be reduced or even lost, and often the anterior adductor muscle is also reduced or lost (Fig. 8.32a). Since sedimentation is not a major problem, the mantle margin does not fuse with that on the opposite side.

Finally, there are a few bivalves like scallops (*Pecten*) and file shells (*Lima*) that live on hard surfaces and have the ability to swim. The single posterior adductor muscle has migrated centrally and is composed of sections of quick and catch muscle fibers. Quick fibers rapidly close the valves, which are then just as quickly opened by the elastic hinge. The rapid clapping of the valves forces water, at least in scallops, to exit on either side of the hinge or at the ventral gape, thus driving the animal by jet propulsion. In keeping with the occasionally pelagic life-style novel to bivalves, the sensory mantle lobe of *Pecten* and *Argopecten* is equipped with numerous short tentacles and small, complex, exquisitely beautiful blue eyes that are used in conjunction with swimming (Fig. 8.28d).

Mantle cavity, ctenidia, and feeding

The mantle cavity of bivalves, particularly large and capacious for molluscs, is located on either side between the mantle and the visceral mass and foot. Primitively, there is communication between the mantle cavity and the exterior all along the distal edges of the mantle. In more advanced bivalves, especially the burrowers, the mantle lobes unite and fuse with each other except at certain points such as the inhalent and exhalent channels and a ventral pedal aperture. Mantle fusions effectively reduce the possibilities of sediments entering the mantle cavity and fouling the ctenidia.

On either side of the spacious mantle cavity is a large bipectinate **ctenidium** suspended from the roof of the cavity. Lamellibranchs became successful in utilizing plankton in water brought into the mantle cavity for gaseous exchange as food. Thus the ctenidia of all lamellibranchs became particularly adapted as filters of suspended particles and planktonic organisms. This filter-feeding habit is ubiquitous in all lamellibranchs and is considered to be one of their most diagnostic features. The size of the ctenidia, many times that necessary for gaseous exchange alone, is primarily correlated with that necessary for food procurement.

A number of important ctenidial modifications have made these organs efficient sieves while at the same time retaining their gas exchange functions. Each ctenidial axis has been lengthened anteriorly, and numerous new filaments have been added. The primitive ctenidial arrangement was for filaments to project lateroventrally (Fig. 8.33b). As the filaments continued to elongate, each was flexed about midway along its length (Fig. 8.33c), and the distal ends were turned dorsally so that each filament became V-shaped or was reflected (Fig. 8.33d). When seen in cross-sectional view, the proximal part of each filament, attached to the axis, is called the **descending limb**, and the distal part is the **ascending limb** (Fig. 8.33e). Since filaments on both sides of the axis have been reflected, what results is a W-shaped ctenidium composed of two V-shaped **demibranchs** (Fig. 8.33d, e).

Adjacent filaments along the length of the ctenidium tend to be united structurally with each other to form a thin plate. Viewed laterally, each ctenidium is a folded, often pleated sheet, hence their common name (lamelli = sheet; branch = ctenidium). Each demibranch is constructed of two sheets or plates, the **outer** and **inner lamellae**, formed corporately by all the ascending and descending limbs, respectively (Fig. 8.33e). Thus each ctenidium has four broad, sheetlike outer surfaces: outer and inner lamellae of outer and inner demibranchs. The positions of the various ctenidial parts divide

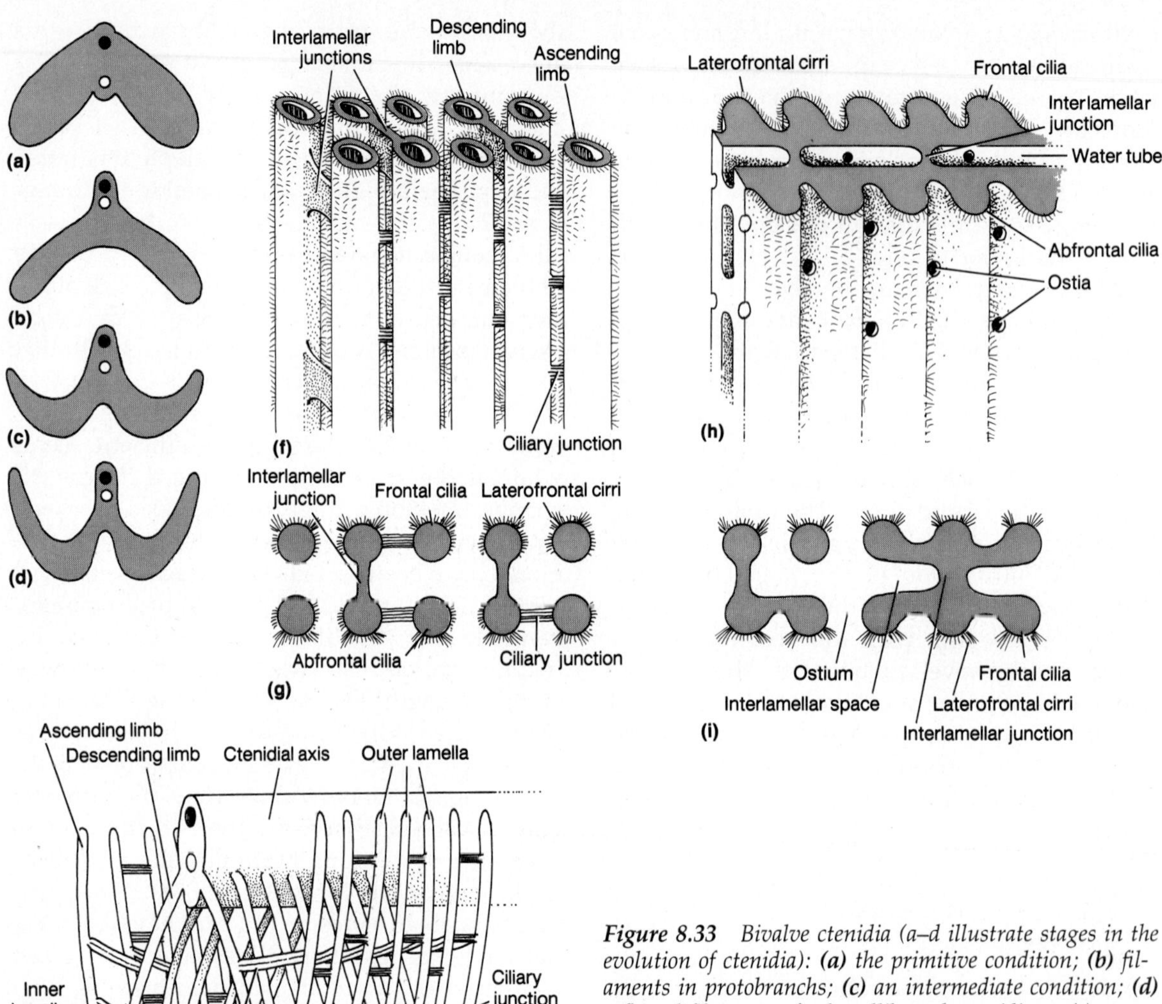

Figure 8.33 *Bivalve ctenidia (a–d illustrate stages in the evolution of ctenidia):* ***(a)*** *the primitive condition;* ***(b)*** *filaments in protobranchs;* ***(c)*** *an intermediate condition;* ***(d)*** *reflected filaments of a lamellibranch ctenidium;* ***(e)*** *stereogram of a primitive filibranch ctenidium;* ***(f)*** *frontal view of a filibranch ctenidium;* ***(g)*** *section through the same filibranch ctenidium;* ***(h)*** *frontal view of an eulamellibranch ctenidium;* ***(i)*** *section through the same eulamellibranch ctenidium.*

the mantle cavity on each side horizontally into a ventral **inhalent chamber** and a dorsal **exhalent** or **suprabranchial chamber**. For water to enter the dorsal exhalent chamber from the ventral inhalent area it must pass between adjacent filaments.

Because of modifications in the ctenidial filaments, their ciliation has also been substantially altered. **Frontal cilia**, present on the outer surfaces of both lamellae, sweep food ventrally to a **food groove** located at the junction of the two lamellae (Fig. 8.33e). Abfrontal cilia, which would have been on the exhalent surface of both lamellae, are lacking altogether. **Lateral cilia**, located near the frontal edge, are primarily responsible for the water flow in the mantle cavity. Situated on the angle of filaments between frontal and lateral cilia are **laterofrontal tracts**, which are fused cilia peculiar to bivalves that form **cirri** (Fig. 8.33g, i). The laterofrontal cirri of adjacent filaments function as a very fine strainer.

Adjacent filaments are joined together by several fundamentally different means. In the primitive ctenidial type, called the **filibranch** and present in the pteriomorphs, adjoining filaments are held together only by the intermeshing of laterofrontal cirri forming **ciliary junctions** (Fig. 8.33e–g). The two lamellae in a demibranch are joined together by infrequent bars of tissue called **interlamellar junctions**. A **pseudolamellibranch** ctenidium, found in palaeoheterodonts, has scattered interfilamentous **tissue junctions** that join adjacent filaments, and scattered interlamellar tissue junctions are also present. The most advanced ctenidium, the **eulamellibranch** form found in the heterodonts and thus in most lamellibranchs, is characterized by extensive interfilamentous tissues. This makes each lamella a continuous, nonpleated sheet perforated by numerous small **ostia**, which in turn are surrounded by laterofrontal cirri (Fig. 8.33h, i). Numbers of interlamellar tissue junctions have increased substantially, resulting in discrete, vertical, interlamellar **water tubes**. Afferent and efferent blood vessels run inside the vertical interlamellar tissues, and respiratory gases are exchanged within these interlamellar tissues. The distal edge of the outer lamella of the outer demibranch fuses to the mantle, while the distal edge of the medial lamella of the inner demibranch fuses to the wall of the visceral mass. This creates an almost perfect ctenidial separation between inhalent and exhalent chambers so that water must pass through the ostia and between the lamellae to enter the suprabranchial chamber. The pathway for water in a bivalve with eulamellibranch ctenidia would be as follows:

> outside → incurrent siphon → inhalent portion of mantle cavity → ostia in lamellae → vertical, interlamellar, water canals → suprabranchial chamber → exhalent siphon → outside.

Particles as small as 1–5 μm are strained out mostly by laterofrontal cirri, which then transfer them to the frontal cilia. Frontal cilia sweep food to a longitudinal **food groove** located ventrally on each demibranch (Figs. 8.33e, 8.35). Some primitive bivalves have additional small food grooves on each ctenidium. Food particles, apparently not entrapped in a mucous sheet produced by the ctenidia as was once thought, are normally transported only by ctenidial cilia toward the mouth. Cilia in the four grooves move particles anteriorly, where they are transferred to large **labial palps**. These organs are modified oral folds whose principal function is to assist in ingestion. Surfaces of the labial palps are heavily ciliated and bear extensive sorting areas in which larger particles are rejected back into the mantle cavity, and small bits are swept into the mouth.

Larger particles that are not to be ingested are sorted out and passed by ciliary tracts from ctenidia and labial palps to the mantle lobes. From time to time, the adductor muscles contract suddenly and force water and the accumulated debris to the outside through the inhalent opening (not surprising when one remembers that accumulated debris in the mantle cavity could not be expelled via the exhalent siphon), thus expelling this material called **pseudofeces**. Most bivalves possess this unusual but effective means for cleansing the entire mantle cavity and discharging nondigestible accumulated debris to the outside.

An interesting nutritional adaptation is found in the giant Pacific clams, *Tridacna*, which live in shallow water attached at the hinge side with the ventral gape uppermost. Within the mantle tissues are found enormous numbers of symbiotic dinoflagellates (zooxanthellae) forming the same type of symbiosis that is also found in corals (see Chapter 5). Even though tridacnids have normal-sized ctenidia and are capable of filter feeding like any other lamellibranch, these clams move their symbionts internally and digest some of them intracellularly as a supplement to their diets. Mantle tissues even contain various pigments, which apparently regulate light intensity ostensibly for the photosynthetic zooxanthellae! These lamellibranchs and their zooxanthellae constitute a remarkable symbiotic relationship.

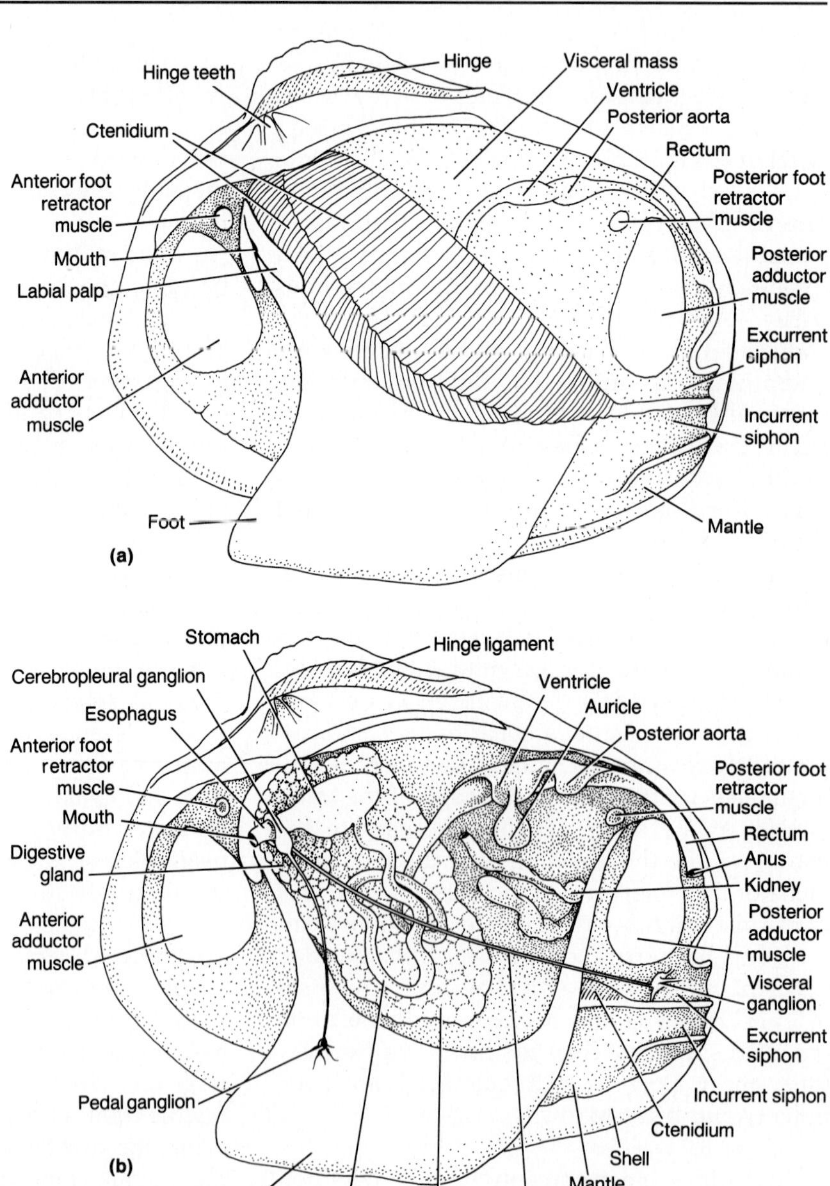

Figure 8.34 *Internal features of* Mercenaria, *a typical eulamellibranch:* ***(a)*** *a lateral view with left valve and mantle lobe removed;* ***(b)*** *a sagittal section of the visceral mass and foot.*

The visceral mass

Digestion, coelom, excretion, and circulation. Notable is the fact that bivalves have neither a head nor a radula as do other molluscs. Food enters the alimentary canal by way of the **mouth**. Mucous secretions from a pair of salivary glands entangle food particles to create a mucus–food string, which is drawn through the **esophagus** into the **stomach** (Fig. 8.34b). A **crystalline style** is usually present that functions in essentially the same way as in continuously feeding gastropods. The stomach, as in some gastropods, bears a number of ciliated sorting areas where particles are sized and graded. Smaller particles enter ducts to the **digestive gland**, where digestion is both extracellular and intracellular. Larger or undigestible particles are swept by cilia into a rather long, coiled **intestine**, which extends dorsally, then turns posteriorly, and becomes the **rectum**. The rectum actually extends through the ventricle of the heart, passes near to the posterior adductor muscle, and opens via the **anus** near or in the excurrent channel (Fig. 8.34). The intestine and rectum function mainly to concentrate fecal materials into pellets or a ribbon that can be voided conveniently in the exhalent stream without fouling the ctenidia or mantle cavity.

The **coelom** is rather small and situated dorsally just beneath the hinge ligament (Figs. 8.34b, 8.35). Waste products, especially water and ammonia, are secreted into the coelom by bordering cells. A pair of U-shaped **kidneys** is located beneath the pericardium (Fig. 8.34b), and each opens into the coelom by a **nephrostome** and into the suprabranchial chamber by a **nephridiopore**. The proximal arm of the kidney is glandular and secretory, while the distal arm functions as a bladder. Copious amounts of water are eliminated by freshwater species.

The dorsal **heart**, bathed by coelomic fluids, is comprised of a pair of auricles and a single ventricle (Figs. 8.34b, 8.35). The muscular **ventricle**, wrapped around the rectum in a unique fashion, forces blood into an **anterior** and a **posterior aorta**. The aortae give rise to smaller arteries that distribute blood to the hemocoel in various regions. Blood is transported to the kidneys and then to the ctenidia before being returned to the thin-walled **auricles** on either side by an efferent blood vessel. Blood in bivalves, often stored temporarily in sinuses within the mantle, is usually colorless; but in a few forms, hemoglobin is present. As was described earlier, blood volume plays an important role in the extension and movement of the foot.

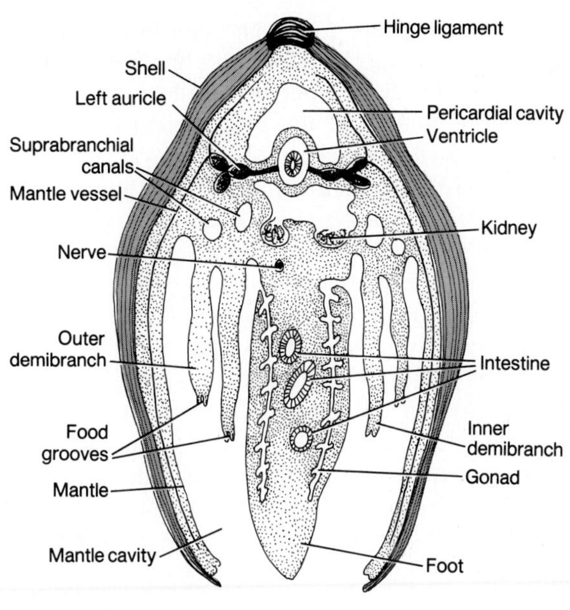

Figure 8.35 *A cross-sectional view through the heart region of* Mercenaria *(compare this illustration with those in Fig. 8.34).*

Nervous system and sense organs. The nervous system, bilaterally symmetrical and relatively simple in keeping with the sedentary existence of bivalves, consists of three pairs of ganglia and two pairs of nerve cords. On either side of the esophagus is a **cerebropleural ganglion** connected to its opposite by a dorsal **cerebropleural commissure** (Fig. 8.34b); these ganglia function principally to coordinate foot and valve movements. Two nerve cords arise from each cerebropleural ganglion: the **pedal cord** passes ventrally to a **pedal ganglion** that innervates the foot, and the **visceral cord** extends posteriorly to a **visceral ganglion** located near the posterior adductor muscle and innervating the visceral mass, posterior adductor muscle, and siphons.

Sensory organs are relatively simple in bivalves and include ocelli (eyes), tactile organs, osphradia, and statocysts. Ocelli and tactile organs are typically located in the sensory lobe of the mantle. **Ocelli** are usually simple structures and function as light-detecting or motion-detecting structures and do not form images. But in a few species such as scallops (*Pecten*), numerous well-developed eyes at the mantle margin are each composed of a cornea, lens, and a retina or photoreceptor cells (see Fig. 8.28d). **Tactile cells** are scattered along the sensory lobe of the mantle and may be located on numerous short tentacles. **Osphradia** are positioned not in the incurrent siphon as one would expect but in the excurrent siphon. Their role in bivalves is not fully known, but they probably function as in other molluscs to monitor the chemical and sedimental qualities of water. A **statocyst** is usually found in or near each pedal ganglion.

Reproduction and development. Most bivalves are dioecious. A pair of almost fused gonads lies in the visceral mass amid the intestinal loops (Fig. 8.34b). In some primitive bivalves, gametes are discharged into the kidneys and are carried to the mantle cavity. In most, however, genital ducts are separate from the renal organs, and the gonopores are near the nephridiopores in the suprabranchial chamber. Hermaphroditic forms may have separate genital ducts or a single duct on either side.

Generally, gametes are carried from the suprabranchial chamber through the excurrent siphon to the outside, where fertilization takes place. In some marine species and in most freshwater forms, eggs are retained by the female, and sperm are brought into the mantle cavity utilizing the normal flow of water. Fertilization then takes place in the suprabranchial chamber, where the young embryos often are brooded. Self-fertilization, while not the rule in monoecious forms, has been reported for a few species, but most hermaphroditic forms function either as males or females at any one time.

Embryonic development proceeds rapidly, and in marine bivalves a **trochophore** is soon produced, followed by a **veliger**. In the veliger, most adult structures appear, both valves develop, the velum is lost, and the adult shape begins to emerge. Soon the veliger, in the process of transforming into a bivalve, sinks to the bottom.

Individuals of the freshwater Order Unionoida are unique in that the veliger is represented by a parasitic glochidial stage. The **glochidium** is a small (up to 0.5 mm) larval stage in which both valves are equipped with hooks or teeth (Fig. 8.36). Ejected through the exhalent siphon, the glochidia in some species are discharged directly into the face or mouth of a fish when the latter contacts the extremities of the excurrent siphon. In other forms, glochidia are ex-

Figure 8.36 *The glochidium of a unionoid bivalve.*

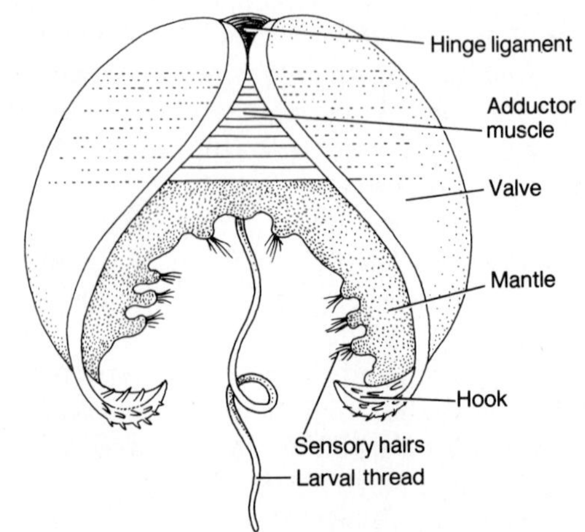

pelled without the contact stimulation of a fish, and they sink to the bottom. No further development takes place unless the glochidium is able to come in contact with a fish within several days. If this chance contact takes place, the glochidium attaches and lives ectoparasitically upon the fish. During this brief but obligatory parasitic existence, fish tissues usually grow over the larval bivalve and form a cyst, and the glochidium loses its larval structures and develops adult features. At the end of the parasitic relationship, the glochidium breaks out of its cyst, sinks to the bottom, and gradually transforms into an adult.

Some individuals may produce up to 3,000,000 glochidia, most of which perish because they do not have an opportunity to parasitize a fish. Fishes are apparently not seriously harmed, since up to 3000 glochidia have been found on an individual fish. The glochidia of some unionoids require a specific species of fish to parasitize, but others have a wide range of tolerances to potential host species.

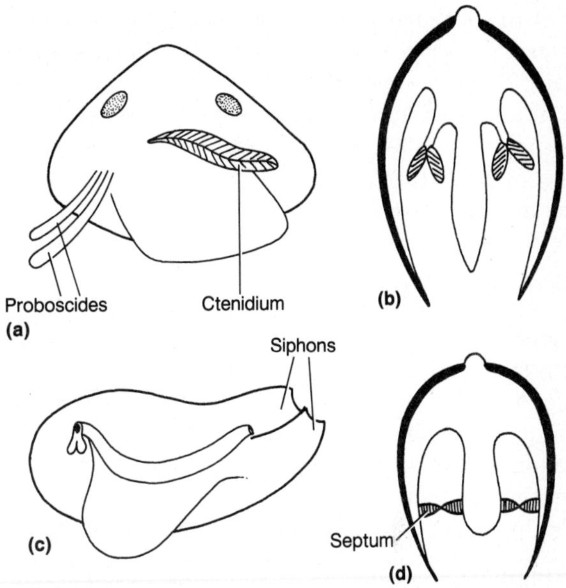

Figure 8.37 *Diagrams of nonlamellibranch bivalves:* ***(a)*** *a protobranch, lateral view;* ***(b)*** *a protobranch, cross-sectional view;* ***(c)*** *a septibranch, lateral view;* ***(d)*** *a septibranch, cross-sectional view.*

Protobranchs

Protobranchs, a group intermediate between the ancestral mollusc and most bivalves, are considered to be the primitive forms from which other bivalves evolved. The approximately 400 living species are all marine and are usually found in sandy or muddy bottoms in cool temperate seas. While some are littoral, most are found on the floor of deep oceans, where they may constitute the majority of all deep-sea bivalves.

The shell is usually thin. The nuculoideans have hinge teeth, but the solemyoideans lack them. The anteroventral foot is small and mostly bladelike, but some can expand the tip to form a flat sole. The head has folds of tissue around the mouth that are enlarged as feeding labial palps, each of whose posterior margins often form a **proboscis** (Fig. 8.37a). Food consists of small organisms or organic debris deposited on the substratum that are swept up by the paired ciliated proboscides. Ciliated sorting areas on these structures enable the animal to ingest conveniently sized particles and to reject the larger ones. The stomach contains a style and gastric shield. Digestion is both extracellular in the stomach and intracellular in the digestive gland. The protobranch ctenidia, similar to those of the ancestral mollusc, are small, bipectinate, nonreflected, and situated toward the posterior mantle cavity (Fig. 8.37a, b). Ctenidial filaments are simple, and in several groups they are used as filters through which suspended particles are strained. The ctenidial axis is supported by a muscular base that contracts and expands rhythmically, thus pumping water into the mantle cavity. Filament cilia function primarily in cleansing the ctenidia.

Since the gonads do not connect to the pericardial cavity, a short gonoduct joins each kidney directly. The cerebral and pleural ganglia are fused rather than being separate. These and other internal and external features reflect the primitive nature of protobranchs. They are commonly known as awning and nut clams.

Septibranchs

Septibranchs or anomalodesmatids, better known as dipper or watering-pot shells, represent a small group of fewer than 100 species of highly specialized marine bivalves living in deep-ocean habitats. Ctenidia are missing altogether, but the ctenidial base on either side remains as a perforated pumping **septum** located between the incurrent chamber and the suprabranchial cavity (Fig. 8.37d). By alternately contracting and relaxing, the septa pump water into the mantle cavity (Fig. 8.37c, d). Septibranchs are carnivores or scavengers; small crustaceans and bits of organic debris are brought into the mantle cavity, where they are captured by small, muscular labial palps and then stuffed into the mouth. The stomach walls are heavily chitinized; this organ is thus an effective grinding and crushing mill. A small style is usually present, and digestion is mostly intracellular in cells of the digestive gland. Gaseous exchange takes place at the surfaces of the mantle cavity. The foot is long, slender, and used for burrowing.

CLASS SCAPHOPODA

Scaphopods, or tusk shells, are marine molluscs that probably arose as an evolutionary offshoot of primitive bivalves. About 300 living species are known in addition to many fossil forms. Scaphopods are mostly sedentary, burrowing animals that are found on the bottoms of shallow littoral zones as well as those of moderately deep water.

External Features

The shell, elongated, cylindrical, and of one piece, is open at both ends with the anterior aperture slightly larger than the posterior one (Figs. 8.4d, 8.38a). The diameter of the shell increases toward the ventral end, and it is slightly concave dorsally, causing it to resemble a tusk or tooth. Shells of most range from 2 to 8 cm in length and 2–10 mm in diameter. They may be sculptured with longitudinal or transverse ribs and are typically white, yellowish, or green in color.

The body of a scaphopod is greatly elongated in an anteroposterior axis with the head and foot protruding through the larger ventral aperture. The animal lives with most of the shell buried in the sand and only the posterior aperture communicating with the water above (Fig. 8.38b). The mantle, fused into a tube, is continued posteriorly as inhalent and exhalent siphons. Gaseous exchange occurs in ciliated folds of the mantle, since ctenidia are lacking altogether.

The reduced head consists of a **proboscis** bearing the mouth and located at the base of the burrowing, cone-shaped **foot** (Figs. 8.4d, 8.38b). The head also bears a number of threadlike prehensile tentacles or **captacula**, each with an adhesive bulb at its tip. These are used to capture particulate food (foraminifera, eggs, etc.), which are conveyed to the mouth by ciliary action.

Internal Features

The mouth opens into a relatively large buccal cavity that contains a well-developed radula and a median **jaw**. A short esophagus is continuous as the stomach, and the latter receives the duct from the large digestive gland, where digestion is extracellular. A convoluted intestine terminates with the ventral anus opening into the mantle cavity. A pair of kidneys also opens into the ventral mantle cavity.

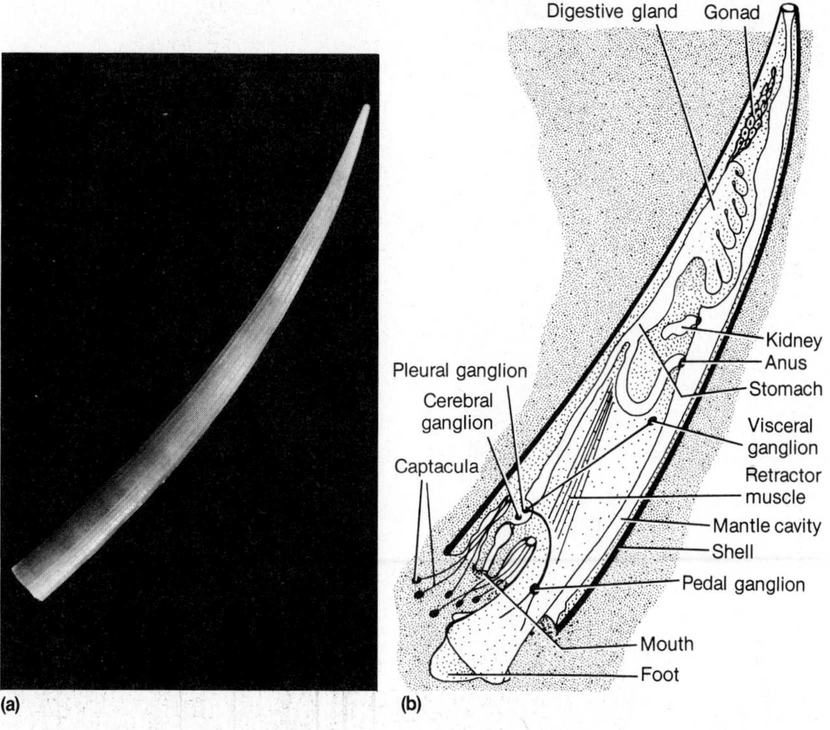

Figure 8.38 *Scaphopod features: **(a)** a photograph of the shell of* Dentalium; ***(b)*** *a longitudinal view of a scaphopod.*

The circulatory system is remarkable in that there is no distinct heart; blood apparently is circulated within a few vessels and hemocoel by forces generated by the alternate protraction and retraction of the foot. The nervous system consists of ganglia (cerebral, pleural, visceral, and pedal) and nerve cords typical of molluscs generally (Fig. 8.38b). Statocysts in the foot and captacula are the sensory organs, but eyes, tentacles, and osphradia are absent.

Scaphopods are dioecious, and gametes are produced by a single gonad lying posterior to the stomach. The right coelomoduct serves as a genital duct to convey gametes to the mantle cavity. Fertilization takes place in the surrounding seawater, and a free-swimming trochophore develops, followed by a veliger stage. Metamorphosis of the veliger into the adult form is gradual and involves the elongation of the body.

Shells of scaphopods are often numerous along some beaches, an indication that they are fairly common. In earlier centuries, shells were used as currency by some Indian tribes in the Pacific Northwest and by islanders in the West Indies. Little is known about scaphopod ecology. Two common genera are *Dentalium* (Fig. 8.38a) and *Cadulus*.

CLASS CEPHALOPODA

The highly specialized and complex cephalopods, including the nautiluses, cuttlefishes, squids, and octopods, are unsurpassed among invertebrates for their neural development, sensory structures, capacities for modified behavior through learning, and efficient locomotion. From both ecological and evolutionary stand-

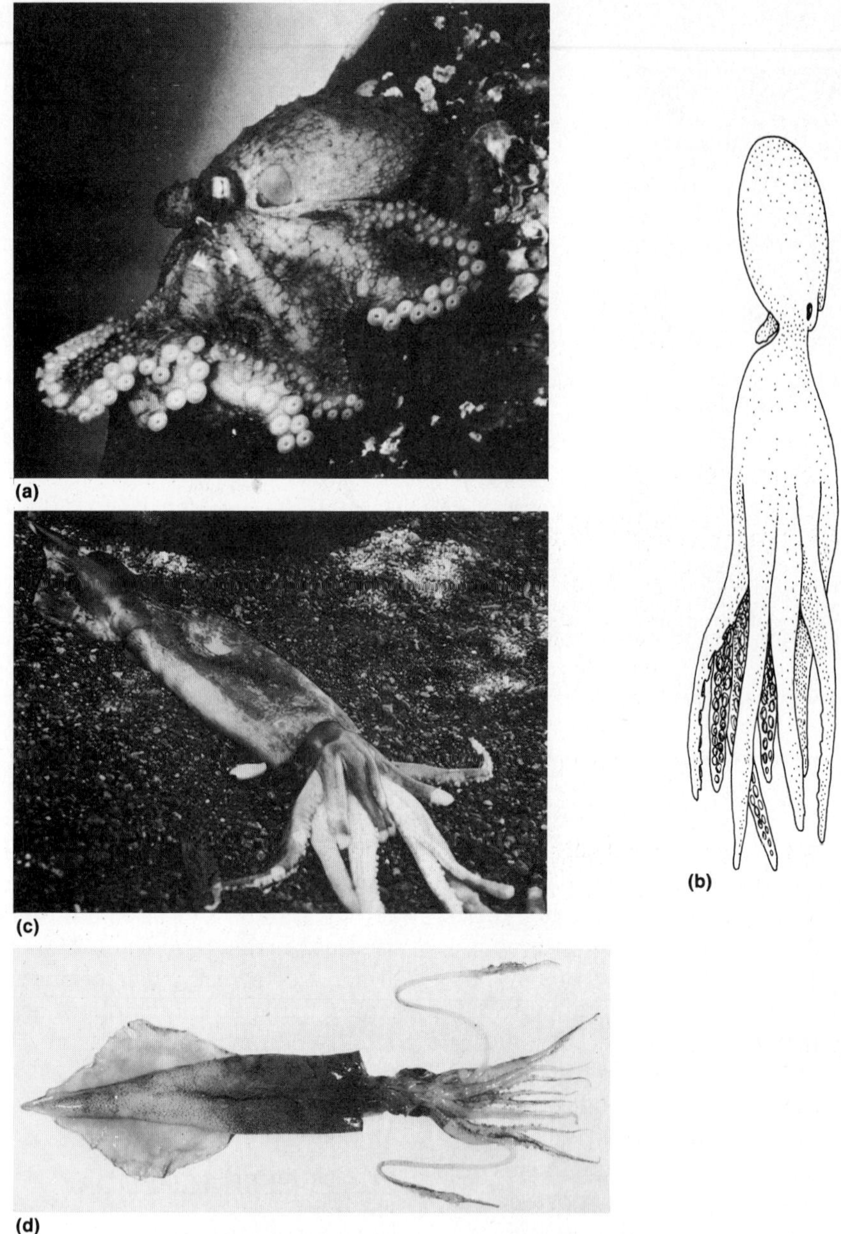

Figure 8.39 *Representatives of the two principal cephalopod taxa:* ***(a)*** *a photograph of* Octopus *(courtesy of Ward's Natural Science, Inc., Rochester, N.Y.);* ***(b)*** Octopus *swimming;* ***(c)*** *a photograph of* Moroteuthis *(courtesy of S.L. Armstrong);* ***(d)*** *a photograph of* Loligo, *a decapod.*

points they are not as successful in numbers of modern species as are the gastropods and bivalves. At present there are only about 1000 species, but species diversity and numerical abundance were much higher during their zenith in the Paleozoic and Mesozoic Eras; about 10,000 fossil species have been discovered to date. They generally range from 6 cm to 1 m, including the arms. One giant squid, *Architeuthis,* is the largest invertebrate known and is reported to have attained a maximum length of 18 m and a diameter of 3 m, while one pigmy squid does not exceed a total length of 15 mm.

Cephalopods are all marine carnivores, and many of their striking adaptations reflect their predaceous habit. Within the class, however, they are not as diversified as are gastropods or bivalves. The major taxa are characterized briefly at the end of this chapter, and the two principal extant cephalopod groups are illustrated in Figure 8.39.

External Features

Although they illustrate some basic molluscan morphological and physiological features, cephalopods have undergone dramatic and significant morphological reorganization from the ancestral plan. Though they are still bilaterally symmetrical, the anteroposterior axis has been greatly shortened and the dorsoventral axis greatly lengthened (Fig. 8.5e). Morphologically, the foot and head are ventral, and the visceral mass is extended dorsally, so a discussion of a functional axis, rather than a morphological one, is more appropriate for cephalopods. The head and arms are considered to be anterior, and the apex of the visceral hump is posterior. Having not lost completely its role in locomotion, the foot has been modified substantially and is represented by the **funnel**, a structure used prominently in locomotion, and a series of prominent grasping **arms** or **tentacles** (Figs. 8.5e, 8.39).

Shell, mantle, and ctenidia

Cephalopods with an external shell reflect the primitive condition. Ancestral cephalopods probably had curved shells that were concave on the ventral side. From these ancient creatures, more recent forms evolved, some of which had straight shells and others coiled; ammonoids and nautiloids all possessed well-developed shells. *Nautilus,* the only extant noncoleoid cephalopod, has a large external shell. But there is a strong tendency among coleoids for the shell to be reduced and internalized or even absent, and the emancipation from the shell is strongly correlated with their more efficient locomotion. The mantle, completely investing the body except for the head and arms, is involved with establishing water currents through the mantle cavity that are vital to locomotion and ctenidial aeration.

Most cephalopods have one pair of ctenidia whose filaments are folded to increase both their surface areas and efficiency. Cephalopod ctenidia lack cilia, and the loss of this ciliation is associated with two other factors: first, there are no longer problems of sedimentation clogging the ctenidia; and second, the function of the cilia has been taken over by the mantle and funnel musculature, which creates water currents through the mantle cavity.

Head and arms

The head and arms of cephalopods are merged into a single morphological unit (Fig. 8.39). The head bears the mouth and a pair of large lateral eyes, details of which will be discussed later with each subclass. The long flexible arms or tentacles, clearly associated with the head and intimately involved in food capture, also represent a portion of the primitive foot. Internal,

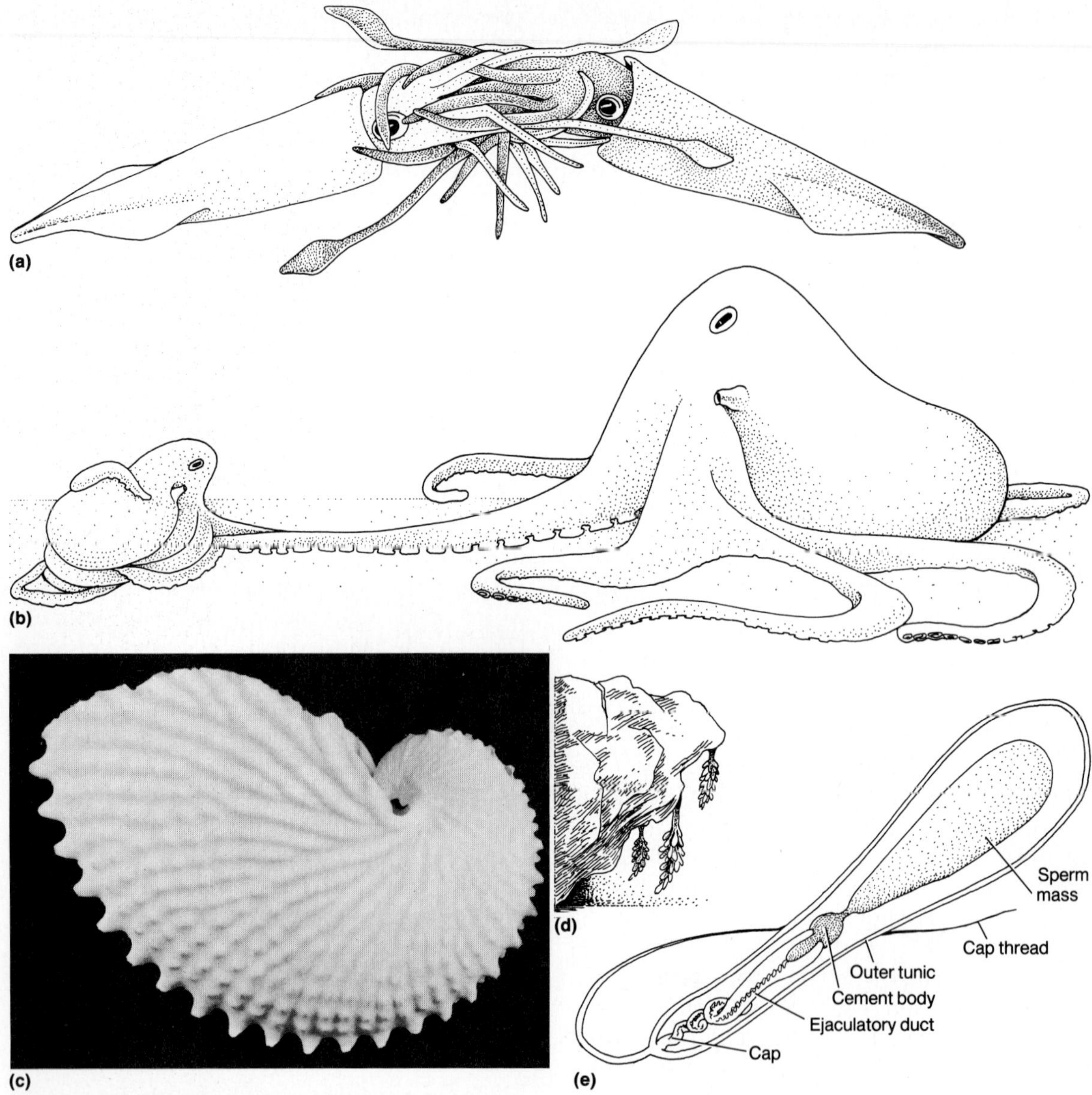

Figure 8.40 *Reproductive adaptations in cephalopods:* ***(a)*** *two* Loligo *individuals copulating;* ***(b)*** *copulation in* Octopus *with larger male using his hectocotylus to transfer sperm to the female;* ***(c)*** *photograph of the egg shell of* Argonauta, *the paper nautilus (octopod) (courtesy of L.Z. Saleeby);* ***(d)*** *several egg masses of an octopod attached to undersurface of a rock;* ***(e)*** *a spermatophore of* Loligo.

strengthening, cartilaginous plates are found in the head region, around the brain, at the bases of the arms, and in the neck region. The **mouth**, surrounded by the bases of the arms, is equipped with a pair of jaws. The chitinous jaws, operated by strong muscles, can crush even hard prey with a single bite.

Internal Features

THE VISCERAL MASS

All cephalopods are raptorial feeders; prey consists of decapod crustaceans, fishes, gastropods, bivalves, and other invertebrates. Once captured, the prey organism is quickly crushed or killed by the powerful beaklike jaws. The buccal cavity contains a **radula** and a pair of **salivary glands** whose secretions lubricate the food. The muscular **esophagus**, part of which may be modified into a crop, opens into the muscular stomach, which is attached to a very large **cecum** extending posteriorly. A complex **digestive gland** (or pancreas) produces enzymes that enter the digestive tract through a duct that opens at the junction of stomach and cecum. Digestion is extracellular and may occur in the digestive gland, but most takes place in the cecum. Cilia lining the cecal diverticula return undigestible particles to the stomach and into the intestine. The **intestine** runs anteriorly from the stomach and ends in the ventral **anus** located in the anterior mantle cavity. Some absorption takes place in the intestine, but most takes place in the cecum.

In primitive cephalopods the coelom is typically molluscan and surrounds only the heart, but in advanced forms it is relatively large. One or two pairs of kidneys are present that usually connect to the coelom. Cephalopod blood contains **hemocyanin** as a respiratory pigment. The nervous system, especially the brain, is remarkably complex for an invertebrate. The brain and behavior have been studied extensively in certain coleoids and will be discussed later.

REPRODUCTION AND DEVELOPMENT

Almost all cephalopods are dioecious, and the two sexes may be of the same size or of different sizes (Fig. 8.40a, b). A single posterior **gonad** is present, and sperm or eggs are liberated into the coelom. Sperm travel from the coelom through a **vas deferens** to a **seminal vesicle**. In this chamber, sperm are compacted and encased into rather complex **spermatophores** (Fig. 8.40e). Spermatophores, treated topically by secretions of accessory glands to harden the capsule, are transferred to a spacious spermatophore reservoir called **Needham's sac**, where they are stored until they are transferred to the female. Eggs are carried from the coelom to an **oviducal gland** by the **oviduct**. Both the seminal vesicle (males) and oviduct (females) open as a gonopore in the anterior mantle cavity.

All cephalopods engage in copulation, which is normally preceded by an elaborate courtship behavior in which both animals employ color changes and body movements. Male cephalopods use one or more arms as a spermatophore-transferring organ (Fig. 8.40b). In squids and cuttlefishes, either the right or left fourth arm is used; in octopods it is the right third arm; and in *Nautilus*, several arms are employed. Each arm is called a **hectocotylus** and is modified by having grooves, smaller suckers, or terminal depressions as modifications for spermatophore transfer. In the process of spermatophore transfer, the hectocotylus of octopods is often detached from the male and even carried for a time partially or wholly within the mantle cavity of the female. The hectocotylus was once erroneously described as a parasitic worm and given the generic name *Hectocotylus*.

Copulation in squids is usually performed with the male and female swimming and head to head (Fig. 8.40a). In octopods it is accomplished with both members either head-on or side to side (Fig. 8.40b). The hectocotylus is first tucked into the mantle cavity of the male, where it attaches to a group of spermatophores. Then

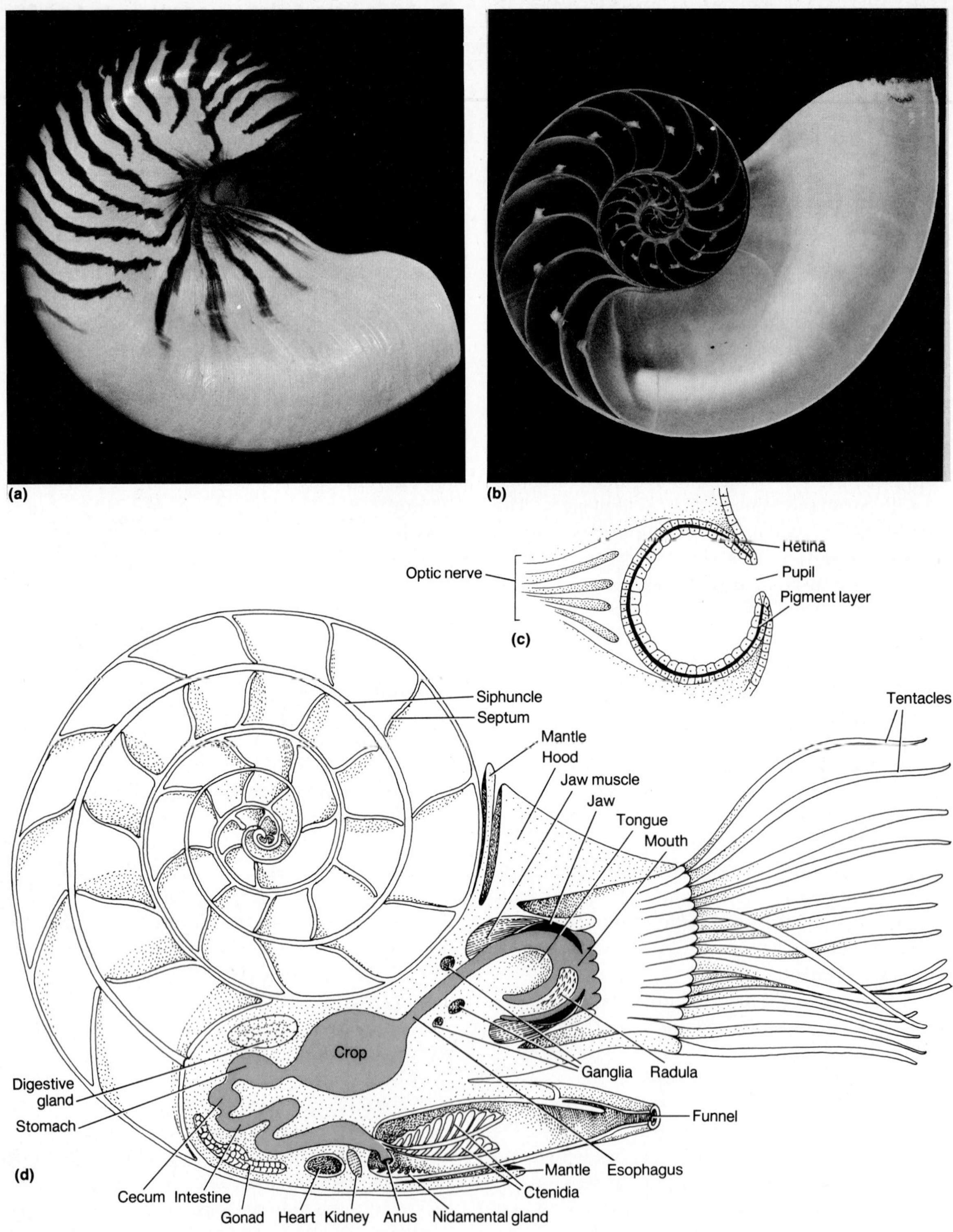

(a)
(b)
Retina
Pupil
Pigment layer
Optic nerve
(c)
Siphuncle
Septum
Mantle
Hood
Jaw muscle
Jaw
Tongue
Mouth
Tentacles
Crop
Digestive gland
Stomach
Ganglia
Radula
Funnel
(d)
Cecum
Intestine
Gonad
Heart
Kidney
Anus
Nidamental gland
Mantle
Ctenidia
Esophagus

it is normally inserted into the mantle cavity of the female, and spermatophores are deposited near the female gonopore. In others, spermatophores are transferred to a seminal receptacle located below the mouth or even into the buccal cavity of the female.

Each **spermatophore** is a complex packet equipped with the sperm mass, a cement body, an ejaculatory organ, and a cap (Fig. 8.40e). Once inside the mantle cavity, the cement body attaches the spermatophore to the inner mantle wall. The cap is removed mechanically either by the hectocotylus or by friction within the mantle cavity of the female. The removal of the cap causes the ejaculatory organ to evert, thus pulling the sperm mass outward and thereby releasing the sperm. Fertilization typically takes place in the mantle cavity as the eggs emerge from the oviduct. Fertilized, yolk-laden eggs are deposited externally either singly or in clutches in strings, capsules, ribbons, or clusters (Fig. 8.40d). Because of the yolk, cleavage is incomplete, and development is direct with no larval stages. The young emerge from their egg cases as small pelagic cephalopods. Some octopod females watch over the eggs until hatching.

In females of the octopod *Argonauta*, a most unusual egg case is formed. A highly modified pair of arms secretes a thin, delicate, calcareous, bivalved-but-united shell in which eggs are deposited (Fig. 8.40c). In this shell the female lives and broods her eggs, and sometimes the small male is a coinhabitant of the shell. This thin shell, although not a true shell, gives representatives of this genus their name, the paper nautiluses.

◀ ***Figure 8.41*** *Features of* Nautilus, *a nautiloid:* ***(a)*** *photograph of the shell from the right side;* ***(b)*** *photograph of the shell seen in sagittal section;* ***(c)*** *section through an eye;* ***(d)*** *semidiagrammatic sagittal section of an individual and its shell.*

Subclass Nautiloidea

Even though many fossil nautiloids are known, there are only three living species of *Nautilus*. The nautiloids live mostly in the Indian and Pacific oceans at depths of 60–600 m. *Nautilus* tends to be a nocturnal benthic feeder, since during the day it uses the arms to cling to the bottom.

The bilaterally symmetrical shell of *Nautilus* is coiled over the head in a two-dimensional, planospiral shape (Fig. 8.41a, b, d). It is composed of two layers: an outer one consisting of calcium carbonate crystals and an inner nacreous layer. The shell is partitioned by **septa** into chambers, each of which is progressively larger than the preceding one; only the last and largest chamber is inhabited by the living animal (Fig. 8.41b, d). Each chamber represents a growth stage. Periodically, the nautilus moves forward, and the mantle secretes a new concave septum behind itself. Septa, joined to the internal surface of the shell by simple sutures, are perforated in the middle by a shelly tube, the **siphuncle**. Within the siphuncle runs a cord of tissue extending from the visceral mass into the shell (Fig. 8.41b, d), and it secretes a gas enriched with nitrogen into the empty chambers. The gas-filled chambers greatly increase the buoyancy to counteract the weight of a rather heavy shell, and the added flotation readily facilitates swimming or resting. Many shells bear alternating bands of orange and white, but other shells may be a uniform pearly white.

Fossil nautiloids had a diversity of shell forms; some had shells coiled in a variety of patterns, and others had straight (uncoiled) shells. The incidence and degree of coiling in the extinct nautiloids followed no general evolutionary pattern; coiled and straight shells were present in both highly advanced and primitive forms. Shell sizes ranged from rather small to some having a diameter of 3 m, and others attained a length greater than 5 m.

The head bears numerous retractable arms, which are arranged in two circles and lack adhesive suckers or hooks. When retracted, each arm fits into a sheath. When feeding, *Nautilus* probably moves forward slowly; but when not feeding, it mostly swims backward slowly. Locomotion is achieved by the muscular contractions in the walls of the bilobed funnel itself. Nautiloid eyes, located on short stalks, are not as complex as those of coleoids. The eye, lacking a cornea or lens and thus open to seawater through the pupil, functions much like a pinhole camera (Fig. 8.41c).

The mantle cavity, formed by the thin mantle, contains two pairs of bipectinate ctenidia (Fig. 8.41d). *Nautilus* lacks an ink sac and pallial chromatophores characteristic of coleoids.

Internally, only the most important differences between *Nautilus* and the coleoids will be mentioned. Correlated with the four ctenidia, there are also two pairs of auricles, kidneys, and osphradia. Each pair of auricles receives oxygenated blood from a pair of ctenidia, and each auricular pair empties into a muscular ventricle, which pumps blood into arteries. *Nautilus* has a typically molluscan open circulatory system in which blood enters hemocoelic spaces. Both pairs of kidneys lack any connection with the coelom. The genital ducts are paired, although the left vas deferens may atrophy. Several different arms are hectocotylized, that is, modified for spermatophore transfer. Little is known about details of copulation or development, and eggs are deposited singly in elaborate cases.

Subclass Ammonoidea

The ammonoids first appeared in the Silurian period, reached their zenith in the Mesozoic Era, and existed into the Cretaceous period, at which point they all became extinct; about 5000 species have been discovered to date. They all had coiled shells with external sculpturing (Fig. 8.42a, b). The septa were variously shaped and joined the inner surface of the shell in complex zigzagging sutures (Fig. 8.42c). Apparently, the siphuncle was located at the margin of the shell rather than medially.

Subclass Coleoidea

Except for *Nautilus*, all living cephalopods belong to this taxon and include squids, cuttlefishes, and octopods. Octopods and some squids are basically benthic dwellers, but the remainder of the coleoids are active swimmers. Historically, coleoids have been the object of both fascination and fear among human beings. There are indeed intriguing molluscs, but much of the folklore about their being dangerous to people should be dispelled. Some of the coleoid diversity is illustrated in Figure 8.39.

External features

Mantle, mantle cavity, locomotion, and ctenidia. The mantle of coleoids is thick and muscular, and its epithelium is equipped with

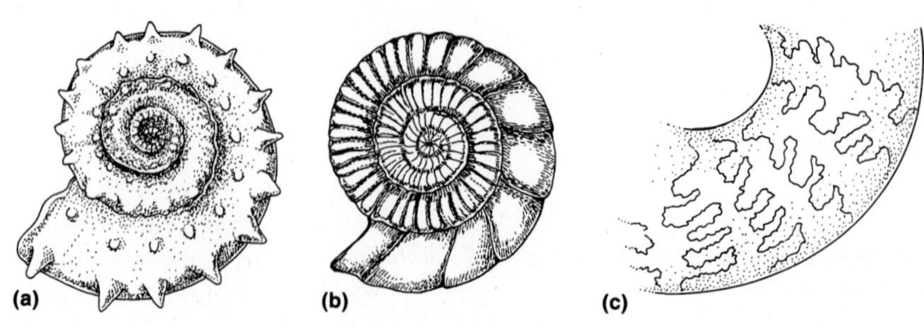

Figure 8.42 *Fossil ammonoids: **(a)*** Eoderoceras, *side view;* ***(b)*** Perisphinctes, *side view;* ***(c)*** *sutures of the shell. (parts a, b from U. Lehmann,* The Ammonites, *Cambridge University Press, 1981.)*

special pigment cells, the **chromatophores** (Fig. 8.43). Several different pigments, including yellow, black, orange, red, brown, and blue, may appear in coleoids, but only one pigment is present in a given chromatophore. Each chromatophore is equipped with minute radial muscles attached to the elastic periphery of the cell. Contraction of these muscles causes the cell with its pigment to be pulled out into a flat, two-dimensional plate, and the coloration becomes quite apparent (Fig. 8.43b). Relaxation of the muscles allows the cell membrane to contract, causing the pigment to be balled up or concentrated, and the animal pales in color (Fig. 8.43a). Deeper cells, the **iridocytes**, differentially reflect light to enhance the chromatophoric effect. Interesting chromatophore patterns result, since the chromatophores are not distributed uniformly in the pallial epithelium (Fig. 8.43c). The chromatophores are controlled by both the nervous system and hormones, and their often dramatic modifications are tied especially to light perceived by the eyes. Total transformation of color can be achieved in some species in less than 1 second! Color changes are associated with courtship, defense when alarmed, and protective coloration in background matching. The chromatophores are also stimulated by general nervous excitation, which suggests these responses may even be, to some extent, emotional.

The mantle cavity is ventral and rather extensive. Coleoids swim by expelling water forcefully from this cavity by contraction of muscles within the mantle itself. With the partial or complete loss of the shell, the entire mantle can contract freely, a feature that greatly enhances the locomotion of coleoids. Incurrent water flows around the head into the mantle cavity because of the contraction of the longitudinal or radial muscles and the concurrent relaxation of the circular muscles, all of which increase the volume of the mantle cavity. In exhalation the reverse sequence of muscular

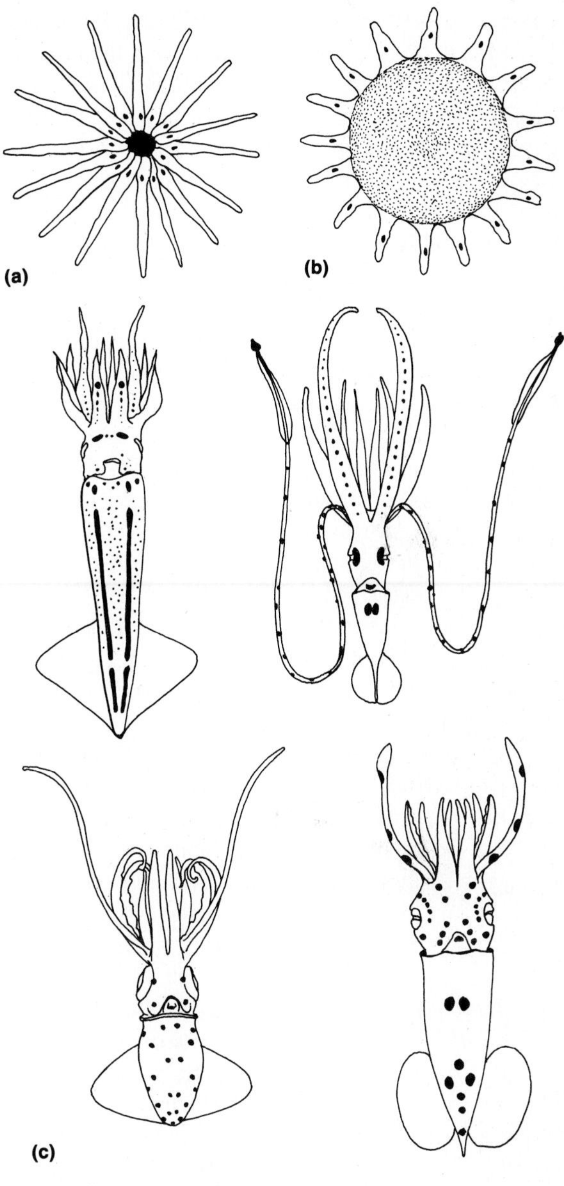

Figure 8.43 *Coleoid chromatophores and luminescent patterns: **(a)** diagram of a contracted chromatophore; **(b)** the same chromatophore expanded; **(c)** some variations in the distribution of luminescent areas (from M. Nixon and J.B. Messinger, eds.,* The Biology of Cephalopods, *1977, reproduced with permission from the Zoological Society of London).*

contractions take place. The anterior portions of the mantle bear cartilages functioning as connectors that fasten the mantle edges to the sides of the head so that water within the mantle cavity can be expelled only through the ventral **funnel**. As the circular muscles contract, the anterior edges of the mantle are snapped tightly around the head, and the jet action of the exhalent water being forced through the funnel propels the animal. The direction of the funnel discharge is usually anterior, thus propelling the animal backward. However, the funnel is very flexible and can also direct the stream posteriorly, which causes the animal to move forward. Squids can hover, glide gently through the water, or swim rapidly; the speed of locomotion is proportional to the force with which water is expelled by the circular muscles. In most squids and cuttlefishes the mantle bears **lateral fins** that function importantly as stabilizers (Fig. 8.39d). Octopods can use jet propulsion, but mostly they use the arms to creep over the bottom.

All coleoids have a single pair of pectinate ctenidia within the mantle cavity. Coleoid ctenidia are unusual among molluscs in that they contain extensive capillaries, thus augmenting their gas exchange efficiency. In a few coleoids, ctenidia are reduced, and gases are exchanged mostly through the general body surface.

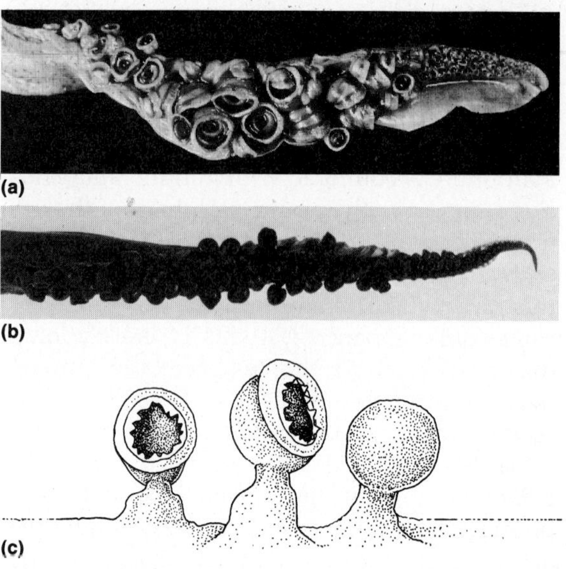

Figure 8.44 *Coleoid arms, tentacles, and suckers:* ***(a)*** *photograph of the distal spatulate end of a decapod tentacle;* ***(b)*** *photograph of a portion of a decapod arm;* ***(c)*** *three stalked suckers.*

Head and arms. The head is well developed and bears eight or ten **arms** on which are borne adhesive **suckers**. Octopods possess eight arms of equal length. All other coleoids, often called decapods, have ten arms, of which eight are of equal length and two, called **tentacles**, are much longer (Figs. 8.39d, 8.40a). It is from the total number of arms present that octopods and decapods derive their names. Octopod arms bear unstalked suction cups, but all other coleoids bear suckers on short stalks situated all along the length of the inner surfaces of the arms and only at the distal spatulate ends of the tentacles (Fig. 8.44). The adhesive suckers, often equipped with a ring of chitinous teeth or hooks, create a strong suction action on the object to which they are applied. Arms and tentacles are used in feeding, grasping, and locomotion (octopods), and one or two hectocotylized arms are employed in spermatophore transfer.

One of the most remarkable adaptations found in coleoids is the pair of complex eyes. Coleoid eyes rival those of vertebrates in structure and function, and the eyes of these two groups are often cited as striking examples of convergent evolution. A transparent **cornea** covers the **lens** suspended in place by **ciliary muscles** (Fig. 8.45). The lens is rigid and therefore has a fixed focal length, but focusing is achieved by moving the lens forward or backward by the ciliary muscles. In front of the lens is an opening, the **pupil**, in the middle of the pigmented **iris**. The pupil is always slitlike with the slit oriented horizontally, and the adjustable iris acts as a diaphragm to regulate amounts of light entering the eye. The **retina** is composed of photoreceptor cells that face outward in the direction of the light (Fig. 8.45). Because of the orientation of the retinal cells outward toward the light, their eyes are termed **direct** in contrast to the indirect eyes of vertebrates. Extrinsic eye muscles, running between the socket and eye, move the eye within the socket. Cephalopods are able to form images, and they can perceive colors. Coleoid vision, however, must be quite different from that of vertebrates, since in both, visual stimuli depend so intimately upon interpretations by the brains.

Other sense organs are also highly developed and specialized in keeping with the advanced nervous system and pelagic habits. Tactile and chemoreceptor cells are present on the head and profusely distributed on the arms. A pair of statocysts is found on either side of the brain, typically embedded within head cartilages.

Figure 8.45 *The coleoid eye.*

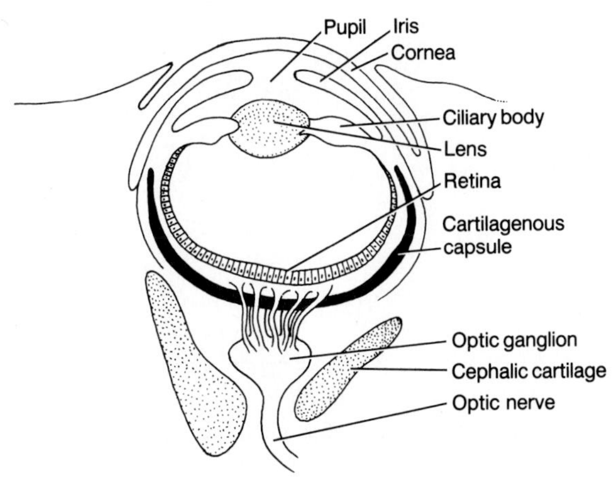

Internal features

Only those internal features which distinguish coleoids in general will be mentioned. Shells of coleoids are either poorly developed (squids) or absent altogether (octopods); when present, they are always internal. The shell is coiled, septate, and posterior in some cuttlefishes like *Spirula* (Fig. 8.46a); it is septate, uncoiled, and highly reduced in *Sepia* and other cuttlefishes (Fig. 8.46b, c); the shell or **pen** is a mere dorsal strip or ribbon of hornlike material in squids like *Loligo*.

The **jaws** of coleoids are particularly well developed. In certain octopods an additional pair of salivary glands, modified as **poison glands**, are present whose neurotoxic poison is quite potent. The bites of a few octopods may be fatal even to human beings.

The paired kidneys are complex, and excretion is carried out in two ways. First, the largest portion of a kidney is a **renal sac** through which passes an afferent branchial vein. Blood enters extensive evaginations called **renal appendages** in the walls of this vein within the renal sac. Wastes are secreted into the renal sac from the vein. Second, some wastes are lost by the circulatory system into the coelom. From there, wastes pass through a **renopericardial canal** to the kidneys, where some reabsorption of water and minerals occurs. Nitrogenous wastes are voided through the nephridiopores into the anterior mantle cavity.

Near the anal opening, the intestine receives the opening of a **rectal gland** specialized into an **ink gland** with an **ink sac** (Fig. 8.47). When the animal is alarmed, the **ink** or **sepia**, composed of melanin, is ejected through the anus, mantle cavity, and funnel to the outside. The ink cloud probably forms a "dummy" animal or creates a screen from behind which the cephalopod escapes. The ink may also be somewhat objectionable to the olfactory senses of potential predators.

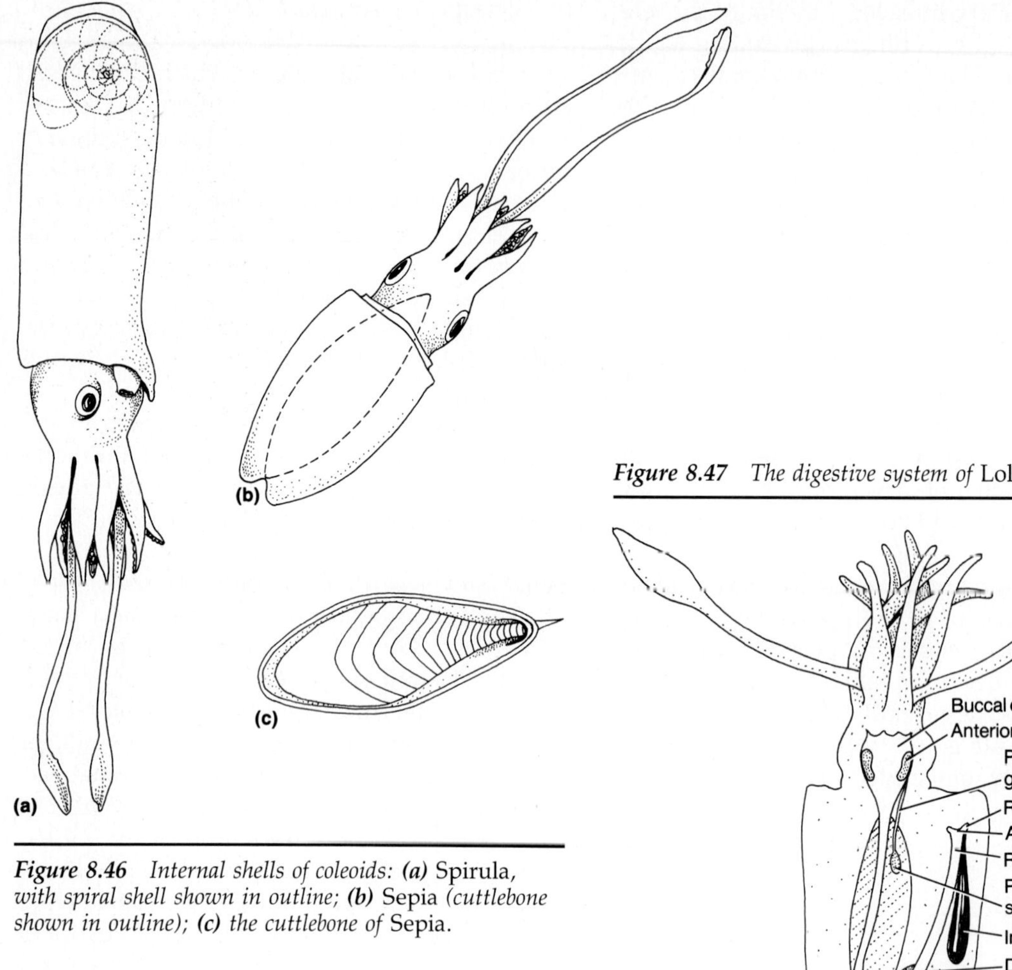

Figure 8.46 *Internal shells of coleoids:* ***(a)*** Spirula, *with spiral shell shown in outline;* ***(b)*** Sepia *(cuttlebone shown in outline);* ***(c)*** *the cuttlebone of* Sepia.

Figure 8.47 *The digestive system of* Loligo, *a coleoid.*

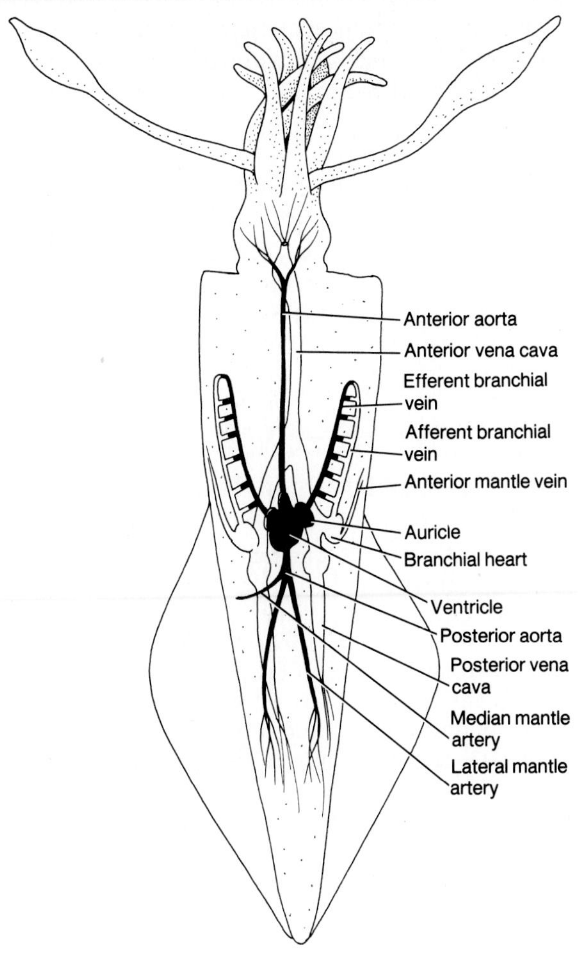

Figure 8.48 *Major parts of the blood vascular system of* Loligo, *a coleoid.*

Circulation in coleoids is more advanced than in other molluscs because theirs is a closed system with extensive arteries, veins, and capillaries. The dorsal **systemic heart** is composed of a median **ventricle** and two lateral **auricles**, each of which receives blood from an **efferent branchial vein** transporting oxygenated blood from a ctenidium (Fig. 8.48). The ventricle pumps blood into an **anterior** and a **posterior aorta**, which ultimately supply all body regions by way of capillaries. Blood from the head and the posterior regions is returned by an **anterior** and a **posterior vena cava**, respectively, and the two vena cavae on either side merge and form an **afferent branchial vein** (Fig. 8.48). Each afferent branchial vein passes through a renal sac, through a muscular **branchial heart**, and into a ctenidium. The branchial hearts increase the pressure and velocity of blood as it enters the ctenidial capillaries. The efferent branchial vein returns blood from each ctenidium to an auricle (Fig. 8.48). Coleoid plasma contains **hemocyanin**.

The nervous system has become highly differentiated and specialized in coleoids, and it has become one of their distinctive characteristics. The nervous system is concentrated in the head region, forming a large **brain** singularly unique in complexity among invertebrates. The brain is composed of pairs of cerebral, brachial, pedal, buccal, and visceral ganglia fused into a massive organ lying dorsal to the esophagus (Fig. 8.49). The large supraesophageal region consists mainly of the **cerebral ganglia**. Anteriorly, each **superior buccal lobe** is connected

Figure 8.49 *Dorsal view of the brain of an octopod (from J.Z. Young,* Biol. Rev. 36:36, *1961, published by Cambridge University Press).*

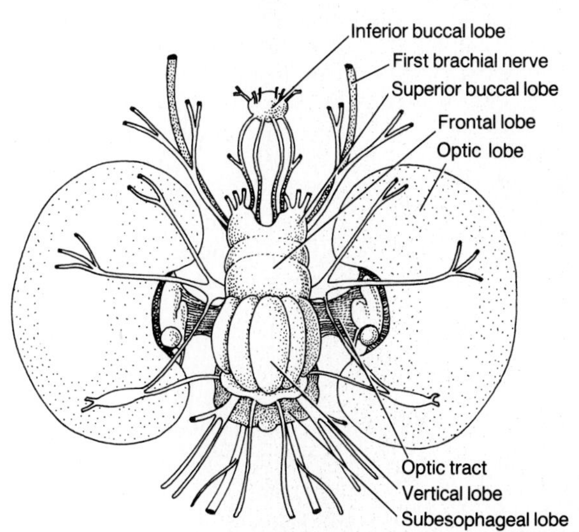

to an **inferior buccal lobe** by a **circumesophageal connective**. Each cerebral ganglion also connects to the large eye by an **optic nerve** whose nerve fibers in some like octopods form an immense **optic lobe** (Fig. 8.49). **Pedal ganglia** supply nerves to the funnel, **brachial ganglia** innervate the arms and tentacles, and **visceral ganglia** supply nerves to the visceral organs and mantle. Brain centers have been located experimentally that control specific functions such as swimming and funnel movements, mantle muscles, and the suckers. Giant neurons found in the brain of cephalopods have been used extensively in neural physiology. It is beyond the scope of this textbook even to mention many of the utterly fascinating studies that have been carried out on the complex behaviors and learning abilities of coleoids, especially those of octopods. A few of the interesting facets of coleoid behavior appear in Box 8.3.

BOX 8.3 □ SOME ASPECTS OF BEHAVIOR AND LEARNING IN COLEOIDS

One of the most distinguishing features of coleoids, and one that separates them from all other molluscs and, indeed, all other invertebrates, is their enormously complex brain and the variety of behaviors made possible by their fantastic nervous system. Coleoids are trainable to a sophisticated degree, and this feature is obviously correlated with their complex brain and the large number of association neurons involved in complex integrative behavior. It has been shown that many activities such as mantle muscle contractions and sucker attachments are purely reflexive. But studies have also shown that coleoids are involved in much higher neural functions such as memory and decision-making.

Coleoids can learn by sight or by touch using the arms. For example, in three trials a day for only two days an *Octopus* learned to avoid a crab offered on a plate electrified with a weak current; however, it would readily devour a crab presented alone. Octopods can be taught to discriminate among many different types of cylinders on the basis of the texture of their surfaces. Other experiments have shown octopods to be able to distinguish between horizontal and vertical rectangles but not oblique rectangles or weights of objects. Squares of varying sizes could be discriminated such as those of 4 and 8 cm. Many different experiments have shown different regions of the brain to be centers of short-term memory of several days, and others are centers for long-term memory of several months.

Octopus has two different anatomical and functional centers for the storage of memory; one is visual, and the other is chemotactile via the suckers. Visual information, stored in the optic lobes, can be activated experimentally to elicit feeding, escape, or attack responses learned much earlier. Chemotactile learning from previous handling of objects or organisms can enable a blinded octopod to pick up the correct item from a battery of objects presented to it. In both visual and chemotactile learning, previously learned information had to be stored within the brain until it was used again in the experiments. Long-term memory implies that the animal retains earlier information and uses it much later in decision-making. Much future research will be devoted to all aspects of coleoid behavior including memory, training, visual and chemotactile recognition, and other basic neural processes. An excellent review of octopod behavior can be found in Wells' (1978) interesting volume, *Octopus*.

PHYLOGENY OF MOLLUSCS

There has been a debate of long standing regarding the evolutionary affinities of the molluscs, and the entire issue of molluscan phylogeny has been reviewed by Stasek (in Florkin and Scheer, 1972). One older theory holds that molluscs and annelids shared a common ancestry; in fact, some biologists hold that molluscs arose from primitive annelids. Evidences to support this theory are mostly derived from embryology, since, in both phyla, patterns of spiral cleavage are very similar, and both develop trochophores that are almost identical. The discovery of *Neopilina*, with repeating structures, was looked upon as a major piece of additional evidence for the theory that molluscs had an annelid as their ancestral stock. A second theory holds that molluscs arose from an acoelomate ancestry, perhaps much like a turbellarian or nemertean; presumably, it was from a similar stock that annelids arose independently. Molluscs are closest structurally to the acoelomates, although acoelomates lack many of the basic molluscan characteristics. Nonetheless, one might theorize methods of how such an ancient animal could evolve into the ancestral mollusc with foot, mantle, shell, etc. Supporting evidence is in the total lack of true metamerism in any molluscan group. The multiple pairs of certain organs in *Neopilina* would then be considered as serially repetitive rather than as any form of metamerism.

It would appear that the best hypothesis for the origins of the molluscs is to be found in an ancient stock from which molluscs and annelids probably diverged early in their phylogenetic history. It is likely that a primitive turbellarian like an acoel served as a plastic adaptable stock that eventually gave rise to both annelids and molluscs. Such a theory must hold that these evolutionary paths diverged before metamerism was developed as an important adaptation in primitive annelids. Thus both the embryological similarities between annelids and molluscs and the lack of metamerism in molluscs can be handled conveniently.

An ancestral mollusc, tentatively characterized in Box 8.1 and Fig. 8.1, is a hypothetical but tenable primitive mollusc. From the ancestral molluscan stock the evolutionary pathways within the phylum are still not definitely understood. There appears to be good theoretical evidence that all gastropods and cephalopods evolved together from monoplacophorans, and bivalves and scaphopods, probably similar phylogenetically, also evolved in another line from monoplacophorans. If these ideas are correct, the suggestion is that aplacophorans and polyplacophorans evolved early in the molluscan line before the development of a univalve. Thus the eight valves of polyplacophorans represent one adaptive line that led to no other molluscs, and the shell-less aplacophorans are therefore primitive rather than degenerate (Fig. 8.50). In fact, there is increasing conviction among many workers that the aplacophorans may indeed be much closer to the ancestral stem mollusc than was previously thought. The monoplacophorans, developed from the primitive aplacophorans, represent a later stem group.

From the monoplacophorans arose two principal evolutionary lines. One line gave rise to gastropods and cephalopods, and these two large groups must have diverged early before torsion and coiling evolved in gastropods (Fig. 8.50). Among gastropods the prosobranchs and more specifically the archeogastropods are considered to be the primitive group, and from them the other prosobranchs, opisthobranchs, and pulmonates were derived. Among cephalopods, two lines of evolution resulted. One produced the ammonoids, which in turn gave rise to the nautiloids, and the other line led to the coleoids. The second evolutionary line from the monoplacophorans ultimately resulted in bivalves and scaphopods. Both lines diverged early and developed independently of each other. The most primitive bivalves are clearly the protobranchs, and from them the septibranchs and lamellibranchs arose (Fig. 8.50).

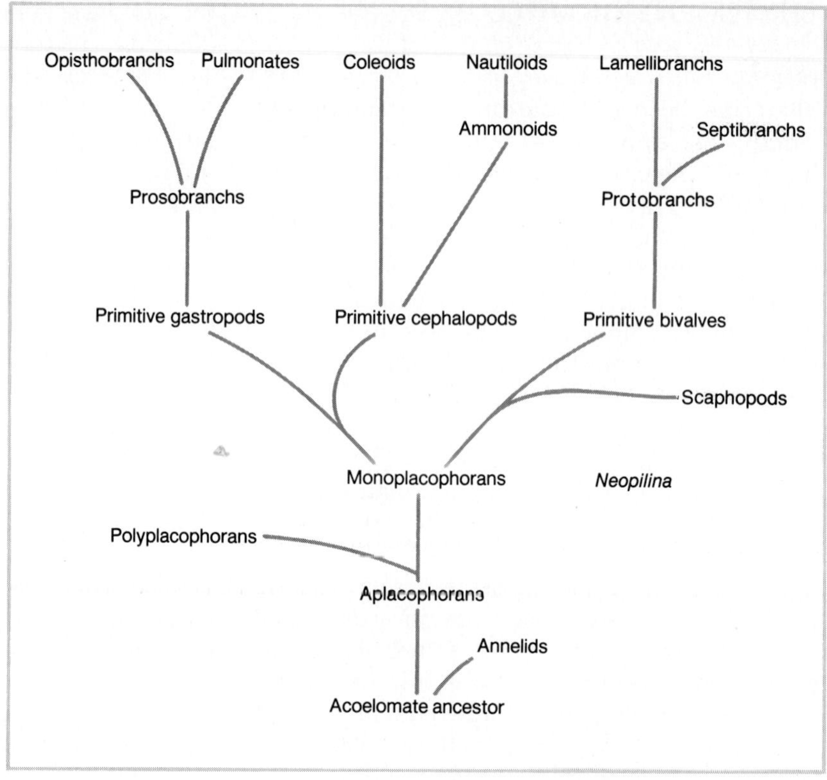

Figure 8.50 *Probable phylogenetic pathways in the Phylum Mollusca.*

SYSTEMATIC RESUME

The Major Molluscan Taxa

CLASS APLACOPHORA—the solenogasters; primitive, marine, wormlike, shell-less molluscs with a simple radula and small posterior mantle cavity.

Subclass Caudofoveata—lack head, foot, gonoducts, kidneys; mantle present; fertilization is external; *Chaetoderma* is the best-known genus.

Subclass Ventroplicida—adults lack true ctenidia and gonoducts; foot is present as a ventral longitudinal furrow; hermaphroditic; *Neomenia* is perhaps the best-known genus.

CLASS MONOPLACOPHORA—mostly extinct; one-piece shell and repeating kidneys, pedal retractor muscles, auricles, gills; *Neopilina* (Fig. 8.13) is the better-known extant genus.

CLASS POLYPLACOPHORA—chitons; shell of eight dorsal plates set in thick girdle; have a creeping foot, radula.

Order Lepidopleurida—small number of posterior ctenidia; lack insertion plates (specialized areas on valves); small group characterized by *Lepidopleurus*.

Order Ischnochitonida—most chitons; numerous pairs of ctenidia; valves have insertion plates that anchor lateral aspects of valves into mantle; includes *Chaetopleura* (Fig. 8.14a, b), *Chiton*, and *Mopalia*.

Order Acanthochitonida—shell plates usually covered by girdle; range in size from small to the largest chitons; *Cryptochiton* and *Acanthochitona* are well-known genera.

CLASS GASTROPODA—snails; all undergo torsion; most are spiraled; asymmetric shell of one piece.

Subclass Prosobranchia—includes most gastropods; mostly marine; effects of torsion are evident, with mantle cavity and associated structures anterior to visceral mass and dorsal to the head; characteristically bear one or two ctenidia; most have an operculum; most are dioecious.

Order Archeogastropoda—primitive; ctenidia are bipectinate; usually two ctenidia, auricles, kidneys, and osphradia are present; includes slit shells, abalones (*Haliotis* [Fig. 8.16f]), limpets with or without apical hole in shell; top shells (*Margarites*, *Trochus* [Fig. 8.25a], turban shells (*Turbo*), and *Nerita*.

Order Mesogastropoda—most prosobranchs; chiefly marine, but some are freshwater or terrestrial; one monopectinate ctenidium, one auricle and kidney; renal and gonadal ducts are separate; most have seven radular teeth per row; includes slipper shells (*Crepidula* [Fig. 8.24b]), carrier shell, conchs (*Strombus*), cowries (*Trivia*), *Littorina*, *Pomatias*, moon shells (*Polinices* [Fig. 8.25l]), helmet shells, tuns, heteropods (*Atlanta*).

Order Neogastropoda—most advanced prosobranchs; all are marine carnivores; radula with three or fewer teeth per row; one monopectinate ctenidium, one auricle, one kidney separate from complex reproductive tract; long proboscis; includes drills (*Murex*, *Urosalpinx* [Fig. 8.19f]), whelks (*Buccinum*, *Busycon* [Figs. 8.16c, d; 8.20, 8.25f]), tulip shells (*Fasciolaria* [Fig. 8.25m]), mud snails (*Ilyanassa*), olive shells (*Oliva* [Fig. 8.25h]), miter and harp shells, volutes, and cone shells (*Conus* [Fig. 8.25i]).

Subclass Opisthobranchia—sea slugs; marine; have undergone detorsion; shell and mantle cavity reduced or absent; secondarily acquired bilateral symmetry; most have lost basic ctenidia; have one auricle and kidney; head with two pairs of tentacles; monoecious.

Order Cephalaspidea—bubble shells; shell is reduced, some have external shell, but in others it is partially or wholly enclosed by mantle; have a head shield used for burrowing; includes *Acteon* and *Bulla* (Fig. 8.26a).

Order Runcinoidea—small (up to 8 mm); marine; shell is very reduced or absent; no tentacles; two eyes; *Runcina* is the type genus.

Order Thecosomata—sea butterflies or shelled pteropods; pelagic, with swimming parapodia or wings; most have an external shell; *Clio* and *Spiratella* are well known.

Order Gymnosomata—naked pteropods with lateral swimming parapodia; planktonic and found in open water; lack a shell and mantle cavity; includes *Cliopsis* and *Spongiobranchaea*.

Order Anaspidea—sea hares; often large (up to 30 cm) and conspicuous; feed on algae; shell reduced and buried in mantle or absent; ctenidium and mantle cavity present; foot with lateral swimming parapodia; best-known genera are *Aplysia*, *Tethys* (Fig. 8.26c), and *Petalifera*.

Order Notaspidea—shell is external, internal, or absent; ctenidium present, but they lack parapodia; *Tylodina* is typical.

Order Acochlidioidea—small interstitial forms with no shell; visceral mass and foot sharply delimited from each other; visceral mass often much longer than foot; some inhabit fresh water; some are dioecious; *Acochlidium* is representative.

Order Sacoglossa—herbivorous; radula with single row of teeth; discarded teeth accumulate in a ventral buccal sac; some forms have a shell; several genera are *Oxynoe* and *Elysia*.

Order Nudibranchia—nudibranchs or sea slugs; lack shell, mantle cavity, and osphradium; body is bilaterally symmetrical and often bears cerata; many have secondary lateral or perianal gills; some of the more familiar genera are *Doris*, *Cadlina*, *Polycera*, and *Eolidia* (Fig. 8.26d).

Order Entoconchida—wormlike endoparasites of holothuroidean echinoderms; greatly modified for parasitic existence; adults lacking ctenidium, radula, shell, and most organ systems; includes *Entoconcha*.

Subclass Pulmonata—mostly terrestrial; have undergone detorsion; mantle cavity modified as "lung"; pneumostome present; no ctenidia; most have a shell; most lack an operculum; monoecious.

Order Archaeopulmonata—primitive pulmonates; shells mostly dextral; hermaphroditic; penis present as a stylet; live in coastal areas; *Ellobium* is a representative genus.

Order Systellommatophora—peculiar, elongated slugs; two pairs of noninvaginable tentacles; foot reduced to median strip; no shell; anus at posterior end; mostly tropical and subtropical species; *Veronicella* is found in Florida.

Order Basommatophora—shelled, aquatic, mostly freshwater pulmonates; one pair of tentacles with an eye at base of each tentacle; a few possess an operculum; *Lymnaea* (Fig. 8.27a), *Planorbis, Helisoma* (Fig. 8.27e), and *Physa* (Fig. 8.27g) are familiar genera.

Order Stylommatophora—slugs and terrestrial snails; many with shell, but many others with reduced shell; two pairs of tentacles, each of posterior pair is tipped with an eye; many are tropical; some of the best-known snail genera are *Partula, Helix,* and *Helicella*; familiar slug genera are *Limax* (Fig. 8.27b), *Arion* (Fig. 8.24a), and *Testacella*.

CLASS BIVALVIA—clams; laterally compressed; shell of two valves; mostly filter-feeders using enlarged ctenidia; no radula.

Subclass Palaeotaxodonta—primitive, small, marine bivalves; mantle cavity largely open ventrally; protobranch ctenidia; enlarged labial palps with feeding proboscides; valves with hinge teeth; *Nucula, Nuculana,* and *Yoldia* are typical genera.

Subclass Cryptodonta—primitive, small, marine bivalves; simple, nonreflected protobranch ctenidia; valves without hinge teeth; reduced labial palps with no proboscides; *Solemya* (Fig. 8.28a) is a representative genus.

Subclass Pteriomorpha—bivalves attached to surfaces either by a byssus or by cementation; mantle margins are not fused; filibranch ctenidia.

Order Arcoida—ark shells; hinge straight, with hinge teeth; adductor muscles of equal size; mantle edges do not form siphons; *Arca* is a common genus.

Order Mytiloida—valves of equal size, but each not equilateral; anterior adductor muscle reduced or absent; common mussel genera are *Mytilus* (Figs. 8.28e, 8.32a), *Modiolus* (Fig. 8.28c), and *Lithophaga*; familiar pen shell genera are *Pinna* and *Atrina* (Figs. 8.28b, 8.32b).

Order Pterioida—valves are often dissimiliar and sculptured; few or no hinge teeth; anterior adductor muscle is small or absent; most are permanently sessile, but some can swim; includes scallops (*Pecten, Argopecten* [Fig. 8.28d]), oysters (*Crassostrea, Ostrea*), and file shells (*Lima* [Fig. 8.28f]).

Subclass Palaeoheterodonta—mostly freshwater forms; valves are similar; hinge teeth are present; pseudolamellibranch ctenidia.

Order Unionoida—large, freshwater bivalves; mostly dioecious; develop a glochidium that is a temporary, obligatory ectoparasite on fishes; *Anodonta, Unio,* and *Margaritifera* are well-known genera.

Order Trigonioida—small; triangular-shaped, oval shells with strong ribs; anterior adductor muscle is small; *Neotrigonia* and *Trigonia* are best known.

Subclass Heterodonta—a large group of mostly burrowing clams; both hinge and lateral teeth are present; valves without nacreous layer; siphons are present; eulamellibranch ctenidia.

Order Veneroida—valve beaks are often toward anterior end; both adductor muscles are of equal size; includes cockles (*Cerastoderma* [Fig. 8.28j]), giant clams (*Tridacna*), razor shells (*Ensis* [Fig. 8.28i]), coquina shells (*Donax* [Fig. 8.28k]), Venus clams (*Mercenaria* [Figs. 8.28h, 8.29a, 8.34]), freshwater fingernail clams (*Sphaerium, Musculium*).

Order Myoida—thin-shelled valves modified for burrowing; mantle edges fused except for foot opening and well-developed siphons; includes geoducks (*Panopea*), wood borers (*Xylophaga* [Fig. 8.31a], *Teredo* [Fig.

8.31b]), soft-shelled clams (*Mya*), and piddocks (*Pholas*).

Order Hippuritoida—rock oysters; valves strongly inequilateral with one cemented to substratum; unequal adductor muscles; *Chama* and *Arcinella* are among the better-known genera.

Subclass Anomalodesmata—the septibranchs; valves variously shaped and sized; hinge teeth reduced or absent; carnivorous; no ctenidia, but with muscular septa for pumping water and food into mantle cavity; mantle margins fused; two of the better-known genera are *Poromya* and *Cuspidaria*.

CLASS SCAPHOPODA—tusk shells; marine; one-piece shell is tubular; have adhesive captacula, radula, but no heart; *Dentalium* (Fig. 8.38a) is common.

CLASS CEPHALOPODA—squids, octopods; foot is locomotive funnel; have eight or ten arms around a well-developed head.

Subclass Nautiloidea—the nautiluses; have an external, many-chambered shell with internal septa; have two pairs of auricles, ctenidia, kidneys, and osphradia; have numerous, suckerless arms; *Nautilus* is the only extant genus (Fig. 8.41).

Subclass Ammonoidea—all extinct; had coiled shells with elaborate external sculpturing; septa joined shell in complex sutural patterns.

Subclass Coleoidea—shell internal or absent; one pair of ctenidia, auricles, and kidneys; osphradia are absent; eight or ten arms with adhesive suckers; a pair of highly developed eyes.

Order Sepioidea—cuttlefishes and sepiolas; internal, calcareous, septate shell may be present, reduced, or missing; eight arms and two tentacles; body short and stocky; familiar genera include *Sepia* (common cuttlefish [Fig. 8.46b]), *Spirula* (Fig. 8.46a), and *Sepiola*.

Order Teuthoidea—squids; shell present only as a horny, ribbonlike pen; eight arms and two tentacles; body tubular, torpedo shaped, with lateral fins; common genera are *Loligo* (Figs. 8.39d, 8.40a, 8.47, 8.48), *Architeuthis*, and *Histioteuthis*.

Order Vampyromorpha—vampire squids; small, deep-water, black-pigmented octopods with eight arms joined by a web; two threadlike tentacles present; no ink sac; *Vampyroteuthis* is best known.

Order Octopoda—the octopods; shell absent; body globular with no fins; eight arms with sessile suckers; best-known genera are *Octopus* (Figs. 8.39a, b, 8.40b), *Argonauta* (the paper nautilus [Fig. 8.40c]), and *Eledone*.

ADDITIONAL READINGS

Abbott, R.T. 1974. *American Seashells*, 2nd ed., 598 pp. New York: Van Nostrand Reinhold Co.

Aldridge, D.W. 1982. Reproductive tactics in relation to life-cycle bioenergetics in three natural populations of the freshwater snail, *Leptaxis carinata*. *Ecology* 63:196–208.

Alkon, D.L. 1983. Learning in a marine snail. *Sci. Amer.* *249*(1):70–84.

Bayne, B.L., ed. 1976. *Marine Mussels: Their Ecology and Physiology*, 494 pp. Cambridge, England: Cambridge University Press.

Boycott, B.B. 1965. Learning in the octopus. *Sci. Amer.* *212*(3):42–50.

Boyle, P.R., ed. 1983. *Cephalopod Life Cycle*. Vol. 1: *Species Accounts*, 440 pp. New York: Academic Press.

Brown, F.A., Jr. 1950. *Selected Invertebrate Types*, pp. 318–357. New York: John Wiley & Sons.

Brown, K.M. 1982. Resource overlaps and competition in pond snails: An experimental analysis. *Ecology* 63:412–422.

Cheng, T.C. 1973. *General Parasitology*, pp. 903–911. New York: Academic Press.

Florkin, M. and B.T. Scheer, eds. 1972. *Chemical Zoology*. Vol. VII: *Mollusca*, 499 pp. New York: Academic Press.

Fretter, V. and J. Peake, eds. *Pulmonates*. Vol. 1: *Functional Anatomy and Physiology*, 400 pp., 1975. Vol. 2A: *Systematics, Evolution and Ecology*, 526 pp., 1978. New York: Academic Press.

Giese, A.C. and J.S. Pearse, eds. 1979. *Reproduction of Marine Invertebrates*. Vol. V: *Molluscs: Pelecypods and Lesser Classes*, 369 pp. New York: Academic Press.

Hochachka, P.W., ed. 1983. *The Mollusca*. Vol. 1: *Metabolic Biochemistry and Molecular Biomechanics*, 504 pp. Vol. 2: *Environmental Biochemistry and Physiology*, 343 pp. New York: Academic Press.

House, M.R. and J.R. Senior. 1981. *The Ammonoidea: The Evolution, Classification, Mode of Life and Geological Usefulness of a Major Fossil Group*, 594 pp. Systematics Association, Special Vol. 18. New York: Academic Press.

Hyman, L.H. 1967. *The Invertebrates*. Vol. VI: *Mollusca I*, 769 pp. New York: McGraw-Hill Book Co.

Jones, D.S. 1983. Sclerochronology: Reading the record of the molluscan shell. *Amer. Sci. 71*:384–391.

Lane, C.E. 1961. The teredo. *Sci. Amer. 204*(2):132–140.

Lehmann, U. 1981. *The Ammonites: Their Life and Their World*, 246 pp. New York: Cambridge University Press.

Lemche, H. 1957. A new living deep-sea mollusc of the Cambro-Devonian class Monoplacophora. *Nature* (London) *179*:413–416.

MacGinitie, G.E. and N. MacGinitie. 1968. *Natural History of Marine Animals*, 2nd ed., pp. 327–401. New York: McGraw-Hill Book Co.

Moore, R.C., ed. 1960–1971. *Treatise on Invertebrate Paleontology. Mollusca*, Parts 1–6 (Parts I–N). Lawrence, Kansas: University of Kansas Press and the Geological Society of America.

Morton, J.E. 1979. *Molluscs*, 5th ed., 254 pp. London: Hutchinson.

Nixon, M. and J.B. Messenger, eds. 1977. *The Biology of Cephalopods*, 567 pp. Symposium of the Zoological Society of London, No. 38. New York: Academic Press.

Pennak, R.W. 1978. *Fresh-water Invertebrates of the United States*, 2nd ed., pp. 710–768. New York: John Wiley & Sons.

Purchon, R.D. 1977. *The Biology of the Mollusca*, 2nd ed., 543 pp., New York: Pergamon Press.

Roper, C.F.E. and K.J. Boss. 1982. The giant squid. *Sci. Amer. 246*(4):96–104.

Russell-Hunter, W.D., ed. 1983. *The Mollusca*. Vol. 6: *Ecology*, 708 pp. New York: Academic Press.

Saleuddin, A.S.M. and K.M. Wilbur, eds. 1983. *The Mollusca*. Vol. 4: *Physiology, Part 1*, 515 pp. Vol. 5: *Physiology, Part 2*, 486 pp. New York: Academic Press.

Thompson, T.E. 1976. *Biology of Opisthobranch Molluscs*. Vol. I, 197 pp. London: The Ray Society.

Verdonk, N.H., J.A.M. van den Biggelaar, and A.S. Tompa, eds. 1983. *The Mollusca*. Vol. 3: *Development*, 343 pp. New York: Academic Press.

Ward, P. 1983. The extinction of the ammonites. *Sci. Amer. 249*(4):136–146.

Ward, R., L. Greenwald, and O.E. Greenwald. 1980. The buoyancy of the chambered nautilus. *Sci. Amer. 243*(4):190–204.

Wells, M.J. 1978. Octopus: *Physiology and Behavior of an Advanced Invertebrate*. 398 pp. London: Chapman and Hall.

Willows, A.O.D. 1971. Giant brain cells in mollusks. *Sci. Amer. 224*(2):68–75.

Yonge, C.M. 1975. Giant clams. *Sci. Amer. 232*(4):96–105.

Yonge, C.M. and T.E. Thompson. 1976. *Living Marine Molluscs*, 273 pp. London: Collins.

9

The Annelids and Small Related Phyla

OVERVIEW

Annelids, a rather large and diverse group of segmented bristleworms, earthworms, and leeches, are particularly common in rich soils and shallow marine habitats. Their two most important features are their metamerism and their coelom. An annelid's body consists of an anterior acron, a posterior pygidium, and a long vermiform trunk composed of a variable number of segments. A spacious, fluid-filled, perivisceral coelom is comparted by internal metamerically arranged septa. Body wall muscles exert hydraulic pressures on the coelomic fluid to achieve various forms of mobility. Locomotion and burrowing are very primal annelidan activities whose efficiencies are directly correlated with metamerism, a hydrodynamic skeleton, and a variable number of slender chaetae on each segment. Other annelidan features include a pair of excretory nephridia in most segments, an unspecialized digestive tract, a ventral nerve cord with metameric ganglia, and diverse reproductive and developmental patterns.

Polychaetes are sedentary or wandering worms that are often very numerous and ecologically important in littoral marine habitats. Their most distinguishing feature is that on each segment is a pair of fleshy, chaetae-bearing, lateral outgrowths called parapodia, which function in locomotion, tube-building, and irrigation of tubes and burrows. These worms also are usually dioecious, lack distinct gonads, bear cephalic sense organs, and have a planktonic larval stage.

Clitellates are mostly freshwater or terrestrial, lack parapodia and gills, are monoecious, bear distinct gonads, and possess a clitellum that produces a cocoon in which embryonic development takes place; no larval stages are developed. Oligochaetes, the earthworms and many freshwater forms, have few chaetae per segment, glands to eliminate excess calcium, and rather complicated reproductive systems. Hirudineans or leeches, basically a parasitic freshwater group, have a body of 34 segments, anterior and posterior suckers, prominent dorsoventral muscles, paired ceca for food storage, and intrusive connective tissue that obliterates the coelom; they lack septa, mesenteries, and a true blood vascular system. Two annelidan groups, branchiobdellids and acanthobdellids are composed of small numbers of species and are ectosymbiotic on crustaceans and fishes.

Four small phyla of worms are distantly related to annelids in different ways. Pogonophorans are marine, and their bodies consist of a forepart with tentacles, a trunk with metameric papillae, and a posterior metameric opisthosoma; no alimentary system is present. Sipunculids and echiurans are marine, benthic, and unsegmented and have an extensible food-gathering organ and an annelidlike body wall; they differ from each other in subtle ways. Onychophorans are tropical terrestrial worms whose annelidan features include a segmented trunk, paired but unjointed legs, body wall construction, and ciliated metanephridia; other arthropodlike features include a chitinous cuticle, an open circulatory system with a hemocoel, and a tracheal system.

Annelids very likely evolved from an acoelomate marine worm. During the course of their early phylogenetic history a coelom and metamerism evolved, which were to be of immense significance to their future success. Within the phylum, two distinct lines of evolution have produced the polychaetes and clitellates. The evolutionary affinities of the four other worm phyla are tenuous, and each is probably only distantly related to annelids.

INTRODUCTION

This chapter is devoted primarily to an understanding of the form, function, and characteristics of the Phylum Annelida. Also included are discussions of four, small, protostomate phyla—Pogonophora, Sipuncula, Echiura, and Onychophora—that are distantly related to annelids. It is probable that all five phyla shared a very remote common ancestry; but if so, the smaller groups diverged quite early before the ontogeny of most annelidan features. This would account for both their few similarities to annelids and their very special differences. Since each is similar to annelids in ways fundamentally different from the other three, there is no one feature common to all five groups. Therefore it is best to have a full treatment of annelids first and to follow it with a much briefer discussion of the four other groups; care will be taken to point out prominently how each is related to annelids. All five phyla are characterized in Table 9.1

PHYLUM ANNELIDA

The segmented soft-bodied worms, including earthworms, leeches, bristleworms, tubeworms, and others, all belong to the Phylum Annelida. A diverse group of about 10,000 species, they are found in the oceans, fresh waters, and moist soils. Often, annelids make a major ecological impact in soils and on many ecosystems, especially those of estuarine or littoral marine habitats. Typically, annelids are the largest of all the worms or wormlike invertebrates and range in length from about 1 mm to 3 m. Some annelids are dark brown in color, but many others are among the most beautiful of all invertebrates and have brilliant shades of red, green, pink, yellow, and blue. As with all other active motile invertebrates, annelids are bilaterally symmetrical.

There are three principal annelidan groups to which frequent references will be made in the introductory discussions: the polychaetes (Class Polychaeta) and two groups of the Class Clitellata, the earthworms (Subclass Oligochaeta) and the leeches (Subclass Hirudinea). All three groups are briefly characterized in Table 9.1.

Representatives of this phylum exhibit several extremely important features that have contributed to both their success and their importance. However, the two characteristics that have had the greatest impact on the functional

TABLE 9.1. □ **CHARACTERISTICS OF THE PRINCIPAL TAXA OF ANNELIDS AND OF FOUR RELATED PHYLA**

Taxon	*Common Name(s)*	*Setae*	*Metamerism*	*Other Features*
PHYLUM ANNELIDA				
Class Polychaeta	Tubeworms, sandworms, bristleworms	Many	Present	One pair of parapodia per segment; lack clitellum
Class Clitellata	Earthworms, leeches	Fewer or none	Present	Possess an enlarged glandular clitellum used in reproduction; lack parapodia
Subclass Oligochaeta	Earthworms	Few per segment	Present	Very little cephalization; most have calciferous glands
Subclass Hirudinea	Leeches	Absent	Present as 34 metameres	Anterior and posterior suckers usually present
PHYLUM POGONOPHORA	Beardworms	Present	Opisthosoma only	Lack digestive system
PHYLUM SIPUNCULA	Peanutworms	Absent	Absent	Feeding introvert; peritoneal urn-shaped cells
PHYLUM ECHIURA	ProbosciswIorms	1 to several pairs	Absent	Proboscis used in feeding
PHYLUM ONYCHOPHORA	Walkingworms	Absent	Present	Paired nonjointed legs; tracheal system

morphology of annelids are that they are segmented and that they possess a coelom, and these immensely important features have fundamentally influenced the process and mechanics of locomotion more than any other function.

Metamerism and the Coelom

Of all annelidan features, none is more important or has more far-reaching consequences than segmentation. In **metamerism** or **segmentation** the body is divided into a series of similar **metameres** or **segments** arranged linearly in an anteroposterior series. The youngest segments are always situated at the posterior end, since that is where new metameres are developed. A quick review of the significance of and details about metamerism, found in Chapter 3, will prove to be most helpful at this point. Metamerism is most pronounced in the least-specialized metameric animals, such as annelids, in which all of the body segments are essentially alike; such a condition is called **homononomous** segmentation. Since annelids possess a generalized metameric plan, it is in this taxon that the basic features of segmentation can best be seen.

Metamerism in annelids is a feature that is manifestly evident both externally and internally (Fig. 9.1). Skin, muscles, peripheral nerves and ganglia, and visceral organs such as those associated with circulation, reproduction, and excretion all reflect segmental influences. Basically, segmentation is a property of the mesoderm, and it impacts directly on all mesodermally derived structures and on most

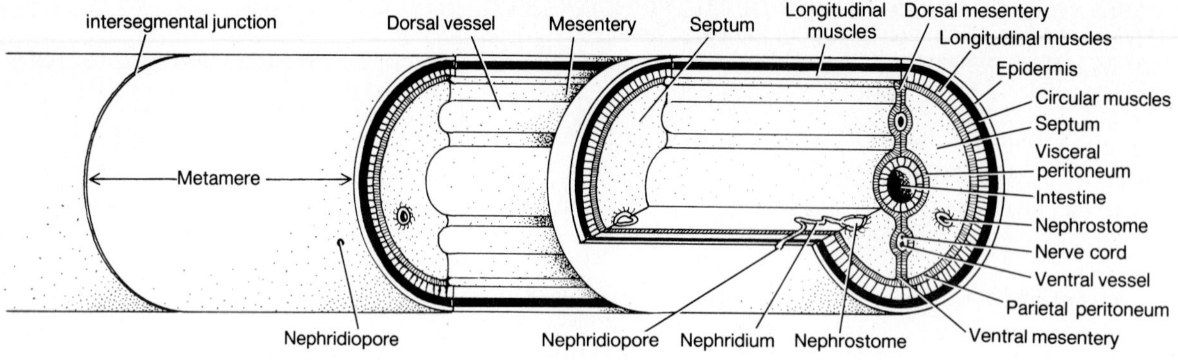

Figure 9.1 *Diagrammatic longitudinal and cross-sectional views of an annelid to show metamerism and its influences on annelidan systems.*

ectodermal derivatives as well. Metamerism is evident quite early in embryonic development with the appearance of the mesoderm and its organization into bilateral blocks or somites. From this internal genesis the effects of metamerism proceed outward and eventually include the obvious ectodermally derived features such as the appendages and the external partitioning of the body wall (Fig. 9.1). Only the endodermally derived tissues, that is, those lining the digestive tract, are unaffected by metamerism.

Is metamerism a unique feature of annelids, or does it appear in other animal groups? Metamerism clearly originated independently in animals at least twice; it appeared in one line of deuterostomates—the vertebrates—and in two large, phylogenetically related protostomate groups—the annelids and arthropods. Each of these three great taxa has capitalized on metamerism in fundamentally different ways. In primitive vertebrates, metamerism is an adaptation for swimming; and in arthropods, segmentation made possible the development of a series of paired appendages that function in a variety of ways including locomotion, mastication, and sensation. In annelids, however, segmentation is an adaptation basically used in burrowing and crawling.

THE COELOM

A second immensely important feature of annelids is that they have a large, capacious, fluid-filled **coelom** or body cavity situated between the digestive tract and the body wall (Fig. 9.1). The reader will find a fuller treatment of the types of body cavities, including a coelom, in Chapter 3. While molluscs do have a small reduced coelom, annelids are the first group discussed in which the coelom is present as a large conspicuous chamber. In annelids and in all coelomate protostomates the coelom originates as a split within the paired mesodermal somites in each segment and is termed a **schizocoelom.** After its ontogeny the coelom is lined on the outside by the **parietal peritoneum,** the innermost layer of the body wall. The innermost coelomic lining or the outermost layer of the alimentary canal wall is the **visceral peritoneum** (Fig. 9.1). Since the coelom is lined both externally and internally by peritoneal layers, it is also called a **peritoneal cavity.** The two peritoneal layers become appressed to each other in the dorsal and ventral midlines to form the **dorsal and ventral mesenteries** (Fig. 9.1). These longitudinal mesenteries, suspending the alimentary canal between them, thus form right and left coelomic spaces. Similarly, the anterior peritoneum of one segment becomes appressed to the posterior peritoneum of the segment immediately anterior to it to form a **septum,** which divides the schizocoelom transversely into a pair of metameric compartments (Fig. 9.1). Thus an annelid with 50 segments would have 100 coelomic compartments. However, many annelids do not exhibit such perfect compartion, since the septa and mesenteries are often perforated or have even partially or completely disappeared.

MOBILITY

Metamerism tends to affect most systems and functions of an annelid, and the coelom is likewise intimately involved in several different activities. But both metamerism and the coelomate condition have the most dramatic impact on one important annelidan activity—mobility or locomotion. Mobility in these worms is a very fundamental activity and involves burrowing, crawling, walking, and even swimming. We shall explore first those annelidan features that contribute directly to this preeminently important activity and then mobility itself.

The body of an annelid is constructed of a small anterior acron, a segmented trunk, and a small posterior pygidium (Fig. 9.2a). The unsegmented **acron** or **prostomium,** technically the head, usually overhangs the mouth and forms an upper lip. In most nonburrowers the

Figure 9.2 *Some external features of annelids:* ***(a)*** *photograph of an earthworm (oligochaete);* ***(b)*** *scanning electron micrograph of the anterior end of* Nais communis *(oligochaete) showing prostomium, mouth, and chaetae (courtesy of T.L. Boullion);* ***(c)*** *some variations in annelidan chaetae.*

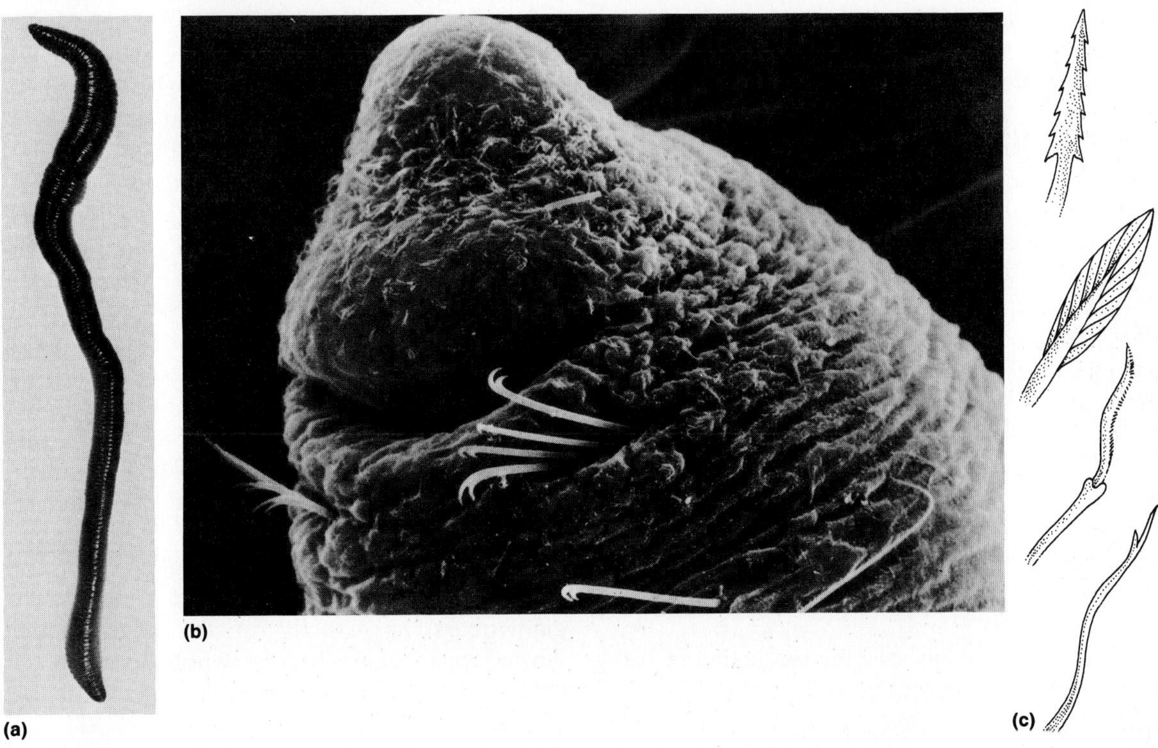

prostomium bears a number of sensory structures including tentacles and eyes. Behind the prostomium is an elongate metamerically arranged **trunk** containing usually from ten to several hundred segments. The first segment, the **peristomium,** surrounds the mouth ventrally and laterally (Fig. 9.2b). The peristomium and several additional anterior segments are often fused with the acron to form a **compound** or **secondary head.** In some annelids, groups of similar segments are associated together functionally to form body regions or **tagmata;** however, tagmatization in annelids is not as pronounced as it is in arthropods. At the posterior end is the unsegmented **pygidium** bearing the anus; new segments are formed immediately anterior to the pygidium.

The body wall of an annelid consists of several different layers, two of which are directly involved in locomotion (Fig. 9.1). The outermost living layer, an **epidermis** composed of columnar epithelium, is responsible for the production of a thin, outer, delicate, nonliving **cuticle.** In terrestrial species, epidermal glands produce mucus, an important material that enables the body wall to remain moist so as to facilitate gaseous exchange across the body surface. The numerous sensory structures and cilia (in some) are also derived from epidermal cells. The epidermal cells rest on a connective tissue **basement membrane** inside of which is a thin layer of **circular muscles** surrounding a much thicker layer of **longitudinal muscles** (Fig. 9.1). The innermost layer of the body wall is a thin **parietal peritoneum.** In earthworms the longitudinal layer is the same thickness throughout, but in most polychaetes these muscles are present as four bundles running the length of the worm. Oblique muscles, originating in the midventral line and extending laterally to the body wall, may also be present. It is the circular and longitudinal muscles that are the driving force for locomotion.

Each segment of annelids (exclusive of hirudineans) bears several to many spinelike, chitinous **chaetae** or **setae,** each of which is secreted by a cell in an epidermal follicle. Composed of protein and chitin, hardened by a chemical process called sclerotization, and readily regenerated, chaetae are present in a variety of sizes and shapes (Fig. 9.2b, c). Setae are directly employed in locomotion, providing temporary **traction points** against the substratum. As we shall see, these traction points are of tremendous importance in locomotion or burrowing in most annelids, and they represent a substantial improvement in the mechanical efficiency of locomotion compared to that of acoelomates or aschelminths.

In polychaetes, each segment bears a pair of flattened lobed structures, the parapodia (see Fig. 9.5). **Parapodia,** each typically bearing two bundles of setae, are important locomotive organs. Specific parapodial details will appear later in the section on polychaetes, but mention is made of them here because they are intimately involved in locomotion.

Some introductory comments should also be made at this point regarding the nervous system, since nerve impulses are required for both layers of body wall muscles to contract. There is a ventral nerve cord that bears a ganglion in each segment. A great deal of autonomy characterizes the annelidan nervous system, since even a single ganglion and its innervations to muscles in that segment can often function independently of adjacent segments. Often, nerve fibers run through several ganglia and tie these metameres together neurologically. An excised portion of a dozen or more segments can generate a coordinated pattern of activity similar to that of an entire animal. Therefore coordination of various annelidan activities such as swimming, crawling, or burrowing is not dependent upon the brain but rather upon these metameric ganglia in the ventral nerve cord. Present in most annelids are giant fibers that usually run the length of the worm; these fibers initiate whole worm reactions like those in escape. More details on the structure and function of the annelidan nervous system will be given later.

Annelids are soft bodied with two layers of body wall muscles—circular and longitudinal—surrounding the comparted, fluid-filled coelom. Such an arrangement of antagonistic muscles lying outside a fluid-filled cavity is referred to as a **hydrostatic skeleton** (Fig. 9.3a). Such a skeleton is also found in several other groups, including the aschelminths. Contraction of the circular muscles forces the coelomic fluid in both anterior and posterior directions, enabling the body to elongate. Contraction of the longitudinal muscles pull the protracted ends of the worm back and causes the coelomic fluid to be extended laterally as the body widens and shortens (Fig. 9.3a). Since the coelomic fluid is incompressible and thus its volume remains constant, the contractions of one set of muscles results in a stretching of the other, which is a stimulus for its contraction. Also important in locomotion is that when longitudinal muscles contract, the setae are fully protruded; and as the circular muscles contract, the setae are fully withdrawn.

But how is the locomotion of an annelid so much better than that of a wormlike aschelminth? The substantial increase in efficiency and effectiveness of the annelidan hydrostatic skeleton over that of a pseudocoelomate is that the body wall muscles are metameric and the coelom is comparted by septa (Fig. 9.3). In effect, the septa create a single hydrostatic unit in each metamere, an arrangement that not only improves the efficiency of these locomotive muscles but also makes possible local changes in shape that are independent of those of other metameres. For example, one segment might be elongated by its circular muscles, and an adjacent segment might be shortened concurrently by the contraction of its longitudinal muscles. Segments in which the circular muscles are contracted are at their minimal diameter and elongated maximally, their chaetae are retracted, and they are not in contact with the substratum. Where the longitudinal muscles are fully contracted, the segments are at their largest diameter and shortened maximally, their chaetae are fully extended, and they are in contact with the substratum (Fig. 9.3). Such modifications in the shape of individual metameres are of immense importance to annelids in locomotion and burrowing. Metamerism enables annelids to move faster, burrow more efficiently, and respond in a far more sophisticated way to environmental conditions than is possible in any other worm group. Metamerism and the coelomate condition plus their related features are the two most important adaptations of annelids, and they are intimately involved in every facet of annelidan movement.

Burrowing and crawling. Most oligochaetes and a great many polychaetes burrow into soft substrata in essentially the same fashion. The anterior end of the worm, often in the form of a truncated cone, initially penetrates the substratum and pushes material aside. Once initiated, burrowing is achieved by alternately employing two different types of anchors. A **terminal anchor** is created by the contraction of the anterior longitudinal muscles, which distends the anterior end and forces the worm's body radially against the walls of the new shallow burrow (Fig. 9.4a). This terminal distension firmly anchors the anterior end and allows the worm to pull its posterior end toward the burrow by contracting the longitudinal muscles. A **penetration** or **flange anchor** is then formed by the dilation of a group of segments toward the anterior end. These segments and their setae are used as flanges that serve as temporary traction points against which the contraction of the more anterior circular muscles forces the head forward and further into the substratum (Fig. 9.4b). The worm then crawls forward, forms a terminal anchor, pulls the posterior end farther into the burrow, and forms another penetration anchor to repeat the cycle (Fig. 9.4c, d). Many annelids are quite facile in burrowing; *Arenicola,*

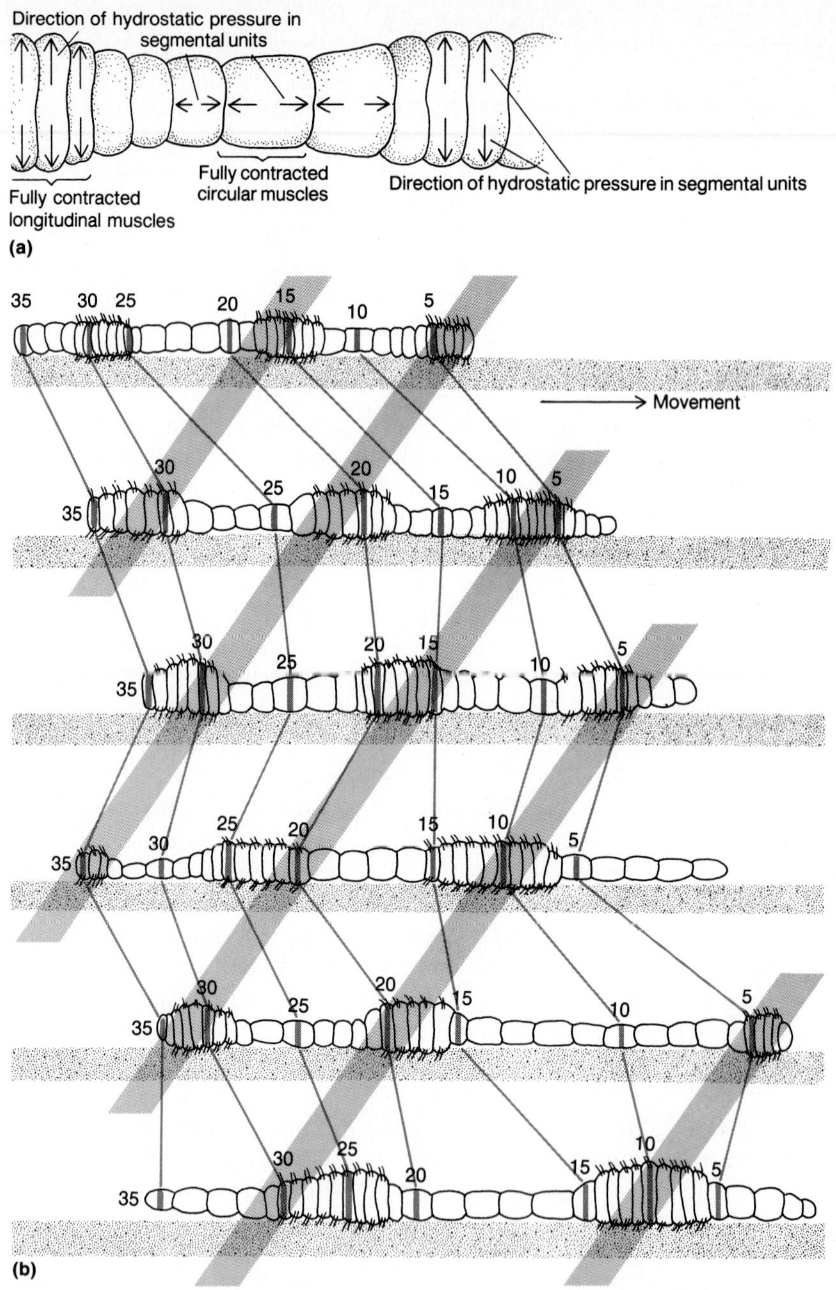

Figure 9.3 *The annelidan hydrostatic skeleton and how it is employed in crawling:* ***(a)*** *diagram of a portion of an annelid in which in one region (center) the circular muscles in each segment are contracting and forcing the hydrostatic skeletal fluid toward either end of the worm, thus elongating these segments; in other regions (right and left) the longitudinal muscles in each segment are contracting and forcing the hydrostatic fluid peripherally, thus shortening and thickening these segments;* ***(b)*** *sequential stages in crawling in an oligochaete (each solid line connects the same segment; each shaded bar connects the temporary regions of contact between the worm and substratum) (adapted from J. Gray and H.W. Lissman, 1938,* J. Exper. Biol. *15:507).*

a large marine polychaete, can burrow more than a meter into soft sand in less than 30 seconds! Burrowing forms have several adaptations to facilitate burrowing, including the absence of cephalic eyes and tentacles, an extensible proboscis in some polychaetes, and a reduced prostomium.

Many oligochaetes and polychaetes crawl, an activity associated with horizontal movement over a surface. Waves of muscular contraction either pass in the same direction as that of the worm's locomotion and are called **direct waves** or pass along the length of the worm in a direction opposite that of the worm and are called **retrograde waves.** Burrowers use one or both types of waves in penetrating the substratum, but waves are best understood in crawling. For a worm like *Lumbricus* that is crawling with direct muscular waves, the longitudinal muscles in a given segment contract and shorten the segment, and chaetae are extended to make temporary traction points against the substratum. Then the segment immediately anterior to the first repeats this process. A number of contiguous segments are in contact with the substratum to form a **region of contact;** several such contact regions are present simultaneously in a given worm (Fig. 9.3b). Direct muscular waves mean that segments are added to the posterior of each contact region and removed

Figure 9.4 *Sequential stages in the burrowing of annelids:* ***(a)*** *the formation of an initial terminal anchor;* ***(b)*** *the formation of the first penetration anchor;* ***(c)*** *the formation of a second terminal anchor;* ***(d)*** *the formation of a second penetration anchor (adapted from E.R. Trueman, 1956,* Biol. Bull. 131:372).

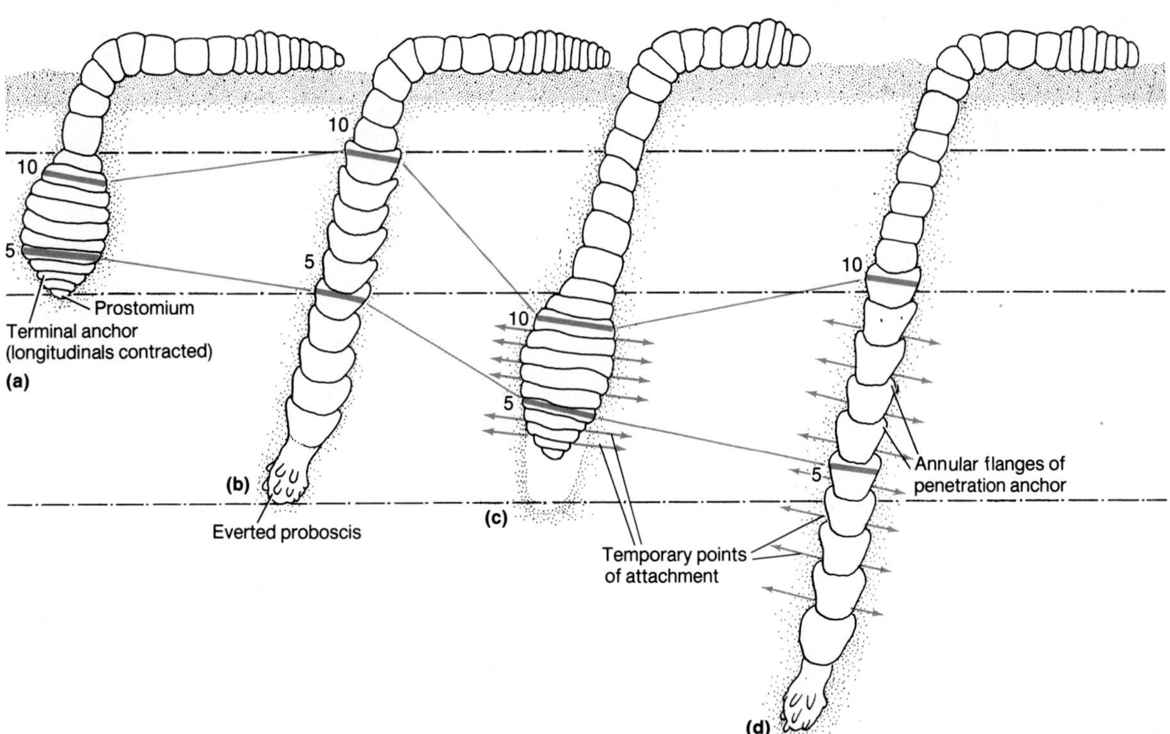

anteriorly. As a segment is removed from the contact region, its circular muscles contract and extend that segment; this action withdraws the setae and lifts that segment slightly off of the surface until it is added to the next contact region situated more anteriorly (Fig. 9.3b). However, the contact regions progress posteriorly in a retrograde wave relative to the substratum; this is reflected as bulges of a worm that move backward during forward motion (Fig. 9.3b). The body moves forward more rapidly than the backward progression of the contact regions, so there is a net forward movement. Most oligochaetes can crawl backward by simply reversing the above process.

Polychaete walking and swimming. In polychaetes there are three basic locomotive patterns unique to them: slow walking, rapid crawling, and swimming; combinations of these are encountered frequently even in the same worm. In **slow walking** or parapodial stepping, a worm like *Nereis* uses the parapodia as a series of levers, and the longitudinal muscle bundles and coelomic fluid are of little importance (Fig. 9.5b). Movement of the parapodia is due to extrinsic oblique muscles and intrinsic parapodial muscles. The effective or propulsive stroke is always backward with the parapodia and chaetae pressed against the surface and lifting the body slightly off of the surface. In the recovery stroke, the parapodium is lifted from the surface, brought forward, and put down a short distance ahead of its previous point of contact. In *Nereis,* every fifth or sixth parapodium on a given side is precisely at the same stage in the cycle of forward recovery and backward effective strokes; and in *Aphrodita,* every tenth or eleventh parapodium is at the same point in the cycle. These metachronal parapodial waves are direct waves. Numbers of waves at any one time are dependent upon the length of the worm. Further, a parapodium on one side of the worm is precisely out of phase with its opposite on the other side.

Rapid crawling involves parapodial movement coordinated with regional alternate contractions of the longitudinal muscle bundles so that the body is thrown into undulating lateral waves (Fig. 9.5c). Rather than the alternate contractions of circular and longitudinal muscles, in rapid crawling or swimming the contractions of powerful longitudinal bands on one side alternate with those on the other side. Therefore right and left longitudinal muscle bundles are antagonistic to each other (Fig. 9.5a). As the longitudinal muscles in one region on one side contract, the parapodia on that side of that re-

*Figure 9.5 Locomotion in polychaetes: **(a)** diagram of a portion of the body of a polychaete showing the relative degrees of contraction of the longitudinal muscles, strokes of the parapodia, and positions of the setae as they form traction points against the substratum; **(b)** slow "walking"; **(c)** rapid crawling; **(d)** swimming.*

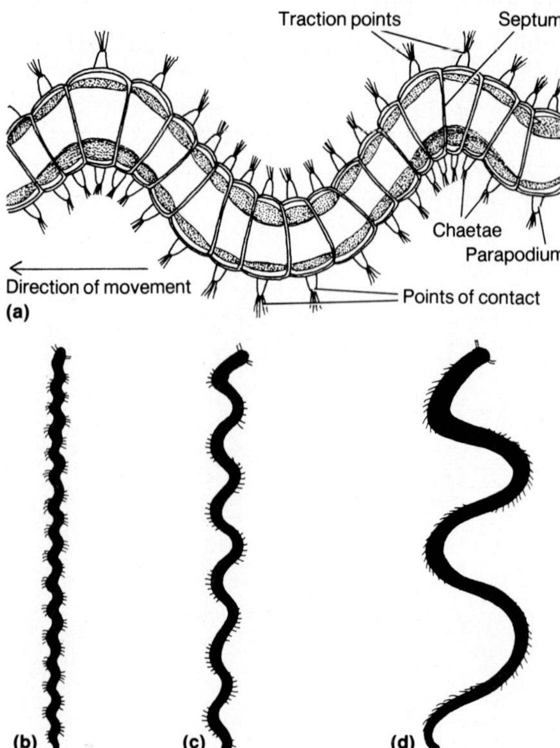

gion are in their effective phase; as those same longitudinal muscles relax, the parapodia are in their recovery stroke. The propulsive or driving force is generated by the longitudinal muscles forcing the body ahead from traction points made by the parapodia and chaetae (Fig. 9.5a). The interval between segments in the same stage of contraction is from 11 to 15 in *Nereis*. As in parapodial stepping, these metachronal waves move anteriorly.

Swimming, a form of locomotion found in a number of species, is essentially like rapid crawling, but the direct lateral waves of muscular contraction are fewer, and their amplitudes are greater (Fig. 9.5d). Swimming, dependent upon the paddlelike power strokes of the parapodia, is a rather inefficient form of locomotion for most polychaetes.

Hirudinean looping. From a mechanical standpoint, leeches have a comparatively simple locomotive pattern. Septa are missing, and the coelom is represented by a series of interconnecting channels. Therefore a leech is, in reality, a noncomparted sac filled with connective tissue, but its hydrostatic skeleton is affected, as in all other annelids, by antagonistic body wall muscles. Leeches are unique among annelids in possessing two muscular suckers: an **oral** or **anterior sucker** surrounding the mouth and a **caudal** or **posterior sucker** at the posterior end. Hirudineans use both suckers in their locomotion over a surface and in feeding; and both serve, in lieu of chaetae, as temporary traction points. While leeches can swim for a short period of time by producing dorsoventral, undulatory, retrograde waves, all hirudineans move mostly by a process called **looping,** in which both suckers attach alternately to a surface. If the posterior sucker is attached and the anterior sucker is free, the circular muscles then contract, causing the worm to elongate anteriorly (Fig. 9.6). The oral sucker then attaches to the surface, the caudal sucker detaches, the longitudinal muscles contract, and the worm becomes shorter as the posterior end is drawn near to the anterior sucker; the posterior sucker reattaches, and the process is repeated (Fig. 9.6). Apparently, the contraction of the longitudinal muscles normally takes place only when the posterior sucker is unattached, and the circular muscles can contract only when the anterior sucker is unattached. In a healthy leech that is looping, there is an obvious, predictable, and elegant rhythm in the alternation of attachment of both suckers and in the alternate contractions of circular and longitudinal muscles.

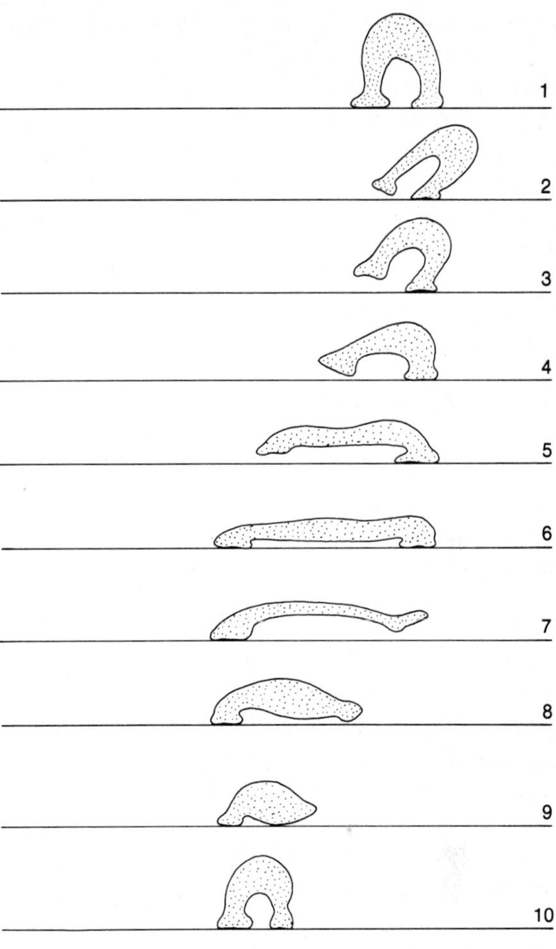

Figure 9.6 *Sequential stages in looping in leeches, in which both suckers serve as traction points.*

This unswerving rhythm, obviously mediated by nerves, does not originate in the central nervous system. Rather, these are reflexive actions initiated by tactile cells in the suckers and proprioceptors (stretch receptors) in the muscle layers. This coordinated rhythm between muscles and suckers is an extremely important dimension of leech locomotion, and the beauty in rhythm of a leech looping is something everyone should experience.

Internal Features

Digestion and circulation

Annelids exhibit considerable diversity in the types of food ingested; predators, herbivores, omnivores, parasites, and scavengers are all present among annelids. They also obtain their food in equally diverse methods including raptorial, filter, and deposit feeding. Many annelids have various accessory structures like jaws, tentacles, and suckers to aid in food acquisition.

The alimentary canal is typically a straight tube running from the anterior **mouth** to the posterior **anus.** Regional specializations are usually present and include a buccal cavity or pharynx, esophagus, stomach, intestine, and rectum. One or more of these digestive parts may be absent or not clearly differentiated in a given species. The wall of the alimentary canal contains both circular and longitudinal muscles that are responsible for the muscular waves of contraction or peristalsis that move food materials posteriorly through the canal.

The annelidan circulatory system typically is a closed one with blood completely contained within vessels. Above the alimentary canal there is a long **dorsal vessel** running the length of the worm, the wall of which is contractile and pumps blood anteriorly. Near the anterior end, the dorsal vessel is connected to a longitudinal **ventral vessel** by one to several pairs of circumenteric commissural vessels ("hearts" in some) or by a network of smaller vessels passing laterally around the gut. The ventral vessel carries blood posteriorly, and in each segment, paired vessels arise from it to supply the body wall, excretory organs, and digestive tract. After blood passes through capillaries, pairs of lateral and intestinal vessels in each segment serve as tributaries for the dorsal vessel.

This basic plan has been modified in various annelidan groups. In some a sinus replaces much of the dorsal vessel, while in others, most blood vessels are contractile. In some polychaetes, accessory hearts are present at various locations, and some worms even have contractile vessels in which blood simply is pushed back and forth rather than moving in a circular flow path. In most species of leeches the closed circulatory system has been replaced by an open system of coelomic sinuses and channels that function as an internal transport system. The blood of annelids generally contains a respiratory pigment dissolved in the plasma; **erythrocruorin** (a hemoglobin), **chlorocruorin,** and **hemerythrin** are found in annelids, and all combine readily with oxygen. In most annelids the only cellular components of blood are colorless phagocytic amebocytes.

Excretion

In coelomate invertebrates there are two different types of tubular organs connecting the coelom and the exterior. **Coelomoducts** are mesodermal in origin and grow outward from the coelom through the body wall. Coelomoducts function primitively in transport of gametes to the exterior from the coelom, into which they are shed by the gonads. **Nephridia,** on the other hand, are ectodermal in origin and grow from the exterior inward to the coelom. Nephridia function primarily in excretion by eliminating nitrogenous wastes and by regulating ionic and osmotic balances.

In most annelids a pair of nephridia is situated in each segment. Nitrogenous wastes (ammonia, urea, uric acid) are shed into the coelom, and the nephridia are primarily concerned with filtering the coelomic fluid, concentrating the wastes, and eliminating them to the outside. Each nephridial pair filters or absorbs wastes from the coelomic spaces in the segment anterior to the segment where wastes are finally discharged. Thus a nephridium consists of a preseptal and a postseptal portion (Fig. 9.7b).

Two types of nephridia are encountered in annelids—protonephridia and metanephridia. **Protonephridia,** found in certain polychaete groups, are more primitive, and each has a blind, multibranched, preseptal or inner end (Fig. 9.7a). At the apex of each branch are groups of specialized cells called **solenocytes,** which are very much like and probably homologous to flame cells of acoelomates. This homology is an important link in relating the evolution of annelids from an acoelomate ancestor. Each solenocyte is perched atop a short tubule that houses a single flagellum (Fig. 9.7a). The solenocytes, bathed in coelomic fluid, absorb fluids and wastes that are driven into the protonephridial lumen by action of the solenocytic flagella. The lumen is ciliated, and fluids are swept down the protonephridial tubule until it opens to the outside at the **nephridiopore.**

Metanephridia are present in most polychaetes and all clitellates and represent a more advanced excretory organ. Each metanephridium has a preseptal end that is an open ciliated funnel, the **nephrostome,** and a postseptal portion that is usually highly coiled and empties wastes to the outside via a **nephridiopore** (Fig. 9.7b). Other tissues may have some excretory functions, the principal one being the **chlorogogen tissue.** This tissue, the site of many metabolic processes, is found on the outer wall of

Figure 9.7 Annelidan nephridia: (a) preseptal end of a protonephridium, and a magnified portion showing individual solenocytes; (b) a metanephridium (from T.I. Storer and R.L. Usinger, 1965, General Zoology, *4th ed., McGraw-Hill Book Co., used with permission).*

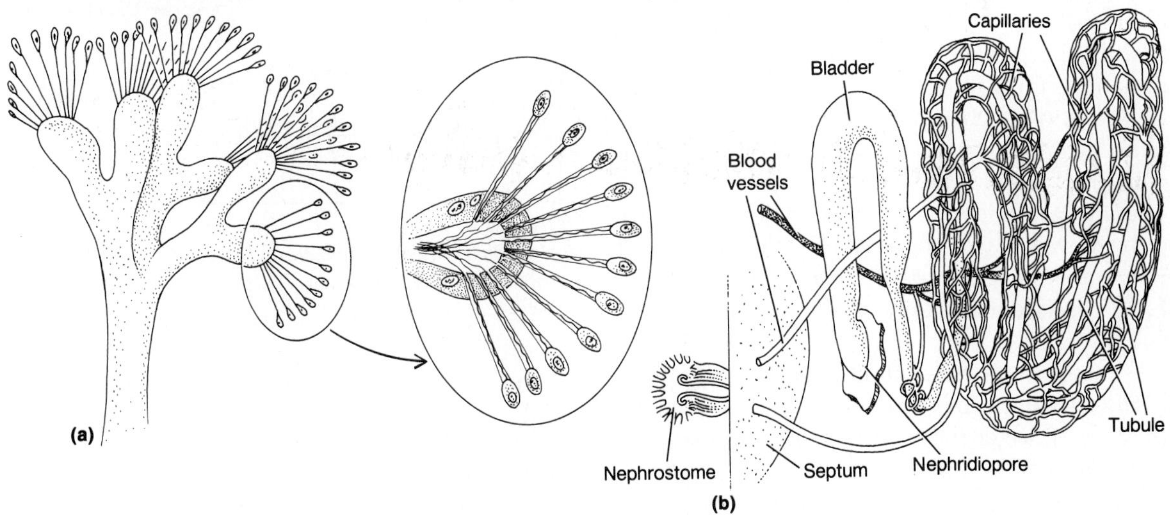

the intestine and produces excretory materials, which are dumped into the coelomic fluid.

Wastes enter the nephridia by either absorption (protonephridia), direct filtration (metanephridia), or direct secretion from the blood vessels. In direct secretion the coiled postseptal tubule is usually supplied with an extensive capillary system that can secrete wastes into the nephridium or absorb salts and water. Aquatic forms normally excrete urine composed of ammonia dissolved in water and isotonic with the environmental water, but some produce a hypotonic urine indicative of salt reabsorption. Terrestrial annelids (oligochaetes) excrete urea or uric acid, and there is considerable reabsorption of salts to produce a hyposmotic urine, an important dimension in their efforts to conserve water and minerals.

An interesting arrangement is present in many polychaetes in which there are various degrees of fusion between the nephridia and coelomoducts, both of which are present in most segments. This fusion often results in a compound organ functioning in both excretion and reproduction. A **protonephromixium** is a combination of a protonephridium and a coelomoduct into one tubule (Fig. 9.8a). A few worms have metanephromixia; a **metanephromixium** is a combined metanephridium and coelomoduct that has a single opening to the outside (Fig. 9.8b). Other worms have **mixonephria,** in which a metanephridium and a coelomoduct have fused completely into a single combined tubule that functions in both excretion and gametic transport (Fig. 9.8c). But many polychaetes have regional specializations so that in some segments these combined organs function as either nephridia or gonoducts but not both.

Excretory organs in leeches, considerably different from those in other annelids because of the absence of septa, a reduced and modified coelom, and the presence of extensive connective tissues, will be included later with a discussion of leeches.

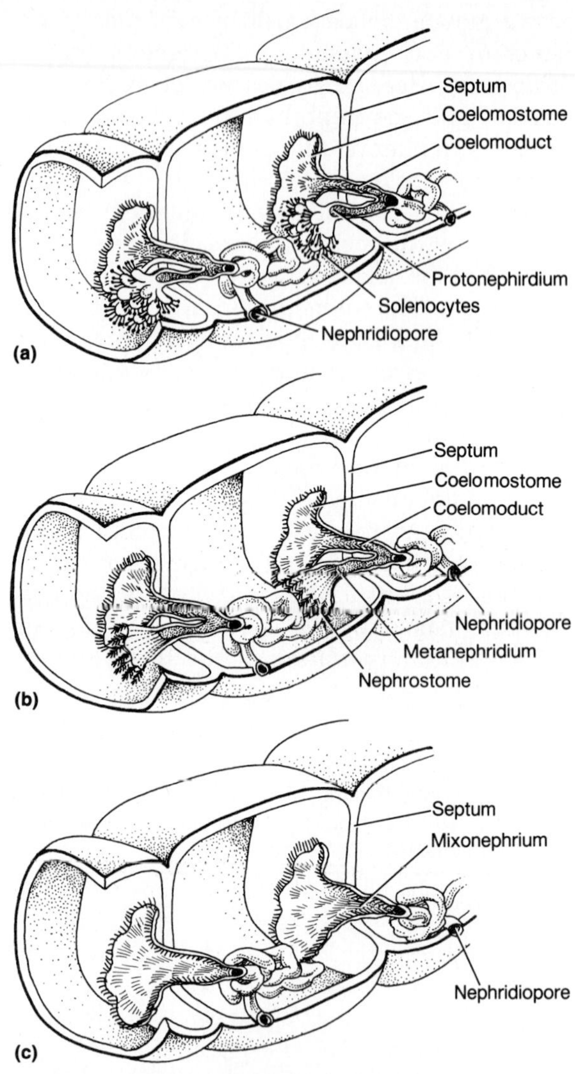

Figure 9.8 *Combinations of nephridia and coelomoducts in polychaetes:* ***(a)*** *protonephromixium* (Phyllodoce); ***(b)*** *metanephromixium* (Hessione); ***(c)*** *mixonephrium* (Arenicola).

Nervous system and sense organs

The central nervous system consists of an anterior brain, a pair of circumenteric connectives, and a ventral nerve cord. The bilobed **brain** is represented by a pair of fused ganglia located above the digestive tract and either in the acron or in one or several anterior metameres. Nerves arising from the brain innervate structures associated with the acron and mouth. **Circumenteric connectives** pass around either side of the pharynx or esophagus and connect brain and nerve cord at the **subenteric ganglia.** The **ventral nerve cord,** extending the length of the worm, is doubled in primitive forms, but in more advanced worms the two cords have fused medially into a single cord. Typically, the nerve cord bears a ganglion (two if the cords are not fused) in each segment. From each ganglionic mass arise three to five pairs of lateral mixed nerves, which innervate portions of each segment. The nervous systems of most annelids that have been carefully studied contain neurosecretory or **chromaffin cells** located in the brain and subenteric ganglion. Secretions of these cells contain several different chemicals that behave like hormones and apparently aid in the regulation of reproduction.

In the nerve cord of most annelids are one to many giant fibers. These fibers equip the entire worm to contract suddenly, and this is an important adaptive mechanism that enables worms to move away rapidly from potential predators. A fiber may be a single giant neuron or a syncytium of many neurons. Usually, there are more giant fibers conducting impulses posteriorly than anteriorly conducting ones. Such an arrangement enables the anterior end to be withdrawn rapidly. More details of this interesting neurological adaptation appear in Box 9.1.

Sense organs such as eyes, nuchal organs, and statocysts, especially prevalent in polychaetes in keeping with their more active lifestyle, are generally absent in other annelids. These specialized organs will be discussed in the treatment of polychaetes. The external surfaces of all annelids, especially near the anterior end, are supplied with tactile cells and chemoreceptors, which provide vital information to a mobile worm.

Reproduction and Development

Most polychaetes are dioecious, but all clitellates are monoecious, and a few also reproduce asexually. In polychaetes, distinct gonads are absent, and gametes are produced by swellings on the parietal peritoneum in a few, some, or most segments. Immature gametes are released into the coelom, where further maturation takes place. In clitellates, discrete gonads develop in only several anterior segments. Gametes in all annelids then pass to the outside through rupture of the body wall (polychaetes), through metanephridia (polychaetes), or by coelomoducts (clitellates). In some polychaetes there is a tendency for segments only in a certain region to produce gametes. Therefore the metanephridia in these fertile segments function only in gametic transport, and those in sterile segments function only in excretion.

Quite diverse patterns of fertilization are found in annelids. In hermaphroditic forms, two worms simply come together, swap sperm, and store the male gametes in seminal receptacles. In many polychaetes, swarming is an event in which worms of both sexes congregate so as to increase the likelihood of fertilization; strict diurnal or lunar periodicities usually dictate this absolutely fascinating activity. In other annelids, fertilization is more problematic as planktonic gametes are shed into the water near other worms. Some leeches have a penis with which sperm are transferred between copulating worms.

Following fertilization, some worms deposit fertilized eggs in masses or in cocoons; a few may brood the embryos internally. In most species, however, the developing embryos exist independently of the mother. Typically, annelid embryos demonstrate spiral cleavage. In most polychaetes a larval stage, the trochophore, is developed; more details of this young larval stage appear later in a discussion of polychaetes. Since there is no larval stage in any

BOX 9.1 □ THE REMARKABLE GIANT NEURONS OF ANNELIDS

One of the most important, and certainly one of the most intriguing, aspects of the annelidan nervous system is the presence of one or more giant neurons in the ventral nerve cord of polychaetes and oligochaetes. They are unusually large in diameter and often extend the entire length of the nerve cord. Since the speed of nervous conduction is directly correlated with the diameter of the nerve fiber, giant neurons are extremely rapid impulse conductors. For example, in earthworms, ordinary fibers 4 μm in diameter conduct at a rate of about 50 cm/sec; but giant fibers, up to about 50 μm in diameter, conduct impulses at speeds of up to 30 m/sec! Such a remarkably faster system enables a worm to react almost instantaneously to adverse stimuli. Further, giant neurons in many worms are syncytial and typically innervate longitudinal muscles in every segment so that the motor response elicited by the giant fiber system is an immediate and almost convulsive shortening of the entire body.

In earthworms an appropriate stimulation of the head generates an impulse in the median giant fiber that causes all the longitudinal muscles to contract; but since the setae of the tail region are protruded and form temporary anchors, the head is rapidly withdrawn. Appropriate stimulation of the tail region creates an impulse in the two lateral giant fibers and a protrusion of the setae in the head region, and the contraction of the longitudinal muscles results in a sudden withdrawal of the tail.

Giant fibers are perhaps most highly developed in tubicolous polychaetes, in which immediate retraction of the protruding anterior end is often a prerequisite for survival. The largest nerve fibers known are found in the sabellids, in which the several fibers are syncytial and interconnected; in *Sabella* the single giant neuron may be as large as 1.5 mm in diameter and conduct at incredible speeds of up to 600 m/sec. One interesting feature of the giant fiber system is that they can be rapidly habituated; after several fast responses in a short period of time, further stimulation is often simply ignored. The failure to respond is due to some temporary inhibition of neural transmission at the junction of neuron and muscles. An antagonistic action to the rapid withdrawal response often removes the inhibition and "cocks" the system for the next stimulus.

The next time you see an earthworm or polychaete, test its giant fiber system so that you can more fully appreciate this dramatic but effective escape mechanism; it is a most important adaptation for most annelids.

clitellate, the young hatch as miniature immature worms and increase in length by forming new segments immediately anterior to the pygidium.

Class Polychaeta

Polychaetes, among the most beautiful and interesting of all invertebrates, exhibit the greatest diversity in form and function among the annelids. They are abundant in marine habitats, particularly in intertidal and subtidal areas, to a depth of about 50 m, where both numerically and ecologically they are among the dominant animals. They range in length from 2 mm to that in *Eunice viridis,* the Pacific palolo worm, which may be up to 3 m. Polychaetes burrow in mud or sand or live in crevices in rocks and shells, in tubes they construct, or commensally with other animals. About 6000 species have been described, and densities of several thousand per square meter in mud flats have been reported. About 50 species are found that are restricted to freshwater habitats mostly near oceans, and others are found in brackish water, but the vast majority are marine. Most polychaetes are exquisitely colored with reds, greens, yellows, and blues predominating; some are strikingly iridescent, and others are luminescent. Polychaetes are a very important ecological group in marine littoral habitats, where they constitute a major link in many food chains. Some examples of polychaete diversity are presented in Fig. 9.9.

The subordinate classificational framework for polychaetes is currently in a state of flux, and there is no general agreement as to subclasses, orders, or suborders. The taxonomic scheme of Fauchald (1977) is a recent and comprehensive plan, and brief characterizations of the orders appear at the end of this chapter. An ecological classification divides them into two groups: Errantia and Sedentaria. The Errantia includes the free-moving active burrowers and pelagic forms; it also includes some crawlers and tube-dwellers that leave their tubes or crevices to feed and reproduce. The Sedentaria are permanent tube-dwellers or burrowers and usually do not expose much more than their anterior ends to the environment. Some biologists consider these two groups to be subclasses, but the basis for this distinction is more ecological than phylogenetic. Even though these two terms are not valid taxonomically, they are very useful designations with which one can refer to a large group; we shall use them for the sake of convenience.

External features

The polychaete body consists principally of a trunk plus the prostomium and pygidium. The trunk is composed of a variable number of segments ranging from seven in *Dinophilus* to over 700 in *Eunice*. Numbers of segments are not constant even for members of the same species. In the least specialized polychaetes, all the segments are homononomous or essentially alike; but in many sedentary forms like the fanworms and featherdusters, some tagmatization is readily apparent. A **compound head** often results from the fusion of the acron and the first one or several segments. The **prostomium** is well developed and equipped with palps, tentacles, and various well-developed sensory structures including tactile cells, eyes, statocysts, and nuchal organs (Fig. 9.10a, b). The stoutly constructed ventrolateral **palps** and filamentous slender **tentacles,** often called antennae, are both major sensory structures. Two to four pairs of **eyes** are present in most errant polychaetes (Fig. 9.10a), but simple eyes are more common and include an inner layer of retinal cells containing a photosensitive pigment and refractive bodies including a cornea and lens. Such eyes, termed direct because the retinal cells face outward toward the source of light, function only in light detection, and most polychaetes are negatively phototactic. Some have eyespots or simple photoreceptor cells on the tentacular crown or along the lateral body wall, and other specialized polychaetes have highly developed image-forming eyes that are equipped with a mechanism for accommodation or focusing. **Statocysts,** present in some burrowers and tube-dwellers and often associated with a part of the nervous system called the circumenteric connectives, are minute bodies that assist the worm in orienting properly when burrowing. **Nuchal organs,** ciliated sensory pits containing chemoreceptors and situated between the prostomium and the first segment, are well developed in predaceous forms but are absent in filter-feeders.

The first adult segment, sometimes representing several fused embryonic segments, is the **peristomium,** which forms the lateral and ventral margins of the mouth and dorsally bears several different filamentous structures. In errant forms like *Nereis,* there are four pairs of

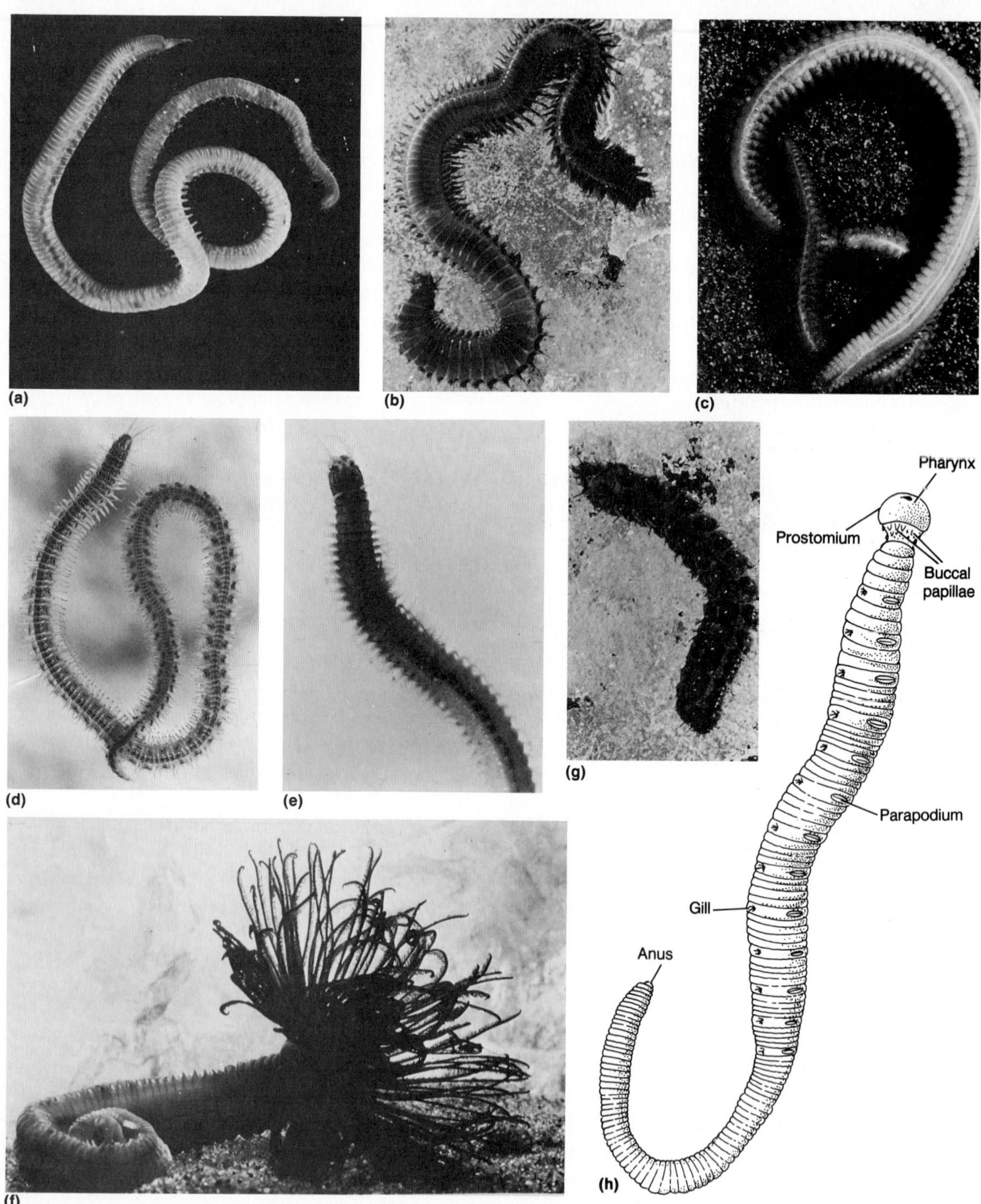
(a)
(b)
(c)
(d)
(e)
(g)
Pharynx
Prostomium
Buccal papillae
Parapodium
Gill
Anus
(f)
(h)

◀ *Figure 9.9 A collage of photographs (a–g) and a drawing (h) of various polychaetes:* ***(a)*** Glycera, *the beak thrower (phyllodocid);* ***(b)*** Nereis, *a sandworm (phyllodocid);* ***(c)*** Nephthys, *a sandworm (phyllodocid);* ***(d)*** Lepidothenia *(phyllodocid);* ***(e)*** *the anterior end of* Eunice rubra *(eunicid);* ***(f)*** Megalomma *(sabellid);* ***(g)*** Halosydna, *a scaleworm (phyllodocid);* ***(h)*** Arenicola, *the lugworm (capitellid). (parts a–c, f, g courtesy of Ward's Natural Science, Inc., Rochester, N.Y.; parts d, e courtesy of C.E. Jenner.)*

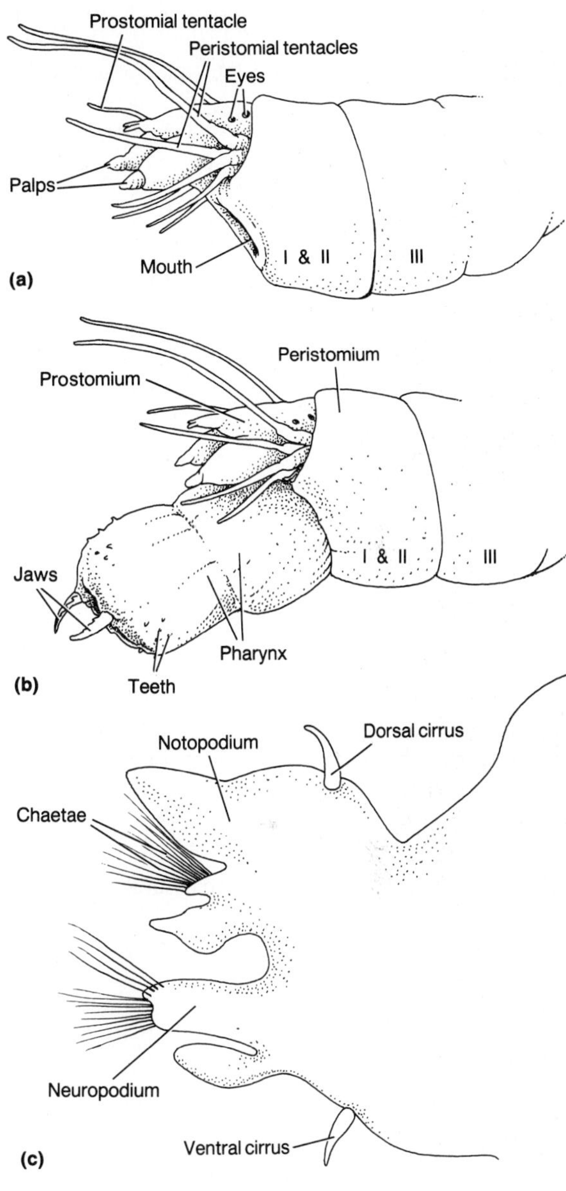

Figure 9.10 Some external features of polychaetes: ***(a)*** *the head of* Nereis *with proboscis retracted;* ***(b)*** *the head of* Nereis *with proboscis extended;* ***(c)*** *an anterior view of a parapodium from the right side. (parts a, b from T.I. Storer and R.L. Usinger, 1965,* General Zoology, *4th ed., McGraw-Hill Book Co., used with permission.)*

cirri (Fig. 9.10a, b). In many sedentary polychaetes there are numerous **tentacles,** and in others there are pinnate **radioles,** both of which are used in particle food gathering and in building permanent tubes or burrows. Cirri, tentacles, and radioles contain numerous tactile cells and chemoreceptors.

Probably the most distinguishing characteristic of polychaetes is that each segment, except for the peristomium, bears a pair of parapodia. A **parapodium** is a fleshy, flattened, lobed appendage projecting laterally from the body wall (Fig. 9.10c). Basically, parapodia are locomotive appendages used in crawling, burrowing, and swimming, but in sedentary polychaetes they are often modified for water circulation, food capture, and ventilation. Some parapodia are simple flaplike lobes; but in most polychaetes, each parapodium is a rather complex organ composed of a dorsal lobe, the **notopodium,** and a ventral lobe, the **neuropodium** (Fig. 9.10c). Often these two lobes are of different sizes and shapes, and sometimes one lobe is highly reduced or even absent. In many different polychaetes, additional lobes function as gills, which are conspicuous because they are filled with either green or red blood. A sensory **dorsal** and **ventral cirrus** are borne by the notopodium and neuropodium, respectively (Fig. 9.10c). Each of the two lobes is supported internally by one or more stiff chitinous rods or **acicula,** which provide mechanical support for both lobes. Acicula are modified chaetae, and they are usually stout and dark and have a pointed apex. At the distal ends of the two

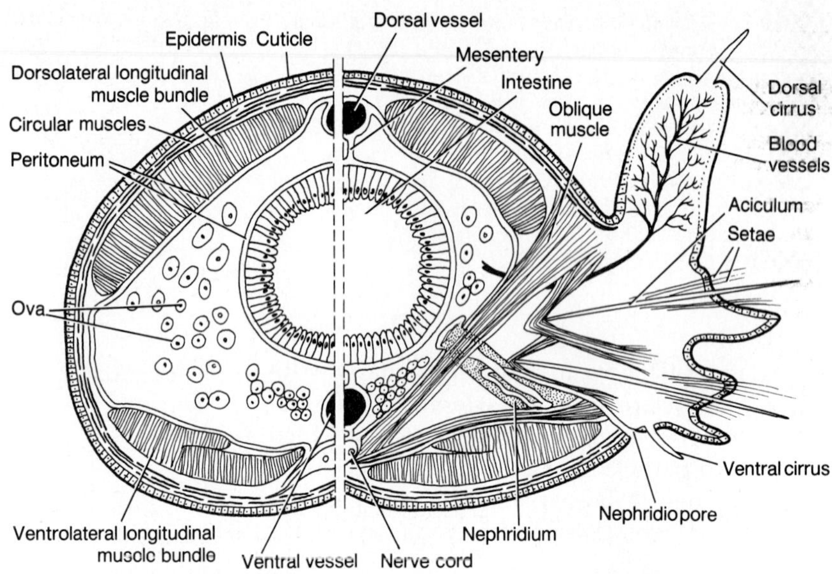

Figure 9.11 *A cross-sectional view of a polychaete (right half is through a parapodium) (from T.I. Storer and R.L. Usinger, 1965,* General Zoology, *4th ed., McGraw-Hill Book Co., adapted with permission).*

parapodial lobes are invaginated ectodermal pockets in which are located fanlike bundles of chaetae from which this taxon derives its name (poly = many; chaetae = bristles) (Fig. 9.10c). In some polychaetes like the fireworms the setae contain a poison, and when a seta is broken, the poison is released that may kill a small predator or sting a larger animal. Setae are absent in certain pelagic worms.

Body wall and coelom

The body wall layers present in polychaetes differ from those of clitellates, since the longitudinal muscles are concentrated into two pairs of **bundles;** one bundle is dorsolateral and the other ventrolateral on either side (Fig. 9.11). The dorsolateral pair is more massive, powerful, and important than the ventrolateral pair. That these muscles play a very important role in various forms of locomotion has already been elucidated. Oblique muscles, extending from the ventral midline and inserting into the bases of the parapodia, are important in parapodial movements (Fig. 9.11).

Medial to the body wall is the coelom, which is sometimes incompletely comparted by septa and mesenteries so that the coelomic fluid circulates more freely between compartments (Fig. 9.11). In most polychaetes that have metanephridia there are paired **dorsal ciliated organs** in each segment situated on the parietal peritoneum just inside the dorsolateral longitudinal muscle bundles. These organs, perhaps representing the ciliated funnels of coelomoducts, aid in circulation of coelomic fluid.

Feeding and digestion

The methods of feeding and types of food are quite varied among polychaetes. Many errant polychaetes are raptorial, preying upon a variety of small invertebrates by using an eversible pharynx or **proboscis** (Fig. 9.10b). Contraction of the body wall muscles exerts pressure on the coelomic fluid so that the pharynx is suddenly everted hydrostatically. The proboscis is equipped with one or several pairs of hardened chitinous **jaws** situated at the tip of the fully extended proboscis (Fig. 9.10b). Prey is seized by the jaws, and the proboscis is withdrawn by retractor muscles extending from pharynx to the body wall. The pharynx usually lies within the first five to eight body segments; but the beak thrower, *Glycera*, has a proboscis about 20% as long as the entire organism and housed in the anterior 20 segments (Figs. 9.9a, 9.12a). As well as in the carnivores, jaws are also present in many herbivores and scavengers and are used to tear off bits of food. Many other worms feed

Figure 9.12 *Organs of food acquisition in polychaetes;* ***(a)*** *the extended beak of* Glycera, *the beak thrower, with about 100 middle segments omitted (compare with Fig. 9.9a) (from G.E. and N. MacGinitie, 1968,* Natural History of Marine Animals, *2nd ed., McGraw-Hill Book Co., used by permission);* ***(b)*** *photograph of* Amphitrita, *a terebellid with tentacles retracted;* ***(c)*** *photograph of a sabellid (probably* Hydroides*) with ciliated radioles;* ***(d)*** *diagram of the anterior end of a sabellid showing pathways of water flow through the radioles.*

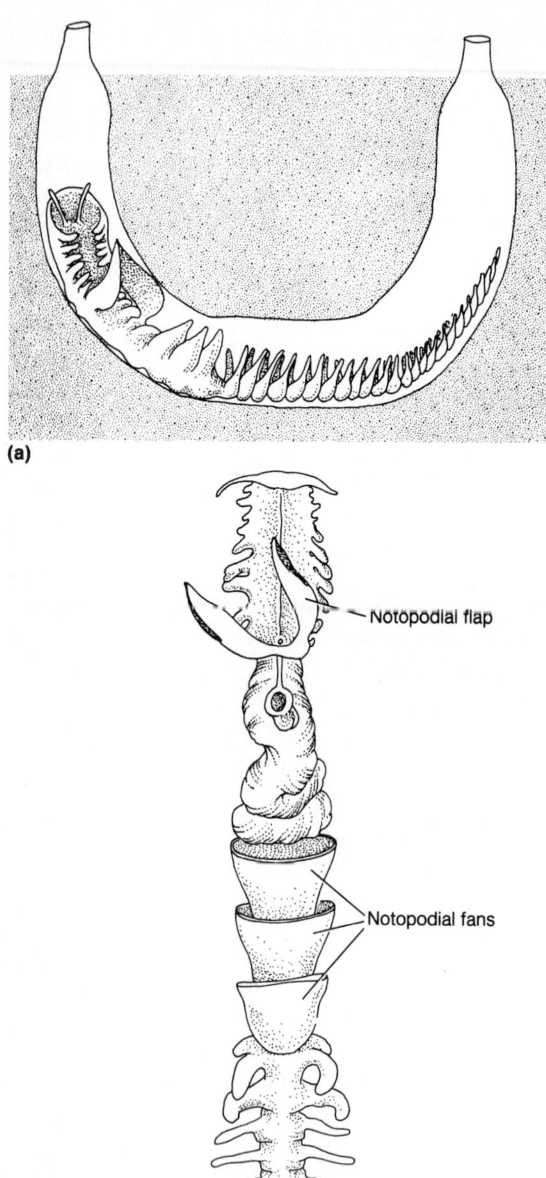

Figure 9.13 Chaetopterus, *the parchment worm (spionid): **(a)** an animal in its U-shaped tube; **(b)** the anterior end of the worm, dorsal view.*

by everting the pharynx to form a short proboscis that sucks up sand, mud, and detritus. Some, like lugworms and bambooworms, ingest copious amounts of mud and detritus with the proboscis as they actively burrow. The passive tube-dwellers merely extend the proboscis to ingest food particles deposited on a substratum. Jaws are usually absent in these forms, although a ring of small teeth is often present.

Most sedentary or tube-dwelling polychaetes, lacking a proboscis, are filter-feeders or ciliary feeders and use ciliated tentacles for food capture. Representatives of a number of taxa have tentacles that gather food as well as assist in the construction of permanent tubes or burrows (Fig. 9.12b). Some ciliary feeders have tentacles that contain a ciliated groove on one side, and the ciliary beat sweeps small particles toward the mouth. Larger particles are moved toward the mouth cooperatively by the cilia and by muscular contractions of the walls of the tentacular groove. Very large bits are brought to the mouth by retraction of the tentacles.

Another type of filter feeding is seen in the fanworms, which possess a large number of bipinnate **radioles** (Fig. 9.12c, d). The radioles form a funnel with the mouth located at its vertex. The radioles and their pinnules are heavily ciliated, and the beat of the cilia drives water centripetally into the funnel and out the top (Fig. 9.12d). Food particles are trapped by the cilia, passed to the base of the radioles, and sorted. The largest particles are rejected, intermediate-sized particles are used in tube construction, and the smallest particles are ingested.

Still another type of filter feeding is found in the unusual parchment worm, *Chaetopterus*, which lives in a U-shaped parchment tube buried in mud (Fig. 9.13a). It has an extremely modified body adapted for both feeding and its tubicolous habit. Segments 14–16 bear notopodial fans, each of which functions as a piston in the cylindrical tube (Fig. 9.13). The rhythmical beating of the fans creates a water current in one end of the tube, over the worm, and out the other end. Segment 12 has modified notopodia that secrete a mucous bag held between the two notopodial flaps (Fig. 9.13). Incurrent water is circulated through the mucous bag, which strains out detritus, food particles, and plankton as small as 1 μm. Periodically, the food bag is removed from the twelfth notopodia, passed to a middorsal groove, carried by ciliary action to the mouth, and ingested.

Those polychaetes that ingest rather large volumes of undigestible sand and mud must therefore produce rather extensive castings. Worms living in tubes with two openings have wastes and castings flushed out with the excurrent water flow. Others that live buried in the mud must periodically bring the pygidium near to the mud–water interface to defecate. Some worms live headdown in the mud, and to defecate they simply back up to the surface of the substrate. Others apparently turn around in the tubes to bring the posterior end near to the mud surface to void the castings.

CIRCULATION, BLOOD, AND GASEOUS EXCHANGE

The circulatory system of polychaetes is well developed and typically annelidan. The principal vessels are a contractile dorsal and ventral vessel, and bulbous contractile swellings or "hearts" are located on the dorsal vessel or on lateral vessels. Variations on this circulatory plan include "hearts" that cause the blood to ebb and flow within a single vessel, the closed system being replaced by one of open sinuses, and the blood–vascular system being lost, the coelomic fluid becoming the only means of transport.

Polychaete blood is usually either red or green owing to the presence of an oxygen-transporting, iron-containing pigment dissolved in the plasma. Occasionally, there are pigments contained within small corpuscles, but these pigments are always of considerably smaller molecular size. Erythrocruorin, a type of red hemoglobin, is frequently encountered among the errant polychaetes, while a similarly constructed green pigment, chlorocruorin, is found in many sedentary forms. Hemerythrin, another red pigment, is encountered only rarely. Other worms have colorless blood, no pigment being present, and a few worms have both hemoglobin and chlorocruorin.

In primitive polychaetes, gaseous exchange takes place over the general body surface, a mode that is particularly effective in small worms or those that are long and slender. In a large number of species and especially in the larger worms, however, gills have developed for gaseous exchange. Most often, gills are outgrowths from the parapodia and particularly the notopodium. The dorsal cirrus is frequently modified as a gill in the form of a cone or lobe or as branched filaments. Gills may also be found along the dorsal surface or at the anterior end as in *Amphitrita* (Fig. 9.12b). In scaleworms the dorsal surface of most segments forms a vascularized gas exchange surface, and dorsal overlapping plates or elytra cover this area like a roof. Ventilating currents of water pass between these dorsal protective layers and the exchange surfaces of the dorsal body wall. Fanworms likewise lack gills, but they use the radioles as gas exchange organs. Many tube-dwellers and burrowers, lacking special gas exchange organs, irrigate oxygenated water through their tubes and burrows, and gases are exchanged across the general body surface. Water is then dispelled from the tube into the surrounding mud or sand or out the end of the tube.

NERVOUS SYSTEM

The nervous system in polychaetes is typically annelidan with a dorsal bilobed **cerebral ganglion** located in the prostomium. In some errant

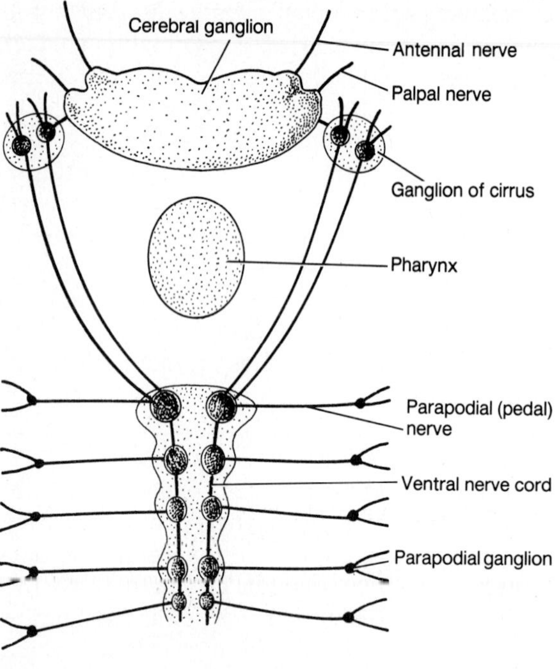

Figure 9.14 *The brain and anterior ventral nerve cord of a polychaete.*

worms, forebrain, midbrain, and hindbrain regions can be identified, each innervating different prostomial sense organs (palps, tentacles, eyes), which are particularly useful in their predatory habits. The brain is connected to the ventral nerve cord at the **subesophageal ganglion** by a pair of lateral **circumesophageal connectives** (Fig. 9.14). Occasionally, the connectives are doubled, they often contain ganglia, and the close association between them and statocysts has already been noted. Primitively, the **ventral nerve cord** is doubled with each cord bearing a ganglion in each segment. In most polychaetes, however, the two cords and their segmental ganglia have fused medially. Each ganglion gives off two to five pairs of lateral mixed nerves, which supply the body wall, parapodia, digestive tract, and nephridia (Fig. 9.14). The ventral cord may be either embedded in the epidermis or muscle layers or situated in the coelom. Characteristically, the nerve cord contains three giant neurons, but some worms contain as few as one, and others have many. These giant neurons are important in rapid retraction of the anterior end (see Box 9.1).

Reproduction and Development

Some polychaetes reproduce asexually by budding or by repeated transverse fissions of the body. Accidental fragmentation can result in many new individuals, since polychaetes have considerable regenerative powers. A few other worms are monoecious, and some reproduce parthenogenetically. But the vast majority of all polychaetes reproduce sexually and are dioecious. Distinct gonads are absent, but germinal patches on various locations of the parietal peritoneum produce gametes. Typically, each segment is fertile; but in those worms in which tagmatization is present, gamete-producing segments are usually restricted to the posterior region; for example, in the lugworm, *Arenicola,* there are only six fertile segments. Immature gametes are shed into the coelom, mature in the coelomic fluids, and are often nourished by special **nurse cells.** Most worms produce a prodigious number of gametes that completely fill the coelom and make the gravid condition of the worm readily apparent. Often, the posterior end of a gravid worm takes on the color of the gametes inside.

Gametes sometimes escape to the outside through gonoducts or by rupture of the posterior body wall, but the most common means is for them to be discharged to the outside via metanephridia. Metanephridia in some worms function both in gamete transmission and in excretion; in others, metanephridia in fertile segments serve only to transmit gametes. The body wall ruptures in several groups of worms pelagic at the time of reproduction. While fertilization usually takes place in the sea, in some worms it is within the tube or burrow of the female. Many tube-dwelling polychaetes brood egg masses within their tubes, under the elytra, in external mucous bags or strings, or beneath the parapodia. A few species retain the eggs and developing embryos internally and are ovoviviparous, and one instance of viviparity has been reported.

The telolecithal egg undergoes spiral determinate cleavage to produce a stereoblastula followed shortly by the development of **trochophore** larva (Fig. 9.15a). A similar, planktonic, pelagic, larval stage is found in certain other marine protostomates, including most marine gastropods and bivalves. The fact that it is developed in both marine annelids and molluscs is strong evidence that both of these large groups shared a common remote ancestry. The trochophore stage is characterized by a ring of cilia, the **prototroch,** situated just above the larval equator and mouth. Later, two additional rings of cilia appear with the **telotroch** surrounding the anal opening and the **metatroch** located below the mouth. An apical tuft of cilia usually appears marking the location of the **apical organ,** which is the forerunner of the adult prostomium (Fig. 9.15b). A trochophore can be divided into three regions: prototroch, growth zone, and pygidium. The **prototrochal region** includes the mouth area, prototroch, apical organ, and all structures dorsal (apical) to the prototroch. The **pygidium** consists of the telotroch and the region around the anus. The **growth zone** is that area between the other two. At the same time as the planktonic larva metamorphoses into a benthic adult, the growth zone is very active as new segments develop immediately anterior to the pygidium (Fig. 9.15b). In metamorphosis, many internal and external changes can be noted as the trunk is gradually

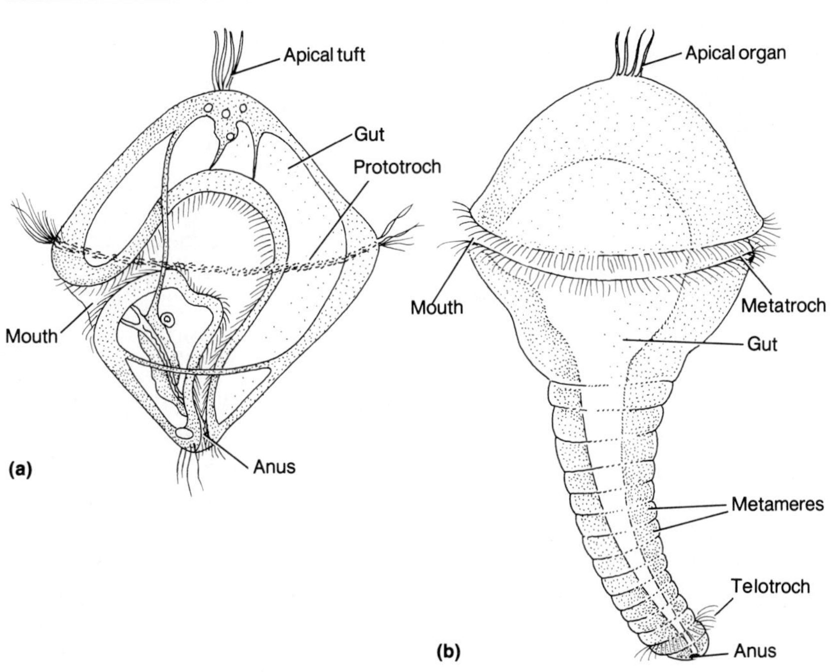

Figure 9.15 *Two larval stages present in polychaetes:* ***(a)*** *an early trochophore (from Shearer in L.H. Hyman,* The Invertebrates. *Vol. II:* Platyhelminthes and Rhynchocoela, *1951, McGraw-Hill Book Co., adapted with permission);* ***(b)*** *a later trochophore with metamerism beginning.*

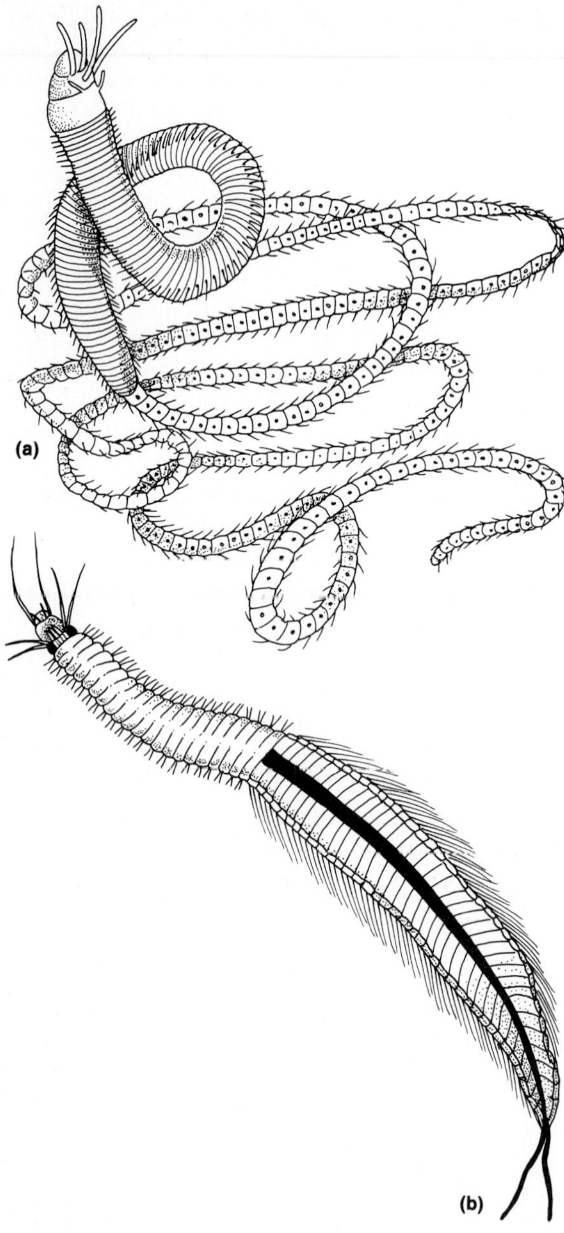

Figure 9.16 *Epitoky in polychaetes:* ***(a)*** Eunice viridis, *the Pacific palolo worm, which has a dramatic swarming of epitokes during the autumn;* ***(b)*** *an epitoke of a nereid.*

lengthened finally to produce a mature worm. Some polychaetes do not have a separate trochophore stage, and in these forms a later larval stage or a miniature worm is the hatching stage. In other worms the trochophore, rather than being planktonic, is a benthic larva.

Epitoky, swarming, and regeneration. At sexual maturity, some worms undergo radical external changes such as alteration of body texture and color and changes in the morphology of parapodia, chaetae, and head. The body of a sexually mature worm appears to be clearly divisible into two regions: the **epitoke,** which is usually posterior, is distended with gametes, and has greatly enlarged parapodial lobes with long swimming setae, and the **atoke,** which is usually anterior and is not a gamete-bearing region. At sexual maturity the parent worm actually divides into two separate individuals. A reproductive individual or epitoke then assumes many other morphological features that make it decidedly different from the atoke (Fig. 9.16b). Epitokes, usually pelagic whereas atokes are nonpelagic, often develop reduced guts and enlarged eyes and are considered by some to be separate individuals produced asexually by budding. Interestingly, some of the epitokal metamorphic changes take place only after separation from the parent worm. Neurosecretory cells in the brain of the parent worm apparently secrete a hormone that inhibits the development of these secondary sexual characteristics, and only after this hormonal and structural separation takes place can metamorphosis of the epitoke be completed.

The best-known example of epitoky is found in *Eunice viridis,* the Pacific palolo worm (Fig. 9.16a), in which only once a year at the full moon in late autumn, the posterior, gamete-bearing portion or epitoke breaks away from the atoke, which remains in its coral burrow. Millions of epitokes swarm to the surface and rupture, and the water actually becomes cloudy from the innumerable teeming gametes from both sexes. People who have seen this dramatic synchronous reproductive activity say that it is an event they will never forget. Other examples of epitoky are known and usually involve sexual dimorphism, swarming, or a high degree of periodicity synchronized to lunar or tidal events. Epitokes die shortly after releasing their gametes.

Some polychaetes demonstrate precisely synchronized periods of spawning in which entire adults, not just epitokes, congregate at or near the surface in a phenomenon called **swarming.** Environmental stimuli such as phases of the moon, sudden changes in temperature, or diurnal changes in light intensities are factors in initiating swarming. In a few worms studied, spawning is initiated by the release of maturation-stimulating hormone by females. Swarming is an adaptation that obviously enhances the chances of fertilization and fecundity.

Polychaetes have an unusual capacity for regeneration of lost parts; parapodia, tentacles, and radioles are quickly regenerated, and even entire regions of the body can be replaced. Regeneration of either end is possible, but heads are more difficult to regenerate than are posterior trunk segments. Even highly specialized worms like *Chaetopterus* are capable of some restorative regeneration.

TUBE BUILDING

While more primitive sedentary polychaetes live in burrows, the more advanced sessile forms construct tubes, a habit found in many diverse groups. Some tubes are simple sacs, others are straight and vertical, and still others take the form of a gallery of interconnecting tubes. A tube serves as a retreat for the worm, to enable the animal to adhere to hard surfaces like corals or rocks, and to aid a mud-inhabiting worm in obtaining a source of clean, oxygenated, plankton-rich water.

Polychaetes possess many different adaptations for their tubicolous habit. Prostomial appendages are usually absent, and special feeding structures such as radioles, tentacles, cilia, and modified chaetae are present. Often, parapodia are reduced or highly modified (for example, in *Chaetopterus* [Fig. 9.13]), and the chaetae are usually modified for gripping the walls of their tubes. Tagmatization is usually evident in tubicolous worms; and the giant fiber system, enabling quick retraction of the anterior end, is extremely well developed.

Tubes are constructed of an organic matrix of materials secreted by tubicolous glands located on the ventral surface, at the anterior end, or over the entire surface. Most tube materials, a mucous complex of proteins and polysaccharides, harden in seawater and have the consistency of soft slime, parchment, or even horn. In many worms, environmental materials such as sand, calcium carbonate crystals, mud, or shell fragments are incorporated into the mucus, adding considerable hardness to the tube and substantially enhancing the protection the tube affords the worm (Fig. 9.17). Some of the most beautiful and common of the tubeworms, the fanworms or featherdusters (Order Sabellida), are exquisite in their color, possess an interesting morphology, and have an intriguing behavior. They are extraordinarily numerous in almost every littoral or intertidal area.

Most often, polychaete tubes are cylindrical in cross section. Some species construct a chimney that projects from the mud surface upward into the water as much as 8 cm. The walls of the chimney are covered with shell fragments, seaweed, or debris the worm has positioned with its jaws (Fig. 9.17b, c). *Chaetopterus* has a U-shaped tube with a chimney at either end (Fig. 9.13a). Some polychaetes even have an unattached tube that is carried around by the worm. Most tube-dwelling polychaetes, however, have tubes that are affixed on or in the bottom materials. Usually, the completely tubicolous polychaetes never leave their tube; only

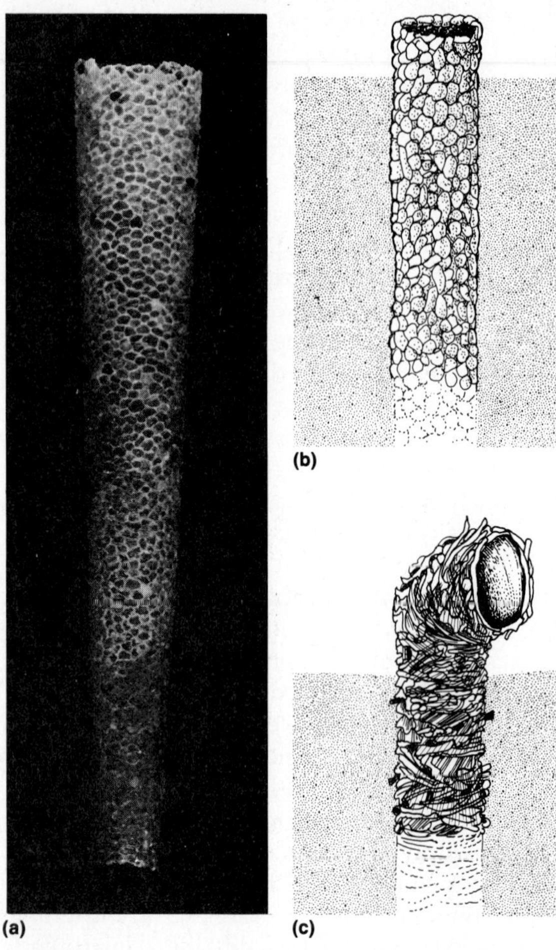

Figure 9.17 *Some types of polychaete tubes: **(a)** a photograph of the sand-grain tube of* Cistenides; ***(b)** the upper end of the tube of* Clymenella; ***(c)** the upper end of the tube of* Diopatra.

the head is extended from the tube opening for feeding. Most worms have evolved some mechanism for closing the tube when the animal retracts. In some an operculum is formed from a modified tentacle or radiole, while in others the tube mouth simply collapses as the worm withdraws, only to be opened again as the head emerges. Tube building is a more or less continuous process, since an enlarging tube is necessary to accommodate a worm growing in length. Building is carried out typically by the peristomium, tentacles, and radioles that sort and size particles and aid the worm in siting them at the tube mouth, where they are held in place by a peristomial collar until the mucous matrix hardens.

Class Clitellata

The Class Clitellata includes the common earthworms, intriguing leeches, and two smaller groups. All four groups share some features, probably had a common ancestry, and thus are considered here as belonging to a single class. Clitellates are mostly inhabitants of soils, fresh waters, or damp places, but a few are marine. They lack parapodia and either have fewer setae per segment than do polychaetes or lack them altogether. Clitellates also have a reduced head usually devoid of tentacles or other projections and generally lack gills. Excretion is always by metanephridia.

From a functional standpoint, clitellates are fundamentally different from polychaetes in the many aspects of their reproductive patterns. All clitellates are hermaphroditic, and copulation involves a pair of worms mutually inseminating each other. Unlike polychaetes, distinct paired gonads are present, and only a few segments bear them. Typically, one or two segments contain testes, and only one segment houses ovaries. A unique characteristic of the clitellates is the presence of a swollen area, the **clitellum,** which is located in the anterior half of the worm and involved with the formation of an egg case or **cocoon.** From two to as many as 50 segments are included in the clitellum, but most species have from two to eight clitellar segments. At the time of reproduction, numerous unicellular mucous glands form a mucous ring that becomes the cocoon. While specific details of reproduction vary among the various taxa, this basic function always involves reciprocal copulation, cross-fertilization, and direct development without any larval stages.

There are two principal subordinate clitellate taxa, the oligochaetes and hirudineans, and two groups that contain only a very few species, the branchiobdellids and acanthobdellids. A discussion of each follows pointing out their individual characteristics and features.

Subclass Oligochaeta

The oligochaetes, with about 3400 species, include the familiar earthworms plus many freshwater species. Earthworms burrow in moist soils from near the surface to a depth of several meters, and they are an important agent in the fertility of the soil; Box 9.2 explores this interesting beneficial activity. Freshwater forms burrow in the bottom debris or live on submerged vegetation. Relatively small numbers of species are found in the tropics living on tree trunks, in estuaries, or in intertidal zones. Sometimes densities of oligochaetes are quite high; as many as 400 earthworms and up to 10,000 soil worms (enchytraeids) per cubic meter have been reported. They range in size from 1 mm to 3 m

BOX 9.2 □ THE ECONOMIC VALUE OF EARTHWORMS TO SOILS

Almost a century and a half ago, Charles Darwin studied extensively the value of earthworms to soil fertility. He found that soils containing earthworms were substantially improved in comparison to soils without them. Since Darwin's time, a number of studies have confirmed the great economic benefit earthworms impart to soils and to soil fertility.

Earthworms significantly affect the structure of soils by actually breaking down mineral particles into smaller bits, mixing soils and organic materials, and bringing subsoils and their minerals to the surface. Thus the soil is mixed and aerated, and its water-holding capacity, porosity, and drainage are greatly improved. One of the most important mechanisms for great soil improvement is in the formation of soil aggregates, which are masses of mineral granules joined together temporarily for weeks or months. Soil fertility is directly related to amounts of soil aggregates present; they promote drainage and allow better aeration. While earthworms are not essential for aggregates to form, their presence greatly enhances the development of soil aggregates by a yet-unknown mechanism.

Earthworms are known to be highly instrumental in turning over soils from deeper areas, a process that directly enhances the fertility of the soil and the productivity of plants grown there. Earthworms can effect a turnover of up to 250,000 kg/hectare, or the equivalent of adding a layer 5 cm thick to the surface each year. The net effect of earthworms, their burrows, and their castings is to provide a stone-free layer at the surface up to 0.2 m thick, and this is manifestly evident in most soil profiles. Earthworm activity enhances the soil-air volume by as much as 30%, which in turn promotes drainage and increases the soil's water-holding capacity.

Greatly improved crop yields have been noted in plants grown in soils containing earthworms. For example, the dry matter of spring wheat was doubled in one study, and clover yields have been increased tenfold. Increased yields have been noted in many other crops and in certain orchard and forest trees.

While studies on earthworms may not lead directly to a cure for cancer or heart disease, these annelids are extraordinarily important to farmers and gardeners and to us all, since we benefit by increased plant yields.

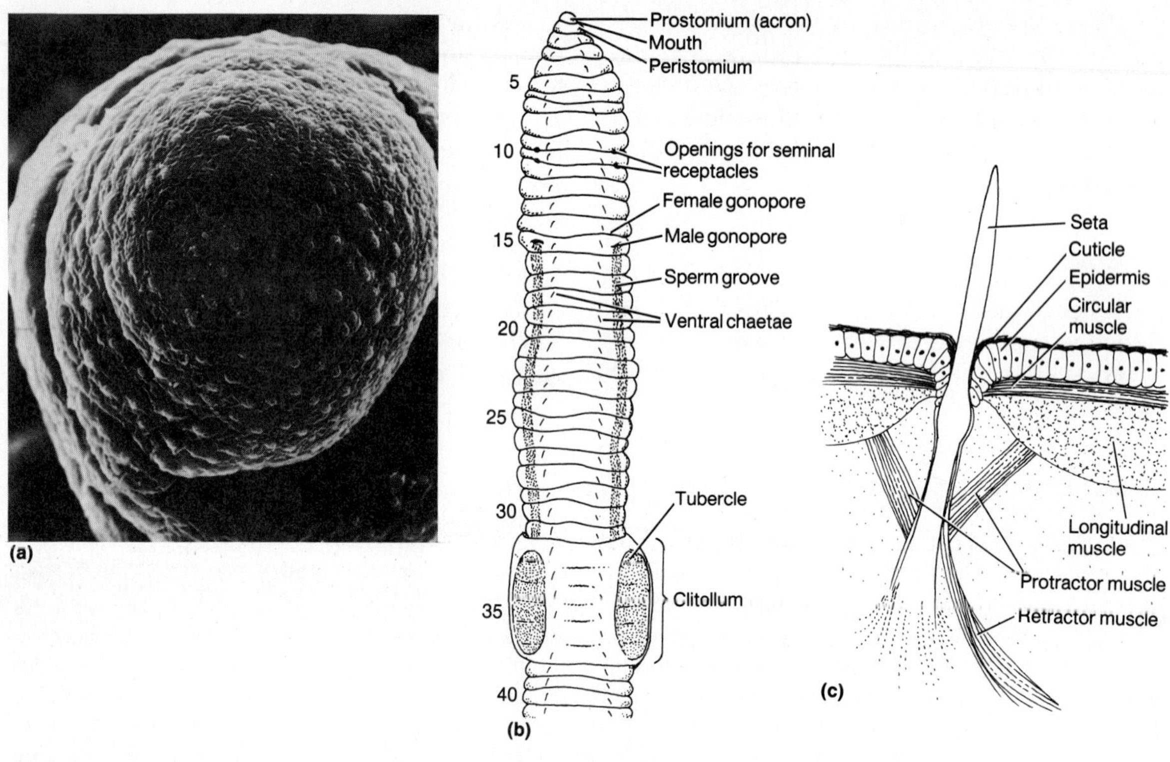

Figure 9.18 Some external features of oligochaetes: ***(a)*** *scanning electron micrograph of the prostomium (acron) and mouth of* Enchytraeus albidus *(note numerous sensory papillae on prostomium) (courtesy of T.L. Boullion);* ***(b)*** *ventral view of the anterior 40 segments of* Lumbricus; ***(c)*** *section through a chaetal sac.*

in length (giant Australian earthworms). Earthworms are basically reddish brown, but freshwater oligochaetes are transparent or red owing to the presence of erythrocruorin, yellow-green because of internal chloragogen cells, or variously colored owing to epidermal pigments.

Most oligochaetes are scavengers and feed on dead or decaying vegetation or, occasionally, animals such as protozoans, insect and fish eggs, other oligochaetes, and miscellaneous invertebrates. Only a few oligochaetes are parasitic. Aquatic species and most earthworms consume rather large quantities of mud and soil and digest the organic material contained within it. Additionally, earthworms collect leaves, pull them into their burrows, and ingest fragments. Aquatic tube-builders must depend upon water-borne food flowing over them, and a few worms use their setae as filters.

External features. The body of an oligochaete is usually cylindrical, although it is somewhat flattened in many earthworms. The **acron** is small, bears no tentacles or eyes, overhangs the mouth (Fig. 9.18a), and, in a few species, is drawn out into a filiform sense organ. The trunk consists of six to over 600 similar segments, and the clitellum is the only externally evident tagma. Each trunk segment is equipped with relatively fewer **chaetae** than in polychaetes, a feature from which oligochaetes derive their name (oligo = few; chaetae = bristles). A surprising variety of setal sizes and shapes is present in oligochaetes.

Typically, chaetae are found in bundles of one to several chaetae each, and two ventral bundles and two ventrolateral bundles are often present (Fig. 9.18b). In some earthworms, however, numerous chaetae (50–125) are distributed uniformly around each segment like a girdle. Representatives of one genus lack setae altogether, while others have many setae per bundle. Aquatic species generally have relatively longer setae. Manipulated by extrinsic protractor and retractor muscles (Fig. 9.18c), the chaetae are useful in crawling and burrowing.

A prominent **clitellum**, found in all oligochaetes, generally involves two segments in aquatic species, seven in *Lumbricus*, and more than 50 in some forms (Fig. 9.18b). Most trunk segments bear a pair of ventral **nephridiopores**, and one or several anterior segments bear reproductive openings or **gonopores** (Fig. 9.18b). Many earthworms have minute **dorsal pores** in the intersegmental furrows through which coelomic fluid can be exuded to the exterior. The opening of each dorsal pore is regulated by a small sphincter muscle. Coelomic fluid lost via the dorsal pores helps to keep the integument moist, and the highly malodorous fluid might also function to dissuade some potential predators.

Internal features. Since internal features have already been discussed both for the phylum and for polychaetes, only the principal differences found in oligochaete systems will be mentioned here. The longitudinal muscle layer of the body wall is of uniform thickness rather than in longitudinal bundles as in polychaetes, and it is usually present as a syncytial cylinder around the coelom (Fig. 9.19). The coelom is comparted by septa, which in many species contain septal openings, each of which is regulated by a sphincter muscle. These openings permit some intersegmental flow of coelomic fluid. Septa often contain muscle fibers whose contractions aid in controlling coelomic fluid flow.

Visceral systems. The mouth, devoid of jaws or teeth, opens into a small buccal cavity, which in turn becomes a rather large muscular

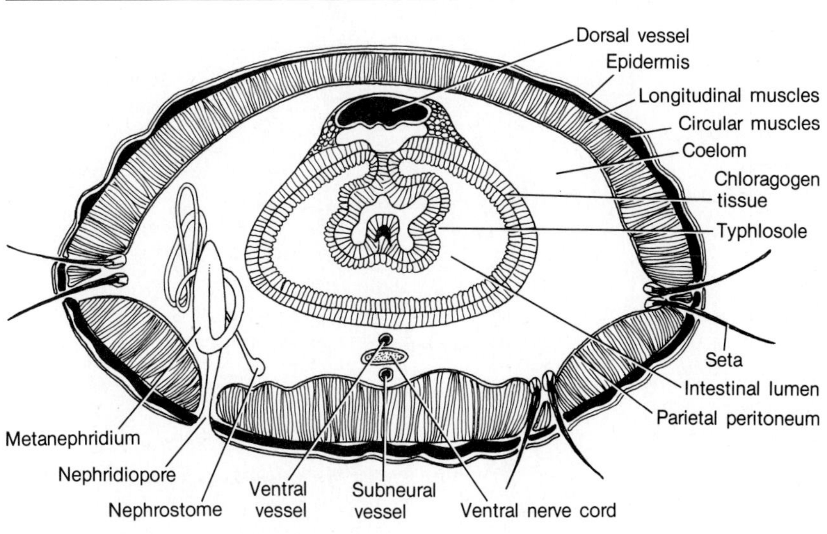

Figure 9.19 *A cross-sectional view through the intestinal region of an earthworm.*

pharynx (Fig. 9.20a). The pharynx, equipped with both extrinsic muscles extending from pharynx to body wall and intrinsic muscles, is an effective organ for pulling food into the alimentary tract. From the pharynx, food enters a tubular esophagus a part of whose length is specialized as a crop, stomach, or gizzard (Fig. 9.20a). A **crop** has thin walls and functions as a storage area, as does a thicker-walled **stomach** when present. Most terrestrial oligochaetes have a **gizzard** that has muscular walls and a chitinous lining; it functions in grinding food. A characteristic feature of oligochaetes, especially the earthworms, is that the esophagus contains a number of **calciferous glands** (Fig. 9.20a). These glands, supplied with a copious vascular supply, eliminate calcium and carbonates as calcite into the esophageal lumen and thus are important adaptations in regulating the pH of the blood and coelomic fluids. A long, straight intestine constitutes the remainder of the digestive canal. Present in all terrestrial oligochaetes, but absent in freshwater forms, is an internal, middorsal, longitudinal fold of the intestine, the **typhlosole** (Fig. 9.19). This ridge greatly increases the surface area of the intes-

Figure 9.20 *Internal features in the anterior 20 segments of* Lumbricus: ***(a)*** *dorsal view of the viscera (dorsal body wall removed);* ***(b)*** *dorsal view of the reproductive organs and anterior nervous system (alimentary canal removed).*

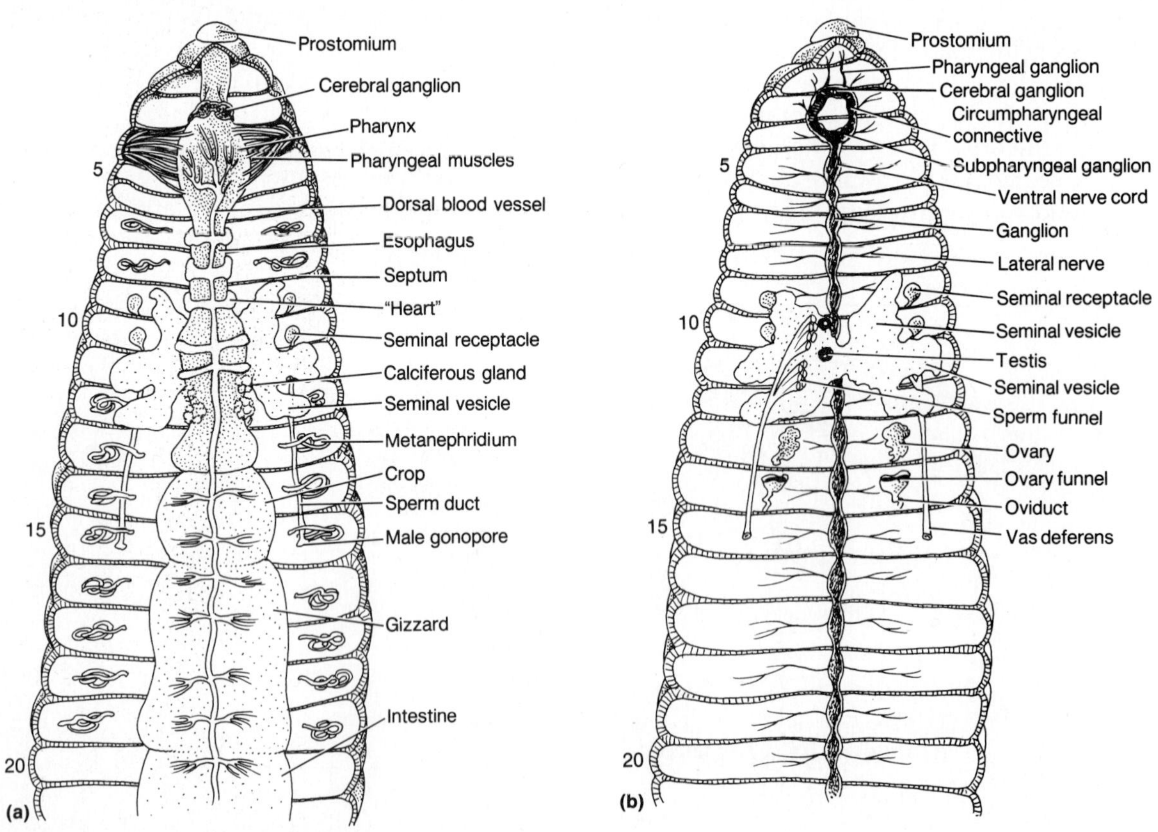

tine, thus facilitating digestion and absorption. Derived from the intestinal visceral peritoneum is the yellowish-green **chlorogogen tissue** (Fig. 9.19) whose cells synthesize and store fats and glycogen and form nitrogenous wastes.

The circulatory system of oligochaetes is typically annelidan. In many species, lateral vessels near the anterior end are especially enlarged and contractile and are termed **"hearts."** Their numbers range from one pair in aquatic forms to five pairs in *Lumbricus* (Fig. 9.20a). Oligochaete blood is colorless if it lacks a respiratory pigment, but more commonly it contains erythrocruorin and thus is bright red. Phagocytic amebocytes are also present. Most oligochaetes use the vascularized, moist, body surface for gaseous exchange. A pair of **metanephridia** typically is located in each segment, where considerable reabsorption of salts enables the worms to excrete hyposmotic urine. Some oligochaetes have multiple nephrostomes for each nephridium, many separate nephridia per segment, or nephrostomes that open into the pharynx or intestine.

The bilobed brain or cerebral ganglion has shifted posteriorly to segment 3 (Fig. 9.20b). Except for the aelosomatids, all oligochaetes have a single ventral nerve cord, which often lies within muscle layers of the body wall. Ganglia in each segment are subordinate in function to the subpharyngeal ganglion in the anterior 1-4 segments, which means that if this ganglion is destroyed, all movements cease. Five giant fibers characterize the nerve cord of oligochaetes, which are instrumental in the rapid sudden contraction of either end of the body (see Box 9.1).

Eyes are absent, although some aquatic forms have simple ocelli or prostomial sense organs that function as photoreceptors; oligochaetes are negatively phototactic to strong light but positively phototactic to weak light. Numerous epidermal and subepidermal chemoreceptors, proprioceptors, and tactile receptors are found aggregated near the anterior end and generally over the integument. Free nerve endings are common in the skin and provide the worm with additional sensory information.

Reproduction. The male reproductive system consists of paired **testes** in one (freshwater forms) or two (earthworms) segments that are always in a segment anterior to the single ovarian segment. The small testes, typically found in the anterior half of the worm, are attached to the posterior surface of a septum and project into the coelom (Fig. 9.20b). They may be simple sacs or lobes or bear fingerlike diverticula. Immature sperm (spermatogonia) are released into the coelom or, as in most oligochaetes, into special coelomic sacs called **seminal vesicles**, where maturation takes place (Fig. 9.20). The one to three pairs of seminal vesicles, lacking in some semiterrestrial species, are usually present. In *Lumbricus* there are two pairs of testes in segments 10 and 11 and three pairs of seminal vesicles in segments 9, 11, and 12 (Fig. 9.20). Each testis is equipped with a **vas deferens** or **sperm duct** whose coelomostome is situated near the posterior septum of the testicular segment. From the funnel, each vas deferens extends posteriorly through several segments; in earthworms the two vasa deferentia on a side fuse to form a single **sperm duct**, which opens to the outside as a ventrally situated **male gonopore** (Figs. 9.18b, 9.20b). Associated with the male gonopore in certain species is a **penis** or an eversible protuberance of the gonopore atrium or body wall called a **pseudopenis**. Accessory **prostate glands** secrete a seminal fluid that enhances the motility and viability of the sperm.

The single pair of **ovaries** is also located on the posterior aspect of a septum and projects into the coelom (Fig. 9.20b). The cells released from the ovaries into the coelom are oogonia, oocytes, or mature ova. Commonly, these sex cells are ovulated into a pair of specialized coelomic **ovisacs** that resemble seminal vesicles except that they are smaller. Mature ova are conducted from the ovisacs to the outside via paired coelomoducts or **oviducts**. Each oviduct passes posteriorly, penetrates one septum, and opens to the outside as a ventral **female gonopore** on the segment immediately posterior to that containing the ovaries (Figs. 9.18b, 9.20b).

Associated with the female reproductive system are one to five pairs of seminal receptacles. **Seminal receptacles** or **spermathecae** are small, rounded, saclike structures that open on the ventral surface of the worm and store sperm provided by the other worm during copulation. Typically, the segmentally arranged seminal receptacles are located in segments anterior to the ovarian segment. In *Lumbricus* the two pairs of seminal receptacles are located in the intersegmental furrows between segments 9 and 10 and between segments 10 and 11 (Figs. 9.18b, 9.20). The exact segmental positions of the reproductive organs, variable in the various taxa, are often used as taxonomic characters.

Figure 9.21 Reproduction in oligochaetes using Pheretima *as an example: **(a)** two worms in copulation; **(b)** sequence of stages in cocoon formation and its subsequent release (parts a, b adapted from V.B. Tembe and P.J. Dubash, 1961,* J. Bombay Natural History Society 58:181).

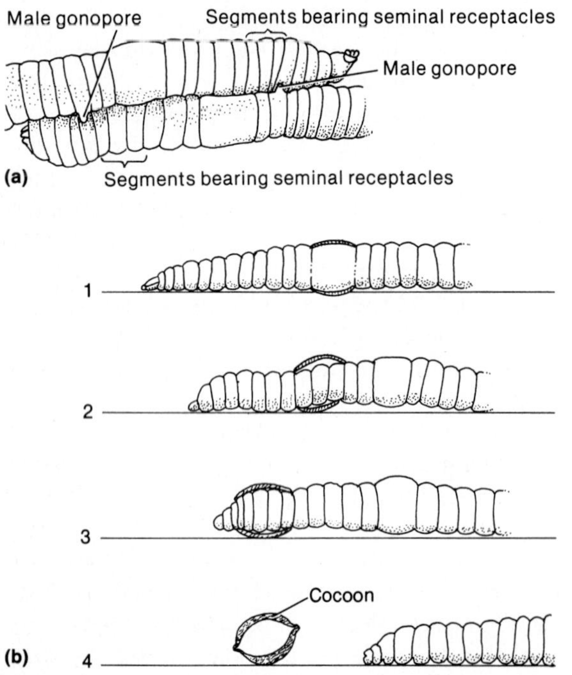

At copulation, two worms come together with the anterior end of each worm pointing toward the posterior end of the other worm (Fig. 9.21a). The ventral surfaces of the copulating pair are appressed and are held together by adhesive mucus from clitellar glands and by modified ventral **genital setae**. In many worms the sperm duct openings of one worm are directly opposite those of the seminal receptacles of the other; but in earthworms like *Lumbricus* and its relatives, sperm have a considerable distance to travel between the male gonopores and the seminal receptacle openings. The intervening segments of both worms contract in such a way as to form two ventral, longitudinal **sperm grooves** (Fig. 9.18b). Sperm movement along the sperm grooves is achieved by the flagellar action of the sperm and by peristaltic contractions of the body wall muscles forming the grooves. Copulation usually lasts from one to three hours, after which the worms separate.

Several days later, the clitellum secretes the mucous band or sleeve, which becomes detached from the clitellum and slides anteriorly as the worm backs out of it (Fig. 9.21b). Albumin glands in the clitellum secrete a soft albuminous material, which will serve as one source of nutrients for the future embryos. As the sleeve passes the female gonopore, ova are deposited in it, and in a similar way, sperm are released subsequently by the seminal receptacles. As the band slides off the anterior end, its ends constrict and seal off the gametes from the outside, its walls become tough and chitinlike, and it forms a cocoon (Fig. 9.21b). Ova are fertilized shortly after sperm enter the mucous ring. Formed singly, a number of cocoons may be produced within a few days after copulation.

Cocoons are yellowish, ovoid or lemon shaped, and range in length from 1.5 to 7.5 mm,

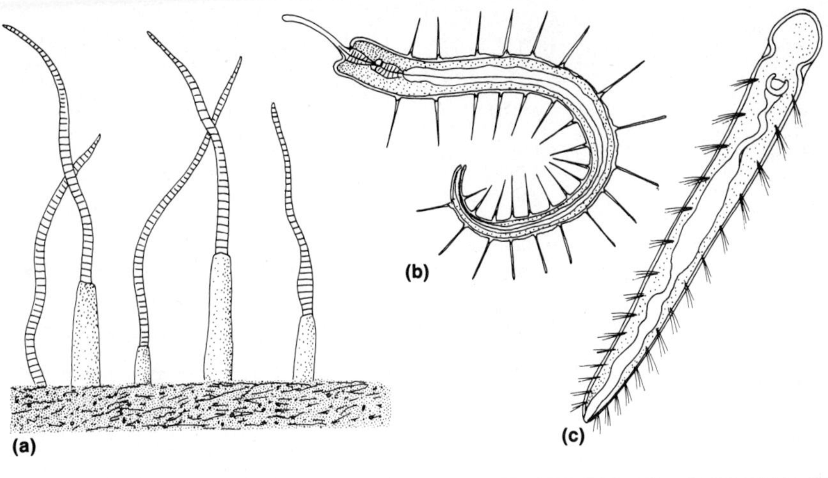

Figure 9.22 *Some freshwater oligochaetes:* ***(a)*** Tubifex, *a tube-builder often found in polluted streams;* ***(b)*** Stylaria, *a predaceous form;* ***(c)*** Aeolosoma, *a questionable oligochaete (see note in Systematic Resume at the end of this chapter). (parts a, c from R.W. Pennak, 1978,* Fresh-water Invertebrates of the United States, *2nd ed., © John Wiley & Sons.)*

and each contains up to 25 embryos. They are left by earthworms in the soil and by aquatic forms in bottom debris or on vegetation. Eggs of aquatic species have relatively larger amounts of yolk than those of terrestrial oligochaetes, and additional nutritive materials are supplied by the albumin inside the cocoon. Specific developmental details vary among various oligochaetes, but the eggs undergo spiral determinate cleavage, and there are no larval stages. Juvenile worms hatch from the cocoon in one to three weeks depending upon environmental conditions. Parthenogenesis and self-fertilization have both been reported for a few species.

Asexual reproduction is common in some freshwater forms (Fig. 9.22) as mature worms undergo transverse fissions to form two to many new individuals. Some species produce a chain of individuals in a process called strobilation; later, each separates to become a new individual. Fragmentation of a worm into several pieces is known, and each piece is capable of regenerating lost parts. The regenerative capacity of oligochaetes is considerable whether in regenerating parts lost in fragmentation or if a worm has been severed accidentally.

Subclass Branchiobdellida

This puzzling group contains over 120 species that are morphologically intermediate between typical oligochaetes and hirudineans. They were at one time grouped with the leeches and have more recently been grouped with oligochaetes; but since they probably diverged early from the stock that gave rise to oligochaetes and hirudineans, they should be considered as a separate subclass. While most are ectocommensal on crayfishes, especially in the northern hemisphere, gill-inhabiting branchiobdellids parasitize the blood and tissue juices from their decapod hosts. A few species are also found on other freshwater crustaceans including crabs, shrimp, and isopods. Some of the better-known genera include *Cambarincola* (Fig. 9.23) and *Sathodrilus*.

Branchiobdellids are from 1 to 12 mm in length and have a head, trunk, and caudal sucker. The head, lacking a prostomium, consists of four fused segments, and the suctorial mouth bears a pair of hard jaws (Fig. 9.23). The trunk contains 11 segments, the last three of which are somewhat fused, indistinct, and form

5 Testes
1
Opening of seminal receptacle
10
Head
Male gonopore
Jaws
Anus
Caudal sucker

Figure 9.23 Cambarincola, *a branchiobdellid, lateral view (from R.W. Pennak, 1978,* Fresh-water Invertebrates of the United States, *2nd ed., © John Wiley & Sons).*

Figure 9.24 Some various hirudineans: ***(a)*** *photograph of a freshwater leech;* ***(b)*** Acinobdella annectens *(rhynchobdellid);* ***(c)*** Placobdella ornata *(rhynchobdellid);* ***(d)*** Theromyzon tessulatum *(rhynchobdellid);* ***(e)*** Mollibdella grandis *(gnathobdellid). (parts b–c from United States Environmental Protection Agency [IERL] Office of Research and Development.)*

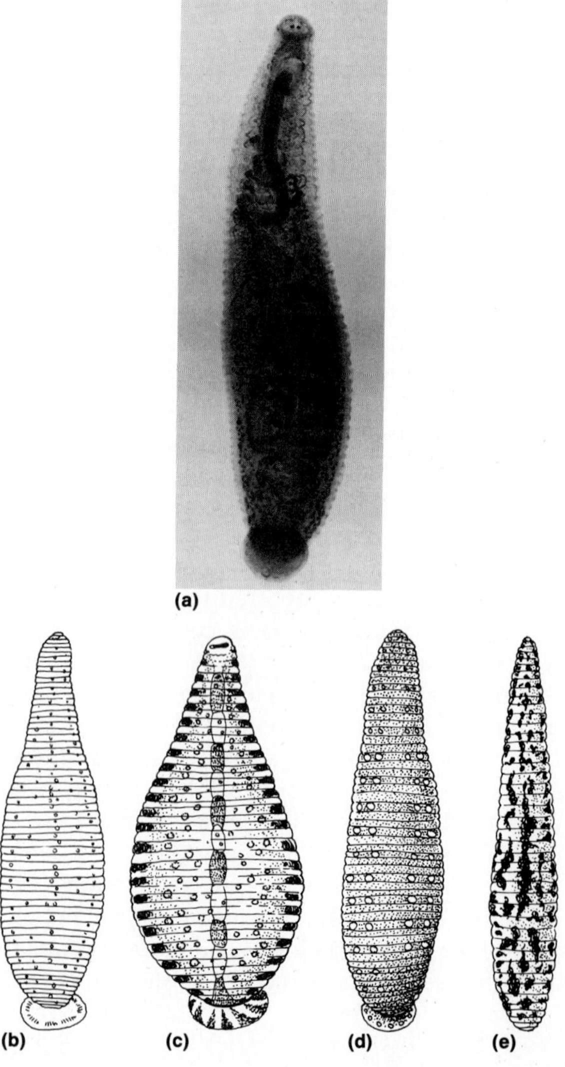

a **caudal sucker**. The anus is present on the dorsal side of trunk segment 10, trunk segments 5–7 constitute the clitellum, and chaetae are typically absent.

Internally, only two pairs of metanephridia are present. Two pairs of testes are found in trunk segments 5 and 6 (one pair in segment 5 in some), and the single pair of ovaries is situated in segment 7. A single male gonopore is located on the ventral surface of trunk segment 6, whereas paired female gonopores are found on trunk segment 8. The one seminal receptacle is located in trunk segment 5 (Fig. 9.23).

Subclass Acanthobdellida

A single species, *Acanthobdella peledina*, comprises this subclass. It, too, has been variously classified and is often grouped with the leeches. Most likely, it is primitive, intermediate between oligochaetes and hirudineans, represents a separate evolutionary line, and should be treated as a subclass.

A. peledina is 2–4 cm long and is an ectoparasite on European salmon. The body consists of a prostomium and 30 segments. A well-developed posterior sucker is present, but there is no anterior sucker. Each of segments 1–5 bears two pairs of chaetae. Internally, in the first five segments there is a circumenteric coelom comparted by septa. The stomach lacks lateral ceca.

Subclass Hirudinea

That leeches are generally repulsive to many people is well known, but it is probably an unfair judgment against them, since they are really quite interesting creatures. The approximately 600 species are basically a freshwater group, but some have invaded the oceans, and others live in semiterrestrial damp places. Most leeches are found in vegetation-choked ponds or shallow lakes, where, incredibly, numbers in excess of 10,000 individuals in 1 m^2 of bottom debris have been reported. Leeches have great powers to contract and extend the body, which sometimes make their true sizes difficult to determine. Most hirudineans are from 1 to 6 cm in length when fully extended, but the medicinal leech, *Hirudo*, reaches a length of 20 cm, and one Amazon species is known to be 30 cm long. Typically, leeches are ectoparasitic on vertebrates on whose blood they feed, but about one-fourth of all leech species, being nonparasitic, prey on small invertebrates and ingest them whole. Many leeches are commonly black or brown, others are brightly colored, and many are striped or patterned with various colors (Fig. 9.24).

External features. The basic external features are much the same in most leech species. The body is elongate and dorsoventrally flattened, and at either end is a sucker that is used in both feeding and locomotion. Characteristic of hirudineans is that the body is invariably composed of 34 segments, but each segment is subdivided into a definite number of **annuli** (Fig. 9.25a). Numbers of annuli vary from species to species and even between segments on the same worm. Some segments may have as many as 14 annuli, but more commonly there are two to five per segment. Chaetae are absent in hirudineans.

Some experts distinguish five major body regions in leeches: head, preclitellar, clitellar, postclitellar, and terminal (see Pennak, 1978). The head, tapered anteriorly, is composed of the prostomium plus the anterior six segments. Dorsally, the head bears a number of paired **eyes** or ocelli; ventrally, it bears the **mouth** surrounded by the **anterior** or **oral sucker** (Fig. 9.25a). In many bloodsuckers the mouth is bordered by teeth and jaws, which are used in feeding. Behind the head is a **preclitellar region** of segments 7–9 followed by the four-segmented **clitellum**, quite an indistinct region except at the time of reproduction. Both gonopores are situated midventrally on a clitellar segment; the male opening is usually on segment 10 or 11, and the female opening is usually

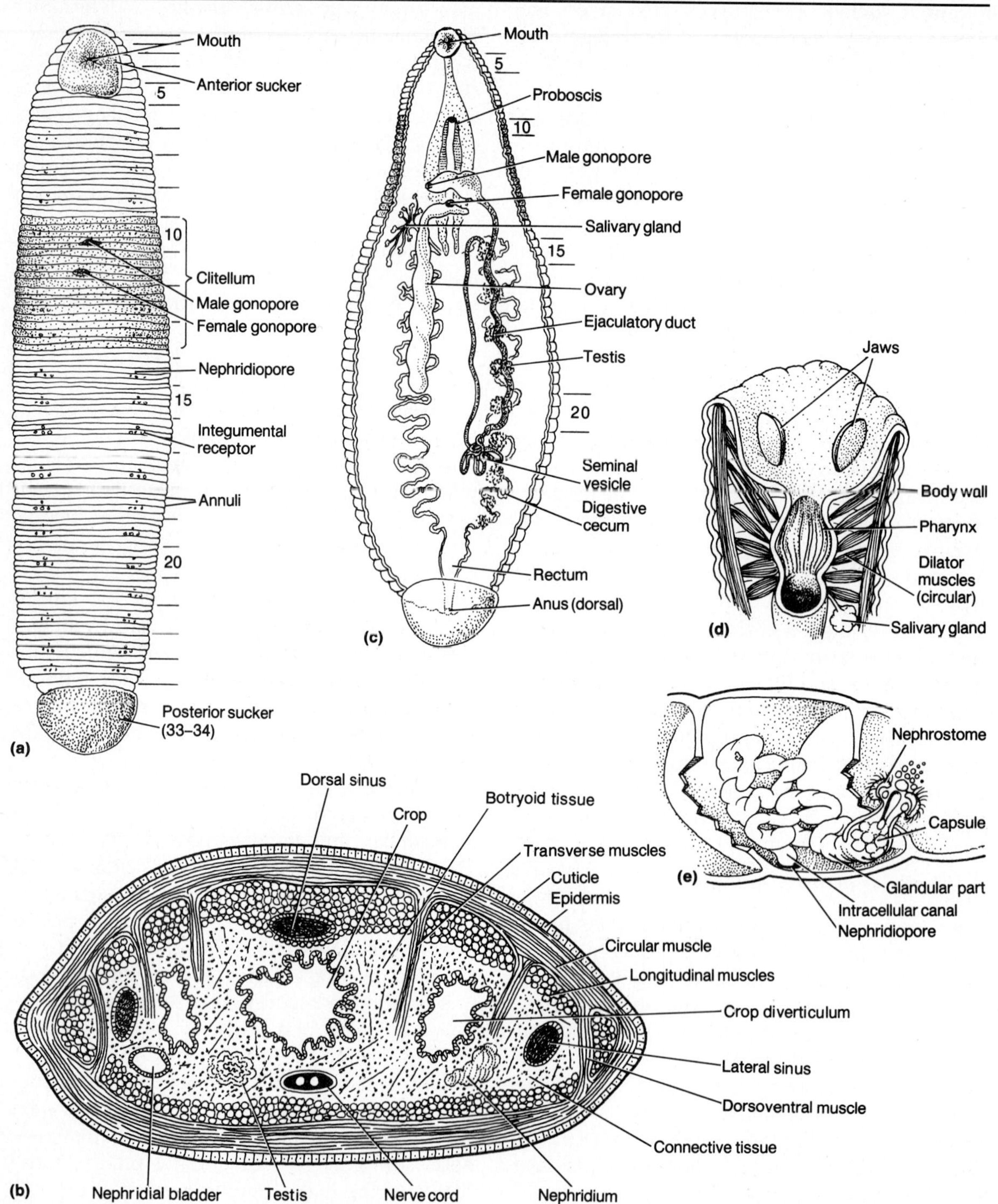
Mouth
Anterior sucker
5
10
Clitellum
Male gonopore
Female gonopore
Nephridiopore
15
Integumental receptor
Annuli
20
Posterior sucker (33–34)
(a)
Mouth
5
Proboscis
10
Male gonopore
Female gonopore
Salivary gland
15
Ovary
Ejaculatory duct
Testis
20
Seminal vesicle
Digestive cecum
Rectum
Anus (dorsal)
(c)
Jaws
Body wall
Pharynx
Dilator muscles (circular)
Salivary gland
(d)
Nephrostome
Capsule
Glandular part
Intracellular canal
Nephridiopore
(e)
Dorsal sinus
Botryoid tissue
Crop
Transverse muscles
Cuticle
Epidermis
Circular muscle
Longitudinal muscles
Crop diverticulum
Lateral sinus
Dorsoventral muscle
Connective tissue
(b)
Nephridial bladder
Testis
Nerve cord
Nephridium

◀ ***Figure 9.25*** *Hirudinean features:* ***(a)*** *the external features of* Hirudo, *ventral view;* ***(b)*** *cross-sectional view through the midbody of* Hirudo; ***(c)*** *the reproductive systems of a typical leech;* ***(d)*** *the anterior end of* Hirudo; ***(e)*** *a nephridium of* Glossiphonia.

on segment 11 or 12 (Fig. 9.25a). The **postclitellar region**, consisting of 11 segments, occupies the majority of the trunk region. The **terminal region** is constructed of 10 segments plus the pygidium and is often subdivided into an anal region (25–32) and a **caudal** or **posterior sucker** (33–34). The anal region bears the dorsal **anus** situated just in front of the large caudal sucker.

Body wall and coelom. The important differences between the body wall of a leech and that of other annelids are found in the musculature. First, there is an oblique muscle layer situated between the circular and longitudinal muscles. Second, prominent dorsoventral muscle strands are present whose contractions cause the body to be flattened dorsoventrally even more than it normally is. Third, both circular and longitudinal muscles are specialized at the suckers to form powerful, strong, radial and circular sucker muscles.

In leeches, all mesenteries, septa, and coelomic compartments, characteristic of other annelids, have disappeared entirely. Rather than the coelom being present as a distinct chamber around the alimentary tract, it is reduced to a series of channels and sinuses as a result of the extensive invasion of the body cavity by intrusive **botryoid tissues** (shaped like a bunch of grapes) (Fig. 9.25b). The coelomic sinuses, often connected by lateral branches and capillaries, usually function as an internal transport system.

Feeding and digestion. A majority of leeches are ectoparasitic on many different vertebrates, and the remainder ingest small invertebrates such as small molluscs, worms, and insects. Parasitic leeches do not exhibit a high degree of host specificity; a given leech will usually feed on any available host within a given class of vertebrates. Two basic types of ingestion are found in leeches, but these do not correspond exactly to the predatory or parasitic habits. In one method, utilized by the rhynchobdellids, the muscular pharynx is everted through the mouth as a **proboscis**, which is forced into the host's tissues. Jaws are absent, but penetration of the host may be facilitated by enzymatic secretions of proboscidian glandular cells. A second method of ingestion, found in most leeches, involves several extremely important adaptations for a leech to feed quickly and unobtrusively. How unobtrusive, you might ask, can a leech be that is 20 cm in length and will suck about 100 ml of blood from your leg? Quite unobtrusive, the answer would be, because of several adaptations for feeding by a leech. First, there are three large, sharp, serrated jaws surrounding the mouth, and the rocking action of the jaws, effected by small muscles, produces a Y-shaped incision (Fig. 9.25d). Second, feeding is facilitated by three different types of pharyngeal glands: one secretes a local anesthetic that masks the pain of the incision, another results in vasodilation of the host's capillaries surrounding the incision, and a third produces **hirudin**, an anticoagulant that causes blood to flow freely from host to leech. Just inside the mouth is a muscular pumping **pharynx**, which is not everted as a proboscis (Fig. 9.25d); alternating contractions of longitudinal and circular pharyngeal muscles make it an effective sucking organ. In leeches that are not bloodsuckers the jaws are reduced to muscular ridges, which are employed in grasping the prey. Thus feeding by a bloodsucking leech is rather quick, painless, and really not debilitating to the vertebrate host.

Posterior to the pharynx is a short esophagus that empties into an elongated stomach often referred to as the crop. In predaceous forms, the stomach is a rather simple tubular organ, but in bloodsuckers it is equipped with one to several pairs of lateral diverticula or **ceca** that store blood until it is digested. The intestine, also often provided with pairs of lateral ceca and acting as the site for most digestion and absorption, is continuous posteriorly as the rectum, which leads to the dorsal anus located anterior to the caudal sucker.

In *Hirudo*, the medicinal leech whose feeding details are best known, an amount of blood up to five times the initial weight of the leech is ingested. The abundant meal is stored in the lateral ceca of the stomach, where the hirudin keeps the blood in liquid form. Shortly after feeding, the leech removes most of the water from the ingested blood and excretes it via the metanephridia. Then the remaining concentrated blood is slowly digested; the globin component of hemoglobin appears to be the chief nutritional source. A single blood meal may last up to six months, and most leeches can survive for up to 18 months between meals.

Other visceral systems. In some of the less-advanced rhynchobdellids a typical annelidan circulatory system has been retained along with the coelomic sinuses, which function as an accessory circulatory system. But in all other leeches the blood–vascular system has been completely replaced by an extensive system of coelomic channels and sinuses that function as a transporting system. A longitudinal **ventral sinus** is the principal channel; typically, a **dorsal** and paired **lateral longitudinal channels** are also present. Transverse vessels, connecting the dorsal and lateral channels to the ventral sinus, are sometimes equipped with contractile **ampullae**. The ampullae may aid in circulation, but the principal contractile channels are the lateral longitudinal ones. The circulating coelomic fluid contains certain cells such as coelomocytes and amebocytes. In some leeches it is red because it contains dissolved hemoglobin.

Most leeches exchange gases through the general body surface, and undulations of the body create ventilatory currents of water over the worm. Other leeches (certain rhynchobdellids) have paired lateral outgrowths or **gills**, which may be simple, leaflike, or branching. Gills are filled with coelomic fluid, which pulsates through the filaments to enhance gaseous exchange.

Leeches have a pair of metanephridia in each of the middle 10–17 body segments. Each metanephridium is embedded within the intrusive connective tissue, and its ciliated nephrostome is situated in a coelomic channel (Fig. 9.25e). The nephrostome opens directly into a peculiar nonciliated **capsule** whose functions may be to produce coelomic fluid, manufacture coelomocytes, and store phagocytes. Particulate wastes, filtered from the coelomic fluid by the nephrostome, enter the capsule, where they are engulfed by phagocytes. Waste-laden phagocytic cells are then transported to the epidermis or intestine for removal.

The remaining **nephridial tubule** is quite different from that in other annelids. The tubule and capsule connect in primitive leeches, but in most forms, the structural connection between capsule and nephridial tubule has been severed. This tubule consists of a strand of cells through which runs a nonciliated intracellular canal that may drain a number of finer intracellular canals. Near the external end the canal is often continuous with a short ciliated **bladder**, which in turn opens to the outside via a ventrolateral **nephridiopore** (Fig. 9.25e). The nephridial tubule is concerned primarily with excretion of ammonia and water and with maintaining an osmotic balance; a hyposmotic urine is usually produced. Variations on this hirudinean nephridial plan are found in various taxa.

The nervous system of leeches, typically annelidan, has some modifications that reflect their functional morphology. The single ventral nerve cord contains segmentally arranged ganglia. The posterior seven ganglia are fused beneath the posterior sucker, and paired nerves innervate the sucker. Anteriorly, the first six ganglia are fused in segment 6 to form a **subpharyngeal ganglion** connected to a pair of **suprapharyngeal ganglia** or the brain by two circumpharyngeal connectives. Nerves from the subpharyngeal ganglion innervate the anterior sucker, mouth, prostomium, and anterior end of the worm.

Leech sensory structures consist of specialized organs including sensory papillae and eyes and dispersed sensory cells. **Sensory papillae** are groups of small protuberances located dorsally or encircling each segment on one annulus, and they function as both tactile and chemical receptors. From one to five pairs of simple dorsal **eyes** are found at the anterior end. Most leeches are photonegative except when they are especially hungry. As in oligochaetes, numerous free nerve endings receive stimuli including those related to temperature, vibrations, chemicals, and touch.

Reproduction and development. Like all other clitellates, leeches are monoecious, but unlike other annelids, leeches do not reproduce asexually, nor do they regenerate lost parts. Immature sperm and eggs are both released from the gonads into surrounding coelomic sacs, where they mature in a manner similar to that of all other annelids. A pair of spherical **testes** is located in each of four to twelve segments beginning with segment 13 (Fig. 9.25c). The testicular sacs on each side of the body are connected to a single vas deferens by means of short vasa efferentia. Both vasa deferentia run anteriorly, and each becomes enlarged or coiled to form a seminal vesicle followed by an ejaculatory duct. The two ejaculatory ducts empty into a single medial atrium, which opens to the outside by the single **male gonopore** located on segment 10 or 11 (Fig. 9.25a). In some leeches the muscular atrium can be everted through the gonopore to form a bursa or penis, but in others it is not eversible but is involved with the formation and expulsion of spermatophores.

A pair of **ovaries** is found usually in segment 12 or in the segment between the anterior testes and that containing the male gonopore (Fig. 9.25c). A short oviduct extends from each ovarian sac and joins its member from the other side to form a vagina. The vagina opens to the outside via the single **female gonopore** located on segment 11 or 12 or the segment just posterior to that containing the male gonopore (Fig. 9.25a).

Incapable of self-fertilization, leeches, like the oligochaetes, copulate and swap sperm; but unlike other clitellates, hirudineans have internal fertilization. At copulation the two worms twine about each other, and their ventral surfaces are brought together with the anterior sucker of one worm grasping the posterior body of the other worm (Fig. 9.26a). In leeches with a penis the male gonopore of one worm is opposite the female gonopore of the other; the penis is simply everted, and sperm are introduced into the vagina. In leeches lacking a penis, spermatophores are expelled, and the sperm hypodermically penetrate the integument of the other worm and migrate through the coelomic channels to the ovaries, where fertilization takes place. Some leeches deposit spermatophores on the clitellar region of the other worm, and the sperm later penetrate the integument (Fig. 9.26b).

The **clitellum**, normally very indistinct, becomes more conspicuous at the time of reproduction. As the mucous ring, secreted by glandular cells, slides anteriorly, fertilized eggs are deposited in it. Each cocoon contains from one to many eggs, depending upon the species. A given leech often produces multiple cocoons during the spring and summer. Cocoons are left in damp soil or on hosts, submerged objects, and bodies of worms themselves. Representatives of one family brood their eggs. A young

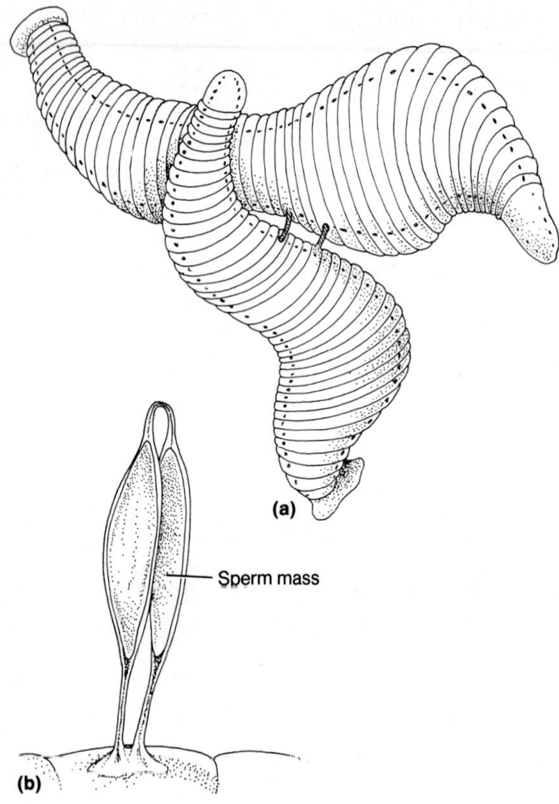

Figure 9.26 *Reproduction in leeches; **(a)** two worms in copulation; **(b)** a spermatophore.*

immature worm, equipped with both suckers, hatches in several weeks.

PHYLUM POGONOPHORA

Pogonophorans, or beardworms, were not discovered until 1900, when living specimens were dredged up near Indonesia. Since that time, almost 100 species have been described, mostly from the northern Pacific Ocean and chiefly by Russian zoologists, but they have also been collected off both coasts of North America and from the Gulf of Mexico. All pogonophorans are marine and live in water whose depth exceeds 500 m. They are sessile and live in long upright tubes embedded within the bottom ooze. Secreted by the worm, the stiff chitinous tubes may be smooth or annulated and may attain a length of up to several meters. Pogonophorans are extraordinarily thin; the body length is generally never less than 100 times its diameter. Two of the more common genera are *Seboglinum* and *Zenkevitchiana*.

The body is divided into an anterior forepart, a middle trunk, and a short posterior opisthosoma (Fig. 9.27). The **forepart** consists of a cephalic lobe, a glandular region behind the cephalic lobe, and a group of anterior tentacles. The long filamentous tentacles number from one to more than 260, and it is from this feature that the group derives its name (pogono = beard; phora = bearer). Each ciliated tentacle possesses a double row of fine pinnules, which are extensions of single tentacular epithelial cells. The tentacles are probably ventral to the cephalic lobe, but because of an incomplete understanding of pogonophoran development, there is disagreement as to whether the tentacles are morphologically ventral or dorsal.

The anterior region of the elongate **trunk** bears paired **papillae** that probably are metamerically arranged, but the more posterior papillae are unpaired (Fig. 9.27). The papillae are perhaps used in movement within the tube. One to several transverse annuli or girdles bearing a number of short setae are found on the posterior trunk. The terminal **opisthosoma** or **anchor** has more recently been discovered, since apparently it is easily broken off by dredging equipment. It consists of up to 25 segments, each equipped with **chaetae** (Fig. 9.27).

Internally, the most striking feature of pogonophorans is the complete lack of an alimentary canal or mouth. How is it possible for a free-living worm like a pogonophoran to exist and thrive without any vestige whatsoever of a mouth or alimentary tract? Apparently, either nutritive materials are absorbed by direct uptake from the water or bottom ooze by the general body surface, especially that of the tentacles, or digestion is extraorganismic amid the

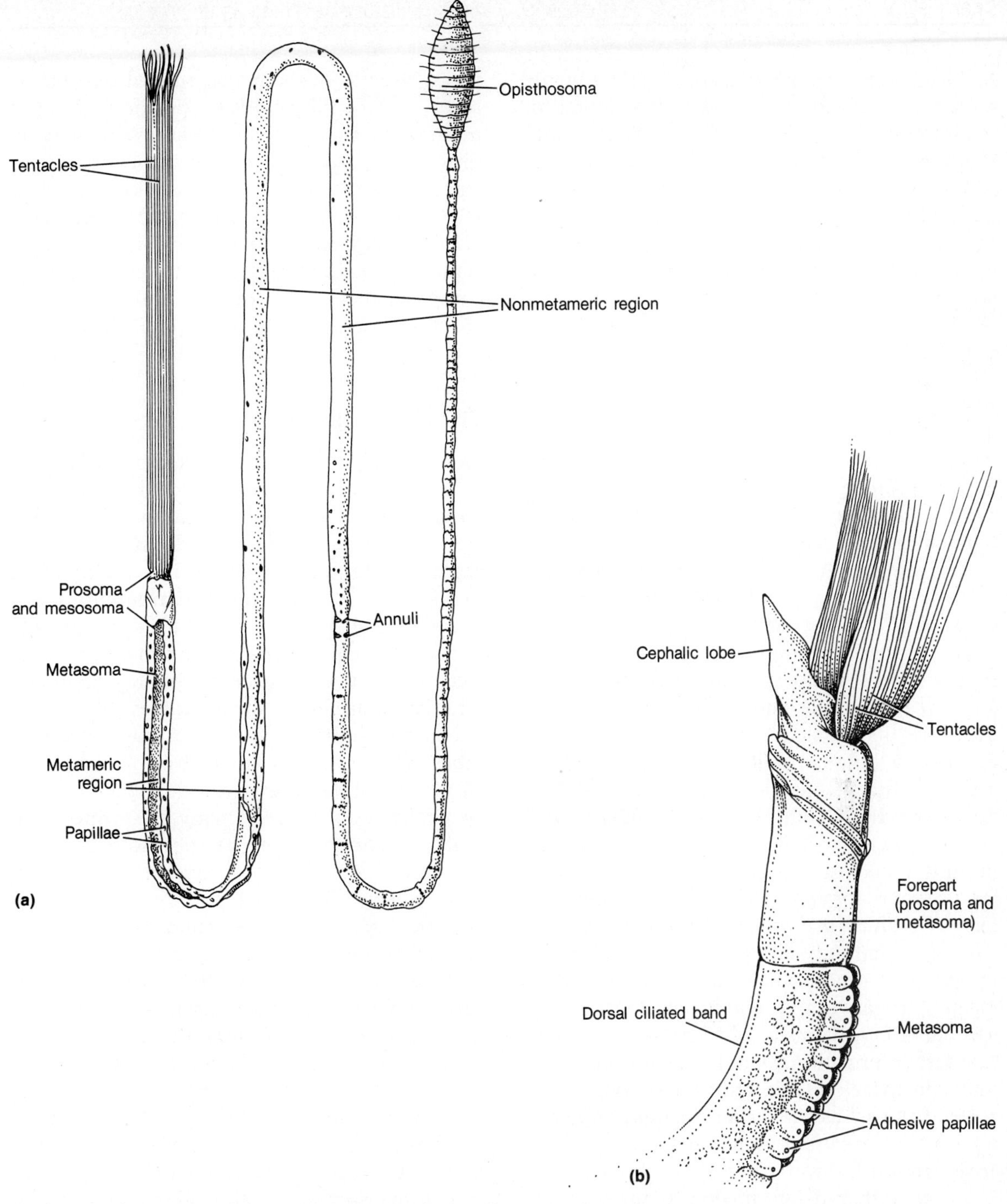

Figure 9.27 *External features of pogonophorans:* ***(a)*** Lambellisabella, *ventral view;* ***(b)*** *anterior end of a female* Spirobrachia *(parts a (modified), b from A. V. Ivanov, 1963,* Pogonophora, *Academic Press, Ltd., London, used with permission).*

tentacles and the foodstuffs are absorbed by the tentacular epithelium. A large coelom is present in all three body regions and even extends into the tentacles; but in the opisthosoma, internal septa divide the coelom into metameric compartments. Gases are undoubtedly exchanged through the tentacular surfaces. A well-developed closed circulatory system is present and supplies all parts of the body, including the tentacles. A pair of coelomoducts functioning as nephridia is present in the trunk.

Pogonophorans are dioecious, and a single pair of gonads lies in the trunk. Sperm are released through a single male gonopore as spermatophores. In some the embryos are brooded within the tube, but details of fertilization and development are poorly known. It is not known whether a larval stage is present.

In the spring of 1979, marine zoologists, collecting near some hot water vents near the Galapagos Islands, dredged up enormously elongate pogonophorans over 2.5 m long. The pink worms produce a white nylon-type tube, and from the upper end of the tube the worm protrudes its tentacles, which are bright red, apparently because of the presence of blood hemoglobin. No means of locomotion is evident, and the worms lack eyes, mouth, and alimentary canal. This entire thermal vent community is a fascinating one, since the base of its food chain is apparently chemoautotrophic bacteria. We know of no other large naturally occurring community in which chemosynthesis rather than photosynthesis is the principal means of converting inanimate energy into foodstuffs by organisms. The future will bring far more information about the pogonophorans found in these odd conditions and about their ecology as well as similar information on many other weird creatures that inhabit these deep areas around hot water vents.

Pogonophorans are related to annelids, although lacking many typical annelidan features. They were once considered to be deuterostomates, but the fairly recent discovery of the posterior opisthosoma has caused a reexamination of the phylogenetic position of pogonophorans. Details of the opisthosoma now indicate clearly that beardworms are protostomates, have a schizocoelom, and are related to annelids. The segmental nature of the papillae and the opisthosoma with internal septa, metamerically arranged schizocoelomic compartments, and metamerically arranged chaetae strongly indicate annelidan affinities.

PHYLUM SIPUNCULA

The sipunculids, called peanutworms because of their shape, are represented by about 300 species of exclusively marine, wormlike, benthic creatures. They are found mostly in shallow water but range from intertidal regions to depths of 5000 m. They live in mud or sand, in mollusc shells or annelid tubes, in crevices in rocks or coral reefs, or in almost any protected retreat; one species lives in abandoned foraminiferan shells. Burrows may be mucus-lined, but sipunculids do not construct true tubes. They range in length from 2 mm up to 75 cm, but most are from 10 to 35 cm long. Generally drab colored, many are white, light pink, or brown. Some of the more common genera are *Phascolosoma, Sipunculus* (Fig. 9.28a), and *Golfingia* (Fig. 9.28b).

The body of a sipunculid is never segmented but is composed of an anterior introvert and a posterior trunk (Fig. 9.28a, b). The **introvert** is a narrow cylindrical structure that can be either shorter or longer than the trunk and can be retracted or invaginated into the trunk. When extended, the anterior tip of the introvert contains the **mouth**, often surrounded by ciliated tentacles, lobes, or a ciliated scalloped fringe. All tentacles, lobes, and fringes contain grooves whose cilia sweep food particles toward the mouth. Food consists mostly of small invertebrates such as protozoans, crustaceans, annelids, and platyhelminths or organic material

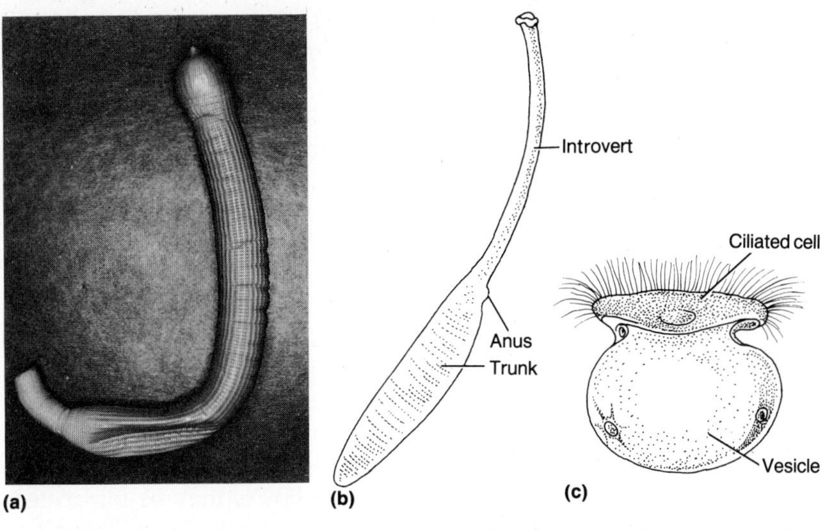

Figure 9.28 *Sipunculids:* ***(a)*** *photograph of* Sipunculus; ***(b)*** Golfingia; ***(c)*** *a free urn cell of* Sipunculus. *(parts b, c [from Selensky] in L.H. Hyman,* The Invertebrates. *Vol. V:* Smaller Coelomate Groups, *1959, McGraw-Hill Book Co., used with permission.)*

contained within ingested mud. The surface of the introvert bears a number of various tubercles, spines, hooks, plates, and other rough projections. The **trunk** may be a short sac or a long cylinder, and its surface lacks spines and hooks but often bears papillae that function as holdfast structures. In most crevice-dwellers a thickened collar is present near the anterior trunk, where it is probably used to block the crevice opening. The **anus** is a middorsal opening near the anterior end of the trunk.

The body wall is annelidlike, having all layers found in any typical annelid. A large fluid-filled coelom runs the length of the trunk, and contraction of body wall muscles against the coelomic fluid causes the hydraulic extension of the introvert. Special retractor muscles, extending from the body wall to the esophagus, pull the introvert back and may invaginate it within the trunk. The coelomic fluid contains many kinds of materials (proteins and ions) and red corpuscles (up to 100,000 per cubic millimeter) containing the respiratory pigment hemerythrin. Peculiar cellular bodies called **urns** are also present and are fixed or free swimming in the coelomic fluid. Fixed urns are clusters of peritoneal cells shaped like a vase and are found on the peritoneum or in compensation sacs (mentioned below). Detached free urns move about in the coelom gathering waste products and degenerating cells (Fig. 9.28c), and this debris is carried to the nephridia for elimination.

The tentacles are hollow and fluid filled, but these spaces, rather than being connected to the coelom, are connected to a series of anterior channels, which in turn join a circumesophageal ring canal. Extending from the ring canal are one to four blind, muscular **compensation sacs**; their muscular contractions force fluid into the tentacles and thus extend them hydrostatically. When the tentacles contract, the compensation sacs receive fluid from the tentacular spaces.

Food, gathered by the tentacles and introvert and swept into the mouth, passes into the esophagus, which in turn continues as a long, U-shaped intestine arranged in a single spiral coil. As the intestine proceeds anteriorly, it becomes a rectum that opens to the outside as the anus.

There are no specialized organs for gaseous exchange or circulation as the function of the latter is assumed by the coelomic fluid and corpuscles. A pair of large metanephridia is situated in the anterior trunk, and each metanephridium opens as a nephridiopore in the anterior trunk. The anterior nervous system is very much annelidlike with a supraesophageal brain, a pair of circumesophageal connectives, and a ventral nerve cord that runs the length of the trunk but lacks ganglia and therefore is not metameric. Chemoreceptors and tactile sensory cells are prevalent especially on the anterior tip of the introvert and on the tentacles. Some sipunculids are equipped with frontal, cerebral, or nuchal organs that are ciliated chemoreceptors, and some have ocelli embedded within the brain itself.

Sipunculids are dioecoius, but there is no external sexual dimorphism. Gonads, attached to the coelomic wall, liberate immature gametes into the coelom to mature; later they escape to the outside via the metanephridia. The advent of sperm into the water is a stimulant for the female to shed her eggs. Fertilization is external, and development may be direct or it may produce a trochophore larva.

Even though metamerism and chaetae are completely missing, sipunculids are considered to be related to annelids. Annelidan similarities are in the construction of the body wall, nature of the anterior nervous system, and certain details of their embryology, including the development of a trochophore larva in some.

PHYLUM ECHIURA

Echiurans are represented today by about 110 species of benthic marine worms. Found in habitats similar to those of sipunculids, they live mostly in shallow water but are sometimes found at considerable depths. They live buried in muddy or sandy bottom material or in coral or rock crevices, and one species is found in the tests of dead sand dollars. Body lengths range from 15 mm to 50 cm, but an extensible proboscis may increase their length up to 2 m; most are from 2 to 8 cm long excluding the proboscis. Generally, they are brown in color, but some may be green or pinkish red. Some of the more frequently encountered genera are *Thalassema* (Fig. 9.29a), *Urechis* (Fig. 9.29d), and *Echiurus*.

The body of an echiuran is divided into a trunk and an anterior proboscis (Fig. 9.29a, c). The **proboscis**, a greatly elongated prostomium, is a large flattened projection that cannot be retracted into the trunk. The proboscis is variable in length depending upon the species; *Ikeda*, a Japanese form, has a trunk of up to 50 cm and a proboscis of 1.5 m, but in *Echiurus* the proboscis is always less than one-half the length of the trunk. The proboscis may be bifurcated, its distal end flared, or it may be cylindrical or ribbon shaped. The lateral edges of the extended proboscis are folded ventrally to form a **ciliated groove** or **gutter** used to transport food, gathered by the proboscis, to the mouth (Fig. 9.29a, c). Echiurans are often called spoonworms because of the spatulate nature of the proboscis. The **mouth** is situated near the base of the proboscis, and just behind the mouth and located near the proboscidian base is a pair of closely placed chitinous **chaetae**, which are employed in burrowing. The chaetae, as in annelids, are secreted by a setal sac, and each has extrinsic muscles enabling it to move. The **trunk** is cylindrical and sausage shaped and may have a smooth surface or one with papillae arranged irregularly or in transverse rings. The papillae contain glandular cells that secrete mucus, which lines the burrow. Some species have additional setae in one or two rows near the posterior end. An **anal opening** is found at the posterior end of the trunk.

The echiuran body wall is very similar to that of annelids and sipunculids. A large, nonsegmented, fluid-filled coelom is present within the trunk. Noncoelomic canals and lacunae within the proboscis are connected to the coelom, and the coelomic fluid circulates within the proboscidian cavities. Coelomic fluid contains

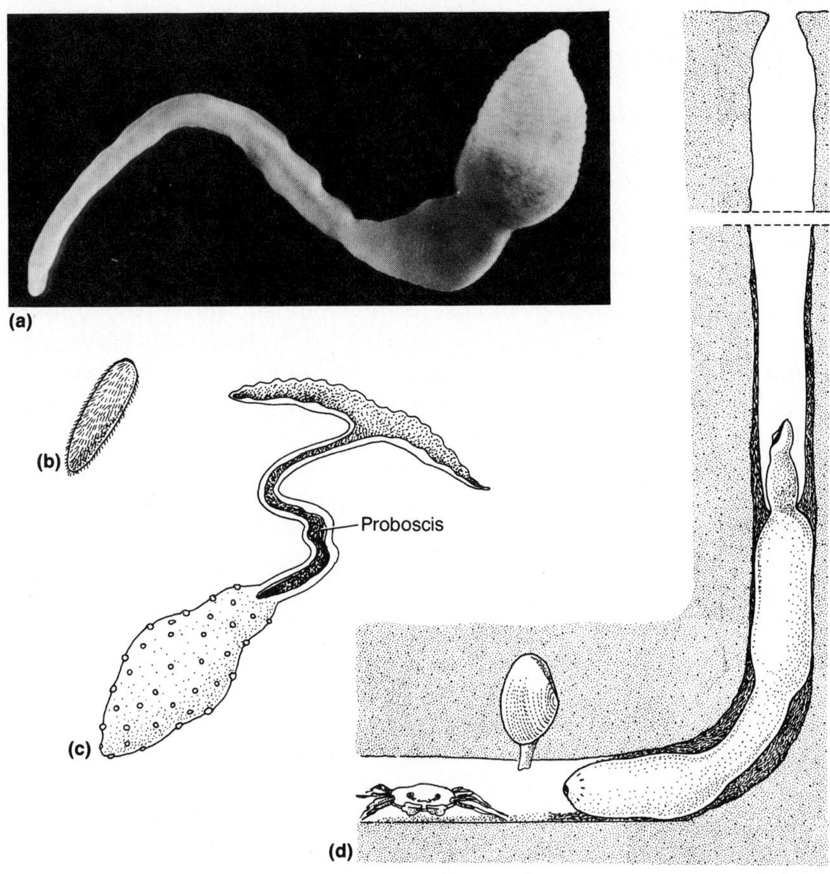

Figure 9.29 *Echiurans:* ***(a)*** *a photograph of* Thalassema hertmani *(courtesy of the Smithsonian Marine Station, Ft. Pierce, FL);* ***(b)*** *a male* Bonellia, *greatly enlarged;* ***(c)*** *a female* Bonellia, *greatly reduced;* ***(d)*** Urechis *in its tube. (parts b–d from G.E. and N. MacGinitie, 1968,* Natural History of Marine Animals, *2nd ed., McGraw-Hill Book Co., adapted with permission.)*

phagocytic amebocytes, erythrocytes containing hemoglobin, and, in several species, coelomic urns like those of sipunculids.

Feeding is accomplished by the remarkable proboscis, which can be extended for a considerable distance and which sweeps detritus, deposited materials, and small invertebrates into the proboscidian ciliated groove and into the ventral mouth. *Urechis*, a Pacific coast form, lives in a U-shaped burrow (Fig. 9.29d) and feeds in a manner similar to that of the polychaete *Chaetopterus*. The alimentary canal consists of a short muscular pharynx, a long coiled esophagus, gizzard and stomach in some, a long and greatly coiled intestine, a short rectum, and a posterior anus. A parallel ciliated tube, the **siphon**, rises ventrally from the anterior end of the intestine and joins the alimentary canal again toward the posterior intestine. The precise function of the siphon is unclear, but it apparently serves as a shunt for some intestinal contents.

A closed circulatory system similar to that of polychaetes is present in most forms. A dorsal vessel carries blood forward to the anterior trunk, where it divides to form two circumpharyngeal branches. They unite ventrally to form a ventral vessel that carries blood posteriorly beneath the alimentary canal. One to several pairs of contractile circumintestinal vessels transport blood dorsally to the dorsal vessel. Blood is colorless but contains some amebocytes.

Some echiurans contain a single metanephridium, but others have up to several hundred pairs; in some forms, females have fewer excretory organs than do males. Echiurans possess an unusual pair of accessory excretory organs called **anal sacs** or **vesicles**. These sacs are often highly branched, and the inner ends are ciliated nephrostomes located in the coelom. The other ends of the tubules open into the posterior intestine or rectum; these organs function primarily in excretion, but the products exit to the outside via the anus. There is a nervous system composed of small cerebral ganglia, a pair of circumesophageal connectives, and a nonsegmented, nonganglionated ventral nerve cord. These are no special sense organs, but the proboscidian surface is well equipped with tactile and chemical receptors.

All echiurans are dioecious, and sexual dimorphism is dramatic in some. Produced by the peritoneum in the posterior of the worm, immature gametes are shed into the coelom, where they mature as in annelids. The metanephridia, never the anal sacs, serve as gonoducts, and gametes are released into the seawater, where fertilization ensues. Species of *Bonellia* are unusual in that females are up to 1 m long while males are only 1–2 mm long (Fig. 9.29b, c). Males live symbiotically in the metanephridia or coelom of the female, and fertilization takes place in the metanephridia. *Bonellia* is also unique in the method of sex determination. All are potentially bisexual, and when a larva develops independently, it becomes a female. But when a larva comes in contact with a female, it become transformed into a male probably by female hormones interacting on the larval genome to produce a male.

Early embryonic development is remarkably similar to that in annelids. Cleavage is spiral, and a trochophore larva is produced. In subsequent development there is a hint of segmentation as a series of mesodermal somites and coelomic pouches develop in several species, but traces of metamerism are soon lost in development, and adults are never metamerically arranged.

It is generally held that echiurans are related to annelids. Details of the circulatory system (except in sipunculids), presence of chaetae, repetitive series of metanephridia, developmental appearance of metamerism and a trochophore larva, and structure of the body wall all suggest close phylogenetic affinities between echiurans, annelids, and sipunculids.

PHYLUM ONYCHOPHORA

Onychophorans are a very interesting group of organisms that share some features with both annelids and arthropods. Earlier writers often referred to them as the "missing link" between these two great phyla, but onychophorans have distinctive characteristics of their own and should be considered as constituting a separate phylum. The ancestral stock from which onychophorans, annelids, and arthropods arose was remarkably plastic and adaptable; onychophorans capitalized on this stock to produce a group of animals with an unusual functional morphology.

Onychophorans are terrestrial and wormlike and range in length from 1.5 to 15 cm. They are found in restricted areas within tropical and subtropical regions such as Australia, New Zealand, East and West Indies, the Congo, South Africa, Mexico, and northern South America. Their discontinuous distribution is explained both by the ideas of continental drift and by the likelihood that onychophorans were at one time more widely distributed but have contracted because of competitive pressures. Invariably, they live in humid habitats mostly in tropical rain forests beneath logs or stone, in tangled vegetation, or near streams. Being nocturnal and susceptible to desiccation, they are rarely seen in the daytime or during dry periods. Body colorations are gray, brown, black, or olive-green, and some have tints of blue, red, or orange. About 80 species are known, of which the most typical and best-known genus is *Peripatus* (Fig. 9.30a).

External Features

The body is elongate, cylindrical, and composed of a head and trunk. The **head**, not distinctly separated from the body, is comprised of the **prostomium** or **acron** plus several fused anterior segments (Fig. 9.30b). The prostomium bears a pair of large annulated **antennae** and a pair of simple **ocelli**. In adults the antennae are preoral, but there is embryological evidence that they are paired appendages of a rudimentary first segment that is postoral. Posterior and lateral to each antenna is a short conical **oral papilla** that flanks the ventral **mouth** (Fig. 9.30b). A gland opens at the tip of each oral papilla and produces a milky adhesive material that can be squirted up to a distance of 50 cm and hardens rapidly upon contact with air to immobilize effectively small would-be predators. On either side of the mouth is a hard two-pronged **jaw**, which represents paired segmental appendages. The **trunk** is segmented, but its metamerism is evident externally only from the presence of 14 to 43 pairs of legs. Each **leg** is a stubby, conical, nonjointed protuberance extending ventrolaterally from the body wall and terminating in a pair of curved **claws** (Fig. 9.30). Proximal to the terminal claws and on each leg are three to six ventral, transverse pads. Surfaces of the body and legs contain numerous transverse rings of sensory tubercles that are covered with minute scales. The **anal opening** is situated at the extreme posterior end.

Locomotion is slow in onychophorans, and retrograde muscular waves propel the body forward methodically. When a segment elongates, the legs of that segment are raised and brought

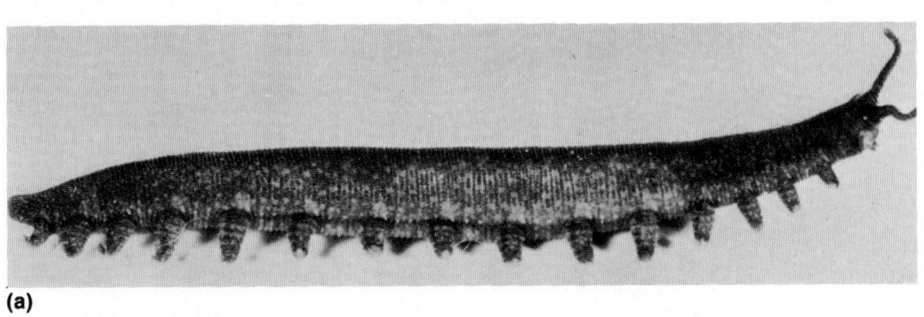

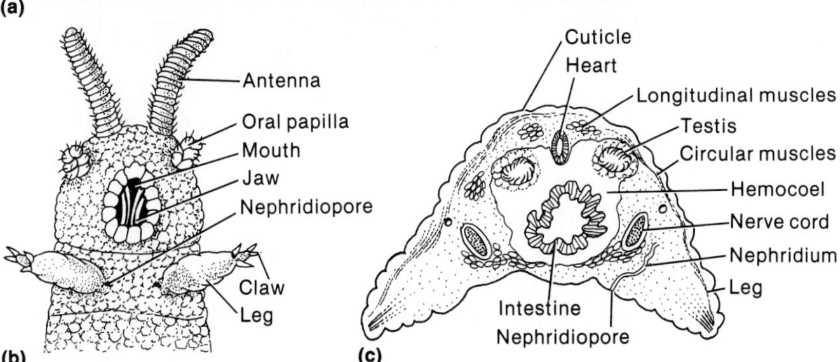

Figure 9.30 *Onychophoran features; **(a)** a photograph of* Peripatus *(from Carolina Biological Supply Co.); **(b)** a ventral view of the anterior end; **(c)** a cross-sectional view through a trunk segment of a male.*

forward in a recovery stroke, and in the effective stroke the segment is pushed forward. The legs of only a few segments are in contact with the substratum at any one time.

Internal Features

The body wall is mostly annelidlike and is composed of the following layers, in order, beginning with the outermost: an exceedingly thin chitinous cuticle, epidermis, connective tissue dermis, and circular, diagonal, and longitudinal muscles (Fig. 9.30c). No parietal peritoneum is widely present, however, because the coelom is reduced to small sacs associated with the gonads and nephridia. Instead, a large, ill-defined **hemocoel** reminiscent of that in arthropods or molluscs is present (Fig. 9.30c).

Onychophorans are predaceous and feed on snails, insects, annelids, and other small invertebrates, but they may ingest plant material as well. Prey is caught by the jaws, maceration is aided by the secretions of a pair of nephridia that function as salivary glands, and the partially digested food is sucked into the foregut consisting of a phayrnx and esophagus. A long straight intestine or midgut follows immediately behind the esophagus, and it is here that digestion is completed and absorption takes place. A short hindgut or rectum opens to the outside as the anus.

The circulatory system consists of a long, dorsal, tubular **heart** whose walls bear a pair of **ostia** in each segment, which are points of entry for the blood into the heart. Blood is pushed anteriorly from the heart into the hemocoel. The hemocoel contains several longitudinal, horizontal partitions or septa, which form several sinuses; the partitions are perforate so that blood moves fully between the sinuses. After somatic circulation the blood, which is colorless and contains amebocytes, is returned to the heart to complete the circuit.

Excretion is accomplished by means of a pair of metanephridia present in each segment; each nephridium is composed of a nephrostome that lies in a small coelomic end sac, a coiled nephridial tubule that is dilated as a contractile bladder, and a nephridiopore located on the medial base of each leg (Fig. 9.30c). Uric acid is the chief excretory product. It is instructive to note that the nephrostomes of onychophorans are ciliated, and cilia are characteristically absent in arthropods.

Gaseous exchange in onychophorans is by means of tracheae similar to those in most arthropods. **Tracheae** are simple, small, straight tubules that permeate the interior of the animal, and each tracheal group arises from a common **atrium** that opens to the outside via a **spiracle**. Ventilation of the tracheal system is achieved by contraction of body wall muscles. Because there are no ways to close the spiracles effectively and because onychophorans potentially lose considerable volumes of water through these pores, onychophorans reduce the possibility of desiccation by living in humid habitats or by being nocturnal.

The nervous system consists of a large pair of suprapharyngeal ganglia or brain, a pair of circumpharyngeal connectives, and a double ventrolateral nerve cord (Fig. 9.30c). Each nerve cord bears a ganglion in each segment, from which arises a number of lateral nerves for each segment. Sensory organs include pairs of ocelli, antennae, oral papillae, numerous tubercles on the body and legs, and hygroscopic and tactile receptors found generally over the body surface.

All onychophorans are dioecious, and males are often somewhat smaller and have fewer legs than females. The elongate gonads are paired and lie in the posterior body above the intestine, and the single gonopore is situated ventrally just anterior to the anus. The ovaries are fused, but each ovary is connected to a nephridium that serves as a gonoduct, which is specialized regionally to form a seminal receptacle and uterus; the two uteri join near the posterior end and open to the outside by the female gonopore. The testes are separate, and, as in females, a nephridium serves as a genital or sperm duct. The two sperm ducts join into a single tube where spermatophores are transferred to the female, and fertilization is internal. In a few species, spermatophores are formed; at copulation the spermatophores are deposited on the body surface of the female, and sperm penetrate her integument, get into the hemolymph, and eventually find their way to the ovaries. Most onychophorans retain the young within the uteri of the female, and they are either viviparous or ovoviviparous, but a few lay eggs with relatively large amounts of yolk and a chitinous shell. Embryonic development is similar to that in arthropods with meroblastic (partial) cleavage, but no larval stage is developed. The young are born after 6–13 months of development, and postnatal maturation takes somewhat less than a year.

Many onychophoran features are clearly annelidan and include details of the body wall, a very thin cuticle, structure of the nephridia, presence of cilia, and the unjointed appendages. Yet other features are decidely arthropodan, including the chitinous cuticle, reduced coelom associated with an open circulatory system and a hemocoel, and modified appendages (jaws) used for feeding. Since onychophorans share features with both great phyla but possess some unique features, they probably arose very early from an ancient worm from which annelids and arthropods also arose. Perhaps it is strategic that onychophorans appear at this point in the text—precisely intermediate between annelids and arthropods.

PHYLOGENY OF ANNELIDS AND SMALL RELATED PHYLA

At the beginning of the Paleozoic period some 600 million years ago, most phyla, including annelids, had already been established with scant fossil evidence of earlier ancestors. Therefore there is no clear paleontological or developmental evidence to describe the ancestral annelid. Yet we can logically speculate that it was small, wormlike, and bilaterally symmetrical and lived on the bottom in littoral marine habitats. The ancestral stock may have been a platyhelminth or nemertean worm that burrowed in the soft benthic material. Strong corroborative evidence comes from several sources, the most important being the close similarities between flame cells of acoelomates and the solenocytes of primitive polychaetes. The presence of a trochopore larva in marine molluscs and annelids strongly suggests that both groups shared a common ancestry in their remote past. At some point in the evolution of the acoelomate ancestor, two adaptations were evolved that were to be of immense importance to subsequent evolution of annelids—a coelom and metamerism (Fig. 9.31). Theories regarding the origin of a coleom were discussed briefly in Chapter 3, but the development of a coelom was most likely an adaptation to a burrowing existence. A fluid-filled coelom could be employed efficiently in peristaltic burrowing; and once it evolved, metamerism of that coelomic space into compartments for more efficient burrowing was a most important subsequent adaptation (Fig. 9.31). However they evolved, the development of a coleom and metamerism are the two most important new adaptations in annelids.

There are two theories as to the origins of the major annelidan taxa. One theory holds that polychaetes were the ancestral group and that clitellates arose from them, and the other states both polychaetes and clitellates arose independently from the ancestral annelid. In either case, both polychaetes and clitellates have retained many of the basic ancestral features but have adapted in other unique ways such as the parapodia in polychaetes and hermaphroditism in clitellates. The archiannelids, considered by some earlier workers to be primitive and approximating the ancestral annelid, are herein

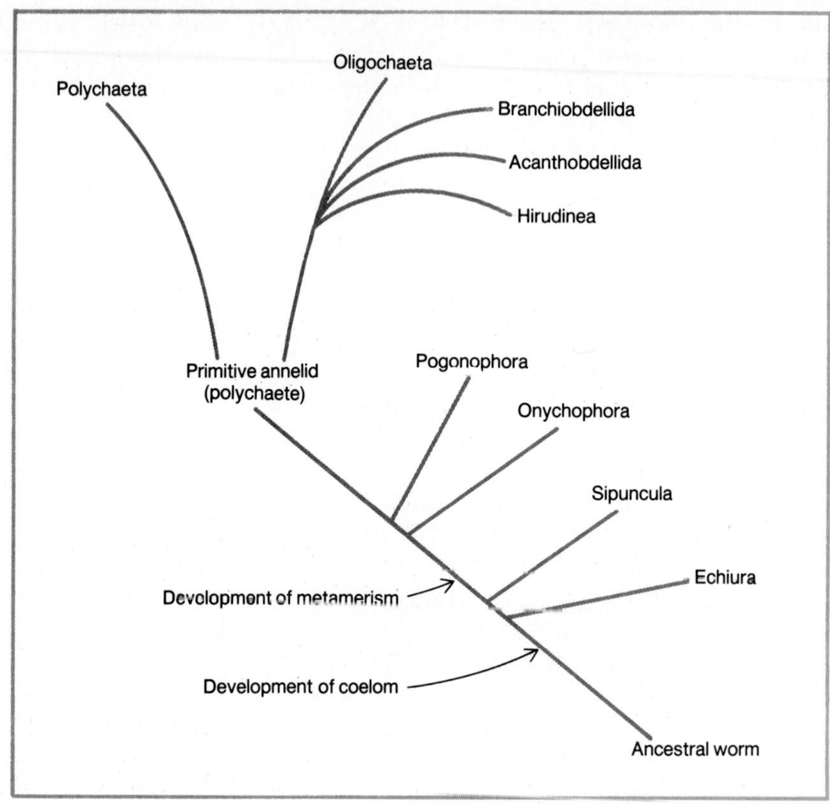

Figure 9.31 *Probable phylogenetic affinities and evolutionary pathways in the Phylum Annelida and four small related phyla.*

treated as a polychaete group that has secondarily become less specialized.

Once established, clitellates and especially polychaetes underwent adaptive radiation to form two quite diverse groups of worms. There seems to be little disagreement with the idea that hirudineans are closely related to oligochaetes and that they arose from a freshwater oligochaete stock (Fig. 9.31). The puzzling branchiobdellids and *Acanthobdella* are intermediate between oligochaetes and hirudineans and probably arose from an ancient oligochaete that may have given rise to hirudineans.

The phylogenetic positions of the remaining phyla distantly related to annelids—Pogonophora, Sipuncula, Echiura, and Onychophora—are tenuous and open to debate. Each of these taxa shares some annelid features, and for reasons stated earlier they are considered to be related phyla. Undoubtedly, each arose very early from the ancestral annelidan stock, and each group became adapted in rather unique ways (Fig. 9.31).

SYSTEMATIC RESUME

The Major Taxa of Annelids

CLASS POLYCHAETA—bristleworms, tubeworms; numerous metameres, each bearing a pair of lateral parapodia; each parapodium bears two bundles of chaetae; head well developed, often with tentacles or radioles; mostly dioecious with a trochophore larva; mostly marine.

Order Orbiniida—acron without appendages; parapodia biramous; simple chaetae; includes *Orbiniella* and *Protoaricia*.

Order Ctenodrilida—prostomium without appendages; proboscis a ventral pad; parapodia without lobes; small, grublike; *Ctenodrilus* is the type genus.

Order Psammodrilida—prostomium lacking appendages; palps absent; setae simple; often interstitial; *Psammodrilus* may be common.

Order Cossurida—acron lacking appendages; biramous parapodia; simple setae; only one genus known, *Cossura*.

Order Spionida—prostomium distinct but without appendages; setae all simple; proboscis often well developed; *Magelona, Polydora, Boccardia,* and the parchment worms (*Chaetopterus* [Fig. 9.13]) are among the best known.

Order Capitellida—prostomium lacking appendages; simple setae; parapodia biramous; includes many common worms such as *Capitella,* the lugworms (*Arenicola* [Fig. 9.9h]), and the bambooworms (*Maldane, Clymenella*).

Order Opheliida—palps absent; prostomium lacking appendages; parapodia usually biramous; simple setae; includes *Ophelia* and *Scalibregma*.

Order Phyllodocida—prostomium with one or more pairs of antennae; one or two pairs of jaws, absent in some; usually biramous parapodia; eversible proboscis; includes most of the errant polychaetes such as the sea mice or scaleworms (*Aphrodita*), *Phyllodoce, Polynoe, Lepidonotus, Syllis,* the sandworms (*Nereis* [Figs. 9.9b, 9.10a, b]), *Neanthes,* beak throwers (*Glycera* [Figs. 9.9a, 9.12a]), and *Nephthys* (Fig. 9.9c).

Order Amphinomida—acron distinct and bearing one or more antennae; eversible proboscis but lacking jaws; includes fireworms (*Hermodice*) and *Euphrosine*.

Order Spintherida—ectoparasitic on sponges; prostomium with a median antenna; body ovate; one genus, *Spinther*.

Order Eunicida—up to five prostomial antennae, absent in some; muscular ventrolateral pharynx; one or more pairs of jaws; some of the more familiar genera are *Diopatra,* palolo worms (*Eunice* [Fig. 9.16a]), and *Arabella*.

Order Sternaspida—dark red or yellow chitinized shield; burrowers; *Sternaspis* is typical.

Order Oweniida—head of fused acron and anterior segments; proboscis is a ventral pad, includes *Owenia* and *Flabelligera*.

Order Fauveliopsida—mostly deep-water forms; acron and peristomium without appendages; *Fauveliopsis* is the only genus.

Order Terebellida—tubicolous worms; peristomium with feeding appendages; includes *Monorchos, Amphitrita* (Fig. 9.12b), *Terebella,* and *Cistenides*.

Order Sabellida—crown of peristomial radioles; smooth, tapering body; almost all sedentary; have companion setae in thoracic region; includes the featherdusters (*Sabella, Myxicola*) and the fanworms (*Serpula, Spirorbis, Hydroides* [Fig. 9.12c]).

Order Archiannelida—heterogeneous, interstitial group; have reduced parapodia, reduced chaetae, and a muscular feeding buccal organ; two of the better-known genera are *Nerilla* and *Polygordius*.

Order Myzostomida—strange group; live symbiotically on or in echinoderms; flattened body; five pairs of reduced parapodia; hermaphroditic; *Myzostoma* is perhaps the best known.

CLASS CLITELLATA—earthworms, leeches; found in terrestrial, freshwater, or marine habitats; hermaphroditic; clitellum present and produces a cocoon; gonads distinct and found only in a few segments; no parapodia; no larval stage.

Subclass Oligochaeta—mostly freshwater, and earthworms are terrestrial; segmentation is conspicuous with large numbers of segments.

Order Lumbriculida—freshwater worms; each segment has four setae; male gonopore is in same segment as one pair of testes; clitellum includes gonopores of both sexes; *Lumbriculus* is the best-known genus.

Order Moniligastrida—tropical, terrestrial, found in Asia; each segment has four chaetae; clitellum includes gonopores of both sexes; best-known genus is *Moniligaster*.

Order Haplotaxida—most oligochaetes; includes aquatic forms and all earthworms; four or eight chaetae present per segment; testes in at least one segment anterior to male gonopores; familiar genera include *Enchytraeus* (Fig. 9.18a), *Tubifex* (Fig. 9.22a), *Dero,* the common nightcrawler (*Lumbricus* [Figs. 9.18b, 9.20]), *Ei-*

senia, and the Australian giant earthworm (*Megascolides*).

(Note: The status of the aeolosomatids is unclear at present (Fig. 9.22c). Traditionally, they have been considered to be a primitive family of the Order Haplotaxida. But many now consider them to be unrelated to oligochaetes on the basis of reproductive details, location of brain, and the absence of blood pigments. This note will not resolve the controversy, only highlight it.)

Subclass Branchiobdellida—leechlike; commensal on crustaceans; lack prostomium and chaetae; trunk of 11 segments, *Cambarincola* (Fig. 9.23) is representative.

Subclass Acanthobdellida—primitive; 30 segments; chaetae and septate coelom present in anterior five segments; one species parasitic on European salmon.

Subclass Hirudinea—leeches; body with 34 segments; anterior and posterior suckers usually present; no chaetae; many are bloodsuckers.

Order Rhynchobdellida—all aquatic in fresh water or oceans; small porelike mouth in anterior sucker; no jaws, but pharnyx can be everted through mouth as proboscis; three or more annuli per segment; blood colorless; circulatory system separate from coelom; *Glossiphonia, Haementeria, Piscicola,* and *Illinobdella* are among the best-known genera.

Order Gnathobdellida—aquatic or terrestrial; medium to large mouth; no proboscis; three jaws; five annuli per segment; stomach ceca but no intestinal ceca; includes the medicinal leech (*Hirudo* [Fig. 9.25a]), *Haemopsis, Macrobdella,* and *Xerobdella*.

Order Pharyngobdellida—mostly freshwater and semiterrestrial; medium to large mouth; no proboscis or teeth; numerous testes; some of the more common genera are *Erpobdella* and *Nephelopsis*.

ADDITIONAL READINGS

Boudreaux, H.B. 1979. *Arthropod Phylogeny with Special Reference to Insects*, pp. 1–41. New York: John Wiley & Sons.

Brinkhurst, R.O. 1982. *British and Other Marine and Estuarine Oligochaetes*, 476 pp. New York: Cambridge University Press.

Brinkhurst, R.O. and D.G. Cook, eds. 1980. *Aquatic Oligochaete Biology*, 539 pp. New York: Plenum Publishing Corp.

Cheng, T.C. 1973. *General Parasitology*, pp. 894–903. New York: Academic Press.

Dales, R.P. 1963. *Annelids*, 194 pp. London: Hutchinson.

Edwards, C.S. and J.R. Lofty. 1977. *Biology of Earthworms*, 2nd ed., 300 pp. New York: Halsted Press.

Fauchald, K. 1977. *The Polychaete Worms—Definitions and Keys to the Orders, Families and Genera*, 179 pp. Science Series 28. Los Angeles: Natural History Museum of Los Angeles County.

Foster, N. 1972. *Freshwater Polychaetes (Annelida) of North America*, 15 pp. Identification Manual 4, Environmental Protection Agency. Washington, D.C.: U.S. Government Printing Office.

Hyman, L.H. 1959. *The Invertebrates: Smaller Coelomate Groups*. Vol. V, pp. 208–225 (Pogonophora), 610–690 (Sipuncula). New York: McGraw-Hill Book Co.

Ivanov, A.V. 1963. *Pogonophora*, 461 pp. New York: Academic Press.

Kaestner, A. 1967. *Invertebrate Zoology*. Vol. I, pp. 425–432 (Sipuncula), 433–441 (Echirua), 442–564 (Annelida); Vol. II, pp. 1–15 (Onychophora). New York: John Wiley & Sons.

Klemm, D.J. 1972. *Freshwater Leeches (Annelida: Hirudinea) of North America*, 53 pp. Identification Manual 8, Environmental Protection Agency. Washington, D.C.: U.S. Government Printing Office.

Laverack, M.S. 1963. *The Physiology of Earthworms*, 183 pp. New York: Macmillan Publishing Co.

Mill, P.J. 1978. *Physiology of Annelids*, 658 pp. New York: Academic Press.

Morris, R.H., D.A. Abbott, and E.C. Haderlie. 1980. *Intertidal Invertebrates of California*, pp. 448–498. Stanford, California: Stanford University Press.

Muller, K.J., J.G. Nicholls, and G.S. Stent. 1981. *Neurobiology of the Leech*, 320 pp. Cold Spring Harbor, N.Y.: Cold Spring Harbor Laboratory.

Nicholls, J.G. and D. Van Essen. 1974. The nervous system of the leech. *Sci. Amer. 230*(1):38–48.

Pennak, R.W. 1978. *Fresh-water Invertebrates of the United States*, 2nd ed., pp. 710–768. New York: John Wiley & Sons.

Reish, D.J. and K. Fauchald, eds. 1977. *Essays on Polychaetous Annelids*, 604 pp. Los Angeles: Allan Hancock Foundation, University of Southern California.

Rice, M.E. and M. Todorovic, eds. 1975. *Proceedings of the International Symposium on the Biology of the Sipuncula and Echiura, 1970*. Vol. I, 355 pp.; Vol. II, 204 pp. Washington, D.C.: National Museum of Natural History, Smithsonian Institution.

Satchell, J.E., ed. 1983. *Earthworm Ecology from Darwin to Vermiculture*, 487 pp. New York: Chapman and Hall.

Sawyer, R.T. 1972. *North American Freshwater Leeches, Exclusive of the Piscicolidae, with a Key to All Species*, 147 pp. Illinois Biological Monograph No. 46, Urbana: University of Illinois Press.

Stephen, A.C. and S.J. Edmonds. 1972. *The Phyla Sipuncula and Echiura*, 474 pp. London: The British Museum (Natural History).

10

Introduction to the Arthropods

OVERVIEW

Many adjectives can be used in attempting to describe arthropods—diverse, ubiquitous, preeminently successful, adaptively versatile, heterogeneous—but none does justice to this largest of all phyla. Arthropods share many features with annelids, but several other extremely important adaptations are unique to arthropods. The exoskeleton or cuticle has had a singular and universal influence on the functional morphology and success of arthropods. Composed of chitin and proteins, the protective, supportive, laminated cuticle consists of hardened regions or plates interspersed with articular points where the cuticle is quite thin and flexible. Muscles insert on the rigid parts, and flexion of the skeleton takes place at the points of articulation. Each segment often bears paired legs that are jointed and mobile because of articular points in the appendicular cuticle. In order for an arthropod to grow, it must secrete a new cuticle beneath the old, molt the old, and expand abruptly before the new cuticle hardens. The presence of a cuticle has affected the nature of sense organs, has altered many visceral systems, and has resulted in severe reduction of the coelom and modifications of muscles essential for movement. Other important arthropod characteristics include a high degree of tagmatization and cephalization, the development of compound eyes, and the presence of an open circulatory system with a hemocoel. Generally, arthropods are dioecious, females produce yolky eggs that undergo meroblastic cleavage, and an embryo usually hatches into some sort of feeding larva that molts repeatedly to form the adult.

Four principal lines of evolution have probably taken place in arthropods, producing the Trilobitomorpha, Chelicerata, Crustacea, and Uniramia. The trilobites, now all extinct, are of great interest because they most closely resemble the ancestors of modern-day arthropods. The trilobite body had three longitudinal lobes, and each of the segments constituting the three tagmata or regions bore biramous limbs. Even several developmental stages are known from the fossil record.

A great deal of debate still swirls around the many questions raised about the phylogeny of arthropods. Did arthropods evolve from primitive polychaetes, or did they develop independently of annelids? Is the phylum monophyletic or polyphyletic? What are the relationships of trilobites to the other three groups? Arguments on both sides of these important phylogenetic questions are briefly presented in this chapter.

INTRODUCTION

The Phylum Arthropoda is an enormously large taxon that encompasses spiders, crustaceans, and insects as well as a number of smaller groups. By any standard one wishes to use, arthropods are the most successful and dominant of all invertebrate taxa. Since this phylum consists of more than 850,000 described living species, it is, without exception, the most diverse of any animal group; some of this diversity is illustrated in Fig. 10.1. Arthropods are found in almost every conceivable habitat from the oceanic abysses to the highest mountains, from cracks in bathroom walls to endoparasites, from aerial dragonflies to terrestrial spiders to aquatic crabs. The extensive adaptive plasticity of these animals has enabled them to fill countless ecological niches, and they are vital and integral components of our biosphere.

Utilizing the best phylogenetic information currently available, this huge assemblage is subdivided into four subphyla: Trilobitomorpha, Chelicerata, Crustacea, and Uniramia. Representatives of the four groups, characterized in Table 10.1, are distinguished from each other principally by the nature of their appendages. To the Trilobitomorpha belong the trilobites, all of which are extinct; they, more than any other arthropod group, were closer to the ancestral stock for the phylum. These primitive forms will be discussed briefly near the end of this chapter. The Chelicerata, including spiders, ticks, scorpions, and many closely related forms, are basically terrestrial, have two pairs of modified anterior appendages that are employed in feeding, but lack true jaws. The Crustacea, including many small aquatic forms, shrimp, crabs,

TABLE 10.1. □ **BRIEF CHARACTERIZATIONS OF THE FOUR SUBPHYLA OF ARTHROPODS**

Subphylum	*Principal Habitat*	*Locomotor Appendages*	*Pairs of Antennae*	*Distinguishing Features*
Trilobitomorpha	Marine, benthic	Many pairs, biramous	1	All extinct; body had 3 longitudinal lobes; distinct head; abdomen with a variable number of metameres
Chelicerata	Terrestrial	4 pairs, uniramous	0	Body divided into anterior prosoma and posterior opisthosoma; no jaws; 2 pairs of feeding appendages; prosoma with legs; compound eyes generally absent
Crustacea	Aquatic, mostly marine	Many pairs, biramous	2	Gills but no tracheae; compound eyes usually present; paired jaws and 2 pairs of maxillae
Uniramia	Terrestrial	3 pairs (insects) or many pairs (myriapods), uniramous	1	Malpighian tubules; tracheae in adults; paired jaws and 2 pairs of maxillae; compound eyes

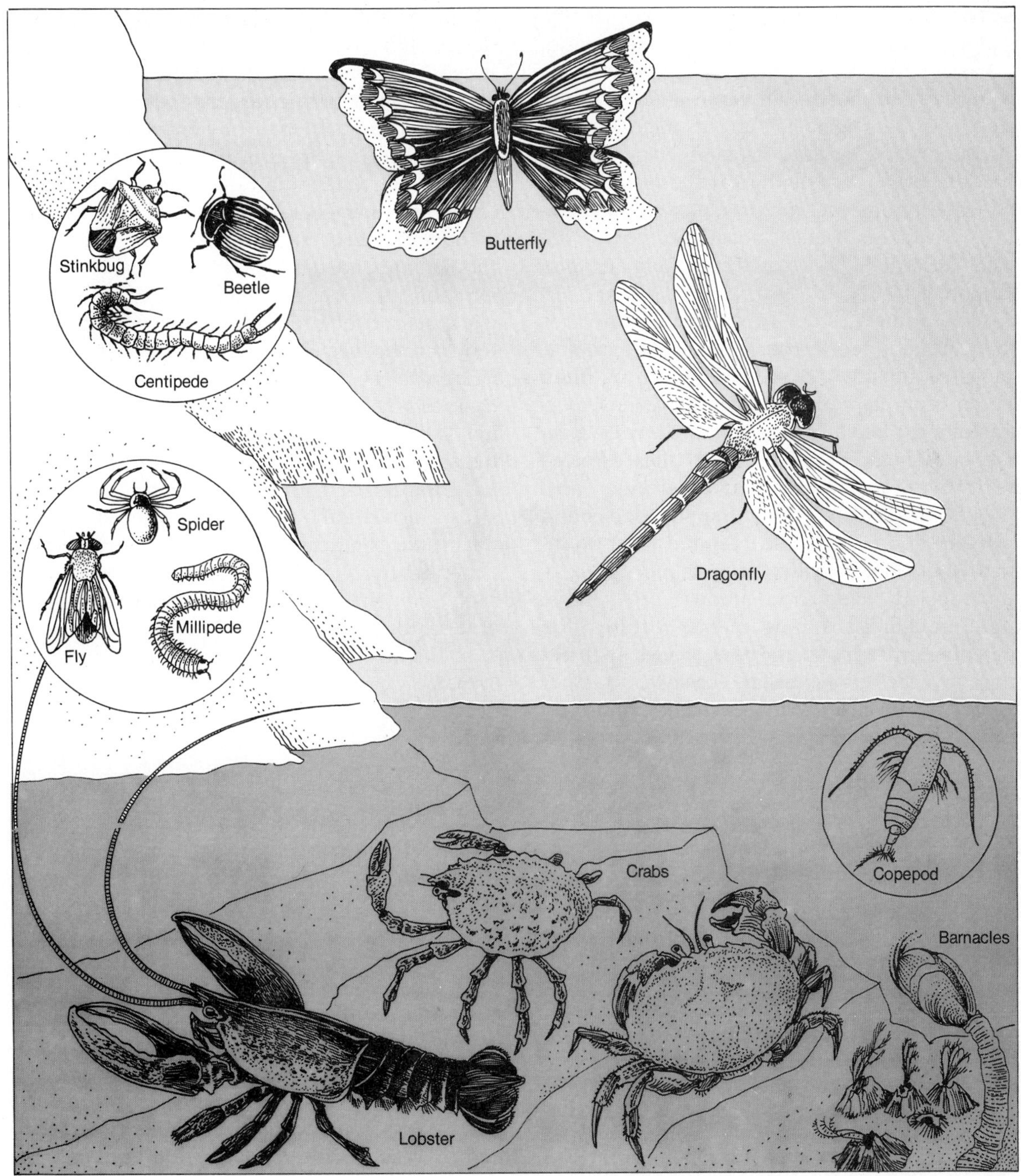

Figure 10.1 *Examples of various arthropods in the oceans, on land, and in the air to illustrate some of the enormous diversity present in this phylum, the largest of all phyla.*

and crayfishes, are mostly aquatic and have biramous appendages. To the Uniramia belongs a massive number of arthropods whose representatives are mostly terrestrial and have uniramous appendages; this group includes both the insects and myriapods (centipedes and millipedes). Because of their great species diversity and their numerical preponderance, chelicerates, crustaceans, insects, and myriapods are each accorded their own separate chapter (Chapters 11–14).

The primeval stock from which arthropods arose undoubtedly possessed a remarkable range of adaptive possibilities. During their long evolutionary history, arthropods have successfully exploited many of these latent potentialities. Their evolution clearly has affected nearly every basic system, but the development of the exoskeleton, paired appendages, and cephalization perhaps have figured most prominently in their success. Even though there is unparalleled variability among arthropods, there are still many fundamental features that are common to all taxa. Therefore we shall explore first these universal and basic features, for by doing so we can then appreciate more fully the extent of arthropodan diversity.

BASIC ARTHROPODAN FEATURES

Arthropods probably evolved from some primitive, marine, segmented worm that also gave rise to the annelids. There are some very important features shared by both phyla. First, arthropods are segmented, even though their metamerism is not as obvious as that in annelids. The presence of metamerism in both phyla is generally interpreted as strong evidence for a common ancestry. In arthropods the functional and often structural association of groups of metameres into regions or **tagmata** is much more pronounced than in annelids. Second, each of the segments of arthropods, like those of polychaetous annelids, bears a pair of appendages, but the arthropodan appendages are much more varied in function and far more complex in construction than are parapodia. There is a fundamental phylogenetic question, still not resolved at present, as to whether the polychaete parapodia and the arthropodan limbs are, in fact, homologous. Third, the arthropodan nervous system is very much like that of annelids; the brain lies dorsal to the anterior gut, a ventral chain of segmented ganglia runs the length of the body, and circumenteric connectives join the brain and nerve cord. Finally, there are some other features that arthropods share, but not exclusively, with annelids; these include their being coelomate, protostomate, and bilaterally symmetrical and having a tubular gut extending from mouth to anus. But several other extremely important arthropodan characteristics have been evolved independently of annelids. Some features—exoskeleton, jointed limbs, compound eyes—are unique to arthropods, and others—cephalization, distinct muscle bundles, open circulatory system—clearly set arthropods apart from annelids. These distinctive arthropodan features are discussed below.

The Exoskeleton

The exoskeleton is clearly the one arthropodan feature that is most important to the overall success of arthropods. The **exoskeleton** or **cuticle** is a protective covering over the entire body and all appendages; it also lines the anterior and posterior parts of the alimentary tract, the outer portions of the reproductive and excretory ducts, and even gas exchange tubules when present. The cuticle is of immense adaptive significance, since its many properties are primarily responsible for the preponderance and success of arthropods. As a tough, continuous, mechanical layer, the outer skeleton effectively affords the animal considerable protection. The cuticle is often folded internally, and

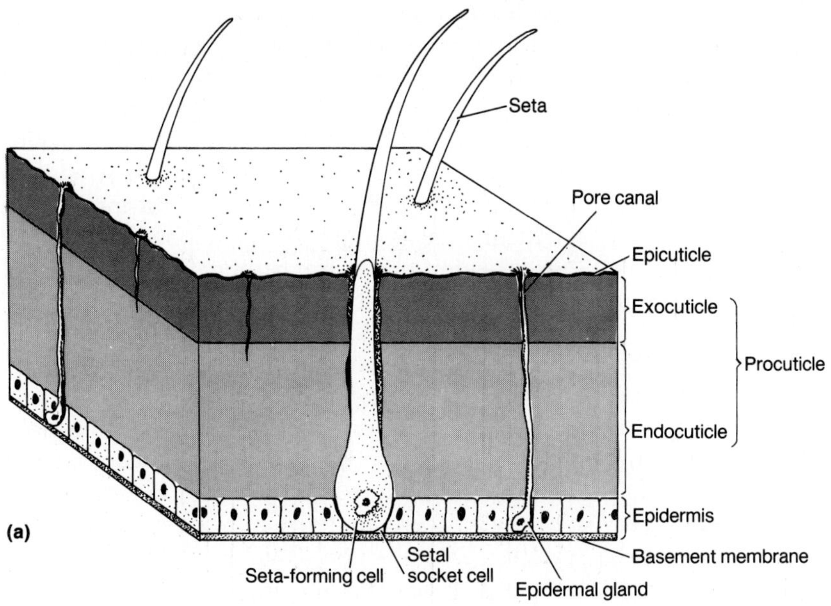

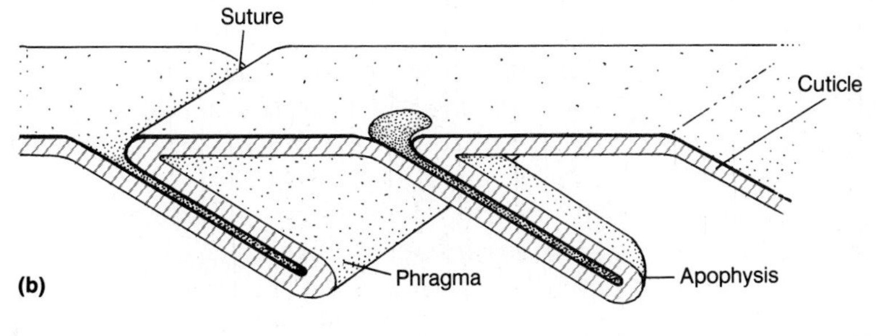

Figure 10.2 *Diagrams of the arthropodan integument:* ***(a)*** *a portion of a typical integument showing cuticular laminae, epidermis, basement membrane, and setae;* ***(b)*** *a section showing an internal strengthening fold.*

these ridges or folds (apophyses, apodemes, phragmata) provide additional strength to the exoskeleton (Fig.10.2b). With many areas being nondeformable the exoskeleton serves as insertion points for muscles essential for movement and locomotion; without this skeletal system, locomotor muscles would be completely ineffective. The impervious cuticle serves as a most important chemical barrier to desiccation and to environmental chemicals. The exoskeleton is a part of the arthropodan **integument,** which, in addition to the cuticle, is composed of an inner **epidermis** that secretes the cuticle and a **basement membrane** on which the epidermal layer rests (Fig. 10.2a). But the many and varied properties of the cuticle make it the arthropods' most important adaptation.

Construction

The cuticle is a noncellular, nonliving, organic complex produced by the epidermis. Chemically, it is composed principally of chitin and proteins. **Chitin,** or N-acetyl-D-glucosamine, can be easily polymerized into long, unbranched, polysaccharide chains similar to cellulose but containing nitrogen. Like cellulose, chitin is a structural molecule, is well suited for the protection and support of the arthropod, and lends strength to the exoskeleton, although it is not responsible for its hardness. Chemically, chitin is very recalcitrant, since it is neither attacked by digestive enzymes of most animals nor soluble in water or most dilute solvents. Although it is an important constituent of the arthropod cuticle, chitin is certainly not restricted to this phylum; it is present in certain protozoans, radiates, and a wide variety of protostomates. While chitin is found in many different annelids, it is absent in their integument and thus is not present as a continuous layer. The striking absence of chitin in deuterostomates is thought to be due perhaps to gene repression somewhere near the base of deuterostomate evolution.

There are two principal proteins often enmeshed within the chitin, and their physicochemical properties give the cuticle some unique attributes. **Arthropodin** and **resilin** are soft flexible proteins that form the bulk of the cuticle. Additionally, the cuticle often contains lipids, waxes, calcium carbonate, polyphenols, and pigments. Melanin pigments, deposited in the cuticle, impart a wide variety of colors, and some individuals are strikingly beautiful. However, the iridescent and metallic colors of many species are due to exceedingly fine striations on the outer surface of the cuticle, which produce refractions in the light and, in effect, make the exoskeleton appear to have color.

Several references have already been made to the hardness of the exoskeleton. What makes this chitinoprotein complex become hardened given that neither chitin nor the proteins are inherently rigid? Hardening of the cuticle is accomplished primarily in two different ways. First, changes are usually effected in the molecular organization of the proteins by the establishment of polyphenolic cross-linkages in their chains. These linkages between chains convert both arthropodin and resilin into a tough, rigid, insoluble protein called **sclerotin;** this important process is termed **sclerotization** or **tanning.** A sclerotized cuticle is therefore quite hard, somewhat elastic but not permanently deformable, and mechanically much stronger than an untanned cuticle would be. A second method of cuticular hardening is by the deposition into the exoskeleton of calcium carbonate in the form of calcite. **Calcification,** also conferring both rigidity and inflexibility upon the exoskeleton, is especially prevalent in higher crustaceans and others like millipedes.

The arthropodan cuticle varies in thickness from 1 μm to perhaps 1 cm and is constructed of many different laminae much like plywood. Basically, there are two principal layers, the epicuticle and procuticle, whose composition and functional significance are quite different. The outermost **epicuticle** is a very thin layer that functionally separates the arthropod from its environment and provides a great deal of chemical independence to the animal (Fig. 10.2a). An effective mechanical barrier to microorganisms, the epicuticle is physicochemically impermeable to most environmental molecules. The epicuticle consists of three thin substrata, which are an outermost protective layer of lipoprotein, a middle waxy layer that imparts to the cuticle a nonwettable property, and an inner layer of proteins and polyphenols. The absence of chitin in the epicuticle is notable. Beneath the epicuticle is the much thicker laminated **procuticle,** a complex of chitin and proteins consisting of an outer exocuticle and an inner endocuticle (Fig. 10.2a); in crustaceans a slightly different terminology is used for the procuticle (see Chapter 12). The **exocuticle** and **endocuticle** are much thicker than the epicuticle, and it is in these layers that both sclerotization and calcification take place, conferring upon the cuticle its hard, inflexible, rigid properties. The complex **epidermis,** the only cellular layer in the integument, contains many types of glands that synthesize various products such as silk, scents, waxes, defensive substances, and, of course, the cuticle. Interestingly, the epidermis secretes the outermost laminae of the epicuticle last after the procuticle is formed. How, you may wonder, is it possible to produce epicuticular layers through the already formed procuticle? There are numerous **pore canals** that extend through the procuticle to the external surface, and lipids and waxes are carried by these collective pore canals to the cuticular surface and deposited there in an orderly and systematic way (Fig. 10.2a).

APPENDAGES, JOINTS, AND MUSCLES

The above discussions have explained how an arthropod is completely encased in a tough, inflexible, chitinoprotein cuticle that is similar to a suit of armor. Then how is it possible that an arthropod is free to move? The secret is that at various strategic points the exoskeleton is not sclerotized, and, as in a real suit of armor, there are joints or points of bending or flexion. The exocuticle is composed of resilient, hardened, cuticular plates, the **sclerites,** between which are areas where only the epicuticle and endocuticle are present. At these points, both cuticular layers are thin, flexible, and nonhardened and form **articular membranes,** which are the joints in the exoskeleton (Fig. 10.3a–c, e). Each segment of the arthropodan body is surrounded by four sclerites: a dorsal **tergum,** a ventral **sternum,** and a lateral **pleuron** on each side (Fig.

Figure 10.3 *Diagrams of the cuticular joints in arthropods:* ***(a)*** *an articular joint;* ***(b)*** *a cross-sectional, three-dimensional view of a segment showing the four sclerites;* ***(c)*** *a longitudinal view through three metameres showing the dorsal extensor and ventral flexor muscles;* ***(d)*** *a leg of seven podomeres;* ***(e)*** *a joint between two podomeres.*

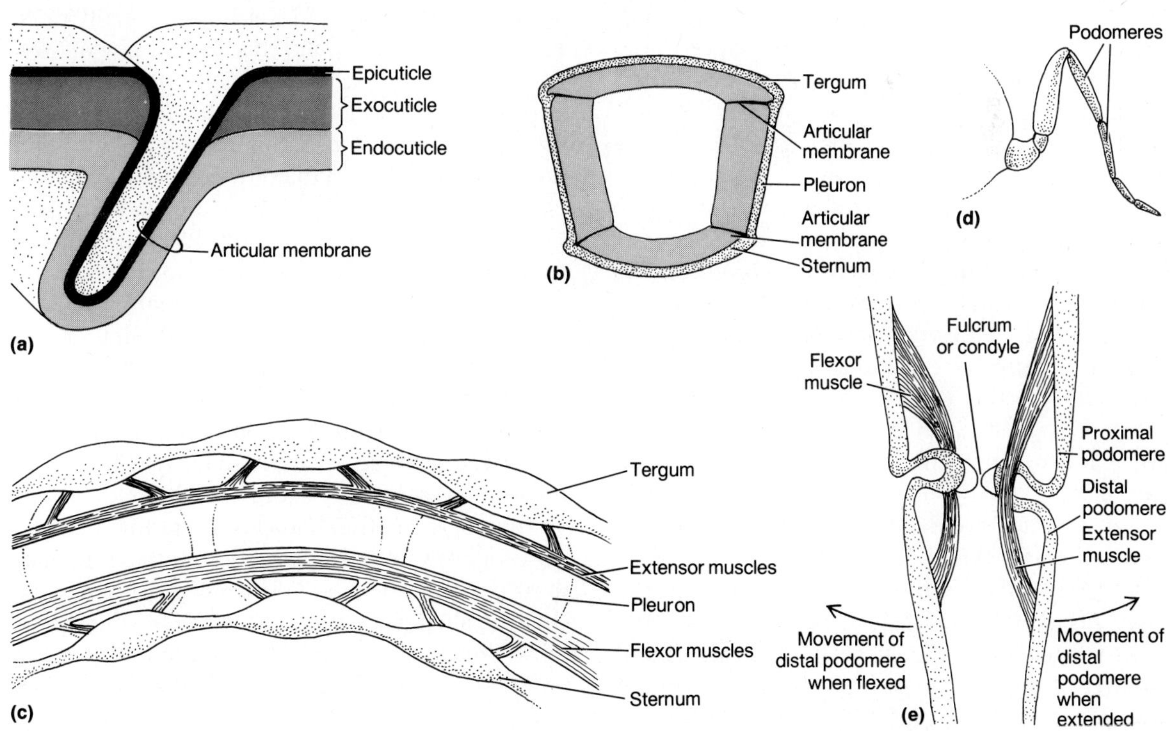

10.3b). At the points where two sclerites adjoin each other are articular membranes. Functionally more important are the articular membranes between similar adjacent sclerites (between sterna, for example), which make intermetameric movement possible in some arthropods (Fig. 10.2c). However, there has been a general tendency in many higher arthropods to increase the rigidity of the body, especially of those segments or regions that bear legs.

One of the characteristic features of arthropods is that each segment typically bears a pair of **jointed appendages,** a diagnostic feature from which the phylum name was coined (arthro = joint; pod = foot) (Fig. 10.3d). Originally evolved as locomotive organs, certain appendages on most arthropods have become specialized for mastication and food-gathering, sensory reception, gaseous exchange, and even reproduction. However, some arthropod appendages always retain their locomotive functions and are used in creeping, walking, swimming, and even jumping. Compared to polychaete parapodia, the locomotive appendages of arthropods are generally longer, more slender, and positioned more toward the ventral surface. Legs are of great adaptive significance to arthropods, since they exploit the principle of leverage. In legs that are basically stiff a rather large amount of movement can take place with a minimal expenditure of muscle contraction energy. Multiple pairs of legs also make possible sustained locomotion, which is far more energy efficient than stop-and-go locomotion, in which a great deal of inertia imposed by those periods of no motion must be overcome. It is very clear that arthropod legs make for a much more efficient system than that in annelids.

Since all appendages are covered by the exoskeleton, appendicular movement is possible only if cuticular joints are present. The cuticle covering each appendage is present as a linear series of two to seven rigid cylinders called **podomeres** or appendicular segments between each of which are the joints (Fig. 10.3d). Mechanically, each podomere is a stiff, inflexible tube joined to another podomere by an articular membrane. Thus the appendages, whose jointedness is made possible by the exoskeleton, are among the arthropods' most obvious and important characteristics.

The cuticle, because it is an exoskeleton and is jointed, has dramatically affected the arrangement of muscles for locomotion and movement. Rather than forming concentric cylinders around the body cavity as in annelids, arthropod effectors are all discrete bundles of striated muscles. In arthropods, all muscles must originate and insert into the inner surface of the integument. In fact, muscles extend through the basement membrane and epidermis as fine **tonofibrils,** which then are attached into the procuticle. In most arthropods the cuticle of sclerites is invaginated at strategic points to form internal cuticular processes into which muscles commonly insert; these processes in effect enhance the leverage effect muscles have on the sclerites. Arthropodan muscles are usually present as antagonistic pairs that span a given joint. Antagonistic muscles are well understood in vertebrates, in which muscles have the opposite effect on the jointed endoskeleton, or in annelids, in which the circular and longitudinal muscles act on an internal hydraulic skeleton (see Chapter 9). A typical arthropodan joint has a **flexor muscle** and an **extensor muscle** that pass on opposite sides of a joint and that are antagonists (Fig. 10.3e). Flexors tend to be larger and more massive and to do more work than extensors, and they are positioned on the side where the greatest movement is possible and where the effect of muscular contractions will be maximized. In general, flexors are situated on the ventral side of the body and on the ventral, posterior, or medial side of a given appendicular joint (Fig. 10.3c, e). It is important to note here that the skeletal or articular condyle, that is, the joint's fulcrum, for muscle contraction must be an apodeme or articular condyle that lies medial to the muscles or near the axis of the limb (Fig. 10.3e). Since a given muscle must be stretched to return to its former, noncontracted state, arthropodan flexors and extensors reciprocally perform that essential task. However, extension of certain appendages in some arthropods is achieved hydrostatically by forcing blood into that part, but flexion is always achieved by muscle contractions.

Arthropod muscles are most unusual in that each is constructed of very few fibers and is innervated by a correspondingly small number of neurons. Several different axon terminals are present in a given muscle fiber, and a given axon usually supplies more than one such fiber. In fact, several functionally different neurons typically innervate a given muscle fiber, including those that cause the fiber to contract rapidly but briefly (phasic), those that contract slowly and for a protracted period of time (tonic), and those that prevent contractions (inhibitory). In arthropods, graded muscular contractions are dependent on the type of muscle fiber involved, the type of neuron employed, and the combinations of different innervating neurons. Thus a given muscle fiber contracts more or less rapidly and more or less powerfully depending upon the stimulus provided to it by its several neurons.

MOLTING

One of the distinct liabilities of the arthropodan cuticle is that after sclerotization and calcification have taken place, any increase in the animal's volume is quite impossible. In order for the arthropod to grow, the entire cuticle—that over the entire body and appendages as well as that lining any internal spaces—must be shed periodically. This entire complicated process is termed **molting** (Fig. 10.4). The actual shedding of the old cuticle, however, must take place after a new cuticle is secreted but before the new exoskeleton is hardened, and this creates a most interesting logistical problem.

At some time prior to a molt, the epidermis becomes quite active with an increase both in the size of epidermal cells and in the incidence of mitotic activity. Then the old overlying cuticle becomes detached from the epidermis by enzymatic action in a process called **apolysis** (Fig. 10.4b). Shortly after apolysis the epidermis begins to secrete a new epicuticle (Fig. 10.4c). Chitinase and proteases, enzymes produced by the epidermis, pass through the new epicuticle and begin to digest chemically the old endocuticle and the inner parts of the exocuticle; up to 90% of the old cuticle may be destroyed. The digested endocuticular chitin and proteins are absorbed by the body, reassimilated in the epidermis, and reused in the secretion of the new procuticle. At this juncture the arthropod is surrounded by portions of the old cuticle (epicuticle, some exocuticle) and the newly formed cuticle (Fig. 10.4d). Interestingly, the animal at this point is still perfectly free to move, since the tonofibrils of most muscles are attached to both the old and new cuticles. The old skeleton is now shed in a process called **ecdysis,** which begins by a splitting of the old cuticle along certain ecdysial lines. At these easily ruptured points, only the epicuticle remains, since there is no exocuticle and the endocuticle has been digested. The animal pulls itself free from the old exoskeleton or **exuvium,** which is either left behind (Fig. 10.4e) or ingested by the arthropod. Therefore the molting process is represented by apolysis and ecdysis, which are separated by several days to a week or more, but ecdysis itself usually lasts from 1 to 3 hours.

The cuticle of an arthropod that has just completed ecdysis is rather soft and pliable and can easily be stretched, at least for up to several hours. To capitalize on this brief period in which the new exoskeleton is expandable, the newly molted or teneral animal increases its size abruptly because of the uptake of water or air and also because of the increased size of cells that were formed before the molt. The important processes of sclerotization and calcification

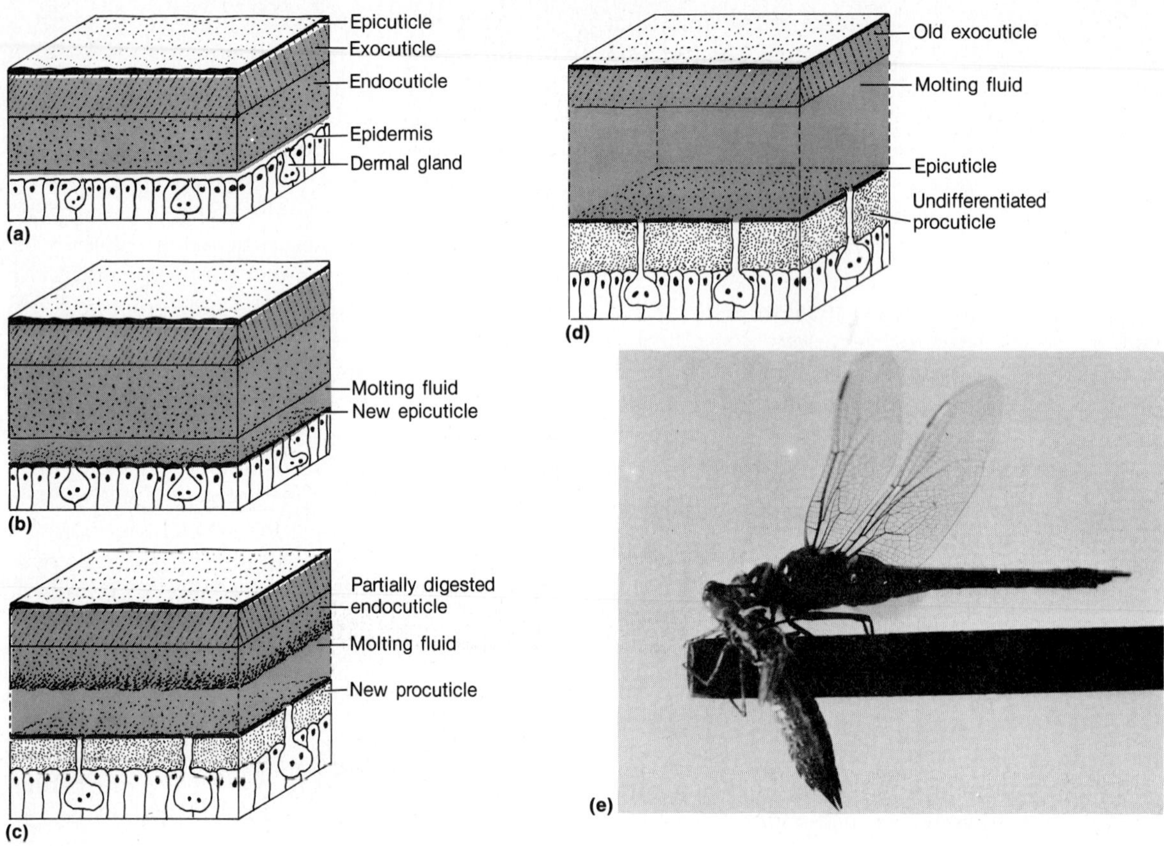

Figure 10.4 *A diagrammatic section of a portion of the cuticle in arthropods to show the events at molting:* ***(a)*** *fully-formed cuticle;* ***(b)*** *apolysis;* ***(c)*** *digestion of the internal portions of the old cuticle and early stages in the formation of a new cuticle;* ***(d)*** *immediately prior to ecdysis with old and new exoskeletons present;* ***(e)*** *photograph of a newly emerged arthropod (dragonfly).*

now take place to harden the new cuticle. Some additional procuticular materials may be added directly onto the inner surface of this layer after tanning has occurred, and in many arthropods and especially the insects and arachnids, secretions via the pore canals add the outermost layers to the epicuticle shortly after ecdysis.

Molting, an exceedingly complex process, is mediated by several hormones and environmental cues, which operate in a most complicated neurohormonal manner; these systems will be discussed later for both crustaceans and insects. In the period of time between molts, an arthropod is termed an **instar.** Some arthropods molt throughout their entire life and thus have 30 or more instars. But many others, including the insects, have 5–15 instars until sexual maturity is reached, after which the final or adult instar does not molt.

Basic Body Plan

The body of a typical arthropod is elongate and segmented, and each of some or all segments bears a pair of appendages. Sizes range from microscopic mites 0.02 mm long to the giant Japanese spider crab, *Macrocheira,* whose leg tips may span more than 3 m. But definite size limitations are imposed by the exoskeleton, since at some point the cuticle cannot adequately support the body. Terrestrial arthropods are generally smaller than aquatic forms because the latter have the benefit of the buoyant properties of their watery medium.

Two exceedingly important trends—tagmatization and cephalization—are especially prevalent in arthropods and are instrumental in the overall success of this phylum. **Tagmatization,** a feature mentioned earlier in connection with annelids, refers to the grouping or fusion of adjacent segments whose appendages perform similar tasks. In arthropods there are usually about 10–25 segments comprising the body. They are normally grouped into two or three tagmata, which are termed the prosoma and opisthosoma (chelicerates); head, thorax, and abdomen (most crustaceans, all insects); or head and trunk (some crustaceans, all myriapods). Regional groups of segments enable the appendages on all the included segments to perform more efficiently and to operate often as a unit.

A second important feature in all arthropods is **cephalization,** which refers to the concentration of several vital functions into the anterior end or head. Arthropods illustrate a very high degree of cephalization that can be matched among invertebrates only by that in cephalopod molluscs (see Chapter 8). Cephalization almost invariably involves the location of the mouth and associated feeding structures, the brain, and most sensory receptors near the anterior end. The arthropodan head, substantially more complex than that of annelids, is constructed of an anterior, nonsegmented **acron** plus two to six anterior segments. The acron is thought to be homologous to the annelidan prostomium. The anteriormost segments have migrated forward, and thus the relative position of the mouth has moved ventrally and subterminally. This means that the first one or two anterior segments lie preorally, and their appendages, the **antennae,** function in sensory reception. The appendages of the segment bearing the mouth are **mandibles** or jaws (absent in chelicerates), while those of the two postoral segments are basically involved in food acquisition (Table 10.2).

TABLE 10.2. □ **THE PAIRED APPENDAGES BORNE BY THE ACRON AND THE FIRST SIX SEGMENTS IN THE MAJOR GROUPS OF ARTHROPODS.**

	Arthropod Group				
	Trilobite	**Chelicerate**	**Crustacean**	**Insect**	**Myriapod**
Acron	None	None	None	None	None
1	Antennae	Chelicerae	Antennules	Antennae	Antennae
		*			
Segment 2	Biramous limbs	Pedipalps	Antennae	None (embryonic)	None (embryonic)
3	Biramous limbs	Legs	Mandibles	Mandibles	Mandibles
4	Biramous limbs	Legs	Maxillae	Maxillae	Maxillae
5	Biramous limbs	Legs	Maxillae	Maxillae (labium)	Maxillae
	*		*	*	*
6	Biramous limbs	Legs	Legs	Legs	Legs

The asterisk denotes the posterior margin of the head.

The midbody of an arthropod contains a variable number of segments whose appendages are basically involved with locomotion either as legs or as paddles (many aquatic forms). Midbody appendages often also function in gaseous exchange and reproduction. At the posterior end is a nonsegmented, appendageless **pygidium** that bears the anus; the arthropodan and annelidan pygidia are thought to be homologous. In chelicerates and many crustaceans the pygidium extends postanally as a **telson.**

Nervous System and Sense Organs

The arthropodan nervous system, built along a plan like that of annelids, has a far greater level of cephalization and neural complexity. The arthropodan nervous system consists of a rather complex anterior brain, a doubled ventral nerve cord, and anterior nerve tracts connecting the brain and cord. The **brain,** comparatively much larger than that of annelids and correlated with much more sophisticated sense organs, integrates the activities of the entire animal. It is constructed of an anterior protocerebrum, a middle deuterocerebrum, and a posterior tritocerebrum. The **protocerebrum** or forebrain is the sensory center for the optic nerves, and into this part come the many nerve impulses that provide the animal with visual information about its environment (Fig. 10.5). The eyes are certainly among the most important sensory organs, and the protocerebrum is correspondingly large and complex. The protocerebrum also has a multitude of motor control functions, and its paired "mushroom bodies" are the center for much of the coordination of arthropods' often complex behavior. The **deuterocerebrum** or midbrain is an important integrative center, since it receives vital sensory information from the paired antennae (Fig. 10.5). The absence of the deuterocerebrum in chelicerates is clearly correlated with the absence of antennae in this group. The **tritocerebrum** or hindbrain lies postorally and contains important motor centers whose neurons innervate the several paired mouthparts, the anterior part of the alimentary canal, and, in crustaceans, the second pair of antennae (Fig. 10.5). Since the mouthparts are copiously equipped with many receptors, the tritocerebrum is the neural center for these tactile and chemical receptors.

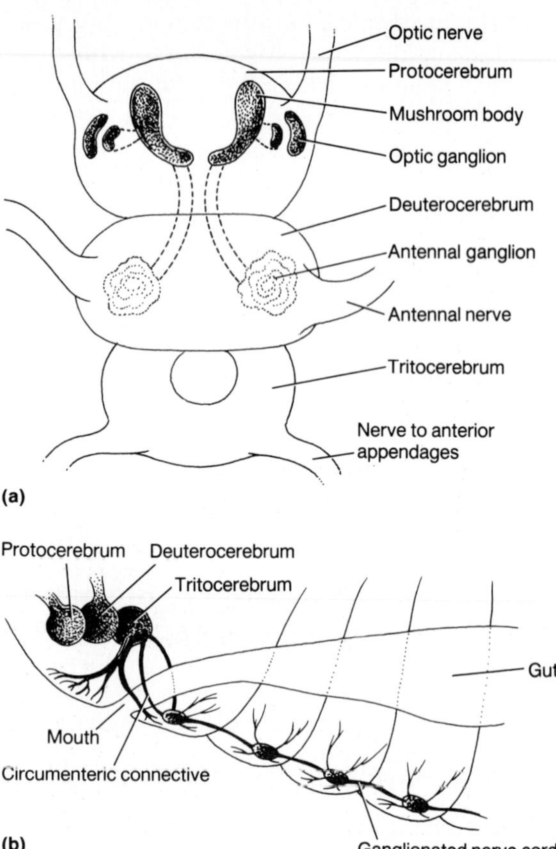

Figure 10.5 *Diagrams of the arthropodan nervous system: **(a)** a dorsal view of a generalized brain; **(b)** a lateral view of the brain and anterior nerve cord.*

There are two contrasting theories, each strongly supported by its adherents, as to the segmental nature of the two anterior brain lobes. One position is that the protocerebrum and deuterocerebrum are really a single unsegmented ganglion that originally evolved in the unsegmented acron, and its division into two lobes is associated with the development of complex sensory centers. The other theory is that the protocerebrum and deuterocerebrum are both paired and segmented ganglia that have migrated forward to a position anterior to the mouth. Accepted by supporters of both positions is that the tritocerebrum is a paired segmental ganglion that has moved anteriorly to a position just behind the mouth.

The **circumenteric connectives,** arising from the tritocerebrum, pass around the anterior gut and connect to the anterior end of the nerve cord (Fig. 10.5b). The ventral **nerve cord** consists of two separate cords in some primitive arthropods, but both cords and their ganglia have fused medially in all other arthropods. Segmentally arranged ganglia on the nerve cord control certain activities within that segment (Fig. 10.5b). There has also been a tendency among many different higher arthropods for cord ganglia to migrate anteriorly and to fuse with each other. Both trends are clear indicators of cephalization.

Sensory receptors

The exoskeleton has markedly affected the nature and arrangement of sense organs, since it serves as a rather efficient shield against most environmental stimuli. Yet data from these stimuli are crucially important for an arthropod to survive and be successful. Therefore sensory receptors are quite numerous but are often restricted to certain areas like eyes, antennae, and mouth appendages. Other sensory structures such as bristles, setae, hairs, or cuticular canals have evolved to transmit sensations through the cuticle to the interior (Fig. 10.2a). Many of these structures are tactile or mechanical receptors and transmit the effects of motion or movement to basal receptor cells. Other cuticular structures include thermoreceptors, simple photoreceptors, chemoreceptors (smell, taste), and proprioceptors, all of which are concentrated on the appendicular articular membranes. The antennae contain a high concentration of mechanoreceptors and chemoreceptors.

Eyes

All arthropods have photoreceptors of some sort, ranging in complexity from simple eyes composed of only a few photoreceptors to very complex eyes with thousands of visual units that collectively form an image. In simple eyes or **ocelli** the photoreceptor cells are arranged in a shallow cup with the receptor surfaces facing toward the light source (direct eyes) or away from the light (indirect eyes). The most intricate eyes are the paired **compound eyes** which are present in insects and higher crustaceans. Each compound eye is a collection of several to over 30,000 individual photoreceptor units, the ommatidia (Fig. 10.6a, c). Each **ommatidium** is a thin, cylindrical structure consisting of an outer cornea, a crystalline cone, and the retinula. The **cornea** is a transparent portion of the cuticle covering the ommatidium; it functions as a lens to focus light rays on the inner receptor surfaces. The external surface of the cornea, the **facet,** is usually round, square, or hexagonal in shape (Fig. 10.6b). Beneath the cornea is the long, cylindrical **crystalline cone,** which is usually formed by four corneal cells, is tapered proximally, and serves as a second lens (Fig. 10.6c). No accommodation is possible in an ommatidium, since both lenses (cornea and crystalline cone) are immobile and have a fixed focal length. At the base of an ommatidium is the rather complex receptor apparatus, the **retinula,** constructed of 7–12 photoreceptor **retinular cells** surrounding a central fluid-filled space (Fig. 10.6d). Retinular cells, as well as distal pigment cells, contain black or brown pigments, and these cells can fit around the ommatidium like a lightproof sleeve and thus photically isolate one ommatidium from adjacent ommatidia.

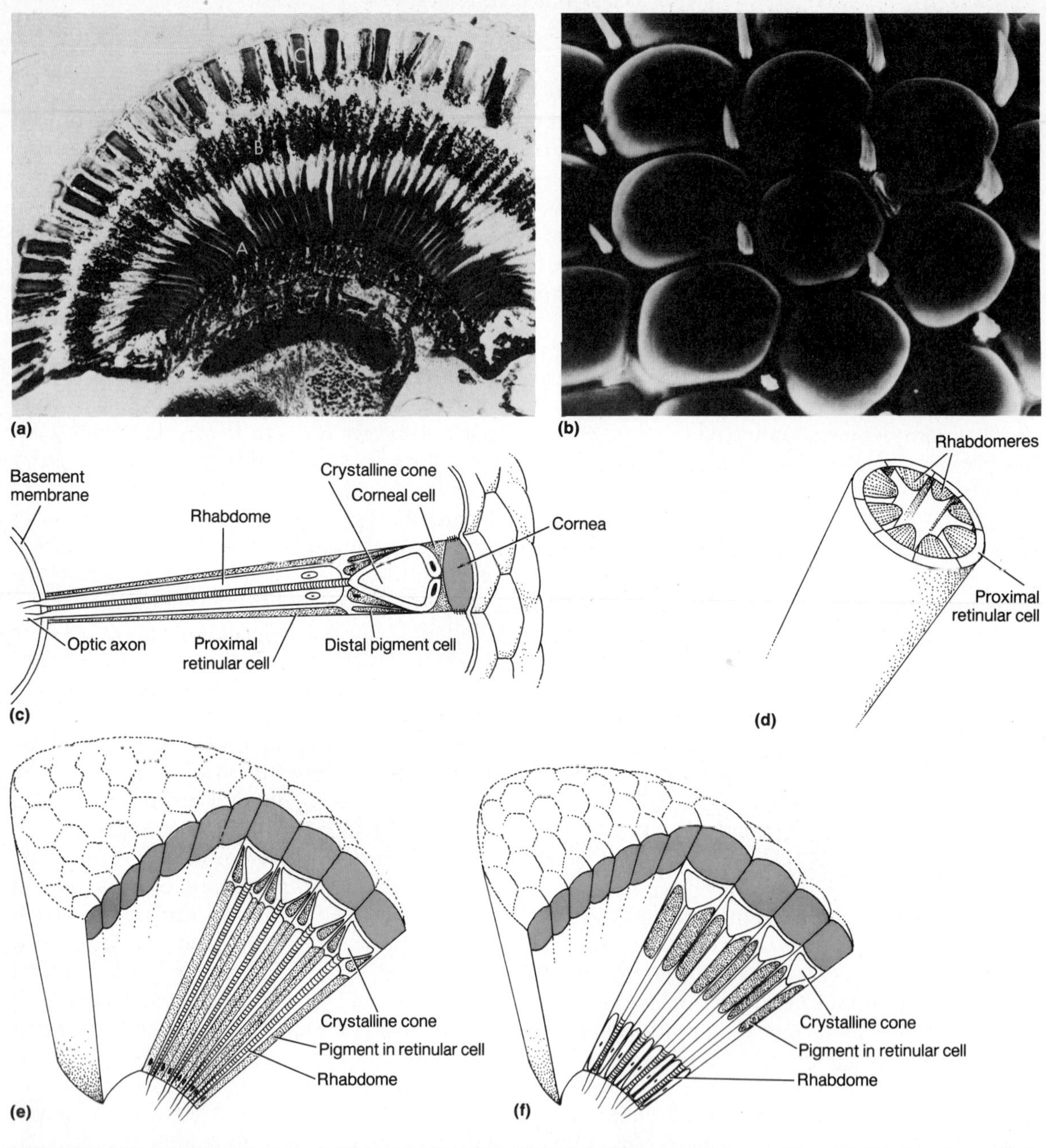

Figure 10.6 *The arthropodan compound eye:* ***(a)*** *photomicrograph of the compound eye of an insect (A = retinular cells; B = area of cytoplasmic filaments; C = crystalline cones);* ***(b)*** *scanning electron micrograph of a small number of facets of the compound eye of an insect (beetle); the function of the hairs between the facets is unknown;* ***(c)*** *an individual ommatidium;* ***(d)*** *diagrammatic, cross-sectional view through a rhabdome;* ***(e)*** *diagram of several ommatidia in an appositional eye;* ***(f)*** *diagram of several ommatidia in a superpositional eye. (parts a, b [from U.S. Department of Agriculture] by courtesy of P.S. Callahan, 1971,* Insects and How They Function, *Holiday House, New York, N.Y.)*

Each of the retinular cells consists of numerous small **microtubles** or microvilli that are perpendicular to the ommatidial axis, and groups of these fibrils, the **rhabdomeres,** are arranged much like the sections of an orange as they project into the central space (Fig. 10.6d). The rhabdomeres and central cavity constitute the **rhabdome,** where actual photoreception and photostimulation occur. Even though each rhabdomere has its own separate sensory neuron, the neurons from each retinular cell of an ommatidium are bundled together, and they collectively stimulate the brain as a single point of light. Therefore an ommatidium functions as a single visual unit.

What sorts of information do the compound eyes provide to the arthropod? In forms with only a few ommatidia, perhaps little more than light intensity and evidence of some motion can be discerned. But in many other arthropods with thousands of ommatidia, images are formed, although they undoubtedly are indistinct and "grainy." The image formed by a compound eye is a **mosaic** with the neuron from each ommatidium contributing a small mosaic of the total brain picture. The outstanding virtue of a compound eye is that is is admirably suited to detect motion. Only a slight shift of an object against a background is noted by many ommatidia, whose impulses stimulate the brain to effect appropriate actions. The ability to discriminate the plane of polarized light and to discern different wavelengths of light (color vision) is found in many higher crustaceans and in most insects. Many insects are especially sensitive to ultraviolet wavelengths.

Even though an ommatidium is structurally a separate, self-contained unit, there are degrees of functional independence that affect the ways in which ommatidia respond to light. **Appositional eyes** are found mostly in terrestrial, shallow-water, and diurnal arthropods. The receptor cells are greatly elongated and often extend distally to the crystalline cone, and the pigments of the retinular and distal pigment cells form an effective light screen between adjacent ommatidia (Fig. 10.6e). Appositional eyes can receive only light rays that are approximately perpendicular to the facet of each ommatidium, they form clear mosaic images, and their visual acuity is quite sharp. In **superpositional eyes,** found in most deep-water or nocturnal forms or where the habitat is poorly lighted, the distal pigment and retinular cells are widely separated and do not form a complete light screen around a given ommatidium. Therefore light often enters a given rhabdome through a number of ommatidial lenses (Fig. 10.6f). Superpositional eyes usually do not form images, or if they do, the images are extremely crude and lack resolution, and the visual acuity of these eyes is very low. Superpositional eyes are much better adapted for night vision or for determining weak light intensities as they gather and concentrate dim light. Mosaic images are almost nonexistent in superpositional eyes, since there is considerable overlapping of images. Many arthropods have modified the degree of ommatidial isolation and, in so doing, possess eyes that are somewhat intermediate between appositional and superpositional.

Visceral Systems

Even though arthropods possess most of the visceral systems typical of other higher animals, there are some important modifications. The digestive system is composed of an anterior **stomodeum** or foregut that extends posteriorly from the mouth, a **midgut,** and a posterior **proctodeum** or hindgut that extends forward from the anus. Both stomodeal and proctodeal regions are derived from ectoderm and are lined with cuticle that must be shed at each ecdysis. The stomodeum is concerned primarily with ingestion, mechanical breakdown, and storage of food, and regions of the foregut may be variously modified as pharynx, esophagus, and storing and grinding organs. The midgut, originating from the endoderm and therefore not lined with cuticle, is the principal site of enzyme production, chemical digestion, and absorption;

midgut outpocketing or digestive glands are frequently present and secrete digestive enzymes. The proctodeum functions primarily in salt and water absorption and feces formation and may be differentiated into an intestine and a rectum.

A spacious, metamerically comparted coelom like that found in annelids is simply not present in arthropods. The **coelom** is represented only by small spaces within each gonad and, in chelicerates and crustaceans, spaces within the excretory organs. Why has the coelom all but disappeared in arthropods? The answer to this important question lies indirectly with the ontogeny of the exoskeleton and the muscles that are necessary to move cuticular sclerites and podomeres. Since distinct muscle bundles are required, there was a concomitant loss of metameric circular and longitudinal muscles and a reduction in the coelom because a hydrostatic skeleton was no longer employed for locomotion. In place of a capacious coelom there is a poorly defined but extensive hemocoel, a circulatory adaptation tied indirectly to the development of the cuticle.

The principal features of the arthropodan circulatory system are the hemocoel, heart, arteries, and blood. The **hemocoel** is a complex series of spaces, sinuses, and channels through which blood is transported as its bathes tissues directly. As in molluscs but contrasting sharply with the annelidan condition, arthropods have an **open circulatory system,** since blood enters the hemocoel and is not contained within vessels. Capillaries and veins are never present, but blood is eventually returned in hemocoelic channels to the **pericardial sinus,** from which it enters the heart again by way of **ostia.** Arthropodan blood contains various amebocytes and often an oxygen-transporting pigment, either hemocyanin or hemoglobin, dissolved in the plasma. Blood composition varies widely between individuals or within the same arthropod over time, and these variations are correlated with growth, molting, and feeding.

Very small aquatic arthropods have very thin cuticles, and these animals carry out both excretion and gaseous exchange through their general body surfaces. However, in larger forms the cuticle is thick and impermeable, and areas for excretory and gas exchange functions are necessarily restricted to certain permeable surfaces. A variety of excretory organs are found, but there is a complete absence of nephridia, a feature that is directly associated with the severe reduction in the coelom. Present in many arthropods are tubular mesodermal structures that are coelomoducts, not nephridia (see Chapter 9). In aquatic chelicerates and crustaceans, coelomoducts are modified into glandular structures (coxal, antennal, maxillary glands) that function in excretion by eliminating ammonia or urea. In terrestrial arthropods (insects, myriapods, most chelicerates), outgrowths of the alimentary canal (Malpighian tubules) are novel structures that eliminate uric acid or guanine and thus conserve precious body water. In order to facilitate gaseous exchange, many aquatic forms clearly augment this function through their thin appendages. Higher crustaceans have developed foliaceous outgrowths or gills at the base of certain appendages, and some even have branchial chambers whose walls are vascularized. Primitive chelicerates often have a peculiar type of gill that has numerous branchial leaves; such gills are appropriately called gill books. But most chelicerates, being terrestrial, have internal chambers called book lungs with many gas exchange lamellae. Insects, myriapods, and terrestrial chelicerates have developed a tracheal system, a series of ectodermal tubules that transmit gaseous oxygen into a vast system of smaller tubules, each of which ultimately leads to a specific local region of the body. These various excretory and gas exchange organs and their functions will be more fully covered when each group is discussed.

Reproduction and Development

Almost all arthropods are dioecious, although members of a few taxa (such as barnacles) are hermaphroditic. A single pair of ovaries or

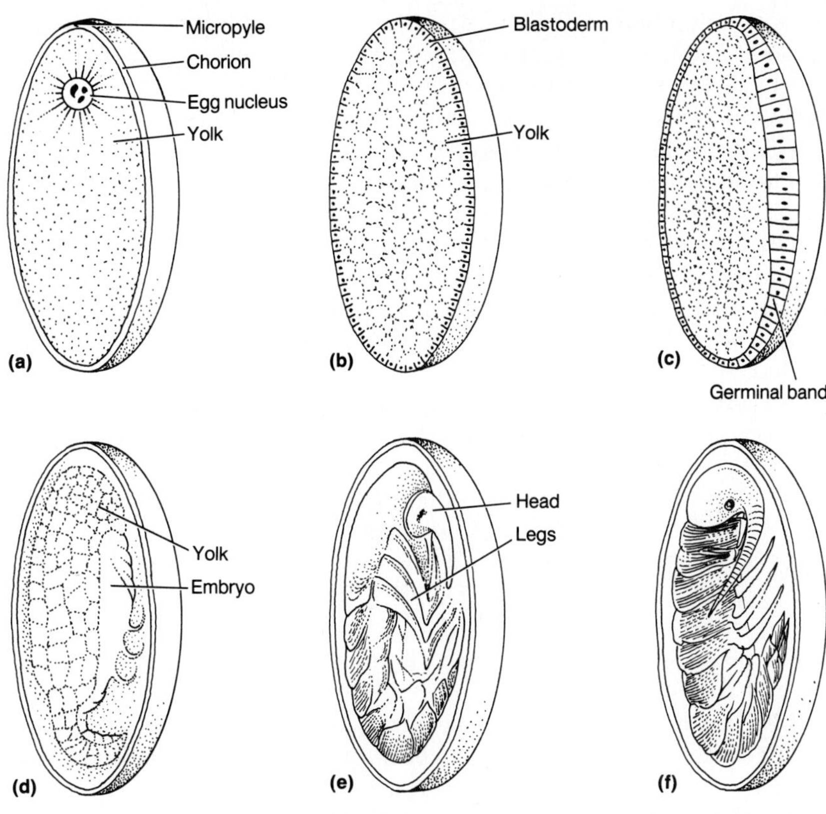

Figure 10.7 *Sequential stages in the embryonic development of a generalized arthropod:* ***(a)*** *a fertilized egg prior to cleavage;* ***(b)*** *nuclei migrate to the periphery of the embryo, and cell membranes form around each nucleus;* ***(c)*** *the formation of the germinal band;* ***(d–e)*** *successively later stages as appendages are formed and the yolk is absorbed by the embryo;* ***(f)*** *the embryo just prior to hatching.*

testes is present at various locations but usually near to the gonopores. The single or paired gonopores may be on anterior (millipedes), midbody (chelicerates, crustaceans), or posterior (centipedes, insects) segments. Although some aquatic forms have external fertilization, most arthropods copulate and have internal fertilization. In many arthropods, sperm enclosed in spermatophores are transferred to the female by the male using a modified appendage. Most females lay fertilized eggs in the environment, so embryonic development is completely independent of the adults.

The typical arthropod egg contains a rather large amount of yolk with the nucleus centrally located in a small area that contains no yolk. Around the nucleus is the thick layer of yolk, and another layer of nonyolky cytoplasm envelops the egg (Fig. 10.7a). Following fertilization the nucleus undergoes mitotic divisions but with no cytokinesis or division of the yolk, so a syncytium is formed in the center. The nuclei then migrate to the peripheral nonyolky cytoplasm, where cell membranes form around each nucleus (Fig. 10.7b). Therefore the egg, with the yolk now inside of the blastomeres, is said to be **centrolecithal,** cleavage is now termed **meroblastic** or superficial, and a solid embryo or **stereoblastula** is produced. Eggs of some arthropods have very little yolk and undergo holoblastic cleavage initially, but later a form of meroblastic cleavage ensues. Spiral cleavage, so characteristic of all other protostomates, occurs only in a very few arthropods. Gastrulation is achieved in a variety of ways, and a germinal band or disc develops on one side of the egg that shortly forms the three germ layers (Fig. 10.7c). As the embryo grows, it wraps around and slowly absorbs the yolk (Fig. 10.7d–f).

Typically, the arthropod embryo hatches as some sort of immature or larval stage that usually lacks many of the adult characters. The larva is a feeding stage that molts and adds new metameres at the posterior end at each molt. The larval stage either gradually comes to resemble the adult through repeated molts or is suddenly metamorphosed into an adult at the last molt.

The Absence of Cilia

Perhaps it is appropriate that we conclude our general treatment of arthropods by noting that cilia are universally absent from all members of this entire phylum. The presence of the cuticle militates against any external cilia, but no ciliated internal surfaces are present either. Perhaps the extensive hemocoel is a partial reason, but cilia are common in molluscs that also have an open circulatory system. Members of the Phylum Onychophora, often considered by some biologists as closely related to arthropods, have ciliated gonoducts, and it was for this and other reasons that onychophorans were considered as being more closely related to annelids (see Chapter 9).

SUBPHYLUM TRILOBITOMORPHA

The trilobites, now all extinct, are of particular interest to zoologists, since we believe that they were the most primitive of all arthropods and were nearest to the ancestral stock from which all other taxa evolved. Because of their important position in the evolution of arthropods, a brief description is included at this point. Trilobites, abundant and widely distributed in marine environments, reached their zenith about 500 million years ago (see Table 1.1 in Chapter 1). They all became extinct near the end of the Paleozoic Era, but they left an extensive fossil record. About 4000 species have been described. Although most trilobites were coastal and benthic forms, some species were pelagic, and others were planktonic. They ranged in size from 0.05 to 70 cm, but most were 3–10 cm in length.

The trilobite body was oval, was flattened dorsoventrally, and bore a dorsal exoskeleton that was much thicker and more heavily calcified than that of the ventral surface. Because of the thick, resistant dorsal exoskeleton, only that part has been preserved in a great many instances. Dorsally, the trilobite body has two anteroposterior furrows that divided the body into a **median axial lobe** and a **lateral lobe** on either side (Fig. 10.8a, b). Trilobites get their name from these three longitudinal lobes (Trilobitomorpha = bearing three lobes) The body was also divided into three tagmata: an anterior cephalon, a middle trunk or thorax, and a posterior pygidium. The **cephalon,** composed of the acron and five segments, all of which were fused together, was covered by a dorsal carapace. Paired eyes, which in some trilobites were compound eyes, were situated on the lateral aspect of the cephalon (Fig. 10.8a). Ventrally, the cephalon bore the mouth located just posterior to a fleshy lobe or labrum. From the first segment arose a pair of appendages, the **antennae;** they are considered to have been homologous to the antennae of uniramians and the first antennae of crustaceans (Fig. 10.8b, Table 10.2). The **trunk** consisted of a variable number of segments, each of which bore a pair of appendages. The posterior **pygidium,** certainly not homologous to that structure of the same name in all modern-day forms, also was composed of several fused segments, each of which had appendages (Fig. 10.8a, b).

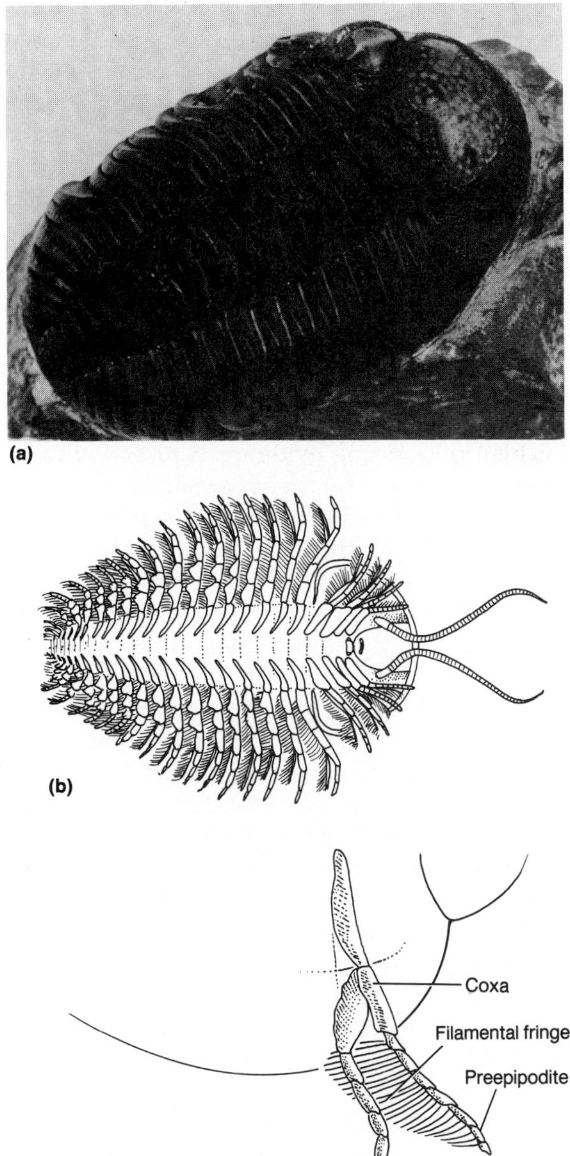

Figure 10.8 *Trilobites:* ***(a)*** *photograph of a trilobite, dorsal view (courtesy of C.F. Lytle);* ***(b)*** *ventral view of a trilobite (from A.S. Packard, 1898,* A Text-book of Entomology, *Macmillan and Co., Ltd., London);* ***(c)*** *a typical trilobite appendage.*

All segments, except the first, which bore the antennae, had paired appendages that were essentially similar in size, shape, and function. Each appendage consisted of a basal **coxa** and was biramous with an inner branch, the **telopodite,** and an outer **preepipodite** (Fig. 10.8c). The telopodite had seven podomeres and was used as a walking limb. The preepipodite was likewise composed of numerous podomeres, and some or all bore a fringe of posterior filaments (Fig. 10.8c). First thought to be gills, these preepipodite filaments may have been important in burrowing, swimming, and food gathering, but they may also have functioned as gills. The fringe of one limb overlapped the field of that of the limb immediately posterior to it; this arrangement must have increased the limbs' efficiency. Certain appendages in some forms were modified around the mouth as feeding limbs. In most trilobites the appendages of the pygidium were progressively smaller toward the posterior end.

It is indeed incredible that paleontologists have been able to determine from the fossil record the developmental stages for certain trilobite species, and these data probably can be safely extrapolated to include those of all trilobites! There were three developmental or larval stages—protaspis, meraspis, and holaspis—that were produced by successive moltings. The **protaspis,** the newly hatched trilobite, was planktonic and consisted only of the acron and the first four or five segments that would eventually form the head (Fig. 10.9a). After several protaspis instars the developing trilobite became a **meraspis** larva with a true pygidium that began to form new metameres on its anterior edge (Fig. 10.9b). After several instars as a meraspis larva a **holaspis** larva was produced that resembled an adult in basic form and outline (Fig. 10.9c). At successive molts, new segments were added to the trunk and adult pygidium until the adult size was reached and sexual maturity was attained. Trilobites may have continued to molt throughout their lives.

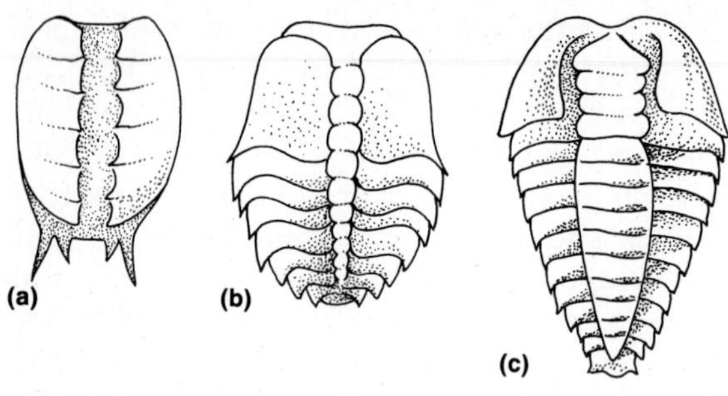

Figure 10.9 *Larval stages in trilobites: (a) protaspis; (b) meraspis; (c) holaspis.*

PHYLOGENY OF THE ARTHROPODS

Arthropods probably arose from some ancient marine worm that was elongate, bilaterally symmetrical, and segmented and possessed a coelom. This ancient ancestral beast perhaps lived as many as a billion years ago but, unfortunately, left no fossil record. By the beginning of the Paleozoic Era, some 600 million years ago, the evolutionary pathways for arthropods, along with those for most other phyla, had already become distinctively different and separate from each other. Therefore we can only hypothesize about the primeval stock from which this great phylum arose. There can be little doubt that annelids and arthropods are related, as is evidenced by both groups being metameric and schizocoelomate, having an unsegmented acron and pygidium, and possessing similar nervous systems. But there are two different schools of thought on the ancestral relationship between arthropods and annelids. Did the arthropods arise independently from annelids, or did arthropods evolve from a primitive annelid? Evidence for an independent evolutionary lineage for arthropods is to be found in their cuticle, their jointed appendages, a hemocoel, a highly reduced coelom, and the total absence of both nephridia and cilia; none of these features appears to be basic in annelids. Evidence for a polychaete ancestry for arthropods is based on the similarities between the two phyla and the probable homologies of the parapodia and segmented appendages. Since the fossil record is of little help, this important phylogenetic question remains unresolved.

But lest the reader think that this is the only major disagreement about the phylogeny of arthropods, it is in fact only the beginning. One general theory, espoused by many but clearly articulated by Boudreaux (1979), maintains that whatever the arthropod ancestors were, all arthropods had a common early phylogeny and thus the entire phylum is monophyletic (Fig. 10.10a). Supporting evidence is to be found in the many ways in which all are similar, the most important being the universal presence in all arthropods of (1) muscular, jointed, paired appendages; (2) a cuticle of chitin and sclerotized protein; (3) extensive tagmatization and cephalization; (4) an open circulatory system with a hemocoel and heart ostia; (5) a highly reduced coelom; and (6) meroblastic cleavage. Another general theory, equally promoted by a large number of invertebrate zoologists and forcefully articulated by Manton (1977), holds that the arthropods are definitely polyphyletic. Manton believes that at least three and probably four different lines of parallel and convergent evolution have occurred in arthropods to produce the uniramians, crustaceans, and chelicerates, and probably the trilobites (Fig. 10.10b). Supporting evidence for the polyphyletic theory is found chiefly in (1) the apparent independent evolution of the cuticle in the three extant groups; (2) the unequal distributions of compound eyes; (3) the differences in patterns of embryonic development; (4) the development of fundamentally different appendages in the three groups; (5) the variations in the ontogeny and morphology of the mandibles; and (6) the basic differences in the tagmata of the three groups.

Either of these theories will support the idea that the similarities between various arthro-

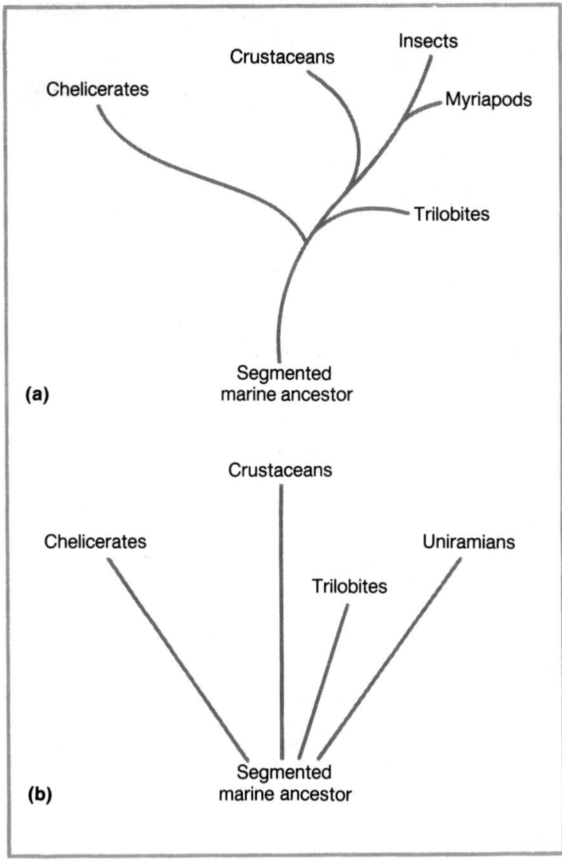

Figure 10.10 *Two different interpretations of the phylogeny of arthropods:* ***(a)*** *a scheme proposed by Boudreaux (1979);* ***(b)*** *a scheme proposed by Manton (1977).*

pods, as elucidated by the opposing theory, are due only to convergent evolution and the vagaries of specific adaptations to local habitats or conditions. Both theories have a weighty mass of supporting data, and the theorists often interpret the exact same data in diametrically opposite ways. Different interpretations of the very same characteristics may be frustrating to the reader, but this is an effective process of science that eventually should arrive at the truth.

The issue of where to place the primitive trilobites generates still another controversy that has produced a trio of possible answers. One hypothesis is that the trilobites were perhaps most closely related to the chelicerates and represent an offshoot of that phylogenetic line. A second hypothesis places the trilobites as an early branch from a main evolutionary line that eventually led to crustaceans (Fig. 10.10a). A third position, one that seems to be attracting adherents currently, is that the trilobites were evolved as an independent line separate from chelicerates and crustaceans and therefore should be accorded their own subphylum status (Fig. 10.10b). I have followed this last plan, but new evidence could alter this scheme substantially.

It is very clear that there were at least three principal evolutionary lines in this enormous group—Chelicerata, Crustacea, and Uniramia—and the inclusion of a fourth line, the Trilobitomorpha, is probably justified. It is very likely that in the not too distant future the term "Arthropoda" may be used to denote a superphylum; the four subordinate groups would then be elevated to the rank of phylum. However, this text treats the arthropods in the more conventional way by considering all taxa as belonging to the same phylum but recognizing four subphyla as representing rather distinct, separate evolutionary lines.

ADDITIONAL READINGS

Austin, P.R., C.J. Brine, J.E. Castle, and J.P. Zikakis. 1981. Chitin: New facets of research. *Science 212:*749–753.

Boudreaux, H.B. 1979. *Arthropod Phylogeny with Special Reference to Insects,* 306 pp. New York: John Wiley & Sons.

Cisne, J.L. 1974. Trilobites and the origin of arthropods. *Science 186:*13–18.

Clarke, K.U. 1973. *The Biology of the Arthropoda,* 259 pp. New York: American Elsevier.

Gupta, A.P., ed. 1979. *Arthropod Phylogeny,* 735 pp. New York: Van Nostrand Reinhold Co.

Herreid, C.F., II and C.R. Fourtner. 1981. *Locomotion and Energetics in Arthropods,* 554 pp. New York: Plenum Publishing Co.

Manton, S. M. 1977. *The Arthropoda: Habits, Functional Morphology, and Evolution,* 514 pp. Oxford, England: Clarendon Press.

Richard, K.S., J. Bereiter-Hahn, and A.G. Matoltsky, eds. 1984. *Biology of the Integument.* Vol. 1: *Invertebrates,* 461 pp. New York: Springer.

Sharov, A.G. 1966. *Basic Arthropodan Stock,* 271 pp. New York: Pergamon Press.

Snodgrass, R.E. 1952. *A Textbook of Arthropod Anatomy,* 363 pp. Ithaca, N.Y.: Cornell University Press.

Symposium on Arthropod Molting. 1972. *Amer. Zool. 12*:341–384.

Wolken, J.J. 1971. *Invertebrate Photoreceptors,* 179 pp. New York: Academic Press.

11

The Chelicerates

OVERVIEW

Chelicerates represent a highly successful group of arthropods. Their body is uniquely divided into an anterior prosoma and a posterior opisthosoma. The segmental prosomal appendages are the paired chelicerae, pedipalps, and four pairs of walking legs. The opisthosoma consists of 12 or fewer segments that may bear gas exchange organs but never walking legs. Often a prosomal carapace and sometimes a posterior telson are present. All chelicerates lack antennae, mandibles, and a deuterocerebrum.

Merostomates, once a large aquatic group but represented by only several extant genera of horseshoe crabs, including *Limulus,* have a large prosoma, smaller opisthosoma, compound eyes, and a prominent telson. Eurypterids, a long extinct group of giant water scorpions, were quite large and had a venomous telson.

Arachnids, the great majority of all chelicerates, are small animals that have adapted superbly to the land. Because of their diversity, there are many variations in body shape, degrees of fusion between the two tagmata, sense organs, and functions of chelicerae and pedipalps. Arachnids lack compound eyes, have a pharyngeal pump, excrete by means of coxal glands and/or Malpighian tubules, and carry out gaseous exchange by book lungs and/or tracheae. Following a rather elaborate precopulatory courtship, fertilization is internal, and the young hatch from cocoons as miniature arachnids without a larval stage.

The Order Araneae includes the versatile and eminently successful spiders. A nonsegmented prosoma and opisthosoma are united by a slender pedicel. There are four pairs of simple eyes, but hunting spiders can form sharp images. Chelicerae are modified as poisonous feeding fangs, and pedipalps in males are adapted as copulatory devices. All spiders produce silk from paired silk glands and three pairs of opisthosomal spinnerets; it is widely used in dispersal and to construct webs, cocoons, and nests. The Order Acarina includes the mites and ticks, whose body consists of a gnathosoma with chelicerae and pedipalps and an idiosoma with the eight legs. With diverse feeding methods, many mites are either symbiotic or parasitic on a variety of hosts, and all ticks feed on vertebrate blood.

Scorpions are primitive arachnids that use the chelate pedipalps and a posterior sting in feeding. They also have opisthosomal pectines and indirect sperm transfer, and females brood the young. Harvestmen have a very small body and four pairs of disproportionately long legs. They are also characterized by having odoriferous glands and preliminary digestion within the gut, and they lack Malpighian tubules and book lungs. The pseudoscorpions resemble scorpions but lack opisthosomal subtagmata and a sting. They have cheliceral glands that produce silk, which is used in constructing nests. Sun spiders have enlarged chelicerae and curious racquet organs on the last legs, and males use their chelicerae to transfer sperm. Six other orders are composed of minute cryptozoic forms.

Pycnogonids or sea spiders have a small modified body with an anterior cephalon and a trunk. There are four (sometimes more) pairs of elongate legs; in males there is an additional anterior pair of brooding legs.

Chelicerates probably arose independently of other arthropod taxa from an ancient segmented marine worm, and merostomates and pycnogonids represent two early evolutionary branches from the chelicerate stock. An early protoarachnid became terrestrial and underwent adaptive radiation to produce four principal arachnid lines: scorpions, spiders, acarines, and pseudoscorpions.

INTRODUCTION

Chelicerates, comprising the second largest subphylum of arthropods, are an important and diversified group of more than 65,000 species. Everyone is familiar with spiders and ticks, and most everyone has seen or heard about scorpions and horseshoe crabs. But there are many different chelicerates that are unfamiliar and unknown to most people. Most chelicerates are small and range from microscopic to about 5 cm in length, although a few are larger. Ancestral chelicerates were marine, as are some primitive extant forms, but most modern-day species are terrestrial, even though some have returned to aquatic habitats. Chelicerates represent a rather distinct phylogenetic line of arthropods that evolved some important and unusual adaptations.

This very large taxon is represented by three subordinate taxa: Merostomata, Arachnida, and Pycnogonida, which are characterized in Table 11.1. The Class Merostomata includes the several contemporary species of xiphosurans called horseshoe crabs and their extinct relatives, the eurypterids. The mostly terrestrial Class Arachnida, by far the largest chelicerate group, includes the very familiar spiders, ticks and mites, scorpions, and many others. The Class Pycnogonida is a small group of marine sea spiders; since they are rather unusual forms, the discussion of general chelicerate features will not necessarily apply to them.

GENERAL CHELICERATE FEATURES

The body of a chelicerate is uniquely divided into two tagmata, the prosoma and opisthosoma. The anterior **prosoma,** sometimes called the cephalothorax, represents the acron plus the anterior six or seven segments, all of which are fused together. Often this entire tagma is covered dorsally by a **carapace** that represents the fusion of all prosomal terga. The first prosomal segment is preoral in position and bears a most unusual pair of appendages, the **chelicerae,** from which the subphylum name was derived. Each chelicera is constructed of a basal coxa and one or two additional podomeres, and this limb is often chelate with the distal two podomeres opposing each other in forming a pincerlike grasping claw. The second prosomal segment is postoral in position and bears a pair of limbs,

TABLE 11.1. □ **BRIEF CHARACTERIZATIONS OF THE CLASSES AND SUBCLASSES OF CHELICERATES**

Taxon	*Habitat*	*Body Features*	*Gas Exchange Organs*	*Other Features*
Class Merostomata				
Subclass Xiphosura	Marine, benthic	Large prosoma with carapace, long telson	Book gills	Have paired compound eyes, legs with gnathobases
Subclass Eurypterida	Aquatic	Elongate, two opisthosomal subtagmata	Book gills	All extinct; had paired compound eyes, piercing poisonous telson
Class Arachnida	Mostly terrestrial	Body segments often obscure	Book lungs and/or tracheae	Pumping chamber in anterior digestive tract
Class Pycnogonida	Marine, benthic	Prosoma of cephalon and thorax; very reduced opisthosoma	None	4–6 pairs of unusually long legs; males have ovigerous legs

the **pedipalps,** one on either side of the mouth. The medial surface of each pedipalpal coxa is formed into a toothed plate, the **gnathobase,** and the two opposing gnathobases are used to macerate food. The pedipalps, composed of six or seven podomeres, may closely resemble the walking legs, or they may be specialized for food capture, defense, or even sperm transfer. Each of the next four prosomal segments bears a pair of **walking legs,** each with usually six or seven podomeres; four pairs of legs is a diagnostic feature for chelicerates. The coxae of the legs often form gnathobases that also crush food particles and pass them forward toward the mouth. Typically, one or more of the anterior legs is chelate. A seventh prosomal segment is present in merostomates but absent in arachnids. The acron and probably only the first segment formed the ancient head, and the remaining prosomal segments formed the thorax (see Table 10.2 in Chapter 10). However, in all modern-day chelicerates there is never a perceptible division between head and thorax so that the prosoma is a single unified tagma. It is indeed notable that the prosoma of all chelicerates lacks both the mandibles and antennae that are so characteristic of all other arthropods. Correlated with the absence of antennae, all chelicerates also lack the deuterocerebrum, that lobe of the brain in all other arthropods that is the center for antennal sensations.

The **opisthosoma** or abdomen consists primitively of 12 distinct segments, none of which ever bears walking legs. In most chelicerates there have been fusions in opisthosomal segments so that this tagma often loses visible evidences of metamerism. If appendages are present on any opisthosomal segment, they are either highly reduced or modified as gills or other gas exchange structures, and no limbs are ever present on the posterior five segments. The first opisthosomal or **pregenital segment** is usually reduced or modified in some manner, and opisthosomal segment II is the **genital segment,** since it bears the gonopores. The pygidium may be present as a postanal **telson** that in some chelicerates is elongate or modified as a sting.

In most chelicerates, mechanical food breakdown and initial enzymatic digestion take place outside of the animal. Epithelial cells in the midgut or its diverticula phagocytize food particles or molecules so that final digestion is intracellular. Excretion is by either coxal glands or Malpighian tubules, and gaseous exchange is achieved by book gills, book lungs, or tracheae. All chelicerates are dioecious, and a great many have internal fertilization.

CLASS MEROSTOMATA, SUBCLASS XIPHOSURA

The merostomates, an ancient group whose fossil record extends back into the Ordovician period, were very likely both quite numerous and the dominant organisms about 500 million years ago. But most have become extinct, and there are only four living marine species, all of which are in the Subclass Xiphosura. Most fossil merostomates looked very much like *Limulus polyphemus,* the common horseshoe crab of the Atlantic coast of the United States and the Gulf of Mexico. *Limulus,* along with two other genera that are found along the coasts of the far East, are the only remnants of what was at one time a very large group. *Limulus* may reach an overall length of 70 cm, and they are usually found plowing through the upper stratum of muddy or sandy bottoms in shallow littoral habitats.

External Features

The somewhat flattened body of *Limulus* consists of a large prosoma, a smaller opisthosoma, and a long prominent telson (Fig. 11.1). The body is covered dorsally by a hard, dark brown, exoskeletal **carapace,** which is almost circular in front and is in the shape of a horseshoe, a feature that is the basis of their common name. The carapace extends laterally and posteriorly to cover partially the sides of the opisthosoma. On the dorsal surface of the convex carapace is a medial ridge and two lateral longitudinal ridges; located just lateral to each ridge is a **compound eye.** Farther anteriorly and located between the medial and each lateral ridge is a **medial eye** (Fig. 11.1a).

The segments and the seven pairs of appendages of the prosoma fit into the ventral concavity created by the carapace (Fig. 11.1b). The **mouth** is located anteriorly and just behind the fleshy labrum, and the chelate **chelicerae** are used in feeding. There are five pairs of **walking legs,** the first of which are the **pedipalps.** Each leg of the anterior four pairs consists of a **coxa, prefemur, femur, tibiopatella,** and two **tarsi** that form a chela. *Limulus* feeds primarily on polychaete worms but may also eat other benthic organisms like molluscs, nemertean worms, and algae. Food organisms are crushed and macerated by the gnathobases located on the coxa of each leg (Figs. 11.1b, 11.2b). The fifth pair of legs is similar to the anterior four pairs except that this pair is not chelate, it lacks gnathobases, and the distal end of the penultimate (next-to-last) podomere bears four flattened, leaflike processes used for pushing the animal out of the mud and for moving mud or sand during burrowing (Fig. 11.1b). Each coxa of the fifth pair is equipped with a spatulate gill-cleansing process, the **flabellum.** The seventh or last prosomal appendages are the **chilaria,** which are short, blunt, and unsegmented and may function as a gnathobase (Fig. 11.1b). It is thought that this posteriormost prosomal segment may be the first opisthosomal (pregenital) segment, which in the course of evolution became a part of the prosoma, and its appendages have been correspondingly reduced.

The polygonally shaped **opisthosoma** joins the prosoma all along the posterior border of the latter, and the two tagmata are articulated by strong ligaments and muscles so that the opisthosoma can be flexed ventrally toward the prosoma. The opisthosoma bears six movable spines on each lateral surface that function in

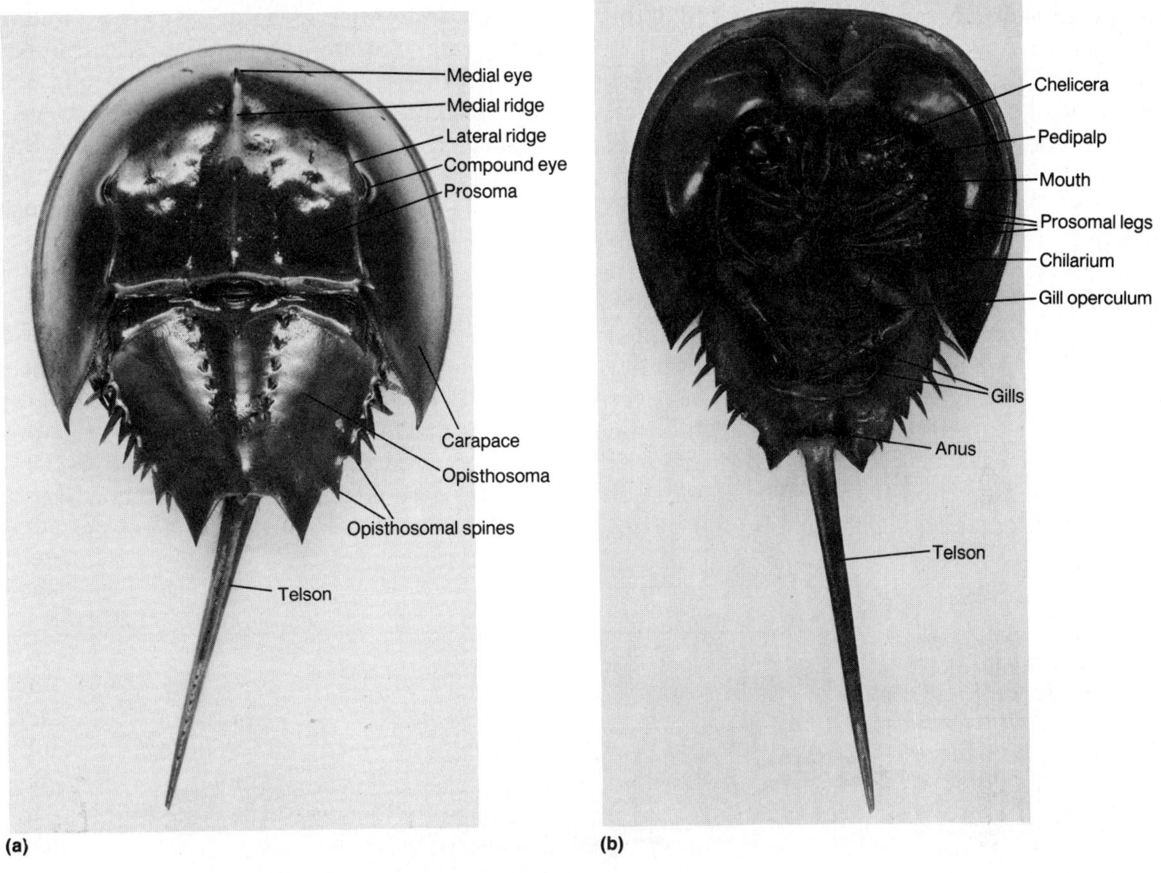

Figure 11.1 Photographs of Limulus polyphemus, *an extant merostomatan, to show external features: (a) dorsal view; (b) ventral view.*

sensory reception and probably assist in burrowing. Ventrally, each of the six opisthosomal segments bears a pair of modified appendages. The appendages of the first opisthosomal segment have fused medially to form a **gill** or **genital operculum** or flap that partially covers the gills and protects the paired **gonopores,** also located on this segment (Fig. 11.1b). The remaining opisthosomal appendages are fused medially to form five single flaplike **gills.** They are often termed **book gills,** since from their undersurfaces arise many leaflike **lamellae** that facilitate gaseous exchange by increasing the exchange surfaces substantially.

The long, spikelike, posterior **telson,** not homologous to the telson of other arthropods, represents the fusion of some opisthosomal terga (Fig. 11.1). The telson is highly mobile where it articulates with the opisthosoma and is used both to right the body when overturned and to push the animal forward when burrowing. Regardless of the perhaps ominous appearance of the telson, it is completely harmless, and horseshoe crabs can be easily manipulated by using it as a handle. The **anus** is located ventrally near the posterior end of the opisthosoma.

Principal sensory organs are the compound and medial eyes and the frontal organ. The lateral **compound eyes** are constructed of about 1000 visual units or ommatidia. An ommatidium in *Limulus* differs from those in other arthropods by having a somewhat different cornea and crystalline cone, by not being shielded from light entering adjacent ommatidia, and by usually having 12 (sometimes 8–15) retinula cells in the rhabdome. *Limulus* cannot form images with so few receptor units, but its eyes may be able to detect movements. Each of the two **medial eyes** or ocelli is a hemispherical invaginated cup lined with a layer of receptor cells, and a crude lens is formed by the cornea overlying the ocellus; the ocelli are sensitive to different intensities of light. A medial **frontal organ,** rather prominent in larvae as an olfactory organ, is a degenerate mass in adults but may continue to function in chemoreception. Sensory receptors are also present on the gnathobases, spatulate processes of the sixth prosomal appendages, and opisthosomal spines.

Visceral Systems

The mouth opens into the stomodeum, which is subdivided into an **esophagus, crop,** and **gizzard** (Fig. 11.2a). The gizzard, whose walls are equipped with denticles or teeth and strong muscles, is an effective grinding organ. Usable food particles pass into the stomach, while un-

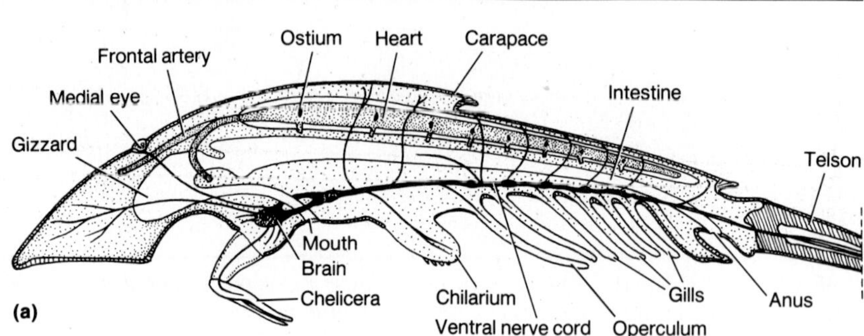

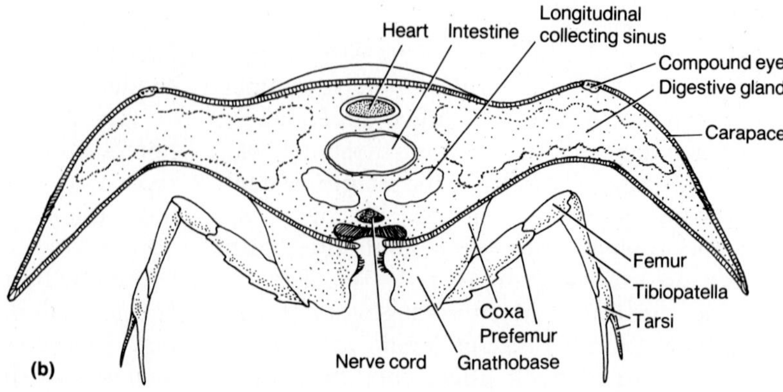

Figure 11.2 *Some internal features of* Limulus polyphemus: ***(a)*** *longitudinal sectional view through the prosoma and opisthosoma (from A.E. Shipley, 1909,* Cambridge Natural History, *Macmillan and Co., Ltd., London);* ***(b)*** *cross-sectional view through the prosoma at the level of the compound eyes.*

digestible bits are regurgitated through the mouth. Arising from the enlarged **midgut** or stomach are the paired **digestive glands,** which ramify extensively into most areas of the prosoma and opisthosoma (Fig. 11.2b). Enzymes, secreted by the digestive glands, empty into the stomach, where initial chemical breakdown takes place. The partially hydrolyzed food then enters the digestive glands, where digestion continues and phagocytosis takes place. Undigested particles pass into the **intestine,** which is continued as a short proctodeum or **rectum** that terminates as the **anus.**

There is a well-developed circulatory system with a heart and a number of arteries. The tubular muscular **heart** lies in the dorsal midline with about half of its length in each tagma and suspended within the **pericardial chamber** by nine pairs of elastic ligaments (Fig. 11.2a). Eight pairs of valvular **ostia,** located segmentally in the heart wall, allow blood to enter the heart from the pericardium. Arteries include four pairs of lateral arteries, a pair of aortae, and a single medial frontal artery. Blood passes from the extensive arterial system into the **hemocoel,** is gathered ventrally into a pair of hemocoelic, longitudinal, collecting sinuses (Fig. 11.2b), and passes into efferent channels, which carry it to the book gills for gaseous exchange. Afferent channels return blood from the gills to the pericardial chamber. The blood of *Limulus* contains amebocytes that function in both phagocytosis and clotting. Present in the plasma is **hemocyanin,** a copper-containing, oxygen-transporting pigment that, when oxygenated, imparts a blue color to blood.

Several adaptations in *Limulus* facilitate gaseous exchange at the gills' surfaces. Many **endochondrites** or small transverse bars are found on the floor of the opisthosoma at the bases of the gills, and many **branchial cartilages** extend into the operculum and gills. Extrinsic muscles, inserting into these endoskeletal parts are responsible for movements of the gills, which enhance water circulation over the lamellae. Movements of the gills also assist in vascular circulation; as the gills are pulled anteriorly, the lamellae become engorged with blood, and as they move posteriorly, blood is expelled from the lamellae. Because of the rhythmic paddling of the gills, very young specimens can even swim with the ventral side uppermost.

Nitrogenous wastes are eliminated by four pairs of modified coelomoducts, the **coxal glands,** with a gland located near the base of each of the prosomal appendages II–V. Supplied with a copious blood supply, these reddish glands filter out and concentrate wastes, which are transported down a common **excretory duct** on each side. The ducts open to the exterior via a pore located on the coxae of the last pair of walking legs.

Illustrating a rather high degree of cephalization, the anterior nervous system or **brain** is a circumenteric collar with the protocerebrum forming the anterior portions and the tritocerebrum and all the prosomal ganglia forming the lateral aspects (Fig. 11.2a). Each ganglion provides lateral nerves to the various parts of an opisthosomal segment and its appendages.

The basic plan of the reproductive systems is essentially the same for both sexes. A single **testis** or **ovary** lies beneath the intestine and extends throughout the opisthosoma and into the prosoma. A short duct (vas deferens, oviduct) opens to the outside on the ventral surface of the genital segment. Mating is synchronized to lunar and tidal cycles and normally takes place in very shallow water or even on the beach at low tide. Sometimes masses of a dozen or more individuals may aggregate into a teeming copulating mass. To copulate, the smaller male climbs on top of the female and holds onto her anterior carapace by the hooks on the distal tarsi of his pedipalps. The female scoops out one or more depressions in the sand into which she deposits several hundred eggs. After the male discharges sperm over them, the fertilized eggs are covered with sand and left by the pair.

There is no parental brooding or care of the young.

Cleavage is total and produces a stereogastrula with two germinal centers. The anterior center forms the future first four prosomal segments, while the posterior center forms all remaining metameres. Interestingly, these two centers correspond to the trilobite cephalon and pygidium, a feature that suggests a close phylogenetic relationship between trilobites and merostomates. Embryonic development is concluded in about two weeks. Since the prosoma of the newly hatched *Limulus* superficially resembles a trilobite cephalon, the juvenile is called a **trilobite larva** (Fig. 11.3). Most appendages are present except for some opisthosomal gills, which are added at succeeding molts. Sexual maturity is achieved only after more than ten molts, requiring about three years.

Subclass Eurypterida

The eurypterids, commonly called the giant water scorpions, are known only from the fossil record, since they became extinct near the end

Figure 11.3 *Photograph of the trilobite larva of* Limulus.

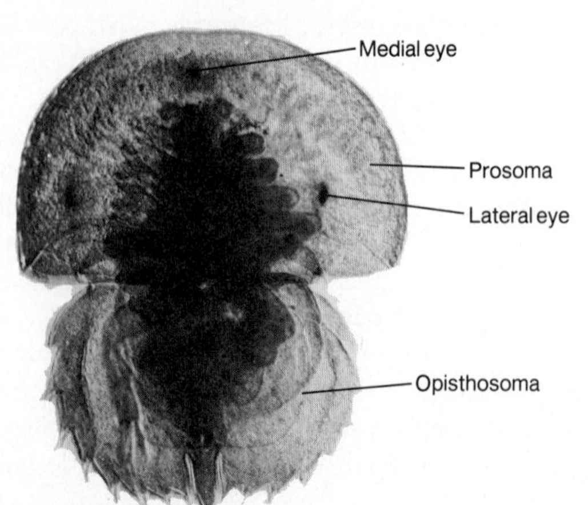

(a)

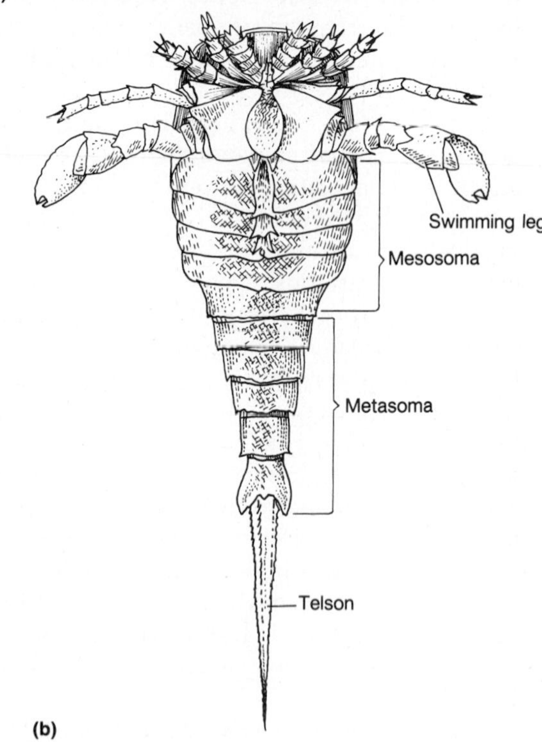

(b)

Figure 11.4 *Eurypterids:* ***(a)*** *photograph of two partial fossilized eurypterids (courtesy of C.F. Lytle);* ***(b)*** *diagram of the external features (from H. Woods, 1909,* Cambridge Natural History, *Macmillan and Co., Ltd., London).*

of the Permian period. The eurypterids, like the trilobites, represented an important group ecologically, since they were large creatures and perhaps were dominant in ancient seas.

Eurypterids both swam in and crawled over the bottom of freshwater and marine habitats. With some attaining a length of almost 3 m these giant water scorpions must have indeed looked formidable because of their size and their venomous telson. They were elongate organisms built much like xiphosurans with a prosoma, opisthosoma, and telson (Fig. 11.4). The prosoma was smaller than that in xiphosurans, and the carapace lacked the lateral extensions. Paired compound and medial eyes were located dorsally on the carapace. Prosomal appendages included the usual paired chelicerae, pedipalps, and walking legs, and the posterior pair of legs was modified into paddlelike appendages utilized in swimming (Fig. 11.4).

The opisthosoma was composed of 12 distinct segments, but they were grouped into two subtagmata, differing from the opisthosoma in xiphosurans. The anterior seven segments formed a mesosoma or preabdomen bearing gill-like appendages, and the posterior five comprised a more narrow metasoma or postabdomen, which was without appendages. A telson, attached to the posterior end of the opisthosoma, probably was modified as a poisonous sting used in defense and feeding (Fig. 11.4).

CLASS ARACHNIDA

This class is a very large and important group of organisms that has become one of the most successful arthropod taxa. Evolving from aquatic ancestors, arachnids were very adept in colonizing the land, and most modern-day species of this diversified group are terrestrial. Arachnids are basically small animals that range in size from 0.02 mm up to 18 cm long, though most are from 0.1 to 5 cm in length. There are about 65,000 known species, which represent about 99% of all chelicerates; because of their small sizes and cryptozoic habits, many other arachnids perhaps remain undiscovered or undescribed. Arachnids are some of the best-known arthropods, since this taxon includes spiders, ticks, mites, scorpions, daddy longlegs, and their many relatives. Perhaps more than any other invertebrate group, many arachnids strike fear among most people, but this phobia is really not justified, since only a very few species are dangerous to human beings. Arachnids are present in almost every terrestrial ecosystem from high mountains to tidal marshes. They favor areas around human habitations, soils, leaf litter, meadows, and woodlands. Some species are highly destructive to domesticated plants, others are ectosymbiotic on a wide range of animals, but most are free living and innocuous to human beings. Many outstanding adaptations have enabled arachnids to be successful on land and fill many terrestrial niches. The most important of these are summarized in Box 11.1.

The heterogeneity of the Class Arachnida is reflected in the fact that there are 12 orders of various degrees of species diversity and importance. Two orders, Araneae (spiders) and Acarina (mites, ticks), have in excess of 20,000 species each, and their representatives are important economically and to their ecosystems. Each of four other orders—Scorpiones (scorpions), Opiliones (harvestmen or daddy longlegs), Pseudoscorpiones (false scorpions), and Solifugae (sun spiders)—has from 750 to 4000 species and is of moderate ecological importance. These six major orders are characterized in Table 11.2. The remaining six orders are composed of minute creatures with very low species diversity; they are comparatively unimportant ecologically. The major features of all 12 orders are listed at the end of this chapter.

External Features

Segmentation in both prosoma and opisthosoma is often obscured externally, especially on the dorsal surface, where the collective terga are

BOX 11.1 □ SOME IMPORTANT ARACHNID ADAPTATIONS NECESSARY FOR A TERRESTRIAL EXISTENCE

Ancient chelicerates were all aquatic and most likely were marine. However, almost all contemporary arachnids are terrestrial, although some species have become aquatic again. Among the invertebrates, only the arachnids and insects as entire large groups have been able to adapt to life on land. What kinds of special adaptations had to be evolved before arachnids could successfully colonize the land?

Much of the success of both arachnids and insects on land is due to the physicochemical and mechanical properties of the cuticle. Arachnids have an epicuticle that contains a special waterproofing layer of lipids and other waxes; this has proved to be a most important water-conserving stratum. The cuticle is sufficiently strong that the arachnid body is adequately supported in an aerial medium and sufficiently strong that powerful appendicular muscles can insert into it to provide the crucial levering leg motions. Becoming terrestrial manifestly affected the feeding habits of arachnids; the majority are carnivorous and have developed a plethora of ways for stalking, capturing, disabling, and ingesting prey.

Arachnids have a number of functional and morphological features that are very instrumental in water conservation. There are three principal physiological processes that potentially could result in great loss of water: egestion, excretion, and gaseous exchange. Most arachnids reabsorb much of the water in the digestive tract so that dry consolidated feces are produced. Excretion in both Malpighian tubules and coxal glands follows basically the same plan whereby excretory products absorbed from the blood are converted into guanine as a semisolid material or as crystals that contain almost no water. Gas exchange organs require extremely large surface areas that obviously are the locus for the loss of great volumes of water. But if these gas exchange organs are internal and their cavities have a high relative humidity, then inadvertent water loss can be kept to a minimum. In arachnids, both book lungs and tracheae have large but internalized surface areas, and muscular spiracles close off these chambers or tubes except at times of actual inhalation or exhalation of air.

Terrestrial existence has a profound impact on reproduction and embryonic development. A terrestrial arachnid must either package sperm in desiccation-resistant spermatophores or transfer sperm directly to the female's reproductive tract by a copulatory organ. Once the eggs are fertilized, an arachnid must either retain the embryos within her reproductive tract or lay them surrounded by a water-conserving shell or silken case. The mechanisms by which sperm are transferred and embryos are protected have emancipated arachnids from moist habitats and permitted much wider geographical and ecological distributions.

To early arachnids the new terrestrial habitat was a mixed blessing. There were many problems associated with living out of water, and these formidable obstacles had to be overcome before a terrestrial life-style was even possible. But at that time there was very little competition on land, and the huge number of unfilled ecological niches permitted extensive diversification. Arachnids had enough adaptive plasticity to overcome these great physiological barriers and to become preeminently successful on land.

completely fused to form a **carapace.** Ventrally, segmentation is somewhat more evident, since the sterna often are not completely fused; but in many higher arachnids the presence of the paired appendages provides the only external clue to metamerism.

The **chelicerae** are very versatile organs in arachnids and are used for feeding, digging, and creating sounds (scorpions) and as poisonous fangs (spiders). If a chelicera is chelate, it usually has three podomeres, but there are only two podomeres if this limb is used in piercing. Each **pedipalp** has six podomeres (the metatarsus is missing), and the coxa has a gnathobase. The pedipalps may be chelate (scorpions) or more leglike, and spiders and solifugids even employ them in copulation. The four pairs of **walking legs** are similarly constructed of seven podomeres: **coxa, trochanter, femur, patella, tibia, metatarsus,** and **tarsus.** Gnathobases may be present on all legs (harvestmen), restricted to anterior pairs (scorpions), or absent altogether (spiders, acarines). Most legs end in one to three chitinous **claws** that may have minute teeth. All prosomal appendages are richly provided with sense organs.

If the point of union between prosoma and opisthosoma is a broad area as in scorpions, the

TABLE 11.2. □ **THE CHARACTERISTICS OF THE SIX MAJOR ORDERS OF ARACHNIDS**

Order	*Common Name(s)*	*Chelicerae*	*Pedipalps*	*Other Features*
Scorpiones	Scorpions	Small, chelate	Large, chelate	Have poisonous telson, opisthosomal pectines; no tracheae; form spermatophores
Araneae	Spiders	Poisonous fangs	Copulatory organs in males	Have 8 eyes, silk glands, and opisthosomal spinnerets; engage in elaborate precopulatory courtship
Opiliones	Harvestmen, daddy longlegs	Small, chelate	With gnathobases	Have very long legs, odoriferous glands; little preingestive digestion; are omnivores
Acarina	Mites, ticks	Variously modified	Variously modified	Body of gnathosoma and idiosoma; various nutritional patterns; many are symbionts
Pseudoscorpiones	False scorpions	Chelate, have silk glands	Enlarged, chelate, have poisonous glands	No book lungs, metasoma, or sting; males produce spermatophores
Solifugae	Sun spiders, camel spiders	Enlarged, chelate	Leglike, tarsal suckers	Live in arid habitats; last pair of legs bears racquet organs

first opisthosomal segment is absent. But if the prosoma and opisthosoma are united by a narrow "waist" as in spiders, then this **pedicel** represents opisthosomal segment I. Usually, the opisthosoma has no appendages, but scorpions have a pair of modified appendages (pectines) on opisthosomal segment III, and spiders have paired spinnerets on opisthosomal segments III and IV. The last opisthosomal segment bears the ventral anus, and a postanal telson is not common in arachnids. The opisthosoma is divided into two subtagmata in scorpions and some other primitive arachnids, but in all others this tagma is smooth in outline and has lost almost all external signs of metamerism. In the Acarina the prosoma and opisthosoma have completely fused together so that the body is a single unit.

CUTICLE AND SENSE ORGANS

The arachnid cuticle is essentially like that of other arthropods generally except that the epicuticle contains an additional layer of waxy materials that greatly retards water loss. This water-conserving cuticular layer is one of the paramount adaptations necessary for life on land.

Because of the nature of the cuticle, sense organs are necessarily modified in arachnids as in all arthropods and include simple eyes, several different sorts of sensory hairs, and slit sense organs. Most arachnids have one to six pairs of simple eyes or **ocelli.** The outermost part of an arachnid ocellus is a transparent cuticle or cornea that also functions as a lens. Beneath the lens are several layers, including a layer of epidermal cells called the **vitreous body,** a layer of **retinal cells,** which are the photoreceptors, and an inner **retinal membrane** (Fig. 11.5a). If the retinal cells are oriented outward toward the light source, they are called direct eyes; if they are directed inward and away from the light, they are indirect eyes. Both direct and indirect eyes are present in arachnids, and most spiders have both types. The eyes basically are sensitive to movements and to light intensities and do not normally form images; however, the eyes of hunting spiders are able to form rather good images. The characteristic compound eyes of crustaceans and insects are absent in all arachnids.

Several different types of sensory hairs are found all over the body but are concentrated on the appendages. Solid, black, movable **spines** are located on the legs of most arachnids and function as tactile receptors. Finer **setae,** distributed in nonuniform patches, account for the proverbial hairiness of the legs and function as both chemoreceptors and tactile organs (Fig. 11.5b). **Trichobothria** are long, exceedingly fine hairs that are movable owing to a ball-and-socket arrangement at their bases (Fig. 11.5c). They have been shown to be extraordinarily sensitive to slight vibrations, sounds, and air currents and have sometimes been called "touch at a distance" receptors. These remarkable structures enable arachnids to capture prey without necessarily seeing it, to sense the approach of another animal, and, in those arachnids without eyes, to overcome the apparent liabilities of blindness. Trichobothria also enable the organism to orient its body to sounds (sonotaxis) or even to the light of the sun and moon (astrotaxis). Trichobothria are absent in individuals of several different orders such as Opiliones and Solifugae.

Slit sense organs are found in the carapace of most arachnids and function as proprioceptors and kinesthetic and mechanical receptors. These organs consist of a slitlike pit in the cuticle, covered by a thin membrane, and a hairlike process that extends from a sensory cell to the membrane. Slit sense organs may be present singly or in groups of cells called **lyriform organs** (Fig. 11.5d).

Visceral Features

Since the basic arthropodan and chelicerate features for most visceral systems have already been mentioned, only those characteristics that

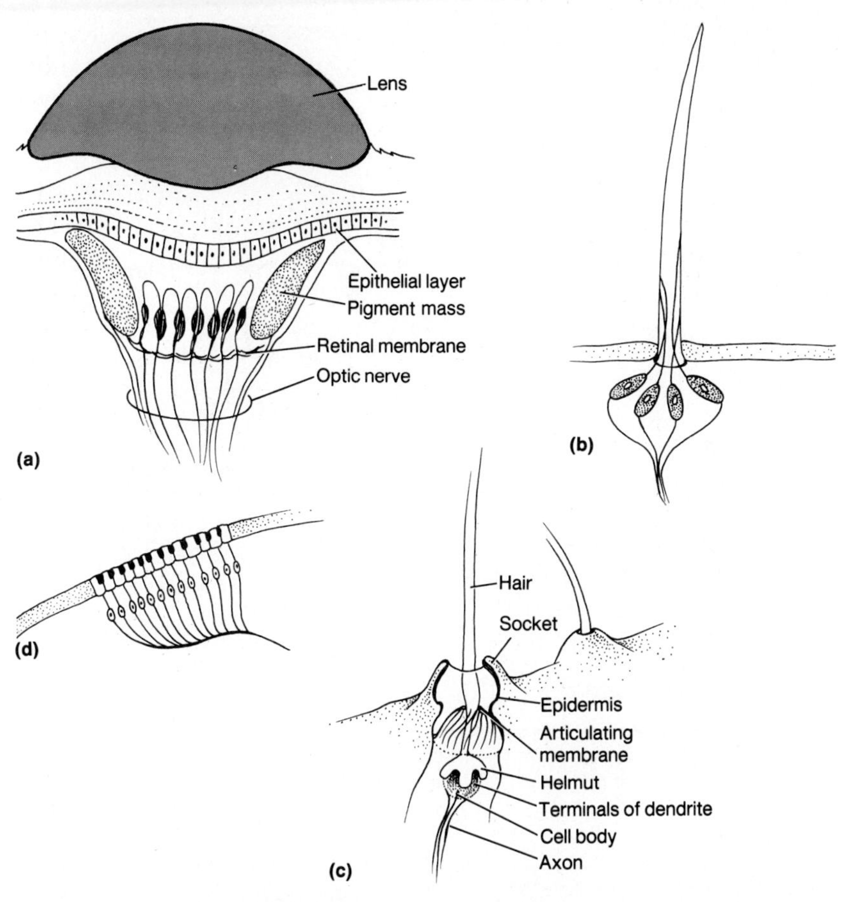

Figure 11.5 *Some arachnid sense organs: **(a)** an ocellus (indirect eye); **(b)** a seta; **(c)** a trichobothrium; **(d)** a lyriform organ.*

are unique to arachnids or those that are particularly prominent will be mentioned here. Many different types of food are used by arachnids; many spiders and scorpions are carnivores, many mites are herbivores, many other arachnids are omnivores or scavengers, and ticks feed on vertebrate blood. Chelicerae, pedipalps, and gnathobases are all involved in the mechanical breakdown of food. Digestive enzymes are usually released through the mouth onto the prey, and the partially digested slurry is then drawn into the anterior digestive tract. An almost universal feature of arachnids is a muscular stomodeal organ (pharynx or stomach) that pumps or sucks soluble food into the anterior alimentary canal. Numerous branching diverticula arise from the midgut and serve as loci for further digestion and absorption. Considerable amounts of food resources are also stored in these diverticula so that most arachnids can go without feeding for a very long time.

Excretory organs in arachnids are of two types: Malpighian tubules and coxal glands. One or two pairs of **Malpighian tubules** are the principal excretory organs; these tubules are not homologous to those of insects, since in arachnids they arise from the midgut and are of endodermal origin. The proximal end of each tubule is attached to the midgut, and its branching distal ends lie free in the opisthosomal hemocoel (Fig. 11.6a). Epithelial cells of the tubules absorb nitrogenous wastes from the hemocoelic

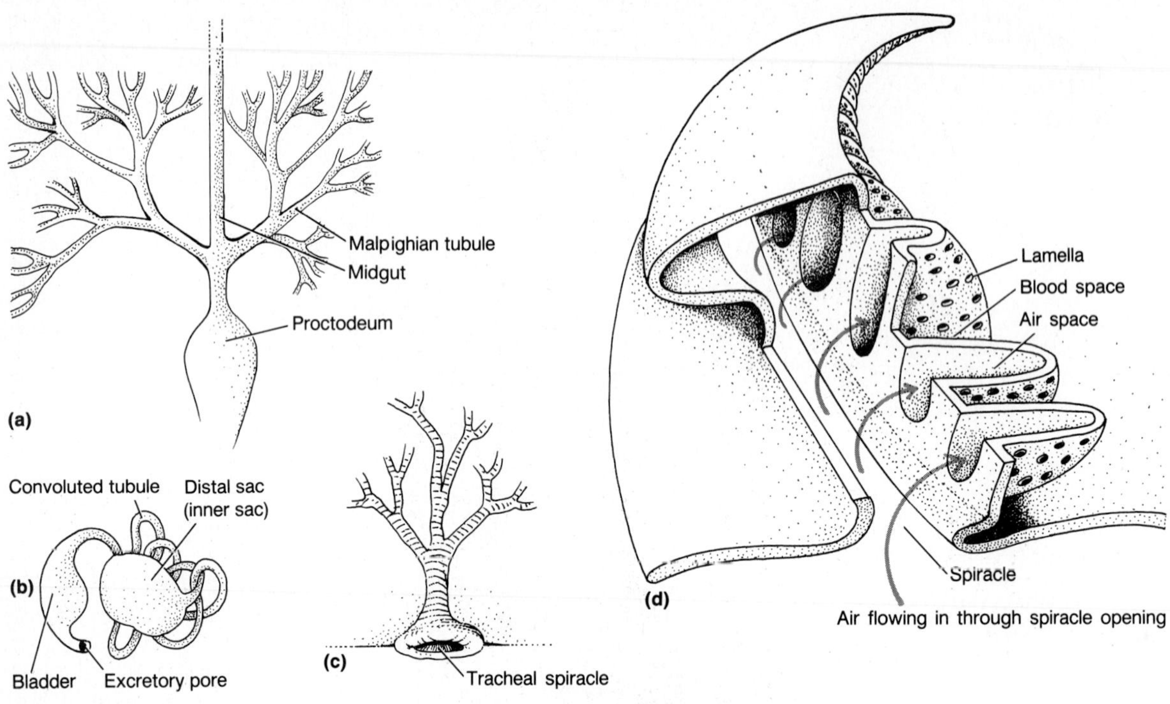

Figure 11.6 Diagrams of special excretory and gas-exchange organs in arachnids: ***(a)*** *Malpighian tubules;* ***(b)*** *a coxal gland;* ***(c)*** *a portion of a trachea and its branches;* ***(d)*** *a book lung.*

blood, and wastes are chemically transformed into guanine crystals, which are carried to the digestive tract and eliminated with the fecal materials. The elimination of crystalline wastes is of great significance, since little or no precious body water is lost in excretion. **Coxal glands** are modified coelomoducts; each consists of a thin-walled sac into which wastes are absorbed from the blood and converted into guanine. From the sac, excretory products enter into a convoluted tubule with an enlarged bladder, and from there a short tubule leads to the exterior and opens through a small pore located on the coxa of an appendage (Fig. 11.6b). From one to four pairs of coxal glands are present in various groups, but one or two pairs are perhaps the most common number.

There are two types of gas exchange structures in arachnids, book lungs and tracheae, and both open to the exterior as **spiracles.** Book lungs develop as invaginations from the anteroventral opisthosomal surface. As invaginations take place, each cavity develops from 5 to 150 folds on its anterior surface that project into the lung cavity like pages in a book (Fig. 11.6d). Scorpions have four pairs of book lungs, but most arachnids have fewer pairs, and the number of book lungs is not in any way correlated with how active the animal is. **Tracheae,** analogous to the tracheae of insects, are apparently derived either from modified book lungs or from the ectodermal invaginations that also produce book lungs (Fig. 11.6c). **Sieve tracheae,** found in pseudoscorpions and some spiders, arise as bundles from a paired tubular origin on the first opisthosomal segment, where the anterior book lungs would have originated. The main tracheae divide and subdivide into progressively smaller tubles, which eventually convey air to tracheoles. The tracheoles are fluid filled and transfer oxygen directly to cells or tissues. A second type, called **tube tracheae,** arise singly from the spiracles, are either branched or unbranched, and also end as fluid-filled tracheoles. Tube tracheae are found in most spiders and in several other taxa. The main tracheae, either sieve or tube, are lined with chitin, often in spiral strips. Pseudoscorpions, harvestmen, solifugids, and acarines have only tracheae; scorpions and others have only book lungs; and some very small forms with efficient cutaneous exchange have neither. But the remaining arachnids, including most spiders, have both book lungs and tracheae.

The arachnid nervous system illustrates the basic arthropodan plan except that in most arachnids the entire ventral nerve cord is fused with the subenteric ganglion. This **ventral nerve mass** lies in the second prosomal segment, and from it arise paired lateral nerves supplying most appendages and a single posterior nerve supplying the opisthosoma. The protocerebrum is small, a feature undoubtedly associated with the lack of compound eyes, and the chelicerae are innervated from the tritocerebrum. In most arachnids that have been carefully studied, hormone-producing neurosecretory cells are present in the brain.

Reproduction and Development

All arachnids are dioecious, and sexual dimorphism is often quite pronounced. Males tend to be smaller and to have different color marking from those of females. There is a rather wide range of differences in the number, shape, and arrangement of gonads and gonoducts. One important and essential factor in terrestrial animals like arachnids is that sperm transfer must be either direct as in copulation or indirect by encapsulating sperm into desiccation-resistant spermatophores. Primitively, indirect sperm transfer was undoubtedly practiced in which spermatophores, deposited in the environment, would chemically attract females to them. In primitive modern-day arachnids like scorpions, a female is maneuvered to the spermatophores by the courtship behavior of the male. More advanced species have direct sperm transfer by employing either modified chelicerae or pedipalps or a specialized intromittent organ, a **penis.** A rather elaborate precopulatory courtship is practiced by most arachnids, and complex patterns of behavior are manifested by both male and female. Fertilization is internal, and female arachnids subsequently lay their eggs. However, female scorpions and some mites and spiders retain the young within the reproductive tract for most or all of the period of embryonic development and are viviparous if the female broods and nourishes the young or ovoviviparous if yolky eggs develop there.

Eggs are deposited in the environment singly or in masses, and most are equipped with coverings that retard desiccation. The eggs of many spiders are wrapped in silken threads, and the entire mass is called a **cocoon.** Eggs normally hatch within several days as miniature arachnids with no true larval stage. The young of ticks and scorpions may hatch as juveniles, which are often referred to as larvae, but this title is misleading. Many females provide a modicum of postnatal care for their young. Over a period of several weeks or months the young arachnids feed, grow, molt, gradually attain their adult size, and become sexually mature.

Order Scorpiones

There are almost 800 species of true scorpions, which are found in most of the warmer parts of the world. In North America their distribution includes the southern and western portions of the United States as well as Mexico and western Canada. Most range in length from 3 to 10 cm, but an African species may be up to 18 cm long, and, incredibly, one fossil species was twice that long. Living in both arid and humid habitats, scorpions are nocturnal animals and hide during the day under logs or stones or in crevices or burrows. Scorpions are frequently found near human habitations and often enter homes or even crawl into empty shoes. Since all are poisonous, they should be treated with considerable care. Scorpions are thought to be the most primitive of all contemporary arachnids. They are of particular importance because they, or some of their ancient ancestors, were ecological mavericks, being among the first animals to become terrestrial.

On the carapace are a pair of large medial eyes located on tubercles and two to five pairs of lateral ocelli. The small chelate chelicerae project anteriorly, and the greatly enlarged chelate pedipalps are effective weapons for defense or food capture. Each of the eight walking legs bears many tactile receptors especially trichobothria, and each leg terminates in a pair of claws (Fig. 11.7b).

There are two opisthosomal subtagmata: an anterior mesosoma or preabdomen and a posterior metasoma or postabdomen. The **mesosoma** has seven segments and is as wide as the prosoma. Mesosomal segment I appears only in embryonic development and is absent in adults. Mesosomal segment II or the genital segment bears the paired ventral **gonopores** and a pair of **opercula** that cover these apertures. Mesosomal segment III has a most unusual pair of sensory appendages, the **pectines,** that have been derived from the first pair of book gills of ancestral scorpions. Each pectine consists of three rows of chitinous plates, and each plate bears numerous toothlike projections and resembles a comb (Fig. 11.7b). Sensory cells on the pectines detect degrees of humidity, sense subtle vibrations in the ground, and, in males, determine the suitability of the soil for depositing spermatophores. Each of mesosomal segments III–VI bears a pair of ventral slitlike **spiracles** that open into book lungs. There are five cylindrical segments in the **metasoma,** postabdomen, or tail, and it is more narrow than the mesosoma. The last metamere bears the ventral anus as well as the telson. Since all opisthosomal segments are so clearly distinct from each other, this is taken as another primitive characteristic.

The **sting** or **telson** is composed of a bulbous portion containing the **poison glands** and a sharp curved point near the apex on which are located the openings of these glands. The metasoma and telson are carried dorsally over the body, and the scorpion's sting is in a forward or anterior stabbing motion. Members of *Centruroides* (Fig. 11.7b), found mostly in the southwestern United States and Mexico, are the most dangerous of the North American forms, but their stings are usually not fatal to human beings. The North African *Androctonus* is quite venomous, and its poison can kill human victims within hours. Other scorpions have a painful sting but are not dangerous or fatal to people.

Scorpions are carnivores and feed on insects or other arthropods or miscellaneous invertebrates. The prey is seized by the pedipalps, stung with the telson, the neurotoxic poison of which is quickly lethal to most invertebrate prey, and torn into bits by the chelicerae. Then enzymes are secreted over the victim, and the digested food is sucked into the alimentary tract by the muscular pharynx.

The heart has seven segmental chambers and extends the length of the mesosoma. Gaseous exchange is carried out by four pairs of book lungs, and tracheae are completely absent. Excretion is by coxal glands that open at the bases of the third legs and by Malpighian tubules. There is a distinct nerve cord with seven unfused ganglia; this feature further characterizes scorpions as primitive or lower arachnids.

Males are more slender and have a longer metasoma than females and have a hook on each opercular plate. In most scorpions there is

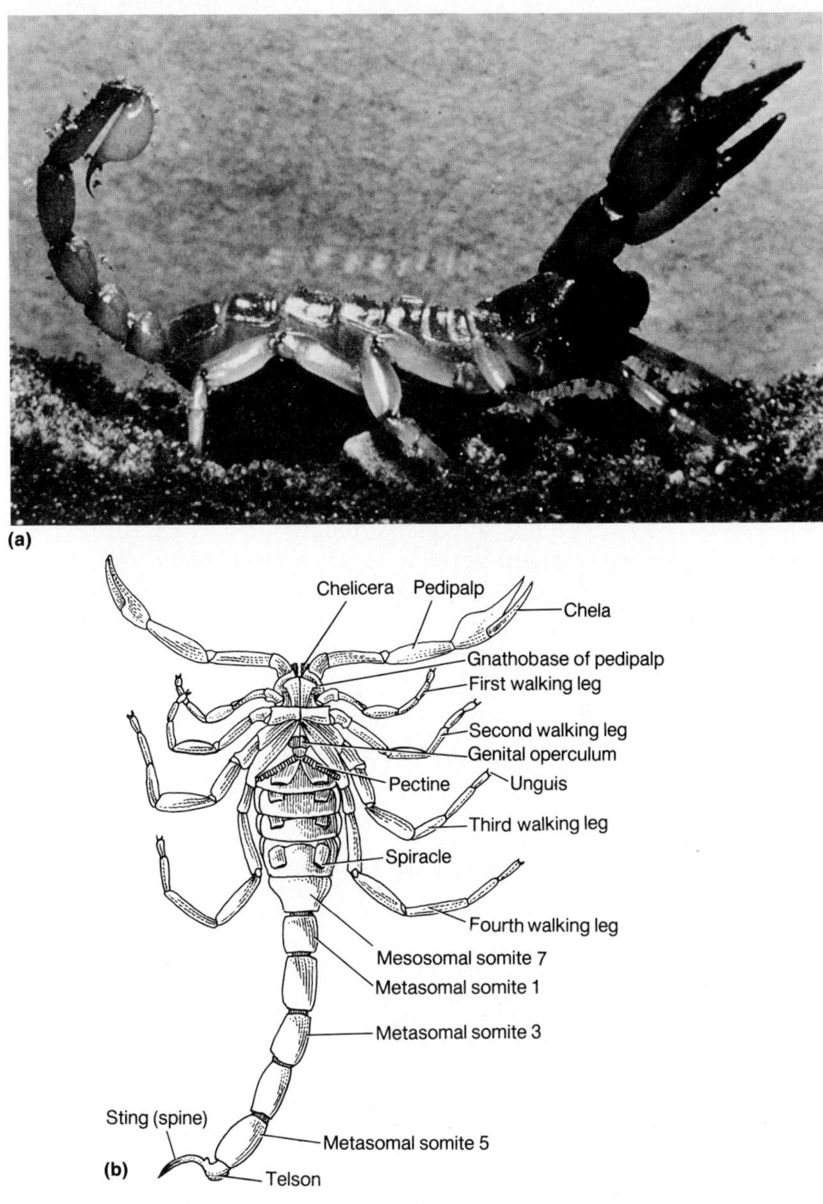

Figure 11.7 *Scorpions:* ***(a)*** *photograph of* Urotocnus *(courtesy of Ward's Natural Science, Inc., Rochester, N.Y.);* ***(b)*** Centruroides, *ventral view.*

an elaborate complicated courtship involving dances or routines followed by each member using its pedipalps to grasp those of the other (Fig. 11.8b). Together they move back and forth, sometimes for a day or more. The male maneuvers the female so that her genital atrium is over the spermatophore he had previously deposited there. A spermatophore is equipped with a lever, which, when moved, releases the sperm into the female's reproductive tract, where fertilization takes place (Fig. 11.8a). In some scorpions the female devours the male

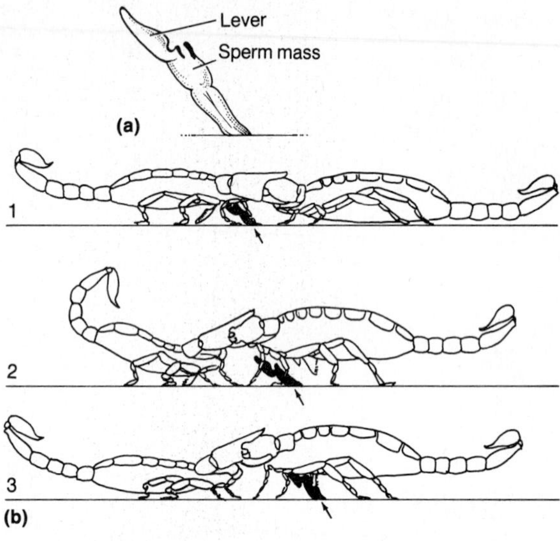

Figure 11.8 *Some reproductive features of scorpions:* ***(a)*** *a spermatophore;* ***(b)*** *three sequential stages in the courtship dance: (1) the male deposits a spermatophore, (2) the male manuevers the female over the spermatophore, and (3) the female picks up the spermatophore. (parts a, b from H. Angermann, 1957,* Zeit. f. Tierpsychol. *14:283, 285, Verlag Paul Parey, Hamburg/Berlin.)*

after the ritual dance. Scorpions are unusual in that they are either viviparous or ovoviviparous. Prenatal development requires about three weeks. After birth the very small scorpions of a typically sized brood of about 50 climb on the dorsal surface of the mother until they molt once or twice. Then they are on the ground but around the mother for several molts until they gradually become independent of her.

Order Araneae

The araneans or spiders represent over half of all arachnids with approximately 36,000 known species; this number will increase substantially as new species are described. Spiders are probably the most intriguing of all chelicerates because of three important adaptations: web building, their diverse and wily means of food capture, and the use of poison. Most spiders range in body length from 2 to 10 mm, but leg spans always add greatly to their overall size. They are found worldwide in almost every type of terrestrial habitat from arctic islands to dry desert regions. Spiders have undergone a remarkable degree of adaptive radiation so that they fill a vast number of ecological niches. They are particularly abundant in areas of rich vegetation; for example, 0.4 hectare (1 acre) of meadow may contain over 2 million individuals! Some are basically drab or dark colored, but many others have bright body markings of yellow, red, green, or blue. Spiders exhibit wide variations in body shapes, sizes, and markings, a few of which are illustrated in Figure 11.9. To many persons, spiders are sinister, repugnant, and to be assiduously avoided, but this is certainly an unjustified reputation. The discussion below should dispel many misconceptions about spiders and instill a more accurate image of these arachnids as being clever, interesting, and possessing a number of important adaptations essential to their success.

External features

The body of a spider is divided into a prosoma or cephalothorax and an opisthosoma or abdomen connected by a slender pedicel (Fig. 11.10a, b). The prosoma has lost most external evidences of metamerism, since it is covered by an unsegmented dorsal **carapace** and a ventral **sternum.** The carapace bears a rather prominent longitudinal indentation of the cuticle; this infolding or apodeme serves as the attachment site for extrinsic muscles operating the sucking stomach.

On the anterior carapace are borne the eight **eyes,** typically lying in two transverse rows, and the pairs are called the anterior lateral, anterior medial, posterior lateral, and posterior medial eyes (Fig. 11.10c). All eyes basically are con-

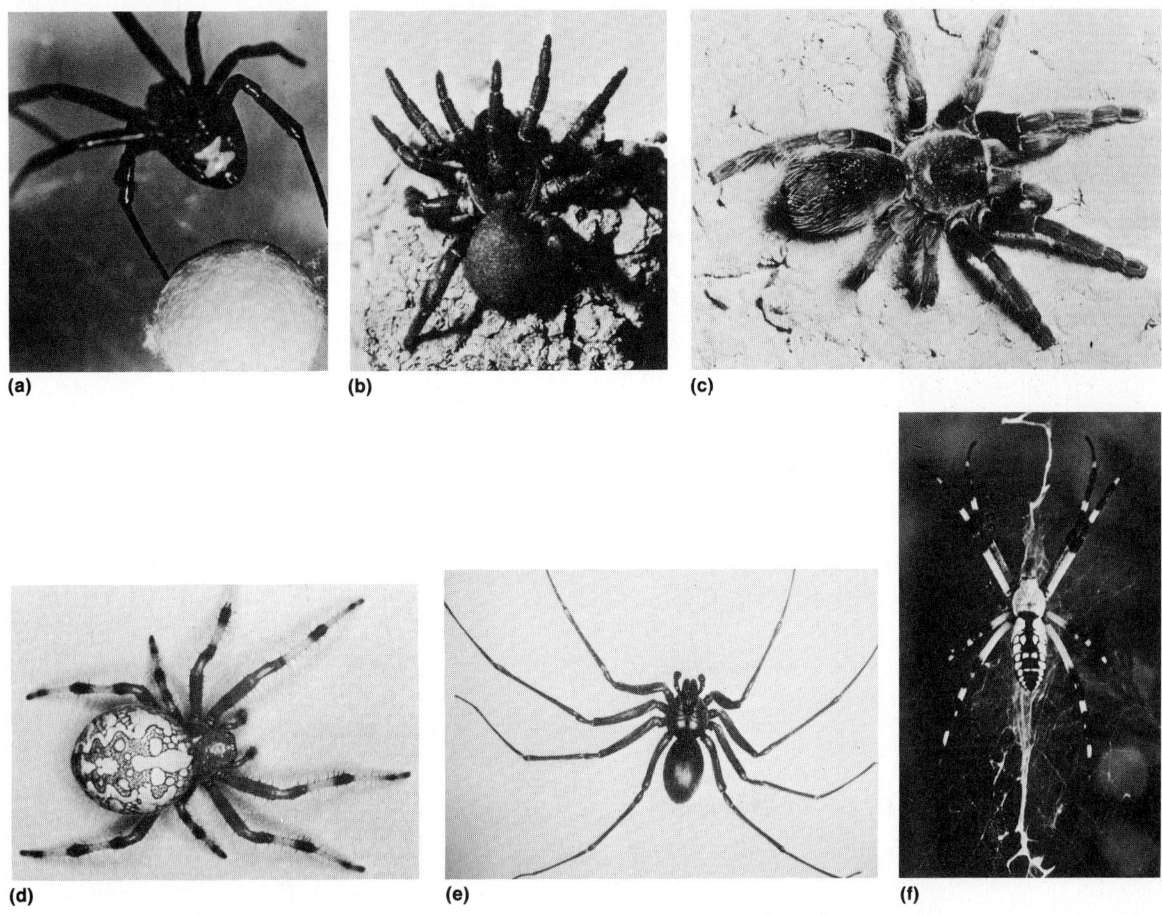

Figure 11.9 *A photographic collage of some types of spiders:* ***(a)*** *female* Latrodectus mactans, *the black widow, with her egg mass;* ***(b)*** Bothriocyrtum, *a trapdoor spider;* ***(c)*** Eurypelma, *a tarantula;* ***(d)*** *orbweaver;* ***(e)*** Loxosceles reclusa, *the brown recluse;* ***(f)*** Argiope, *a common garden spider (courtesy of J. C. Killian); (parts a, e courtesy of the Centers for Disease Control, Atlanta, Ga. 30333; parts b–d courtesy of Ward's Natural Science, Inc., Rochester, N.Y.)*

structed of a large lens, a vitreous body, and a laminated retina of a variable number of visual receptors, which point either outward (direct eyes) or inward (indirect eyes). The anterior medial eyes are considered as **primary eyes** and are always direct, while all other eyes are **secondary** and are indirect. In most spiders, eyes are important in detecting subtle changes in light intensity and perhaps motion, but vision plays a minor role in behavior, since spiders depend much more on tactile and chemical clues than on vision. Some spiders have only one to three pairs of eyes, and some cave-dwelling spiders are eyeless. The arrangement and

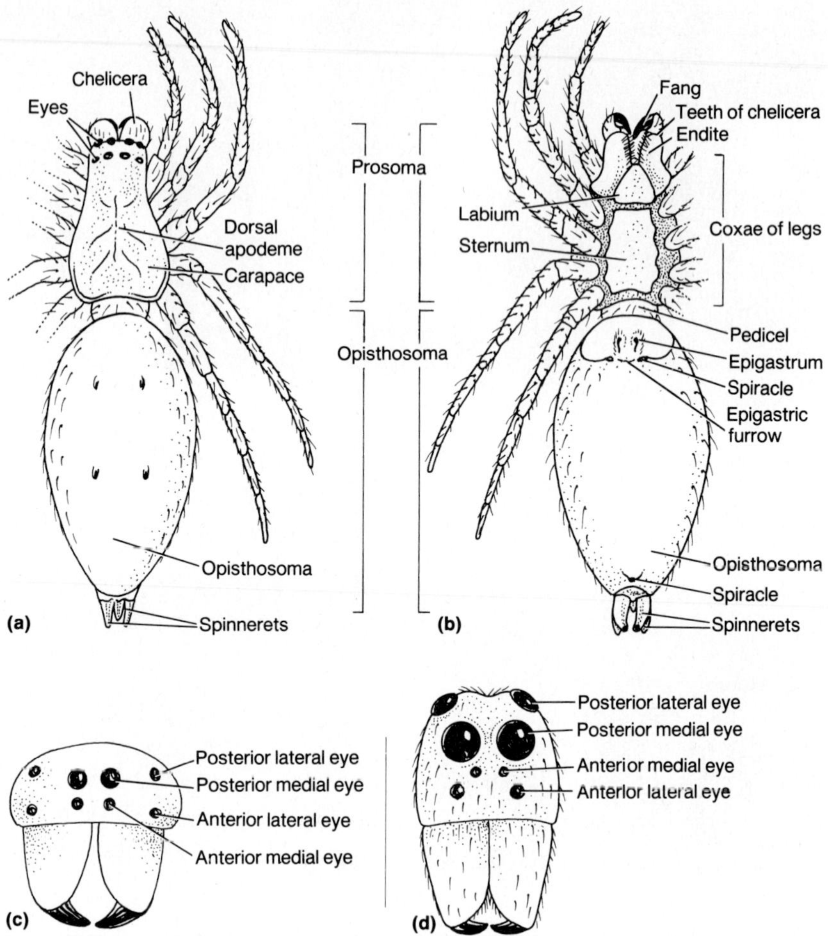

*Figure 11.10 The main external features of a spider: **(a)** dorsal view; **(b)** ventral view; **(c)** "facial" view showing the typical arrangement of eyes; **(d)** "facial" view showing the arrangement of eyes in a wolf spider.*

position of the eyes are taxonomically important in the classification of spiders.

However, the eyes of most hunting spiders (wolf, crab, and jumping spiders) are capable of perceiving motion and form, and such eyes are essential for successful hunting, stalking, and feeding. The secondary eyes are most sensitive in detecting and locating motions as exemplified by the wolf spiders (Fig. 11.10d), but in hunting spiders the primary eyes in particular are well developed and can form quite detailed images. They are not compound eyes, but the retina of each has about 1000 visual receptors. Only if light rays fall on the central portion of the retina where the receptors are concentrated can the animal discern form and thereby create sharp images. The primary eyes of jumping spiders may be as sophisticated as the compound eyes of insects in terms of perception of both motion and form. There is a possibility that these spiders can even see colors.

Prosomal appendages and their functions. The chelicerae of all spiders are the characteristic biting–piercing organs, and each is composed of two podomeres of which the distal one is the **fang** or **unguis** (Fig. 11.11a). At the tip of the fang is the opening of the **venom gland,** which is located either in the basal podomere or in the prosoma. When ready for use, the very hard, sharp fang is extended into the strike position; when not in use, the fang is folded back into a groove on the basal podomere. In about 5% of the species of spiders (Suborder Orthognatha) the chelicerae strike downward and parallel to each other. But in all other spiders the chelicerae strike transversely, that is, inward,

Figure 11.11 *Certain prosomal appendages of spiders: (a) a chelicera of* Argiope; *(b) a pedipalp of a female* Argiope; *(c) a pedipalp of a typical male with tarsal organ extended; (d) a walking leg of* Argiope.

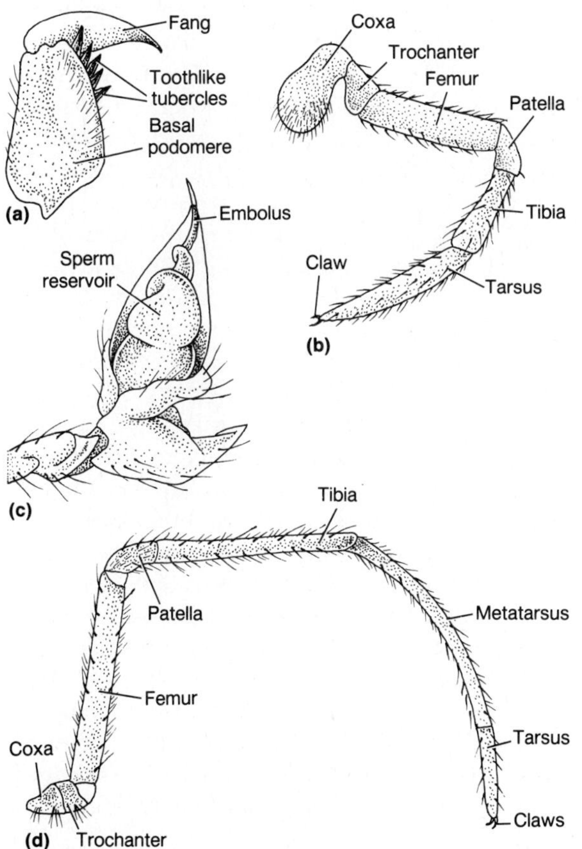

and meet in the midline of the victim. The venom, while poisonous and deadly to small invertebrates, is not usually toxic to people. The bite of the well-known black widow spider, *Latrodectus mactans* (Fig. 11.9a), can be extremely painful, and its neurotoxic poison induces severe muscular aches, nausea, some breathing difficulties, and paralysis. The brown recluse spider, *Loxosceles reclusa* (Fig. 11.9e), produces a hemolytic poison that results in a spreading necrosis of tissues around the bite. Antivenom medications for the bites of both species reduce the danger of these spiders appreciably; therefore their bites are fatal in only a small percentage of cases. Probably fewer than ten other species around the world are also considered as dangerous, and the bite of the North American tarantulas, the mygalomorphs, commonly thought to be very venomous, is no worse than a wasp sting.

The pedipalps generally resemble walking legs and, in females, end in a pair of claws (Fig. 11.11b). But in males the distal podomere is modified into a remarkable structure, the **tarsal organ,** for transferring sperm to the female (Fig. 11.11c). The functional morphology of this organ will be described in connection with reproduction.

Of the four pairs of walking legs the anterior two pairs are usually longer than the others and are used to feel or probe the environment. The tarsus on each leg bears two to three bent **claws** (Fig. 11.11d), and between the claws of many hunting spiders are dense tufts of hair, the **scopulae.** The distal surface of each scopula splits into thousands of minute extensions or end feet. The physical adhesion of the innumerable end feet provides sure traction for the animal to walk up a slick vertical surface or upside down on smooth surfaces such as glass.

Opisthosoma. The opisthosoma is normally a cylindrical or oval sac without external signs of segmentation except in one primitive family. All spiders have a **pedicel,** which is the pregenital segment and is a narrow constriction that unites the prosoma and opisthosoma (Fig. 11.10a, b). On the ventral surface near the pedicel is a heavily chitinized region called the **epigastrum** demarcated posteriorly by an **epigastric furrow.** In this furrow lie the gonopores and book lung spiracles. Toward the posterior end of the opisthosoma are one to two pairs of spiracles, which open into the tracheae (Fig. 11.10b). The posteriormost segment bears a small tubercle with the anus at its apex.

Spinning organs, silk, and webs

By far the most dramatic characteristic, and certainly one of the most diagnostic and universal features, of all spiders is their ability to produce long silken threads. Silk production is a complex function involving secretory glands and organs for blending the product into a variety of forms. A spider typically has three pairs of modified appendages, the **spinnerets,** located on the posterior opisthosoma through which silk is extruded (Fig. 11.12). Originally, spiders had four pairs of spinnerets that represented the divided appendages of opisthosomal segments IV–V. This number has been reduced to three pairs of spinnerets in most, and some spiders have only one or two pairs. Because of their well-developed muscles, the spinnerets are mobile, a feature that is essential to orienting the silk as it is produced by the spider. Located in the opisthosoma are a number of **silk glands;** the duct of each silk gland opens into one or more spinnerets by means of many tiny **spigots.** Six different glandular types have been noted in some spiders, and each secretes a different type of silk, usually with varying diameters for use by the spider in different ways. In most spiders the amount of silk leaving the gland is regulated by a valve in the duct. Cribellate spiders bear an oval plate, the **cribellum,** located in front of the three pairs of spinnerets and containing up to 40,000 spigots (Fig. 11.12). Silk exuded through the cribellum is in the form of a broad, woolly-appearing band containing thousands of fibers.

Figure 11.12 *Ventral view of the posterior opisthosoma of a cribellate spider showing the arrangement of spinnerets.*

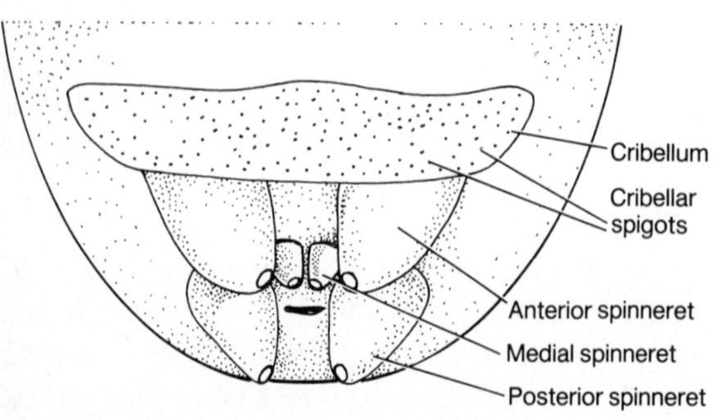

Spider silk is a proteinaceous liquid secretion of the silk glands. As it is exuded from the spigots, liquid silk is rapidly transformed into an insoluble solid thread. Once thought to harden on contact with air, silk is now known to be converted into a strong, tough material by the spider's actually pulling on the thread, a physical process that changes the molecular configuration of silk. With the strength generally of nylon, silk threads are usually no more than several micrometers in diameter, and many are even less than 1 μm. It is important to note that abandoned spider webs are not readily attacked by decomposers, probably owing to the low pH (~4) of silk.

Silk is an extremely important and versatile product of spiders. It is used to construct webs for prey capture, to build nests, to clothe newly laid eggs in cocoons, and for hinges on trapdoors. Commonly, as a spider wanders, it produces a drag line, which is attached at various points to the substratum, and this line functions as a safety line. Silk is important in dispersal, since young spiders produce silken threads that, when picked up by the wind, enable the spiderlings to "balloon" considerable distances.

Certainly the most familiar use of silk by a spider is in the construction of an intricate web.

Web-building behavior appears to be innate, since newlyhatched spiders can build complex webs. Primitively, spiders probably did not build webs, but the drag lines produced by their wanderings must have served as a crude trap for insects. Such a protoweb of drag lines serving as stumble threads is found in some modern primitive forms like certain hunting spiders. From this simple beginning, many spiders have evolved complex webs that have a variety of forms. Webs undoubtedly became adaptively significant, since better-constructed webs would catch proportionately more prey.

The orb web, produced by garden spiders, is the most familiar, and it is an amazing piece of construction in its beauty and symmetry (Fig. 11.13). In the initial phases of construction the spider spans various objects that are located near to each other by **framing,** bridge, or spanning **threads.** Usually, the initial framing threads are carried by the wind from the point of production to a distant object to which they adhere. The spider then pulls each thread taut and tests its strength. By using the bridge threads as a line the spider subsequently lays down heavier frame or **foundation lines** that form the basic scaffolding (Fig. 11.13b); some of the earlier framing lines often are consumed by the spider at the same time as it secretes the foundation lines. After the scaffolding is in place, from 20 to 60 **radial** or spoke **threads** that converge into a central hub are produced (Fig. 11.13c). Then a nonviscous **temporary spiral line** is laid down beginning at the center and progressing to the outside (Fig. 11.13d). This is followed by the production of a quite viscous or adhesive **permanent spiral thread** laid down from the outside toward the center. As the viscous spiral is produced, the spider removes and often consumes the temporary lines including the nonviscous spiral and some scaffolding lines and spiral lines near the hub. By consuming superfluous threads the spider quickly metabolizes the used silk, converts it into new silk, and uses it subsequently in producing new lines. In the web's final form, there is typically a **free zone** and the outer spiral **capturing zone** (Fig. 11.13e, f). The viscous spiral threads are especially effective in capturing prey insects that cannot free themselves from the adhesive material. Apparently, web size is determined by the size of the spider, and mesh size is a function of the span of legs. A common structural variation in the orb web is the production of **stabilimenta,** which are zigzag silken bands produced radially. Once thought to be stabilizers, these threads are now believed to be a form of camouflaging for the spiders that sit at the hub of their web. Since the viscous nature of the permanent spiral lines lasts only for a day or two, portions of the web constantly must be destroyed and new lines established. Many garden spiders replace their entire web each night!

Different families or groups of spiders have evolved quite different webs suitable for the habitats in which they live. For example, spiders living among vegetation, on the exposed surfaces of rocks, and beneath the soil surface would all have quite different web forms. Some webs are in the form of a funnel, a sheet, or an irregular mass such as the familiar cobwebs. Cribellate spiders form woolly bands of silk without any adhesive properties. Certain aquatic spiders like *Argyronecta* construct an underwater silken diving bell filled with air, and to this bell the spider returns frequently for gaseous exchange and feeding.

Feeding

All spiders are carnivores, and insects are the preferred prey probably because of the abundance of insects and the fact that they are the only invertebrates that can be captured in an aerial net. Ecologically, spiders can be divided into the wanderers and web-builders, and methods of food capture are different in the two groups. Wandering spiders do not entrap prey but rather must locate and stalk or overpower

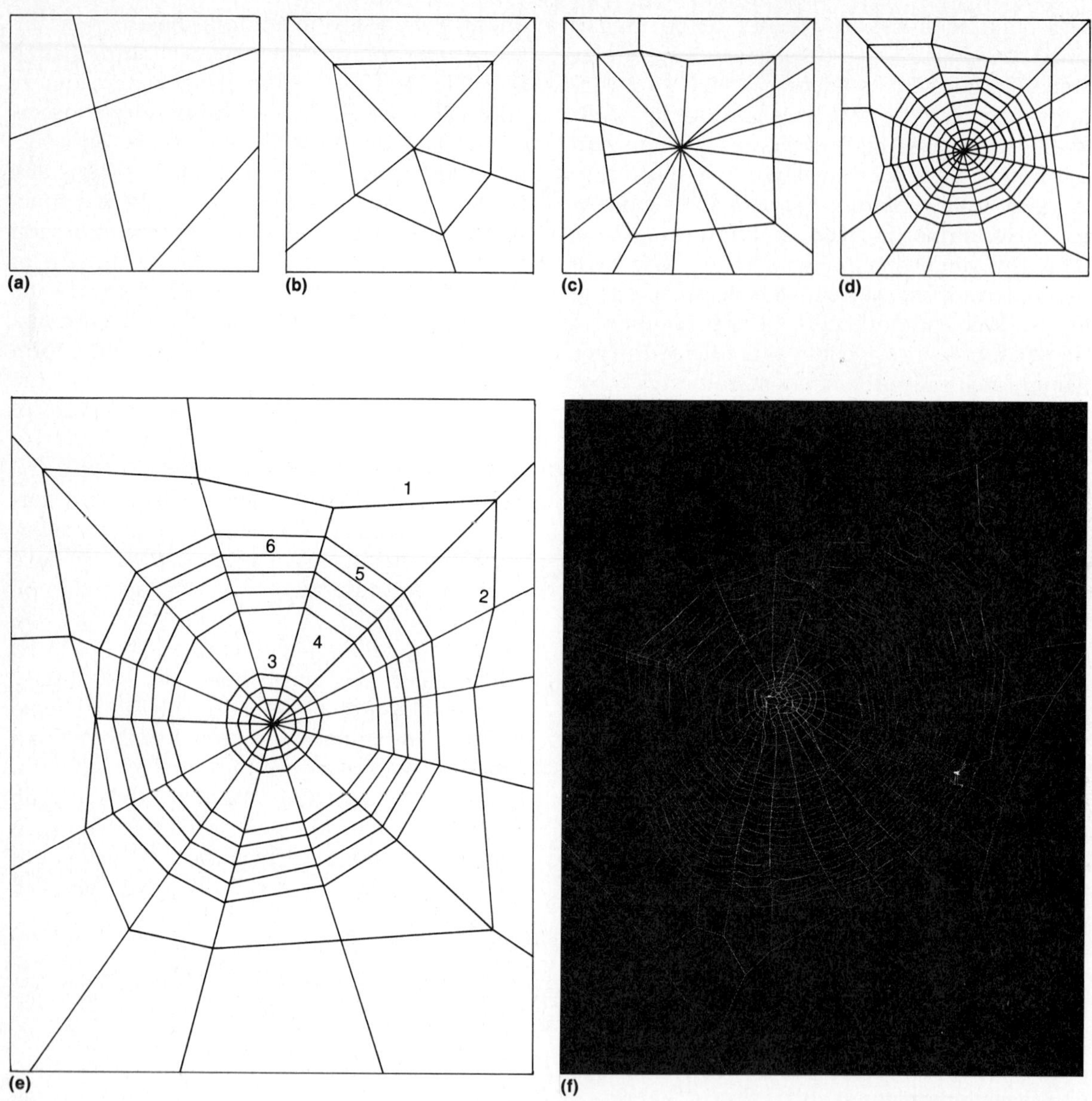

Figure 11.13 *Stages of construction of an orb web:* ***(a)*** *framing threads are initially formed;* ***(b)*** *foundation lines form the scaffolding;* ***(c)*** *radial threads are produced;* ***(d)*** *the temporary spiral thread is formed;* ***(e)*** *completed web (1 = foundation threads, 2 = radial thread, 3 = strengthening zone; 4 = free zone, 5 = capturing zone, 6 = permanent viscous spiral thread);* ***(f)*** *photograph of a completed orb web of* Aranea *(Neg. No. 37858 [Photo: K.C. Lenskjold] courtesy Department Library Services, American Museum of Natural History).*

an insect. Some lie in ambush, while others are active hunters. Visual, tactile, and vibrational cues are quite important. Prey normally is grasped by the front legs, pulled toward the spider's body, and pierced with the fangs. The prey, now held only by the chelicerae, is quickly immobilized by the poison and often is secured by silk threads before feeding begins. Digestive enzymes pour over and into the prey, and this extraorganismic digestion produces an organic broth that is then sucked into the alimentary canal. Variations on this theme are found in tarantulas (Fig. 11.9c), which leap upon passing prey, in trap-door spiders (Fig. 11.9b), which hide beneath a hinged trapdoor, or by a few other forms that rush out to grab an insect that has "tripped" on the stumble threads.

Web-building spiders obviously use the web as a trap or snare for insects. Once a prey touches the web, the spider rushes over to bite it. Some web-builders use the chelicerae to immobilize the prey first, but others wrap it in silken threads before biting it. Prey organisms are then carried to a safer place such as into the hub of an orb web or into a retreat for actual ingestion.

There are a number of highly specialized methods of predation and feeding. The bolas spider *(Mastophora)* throws a sticky droplet at an insect flying nearby and then grabs the stunned prey. Pirate spiders such as *Ero* prey only on other spiders, and other forms like *Zodarium* eat only ants. The ogre-faced spider, *Dinopis,* has enormously enlarged posterior medial eyes used to locate prey. Upon locating a prey organism, *Dinopis* throws an entangling rectangular web over the victim, and the prey is then bitten and consumed.

Internal features

The stomodeum of the alimentary tract consists of the pharynx, esophagus, and stomach. Liquified food is strained by bristles around the mouth and again by cuticular platelets in the pharynx that, unbelievably, remove particles as small as 1 μm in diameter. The stomach serves as a pump for food intake, and strong extrinsic muscles originating from the dorsal apodeme of the cuticle and inserting into the stomach facilitate this sucking action (Fig. 11.14). The midgut lies for the most part in the opisthosoma; from

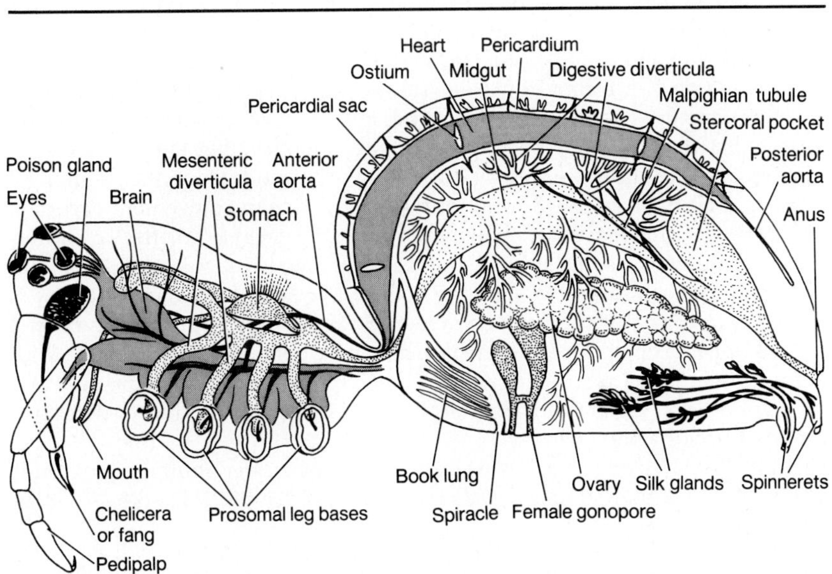

Figure 11.14 *Internal features of a female spider seen in sagittal section (from J.H. Comstock,* The Spider Book, *revised by W.J. Gertsch, © 1940, Doubleday, Doran & Co., Inc.; copyright assigned to Comstock Publishing Co., Inc., 1940, used by permission of the publisher, Cornell University Press).*

it arise numerous digestive diverticula. The posterior midgut forms an enlarged **stercoral** or cloacal **chamber** that stores excretory and fecal materials until they are eliminated. The stercoral chamber empties into the proctodeum, a short tube that opens to the exterior as the **anus.**

Excretion is accomplished by one or two pairs of coxal glands located in the prosoma and by Malpighian tubules, the principal excretory organs, which arise from the stercoral chamber and branch extensively into the hemocoel of the opisthosoma (Fig. 11.14). Gaseous exchange is achieved by a single pair of book lungs and one or two pairs of tracheae. Book lungs lie in the anterior ventral opisthosoma, and their spiracles are situated in the epigastric furrow. Tracheae, which are modified book lungs, are in opisthosomal segment III, and most spiders have a single spiracle that opens into the two tubular tracheae. However, there are variations on this basic gas exchange plan; some spiders have one or two pairs of book lungs only, and some have one or two pairs of tracheae only.

The heart lies dorsally in the anterior opis-

BOX 11.2 □ SOME OTHER INTERESTING DIMENSIONS OF SPIDER BEHAVIOR

That spiders have complex, intricate, and interesting behavioral patterns is well known, and much of our knowledge about spider behavior involves mating, web-building, and ballooning. But there are several other aspects of spider behavior that are equally provocative and interesting. We shall look briefly at the behavior of spiders when threatened and as they produce sounds and their social behavior.

If some spiders are startled, they respond with equal suddenness. A rather common response is to drop suddenly along a thread. The tropical *Cyrtophora* immediately drops from its web, and the color of its abdomen changes abruptly to blend with the background. Some spiders "play dead" by falling to the ground and pulling their legs in close to the body. *Argiope,* a common garden spider, will shake its web vigorously if disturbed by vibrations or shadows. Some hunting spiders display a defensive posture involving raising the front legs, opening the chelicerae, and elevating the prosoma.

Some spiders produce sounds either by drumming or by stridulation. "Drumming," practiced by many male wolf spiders during courtship, involves the movement of certain parts of the body (limbs, opisthosoma) against the substratum, which, in effect, indicates the presence of a male. But in some spiders, one part, the file, is rubbed against another part, the teeth, to produce a distinctive grating sound; this sound-producing activity is called stridulation. In males of the cobweb spider, *Steatoda,* a prosomal file is rubbed against a toothed opisthosomal area, and the resulting sound stimulates the female prior to copulation. Stridulating parts in other spiders involve the chelicerae, pedipalps, and one or more pairs of legs. Certain mygalomorphs or "tarantulas" produce a defensive hissing sound that drives away enemies.

Most all adult spiders lead solitary lives, and if they encounter another spider, it is

thosoma and has two or three pairs of ostia (Fig. 11.14). A rather distinct arterial system conveys blood to the various body parts and even to the tips of the legs, and blood is returned to the heart through the extensive hemocoel. Spider blood contains a variety of hemocytes that function in clotting, phagocytosis, wound healing, and even sclerotization of the cuticle.

Spiders have a well-developed and condensed central nervous system illustrating a high degree of cephalization. There are two prominent ganglia, the **supraesophageal ganglion** representing the basic chelicerate **brain** and the **subesophageal ganglion,** which is the fused prosomal and opisthosomal segmental ganglia (Fig. 11.14). Several different lobes of the brain may be evident and usually function as visual centers as in wandering spiders. Neurosecretory cells are found in the brain, and their secretions correlate life-cycle events like molting and the attainment of sexual maturity. Spiders have elaborate and complicated behavioral patterns. A few details of their versatile activities are explored in Box 11.2.

usually a potential mate or prey. But there are perhaps 20 species that are highly gregarious and sometimes build large communal nests. Two of the most-studied social spiders are the African *Agelena consociata* and the Mexican *Mallos gregalis.* It is indeed remarkable that since spiders are so aggressive, a few forms not only have developed a tolerance for each other but actually cooperate in hunting prey and in parental care of the young. Colonies perhaps began with young spiderlings that, rather than dispersing, aggregated and joined in web-building, a primitive degree of sociability practiced by several contemporary species. By combining their labors these spiders have evolved the ability to construct a web that is both larger and much more elaborate than that of a single spider. Also, the larger web permits the capture of larger-sized prey, which, after capture, is even fed upon communally.

How are these spiders so tolerant of one another? Undoubtedly, there are chemical cues or forms of recognition that are intimately involved. In fact, members of one colony can be moved to another colony of the same species without disrupting the community's activities. But, interestingly, males still go through an elaborate precopulatory behavior to prevent possible predation by the female. Then if spiders are so tolerant of each other in feeding, web-building, and care of the young, why does all this change at the time of mating? Could it be that mating is such a crucial event that it somehow lowers the tolerance levels of the female? Conversely, could some precopulatory pheromone promote an elevated level of aggressiveness in the female?

It is clear that spider sociality is not nearly so pronounced or complicated as that in bees and termites, but the very fact of communality in any spider activity is remarkable and merits further study.

REPRODUCTION AND DEVELOPMENT

Sexual dimorphism is evident in most spiders. Males are usually dwarfs, molt fewer times than females, and have conspicuously enlarged, modified pedipalpal tarsi. The paired gonads of either sex lie in the opisthosoma, and the ductal system, at least in males, is not complicated. A drop of mature sperm is produced onto a special small sperm web produced by the male. Then the male transfers the drop of sperm from this web to his tarsal organs. Each **tarsal** or palpal **organ** contains a bulb or reservoir of sperm and a narrow tip, the **embolus** (Fig. 11.11c).

As a male attains sexual maturity and goes through his final molt, he stops feeding and becomes nomadic in search of a female. When he locates a female, there ensues a very complex, ritualized, precopulatory courtship that, in effect, accomplishes two things. First, it stimulates the female sexually; second, it is a means by which the male can be recognized for what he is and not mistaken for just another prey organism. Courtship behavioral patterns are very species specific so as to prevent any interspecific mating.

Without going into much detail, the following are events that take place in the courtship of various spiders: interplay of front legs, tying the female up with silk, drumming on the web, nest-building by both sexes, secretion of a mating thread by the male, raising the front legs, opisthosomal vibrating up and down, pelipalp waving, presentation of a fly as a bridal gift, zigzagging dances, and other body motions (Fig. 11.15). Apparently, various chemicals such as pheromones play an important role in concert with other body signals from both members of the pair.

After a successful courtship the tarsal organ of the male, loaded with sperm prior to the courtship, is extended by hemocoelic pressures. The embolus or entire tarsal organ is inserted into the female gonopore and into her **uterus.** Copulation is achieved by both members meeting head-to-head, hanging upside down from the web, or the male on top of the female. Sperm are stored by the female in a pair of **seminal receptacles** until they are used in fertilization. Contrary to popular belief, the aggressiveness of the female toward her mate immediately following copulation is a myth, and males of only a few species are in great peril from the female after mating.

In one to three weeks after copulation the female constructs a silken cocoon into which she lays the fertilized eggs, but some spiders are ovoviviparous. Numbers of eggs per cocoon range from 20 to 2500. The cocoon, tightly closed by the female after egg laying is completed, is usually located in a retreat, on the web or in vegetation, or even carried ventrally by the female. Females generally protect their

Figure 11.15 *Diagrams illustrating some of the courtship postures and behavior in male wolf and jumping spiders:* ***(a)*** *first legs raised;* ***(b)*** *first legs raised, second legs extended laterally;* ***(c)*** *first legs extended laterally, pedipalps waving;* ***(d)*** *anterior end of spider raised;* ***(e)*** *opisthosoma raised.*

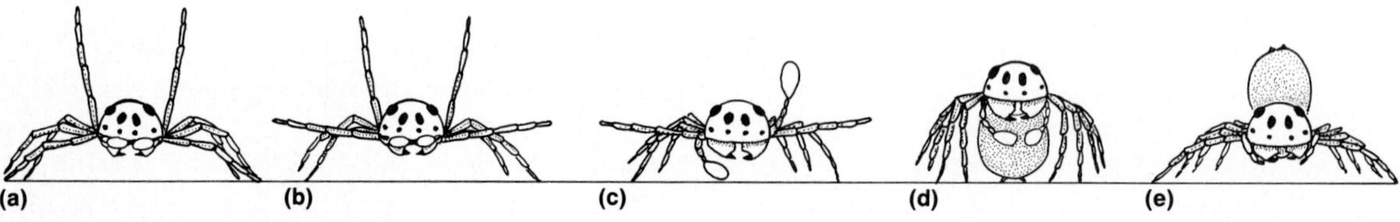

cocoons by aggressive behavior. After several weeks of embryonic development the spiderlings hatch and immediately crawl onto the dorsal surface of the female. During their 7- to 30-day stay with the mother the young of some spiders are fed by the mother; in some cases the female even regurgitates food directly into the mouths of the young. After a period of feeding and molting, young spiders of a great many species climb to the top of vegetation, produce silken threads, and balloon away.

Order Opiliones

There are about 3500 species of opiliones, commonly known as harvestmen or daddy longlegs. The body of an opilione, measuring 5–10 mm in length, is mounted on four pairs of disproportionately elongated legs. One tropical form may have a body length of 20 mm and a leg span of 16 cm! Though they have a worldwide distribution, harvestmen are most abundant in moist tropical regions. They are found amid vegetation and on the forest floor in humus and under fallen trees. Often they live around houses, barns, and other buildings, and they are usually most active at twilight. Most are cryptically colored, drab brown and black being the prevalent colors.

The prosoma and opisthosoma are broadly joined together, and the absence of a pedicel connecting the two is a feature that clearly differentiates harvestmen from spiders. A dorsal carapace covering the prosoma is continuous with a shield covering the opisthosoma. The carapace bears a pair of direct eyes located laterally on a prominent medial **ocular tubercle** (Fig. 11.16b). Located on each lateral margin of the prosoma is an **odoriferous** or scent **gland,** which is characteristic for all harvestmen. The odor of the glandular secretions—variously described as aromatic, sweet, acrid, like horseradish, offensive, and repulsive—is an effective repellant to would-be predators. The chelicerae are chelate, the pedipalps are leglike and have coxal gnathobases, and some harvestmen have distal claws that function as chelae. The four pairs of walking legs are inordinately long, sometimes as much as 40 times as long as the body (Fig. 11.16a). Someone has quipped that a study of opiliones is a study of legs! The legs not only serve for locomotion, but also are very important because they bear numerous sensory

(a)

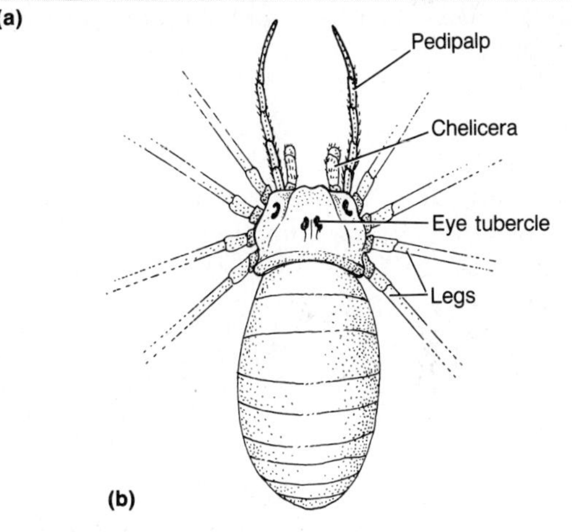

(b)

Figure 11.16 *Opilionids, harvestmen, or daddy longlegs:* ***(a)*** *a photograph of an opilionid;* ***(b)*** *a dorsal view of the body.*

organs. In most common harvestmen (Phalangidae) the second pair is so indispensable that loss of one is a serious handicap, and the loss of both is fatal. Some harvestmen can autotomize (i.e., self-amputate) one or more legs, but legs are not regenerated. The first one or two pairs of legs bear gnathobases. The long legs and the odoriferous glands are the distinctive features of the opiliones.

The opisthosoma, almost imperceptibly fused to the prosoma, is visibly segmented and composed of ten segments, but fusion and some displacement of sterna and terga complicate the segmental pattern. Usually, the opisthosoma is small, oval or globular, and ventrally has an anterior operculum on which are found the paired gonopores. Also on the ventral surface of females is a long, tubular **ovipositor,** which is housed in a sheath when not in use.

Opiliones are unusual among arachnids, since they macerate the food and ingest it before much digestion has taken place. Some harvestmen are predators and consume small invertebrates and especially aphids, but most opiliones are omnivores or scavengers and consume dead organic material. The pedipalps grasp food and pass it along to the chelicerae, which crush and fragment it. Solid particles are then sucked into the alimentary canal. A pair of coxal glands, whose ducts open between the coxae of the third and fourth legs, are the only excretory organs. Gaseous exchange is accomplished by tracheae that may not be homologous to those of other arachnids. Spiracles are on opisthosomal segment II; some accessory spiracles are even located on the legs. Malpighian tubules and book lungs are absent in harvestmen.

Opiliones do not practice precopulatory courtships. Males of many species possess a long protrusible **penis** with which sperm are transferred to females. Parthenogenesis is known to occur in a large number of cases. Females use the ventral ovipositor, which can be extended for a great distance, to deposit eggs into damp soil, humus, or rotting wood. Often hundreds of eggs are laid at one time, and several clutches are typically produced each year. Temperate species live only one summer, but tropical species may have a life span of several years.

Order Acarina

The acarines, or mites and ticks, are a very important group of arachnids, and they possess a number of adaptations that have contributed to their numerical success. There are more than 20,000 known species, but estimates of total numbers of acarines (including those not yet described) often range well over 100,000. Acarines, especially the free-living mites, are found in almost every conceivable terrestrial habitat wherever adequate moisture is available, and many live in freshwater and marine environments as well. Mites are especially abundant in tropical or temperate regions in mosses, leaf mold, humus, and soils. In excess of 100,000 mites per square meter of forest humus have been reported. A great many mites live symbiotically with domesticated plants and animals; many others are predaceous on crop plants and dry stored food and therefore are economically important. Mites are usually quite small and range from 0.02 to 3 mm in length, but ticks usually measure from 0.1 to 1.5 cm in length. Because all ticks and some mites are parasitic on higher vertebrates, they are of great medical and veterinary importance. Since acarines are very small, many are symbionts, and they have high reproductive potentials, acarines are of unusual significance to the biosphere and to human cultures.

External features

The acarines show no external signs of segmentation; prosoma and opisthosoma have fused, and a single shieldlike dorsal **carapace** covers the entire body. The body is usually ovoid, but

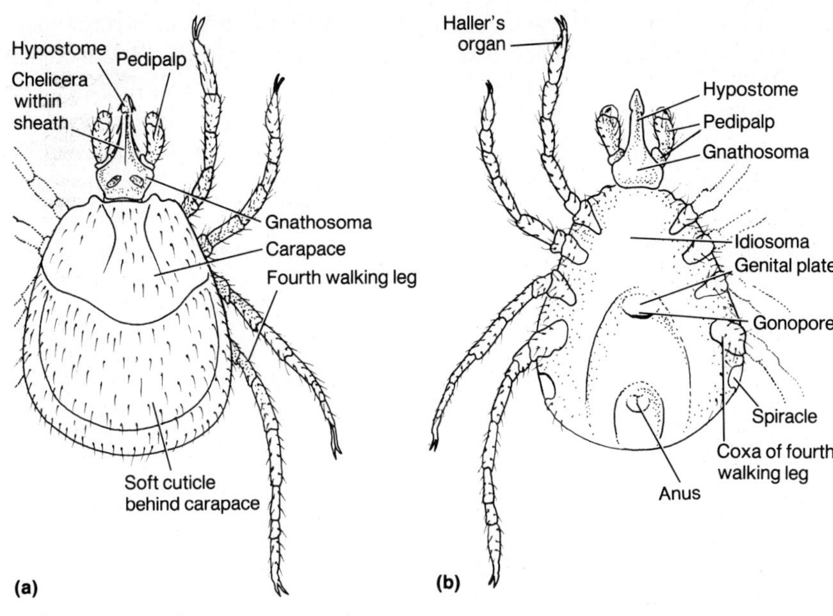

Figure 11.17 *Some external features of ticks (acarines):* ***(a)*** *a dorsal view;* ***(b)*** *a ventral view.*

some acarines are more elongated. Since body regions do not conform to the pattern in other arachnids, a different system of nomenclature for their tagmata has been developed in which the body is divided into an anterior gnathosoma and a posterior idiosoma (Fig. 11.17). The **gnathosoma** or **capitulum** includes the anterior two segments and bears one or more pairs of simple eyes, chelicerae, and pedipalps. Chelicerae, typically two-segmented, are either chelate or needlelike. The pedipalps, constructed of five podomeres, are variable in function; in some acarines they are tactile, while in others they are modified for grasping and feeding. Anteriorly, the gnathosoma bears a **rostrum** or beak composed of the dorsal body wall, the coxae of pedipalps, and the labrum. Ticks often have an organ, the **hypostome,** that is used to penetrate the skin of the host and as a holdfast organ (Fig. 11.17).

The **idiosoma** is the remainder of the body including the four thoracic leg-bearing segments. Each leg is well developed and has six podomeres, and the tarsus bears a pair of claws (Fig. 11.17). The legs have become variously modified for specific modes of life. A **genital plate** bearing the single or paired gonopores is located ventrally between the last pair of legs, and the anus is located on the ventral side of the posterior idiosoma. From one to four pairs of spiracles are present, usually near the bases of the legs, but their positions are variable.

Feeding

Feeding habits in acarines are unusually varied; almost every type of heterotrophic nutrition is present in this order. This diversity is often reflected in the gnathosomal appendages, digestive tract, and even the general body form. Among the herbivores are the spider mites and gall mites, which are serious pests on many fruit trees, house plants, cotton, clover, and other crops. The chelicerae in these mites are piercing stylets that puncture plant tissues and suck out their juices. Gall mites, perhaps more accurately designated as plant parasites, burrow into plants and suck their cell juices.

Carnivorous mites will use their chelate chelicerae to grasp and feed on almost any appropriately sized invertebrate such as minute crustaceans, nematodes, other acarines, and insect eggs. Fluids are then drawn into the anterior alimentary canal, where most of the enzymatic digestion takes place. Scavenger mites ingest almost any form of organic material present in their environment. A vast number of mites live in soils or humic layers; others prefer dead insects, flour, stored grain, old mattresses, household dust, and many other nutrient sources. The gnathosomal appendages of scavengers are principally of the grinding and chewing type.

Many mites are symbiotic and as such are commensals or parasites. A large number of plants and animals have mites as ectocommensals, and most do no damage to their hosts. But some, like the feather mites, may irritate the skin of birds as they feed on feather fragments, dead skin, and integumental oil. Parasitism is a very common life-style in acarines, and a discussion of the adaptations for feeding by parasites appears later.

Visceral systems

The acarine digestive tract contains a pumping pharynx and food-storing midgut ceca that are greatly distensible in ticks. The absence of an anal opening in some ticks is a measure of how completely they utilize blood as food. In most acarines a circulatory heart is lacking, and the hemolymph is colorless. Excretion in acarines can be achieved by epithelial cells of the digestive glands absorbing excretory products and secreting them into the alimentary lumen, by one to four coxal glands in the idiosoma, and by Malpighian tubules. In those mites with an incomplete digestive tract, a unique posterior pouch, representing the modified proctodeum, serves as an excretory vesicle that opens to the exterior as a posterior medial pore.

Gaseous exchange is by means of a tracheal system whose spiracles are located on the coxae of the legs. This bizarre arrangement suggests that acarine tracheae are not homologous to those of other arachnids. Book lungs are absent, but many acarines are small and have thin cuticles so that gaseous exchange takes place at the general body surface; however, this respiratory method restricts these mites to humid areas like moist soils.

The brain is like that of arachnids generally and represents considerable fusion of body ganglia. Tactile and chemical sense organs are well developed and found all over the body but appear to be concentrated on the first pair of legs. Trichobothria are abundant on the anterior body. A chemical receptor area called **Haller's organ** is found on each of the first legs. Simple eyes or ocelli are found on some acarines, but many mites and all ticks lack eyes altogether.

Reproduction and development

The Acarina are all dioecious, and fertilization is always internal. The male reproductive system consists of a pair of testes, vasa deferentia, and accessory glands. A single ovary, an oviduct, and a gonopore characterize the female system, and the females of some species have seminal receptacles. Ticks have indirect sperm transmission; spermatophores are attached to a substratum, and the female later contacts them and transfers the sperm to her gonopore. In some mites, one of the appendages is used to transfer spermatophores directly to the female's gonopore. In still other species, males have a chitinous **penis** that can be projected through the gonopore located medially between the first and second pair of legs; it is used to transfer sperm directly to the female's reproductive tract. During copulation, male appendages are often used to hold the pair together. Eggs are laid in a nest singly or in clutches, but a few species are ovoviviparous. After a short incubation period a larva hatches; it differs from an adult principally by having only three pairs of legs. The fourth pair of legs is added by the succeeding stage, the **nymph.** Several nymphal molts gradually produce all adult features.

PARASITIC ACARINES

Parasitism is a common form of symbiosis in acarines. While ectoparasitism is the rule, some mites have even become endoparasites. A parasitic existence has resulted in rather severe modifications of the appendages and body form. Appendages have been either reduced or modified into burrowing structures, and the body shape is flattened, streamlined, and reduced so as to make the body relatively obscure on the surface of the host. In ticks the hypostome or chelicerae pierce the skin of the host, and lateral cheliceral teeth widen the puncture for insertion of the pedipalps, which form a suction tube. As blood is sucked into the tick, extensive midgut ceca become enormously engorged with the meal. All ticks are parasitic on a variety of vertebrates such as reptiles, birds, and a great many mammals. Mites are parasitic on a variety of invertebrates, vertebrates, and plants. Ticks are ectoparasites and attach to a suitable vertebrate host temporarily while they are feeding; once satiated, they drop from the host. The effects of tick bites fall into two categories: the bite itself and diseases that are transmitted with the bite. Tick bites often cause local inflammation, itching, and some traumatic damage, and the anticoagulant secreted by the tick can cause paralysis and, in severe cases, death in sheep and calves. Ticks are the vectors for Texas cattle fever, tick-borne typhus, encephalitides, tularemia, relapsing fever, and several other diseases. The best-known disease for which ticks are vectors is Rocky Mountain spotted fever, which is carried by several species of *Dermacentor* (Fig. 11.18a).

Mites are either permanent or temporary parasites and either ectoparasites or endoparasites. Some permanent parasitic forms are follicle mites *(Demodex)* and human-itch and mange-producing mites *(Sarcoptes* [Fig. 11. 18e). Mites, like ticks, are vectors for some microorganismic diseases like scrub typhus and rickettsial pox, and other mites are intermediate hosts for several different cestodes. Some mites are endoparasitic in the air passageways of vertebrates and in tracheal systems of other arthropods. The juvenile stages may be temporary ectoparasites on vertebrates such as the larvae of *Trombicula,* the harvest mite, which are the familiar common chiggers or redbugs (Fig. 11.18d). Larvae feed on dermal host tissues broken down by digestive enzymes, and an intense, aggravating itch and dermatitis result from the oral secretions of the larval mites. Larvae drop off the host after about a ten-day feeding period and become free-living nymphs and, later, adults. Other mites cause a variety of skin disorders in birds, many mammals, and some insects.

Order Pseudoscorpiones

Pseudoscorpions are small arachnids, most being from 4 to 8 mm long, and they resemble scorpions somewhat but have neither a metasoma nor a sting. Because of their minute size and cryptic habits, they are not frequently seen, but they are rather common in leaf mold, beneath bark, under stones, and in the nests of birds, mammals, and insects. About 2000 species have been described to date.

The prosoma is covered dorsally by a shield-like carapace bearing one or two pairs of lateral sessile eyes, but some are eyeless. Covering the entire ventral surface of the prosoma are the coxae of the pedipalps and walking legs. The chelicerae are chelate, and the movable podomere bears a grooming organ (Fig. 11.19). The chelicerae also bear the openings of prosomal

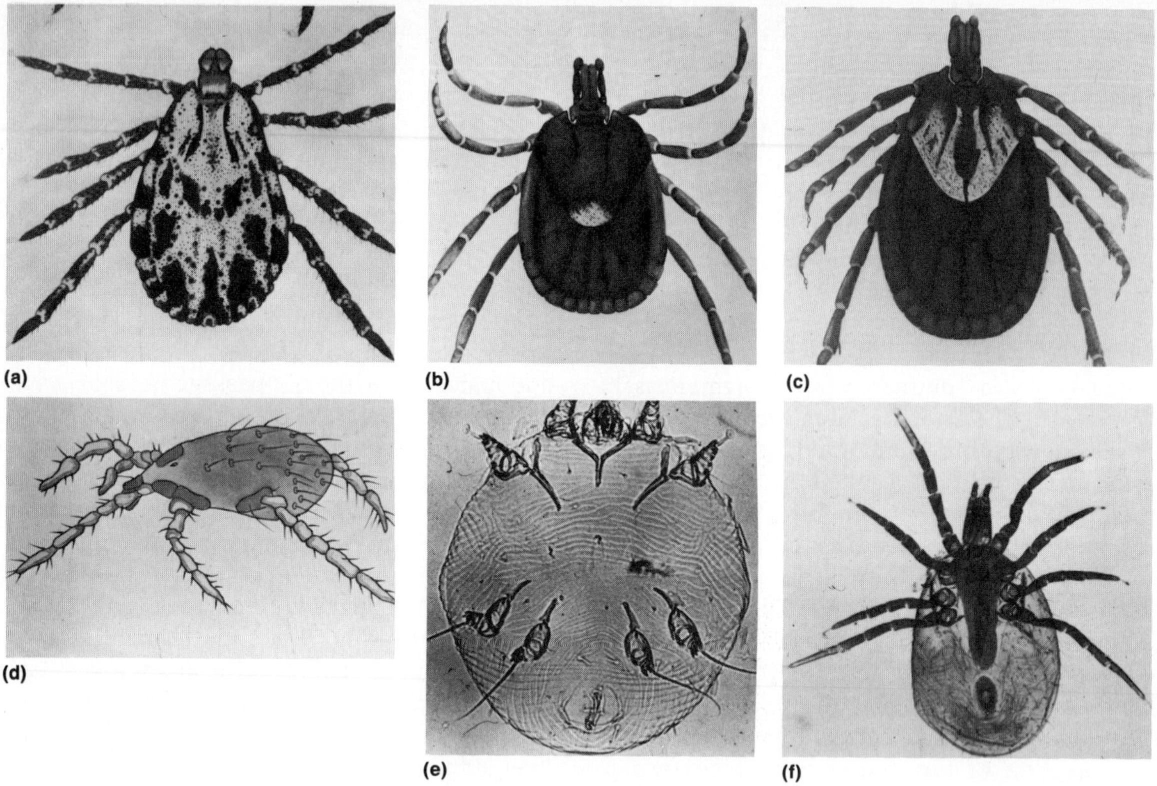

Figure 11.18 A photographic collage of types of acarines: ***(a)*** Dermacentor andersoni *male, a tick;* ***(b)*** Amblyomma americanum, *the lone-star tick;* ***(c)*** Amblyomma maculatum, the Gulf-coast tick; ***(d)*** *diagram of a larval* Trombicula *mite, better known as a chigger;* ***(e)*** Sarcoptes scabiei, *an itch mite of human beings (courtesy of the World Health Organization);* ***(f)*** Allodermanyssus, *a house mouse mite and carrier of rickettsial pox. (part a [from G. Anastos], b–d, f courtesy of the Centers for Disease Control, Atlanta, Ga. 30333.)*

Figure 11.19 Pseudoscorpions: ***(a)*** *dorsal view;* ***(b)*** *ventral view. (From C. Warburton, 1909,* Cambridge Natural History, *Macmillan and Co., Ltd., London.)*

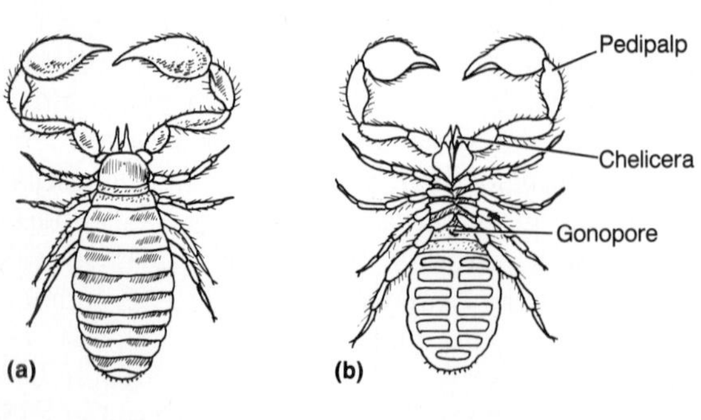

silk glands, which are homologous to the venom glands of spiders and produce a silken material used to build nests. On the inner cheliceral margins are groups of specialized setae, the **flagella,** which are used to discard undigested food while feeding. The pedipalps, resembling those of scorpions in that they are large and chelate, are unusual because each contains one or more poison glands. Used in capturing prey, mating, and defense, the pedipalps also carry most of the sensory organs. Each walking leg consists usually of seven podomeres, although this number has been reduced to six or even five in some families, and each leg terminates with a segment bearing two claws and an adhesive pad (Fig. 11.19).

The opisthosoma is externally segmented but bears no appendages. The sterna of segments II and III form **genital opercula** covering the gonopores situated on the genital segment (Fig. 11.19b). Paired tracheal spiracles are located ventrally on segments III and IV, but book lungs are absent. The last segment is reduced to a small cone bearing the anus.

Pseudoscorpions prey chiefly upon other arthropods, which are caught by the raptorial pedipalps and killed with poison. The victim is passed to the chelicerae, which tear it open; partial digestion takes place before ingestion.

Male pseudoscorpions deposit stalked spermatophores on the substratum, and the sperm are later taken up by the female. In several species, no courtship or mating takes place; in others the male deposits spermatophores only after encountering a female. In both cases the females are attracted chemically to the spermatophores. In many pseudoscorpions a mating dance ensues as the couple dances back and forth with their pedipalps attached; others perform much the same dance but without body contact. A male deposits a spermatophore and either manuevers the female to it, guides her to it using silken threads, or forcefully pulls her to it. Once she is over the spermatophore, it attaches to her ventral surface, the sperm are released and enter her reproductive tract, and internal fertilization soon follows.

A brood nest, built by the female, is often constructed by using small particles of wood, sand grains, and other debris to reinforce the silk. The embryos of all pseudoscorpions are attached externally to the undersurface of the female. Several broods, each usually averaging about 50 young, are produced annually. Nutritive materials are supplied to the developing embryos by the mother. The young hatch and are fed in the nest until they become independent of the female.

◀ ***Figure 11.20*** *Photograph of a solifugid (courtesy of Ward's Natural Science, Inc., Rochester, N.Y.).*

Order Solifugae

The solifugids, commonly called sun or camel spiders, live mostly in arid tropical or temperate areas and are found in the southwestern quadrant of the United States, Central America, and some other desert locations. About 800 species are known, and the animals range in size from 1 to 5 cm. They are cryptozoic and are usually found under stones, in crevices, or in burrows.

The prosoma is characterized by a conspicuously swollen anterior "head" that contains large muscles for the chelicerae. Dorsally, the prosoma is covered by a three-piece carapace of which the anterior part covers the head and bears one pair of medial eyes. The most striking feature of the solifugids is the presence of a pair of greatly enlarged chelicerae (Fig. 11.20). Each chelicera is two-jointed, chelate, and lacks poisonous glands; and these appendages rip, crush, and tear the prey. The pedipalps are leglike and nonchelate, and the distal tarsus is equipped with an adhesive organ used to capture prey such as termites or other small animals. Additionally, the pedipalps are used in defense and mating. The first pair of walking legs is long, nonambulatory, and held forward as tactile organs; the remaining legs are used in running. The coxae and trochanters of the last pair of legs bear three to five curious **racquet organs,** the function of which is uncertain. A pair of prosomal spiracles is located between the bases of the second and third legs.

The opisthosoma is without a pedicel and is constructed of 11 clearly defined metameres (Fig. 11.20). Ventrally, segment II bears the gonopores, and the anal opening is on the last segment. Segments II–V bear tracheal spiracles of which the last pair may be fused into a single opening. Both prosoma and opisthosoma, along with the appendages, are profusely covered with long sensory setae.

During the brief courtship the female becomes passive and enters a trancelike state with her ventral side up. The male secretes a globule of semen on the substrate, picks it up with his chelicerae, and transfers it to the upturned gonopores of the female. A clutch of 100–250 eggs is laid by the female in a previously prepared burrow, and the eggs hatch into juveniles, which attain the adult size after several molts.

BOX 11.3 □ BRIEF CHARACTERIZATIONS OF THE MINOR ORDERS OF ARACHNIDS

Palpigrades live in ground litter in warm tropical zones, are eyeless, and are never more than 3 mm in length. They have a carapace of three pieces and a pedicel, and the opisthosoma ends in a long jointed telson or flagellum (Fig. 11.21a). There are no gnathobases, and the segmented sterna are distinct from each other.

Uropygians, called whip scorpions or vinegaroons, usually tunnel in moist soils. Their bodies have a narrow prosoma, and the opisthosoma has a terminal telson or flagellum (Fig. 11.21b). Their pedipalps are enlarged and raptorial. A unique defensive mechanism is present: paired anal glands secrete acetic acid, which is sprayed on enemies to dissuade or chemically injure them.

Schizomides are minute, eyeless, tropical arachnids. A tripartite carapace, chelate chelicerae, a very short telson, and their having the first pair of legs as sensory limbs characterize this group (Fig. 11.21c).

Amblypygians, the tailless whip scorpions or whip spiders, have a flattened body and no telson. Their first legs are elongate, many segmented, nonambulatory, and whiplike in appearance (Fig. 11.21d). Raptorial pedipalps, book lungs, precopulatory courtship, and indirect sperm transfer characterize the amblypygians.

Ricinuleids are eyeless and live in moist dark places like caves. They have an unsegmented carapace on whose anterior surface is borne a hood or **cucullus** that protects the mouth and raptorial chelicerae (Fig. 11.21f). The opisthosoma is rounded and ticklike. They are rather small and rare creatures and not studied as thoroughly as most other arachnids.

Cyphophthalmians, similar to harvestmen, have a scattered global distribution. They have an unsegmented carapace, chelate chelicerae, and odoriferous glands whose ducts open between the second and third legs and whose secretions are a repellant toward enemies (Fig. 11.21e). The tarsi of the fourth legs in males are distinctively swollen and glandular.

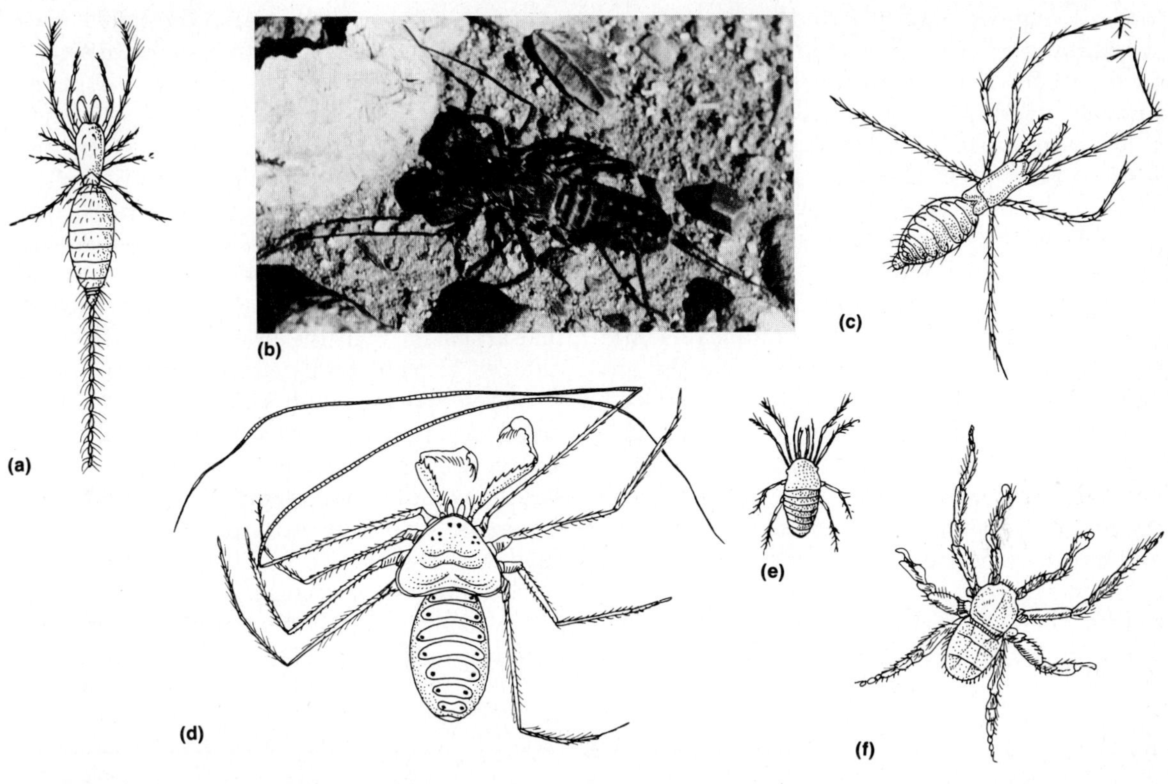

Figure 11.21 *A representative of each of the six minor orders of arachnids:* ***(a)*** *a palpigrade;* ***(b)*** *a photograph of* Tithyreus, *a whip scorpion (courtesy of Ward's Natural Science, Inc., Rochester, N.Y.);* ***(c)*** *a schizomide;* ***(d)*** *an amblypygid;* ***(e)*** *a cyphophthalmian;* ***(f)*** *a ricinuleid. (parts a, e, f from C. Warburton, 1909,* Cambridge Natural History, *Macmillan and Co., Ltd., London.)*

Minor Arachnid Orders

There are six orders of arachnids: Palpigradi, Uropygi, Schizomida, Amblypygi, Ricinulei, and Cyphophthalmi, which collectively contain fewer than 350 species. The representatives of all six orders share some common features that include their being very small (always less than 7 cm, and most are smaller than 1 cm), cryptozoic, and mostly nocturnal and living in damp or moist habitats like soils and leaf litter. Brief characterizations of these six orders appear in Box 11.3, and a representative of each is illustrated in Figure 11.21.

CLASS PYCNOGONIDA

Pycnogonids are a group of about 600 species of bizarre chelicerates called sea spiders. They are characterized by having four pairs of unusually long legs, which account for their alternate name, Pantopoda (= all legs). In some cases they are common animals in the benthic marine community, where they crawl over corals, bryozoans, and other encrusting organisms, but some swim or are planktonic. Most do not exceed a length of 10 mm, although some deep-sea forms are reported to have an incredible leg span of 75 cm. While some are red or green, generally they are drably colored.

The pycnogonid body is greatly reduced, having two prosomal subtagmata, the cephalon and trunk, and a reduced opisthosoma (Fig. 11.22). The **cephalon,** composed of the acron and the anterior four segments, bears a medial, anteriorly directed **proboscis** that is a suction tube for feeding on liquids and fine particles. The appendages of the cephalon are the paired chelicerae, pedipalps, ovigerous legs, and first walking legs. The chelicerae, attached to the junction of proboscis and cephalon, are reduced or even absent in some (Fig. 11.22). The pedipalps are leglike, multisegmented, and supplied with sensory setae. **Ovigerous legs,** unique to pycnogonids and usually absent in females, are employed by males to carry eggs and for grooming. The last segment is really a leg-bearing trunk segment that has become fused to the cephalon. The posterior part of the cephalon is narrow and forms a **neck** that, quaintly enough, bears two pairs of simple eyes. The **trunk** consists of three to five segments, each of which has two lateral extensions that articulate with the base of a leg. Each leg consists of eight podomeres, the distal one of which is equipped with a claw (Fig. 11.22). Individuals of some species have slender bodies and very long legs, but in other species with stouter bodies the legs are shorter. Most pycnogonids have four pairs of legs, but others have five or six pairs. The opisthosoma is greatly reduced or vestigial, and in most forms it is present only as a small protuberance bearing the anus (Fig. 11.22).

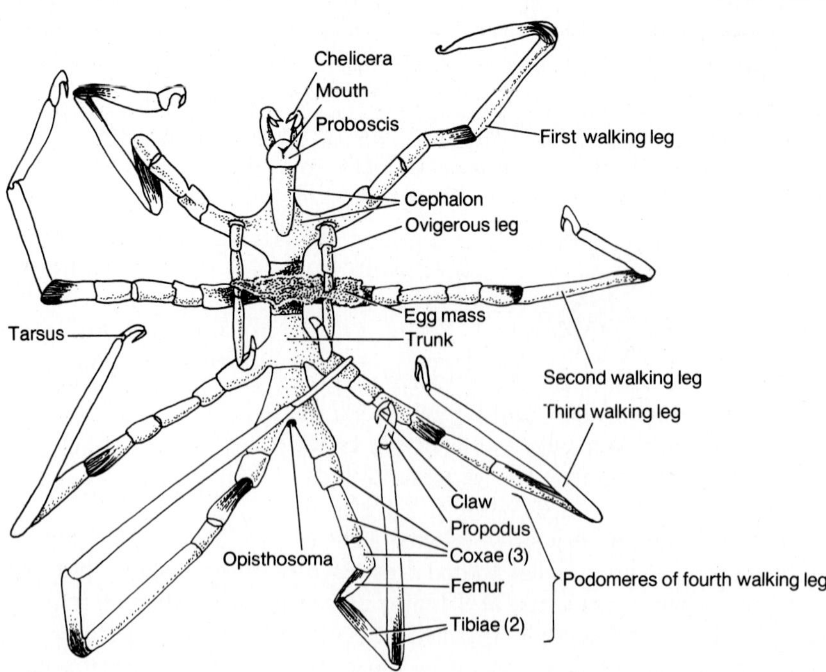

Figure 11.22 *A ventral view of a male* Phoxichildium maxillae, *a pycnogonid.*

Pycnogonids are carnivores, and they use the chelicerae to grasp prey and to grind particles. Strange as it seems, digestive and reproductive systems extend into the legs. The male fertilizes the eggs as they exit from the female, and he collects them into a mass of up to 1000 eggs on his ovigerous legs and broods them there until they hatch (Fig. 11.22). Eggs stick to the legs as a result of appendicular bristles and cement secretions from glands on the legs. The juvenile or larval stage, the **pronymphon,** has only the three anteriormost appendages. A trunk segment and a pair of walking legs are added at each of the next several molts. Larvae may remain on the ovigerous legs, or they may be temporarily parasitic on cnidarians, molluscs, or other marine invertebrates until metamorphosis is complete. Adults are mostly free living, but a few are ectoparasitic on anthozoans.

The phylogenetic status of pycnogonids is really in limbo. The chelicerae, the usual four pairs of walking legs, and some internal features strongly suggest a chelicerate affinity. But the presence of a segmented trunk, ovigerous legs, and more than four pairs of walking legs in some argue for a more distant relationship between pycnogonids and chelicerates. Their aberrancy is much more evident than are their phylogenetic affinities.

CHELICERATE PHYLOGENY

Most biologists today hold that chelicerates represent a distinct evolutionary group that arose from some primitive, segmented, marine worm that also produced the other three lines of arthropods. Unfortunately, there are no known fossils in the paleontological record that reflect or even suggest the sequence of early stages in the evolution of chelicerates.

Chelicerate evolution initially produced a hypothetical protochelicerate that most likely had a ventral mouth, an opisthosoma of 12 segments, and a brain that had lost the deuterocerebrum. This protochelicerate could also have been ancestral to trilobites, since trilobites and primitive chelicerates were very similar morphologically. A more plausible idea is that both chelicerates and trilobites arose from a common ancestor, and in the evolution or both groups there must have been stages that were quite similar but probably not identical. From this protochelicerate, one early line of evolution produced the aberrant pycnogonids. The evolution of eurypterids and xiphosurans must have occurred in the Ordovician and Silurian periods, since their fossil record from these periods in extensive. Eurypterids and xiphosurans probably evolved independently of each other after having originated from a common or similar ancestor (Fig. 11.23).

Arachnids evolved from a hypothetical ancient protoarachnid, which presumably was aquatic at first, had a prosoma and opisthosoma, and possessed six pairs of prosomal appendages. These early arachnids were probably very common in the Silurian seas. Perhaps because of overcrowding in their shallow marine habitats, they become progressively more terrestrial. In its early evolution the protoarachnids underwent adaptive radiation to produce four rather distinct evolutionary lines. One very early branch produced the scorpions, which first became fully adapted to a terrestrial existence and thus became the first land animals. The other three lines also evolved in the Carboniferous period from the terrestrial protoarachnids. One line led eventually to the primitive Palpigrada, the versatile Araneae, and several minor groups. Another evolutionary branch resulted in the Acarina, the Opiliones, and two small taxa. The Pseudoscorpiones and Solifugae represent still another line (Fig. 11.23).

There is still considerable debate about the possible relationships between scorpions and eurypterids, which shared a number of morphological features. If eurypterids were the ancestors of scorpions, then either scorpions

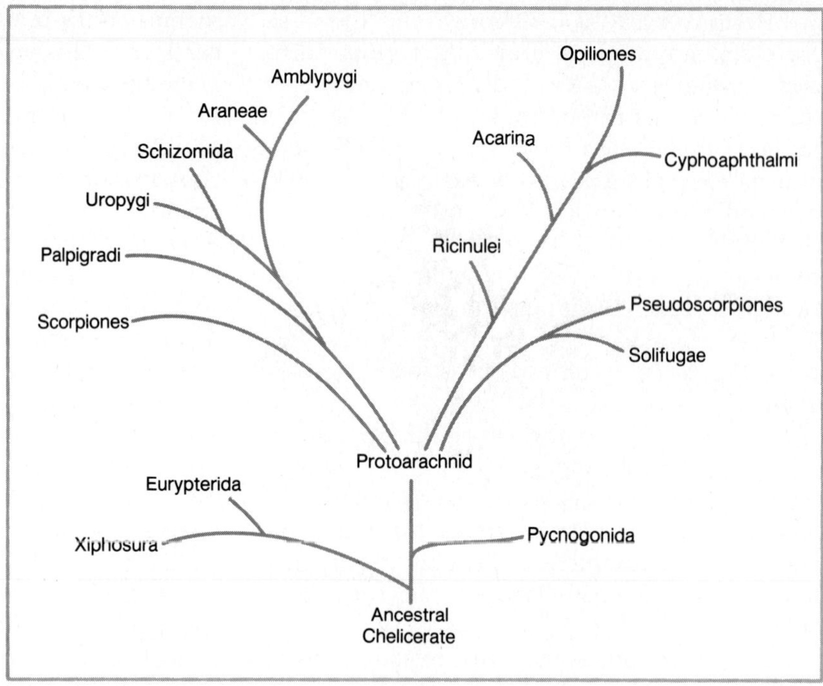

Figure 11.23 *Probable phylogenetic relationships among the chelicerates.*

would have to be classified as merostomates or eurypterids would then be treated as arachnids. A more conventional interpretation of the similarities between eurypterids and scorpions would indicate that this is a case of convergent evolution and their presumed relationship is not based on paleontological evidence.

SYSTEMATIC RESUME

The Chelicerates

Subphylum Chelicerata

Class Merostomata—all aquatic; opisthosomal book gills and long posterior telson; mostly marine and mostly extinct.

Subclass Xiphosura—horseshoe crabs; large prosoma with dorsal carapace; seventh prosomal segment bears chilaria; mostly extinct; *Limulus polyphemus* (Figs. 11.1, 11.2) is common on Atlantic coasts.

Subclass Eurypterida—giant water scorpions; all extinct; had a narrow prosoma, opisthosoma with two subtagmata, and long poisonous telson; marine and freshwater.

Class Arachnida—large group and mostly terrestrial; body segments mostly fused; have book lungs and/or tracheae, Malpighian tubules and/or coxal glands; lack compound eyes, but have many hairlike sensory receptors.

Order Scorpiones—scorpions; book lungs but no tracheae; long metasoma wth poisonous telson; have complex precopulatory behavior; *Centruroides* (Fig. 11.7b), *Isometrus*, and *Diplocentrus* are often encountered.

Order Palpigradi—very primitive, minute, eyeless; similar to uropygians; have opisthosomal flagellum; first pair of legs are sensory; *Koenenia* is typical.

Order Uropygi—whip scorpions or vinegaroons; have terminal flagellum, raptorial pedipalps, and defensive anal glands that secrete acetic acid as a repellant; *Thelyphonus* is often encountered.

Order Schizomida—minute, eyeless; carapace of three segments; short telson; first legs are sensory and nonambulatory; *Nyctalops* is a typical form.

Order Amblypygi—tailless whip scorpions or whip spiders; first legs are elongate and whiplike; have raptorial pedipalps; *Stegophrynus* is relatively common.

Order Araneae—spiders; have four pairs of eyes, poisonous chelicerae, copulatory pedipalps in males; form silk through spinnerets; have diverse means of food capture; includes:

- Mygalomorphs—*Eurypelma* (a "tarantula" [Fig. 11.9c])
- Orb-weavers—*Araneus* and *Argiope* (garden spiders [Fig. 11.9f]); *Cyclosa, Micrathena*
- Cobweb spiders—*Achaearanea* and *Latrodectus* (black widow [Fig. 11.9a])
- Funnel-web or grass spiders—*Cybaeus, Agelena, Tegenaria*
- Wolf spiders—*Pirata, Arctosa, Lycosa, Hogna,* and *Bothriocyrtum* (trap-door spider [Fig. 11.9b])
- Jumping spiders—*Salticus, Phidippus*
- Nursery-web spiders—*Dolomedes*

Order Ricinulei—thick cuticle; anterior cucullus over mouth; short ticklike opisthosoma; eyeless; raptorial chelicerae; *Ricinoides* is the type genus.

Order Opiliones—harvestmen or daddy longlegs; elongate walking legs and very small body; have odoriferous glands with repugnant scent; lack Malpighian tubules and book lungs; common genera include *Phalangium, Leiobunum,* and *Phalangodes.*

Order Cyphophthalmi—similar to harvestmen; have unsegmented carapace, pincer chelicerae, and odoriferous glands whose secretion is repellant; *Siro* is typical.

Order Acarina—mites, ticks; prosoma and opisthosoma fused; body has an anterior gnathosoma and an idiosoma with walking legs; variable feeding patterns; some typical mite genera are *Demodex, Trombicula* (Fig. 11.18d), *Arrenurus,* and *Hydrachna,* and some tick genera are *Ixodes, Dermacentor* (Fig. 11.18a), and *Amblyomma* (Fig. 11.18b, c).

Order Pseudoscorpiones—superficially resemble scorpions, but much smaller; have silk glands; produce spermatophores and have mating dance; *Chthonius, Neobisium,* and *Chelifer* are frequently encountered.

Order Solifugae—sun or camel spiders; have very large chelicerae, copulatory pedipalps, sensory first legs, and racquet organs on last legs; some common genera are *Eremobates, Galeodes,* and *Solpuga.*

Class Pycnogonida—sea spiders; body of cephalon and trunk; males have cephalonic ovigerous legs that are reduced or absent in females; four to six pairs of long legs; feed on small invertebrates; *Nymphon* and *Pycnogonum* are two common genera.

ADDITIONAL READINGS

Bonaventura, J., C. Bonaventura, and S. Tesh, eds. 1982. *Physiology and Biology of Horseshoe Crabs; Studies on Normal and Environmentally Stressed Animals,* 306 pp. New York: Alan R. Liss, Inc.

Boudreaux, H.B. 1979. *Arthropod Phylogeny with Special Reference to Insects,* 306 pp. New York: John Wiley & Sons.

Brownell, P.H. 1984. Prey detection by the sand scorpion. *Sci. Amer. 251:*86–97.

Burgess, J.W. 1976. Social spiders. *Sci. Amer. 234:*100–106.

Cheng, T.C. 1973. *General Parasitology,* pp. 725–775. New York: Academic Press.

Cloudsley-Thompson, J.L. 1968. *Spiders, Scorpions, Centipedes and Mites,* pp. 85–252. New York: Pergamon Press.

Foelix, R.F. 1982. *Biology of Spiders,* 297 pp. Cambridge, Mass.: Harvard University Press.

Forster, L. 1982. Vision and prey-catching strategies in jumping spiders. *Amer. Sci. 70*:165–175.

Gertsch, W.J. 1979. *American Spiders*, 2nd ed., 260 pp. New York: Van Nostrand Reinhold Co.

Gupta, A.P., ed. 1979. *Arthropod Phylogeny*, 735 pp. New York: Van Nostrand Reinhold Co.

Headstrom, R. 1973. *Spiders of the United States*, 257 pp. New York: A.S. Barnes.

King, P.E. 1975. *Pycnogonids*, 144 pp. London: Hutchinson Publishing Group.

Manton, S.M. 1977. *The Arthropods: Habits, Functional Morphology, and Evolution*, 514 pp. New York: Oxford University Press.

Merrett, P., ed. 1978. *Arachnology*, 509 pp. Symposium No. 42 of the Zoological Society of London. New York: Academic Press.

Pennak, R.W. 1978. *Fresh-water Invertebrates of the United States*, 2nd ed., pp. 488–511. New York: John Wiley & Sons.

Savory, T.H. 1962. Daddy longlegs. *Sci. Amer. 207*:119–128.

Savory, T.H. 1966. False scorpions. *Sci. Amer. 214*:95–100.

Savory, T.H. 1971. *Evolution in the Arachnida*, 41 pp. Watford, England: Merrow Publishing Co.

Savory, T.H. 1977. *Arachnida*, 2nd ed., 334 pp. New York: Academic Press.

Snow, K.R. 1970. *The Arachnids: An Introduction*, 81 pp. New York: Columbia University Press.

Weygoldt, P. 1969. *The Biology of Pseudoscorpions*, 138 pp. Cambridge, Mass.: Harvard University Press.

Witt, P.N. and J.S. Rovner, eds. 1982. *Spider Communication*, 432 pp. Princeton, N.J.: Princeton University Press.

12

The Crustaceans

OVERVIEW

Crustaceans are an unusually varied taxon of aquatic and mostly marine arthropods. All crustaceans have a head and one or two postcephalic tagmata. Each segment in a crustacean typically bears a pair of unique biramous appendages, and these limbs are specialized for a variety of functions. Head appendages are two pairs of antennae, mandibles, and two pairs of maxillae, which are used in feeding and sensory reception. The second pair of antennae, borne by head segment II, is unique to crustaceans. Appendages of the remaining segments are specialized for locomotion, reproduction, brooding the young, and creating water currents. A carapace, paired compound eyes, and numerous sensory structures are present.

The digestive tract often contains a triturating gastric mill and a midgut gland. There is an open circulatory system, and in higher crustaceans the blood plasma contains hemocyanin. Gills are modified, lateral, appendicular processes and are efficient gas exchange organs. Paired coelomoducts have evolved as excretory organs. The nervous system is typically arthropodan, and there are many degrees of fusion of the cord ganglia. Most crustaceans are dioecious and have internal fertilization; the eggs hatch as free-living, planktonic larvae called nauplii. Many higher crustaceans also have other, postnaupliar, larval stages.

There are six classes of crustaceans with varying degrees of species diversity. Cephalocarids are extremely primitive marine crustaceans with a horseshoe-shaped carapace and a 19-segment trunk with nine pairs of triramous appendages. Branchiopods have foliaceous appendages, are small, mostly live in small bodies of fresh water, and often are quite numerous. Remipedians, discovered in 1980 in the Bahamas, are small raptorial feeders.

The maxillopods, including copepods, barnacles, and other forms, have maxillae specialized for feeding, a reduced trunk, and uniramous thoracopods. Many maxillopods are planktonic, others are sedentary, and a rather large number are parasitic or commensal. The ostracods are mostly small benthic creatures with a bivalved carapace and a reduced trunk with zero to two pairs of appendages, and locomotion is usually by means of one or both pairs of antennae.

The malacostracans, whose members constitute the largest crustacean subclass, always have three tagmata with eight thoracic and six to eight abdominal segments and equal numbers of pairs of appendages. The first antennae are usually biramous, they have a cephalothorax composed of the head and one to three thoracic segments, abdominal appendages are pleopods, and there is a large tail fan. This subclass includes the amphipods, isopods, mysids, euphausiids, and decapods. The decapods, the largest order of crustaceans, have ten pereiopods with gills and three pairs of maxillipeds and are predaceous, and their young are brooded on the pleopods. Internally, they have paired antennal glands, blood with hemocyanin, and a condensed nerve cord. Decapods include shrimp, hermit and true crabs, crayfishes, and lobsters.

Crustaceans probably evolved independently of other arthropods and are not closely related to either trilobites or insects. The protocrustaceans were small, marine, and wormlike, and each segment bore a pair of biramous or triramous appendages. From this ancestral crustacean stock the various classes arose. Higher crustaceans evolved along two basic lines: one line led to the diverse maxillopods, and the second eventually culminated in the malacostracans, in which species diversity and adaptive radiation are the highest among this subphylum.

INTRODUCTION TO THE CRUSTACEANS

Crustaceans, represented by crabs, shrimp, lobsters, barnacles, and a great many small forms like copepods, ostracods, and cladocerans, perhaps illustrate a greater range of variations in their functional morphology than do representatives of either of the other two arthropod subphyla. Because of their variability in structure, function, appearance, and habits, it is difficult to characterize this group as a single taxon. Basically an aquatic group, the vast majority of the almost 40,000 crustacean species are marine. There are many freshwater crustaceans that are found mostly in lentic habitats, a few species live in moist terrestrial habitats, and a surprisingly large number of representatives of several different taxa live symbiotically with other animals. Although most crustaceans are from 0.1 to 20 cm in length, they range in size from microscopic to the American lobster *(Homarus)*, which attains a length of up to 60 cm, and the Japanese spider crab *(Macrocheira)*, whose legs span a distance of up to 3 m. Small crustaceans occupy a basic and fundamental position in every aquatic food chain, since they serve as prey for myriads of larger animals, and the larger crustaceans are often ecological dominants in many ecosystems, where they are predaceous on a wide variety of animals.

Many crustaceans are of great direct economic importance. Vast quantities of seafood such as shrimp, lobsters, and crabs are consumed daily, and freshwater crabs and crayfishes are part of the human diet in many parts of the world. Until recently, the efficiency of many ocean-going ships was substantially lowered by barnacles encrusting their underwater surfaces. Crabs severely damage irrigation dikes and fields of rice and cotton.

There are many different taxonomic schemes for grouping the diverse crustaceans. We shall use the taxonomic framework of Bowman and Abele in Abele (1982), who recognize six classes of crustaceans, which are briefly characterized in Table 12.1. Two classes, Cephalocarida and Remipedia, contain a total of only about ten species, but the Cephalocarida is a particularly important group in the understanding of crustacean evolution. Two other classes, Branchiopoda and Cirripedia, are moderately sized taxa whose representatives are common in aquatic habitats. Representatives of the remaining two classes, Maxillopoda and Malacostraca, have very high species diversity and are

TABLE 12.1. □ **BRIEF CHARACTERIZATIONS OF THE CLASSES OF CRUSTACEANS**

Class	*Common Name(s)*	*Habitat*	*Chief Locomotor Appendages*	*Other Features*
Cephalocarida	—	Marine, benthic	9 pairs of trunk limbs	Small, primitive; lack compound eyes; have horseshoe-shaped carapace
Branchiopoda	Daphnids, brine and clam shrimp	Temporary bodies of water mostly	Antennae or trunk appendages	Epipods of trunk limbs function as gas exchange organs
Remipedia	—	Marine cave	Swimming trunk limbs	Lack eyes; raptorial uniramous maxillae and maxillipeds
Maxillopoda	Copepods, barnacles	Planktonic, sessile, benthic in all aquatic habitats	Abdominal appendages or antennae	Large maxillae; reduced trunk; thoracopods lack endites; flagellated sperm
Ostracoda	Ostracods or seed shrimp	Mostly benthic in all aquatic habitats	Second antennae	Bivalved carapace; reduced trunk that terminates as 2 caudal rami
Malacostraca	Isopods, amphipods, shrimp, crabs, crayfishes	All aquatic habitats; all life-styles	Pereiopods, pleopods	Thorax of 8 segments, abdomen of 6 segments plus telson

ecologically important in most aquatic ecosystems. Copepods (Maxillopoda) and decapods (Malacostraca) are of particular importance and will be mentioned frequently in the general discussions below. The nonmalacostracans are sometimes collectively called the "entomostracans," a nonvalid taxonomic term but one that is conveniently used to refer collectively to these small but often incredibly numerous crustaceans.

GENERAL CHARACTERISTICS

Each segment of a crustacean typically bears a pair of biramous appendages. These appendages are characteristic for members of this entire subphylum. Cephalization and tagmatization are phenomena that are evident in crustaceans, even though the tagmata are not the same in all groups. There is always a **head** with five segments and five pairs of appendages, including those principally concerned with sensory reception and feeding. The entire postcephalic body is a **trunk** if all the segments are similar to each other, but tagmatization of the trunk more commonly results in an anterior **thorax** and a posterior **abdomen.** Often one or several thoracic metameres are fused to the head to form a **cephalothorax.** The appendages of the trunk or thorax and abdomen are concerned principally with locomotion.

External Features

The crustacean cuticle differs in several important ways from that in other arthropods generally. In crustaceans the entire procuticle is termed the **endocuticle,** and it consists of three strata: an outer calcified **pigmented layer** with tanned proteins; a relatively thin middle **calcified layer** that is untanned and unpigmented; and a thin inner untanned **uncalcified layer** (Fig. 12.1). In most crustaceans except for some primitive entomostracans there is considerable deposition of calcium carbonate into the endocuticular pigmented and calcified layers and even in the epicuticle. This cuticular **calcification,** along with sclerotization, results in rigid, armored, integumental sclerites. Beneath the cuticle secreting **epidermis** are clusters of **tegumental glands** that open via ducts onto the surface of the exoskeleton (Fig. 12.1); their secretions are involved in the production of the epicuticle.

Crustaceans are often colored with various shades of red, orange, blue, and green. Coloration is due to the calcified salts or pigments deposited within the endocuticle, the coloration of the blood or that of ingested food, or various chromatophores located beneath the epidermis. Chromatophores are present in a great many malacostracans and are responsible for their colors and their abilities to change colors. The interesting ways in which chromatophores function is the subject of Box 12.2, which appears later in this chapter.

Characteristic of most crustaceans is the presence of a **carapace,** a dorsal fold of the integument. Arising from head segment V, it usually extends posteriorly, and it most often is fused with a variable number of terga behind it, although it remains free from thoracic terga in some forms. The lateral margins of the carapace frequently extend ventrally to cover the lateral aspects of the anterior trunk. In some taxa the carapace even encloses the entire body as a bivalved shell, and the two valves may be closed by an adductor muscle. Between the lateral carapace and the body wall is the **branchial chamber,** in which gills may be found (see Fig. 12.5b).

Figure 12.1 *Diagram of a section of crustacean cuticle.*

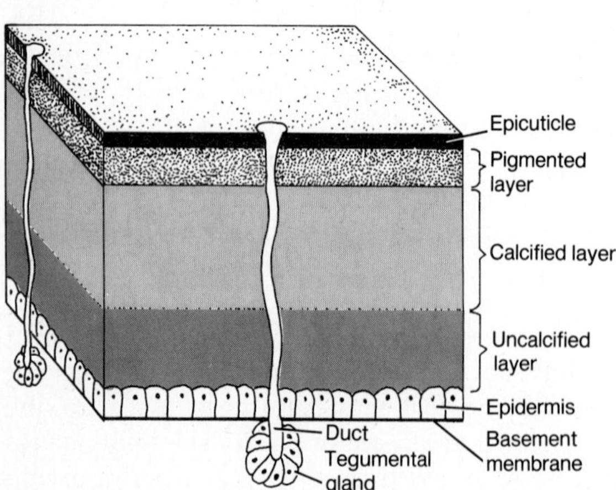

The basic crustacean appendage

Perhaps the premier feature that has contributed most significantly to the great success of crustaceans is their paired appendages. It is the evolutionary plasticity of the appendages that is primarily responsible for the high degree of adaptive radiation exemplified by crustaceans. A great measure of appendicular variability is often present on a given animal, and these paired limbs function in a variety of ways to enhance the success of the organism.

With the possible exception of the first antennae, all crustacean appendages are basically biramous; each consists of a basal **protopod** and two distal rami, an outer **exopod** and a medial **endopod** (Fig. 12.2a). Thus the biramous appendage takes the form of an inverted Y with the protopod at the base and the exopod and endopod as branches. The protopod most often has two podomeres, a basal **coxa** that joins the

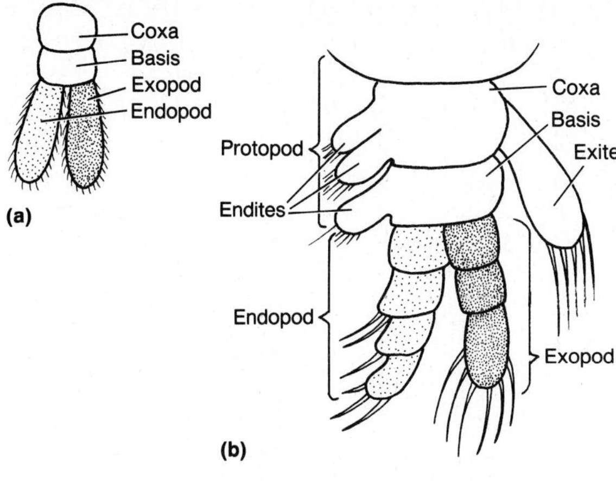

Figure 12.2 *The basic crustacean appendage: **(a)** a generalized and simple appendage; **(b)** a generalized appendage with an exite and endites.*

appendage to the body and a distal **basis** from which arise the two rami (Fig. 12.2). The protopod often gives rise to two other protrusions, lateral exites and medial endites (Fig. 12.2b). From one to three **exites** or epipods usually develop from the lateral margins of the coxa, serve as the principal gas exchange organs, and, in many crustaceans, form rather specialized **gills. Endites** are the numerous teeth, spines, or bristles that arise from the medial face of the protopod and function basically in the movement of food and in mastication. Endites of appendages around the mouth function as gnathobases used to fragment food particles.

The exopod normally consists of two podomeres, but often this entire ramus is missing on those appendages that are specialized for walking or burrowing; therefore these appendages are secondarily uniramous. The exopod functions primarily as a swimming limb or in feeding and water movement. An endopod usually is well developed and consists of three to five podomeres. In the malacostracans a typical walking leg endopod is constructed of five podomeres, which are, from proximal to distal, the **ischium, merus, carpus, prodopus,** and **dactyl** (Fig. 12.3h). Endopods often are specialized into sensory, feeding, locomotive, and burrowing limbs.

Crustaceans have great powers of regeneration and can usually regenerate a missing leg with one or several molts. Many crustaceans can voluntarily break off or autotomize a trapped or damaged appendage. Autotomous breakage usually occurs at the basis–ischium joint, and an integumentary diaphragm both prevents the loss of blood and enhances the clotting process.

THE HEAD

The crustacean head invariably consists of the acron plus five segments fused together as a tagma. Embryologically, only the acron and segment I are preoral in position; but during development, segment II also moves to a preoral position. Segment I bears the pair of **first antennae** or **antennules,** which in entomostracans are always uniramous. In malacostracans, however, two **flagella** or rami arise from the three-segmented **peduncle,** and there is currently considerable debate as to whether these two flagella are homologous to the endopods and exopods of other appendages (Fig. 12.3a). Segment II bears the paired **second antennae** or simply the **antennae** (Fig. 12.3b). These appendages, quite variable both in morphology and in function, are basically biramous, but in adults of some taxa they are secondarily uniramous. The antennae are variously modified for swimming, copulation (males), crawling, feeding, and other functions. Both pairs of antennae contain numerous sensory receptors, especially those concerned with tactile and chemical sensations. The two pairs of antennae are a diagnostic feature for crustaceans, and they are the only arthropods with two pairs of antennae.

Head segments III–V bear paired appendages that are collectively called mouthparts.

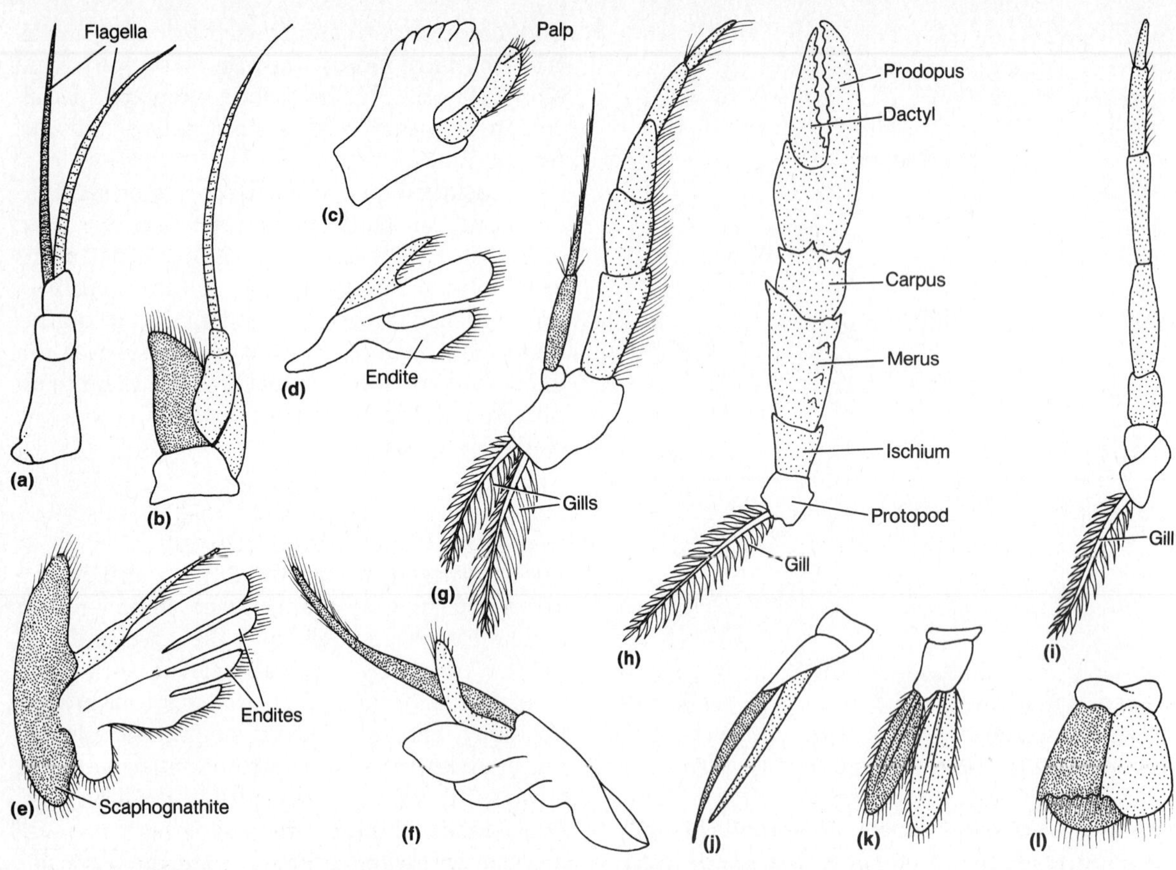

Figure 12.3 Examples of the various types of appendages of a decapod (crayfish) to illustrate appendicular diversity in crustaceans: ***(a)*** *first antenna;* ***(b)*** *second antenna;* ***(c)*** *mandible;* ***(d)*** *first maxilla;* ***(e)*** *second maxilla;* ***(f)*** *first maxilliped;* ***(g)*** *third maxilliped;* ***(h)*** *first walking leg (cheliped);* ***(i)*** *fourth walking leg;* ***(j)*** *first pleopod of a male;* ***(k)*** *a typical pleopod;* ***(l)*** *uropod.*

Flanking the mouth, the **mandibles** are composed basically of the protopod and a well-developed gnathobase or **molar process** that extends toward the midline to meet its opposite (Fig. 12.3c). Both rami may be reduced to a single **palp,** or they often disappear completely. The shape and construction of the mandibles reflect the types of food ingested and the modes of ingestion. The appendages of segment IV are the **first maxillae** or **maxillulae;** those of segment V are the **second maxillae** or simply the **maxillae** (Fig. 12.3d, e). The maxillae of both pairs are relatively similar to each other; they have a protopod of one or two podomeres with well-developed endites or gnathobases, and exites often are present. But one or both rami are often reduced or even absent, and the reduced endopods are the **palps.** The functions of the maxillae are to assist in feeding and to manipulate food between the masticating mandibles. Associated with the mouth are two unpaired structures: the **labrum,** a fleshy "upper lip" lying anterior to the mouth, and a bilobed **labium** or "lower lip" lying posterior to the mandibles. The labium of crustaceans is in no way homologous to the labium of insects, in which it represents a pair of fused appendages.

The head of a crustacean bears a pair of compound eyes and often an unpaired medial eye. The functional morphology of these visual receptors appears in a subsequent section on sense organs.

THE POSTCEPHALIC BODY

The remainder of the body is much more variable than the head in number of segments, details of tagmatization, and nature of the appendages. Primitively, the postcephalic body lacks regionalizations and is termed simply the **trunk.** But in the vast majority of all crustaceans there are a thorax and abdomen as rather distinct tagmata. The thorax consists of from 1 to 19 segments, and each bears a pair of appendages, the **thoracopods,** which are basically involved in locomotion. Several important trends concerning the thoracopods are evident. First, there has been a tendency for the anterior thoracopods to be specialized and the posterior ones to be the least specialized. In the malacostracans, one to three thoracic segments are fused to the head and they and the head constitute a **cephalothorax;** the thoracic appendages of the cephalothorax, termed **maxillipeds,** are turned forward and assist the mouthparts in feeding (Fig. 12.3f, g). Second, one or more of the thoracopods are chelate, that is, the prodopus and dactyl oppose each other, forming a pincerlike mechanism (Fig. 12.3h). Third, the thoracopods generally have become more slender, and they have lost their exopod and become secondarily uniramous.

The **abdomen** is also constructed of a variable number of segments with six metameres in this tagma in malacostracans. Abdominal appendages are absent in all nonmalacostracans with the exception of certain branchiopods (Notostraca). In the malacostracans the anterior five pairs of appendages are specialized as swimming **pleopods,** and the last abdominal appendages are the more specialized **uropods,** which are used in various forms of locomotion (Fig. 12.3j–l). Pleopods may be variously modified as copulatory limbs in males or as brooding appendages in females. The abdomen typically terminates in a **telson,** although in many primitive crustaceans the telson is missing. When present in entomostracans, the telson resembles the abdominal segments lying immediately anterior to it; since its functional morphology is similar to that of adjacent segments, it is considered to be a true body segment. But in malacostracans, in which the telson is always present, it is quite different from other abdominal segments and is not considered as a true body segment. The **anal opening** is present at the posterior end of the abdomen.

SENSE ORGANS

A number of different types of sensory receptors have been very instrumental in the success and diversity of crustaceans. Sense organs in crustaceans include visual receptors, statocysts, chemoreceptors, proprioceptors, and tactile receptors.

Visual receptors are found on the head of a crustacean and include a pair of **compound eyes** and often an unpaired **medial** or **naupliar eye.** A compound eye, previously discussed in Chapter 10, varies from being very simple with fewer than 20 ommatidia to being very complex with up to 15,000 visual units. The eyes of crustaceans are located on movable stalks or **peduncles,** are sessile, or are fused medially into a single eye; some crustaceans are even blind. The movable stalk and the highly convex corneal surface enable each eye to have an exceptionally wide visual field. Visual acuity is not nearly as good as it is in insects, and crustaceans living in deep water or those active at night have ommatidia that are particularly adapted to dim light. Other crustaceans living in habitats with bright light have ommatidia adapted for light, which enhance their visual acuity (see Chapter 10). A few decapods can discriminate between several colors; selected aspects of their color vision ability appear later in the chapter in Box 12.2.

The medial eye, characteristic of the nauplius larva, persists in adults of some species, but in the adults of most higher crustaceans it is absent. The naupliar eye is constructed of three fused ocelli with only a few photoreceptor cells but with no lens. This eye probably functions in orienting the animal to polarized light, in perception of movement, and in distinguishing degrees of light intensity.

Paired **statocysts** are found in some malacostracans and function in the maintenance of equilibrium. Located at the bases of the first antennae, uropods, or telson or at the junction of the thorax and abdomen, they are epidermal invaginations lined with sensory hairs that usually retain their openings to the outside. Each statocyst contains a statolith, which may be secreted by the animal but more likely is composed of grains of agglutinated sand that the animal picks up from the environment. Displacement of the statolith as the body is disoriented stimulates sensory hairs of the sac to send impulses to the brain, and corrective motor action is taken.

Chemoreceptors are numerous on certain appendages, especially the five pairs of head appendages. The most common type of chemoreceptor, the **esthetascs,** are in the form of sensory plates, sacs, or hairs. Canals, funnels, and pores in the cuticle may also be involved in chemoreception.

Proprioceptors are special receptor organs located in the dorsal musculature of each of the last two thoracic and all abdominal segments. They are best known in decapods, but they may be present in all crustaceans. All proprioceptors supply the animal with vital information about various muscle contractions. Each organ, made up of a modified muscle cell, is stimulated either by the contraction of the dorsal extensor muscles or by the stretching of the proprioceptor cells when the ventral flexor muscles contract. Other proprioceptors, located in the joints of walking legs, are stimulated by muscular movements around the joint. A special type of proprioceptor found in decapods is the **chordotonal organ** found in the third segment of each walking leg.

Tactile organs are common in crustaceans as in all arthropods and are in the forms of hairs or setae. They are especially numerous on those appendages such as antennae and legs that come in contact most directly with the environment. The hairs or setae, stimulated either by an object or when moved by water currents, convey this motion through the cuticle to a cup-like base supplied with sensory dendrites. Probably, the crustacean responses to noise are due to sensations transmitted to them via surface vibrations, which they detect by tactile hairs and by the statocysts.

Feeding and Locomotion

Both feeding and locomotion in crustaceans are complex and varied functions that basically involve the appendages. Further, among the entomostracans and certain malacostracans, feeding and locomotion are so intimately interrelated that both functions have to be seen in concert with each other.

Considerable variety exists among crustaceans in their sources of food and their specializations for feeding. The most common forms of feeding are filter- or suspension-feeding, predation, parasitism, and scavenging. Even though primitive modern crustaceans are for the most part filter-feeders, the earliest crustaceans may have been raptorial-feeders. From this method of grabbing and ingesting small prey, more specialized forms of filter-feeding, predation, and parasitism evolved. Each method of feeding requires a rather high degree of appendicular differentiation and specialization.

Filter-feeding is prevalent among the entomostracans as many locomotor appendages create currents of water over specialized feeding appendages. On various head and trunk appendages, filtration of particulate material is achieved by the use of finely spaced setae. In many cases the setae themselves are equipped with more finely spaced plumosities that may create a mesh size as small as several micrometers. The minute food particles, such as plankton and finely divided detritus, are then transferred to the mouth by food grooves or collected by setae on mouthparts.

Predators, found among higher crustaceans and some copepods and other entomostracans, require some means for grasping or capturing prey. Some mouth appendages or thoracopods often have spines or teeth that are used to impale small organisms or to tear off bits of food. A great many malacostracans are predaceous, and the maxillae, maxillipeds, or thoracopods are used as cutters, shredders, spears, or smashers. Many decapods are predators, and one or more pairs of legs are chelate and used to grasp, crush, tear, or stun prey organisms.

A great many crustaceans are deposit-feeders, scavengers, herbivores, or omnivores. Straining setae and scraping spines on thoracopods, maxillipeds, and mouth appendages are frequently present. Parasitic crustaceans, represented mostly by certain taxa of maxillopods, show a great range of modifications in their modes of feeding from ectocommensal to endoparasitic. Parasitic crustaceans feed on tissues or body fluids of polychaetes, echinoderms, bivalves, tunicates, and fishes. In parasitic crustaceans, appendages are modified as holdfast, piercing, and sucking organs, and in many the mouthparts have been lost altogether.

Locomotion is also an extremely varied activity in crustaceans; a great many are planktonic, pelagic, benthic crawlers or walkers, periphytonic, or burrowers. Virtually every appendage is used by one or another crustacean as a locomotor appendage. Planktonic species are inherently small, and a great many copepods use their antennae as flotation devices. Many different entomostracans are weak swimmers, and they use antennae, thoracopods, or other appendages to move through the water. Other entomostracans creep, crawl, or glide over surfaces using spines, claws, bristles, and teeth on the rami of a wide variety of appendages but especially those of the antennae, maxillae, and thoracopods. Locomotion in the malacostracans has led to the development of two basic types of locomotion: walking or crawling and swimming. In walking or crawling over the bottom, thoracic appendages have lost their exopods and become slender thin stenopods. Some stenopods have become modified as **pereiopods** or walking legs (Fig. 12.3h, i), and others have been specialized as food-gathering appendages. In the pelagic malacostracans the abdominal appendages are modified as swimming limbs and are called **pleopods** (Fig. 12.3k).

Visceral Systems

The digestive tract is a fairly straight tube without excessive modifications (Fig. 12.4). The foregut or stomodeum is divided into the **esophagus** and the **stomach.** If the stomach is equipped with ridges, teeth, or ossicles that enable it to grind or triturate food, it is called a **gastric mill.** Several pouchlike **midgut glands** originate from the midgut or mesenteron and are primarily responsible for secreting digestive enzymes. Typically, food enters the ducts of the midgut glands, where nutrients are absorbed. The hindgut, proctodeum, or **intestine** terminates in the ventroposterior **anus.**

The circulatory system consists principally of the heart, arteries, and hemocoel (Fig. 12.5a). The **heart,** a muscular organ located dorsally in the thorax and surrounded by the peritoneal **pericardium,** may be elongated and extend most of the length of the body, or may be com-

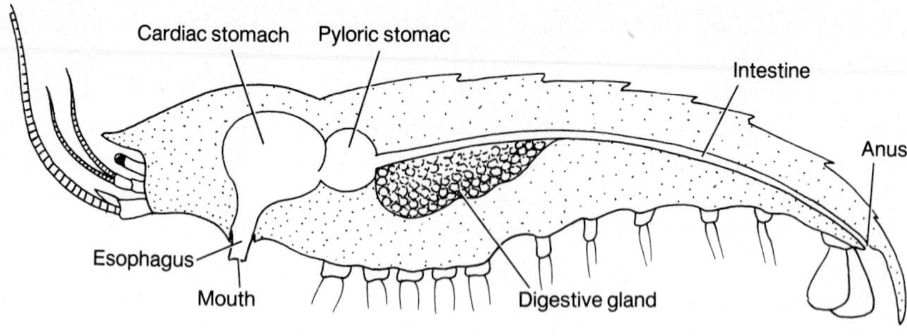

Figure 12.4 Generalized plan of the digestive system of a malacostracan, lateral view.

pact and often spherical; it is completely absent in some smaller forms. Blood enters the heart from the pericardial sac by way of one to several pairs of valvular **ostia,** and it exits through a **medial anterior artery.** Additionally, there may be other arteries (posterior, ventral, lateral) that convey blood to specific regions. In advanced forms, the arterial system may be branched extensively. Blood inevitably passes from arteries into ill-defined blood sinuses constituting the **hemocoel,** where it bathes the tissues directly. In malacostracans and other forms that possess gills, blood is cycled through these gas exchange organs before being returned to the per-

Figure 12.5 Generalized plan of the circulatory system of crustaceans: ***(a)*** diagrammatic lateral view; ***(b)*** cross-sectional view through the thoracic region of a malacostracan.

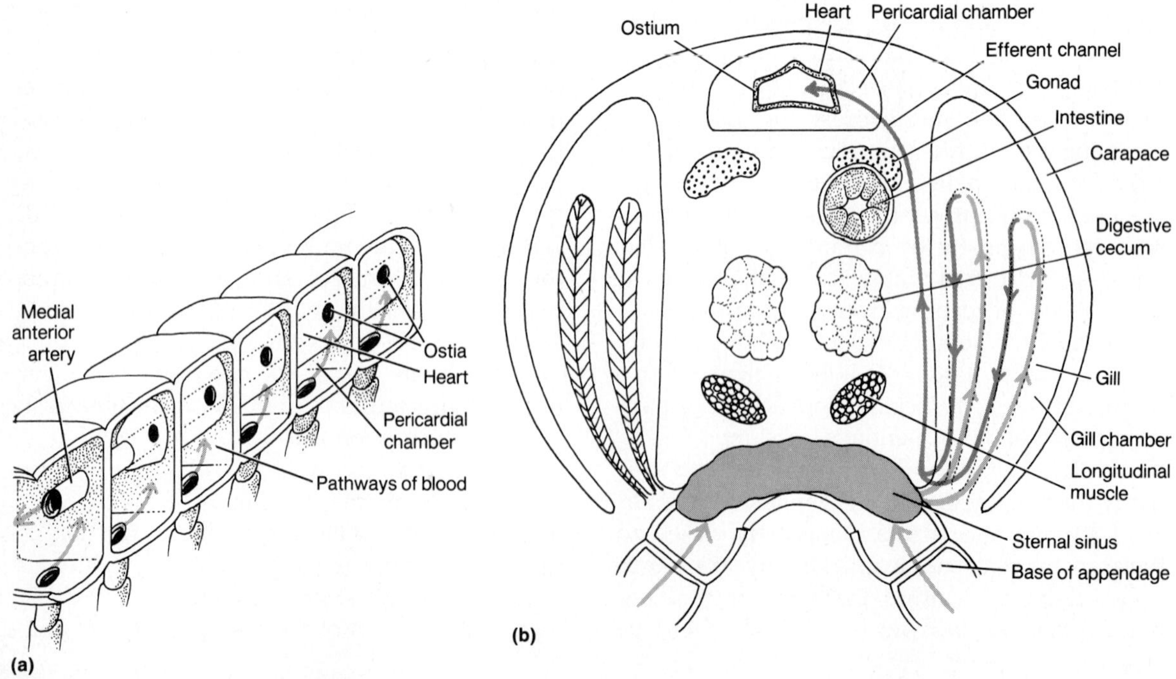

icardial chamber (Fig. 12.5b). In some crustaceans there are **accessory hearts** located either at the bases of certain appendages or along the lengths of arteries that aid in pumping; body movements also augment blood flow.

The blood of crustaceans often has one or several respiratory pigments dissolved in the plasma. **Hemoglobin** is present in many of the lower entomostracans, especially those living in water with low oxygen concentrations; **hemocyanin,** a copper-containing pigment, is present in malacostracans. The blood also contains one to several different types of cells that are involved with clotting and phagocytosis. Blood cell counts vary widely depending upon the animal's physiological state.

In many groups of small Crustacea, gases are exchanged through the general body surface or by way of foliaceous appendages (Fig. 12.6a) But in a vast majority of crustaceans, **gills** are developed, and water currents are created by a variety of appendages to facilitate gaseous exchange. Gills generally are intimately associated with appendages and often are lamellar modifications of exites (Fig. 12.3g–i). Gills are usually found associated with thoracic legs, but in certain malacostracans there are abdominal gills. Primitively, a gill is a broad platelike process, but those of most species are complicated by the development of lateral filaments or lamellae that arise from a central axis (Fig. 12.6b). Such filaments greatly increase the surface area and thus the efficiency of gills. The axis and filaments have afferent and efferent blood channels through which blood courses to augment gaseous exchange between the water and blood. Terrestrial crustaceans may have gills or vascularized branchial chambers enlarged as "lungs."

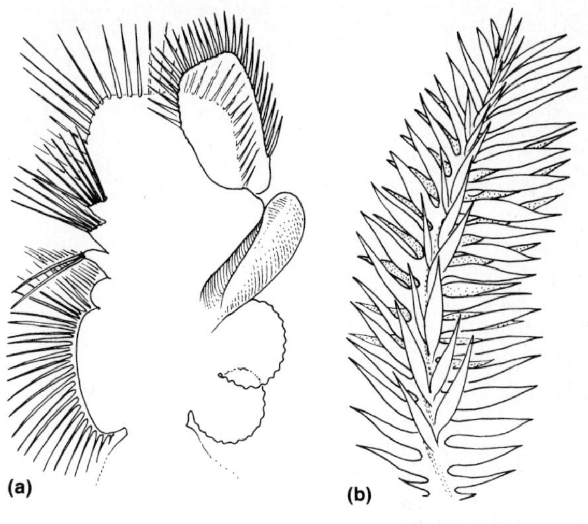

Figure 12.6 *Crustacean gills:* ***(a)*** *a simple lamellar gill (branchiopod) (from G. Smith, 1909,* Cambridge Natural History, *Macmillan and Co., Ltd., London);* ***(b)*** *a complex gill (decapod).*

Crustacean excretory organs are paired structures that are modified coelomoducts located either in the second antennal segment and called **antennal glands** or in the second maxillary segment and called the **maxillary glands.** While generally only one pair persists in mature individuals, both pairs are usually found in larvae as well as in a few adults. Both antennal and maxillary glands open to the outside by **excretory pores** located at the bases of the second antennae and the second maxillae, respectively. Each organ consists of an end sac or saccule, labyrinth (in some), tubule, bladder, and urinary duct (Fig. 12.7). Blood bathes the end sac, and nitrogenous wastes are filtered into the excretory gland. Crustaceans excrete traces of amines, urea, and uric acid, but the principal waste product is ammonia. Apparently, the excretory organs of most crustaceans do not function in osmoregulation, since the urine produced is isosmotic with the blood. However, a few forms, including some freshwater crayfishes, excrete a hyposmotic urine indicative of some ionic reabsorption. The gills usually are the principal osmoregulatory organs, since they either secrete salts into the water (marine species) or absorb salts from the water (freshwater species). Waste-accumulating **nephrocytes** are particularly common in the gills and appendicular bases.

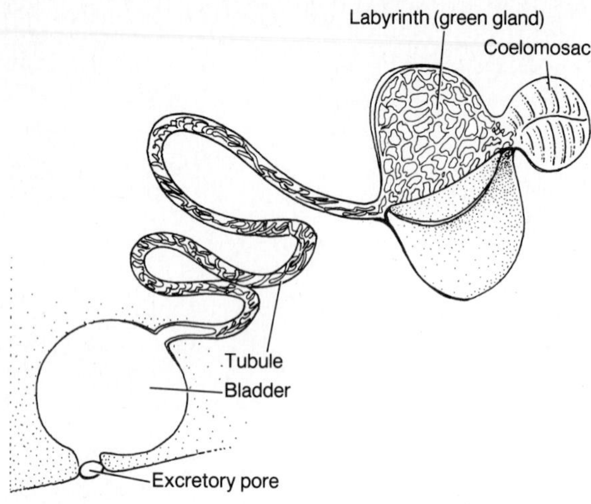

Figure 12.7 *An excretory organ in crustaceans.*

The nervous system of crustaceans is typically arthropodan both in basic construction and in the tendency toward fusion of ganglia. The **brain or supraesophageal ganglion** represents the fusion of the protocerebrum, deuterocerebrum, and tritocerebrum, and paired nerves from each of the three parts of the brain innervate the two pairs of antennae and the paired compound eyes, respectively. When the naupliar eye is present, a medial nerve innervates it. Paired **circumesophageal commissures** arise from the tritocerebrum and connect it with the paired **subesophageal ganglia,** the anteriormost ganglia of the **ventral nerve cord.** In the primitive condition there are two cords that are spatially separated but connected by interganglionic commissures, and each bears segmental ganglia (Fig. 12.8a). Each ganglion in each nerve cord gives off three pairs of lateral nerves to various parts of that segment. The anterior two nerves are mixed and innervate appendicular and segmental muscles, while the posterior nerves are mostly motor and innervate the flexor muscles. But in most advanced forms there is a tendency for both medial and longitudinal fusion of ganglia, since the two cords usually are fused into one, and the cord may be shortened and condensed (Fig. 12.8b). In true crabs (Order Decapoda) the ventral cords are completely consolidated to form a single ventral mass (Fig. 12.8c).

Located in the ventral nerve cord of higher crustaceans are two pairs of dorsally located giant fibers. Two dorsomedial fibers arise in the brain and run the length of the cord; each represents one very long neuron. Two dorsolateral fibers, which are intrinsic in the cord, are composed of serially arranged neurons that can conduct impulses both anteriorly and posteriorly.

Figure 12.8 *Variations in the central nervous system of crustaceans: **(a)** a primitive arrangement (branchiopod); **(b)** an advanced arrangement (crayfish) illustrating medial fusion of the nerve cords; **(c)** an advanced arrangement (crab) illustrating both medial and longitudinal fusion of the nerve cords.*

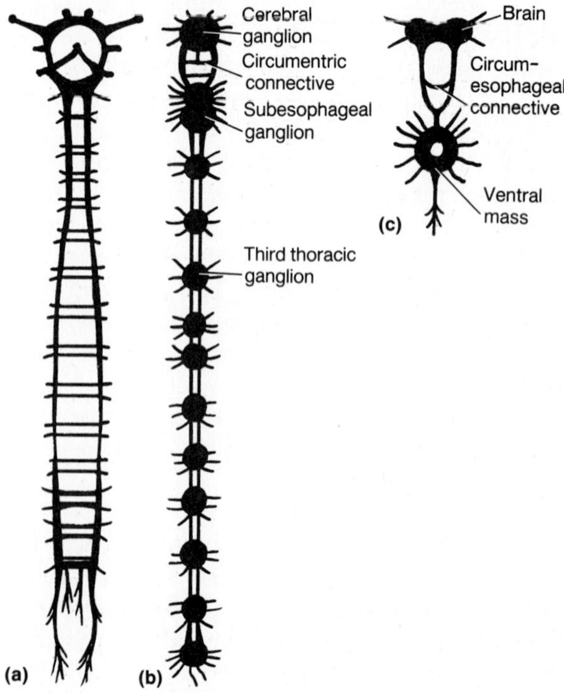

Both pairs of nerve fibers are noteworthy for their unusually large diameter and for their ability to have a single impulse evoke a complex muscular response. In malacostracans the reflexive contraction of the ventral abdominal muscles flexes the tail fan ventrally, and this constitutes an effective, rapid, escape reflex. The brain contains some neurosecretory cells whose function will be discussed later in this chapter.

Reproduction and Development

Even though some groups like barnacles are hermaphroditic, most crustaceans are dioecious. Sexual dimorphism is evident in many crustaceans; either sex can be larger than the other, often there are color differences in the body or differences in the degree of development of certain appendages, and males often have various appendages modified as clasping organs. In male decapods the first and second abdominal appendages are modified as copulatory organs, and a **penis** (or paired penes) is found in some. Parthenogenesis is common in many lower crustaceans.

The **gonads** are paired organs lying in the dorsal thorax and abdomen. The **gonoducts** (oviducts or sperm ducts) are simple tubules that end in **gonopores** located either on a thoracic sternite or at the base of thoracopods. In copulation the male usually clasps the female by using modified antennae, mouthparts, or thoracic appendages; by means of a penis or modified abdominal appendages, sperm are deposited within the seminal receptacle of the female or on the surface near the female gonopore. Spermatophores are sometimes transferred to the ventral thoracic surface of the female. The sperm of most crustaceans are distinctive, since they lack flagella and are therefore nonmotile.

Fertilization is either internal if sperm are inserted into the seminal receptacle or external as the eggs are extruded from the female. Characteristically, crustaceans brood their eggs or young for varying lengths of time either within a specialized chamber or sac produced as the eggs are released or on various appendages. Some shrimp and many copepods shed their eggs directly into seawater. A wide variety of cleavage patterns is found in this group, and some crustaceans even have indeterminate cleavage.

Most crustaceans have indirect development in that a larval stage is developed that usually is free living and planktonic. The most basic and primitive larval stage is the **nauplius** (Fig. 12.9a). The nauplius has an unsegmented oval body, a median naupliar eye on the acron, and three pairs of head appendages of which the second antennae and mandibles are used as swimming limbs. In successive molts, additional pairs of appendages are added immediately anterior to the telson. Trunk segmentation becomes evident in some of the earlier naupliar instars. Formation of the carapace, development of the compound eyes, and other morphological changes soon follow. In higher malacostracans, when development finally produces a stage in which all eight paired thoracopods have developed, the larva is termed a **zoea** (Fig. 12.9b). With additional molts, the zoea acquires the remainder of its segments, and when the appendages all become functional, it is called a **postlarva** (Fig. 12.9c). The postlarva often is quite similar to the adult in its general appearance, but in some malacostracans it is fundamentally different from the adult. Successive molts increase its size until the animal attains sexual maturity. This basic sequence of nauplius, zoea, and postlarva is often modified in various groups. Malacostracan larvae do not feed, and the young of some crustaceans like crayfishes hatch as postlarvae and therefore do not have a naupliar or zoeal stage. Others, like barnacles, experience a high degree of metamorphosis from a swimming larval stage to a sessile adult form. Postlarvae in certain well-known groups are given specific names

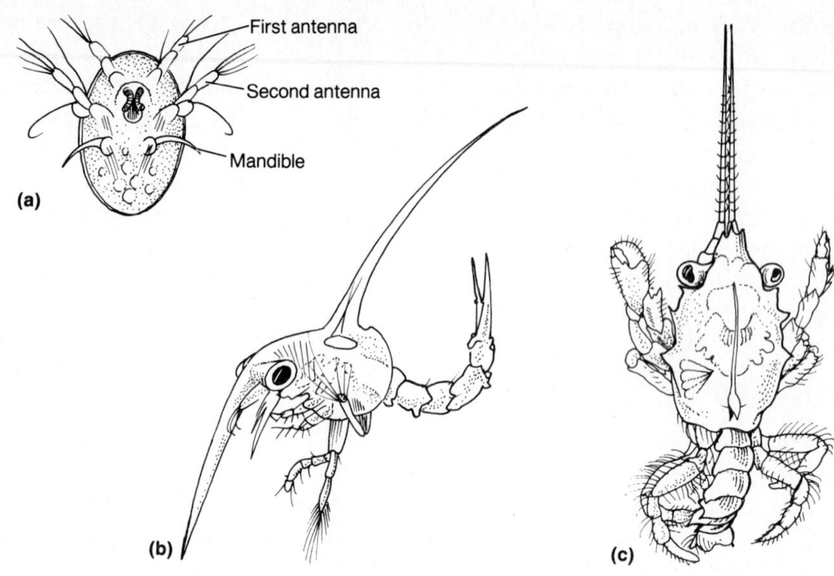

Figure 12.9 *Larval stages of crustaceans;* ***(a)*** *a nauplius;* ***(b)*** *a zoea (crab);* ***(c)*** *a postlarva (megalops of a crab). (Parts a-c from G. Smith, 1909,* Cambridge Natural History, *Macmillan and Co., Ltd., London.)*

such as cypris (barnacles), copepodid (copepods), phyllosoma (spiny lobsters), megalops (crabs), manca (isopods), and mysis (lobsters).

CRUSTACEAN DIVERSITY

The following pages represent a brief treatment of the principal subordinate crustacean groups. Each of the six classes (see Table 12.1) is discussed, more detailed treatment being given to the classes with greatest species diversity. Subordinate taxa are discussed separately if they have high species diversity, are economically important, or are commonly encountered. In most instances except for decapods the descriptions for each taxon is restricted to external features, and special attention is given to appendages, since they are often the chief taxonomic indicators. Brief characterizations of all crustacean taxa down to order appear at the conclusion of this chapter.

Class Cephalocarida

Members of this taxon, first discovered in 1954 in Long Island Sound, have also been found in various other parts of the world including California, the West Indies, Japan, and New Zealand. They are the most primitive of all modern crustaceans, and it is very likely that the earliest crustaceans resembled cephalocarids in their functional morphology. Cephalocarids are small (up to 4 mm), transparent, shrimplike forms that live in marine bottom oozes. The head is horseshoe shaped and covered by a carapace (Fig. 12.10a). The two pairs of antennae are short, and compound and naupliar eyes are missing. Behind the head is a trunk of 19 segments of which the anterior nine bear appendages. The trunk appendages are similar to each other as well as to the second pair of maxillae. Each trunk appendage is broad, flattened, and essentially triramous, since the basal portion bears a large, flat outer ramus, the **pseudepipod,** in addition to the exopod and endopod

(Fig. 12.9b). Each appendage also bears gnathobases that collectively form a food groove along which detrital particles move to the mouth. The telson bears two long, spinelike **caudal rami.** Cephalocarids, unlike most crustaceans, are monoecious.

Class Branchiopoda

Branchiopods are small crustaceans that are restricted mostly to freshwater habitats. Only one order (Cladocera) has both marine forms and representatives in large ponds and lakes; another branchiopod, *Artemia,* lives in salt lakes throughout the world. The remaining branchiopods are found in temporary bodies of water that periodically dry up or that form briefly from melting snow or from rains, and branchiopods develop an egg stage that is adapted to withstand long periods of desiccation. They range in size from 0.02 to 12 cm, and about 900 species have been described. They are mostly pale and transparent, but some have red hues owing to the presence of hemoglobin, and some marine forms are dark brown or black.

Figure 12.10 *Cephalocarids:* ***(a)*** Hutchinsoniella, *ventral view;* ***(b)*** *a thoracic appendage. (from H.L. Saunders, 1963, Mem. Connecticut Academy Arts and Letters 15:47, 48).*

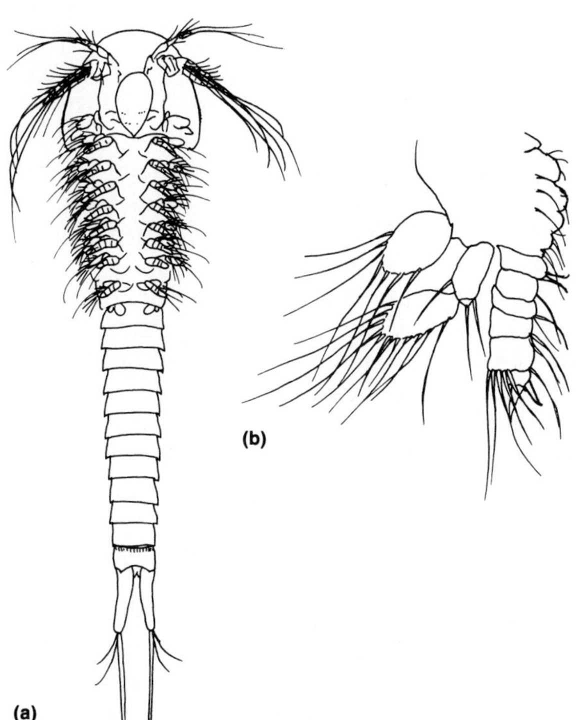

Even though there is considerable diversity within this taxon, a number of common characteristics are found in most forms. They all have flattened, leaflike trunk appendages, each of whose endopod and exopod form a platelike lobe heavily equipped with setae. Each epipod, along with the entire appendage, functions as a gill, which gives the group its name (branchio = gill; pod = foot). Branchiopods (except anostracans) have a carapace that is either a dorsal shield or a bivalved shell covering most or all of the animal (Figs. 12.11c, d, 12.12a, b). The head bears compound eyes, usually the first antennae and second maxillae are reduced or vestigial, and the trunk appendages are used for feeding, locomotion, and gaseous exchange.

With some exceptions, branchiopods are filter-feeders. Food-gathering involves the movement of trunk appendages creating currents of water in an anteroposterior direction. Appendicular setae and setules filter out plankton, detritus, bacteria, and other small bits of food, which are passed to a midventral **food groove** where they are entangled with mucus. The first maxillae push the food to the mouth, where it is crushed by the mandibles.

Branchiopods are dioecious and possess a rather simple reproductive system. Fertilized eggs, produced in clutches ranging up to several hundred, are always brooded by branchiopods for varying periods in special sacs or beneath the carapace or trunk appendages. Development in some is direct, but in most a

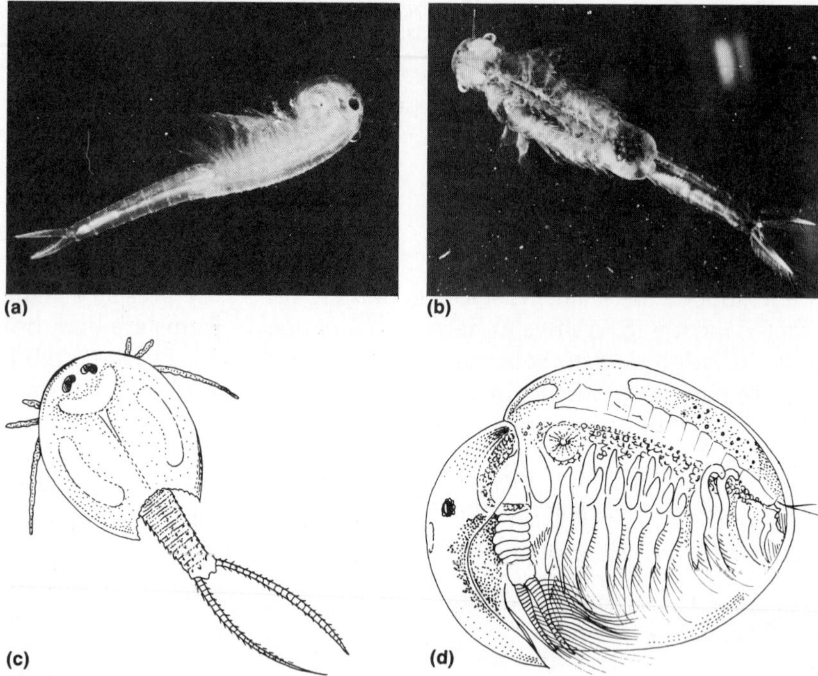

Figure 12.11 *Some examples of branchiopods:* ***(a)*** *photograph of an anostracan,* Streptocephalus, *lateral view;* ***(b)*** *photograph of an anostracan,* Eubranchipus, *ventral view;* ***(c)*** *a notostracan* (Triops); ***(d)*** *a conchostracan* (Lynceus). *(parts a, b courtesy of D.L. Belk; parts c, d from R.W. Pennak,* Freshwater Invertebrates of the United States, *2nd ed., © 1978, John Wiley & Sons, Inc., used with permission.)*

nauplius is developed. Parthenogenesis is common, and males are rare or have not been discovered in some species. Many branchiopods produce two types of eggs: **thin-shelled** eggs in which development is rapid and **thick-shelled** or **dormant** eggs. Dormant eggs are produced as a result of a variety of environmental conditions, including decreasing photoperiods and temperatures. Imminent desiccation, as in the drying up of temporary pools, also induces the formation of dormant eggs. The return of more favorable environmental conditions stimulates rapid egg development and early hatching.

Four orders of branchiopods are usually recognized: Notostraca, Conchostraca, Cladocera, and Anostraca. Representatives of the last three orders are rather familiar, common, and distinctively different from each other; a brief treatment of each follows.

Conchostracans are called clam shrimp because of the presence of a bivalved carapace completely enclosing the animal (Fig. 12.11d). The animal is laterally compressed, and the dorsally folded carapace is closed by an adductor muscle located in the fifth head segment. All these features make a conchostracan look much like a miniature clam. The second antennae are well developed and used for swimming. There are 10–32 trunk somites, each bearing a pair of foliaceous appendages that are also used in swimming or crawling through the bottom mud. Eggs, brooded beneath the carapace, hatch into nauplii. Typically, one generation is produced each year. The 30 or so North American species generally are found in muddy alkaline lakes.

Cladocerans are small (0.1–3 mm) familiar water fleas that are found in most ponds and lakes, where they often are present in great numbers; they are a fundamental link in the food chains of many aquatic ecosystems. While most of the 450 species live in fresh water, some are marine, where they may be present in tremendous numbers. *Daphnia* is a very common taxon and is frequently used as a multipurpose laboratory animal (Fig. 12.12b).

The unhinged carapace is folded dorsally into two halves that enclose the entire body except for the head and antennae (Fig. 12.12b). The carapace often terminates posteriorly in an **apical spine.** The antennules are small, but the antennae are greatly enlarged, and both rami bear several branches. These powerful antennae are used for swimming mostly in a vertical direction as the swimming stroke usually produces a jerky motion. The head bears a single

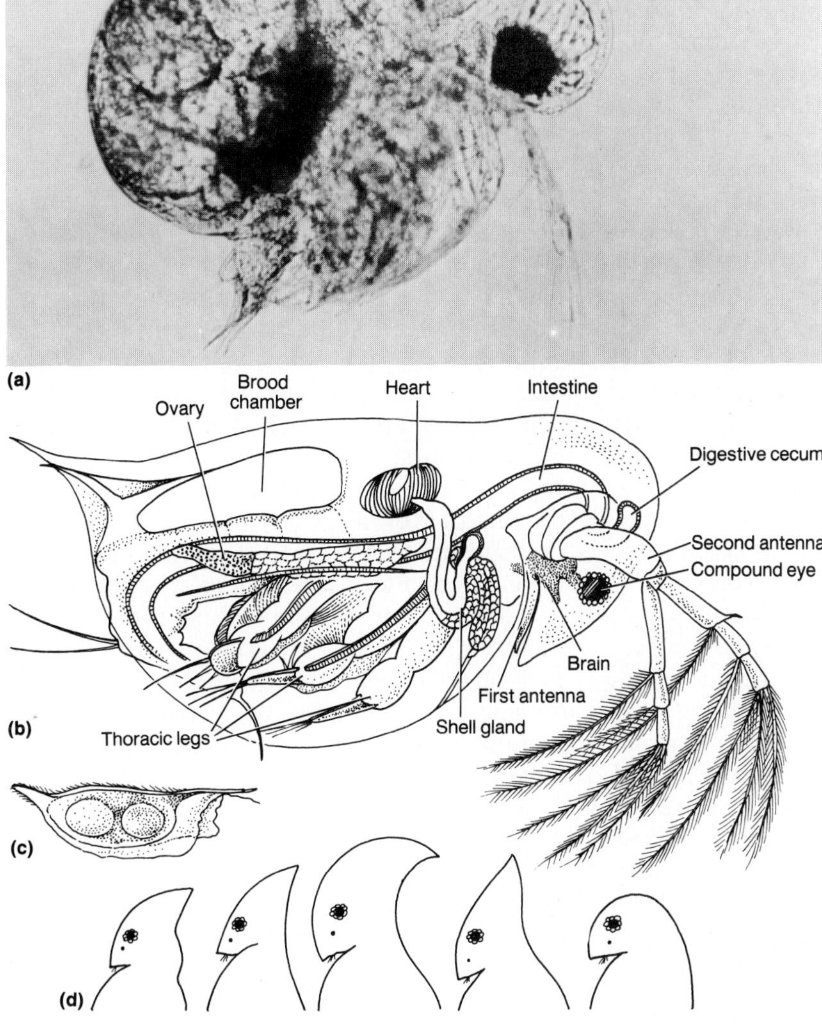

*Figure 12.12 Cladoceran features: **(a)** photograph of* Podon, *a marine planktonic form (courtesy of Ward's Natural Science, Inc., Rochester,N.Y.); **(b)** lateral view of a female* Daphnia *to show the principal internal features (from O. Storch, 1925,* Biol. der Tierre Deutschl. 15*(14): 14.30, used by permission of Gebruder Borntraeger Verlagsbuchhandlung); **(c)** an ephippium with two winter eggs (from G. Smith, 1909,* Cambridge Natural History, *Macmilllan and Co., Ltd., London); **(d)** some cyclomorphic variations in the shape of the head of* Daphnia retrocurva *(from R.W. Pennak,* Fresh-water Invertebrates of the United States, *2nd ed., © 1978, John Wiley & Sons, Inc., used with permission).*

medial, compound eye that is used in orientation, and a naupliar eye is often present. Behind the head is a trunk bearing five or six pairs of flattened appendages (Fig. 12.12b). The movement of these limbs creates feeding water currents. Posterior to the trunk is an appendageless postabdomen that is bent downward and anteriorly; it bears two claws and a pair of abdominal setae.

From 10 to 20 parthenogenetically produced eggs per clutch are released from the oviducts into a dorsal **brood chamber,** where they hatch into miniature first instars and where the young are brooded (Fig. 12.12b). After being released to the outside they molt several times and become sexually mature. Certain environmental conditions like decreasing photoperiods, temperature, or food supplies will induce the appearance of males. With males in the population, most subsequent eggs are fertilized. Only a few fertilized eggs are produced in a clutch, and they are often called winter or dormant eggs. The walls of the brood chamber containing fertilized eggs thicken, and the chamber is transformed into a tough, thick, saddlelike capsule called the **ephippium,** which is shed at the next molt (Fig. 12.12c). Ephippia are often inadvertently picked up and dispersed to other bodies of water by aquatic birds. With the return of favorable conditions the eggs hatch promptly. Populations of cladocerans may increase in numbers dramatically as a pulse or bloom in the spring and autumn. Some cladocerans are striking in that their gross morphology, especially that of the head, changes seasonally. This phenomenon is called **cyclomorphosis** (Fig. 12.12d).

Anostracans, including the fairy shrimp and brine shrimp, are represented by 180 species of which about 30 are found in North America. They are mostly transparent elongated animals with some blue, green, and red tones. Characteristically, they swim upside down with their leaflike trunk appendages serving as locomotor organs. They lack a carapace (Anostraca = without a shell) (Fig. 12.11a, b). The antennules are small, while the antennae are well developed; in males the second antennae are greatly enlarged and are used as strong grasping organs in copulation. Compound eyes are on short stalks. The trunk consists of 20 or more segments of which the anteriormost 11–19 segments bear paired appendages. In addition to their locomotive function, the setae of these trunk appendages form a filter chamber for feeding. The first trunk segment that does not bear appendages has the **penis** or **ovisac.** Eggs either hatch promptly or can withstand periods of desiccation before hatching. The trunk terminates with an anal segment that bears a pair of short **caudal rami** (Fig. 12.11a, b).

Class Remipedia

Representatives of this class were discovered as recently as 1980 in a submarine cave in the Bahamas. The single species *Speleonectes lucayensis* (Fig. 12.13) does not appear to be closely related to any other crustacean taxon, since most appendages on the head and trunk make this species (and several other closely related species currently being described) quite distinct. These individuals are approximately 2.4 cm long, are pelagic, and lack eyes. The head bears a pair of biramous first antennae similar to those of higher crustaceans. Both pairs of maxillae and the one pair of maxillipeds are uniramous and used in raptorial feeding. The biramous trunk appendages are directed laterally, lack endites, and are apparently used exclusively for swimming (Fig. 12.13b). There is no fossil record, and nothing is known about their embryonic development. Further studies will certainly elucidate more fully the phylogenetic significance of this remarkable recent discovery.

Class Maxillopoda

This very large class of almost 10,000 species includes the copepods, barnacles, and two small groups. Maxillopods are small crustaceans and are found in both marine and freshwater

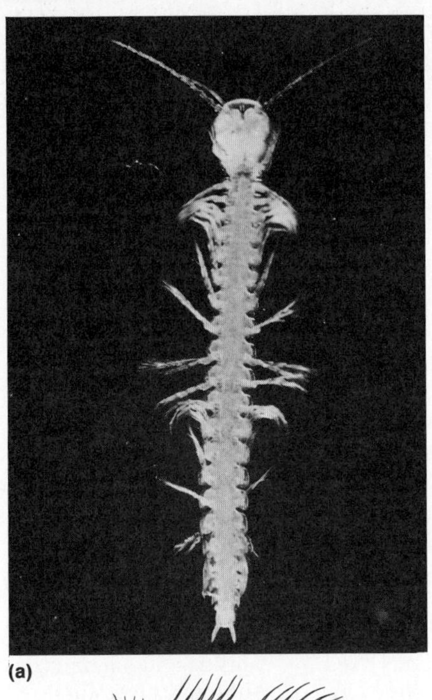

(a)

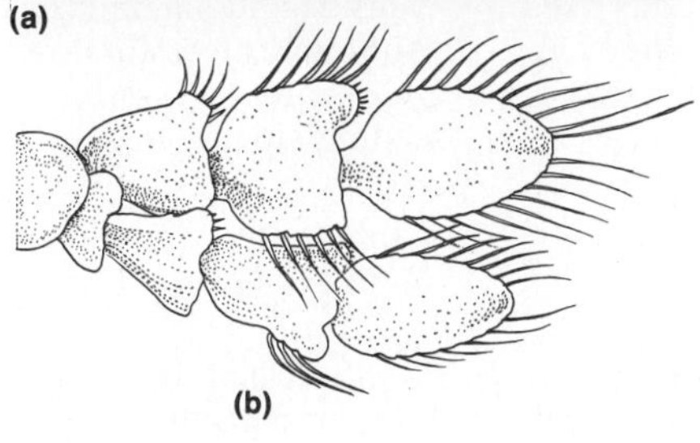

(b)

Figure 12.13 Speleonectes lucayensis, *a remipedian: **(a)** photograph of a living, swimming remipedian (courtesy of J. Yager); **(b)** a trunk appendage (courtesy of L. Ohman and J. Yager).*

habitats. Copepods in particular are often present in very high densities.

Maxillopods are characterized by several important common features: the maxillae are large and are used in food-gathering, the trunk is reduced and usually consists of 11 segments including the telson, and the thoracopods lack endites. Excluding copepods, maxillopods are unique among crustaceans in having flagellated sperm. Copepods and barnacles have six naupliar instars. A great many maxillopods are parasitic, and they have exploited the parasitic lifestyle more than any other crustacean group has.

This large class is represented by two subclasses, Mystacocarida and Branchiura with very few species, and two other taxa, Copepoda and Cirripedia (barnacles) with high species diversity. The four subclasses and the principal orders are characterized in Table 12.2.

Subclass Mystacocarida

The mystacocarids are very small primitive crustaceans found in the interstices of sand grains in intertidal locations. They were first discovered in 1943 in Woods Hole, Massachusetts, and only nine species have been reported. Their body, no more than 0.5 mm long, is cylindrical with divisions similar to those of copepods, to which they are related (Fig. 12.14). The head

Figure 12.14 *A mystacocarid (from R.R. Hessler and H.L. Saunders, 1966,* Crustaceana *11:142).*

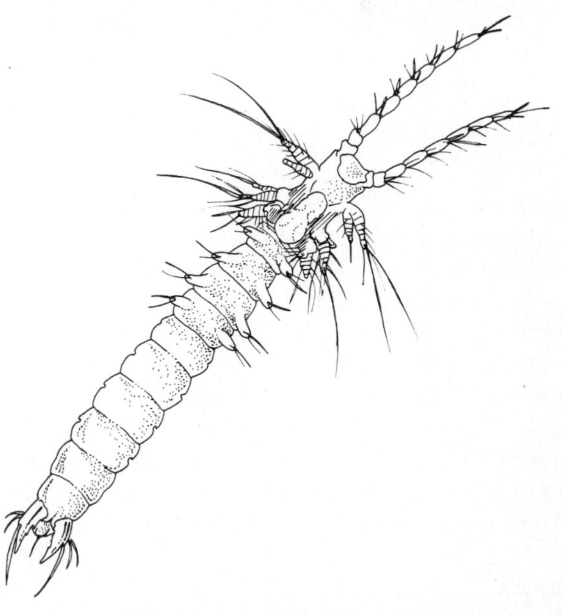

TABLE 12.2. □ **BRIEF CHARACTERIZATIONS OF THE SUBCLASSES AND PRINCIPAL ORDERS OF MAXILLOPODS**

Taxon	*Habitat*	*Unique Appendages*	*Other Features*
Subclass Mystacocarida	Marine, interstitial	1 pair of maxillipeds	Naupliar eye; 5 abdominal segments that lack appendages
Subclass Branchiura	Ectoparasitic mostly on freshwater fishes	Antennae I, maxillae I and II form holdfast devices	Small bilobed abdomen
Subclass Copepoda			
Order Calanoida	Mostly marine, pelagic	Very long first antennae	Herbivorous; one ovisac
Order Harpacticoida	Mostly marine, benthic	First antennae very short	Wormlike body
Order Cyclopoida	Marine and freshwater	Relatively short first antennae	2 ovisacs; omnivores; often predaceous
Order Poecilostomatoida	Marine	Thoracopods used to grasp and cling to host	Parasitic and commensal on fishes and invertebrates
Order Siphonostomatoida	Marine	Thoracopods used to grasp and cling to host	Parasitic and commensal on fishes and invertebrates; labium and labrum form long feeding siphon
Subclass Cirripedia			
Order Thoracica	Sessile and commensal in marine habitats	6 pairs of well-developed cirri	Mantle with calcareous plates; stalked or sessile
Order Rhizocephala	Endoparasitic in marine decapods	All appendages are lost	Form nutritive rhizoids within host; no digestive tract; most other visceral systems degenerate

bears a naupliar eye, and both pairs of antennae are large and used for locomotion. The first thoracic segment, partially fused to the head to form a cephalothorax, bears a pair of maxillipeds. The thorax proper is composed of five segments of which the anterior four bear appendages. The five abdominal segments lack appendages. Mystacocarids develop nauplii, which are detritus feeders, and employ the second antennae for food-gathering.

Subclass Branchiura

Representatives of this group of 150 species, of which *Argulus* (Fig. 12.15) is the most common genus, are ectoparasitic on the skin or gills of marine and freshwater fishes and some amphibians. They are from 5 to 25 mm long, are called fish lice, and are variously adapted for attachment and blood sucking. Branchiurans have a flattened, disc-shaped cephalothorax

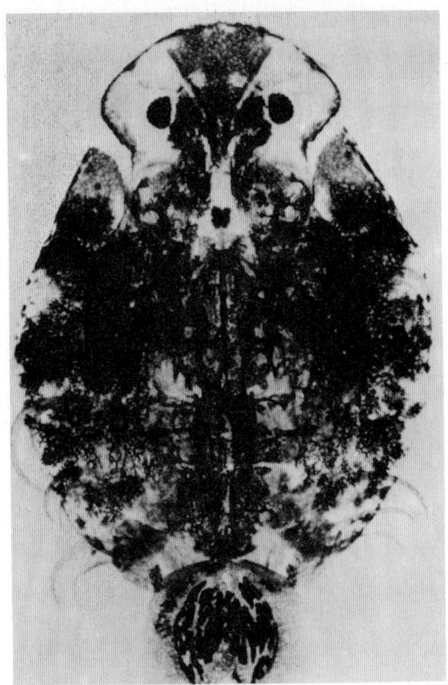

(a)

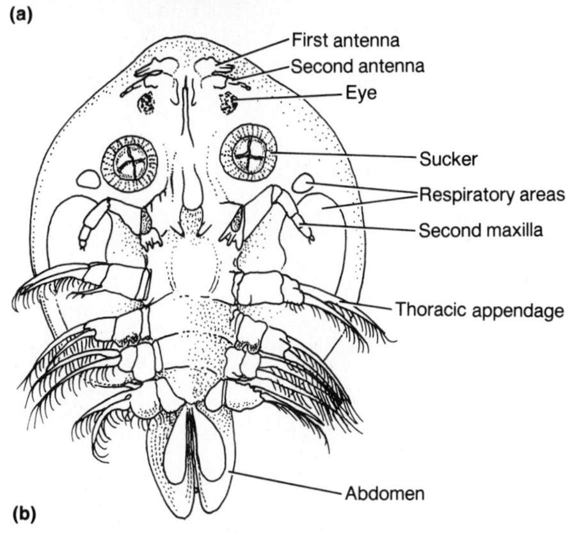

(b)

Figure 12.15 Argulus, *a branchiuran: **(a)** photograph of an individual, dorsal view; **(b)** ventral view of a male (by permission of Smithsonian Institution Press from "A review of the parasitic Crustacea of the genus* Argulus *in the collections of the United States National Museum," Figure 21a.* Proceedings of the United States National Museum, Volume 88, Number 3087, *by O. Lloyd Meehan. Smithsonian Institution, Washington, D.C., 1940).*

covered by a carapace and bearing a pair of sessile compound eyes (Fig. 12.15). Both pairs of antennae are short. The first maxillae are modified to form two large **suckers,** and the second maxillae and the second antennae both terminate in holdfast **claws.** The four pairs of biramous thoracic appendages are large and used for copulation and for swimming, since branchiurans often swim from one host to another. The abdomen is small, unsegmented, and bilobed (Fig. 12.15).

Branchiurans mate on the host, but eggs are often deposited on some object on the bottom. No naupliar stage develops, but rather the egg hatches into a juvenile stage, which is also parasitic. Fish lice continue to molt after attaining sexual maturity.

Subclass Copepoda

The Copepoda is by far the largest and most successful of the nonmalacostracan taxa with almost 8500 known species. Most copepods are marine, but many species are found in fresh water, and they often have unbelievably high densities. Several other copepods are found in moist terrestrial habitats such as in moss mats or humus. Most copepods are free living, but a great many species and representatives of several entire orders are parasitic or commensal on a variety of aquatic animals (see Table 12.2). Many copepods are planktonic, and the calanoids in particular feed on phytoplankton; these copepods, in turn, are a vital link in the food chains of many aquatic animals. Copepods are usually small crustaceans ranging in length from 0.5 to 2 mm. Parasitic forms tend to be longer, and the length of one species parasitic on whales may exceed 30 cm. Some are grayish or pale in color, but others are colored brilliant red, orange, blue, or purple owing to altitude of the habitat, food recently eaten, or oil droplets located beneath the skin.

The body of free-living copepods is cylindrical, rounded anteriorly, and tapered posteriorly (Fig. 12.16a). The first (and occasionally the second) thoracic segment is fused to the head to form a cephalothorax that bears little evidence of segmentation. A carapace and compound eyes are absent, but typical of copepods is the naupliar eye. The long uniramous first antennae are used in copulation and locomotion, and the second antennae are usually biramous and used in locomotion. The mandibles, maxillae, and a pair of maxillipeds are

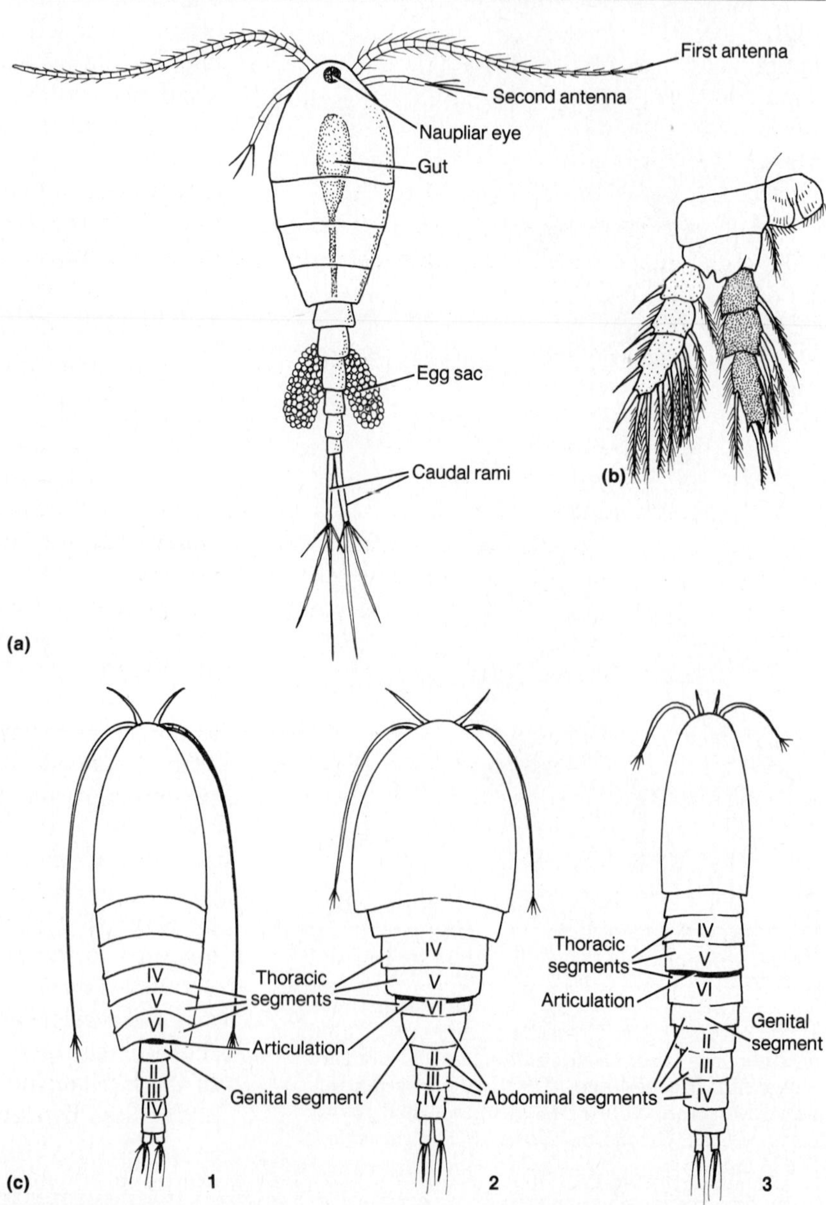

Figure 12.16 *Copepod features:* ***(a)*** *a female cyclopoid illustrating basic copepod features;* ***(b)*** *a typical thoracic swimming appendage (from R. Gurney, 1933,* British Freshwater Copepoda, *Vol. III,* The Ray Society*);* ***(c)*** *points of articulation between the prosome and urosome in a calanoid (1), cyclopoid (2), and harpacticoid (3).*

variously modified for feeding. The trunk consists of 11 segments of which six are usually considered thoracic and five abdominal. Numbers of trunk segments are sometimes complicated owing to fusion of segments; often thoracic segments IV and V are fused, as are thoracic segment VI and abdominal segment I. Each of thoracic segments II–V bears a pair of biramous swimming legs. Appendages of the sixth thoracic segment are typically absent in females, and in males they are modified for copulation. The narrow cylindrical abdomen is composed of four or five appendageless segments, and the telson often bears a pair of **caudal rami** (Figs. 12.16a, 12.17a–c, e).

In most copepods the trunk has a major point of articulation or flexion where the anterior and posterior parts of the body can be bent or moved in relation to each other. This point of articulation is located either between thoracic segment VI and abdominal segment I or between thoracic segments V and VI (Fig. 12.16c). Copepod specialists term the anterior part the **prosome** and the posterior part the **urosome.**

Locomotion in copepods is achieved mainly by the biramous second antennae and the thoracopods, which are moved in rapid oarlike motion and from which this group gets its name (cope = oar; pod = foot) (Fig. 12.16b). When the thoracic appendages or antennae are in motion, the long first antennae are laid back against the body. But when the movements of these limbs cease, the first antennae are extended laterally, which creates a parachute effect and slows the rate of sinking. The sudden jerky motion of many copepods is due to movements of alternating swimming and floating. Some calanoid copepods move slowly through the water using the second antennae as they graze on plankton. The harpacticoids crawl over the substratum using the thoracic appendages as legs.

Filter-feeding, common in planktonic copepods, is achieved by the second maxillae, which mainly strain out phytoplankton. Water is brought over the maxillae by currents created by the second antennae, mandibular palps, and maxillulae, and filtered particles are passed on to the mouth. Predaceous copepods, that is, the harpacticoids and cyclopoids, capture prey with the second maxillae or maxillipeds, and the maxillae sometimes serve as a scoop to catch food.

The **gonopores** for both sexes are on the **genital** or first abdominal **segment** (Fig. 12.16c). The antennae, and sometimes the last thoracic appendages, are used by the male for clasping the female during copulation. During sperm transfer, spermatophores are attached to the ventral surface, sperm enter the seminal receptacle of the female, and fertilization is internal. The fertilized eggs are usually brooded in one or two **ovisacs,** but ovisacs are absent in those species in which eggs are laid directly. Broods can be produced at varying intervals ranging from every two weeks to one or two a year depending upon the species. The copepod egg hatches into a typical nauplius, and appendages are added at each subsequent molt. When the sixth naupliar instar molts, it resembles an adult; this postlarva is called a **copepodid** stage. After five copepodid instars the adult instar is reached. Development may be completed in as little as one week or as long as a year.

Parasitic copepods. Most of the parasitic crustaceans are copepods, and some copepod orders are even exclusively parasitic. Many species, often called fish lice, are ectoparasitic on freshwater or marine fishes. Other copepods are endoparasitic or commensal in cnidarians, polychaetes, certain echinoderms, tunicates, bivalves, and other crustaceans. Compared to their free-living relatives, some parasitic copepods are only slightly modified for their habit, but others are so bizarre that they have lost any resemblance to free-living forms (Fig. 12.17f–h). Interestingly, females of a species generally have more morphological adaptations as a parasite than do males. Adaptations present in parasitic species include reduction or loss of segmentation and most appendages and the

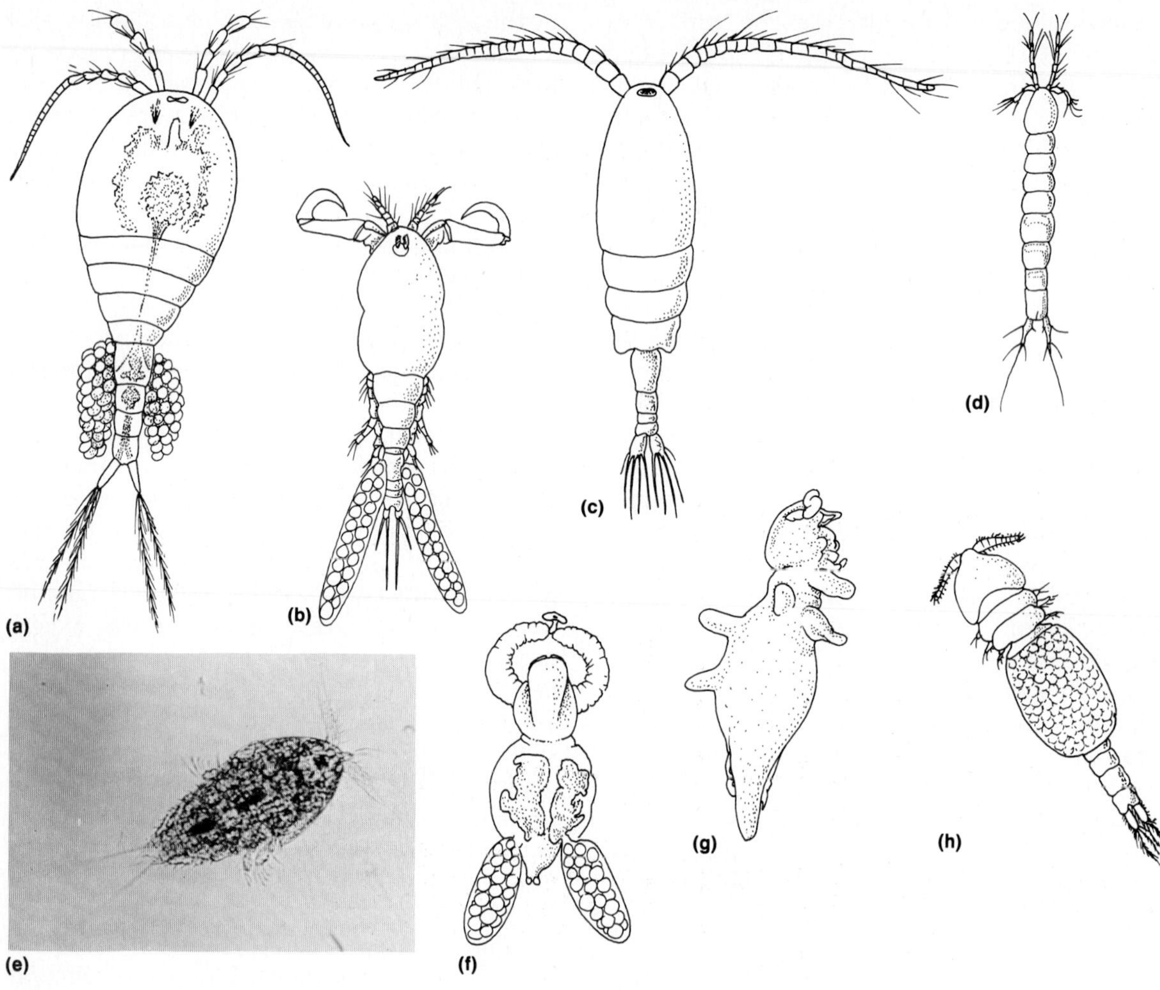

Figure 12.17 *Some types of copepods:* ***(a)*** *female cyclopoid* (Cyclops); ***(b)*** Ergasilus, *a cyclopoid parasitic on the gills of freshwater fishes;* ***(c)*** *female calanoid* (Senecella) *(by permission of Smithsonian Institution Press from "*Senecella calanoides, *a recently described freshwater copepod," Plate 1, Figure 1.* Proceedings of the United States National Museum, Volume 66, Number 2541, *by Chancey Juday. Smithsonian Institution, Washington, D.C., 1925);* ***(d)*** *harpacticoid;* ***(e)*** *photomicrograph of a marine harpacticoid (courtesy of Ward's Natural Science, Inc., Rochester, N.Y.);* ***(f)*** Achtheres, *a parasite on fishes (by permission of Smithsonian Institution Press from "North American parasitic copepods belonging to the Lernaeopodidae, with a revision of the entire family," Figure 86.* Proceedings of the United States National Museum, Volume 47, Number 2063, *by Charles Branch Wilson. Smithsonian Institution, Washington, D.C., 1915);* ***(g)*** Chondracanthus, *a parasitic form, lateral view (from J-S. Ho, 1972,* Proc. Biol. Soc. Washington 85:*528);* ***(h)*** Notodelphoida, *a female, lateral view (by permission of Smithsonian Institution Press from "North American Copepods of the Family Notodelphyidae," Figure 2a.* Proceedings of the United States National Museum, Volume 108, Number 3390, *by Paul L. Illg. Smithsonian Institution, Washington, D.C., 1958). (parts c, e [from Wilson] in R.W. Pennak, 1978,* Fresh-water Invertebrates of the United States, *2nd ed., © 1978, John Wiley & Sons, Inc., used with permission.)*

development of a wormlike body (Fig. 12.17e–h). In ectoparasites, certain appendages terminate in holdfast **claws,** others have a **frontal gland** that becomes a special attachment button, and mouthparts have evolved for piercing and sucking. Endoparasites usually lose their mouthparts and absorb nutrients from the host directly. Often adults are parasitic, while the larval stages are free living.

Subclass Cirripedia

Cirripedes, including the familar barnacles along with some smaller parasitic groups, are different from other crustaceans because as adults they are all sessile and possess a number of unique characteristics associated with their being sedentary. They are exclusively marine, and about 70% of the more than 1000 known species are nonparasitic and live attached to rocks, shells, coral, ships, timber, and many other inanimate surfaces. Others are commensal on whales, turtles, jellyfishes, fishes, and on seaweed. One large group of cirripedes is endoparasitic mostly in decapod crustaceans. Some cirripedes live in deep water, but the group as a whole is typified by being intertidal or occurring in shallow water. Cirripedes are both stalked and sessile, free living and symbiotic, and some of their diversity is illustrated in Figure 12.18. Most cirripedes are small and range in diameter or length from 1 to 4 cm, but there are a few enormously large species; one stalked barnacle *(Lepas anatifera)* may attain a length of 70 cm, and one sessile species is nearly 9 cm in diameter. When not encrusted with lots of other organisms, barnacles are often quite colorful with both solid and striped patterns. Barnacles are of considerable economic importance, since they foul the underwater surfaces of ships and pilings in all seas. Since the 1940s, plastic coatings have been applied to the hulls of most ships, reducing this problem substantially.

(a)

(b)

(c)

Figure 12.18 *Collage of photographs of several types of free-living cirripedes:* ***(a)*** Chthamalus, *a sessile form;* ***(b)*** Mitella, *a pedunculate goose barnacle;* ***(c)*** *closeup of a sessile barnacle, probably* Balanus *(courtesy of C.F. Lytle). (parts a, b courtesy of Ward's Natural Science, Inc., Rochester, N.Y.)*

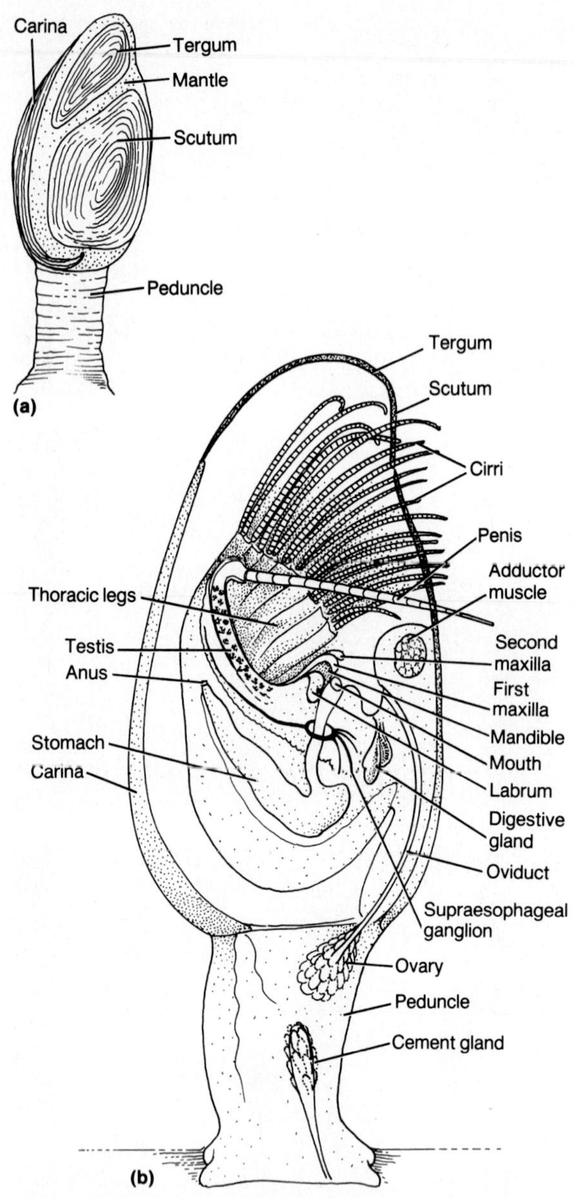

Figure 12.19 *Some features of* Lepas, *a stalked or pedunculate barnacle;* ***(a)*** *valve arrangements seen from the left side of the capitulum;* ***(b)*** *diagrammatic view of the internal features (left carapace and appendages removed).*

Since most cirripedes are typical barnacles, the following discussion of the general characteristics applies to the free-living and commensal representatives (Order Thoracica); the aberrant parasitic forms will be covered later. The chief characteristics of the two principal orders of barnacles appeared in Table 12.2.

The last larval stage of barnacles is called a **cypris** larva, since it looks remarkably like the ostracod *Cypris*. The larva swims for a time, settles on a substratum, and attaches by means of secretions from **cement glands** located near the base of the first antennae. This cement, a polysaccharide, is probably the toughest, strongest cement known. It has been the object of considerable research in the hopes of finding a more perfect adhesive for dental fillings. The larval carapace persists in adult barnacles as a fleshy **mantle,** which encloses the body and is attached only at the animal's dorsal side. External calcareous plates develop in the mantle of most barnacles, giving them the superficial appearance of molluscs; in fact, barnacles were considered to be molluscs until about 1830, when their true affinities with crustaceans were established. The cavity between the mantle and the body is the **mantle cavity.**

Barnacles are divided into the pedunculate (stalked) and the sessile (nonstalked) forms (Figs. 12.18, 12.19, 12.20). In the stalked or goose barnacles there is a long **peduncle** (stalk) attached to the substratum on one end and bearing the major part of the body, the **capitulum,** on the other (Fig. 12.19). The peduncle, representing the greatly enlarged preoral end, is equipped with longitudinal, circular, and oblique muscles, which make it quite movable. The capitulum represents all but the preoral part and is enclosed by the mantle. The external surface of the mantle bears five basic plates: a dorsal, keellike, unpaired **carina,** paired posterior **terga,** and paired anterior **scuta** (Fig. 12.19). The opposing upper margins of the terga and scuta can be pulled together, closing the mantle

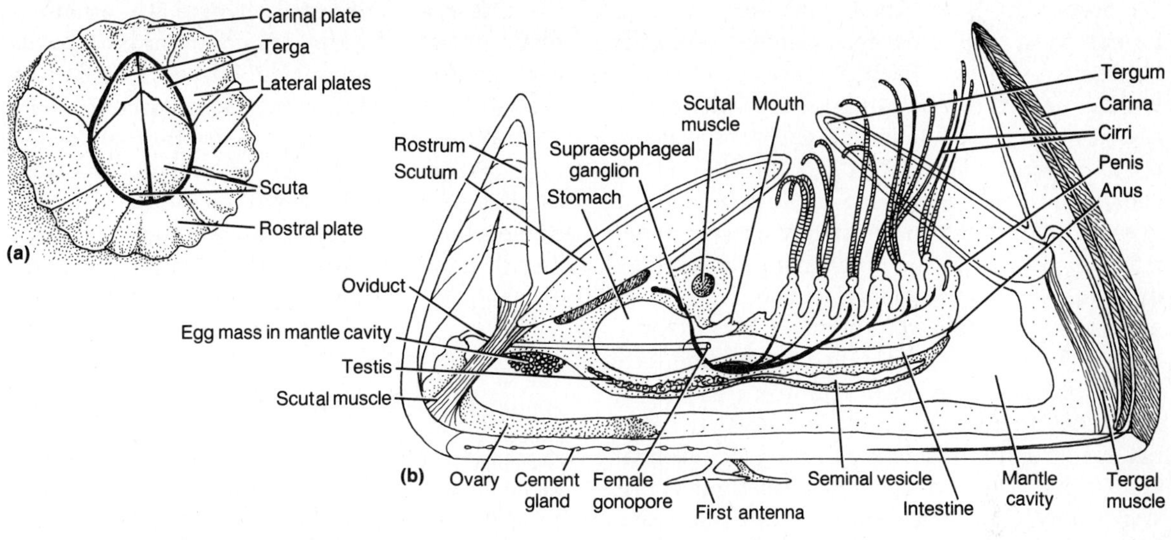

Figure 12.20 *Some features of sessile barnacles:* ***(a)*** *the external plates seen from above;* ***(b)*** *diagrammatic longitudinal view of the internal features (left carapace and appendages removed).*

cavity, or they can be pulled apart, thus enabling the thoracic appendages to be extended.

In sessile barnacles a peduncle is absent, and the preoral region is represented by a flat membranous or calcareous surface permanently attached to the substratum (Fig. 12.20b). The mantle of sessile barnacles also bears calcareous plates that form both the walls and the **operculum** or lid situated on top of the animal. Primitively, sessile barnacles have eight mural (wall) plates, which are the unpaired **rostral** and **carinal plates** and three pairs of lateral plates, the **rostrolateral, lateral,** and **carinolateral plates.** The operculum is composed of the paired **terga** and **scuta** (Fig. 12.20a). The mobility of these four opercular plates allows the thoracopods to be extended and permits water to enter the mantle cavity. In both the stalked and nonstalked barnacles there has been an evolutionary tendency for the plates to be reduced both in number and in size.

The body is composed of a head and a thorax, and the abdomen is absent or is represented only by a pair of **caudal rami.** Evidences of external segmentation are all but absent. The second antennae, present in the larval stage, are absent in adults, and the usual mouth appendages are present. The thorax bears six pairs of long biramous appendages called **cirri** from which the group gets its name (cirrus = filamentous; ped = feet) (Figs. 12.19b, 12.20b). Both rami of each appendage are segmented and equipped with setae. When the mantle cavity is opened, the cirri are unrolled and extend into the surrounding water. The six paired cirri form a basket that sweeps through the water, trapping small particles of animal or plant material. The thoracic legs beat rhythmically on an average of 20–35 times per minute, but several environmental conditions can alter the rate. The food particles are passed to the first maxillae and on to the mouth.

Free-living barnacles are for the most part monoecious, and cross-fertilization is usually between neighbors. Each barnacle possesses a long penis that can be extended into the mantle cavity of an adjacent individual, where it deposits sperm (Figs. 12.19b, 12.20b). Some stalked and sessile barnacles exhibit extreme sexual dimorphism in which tiny dwarf males may be attached to the inner mantle of females or of hermaphrodites. Such dwarfs are usually highly degenerate and, when attached to monoecious individuals, are called **complementary males.** Eggs are brooded within an ovisac located within the mantle cavity, and the eggs hatch into nauplii. The six planktonic naupliar instars are followed by the nonfeeding cypris stage, which is the settling stage. Barnacles exhibit a high degree of aggregrating behavior as they settle, but the underlying causes are not fully understood. Following attachment to some suitable substratum, the mantle infolds, the cirri elongate, the plates develop, and the body undergoes flexion to attain the adult position, which is, in effect, the animal standing on its head with its feet uppermost. Barnacles molt periodically and shed the chitinous cuticle but not the calcareous plates. The plates grow by accretion, new calcareous layers being added continuously to their margins and inner surfaces.

Parasitic cirripedes. Representatives of two orders of cirripedes are parasitic. All parasitic forms have the expected reduction and modification of structures along with adaptations for a parasitic existence. Usually, parasitic species are dioecious and males are smaller than females.

Not only are members of the Order Rhizocephala the largest parasitic taxon, they have worldwide distributions, are structurally bizarre, and are parasitic in other crustaceans. The female cypris larva settles on a crab or other decapod, attaches by its first antennae, and injects a mass of undifferentiated parasitic cells into the crab. The cellular mass finds its way to the intestine of the crab, where its growth is made possible by the development of rhizoids that ramify throughout the host and supply the growing parasite with nourishment (Fig. 12.21a). A **visceral mass** develops that, at the next crab molt, becomes an **external mass** located near the ventral base of the decapod abdomen. The visceral mass contains a large brood chamber to which a male cypris larva can

Figure 12.21 *Two examples of parasitic cirripedes:* ***(a)*** Sacculina, *a rhizocephalid barnacle endoparasitic in decapods (crab is shown as transparent and from the ventral surface; rhizoidal system of the parasitic barnacle is shown only on the crab's left side);* ***(b)*** *photograph of* Peltogaster, *a barnacle parasitic on the abdomen of a hermit crab (courtesy of Ward's Natural Science, Inc., Rochester, N.Y.).*

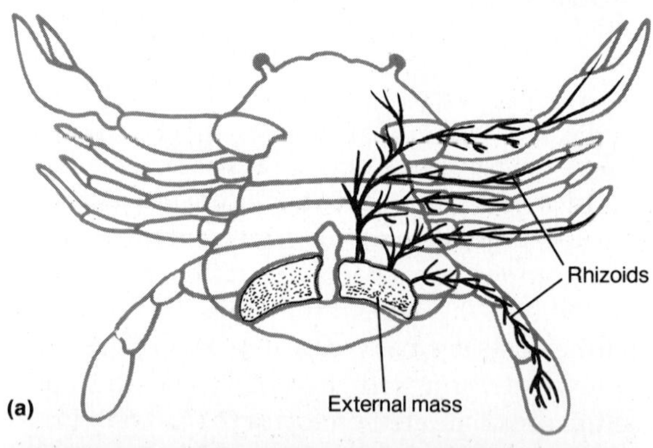

attach. The male discharges a mass of undifferentiated cells, all of which migrate into the brood chamber. Once inside, these cells differentiate into testicular tissue, and the parasitic female becomes a functional hermaphrodite. Self-fertilization follows, and the nauplii exit into the water.

Parasitism by rhizocephalids produces some strange effects within the host. The parasite usually causes castration of the infected crab by retarding testicular development. Associated with gonadal atrophy are some sex changes; in male crabs the abdomen becomes broader as in females, and in females the parasite causes the abdomen of juvenile crabs to develop precociously. Probably, these changes in the host are caused by the parasite digesting away the androgenic gland, which induces maleness in crustaceans. In some cases the host decapod is literally eaten away from the inside, and both host and parasite die, but only after the parasite has reproduced.

Members of the Order Ascothoracica are ectoparasitic or endoparasitic on various echinoderms and soft corals. Their functional morphology is a combination of primitive features and adaptations to a parasitic existence. The first antennae are prehensile, and the second antennae are absent. The mouthparts are adapted for sucking or absorbing host fluids. The six pairs of thoracopods are often reduced or even lost, and there are no abdominal appendages.

Class Ostracoda

Ostracods are found in all types of marine and freshwater habitats. Typically, they are found on the surface of rooted plants, in algal mats, and in the bottom debris and mud. A few species are active swimmers, and others live as commensals on the gills of other crustaceans. Several species are found in the humus of tropical forests. They are small animals typically ranging in size from 1 to 5 mm, but one giant form *(Gigantocypris)* attains a length of up to 25 mm. There are over 5500 described living species, but numbers of fossil species exceed 10,000, dating back to the Cambrian period (see Table 1.1 in Chapter 1). Brief characterizations of the subclasses and principal orders of ostracods appear in Table 12.3.

TABLE 12.3. ☐ **BRIEF CHARACTERIZATIONS OF THE SUBCLASSES AND MAJOR ORDERS OF OSTRACODS**

Taxon	*Habitat*	*Habits*	*Nature of Antennae II*	*Other Features*
Subclass Myodocopa				
Order Myodocopida	Marine, few brackish water	Benthic, meroplanktonic, or pelagic	Biramous	Carapace with antennal notches
Order Halocyprida	Marine	Benthic or pelagic	Biramous	No trunk appendages; no antennal notches
Subclass Podocopa				
Order Podocopida	Marine and freshwater	Benthic	Uniramous	2 pairs of trunk appendages; no antennal notch in carapace
Subclass Palaeocopa	Marine			(Only shells found)

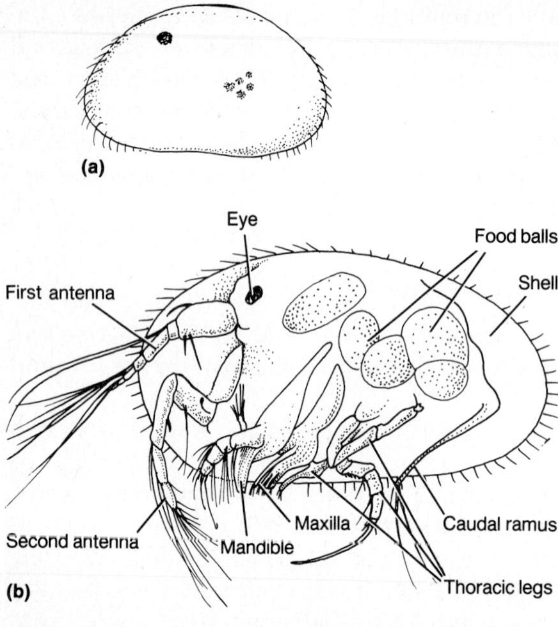

Figure 12.22 *Some features of ostracods:* ***(a)*** *lateral view of the left valve;* ***(b)*** *lateral view of a typical marine ostracod with the left valve and appendages removed (from C.C. Hoff, 1942,* Univ. Ill. Biol. Monogr. *19:177).*

The body is completely enclosed in a bivalved carapace equipped with a dorsal hinge. The valves are oval in shape, giving them a superficial appearance of a seed and the source of their common name, seed shrimp (Fig. 12.22). The valves are impregnated with calcium salts and may be variously equipped with notches, pits, tubercles, and teeth. The body, not visibly segmented, is composed of a head making up about half of the body and a reduced trunk (Fig. 12.22b). The locomotive appendages are the second antennae, and often the antennules are well developed for crawling or swimming. There is some debate about the nature of the pair of appendages immediately behind the first maxillae. Some specialists maintain that the head lacks a second pair of maxillae and thus the next pair are the first trunk appendages. Others believe that these appendages are indeed a second pair of maxillae homologous to the second maxillae of other crustaceans. The latter theory has obvious advantages, since it conforms these appendages to those in other crustaceans. The maxillae, along with the trunk appendages, are cylindrical and leglike and are also used for swimming, crawling, feeding, and grasping. Some species lack trunk appendages altogether, while others have one or two pairs. The trunk terminates in two long **caudal rami** that project anteroventrally (Fig. 12.22b). A naupliar eye is present in all ostracods, and some species also possess a compound eye.

Ostracods are quite diverse in their food preferences as carnivorous, herbivorous, scavenger, and filter-feeding forms are known. Most appendages are partially responsible for stirring up detritus, mud, bacteria, and other small benthic organisms.

Most ostracods are dioecious, and the male uses his paired penes to introduce sperm into the paired seminal receptacles of the female. The fertilized eggs may be released promptly, or they may be brooded for a time by the female. Parthenogenesis is common among freshwater species. Some ostracod eggs that were desiccated experimentally for 20 years were shown to be viable when water was added! Hatching produces a nauplius with a bivalved shell; eight or nine molts are required before the ostracod reaches sexual maturity. Several species are luminescent. The bluish light they produce is from secretions of a gland located in the labrum.

Class Malacostraca

This class contains slightly more than one half of all crustacean species and includes most of the larger, better-known forms such as crabs, shrimp, and crayfishes. They are mostly marine, but many species live in fresh water, and a few are found on land. Even though a high degree of diversity is found among malacostracans, they all have some common features.

The malacostracan body is divided into three tagmata; head, thorax, and abdomen (Fig. 12.23a). The first antennae are usually biramous, and the second antennae often have a scalelike exopod. The mandibles and both pairs of maxillae are each equipped with a palp, and the compound eyes are often located on movable stalks. The anterior eight postcephalic segments constitute the thorax, and the posterior six make up the abdomen. An exception to this basic plan occurs in one small primitive group (Subclass Phyllocarida) in which there are 16 trunk segments of which eight are abdominal. In malacostracans the anterior one to three thoracic segments are fused with the head, and their appendages are maxillipeds. The remaining thoracic appendages are generally walking legs or pereiopods, which usually are secondarily uniramous, since they have lost their exopods. The anterior pair (or several pairs) of pereiopods is chelate, and each is called a **chela** or **cheliped.** Also, each pereiopod characteristically bears one or more modified epipods or gills. A carapace is usually present and often covers the entire head and thorax.

The abdomen is flexed ventrally to some degree and bears six pairs of appendages (Fig. 12.23a). The first five pairs are pleopods and are used primarily in swimming. Pleopods sometimes function in gaseous exchange, and eggs may be brooded on them by females. In males the first one or two pairs of pleopods are used as copulatory organs. The last abdominal segment bears the paired uropods, each of which is composed of two narrow or broad, flat or cylindrical rami. The uropods, along with the telson, may make a large tail fan (Fig. 12.23a). Gonopores are always located at the bases of the sixth thoracic appendages in females and those of the male are on the eighth thoracic segment. There may be a nauplius stage, but most often this larval stage is passed within the egg.

Considerable diversity exists among the various malacostracan groups; they have modified various facets of the generalized pattern

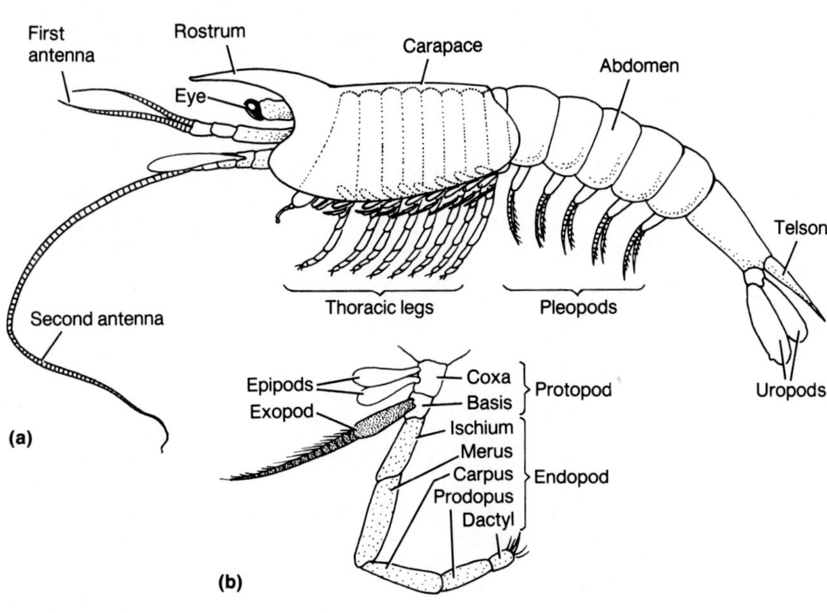

Figure 12.23 *Generalized malacostracan: **(a)** lateral view; **(b)** thoracic appendage (parts a, b from Calman).*

TABLE 12.4. □ **BRIEF CHARACTERIZATIONS OF THE PRINCIPAL TAXA OF MALACOSTRACANS**

Taxon	*Habitat*	*Carapace*	*Food-gathering Mechanisms*	*Other Features*
Subclass Phyllocarida	Marine, benthic	Bivalved, unhinged	Maxillary filtration	Abdomen of 8 segments
Subclass Hoplocarida	Marine, benthic	Shieldlike, over head and anterior thorax	Subchelate chelae	Body flattened dorsoventrally; triramous antennae I
Subclass Eumalacostraca				
Superorder Syncarida	Freshwater, benthic	Absent	Filtering thoracopods and maxillae	Biramous thoracopods
Superorder Pancarida	Brackish water	Present	Filtering maxillae	Small; disjunctive distributions
Superorder Peracarida				
Order Mysidacea	Mostly marine	Free from last 4 thoracic segments	Filtering by various limbs	Rostrum covers movable eyes
Order Cumacea	Marine, benthic burrowers	Forms lateral branchial chambers	Filtering by mouthparts	Eyes are sessile; male antennae very long
Order Amphipoda	Mostly marine, benthic	Absent	Gnathopods	Body laterally compressed; have thoracic gills
Order Isopoda	Mostly marine, benthic	Absent	Filtering by mouthparts	Body dorsoventrally flattened; have abdominal gills
Order Tanaidacea	Mostly marine, benthic	Reduced, covers head	Filtering mouthparts; chelate thoracopods II	Flattened dorsoventrally; thoracopods I are maxillae; many are eyeless
Superorder Eucarida				
Order Euphausiacea	Marine, pelagic	Does not cover gills completely	Filtering by anterior 6 pairs of thoracic limbs	Luminesce by photophores
Order Decapoda	Mostly marine	Large, covers gills completely	Chelipeds	Scaphognathite; many gills; movable eyes

just discussed. A discussion of each of the main taxa follows, the Table 12.4 gives a brief synopsis of the important features of the principal malacostracan taxa.

SUBCLASS PHYLLOCARIDA

This group of fewer than 20 species has a rather cosmopolitan marine distribution. The phyllocarids are found in littoral benthic mud mostly amid seaweed, range in length from 4 to 12 mm, and are considered to be the most primitive of the malacostracans. The fossil record indicates this group had a great many representatives in the Proterozoic Era. The thorax and part of the abdomen are enclosed by an unhinged bivalved carapace, and a small rostrum covers the head (Fig. 12.24a). The eight pairs of similar thoracic appendages create water currents that bring small particles of food ventrally, where maxillary filtration occurs. Thoracic epipods are flattened lamellar gills. The abdomen consists of eight rather than six somites; the first six bear pleopods, but the posterior two segments are very much reduced and lack appendages. The eighth abdominal segment is the telson; unlike that in other malacostracans, it represents a true segment. Brooded on the thoracic legs, eggs hatch into postlarvae.

SUBCLASS HOPLOCARIDA

The hoplocarids, often called the stomatopods or mantid shrimp, comprise about 350 species. They range in length from 1 to 36 cm; a majority are brightly colored with striped or solid patterns of red, green or blue; and they are all marine and live in coral crevices, rock fissures, or burrows in bottom mud.

The body is elongated and flattened dorsoventrally. A shieldlike carapace covers the anterior two thoracic segments and most of the head except for the eyes and first antennae, which are located beneath a short rostrum (Fig. 12.25b). Each first antenna has three rami, is larger than the second antennae, and is equipped with a prominent **scale.** The mandibles and both pairs of maxillae are typical of those of other crustaceans. The eyes are large and on movable stalks.

Figure 12.24 *Some primitive malacostracans: (a)* Nebalia, *a phyllocarid; (b)* Anaspides, *a syncarid (parts a [after Claus], b in Calman).*

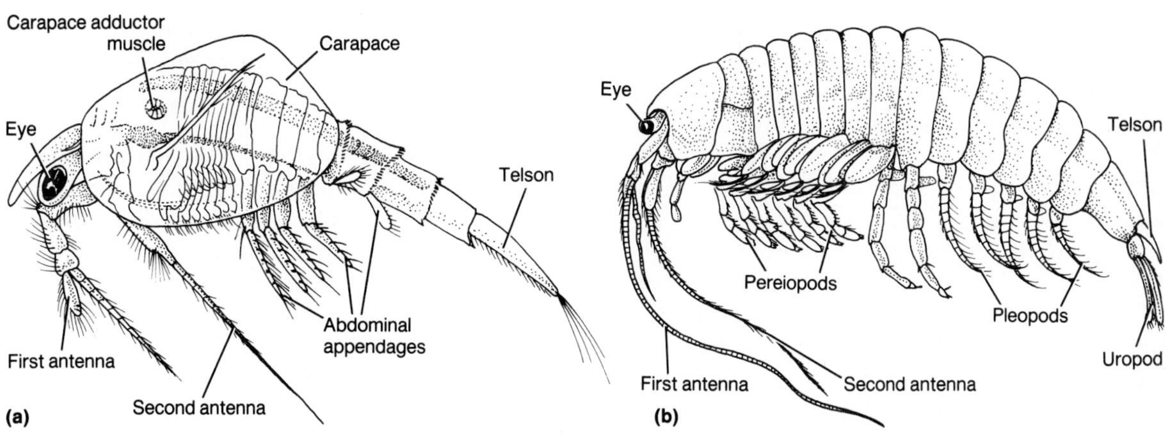

(a)

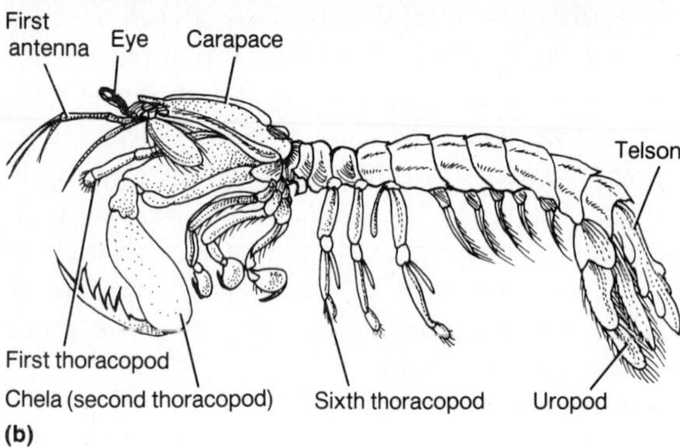

(b)

Figure 12.25 Squilla, *a typical stomatopod:* ***(a)*** *photograph of an individual, dorsal view;* ***(b)*** *lateral view of a male (from Calman).*

All thoracic limbs are uniramous, and the first five pairs are distinctively subchelate, that is, the dactyl folds back into the prodopus. The first pair of thoracic limbs is slender and used for cleaning, but the second thoracopods or **chelae** are enormously enlarged and used in raptorial feeding. The inner surface of the dactyl of the chela bears long sharp spines that impale the victim when the dactyl is flexed back into a deep groove in the prodopus (Fig. 12.25b). Chelae can be extended rapidly to capture or stun fishes, molluscs, crustaceans, and other invertebrates. Some stomatopods actively hunt for food; others lie in wait for prey to pass near the burrow's aperture. Thoracic segments III and IV are very much reduced in size, even though they bear appendages. Thoracopods III–V are used as accessory feeding limbs, while legs VI–VII are nonchelate and function in walking or crawling.

The abdomen has five pairs of well-developed, biramous pleopods, and each limb bears an epipodous filamentous gill. The uropods and telson form a large tail fan (Fig. 12.25a).

In males a pair of long slender penes, located at the base of the last pair of thoracic legs, transfers sperm to the female. As fertilized eggs are discharged, they often agglutinate to form an egg mass sometimes up to 5 cm in diameter and containing up to 50,000 eggs. The female broods the egg mass amid her anterior thoracic legs until hatching. A planktonic zoea stage is the hatching stage, and many molts ensue until the adult stage is attained.

Subclass Eumalacostraca

This subclass includes all the remaining malacostracans. They all have an abdomen of six metameres, each of which bears limbs; they almost never use the thoracopods as straining devices; they have biramous first antennae; the carapace encloses the entire thorax; and eyes are on movable stalks. Four superorders are recognized (see Table 12.4).

Superorder Syncarida. The syncarids represent about 120 species of primitive freshwater crustaceans. They are benthic and are found mostly in caves, springs, and mountain lakes. Their primitive features are their small size (1–50 mm), absence of a carapace, unspecialized thoracic limbs with exopods, absence of chelate legs, and similar thoracic and abdominal legs (Fig. 12.24b). Eggs are not brooded, and there is no free-swimming larval stage.

Superorder Pancarida. This is a very small group of small crustaceans found in brackish water in Texas, Italy, Israel, the West Indies, and Yugoslavia. A dorsal brood pouch is present beneath the carapace. There is no external division between the thorax and abdomen, but there are two pairs of abdominal appendages.

Superorder Peracarida. This large superorder contains small shrimplike crustaceans including the sow bugs and sideswimmers. A carapace may be present, but in most species it is absent. The first thoracic segment is always fused to the head and bears the maxillipeds. The last four thoracic segments are distinct and not fused to the carapace. Characteristic of many females of this group is the presence of a ventral brood pouch or **marsupium** in which eggs are maintained until they hatch. Flat plates or **oostegites** extend medially from the thoracic coxae and meet in the ventral midline to form the floor; the thoracic sternites form the roof of the marsupium. Development of the oostegites is apparently controlled by hormones, and the oostegites do not develop completely until sexual maturity is achieved. Upon release from the brood chamber the planktonic postlarvae undergo numerous molts until the final instar is reached.

Species diversity in this superorder is very high in two orders (Isopoda and Amphipoda) and moderate in the other three (Mysidacea, Cumacea, and Tanaidacea).

Order Mysidacea. This group of almost 800 species contains members that are mostly marine. Mysids range in length up to 35 cm, but most are from 1.5 to 3 cm long. They are often found in large swarms or schools and are important in the diet of many fishes. Most mysids prefer littoral habitats; other forms migrate shoreward only at night.

A carapace covers the stalked movable eyes and the thorax but is free from the last four thoracic segments (Fig. 12.26a). Both pairs of antennae are long. The first pair of thoracic appendages (and occasionally the second pair also) are the biramous, slender, setose maxillipeds. All remaining thoracic legs are similar and are used for walking, crawling, and swimming. The mysids are commonly known as opossum shrimp because of the prominent ventral **marsupium** located at the posterior thoracic region. The five pairs of pleopods may be reduced in females, and the fourth pair is often elongated in males.

Most mysids filter food particles as they swim. Swimming, ventilating, and feeding currents may be carried out by different appendages and may be independent of each other.

Figure 12.26 *Some representative peracarids:* ***(a)*** *a freshwater mysid (from R.W. Pennak,* Fresh-water Invertebrates of the United States, *2nd ed., © 1978, John Wiley & Sons, used with permission);* ***(b)*** *photograph of a cumacean (courtesy of A.B. McCrary);* ***(c)*** *a female* Apseudes, *a tanaidacean (from Sars in Calman).*

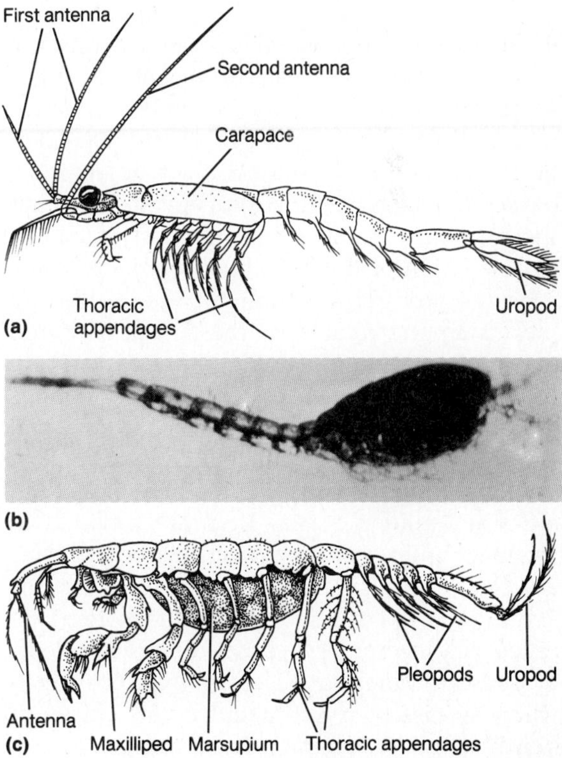

Some mysids may be scavengers, and representatives of one small group are predaceous.

Order Cumacea. All cumaceans live tail first in burrows or tubes in the bottom mud and sand of marine habitats. Although some are deep-water forms, most are numerous in littoral zones. They are small and range in length from 1 to 35 mm, and about 800 species have been described.

The body shape is quite distinctive with a much enlarged head and thorax and a narrow slender abdomen (Fig. 12.26b). The carapace, fused to the first three or four thoracic segments, overhangs each side to form a branchial chamber containing the gills. Two unusual extensions of the carapace form a ventroanterior **rostrum.** When present, the eyes are sessile and often fused into a single median eye. Antennae in males are very long and often folded back along the sides, but in females they are vestigial.

The epipods of the three pairs of maxillipeds bear filamentous gills located within the branchial chamber. Most cumaceans are filter-feeders; they strain water entering the branchial chamber. An incurrent water stream is created by the action of the anterior thoracic epipods, passes over the mouthparts, upward through the branchial chamber, and then is forced out by the action of the maxilliped epipods. Some species scrape detritus from sand grains that are picked up and rotated by the first and second maxillipeds. The remaining thoracic legs are used for burrowing or for swimming. Pleopods are absent in females, but males have several pairs.

Swarming is common at the time of mating, especially among males. Eggs are brooded in the marsupium until they hatch as postlarvae.

Order Amphipoda. This order contains about 6000 species of crustaceans commonly called side swimmers, scuds, and beach fleas. Most are marine, but there are over 600 freshwater species of which about 60 are found in North America. A small number are semiterrestrial, a few species are even found in terrestrial habitats such as in humus of tropical forests, and a few others are ectoparasitic on fishes. They range from 5 to 20 mm in length, and one species is reported to be 28 cm long. They tend to be either semitransparent or brownish in color, but some are red, green, or bluish.

Amphipods and isopods share a number of similar features including sessile eyes, absence of a carapace, fusion of the first (and frequently the second) thoracic segment to the head, first thoracic appendages being maxillipeds, and uniramous thoracopods; the thorax and abdomen are not sharply delimited from each other and are about the same size. However, two amphipod features distinguish them from isopods: the amphipod body is laterally compressed rather than dorsoventrally flattened, and the gills are thoracic rather than abdominal.

The amphipod head bears a pair of unstalked eyes (although a few are blind) and the usual head appendages. Both pairs of antennae are well developed and uniramous, having lost their exopods (Fig. 12.27). The second and third maxillipeds are chelate or subchelate, are called **gnathopods,** and are used in feeding and in prehension. Most amphipods are scavengers and feed on organismic remains. Filter-feeding, predation on smaller animals, and even parasitism are other methods of food gathering. Except for the first pair of gnathopods, the remaining thoracic appendages bear two to six pairs of simple lamellar gills. The anterior three pairs of abdominal pleopods are used in swimming and in creating a current of water that is pulled posteriorly over the gills, and the posterior three pairs of pleopods are uropodlike and are used in various types of locomotion.

Amphipods generally are bottom-dwellers, even though most can swim. The thoracic legs and the posterior abdominal uropods are utilized for walking and crawling. Some, like the caprellids, even use the gnathopods as grasping devices to climb on vegetation (Fig. 12.27c). They burrow in the benthic debris and may create temporary burrows or permanent tunnels and tubes. Cementing sand grains, shell frag-

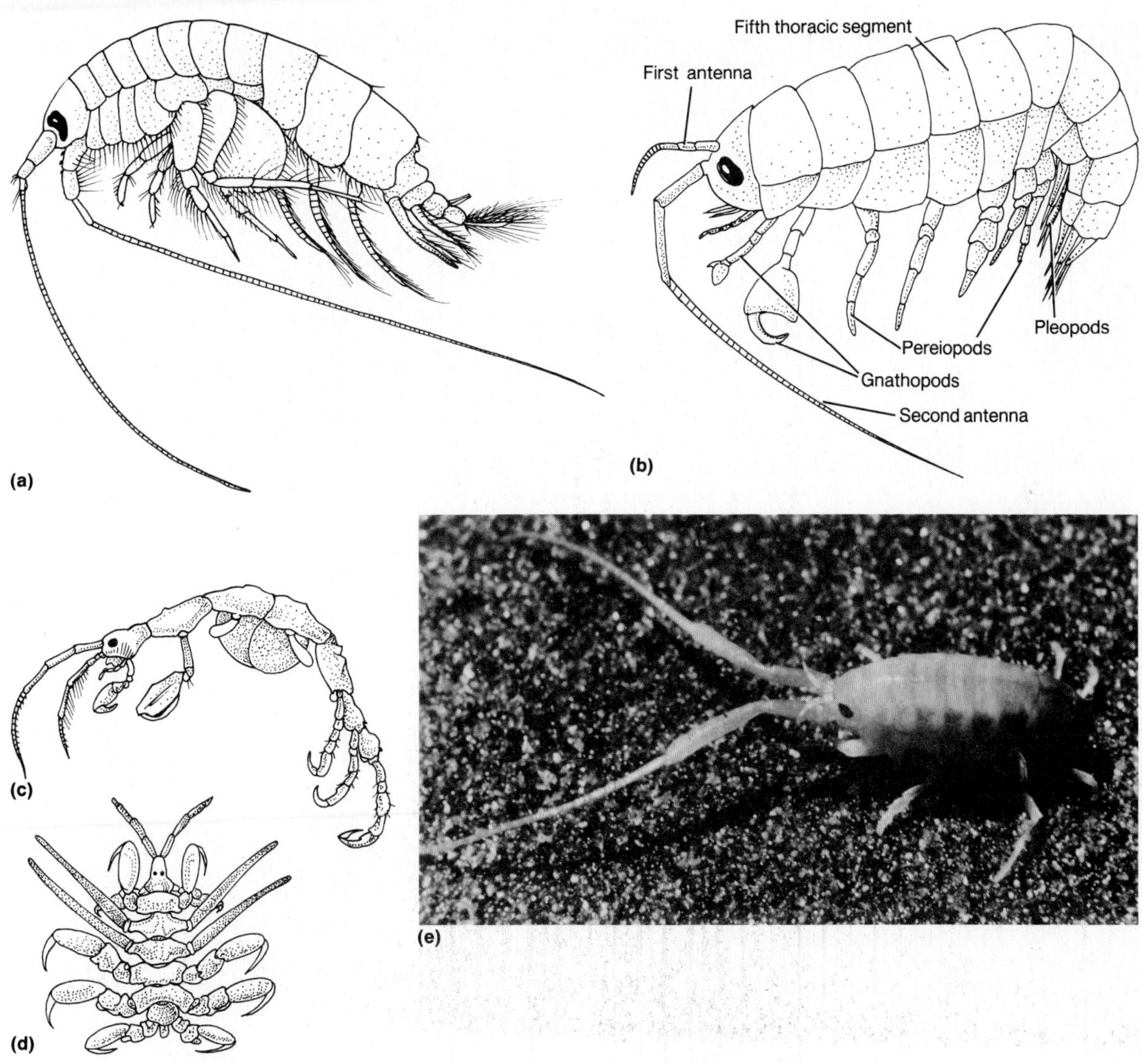

Figure 12.27 *Amphipods: **(a)** a male* Pontoporeia, *a freshwater form illustrating basic amphipodan features (from S.G. Segerstrale, 1937,* Societas Scientiarum Fennica 7: *Taf I, adapted with permission); **(b)*** Orchestia, *a marine beach flea; **(c)*** Caprella, *a marine skeleton shrimp (after Sars); **(d)*** Cyamus, *the whale louse (from Calman); **(e)** photograph of* Orchestoidea, *a beach flea (courtesy of Ward's Natural Science, Inc., Rochester, N.Y.).*

ments, and mud together, some amphipods construct rather elaborate tubes or cases, in a few species the cases are carried along with the animal. Most amphipods, being compressed, move along on one side of their body by a great deal of flexing and squirming. One group, the beach fleas (Fig. 12.27b, e), can hop vigorously for great distances. By suddenly extending the abdomen and telson against the sand, some can leap a distance of up to 50 times their length.

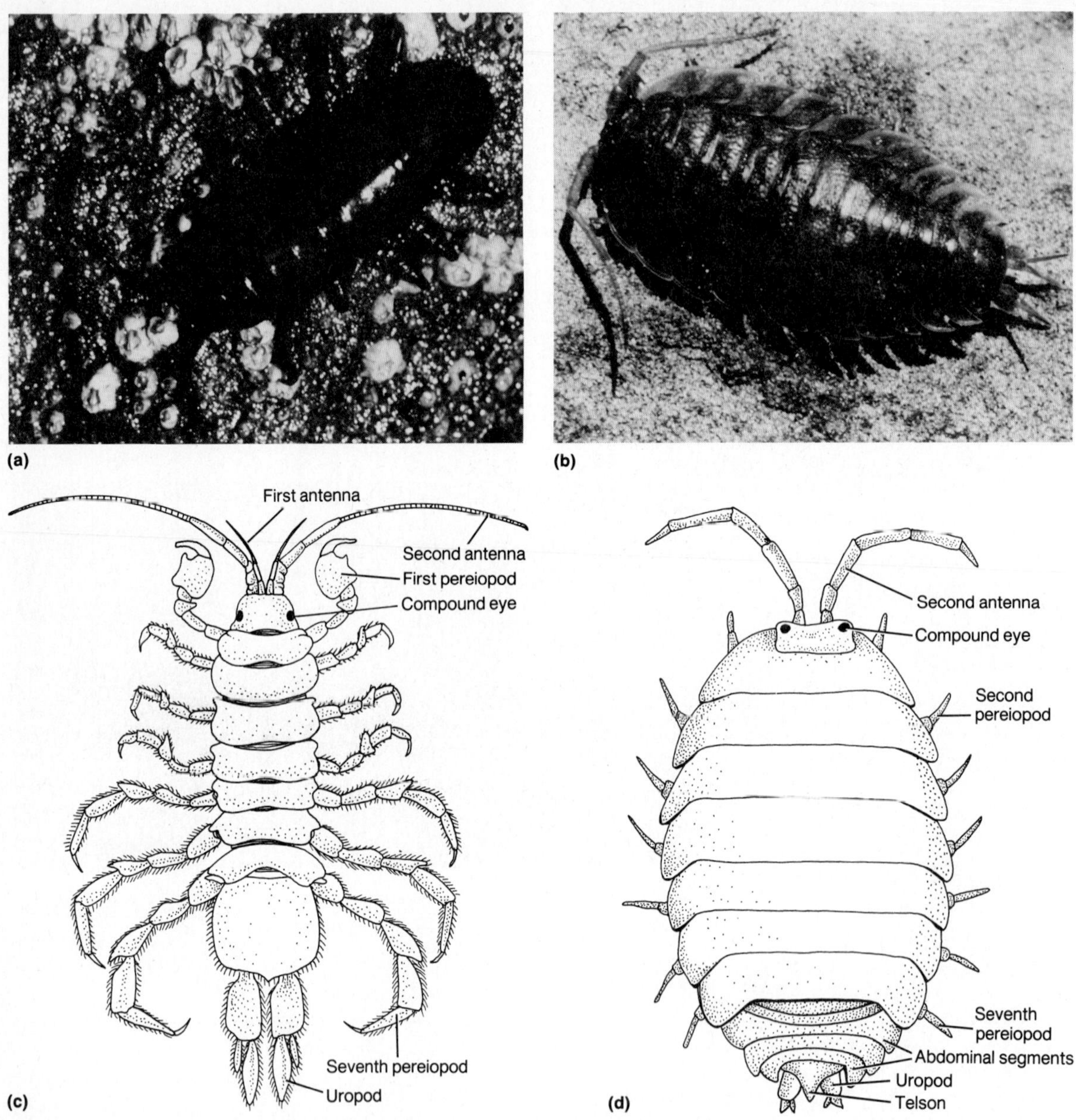

Figure 12.28 *Isopods:* ***(a)*** *photograph of* Idothea, *a marine form;* ***(b)*** *photograph of* Ligia, *a marine form;* ***(c)*** Asellus, *a common, freshwater form (from R.W. Pennak,* Fresh-water Invertebrates of the United States, *2nd ed., © 1978, John Wiley & Sons, Inc., used with permission);* ***(d)*** Oniscus, *a pill bug, common in terrestrial habitats. (parts a, b courtesy of Ward's Natural Science, Inc., Rochester, N.Y.)*

Sperm transfer is accomplished by the twisting of the male body so that his uropods are in contact with the female marsupium. Sperm are swept into the marsupium by water currents; and once ova are discharged into the brood chamber, they are fertilized. Amphipods develop directly without a larval stage. One to several broods are produced each year with up to several hundred eggs per brood.

Order Isopoda. This is a large order of about 4000 species of which a majority are marine, but a large number of species lives in fresh water, and one rather large group is completely terrestrial. Although some are parasitic, the vast majority are free living and are benthic burrowers. Marine forms most often are found in littoral zones, but some have been collected from abyssal and hadal regions. They scurry over wharf pilings, rocks, jetties, and shell beds either above tide levels or in shallow water. Freshwater isopods are found under rocks and in the benthic debris in streams and ponds. Terrestrial forms usually live in humid regions, normally under rocks, bark, and logs. Isopods range in length from 1 mm to over 40 cm, but most are from 5 to 20 mm.

The body of most isopods is distinctively flattened dorsoventrally (Fig. 12.28). Body segments generally are distinct, but there is a tendency for abdominal segments to be fused. Eyes are sessile or even absent in some. When viewed from the dorsal side, most isopods are ovate with the shapes of the cephalic, thoracic, and abdominal regions blending into each other.

The first antennae are short and uniramous, and in terrestrial forms they are highly reduced. The second antennae are large and often uniramous (Fig. 12.28). The two maxillipeds are often hooked together by specialized setae. The remaining seven pairs of thoracopods are uniramous and are used for walking or crawling. The five pairs of broad biramous pleopods are used in swimming and for gaseous exchange. These abdominal gills are typically protected by some type of **operculum** representing the modified first pair of pleopods or the uropods. The last abdominal segment, bearing the uropods, is almost always fused to the telson.

Isopods exhibit a variety of means of locomotion. Basically, they are adapted for crawling, but most of the supralittoral forms can run rapidly over rocks. Aquatic isopods can swim and also burrow in bottom debris. Isopods are either herbivorous, scavengers, or omnivorous but are never filter-feeders. Some groups of isopods are ectoparasites on fishes or other crustaceans, and parasitic adaptations include modifications of appendages, body shape, and life cycles.

Land isopods have evolved a number of features that have enabled them to be truly terrestrial. Some adaptations reduce water loss, but these species are not nearly as well adapted in this regard as are other terrestrial arthropods. For example, they lack a waxy cuticle that would retard water loss, and they still rely upon gills for gas exchange. Their usual location in moist, humid places, their being nocturnal, and the ability of some terrestrial isopods to roll up in a ball (hence the name "pill bug") are other water-conserving mechanisms. The abdominal gills must be kept moist by water derived from food or from the damp soil in which they live. Some terrestrial forms have developed small ventral invaginations that function like primitive lungs, a feature that enables these isopods to survive in much drier habitats than would otherwise be possible.

In copulation the male employs pleopods I and II to transfer sperm to each of the paired female gonopores, and fertilization takes place in the oviducts. Eggs are brooded in the marsupium until hatching, and the young hatch as postlarvae.

Order Tanaidacea. About 500 of these small, mostly marine species have been described. They rarely exceed 2 mm in length and are found buried in bottom mud or coral rubble, but numerous deep-sea species are known. They are morphologically intermediate between cumaceans and isopods. The body is elongated and flattened dorsoventrally. A reduced carapace, the inner surface of which functions in gaseous exchange, covers most of the head and is fused to the first two thoracic segments (see Fig. 12.26c). Many forms are blind, and the remainder have nonmovable eyes located on short stalks. The first pair of thoracopods is the maxillipeds, and the second pair is uniramous, large, and chelate. Thoracopods III are biramous and used for burrowing, and the remaining five pairs are uniramous and employed in swimming. Pleopods are often reduced or absent, especially in females of some species.

Tanaidaceans are filter-feeders, but some forms are raptorial. Most species are dioecious, eggs are brooded in the marsupium, and they hatch as postlarvae.

Superorder Eucarida. This taxon includes the most advanced crustaceans. The carapace is prominent and fused with all thoracic segments, and the eyes are large, stalked, and movable. They do not have a brood chamber, development is indirect, and there is usually a zoea stage. Of the three orders of eucarids, one contains only a single species, another (Euphausiacea) is rather small, and the third (Decapoda) has the greatest species diversity of all crustacean orders.

Order Euphausiacea. The euphausiids represent about 100 species of exclusively marine, shrimplike crustaceans. Found from the surface to depths of more than 3500 m, they are pelagic and often live in huge swarms that may cover several hundred hectares and in which densities near the surface may exceed 60,000 individuals per cubic meter. They range in length from 2 to 6 cm and are either transparent or bright red in color. They often are called "krill" and constitute the main component in the diet of several species of baleen whales including blue whales. Most euphausiids undergo dramatic diurnal migrations in which they move from depths as great as 1000 m in the daytime to near the surface at night. Most are luminescent, and the bluish-green light is probably useful for forming aggregations and for reproduction. Their light-producing abdominal organs are called **photophores** (Fig. 12.29); light is produced intracellularly in the photophore cells.

The sides of the carapace do not extend far enough ventrally to cover the gills completely (Fig. 12.29). All thoracic appendages are biramous and similar, and the first pair is not specialized as maxillipeds. Since most euphausiids are filter-feeders, the first six pairs of thoracic legs are modified somewhat as a filtering mech-

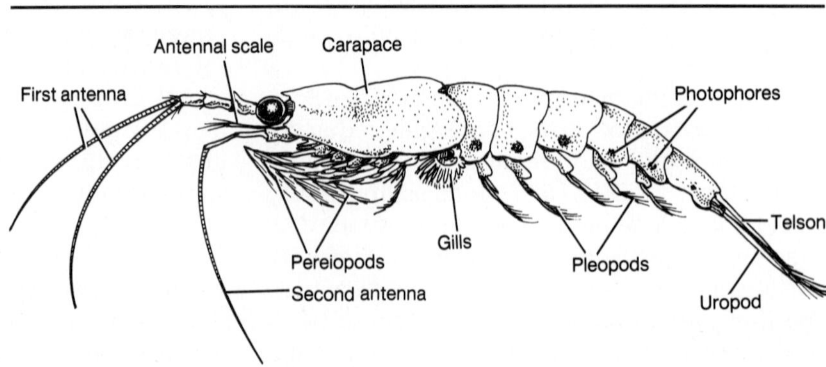

Figure 12.29 Meganyctiphanes, *a representative euphausiid (from Calman).*

anism. Large filamentous gills arise from all thoracic legs. The well-developed abdomen bears five pairs of setose swimming pleopods, and the telson is equipped with a pair of movable spines.

Spermatophores are transferred to the female, and after fertilization the eggs are either released or retained temporarily beneath the body of the female. A nauplius is the hatching stage; it is succeeded by a number of other stages, which are given specific names (calyptopis, furcilia, and micrytopia).

Order Decapoda. This order contains most of the familiar crustacean species such as crabs, lobsters, crayfishes, and shrimp. It is a heterogeneous group in which high degrees of specialization are evident (Fig. 12.30). To date, about 10,000 species have been described of which the vast preponderance are marine, some are adapted to fresh water, and a few are either terrestrial or amphibious. While some commensal crabs are less than 1 cm in diameter, most decapods range from 3 to 25 cm in length. Most decapods are free living and benthic, but some are commensal, and others are pelagic. Most decapods live in shallow littoral zones, although some have been captured from abyssal depths.

Decapods can conveniently be subdivided into two suborders: Dendrobranchiata and Pleocyemata. The dendrobranchiates include the swimming shrimp possessing three principal adaptations that aid their pelagic existence. First, the abdominal pleopods are large, setose, and efficient swimming limbs. Second, the body is usually laterally compressed and streamlined. Third, each second antenna has a large platelike **scale** that is used as a rudder. Dendrobranchiates are also characterized by having a well-developed abdomen, feathery gills, and thoracic legs that are slender and mostly uniramous. The first two or three pairs of pereiopods are chelate. The pleocyemates are typically benthic, even though some do swim, but in doing so they do not use their pleopods. The body is either dorsoventrally flattened or cylindrical in shape, thoracic legs are always uniramous and more massive than in dendrobranchiates, anterior pereiopods are usually powerful strong chelipeds, and young are always brooded on the pleopods of females. A brief characterization of the principal decapod groups appears in Box 12.1.

External Features. In decapods there is a conspicuous cephalothoracic region covered by a prominent carapace that extends laterally and ventrally to enclose the gills completely, thus forming a **branchial chamber** on each side. Each of the 19 segments bears a pair of appendages (Figs. 12.3, 12.31). The paired biramous first antennae and stalked compound eyes are located on the first segment. Each of the second antennae typically has a very long endopod, the **flagellum,** and a scalelike exopod that aids in swimming. At the base of each of the second antennae is the **excretory pore** for the antennal or excretory gland. Both pairs of maxillae and the first maxillipeds are broadly lobate. Each second maxilla bears a broad, flaplike **scaphognathite** or water bailer (see Fig. 12.3e) whose motion pulls a water current anteriorly over the gills. The second and third maxillipeds are either broad or slender, and they may be modified as a cover or operculum over the other mouthparts. There are five pairs of pereiopods, from which this group derives its ordinal name (decapod = 10 feet). The first pair, the **chelipeds,** is usually much more developed than the remaining four (Figs. 12.3h, i, 12.31). Pereiopods II and III may also be chelate. The pereiopods in most decapods contain seven podomeres, but fusion of adjacent podomeres sometimes reduces this number. Pleopods in females of most decapods are used for brooding the young; in males the first one or two pairs are used in copulation. Uropods may be well developed and, with the telson, form the tail fan, or they may be reduced or even absent. When the tail fan is suddenly flexed ventrally, it is especially

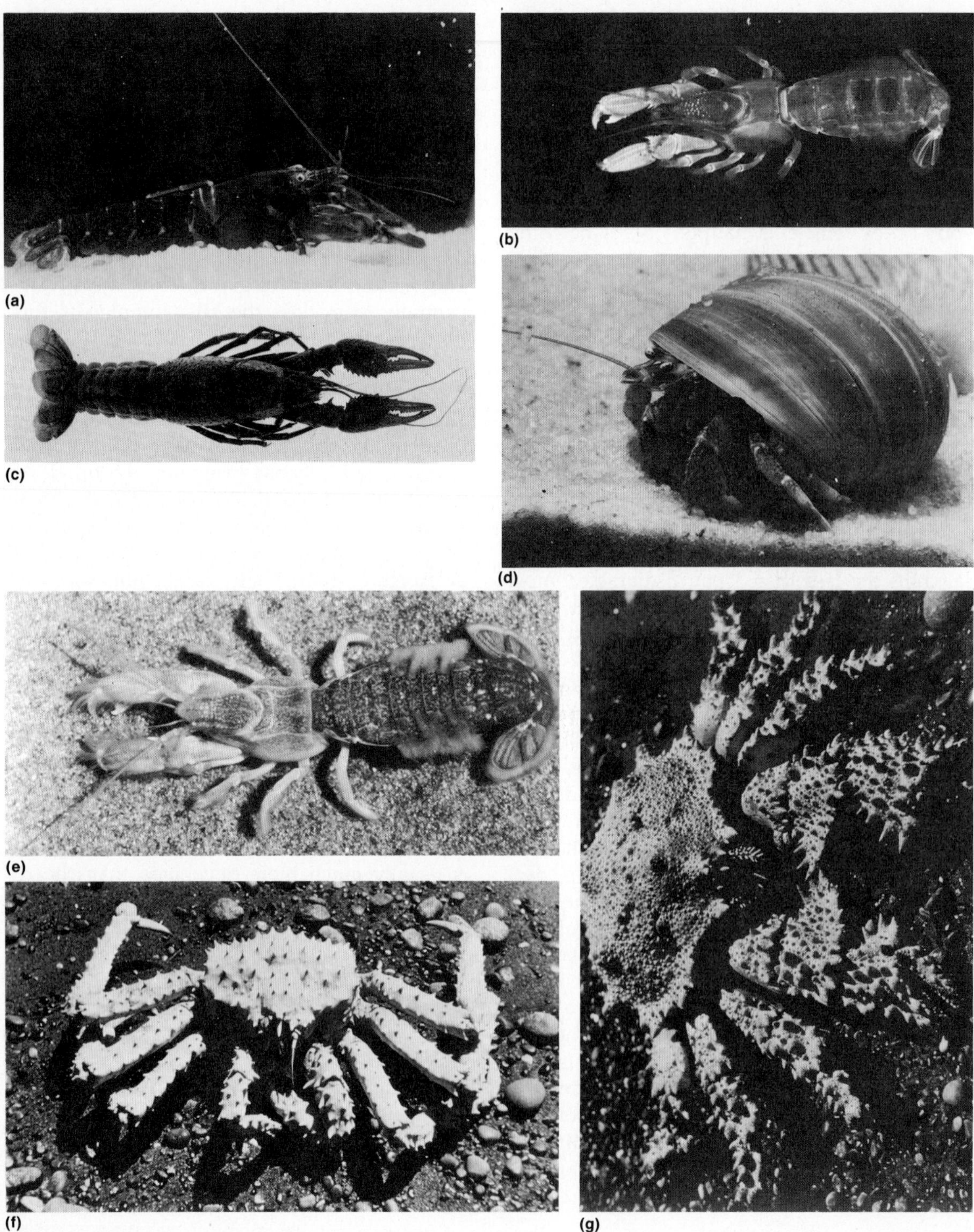
(a)
(b)
(c)
(d)
(e)
(f)
(g)

▲
◀ ***Figure 12.30*** *Photographic collage of some decapods to illustrate diversity:* ***(a)*** Penaeus, *an edible shrimp (dendrobranchiatan);* ***(b)*** Alpheus, *a snapping shrimp (caridean);* ***(c)*** Cambarus, *a crayfish (astacidean);* ***(d)*** *a hermit crab, probably* Pagurus *(anomuran) in a* Polinices *shell;* ***(e)*** Upogebia, *the blue mud shrimp (thalassinidean);* ***(f)*** Lithodes, *a spider crab (brachyuran);* ***(g)*** Lopholithodes, *a box crab (brachyuran);* ***(h)*** Callinectes, *the blue crab (brachyuran);* ***(i)*** Menippe, *a stone crab (brachyuran);* ***(j)*** Uca, *a fiddler crab (brachyuran);* ***(k)*** *a deep-sea crab (brachyuran) (courtesy of B. Best). (parts e–g courtesy of Ward's Natural Science, Inc., Rochester, N.Y.)*

BOX 12.1 □ AN ANALYSIS OF THE MAJOR TYPES OF DECAPOD CRUSTACEANS

There are lots of different types of decapod crustaceans, and the reader may wish a further breakdown of the types frequently encountered. Mention has been made that there are two suborders of decapods: Dendrobranchiata, which includes many laterally compressed shrimp, all with dendrobranchiate gills, and Pleocyemata, which includes all other decapods. Suborder Pleocyemata is composed of six subordinate infraorders, the most important of which are briefly characterized below.

Carideans also include many shrimp in which the first three pairs of legs are chelate, the cephalothorax is more or less cylindrical, and the pleuron of abdominal segment II overlaps those of segments I and III. This group includes the sand, snapping, and cleaning shrimp and many freshwater genera (Fig. 12.30b).

Astacideans are all the crayfishes and those lobsters with large claws; the cephalothorax is more or less cylindrical, the first three pairs of pereiopods are chelate, the first pair is greatly enlarged, and there is a prominent tail fan (Fig. 12.30c).

Palinurans have a cylindrical cephalothorax, the abdomen is well developed, and legs may be chelate but none is enlarged. To this group belong certain lobsters including the spiny and Spanish lobsters.

Anomurans have a dorsoventrally flattened cephalothorax, the third pair of pereiopods is never chelate, the fifth pereiopods are reduced or turned upward, and the abdomen is often asymmetrical. This taxon includes burrowing shrimp and hermit, porcelain, king, and sand or mole crabs (Fig. 12.30d).

The brachyurans include all the true crabs that have a broad flattened carapace, first pereiopods that are strong and enlarged chelipeds, third pereiopods that are never chelate, and an abdomen that is reduced and flexed beneath the cephalothorax. There is considerably diversity among crabs as exemplified by their common names, which include box, decorator, spider, cancer, swimming, mud, stone, freshwater, ghost, fiddler, soldier, and land crabs (Fig. 12.30f–k).

effective in aiding the animal to dart backward abruptly. In hermit crabs the uropods are used to hook around the columella of the gastropod shell that the crab occupies. In crabs the abdomen is highly reduced and tucked ventrally beneath the large thorax (Fig. 12.31b).

Decapods generally are predaceous, and prey is captured by the chelipeds and passed to the third maxillipeds. From there the food is pushed between the other mouthparts and finally crushed or parts bitten off by the mandibles. Some decapods are herbivorous, and others are scavengers, eating dead organic remains. In other decapods, filter-feeding is an important means of acquiring food with the maxillipeds usually functioning as a strainer. The mole crab, *Emerita*, buries itself backward in the sand in the surf zone but permits its densely setose second antennae to project above the sand, forming a familiar V shape. As a wave recedes, the antennae filter particulate material out of the water, and this food is conveyed to the mouth as the antennae are rolled up.

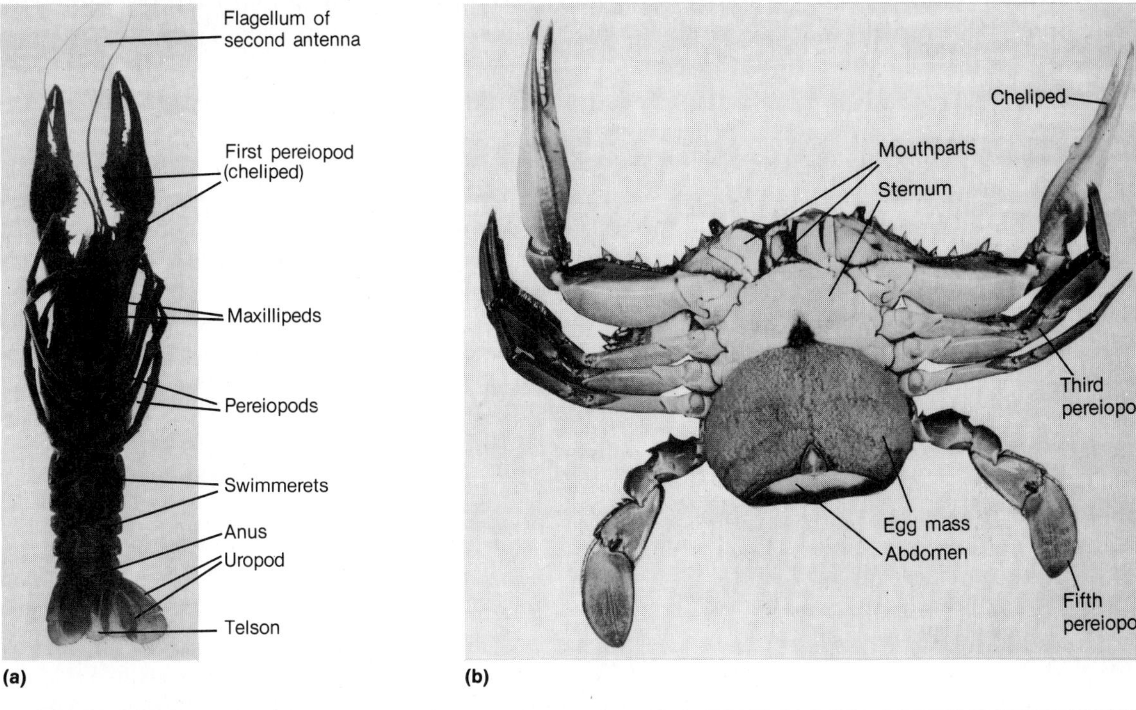

Figure 12.31 *Photographs of the ventral view of two decapods to illustrate external features:* ***(a)*** Cambarus, *a crayfish;* ***(b)*** *a female* Callinectes, *the blue crab with her egg mass.*

Internal Features. The stomodeum consists of a short **esophagus,** the **cardiac stomach,** and the anterior half of the **pyloric stomach** (Fig. 12.32). The walls of the cardiac stomach and the anterior half of the pyloric stomach are strengthened with a series of movable calcareous ossicles controlled by a complex series of muscles to form a triturating **gastric mill** (Fig. 12.32b). Digestive enzymes from the midgut glands are secreted into the mill, where food is reduced to a liquid pulp. The mill reaches its greatest development in the brachyurans.

The mesenteron is composed of the posterior part of the pyloric stomach and, in some, the anterior part of the **intestine.** The pyloric stomach has folded walls heavily equipped with setae that squeeze, knead, and filter the partially digested food. The fine particles are sorted into a duct leading to the midgut glands, an organ that represents a pair of diverticula that have coalesced to form a very large glandular mass in which absorption and food storage occur (Fig. 12.32a). The larger, undigestible particles are shunted into a ventral groove leading to the intestine. The proctodeum consists of the posterior portion of the **intestine,** which terminates posteriorly in the **anus** located at the base of the telson.

The **heart** is a polygonally shaped short, muscular organ located in the dorsal thoracic region (Fig. 12.32a). It lies in a **pericardial chamber,** and blood passes from this chamber into the lumen of the heart through two or three pairs of ostia. Seven arteries conduct blood from the heart to various somatic regions; they are the unpaired median anterior (ophthalmic), median posterior (abdominal), and sternal arteries and the paired lateral cephalic (antennal) and

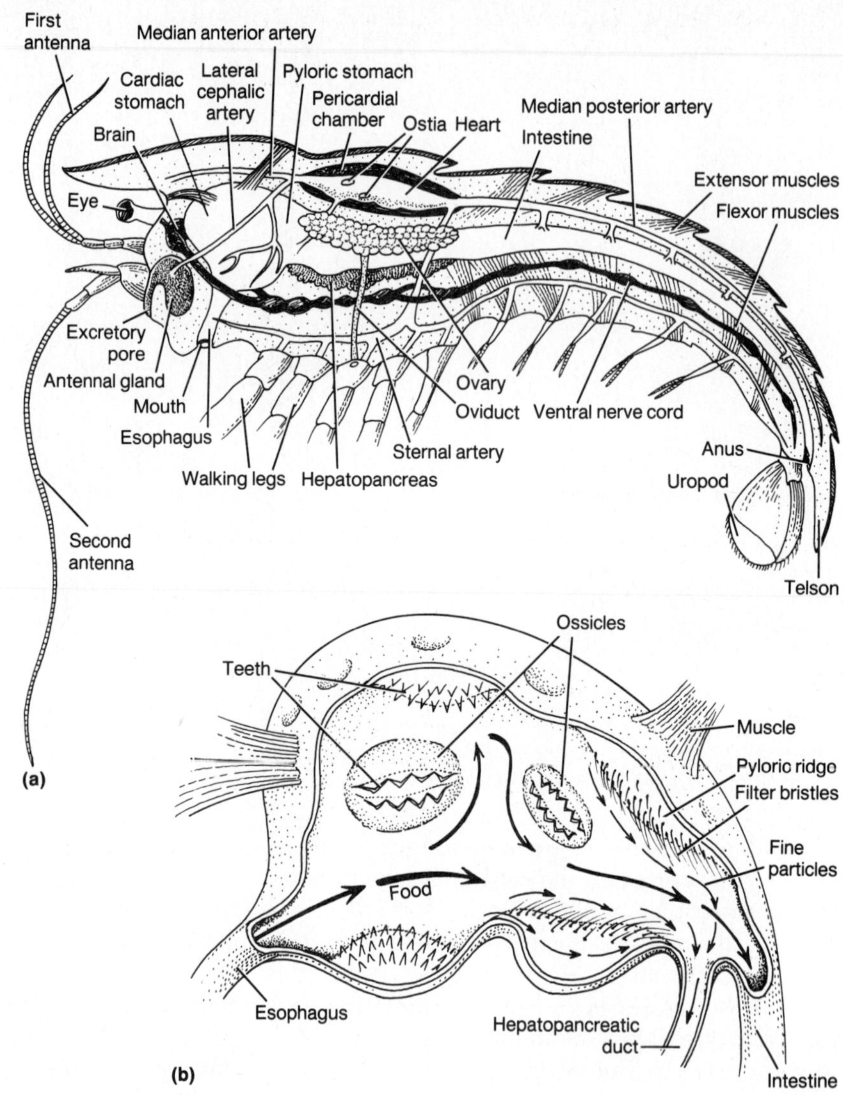

Figure 12.32 *Some internal features of the crayfish* (Cambarus): ***(a)*** *lateral view of a female (from T.I. Storer and R.L. Usinger, 1965,* General Zoology, *4th ed., McGraw-Hill Book Co., used with permission);* ***(b)*** *the gastric mill (path of fine particles shown with short arrows, that of larger particles shown with longer arrows).*

hepatic arteries. Each branches extensively, and the smaller arteries ultimately end in the hemocoel. Venous blood drains into a **sternal sinus,** where it is conveyed by channels to the gills for aeration and returned again to the pericardial sinus (Fig. 12.33a). The plasma of decapods contains the gas-transporting pigment **hemocyanin.** Various types of leucocytes are present and function in clotting and phagocytosis.

Characteristically, decapods have gills standing vertically in the branchial chamber and associated with some or all of the thoracic segments and their appendages. On each side of each thoracic metamere there is typically a set of four gills (Fig. 12.33a). Two gills, called **arthrobranchs,** are attached to the articular membrane between the segment and the appendicular coxa. One gill, a **pleurobranch,** is attached to the lateral thoracic wall; another, a **podobranch,** is attached to the coxa and represents a modified epipod. Maximally, decapods could have 32 gills on a side, but this number is never present, since there are omissions and modifications. For example, the gills of the first thoracic segment are almost always absent, and in crayfishes, most pleurobranchs are either vestigial or absent. Actual numbers of gills per side range from 3 to 24 with a more common range of 9–20 per side.

Basically, a decapod gill has a main axis with a large number of lateral filaments or lamellae. Based on the arrangement of filaments, three gill types are recognized regardless of their position on the segment (Fig. 12.33b–d). A **trichobranchiate** gill has simple unbranched filaments that are arranged in several series around the axis. A **dendrobranchiate** gill has two series of filaments that, in turn, have secondary branches. A **phyllobranchiate** gill has two series of flattened, platelike branches arranged on the axis. Dendrobranchiate gills are characteristic of pelagic shrimp, and the other two types are found in the remaining decapods. The gills of all individuals of a given species are all of one type regardless of their position on the thorax or leg. Blood always enters the gill axis through the afferent channel, passes into each filament through a fine afferent canal to the gill tip, returns by way of a minute efferent canal to the efferent channel, and passes to the pericardial chamber. Gases are exchanged in the numerous filaments, and oxygen absorption is especially facilitated by hemocyanin. Water is drawn anteriorly over the gills and expelled from the branchial cavity beside the head by the rapid back-and-forth movements of the scaphognathite. Water enters the branchial chamber from different points in the various groups. In shrimp it enters all along the ventral and posterior borders of the carapace; but in many others, entry is restricted mostly to the posterior margins. In true crabs, however, the carapace fits snugly all around the animal except for an opening near the bases of the chelipeds. Thus

Figure 12.33 *Decapod gills:* ***(a)*** *cross-sectional view of a thoracic segment showing the set of four gills on one side;* ***(b)*** *a trichobranchiate gill;* ***(c)*** *a dendrobranchiate gill;* ***(d)*** *a phyllobranchiate gill.*

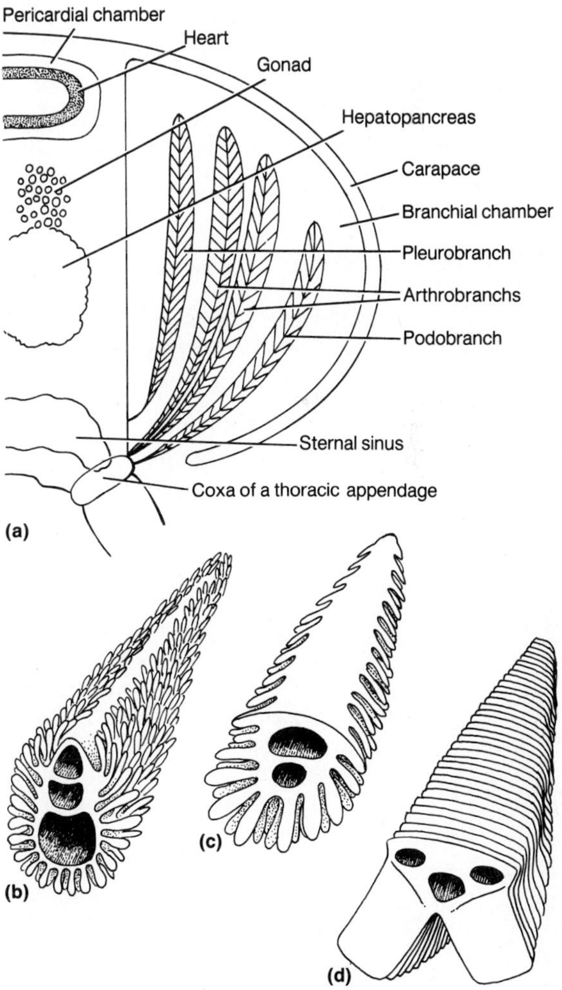

water follows a U-shaped course, passing posteriorly from the incurrent opening, dorsally over the gills, and anteriorly to exit beside the head.

Burrowing decapods are confronted with problems of sediments clogging the gills. In some this potential problem is overcome by filtering the water with dense setae lining the incurrent channels. Others may reverse the water flow, and the inhalent channel may be extended by tubelike siphons or siphons created by the antennae and the third maxillipeds. Also, many burrowing decapods have maxilliped epipods that are especially adapted for sweeping the gills, thus clearing them of debris.

Terrestrial or amphibious decapods have some features associated with gas exchange that permit them to survive in these habitats. Gills must always be kept moist, and this is achieved by periodic entries of the animal into the water, by filling both branchial chambers with water before the decapod emerges on land, or by keeping the surfaces moist by simply reducing evaporation. Some gills have filaments strengthened by sclerotized areas that hold them apart and prevent them from collapsing against each other, which otherwise would greatly lower their efficiency. A few like *Birgus*, the coconut crab, have vascularized folded epithelia in the branchial chamber through which gases are exchanged. This gill chamber converted into a "lung" is analogous to that developed in pulmonate snails.

The excretory organs are **antennal** or **green glands,** and maxillary glands are present only in larval decapods (Figs. 12.7, 12.32a). Some

BOX 12.2 □ SOME INTERESTING DIMENSIONS INVOLVING THE INTERACTIONS OF CRUSTACEANS AND COLOR

Some of the most intriguing aspects of all of crustacean biology involve the responses of these arthropods to color. Some have color vision, and many others utilize pigments within chromatophores to alter abruptly their colors or patterns to blend in with their backgrounds.

There is an expanding list of more than 20 species of mostly shallow-water or terrestrial decapod species that are known to be able to discriminate colors. In other arthropods with color vision (some flies, cockroaches, bees) the color receptors are within a given ommatidium, and there is growing evidence that the same mechanisms are present in these few crustaceans.

In several species of the crayfishes *Procambarus* and *Orconectes* the capabilities of perceiving orange and blue wavelengths were demonstrated. Similar results were discovered in several species of prawns belonging to *Palaemonetes* and *Crangon*. But why these two colors and not others? What is the ecological and survival significance of being able to sense orange and blue? Very little is known about the correlation of the behavior of these crustaceans with their capability for color vision.

That crustaceans are diverse in their color changes is well known. Some slow morphological color changes are in response to relative celestial positions of the sun, moon, and earth. Other more rapid changes are in response to the degree of illumination and changes in background coloration, and these are termed physiological color changes. Physiological color changes are of considerable importance in protective coloration,

nitrogenous wastes are eliminated by these glands, but ammonia, the chief nitrogenous waste, diffuses into the water from the general body surface. The extensive gill surface are especially efficient in disposing of ammonia. The urine of most decapods, like that of other crustaceans, is isosmotic with the blood; but in some, ions are reabsorbed to make the urine hyposmotic.

The nature of the ventral nerve cord of decapods is variable. In some like lobsters and crayfishes the ventral cords are fused into a medial cord containing 12 ganglia, which are the subesophageal ganglion and five thoracic and six abdominal ganglia (see Fig. 12.8b). In brachyurans (true crabs), all thoracic and abdominal ganglia have completely fused to form a single **ventral mass** located in the thorax (see Fig. 12.8c). Between these two extremes, various degrees of ganglionic fusion are found in other decapods.

All the usual crustacean sense organs are present in decapods, in which they reach their highest development. Well-developed, stalked, compound eyes are present containing large numbers of ommatidia; a few decapods are blind, and a few even have color vision. An important characteristic of decapods is their ability to blend in with their background in both color and patterns. This important feature, clearly mediated by the compound eyes and influenced by subepidermal pigments, is briefly discussed in Box 12.2 along with a short treatment of color vision. The naupliar eye generally does not persist in adults. The base of each antennule contains a statocyst, and chemore-

---

thermoregulation, courtship and mating, and parental behavior. Some species undergo a rhythmic diurnal change, while other color changes are arhythmic but take place in response to changing environmental or behavioral conditions. These latter responses are under hormonal control.

Characteristic of higher crustaceans is the presence of pigment cells called chromatophores. A **chromatophore** is a cell with branching, stellate, nonretracting processes usually located just beneath the epidermis. White, black, blue, yellow, and red pigments are usually present. Each chromatophore may contain one (monochromatic), two (dichromatic), or several (polychromatic) pigments. When the pigments are dispersed in the radiating processes, the darker tones of the color(s) result. But when the pigments are concentrated in the central chromatophore, the animal appears pale or blanched. In dichromatic and polychromatic chromatophores the several pigments are dispersed or concentrated independently of each other.

Color changes manifested by the chromatophores are controlled by hormones called chromatophorotropins produced in the X-organ–sinus gland complex and in other parts of the central nervous system. Antagonistic pairs of hormones are involved in which one induces pigment dispersion and another promotes pigment concentration. The "multiple-hormone principle" is well established, but the complete system has not yet been elucidated fully for any one species.

ceptors are prevalent especially on the head appendages. Proprioceptors and tactile receptors are particularly well developed and numerous in decapods. Some decapods like ghost crabs *(Oxypode),* fiddler crabs *(Uca),* spiny lobsters *(Panulirus),* and snapping shrimp *(Alpheus)* stridulate or produce various sounds.

Reproduction and Development. With the exception of a few shrimp, all decapods are dioecious, and sexual dimorphism is especially prevalent. Often one cheliped (or both) is greatly enlarged in males, and copulatory organs (penis or modified pleopods) also make males externally distinct. One sex is often larger than the other for a given species, and often there are color or pattern differences between the sexes.

The paired testes or ovaries lie in the thoracic region, but in hermit crabs they may be entirely within the abdominal region. Testes are often connected and may even be fused into a single organ. The paired oviducts are simple and open by pores on or near the bases of the third pereiopods (Fig. 12.32a). In brachyurans the distal part of each oviduct is expanded to form a seminal receptacle; in other decapods there is a single median **seminal receptacle** developed from parts of the last two thoracic sterna. One or two sperm ducts open at the base of pereiopods V.

Some form of precopulatory courtship is typical of many decapods. The fiddler crab ritual is perhaps the most elaborate, the male signaling with his enlarged claw, and tapping it on the ground. Body movements are also significant in attracting a female. In many crabs the male attends the female by carrying her either on top of or beneath him until she attains sexual maturity at the next molt, which is then followed by copulation. Dances, signals, and other behavioral patterns precede copulation in most species. Pheromones have been shown to be important in sexual attraction in some species.

A variety of copulatory positions are taken by both animals, but most involve the ventral surfaces of both decapods opposing each other. In some decapods the copulating members are oriented at right angles to each other with thoracic surfaces appressed. The male often strokes and caresses the female immediately prior to or during mating. Copulation may take place while both members are swimming, on the bottom, in burrows, or on land (terrestrial crabs).

Decapods usually transfer sperm using the first two pairs of pleopods, but male hermit crabs have a single penis, and true crabs have a pair of penes. Some have internal fertilization, but the eggs are generally fertilized as they are laid.

Penaeid shrimp release eggs into the water, but females of the other decapods attach or cement eggs to their pleopods. Most females flex the abdomen so that as the eggs are extruded, they are swept over the pleopods, to which they adhere. In crabs the brooded egg mass usually turns orange and is called a "sponge" (see Fig. 12.31b).

The hatching stage and several larval stages vary among members of this order. In the nonbrooding shrimp the egg hatches as a nauplius, and it is succeeded by a metanauplius, protozoea, zoea, and mysis stages. In many other decapods the nauplius is completed in the egg, and the animal hatches as a **zoea** (see Fig. 12.9b). The zoeal stage of crabs is distinctive because of the presence of the very long **rostral spine.** The crab zoea is succeeded by a postlarval **megalops** (see Fig. 12.9c), a stage that is characterized by a large unflexed abdomen and five pairs of functional pleopods. Since most freshwater forms do not have separate larval stages, the young hatch as juveniles.

SOME ASPECTS OF CRUSTACEAN ENDOCRINOLOGY

In recent years a tremendous amount of research has been carried out on crustacean endocrinology, but many questions still remain

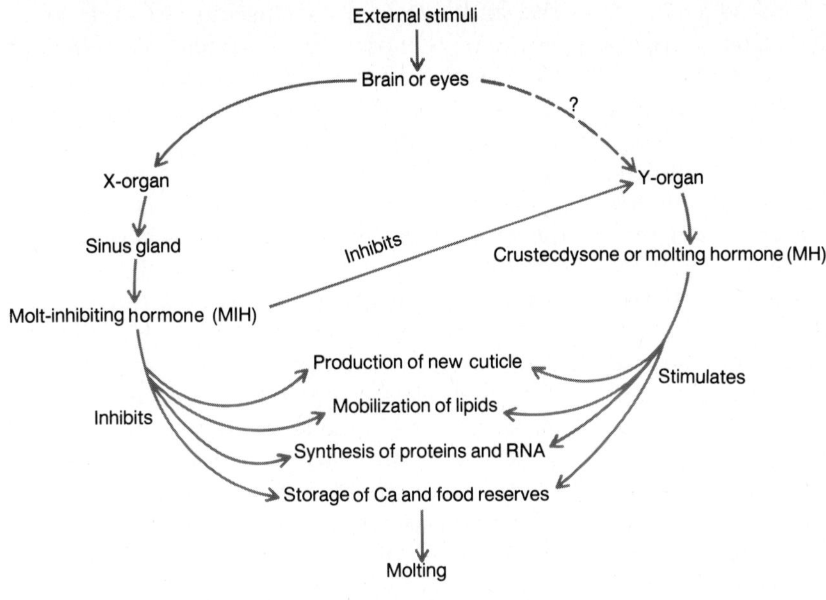

Figure 12.34 *Word diagram of the hormonal control of molting in crustaceans.*

unanswered. Most of these physiological studies have been made on malacostracans, in particular on decapods, and this should be kept in mind when general conclusions are made about hormonal systems. The principal endocrine functions include the control of molting, sex differentiation and reproduction, color changes, movement of retinal pigments, and other basic physiological processes.

The principal endocrine organs are the X-organ, sinus gland, Y-organ, androgenic gland, and ovary. The **X-organ** is in reality composed of at least two pairs of distinct groups of neurosecretory cells located either in the lateral aspects of the brain or, more commonly, in the eyestalks. Intimately associated with each X-organ is a **sinus gland,** which represents the swollen endings of the nerve fibers that originate in neurosecretory cells of the X-organ and other parts of the brain. The sinus gland may produce hormones itself or it may serve as a reservoir for those produced by the various neurosecretory cells principally of the X-organ. The paired **Y-organs** are masses of secretory cells located at the base of either the second maxillae or the second antennae. If the excretory glands are at the base of the second antennae, the Y-organs are at the base of the second maxillae and vice versa. The paired **androgenic glands** are found in males on or near the testes, and the paired **ovaries** lie in the thoracic region.

Molting in crustaceans is a complex process involving the nervous system, hormones, X-organ, Y-gland, and epidermis (Fig. 12.34). Molting is induced by a hormone produced by the Y-organs. This hormone, called **crustecdysone** or **molt-inducing hormone** (MH), is very similar to ecdysone in insects, and even injections of insect ecdysone promote crustacean molting. The Y-organs are analogous to the ecdysial glands of insects (see Chapter 13). The principal sites of action of MH are the epidermal cells, which are responsible for the production of the new cuticle, and the midgut gland, which functions in mobilization of lipids and as a site for synthesis of proteins and RNA. The X-organ–sinus gland complex is also involved in the molting process by producing a **molt-inhibiting hormone** (MIH). The MIH is stored for a time in the sinus gland and then released into the hemolymph. Its principal action is to inhibit the production of MH by the Y-organs. MIH may also act by inhibiting events secondary to molting, including inhibition of storage of calcium and food reserves necessary for a subsequent molt and the regulation of water taken up at ecdysis. Social contacts and environmental conditions, especially photoperiod and temperature, are important external cues that influence this hormonal system and thus regulate the molting cycle.

A complex of hormones is directly responsible for controlling sexual differentiation (Fig. 12.35). A pair of androgenic glands is present in males but is rudimentary in females. In males the androgenic glands secrete an **androgenic hormone** that promotes testicular functions and the development of all secondary male sexual characteristics. There is a hormonal relationship between the X-organ–sinus gland complex and gonads similar to that between the pituitary and gonads in vertebrates. In males the X-organ produces a hormone that inhibits the development of the androgenic glands and their hormonal production. In females the X-organ produces a **gonad-inhibiting hormone** (GIH) that prevents egg development and maturation, especially vitellogenesis or yolk deposition. Apparently, oogenesis is dependent upon crustecdysone produced by the Y-organ, although it is not, per se, an ovary-stimulating hormone. Stimulation of gonadal function is undoubtedly not a response solely to the absence of inhibiting hormones of the X-organ. Recent evidence suggests a **gonad-stimulating hormone** (GSH) perhaps produced by neurosecretory cells of the brain

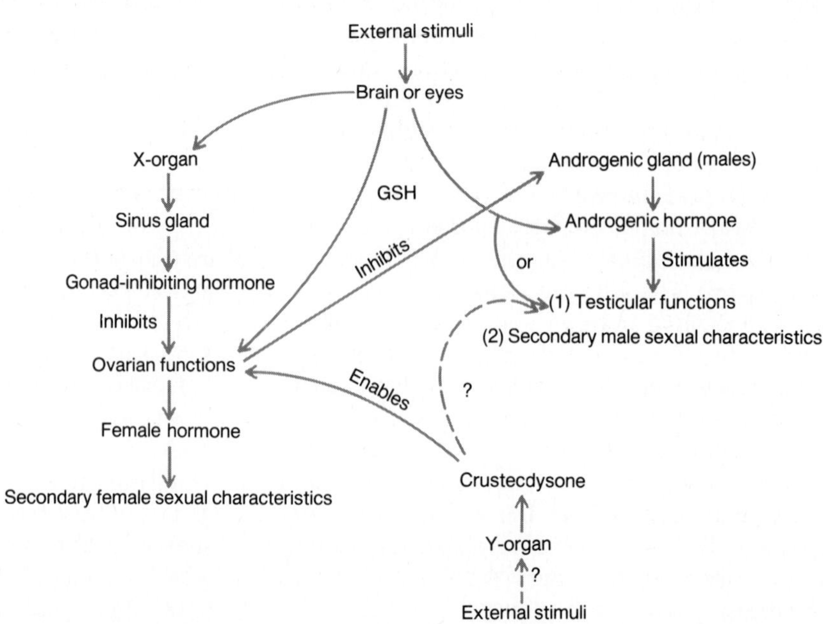

Figure 12.35 *Word diagram of the hormonal control over reproduction in crustaceans (GSH = gonad-stimulating hormone).*

that stimulates ovarian and testicular function (Fig. 12.35).

Recent studies suggest that the heartbeat may be controlled hormonally by substances produced by neurosecretory cells as one or more hormones have been shown to be cardiostimulators. Other metabolic functions including oxygen consumption, calcium deposition, water balance, blood sugar levels, and color changes (see Box 12.2) are also controlled by the endocrine system.

PHYLOGENY OF THE CRUSTACEANS

Even though crustacean fossils have been found in sediments deposited as early as the Proterozoic Era (see Table 1.1 in Chapter 1), there are few clues as to the origins of this group. In Chapter 10, mention was made of two conflicting theories of the origins of various arthropodan groups. One theory holds that arthropods are polyphyletic; this would mean that crustaceans arose independently of any other arthropodan group. This position is predicated on fundamental differences in mandibles, other appendages, and embryonic details between crustaceans and representatives of the other subphyla. But a second theory, that of monophyletism, holds that all arthropods, especially crustaceans and trilobites, may have arisen from a common ancestor. This theory holds that patterns of feeding by trilobites and branchiopods were quite similar, even though there are fundamental differences in the limbs of these two groups. Unfortunately, the fossil record tells us very little about the origins of this group. While paleontology cannot convincingly resolve these different theories at this time, most specialists probably now support the theory that suggests that crustaceans evolved independently of other arthropods.

The earliest protocrustaceans were probably small, benthic, marine, wormlike, segmented, and capable of swimming. The body presumably was composed of a head of five segments and a trunk, and each segment bore a pair of similar foliaceous appendages that were either triramous or biramous. The resemblances between the protocrustaceans and cephalocarids are quite strong and lend great credence to the belief that these modern-day forms are the most primitive of all crustaceans. But an intangible feature of this protocrustacean was that it must have possessed a great capacity for modification and adaptability. Therefore in their long and illustrious history, crustaceans have adaptively radiated to fill a great many ecological niches.

In the evolution of early crustaceans, three relatively primitive classes diverged and began their own separate evolution: the primitive cephalocarids, the branchiopods with foliaceous appendages, and the remipedes (Fig. 12.36). With a further evolution of filter-feeding, one phylogenetic line led to the very numerous ostracods and the diverse maxillopods. The maxillopods evolved initially as a group exploiting filter-feeding and modifications of the maxillae and other head appendages; the ontogenies of various life-styles resulted in the copepods, barnacles, and others. The remaining great crustacean group, the malacostracans, capitalized on evolving a high degree of tagmatization and cephalization, development of a telson, and the modification of the thoracic and abdominal appendages for walking, crawling, swimming, and food-gathering. The phyllocarids and hoplocarids must be considered to be rather primitive malacostracans. In the eumalacostracans, three evolutionary lines have produced the syncarids, the peracarids including isopods and amphipods, and the eucarids including euphausiids and decapods. Their larger sizes, raptorial feeding, and abilities to tolerate a myriad of environmental conditions are significant in the success and prominance of decapods today (Fig. 12.36).

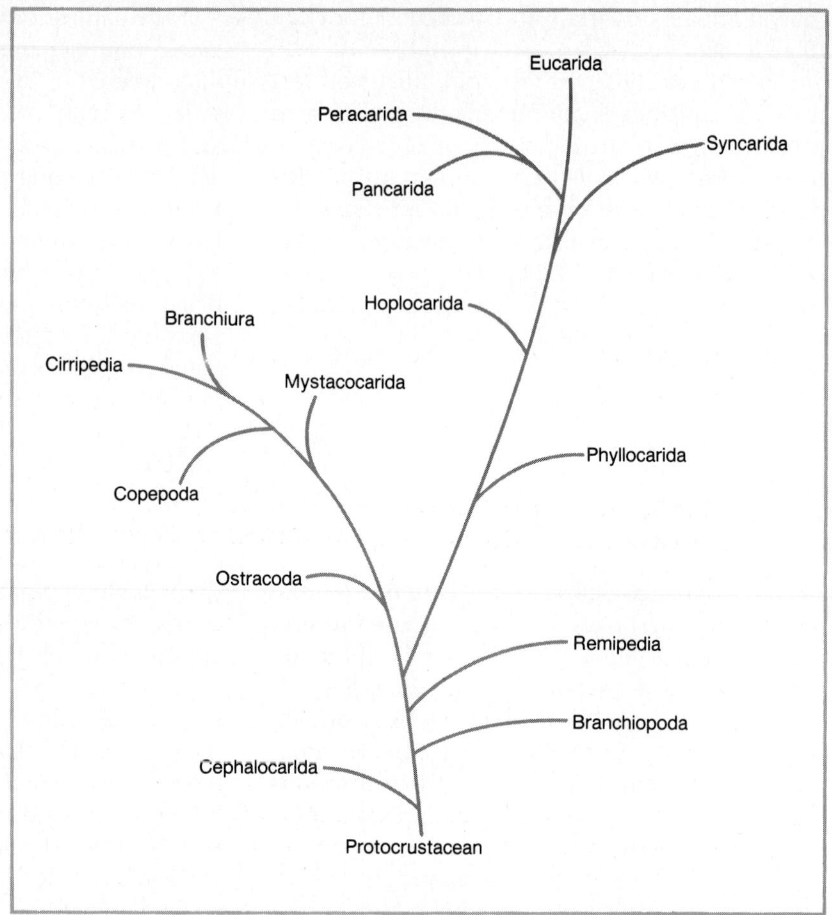

Figure 12.36 *Probable phylogenetic relationships within the crustaceans.*

SYSTEMATIC RESUME

The Principal Taxa of Crustaceans

Class Cephalocarida—the most primitive of all contemporary crustaceans; small; horsehoe-shaped head; trunk of 19 segments; trunk appendages are all similar and have a third ramus or pseudepipod; *Hutchinsoniella* (Fig. 12.10) is the best known.

Class Branchiopoda—foliaceous appendages; mostly filter-feeding; most inhabit temporary bodies of water.

Order Notostraca—tadpole shrimp; large dorsal carapace; trunk with 35–71 pairs of appendages; telson with two long caudal cercopods; *Triops* (Fig. 12.11c) occurs in western North America.

Order Conchostraca—clam shrimp; bivalved carapace; body laterally compressed; second antennae natatorial and well developed; 10–32 trunk segments; common genera include *Lynceus* (Fig. 12.11d) and *Limnadia.*

Order Cladocera—water fleas; swim by enormously enlarged, biramous, second antennae; carapace large but does not enclose the head; five to six pairs of trunk appendages; mostly parthenogenetic; *Daphnia* (Fig. 12.12b), *Leptodora, Podon* (Fig. 12.12a), and *Bosmina* are frequently encountered.

Order Anostraca—fairy and brine shrimp; swim upside down using foliaceous trunk appendages; lack carapace; *Artemia, Eubranchipus* (Fig. 12.11b), and *Branchinecta* are typical genera.

Class Remipedia—found in a Bahamian marine cave; biramous first antennae; the maxillae and maxillipeds are uniramous and raptorial; trunk appendages are biramous; *Speleonectes lucayensis* (Fig. 12.13) is the only species described to date.

Class Maxillopoda—maxillae large and used in feeding; trunk is reduced usually to 11 segments; thoracopods are uniramous; many are parasitic.

Subclass Mystacocarida—primitive, small, and live in intertidal interstices; thorax and abdomen each with five segments; *Derocheilocaris* is the type genus.

Subclass Branchiura—fish lice; ectoparasitic on marine and freshwater fishes; carapace covers cephalothorax; maxillae I form holdfast suckers, and maxillae II and antennae II have holdfast claws; *Argulus* (Fig. 12.15) is a common genus.

Subclass Copepoda—copepods; mostly marine, but some live in fresh water, and many are parasitic; locomotion by oarlike antennae and thoracopods; great versatility in foods and food-gathering adaptations.

Order Calanoida—mostly planktonic; freshwater and marine; chiefly herbivorous and filter-feeders; prosome/urosome articulation is between thoracic segment VI and abdominal segment I; antennae I are very long; females usually with only one ovisac; *Diaptomus, Calanus,* and *Senecella* (Fig. 12.17b) are common genera.

Order Harpacticoida—mostly benthic in fresh and salt water; prosome/urosome articulation between thoracic segments V and VI; antennules are very short; body is wormlike; a few species are parasitic; some better-known genera include *Harpacticus* and *Parastenocaris.*

Order Cyclopoida—planktonic and benthic forms in freshwater and marine habitats; omnivorous, often raptorial feeders; prosome/urosome articulation between thoracic segments V and VI; relatively short antennules; females with two ovisacs; some are parasitic; *Cyclops* (Fig. 12.17a), *Macrocyclops,* and *Sapphirina* are frequently encountered.

Order Poecilostomatoida—all marine; parasitic and commensal on fishes and invertebrates; thoracopods modified for grasping and clinging to host; *Pseudanthessius* and *Chondracanthus* (Fig. 12.17g) are representative.

Order Siphonostomatoida—adults parasitic on or commensal with marine and freshwater fishes and invertebrates; thoracopods modified for attaching to host; labium and labrum prolonged into siphon or tube; *Pontoeciella, Nemesis,* and *Clavella* are often collected.

Order Monstrilloida—adults lack antennae II and mouthparts; nauplius is free living, but later larval stages are parasitic in polychaetes; adults are free living in marine habitats; the best-known genus is *Monstrilla.*

Order Misophryoida—all marine and pelagic; thoracopods with swimming plumose setae; *Misophria* is the type genus.

Order Mormonilloida—marine and pelagic; first antennae of three to four long podomeres; lack a heart and the fifth thoracopods; *Mormonilla* is the type genus.

Subclass Cirripedia—barnacles; adults all sessile and encrusting or parasitic; marine; free-living forms have six pairs of feeding thoracopods or cirri.

Order Ascothoracica—parasites in echinoderms and corals; bivalved saccular mantle and six or fewer pairs of thoracic legs; dioecious; about 45 species; best-known genera include *Baccalaureus, Laura,* and *Dendrogaster.*

Order Acrothoracica—naked barnacles boring holes in calcareous shells and corals; calcareous plates are absent, and usually there are only four pairs of cirri; typically dioecious; about 50 species; genera include *Kochlorine* and *Trypetesa.*

Order Thoracica—free-living and commensal barnacles; six pairs of well-developed thoracic cirri; mantle usually with calcareous plates; stalked or sessile; hermaphroditic; *Lepas* (Fig. 12.19), *Balanus* (Fig. 12.18c), *Pollicipes, Scalpellum,* and *Chthamalus* (Fig. 12.18a) are common genera.

Order Rhizocephala—highly specialized endoparasites of decapod crustaceans; appendages and digestive tract are absent; form nutritive rhizoids within host; dioecious; *Sacculina* (Fig. 12.21a), *Peltogasterella*, and *Peltogaster* (Fig. 12.21b) are among the best known genera.

Class Ostracoda—ostracods or seed shrimp; body completely enclosed in bivalved carapace with adductor muscle; second antennae are locomotive limbs; trunk reduced; mostly scavengers.

Subclass Myodocopa—exclusively marine and mostly benthic; second antennae biramous; zero to two pairs of thoracopods.

Order Myodocopida—notch in carapace for the extension of the first antennae when valves are closed; second antennae biramous with enlarged basal segment; two pairs of trunk appendages; *Cypridina*, *Philomedes*, and *Vargula* are often encountered.

Order Halocyprida—no notch in carapace for antennae; flattened, biramous second antennae used for locomotion; trunk appendages are lacking; unpaired genitalia; *Polycope* is typical.

Subclass Podocopa—benthic or ectocommensal; strongly calcified carapace; found in a variety of aquatic and terrestrial habitats; no antennal notch in carapace for antennae.

Order Platycopida—biramous second antennae; locomotion by maxillae and the one pair of trunk appendages, but not by antennae I or II; exclusively marine; *Cytherella* is a common genus.

Order Podocopida—antennae II uniramous; two pairs of trunk appendages; compound eyes are missing; marine, freshwater, or terrestrial (a few); *Cypris*, *Darwinula*, and *Candona* are typical genera.

Subclass Palaeocopa—mostly extinct; contemporary forms known only from carapaces; marine; two modern-day species have been discovered.

Class Malacostraca—higher crustaceans; thorax usually of eight segments; one to three anterior thoracopods are maxillipeds; most thoracopods are pereiopods, and some are chelate; abdominal pleopods are well developed.

Subclass Phyllocarida—primitive; unhinged bivalved carapace; abdomen of eight segments of which the last two are reduced; *Nebalia* (Fig. 12.24a) and *Nebaliopsis* are well known.

Subclass Hoplocarida—the mantid shrimp; raptorial feeders using the second, large, subchelate thoracopods; body flattened dorsoventrally; triramous antennules.

Order Stomatopoda—features same as those of subclass; common genera include *Squilla* (Fig. 12.25), *Gonodactylus*, and *Pseudosquilla*.

Subclass Eumalacostraca—all remaining malacostracans; thoracic limbs usually do not function as strainers for filter-feeding; mostly predaceous and often raptorial; abdomen with six segments and usually five pairs of pleopods.

Superorder Syncarida—primitive; freshwater; no carapace or chelae.

Order Anaspidacea—live in Tasmania and Australia; slender body with homononomous segmentation; fringed pleopods; *Anaspides* (Fig. 12.24b) is representative.

Order Bathynellacea—blind, small, wormlike; found in interstices of sand grains; reduced pleopods; telson is absent; includes *Bathynella*.

Superorder Pancarida—very small group; dorsal brood pouch beneath carapace; only two pairs of abdominal appendages.

Order Thermosbaenacea—features same as those of superorder; *Monodella* is representative.

Superorder Peracarida—one pair of maxillipeds; ventral marsupium in females; carapace generally absent.

Order Mysidacea—mysids; mostly marine; marsupium prominent; pleopods reduced in females; filter-feeders; *Mysis*, *Neomysis*, and *Gnathophausia* are often common genera.

Order Cumacea—marine; live in burrows or tubes; anterior rostrum; filter-feeders; pleopods absent in females; common genera include *Diastylis*, *Cumopsis*, and *Leptocuma*.

Order Amphipoda—amphipods; mostly marine; body laterally compressed; thoracic gills; maxillipeds II and III are gnathopods used in feeding; some are parasitic; *Gammarus, Orchestia* (Fig. 12.27b), *Hyallela, Talorchestia, Caprella* (Fig. 12.27c), and *Ingolfiella* are very common.

Order Isopoda—isopods; marine, freshwater, or terrestrial; body flattened dorsoventrally; pleopods function as gills and are covered by an operculum; some are parasitic; familiar genera include *Oniscus* (Fig. 12.28d), *Asellus* (Fig. 12.28c), *Porcellio, Ligia* (Fig. 12.28b), and *Idothea* (Fig. 12.28a).

Order Tanaidacea—small, marine; body flattened; carapace functions in gaseous exchange; most are filter-feeders; representative genera include *Apseudes* (Fig. 12.26c), *Tanais,* and *Sphyrapus.*

Superorder Eucarida—prominent carapace fused to all thoracic segments; lack brood chamber.

Order Euphausiacea—krill; marine; shrimplike; often present in large numbers; many luminesce because of light-producing organs; *Euphausia, Meganyctiphanes* (Fig. 12.29), and *Pentheuphausia* are common genera.

Order Decapoda—crabs, shrimp, lobsters; have branchial chamber, scaphognathite, three pairs of maxillipeds, five pairs of pereiopods; generally predaceous; antennal glands; internal fertilization; includes *Penaeus* (Fig. 12.30a) and *Crangon* (shrimp), *Homarus* and *Astacus* (lobsters, crayfishes), *Pagurus* (Fig. 12.30d) and *Clibanarius* (hermit crabs), and *Callinectes* (Fig. 12.30h,12.31b), *Uca* (Fig. 12.30j), and *Ocypode* (crabs).

ADDITIONAL READINGS

Abele, L.G., ed. 1982. *The Biology of Crustacea.* Vol. 1: *Systematics, the Fossil Record, and Biogeography,* 304 pp. Vol. 2: *Embryology, Morphology, and Genetics,* 403 pp. New York: Academic Press.

Atwood, H.L. and D.C. Sandeman, eds. 1982. *The Biology of Crustacea.* Vol. 3: *Neurobiology: Structure and Function,* 468 pp. New York: Academic Press.

Bate, R.H., E. Robinson, and L.M. Sheppard, eds. 1982. *Fossil and Recent Ostracods,* 494 pp. Chichester, England: Horwood.

Browne, R.A. 1982. The costs of reproduction in brine shrimp. *Ecology 63:*43–47.

Caldwell, R.L. and H. Dingle. 1976. Stomatopods. *Sci. Amer. 234:*81–89.

Calman, W.T. 1909. *A Treatise on Zoology.* Part VII: 3rd Fascicle, *Crustacea,* 332 pp. London: Adam and Charles Black.

Cheng, T.C. 1973. *General Parasitology,* pp. 695–719. New York: Academic Press.

Gosner, K.L. 1971. *Guide to Identification of Marine and Estuarine Invertebrates,* pp. 423–555. New York: John Wiley & Sons.

Green, J. 1961. *A Biology of Crustacea,* 165 pp. Chicago: Quadrangle Books.

Grosberg, R.K. 1982. Intertidal zonation of barnacles: The influence of planktonic zonation of larvae on vertical distribution of adults. *Ecology 63:*894–899.

Mantel, L.H., ed. 1983. *The Biology of Crustacea.* Vol. 5: *Internal Anatomy and Physiological Regulation,* 457 pp. New York: Academic Press.

Manton, S.M. 1977. *The Arthropoda: Habits, Functional Morphology, and Evolution,* 514 pp. Oxford, England: Clarendon Press.

Mauchline, J. 1980. *Advances in Marine Biology.* Vol. 18: *The Biology of Mysids and Euphausiids,* 682 pp. New York: Academic Press.

McLaughlin, P.A. 1980. *Comparative Morphology of Recent Crustacea,* 173 pp. San Francisco: W.H. Freeman and Co.

Minor, R.W. 1950. *Field Book of Seashore Life,* pp. 397–532. New York: G.P. Putnam's Sons.

Moore, R.C., ed. 1969. *Treatise on Invertebrate Paleontology.* Part R, Vols. I and II, 629 pp. Boulder, CO.: The Geological Society of America.

Pennak, R.W. 1978. *Fresh-water Invertebrates of the United States,* 2nd ed., pp. 318–487. New York: Ronald Press Co.

Provenzana, A.J., ed. 1983. *The Biology of Crustacea.*

Vol. 6: *Pathobiology,* 273 pp. New York: Academic Press.

Rebach, S. and D.W. Dunham, eds. 1983. *Studies in Adaptation: The Behavior of Higher Crustacea,* 264 pp. New York: John Wiley & Sons.

Sandeman, D.C. and H.L. Atwood, eds. 1982. *The Biology of Crustacea.* Vol. 4: Neural Integration and Behavior, 319 pp. New York: Academic Press.

Saunders, H.L. 1957. The cephalocarids and crustacean phylogeny. *Syst. Zool.* 6:112–128

Schmitt, W.L. 1965. *Crustaceans,* 194 pp. Ann Arbor: University of Michigan Press.

Smith, D.W. and S.D. Cooper. 1982. Competition among Cladocera. *Ecology 63:*1004–1015.

Vernberg, F.J. and W.B. Vernberg, eds. 1983. *The Biology of Crustacea.* Vol. 7: *Behavior and Ecology,* 319 pp. Vol. 8: *Environmental Adaptations,* 363 pp. New York: Academic Press.

Warner, G.F. 1977. *The Biology of Crabs,* 197 pp. New York: Van Nostrand Reinhold Co.

Wicksten, M.K. 1980. Decorator crabs. *Sci. Amer. 242:*146–154.

Yager, J. 1981. Remipedia, a new class of Crustacea from a marine cave in the Bahamas. *J. Crustacean Biol. 1:*328–333.

13

The Insects

OVERVIEW

Insects are the most successful of all invertebrates, having unparalleled species diversity, numerical predominance, and importance in terrestrial habitats. But they have a surprisingly stereotypic body plan. The insect head is enclosed by rigid cuticular sclerites and strong internal apodemes. The four adult head segments bear the sensory antennae and three pairs of variously modified mouthparts: mandibles, maxillae, and a second pair of fused maxillae that forms the labium. The head bears three ocelli and a pair of compound eyes with high visual acuity. The thorax always has three segments, each of which bears a pair of legs. Each of the two posteriormost thoracic segments usually bears a pair of membranous wings that develop from outfoldings of the thoracic pleura. Wings are driven both by direct muscles that attach to the wing bases and by indirect muscles that alter the shape of the thorax. The abdomen consists of 11 or fewer segments, never bears legs in adults, has posterior gonopores, and in females terminates in an ovipositor. The external body surface contains a plethora of sensory receptors that transmit mechanical, chemical, and tactile stimuli to the central nervous system.

The alimentary canal contains a pumping pharynx, a ventriculus where most digestion takes place, and a hindgut where water and minerals are absorbed. A dorsal vessel containing the heart, a comparted hemocoel, and several kinds of blood cells are circulatory features. Excretion, accomplished by means of Malpighian tubules that arise from the hindgut, involves the elimination of uric acid crystals. An extensively branched ventilating tracheal system contains as its functional units exceedingly fine tracheoles that end within tissues. The nervous system contains a large brain, a ventral nerve cord, peripheral nerves, and a visceral nervous system innervating internal organs.

Insects are dioecious, most males have a copulatory organ, and fertilization is internal. Both male and female systems are relatively complicated with various ducts and glands specialized for particular functions. Eggs are laid in the environment, and development begins without delay. Cleavage is superficial and regulative, soon producing a stereoblastula. Usually after several weeks of development the embryo hatches as a very small immature insect. Immatures of primitive insects resemble adults and have no metamorphosis. But the vast majority of all insects undergo some metamorphosis or transformation from the immature stages to the adult. In some orders, the immatures or nymphs gradually come to resemble the adults; they experience incomplete metamorphosis. The most advanced insects have a radical, complete metamorphosis between the larval and adult stages. To accommodate complete metamorphosis, juxtaposed between the larval and adult stages is a pupal stage, a nonfeeding, nonambulatory stage in which these dramatic alterations take place.

Taxonomically, insects are apterygotes if they are primitively wingless or pterygotes if they have wings. Pterygotes are placed in two divisions: Exopterygota, the insects in which the wings develop externally in nymphs and which have incomplete metamorphosis, and Endopterygota, the insects in which the wings develop internally in larvae and pupae and which have complete metamorphosis. Twenty of the largest and most important orders are briefly characterized.

Insects, evolving from an ancient arthropod that had uniramous appendages, were at first undoubtedly small and apterous. Important developments in insect evolution were the ontogeny of wings, wing-flexing mechanisms, and immature stages that could utilize different habitats and food resources from those of adults; the last of these trends led to the phenomenon of metamorphosis.

INTRODUCTION TO THE INSECTS

Insects are extraordinarily, preeminently, singularly successful. Using as standards such important criteria as species diversity, numbers of individuals, ecological importance, and economic effects on human cultures, insects are the dominant invertebrates today. While estimates of numbers of species vary widely, there are probably about 750,000 described species. This means that about three-fourths of all animal species and over half of all living species are insects. Many more insects are certainly yet undescribed and unidentified, and the total number of species could eventually be as high as 10 million! Box 13.1 presents a fairly good case that species diversity in insects may, in fact, be much higher than anyone thought possible until now. Not only are insects a huge assemblage of different species, they are also incredibly numerous as individuals. An area of 0.4 hectare (1 acre) of backyard, woods, or meadow might be home for several million individuals!

Ecologically, insects are clearly the dominant invertebrates in their habitats. Insects are basically land animals, and their ability to fly has contributed markedly to their terrestrial success. Insects are found from polar regions to arid deserts, from alpine areas with an altitude in excess of 6400 m to estuarine marshes, from deep caves to hot springs. Insects have successfully filled innumerable ecological niches in almost every known habitat on land. A relatively small number of species (~2% of the total) live some stage in their life cycle in fresh water. But insects are almost totally absent from the marine environment, a rather remarkable fact considering the adaptability of this enormous group. Insects are especially important in ecosystem energetics, since they ingest so many different types of food. There are almost no natural terrestrial food chains that do not involve insects to a substantial degree. In fact, an overwhelming majority of the products of photosynthesis on land and in fresh water is directly ingested by insects.

The importance of insects to human societies cannot be overstated. Even though the number of pests is small in comparison to the total number of species, certain insects cause

BOX 13.1 □ HOW MANY SPECIES OF INSECTS ARE THERE?

This text states that there are about 750,000 species of insects—a fairly common estimate of the species diversity of this enormous taxon. Other recent estimates of numbers of insect species have dared to range from one to five million. But now the incredible estimate of 30 million different species of insects has been advanced! To what do we attribute such a monumental increase? Biologists studying tropical rain forests are the source of this estimate—an unimaginable figure only a few years ago.

Biologists with the Smithsonian Institution and with several universities have been studying the biota of the upper canopy in jungle forests in Puerto Rico and in Central America. The upper canopy is a vast complex of tree limbs, leaves, and vines about 30–70 m above the ground. These studies have shown that the upper canopy is a prodigious, productive, verdant mat with a plethora of different plants and animals. Both the profusion and the diversity of life are far greater than anyone had ever imagined. In one instance a pesticide was experimentally sprayed by remote control into a small section of the canopy in a Panamanian rain forest. So many insect specimens, many of which were simply unknown to that point, were collected that the estimate of insect species was raised by this research group to 30 million.

There is one very disconcerting dimension to this entire issue, and that concerns the rate at which tropical rain forests are being destroyed. By clear-cutting, slash-and-burn agriculture, and other human activities we are destroying over 500,000 km^2 of tropical rain forests per year. In short, this means that tropical rain forests and most of the heretofore unknown plants and animals associated with the canopies of these ecosystems will be soon lost forever. Most will probably have disappeared by the end of this century. Science and society will be the decided losers.

How many species of insects are there? Lots—perhaps 30 million or more, but that number, sadly enough, will surely decline as more jungle areas are destroyed in the name of progress.

trouble to human cultures in astonishing proportions. Several billion dollars worth of damage is wrought each year by insects to crops, forests, ornamental plants, and structures inhabited by human beings. In many parts of the world, millions of people die or are seriously debilitated by insect-borne diseases. Domestic animals are also susceptible to the ravages of insects and the diseases they transmit. Yet far more insect species are beneficial directly to human beings and indirectly to the natural world. Insect products like honey, beeswax, silk, certain dyes, and shellac contribute to our well-being. One of the outstanding contributions of insects is their role in pollination, including that of many important crops and most all fruits; in fact, almost any angiosperm with a colorful, showy, or strongly scented flower is pollinated by insects. Many other insects are predaceous or parasitic on other insects and thus are important biological control agents.

GENERAL CHARACTERISTICS

Amid the great species diversity epitomized by insects, their functional morphology is surprisingly stereotypic. Apparently, in the evolution

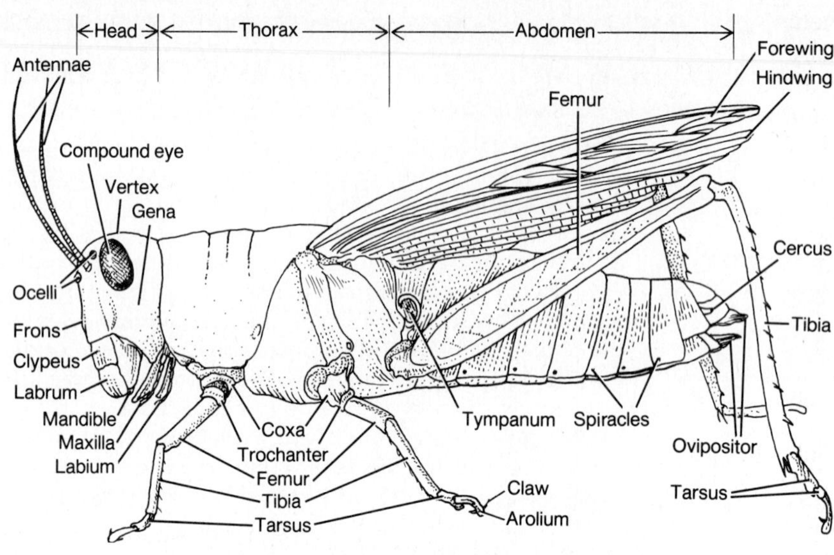

Figure 13.1 *The principal external features of a grasshopper, a generalized insect (from T.I. Storer and R.L. Usinger, 1965,* General Zoology, *4th ed., McGraw-Hill Book Co., used with permission).*

of insects a remarkably successful, utilitarian plan was developed—one that allowed success in almost any terrestrial environment and one that could be modified in very specific and subtle ways to adapt to innumerably different conditions and niches. This basic stereotypic plan is present with great fidelity in such varied insects as grasshoppers, hornets, dragonflies, and moths. Figure 13.1 illustrates the principal external features present generally in insects. This fundamental body plan has three tagmata, a water-conserving cuticle, one pair of antennae, three pairs of mouthparts, three pairs of legs, a tracheal system for gaseous exchange, Malpighian tubules for excretion, and gonoducts at the posterior end. Additionally, most insects have a pair of compound eyes and two pairs of wings as adults (Fig. 13.1). All insects are relatively small, and though some Mesozoic dragonflies had a wingspan of nearly 70 cm, most contemporary insects range in length from 0.2 to 4 cm. Variations on this basic plan are the bases for distinguishing between the various subordinate taxa of insects.

An Initial Taxonomic Framework for Insects

Obviously, a taxon with three-quarters of a million species requires a rather elaborate taxonomic scheme. While there are many different plans for grouping insects taxonomically, we shall follow the outline of Borror, De Long, and Triplehorn (1981), whose basic scheme appears in Table 13.1. Three features are of particular importance in separating the major subtaxa: (1) the presence or absence of wings, (2) whether wings develop externally or internally in the immature forms, and (3) the degree of metamorphosis by the immature stages as they become adults. Initially, insects can be divided into two subclasses: the Apterygota, or the insects that primitively do not have wings, and the Pterygota, or insects that have wings or that have lost them secondarily. Among the pterygotes there are two subordinate taxa. Insects in the Division Exopterygota have immature stages called **nymphs** that gradually become adults, which means that they have incomplete metamorphosis, and their wings develop externally in later nymphal instars. Insects belonging to the Division Endopterygota have immature stages called **larvae** and **pupae** which metamorphose dramatically and completely into adults and thus have complete or total metamorphosis; also, wings develop internally in the larvae (Table 13.1).

Once insects have been subdivided into the apterygotes, exopterygotes, and endopterygotes, there are 28 groups of insects at the ordinal level. The principal distinctions between

TABLE 13.1. □ **THE PRINCIPAL CHARACTERISTICS OF THE MAJOR ORDERS OF INSECTS***

Taxon	*Common Name(s)*	*Principal Food*	*Adult Mouthparts*	*Other Features*
Subclass Apterygota				
Subclass Pterygota				
Division Exopterygota				
Order Odonata	Dragonflies, damselflies	Flying insects	Chewing	Very large compound eyes; aquatic nymphs; extensible nymphal labium
Order Orthoptera	Crickets, grasshoppers	Vegetation	Chewing	Parchment forewings, membranous hindwings; most stridulate
Order Hemiptera	True bugs	Variable	Piercing-sucking; form a beak	Forewings are thickened at base, overlap distally when at rest
Order Homoptera	Aphids, hoppers, cicadas	Vegetation	Piercing-sucking; form a beak	Many are pests on plants; wings held rooflike over abdomen when at rest
Division Endopterygota				
Order Coleoptera	Beetles, weevils	Variable	Chewing	Forewings are thickened, meet in midline when at rest
Order Trichoptera	Caddisflies	Variable	Chewing, but often not functional	Wings hairy with few crossveins
Order Lepidoptera	Butterflies, moths	Plant fluids	Sucking proboscis	Herbivorous larvae; wings covered with scales
Order Diptera	Flies	Variable	Piercing, lapping	Only mesothoracic wings functional; metathoracic wings as halteres
Order Hymenoptera	Bees, wasps	Plant material (nectar)	Chewing	Complex societies; wings on either side hooked together

* Major is defined here as having 5000 or more species.

these orders, in addition to the three features mentioned above, involve the types of food ingested, specializations of mouthparts, and patterns of veins in the wings. Also listed in Table 13.1 are the nine largest orders with some of their principal characteristics; references will be made to these in the general discussions.

External Features

The entire external surface of an insect, like that of all arthropods, is covered by a tough, nonliving, chitinoprotein **cuticle.** As in arachnids, the insect epicuticle contains an additional layer of waxes and other lipids. This layer is particularly significant in water conservation, an important feature that has permitted insects to be highly successful as terrestrial arthropods. The cuticle does not have a uniform thickness, since there are many thick sclerotized plates or **sclerites** and thinner sutures or quite thin articular membranes between the sclerites.

The body of every insect consists of an anterior acron, a posterior pygidium, and 18 or fewer segments. These segments are invariably grouped into three distinct tagmata: head, thorax, and abdomen (Fig. 13.1).

Head

The anteriormost tagma of an insect is the **head** composed of four adult segments and the preoral **acron,** a structure that is homologous to the acron of annelids. In addition to the preoral segment I there is another preoral segment, the **intercalary segment,** which appears briefly in embryonic development, does not bear appendages, is absent altogether in adults, and may be homologous to that crustacean segment that bears the second antennae. Adult segments II–IV are all postoral; each bears a pair of appendages involved with mastication. The acron and all four head segments have imperceptibly fused to form a single unit.

The exoskeleton of the head is a hardened capsule or **cranium** that represents a number of fused sclerites. It is beyond the scope of this text to treat the nomenclature of the individual head sclerites and the lines of fusion or sulci that mark the articulation of adjacent sclerites, but Figure 13.2 illustrates some of the more important cranial features of the insect head. Externally, head segmentation can be discerned only by the presence of metameric paired appendages. The head is attached to the thorax by a flexible membranous neck or **cervix.** At the posterior aspect of the head is an opening in the head capsule, the **foramen magnum,** through which several internal systems pass between the head and thorax (Fig. 13.2c). Internally, an integumental framework or **tentorium** provides additional strengthening to the head capsule and attachment points for muscles that operate the mouth appendages (Fig. 13.2d). The tentorium consists of the paired **anterior** and **posterior apodemes,** which fuse to form a rigid internal endoskeleton for the cranium. Often, a third pair of apodemes, the **dorsal arms,** is present as additional internal braces.

The **antennae** are paired appendages borne by the first head segment and are usually anteroventral to the compound eyes. Often reduced in immature insects, the antennae in adults take many different forms or shapes. An antenna is composed of a basal **scape,** a middle **pedicel,** and a long **flagellum** composed of many podomeres (Fig. 13.2a, b). Adult antennae are equipped with chemical, olfactory, tactile, and auditory receptors; in most insects the pedicel contains **Johnston's organ** which senses vibrations.

Eyes. On either side of the head are the **compound eyes,** usually with a large number of ommatidia (up to 30,000 in some dragonflies) (Fig. 13.2a, b). Generally, the fast fliers who depend upon a high degree of depth perception for feeding or locating mates have the largest

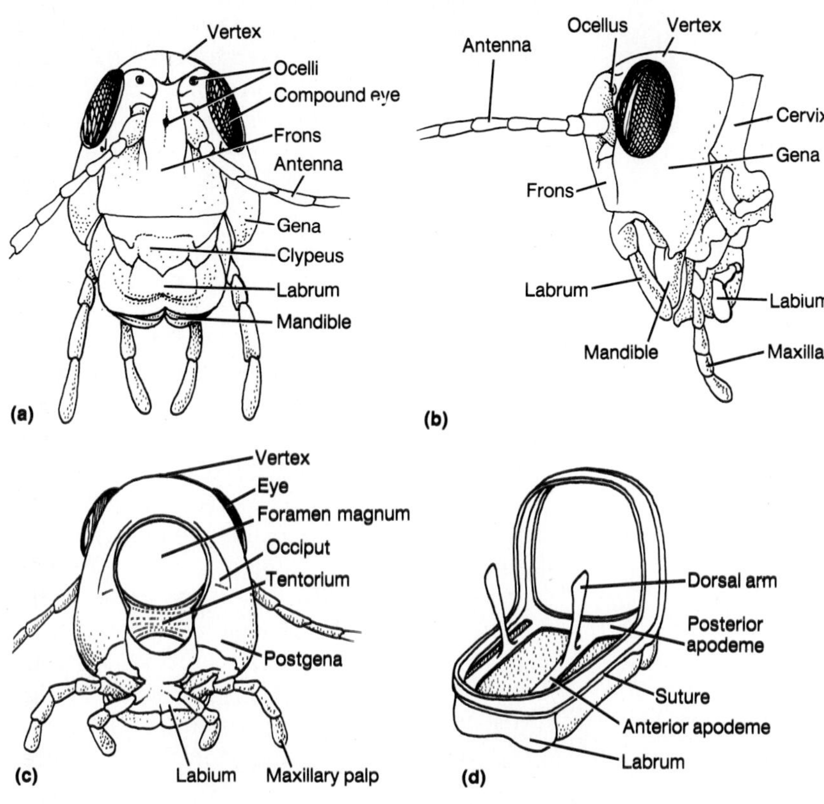

Figure 13.2 *The most important features of the head of a generalized insect: **(a)** anterior view; **(b)** lateral view; **(c)** posterior view; **(d)** the tentorium (parts a-d from R.E. Snodgrass, 1935,* Principles of Insect Morphology, *McGraw-Hill Book Co., used with permission).*

numbers of ommatidia. Nocturnal insects have dark-adapted compound eyes, and many insects can perceive different wavelengths of light, especially ultraviolet, blue-green, and red. The reader is encouraged to refer to the treatment of the functional morphology of compound eyes found in Chapter 10.

In addition to the compound eyes there are normally three dorsal **ocelli** or simple eyes on the anterior cranial surface (Fig. 13.2a, b). Sudden changes in light intensity, apparently perceived by the ocelli, result in an increase in muscle tonus and preparation of the insect to fly, jump, or run. Some immature insects have photoreceptors called **stemmata** or lateral ocelli that are intermediate in structure and function between compound eyes and ocelli. As the adult stage is reached, the stemmata are generally lost, but in some primitive insects the stemmata persist as the only eyes of the adult.

Mouthparts and feeding. In association with the mouth are the three pairs of mouthparts that are the appendages of the adult head segments II–IV plus several unpaired structures. The mouthparts are the unpaired labrum and hypopharynx, the paired mandibles and maxillae, and a second pair of maxillae that is fused to form a labium. While there are a number of important modifications in mouthparts that enable a given insect to feed on a specific type of food, mouthparts can be reduced to two general types: mandibulate and haustellate. The **mandibulate** mouthparts are less specialized

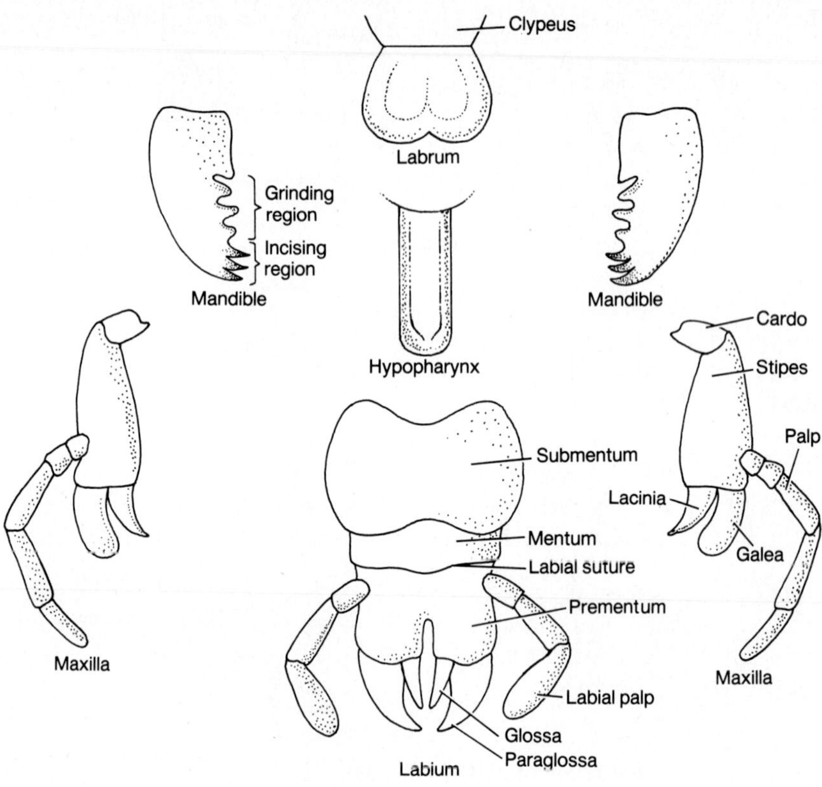

Figure 13.3 The mandibulate mouthparts of a generalized insect.

and are principally adapted for biting off and chewing discrete particles of food; we shall consider this type first. The paired mouthparts, in effect, create a **preoral cavity** through which food passes before it enters the **mouth.** The **labrum** or "upper lip" is an unpaired movable part suspended from the head (clypeus) and forming the anterior border of the preoral cavity (Figs. 13.2a, 13.3). The **mandibles,** representing the paired appendages of segment II, are heavily sclerotized unsegmented jaws that are opened and closed by powerful muscles. Each mandible has a distal **incisor** or cutting **region** and a proximal **molar** or grinding **region** (Fig. 13.3). The mandibles of certain beetles can even bite through metals such as lead, tin, and copper, a feat that certainly attests to the hardness of the mandibles and the force that mandibular muscles can generate.

Posterior to the mandibles lies a pair of jointed **maxillae,** the appendages of segment III, that forms part of the sides of the preoral cavity (Fig. 13.3). The maxillae manipulate the food by keeping it properly placed against the mandibles, function in tasting, and are used in cleaning antennae and legs. Each maxilla consists of a proximal podomere, the **cardo,** and a more distal podomere, the **stipes.** The stipes, in turn, bears three distal lobes: a **galea,** a toothed **lacinia,** and a lateral **palp.** The palps bear many fingerlike projections that contain a prodigious number of chemical and tactile receptors. The **labium** or "lower lip" is a single structure but represents the fusion of the paired **second max-**

illae or the limbs of segment IV (Fig. 13.3). Forming the posterior and some lateral boundaries for the preoral cavity, the labium consists of the proximal **submentum,** a middle **mentum,** and a distal **prementum** that bears a pair of **labial palps.** Like the first maxillae, the labium manipulates and tastes food.

The other unpaired mouthpart is the **hypopharynx,** which lies in the middle of the preoral cavity in much the same way as does a tongue (Figs. 13.3, 13.4). Between the hypopharynx and labium is the **salivarium,** a region into which salivary glands empty **saliva,** a watery fluid that moistens food and initiates digestion. The salivary glands of some lepidopteran larvae secrete natural silk, which is used to construct cocoons. Between the hypopharynx and the labrum is the **cibarium,** a cavity through which food passes from the preoral cavity to the mouth (Fig. 13.4). The cibarium is often modified as a pump used to bring food into the mouth.

Figure 13.4 Diagrammatic sagittal section of the oral region of a generalized insect with mandibulate mouthparts (from R.E. Snodgrass, 1935, Principles of Insect Morphology, *McGraw-Hill Book Co., used with permission).*

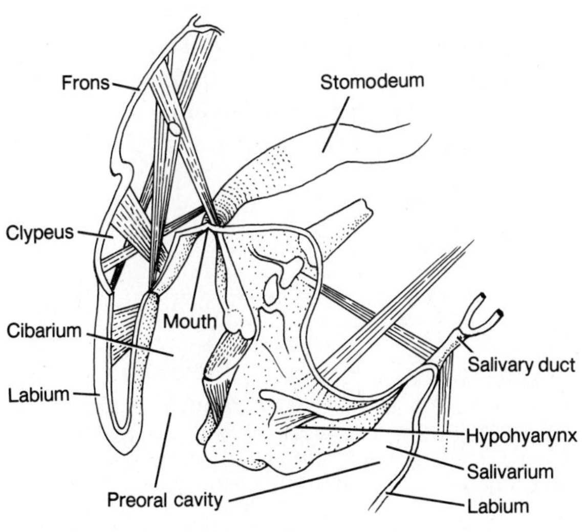

Insects have developed many modifications on this mandibulate mouthpart plan. In various insects, chewing mouthparts are adapted as elongated structures for holding prey, as devices for holding females during copulation, as a tube through which the body fluids of prey are sucked, as flattened structures used to mold and shape balls of dung or pollen, and as grossly enlarged defensive weapons. In some insects the mandibulate mouthparts have even become so specialized that they have completely lost their basic function as chewing appendages.

Many advanced insects have mouthparts of a second type called **haustellate,** which are adapted for sucking liquids. Usually, one or more of the mouthparts have become stylets or needlelike blades used to pierce or abrade prey tissues (Fig. 13.5a), and some of the mouthparts may be severely reduced or even absent. Such piercing-sucking mouthparts are used to penetrate the prey by female mosquitoes, most biting flies, and hemipterans and homopterans. Others, like honey bees, have lapping mouthparts used to suck nectar from flowers. Lepidopterans possess a coiled proboscis that, when uncoiled and inserted into the throat of a flower, functions as a nectar siphon (Fig. 13.5c). House flies and some other dipterans have a sponging method of feeding in which the enlarged distal portion of the labium, the **labella,** soaks up fluids through minute canals (Fig. 13.5b). Often, saliva is secreted onto solid food, which is quickly dissolved, and the broth is drawn into the mouth.

Adult mayflies, most caddisflies, and some moths do not feed, and the mouthparts are nonfunctional. Most immature forms (nymphs, larvae) have mandibulate mouthparts regardless of the type found in adults.

The mouthparts are variable in their orientation relative to the head. The ancestors of insects probably had mouthparts that were directed downward from the head. This primitive **hypognathous** position enabled early insects to

(a)

(b)

(c)

consume food on the substratum directly beneath the head; this position characterizes orthopterans and many others. Many insects such as hemipterans and homopterans have evolved a more advanced **opisthognathous** position whereby the long stylet mouthparts project backward between the legs when not in use. Another modification is the **prognathous** position whereby the mouthparts project anteriorly from the head; this arrangement is better suited for insects that tunnel or actively pursue prey such as many beetles.

THORAX

This tagma is always composed of three segments: an anterior **prothorax,** a middle **mesothorax,** and a posterior **metathorax** (Fig. 13.6a). These three segments are most specifically associated with locomotion, since each one invariably bears a pair of legs, and each of the posterior two metameres typically bears a pair of wings. The three pairs of legs and, in most, two pairs of wings are the two most distinctive characteristics of insects. The sclerites of the thorax (especially the terga and sterna) are exceptionally thickened to support the legs and wings and to provide adequate attachment sites for the extrinsic muscles that operate them. Immature stages never have functional wings, and some lack legs. Usually the mesothoracic and metathoracic segments bear a single **spiracle** on either side, which is the opening associated with the gas exchange system (Fig. 13.6a).

◀ ***Figure 13.5*** *Scanning electron micrographs of the haustellate mouthparts of several different insects:* ***(a)*** *the piercing or cutting mouthparts of a horse fly;* ***(b)*** *the spongelike labella of a house fly;* ***(c)*** *the coiled sucking proboscis of a moth. (parts a, b [courtesy of T. Carlysle and P.S. Callahan], c [courtesy of the U.S. Department of Agriculture] in P.S. Callahan, 1971,* Insects and How They Function, *Holiday House, New York, N.Y.)*

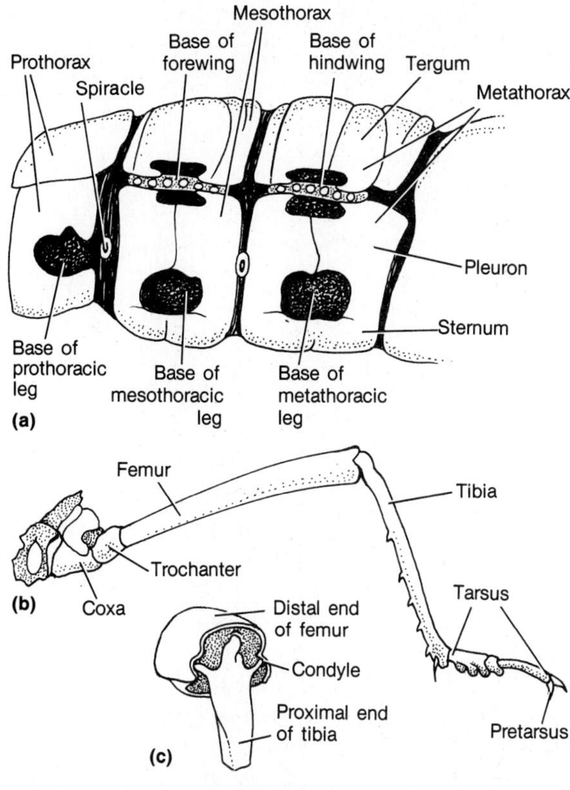

Figure 13.6 *The thorax and leg of a generalized insect: **(a)** a lateral view of the left side of the three thoracic segments; **(b)** a typical leg; **(c)** the articulation between the femur and tibia. (parts b, c from R.E. Snodgrass, 1935,* Principles of Insect Morphology, *McGraw-Hill Book Co., used with permission)*

Legs. The three legs on a side are termed **prothoracic, mesothoracic,** and **metathoracic** to indicate the thoracic segment to which they are attached. A generalized insect leg is composed of six podomeres, which are, in order from proximal to distal: **coxa, trochanter, femur, tibia, tarsus,** and **pretarsus** (Fig. 13.6b). The short sturdy coxa articulates with a pleuron of the thoracic segment. The femur is typically the largest segment of a leg and usually contains the greatest musculature. The pretarsus consists of one or two claws and an adhesive pad that permits some insects like flies to walk upside down or on smooth vertical surfaces. One or two condyles are present at each leg joint and act as a fulcrum or pivot for movement of the more distal podomere (Fig. 13.6c). A number of intrinsic appendicular muscles are usually present so that a wide range of leg movements is possible.

The least specialized of leg types that has been described above is for legs that are modified for walking. But variations on this basic plan are present in many insects. In some the legs are elongated and slim and function in running, and grasshoppers have greatly enlarged metathoracic femora, which enable these insects to jump for considerable distances. Sometimes the prothoracic legs are modified as raptorial appendages for grasping and holding prey as in the praying mantids. Many aquatic insects have legs that are flattened, equipped with "swimming hairs," and modified in other ways to function as oars or paddles. A few insects have legs that are heavily sclerotized and that are modified to enable the animal to dig in the soil. Still other legs on various male insects may be modified for grasping the female during copulation. In honey bees the legs are profusely equipped with hairs that form baskets for transporting pollen to the hive.

Wings and flight. Insects are singularly different from all other invertebrates because they have wings and can fly. Unlike the wings of birds or bats, insect wings are not modified paired appendages but rather are novel lateral outgrowths or outfoldings of the integument. An insect wing is anchored to its thoracic segment in a horizontal orientation between tergum and pleuron. Both upper and lower surfaces of mature wings consist primarily of cuticular layers. As wings develop, the space between the two layers is occupied by extensions of the hemocoel and wing tracheae. As

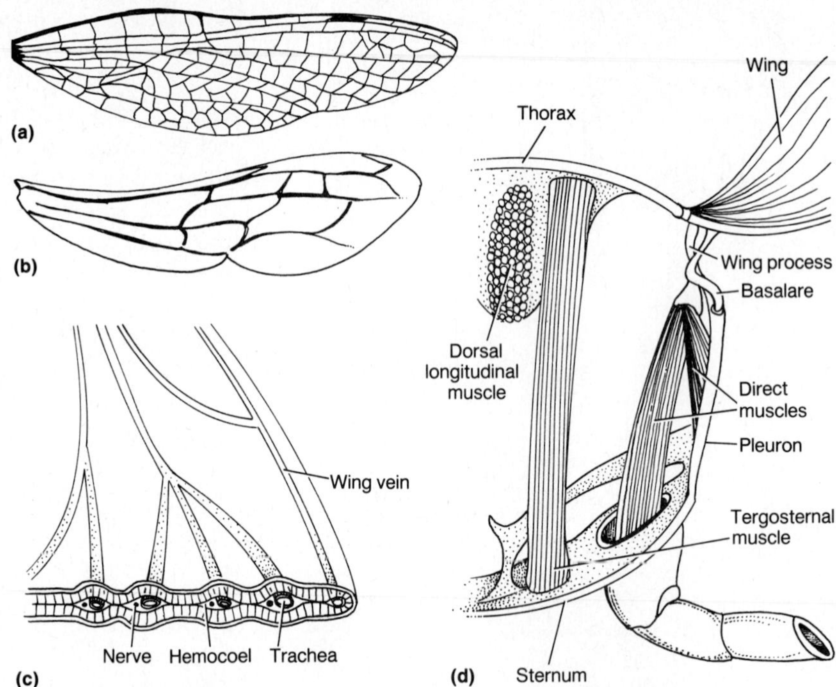

Figure 13.7 *Insectan wings: **(a)** from a primitive insect (dragonfly); **(b)** from an advanced insect (wasp); **(c)** diagrammatic cross-sectional view of a portion of a wing; **(d)** the direct and indirect flight muscles (from R.E. Snodgrass, 1935,* Principles of Insect Morphology, *McGraw-Hill Book Co., used with permission).*

the wing becomes fully developed, the two cuticular layers become closely appressed to one another to obliterate most of the hemocoelic spaces (Fig. 13.7c). The tracheae, along with thin strands of hemocoel persisting around the tracheae, constitute the wing **veins** (Fig. 13.7a, b). The cuticle around the tracheae sclerotizes, resulting in strong braces for the delicate wings. Both tracheae and blood spaces are essential for supplying nourishment to the living integumental cells of the wings; cyclic circulation of hemolymph is made possible by many crossveins. Wing venation is very characteristic for all insects within an order and for all subordinate taxa, and venational patterns are used extensively by entomologists in demonstrating evolutionary relationships and for classifying insects.

Primitive insects (Apterygota) never develop wings, but in higher insects (Pterygota) there are two pairs of wings with one pair, the **forewings** or **mesothoracic wings** borne by the mesothoracic segment and the **hindwings** or **metathoracic wings** arising from the metathoracic segment. When at rest, wings are primitively held horizontally from or vertically over the thorax. But in most insects the wings are flexed so that they "lie down" over the abdomen. Some advanced insects like certain termites and ants lose both pairs of wings and thus are secondarily wingless.

The wings are mounted so that the tergum provides the downward force against the base; the edge of the pleuron serves as a fixed fulcrum for the wing motion (Fig. 13.7d.) To move the wings, the two wing-bearing segments are equipped with two basic types of muscles. One set of muscles is called **direct muscles,** since they run from thoracic sclerites (usually sterna) directly to sclerites that in turn are attached to a point near the base of each wing but lateral to the fulcrum (Fig. 13.7d). A second set of muscles is called **indirect** or **tergosternal muscles;** these do not attach directly to the wings but extend from sternum to tergum; these muscles alter the thorax and, indirectly, move the wing. The contraction of the direct muscles depresses the wing tip, since their point of insertion is lateral to the fulcrum (Fig. 13.7d). The wing tips are elevated by the contraction of the tergosternal muscles, which depresses the tergum and causes the wing bases to be depressed; this results in the wing tips being raised. This arrangement of both direct and indirect muscles is characteristic of some insects like Orthoptera, Odonata, Ephemeroptera, and Lepidoptera, in which both sets of muscles provide the force for the up-and-down movements of the wings. Direct muscles are also **synchronous** since they are innervated directly, and a single nerve impulse produces a single muscle twitch. Thus synchronous muscles contract only as fast as impulses reach them. Wing beats in insects with direct muscles are usually fewer than 25 per second.

For most insects the main force for wing movements is supplied by two different indirect muscles (Fig. 13.7d). The thorax, especially the mesothorax and metathorax, is a flexible but resilient box whose shape can be altered by the action of indirect muscles, which are of two types: **dorsal longitudinal** and **tergosternal** (Fig. 13.7d; 13.8). When the dorsal longitudinal muscles contract, the tergum is vaulted dorsally; this causes the wing base to be elevated in relation to the fulcrum and the wing tips to be lowered (Fig. 13.8b, d). The tergosternal muscles then contract, pulling the tergum downward and thus raising the wing tips (Fig. 13.8a, c). Alternate contractions of the longitudinal and tergosternal muscles cause the up-and-down movement of the wings. But as one

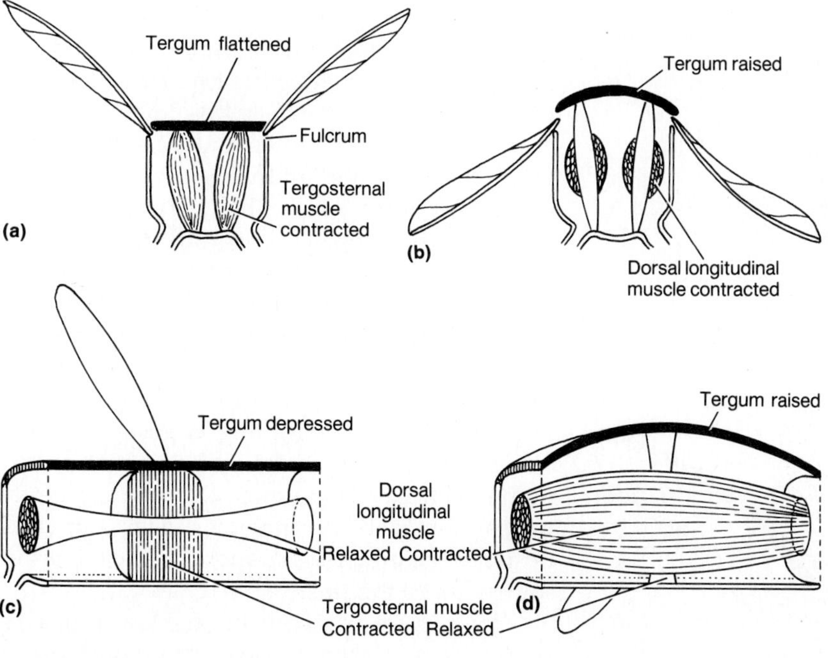

Figure 13.8 *Diagrams of the events during flying in an insect:* ***(a)*** *a cross-sectional view through the thorax with wings elevated;* ***(b)*** *a cross-sectional view through the thorax with wings depressed;* ***(c)*** *a longitudinal-sectional view of the thorax with wings elevated;* ***(d)*** *a longitudinal-sectional view of the thorax with wings depressed.*

set of muscles contracts, the other set is stretched, and the stretching is a stimulant for its contraction. Therefore such muscles may contract repeatedly in response to a single nerve impulse. Such muscles are considered to be **asynchronous,** since they contract on indirect stimuli (stretching). In insects that employ indirect muscles as the chief wing movers, wing beats of up to 1000 per second are possible. Direct muscles are also present and are used to twist the wings in various ways to produce an ellipse or a figure eight, which the wing tips trace out during flight (Fig. 13.9a, b). Many insects are able to hover, and some can even fly backward by altering the inclination of the figure eight (Fig. 13.9c, d).

Figure 13.9 *Pathways of the wing tip during:* ***(a)*** *forward flight (grasshopper);* ***(b)*** *forward flight (bee);* ***(c)*** *hovering (bee);* ***(d)*** *backward flight (bee).*

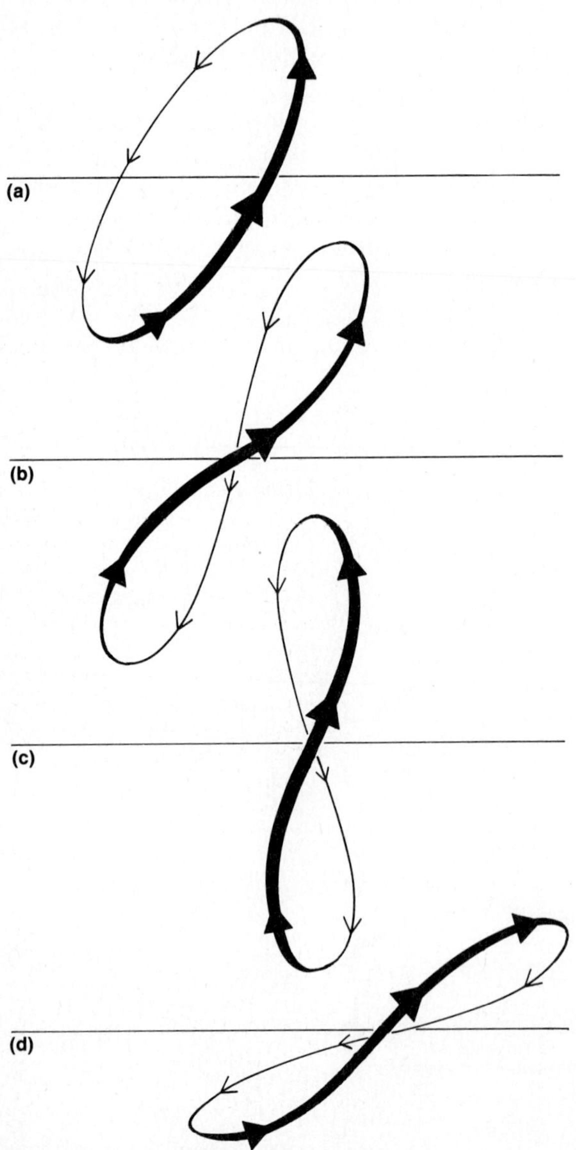

An important modification of the indirect muscle system is found in some Diptera and Coleoptera, in which the wings have two stable positions—elevated and depressed (Fig. 13.10). Movement away from one of these two stable positions is resisted by the rigidity of the thorax and must be overcome by muscle exertion. This resistance is exerted up to that point at which the wing is suddenly driven by the elasticity of the thorax into the other stable position, a mechanism aided by the arrangement of sclerites at the base of the wing (Fig. 13.10b, e). Thus the wing snaps into the opposite position; this is called the "click" mechanism. As the wing snaps from wings up to wings down, for example, the tergosternal muscles are suddenly stretched; the stretching in turn stimulates them to contract spontaneously. As they contract, they pull the thorax down and thus elevate the wing tips. At the midpoint of this upward movement of the wings, they snap upward into that stable position. This upward click stretches the longitudinal muscles, which then contract, and the cycle is repeated (Fig. 13.10). The thorax may be distorted and the wings flap many times for each nerve impulse arriving at the muscles. Moreover, direct flight muscles usually twist the wings so that their leading edges are pulled downward during the downstroke and upward in the upstroke. This almost 180° rotation of the wings substantially increases their lift and thrust in the downstroke and increases their lift somewhat in the upstroke (Fig. 13.10).

Wing pairs in some of the lower insects work independently of each other, but in many

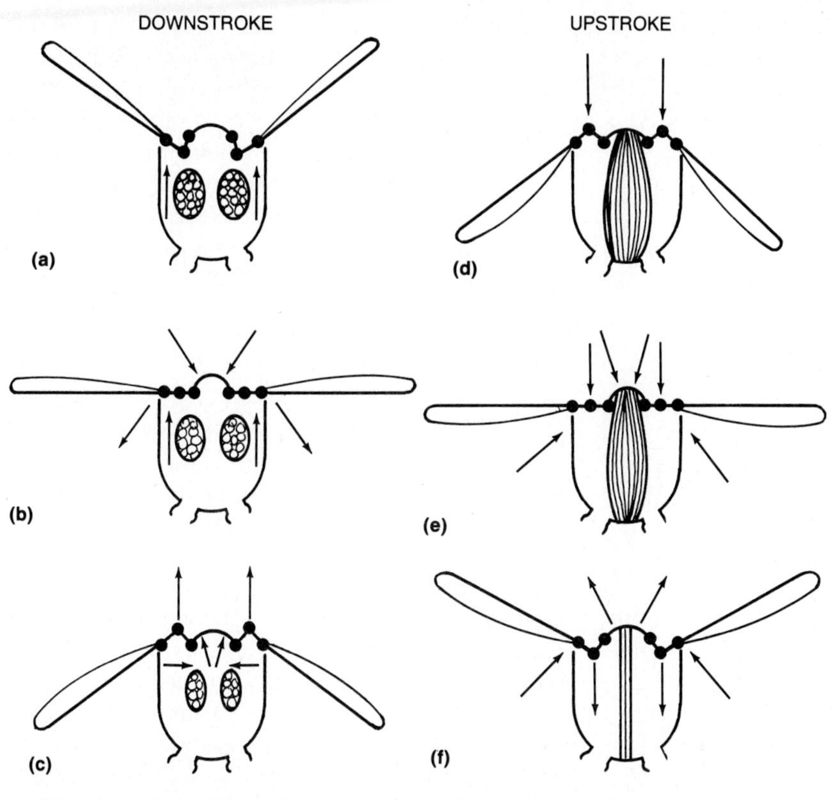

Figure 13.10 *The "click" mechanism found in certain insects like beetles and flies:* ***(a)*** *dorsal longitudinal muscles contract, scutellar lever (arrow) is forced upward;* ***(b)*** *tergum is sprung, lever passes the midpoint;* ***(c)*** *dorsal longitudinal muscles relax abruptly, tergum is vaulted dorsally;* ***(d)*** *tergosternal muscles contract, scutellar lever is forced downward;* ***(e)*** *tergum is sprung, lever passes the midpoint;* ***(f)*** *tergosternal muscles relax abruptly, tergum springs dorsally.*

insect groups there is a tendency for both wings on a side to be coupled together so that they can function as a single unit. In hymenopterans, for example, the flight muscles mostly affect the metathoracic wings, which are hooked to the mesothoracic wings with numerous small hooks or **hamuli** so that both are driven as a unit. In certain insects there is a tendency for the hindwings to be reduced or severely modified. This tendency culminates in the Diptera, in which the metathoracic wings, called **halteres,** do not function as wings but are modified as stabilizers. In flight the halteres vibrate up and down, usually with the same frequency as the mesothoracic wings. Halteres are especially sensitive to subtle changes in orientation of the body, and nerve impulses from them to the brain enable the fly to correct its course by altering the mesothoracic wing strokes. Exactly the opposite trend has occurred in the Coleoptera, in which the principal flying organs are the metathoracic wings, and the mesothoracic wings, or **elytra** are hardened and nonfunctional in flight and protect the membranous hindwings when at rest.

Abdomen

The posterior insect tagma is the **abdomen,** comprised primitively of 11 segments and a highly reduced pygidium. Terminal segments may be reduced in number or size or may be fused with each other, so the actual number is usually ten or fewer. The tergum and sternum of each segment are hard sclerites, and the pleural regions are typically membranous and usually not sclerotized (Fig. 13.11). In each of the anterior eight segments the tergum bears a

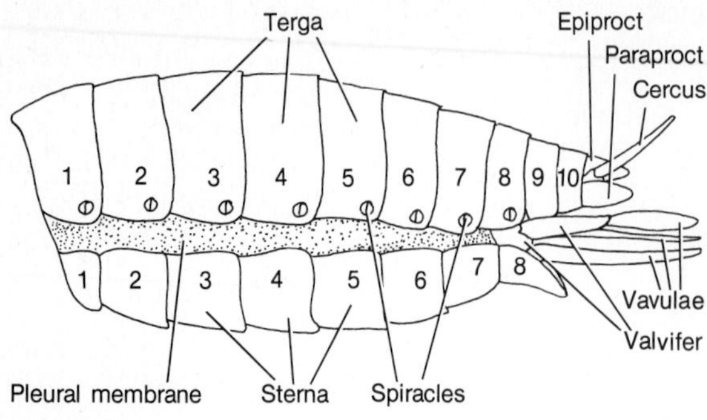

Figure 13.11 A lateral view of the genralized abdomen of a female valvifers and valvulae comprise the ovipositor) (from E.M. Duporte, 1959, Manual of Insect Morphology, *Reinhold Publishing Corp., New York, NY)*

pair of **spiracles.** The terminal segment may bear a pair of **cerci** which are serially homologous to mouthparts and legs and have numerous tactile setae. The anal opening is situated at the posterior end of the abdomen.

Located on the abdomen of adults are the external genitalia, which in some insects are conspicuous but in others concealed. In males an unpaired ventral copulatory organ, the **aedeagus,** is present on the ninth segment. This same segment also often bears a pair of **claspers** used to grasp and orient the female in a proper position during copulation. Females possess paired structures (valvifers, valvulae) located on the eighth and ninth segments that collectively constitute the **ovipositor** (Fig. 13.11). The ovipositor is variously modified to accommodate the diverse ways in which eggs are to be deposited—underground, in plant tissues, in a host, under bark, and so on. In a great many hymenopterans the ovipositor has been modified as a stinger, and its primary function of egg deposition has been lost. In one order (Collembola) a unique springing device is borne by abdominal segments.

The immature stages of many insects develop paired abdominal appendages, which are lost as adulthood is attained. Some aquatic forms may bear abdominal gills, and others living in swift-running water have hooked holdfasts to help maintain their position. Many larvae, including caterpillars, have lobelike **prolegs** with tiny hooklets that aid in locomotion.

Sense organs

The presence of the cuticular exoskeleton has altered to some extent sensory reception in insects. Receptors are generally found all over the integument, but they are especially concentrated on the appendages, on the head, and at the posterior end. Insect receptors fall into three main categories: photoreceptors, mechanoreceptors, and chemoreceptors. **Photoreceptors** include compound eyes, ocelli, and stemmata, which have already been discussed.

Insect **mechanoreceptors** sense various mechanical changes such as stretching, bending, compression, and vibrations. All mechanoreceptors are thought to have been derived from a common structure, and these typically filamentous, bristlelike, or peglike organs are called **setae** or **sensillae** (Fig. 13.12a–c). From a

*Figure 13.12 Several different types of insect sensory receptors: **(a)** diagram of a single sensillum; **(b)** scanning electron micrograph of the anterior end of a cotton boll weevil showing sensory spines (sensilla) on the integument; **(c)** greater magnification of part b showing several sensilla in greater detail; **(d)** scanning electron micrograph of the antennal base of a wasp showing numerous proprioceptors that control the position of the antenna; **(e)** diagram of a tympanic organ (from W.N. Hess, 1917,* Ann. Entomol. Soc. Amer. 10:*Plate VI); **(f)** Johnston's organ (from R.E. Snodgrass, 1935,* Principles of Insect Morphology, *McGraw-Hill Book Co., used with permission). (parts b-d courtesy of T. Carlysle and P.S. Callahan in P.S. Callahan, 1971,* Insects and How They Function, *Holiday House, New York, N.Y.)* ▶

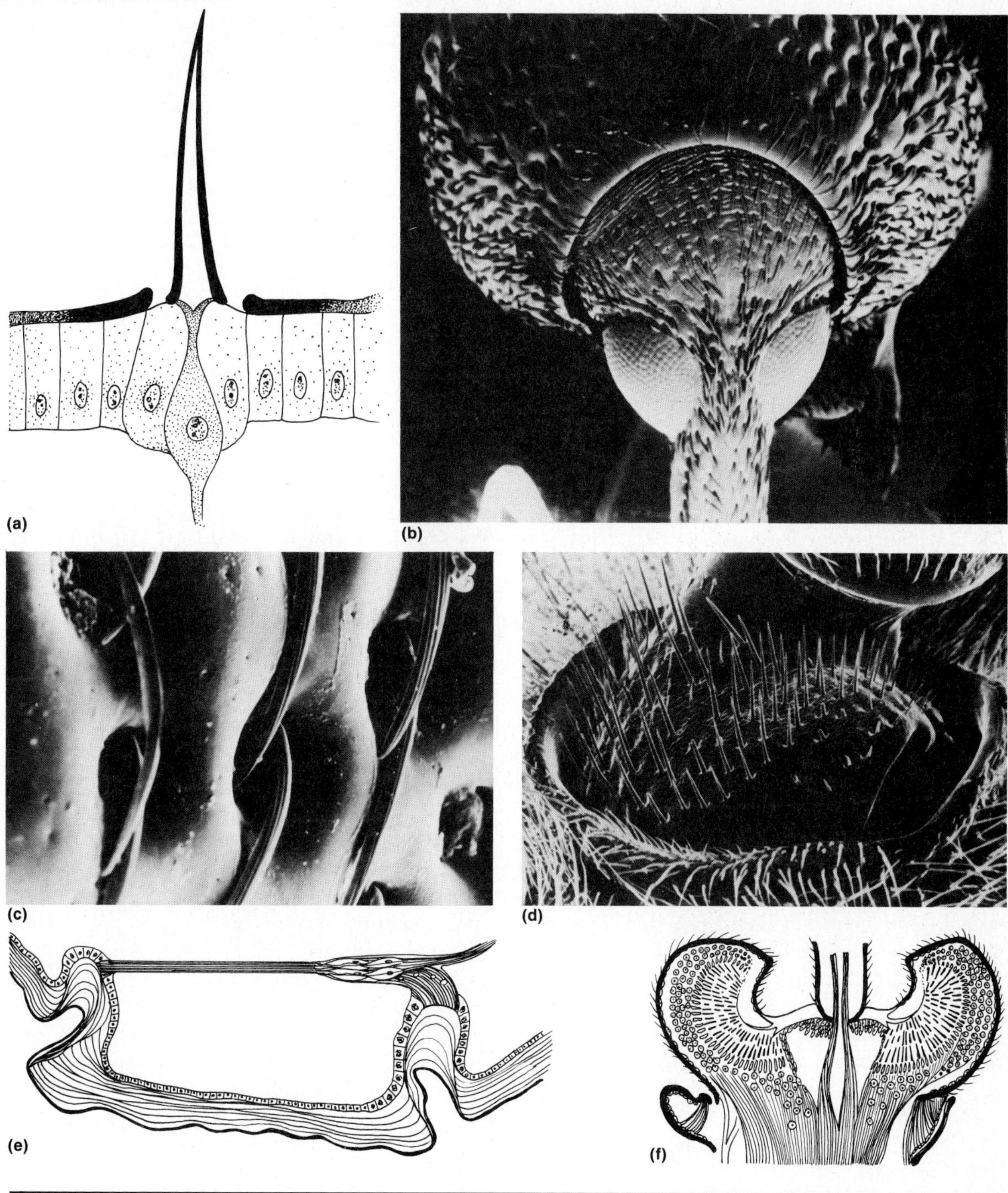
(a)
(b)
(c)
(d)
(e)
(f)

functional standpoint, mechanoreceptors can be divided further into tactile receptors, proprioceptors, and sound receptors. **Tactile receptors** are stimulated when a hair comes in contact with another surface or object. Some tactile setae are sensitive to constantly changing stimuli; in a flying insect, setae are situated on the leading edges of wings and provide information about the velocity of flight. Other tactile sensillae respond when the hair or bristle is bent or deformed by the presence of another object, and they are concentrated on those insect parts most likely to be contacted such as on antennae, legs, wings, mouthparts, and cerci. **Proprioceptors** provide the animal with continuous information regarding the relative positions of parts of the body to each other, and they are stimulated by relative changes such as in length or tension in various parts of the insect (Figs. 13.12d). Since they are particularly sensitive to stresses in the cuticle, these sense organs are important in body orientation, posture, and the relationship of the body to gravity. Proprioceptors, present in a variety of forms and shapes, are found especially on the antennae, head, legs, wings, ovipositor, and mouthparts and within several internal organs. In almost every insect studied, a group of setae found in the pedicel of each antennae constitutes **Johnston's organ,** which functions in proprioception and as a vibration detector (Fig. 13.12f). **Sound** reception involves the detection of sound waves through air or water (hearing) or through the substratum (vibration detection). Most insects have no specific hearing organs as such. The Orthoptera (grasshoppers) and some Homoptera (cicadas) and Lepidoptera (certain moths) have a **tympanic organ,** which involves a thin integumental tympanum and groups of setae attached to its inner surface (Fig. 13.12e). Sound waves cause the vibration of the tympanum, which in turn bends the setae and generates a nerve impulse. Insects with an auditory tympanum also usually stridulate, and the sounds they produce play a major role in their social life. Some other insects possess a group of sound setae in the tibial leg segment (subgenual organ), and others use Johnston's organ as an auditory structure. Perception of vibrations is a well-developed sense in most insects, especially in those with a subgenual organ. In still others, specialized sensillae located on the legs are sensitive to vibrations.

In **chemoreception,** hair setal organs come in contact with chemicals either directly (taste or gustation) or at some distance (smell or olfaction). Both gustatory and olfactory receptors are distributed all over the body but especially on the antennae, mouthparts, legs, and ovipositor. **Gustatory sense organs** respond to direct contract with chemicals in fairly high concentrations and are normally associated with feeding. The tips of most of the mouthparts are profusely equipped with taste organs; the cibarium and even the foregut may also contain contact chemical receptors. The tibiae and tarsi of many Hymenoptera bear chemoreceptors, and other parasitic hymenopterans have contact chemoreceptors on the ovipositor. **Olfactory receptors** enable the insect to respond to chemicals or ions in the atmosphere at exceedingly low concentrations. These receptors, for the most part located on the antennae, show a high degree of specificity to the types of chemical stimuli received. Olfactory senses are used in locating food or specific types of flowering plants and in mate attraction. Often, chemicals called **pheromones** are produced by the insects themselves; these are discussed in Box 13.2. The male silkworm moth reacts to a sex attractant produced by a female in concentrations as low as 100 molecules of the attractant per cubic centimeter of air. Virgin gypsy moth females were able to attract males from as far as 8 km downwind! Insects apparently have a much greater capacity for discriminating between olfactory chemicals than between gustatory chemicals. Light, sounds, and vibrations are all important environmental stimuli to insects, but they are extremely sensitive to chemicals, and their environment is one that is sensed and experienced predominately by taste and smell. Unfortunately, the physiological mechanisms behind the inordinate sensitivity of insects to specific chemicals are not fully understood.

Many insects are also responsive to changes in temperature, humidity, and magnetism. In some insects, thermoreceptors are located on the antennae; in others, temperature sensations are received generally over the body. Responses to differences in humidity have been noted in some insects, but humidity receptors have not been specifically identified except in the human louse, in which antennal "tuft" organs function in this capacity. Honey bees build their nests oriented along the magnetic field, but magnetoreceptors have not yet been discovered.

Functional Morphology of Visceral Systems

While a schizocoelom appears in insects, as in all arthropods, early in embryonic development, it does not persist in the adult insect as a chamber or space. The walls of the embryonic coelomic sacs contribute to the formation of the dorsal vessel, but this cavity disappears in development and is never functional in adults.

Digestive system

Though there are wide variations in types of food ingested and mouthparts employed in feeding, the alimentary systems of all insects have remarkably similar morphological and physiological attributes. This system is composed of three basic regions: a foregut or stomodeum, a midgut or mesenteron, and a hindgut or proctodeum. Both stomodeum and proctodeum are lined by an ectodermally derived epidermis that produces the cuticle or **intima** lining their lumens. Both foregut and hindgut grow inward and join the hollow, noncuticular, endodermally derived mesenteron (Fig. 13.13a, b).

The **foregut** exhibits the most variation of the three regions, owing to the diversity of foods ingested. Food enters the cibarium and is mixed with saliva; some initial digestion begins there (Figs. 13.4, 13.13c). From the mouth, food enters a variably constructed **pharynx;** the cibarium or pharynx or both may have muscular walls and function as food pumps. The pharynx is continuous with the **esophagus** whose posterior portion is enlarged typically as a **crop** where food is stored temporarily. Immediately posterior to the crop is the **gizzard** or **proventriculus,** a highly variable chamber that is often equipped with teeth or denticles that macerate the food further and, in some species, strain or filter the food materials. At the posterior end of the foregut is a **stomodeal valve** that regulates the peristaltic movement of food into the midgut. Because of the intima, there is practically no absorption whatsoever in the stomodeum.

The **midgut,** beginning with the posterior aspects of the stomodeal valve, is also called the **stomach** or **ventriculus** (Fig. 13.13c). At the anterior end of the ventriculus near the stomodeal valve there may be a group of sacs or diverticula, the **gastric ceca,** which temporarily store food. The major digestive enzymes are secreted by epithelial cells of the mesenteron. In the midgut there is no cuticular lining, but there is often a thin chitinous **peritrophic membrane** that surrounds the food bolus (Fig. 13.13c). In some insects this membrane is continuously secreted as a "sleeve" by epithelial cells near the stomodeal valve, and it passes posteriorly with the food until it is evacuated eventually with the feces. In other species, however, the peritrophic membrane results from a delamination of cells along the length of the mesenteron. It acts as a differentially permeable membrane that allows only enzymes and digested food to pass freely through it, and most of the enzymatic digestion

BOX 13.2 □ SOME INTERESTING AND PROVOCATIVE DIMENSIONS OF INSECT COMMUNICATION

One of the most remarkable properties of insects is their ability to communicate with each other. There is some evidence that certain insects can receive sounds of an animal of a different species; for example, certain moths can actually pick up ultra-high frequency sounds generated by a potential predator, a bat, and the moth takes evasive action. But certainly the more obvious and better understood communications involve two or more individuals of the same species. This ability to disseminate information to another individual is a fascinating adaptation that greatly enhances individual survival and effectiveness. Insect communications are usually tactile, chemical, auditory, or visual.

In tactile communications, setae sensitive to pressure in one insect are stimulated by the physical presence of another insect. This certainly occurs in almost all insects at the time of mating and is a major component in sexual recognition. Social insects often touch, feel, and groom each other with tactile mouthparts and antennae. Termite workers can alarm the colony by tapping their mandibles on the adjacent wall of the nest, and these indirect tactile vibrations are detected as sound sensations by receptors on the tarsi of other termites.

One of the most remarkable forms of tactile communication is the waggle dance of the honey bee. A worker bee discovers a good source of nectar, returns to the hive, and communicates to other workers the nectar's exact location, direction, distance from the hive, scents, and richness. All these factors are indicated by different aspects of the dance. One of the dances, the so-called waggle dance, is performed on a vertical surface inside the dark hive. The dance involves a short straight run with the tail waggling; the bee then turns to the right or left thus making a circle, only to redo the dance again. The straight run deviates from the vertical direction in an angle equal to the angle that the food source lies in relation to the direction of the sun. The rapidity of the tail waggling is indicative of the distance of the food source from the hive. The entire dance,

and absorption of food take place in the midgut. The peritrophic membrane probably also serves to protect the mesenteric epithelium from the abrasive action of the food's passage. The ventriculus ends with a **pyloric valve,** a sphincter muscle regulating the passage of materials from the mesenteron into the proctodeum.

The first region of the hindgut is the **pylorus** from which arises a variable number of slender **Malpighian tubules** that function in excretion. The remaining hindgut is differentiated into an anterior **intestine** and a highly muscular enlarged **rectum** (Fig. 13.13c). Digestive activity is substantially reduced in the hindgut, and it is here that feces preparation takes place, involving a considerable reabsorption of water from the hindgut contents. The intima of the hindgut, as well as the surrounding tubular peritrophic membrane, permits continued absorption of water and salts and some food. The fairly dry fecal materials surrounded by the peritrophic membrane are usually egested in the form of small pellets.

In almost all insects are found symbiotic

performed in the dark, is conducted amid a crowd of other bees that closely monitor the dance with their antennae. Exiting from the hive, they can fly directly to the source on the basis of the tactile information communicated by the waggle dance. However, recent studies have shown that audible sounds and chemical pheromones also seem to be involved, so the honey bee dance is not completely tactile.

Many insects communicate with each other by chemical substances called pheromones that are produced by one insect and affect another of the same species. Pheromones are hormones secreted externally that provide important information for other individuals. Pheromones usually volatilize rapidly in the atmosphere, and incredibly small amounts can stimulate activity over long distances. These chemical substances serve to aggregate individuals, mark trails, generate alarm reactions, and attract mates. In the termites and perhaps even in hymenopterans, pheromones are involved in the formation of castes.

Cicadas and many orthopterans regularly communicate auditorily by stridulation; this is primarily to attract mates, although some sounds have the effect of alarms or serve to establish a territory. Other insects are attracted to a member of the opposite sex by the buzzing of their wings or by sounds created while feeding.

Visual communication is probably the most complex, since there are many varieties of patterns and colors, which, when combined with a particular behavioral mode, produce a great many different responses. Many insects are visually attracted to members of the other sex. Males of some insects "stake out" a territory and defend it against all other males of the same species who enter their area. Several species produce light flashes that are detected by other individuals, which respond to them. Fireflies are the best known of the light-emitting insects. Both males and females emit species-specific pulsating codes that serve as mate attractants.

bacteria, protozoa, and fungi that constitute the microbiota of the alimentary canal. The microbiota of many insects digest cellulose, starch, and other compounds and may produce vitamins essential for the nutrition of the insect. Often, there are special diverticula or fermentation chambers that house the symbionts and in which some of the microbiota are even digested as food by some insects. In the alimentary canal of some Isoptera are zooflagellates that have the capabilities of digesting cellulose; this fascinating mutualistic relationship between termites and flagellates is essential for the survival of both partners (see Chapter 2).

Circulatory System and Hemolymph

The open circulatory system has an extensive **hemocoel** that extends into all parts of the insect including the appendages; its basic functions are in transport and storage. The only contractile tube is the **dorsal vessel,** which extends the entire length of the animal. The dorsal vessel is

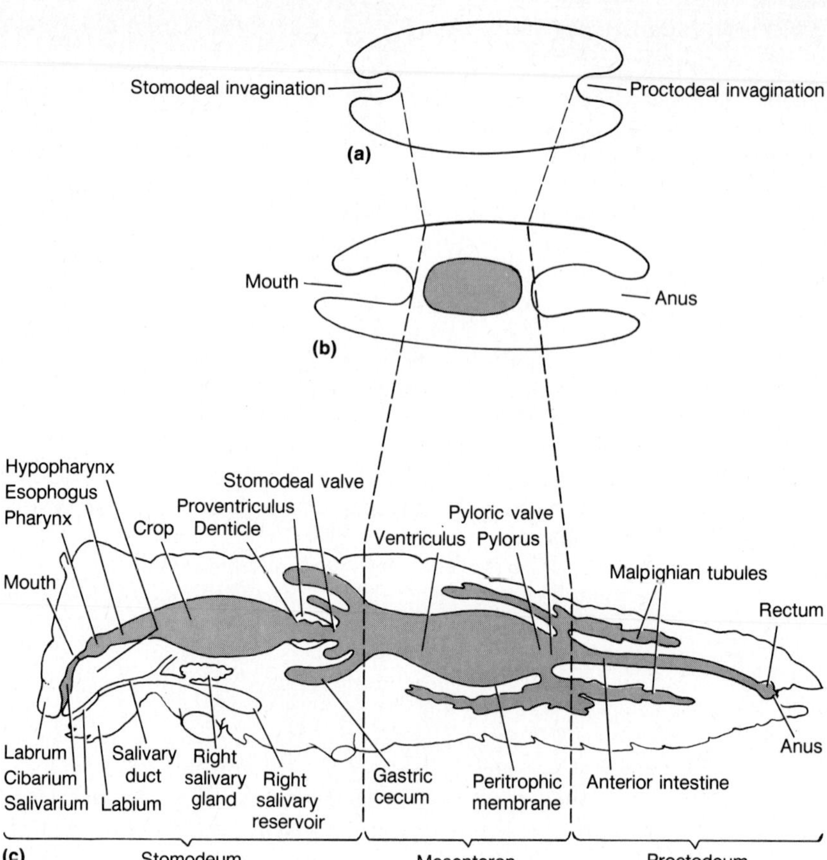

Figure 13.13 *The digestive tract of a generalized insect;* ***(a, b)*** *embryonic stages;* ***(c)*** *a stylized lateral view of the principal parts of the alimentary canal and its derivatives.*

composed of a posterior **heart,** which is mostly restricted to the abdomen, and an anterior **dorsal aorta** that extends nearly to the brain, where it opens into the hemocoel (Fig. 13.14). The heart is closed at the posterior end and is equipped with 1–13 paired, segmentally arranged **ostia** that permit entry of the blood into the heart. On either side of the heart are a number of **alary muscles** resembling wings that extend from the lateral wall of the heart to the body wall. They suspend and support the heart and may be partially responsible for the heartbeat (Fig. 13.14).

The hemocoel is divided by one or two horizontal partitioning septa into cavities or sinuses that substantially affect patterns of circulation. The **dorsal diaphragm,** mostly restricted to the abdominal region and enclosing the alary muscles, divides the hemocoel into a dorsal **pericardial sinus** and a ventral **perivisceral sinus.** The dorsal diaphragm has lateral openings through which hemolymph can pass from the perivisceral sinus dorsally into the pericardial sinus and then into the heart. In many insects a **ventral diaphragm** is situated between the alimentary canal and the ventral nerve cord. This perforate diaphragm partitions off the ventral aspect of the perivisceral sinus into a ventral **perineural sinus** (Fig. 13.14). Intrinsic muscles in both dorsal and ventral diaphragms may produce an undulating motion that facilitates and speeds hemolymph circulation. Most insects have accessory pulsating structures that assist in the circulation of blood into regions most remote from the heart. Located at the bases of the antennae, wings, and legs, these pumping structures aid in circulating blood into and out of the hemocoel extending into the appendages.

The **blood** or **hemolymph** of insects is a relatively clear fluid that contains at least seven different types of **hemocytes.** The most important are the **prohemocytes,** small cells found in all insects that give rise to other hemocytes; phagocytic **plasmatocytes** and **granular hemocytes;** and others that function in various ways including clotting and wound formation. Total hemocyte counts range from 30,000 to 50,000 per cubic millimeter.

In the fluid or plasma component, some insects have the pigment **insectoverdin** that imparts a green color to the hemolymph, and a few insects have the red pigment, **hemoglobin.** Hemolymph is composed principally of water (up to 92%) but contains many organic monomers absorbed from the digestive tract, carbon dioxide, and nitrogenous wastes usually as uric acid. The concentrations of free amino acids in insect hemolymph are much higher than in any other animal group, and this indicates that the hemolymph is a major reservoir for amino acid

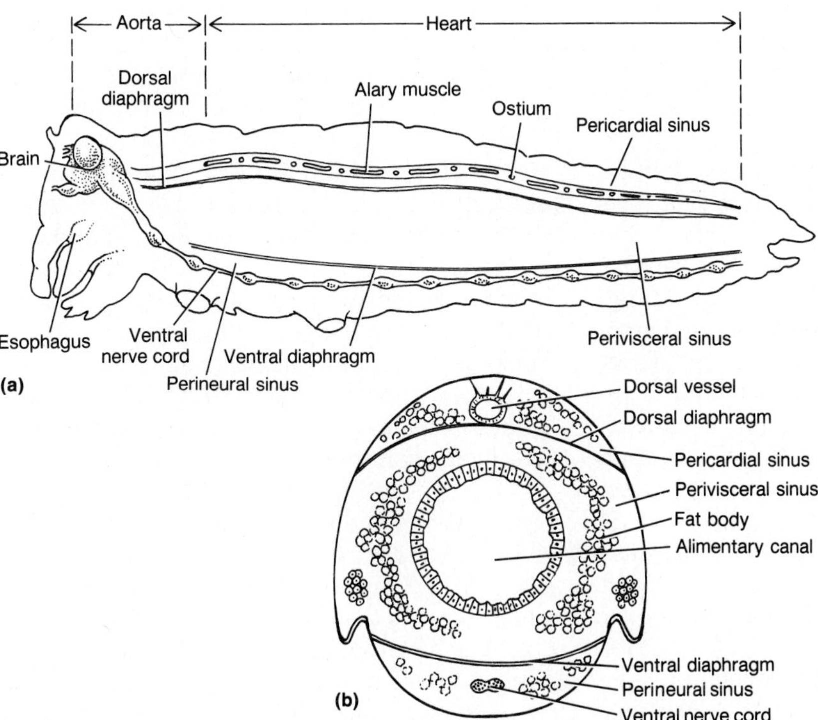

Figure 13.14 *The basic circulatory system in a generalized insect:* ***(a)*** *a stylized lateral view of the heart and hemocoelic spaces;* ***(b)*** *a cross-sectional view through the anterior abdominal region (from R. E. Snodgrass, 1935,* Principles of Insect Morphology, *McGraw-Hill Book Co., used with permission).*

storage. The amino acid concentration in blood probably has an important water-conserving function, since water would move by osmosis into the hypertonic hemolymph. The principal blood sugar is a disaccharide, α-trehalose, that only rarely is found in any other organism. The chemical composition of the blood varies according to the diet, the time elapsed since the last meal, and the stage of development. The open circulatory system is rather inefficient in terms of the speed of hemolymph transport; but because insects do not depend on blood for oxygen transport as do most other animals, there is less dependence on an efficient system that rapidly circulates hemolymph.

An unpaired organ, the **fat body,** lies in the hemocoel and is associated with the circulatory system (Fig. 13. 14b). This variably lobed body functions as a storage depot for glycogen, fats, and proteins, and it contains large amounts of food reserves in larval insects with complete metamorphosis.

VENTILATORY SYSTEM

The ventilatory system in insects is composed of an elaborate system of branching tubules called **tracheae.** Tracheae are found in all insects except for some Collembola (springtails) and larvae of some Hymenoptera, in which gaseous exchange takes place through the general integument. Tracheae originate as segmental ectodermal invaginations from the lateral walls of the posterior two thoracic segments and in up to eight of the anterior abdominal segments (Fig. 13.15a, b). The cuticle-lined tracheae open to the outside by small openings, the **spiracles,** which are the portholes through which air is inhaled. The spiracles in some insects are simple openings, but in others they are equipped with a muscle-operated valve (Fig. 13.15c, d). The valve is closed most of the time except for brief periods of inhalation and exhalation, thus preventing excess loss of water by transpiration. The spiracles are often equipped with comblike hairs that both filter the inhalent air and reduce tracheal water loss. In aquatic insects, glands surrounding the spiracles secrete a water-repelling substance that prevents the tracheae from becoming wet.

The tracheal cuticle, continuous with the integument, is present in a series of folds, the **taenidia,** that run spirally around the tracheal lumen (Fig. 13.15e). The taenidia lend some support to the tracheae; and while the tracheae can be stretched longitudinally, the taenidia prevent the tracheae from collapsing. Typically, each original trachea gives off branches dorsally to the body wall and dorsal vessel, medially to the alimentary canal and other viscera, and ventrally to the body wall and nerve cord (Fig. 13.15a,b). In a few primitive insects the tracheae remain independent of other adjacent tracheae, but in most forms the tracheae become joined by **longitudinal trunks** so that all parts of the tracheal system are interconnected (Fig. 13.15a).

In most flying insects, tracheae are dilated at various intervals as **air sacs.** The walls of the air sacs lack taenidia and therefore can be greatly distended and contracted. The functions of the air sacs are to increase the volume of inspired air, aid in flight by decreasing the animal's specific gravity, increase stability in flight, and facilitate the circulation of hemolymph.

Internally, the tracheae divide repeatedly, forming extensively branched, progressively smaller tubules that penetrate into every tissue. The very small branches of tracheae terminate in **tracheal end cells.** Each cell in turn gives off several smaller extensions, the **tracheoles,** which are continuous with the tracheae (Fig. 13.15f). It is in the tracheoles that gaseous exchange between the ventilatory system and specific tissues takes place. The tracheoles range in size from 0.2 to 1 μm in diameter and bear very small taenidia; they are lined with cuticle, which is not shed at molting. Each tracheole has a blind end filled with fluid. In resting muscle tissue the terminal end of each tracheole is filled with fluid. But at times of intensive muscular

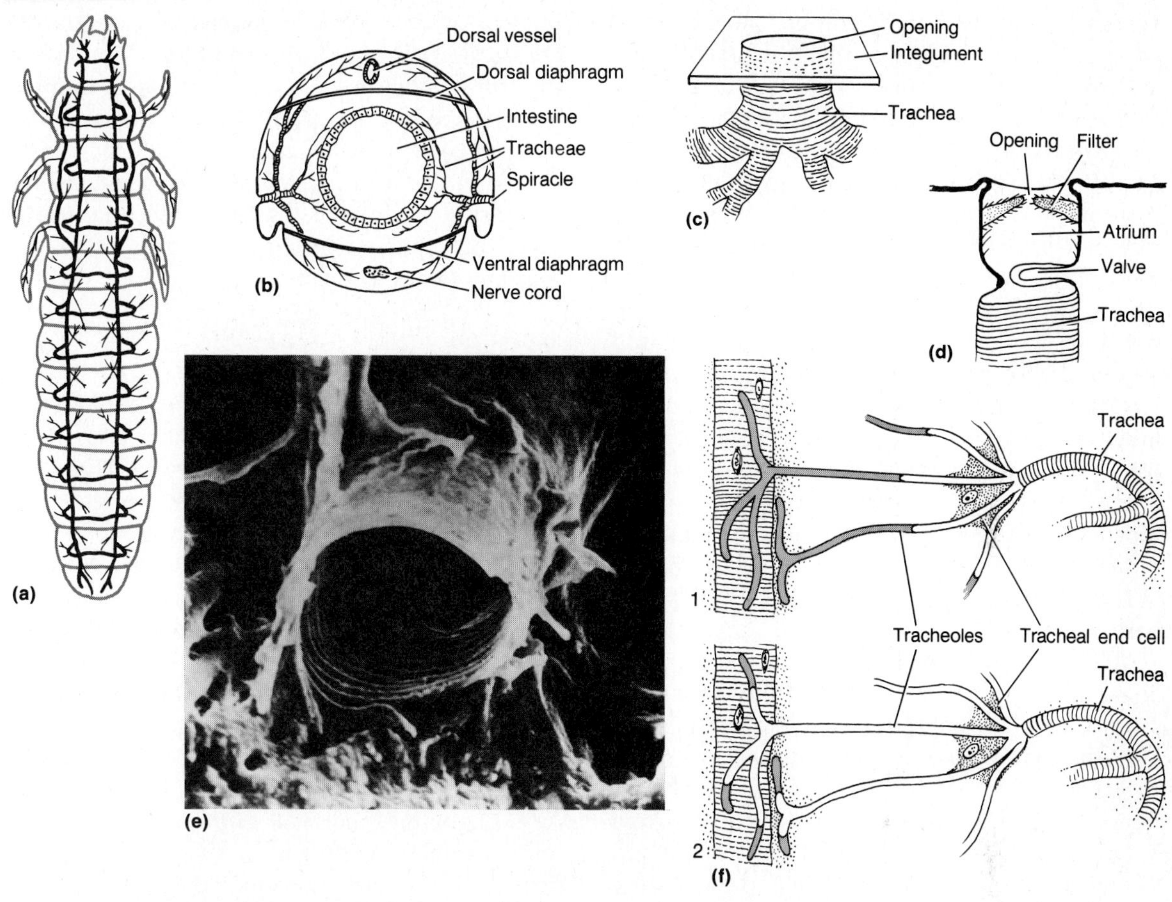

Figure 13.15 *The ventilatory system in insects: **(a)** the distribution of tracheae in the larva of a beetle (from L.F. Henneguy, 1904,* Les Insectes, *Masson et Cie, Paris); **(b)** diagrammatic cross-sectional view of an abdominal segment; **(c)** a simple spiracle without an atrium; **(d)** a complex spiracle with an atrium; **(e)** scanning electron micrograph of a portion of a trachea (note spiral taenidia in the walls) (courtesy of the U.S. Department of Agriculture in P.S. Callahan, 1971,* Insects and How They Function, *Holiday House, New York, N.Y.); **(f)** the inner ends of the tracheal system showing tracheal end cells and tracheoles in (1) a resting muscle and (2) an active muscle (from V.B. Wigglesworth, 1930,* Proc. Roy. Soc. London, B, 106:231*). (parts b–d from R.E. Snodgrass, 1935,* Principles of Insect Morphology, *McGraw-Hill Book Co., used with permission.)*

contraction the tracheoles are surrounded by hypertonic tissue fluid created by the accumulation of lactic acid. This causes the tracheolar water to move by osmosis out into the tissue fluid; tracheal air replaces the fluid (Fig. 13.15f). Since the tracheoles in active muscle tissues are filled with inspired air, the rate of gaseous diffusion into these tissues is greatly enhanced.

Two basic types of ventilatory systems are found in insects: open and closed. An **open**

system, present in most insects, permits free communication between the tracheal system and the exterior. In a **closed system,** gases are exchanged from the outside across the integument into a closed, air-filled, tracheal system, and functional spiracles are absent. The closed system is particularly characteristic of the immature stages of Ephemeroptera, Odonata, and Plecoptera, which are aquatic and possess tracheal gills through whose integument gases diffuse into or out of the tracheal system. A closed system is also present in most parasitic insects.

All aquatic insects with an open system must have some means of acquiring gaseous oxygen for the tracheal system. Some forms must surface periodically to replenish the atmospheric oxygen of the tracheae; others can remain submerged indefinitely and are mostly independent of the atmosphere. Of those that periodically surface, several mechanisms have evolved to facilitate this visit to the surface. Some have nonwettable **hydrofuge** structures surrounding several spiracles so that the surface film can be broken easily and, at the same time, keep the tracheae dry. Others carry a bubble or film of stored air held in place by **hydrofuge hairs.** The bubble often is sufficient to sustain the animal underwater for up to a day or more, at the end of which the insect must surface to trap another air bubble. The bubble acts as a "physical gill" across which oxygen diffuses from the water to the animal and carbon dioxide from the animal into the water. A further modification of the "physical gill" is found in insects that are able to remain submerged indefinitely; they obtain their oxygen by means of a **plastron.** A plastron is a thin film of air like a physical gill except that plastron air does not have to be replenished by periodic surfacing. Some other aquatic insects obtain oxygen directly from submerged plants.

Oxygen enters and carbon dioxide leaves the tracheal system passively by diffusion in resting insects. But when active, most insects must provide a greater degree of gaseous exchange by ventilating movements. Most often the contraction and relaxation of the abdominal muscles are synchronized with spiracular openings and closings to improve significantly the air flow through the ventilatory system. In many species, inhaled air enters by certain spiracles, and exhaled air exits by others, thus further enhancing the flow of air.

EXCRETORY SYSTEM

The excretory system of insects consists of a number of **Malpighian tubules** that arise from the anterior part of the proctodeum. These insectan tubules are of ectodermal origin and therefore are not homologous with tubules by the same name in arachnids that have an endodermal origin. But the Malpighian tubules of insects are functionally similar to those of arachnids. These long, slender, convoluted tubules have closed distal ends that are free in the hemocoel (Fig. 13.13c). Most insects have 2–150 tubules, but aphids (Homoptera) and representatives of a small order (Collembola) lack them. A Malpighian tubule wall may be one or several cells in thickness, and the cells contain muscle fibers whose contractions may cause the tubular tip to move within the hemocoel. The tubules are usually supplied by a large number of tracheoles.

As a result of protein metabolism, **uric acid,** produced within somatic cells, diffuses into the hemolymph and is then combined with potassium bicarbonate and water to form a fairly soluble material, potassium acid urate. This material is actively taken up by Malpighian tubule cells from the hemolymph and into the distal portions of the tubule. As the potassium acid urate passes into the proximal part of the tubule, potassium bicarbonate and water are reabsorbed; this results in the precipitation of crystalline uric acid. These anhydrous crystals then enter the proctodeum and are eliminated with the fecal material as a dry pellet. The elimination of uric acid by the Malpighian tubules is a paramount water-conserving feature that has contributed in no small way to the inordinate success of insects on land.

But insects living in various habitats have different salt and water problems. Some plant-eaters excrete or defecate excess water ingested with their food. Fresh-water insects, because they gain water by osmosis from their environment, must excrete large volumes of water while, at the same time, eliminating nitrogenous wastes as ammonia, allantoin, or even urea. But all insects that excrete considerable volumes of water must concomitantly conserve mineral salts. Thus the cells of the Malpighian tubules selectively or facultatively reabsorb by active transport those salts, ions, and minerals needed by the animal.

Nervous system

The nervous system, reflecting the basic arthropodan plan, has some important modifications in keeping with insects' active life-style. This system consists of the brain, the nerve cord, and a peripheral system of nerves (Fig. 13.16). The **brain** or **supraesophageal ganglion** has three principal parts: the protocerebrum, deuterocerebrum, and tritocerebrum. The **protocerebrum** is the sensory center for the compound eyes and ocelli and the major association center for the brain. Within the protocerebrum are the **"mushroom bodies"** which are the most extensive in insects with complex behavioral patterns. Since these insects can learn, the mushroom bodies may be the centers for storing information as memory and for other forms of complex advanced behavior. The **deuterocerebrum** receives sensory information from the antennae, controls antennal movements, and is important as a coordinating center. The **tritocerebrum,** bilobed and reflecting its paired segmental origin, receives nerves from the labrum and other ganglia. Arising from the tritocerebrum are the two **circumesophageal connectives** that join the brain and subesophageal ganglion (Fig. 13.16a). There are several important bodies associated with the brain and important neurosecretory cells located within the protocerebrum that are intimately involved with a neurohormonal mechanism that will be discussed later.

The **subesophageal ganglion,** representing three fused ganglia from head segments II–IV, innervates the sense organs and muscles associated with the mouthparts and salivary glands (Fig. 13.16a). Extending posteriorly from the subesophageal ganglion is the doubled ventral **nerve cord.** In ancient insects the cords probably were distinct, but in most contemporary insects the cords and their segmental ganglia have fused medially. Posterior to the subesophageal ganglion are three segmentally arranged **thoracic ganglia;** each coordinates sensory and motor functions within that segment, including those of the legs and wings. In primitive insects each of the first eight abdominal segments contains an **abdominal ganglion** although the number of ganglia is usually reduced. The posteriormost abdominal ganglion is the **caudal ganglion,** which is involved directly in the coordination of the genitalia, copulation, and oviposition. Each ganglion, whether in the head, thorax, or abdomen, gives rise to several peripheral nerves supplying portions of that segment or region. There is also a tendency among insects for the ganglia to move anteriorly and fuse longitudinally (Fig. 13.16c, d). For example, the three thoracic ganglia are fused into a longitudinal mass in insects of several orders. Only one abdominal ganglion persists in some coleopterans and dipterans; and in some insects, all abdominal and thoracic ganglia have fused into a single ventral mass (Fig. 13.16d).

Insects have a system of nerves supplying the visceral organs. This **visceral nervous system** is similar in function to the autonomic ner-

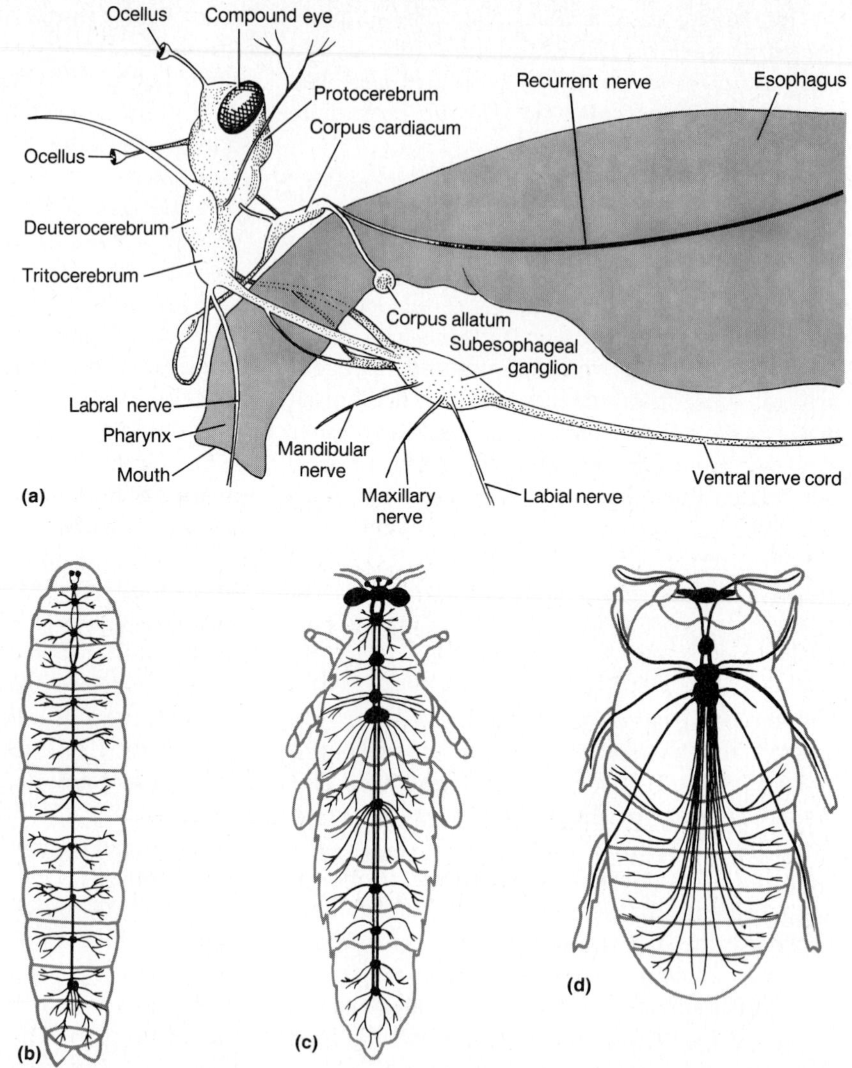

Figure 13.16 *The insectan nervous system: **(a)** lateral view of the brain and anterior ventral nerve cord; **(b)** an unfused nerve cord (caterpillar); **(c)** a nerve cord with some anterior fusions (grasshopper); **(d)** extreme fusion of the nerve cord into a prothoracic mass (from A.S. Packard, 1898,* A Textbook of Entomology, *Macmillan and Co., Ltd., London). (parts a–c from R.E. Snodgrass, 1935,* Principles of Insect Morphology, *McGraw-Hill Book, Co., used with permission.)*

vous system in vertebrates. It regulates such functions as heartbeat, peristalsis, salivation, and hormonal secretion. The visceral nervous system has three components. The anterior stomodeal component connects to the central nervous system via the tritocerebrum and includes a **frontal ganglion** lying dorsoanteriorly to the brain and several nerves to the anterior viscera (Fig. 13.16a). It controls peristalsis in the anterior alimentary canal and sends nerves to the dorsal vessel and to several brain-associated bodies (corpora allata and cardiaca). The ventral sympathetic component is composed of segmental nerves arising from the ventral nerve cord and innervating the spiracles, and the caudal sympathetic component supplies the posterior portions of the alimentary canal and the internal reproductive organs.

REPRODUCTION

Reproduction in insects is nearly always sexual, and all are dioecious except for one species (*Icerya purchasi,* a cottony-cushion scale [homopteran]). The internal reproductive organs of both sexes lie in the posterior abdominal region, and the gonopores of both sexes are situated at the posterior end of the abdomen. In most species, sperm are transferred by the male to the female in copulation, and fertilization is internal. Parthenogenesis occurs in many insects, and in some species, no males are known. Certain adults of the social insects do not reproduce owing to atrophied reproductive organs.

The male system. The male reproductive system consists of a pair of testes connected by ducts to a copulatory organ, the aedeagus or penis, located on the ninth abdominal segment. Each **testis** is composed of a variable number of **testicular follicles** or sperm tubes enclosed in a connective tissue sac (Fig. 13.17b, c). Within these tubes, unspecialized spermatogonia become primary spermatocytes, which meiotically become secondary spermatocytes and in turn transform into **sperm.** These stages are found in zones or regions progressing from follicular tip to base. Each follicle wall is composed of a layer of epithelial cells that nourish the developing sperm. Each follicle has a tiny duct, the **vas efferens,** through which sperm pass from the follicle to the **vas deferens.** Sperm may be stored temporarily in a dilated portion of the vas deferens, the **seminal vesicle.** The two vasa deferentia join to form a common sperm duct or **ejaculatory duct** that opens to the outside through the gonopore located on the **aedeagus** (Fig. 13.17a). The vasa deferentia and ejaculatory duct are of ectodermal origin and hence are lined with a thin cuticle. Usually, a pair of **accessory glands** opens either into the vasa deferentia or into the ejaculatory duct, and their seminal fluid secretion enhances sperm vitality and mobility.

Mating behavior in insects is rather diverse. In most there is some form, albeit brief and simple, of mutual courtship after which copulation ensues. Typically, sperm are transferred to the vagina of the female by the aedeagus, but there are modifications of this general plan. In more primitive insects like the apterygotes

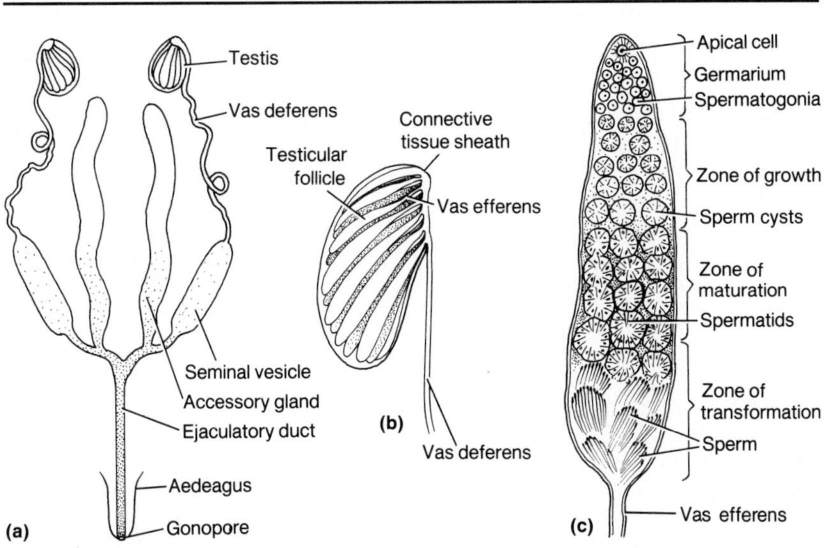

Figure 13.17 *Diagrams of the male reproductive system in insects: **(a)** the principal organs; **(b)** structure of a testis; **(c)** longitudinal section through a sperm tube (parts a–c from R.E. Snodgrass, 1935,* Principles of Insect Morphology, *McGraw-Hill Book Co., used with permission).*

and the Odonata, sperm are produced in packets called **spermatophores,** which are held together by accessory gland secretions. The packets are deposited externally by the male. A male Odonata (dragonflies and damselflies) transfers spermatophores from his gonopores to the female by way of a special ventral copulatory organ on abdominal segment II. Some other insects transfer spermatophores directly to the female after which the sperm are released by digestion and rupture of the spermatophores. A male bed bug (Hemiptera) inseminates a female by injecting sperm into her abdominal hemocoel using his needlelike aedeagus.

The female system. Typically, the female system consists of a pair of **ovaries** composed of a number of functional units called **ovarioles** (Fig. 13.18). There are commonly four to eight ovarioles per ovary, and greater numbers of ovarioles are generally indicative of greater fecundity rates. Each ovariole ends in a terminal thread, and the many threads collectively form the **terminal filament** that attaches the ovary to the dorsal diaphragm. Each ovariole consists of a **germarium,** a mitotic region that produces oocytes and nurse cells, and a **vitellarium,** in which yolk is added to the developing oocytes (Fig. 13.18b, c). Developing oocytes enter the vitellarium and are surrounded by follicular epithelium to form **follicles.** It is interesting to note that in most insects, follicle development is completed in the pupal or last nymphal stage. As an oocyte moves toward the ovariole base, progressively more yolk is incorporated into its cytoplasm. The follicles nearest the ovariolar base are the most mature, and they are forced out of the ovariole into the oviduct at ovulation. Ovum production of females varies widely from fewer than 100 in the lifetime of the coleopteran *Tenebrio* to in excess of 125,000 per year in *Apis*, the common honey bee.

Each ovum is surrounded by a thin **vitelline** or egg **membrane** and a thicker hardened **cho-**

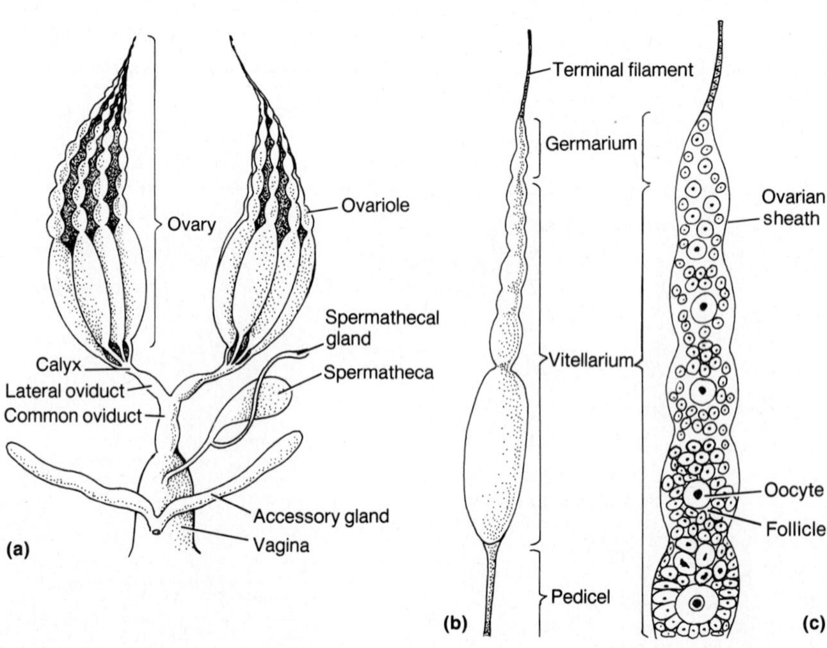

Figure 13.18 *Diagrams of the female reproductive system in insects: **(a)** the principal organs; **(b)** structure of an ovariole; **(c)** longitudinal section through an ovariole. (parts a–c from R.E. Snodgrass, 1935,* Principles of Insect Morphology, *McGraw-Hill Book Co., used with permission.)*

rion or eggshell secreted by the follicular cells. The chorion bears several pores, the **micropyles,** through which sperm enter the ovum. Unfertilized ova pass from the ovariole through a short duct or **pedicel,** an enlarged **calyx,** a **lateral oviduct,** and into a short **common oviduct,** which becomes the **vagina** (Fig. 13.18a). The common oviduct and vagina are of ectodermal origin and are lined with a thin cuticle, whereas the inner parts of the female system are of mesodermal origin. The **seminal receptacle** or **spermatheca,** an outpocketing of the vagina or common oviduct, serves as a storage point for sperm introduced by copulation. In most cases, many sperm enter the ovum via the micropyles, but only one sperm is finally involved in fertilization. The entry of the sperm into the egg triggers the diploid egg nucleus to undergo meiosis, resulting in a haploid nucleus; the two haploid gametes then unite in fertilization to produce a diploid zygote. Oviposition normally follows shortly thereafter. One or two pairs of **accessory glands** are present whose secretions onto the surfaces of the eggs either cement the eggs to a substratum, cause the eggs to adhere to each other, or form an enveloping cocoon around the eggs.

In most female insects, egg-laying is facilitated by an **ovipositor** composed of paired structures of abdominal segments VIII and IX (Fig. 13.11). Ovipositors are variously adapted to deposit eggs in a variety of places including twigs or leaves or stems, insect hosts, water, soil, or within egg cases. Eggs may be laid singly or in groups, and those of some species may be covered with additional layers of froth, silk, or waxes. Usually, eggs are laid near food that is accessible for the future immature stages.

The vast majority of insects lay eggs and are oviparous, but important deviations are not uncommon. The eggs of some species are retained by the female until hatching, and she, in effect, lays nymphs or larvae; such a condition is termed ovoviviparous. A few species are viviparous, since they produce eggs with very little yolk, and the embryos are retained and nourished by the female until the immatures are born. Also, the phenomenon of **polyembryony** has been observed in a few insects in which each of the blastomeres in early cleavage stages develops asexually and independently into a complete embryo.

Development

After oviposition the fertilized eggs usually begin cleavage without delay. The early patterns of cleavage and development follow the general plan of arthropods (see Chapter 10). Cleavage is partial, meroblastic, regulative, and mosaic and produces a stereoblastula with a **blastoderm** or **germinal band** that will differentiate into the embryo proper. Other blastodermal cells become flattened and form the embryonic membranes—an inner **amnion** and an outer **serosa**—that surround the embryo and cushion and protect it. As development proceeds, germinal band cells move inward by invagination in a process that is comparable to gastrulation, and the three germ layers soon develop and become distinct. Ectoderm gives rise to the integument, nervous system, stomodeum and proctodeum, tracheal and excretory systems, oenocytes, and the outermost parts of the reproductive systems. Mesoderm results in the muscles, fat body, circulatory system, reduced and embryonic coelom, and innermost portions of the reproductive systems. Endoderm lines the midgut region. Metamerism of the embryo begins shortly after the formation of the germinal band and is always initiated at the posterior end at the pygidium. Tagmata soon become evident, as do the appendicular buds. Toward the end of egg development a phenomenon called **dorsal closure** takes places in which the sides of the developing embryo meet dorsally and form the dorsal parts of the embryo including the dorsal vessel and hemocoel.

Eclosion is the process of hatching from the egg. Often, amnionic fluid is gulped by the embryo to aid in the hydrostatic rupturing of the tough chorion. Others may chew their way out, digest away the chorion with enzymes, or used specialized hatching spines. The newly hatched immature insect crawls out of its egg case and begins an independent existence.

As each immature insect feeds and grows, the exoskeleton is periodically molted, and a new, somewhat larger one is produced. After passing through 3–30 immature instars, the final instar, or the adult or **imago** stage is achieved, in which the insect develops functional wings and becomes sexually mature. Thysanurans continue to molt after reaching the adult stage, and mayflies have two winged instars. Although biomass is added throughout the immature stages, molting gives the superficial appearance of a series of discontinuous spurts or growth taking place at each ecdysis.

Metamorphosis. In the vast majority of insects, early instars are different morphologically and functionally from the adult. The degree of difference, ranging from slight to dramatic, is a

Figure 13.19 *Three basic types of metamorphosis in insects: **(a)** without metamorphosis (bristletail); **(b)** incomplete (grasshopper); **(c)** complete (mosquito). (parts a–c from R.J. Elzinga, 1981,* Fundamentals of Entomology, *2nd ed., © 1981, p. 105, reprinted by permission of Prentice-Hall, Inc., Englewood Cliffs, N.J.)*

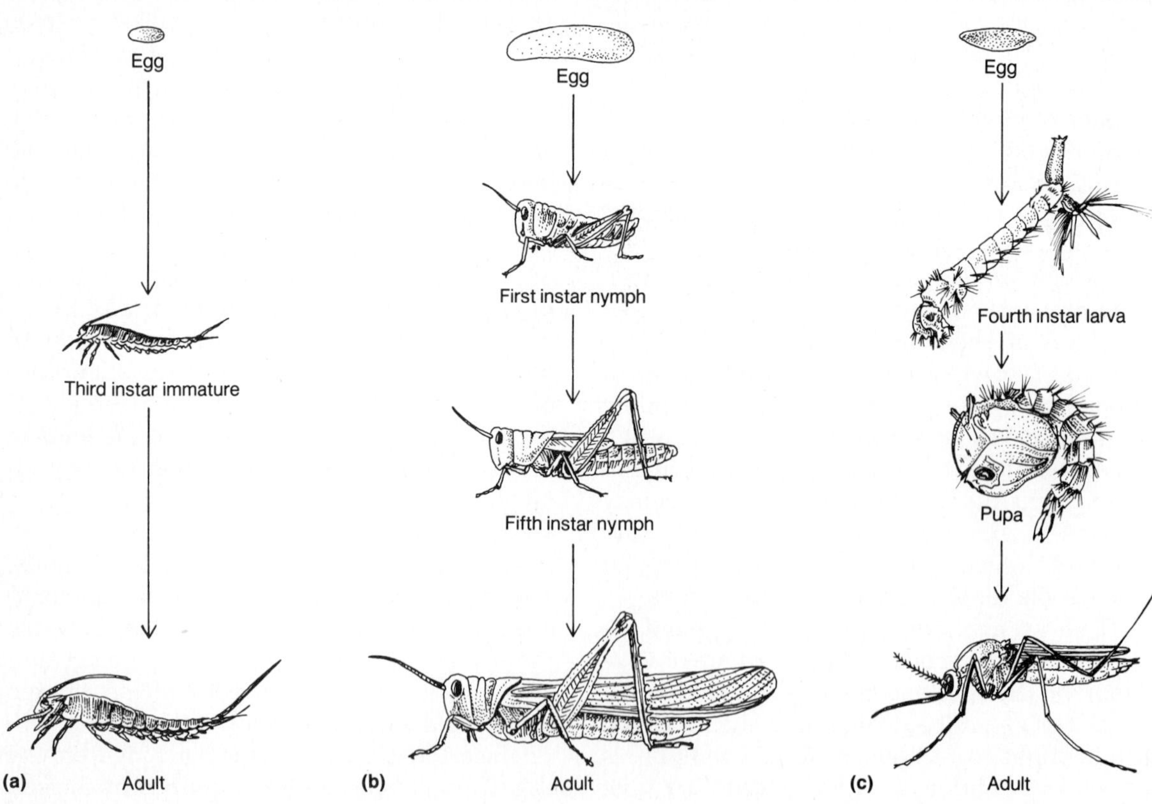

measure of the amount of **metamorphosis** or transformation that takes place. Members of the Subclass Apterygota do not undergo any metamorphosis whatsoever in that the immature instars resemble the adults morphologically and functionally except that they are smaller and are sexually immature; these insects are said to be **ametamorphic** (Fig. 13.19a). All the remaining insects, belonging to the Subclass Pterygota and representing more than 99% of all insect species, undergo some degree of metamorphosis.

The pterygotes can be divided into two groups depending upon the degree of metamorphosis: incomplete and complete. **Incomplete** or simple **metamorphosis,** often referred to as hemimetabolous development, is found in insects whose instars resemble the adults in many ways but lack functional wings and are sexually immature (Fig. 13.19b). Since wings appear as external pads during later immature instars in insects with incomplete development, these insects are called **exopterygotes.** The immature stages of insects with simple metamorphosis are usually termed **nymphs.** Nymphs of some hemimetabolous insects are aquatic, but a majority are terrestrial.

Insects with **complete** or total **metamorphosis** are those in which the immature stages are radically different from the adults, and they have what is sometimes called holometabolous development (Fig. 13.19c). The immature instars of these insects, called **larvae,** lack compound eyes and almost always have mandibulate mouthparts. Most often, larvae utilize food and other resources that are quite different from those of the adult, and they have become the dominant feeding stage in most holometabolous insects. Since wings develop internally during the larval stages of insects with total metamorphosis, these insects are called **endopterygotes.** Small groups of undifferentiated epidermal cells called **imaginal discs** are present in larvae and will eventually give rise to adult parts such as antennae, wings, and legs. The principal larval types are outlined in Box 13.3, and two are illustrated in Fig. 13.20.

To accommodate this radical transformation between the larva and adult, a stage is required in which these dramatic changes in the functional morphology take place. Juxtaposed between the last larval stage and the adult stage is the **pupal stage** (Fig. 13.19c). This nonfeeding and usually nonmotile stage is referred to as a "resting stage," but it is extremely active metabolically. During the pupal stage, most of the larval tissues are histolytically destroyed, and imaginal discs rapidly develop by histogenesis. Gonads mature, wings develop, and the alimentary system and mouthparts are reorganized. The pupa is a preeminently significant stage of transformation from larva to adult. Since it is a vulnerable instar, pupae are often encased in elaborate protective cocoons. Many insects use silk for constructing **cocoons,** others build cells in the soil, and others utilize the last larval exuvium to form a **puparium.** Types of pupae are also described in Box 13.3, and two are illustrated in Figure 13.20.

As the time approaches for a holometabolous insect to emerge, the pupa darkens. At emergence the pupal cuticle splits along the middorsal ecdysial line. Finally emerging from the pupa is a **teneral** adult, that is, a newly emerged, pale, and soft bodied adult. Shortly, the body expands, its cuticle is sclerotized, it assumes its imaginal color, and the animal becomes sexually mature. The adult is, of course, that stage in which dispersal and reproduction take place.

Endocrine control of metamorphosis and growth. One of the most active areas of research in recent years concerns the mechanisms controlling growth and metamorphosis. Space here will permit only a brief summary of the most important hormones involved in this complex problem. At least four separate tissues or organs are involved in the hormonal control of

BOX 13.3 □ THE MAJOR TYPES OF INSECT LARVAE AND PUPAE

A surprisingly large number of different larvae and pupae develop in endopterygotes. Larval types are distinguished primarily by the shape of the body, the presence or absence of abdominal prolegs, and their ecological niches. Pupal types are based primarily on the nature of the puparium. The major larval and pupal types are as follows:

LARVAE

Campodeiform—larvae resembling the dipluran *Campodea;* flattened and elongated body; long legs, antennae, and cerci; found in Neuroptera, Trichoptera, and many Coleoptera.

Carabiform—larvae resembling larval carabid beetles; shorter legs and antennae than campodeiform; occur in some Coleoptera.

Elateriform—wirewormlike; larvae resemble click beetles; cylindrical, elongated bodies with hard shells; small head; short legs; occur in some Coleoptera.

Eruciform—caterpillars; larvae have cylindrical body with well-developed head; short antennae, thoracic legs, and abdominal legs (prolegs) (Fig. 13.20a); are present in Lepidoptera, Mecoptera, and some Hymenoptera.

Platyform—larvae have flattened bodies; thoracic legs may be absent; found in some Diptera, Coleoptera, and Lepidoptera.

Scarabaeiform—grublike; cyclindrical body curved in the shape of a C; well-developed head; no abdominal prolegs; mostly sluggish and inactive (Fig. 13.20b); are found in some Coleoptera.

Vermiform—maggotlike; body elongate, wormlike, and without legs and often with a head; found in certain Diptera, Siphonaptera, Coleoptera, Lepidoptera, and most Hymenoptera.

PUPAE

Coarctate—the hardened, last, larval exuvium covers the pupa and is a puparium (Fig. 13.20c); found in some Diptera.

Exarate—appendages (legs, wings, antennae) are free and not adhering to body; pupa resembles a pale mummified adult, and usually there is no cocoon; found in most holometabolous insects except for the Diptera and most Lepidoptera.

Obtect—appendages more or less fused to body; pupa is often covered by a silken cocoon (Fig. 13.20d); are in most Lepidoptera and some Diptera.

metamorphosis (Fig. 13.21). **Neurosecretory cells,** numbering from five to several thousand, are located in the protocerebrum and secrete the **brain hormone.** This hormone is passed along neurons leading from the brain to the **corpora cardiaca,** paired neuroglandular bodies lying close to the brain, where the hormone is released at specific times into the hemolymph. The target tissues of the brain hormone are the two small **ecdysial glands** (often called prothor-

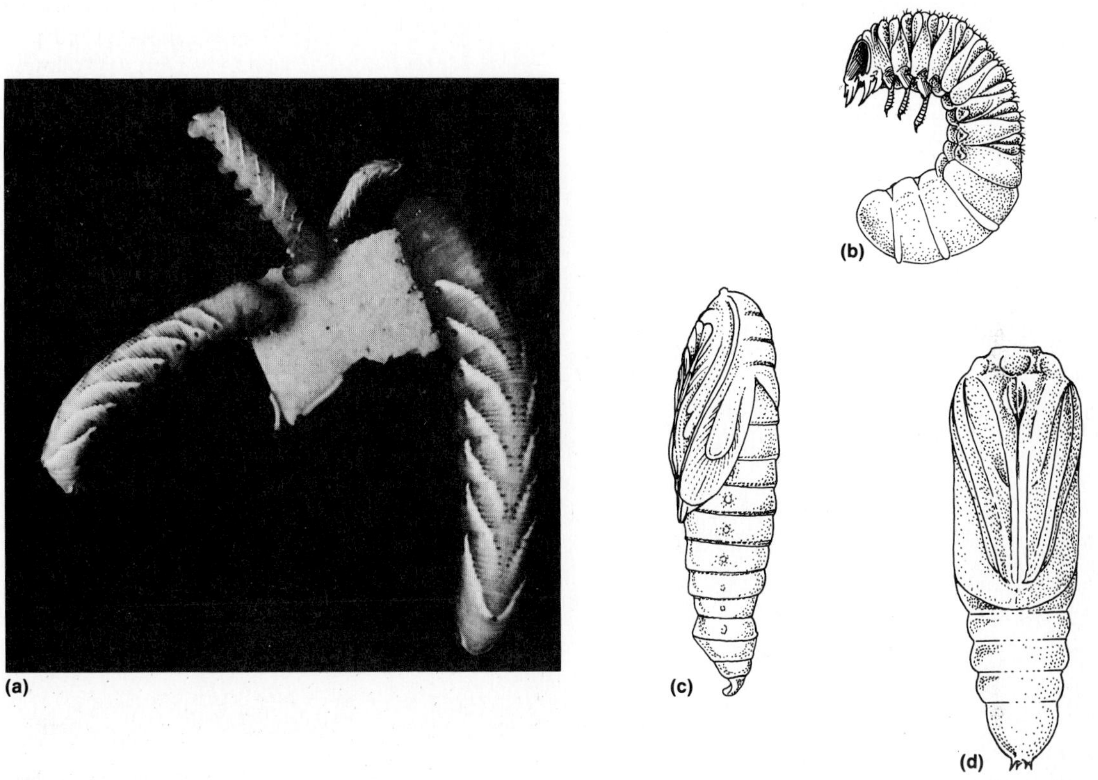

Figure 13.20 *Selected types of larvae and pupae:* ***(a)*** *photograph of eruciform larvae (caterpillars) (courtesy of H. Itagaki);* ***(b)*** *a scarabaeiform larva (grub);* ***(c)*** *a coarctate pupa;* ***(d)*** *an obtect pupa (parts b, c from A.S. Packard, 1898,* A Text-book of Entomology, *Macmillan and Co., Ltd., London.)*

acic glands, since in some insects they are found in this segment), which produce **ecdysone** or its precursor. Ecdysone is often referred to as the molting hormone that activates the integument to produce a new cuticle and molting fluid. In the absence of other inhibiting hormones, ecdysone induces metamorphosis.

Another pair of small bodies, the **corpora allata,** lies on the lower side of the alimentary canal near the circumesophageal connectives.

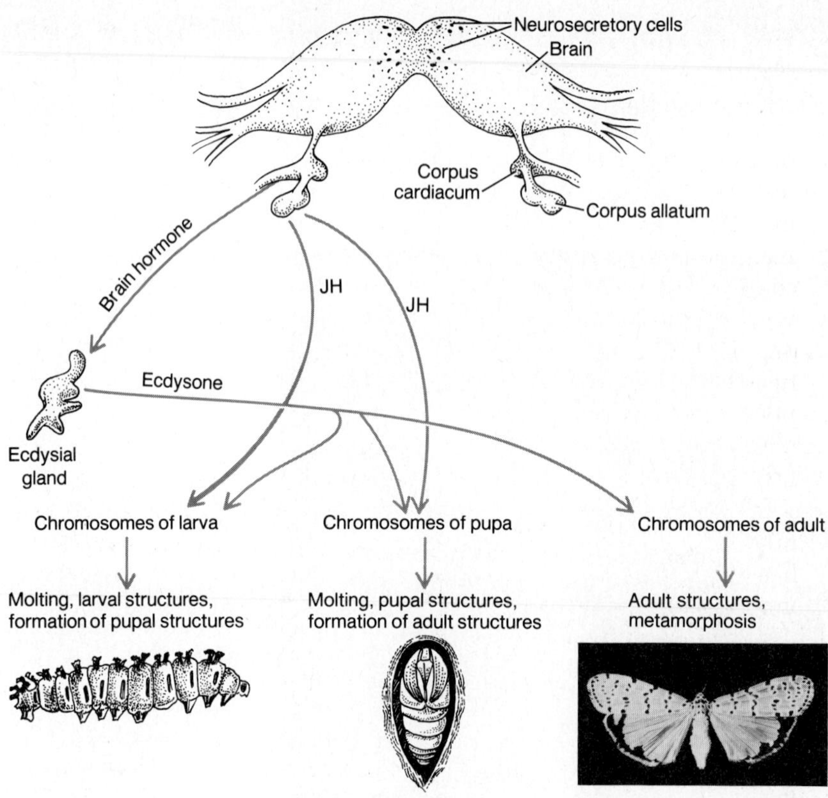

Figure 13.21 *Schematic diagram of the hormonal control of growth and metamorphosis in insects (photograph is an adult* Utetheisa *[courtesy of H. Itagaki]).*

The corpora allata produce **juvenile hormone** (JH), which stimulates larval development and prevents or retards the development of adult features (Fig. 13.21). In ways not yet fully understood, the brain directs the corpora allata to produce JH. During larval or nymphal existence, both JH and ecdysone are present; the former causes larval characteristics to be perpetuated, and the latter promotes molting. As the time for metamorphosis nears, either JH production is completely arrested or the JH concentration drops to a very low level. Without the inhibitory influences of JH, ecdysone now stimulates metamorphosis and emergence. The reduction of JH takes place gradually in the last several nymphal instars in hemimetabolous insects or is confined to the pupal stage in holometabolous forms. It has been shown that the definitive effects of both JH and ecdysone are at the chromosomal level, where specific genes regulate either larval or metamorphic patterns.

In many insects, development is arrested in the anticipation of adverse environmental conditions, a condition known as **diapause.** Diapause typically occurs in the egg or in larval or nymphal instars, although it is a phenomenon that may be present in the pupal or imaginal stages; diapause normally takes place in only one instar for a given species. In some insects, diapause is obligatory in that all individuals in every generation pass through this stage regardless of environmental conditions; but in other insects, diapause is facultative in that individuals in only some generations anticipate adverse environmental conditions and enter diapause.

Many environmental factors may induce diapause, but undoubtedly the most important factor is length of day (photoperiod). Normally, decreasing day lengths in the summer and autumn are the trigger for the onset of diapause in instars that must overwinter. Temperature, in concert with photoperiod, has also been shown to be important in the induction of diapause. During diapause, no morphological changes take place, and many physiological events are reduced or curtailed (lower metabolic rate, reduced incidence of feeding, etc.); however, other physiological changes (diapause development) must take place before diapause is terminated. Various environmental conditions such as temperature, moisture, and other environmental shocks can terminate diapause.

Diapause is obviously an organismic response to anticipated environmental conditions that results in dramatic changes in hormonal concentrations. In animals that have been closely studied, diapause is characterized by a reduction of both JH and ecdysone to very low levels. Experimental injections of high concentrations of ecdysone often abruptly terminate the diapause condition. However, a complete understanding of the specific roles of the brain, endocrine glands, and hormones is not yet known.

SOCIAL INSECTS

Most insects live entirely solitary lives. If two individuals happen to meet in a nonreproductive encounter, competition is almost immediately manifested, and mortal combat sometimes ensues. However, representatives of a great many insect taxa, including web-spinners and certain coleopterans like scarab, burying, and ambrosia beetles, practice a degree of parental care and even some division of labor. These many intermediate grades or levels of organization are termed **presocial.** But true sociality or **eusociality** has evolved in invertebrates only in termites and certain hymenopterans. Eusocial insects are characterized by a colony in which numerous individuals live and generations overlap, cooperative brood care is practiced, and a sharp division of labor is evident. It is clear that eusociality evolved in termites independent of that in hymenopterans, but it is indeed incredible that social organization is so very similar in these two widely unrelated orders.

In any insect colony, individuals need to communicate with fellow colonists about various factors that affect them or the colony. Such things as location of food, sexual recognition, marking trails, creating sounds, and visual cues must be generated by one insect and perceived by another. While communication between insects is not confined to colonies, it is most evident in the social insects. Some interesting facets of insect communication appeared earlier in this chapter in Box 13.2.

In all insect societies, no one individual can survive outside of the colony for very long, and an individual is not permitted to join another established colony. Individuals within a colony have an inflexible, inviolate polymorphism in which different morphological forms are present, each with specific functions. This has led to a **caste system.** In all insect societies there are basically three castes: queens, males, and workers. **Queens** and **males** function mostly in reproduction, and **workers** do all the other chores that are required for the support and maintenance of the colony. Workers forage, build nests, care for the young, and defend the nest and colony against intruders.

Since sex and the caste system in eusocial insects are so intimately entwined, how is it determined that a given zygote or larva will develop into an individual belonging to a given caste? In some termites the castes are determined by extrinsic or environmental influences in addition to intrinsic hormones. Young workers called **pseudergates** have the developmental

BOX 13.4 □ SOCIAL BEHAVIOR IN ANTS

The colonial and social nature of hymenopterans like bees and wasps is well known. But all ants are also social, and they have a very complex, rigid social organization and behavioral patterns that are probably not as well known as those in other hymenopterans. Therefore this box briefly explores sociability and cooperation in ants.

All contemporary ants are social, and most cooperate to construct an underground nest in which to retreat and to rear the young. All ant species have a very rigid caste system with a sharp division of labor among colonists. A given colony contains many males, one or more queens, and a variety of workers. A queen is the colony's founder and mother; in an established colony the queen reproduces and is involved in grooming. Males are winged reproductives that also only mate and are involved in grooming. Workers are sterile females that normally fall into three subcastes: **majors** or **soldiers,** which have oversized heads and mandibles; **media,** which are intermediate in size; and the small **minors.** These various caste functions can clearly be seen in the army ants *(Eciton)* of tropical rain forests. These ants do not build subterranean nests but form temporary overnight bivouacs in fallen tree trunks or in the shelter of a large boulder. In a bivouac at night, several hundred thousand workers are linked to each other by their tarsal claws; one or more queens, hundreds of males, and thousands of larvae and pupae are located in the center of this mass of workers.

At dawn the workers move away from the bivouac in all directions and usually form raiding or exploratory columns in search of food. Workers lay a pheromone trail to guide the workers who bring along the larvae and pupae, and soldiers guard the colony's flanks. Foragers in the front of the advancing columns locate prey, usually other arthropods, which are stung and killed, and the dismembered prey parts are carried rearward in the column to feed the larvae. A colony might move several hundred meters in a day's time. This nomadic nature is associated with reproductive periods in the colony when large numbers of larvae are to be fed. At dusk the workers return, and the bivouac is reformed.

In temperate regions, familiar ants like *Myrmica* and *Formica* nest in the ground. In late summer, winged males and females are produced, sometimes in large numbers. Sometimes these reproductives swarm out of the nest in spectacular nuptial flights during which mating takes place. Queens then locate a new site for a nest, and each sheds her wings, excavates a nest, and seals herself inside. She lays a group of eggs the following spring that become the new group's primary colonists. The colony grows numerically and produces workers, which then forage for the colony.

A number of ant genera, including *Atta,* are fungus-culturers. Workers gather small bits of vegetation, carry them into the underground nest, and wet them with anal secretions, forming a moist pulp. Fungal mycelia are then "inoculated" into the pulp beds, and soon many new hyphae are formed. Hyphal tips are then fed to the ant larvae. The fungal gardens are tended by minors. A very perplexing question is how the ants are able to be monoculturists. Apparently, the ants are able to discriminate between the desired crop and all the "weed" fungi. Fungicidal and bactericidal substances are also secreted by salivary and anal glands. The newly formed queen even carries a wad of hyphae in her mouth cavity as she begins her nuptial flight.

Ants are really remarkable animals in their gregariousness, their colonial division of labor, and their symbiotic gardening of fungi.

potentiality of becoming reproductives or becoming sterile workers. Pheromones, chemicals produced by existing caste members of the colony, are instrumental in determining the future of pseudergates. Queens or female alates produce a pheromone substance that prevents female workers from becoming functional females or queens. Males likewise produce a male-inhibiting substance, and they also produce a pheromone that stimulates female workers to become functional queens. Internal hormones within the pseudergates such as ecdysone and juvenile hormone are also involved; an excess of ecdysone promotes the development of reproductives, and an excess of juvenile hormone relegates pseudergates to permanent worker status.

In hymenopterans, caste determination is mostly genetically controlled, since females develop from fertilized eggs, and males are haploid and develop from unfertilized eggs. In ants, queens are the functional females. Workers are also females, but their attainment of sexual maturity is predicated on an interwoven set of environmental or maternal factors that inhibits the workers from becoming queens. Pheromones, secreted by the queen, inhibit the development of other queens. But environmental influences like the availability of food and autumnal chilling temperatures often result in several workers becoming queens. Some further details of eusociality in ants appear in Box 13.4.

In honey bees, all adults in a hive are winged. Males or drones are haploid and are produced during the spring and summer; after the queen's mating, they die. Usually, there is only one queen per colony, and she never leaves the hive and is the only source of all eggs and a new generation. Fertilized eggs are females that become workers. Then what distinguishes between a queen and a worker? Larvae destined to form workers and drones are fed for the first two days on a mixture of clear hypopharyngeal gland secretions and white mandibular gland secretions from workers that are functioning as nurse bees. On the third day the white secretions are terminated by a still unexplained mechanism, and the larvae are fed "bee bread," a mixture of clear hypopharyngeal fluids, pollen, and honey. But a female larva destined to be a queen is fed for its entire existence on "royal jelly," a mixture of clear and white secretions from the nurse bees. It would appear that the quality and quantity of food and the timing of the feeding serve as the discriminating factors between queens and workers.

While many advertisements extol the benefits to human beings of so-called "royal jelly," which is supposed to be a powerful aphrodisiac, restore sexual vitality, and promote virility, these are simply unfounded claims. Synthetic "royal jelly" will not even convert a female larval honey bee into a queen.

INSECT DIVERSITY

It is certainly beyond the scope of this text to discuss in detail the various orders of insects. Rather, I have chosen to describe 20 orders in some brief detail; the remaining eight orders contain fewer species and are highlighted in Box 13.5 and illustrated in Figure 13.42 later in this chapter. The decision regarding which orders to include was arbitrary; I chose to include a discussion of the 18 orders that contain more than 1500 species each, since they collectively represent more than 99% of all insect species. In addition, the Order Anoplura is included because of its medical importance and the Order Thysanura because it contains very primitive archetypal insects. Brief characterizations of all insect orders appear at the conclusion of this chapter.

Subclass Apterygota

All members of this subclass are wingless and primitively so, since none of their ancestors had wings. The thorax of apterygotes, differing from that in pterygotes, lacks a number of features (sclerites, sutures, muscles) that are associated with the possession of wings and the ability to fly. The apterygotes are very small insects that lack any form of metamorphosis. In many apterygotes, some of the anterior abdominal segments bear paired styluslike appendages that are not homologus to paired appendages of the head or thorax.

A closer inspection of the five small orders comprising this subclass reveals that they are in two rather distinct groups. One group, composed of the Protura, Collembola, and Diplura, have mouthparts withdrawn into the head and are termed **entognathous,** while the other group, the Thysanura and Microcoryphia, have protruding mouthparts and are **ectognathous,** as are all other insects. There is growing support among entomologists for treating the entognaths as a separate evolutionary line; in fact, many entomological taxonomists consider them to be noninsects and in a separate taxon, Class Entognatha. However, we shall follow the more traditional scheme of Borror, De Long, and Triplehorn (1981) and consider all five orders as apterygotes but recognize two different groupings. Two apterygote orders are discussed below, the entognathous Collembola and the ectognathous Thysanura.

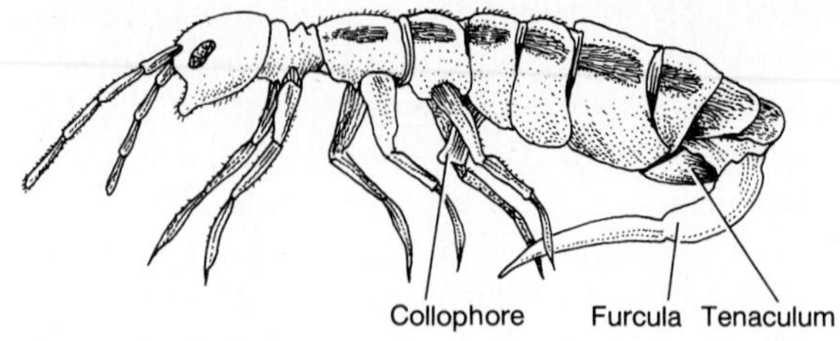

Figure 13.22 *A collembolan or springtail (from L.C. Miall, 1903,* The Natural History of Aquatic Insects, *Macmillan and Co., Ltd., London).*

ORDER COLLEMBOLA

The Collembola are best known as springtails because a unique springing device characterizes this group of about 2000 species. They are small primitive animals usually not over 6 mm in length, but they are incredibly abundant in soil and leaf litter, around damp places, and on surfaces of bodies of fresh water. Estimates in excess of 100,000 springtails per cubic meter of surface soil are common. They are widely distributed in polar, temperate, and tropical habitats.

The head bears entognathous mandibulate mouthparts, one to eight stemmata, and short antennae, but they lack compound eyes and dorsal ocelli (Fig. 13.22). The abdomen consists of only six segments, the first of which bears a ventral lobelike **collophore** that is now thought to be involved with water uptake. The ventral surface of the fourth abdominal segment bears the unique **furcula** or fork, which, when bent anteriorly, is held in place by the **tenaculum** located on the ventral surface of the third segment (Fig. 13.22). When the tenaculum suddenly releases the furcula, the latter springs against the substratum and propels the animal forward and upward; this has given them their common name. Adults lack Malpighian tubules, external genitalia, and tracheae (some), and they often molt after reaching adulthood.

While none are medically important, they can be pests in truck garden, sugar cane, cereal, and forage crops. One species is a serious pest in Australian alfalfa.

ORDER THYSANURA

The Thysanura, or bristletails, are considered by most entomologists to be the most primitive of all living insects. With about 400 species they are elongated, range up to 30 mm in length, and typically are found in grassy or wooded areas, in the soil, under stones or bark, or in dead wood. Several species live in or around human dwellings.

The head of the adult is characterized by rudimentary compound eyes with small numbers of ommatidia, elongated filiform antennae, and visible mandibulate mouthparts. The abdomen is composed of 10 or 11 segments, each of which bears a pair of **styli** or vestigial legs. Characteristic of the thysanurans are three long, bristly filaments at the posterior end of the abdomen; these are a pair of well-developed **cerci** and a median **caudal filament** (Fig. 13.23). The entire body is nearly always covered with scales.

Bristletails are unique in that they continue to molt after reaching sexual maturity, and one study reported that some thysanurans lived for as long as seven years through 60 molts. They are capable of rapid movement, and some species can jump. Males deposit spermatophores on various substrata, and the females later pick them up.

To this order belong several human pests including firebrats *(Thermobia)*, found near furnaces, steam pipes, and other hot dry areas, and silverfish *(Lepisma)*, which are usually found in cool, damp areas (Fig. 13.23). Both species feed on book bindings, flour, wallpaper paste, and starched clothing. None are medically significant.

Figure 13.23 *A thysanuran or bristletail* (Lepisma) *(from D. Sharp, 1909,* Cambridge Natural History, *Macmillan and Co., Ltd., London).*

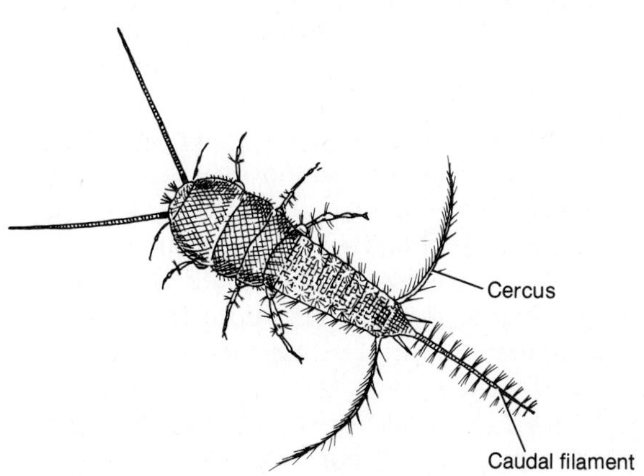

Subclass Pterygota, Division Exopterygota

This division includes a number of orders whose representatives are winged, have incomplete metamorphosis, and have nymphs that develop wings externally. In this very large grouping, three subgroups of orders are recognized. In one group, represented by the Ephemeroptera and Odonata, all representatives have aquatic nymphs, and adults lack a mechanism for flexing the wings over the abdomen at rest. Representatives of a second group, including the Orthoptera, Isoptera, and two minor orders, have mandibulate mouthparts, forewings that are usually more narrow than hindwings, and many crossveins on both pairs of wings. Members of the third group, including the Plecoptera, Psocoptera, Mallophaga, Anoplura, Thysanoptera, Hemiptera, Homoptera, and a minor order, have specialized haustellate mouthparts, lack abdominal cerci, and have equally sized wings with relatively few crossveins.

Order Ephemeroptera

The Ephemeroptera or mayflies are commonly found around streams and ponds, and there are about 2100 species. Mayflies have a unique winged subadult form that lasts usually for only a brief time before it molts to a slightly larger adult.

Adults are small to medium-sized insects with fragile soft bodies (Fig. 13.24a). The head bears a pair of short setaceous antennae and vestigial or nonfunctional mouthparts. Two pairs of wings are present, although the hind pair is frequently reduced and occasionally missing. When at rest, the wings are held vertically over the thorax. The abdominal tip bears a pair of very long filamentous **cerci,** and in most there is an additional median **caudal filament.**

Nymphs are all aquatic, are diverse in appearance, have chewing mouthparts, and are either herbivorous or predaceous. The abdomen

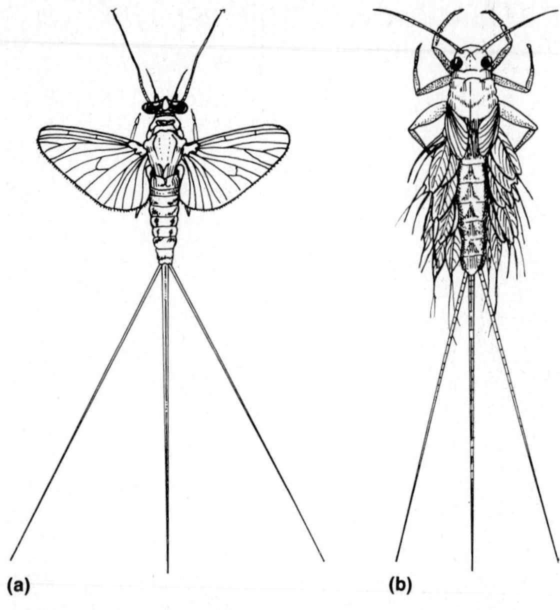

Figure 13.24 *Ephemeropterans or mayflies:* ***(a)*** *an adult of* Caenis diminuta; ***(b)*** *a nymph of* Leptophlebia intermedia *(parts a, b from L. Berner, 1950,* Mayflies of Florida, *University Presses of Florida, used by permission).*

bears four to seven pairs of gills on the anterior abdominal segments, and nymphs have a closed tracheal system. The posterior abdomen bears a pair of **cerci** and usually a median **caudal filament** (Fig. 13.24b). Nymphs live for one to several years, during which time they molt up to 50 times. When the nymph emerges, it gives rise to a nonfeeding, nonreproductive, winged **subimago,** which molts within several hours to produce a nonfeeding reproductive **imago** that usually dies within several days. Mayflies are the only insects in which a winged form molts.

Sometimes mayflies are present in prodigious numbers; they may be a nuisance, and discarded exuviae and dead animals may accumulate in extensive piles or drifts near lakes. Mayflies often fly in huge, over-water swarms in which mating takes place. Adults and nymphs are an important source of food for many freshwater fishes, and many artificial fishing flies are patterned after the winged stages.

ORDER ODONATA

The Odonata, which includes over 5100 species of dragonflies and damselflies, are large, conspicuous, often colorful insects around ponds and streams from early spring to late autumn in temperate regions. They spend a large part of their time on the wing, are easily recognized, and are completely harmless to human beings, even though folklore depicts them as biting or stinging.

Adults are large insects (up to 13 cm in length, with a wingspan of 19 cm), and are often strong fliers. The head bears a pair of enormously enlarged compound eyes with thousands of ommatidia, and the eyes provide odonates with excellent vision used to capture food on the wing. There are two pairs of elongated membranous wings with many veins (Fig. 13.25a, c). At rest, the wings are held either horizontally or vertically over the thorax. The three pairs of legs form an effective basket in which prey is always caught; prey is often eaten in midair. The abdomen is elongated and slender, and in females its posterior tip bears the ovipositor.

Odonates have a very unusual method of indirect sperm transfer. In males there is a copulatory organ on the ventral aspect of the second abdominal segment, and a male must transfer sperm from the gonopores at the abdominal tip to this copulatory organ by bending his abdomen ventrally and anteriorly. In males the posterior abdomen bears **claspers** used to hook behind the head of the female to achieve a tandem position for copulation with the male in front. During the tandem copulatory flight or while perched, the trailing female must bend her abdomen so that its tip comes in contact with the male copulatory organ for sperm transfer. Females then deposit the fertilized eggs in emergent vegetation or in water either singly or in masses.

Nymphs are all aquatic and have a unique, greatly enlarged, protrusible **labium** that is folded beneath the head when not in use, but it can be extended suddenly and rapidly to grasp any moving prey of the proper size. Prey usually consists of small oligochaetes, crustaceans, and other insects, but some nymphs even attack tadpoles and small fishes. Rectal and abdominal gills are used in ventilation of the closed tracheal system (Fig. 13.25b). At the conclusion of the final nymphal instar the nymph crawls out of the water onto some emergent object, undergoes its final ecdysis, and emerges as an adult. Improved flying abilities

Figure 13.25 *Odonates or damselflies and dragonflies:* ***(a)*** *photograph of an adult damselfly (courtesy of J.R. Meyer);* ***(b)*** *dorsal view of a nymphal* Argia *(damselfly) (by permission of Smithsonian Institution Press from "Notes on the life history and ecology of the dragonflies (Odonata) of Washington and Oregon," Figure 36.* Proceedings of the United States National Museum, Volume 49, Number 2107, *by Clarence Hamilton Kennedy. Smithsonian Institution, Washington, D.C. 1915);* ***(c)*** *a photograph of an adult* Celithemis *(dragonfly);* ***(d)*** *a photograph of a nymphal* Hagenius *(dragonfly). (parts c, d courtesy of S.W. Dunkle.)*

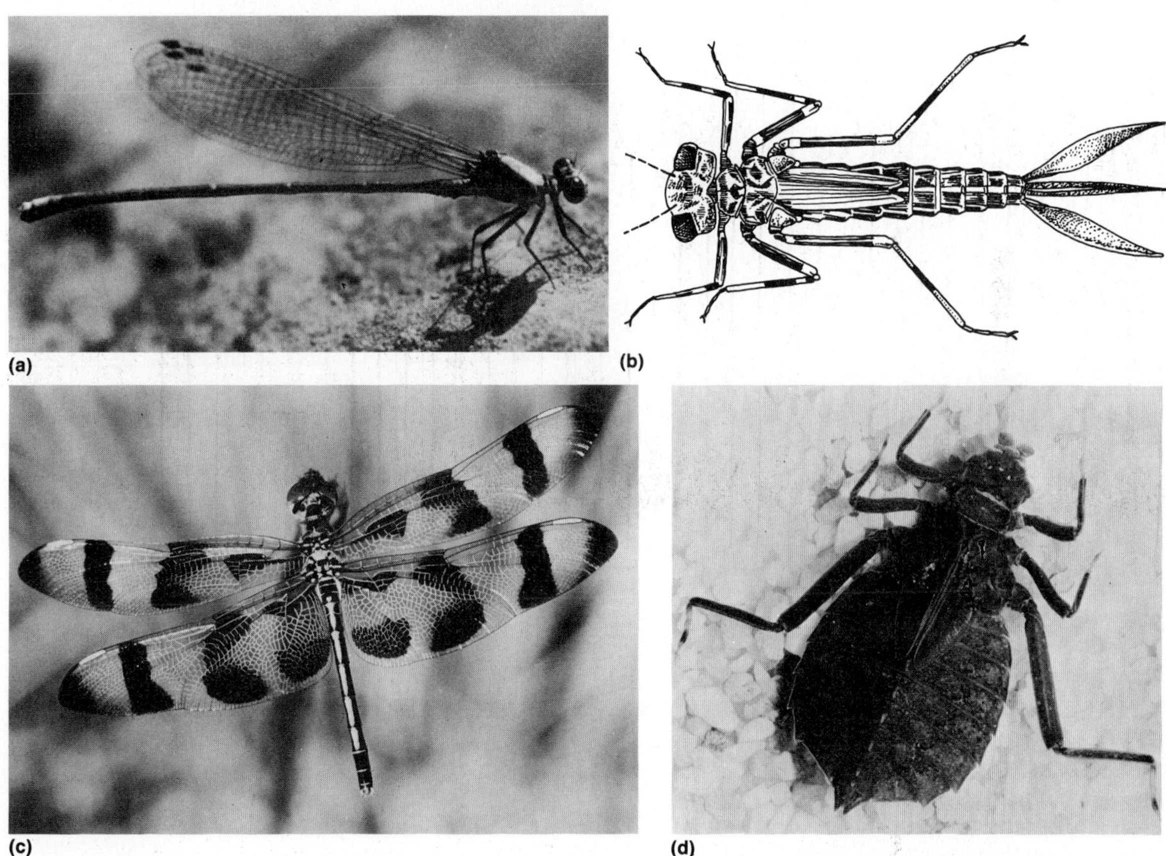

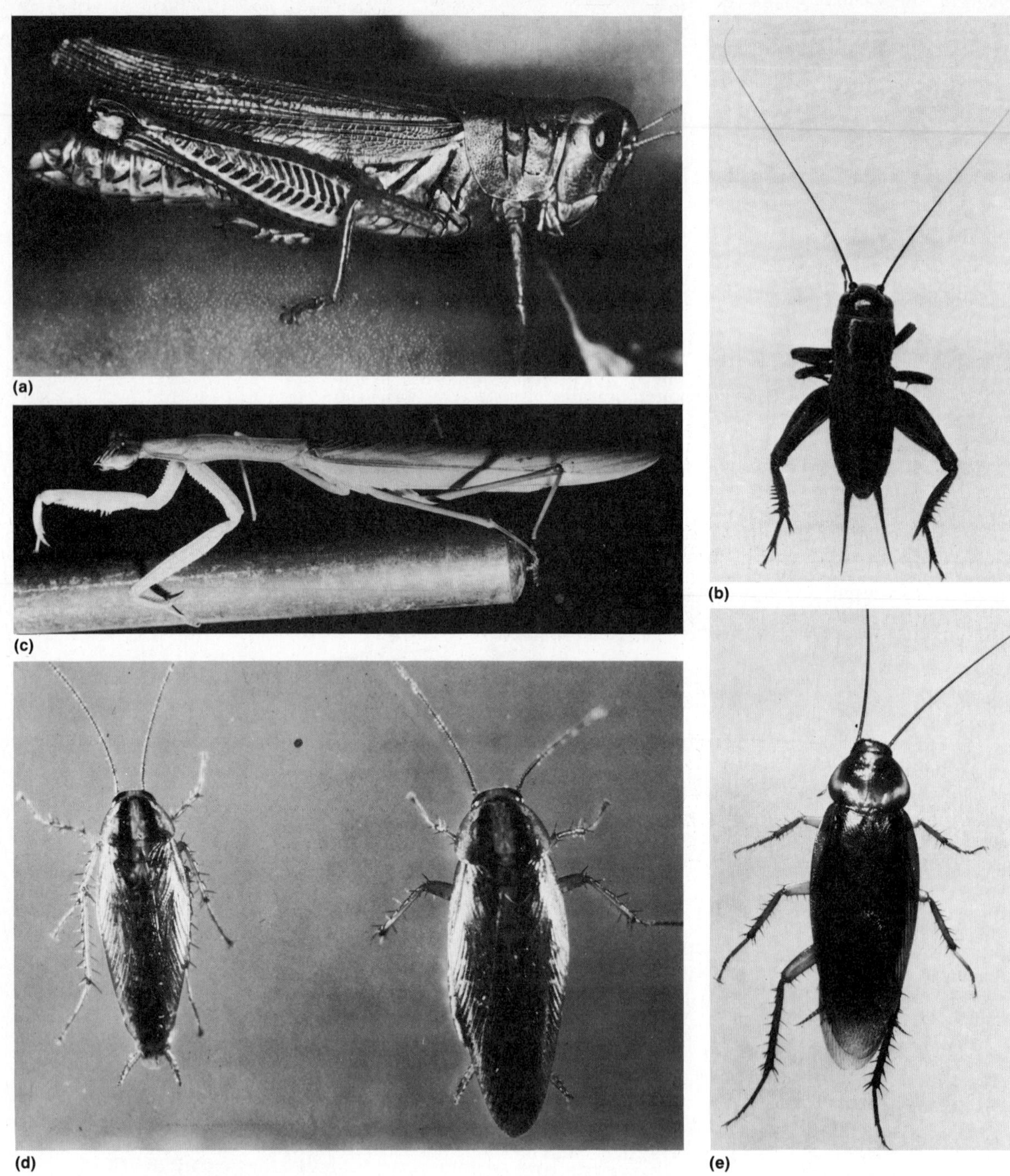

Figure 13.26 *Orthopterans or grasshoppers, crickets, mantids, and roaches:* ***(a)*** *photograph of* Melanoplus, *the differential grasshopper (courtesy of Ward's Natural Science, Inc., Rochester, N.Y.);* ***(b)*** *photograph of* Gryllus, *the common house cricket;* ***(c)*** *photograph of* Tenodera, *a praying mantid (courtesy of H. Itagaki);* ***(d)*** *photograph of* Blatella germanica, *the German cockroach (female on right, male on left);* ***(e)*** *photograph of* Periplaneta americana, *the American cockroach. (parts d, e courtesy of the Centers for Disease Control, Atlanta, Ga. 30333.)*

are soon developed, adult colorations appear, and the adult becomes sexually mature in about one week.

Two groups of odonates are recognized. Dragonflies (Anisoptera) are larger and stouter, wings are held horizontally at rest, and nymphs have rectal gills (Fig. 13.25c, d). The damselflies (Zygoptera) are more slender and dainty appearing, have wings that can be moved vertically over the abdomen, and have nymphs with three external abdominal gills (Fig. 13.25a, b). Odonata are voracious feeders both as nymphs and as adults and often prey upon mosquitoes, gnats, flies, and other nuisance insects.

Order Orthoptera

The Orthoptera includes the grasshoppers, crickets, cockroaches, mantids, and related forms. It is a large, diverse order with almost 30,000 identified species. Because it is a varied assemblage of insects, some taxonomic schemes divide this grouping into several different orders. Adults range in size from 1.2 to 25 cm, and almost all orthopterans are terrestrial.

Adults have mandibulate mouthparts, and most are herbivorous. Large compound eyes are present. The antennae are variable in length but tend to be elongated and multisegmented (Fig. 13.26). Those winged orthopterans usually have two pairs of wings; the forewings or **tegmina** are elongated and parchmentlike, and the membranous broad hindwings are folded pleatlike under the tegmina when at rest. Flight is mainly through the action of the hindwings. Many orthopterans like katydids, crickets, and some walkingsticks are wingless. The variable legs may be adapted for jumping, running, grasping, or digging. In most the ovipositor is well developed, and eggs are typically deposited in the soil or on vegetation. In some species the eggs are retained by the female until eclosion. Orthopterans generally have five to ten nymphal instars.

Many orthopterans are able to produce sounds or to **stridulate** and to perceive sounds made by others. The "songs" are generated by males and are designed to attract females, although "fighting" or "alarm" songs can also be produced. The singing Orthoptera usually possess auditory organs called **tympana** located on the abdomen or at the base of the legs.

This order includes a larger number of economically important forms. Many grasshoppers and some crickets (Acridoidea, Ensifera) are serious pests in field and truck crops, and some shorthorned grasshoppers (called locusts) are migratory and are serious pests in North Africa. Cockroaches (Blattoidea) are especially familiar and cosmopolitan inhabitants of cupboards, pantries, and bathrooms (Fig. 13.26d, e). Though they neither bite nor sting, roaches may be vectors for a larger number of pathogenic microorganisms. Mantids (Mantidae), sometimes very large orthopterans, are all strongly predaceous, and their forelegs are modified as strong grasping hooks for raptorial feeding on other arthropods (Fig. 13.26c). Mantids are especially coveted by gardeners, since they are voracious feeders on pest insects.

Order Isoptera

The Isoptera or termites are small to medium-sized insects that live in colonies and have a well-developed caste system. Most of the 2100 species are tropical and subtropical, but some live in temperate regions.

Reproductive adults or **alates** possess short antennae, and most have compound eyes, although some apterous forms lack them. When present, there are two pairs of identically sized membranous wings that are usually shed after mating (Fig. 13.27b). Termites are sometimes confused with ants and are called "white ants," but termites are distinguished by having wings of equal size from which they get their name (Isoptera = equal wings), being pale and soft bodied, and lacking a "waist" or constriction between thorax and abdomen.

All termites are social and live in highly organized colonies of thousands or millions of

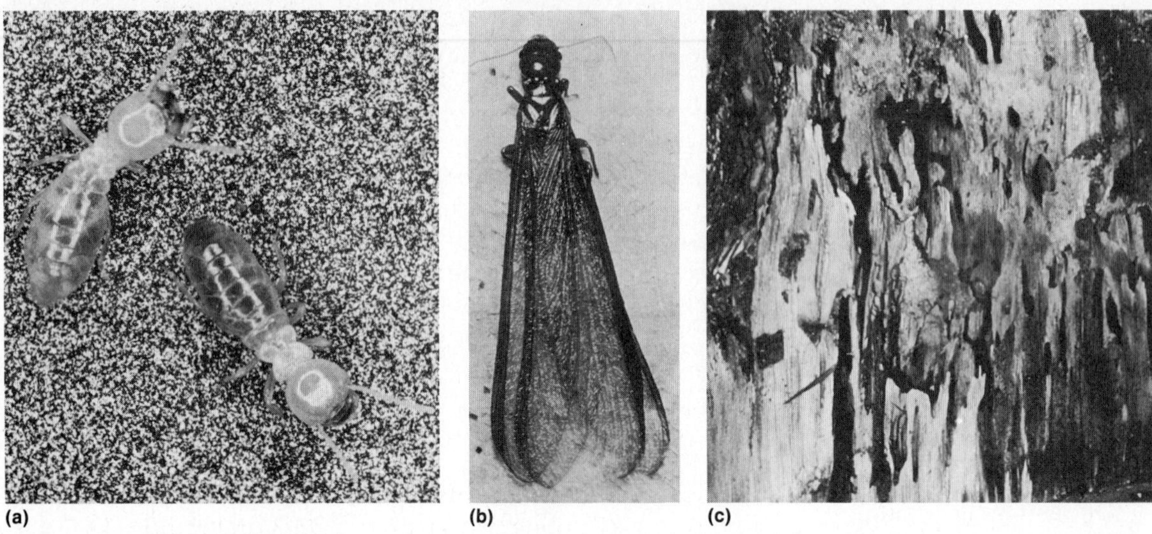

Figure 13.27 Isopterans or termites: *(a)* photograph of two workers; *(b)* photograph of a Zootermopsis winged form; *(c)* photograph of a section of wood showing destruction by termites. (parts b, c courtesy of Ward's Natural Science, Inc., Rochester, N.Y.)

individuals. Each colony has several specialized castes. **Males** and **queens,** the primary reproductives with wings and compound eyes, swarm from the nest, construct a chamber, mate, lose their wings, and start new colonies. Secondary reproductives, having wing pads but not functional wings and having eyes smaller than those of the primary reproductives, often reproduce and aid the primary reproductives in building the colony. **Workers** are nymphs and sterile adults who are wingless, lack compound eyes, and have small mandibles; they are primarily nursemaids, cleaners, builders, and foragers (Fig. 13.27a). **Soldiers** are sterile adults with modifications for defense against intruders in the colony. Many species have forms with enlarged heads and mandibles. **Nausti,** present in a number of species, have heads elongated into a snout through which a sticky secretion is squirted at intruders. Caste determination still is not fully understood but is controlled by pheromones and intrinsic hormones.

Termites do substantial damage each year to wooden structures (timbers, floors, furniture); estimates of $40 million damage per year in the United States alone have been reported (Fig. 13.27c). Most termites eat wood, seeds, and grasses that they forage, and some eat fungi that they cultivate. Termites lack the ability to digest wood; their digestive tracts harbor an assemblage of symbiotic bacteria and zooflagellates that have the necessary cellulases (see Chapter 2). Many termites build elaborate extensive nests either above ground or in subterranean habitats.

Order Plecoptera

The Plecoptera or stoneflies are 12–65 mm in length and are usually drab-colored insects found near streams or rocky lake shores. About 1600 species are known, and all are soft bodied and elongated.

Adults have long antennae, well-developed compound eyes, but sometimes reduced and often nonfunctional chewing mouthparts. The two pairs of similar-sized wings are membranous, and part of the hindwings are folded in pleats beneath the forewings and over the abdomen when at rest (Fig. 13.28a). It is from this folded nature of the hindwings that the ordinal name was coined (Plecoptera = folded wings). Wings are reduced or absent in the males of a few species. The abdomen bears two prominent **caudal cerci.**

Nymphs are all aquatic, flattened, and basically herbivorous. Mandibulate mouthparts are used for feeding on various, small, aquatic plants. The thoracic segments are quite distinct from each other with no segmental fusion (Fig. 13.28b). Each thoracic segment bears a pair of **gill tufts** at the posterolateral aspect, and nymphs have a closed tracheal system. The abdomen bears two long **caudal cerci** similar to those in adults. Nymphal stages may last one to several years; most species emerge in the summer, but others even emerge, feed, and mate during the winter months.

Stoneflies are an important source of food for many freshwater fishes, are only rarely a pest, and have no medical importance.

ORDER PSOCOPTERA

The Psocoptera or psocids are usually less than 6 mm in length and are often not noticed, even though there are more than 1600 species. All have chewing mouthparts, long filiform antennae, and tarsi with two or three segments. They

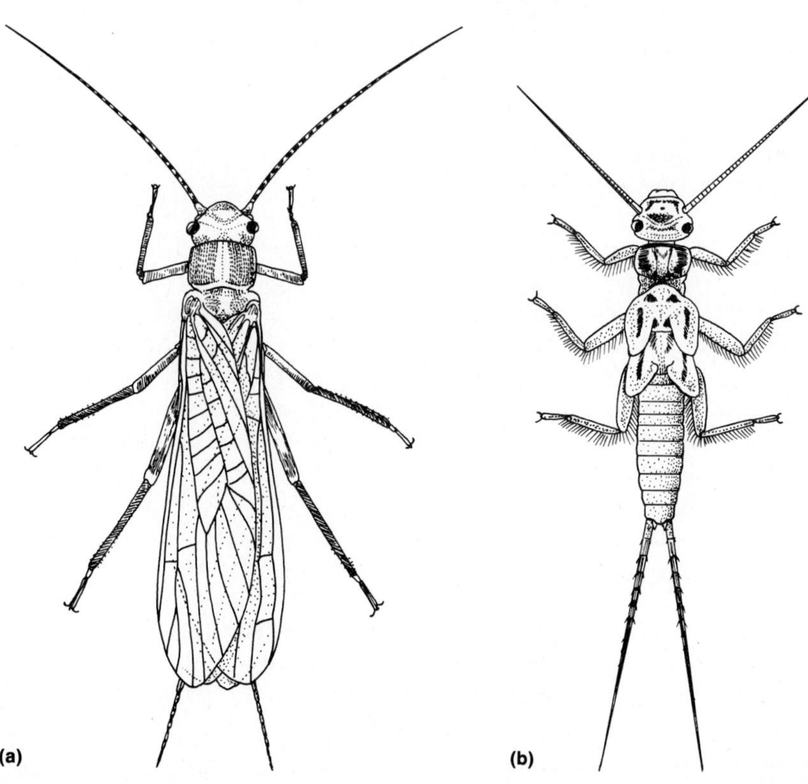

Figure 13.28 *Plecopterans or stoneflies:* ***(a)*** *an adult* Isoperla; ***(b)*** *a nymphal* Isoperla *(parts a, b from T.H. Frison, 1935,* Bull. Illinois Nat. Hist. Sur. *20:283, 442).*

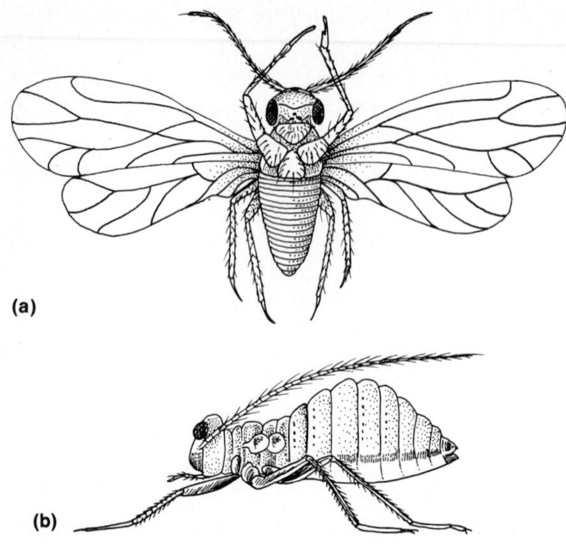

Figure 13.29 *Psocopterans or psocids:* ***(a)*** *an adult bark louse;* ***(b)*** *lateral view of an apterous adult book louse (from D. Sharp, 1909,* Cambridge Natural History, *Macmillan and Co., Ltd., London).*

have either large or reduced compound eyes. Wings are either present or absent; in those that have wings these structures are either long or short. Hindwings tend to be smaller than forewings, and both pairs are flexed into a rooflike position over the abdomen when at rest (Fig. 13.29a).

Most psocids live on the bark or foliage of trees or shrubs, under bark or stones, or in leaf litter. But a few well-known species live in houses, where they often feed on the starchy pastes of bookbindings and are therefore called booklice (Fig. 13.29b). They may also feed on glue, grain and cereal products, and fungi. Rarely do they cause serious damage, but they are frequently a nuisance.

Eggs are laid singly or in clusters and may be covered with silk or debris. Most have six nymphal instars, and nymphs and adults eat the same types of food.

Order Mallophaga

The Mallophaga or chewing lice are represented by about 2700 species. They are small (2–6 mm), flattened, wingless insects with chewing mouthparts. They are ectoparasitic mainly on birds and some mammals and are often referred to as bird lice. They feed on the feathers, skin, or hair of the host.

Adults are minute and dorsoventrally flattened with a triangular-shaped head bearing small compound eyes and no dorsal ocelli (Fig. 13.30). The legs are modified for grasping the feathers or hair of the host. The abdomen lacks cerci or other appendages.

Eggs are attached to hair or feathers, and there are three nymphal instars. The apterous lice, dispersed by physical contact between hosts, are important pests of domestic animals, particularly poultry, in which serious infestations cause the host to be emaciated, thus making it easier prey for other diseases. Cattle, horses, goats, sheep, cats, and dogs are often attacked by chewing lice, but such lice do not live long on people and are of no medical importance.

Order Anoplura

The Anoplura are the sucking lice, which are small (0.5–6.5 mm) wingless insects comprising about 250 species. They differ from the Mallophaga in having sucking mouthparts and a head more narrow than the thorax, and they are always ectoparasitic on mammals.

Adults are dorsoventrally flattened and have piercing-sucking mouthparts that are retracted into the head when not in use. They have greatly reduced eyes and lack wings, dorsal ocelli, and cerci. The tarsus of each short leg is composed of only one segment, which is equipped with a large claw; this makes an effective grasping organ for holding onto the hair of the host (Fig. 13.31). The entire life cycle is spent on a host; eggs (nits) are glued to hair,

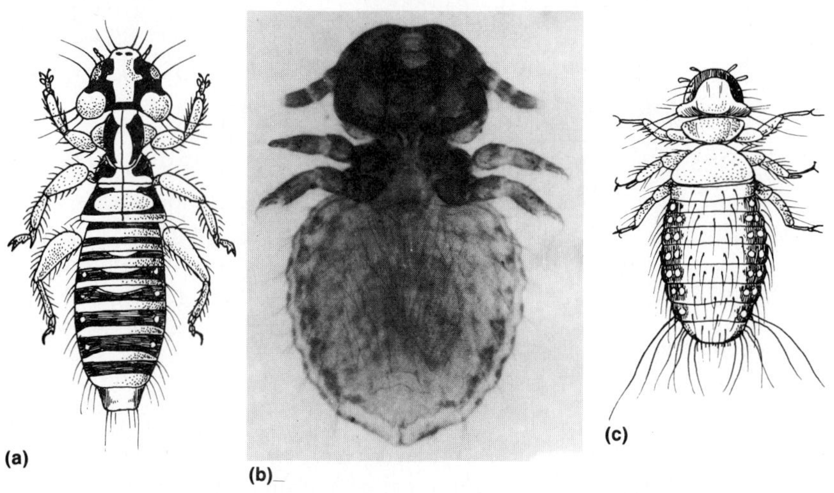

Figure 13.30 *Mallophagans or chewing lice: **(a)** a duck louse; **(b)** photograph of a mallophagan (courtesy of J.R. Meyer); **(c)** a chicken louse. (parts a, c from D. Sharp, 1910,* Cambridge Natural History, *Macmillian and Co., Ltd., London.)*

Figure 13.31 *Anoplurans or sucking lice: **(a)** photograph of* Pthirus pubis, *the human pubic or crab louse; **(b)** photograph of* Pediculus humanus, *another human body louse (parts a, b courtesy of Ward's Natural Science, Inc., Rochester, N.Y.).*

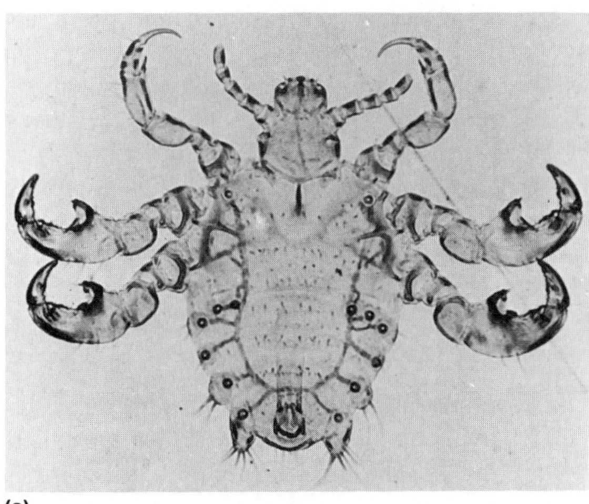

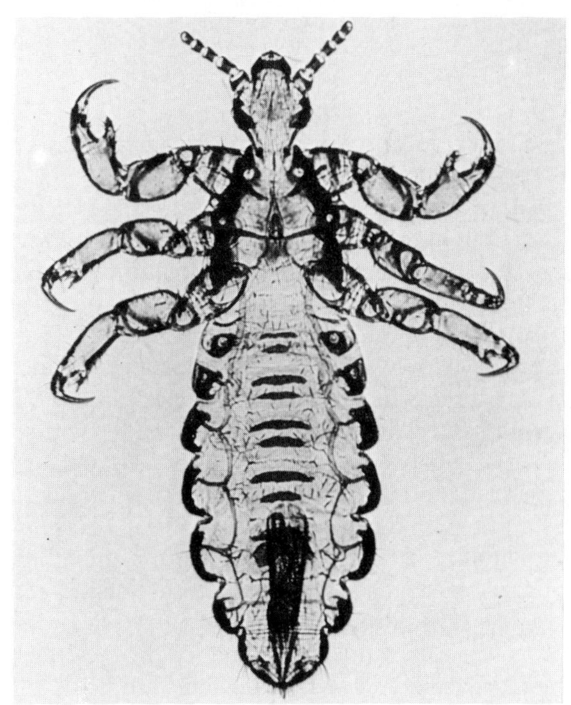

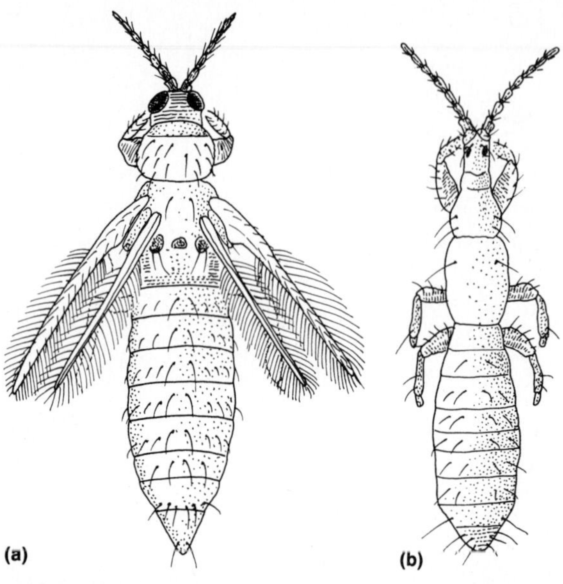

Figure 13.32 *Thysanopterans or thrips: **(a)*** Heterothrips salicus, *a winged form;* ***(b)*** Merothrips morgani, *an apterous form (parts a, b from L.J. Stannard, 1968, The Thrips, or Thysanoptera, of Illinois.* Ill. Nat. Hist. Sur. Bull. 29:*215–552).*

and the young nymphs emerge after a short embryonic period.

Lice exhibit a high degree of host specificity and even specificity to body regions. They are found on a variety of mammals including hogs, horses, cattle, sheep, dogs, cats, seals, and walruses. Two species of sucking lice are ectoparasitic on people. *Pthirus pubis* infests pubic and armpit areas and causes intense itching (Fig. 13.31a). *Pediculus humanus* is a human louse that has two subspecies: *corporus,* which lives on various places on the body, and *capitis,* which usually lives on the head (Fig. 13.31b). *Pediculus* individuals serve as vectors for a number of human pathogens including those causing relapsing fever, trench fever, and epidemic typhus. People who bathe and change clothes regularly seldom remain infested with lice.

Order Thysanoptera

The Thysanoptera are the thrips and are represented by about 4700 species. They are small (0.5–14 mm) elongated insects that feed on a wide range of flowers and cultivated crops. They are especially abundant on daisies and dandelions and may attack tobacco, onions, gladioli, and several citrus crops, in which they can cause considerable damage. Sometimes, thrips are found in enormous numbers. While a few species may occasionally bite humans, they are of little medical importance.

Adults are usually dark colored with small compound eyes, and the sucking mouthparts are in the form of a stout conical beak in which the right mandible is reduced or absent. Wings are absent in some, but in most there are two pairs of very long, narrow wings that are fringed with long hairs (Thysanoptera = fringed wings) (Fig. 13.32). Some possess an ovipositor, while in others it is absent. Some species reproduce parthenogenetically.

Thysanopterans have an unusual life cycle that is intermediate between incomplete and complete metamorphosis. The first two instars are apterous and usually are called **larvae;** in some the wings even develop internally. The third and fourth instars are inactive, are nonfeeding, and have external wing pads; these stages are often termed **prepupa** and **pupa,** respectively. Some thrips may have two prepupal instars, and the pupal stage may even be enclosed in a cocoon. This type of metamorphosis is simple in that more than one immature stage has external wing pads. It resembles complete metamorphosis in that wing development is internal in earlier stages, and there is a quiescent stage preceding the adult.

Order Hemiptera

The Hemiptera is a rather large order to which belong the bugs. While lots of things are called "bugs" by people generally, the true bugs belong to this order. Some authors also group homopterans in this order, but our scheme will consider the two groups to be separate orders. With about 25,000 species the insects in this order are quite variable (Fig. 13.33). They range in size from minute up to 10 cm in length. While most are terrestrial and have wings, many are aquatic.

Adults possess well-developed compound eyes, and most have long antennae with four or five segments, but the ocelli often are absent. They have piercing-sucking mouthparts arising from the anterior part of the head that are associated together to form a slender, often-segmented **beak** or **rostrum** that is carried in an opisthognathous position. The segmented labium serves as a sheath for the mandibles and maxillae, which are modified as sharp piercing stylets. The maxillae, lacking palps, fit together in the beak to form both food and salivary channels. A distinctive feature of most bugs is the structure of the forewings, the basal portions of which are thickened and tough while the distal parts are membranous. This type of wing is called a **hemelytron,** and it is from this feature that they get their name (Hemiptera = half wing). The hindwings are entirely membranous and are usually somewhat shorter than the front wings. When at rest, the wings are folded over the abdomen. The membranous tips of the forewings overlap each other, and the triangular **scutellum** between the folded wings is a rather distinctive feature (Fig. 13.33a, c, e). But many hemipterans are apterous, and others have one pair of rather short wings. Abdominal cerci are absent. Many bugs have thoracic or abdominal **stink glands** that produce characteristic unpleasant odors that repel enemies. Many hemipterans are capable of stridulation.

Most hemipterans (Geocorizae) are terrestrial, and they have conspicuous antennae and lack trichobothria. Many other bugs such as water striders and treaders (Amphibicorizae) are semiaquatic or live on shores and have three pairs of cephalic trichobothria (Fig. 13.33f, g). A great many other bugs such as water boatmen (Fig. 13.33h) and backswimmers (Hydrocorizae) are aquatic as adults, lack trichobothria, and have short concealed antennae. A number of bugs are of considerable economic and medical importance. Many are pests of cultivated plants including the chinch, squash (Fig 13.33a), plant, harlequin (Fig. 13.33c), and stink bugs and cotton fleahoppers. Bed bugs and assassin bugs (Fig. 13.33b, d) cause painful bites to persons, and several species are vectors of Chagas' disease. Some are predaceous on other insects or parasitic on birds and mammals, but most bugs feed on plant juices.

ORDER HOMOPTERA

The Homoptera include the cicadas, hoppers, whiteflies, aphids, and scale insects. They constitute a very large and diverse group, the largest of the exopterygote orders, of about 34,000 species (Fig. 13.34). Homopterans, ranging from 0.03 to 8 cm in length, are quite similar to the hemipterans, but there are also some basic differences. All homopterans are terrestrial plant feeders, and many are serious pests of cultivated plants.

The mouthparts of adults, similar to those of the Hemiptera, are piercing-sucking and in the form of a **beak** or rostrum that arises from the back of the head near the forelegs rather than from the anterior aspect as in bugs. Large compound eyes are often present; ocelli are variable in their occurrence. Antennae range from short and bristlelike to long and filiform. Abdominal cerci are absent, but females often possess an ovipositor of modified appendages. Some are apterous, and scale insects possess only mesothoracic wings (Fig. 13.34d), but most bear two pairs of uniformly membranous wings. At rest, the wings are held rooflike over the body with the distal tips slightly overlapping (Fig. 13.34b, f). In some species such as

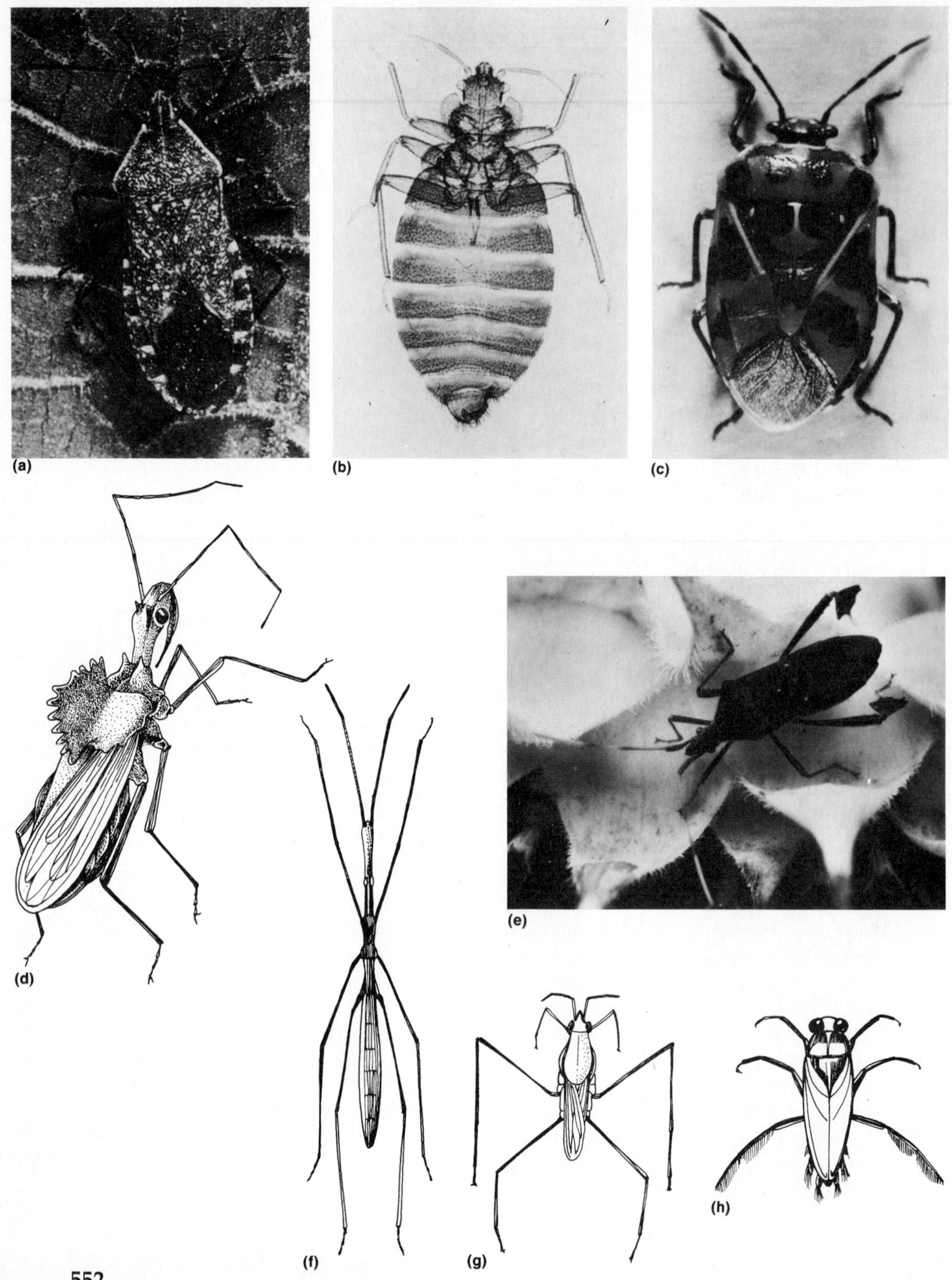
(a)
(b)
(c)
(d)
(e)
(f)
(g)
(h)

◀ *Figure 13.33* *Hemipterans or true bugs:* ***(a)*** *photograph of* Anasa tristis, *the squash bug;* ***(b)*** *photograph of* Cimex, *the bed bug;* ***(c)*** *photograph of* Murgantia histrionica, *the harlequin bug;* ***(d)*** Arilus, *an assassin bug;* ***(e)*** *photograph of* Leptoglossus, *the leaf-footed bug;* ***(f)*** Hydrometra, *an apterous water treader;* ***(g)*** Gerris, *a water strider (from R.W. Pennak,* Fresh-water Invertebrates of the United States, *2nd ed., © 1978, John Wiley & Sons, Inc., New York, N.Y., used with permission);* ***(h)*** Notonecta, *a backswimmer. (parts a, b courtesy of Ward's Natural Science, Inc., Rochester, N.Y.; parts c, e courtesy of J.R. Meyers; parts f, h from L.C. Miall, 1903,* The Natural History of Aquatic Insects, *Macmillan and Co., Ltd., London).*

Figure 13.34 *Homopterans or hoppers, whiteflies, aphids, and scale insects:* ***(a)*** *photograph of a membracidid or tree hopper;* ***(b)*** *photograph of* Trialeurodis, *the strawberry whitefly;* ***(c)*** *photograph of newly hatched aphids;* ***(d)*** *photograph of the white peach scale;* ***(e)*** *photograph of the black peach-tree aphid;* ***(f)*** *photograph of* Magicicada, *the periodical cicada. (parts a, b courtesy of Ward's Natural Science, Inc., Rochester, N.Y.; parts d, e courtesy of J.R. Meyer.)*

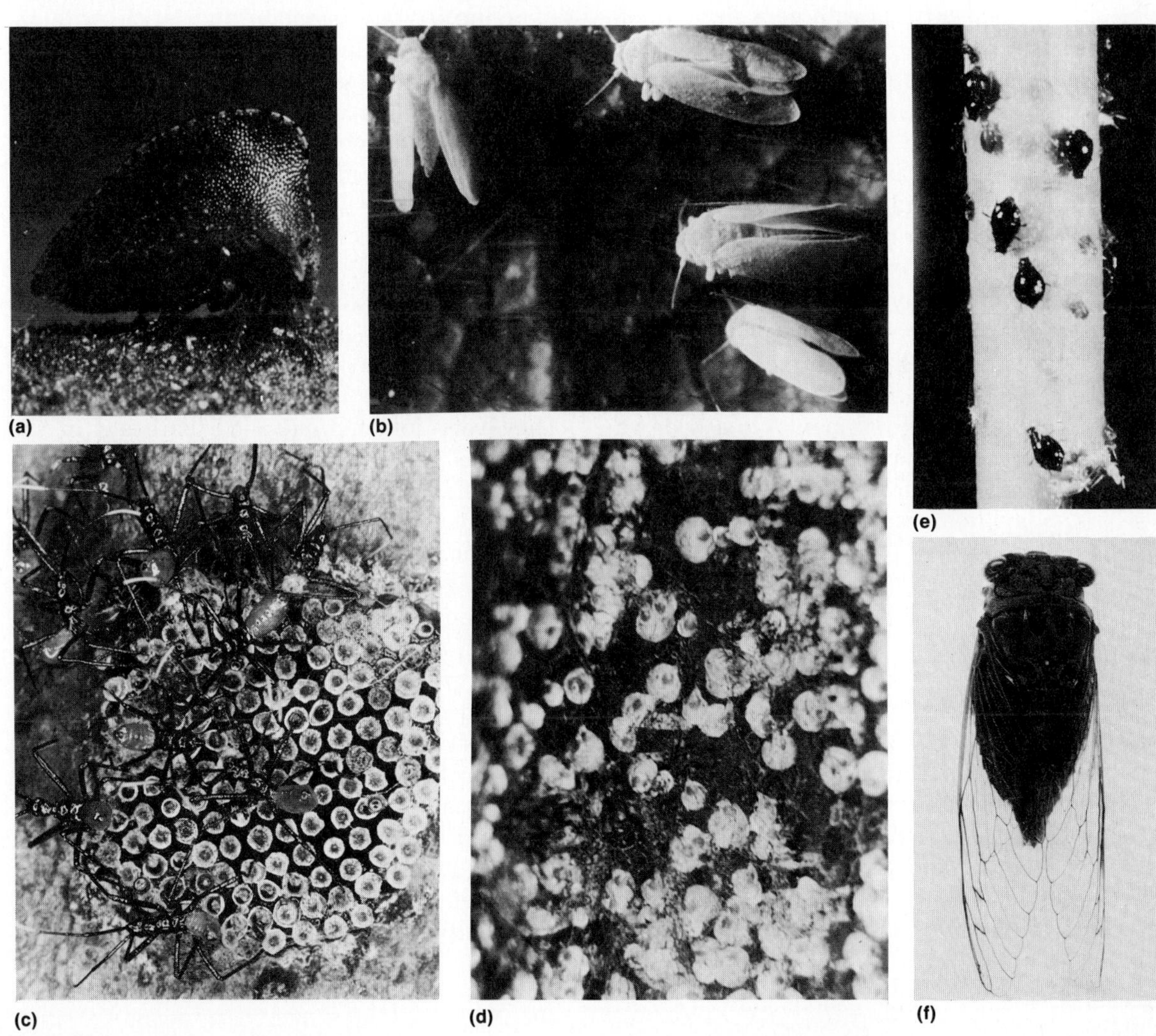

scale insects, one sex may be apterous, and even winged and wingless individuals of the same sex are found.

Members of this order typically undergo simple metamorphosis, but the last nymphal instar of whiteflies and male scale insects is immobile and pupalike. Homopteran life cycles may also be very complex involving winged and wingless stages. Many aphids reproduce parthenogenetically and are viviparous during the summer, but they become oviparous in the autumn. Completion of the life cycle may take from one week to 17 years in some cicadas.

Cicadas and hoppers (Auchenorrhyncha) are active mobile insects with very short antennae. Cicadas are relatively large insects, they have nymphal development in the ground, and males produce shrill stridulating sounds (Fig. 13.34f). Hoppers are much smaller, and their forewings are thickened somewhat (Fig. 13.34a). Whiteflies and aphids (Sternorrhyncha) are not very active or mobile (Fig. 13.34b, c, e), and the scale insects are sedentary (Fig. 13.34d). Antennae are usually long and filiform. No homopterans are important medically, although many are economically important pests. Most hoppers, whiteflies, aphids, and scale insects (Coccoidea) are quite detrimental to a wide variety of crops and ornamental plants. Most viral plant diseases are transmitted by hoppers and aphids. Some homopterans have become symbiotic with ants.

Subclass Pterygota, Division Endopterygota

These insects undergo complete or total metamorphosis, and their wings develop internally in larval and pupal instars. About 88% of all living insect species belong to this division. Endopterygote larvae have become more specialized for feeding, and adults are involved with reproduction and dispersal. Larvae are typically found in quite different habitats from those of adults, and their general appearances are fundamentally different. In the immobile, nonfeeding, pupal instar, larval tissues are fundamentally reconstructed to form adult structures. Insects with complete metamorphosis are said to have holometabolous development.

Three lines of evolution have occurred in the endopterygotes. In one group, including the Neuroptera, Coleoptera, and a minor order, all individuals have chewing mouthparts, and most have modified forewings and lack an ovipositor. A second group, including the Trichoptera, Lepidoptera, Diptera, Siphonaptera, and a minor order, are thought to be related on the basis of reduced wing venation and larvae with reduced antennae and similar mouthparts. Insects in a third group, represented only by the Order Hymenoptera, have more numerous Malpighian tubules and wings with large cells, and the ovipositor is usually modified into a sting.

Order Neuroptera

The Neuroptera consists of about 4800 species of insects called alderflies, dobsonflies, lacewings, and antlions. Neuropterans are usually found in vegetation and around bodies of water. They vary in length (0.3–6.5 cm) and in the nature of the food they ingest. Neuropterans are considered the most primitive of the endopterygote insects and perhaps were ancestral to the higher insects.

Larvae are either aquatic (dobsonflies, alderflies) or terrestrial (lacewings, antlions) (Fig. 13.35a). Aquatic larvae are elongated and strongly predaceous and have paired abdominal gills and strong mandibulate mouthparts. The larvae of dobsonflies, called hellgrammites, are prized as fish bait. Terrestrial larvae often have the mouthparts forming a hollow tube for sucking fluids. Antlion larvae build cone-shaped depressions in loose sand into which fall small prey that are unable to escape. The larvae of one group are parasitic on ground-spider eggs.

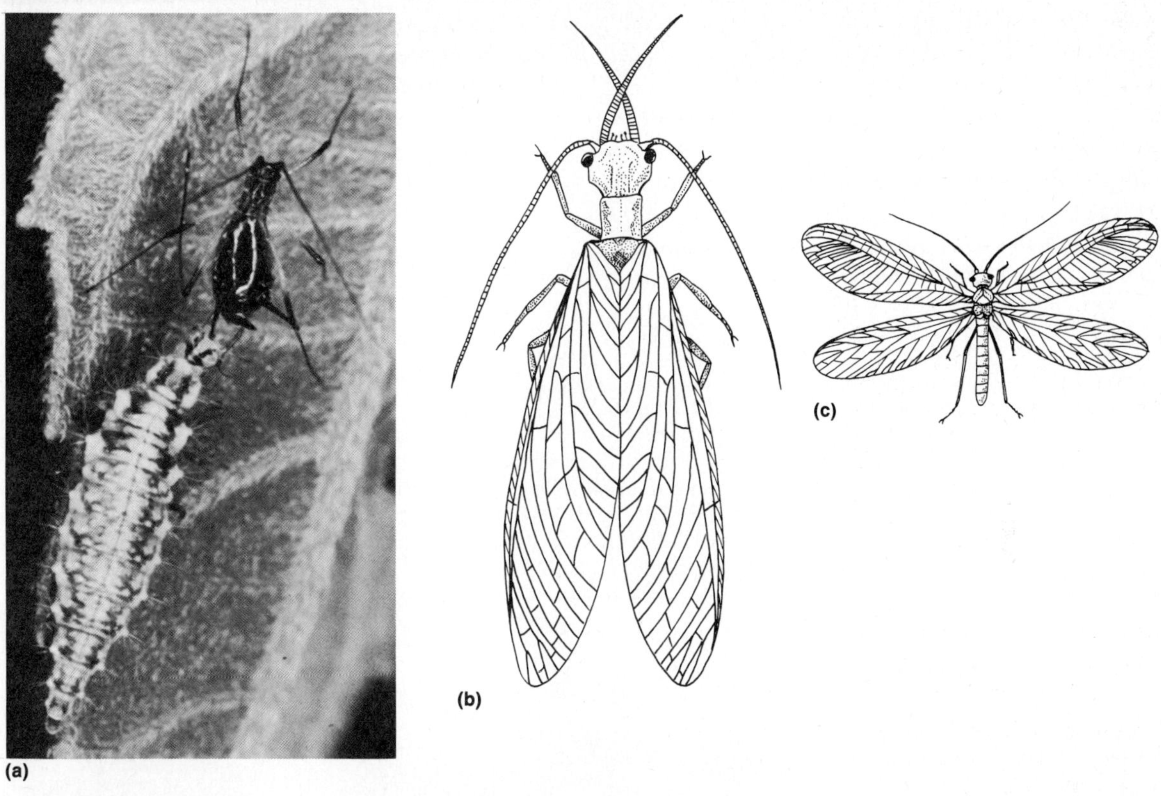

Figure 13.35 *Neuropterans or alderflies and lacewings;* ***(a)*** *photograph of an antlion larva (courtesy of Ward's Natural Science, Inc., Rochester, N.Y.);* ***(b)*** *an adult dobsonfly* (Corydalus) *(from R.W. Pennak,* Fresh-water Invertebrates of the United States, *2nd ed., © 1978, John Wiley & Sons, Inc., New York, N.Y., used with permission);* ***(c)*** *an adult lacewing* (Chrysopa) *(from D. Sharp, 1909,* Cambridge Natural History, *Macmillan and Co., Ltd., London).*

Pupation never takes place underwater but in earthen cells or silken cocoons.

The head of adults bears chewing mouthparts that may be reduced (or nonfunctional as in dobsonflies) as feeding devices (Fig. 13.35b, c). They all bear four equal-sized wings with a netlike pattern of many crossveins, and adults are generally weak fliers. At rest, the wings are usually held rooflike over the body.

Neuropterans have no medical significance. They are often considered to be economically beneficial because of their predatory habits.

Order Coleoptera

The Coleoptera is the most successful of all insect orders; over 300,000 species of beetles have been identified representing about 40% of all insects. Unbelievably, there are more beetle species than there are species of all noninsectan animals combined! They range in length from 0.02 to 15 cm and are found almost everywhere in a wide variety of habitats. Beetles are incredibly varied, all the major larval types being present. Many larval forms are aquatic, but the vast majority are terrestrial. The larval stage, extremely varied in length, is followed by an exarate pupa.

In spite of their enormous species diversity, adults are rather homogeneous in appearance (Fig. 13.36). Their bodies are normally stout and extensively sclerotized. All coleopterans have mandibulate mouthparts, and the mandibles in particular are well developed and can pierce or crush very tough materials. To a degree, the mouthparts reflect the extraordinary variety of foods ingested by beetles, since phytophagous, predaceous, scavenger, and parasitic forms are known.

The most distinctive coleopteran feature is the structure of the wings. The forewings are thick, leathery, and tough and normally are not used in flying, but rather these **elytra** serve as

Figure 13.36 *Coleopterans or beetles and weevils:* ***(a)*** *photograph of* Popilla japonica, *the Japanese beetle;* ***(b)*** *photograph of* Adalia bipunctata, *a ladybird beetle;* ***(c)*** *photograph of* Lucanus elephus, *the stag beetle (courtesy of D. and L.P. Burney);* ***(d)*** *photograph of an aquatic hydrophilid;* ***(e)*** Photuris, *the lightningbug or firefly;* ***(f)*** Hydrophilus, *a water-scavenger beetle (from L.C. Miall, 1903,* The Natural History of Aquatic Insects, *Macmillan and Co., Ltd., London);* ***(g)*** *photograph of* Sitophilus, *a granary weevil (courtesy of the U.S. Department of Agriculture). (parts a, b courtesy of Ward's Natural Science, Inc., Rochester, N.Y.)*

protective sheaths for the hindwings (Fig. 13.36); it is from this feature that their ordinal name was coined (Coleoptera = sheathed wings). The membranous hindwings are the only flight organs, are longer than the forewings, and when not in use fold up and are covered by the elytra, which meet in a straight middorsal line (Fig. 13.36). Coleopteran life cycles vary in length from several generations a year to a single generation in several years. Some beetles produce courtship sounds associated in the course of feeding or flying, by stridulation, or by striking some part of the body against the substratum.

This group is of tremendous economic significance, since many are destructive to agricultural crops; they are especially damaging to cotton, corn, potatoes, and beans. Many infest homes and destroy rice, flour, meal, clothes, and structural wood. Japanese beetles are serious pests on many different types of vegetation (Fig. 13.36a), and some beetles bore into fruit, destroy leaves, and attack roots of grasses and other plants. A few adults are predaceous on other insects like aphids and are therefore beneficial to human beings.

Several of the larger taxa of coleopterans are the snout, ground, rove, leaf, longhorn, and scarab beetles. The snout beetles (Curculionoidea) have the head prolonged into an anterior beak or snout, all feed on plant material, and larvae and adults (many of whom are weevils) are of considerable economic importance, since they are pests. Ground beetles (Carabidae) are common under stones, logs, leaves, and other debris. Most ground beetles are predaceous, run rapidly but seldom fly, and have brightly striated elytra. Rove beetles (Staphylinidae) are slender and elongated, have short elytra, and are predaceous. They are usually found in decaying materials like dung or carrion and in a variety of other habitats. Leaf beetles (Chrysomelidae) are small, often brightly colored beetles that feed on flowers and foliage and therefore are serious pests of cultivated plants. This group includes tortoise and flea beetles and leaf miners. The longhorn beetles (Cerambycidae) all have very long antennae, are brightly colored, and are phytophagous. Many species are wood-borers and are destructive to a great many trees and to newly cut logs and lumber. Scarab beetles (Scarabaeidae) have heavily cuticularized bodies but vary greatly in size and color. Many feed on dung or decomposing plant material, and some are serious pests to lawns and to a variety of agricultural crops.

ORDER TRICHOPTERA

The Trichoptera are the caddisflies, comprising about 5000 species. Adults range in size from 0.15 to 4 cm in length.

Eggs are laid in fresh water or on objects at the water's edge. Larvae are all aquatic and are caterpillarlike with a well-developed head and thoracic legs (Fig. 13.37b). The abdomen bears segmental pairs of filamentous gills; at its posterior end is a pair of hooklike appendages. Most larvae construct cases composed of a silklike base into which they incorporate environmental materials such as pieces of plants, twigs, sand, or pebbles (Fig. 13.37b, c). Each species builds a characteristic case, and case construction is often used as a taxonomic character. The hooked abdominal appendages aid the larva in remaining inside the case. The case-building larvae are all herbivorous. Other larvae inhabiting running water often build underwater nets constructed from silk produced by the salivary glands. Such nets tend to be funnel shaped with the large end upstream and the larva at the smaller, downstream end; larvae feed on plant and animal material caught in the net. Other caddisfly larvae construct neither cases nor nets and are free living and predaceous. Pupation occurs underwater, often within the larval case.

Adults are weak fliers and mostly dull colored. The mandibulate mouthparts are often vestigial and nonfunctional, and caddisflies usually feed only on liquid foods. The head

(a)

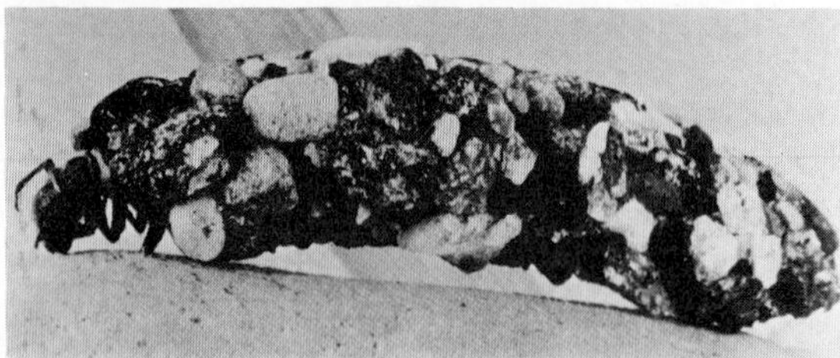
(b)

(c)

*Figure 13.37 Trichopterans or caddisflies: **(a)** photograph of an adult caddisfly; **(b)** photograph of a larva inside a sand-grain case; **(c)** photograph of two larvae and three cases constructed of vegetation of* Limnephilus indivisus *(courtesy of D.B. and B.K. MacLean). (parts a, b courtesy of Mrs. R.E. Hutchins.)*

bears large compound eyes and long filiform antennae (Fig. 13.37a). The wing surfaces are hairy (Trichoptera = hairy wings), and the wings are held rooflike over the abdomen at rest. The membranous wings have few crossveins, and forewings and hindwings are similar in size.

Both aquatic stages and adults are important in many aquatic food chains. Caddisflies are of no medical significance.

Order Lepidoptera

The Lepidoptera are the very familiar butterflies and moths. They constitute a rather large order with about 115,000 known species. They are readily recognized, since a great many are colorful and all have minute overlapping **scales** covering the wings and most of the body. They are medium to large insects having wingspans of up to 28 cm.

Larvae are eruciform and are called caterpillars. All are phytophagous and often are serious pests on many cultivated plants. Many caterpillars have colorations and structures that give them a ferocious or ominous appearance, but most are harmless to people. They have a well-developed head and cylindrical trunk of three thoracic and ten abdominal segments. The head bears three pairs of ocelli and a pair of short antennae. Each thoracic segment has a pair of legs, and some abdominal segments bear pairs of short, fleshy **prolegs,** the distal end of which bears hooks or **crochets;** in some larvae, some or all of the legs may be absent. Using their mandibulate mouthparts, smaller larvae skeletonize leaves, whereas larger larvae consume entire leaves. Most lepidopteran larvae have modified salivary glands that function as silk glands; silk is used for making cocoons, tying leaves together, and building tents or shelters. Many larvae form simple to elaborate cocoons in which to pupate. But the majority of lepidopteran larvae do not form cocoons; rather the naked pupa is a **chrysalis** surrounded by the exuvium of the final larval instar (see Fig. 13.20d).

The body of adults is covered with distinctive scales or macrotricha. In moths the scales are often dull brown or gray; but in butterflies, all colors may be present and in characteristic, contrasting, and usually beautiful patterns. The scales are easily detached when handled, and this feature enables them often to escape predators. The head bears large compound eyes and sucking mouthparts that typically are formed into a long sucking **proboscis** that is usually coiled (see Fig. 13.5c). Since the mandibles are often vestigial or absent, the proboscis is used only to obtain fluids in the form of nectar from flowers. The forewings are larger than the hindwings, and both are profusely covered with scales (Fig. 13.38). Lepidopterans have different joining mechanisms so that the forewings and hindwings can be coupled to synchronize their beat.

Popularly, lepidopterans are subdivided into moths and butterflies. Moths have feathery antennae, a stout body, and wings that are held horizontally or flexed over the abdomen at rest; are mostly dull colored and basically nocturnal; and often form cocoons (Fig. 13.38d–h). Butterflies have clubbed antennae, a slender body, and wings that are held vertically over the body when at rest; are brightly colored and essentially diurnal; and form a naked chrysalis (Fig. 13.38a–c).

Most lepidopterans (Ditrysia) are also subdivided into the microlepidopterans and macrolepidopterans, a distinction based solely on size. The microlepidopterans are mostly small moths such as bagworms, and clearwing, leafroller, and plume moths. The larvae of a great many species are pests to clothes, stored foods, cultivated crops, and lawns. The macrolepidopterans are the much larger forms and include a variety of moths (sphinx or hawk, giant silkworm, tussock, gypsy, tiger, owlet). The noctuid moths are a particularly large group of nocturnal species that are attracted to lights at night. Butterflies (Papilionoidea) and skippers (Hesperioidea) are the most colorful and familiar of the macrolepidopterans. Swallowtails (Papilionidae) have a pair of distinctive slender projections on the hindwings. The whites and sulfurs (Pieridae) are common, moderately sized butterflies whose larvae feed mostly on herbaceous plants. The brush-footed lepidopterans have forelegs that are small and brushlike, and thus they appear to have only four legs.

Natural silk is a product of the highly prized silkworm, *Bombyx mori,* and many other larvae are of considerable economic importance. Plants damaged by larvae include forests and fruit trees, many grain plants, sugarcane, cotton, tobacco, cabbage, and many stored products. Clothing is often ravaged by certain moth larvae. A very few species may cause blindness or may cause blistering of the skin in African cattle.

ORDER DIPTERA

The Diptera or true flies constitute a very large order and are represented by about 120,000 species of small to medium-sized insects with wingspans ranging from 0.1 to 10 cm. Adult dipterans are present sometimes in incredible numbers in a wide variety of ecosystems.

Larvae are found in a plethora of moist or aquatic habitats or anywhere the eggs are deposited. Many larvae are aquatic, others live within plants, and still others are found in decaying materials. The larval body is most commonly cylindrical and soft and always lacks true, segmented, thoracic legs. Larvae are wormlike and have mandibulate mouthparts. Pupae may be active (mosquitoes), but most are sessile within a puparium.

Adult dipterans are characterized by having only the forewings as flight organs, a feature

(a)
(b)
(c)
(d)
(e)
(f)
(g)
(h)

◀ ***Figure 13.38*** *Lepidopterans or butterflies and moths:* ***(a)*** *photograph of* Graphium marcellus, *the zebra swallowtail;* ***(b)*** *photograph of* Battus philenor, *the pipe-vine swallowtail;* ***(c)*** *photograph of* Danaus plexippus, *the monarch butterfly (courtesy of J.R. Meyer);* ***(d)*** *photograph of* Manduca sexta, *the adult of the tobacco hornworm (courtesy of H. Itagaki);* ***(e)*** *photograph of* Antheraea polyphemus, *the polyphemus moth;* ***(f)*** *photograph of* Anisota rubicunda, *the rosy maple moth;* ***(g)*** *photograph of* Actias luna, *the luna moth;* ***(h)*** *photograph of* Celerio lineata, *the white-lined sphinx moth.*

from which they derive their ordinal name (Diptera = two wings) (Fig. 13.39). The mesothoracic wings are membranous and contain few cells and lack many crossveins. Since the forewings are the only functional wings, the mesothorax is disproportionately large and sclerotized. The metathoracic wings are reduced to small knoblike **halteres** that function as gyroscopic organs of equilibrium. The dipteran head is well developed, the antennae are of variable form, and the compound eyes are quite large. The mouthparts are basically suctorial but are highly specialized for cutting, piercing, rasping, and sponging. A substantial number of adults utilize blood or nectar as food.

There are three suborders of dipterans: Nematocera, Brachycera, and Cyclorrhapha. The nematocerans, including crane, moth, and black flies, mosquitoes, midges, and gnats, are the least specialized of the three, and all have long antennae of more than three podomeres (Fig. 13.39a–c). Male nematocerans lack mandibles and therefore do not bite. Nematocerans are among the most serious medical pests; these dipterans are also vectors of important diseases like malaria, yellow fever, encephalitis, filariasis, African sleeping sickness, typhoid fever, and dengue. The brachycerans, including soldier, horse, deer, snipe, flower-loving, robber, bee, and dance flies, are often predatory on other insects (Fig. 13.39d, f, g, i). They have stout bodies, and all have antennae with fewer than five podomeres (usually three). The cyclorrhaphans, including fruit, shore, skipper, vinegar, house, louse, blow, and bot flies, are the most advanced dipterans (Fig. 13.39e, h). They are small, are mostly stout bodied, and have three-segmented antennae.

Dipterans in general are an important group because of the impact they have on human beings. Human blood and that of other mammals is the food for many dipteran adults including sand, black, horse, and deer flies, mosquitoes, and biting midges (Fig. 13.39a-c, f, h). Other flies like the Hessian fly and Mediterranean fruit fly and the larvae of certain other dipterans are especially harmful to plants. Many flies are scavengers, and larvae are found in garbage, excrement, carcasses, and other decaying materials. Still others like crane flies, midges, gnats, and house flies may be a nuisance especially when present in large numbers. A few species are of direct human benefit because they are parasitic on other insect pests.

ORDER SIPHONAPTERA

The Siphonaptera are the fleas represented by about 1600 species. Fleas are minute (0.8–5 mm) hard-bodied, wingless insects that are highly specialized as intermittent ectoparasites on mammals and birds.

Eggs are laid in the soil or the nest of the host and hatch into whitish legless larvae (Fig. 13.40b). The larvae feed on organic material, using their mandibulate mouthparts. A silken cocoon is spun, and pupation occurs within. The newly formed adult can remain quiescent in the cocoon for extended periods and often emerges in response to vibrations generated by an approaching host. This is an important adaptation that prevents adults from appearing when there is little hope of finding a host.

Adults are strongly compressed laterally, and the body bears many spines and bristles.

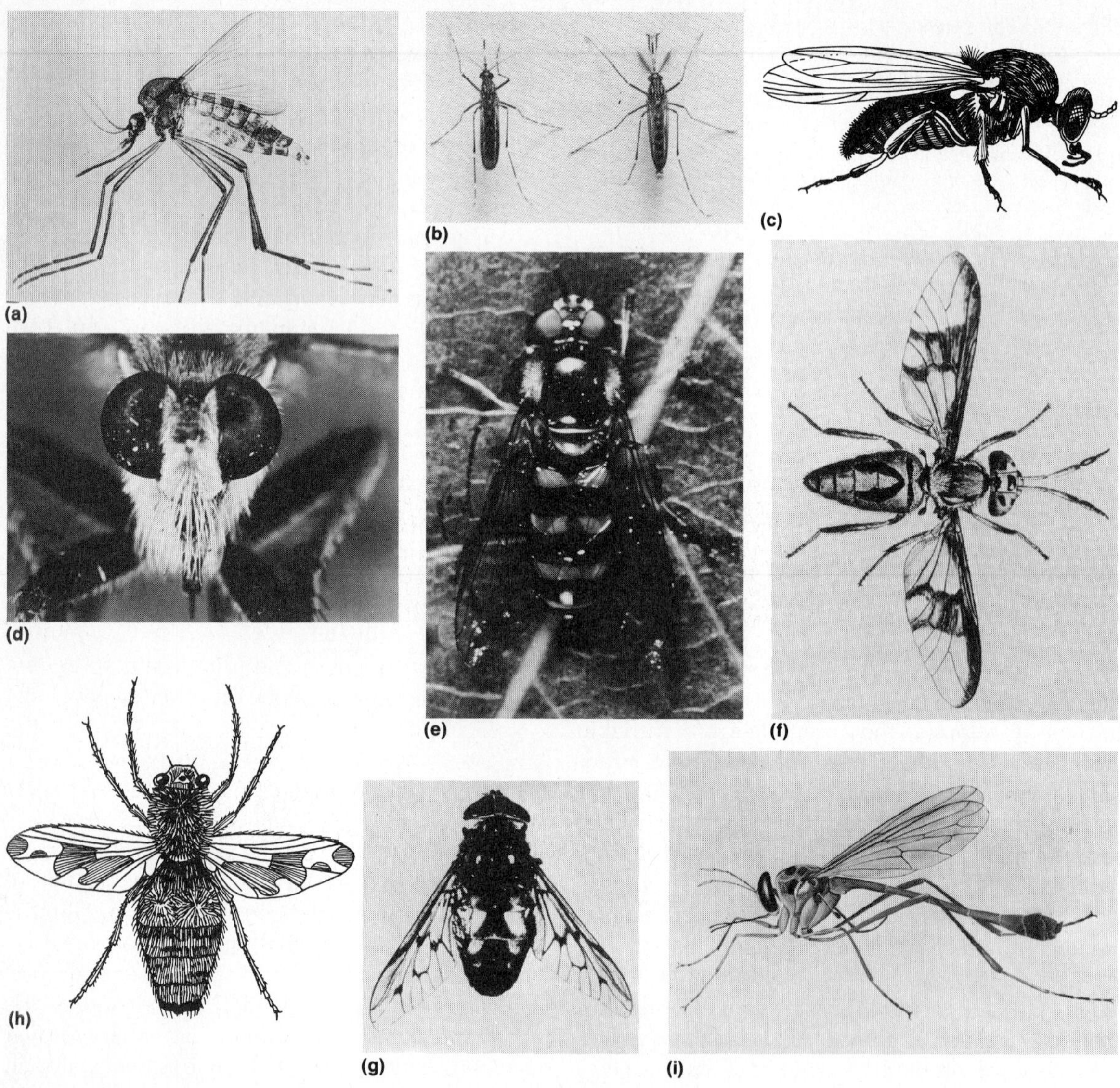

Figure 13.39 *Dipterans or true flies:* ***(a)*** *photograph of* **Aedes aegypti,** *a mosquito;* ***(b)*** *photograph of* Culex, *a mosquito;* ***(c)*** Simulium, *a black fly;* ***(d)*** *photograph of the anterior end of* Erax, *a robber fly;* ***(e)*** *photograph of* Syrphus, *a syrphid fly;* ***(f)*** *photograph of* Chrysops dispa, *a deer fly;* ***(g)*** *photograph of* Tabanus atratus, *a horse fly;* ***(h)*** *horse bot fly;* ***(i)*** *photograph of* Systropus arizonemis, *a wasp mimic (by permission of Smithsonian Institution Press from* Bee Flies of the World: The Genera of the Family Bombyliidae, *United States National Museum Bulletin 286, by Frank M. Hull. "Systropus arizonensis," p. ii. Smithsonian Institution, Washington, D.C., 1973). (parts a, b, d, and e courtesy of Ward's Natural Science, Inc., Rochester, N.Y.; f and h courtesy of the Centers for Disease Control, Atlanta, Ga. 30333.)*

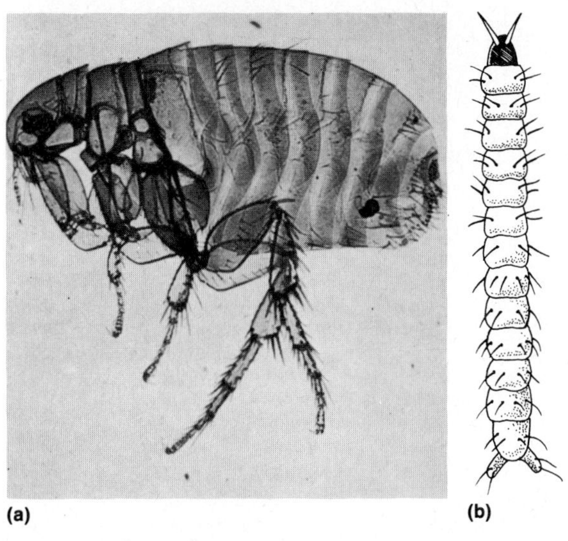

Figure 13.40 *Siphonapterans or fleas:* ***(a)*** *photograph of* Xenopsylla cheopis *(courtesy of the Centers for Disease Control, Atlanta, Ga. 30333);* ***(b)*** *flea larva.*

They have piercing-sucking mouthparts and reduced compound eyes, or eyes may be absent altogether. Adults are apterous but have legs adapted for jumping and running (Fig. 13.40a). Fleas are active and move with ease through the hair of the host. They are not very specific in their selection of hosts and may spend up to several weeks at one time away from a host. They are especially common on cats, dogs, poultry, rats, and sometimes human beings. They create irritations that result in the familiar scratching by the host.

Fleas are a problem because serious diseases are carried by them such as bubonic plague (black death) and endemic fever. Fleas also serve as the intermediate host for several tapeworms. A tropical flea, the chigoe, burrows into the human skin and can cause severe lesions.

Order Hymenoptera

The Hymenoptera are represented by about 110,000 species of ants, bees, wasps, sawflies, and their many relatives. They are mostly small insects ranging from 0.02 to 6.5 cm in length. Members of this order are very diverse and have adapted to countless modes of life.

Larvae of sawflies are caterpillarlike and strongly herbivorous, but all the rest are wormlike and vary in eating habits from being scavengers to parasitic. In many hymenopterans, adults either feed their larvae daily or have previously provided food materials in nests on which larvae feed. In apodous larvae, fecal materials are stored in a **meconium** or chamber until pupation occurs and the midgut–hindgut connection is established. Pupae are of the exarate type and may be enclosed in a cocoon.

The adult mouthparts are basically mandibulate but may be modified to use liquid food. Compound eyes and antennae are well developed (Fig. 13.41). Most adults possess two pairs of membranous wings (Hymenoptera = membranous wings), and the hindwings are usually smaller than the forewings. The hindwings bear a row of tiny hooks or **hamuli** on their anterior margin that hook into a groove on the posterior edge of the forewings, enabling both wings to move as a single unit. Workers of many species are secondarily apterous. The ovipositor in females is usually modified for sawing, piercing, or stinging. The stinger has lost its ovipositing function and is an effective defensive or offensive organ, but only females can sting. In hymenopterans, fertilized eggs usually develop into females, whereas unfertilized eggs parthenogenetically become males.

There are two suborders of hymenopterans, the Symphyta and Apocrita. The symphytans or sawflies are more primitive and have a sawlike ovipositor, adults lack a waist, and the larvae are herbivorous and caterpillarlike. Apocritans include all other hymenopterans (95% of the total) whose larvae are apodous, grublike, and have minute mandibules. Adults are small to medium insects that always have a waist or constriction where the thorax and abdomen join. Larvae of a great many apocritans such as

(a)
(b)
(c)
(d)
(e)
(g)
(f)

◀ ***Figure 13.41*** *Hymenopterans or ants, bees, and wasps:* ***(a)*** *a scanning electron micrograph of a fire ant worker (courtesy of T. Carlysle in P.S. Callahan, 1971,* Insects and How They Function, *Holiday House, New York, N.Y.);* ***(b)*** *a photograph of a carpenter ant, probably* Camponotus; ***(c)*** *a photograph of a worker of* Apis mellifera, *the common honey bee (courtesy of Ward's Natural Science, Inc., Rochester, N.Y.);* ***(d)*** *a photograph of* Polistes, *a wasp;* ***(e)*** *a photograph of* Bombus, *a bumble bee;* ***(f)*** *a photograph of a queen of* Paravespula germanica *(from J.P. Spradbery, 1973,* Wasps, *University of Washington Press, used with permission);* ***(g)*** *a photograph of* Vespula maculfrons, *the yellow jacket. (parts d [from M.V. Smith], e [from F. Perlman], g courtesy of the Centers for Disease Control, Atlanta, Ga. 30333.)*

ichneumonoid, braconid, and chalcidoid wasps plus a number of smaller families are parasitic in other insects. These parasitic hymenopterans are important biological-control agents, since they kill larvae of serious human pests. The aculeate (those that sting) apocritans include the ants (Formicidae), velvet ants (Mutillidae), spider wasps (Pompilidae), common wasps and hornets (Vespidae), and bees (Apidae).

Many of the hymenopterans have evolved complex societies with several castes. Bumble bees (*Bombus* [Fig. 13.41e]) and honey bees (*Apis mellifera* [Fig. 13.41c]) are two familiar taxa that have elaborate societies. For example, in higher bees a colony is composed of only one queen, thousands of sterile female workers, and hundreds of male drones at certain seasons. In ant colonies there are reproductives of both sexes and female workers.

Although many species are considered to be pests, they are only a small percentage of hymenopteran species. The stings of some are painful and can cause death to hypersensitive persons, and some species are destructive to various crops. Even though both honey bees and bumble bees sting, sometimes painfully so, they are enormously beneficial to human agriculture and horticulture, since they are the chief insect pollinators.

Minor orders

There are eight minor orders of insects that represent a relatively few number of species and the members of which are neither very large nor ecologically very important. These eight orders are briefly characterized in Box 13.5, and a representative of each is shown in Figure 13.42.

PHYLOGENY OF INSECTS

The fossil record of ancient insects is of little help in determining the details of the earliest insects. The paleontological record is poor because these early insects were soft bodied and not easily fossilized. While there is some evidence of Devonian fossils that may have been insectan in origin, insect fossils first appeared in abundance in the Mesozoic Era (see Table 1.1 in Chapter 1). By the time these Mesozoic forms were being fossilized, insects had, as a group, become a distinct evolutionary line, and a number of separate orders were already evolved. Therefore the fossil record is not helpful, since it does not shed much light on the ontological stages of uniramians or insects.

Insects and myriapods undoubtedly arose from a common ancestral stock. These ancient wormlike uniramians were probably terrestrial and lived in damp habitats and had some degree of cephalization, modified mouthparts, a long tubular digestive tract without diverticula, and uniramous appendages. From this ancestor protoinsects developed three distinct tagmata, three pairs of thoracic legs, and gonopores located at the posterior end of the abdomen. These first insects were undoubtedly small and apterous and may have resembled contemporary thysanurans. Early in the evolution of insects, a tracheal system was evolved that permitted them to ventilate with air, thus emancipating insects from a watery environment and enabling them to become fully terrestrial.

There is considerable debate regarding the position of the entognathous insects. One

BOX 13.5 □ BRIEF RESUMES OF THE MINOR ORDERS OF INSECTS

SUBCLASS APTERYGOTA

Order Protura—small, whitish insects; no eyes, antennae, or tracheae; live in moist humus (Fig. 13.42a).

Order Diplura—similar to bristletails but lack median caudal filament and scales; small, pale colored; live in damp places (Fig. 13.42b).

Order Microcoryphia—relatively large compound eyes; ocelli and abdominal stylets present; most are nocturnal and are capable of jumping (Fig. 13.42c).

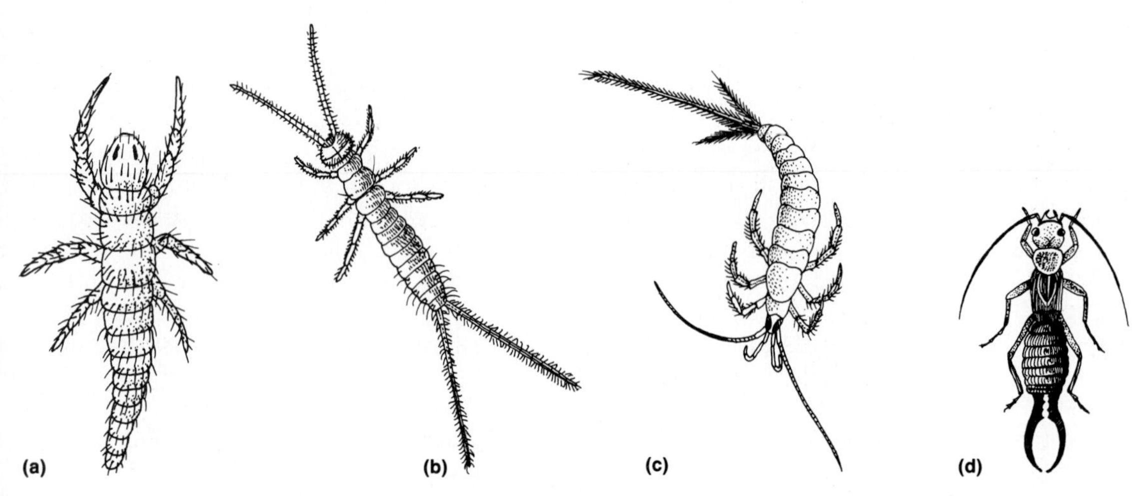

Figure 13.42 *Representatives of the eight smallest orders of insects: **(a)** proturan; **(b)** dipluran; **(c)** microcoryphian; **(d)** dermapteran (earwig); **(e)** embiopteran; **(f)** zorapteran; **(g)** strepsipteran (by permission of Smithsonian Institution Press from "The comparative morphology of the Order Strepsiptera together with records and descriptions of insects," Figure 1.* Proceedings of the

school of thought treats these three primitive orders as noninsects in which the hexapodous condition evolved independently of the other insects. This position is buttressed by the several ways in which the mouthparts, especially the mandibles, differ from those in ectognathous insects. In fact, the entognaths may resemble the myriapods more than they do insects. The other taxonomic position is to consider all hexapodous arthropods as insects; in this case their entognathous mouthparts and other details of their mandibles are simply deviations on the basic plan of the protoinsects. Whether the apterygotes are monophyletic or diphyletic remains an open question.

Certainly one of the most significant steps in the evolution of all other insects was the development of wings. The earliest insect fossils that have been discovered were from the Carboniferous period, and these insects already

SUBCLASS PTERYGOTA

Division Exopterygota

Order Dermaptera—earwigs; have large, forceplike, abdominal cerci; mainly nocturnal and scavengers (Fig. 13.42d).

Order Embioptera—web-spinners; small, slender, mostly tropical forms; live in silken tunnels beneath stones and bark (Fig. 13.42e).

Order Zoraptera—minute; many lose wings as in termites; gregarious; scavengers (Fig. 13.42f).

Division Endopterygota

Order Strepsiptera—minute; mostly endoparasitic on other insects; males with reduced forewings and large hindwings (Fig. 13.42g).

Order Mecoptera—scorpionflies; medium-sized slender body; head peculiarly shaped; male genitalia resemble the stinger of scorpions (Fig. 13.42h).

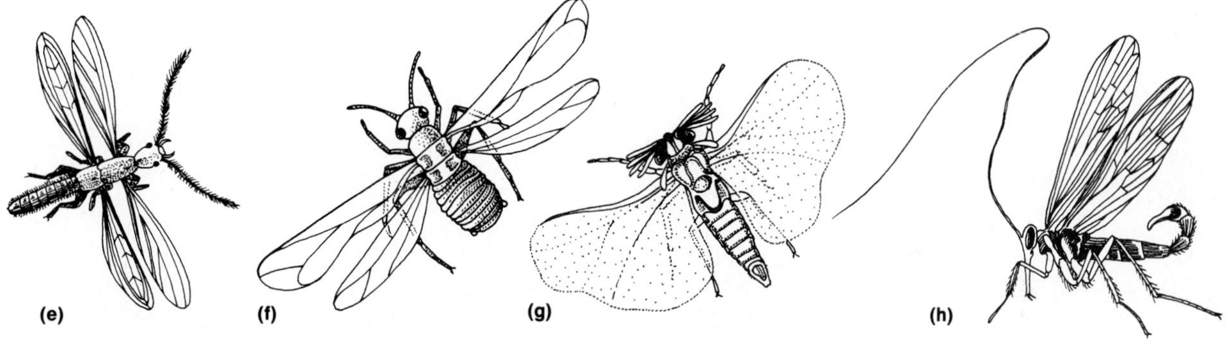

United States National Museum, Volume 54, Number 2242, *by W. Dwight Pierce. Smithsonian Institution, Washington, D.C., 1918);* ***(h)*** *mecopteran. (parts b, d, e, h from D. Sharp, 1909,* Cambridge Natural History, *Macmillan and Co., Ltd., London.)*

had wings. Therefore the stages in the ontogeny of wings are matters mostly of speculation. Entomologists feel that the winged condition evolved only once and thus is a very basic and fundamental characteristic of all pterygote insects.

Insects were the first flying animals, and flight capability was a most important factor as they underwent adaptive radiation. There are two theories as to the origin of wings, neither of which can be totally discounted today. One theory holds that wings first evolved in arboreal insects that utilized lateral expansions of the thoracic terga to stabilize their fall out of trees. From these short, primordial, tergal flaps, longer wings evolved that were of adaptive significance in gliding as an insect could run, jump, and then glide to escape enemies. This ability was later modified into organs that could be flapped rapidly in the air to permit sudden takeoffs and efficient propulsion through the air. The second theory proposes that flight originated in insects with aquatic nymphs that had lateral thoracic gills. Nymphs may have been able to beat the gills in water to facilitate gaseous exchange, and more powerful strokes could propel the animal forward. As these gills evolved into stronger fins and were retained by the adult, they would then be used as aerial wings. Whatever the mechanism by which wings evolved, the earliest insects with wings lacked a flexing mechanism, so the wings were held horizontally or vertically from the thorax. These insects were dominant toward the end of

Figure 13.43 *Probable evolutionary pathways in the insects.*

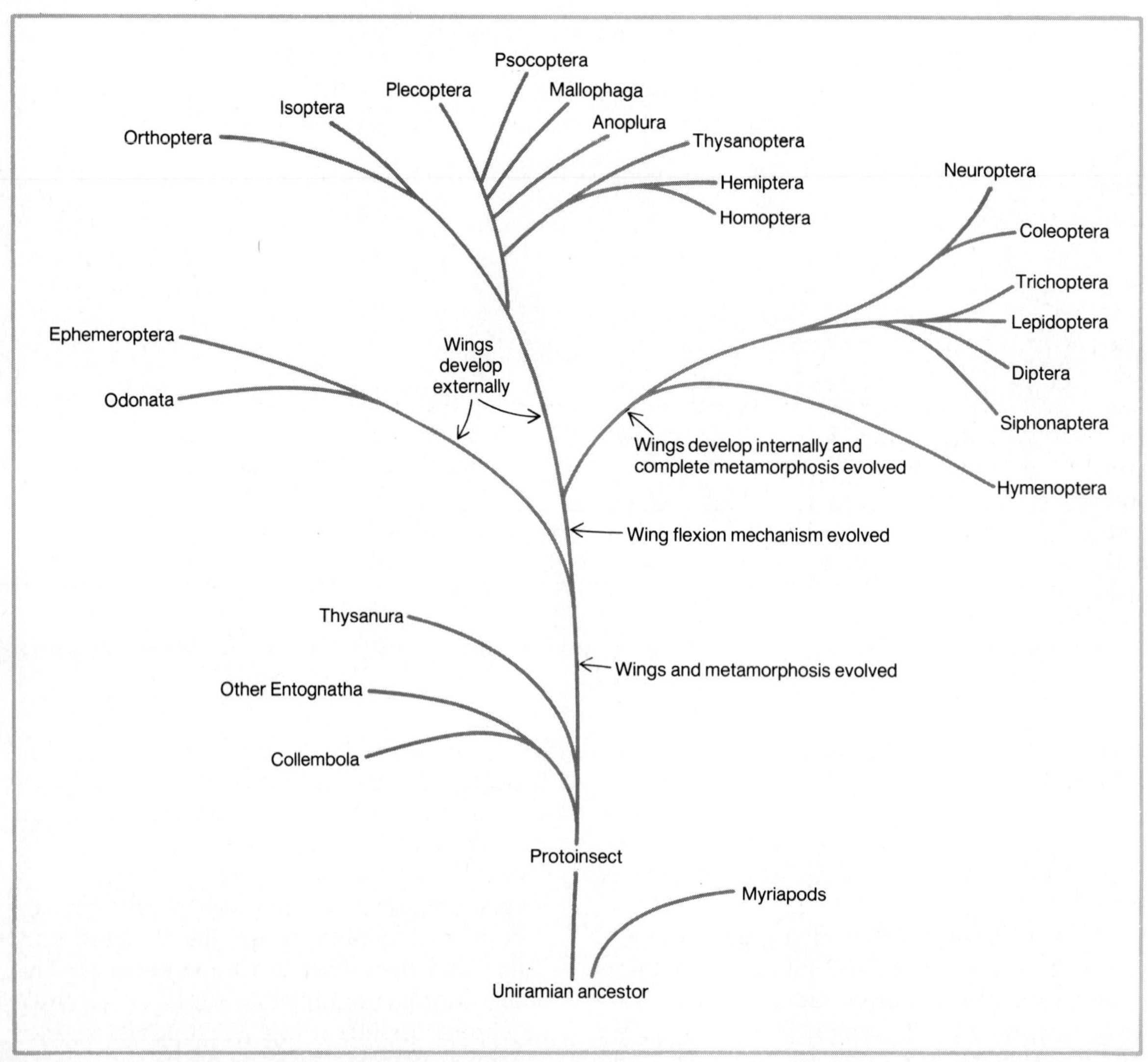

the Proterozoic Era and are represented today by the Ephemeroptera and Odonata.

The next major evolutionary step was the development of a wing-flexing mechanism so that the insect could bend the wings posteriorly over the abdomen. The ability to fold the wings out of the way and thus hide from predators in small crevices was an immensely important survival mechanism. The ability to flex the wings is found in all higher orders of insects (Fig. 13.43).

Another major evolutionary event was the development of metamorphosis. The ability to metamorphose allowed the immature stages to live in habitats and fill ecological niches that were different from those of adults. This type of resource partitioning was a substantial adaptation that tended to separate spatially and temporally the feeding stages from the dispersal and reproductive adults. In a large number of insects the degree of ecological difference is not so great; these insects, exemplified by the exopterygotes, have incomplete metamorphosis. The ultimate manifestation of this trend was in the development of a feeding larval stage that was radically different from the winged adult. Thus a pupal stage was necessary for the reformation of larval features into adult structures, and these insects are represented today as the endopterygotes. The probable evolutionary pathways and phylogenetic groupings are summarized in Fig. 13.43.

SYSTEMATIC RESUME

The Insects to Ordinal Level

Subclass Apterygota—primitive, small, wingless insects with no metamorphosis; some anterior abdominal segments bear styluslike appendages.

Order Protura—minute whitish insects; entognathous; no eyes or antennae; *Acerentulus* is representative.

Order Collembola—springtails; have springing furcula, tenaculum, collophore; entognathous; very abundant; *Podura* and *Tullbergia* are common.

Order Diplura—small; have two caudal filaments; lack compound eyes and ocelli; entognathous; *Campodea* is often encountered.

Order Thysanura—bristletails; three caudal filaments; body covered with scales; abdomen of 11 segments; many primitive features; includes *Lepisma* (silverfish [Fig. 13.23]) and *Thermobia*.

Order Microcoryphia—jumping bristletails; have large compound eyes and ocelli; are active and can jump considerable distances; *Pedetontus* is representative.

Subclass Pterygota—winged insects or those that are secondarily wingless; immature stages undergo some metamorphosis to become adults.

Division Exopterygota—have wings that develop externally in later nymphal instars; have incomplete metamorphosis.

Order Ephemeroptera—mayflies; forewings larger than hindwings; have aquatic nymphs with abdominal gills; have winged subimago and adult stages; *Hexagenia*, *Ephoron*, and *Baetis* are quite common.

Order Odonata—dragonflies and damselflies; adults have large compound eyes and are often strong fliers with four membranous wings; have indirect sperm transfer; nymphs are aquatic and have enlarged feeding labium; *Epitheca*, *Anax*, and *Ischnura* are all common.

Order Orthoptera—grasshoppers, crickets, mantids, cockroaches; common and well known; forewings or tegmina often thickened; most stridulate; most have long ovipositor; *Romalea* and *Melanoplus* (grasshoppers [Fig. 13.26a]), *Gryllus* (crickets), *Stagmomantis* (mantids), and *Periplaneta* (cockroaches [Fig. 13.26e]) are common.

Order Dermaptera—earwigs; adults have posterior forceplike cerci; wings variable; mostly nocturnal and sometimes destructive to plants; *Forficula* is typical.

Order Isoptera—termites; live in social groups with several different castes; colonies may number in the millions of individuals; thorax and abdomen broadly joined; some are destructive to wood; *Reticulitermes* and *Zootermopsis* (Fig. 13.27b) are frequently encountered.

Order Embioptera—web-spinners; mostly tropical, small; females wingless; adults spin silken cryptic retreats using modified salivary glands; *Anisembia* is found in the south and southwest United States.

Order Plecoptera—stoneflies; adults are weak fliers and most do not feed; nymphs are aquatic with two anal cerci and thoracic gills; two of the frequently encountered genera are *Isoperla* (Fig. 13.28) and *Nemoura*.

Order Zoraptera—minute; adults often apterous; winged forms have compound eyes and ocelli; abdomen of ten segments; *Zorotypus* is the only genus.

Order Psocoptera—psocids; small; wings variable; some like *Trogium* (booklice) and *Lepinotus* (in granaries) can be destructive.

Order Mallophaga—chewing lice; adults are apterous and ectoparasitic on birds and mammals; dispersed to new hosts by direct contact; *Menacanthus* and *Cuclotogaster* are found on poultry.

Order Anoplura—sucking lice; adults are apterous and parasitic on mammals; body flattened dorsoventrally; each tarsus has a large claw; *Pthirus* (Fig. 13.31a) and *Pediculus* (Fig. 13.31b) parasitize human beings.

Order Thysanoptera—thrips; slender body; fringed wings; sucking mouthparts; most feed on plants; common genera include *Taeniothrips* and *Heterothrips* (Fig. 13.32a).

Order Hemiptera—bugs; basal portion of forewings is thickened, form hemelytra, and tips overlap at rest; hindwings are membranous; sucking-piercing mouthparts in opisthognathous position; *Murgantia* (harlequin bugs [Fig. 13.33c]), *Anasa* (squash bugs [Fig. 13.33a]), *Reduvius* (assassin bugs), *Cimex* (bed bugs [Fig. 13.33b]), *Sigara* (water boatmen), *Notonecta* (backswimmers [Fig. 13.33h]), and *Ranatra* (water scorpions) are common.

Order Homoptera—cicadas, hoppers, whiteflies, aphids, scale insects; all herbivorous; have mouthparts as beak arising from back of head; usually four membranous wings; includes *Magicicada* (periodical cicadas [Fig. 13.34f]), *Microcentrus* and *Aetalion* (hoppers), *Aleurocanthus* (whiteflies), *Aphis* and *Eriosoma* (aphids), and *Coccus* and *Lepidosaphes* (scale insects) are common pests.

Division Endopterygota—insects with complete metamorphosis; larvae develop internal wing pads; have a quiescent pupa.

Order Neuroptera—alderflies, dobsonflies, antlions; larvae are predaceous; adults have membranous wings, mandibulate mouthparts; *Sialis* and *Corydalus* (Fig. 13.35b) are often present.

Order Coleoptera—beetles; contain about 40% of all insect species; forewings are thickened as elytra and are nonflying organs; hindwings are membranous; have powerful mandibles; some of the more common genera are *Cupes* (reticulated beetles), *Calosoma* (ground beetles), *Gyrinus* (whirligig beetles), *Dytiscus* (diving beetles), *Hydrophilus* (water-scavenger beetles [Fig. 13.36f]), *Lathrobium* (rove beetles), *Copris* (scarab beetles), *Phyllophaga* (June beetles), *Ctenicera* (click beetles), *Photuris* (fireflies [Fig. 13.36e), *Hippodamia* (ladybugs), *Prionus* (longhorn beetles), and *Curculio* (weevils).

Order Strepsiptera—twisted-winged parasites; most females are ectoparasitic on other insects; forewings reduced; hindwings large and membranous; *Triozocera* is representative.

Order Mecoptera—scorpionflies; wings membranous; mouthparts in form of beak; male genitalia form scorpionlike mechanism; *Panorpa* and *Bittacus* are often collected.

Order Trichoptera—caddisflies; larvae all aquatic and often construct larval sacs or webs utilizing silk; adults are small, usually dull colored, feed on liquid; *Hydropsyche*, *Polycentropus*, and *Lepidostoma* are common genera.

Order Lepidoptera—butterflies, moths; forewings larger than hindwings and both covered by minute scales; mouthparts usually in coiled proboscis; larvae are herbivorous caterpillars; moth genera include *Tineola*, *Bombyx*, *Cydia*, *Ostrinia*, and *Sibine*; common butterfly genera are *Papilio*, *Pieris*, *Danaus* (Fig. 13.38c), *Heliconius*, and *Lycaenopsis*.

Order Diptera—flies; forewings membranous; hindwings reduced to halteres; larvae are often called maggots; many feed on plant or animal juices; common dipterans include *Tipula* (crane flies), *Chaoborus* and *Culicoides* (midges), *Culex* and *Aedes* (Fig. 13.39a, b) (mosquitoes), *Tabanus* (horse flies [Fig. 13.39h]), *Drosophila* (small fruit flies), *Musca* (house flies), and *Oestrus* (bot flies).

Order Siphonaptera—fleas; adults are apterous, laterally compressed, have sucking mouthparts, and are parasitic on birds and mammals; includes *Pulex* and *Ctenocephalides*.

Order Hymenoptera—sawflies, ants, wasps, bees; the smaller hindwings are hooked to the forewings by hamuli; wings with few veins, ovipositor well developed and often modified as a sting; fertilized eggs become females and unfertilized eggs develop into males; includes *Xyela* and *Cimbex* (sawflies), *Rhyssella* (ichneumonoids), *Formica* and *Polyergus* (ants), *Vespula* and *Polistes* (wasps, yellowjackets, hornets [Fig. 13.41d, g]), *Chalybion* and *Sphecius* (sphecid wasps), *Apis* (Fig. 13.41c), *Hylaeus*, and *Bombus* (Fig. 13.41e) (bees).

ADDITIONAL READINGS

Atkins, M.D. 1978. *Insects in Perspective*, 500 pp. New York: Macmillan Publishing Co.

Borror, D.J., D.M. De Long, and C.A. Triplehorn. 1981. *An Introduction to the Study of Insects*, 5th ed., 753 pp. New York: Saunders College Publishers.

Boudreaux, H.B. 1979. *Arthropod Phylogeny with Special Reference to Insects*, 306 pp. New York: John Wiley & Sons.

Chapman, R.F. 1982. *The Insects: Structure and Function*, 3rd ed., 885 pp. Cambridge, Mass.: Harvard University Press.

Cheng, T.C. 1973. *General Parasitology*, pp. 786–883. New York: Academic Press.

Daly, H.V., J.T. Doyen, and P.R. Ehrlich. 1978. *Introduction to Insect Biology and Diversity*, 538 pp. New York: McGraw-Hill Book Co.

Downer, R.G.H., ed. 1981. *Energy Metabolism in Insects*, 244 pp. New York: Plenum Press.

Eisner, T. and E.O. Wilson, eds. 1977. *The Insects: Readings from* Scientific American, 328 pp. San Francisco: W.H. Freeman and Co.

Elzinga, R.J. 1981. *Fundamentals of Entomology*, 2nd ed., 410 pp. Englewood Cliffs, N.J.: Prentice-Hall.

Hermann, H.R., ed. *Social Insects*. Vol. I: 413 pp., 1979. Vol. II: 453 pp., 1981. Vol. III: 480 pp., 1982. Vol. IV: 385 pp., 1982. New York: Academic Press.

Horn, D.J. 1976. *Biology of Insects*, 417 pp. Philadelphia: W.B. Saunders Co.

Manton, S.M. 1977. *The Arthropoda: Habits, Functional Morphology, and Evolution*, 514 pp. Oxford, England: Clarendon Press.

Matthews, R.W. and J.R. Matthews. 1978. *Insect Behavior*, 484 pp. New York: John Wiley & Sons.

Novak, V.J.A. 1975. *Insect Hormones*, 2nd ed., 569 pp. New York: John Wiley & Sons.

Pennak, R.W. 1978. *Fresh-water Invertebrates of the United States*, 2nd ed., pp. 512–709. New York: John Wiley & Sons.

Price, P.W. 1975. *Insect Ecology*, 482 pp. New York: John Wiley & Sons.

Richards, O.W. and R.G. Davies. 1978. *Imm's Outlines of Entomology*, 6th ed., 237 pp. London: Chapman and Hall.

Romoser, W.S. 1973. *The Science of Entomology*, 413 pp. New York: Macmillan Publishing Co.

Steinmann, H. and L. Zombori. 1981. *An Atlas of Insect Morphology*, 153 pp. Budapest, Hungary: Akademiai Kiado.

Wigglesworth, V.B. 1972. *The Principles of Insect Physiology*, 747 pp. London: Chapman and Hall.

Wilson, E.O. 1971. *The Insect Societies*, 527 pp. Cambridge, Mass.: Harvard University Press.

14

The Myriapods

OVERVIEW

Myriapods, a group of similarly constructed, uniramous arthropods including centipedes and millipedes, are mostly tropical or temperate and found in moist terrestrial habitats. The wormlike body has a head with an acron plus three or four head segments that bear the antennae, mandibles, and one or two pairs of maxillae, respectively, and a trunk with a variable number of segments, each with a pair of walking legs. Internally, myriapods are quite similar to insects in their tracheal system, Malpighian tubules, dorsal tubular heart and hemocoel, and details of the central nervous system. Myriapods are dioecious, most have indirect sperm transfer, eggs hatch without much delay into miniature forms, and subsequent molts add new segments and appendages.

Chilopods are dorsoventrally flattened, the head has two pairs of maxillae, the first pair of trunk limbs are poisonous maxillipeds used to disable prey, and gonopores are at the posterior end. Symphylans are minute and have a trunk of 12 segments and 12 pairs of legs. Diplopods are circular in cross section and have calcified sclerites, and there is one pair of maxillae. Typically, pairs of trunk segments are fused as diplosegments, each with two pairs of legs, and gonopores and gonopods are on the anterior trunk. Pauropods are small and have biramous antennae and a trunk of 11 segments and nine pairs of legs.

Myriapods and insects undoubtedly arose from a similar ancestry, since many external and internal features are common to both groups. Chilopods and symphylans appear to be related to each other as are diplopods and pauropods. Each of the four groups has had a long and independent evolutionary history.

INTRODUCTION TO THE MYRIAPODS

Myriapods, a most interesting group of centipedes, millipedes, and others, are uniramous arthropods with many pairs of legs. They are all rather similar in their functional morphology and in the habitats in which they live. There are about 11,500 species of myriapods, living in mostly tropical and temperate forest ecosystems. Myriapods are characteristically terrestrial, even though a few species have become secondarily aquatic and a few of these are even intertidal! Myriapods are usually found in moist habitats under stones, wood, or bark and in soil and humus; the most important reasons why myriapods are confined to such areas are summarized in Box 14.1. Some species are found near or actually in human dwellings. Myriapods are wormlike and range in length from 0.5 to nearly 30 cm, but most are from 1 to 8 cm long.

Myriapods and insects are rather closely related as is shown by the fact that both groups are placed in the Subphylum Uniramia. Myriapods and insects share many features, the most important being the similarities of the head segments and appendages, tracheal system, and Malpighian tubules and the close common features of the digestive, circulatory, nervous, and reproductive systems. These similarities will be explicated further in the discussions below.

Even though "myriapod" is an invalid taxonomic term, it is a convenient way to refer collectively to these multilegged uniramians. There are four classes of myriapods, which are characterized in Table 14.1. The Class Diplopoda includes the millipedes and is by far the largest of the four classes. The Class Chilopoda, to which belong the centipedes, has a moderate

BOX 14.1 □ WHY ARE MYRIAPODS RESTRICTED TO HUMID ENVIRONMENTS?

It is well known that myriapods are restricted to moist areas like soils, leaf litter, beneath logs or bark, or other humid areas. Myriapods are much more common in tropical forests, where the relative humidity is usually quite high. It is indeed strange that two groups of terrestrial arthropods, arachnids and insects, are marvelously adapted for life on land. Why then are the myriapods so vulnerable to desiccation?

Several features in the functional morphology of myriapods regulate the kinds of habitats these animals can live in. First, the cuticle of myriapods lacks the waxy waterproofing layer that markedly retards water loss in other terrestrial arthropods. Second, the spiracles of the gas exchange system are nonclosable, so rather large volumes of water could be lost via these openings in arid conditions. Third, myriapods apparently lose a considerable amount of water in defecation, unlike arachnids and insects, in which there is almost total reabsorption of water in the hindgut. Finally, some water loss occurs through the mouth and at times of reproduction.

If myriapods are so sensitive to desiccation, how are they able even to live in terrestrial habitats? Obviously, they live in moist areas where the relative humidity is high, many are active at night when water loss would be minimal, some burrow deep into the soil during dry periods, and the immatures of some forms enter a state of diapause in dry seasons. Many millipedes are known to attack growing crops or seedlings; this may be an attempt to take in water. Some myriapods have increased locomotor activities if the relative humidity drops, and the animals clearly move toward areas of greater moisture.

Why myriapods never evolved some of the water-conserving features that their close relatives, the insects, developed will never be known. But they are eternally bound to moist habitats or those where the humidity is rather high.

TABLE 14.1. □ **BRIEF CHARACTERIZATIONS OF THE FOUR CLASSES OF MYRIAPODS**

Class	*Common Name*	*Type of Food*	*Number of Trunk Segments*	*Other Features*
Chilopoda	Centipedes	Mostly invertebrates	15–181	First trunk appendages are poisonous jaws or maxillipeds; 2 pairs of maxillae; gonopores at posterior end
Symphyla	"Centipedes"	Plant material	12	Second pair of maxillae fused into a labium; gonopores at posterior end
Diplopoda	Millipedes	Mostly plant material	15–200 diplosegments	Maxillae fused into a gnathochilarium; no second maxillae; gonopores and gonopods at anterior end of trunk
Pauropoda	—	Dead organisms, organic material	11	Biramous antennae; one pair of maxillae forms a gnathochilarium; 9 pairs of legs

number of species of which most are important predators. Two other classes, Symphyla and Pauropoda, have low species diversity and consist of very small myriapods.

GENERAL CHARACTERISTICS

External Features

All myriapods are constructed along a similar plan with two tagmata: a head and a trunk. The **head** is composed of the anterior acron and four adult segments. The first segment is preoral in position and bears a pair of many-segmented **antennae.** Immediately posterior to this segment is an intercalary segment that appears only briefly in embryonic development and is never present in adult myriapods; a similar embryonic segment is also present in insects (see Chapter 13). Adult head segment II bears a pair of **jaws** or **mandibles,** and segments III and IV bear paired **maxillae.** However, millipedes and pauropods have neither a fourth head segment nor the second pair of maxillae. In addition to the mandibles and one or two pairs of maxillae, myriapod mouthparts include an unpaired **labrum** or upper lip and a tonguelike **hypopharnyx.**

Behind the head is a **trunk** containing a variable number of similar or homononomous segments; each bears a pair of ambulatory legs. Each leg typically consists of six or seven podomeres, and the full complement of legs is adapted for walking. In millipedes, pairs of segments have fused together so that each apparent trunk segment bears two pairs of legs. Most trunk segments bear lateral **spiracles** that open internally into the tracheal system. The trunk terminates posteriorly with the **pygidium.** The **anal opening** is situated on either the pygidium or the posteriormost segment.

Sense organs in myriapods are numerous and diverse. Photoreceptors are sometimes totally lacking, and in other myriapods there are 2–80 ocelli, usually at the base of the antennae. Some chilopods actually have clusters of ocelli that are organized to form compound eyes of up to 200 units, but these eyes are not capable of forming images. Most myriapods are negatively phototactic. Mechanoreceptors and chemoreceptors are especially concentrated on the antennae; they appear to a lesser degree on the mouthparts, and some are scattered over the entire body. **Organs of Tomosvary** are located at the antennal bases of some millipedes and centipedes; they probably function in the detection of vibrations or sound waves.

Most myriapods bear **repugnatorial** or stink **glands** along either side of most trunk segments. The secretions of these glands are offensive and can effectively repel a great many would-be predators.

Internal Features

Since myriapods are similar to insects and since the basic arthropodan features have already been mentioned, only the most important, salient, internal features will be mentioned. The digestive tract is essentially a straight tube without prominent diverticula. In most a **peritrophic membrane** is secreted by the anterior midgut and surrounds the food as it passes posteriorly. The circulatory system is typically arthropodan with a dorsal tubular heart with 2–20 pairs of segmentally arranged **ostia,** a principal anterior **aorta,** and a rather extensive **hemocoel.** Ventilation is achieved by a rather extensive **tracheal system.** Each segment bears a pair of nonclosable spiracles, each of which opens into a trachea that in turn gives off a number of small tracheal branches. **Malpighian tubules** arise from the anterior part of the hindgut and extend forward, where they lie in the hemocoel. These tubules excrete mostly uric acid and ammonia into the hindgut for nitrogenous waste elimination. Some millipedes and centipedes possess paired tubular organs in the head that also may be involved in excretion.

The nervous system is not particularly concentrated, and the parts of the brain—protocerebrum, deuterocerebrum, and tritocerebrum—are rather distinct and only partially fused. Circumesophageal connectives unite the tritocerebrum and **subesophageal ganglion,** and nerves from this latter ganglion innervate the head segments (except I) and mouthparts. The ventral nerve cords are fused medially into a single **nerve cord,** which bears a ganglion in each segment. Trunk ganglia are of about the same size, and each gives rise to four pairs of nerves that innervate specific parts of that segment.

All myriapods are dioecious, and the paired gonads lie in the trunk either in midbody (diplopods) or at the posterior end (chilopods). Most myriapods have indirect sperm transfer in that males produce spermatophores. Female centipedes typically are chemically attracted to the spermatophores, and they attach the spermatophores to their posterior end. Millipedes have a courtship replete with tactile and chemical signals followed by sperm transfer using specialized trunk appendages or **gonopods** that have previously been charged with sperm. Ova are usually fertilized as they are laid, and 10–300 eggs are produced in a single clutch. Eggs are usually deposited in masses in the soil, leaf litter, or humus or even in nests. No larval stage is developed, and the young hatch in several weeks, often with only three pairs of legs. Subsequent molts add new segments and trunk appendages.

Class Chilopoda

The chilopods, commonly called centipedes or hundred-legged worms, are rather common myriapods. They are found worldwide but are most prevalent in tropical and temperate areas. Over 3000 species are known. Centipedes are routinely found in moist habitats under logs and stones or in leaf litter or soil, and some may burrow in the soil to a depth of 40 cm. A few species are found in intertidal areas, where they live in protected places like empty shells or crevices. Most centipedes range from 3 to 15 cm in length, though one tropical form may be up to 27 cm long. Most are yellow or brown, but parts of the body of many tropical centipedes are red, green, blue, or yellow. Two subclasses of chilopods are usually recognized. The Epimorpha contains centipedes with more than 20 pairs of legs, and the young are brooded and hatch with all adult segments. The Anamorpha are centipedes with 15 pairs of legs, and the young are not brooded but hatch with less than the full complement of adult segments.

External features

Chilopods are slender, elongated, and flattened dorsoventrally (Fig. 14.1a–d). The body is very flexible and deformable, a feature that aids in

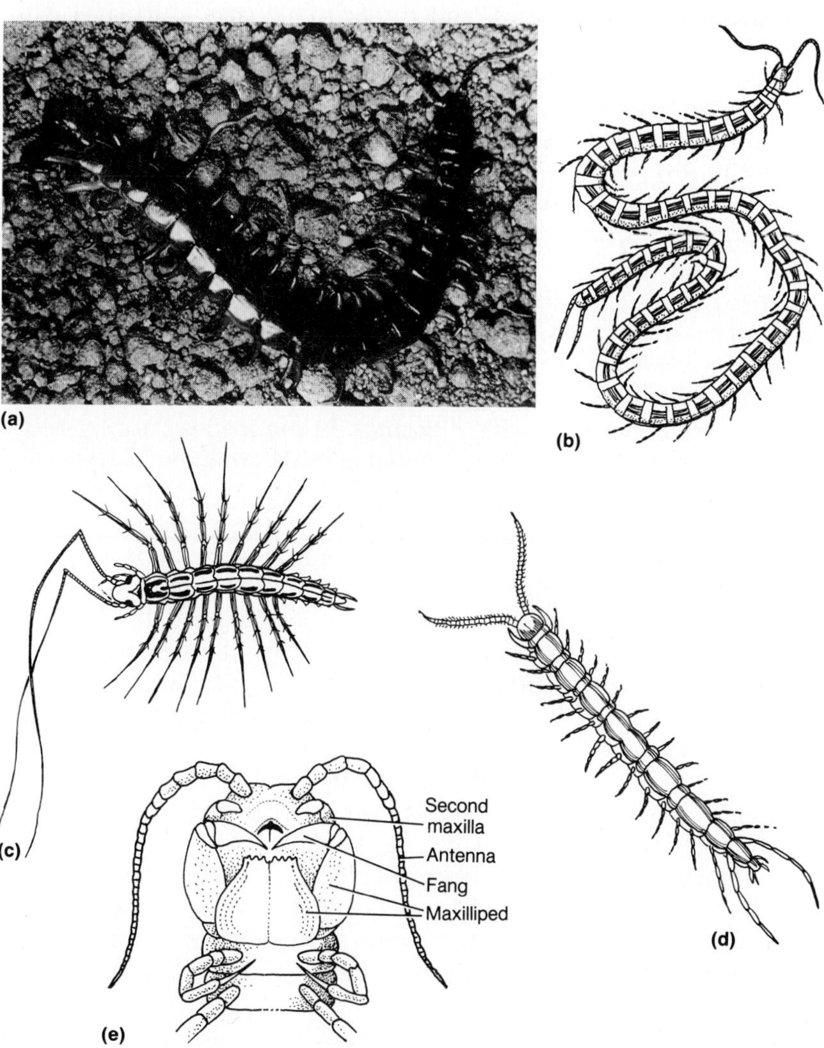

Figure 14.1 *Chilopods:* ***(a)*** *photograph of* Scolopendra *(scolopendromorphan) (courtesy of Ward's Natural Science, Inc., Rochester, N.Y.);* ***(b)*** Geophilus *(geophilomorphan);* ***(c)*** Scutigera, *a common house centipede (scutigeromorphan);* ***(d)*** Lithobius *(lithobiomorphan);* ***(e)*** *ventral view of the anterior end of a typical centipede (from A.S. Packard, 1898,* A Text-book of Entomology, *Macmillan and Co., Ltd., London). (parts b–d from F.G. Sinclair, 1910,* Cambridge Natural History, *Macmillan and Co., Ltd., London.)*

burrowing and hiding in crevices. The head is typically flattened, and mouthparts are located on its ventral surface. The **antennae** arise at the anterolateral margins of the head, and each consists of a variable number of segments. On the dorsal surface are several to many simple eyes. Individuals of one order (Scutigeromorpha) have ocelli clustered together to form compound eyes of up to 200 visual units per eye. Experimental evidence indicates that these eyes do not form images; rather, they, along with the simple eyes, function only in detection of light intensities and perhaps some motion. The **mandibles** in most chilopods are composed of two podomeres, and the distal one bears sharp strong teeth. The foliaceous **first maxillae** lie beneath the mandibles and constitute a lower lip. Each **second maxilla** is represented by a **palp** that is used to hold bits of food. Surrounding these three pairs of mouthparts is a pair of large **poison jaws** or **maxillipeds** (Fig. 14.1e). They are the appendages of the first trunk somite, but they are turned forward and are used in feeding. In many centipedes the coxae of these appendages are fused together to form a ventral plate covering the mouthparts. Each maxilliped is composed of four podomeres, the distal one of which is a sharp pointed **fang.** A poison gland is housed in each maxilliped and opens at or near the tip of the fang.

All chilopods are predaceous and feed on a variety of animals including small arthropods, earthworms, snails, nematodes, or even small toads, snakes, and mice. Prey is located by tactile receptors on the antennae, impaled by the maxillipeds, and quickly killed or stunned by the poison. While held by the maxillipeds and second maxillae, the victim is shredded by the first maxillae and the mandibles. People often fear centipedes, especially the very large tropical forms, since it has been erroneously assumed that their bite could be fatal. A centipede bite is indeed painful, and the hemolytic poison causes local swelling and may cause vomiting and fever. Even though the pain usually disappears within several hours to a day, centipedes should be treated with considerable care.

Behind the maxillipeds the **trunk** consists of 15 or more flattened, legbearing segments (Fig. 14.1a–d); some centipedes may have as many as 181 such segments. Interestingly, the number of leg-bearing segments (and therefore the number of pairs of legs) is always an odd number. The legs are joined to each trunk segment on its lateral aspect rather than ventrally or beneath the animal. Each leg has seven podomeres, and the proximal coxa articulates with the lateral aspects of the sternal plate near the pleural membrane.

Centipedes (except geophilomorphs) are rapid runners and possess a number of structural and functional adaptations that facilitate running. The faster centipedes have longer legs, greater frequency of the stride, and increased angle of leg movement to enhance speed. Legs are progressively longer posteriorly, a feature that reduces the interference between them. During rapid movement, most of the legs are off the ground at any one moment, thus enhancing efficiency. In centipedes the long body has the potential of undulating in a wavelike or serpentine motion, which would slow forward motion. If the body were to undulate, the long legs on each concave side would most certainly interfere with one another. This tendency toward undulation is counteracted by alternate movement of the two legs in a pair and by the rigidness afforded the trunk by overlapping terga or by alternating broad and narrow tergal plates.

At the posterior end of the trunk are two small, legless segments termed the **pregenital** and **genital segments** followed by a terminal rudimentary **pygidium.** The genital segment bears the anal opening and a single **gonopore,** which in males is at the apex of a short **penis.** Some chilopods have a pair of short **gonopods** on the genital segment of both sexes; these function in handling spermatophores.

Each trunk segment typically bears a spiracle located in each pleural membrane. Spiracles are always absent from the head and the first and the last three trunk somites; in some centipedes, other segments also lack spiracles. In the scutigeromorphs, each of the trunk segments II–VIII bears a single dorsal spiracle that opens into a saclike tracheal lung. Most trunk metameres in a majority of centipedes bear effective repugnatorial glands. Additionally, the last pair of legs can be used defensively as pincers, and the powerful maxillipeds are also potent defensive weapons.

Many chilopods engage in a brief courtship before sperm transfer. A spermatophore is deposited in a mat of silk strands produced by the male. The female then picks up the spermatophore, and sperm enter her reproductive tract. In the epimorphs, females externally brood their young, which number from 15 to 70 per brood. A female will roll up into a circle with her ventral side on either the outside or the inside and with the young on the inside of the circle. She broods them until they hatch with their full complement of trunk segments and legs. In the anamorphs the eggs are deposited singly in the soil, no maternal care is provided, and the young gradually acquire more segments and legs at subsequent molts.

Class Symphyla

Symphylans are minute myriapods resembling centipedes and not exceeding 8 mm in length. The approximately 125 species are found in moist habitats in soil or leaf litter; some are often pests in greenhouses. They are vulnerable to desiccation and are usually found in areas of high humidity. They are a very interesting group phylogenetically because many of their features are shared with primitive insects. A few authorities even consider them to be possible ancestors to insects.

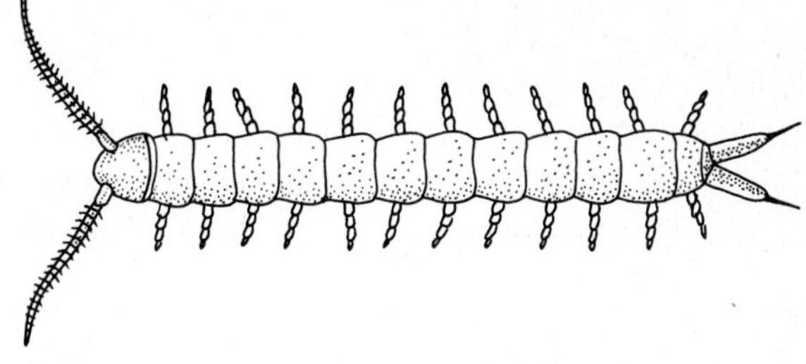

Figure 14.2 Scutigerella, *a symphylan.*

The symphylan body is soft, slender, and composed of a head and trunk (Fig. 14.2). The head is similar to that of centipedes with a pair of many-segmented antennae, a pair of toothed mandibles, and two pairs of maxillae. The second maxillae are fused to form a single nonpoisonous **labium.** There are no eyes, but there are many tactile receptors, especially on the antennae. Symphylans are vegetarians but on occasion will ingest small invertebrates.

Ventral and dorsal trunk segmentations do not correspond; there are more terga than sterna. Each of the 12 trunk segments bears a pair of legs. Each leg, except for the first pair, is composed of six podomeres whose distal segment bears a claw. The base of each leg is equipped with a **stylus,** the function of which is unknown, and a **coxal sac** that absorbs moisture. Posteriorly, there is a preanal segment bearing short unjointed appendages that function as **spinnerets** and a pair of long sensory trichobothria. The trunk terminates in a short pygidium bearing the anus (Fig. 14.2).

A tracheal spiracle is situated on each lateral side of the head. The gonopores are located on the ventral surface of the third trunk segment. Indirect sperm transmission is probably the rule, and the young hatch with only a partial number of legs. Some symphylans are parthenogenetic.

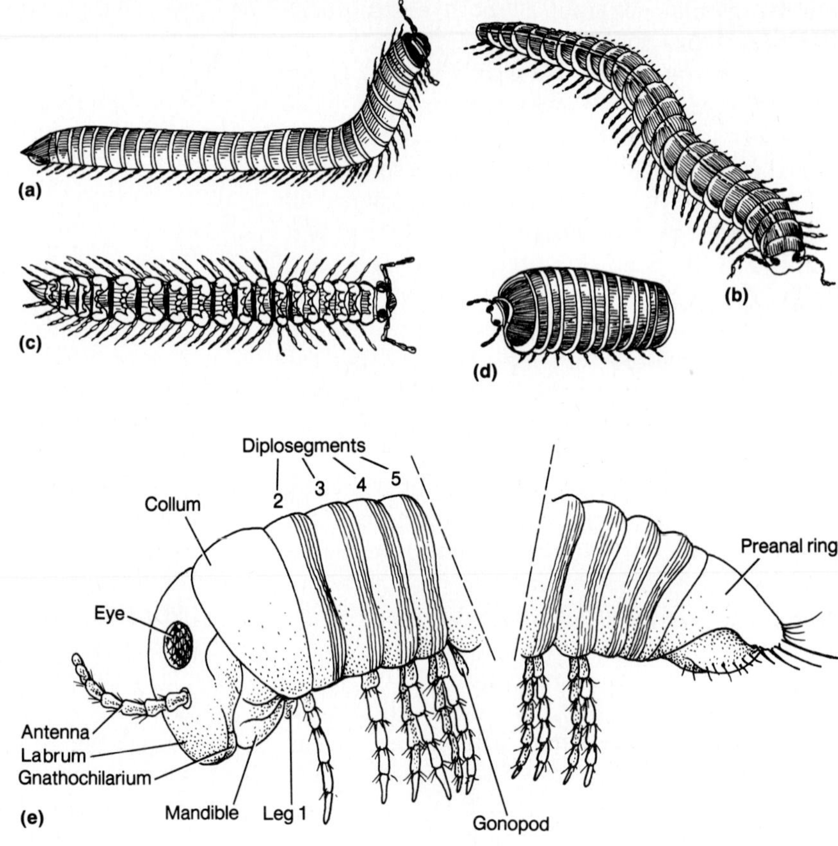

Figure 14.3 *Diplopods:* ***(a)*** Julus *(julidan);* ***(b)*** Chordeuma *(chordeumidan);* ***(c)*** Polydesmus *(polydesmidan);* ***(d)*** Glomeris *(glomeridan);* ***(e)*** *diagrammatic lateral view of the anterior and posterior ends of a typical millipede. (parts a–d from F.G. Sinclair, 1910,* Cambridge Natural History, *Macmillan and Co., Ltd., London.)*

Class Diplopoda

The diplopods, commonly called millipedes or thousand-legged worms, are distributed worldwide but especially in tropical and temperate regions. They are cryptozoic and negatively phototactic myriapods living under stones, bark, or logs or in soil or burrows, and some are troglodytic (cave-dwellers). Approximately 8000 species are known at present, and careful investigations of tropical forest ecosystems may result in the identification of many more species. They mostly are from 3 to 10 cm long, but extremes of 2 mm and 28 cm are known. Most diplopods are dark brown or black, but some species are red or orange spotted.

Millipedes are distinguished by having two pairs of legs per trunk somite, and it is from this arrangement that diplopods derive their name (Fig. 14.3). Ontogenetically, each embryonic segment contains a single pair of legs, but pairs of most trunk segments fuse to form **diplosegments,** each with two pairs of legs. The double nature of each diplosegment is supported by the presence of two pairs of ventral nerve cord ganglia, two pairs of heart ostia, and two pairs of spiracles within each diplosegment.

Millipedes are subdivided into two subclasses: Penicillata, a small group of minute millipedes, and Chilognatha, representing the vast majority of all millipedes. The principal subordinate taxa of the Subclass Chilognatha are characterized in Table 14.2, and resumes of all taxa down to ordinal level appear at the end of this chapter. The chilognaths are the typical familiar diplopods; a general discussion of their external features follows.

External features

Both head and trunk, the two body tagmata, are covered by a tough cuticle that is hardened owing to the deposition of calcium salts into the procuticle. The cuticular calcification is much more evident in the sclerites, and the terga are particularly large, tough, and calcified. Dorsally,

TABLE 14.2. □ **THE PRINCIPAL CHARACTERISTICS OF THE MAJOR TAXA OF CHILOGNATH DIPLOPODS**

Taxon	*Body Features*	*Number of Trunk Segments*	*Gonopods*	*Repugnatorial Glands*	*Other Features*
Superorder Pentazonia					
Order Glomerida	Short trunk, cylindrical, flattened ventrally	15–17	Absent	Absent	Trunk can be rolled up into a sphere
Order Glomeridesmida	Trunk long and flattened	22	Absent	Absent	Body cannot be rolled up
Superorder Colobognatha	Many, flattened, telescopic, trunk segments	30–191	Last legs on VII, first legs on VIII	Present	Mouthparts often in a snout; rod-shaped mandibles
Superorder Helminthomorpha					
Order Chordeumida	Small	26–32	1 or 2 pairs on segment VII	Present	1 or 3 pairs of posterior spinnerets
Order Polydesmida	Ribbon-shaped, flattened	18–22	First pair on segment VII	Present	Terga prominent and keellike
Order Julida	Cylindrical trunk	40 or more	2 pairs on segment VII	Present	Eyes usually present
Order Spirobolida	Cylindrical trunk	40 or more	Housed in pouches	Present	Anterior 5 trunk segments not diplosegments

the **head** is typically rounded and bears a pair of **antennae** composed of seven or eight segments. The dorsal head surface also bears several **ocelli,** but some diplopods are blind. Laterally, the head is covered by large segmented **mandibles,** each bearing a movable serrate lobe. The second pair of mouthparts, the **maxillae,** is fused to form a broad flattened **gnathochilarium** or lower lip (Fig. 14.3e). The gnathochilarium forms the floor of the preoral cavity and imparts a flatness to the ventral head surface. The diplopod head bears neither a second maxillary segment nor second maxillae. Characteristically, millipedes are herbivorous and feed mostly on decaying plant material. Some (colobognaths) suck plant juices, and a few others are omnivores or even consume small invertebrates. Food is moistened by oral secretions and chewed by the mandibles.

The **trunk** is composed of 15–200 diplosegments (Fig. 14.3), each of which is covered by a large convex tergum that may extend laterally and two small pleura and sterna. In many millipedes the sclerites in each diplosegment fuse together to form a rigid ring. The trunk of the flat-backed millipedes (Polydesmida) is dorsoventrally flattened, but in most this tagma is cylindrical in cross section. Each diplosegment bears two pairs of legs with six or seven podomeres in each leg. Each coxa joins the diplosegment at the sternal plate, and thus the legs of millipedes are more ventral in position than are those of centipedes. Two pairs of spiracles are located near the coxae in each diplosegment.

The anterior four trunk segments differ from the typical trunk segments in that they are not diplosegments. The first segment or **collum** is legless, and each of segments II–IV bears only one pair of legs (Fig. 14.3e). Segment II is the genital segment, which bears the **gonopores.** Gonopods, located on trunk segment VII (sometimes on segment VIII also), are extensively used as taxonomic or diagnostic features for various diplopod taxa. In most millipedes the posteriormost one to several metameres are legless, and the pygidium bears the ventral anal opening (Fig. 14.3e).

Millipedes are relatively slow crawlers, but they can force their way through various compact substrates. The pushing force is exerted solely by the legs, which, along with the rounded head and hard rigid body, enable millipedes to burrow through humus and soil. The flat-backed diplopods use the entire dorsal surface to open up cracks in bark and logs.

Millipedes have two principal means of defense. Many forms contort the body into protective spirals. The pill millipedes can roll up in a tight sphere with the terga outermost, a type of behavior that effectively protects them from small predators. Each diplosegment of many diplopods typically bears a pair of repugnatorial glands, but they are absent from certain segments. Each gland opens on the lateral aspects of the tergum, secreting mostly hydrogen cyanide, iodine, or phenol. The reddish or yellowish pungent secretion, offensive to many predators, can be exuded slowly or ejected in squirts. Incredibly, one species can discharge its secretion in a fine jet of more than 80 cm to either side at the same time.

In male millipedes the one or two pairs of gonopods on trunk segments VII and VIII are charged with sperm by bending the anterior body ventrally, thus bringing them in contact with the gonopores on segment III. During copulation the trunk of the male is coiled around that of the female so that his copulatory gonopods are opposite her gonopores. In other millipedes, sperm transfer can even be achieved by using the mandibles. Eggs are fertilized at the time of laying; they may be laid singly in the soil or in clutches in a nest constructed of maternal fecal material. The young of some species are brooded externally by the female. The young hatch in several weeks with only the first three pairs of legs, and subsequent molts produce instars with progressively more legs.

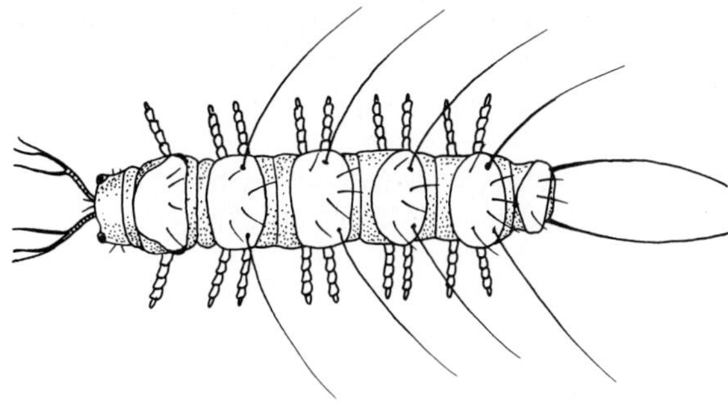

Figure 14.4 Pauropus, *a pauropod.*

Class Pauropoda

The pauropods are represented by about 375 species of soft-bodied, white, eyeless, wormlike animals. They range in size from 0.5 to 2.0 mm. They typically are found in moist soil and leaf mold in temperate and tropical regions. They may be extraordinarily common in forest litter as over two million individuals have been reported per 0.4 hectare (1 acre).

The body is constructed of a head and a trunk (Fig. 14.4). The **head** bears a pair of two-branched antennae; one ramus terminates in a single flagellum and the other in two flagella. On either side of the head is a peculiar disc-shaped sensory organ whose function is not fully understood. The mouthparts are similar to those of millipedes with a pair of mandibles and a lower lip or **gnathochilarium** formed by the single pair of maxillae. Food consists of fungal hyphae, dead organisms, and organic debris.

The **trunk** is composed of 11 segments plus the pygidium. Since segments I and XI are legless, there are nine pairs of legs (Fig. 14.4). Each leg consists of six podomeres, and the distal one bears one or several claws. Dorsally, there are six **diploterga,** each covering a pair of segments. Each of the posterior five diploterga bears a pair of long, lateral sensory **setae.** On segment III or the genital segment are a gonopore and a seminal receptacle opening in females and paired sperm duct openings in males. Spermatophores are transferred to the female, fertilized eggs are laid in the soil, the young hatch with only three pairs of legs, and the remaining legs are added at subsequent molts.

PHYLOGENY OF THE MYRIAPODS

Details of the ancestral forms of myriapods can only be deduced, since there are no early fossils to indicate the evolutionary stages of these arthropods. That myriapods and insects arose from a common or similar ancestry is not in doubt. Both myriapods and insects probably arose from a primitive segmented worm that was already semiterrestrial in its habits. It is even possible that insects arose from a myriapod ancestor such as a symphylan that was already sufficiently different from the primitive segmented ancestors. The similarities between symphylans and primitive insects are numerous, and the possibility that insects originated from a symphylan ancestor is intriguing. The myriapods are considered to be polyphyletic, and the four classes have evolved independently (Fig. 14.5). Chilopods and symphylans

Figure 14.5 *Probable phylogenetic trends in the myriapods.*

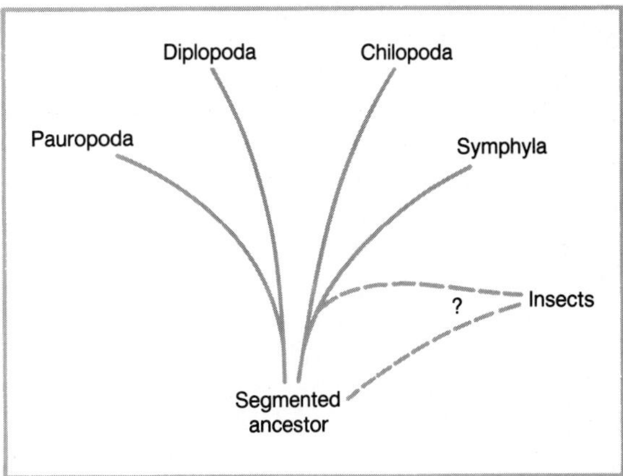

as a group and diplopods and pauropods as another group are more similar to each other than they are to those taxa in the other group. In all four classes there are fundamental similarities in serial trunk segmentation, head construction and appendages, and almost all visceral systems.

SYSTEMATIC RESUME

The Major Taxa of Myriapods

Class Chilopoda—centipedes; head with two pairs of maxillae; first trunk appendages are poisonous maxillipeds; trunk segments each with one pair of legs; gonopores at posterior end; mostly predaceous.

Subclass Epimorpha—more than 20 pairs of legs; young are brooded and hatch with all adult segments present.

Order Scolopendromorpha—abundant especially in tropical areas; some may be up to 27 cm long; 21 or 23 pairs of legs; two pairs of ocelli or eyeless; antennae with 17–31 segments; 9–11 pairs of spiracles; *Scolopendra* (Fig. 14.1a) and *Theatops* are common.

Order Geophilomorpha—slender, burrowing, eyeless; widespread distribution; 31–181 pairs of legs; antennae with 14 segments; spiracles on almost all segments; *Geophilus* (Fig. 14.1b) and *Strigamia* are frequently collected.

Subclass Anamorpha—15 pairs of legs; young are not brooded; young hatch with less than full number of segments.

Order Lithobiomorpha—widely distributed; trunk of 18 segments; 15 pairs of legs; usually many ocelli present; antennae of 19–70 segments; paired lateral spiracles usually on alternate segments; *Lithobius* (Fig. 14.1d) and *Bothropolys* are rather common.

Order Craterostigmomorpha—strange forms intermediate between Lithobiomorpha and Scolopendromorpha; found in Tasmania and New Zealand; two known species of *Craterostigmus*.

Order Scutigeromorpha—worldwide distribution; 18 trunk segments; 15 pairs of long legs; large compound eyes and long antennae present; single row of seven dorsal spiracles; common genera are *Scutigera* (Fig. 14.1c) and *Thereuonema*.

Class Symphyla—minute; resemble centipedes; often pests in greenhouses; trunk of 12 segments, each with a pair of legs; gonopores on third trunk segment; includes *Scutigerella* (Fig. 14.2) and *Hanseniella*.

Class Diplopoda—millipedes; one pair of maxillae; numerous trunk diplosegments, each with two pairs of legs; gonopores on anterior trunk; mostly herbivorous.

Subclass Penicillata—minute; soft cuticle with tufts and rows of setae; 13 or 17 pairs of legs; no gonopods or repugnatorial glands; *Polyxenus* and *Lophoproctus* are typical genera.

Subclass Chilognatha—large; cuticle hard and calcareous; no tufts or rows of setae; toothed mandibles; gonopods and repugnatorial glands often present; contain most diplopods.

Superorder Pentazonia—body mostly circular in cross section, ventral surface flat; posterior one or two pairs of legs in males are modified as gonopods or claspers.

Order Glomerida—pill millipedes; trunk short and can be rolled up into a sphere; 15–17 trunk segments with 17–23 pairs of legs; no gonopods, but posterior legs are claspers; no repugnatorial glands; mostly tropical; *Onomeris* and *Glomeris* (Fig. 14.3d) are typical.

Order Glomeridesmida—body is long and flattened and cannot be rolled up into a sphere; 22 trunk segments with 36–37 pairs of legs; no gonopods, but last one or two pairs of legs used as claspers; no repugnatorial glands; *Glomeridesmus* is typical.

Order Sphaerotheriida—giant pill millipedes, sometimes in excess of 25 cm in length; mostly tropical; trunk with 13 terga; best known genus is *Sphaerotherium*.

Superorder Colobognatha—mandibles are rod shaped, hidden under gnathochilarium; mouthparts often form a snout; trunk of 30–192 flattened segments that can be somewhat telescoped; posterior legs of segment VII and anterior legs of segment VIII are gonopods; repugnatorial glands present; *Brachycybe* and *Polyzonium* are commonly encountered.

Superorder Helminthomorpha—includes most millipedes; one or two pairs of legs on trunk segment VII are modified as gonopods; repugnatorial glands present.

Order Chordeumida—small; widely distributed; 26–32 trunk somites; one or three pairs of spinnerets on posterior tergum; *Conotyla* and *Cleidogono*.

Order Polydesmida—common; flat-backed millipedes; most are ribbon shaped and flat; 18–22 trunk segments with prominent lateral tergal keels; no eyes; anterior legs on segment VII are gonopods; *Polydesmus; Desmonus,* and *Oxidus*.

Order Julida—cylindrical trunk; males have two pairs of gonopods; eyes usually present; *Blaniulus, Nemasoma, Julus* (Fig. 14.3a), and *Diploiulus* are common.

Order Spirobolida—cylindrical trunk; anterior five segments bear one pair of legs; male gonopods housed in pouch; *Narceus, Arinolus,* and *Spirobolus*.

Order Spirosteptida—mostly tropical and African; cylindrical trunk; posterior pair of gonopods in males absent or vestigial; *Scaphiostreptus* and *Orthoporus*.

Class Pauropoda—small; biramous antennae; one pair of maxillae; trunk of 11 segments, legs on segments II–X; diploterga with paired sensory setae; frequently encountered genera are *Pauropus* (Fig. 14.4) and *Allopauropus*.

ADDITIONAL READINGS

Blower, J.G., ed. 1974. *Myriapoda,* 679 pp. Symposium No. 32 of the Zoological Society of London. New York: Academic Press.

Brown, F.A. 1950. *Selected Invertebrate Types,* pp. 462–475. New York: John Wiley & Sons.

Cloudsley-Thompson, J.L. 1968. *Spiders, Scorpions, Centipedes, and Mites,* pp. 20–84. New York: Pergamon Press.

Kaestner, A. 1968. *Invertebrate Zoology,* pp. 302–445. New York: John Wiley & Sons.

Lewis, J.G.E. 1981. *The Biology of Centipedes,* 455 pp. New York: Cambridge University Press.

Manton, S.M. 1977. *The Arthropoda: Habits, Functional Morphology, and Evolution,* pp. 344–397. Oxford, England: Clarendon Press.

15

The Bryozoans and Entoprocts

OVERVIEW

Even though bryozoans and entoprocts have puzzling phylogenetic affinities, they are considered to be protostomates and closely related to each other. The most important common features are that all have a food-gathering lophophore bearing a circle of hollow ciliated tentacles surrounding the mouth and a U-shaped digestive tract. Representatives of both groups are bilateral, aquatic (mostly marine), very small, and sessile.

Bryozoans, a large phylum numerically but not at all dominant ecologically, usually live as colonies on almost any aquatic surface. Most colonies are either flat encrustations or grow in upright arborescent patterns. Individual bryozoan zooids are small, coelomate, and lacking excretory organs, and most have a calcified body wall. The lophophore is hydraulically protruded for feeding by the animal's compressing its coelomic fluid. A great many bryozoans are polymorphic, since certain zooids are modified for special functions. Because zooids are minute and sessile, several conventional body systems are either absent or severely reduced. Asexual reproduction is unusually varied in bryozoans. Most are hermaphroditic, and a zooid broods a single egg that hatches into a small larva; after settling, the larva metamorphoses and produces a primary zooid that buds off other zooids, thus developing a colony of interconnected individuals.

Of the three classes of bryozoans the phylactolaemates live exclusively in fresh water, have body wall muscles that are responsible for protracting the horseshoe-shaped lophophore, and have coelomic connections between zooids. The stenolaemates have cylindrical zooids, a nonmuscular body wall, and some polymorphism. The gymnolaemates, by far the largest class, have a calcified body wall and a circular lophophore that is protruded by deforming the body wall; most are polymorphic.

The entoprocts represent a small number of mostly marine organisms in which individuals are small, are often colonial, and consist of the body atop a stalk. Characteristic of this group is that each has a pseudocoelom, has its anus positioned inside the lophophoric tentacles, and has a pair of excretory protonephridia. Following internal fertilization the embryo hatches into a trochophorelike larva that soon settles, metamorphoses, and develops into an adult.

The phylogenetic affinities of bryozoans and entoprocts with each other and with other protostomates are open to serious questions, but many embryological features and larval and adult morphological similarities are compelling evidence for a protostomate affiliation for both groups. The ancestral form was probably a marine bilateral worm that evolved adaptations for a sessile existence. There are no protostomate groups to which they are closely related.

INTRODUCTION

The phyla Bryozoa and Entoprocta are rather unusual in that they have certain features that strongly link them with protostomates and a few other characteristics that are like those of deuterostomates. This in no way is meant to imply that they are a phylogenetic link between the two great groups of higher metazoans, although some zoologists have speculated that bryozoans might represent a third line of evolution apart from both protostomates and deuterostomates. In the discussions that follow, the reader will soon see the dilemma of trying to relate bryozoans and entoprocts to other invertebrates in either of the supergroups. In fact, two other small taxa, the phoronids and brachiopods, have also traditionally been included in this grouping, but since recent research has shown that they are more decidedly deuterostomate, they are so categorized (see Chapter 16). Future interpretations of bryozoan and entoproct development and morphology may also place one or both of these phyla with the deuterostomates, but for now they are grouped as protostomates.

What are the diagnostic features of the bryozoans, and why are entoprocts associated with them? Representatives of both groups are aquatic and mostly marine, sedentary, and quite small, and they secrete a protective covering around themselves. The most important characteristics of the bryozoan classes and the Phylum Entoprocta appear in Table 15.1. But apart

TABLE 15.1. □ **CHARACTERISTICS OF THE THREE CLASSES OF BRYOZOANS AND THOSE OF THE PHYLUM ENTOPROCTA**

Taxon	*Habitat*	*Nature of Lophophore*	*Nature of Body Wall*	*Other Characteristics*
Phylum Bryozoa				
Class Phylactolaemata	All freshwater	Horseshoe-shaped	Muscle fibers, noncalcareous	Form statoblasts; no polymorphism or brown body formation
Class Stenolaemata	All marine	Circular	No muscles, some calcification	Have an internal membranous sac; undergo polyembryony
Class Gymnolaemata	Mostly marine	Circular	No muscles, often calcified	Cystid pores with cellular plugs; extensive polymorphism; form brown bodies
Phylum Entoprocta	All marine but 1 species	Circular, encloses anus	Some longitudinal fibers	Small, stalked; no exoskeleton; have a pseudocoelom

from important embryological details to be mentioned later, the most basic unifying feature is the presence of a rather remarkable food-gathering adaptation, the lophophore.

A **lophophore,** a simple or complicated fold of the body wall, surrounds the mouth and bears a variable number of ciliated tentacles. The tentacles are simple and unbranched, and their lumens are extensions of the internal body cavity. The lophophoric complex can be protracted for feeding or retracted for protection. When protracted, the tentacles form a funnel, and their ciliary beat drives water down into the funnel and carries small planktonic organisms to the mouth, which is situated in the funnel's smaller end. So universal and constant is the lophophore among the bryozoans and entoprocts that they are often called the lophophorates. However, both phoronids and brachiopods also possess a lophophore; since they apparently are not close relatives of bryozoans, the term "lophophorate" should not be used to imply phylogenetic affinities. Apparently, the presence of a lophophore in these two protostomate phyla and in two deuterostomate taxa is a case in which quite dissimilar groups evolved a similar adaptation to a sedentary, filter-feeding habit.

PHYLUM BRYOZOA

The bryozoans have a high species diversity with about 4000 living species, but they are not a dominant group ecologically. Their extensive fossil record contains about 20,000 species, mainly because most have secreted a hard covering around themselves that often has led to preservation. Living bryozoans are particularly common in littoral, marine, subtidal areas, where they are both numerous and diverse. One *Pinna* shell (a bivalve) bore 30 bryozoan species, and a single dredge haul over a shell bed produced 90 different bryozoan species! But bryozoans are very small animals that usually do not exceed 1.5 mm in length and therefore are easily overlooked. Although almost all bryozoans are colonial, their colony size normally does not exceed 3–8 cm in width or height; thus they are rather inconspicuous even as colonies. Most bryozoans are marine and grow on a variety of surfaces, both inanimate and animate, in shallow coastal zones. Many species form encrustations on shells, rocks, and other reasonably firm surfaces; others have upright branching systems reminiscent of colonial hydroids (Fig. 15.1). They also may appear to be plantlike, and their superficial resemblance to mosses is the basis for their name (bryo = moss; zoa = animal). While some bryozoans are brightly colored, most are drab and inconspicuous.

It will be helpful to the reader to become familiar at the outset with the three subordinate classes of bryozoans: the exclusively freshwater Phylactolaemata, the small and exclusively marine Stenolaemata, and the largest and almost entirely marine Gymnolaemata. All three are characterized in Table 15.1, and examples of bryozoan diversity appear in Fig. 15.1.

The Bryozoan Colony

With only a few exceptions, all bryozoans are colonial. Each small colonist is termed a **zooid.** Zooids, often numbering in the hundreds of thousands or even millions, are associated together in colonies that vary widely in shape, form, and organization (Fig. 15.1). Perhaps not surprisingly, zooids in a colony are connected via pores in their body walls, thus permitting some communication throughout the colony. Perhaps the most primitive colony shape is stoloniferous as exemplified by *Bowerbankia* (Fig. 15.2a). The stolon, either erect or creeping, is composed of modified zooids (kenozooids), and the stolons give rise to unmodified zooids. More commonly, colonies like those of *Electra* or *Membranipora* are in the form of flat encrustations in which an individual zooid is abutted on all four sides by other zooids so that the colony is one layer thick (Fig. 15.1d). Such colonies may be

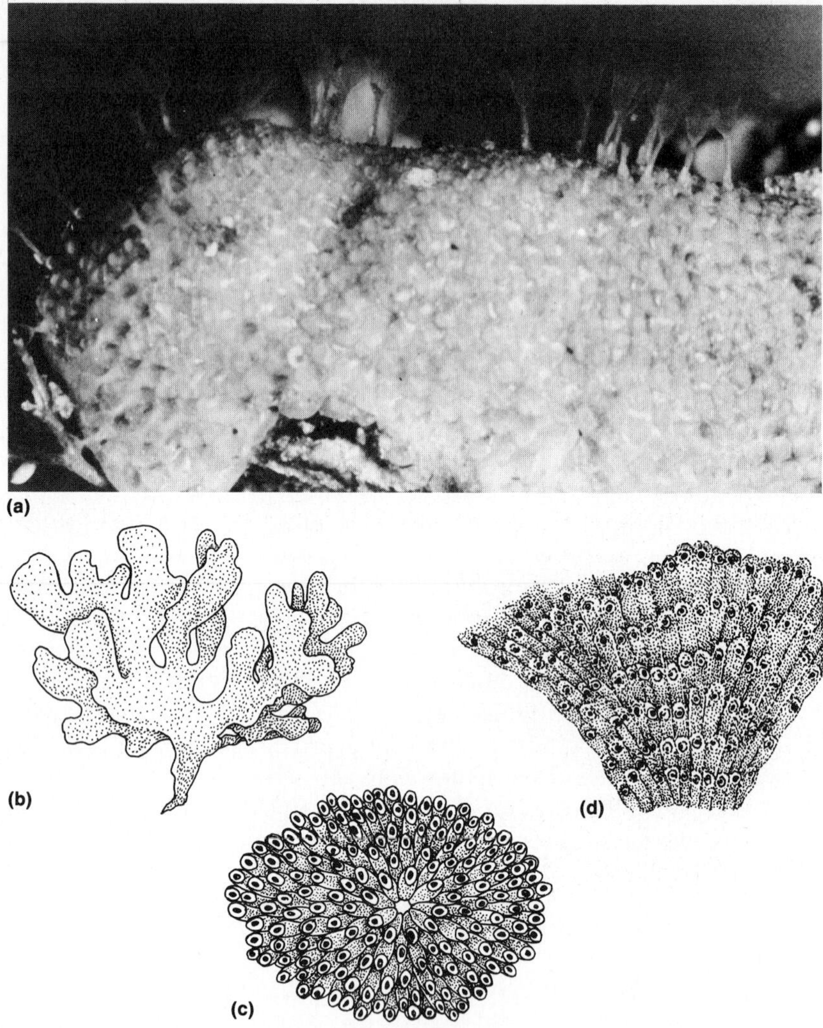

Figure 15.1 *Some types of bryozoan colonies:* ***(a)*** *photograph of a portion of a colony of* Eurystomella, *an encrusting bryozoan (courtesy of Ward's Natural Science, Inc., Rochester, N.Y.);* ***(b)*** *a colony of* Flustra; ***(c)*** *a colony of* Cryptosula; ***(d)*** *a portion of a colony of* Schizoporella. *(parts b–d from L.H. Hyman,* The Invertebrates, *1959, Vol. V:* Smaller Coelomate Groups, *McGraw-Hill Book Co., used with permission.)*

Figure 15.2 *The principal types of gymnolaemate bryozoans:* ***(a)*** *diagram of* Bowerbankia, *a stoloniferous form (from J.S. Ryland, 1970,* Bryozoans, *Hutchinson Publishing Group, used with permission);* ***(b)*** *photomicrograph of an autozooid of* Bugula, *an arborescent form (courtesy of Ward's Natural Science, Inc., Rochester, N.Y.):* ***(c)*** *diagram of an individual autozooid of* Bugula, *an arborescent bryozoan;* ***(d)*** *diagram of* Electra, *an encrusting form (from Marcus in L.H. Hyman,* The Invertebrates, *1959, Vol. V:* Smaller Coelomate Groups, *McGraw-Hill Book Co., used with permission).* ▶

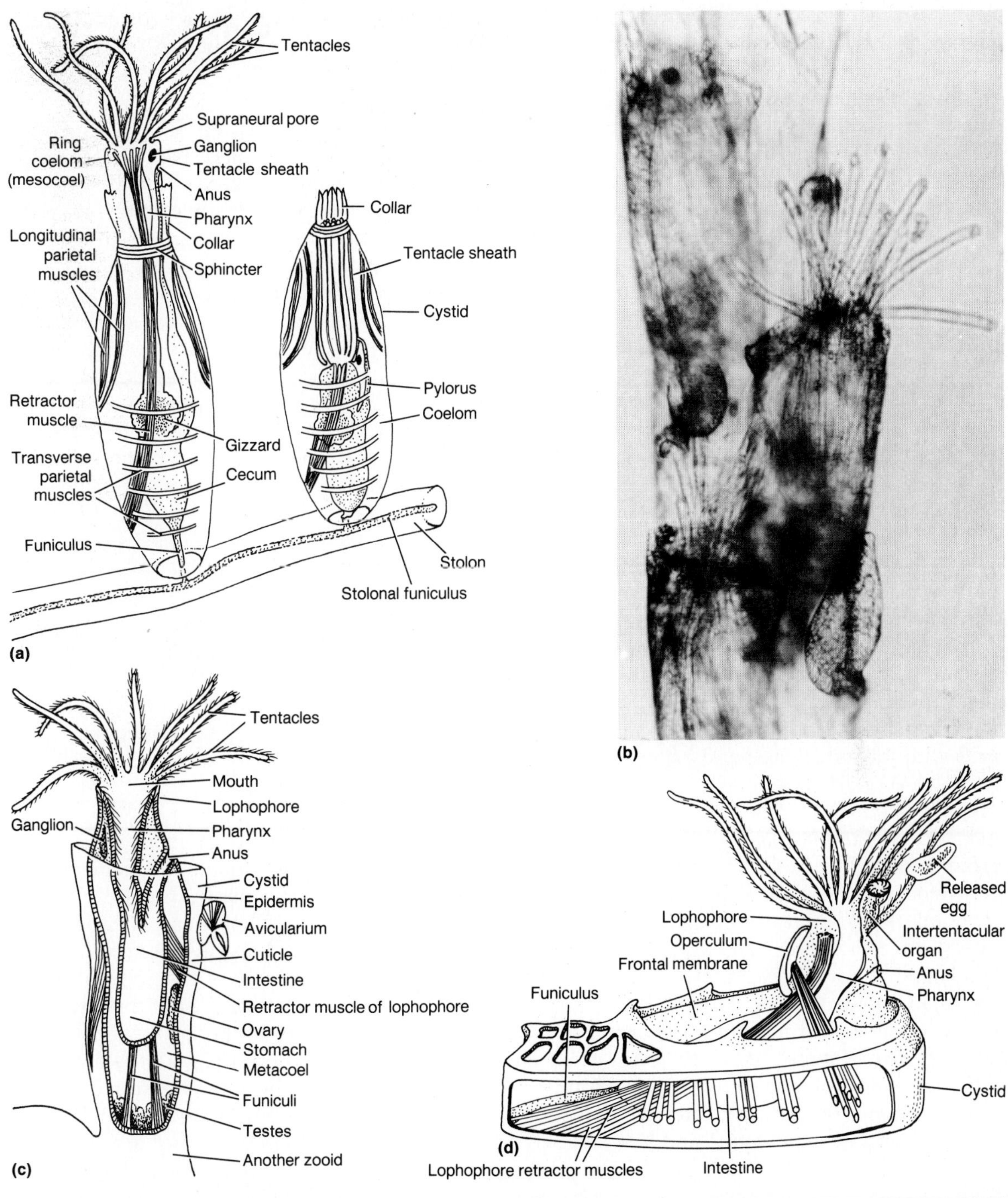
Tentacles
Supraneural pore
Ring coelom (mesocoel)
Ganglion
Tentacle sheath
Anus
Pharynx
Collar
Longitudinal parietal muscles
Sphincter
Collar
Tentacle sheath
Cystid
Retractor muscle
Pylorus
Coelom
Gizzard
Transverse parietal muscles
Cecum
Funiculus
Stolon
Stolonal funiculus
(a)
(b)
Tentacles
Mouth
Lophophore
Ganglion
Pharynx
Anus
Cystid
Epidermis
Avicularium
Cuticle
Intestine
Retractor muscle of lophophore
Ovary
Stomach
Metacoel
Funiculi
Testes
Another zooid
(c)
Released egg
Lophophore
Operculum
Intertentacular organ
Frontal membrane
Anus
Funiculus
Pharynx
Cystid
(d)
Lophophore retractor muscles
Intestine

found on almost any substratum like submerged plants, shells, external surfaces of other animals, and stones. The shape of encrusting colonies may be irregular, rounded spots, or lobed or stellate patches. Some encrusting forms are bilaminar, that is, have two layers back-to-back, and other colonies are mounded and have multiple layers. A great many bryozoans have upright dendritic growths with the colony base anchored into a substratum. The common marine *Bugula* forms upright branching colonies of up to 8–10 cm high in which individual zooids are arranged in a biserial pattern (Fig. 15.2b). A colony of the freshwater *Pectinatella* is a slimy gelatinous mass of up to 2 m in diameter with large numbers of zooids at the surface.

The Bryozoan Zooid

The individual zooid possesses a number of important features that will be treated more fully later including their bilateral symmetry, coelom, and being mostly colonial, sessile, and monoecious. On the other hand, bryozoans lack important organs that are present in most other metazoans; these omissions are clearly associated with their small size and being sedentary.

Each zooid is quite small, usually averaging less than 3 mm in length. The orientation of the bilateral zooid is variable depending on the growth form. In upright colonies the lower or attached surface of the zooid is ventral, and the lophophore and anus are dorsal. In encrusting forms, the surface against the substratum is basal, and the upper surface is frontal. In all bryozoan zooids the end nearest to the origin of the colony is proximal, and the opposite end is distal. While many bryozoans have several different kinds of zooids (polymorphism), a phenomenon to be discussed later, we shall first consider the general structure and function of a typical feeding zooid or **autozooid** of a gymnolaemate. Even though autozooids come in a variety of shapes including the shape of a box, tube, vase, or even an egg, all autozooids of a colony have a uniform structure.

The body wall

The zooid has a body wall or **cystid** (often called the zooecium) enclosing the lophophore and visceral mass collectively called the **polypide.** The outer cystid layer is an organic **cuticle** composed of protein and chitin and secreted by the underlying **epidermis.** In the phylactolaemates the cystid is gelatinous, membranous, or chitinous but always flexible, and it also contains circular and longitudinal muscles. In the stenolaemates and most gymnolaemates there is a thick depositional layer of calcium carbonate between the cuticle and epidermis, and body wall muscles are absent. In other gymnolaemates the body wall is more flexible, lacks muscles, but still may contain calcareous deposits within the cuticle. If a calcareous layer is present, it lies outside of the epidermis and constitutes an **exoskeleton.** Although the degree of calcification varies considerably, in most bryozoans the polypide is encased in a hard, calcareous, rigid cystid. The remainder of the body wall consists of muscles (in some) and the **peritoneum.**

Internal features

The visceral organs of a zooid lie in a rather large coelom that has both an interesting ontogeny and shape. The **coelom** is formed as a split within the mesodermal mass and later develops into a two-chambered space, the mesocoel and metacoel. The **mesocoel** is the coelom of the tentacles and lophophore, and the **metacoel** persists as the cavity of the trunk surrounding the visceral organs (Fig. 15.2). The two parts of the coelom, divided by a transverse septum but with a pore connecting the two, are filled with fluid, contain coelomocytes that are useful in absorption and waste removal, and are the principal internal transport cavities. It is thought by some that a third chamber, the **protocoel,** was consequently lost when bryozoan evolution resulted in the loss of the head. The coelom reflects both protostomate and deuterostomate characteristics; the schizocoelom is decidedly protostomate, but its presumed tripartite construction (assuming the ancestral bryozoan had a head) is a feature characteristic of many deuterostomates like echinoderms and chordates.

The digestive tract is decidedly U-shaped and consists of a mouth, pharynx, esophagus (in some), enlarged stomach, intestine, and the anus, which is always situated outside of the tentacular circle (Fig. 15.2). Food particles are swept into the mouth by cilia both on the tentacles and lining the pharynx. The **stomach** is typically divided into an anterior nonciliated cardia, a middle nonciliated cecum, and a posterior ciliated pylorus; most digestion and absorption take place in the stomach. Food molecules are apparently stored in the epithelial lining of the alimentary tract.

It has already been noted that individual zooids within a colony are not separate entities but are functionally connected to each other. Pores in the abutting cystids of contiguous zooids allow for interzooidal communication. Attached to the undersurface of the stomach and extending to the base or sides of the cystid are one or several distinct mesenchymal cords, the funiculi. Each **funiculus** passes through a pore in the cystid wall and joins funiculi from other zooids; therefore the zooids of a colony are structurally and functionally connected through their collective funiculi (Fig. 15.2). The funiculi are hollow in primitive bryozoans, so there are direct coelomic connections between zooids. In the gymnolaemates, even though the funiculus is solid and the cystid pore is filled with a plug of funicular cells, there is still some interzooidal transport of materials through this cellular plug.

Closely associated with the polypide is a pair of large **retractor muscles** that originates on the proximal end or the base of the cystid and inserts in a ring around the base of the lophophore; these muscles retract the lophophore complex (Fig. 15.2). Because of the high degree of miniaturization, a zooid can easily exchange materials with its aquatic medium, and it is not surprising that separate gas exchange, circulatory, and excretory systems are missing. The nervous system is chiefly represented by a **ganglion** situated near the base of the lophophore (Fig. 15.2a). Nerves arising from the ganglion supply the lophophore and tentacles, retractor muscles, and viscera. Sensory organs are absent, but numerous sensory cells are present on the tentacles.

The Lophophore and Mechanisms for Its Extension

In the phylactolaemates the lophophore is shaped like a doubled horseshoe with the mouth situated inside the two ridges at the bend of the horseshoe; the two ridges bear a total of 16–106 tentacles. In the gymnolaemates and stenolaemates the lophophore is circular, not doubled, and bears 8–40 tentacles. When fully extended, the collective tentacles form a funnel with the mouth at its smaller end (Fig. 15.3a). Each tentacle bears a ciliated tract along both of its lateral borders (those facing either adjacent tentacle) and a third tract of shorter cilia on its medial (inner) surface (Fig. 15.3b). Water currents are created by the lateral cilia, which beat downward and carry particles directly into the mouth. As water enters the top of the funnel and flows centrifugally between the tentacles to the outside, particles are strained out by the lateral cilia and carried to the mouth by the medial cilia. It has been shown recently that the tentacles are held close together so that their tips are usually no more than 110 μm apart, and the lateral cilia can effectively strain out small planktonic forms and bacteria for ingestion. The lophophore complex is, in reality, a very efficient device that not only creates a water current but also acts as a sieve.

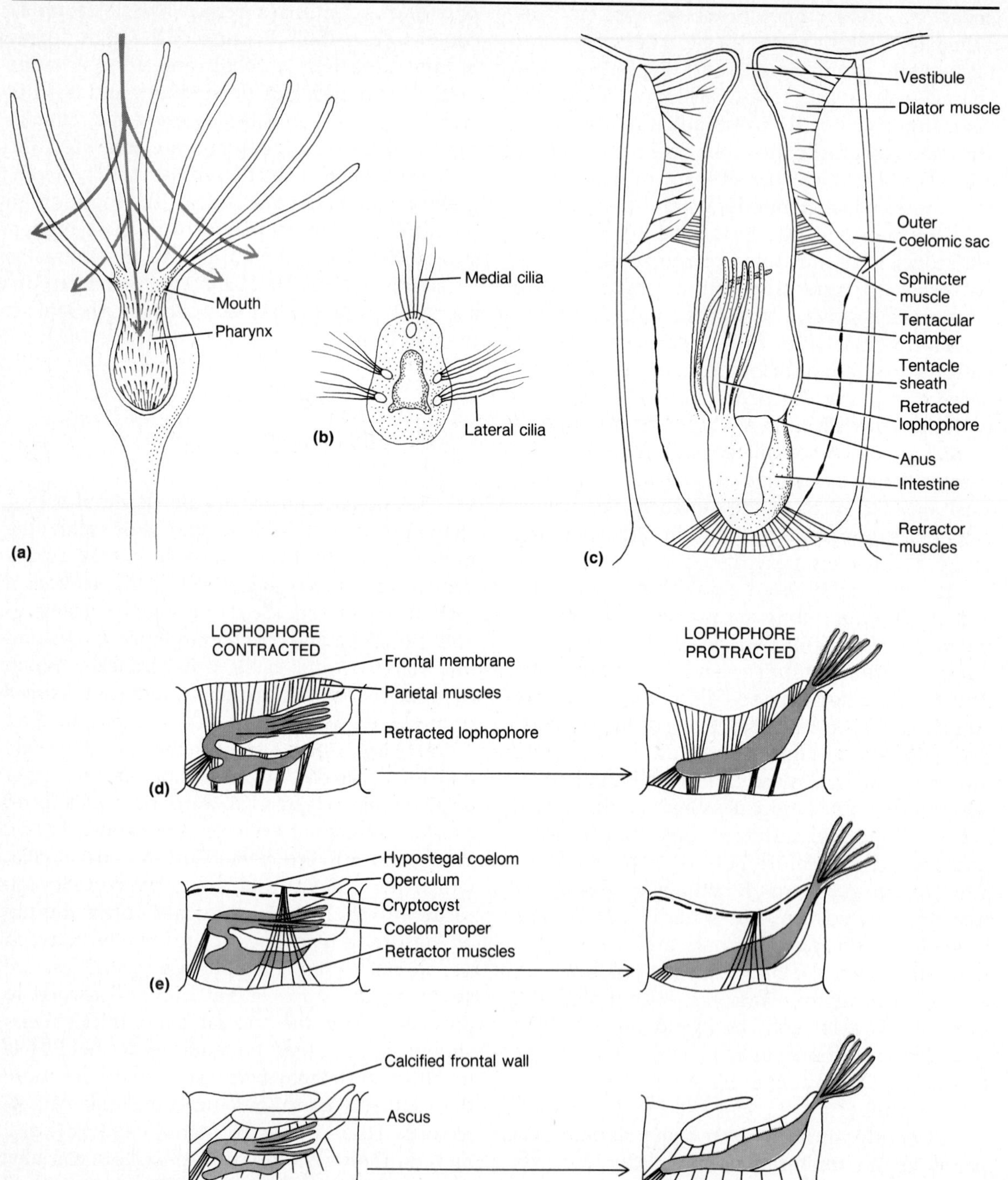
Mouth
Pharynx
(a)
Medial cilia
Lateral cilia
(b)
Vestibule
Dilator muscle
Outer coelomic sac
Sphincter muscle
Tentacular chamber
Tentacle sheath
Retracted lophophore
Anus
Intestine
Retractor muscles
(c)
LOPHOPHORE CONTRACTED
LOPHOPHORE PROTRACTED
Frontal membrane
Parietal muscles
Retracted lophophore
(d)
Hypostegal coelom
Operculum
Cryptocyst
Coelom proper
Retractor muscles
(e)
Calcified frontal wall
Ascus
(f)

◀ *Figure 15.3 The bryozoan lophophore and mechanisms for its hydraulic extension:* ***(a)*** *pathways for water currents through the tentacular funnel;* ***(b)*** *cross-sectional view through a tentacle;* ***(c)*** *the vestibule and tentacular chamber;* ***(d)*** *extension of the lophophore by means of a flexible frontal wall;* ***(e)*** *extension of the lophophore by means of a cryptocyst;* ***(f)*** *extension of the lophophore by means of an ascus.*

The anteriormost alimentary organ, the pharynx, is ciliated and often draws particles into the mouth. Ingestion can even be augmented by pharyngeal muscles functioning as a suction pump or by the animal's retracting the lophophore to force particles into the mouth.

The **anal opening** is situated near the base of the lophophore but never inside the ring of tentacles. On the basis of the position of the anus relative to the lophophore, bryozoans have also been called ectoprocts (ecto = outside; proct = anus).

An opening in the cystid, the **orifice,** enables the lophophore and tentacles to be protracted. In a large number of marine species the orifice is equipped with a protective hinged **operculum** or lid that closes the orifice when the lophophore is retracted (Fig. 15.3e). There is often a two-part space to accommodate the retracted lophophore. The outer part, called the **atrium** or **vestibule,** represents an invagination of the body wall and is continuous as an inner space, the **tentacular chamber;** at the junction of the atrium and tentacular chamber is a thickened **diaphragm** provided with a sphincter muscle (Fig. 15.3c). When the lophophore is retracted, it is pulled through the atrium and comes to rest in the tentacular chamber, and the operculum, diaphragm, and sphincter muscle effectively close off the tentacular chamber from the outside. Bounding the tentacular chamber is a **tentacular sheath,** a portion of the body wall that extends from the diaphragm to the base of the lophophore and extends around the withdrawn lophophore like a sleeve (Fig. 15.3c). When the lophophore is protracted, the sheath everts and covers the base of the lophophore like a collar.

Whereas the lophophore is retracted by retractor muscles, its protrusion is achieved by several quite elegant means that always involve compression of the coelomic fluid first in the metacoel and then in the mesocoel. This forces the lophophore and tentacles out hydraulically. Bryozoans differ only in the mechanisms by which this is accomplished. In the phylactolaemates the contraction of the circular muscles in the body wall compresses the coelomic fluid. Many other bryozoans have a flexible frontal (roof of the box) wall into which protractor muscles are inserted. As these muscles contract and the frontal membrane is depressed, the volume of space within the cystid is decreased, and the coelomic fluid is compressed (Fig. 15.3d). In other bryozoans in which the frontal wall becomes rigid through calcification a circular flexible **frontal membrane** persists that can be altered by muscles in a similar manner. But as the frontal wall or membrane is depressed, there is the possibility of damage to the polypide; therefore several important adaptations have evolved to permit hydraulic protrusion of the lophophore while at the same time providing some protection for the underlying tissues. In many bryozoans there is a narrow, annular, interior wall—the cryptocyst—beneath the frontal membrane. The **cryptocyst** is usually a wide shelf that divides the metacoel into two parts, the **coelom proper,** located beneath the cryptocyst, and the **hypostegal coelom,** located between the cryptocyst and frontal wall (Fig. 15.3e). One crucial functional feature of the cryptocyst is that it is perforated. As the frontal wall is depressed by muscles, the fluid in the hypostegal coelom is compressed, and the compressional effect passes through the perforations of the rigid cryptocyst to the coelom proper and hydraulically forces the lophophore outward. Therefore the cryptocyst provides a great deal of mechanical protection to the polypide and, at the same time, permits compression of coelomic fluids. In still other bryozoans in which the frontal wall is completely calcified, there is a thin-walled compensation sac, or **ascus,** just beneath the rigid frontal wall and opening to the exterior via a pore (Fig. 15.3f). Contraction of muscles inserting into the floor of the ascus causes it to dilate with seawater from the outside. Ascal dilation results in compression of the coelomic fluid and thus indirectly extends the lophophore and tentacles.

Polymorphism

Polymorphism is one of the most striking and unusual features of many different bryozoans. It is particularly pronounced among members of the very large Order Cheilostomata (gymnolaemates). Basically, bryozoan zooids are of two types: autozooids, or the feeding zooids already discussed, and **heterozooids,** which are reduced or modified zooids. While bryozoologists recognize many different types of heterozooids, there are several that merit consideration: avicularia, vibracula, ooecia, and kenozooids.

Avicularia, present in most cheilostomatids, are perhaps the most spectacular and unusual of all heterozooids. An **avicularium** (= small bird) is shaped like a bird's head and is attached to the surface of an autozooid (Figs. 15.2c, 15.4c). It is generally smaller than autozooids and lacks a functional polypide. The most characteristic feature of an avicularium is a movable **mandible** or jaw, which represents its modified operculum. The mandible can be opened or closed in a single plane by several well-developed muscles (Fig. 15.4c). The avicularia of some bryozoans are sessile, but those of many others are stalked. They are generally considered to have a protective function and are analogous to the pedicellariae of many echinoderms (see Chapter 17). Avicularia are effective deterrents against predation by small organisms and against the settling of larvae on the surface of an autozooid. The avicularia commonly grasp small nematodes and tubicolous amphipods and polychaetes, which are either killed or discouraged from building tubes there. Avicularia are perhaps best displayed in *Bugula* and other members of its family.

The operculum of a **vibraculum** is drawn out into a long **seta** that is more freely articulated than a mandible (Fig. 15.4a). It can be moved in many different planes, and can even be swiveled, by muscles located all around its base. The effect of the setal action is to sweep detritus and settling larvae away. Vibracula are present in the common cosmopolitan *Scrupocellaria* and in several other genera like *Cupuladria.*

Another type of heterozooid is the **ooecium** or **ovicell,** which is an external calcified brood chamber (Fig. 15.4b). In its most common form, the ooecium is a hoodlike structure attached to one end of an autozooid. The functional significance of ovicells will be explained below in the discussion of reproduction. Here we note that it is in this chamber that a single embryonic bryozoan develops. **Kenozooids** are individuals that are devoid of internal structures and form stolons, spines, rhizoids, and other types of empty chambers (Fig. 15.2a).

Reproduction and Development

Bryozoans exhibit a rather wide diversity of reproductive methods and patterns. The most prevalent form of asexual reproduction is **budding,** in which a constriction divides a parent zooid into two daughter zooids. The stenolaemates are unusual in that they exhibit **polyembryony,** a process by which multiple embryos are produced asexually from a primary embryo, and the phylactolaemates (freshwater bryozoans) form asexual **statoblasts** that survive the rigors of winter conditions.

The vast majority of bryozoans are hermaphroditic, but dioecious, protandrous, and

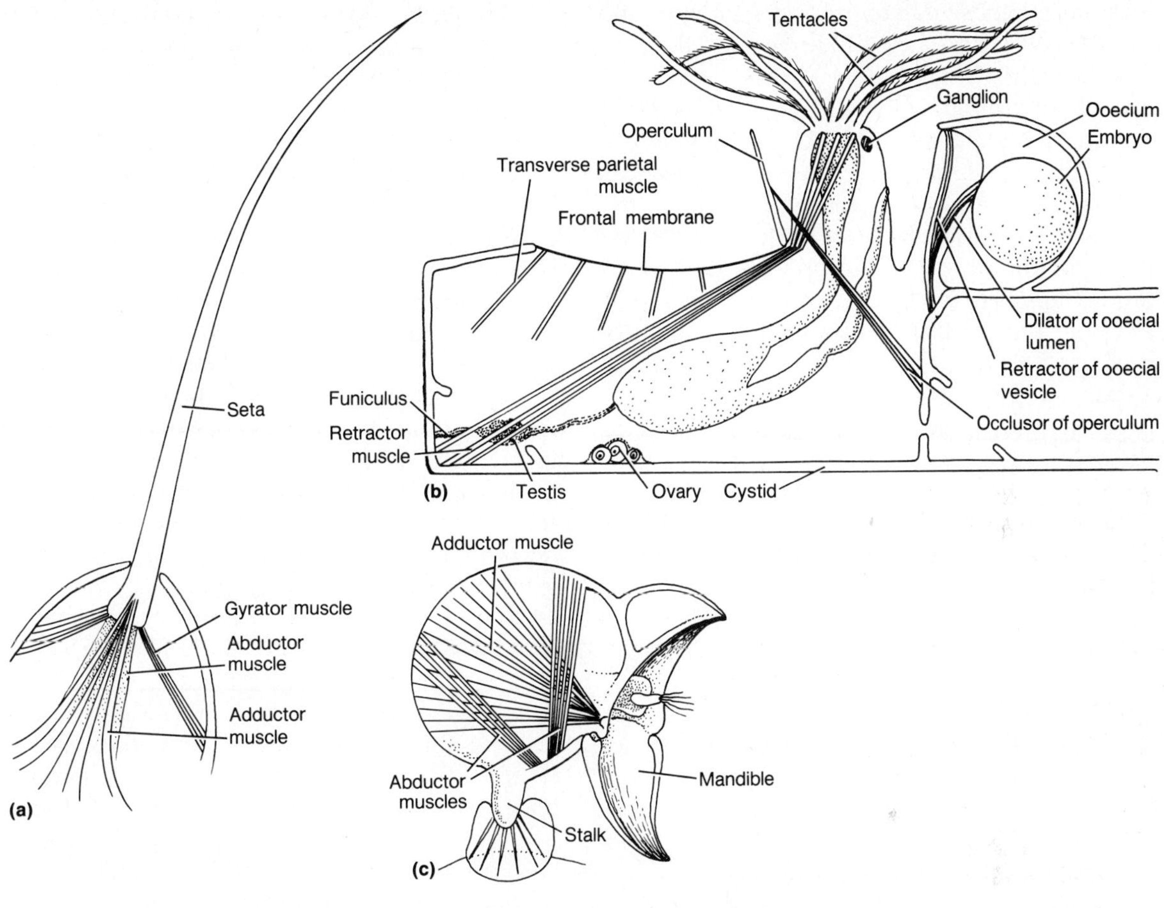

Figure 15.4 *The principal types of heterozooids in gymnolaemates:* ***(a)*** *a vibraculum;* ***(b)*** *an ooecium (from J.S. Ryland, 1970,* Bryozoans, *Hutchinson Publishing Group, used with permission);* ***(c)*** *an avicularium (from Marcus in L.H. Hyman,* The Invertebrates. *1959, Vol. V:* Smaller Coelomate Groups, *McGraw-Hill Book Co., used with permission).*

even sterile individuals are frequently encountered. Germ cells differentiate from the peritoneum and amass together to form **gonads** on the inside of the body wall or on the funiculus. The one or two ovaries generally are more distally located than the one to many testes (Figs. 15.2c, 15.4b). Since gonadal ducts do not develop, mature gametes are shed into the coelom. For a long time the mechanism of sperm release was a mystery; but it is now known, at least in *Electra* and several other genera, that sperm enter the mesocoel of the tentacles and escape via a minute terminal pore in certain tentacles.

The eggs of a few marine bryozoans like *Electra* and *Membranipora* are liberated from the coelom into the sea through a small pore, the **coelomopore,** at the base of the lophophore or atop a small tentacular organ. The vast majority of all bryozoans brood their eggs in a variety of locations; in those that brood the eggs externally the eggs escape to the outside through the coelomopore. Brooding rarely occurs within the coelom proper, but sometimes it takes place within the tentacular chamber as the entire polypide degenerates. In a few other bryozoans the egg develops within an invagination of the atrial wall or in an internal ovisac. By far the most common brooding locus is within an external chamber, the **ovicell** or **ooecium,** a heterozooid formed at the distal end of one autozooid and resting on or embedded in the next autozooid (Fig. 15.4b). A coelomic evagination from the autozooid develops and bulges into the ovicell. Following fertilization, the egg, still within the coelom, is squeezed through the minute connection into the coelomic cavity of the ovicell, where subsequent development takes place.

Both self- and cross-fertilization are thought to occur, but self-fertilization may be the exception. Sperm, released through the tentacles, are carried by water currents to another zooid and adhere to its tentacles. In ectoprocts that spawn their eggs, fertilization takes place as the ovum is released through the coelomopore. Since most bryozoans brood their eggs within the ovicell, sperm enter the second zooid through the coelomopore. Fertilization commonly takes place just before the ovum is extruded into the ooecium.

Early cleavage is total, equal, and usually radial or biradial, but there is considerable variation in cleavage patterns. In bryozoans that have external fertilization the embryo develops into a triangular-shaped larva, the **cyphonautes** (Fig. 15.5a). This larval stage has a bivalved shell, a sensory apical organ, and a lower feed-

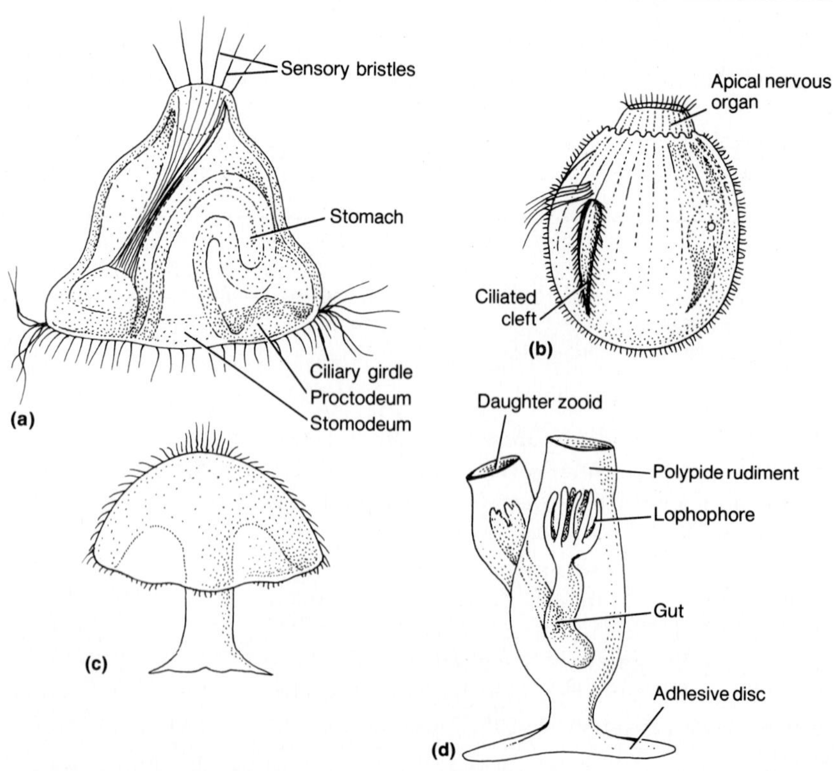

Figure 15.5 *Stages in bryozoan development:* ***(a)*** *a cyphonautes larva of* Farella; ***(b)*** *a larval stage of* Bugula; ***(c)*** *the settling, "umbrella," or "mushroom" stage;* ***(d)*** *a fully formed primary zooid with its first bud. (parts a [from Marcus], b [from Calvet] in L.H. Hyman,* The Invertebrates. *Vol. V,* Smaller Coelomate Groups, *McGraw-Hill Book Co., 1959, used with permission.)*

ing corona surrounding the gape of the valves. The cyphonautes also develops a functional digestive system, is an active feeder on small planktonic organisms, and may exist for several months before finally settling to the bottom.

In those bryozoans (most) that brood their eggs a trochophorelike larval stage is developed that escapes from the ovicell (or other point of brooding) after about two weeks of development (Fig. 15.5b). Each larva of brooding bryozoans also has a locomotor ciliated corona, an anterior tuft of long cilia, and a posterior adhesive sac. These larvae do not feed; therefore they have only a very ephemeral free-swimming existence before they come to rest on a suitable surface. The bryozoan larva settles, and its adhesive disc, everted by muscular action, spreads out over the substratum and anchors the larva (Fig. 15.5c). Within several hours after settling, metamorphosis begins; all larval organs disintegrate while the coelom persists, the ectoderm remains and secretes a cuticle, and the larva becomes an **ancestrula.** The ancestrula at first consists of an undifferentiated cellular mass within its cystid; but very shortly, it is transformed into a complete polypide. The fully formed ancestrula begins to form a colony by budding off daughter zooids (Fig. 15.5d), which in turn bud off new zooids, a pattern that continues during the life of the colony.

The life span of a colony varies immensely from several months to years or even decades. Most freshwater bryozoans and many that are marine and feed on phytoplankton die in the autumn because of decreasing water temperatures and reduced food supplies. Many colonies of marine species live for two to six years; some of these perennial colonies die back to their stolons or bases in the winter only to reform the following spring.

An extremely fascinating dimension of most bryozoans is that the polypide of a given zooid normally does not persist for more than several weeks to a month. Degeneration reduces the polypide to one or two conspicuous dark masses called **brown bodies** (Fig. 15.6). Many different theories have been proposed to account for the formation of brown bodies and their significance. Following regression, a small group of undifferentiated cells completely regenerates a new polypide, and the brown bodies are either retained by the new polypide or defecated by it. Box 15.1 explores several interesting dimensions to brown body formation.

Class Phylactolaemata

The representatives of this class all live in fresh water and are considered by many to be primitive. They lack a high degree of species diversity in that only about 50 species of freshwater bryozoans are known, and only 21 species have been reported from North America (two of which are gymnolaemates). However, they are widely distributed and live in almost every lake, pond, and stream. Lentic habitats, shallow water, and slightly alkaline water appear to be preferred conditions. Most of their special adaptations are correlated with life in fresh water. The lophophore is uniquely horseshoe shaped and doubled so that the tentacles are present in two series. Comparatively speaking, the lophophore in phylactolaemates is larger and bears more tentacles than lophophores in their marine counterparts (Fig. 15.7a). The zooids are mostly cylindrical, the body wall is muscularized and not calcified, and contraction of the circular muscles protrudes the lophophore. Phylactolaemates form neither heterozooids nor brown bodies, since polypide regression is complete, and the polypide does not reform.

Brooding is within an internal sac, and embryonic development culminates in a small active larva that settles and develops into a colony by repeated buddings and bifurcations. Characteristic of the freshwater bryozoans is the asexual production of statoblasts that are environmentally resistant. A **statoblast** is a small cellular mass about 0.7–1.8 mm in diameter. Formed usually as cell aggregates on the funiculus, each statoblast is a mass of undifferentiated germinative cells and stored food reserves

BOX 15.1 □ WHAT ARE BRYOZOAN BROWN BODIES ALL ABOUT?

One of the most characteristic features of marine bryozoans is the cyclic phenomenon of polypide regression and subsequent renewal (Fig. 15.6). It is now clear that brown bodies are the products of polypide deterioration and are neither ova, statoblasts, embryos, incipient larvae, nor excretory wastes as once thought. What are the factors that induce brown body formation? The principal reasons normally given for their inception are water quality (low O_2, extremes of pH and salinity, other chemicals), reproductive activity, and an accumulation of waste materials because of the absence of excretory organs. Recently, however, it has been shown that brown body formation is a very natural phenomenon that may be the final product of aging in bryozoans.

In the bryozoans that have been studied carefully it has been shown that the longevity of polypides is 6–72 days. One of the first signs of regression is the appearance of orange-brown inclusions in the stomach wall. Then the tentacles begin to regress from their tips, and as regression proceeds to the lophophore, significant degeneration, phagocytosis, and cellular death occur. Regression and deterioration of muscles, gut, other visceral organs, and peritoneum soon follow, resulting in one or two brown bodies in a given cystid. These bodies are then defecated by the reformed polypide or left *in situ* until the entire zooid disintegrates later (Fig. 15.6). Remarkably, however, a few viable cells remain, and they aggregate, begin to grow, and shortly restore a complete differentiated polypide to the still-intact cystid. Therefore the new polypide is completely rejuvenated following the almost total regression of the old.

Brown body formation is unique to bryozoans, but the qualitative changes that go on during polypide regression are remarkably similar to those in cells of vertebrates that are regressing or even aging. Electron microscopy and histochemical studies have documented a number of similar characteristic alterations in most cellular organelles in both bryozoans and vertebrates. What happens as these necrotic brownish bodies are formed is really a generalized tissue response that turns out to be rejuvenatory to the bryozoan zooid. Could it be that brown bodies are bryozoans' response to cellular senescence, and, like the mythological Phoenix, a new polypide arises from the "ashes"? If other animals could form brown bodies, could they perhaps be immortal?

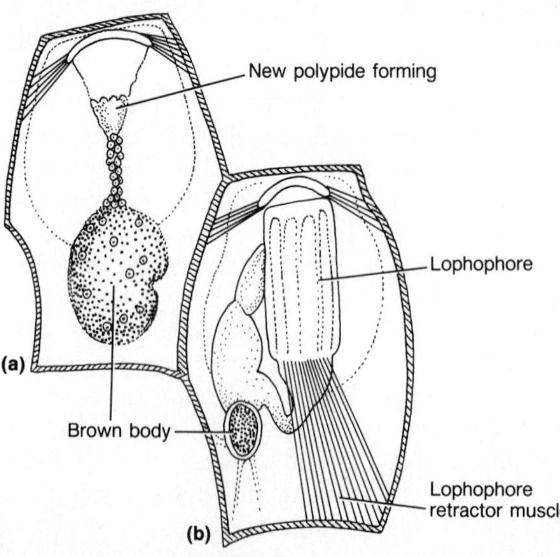

Figure 15.6 *A brown body in a zooid of* Electra: ***(a)*** *zooid is degenerating and a new polypide is forming;* ***(b)*** *the brown body is being taken into the stomach of a new polypide (parts a, b, from Marcus in L.H. Hyman,* The Invertebrates, *1959, Vol. V:* Smaller Coelomate Groups, *McGraw-Hill Book Co., used with permission).*

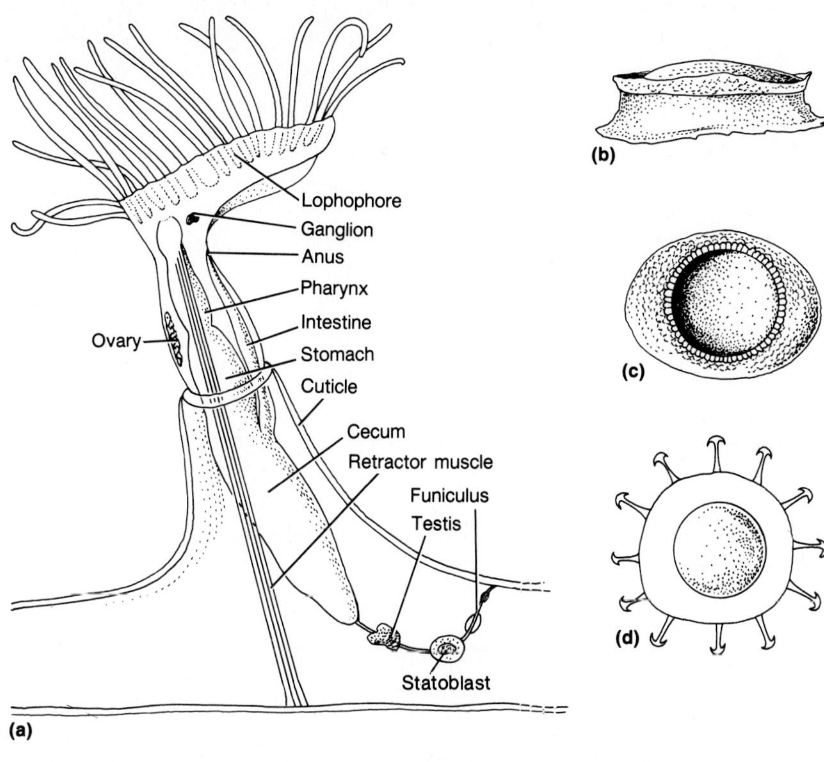

Figure 15.7 Some features of phylactolaemates: ***(a)*** *the structure of a zooid of* Plumatella *(from R.W. Pennak,* Fresh-water Invertebrates of the United States, *2nd ed., © 1978, John Wiley & Sons, Inc., used with permission);* ***(b)*** *an attached statoblast;* ***(c)*** *a floating statoblast;* ***(d)*** *a statoblast with hooks. (parts b, c [from Rogick], d [from Kraepelin] in L.H. Hyman,* The Invertebrates, *1959, Vol. V:* Smaller Coelomate Groups, *McGraw-Hill Book Co., used with permission.)*

and is surrounded by two sclerotized or chitinous nonliving valves (Fig. 15.7b–d). The outer chitinous layer often forms spines or hooks in some species, and the peripheral portion (annulus) frequently consists of air cells, especially in statoblasts that float (Fig. 15.7c, d). Numbers of statoblasts formed by an individual zooid range from one in *Pectinatella* to a dozen or more in *Plumatella,* but two to eight is a more common number. In some species, statoblasts develop only in the autumn, but in others they are formed continuously during the summer and autumn and are either expelled into the water or retained by the parent zooid. With the onset of colder temperatures the zooids die and disintegrate, releasing their statoblasts; some statoblasts float, some sink to the bottom, and others are attached to a surface (Fig. 15.7b–d). With the return of favorable vernal temperatures the valves separate, and the germinal mass expands to produce a single polypide. By repeated budding, the original mass of cells forms a large colony. Because of the ease with which statoblasts are transported, most freshwater bryozoans are cosmopolitan in their distribution and habitats.

Class Stenolaemata

This group contains a very large number of extinct (fossil) forms and perhaps 900 living species, all of which are marine. Their autozooids are cylindrical, and there is a terminal circular orifice for lophophore protrusion. The cystid is cellular, contains some calcification, but lacks muscles. Surrounding the entire polypide is a **membranous sac,** a unique feature of this group located inside the body wall and surrounding

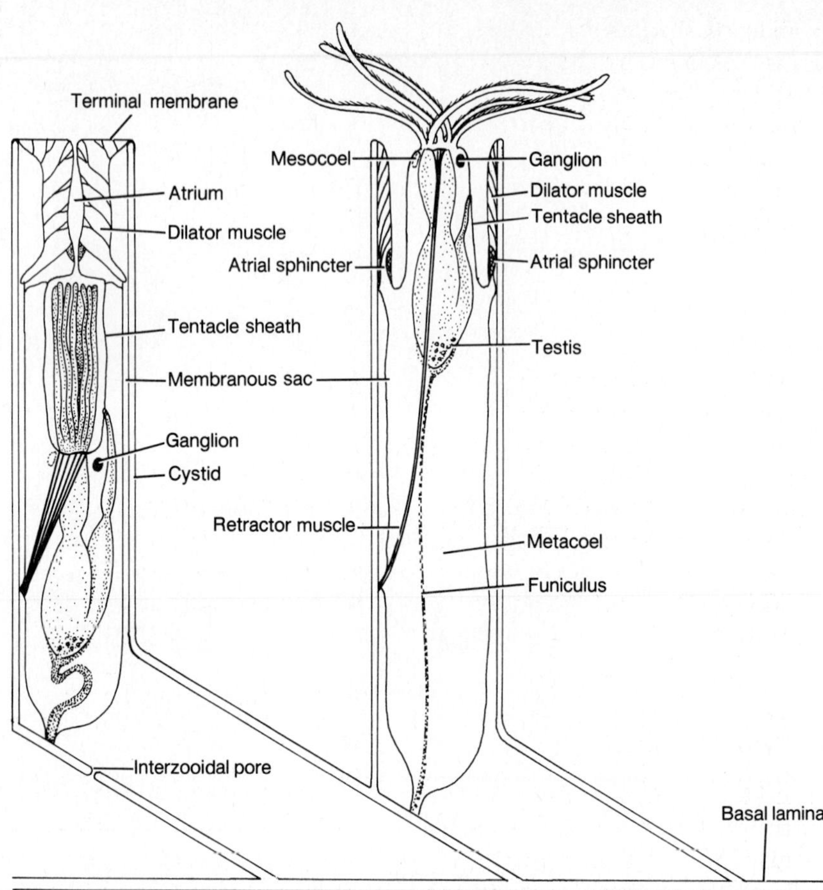

Figure 15.8 *Diagrams of some zooidal features of a retracted (left) and protracted (right) lophophore of a stenolaemate (from J.S. Ryland, 1970,* Bryozoans, *Hutchinson Publishing Group, used with permission).*

part of the coelom (Fig. 15.8); its true function is still obscure. By dilation of the atrium the coelomic fluid is forced proximally, causing the lophophore complex to be protracted hydrostatically. Although avicularia and vibracula are absent, some heterozooids often are produced.

Reproduction in stenolaemates is remarkable, since they undergo a form of **polyembryony** within a special reproductive zooid; a primary embryo buds off secondary embryos, which may in turn form tertiary embryos, resulting in a hundred or more embryos. Each separate embryo then develops into a larva that is free swimming for a short period before it settles.

Class Gymnolaemata

This taxon includes the vast majority of all living bryozoans. They are, with a few exceptions, marine and are numerically successful. Most of the foregoing general discussion has been made with gymnolaemates in mind, and only a quick review of their most important features will be made.

The lophophore is circular and bears fewer tentacles, on the average, than that of freshwater species. The lophophoric complex is hydraulically protruded by muscles that insert into the body wall, which itself lacks intrinsic muscles and often is calcified. Cystids of adjacent zooids bear pores for interzooidal communications, but the pores are plugged with tissues so that free coelomic circulation is not possible. Most are hermaphroditic, and fertilization is in the sea or, more commonly, in the coelom or brood chamber. A typical larval stage is produced.

This class can be divided easily into two orders: Ctenostomata and Cheilostomata. The ctenostomatids have noncalcified cystids that are membranous or gelatinous. The orifice is usually terminal and often closed by a pleated **collar.** Heterozooids are found in the ctenostomatids, but avicularia, vibracula, and ooecia are absent. The cheilostomatids include three-fourths of all bryozoan species. Their cystids generally are shaped like a flattened box with thick calcified walls. Polymorphism is especially pronounced with avicularia, vibracula, and ovicells present.

PHYLUM ENTOPROCTA

The entoprocts constitute a group of about 90 species of very small (5 mm or less) sessile organisms that are similar in appearance to bryozoans. Except for a single small genus (*Urnatella* [Fig. 15.9b]), all entoprocts are marine, and they usually live in shallow water either attached to various inanimate objects or as commensals on various invertebrates. Members of one family are solitary, while the others are colonial (Fig. 15.9a, b). Long considered to be allied with the pseudocoelomate phyla, a close phylogenetic relationship between entoprocts and bryozoans has been strongly suggested recently but not universally accepted. Some authorities even consider entoprocts as a bryozoan subphylum, but we shall still treat them as a separate but related phylum.

The body of an entoproct consists of a stalk bearing the central calyx. The slender **stalk** is commonly composed of short segments, often bears muscular enlargements, and may arise from a horizontal stolon or be anchored singly into the substratum (Fig. 15.9a, b). The elliptical **calyx,** housing the internal organs and supporting the lophophore, is attached to the stalk by its dorsal surface, and therefore its ventral surface is uppermost. The **lophophore** is oval or elliptical in outline and bears 8–30 hollow ciliated tentacles, which are simple outgrowths of the body wall (Fig. 15.9c). Entoprocts are filter-feeders that trap microplankters in mucus secreted by the tentacles; the mucus–food complex is swept by ciliary action to the mouth. That area surrounded by the tentacles is the **vestibule** or **atrium** that bears the **mouth** at the anterior end and the **anus** at the posterior end. The position of the anus within the tentacular circle is the basis for their name (ento = inside; proct = anus). As the lophophore complex is retracted inward into the vestibule, a membrane attached to the lophophore base is pulled medially to protect the lophophore.

The entoprocts do not form a calcareous exoskeleton as in many bryozoans, but a thin protective cuticle is secreted by the underlying epidermis. Longitudinal muscle strands are present only in certain areas of the body wall such as those of tentacles and certain parts of the calyx. Internally, there is a **pseudocoelom** that surrounds the viscera and is filled with mesenchymal cells and fluid (Fig. 15.9c). Then why aren't these creatures grouped with the other animals with a pseudocoelom (see Chapter 7)? As we shall see below, there are other features that more strongly suggest that the entoprocts are more closely related to bryozoans than they are to rotifers and nematodes. The digestive tract is U-shaped with an anterior mouth, an esophagus, a large bulbous stomach, an intestine, a rectum, and a posterior anus. Two **protonephridia,** each equipped with a single flame bulb, eliminate waste from the pseudocoelom and conduct them to a single **nephridiopore** near the mouth. There are no organs for circulation or gaseous exchange. The nervous system consists of a large median **ganglion** located dorsal to the vestibule. From the ganglion arise paired nerves that innervate the stalk, tentacles, lophophore, and calyx walls. Bristles on the tentacles and calyx margins are the principal sensory structures.

Like bryozoans, entoprocts are quite varied in their reproductive patterns. Asexually formed buds are common among entoprocts; new colonies or individuals are readily produced by budding. The two gonads present in dioecious forms or the two pairs of gonads in monoecious individuals are located between the vestibule and stomach. Gonadal ducts merge to a single **gonopore** located just posterior to the nephridiopore. Ova are fertilized in the ovary, and early developmental stages are brooded in a special part of the vestibule. Cleavage is spiral and determinate, and the embryo hatches into a larval stage. This trochophorelike stage is characterized by an anterior apical tuft, frontal organ, girdle of cilia around the ventral margin, and ciliated foot (Fig. 15.9d). After a brief free-

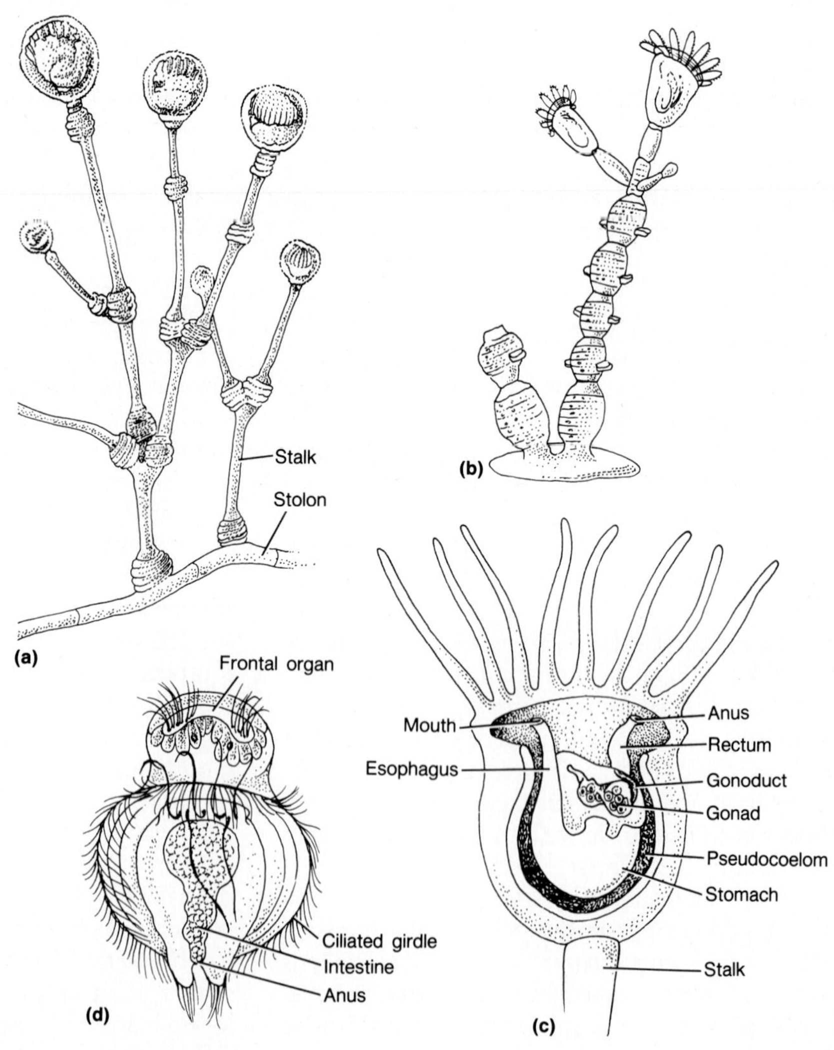

Figure 15.9 *The entoprocts: **(a)** portion of a colony of* Gonypodaria; ***(b)** portion of a colony of* Urnatella, *a freshwater entoproct (from R.W. Pennak,* Fresh-water Invertebrates of the United States, *2nd ed.,* © *1978, John Wiley & Sons, Inc., used with permission); **(c)** diagrammatic section through the calyx of a typical entoproct; **(d)** larva. (parts a [from Robertson], d [from Atkins] in L.H. Hyman,* The Invertebrates. *Vol. III.* Acanthocephala, Aschelminthes and Entoprocta, *McGraw-Hill Book Co., 1951, used with permission.)*

swimming existence, the larva settles, attaches by its frontal organ, and undergoes a metamorphosis to form an adult entoproct.

PHYLOGENY OF BRYOZOANS AND ENTOPROCTS

The exact phylogenetic positions of the bryozoans and entoprocts are still not known. There have been and still are major important questions regarding the affinities of these two groups and the possible relationships between them and the phoronids and brachiopods. All four phyla have been variously grouped together in the past, primarily on the basis of the common appearance of the lophophore and, at the same time, minimizing the importance of certain basic embryological differences. There is a growing conviction among specialists that the lophophore in bryozoans, entoprocts, phoronids, and brachiopods represents similar adaptations to a sessile, filter-feeding life and thus is a prime example of convergent evolution among nonrelated groups. Rather, what the bases of phylogenetic affinities should be are the embryological similarities that generally reflect ancestral linkages. If we examine the embryological details of bryozoans and entoprocts, they appear to be similar to each other and generally like other protostomates in patterns of cleavage, methods of coelom formation (bryozoans), and the development of a trochophorelike larva. Conversely, because the embryology of phoronids and brachiopods is decidedly more like that of the deuterostomates, they are not treated as protostomates.

While there are few definitive facts on their ancestral origins, bryozoans and entoprocts probably arose from an ancient, bilaterally symmetrical, marine, protostomate worm (Fig. 15.10). The development of a coelom in bryozoans and a pseudocoelom in entoprocts is an especially vexing problem that cannot be resolved at the present time. But the other similarities between bryozoans and entoprocts are compelling evidence; they include similarities in life cycles, larval stages, metamorphosis, various means of asexual reproduction, and the adult ganglionic mass. Presumably, this ancestral wormlike bryozoan underwent adaptive radiation to form three groups of bryozoans and the entoprocts. But the entire question of bryozoan–entoproct phylogeny is still largely unresolved.

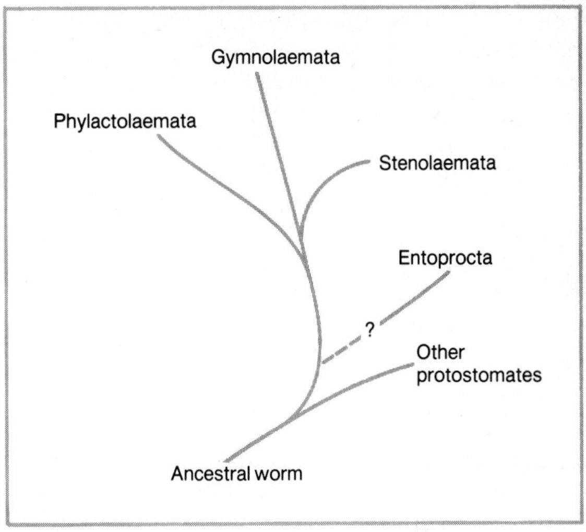

Figure 15.10 *Probable phylogenetic affinities and pathways in the bryozoans and entoprocts.*

SYSTEMATIC RESUME

The Principal Taxa of Bryozoans

Class Phylactolaemata—exclusively freshwater; zooids mostly cylindrical with noncalcareous muscular body wall; have a horseshoe-shaped lophophore; coelom continuous between zooids; produce statoblasts; *Lophopus*, *Plumatella* (Fig. 15.7a), and *Pectinatella* are representative.

Class Stenolaemata—all marine; cylindrical zooids whose body wall is calcified and contains no muscles;

coelom sometimes continuous between zooids via pores.

Order Cyclostomata—only extant order; orifice is circular; adjacent zooids connected by open pores; asexual reproduction involves polyembryony; includes *Crisia, Tubulipora,* and *Stomatopora.*

Class Gymnolaemata—mostly marine; zooids cylindrical or squat and polymorphic; lophophore is circular; body wall either soft or calcified but without muscle fibers; coeloms of adjacent zooids are separated by cellular plugs.

Order Ctenostomata—zooids cylindrical to flat; walls membranous, not calcified; no ooecia, vibracula, or avicularia; orifice closed by pleated collar; typical representatives are *Nolella, Alcyonidium,* and *Paludicella.*

Order Cheilostomata—zooids box-shaped and flat; walls calcified; orifice closed by hinged operculum; avicularia and vibracula commonly present; embryos develop in ovicells; contains a majority of all extant species including those of *Membranipora, Electra* (Fig. 15.2d), *Bugula* (Fig. 15.2b, c), and *Callopora.*

ADDITIONAL READINGS

Hayward, P.J. and J.S. Ryland. 1979. *British Ascophoran Bryozoans,* 306 pp. New York: Academic Press.

Hyman, L.H. *The Invertebrates.* Vol. III: pp. 521–553 (Entoprocts), 1951. Vol. V: pp. 275–501 (Bryozoans), 1959. New York: McGraw Hill Book Co.

Larwood, G.P., ed. 1973. *Living and Fossil Bryozoa,* 606 pp. New York: Academic Press.

Larwood, G.P. and M.B. Abbott, eds. 1979. *Advances in Bryozoology,* 619 pp. Special Vol. No. 13, The Systematics Association. New York: Academic Press.

Larwood, G.P. and C. Nielsen, eds. 1981. *Recent and Fossil Bryozoa,* 334 pp. Fredensborg, Denmark: Olsen and Olsen.

Morris, R.H., D.P. Abbott, and E.C. Haderlie. 1980. *Intertidal Invertebrates of California,* pp. 91–107. Stanford, Calif.: Stanford University Press.

Pennak, R.W. 1978. *Fresh-water Invertebrates of the United States,* 2nd ed., pp. 254–274. New York: John Wiley & Sons.

Ryland, J.S. 1970. *Bryozoans,* 168 pp. London: Hutchinson University Library.

Symposium on the Biology of Lophophorates (14 authors). 1977. *Amer. Zool. 17:*1–150.

Woollacott, R.M. and R.L. Zimmer. 1977. *Biology of Bryozoans,* 541 pp. New York: Academic Press.

Yoshioka, P.M. 1982. Role of planktonic and benthic factors in the population dynamics of the bryozoan *Membranipora membranacea. Ecology 63:*457–468.

16

Some Minor Phyla

OVERVIEW

There are eight minor phyla, each of which has relatively few species and is not closely related to another larger group. These phyla are treated as protostomates or deuterostomates, primarily on the basis of details of their cleavage, early development, larval stages, and ways in which the coelom is formed.

The minor protostomate phyla are the Priapulida, Tardigrada, and Pentastomida. Priapulids are predatory, benthic, marine worms that have a proboscis, trunk, and paired urogenital organs. Both sexes spawn gametes into the sea. By repeated molts, the larva, surrounded by a lorica, develops into an adult. Tardigrades, found mostly in aqueous films on terrestrial mosses, have a stout body, four pairs of legs, feeding stylets, and excretory Malpighian tubules; they are dioecious and have internal fertilization but no larval stage. As endoparasites in carnivorous vertebrates, pentastomids have two pairs of clawed holdfast legs, a hemocoel, and a muscular pharynx. Larvae live in intermediate hosts, have three pairs of jointed legs, and periodically molt their cuticle.

The minor deuterostomate phyla include the Phoronida, Brachiopoda, Chaetognatha, Hemichordata, and the nonvertebrate Chordata. A phoronid lives in a chitinous tube and has a food-gathering lophophore, a closed circulatory system, and a pair of metanephridia. Phoronids are monoecious and develop a planktonic larva that later becomes a benthic adult. Brachiopods are marine and benthic and have a hard outer dorsal and ventral valve secreted by corresponding mantle folds. A complex food-gathering lophophore is situated in the anterior mantle cavity. Brachiopods, being dioecious, spawn gametes, and a free-swimming, feeding, bivalved larva is formed. Chaetognaths are marine, planktonic, very streamlined, and strongly predaceous; they are fertilized internally, but their embryos develop in the sea. Hemichordates are benthic marine worms with a proboscis, collar, and trunk. They have pharyngeal slits, an open circulatory system, and what might be a dorsal nerve cord; their tornaria larva is very similar to the echinoderm larval stage. Chordates are characterized by having pharyngeal slits, a dorsal nerve cord, a skeletal notochord, and a postanal tail. The urochordates secrete an enveloping tunic around themselves; ascidians, the most prevalent urochordates, are mostly sessile and have prominent siphons and a large peripharyngeal atrium for water flow. Adult cephalochordates have a dorsal nerve tube and a notochord and are metameric, as is exemplified by their muscles and nerves. Additionally, cephalochordates are benthic filter-feeders and have an unusual digestive cecum.

Each of these eight phyla has had a very long and separate evolutionary history. Phoronids and brachiopods have similar developmental details and a lophophore. The hemichordates, urochordates, and cephalochordates are also closely related to each other in their embryology and in the presence of pharyngeal slits.

INTRODUCTION

This chapter might just as easily be entitled "miscellaneous invertebrates" or even "odds and ends." About two dozen phyla are considered to be minor numerically, since each has fewer than 2000 species. We have already mentioned a number of these groups—Mesozoa, Sipuncula, Gnathostomulida, and Onychophora, to mention only a few—but each of these is related phylogenetically to another larger phylum, and both were discussed together. Remaining, therefore, are eight small groups that are usually considered to be not closely related to any other taxon. All have a coelom and bilateral symmetry, they are generally marine and lack metamerism, and most are not frequently encountered by novices. They have little else in common.

The eight phyla are separated into two categories, the protostomates and deuterostomates; the phyla in either supergroup share some very basic evolutionary and embryological characteristics (see Chapter 3), even though representatives have very few similarities as adults. The protostomate phyla are the Priapulida, Tardigrada, and Pentastomida. The deuterostomate phyla are the Phoronida, Brachiopoda, Chaetognatha, Hemichordata, and Chordata. Immediately, one might properly challenge the statement that the Chordata is a minor phylum; to be sure, it is not, since it includes about 36,500 species of vertebrates. But we are not concerned with the vertebrates, and the nonvertebrate chordates are a minor group. Each phylum, briefly characterized in Table 16.1, has been variously classified in the past, and their grouping here is solely one of convenience.

THE MINOR PROTOSTOMATE PHYLA

The priapulids, tardigrades, and pentastomids have very little in common. They are found in entirely different habitats, and the functional morphology of each is fundamentally different from that of the other two.

Phylum Priapulida

Priapulids are benthic marine worms that are rather widely distributed in temperate and polar regions. There are only nine known species, and most live buried in mud and sand from intertidal zones to water depths of 3000 m. One species lives in coral sediments, and another is tubicolous. Elongate, wormlike, and ranging in length from 0.05–21 cm, priapulids are not conspicuously colored and usually are solitary, but one report indicated a priapulid density of 250 per square meter! The most commonly encountered species is *Priapulus caudatus* (Fig. 16.1).

The elongated body is cylindrical in cross section, is shaped somewhat like a cucumber, and consists of a proboscis and trunk (Fig. 16.1). The anteriormost **proboscis** or **introvert** constitutes about one-third of the total length. Often

TABLE 16.1. □ **THE MOST IMPORTANT CHARACTERISTICS OF THE EIGHT UNRELATED MINOR PHYLA**

Phylum	*Principal Habitat*	*Feeding Adaptations*	*Reproductive Strategies*	*Other Features*
Protostomates				
Priapulida	Marine, benthic	Eversible pharynx; spines	Dioecious; benthic larva	Body of proboscis and trunk; urogenital organs and ducts
Tardigrada	Watery films on terrestrial mosses	Stylets for procuring tissue juices	Dioecious; internal fertilization; some parthenogenesis	4 pairs of stubby clawed legs; hemocoel; Malpighian tubules
Pentastomida	Nasal passages of vertebrates	Clawed legs; snout; pumping esophagus	Dioecious; larva in intermediate host	Hemocoel; ventral nerve cord with 3 pairs of ganglia
Deuterostomates				
Phoronida	Marine, benthic	Lophophore for filter feeding	Monoecious; actinotroch larva	Secrete a chitinous tube; have metanephridia
Brachiopoda	Marine, benthic	Complex filter-feeding lophophore with 2 brachia	Mostly dioecious; some have larval stage	Dorsal and ventral valves secreted by mantle lobes
Chaetognatha	Marine, planktonic	Predatory spines	Monoecious; internal fertilization	Fusiform body; lateral and caudal fins; no peritoneum
Hemichordata	Marine, benthic	Ciliated proboscis	Dioecious; produces tornaria larva	Body of proboscis, collar, trunk; open circulatory system
Chordata (nonvertebrates)	Marine, benthic	Ciliated pharynx with stigmata	Produces tailed larva	Notochord and dorsal hollow nerve cord

with a diameter greater than that of the rest of the body, the proboscis has numerous longitudinal rows of small spinelike **scalids** on its surface. The mouth is located at the anterior end of the proboscis and usually is surrounded by several concentric circumoral rings of spines that are used in food capture. While the proboscis can be fully retracted into the trunk, it normally is partially extended. The **trunk** is demarcated from the proboscis by a shallow constriction, and its surface contains irregularly spaced spines and tubercles that give it a warty appearance. The trunk is also annulated with 30–120 superficial rings that clearly do not suggest segmentation. At the posterior end the trunk bears the anus, a pair of nephridiopores, and sometimes hollow **caudal appendages** or a **tail** (Fig. 16.1). Although the functions of the caudal appendages are not fully known, they may play a role in gaseous exchange and chemoreception.

The body wall consists of a thin outer cuticle, an underlying epidermis, circular and longitudinal muscle layers, and a thin cellular peritoneum. The nonliving cuticle is periodically molted, and the epidermis secretes another cu-

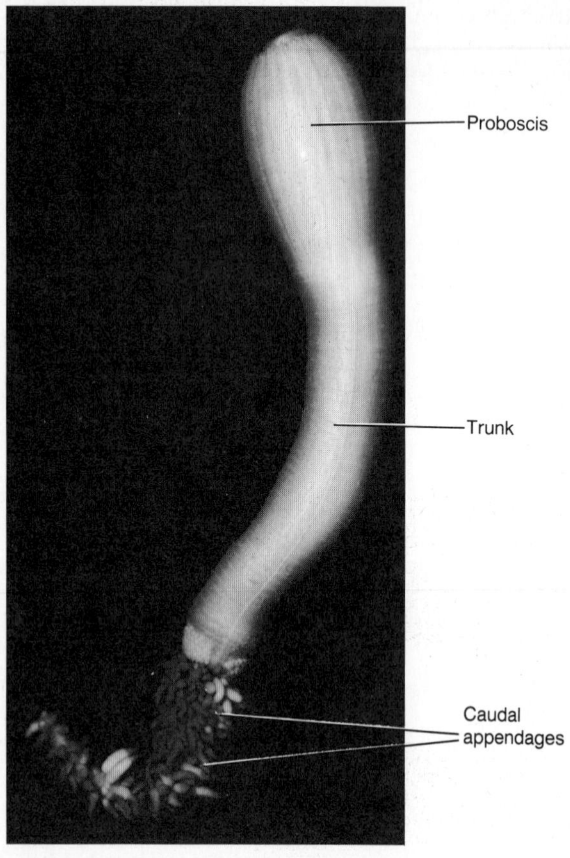

Figure 16.1 *Photograph of* Priapulus caudatus, *a priapulid (courtesy of C.B. Calloway).*

ticular layer inside of the old prior to ecdysis. Alternate contractions of the circular and longitudinal muscles facilitate burrowing. Inside the body wall is a coelom that contains fluid, coelomocytes, and—at least in *Priapulus caudatus*—hemerythrin-containing corpuscles. The coelom extends into the caudal appendages and tail when present.

Priapulids are predatory mostly on polychaetes, but they may eat other softbodied animals like nemerteans and other priapulids, which are seized by the oral spines and ingested whole. The alimentary tract consists of a muscular pharynx, a tubular intestine, and a short rectum held in place by a peritoneal mesentery. Lacking are definite gas exchange and circulatory organs. There is a circumpharyngeal nerve ring; an unpaired, ventral, ganglionated nerve cord; and an extensive epidermal plexus. These are the principal parts of the nervous system.

Within the coelom is a pair of elongated **urogenital organs,** each of which is a urogenital duct that bears a **gonad** on one side and clusters of **solenocytes** on the other. This system is probably a protonephridial tubule bearing solenocytes. The gonads have capitalized on the conveniently placed ducts so that the tubules also function as gonoducts. Each urogenital duct opens to the outside as a posterior nephridiopore. Priapulids are dioecious, and sperm shed into the sea apparently act as a stimulus for the spawning of ova, which are subsequently fertilized externally. Cleavage is apparently radial in some priapulids rather than spiral as in most other protostomates, but many details of cleavage are still not known; development soon results in a benthic larval stage. It is not known whether the coelom is a schizocoelom or an enterocoelom. Each larva is surrounded by a cuticular **lorica** formed by the posterior larval body. By repeated moltings the larva grows, and in about two years it emerges as a juvenile with most adult structures but without a lorica. Priapulids molt throughout their juvenile and adult existence.

The phylogenetic affinities that priapulids might have with other groups are extremely obscure. There are certain aschelminth characteristics like the cuticle, scalids, and radial symmetry around the mouth. But the true coelom, the cellular peritoneum, and the body wall construction argue strongly for more annelidlike affinities. Perhaps the most critical areas, and those about which little is known, are the patterns of cleavage and the method of coelom formation. Definitive embryological information will aid immeasurably in establishing more accurate affinities for the priapulids.

Phylum Tardigrada

Tardigrades are among the most curious of all invertebrates, even though they are mostly no longer than 0.3–0.5 mm. Commonly called water bears because of their body shape and legs, tardigrades live in marine interstitial areas, in detritus in freshwater habitats, and, most commonly, in watery films on the above-ground parts of terrestrial mosses, lichens, and liverworts. Many of the approximately 400 described species have been found in all latitudes from equatorial to polar regions. They are not at all well known because of their extremely small size, but some of the more common genera are *Echiniscus* (Fig. 16.2a), *Macrobiotus*, and *Echiniscoides*.

The body of a water bear is elongated but plump, cylindrical in cross section, and unsegmented, and it bears four pairs of stubby legs. The head, simply a continuation of the trunk, carries the terminal mouth and often bears a pair of simple eyespots. The four short, nonjointed **legs** on the either side extended ventrolaterally from the body; each terminates in four single or two pairs of claws, which are used for climbing or clinging to a surface (Fig. 16.2a). The anus is at the posterior end.

The entire body is covered by a sculptured or smooth cuticle composed of protein but no chitin. The thin cuticle is secreted by an eutelic epidermis (look back to Box 7.1). Periodically, a new cuticle is secreted by the underlying epidermis, and the old cuticle is shed, often mostly intact. Beneath the epidermis are separate muscle strands, not muscular layers, that extend from one subcuticular point to another, especially from trunk to legs, to facilitate movement of the legs and body. Beneath the muscle strands is an ill-defined **hemocoel,** a space in which the viscera lie and through which some of the muscle strands extend. The coelom is confined to an area within the single gonad.

At the anterior end of the alimentary tract is the buccal cavity with an anterior **stylet apparatus** that is similar to that in herbivorous nematodes (Fig. 16.2b). Two pointed stylets project anteriorly toward the mouth, and each is secreted by a large gland. The stylets are supported by several hardened struts and operated by muscles. The entire mechanism is an efficient piercing device (Fig. 16.2b). Tardigrades use the apparatus to pierce plant cell walls, and then they suck out the cell contents by using the muscular pharynx as a pump. Some tardigrades impale small nematodes with the stylets and suck their juices. The pharynx, triradiate in cross section as in most pseudocoelomates, opens into a tubular esophagus. The esophagus is continuous as a large midgut where absorption takes place, and the remainder of the digestive tract is a short rectum. At the junction of the midgut and rectum of some tardigrades are three organs sometimes called **Malpighian glands** and thought to be excretory in function; the reader will recall that insects and some arachnids also have Malpighian tubules. Tardigrades lack cilia and organs for gaseous exchange, osmoregulation, and circulation. Their nervous system is typically like that of protostomates with a multilobed suprapharyngeal brain joined to a subpharyngeal ganglion by connectives. Arising from the subpharyngeal ganglion is a double ventral nerve cord bearing four prominent ganglia in the trunk. (Are these beasts then metameric?) Sensations are received through the eyespots, sensory bristles, and spines located especially on the head.

All tardigrades are dioecious, and the single gonad is saclike and located dorsal to the intestine. A single oviduct or two sperm ducts extend posteriorly and open to the exterior as one or two gonopores. Females are usually more common than males, and since males have not

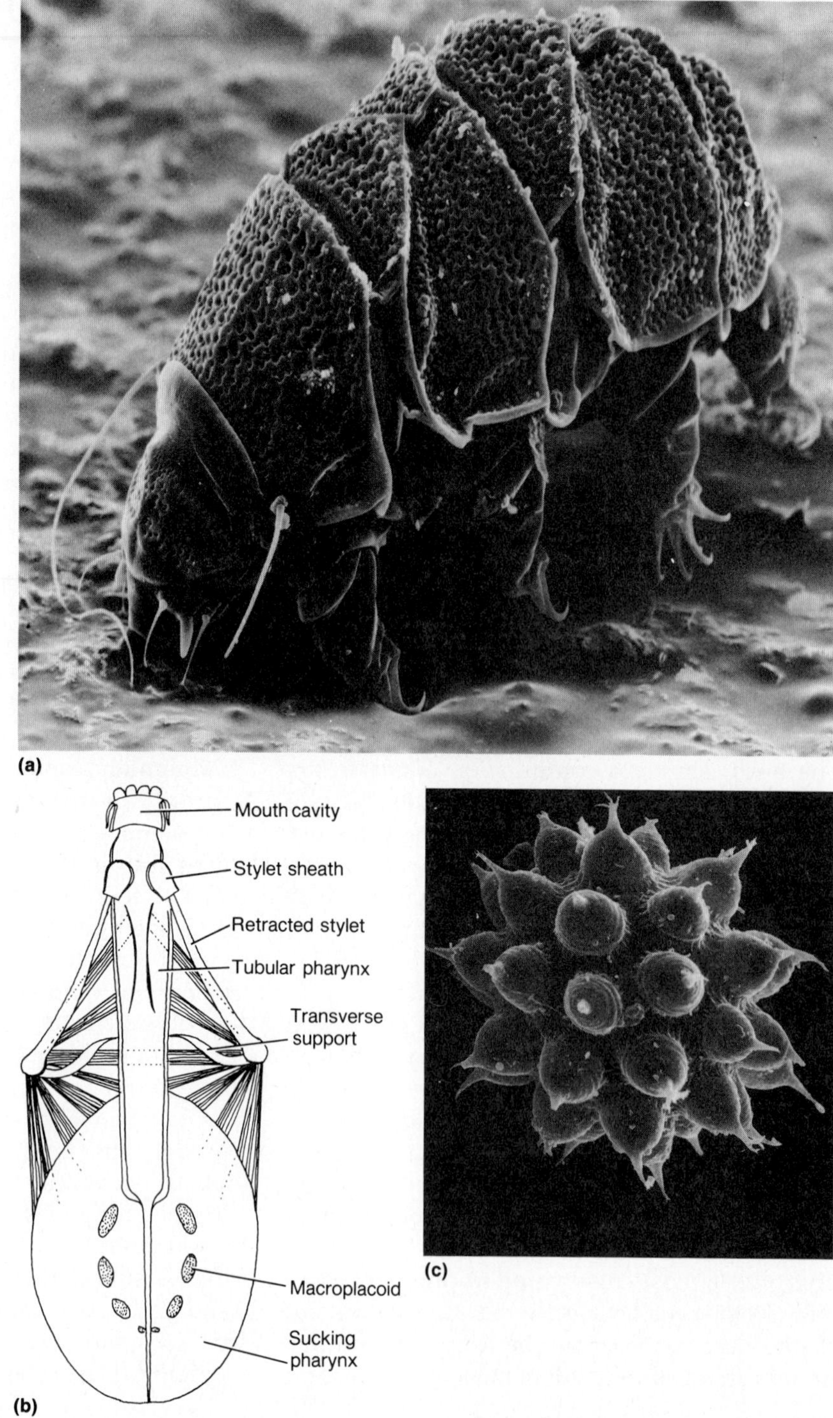

Figure 16.2 *Tardigrades:* ***(a)*** *scanning electron micrograph of* Echiniscus mauccii; ***(b)*** *the stylet apparatus (from R.W. Pennak,* Fresh-water Invertebrates of the United States, *2nd ed., © 1978, John Wiley & Sons, Inc., used with permission);* ***(c)*** *scanning electron micrograph of an egg of* Macrobiotus harmsworthi. *(parts a, c courtesy of D.R. Nelson.)*

been found in some species, parthenogenesis is common. Sperm are transferred internally to the female through either her anus or her gonopore, usually at time of a molt, and fertilization is within the ovary. From one to several dozen fertilized eggs are laid by the female (Fig. 16.2c). Terrestrial species all produce eggs with thick shells that are able to withstand the rigors of the environment. But many of the freshwater tardigrades produce two types of eggs; when conditions are optimal, they produce thin-shelled eggs that develop rapidly, and when adverse conditions are imminent, thick-shelled resistant eggs are formed. A similar condition exists in many rotifers, which produce both thin- and thick-shelled eggs, depending upon environmental conditions. Interestingly, an enterocoelom is formed as five pairs of pouches, but the walls of the four anterior pairs form muscles as their spaces disappear, and the fifth pair persists inside the gonad. Within about two weeks the juvenile hatches from its egg case using the stylets, and it continues to molt even as an adult to accommodate its growth. Tardigrades, along with rotifers and many nematodes, are able to enter into a period of suspended animation when environmental conditions are not conducive to normal existence. This provocative phenomenon, called **cryptobiosis,** is probed in Box 16.1.

The relationship between tardigrades and other groups is most uncertain. Water bears possess several important aschelminth features such as a triradiate pharynx, eutely, parthenogenesis, and variable egg types. But they also exhibit several classical features such as the nervous system and the ontogeny of multiple pairs (metameric?) of coelomic pouches that definitely relate them to annelids and arthropods; further, their cuticle is very much like that of arachnids. But to make matters even more confusing, the coelomic pouches develop as outpockets from the gut—a most distinctive deuterostomate feature. Which set of these features is most important? By virtue of their location in this chapter, I have indicated where I think they belong (see Fig. 16.13)—but I do so with considerable temerity.

Phylum Pentastomida

The pentastomids, commonly called tongue worms, are all endoparasitic in the lungs or nasal passageways of carnivorous vertebrates like reptiles and sometimes birds and mammals. Tropical lizards are apparently the preferred host for many of the 70 described species. Sometimes pentastomids are up to 15 cm in length, but a more common length is 4–8 cm. Two of the better-known genera are *Linguatula* (Fig. 16.3b) and *Raillietiella.*

The pentastomid body is elongate and wormlike, cylindrical or flattened in cross section, and not differentiated into recognizable regions, and it bears a posterior terminal anus. At the anterior end are five small protuberances, each of which was erroneously thought to bear a mouth and from which their phylum name was coined. Four of these oral protuberances are now termed **legs,** and each bears a retractile hook provided with strong muscles; the hooks are important as holdfast structures. The fifth projection is a **snout** bearing the mouth. The legs grasp the host's tissues with their claws, the mouth is applied to the wound created by the claws, and blood is sucked into the worm. The pentastomid surface is covered by a thick chitinous cuticle that typically gives the entire body an annulated appearance (Fig. 16.3a). The cuticle is molted periodically. Beneath the cuticle-secreting epidermis are circular and longitudinal muscles.

Internally, there is a hemocoel as in arthropods. A coelom is thought to develop but persists only within the gonads. The straight digestive tract is especially simple with a muscular sucking pharynx, midgut, and rectum. Tongue worms lack organs for excretion, gaseous exchange, and circulation; these absences are not unexpected, since pentastomids are endoparasitic. The nervous system is annelidlike or ar-

BOX 16.1 □ CRYPTOBIOSIS: A DEATH-DEFYING ACT?

The human species has had a preoccupation with death ever since our ancestors began to recognize their own mortality. Popular literature is replete with articles on death and dying, the afterlife, and possible ways to circumvent death or at least to prolong life. But several different invertebrate groups have been doing exactly this for millennia. Tardigrades, rotifers, and many nematodes can naturally enter into a period of suspended animation and greatly extend their life span by a hundred times or more. This life-prolonging state is called **cryptobiosis** (= hidden life) and is one of the most remarkable and fascinating phenomena present in the entire animal world.

Most tardigrades, rotifers, and nematodes probably have life spans measured in weeks or months rather than in years. But nematodes have been revived after 39 years of cryptobiosis, and estimates for tardigrades range up to 60 years. Incredibly, a piece of moss was moistened after having been dry for 120 years, and even though they did not live long, some cryptobiotic rotifers and tardigrades on the dry moss actually revived! Tardigrades have resumed normal living after having been subjected to a temperature at the brink of absolute zero (0.008°K), after a radiation dosage of 570,000 roentgens (1000 times the dosage that would have killed 50% of exposed human beings), and after being in a severe vacuum for 48 hours, all while in a state of cryptobiosis. How is it possible that living cells could possibly survive such extreme conditions?

As tardigrades begin to dry, they contract into a barrel shape or tun, a phenomenon that is thought to produce an ordered packing of organs and tissues to minimize mechanical damage caused by continued desiccation. This contraction also slows down the initial rate of desiccation, a feature that is apparently very important to the future recovery rate for tardigrades. Do all metabolic activities cease in a cryptobiotic state? No, apparently an incredibly small amount of oxygen is consumed by the cells and tissues, for if there were a total cessation of all metabolism, then we would certainly need to redefine our concepts of death and life. But in cryptobiosis, some irreversible degradation of cells apparently occurs if water, oxygen, and heat are all present. The absence of one of these three is usually characteristic of cryptobiosis; if two or all three factors are absent, then cryptobiosis is very stable, and the likelihood of revival is greatly increased.

Why do these creatures undergo cryptobiosis? Is it of any adaptive significance to them? Tardigrades and other cryptobiotic animals live in habitats where conditions can suddenly become adverse, and this is their way of suspending animation until the return of favorable conditions. So cryptobiosis, with its great survival significance, is a most important adaptation.

Is it possible that human beings have a set of genes somewhere that, when activated, could permit us to suspend almost all metabolic activities for a prolonged period of time and would later ensure our successful revival? Probably not, but it is intriguing that even such a remote possibility would allow us to prolong life or even be immortal. Think about it. Who knows what startling discoveries in suspended animation lie just around the corner?

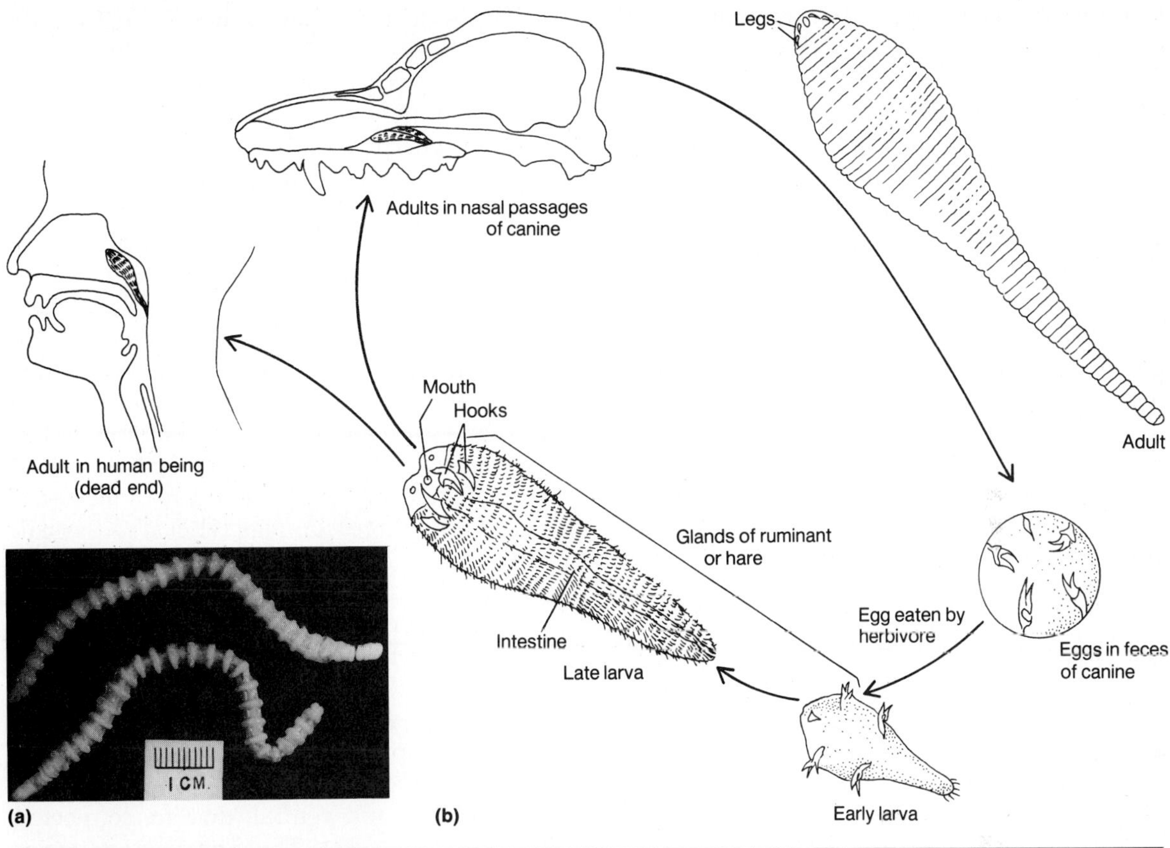

Figure 16.3 *Pentastomes:* ***(a)*** *photograph of two adult pentastomes from a cobra* (Naja) *(courtesy of R.L. Muller);* ***(b)*** *diagram of the life cycle of* Linguatula *(adapted from J.T. Self).*

thropodlike with a brain and a paired ventral nerve cord bearing three pairs of ganglia.

Pentastomids are dioecious, and females are usually much larger than males. Gonads are unpaired, and gonopores are observed in a variety of locations. Following internal fertilization, the shelled eggs are liberated by the female into the alimentary tract of the host and voided with its feces. Members of one genus are known to have direct life cycles, but most pentastomids require an intermediate host in which their larvae can mature. Eggs are eaten by a variety of vertebrates, and the developing larva molts and grows within the lungs, liver, or kidneys of the intermediate host. The larval stage is characterized by having two or three pairs of jointed leglike appendages that are definitely like those of arthropods (Fig. 16.3b). When the intermediate host is eaten by the definitive host, the larvae, freed by the host's digestive processes, migrate through the esophagus to the lungs, trachea, or nasal passages (Fig. 16.3b).

The pentastomids' phylogenetic affiliations are greatly clouded by their many adaptations to a parasitic life. Perhaps they are most closely related to arthropods, as is evidenced by their cuticle and its ecdysis, hemocoel, jointed paired legs in larvae, and lack of cilia. Various ancestral lines within the arthropods have been proposed for a close pentastomid affinity including those of mites, myriapods, and crustaceans. Other biologists hypothesize that they are related more closely to polychaetes or some other annelid group. Most zoologists agree that they are protostomates, but general consensus about their phylogeny stops there.

THE MINOR DEUTEROSTOMATE PHYLA

The phoronids, brachiopods, chaetognaths, hemichordates, and invertebrate chordates are all exclusively marine and, excepting the chaetognaths, are almost all benthic. However, each taxon has evolved a set of characteristics that clearly distinguishes it from the others (see Table 16.1).

Phylum Phoronida

The phoronids, with only about 16 species and belonging to only two genera (*Phoronis* and *Phoronopsis*), are marine and wormlike and live in tubes that are buried in the soft bottom material or attached to various hard substrata. A few phoronids actually bore into calcareous shells and rocks. Phoronids are from 0.5 to 25 cm in length and secrete chitinous tubes in which they live permanently.

At its posterior end, the phoronid body is dilated somewhat to anchor the worm in the tube, and at the anterior end is a lophophore. The phoronid **lophophore** is horseshoe shaped with the mouth inside of two parallel ridges or rami. The bend of the crescent lophophore marks the ventral surface, and the two lophophoric rami, projecting dorsally, may be spiraled. Each of the two ridges supports a large number of ciliated tentacles, which are involved in food gathering (Fig. 16.4a, b). Particulate food, swept toward the groove formed by the two lophophoric ridges, is trapped by mucus and carried to the mouth. The mouth is covered by the flaplike **epistome** and flanked by a pair of nephridiopores (Fig. 16.4b).

The body wall is delicate and consists of the outer epidermis, a thin layer of circular muscles, a thick layer of powerful longitudinal muscles, and the parietal peritoneum. Contraction of the circular muscles extends the lophophore hydrodynamically, and that of the longitudinal muscles enables the animal to withdraw the lophophore quickly. The tripartite coelom consists of a **metacoel** surrounding the viscera, a **mesocoel** in the lophophore complex, and a very small **protocoel** in the epistome; the coelom contains coelomocytes. Small planktonic organisms are swept into the mouth and into a U-shaped digestive tract that is regionally specialized into a descending esophagus, stomach, ascending intestine, and ventral anus lying outside of the lophophoric spirals. The principal parts of the closed circulatory system are two contractile vessels, dorsal and ventral, that are connected by capillary networks. The dorsal vessel carries blood anteriorly to the mouth and lophophore, and the ventral vessel transports blood posteriorly to supply the viscera. Hemoglobin-containing blood corpuscles carry respiratory gases to and from the tentacles, where gaseous exchange takes place. A pair of metanephridia directs nitrogenous wastes from the metacoel to the nephridiopores. An epidermal nerve ring is located at the base of the lophophore, and its nerves supply the tentacles, lophophore, and muscles. No specialized sensory organs are present.

Most phoronids are monoecious and produce both sperm and eggs from the peritoneal layers. Sperm exit by way of the metanephridia, and fertilization is internal in about half of the species. In those that have internal fertilization, spermatophores penetrate the body wall and

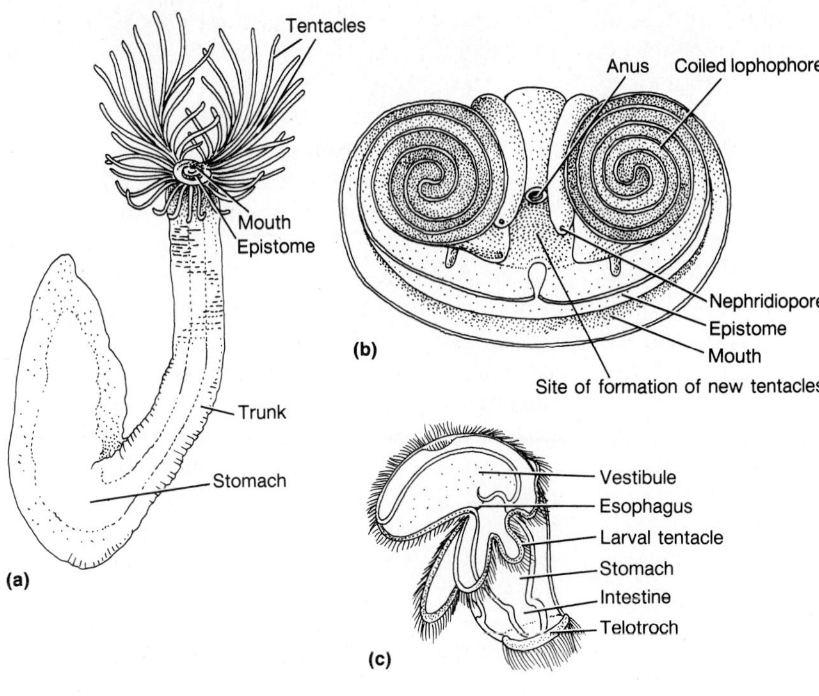

Figure 16.4 *Phoronids:* ***(a)*** Phoronis architecta; ***(b)*** *diagrammatic view looking into the anterior end of the lophophore of a phoronid;* ***(c)*** *an actinotroch larva. (parts a [from Wilson], b [from Benham], c [from Selys Longchamps] in L.H. Hyman,* The Invertebrates, *1959, Vol. V:* Smaller Coelomate Groups, *McGraw-Hill Book Co., used with permission.)*

thus facilitate entry of the sperm into the coelom. Fertilized eggs also exit via the metanephridia and may be brooded on the lophophore. Fertilization is external in the remaining phoronids, and both eggs and sperm are spawned into the sea. Cleavage is total, radial, and indeterminate at least early in development. A larval stage, the **actinotroch,** is characterized by an oral collar of ciliated tentacles, a complete digestive tract, an enterocoelom, and a posterior ring of locomotive cilia (Fig. 16.4c). During its approximately one month of existence the actinotroch feeds, grows, then settles, and secretes a tube.

The phoronids were long considered to be related to bryozoans because of the lophophore, but at least some experts now believe that they are deuterostomate (Emig, 1977, 1979; Nielsen, 1977). Radial indeterminate cleavage and the development of an enterocoelom that originates as three paired chambers probably clearly align the phoronids with deuterostomates, but they are not closely related to any other phylum except for the brachiopods.

Phylum Brachiopoda

The brachiopods are a most interesting group of exclusively marine benthic invertebrates. At first glance a brachiopod might be confused with a mollusc because it has a bivalved shell; but as we shall see below, there are some fundamental differences between brachiopods and bivalves. Most of the approximately 300 species are found in colder waters in depths of 20–200 m, where they are either attached to a firm surface or buried in the benthic material. But modern brachiopods are only a remnant of a once-dominant and extensive group. There are 15,000–30,000 described extinct species whose shells were easily preserved in the fossil record beginning in the early Cambrian period (see Table 1.1 in Chapter 1). Why are there so many fossil brachiopods and so few extant species? Perhaps the answer is that ancient brachiopods could not compete with other organisms like bivalves that fed in quite similar ways.

There are two classes of modern brachiopods to which frequent references will be made. The Class Inarticulata contains the more primitive forms whose valves, lacking a hinge, are held together by muscles only; some of the more common inarticulate genera include *Lingula* (Fig. 16.5b) and *Crania*. Members of the Class Articulata, typified by *Terebratula* and *Laqueus,* have valves that are hinged, but this articulation greatly restricts the gape opening. Other differences between the inarticulates and articulates will be highlighted in the discussions below.

THE SHELL, MANTLE, AND LOPHOPHORE

The brachiopod shell is an important adaptation for protection of the soft underlying tissues; and since it is wholly external, it is an **exoskeleton.** The shell, which is not conspicuously colored but may be variously sculptured, is composed of two **valves** constructed of calcium carbonate in the form of calcite (all articulates, a few inarticulates) or chitinophosphate (most inarticulates). The valves are both bilaterally symmetrical, usually convex, and joined together at their posterior surfaces, and their gape is anterior. In most brachiopods the **dorsal valve** is smaller than the **ventral valve,** the latter being usually attached to a hard surface. In burrowing brachiopods, including *Lingula,* the two valves are the same size. Near the anterior end where the valves are joined, the ventral valve is often drawn out into a spout, and in profile the two valves and spout are in the shape of an ancient lamp—thus their common name, lamp shells (Fig. 16.5c). The valves of some brachiopods are perforated with a series of minute holes, but the significance of these perforations is obscure.

While the valves fit snugly together to maximize the protection they afford the animal, they must be opened periodically in all brachiopods to permit entry of water from the environment. In the more primitive inarticulates in which no hinge of any sort is present, the posterior valvar edges remain almost in contact, and the anterior edges gape slightly. Closing the valves is achieved by two pairs of **adductor muscles** that extend between the valves and through the body. The opening mechanism is still somewhat of a mystery, since there is no opening ligament or spring. Retraction of the body may well cause the valves to gape slightly, and perhaps the elastic recovery of the adductor muscles is sufficient to force the valves apart, if only a few degrees. In articulates there is a posterior **hinge line** marking the point of junction of the valves, and that hinge becomes the fulcrum (Fig. 16.5c). The hinge line is provided with a pair of interlocking teeth on the ventral valve and corresponding sockets in the dorsal valve that effectively restrict the gape usually to no more than 10°. Closure of the valves is achieved by contraction of a pair of **adductor muscles** composed both of striated "quick" and smooth "catch" fibers (Fig. 16.5c). The shell is opened by the contraction of a pair of **diductor muscles** that originate on the ventral valve, pass behind the hinge line, and insert into the posterior edge of the dorsal valve (Fig. 16.5c). For the diductors to open the valves their fibers obviously must pass posterior to the fulcrum. The diductor contractions draw the posterior edges of the valves closer together, and, because of the location of the fulcrum, this leverage action causes the anterior edges to gape. Various evolutionary pathways in brachiopods have involved some modifications of the placement of the hinge and muscle attachments to improve leverage.

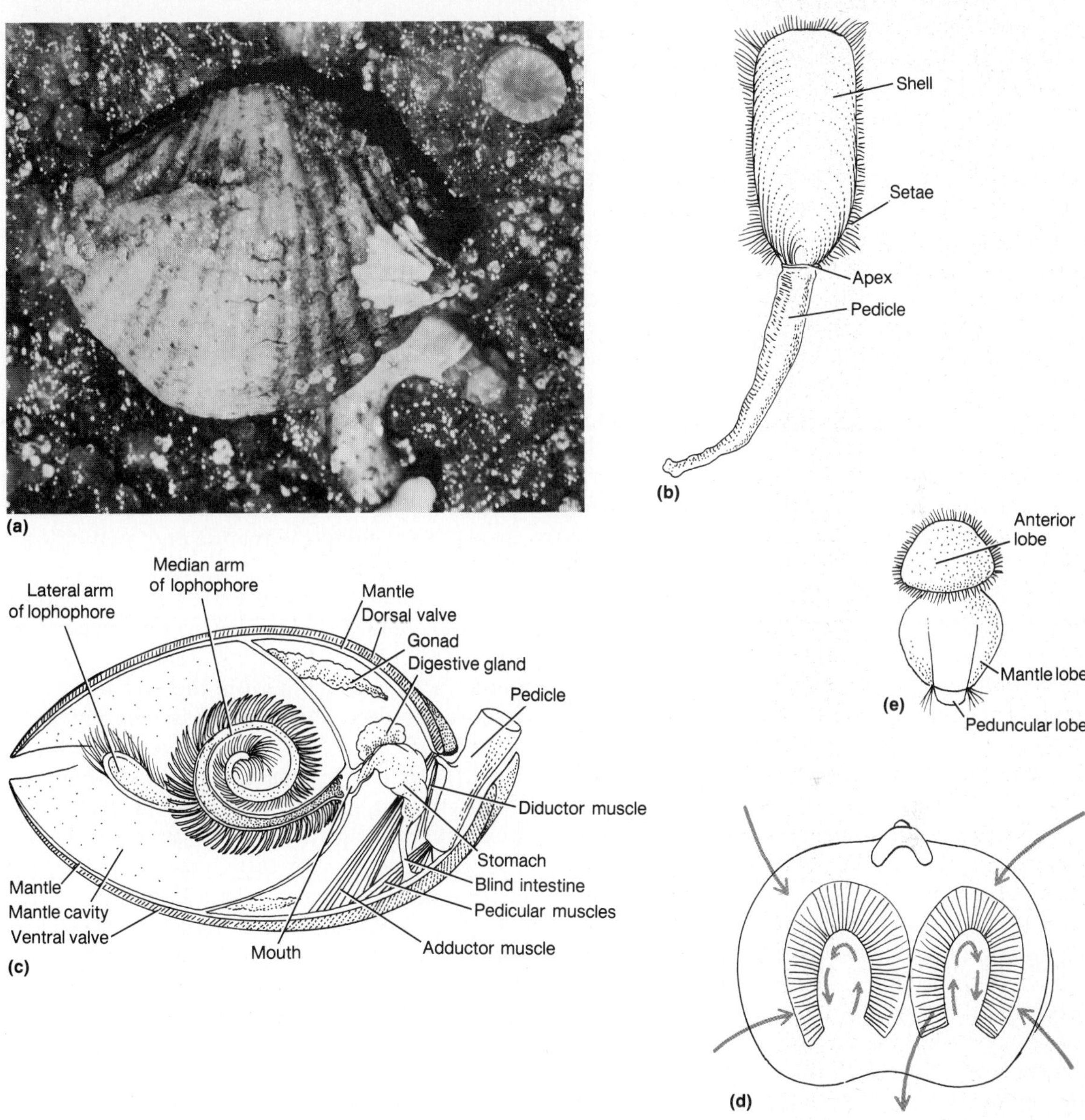

*Figure 16.5 Brachiopods: **(a)** photograph of* Terebratella transversa, *an articulate (courtesy of Ward's Natural Science, Inc., Rochester, N.Y.); **(b)*** Lingula, *an inarticulate; **(c)** diagrammatic sagittal section through a brachiopod to illustrate the positions and relationships of the principal organs; **(d)** diagrammatic dorsal view of the lophophore showing directions of water currents (large arrows) and ciliary currents (small arrows); **(e)** mature larva of* Terebratella. *(parts b, e, [from Percival] in L.H. Hyman,* The Invertebrates. *Vol. V, 1959,* Smaller Coelomate Groups, *McGraw-Hill Book Co., used with permission.)*

Most brachiopods are attached to a hard surface by the ventral valve being attached either directly or by means of a cord or stalk of tissue, the pedicle, which is present in most brachiopods. The **pedicle,** usually cylindrical, short, and constructed mostly of connective tissue, emerges from between the valves at the posterior edge and is attached into the substratum by short rootlike papillae (Fig. 16.5c). Several different muscles insert into the pedicle, and their actions enable the rest of the animal to move in many different ways in relation to this stationary organ. In certain forms like *Lingula* the pedicle is drawn out into an elongated, muscular, ventral, burrowing organ (Fig. 16.5b).

The mantle consists of two fleshy folds arising from the anterior body wall, one fold lying dorsal and the other lying ventral to the body. Each fold underlies its respective valve and is responsible for its secretion. At the anterior or distal edges of the mantle are two lobes—an outer **secretory lobe** and an inner **sensory lobe.** Accretional valvar growth is accomplished by epithelial cells of the secretory lobe depositing materials onto the extreme edges of the valves so that concentric growth lines are produced. Some additional shell materials are also deposited all along the inner surface of the shell by mantle tissues, which increase the thickness of the valves. Internally, the mantle lobes contain a large number of **mantle canals** that are continuous with the coelom. The function of these canals is still obscure, but they probably aid in gaseous exchange and transport. The space enclosed between the mantle lobes is the **mantle cavity,** and the mantle and mantle cavity are analogous but not homologous to those of molluscs.

The **lophophore** is an extremely important organ; it serves as a ciliary pump to bring water currents into the mantle cavity for feeding, gaseous exchange, and waste elimination. The lophophore, attached to the anterior body wall, consists of two long arms or **brachia** that are usually looped or coiled within the mantle cavity; it is from this feature that the phylum name was derived (Fig. 16.5d). Each brachium, strengthened internally by a cartilaginous axis, bears a row of long slender **tentacles** or filaments, each of which is equipped with two rows of lateral cilia and a row of frontal cilia. The position of the lophophore within the mantle cavity effectively divides this space into a dorsal inhalent and ventral exhalent chamber. Water can enter the exhalent chamber only by passing between the tentacles. This system, while reminiscent of that in eulamellibranch bivalves (see Chapter 8), is fundamentally different, since in brachiopods the tentacles are neither attached to each other nor supported internally, but rather are held in place only by their tips touching the mantle and by their own intrinsic strength.

Particulate and suspended food is carried into the mantle cavity by currents created by the lateral cilia, and often the tentacles are staggered in their attachment to the brachium so as to enhance their effectiveness in creating water currents. As water passes between the tentacles, the particles collide with the frontal tentacular surface and are enmeshed in mucus. The mucus–food complex is then carried by frontal cilia to the tentacular base, into a ciliated food groove, and on to the mouth (Fig. 16.5d). Brachiopods feed indiscriminately, since there are no sorting areas, but if conditions become unfavorable as in the water becoming turbid, both frontal and lateral cilia can reverse their beat and eventually eliminate the mucus–particle complex from the mantle chamber.

INTERNAL FEATURES

The brachiopod coelom consists of two parts, a **mesocoel** that surrounds the esophagus and extends into the lophophore complex and a **metacoel** that surrounds most of the viscera and extends into the pedicle and mantle lobes. Coelomocytes are present, some of which even contain hemoglobin.

Food is swept by ciliary action from the mouth into a dorsal esophagus that is continuous with the stomach (Fig. 16.5c). Opening into the stomach are the ducts from two pairs of digestive glands or ceca in which most digestion occurs, and arising from the stomach is an intestine that runs ventrally. In inarticulates the intestine opens into the right side of the mantle cavity as the anus; but in articulates, no anus is present, since the intestine ends blindly. Fecal materials are formed into distinct pellets that are voided into the mantle cavity in the inarticulates; in articulates the pellets are discharged, incredibly, through the mouth by antiperistaltic contractions. Regardless of the opening through which the pellets are voided, they are carried by ciliary action to the mantle's edge and are ejected by a sudden closing of the valves. The "snapping" action of the valves, similar to that phenomenon in bivalves, forcefully ejects any debris in the mantle cavity so that it will not be reingested.

Excretion is accomplished by a pair of metanephridia whose nephrostomes open into the metacoel and whose nephridiopores open into the mantle cavity at a point flanking the mouth. Coelomocytes, which have phagocytized coelomic wastes, are drawn into the nephridia and voided. The open circulatory system, separate from the coelom, has a contractile heart, situated dorsally to the stomach, and dorsal and ventral channels. The role of the circulatory system in brachiopods is not clear but probably is mostly concerned with nutrient transport. Gaseous exchange is achieved by the mantle lobes and not by special organs such as gills. The nervous system has a brain or subenteric ganglion from which arise important nerves to mantle lobes, pedicle, muscles, and the lophophore. Sensory structures are represented by numerous epidermal cells on the tentacles and setae along the sensory lobe of the mantle.

Reproduction and development

The vast majority of all lamp shells are dioecious, but there is little sexual dimorphism. Two pairs of gonads, one dorsal and one ventral, develop from the metacoelomic peritoneum and often spread into the coelomic canals of the mantle. When mature, the gametes enter the coelom proper and are spawned through the metanephridia into the mantle cavity and to the outside, where fertilization takes place. A few species brood the ova and young embryos within the mantle cavity. Cleavage is equal and radial and produces a hollow blastula and, later, a gastrula. Most lamp shells are enterocoelous, although details vary among the species. Soon a larval stage is formed, a miniature brachiopod that both swims and feeds (Fig. 16.5e). This larval stage is characterized by the paired mantle lobes and valves and the ciliated lophophore, which also serves as a locomotor organ. As new shell materials are added to the valves, the larva sinks to the bottom, attaches by its rudimentary pedicle, and becomes an adult brachiopod. While some forms undergo a metamorphosis, others, like *Lingula,* do not.

Brachiopod phylogeny

In most invertebrate zoology textbooks the brachiopods are considered to be protostomates, primarily on the basis of the presence of a lophophore; their radial cleavage and enterocoelous formation are concomitantly ignored. It is clear that these two fundamentally important deuterostomate features should relate them to echinoderms and chordates, and it is time to correct this perpetual error. They are considered here to have deuterostomate affiliations, a position that is also supported by Nielsen (1977).

The brachiopod fossil record is rather extensive and goes back into the lower Cambrian period. Proterozoic strata simply have not produced many brachiopod fossils, although it seems likely that lamp shells lived and thrived in Proterozoic seas. By the early Cambrian, when their fossils were extensively preserved, brachiopods were quite distinct from any other phylum. Their early history is not known, and neither are their evolutionary relationships. Perhaps all that can be said at this point is that brachiopods have had a very long and singular history, are remotely related to phoronids, and arose from some ancient stock that also gave rise to echinoderms and chordates.

The inarticulates, more common and more varied than their articulate cousins during the Cambrian period, are thought to be more primitive. During the Ordovician period the articulate lamp shells underwent extensive adaptive radiation and become one of the dominant benthic organisms. However, from the evidence of their fossil record, brachiopods experienced a massive die-off at the end of the Paleozoic Era in which thousands of species became extinct, and they never recovered from this crisis to assume a dominant position ecologically or to undergo a major adaptive radiation again.

Phylum Chaetognatha

Chaetognaths are extremely interesting organisms, and the approximately 90 species are all marine. Except for the species of *Spadella,* which are benthic, all are planktonic and are quite important in planktonic communities, where densities of up to 1000 per cubic meter have been reported. As voracious carnivores, they feed on other zooplankters and thus are important organisms in the planktonic food chains of most seas. They range in length from 2 to 10 cm, although most are about 3–4 cm long. Most of the species belong to the genus *Sagitta* (Fig. 16.6a), some of which are cosmopolitan in their distribution. In addition to their usual habitat of the euphotic stratum, chaetognaths may also be present in deeper waters of 1000 m or more. Temperature appears to be the most critical factor in determining latitudinal and vertical distributions. Many chaetognaths undergo rather dramatic, vertical, diurnal migrations that apparently coincide with similar migrations of food organisms.

The chaetognath body is mostly transparent and is shaped very much like a feathered dart, a feature from which they get their common name, arrowworms. The fuselage body consists of an anterior head, a middle trunk, and a tail (Fig. 16.6a, b). The head is rounded and bears a pair of dorsal eyes or ocelli, and on the ventral surface is a depression, the **vestibule,** at the bottom of which is the mouth. Flanking the vestibule are 4–14 pairs of long curved **spines** and several rows of shorter spinelike teeth (Fig. 16.6c). These spines and teeth are responsible for food capture and for the phylum name (chaeto = bristle; gnath = jaws). As the animal suddenly darts forward with spines spread, the spines can surround or impale prey, which most commonly are copepods but also can be small fishes or other arrowworms. Several reports have shown that chaetognaths can decimate or almost totally destroy the very young planktonic stages of several commercially important fishes. At the back of the head is a fold or **hood** that can be pulled forward over the head and spines, presumably to enhance the streamline contours of the anterior body. The **trunk** is fusiform and bears one or two *(Sagitta)* pairs of horizontal **lateral fins** on the posterior trunk; they function in providing flotation and stability to the animal. The **tail** bears a part of the horizontal fins and a large rounded **caudal fin,** which, like the lateral fins, is strengthened by rays (Fig. 16.6a, b).

The body is covered by a thin cuticle secreted by an underlying epidermis. Beneath the epidermis is a basement membrane, which is thickened in the fins as rays and in the head as strengthening plates. There are no circular muscles, but paired dorsolateral and ventrolateral muscle bands are present. Rapid alternating contractions of these dorsal and ventral bands propel the animal forward in the water, after

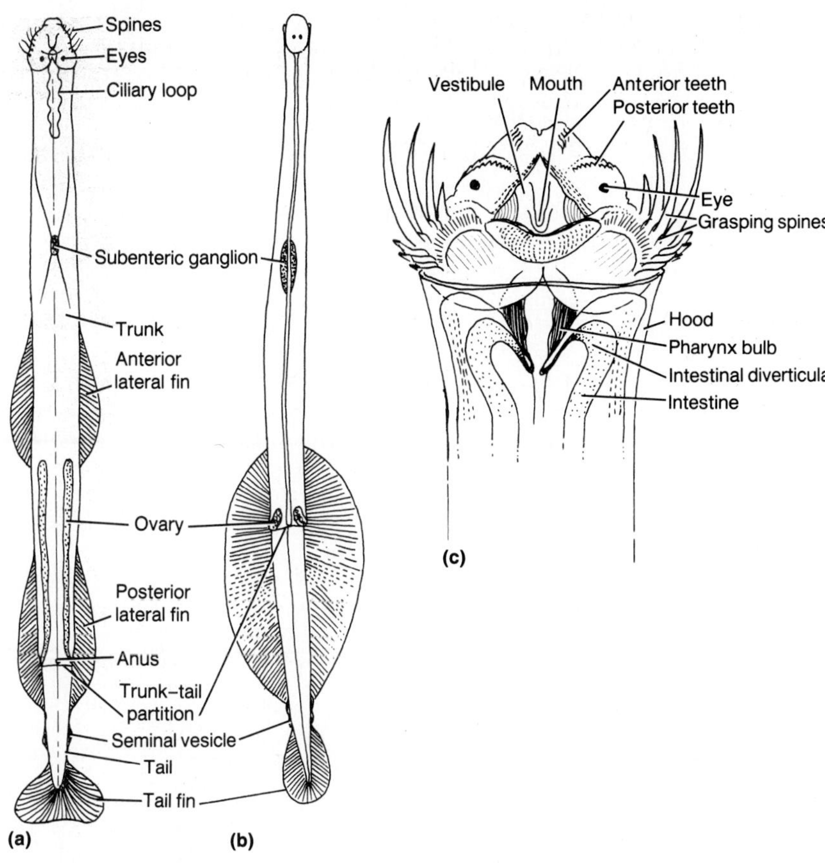

Figure 16.6 *Chaetognaths:* ***(a)*** *ventral view of* Sagitta; ***(b)*** *ventral view of* Krohnitta; ***(c)*** *ventral view of the anterior end of a chaetognath. (parts a, c [from Ritter-Zahony], b [from Tokioka] in L.H. Hyman,* The Invertebrates, *1959, Vol. V:* Smaller Coelomate Groups, *McGraw-Hill Book Co., used with permission.)*

which the animal floats, aided by its fins. Therefore locomotion consists of alternate periods of the chaetognath darting forward and then floating.

Internally, there is a coelom without peritoneal linings and comparted by septa into a single chamber in the head, paired trunk chambers, and one or two tail spaces. The digestive tract is simple with an anterior muscular pharynx and an intestine in the trunk, which, at its anterior end, gives rise to paired lateral diverticula. The intestine terminates in the anus situated at the junction of trunk and tail. There are no special organs for excretion, circulation, or gaseous exchange, and presumably, fluid within the coelom functions in internal transport. The nervous system has a circumpharyngeal ring consisting of a dorsal cerebral ganglion innervating the head, lateral connectives, and a subpharyngeal ganglion whose nerves innervate the trunk and tail. Sense organs include the pair of inverse (i.e., receptor cells point inward) cephalic eyes; numerous vibration-sensitive tactile bristles and hair fans; and a curious cephalic **ciliary loop,** a U-shaped band of cilia extending from head to the anterior trunk (Fig. 16.6a).

Chaetognaths are all monoecious; both gonads are paired, are elongated, and lie in the posterior coelom. The ovaries are situated on

TABLE 16.2. ☐ **CHARACTERISTICS OF THE PRINCIPAL TAXA OF THE PHYLUM HEMICHORDATA AND THE INVERTEBRATES OF THE PHYLUM CHORDATA**

Taxon	*Common Name*	*Pharyngeal Slits*	*Habit*	*Other Features*
Phylum Hemichordata				
Class Enteropneusta	Acorn worms	1–100 pairs	Mostly burrowing, solitary	Body of proboscis, collar, trunk; branchial sacs; form a tornaria larva
Class Pterobranchia	—	0–1 pair	Solitary or colonial	Collar with 1–9 pairs of arms and numerous tentacles similar to a lophophore
Phylum Chordata				
Subphylum Urochordata	Tunicates			
Class Ascidiacea	Sea squirts	Many	Solitary or colonial; sessile	Enlarged pharynx; secreted outer tunic; variety of blood corpuscles
Class Larvacea	—	1 pair	Planktonic	Secrete a surrounding house; no atrium; food trapped on internal net of house; have a tail
Class Thaliacea	Salps	2–many	Planktonic	Barrel shaped; atrial siphon at posterior end; some are polymorphic
Subphylum Cephalochordata	Lancelets	~ 100 pairs	Solitary, benthic	Notochord and dorsal nerve cord present in adults; prominent cecum; protonephridia

the anterior surface of the trunk–tail septum, and the testes lie on the posterior surface of this same partition. Immature spermatogonia are shed into the coelom. After maturing, sperm enter a sperm duct and go to a seminal vesicle, where they are formed into a spermatophore. Upon rupture of the seminal vesicle the spermatophore is released and is deposited on the anterior trunk region of another worm. As the spermatophore deteriorates, sperm stream back to the female gonopore and enter a seminal receptacle; fertilization is internal. Fertilized eggs are transferred along two oviducts to a pair of female gonopores located in the posterior trunk. Early embryology is characterized by total, radial, indeterminate cleavage, and a hollow gastrula is formed. The coelom is enterocoelous but is formed in a curious way by a pair of backward folds or archenteric invaginations that later form two pairs of coelomic chambers. Since no larval stage is found, development is direct, hatching occurs in about 48 hours, and all adult features appear within a week.

There are a number of chaetognath features that resemble those in pseudocoelomates, such as the cuticle, body wall construction, absence of peritoneum, and the anterior oral spines. Yet their cleavage patterns and enterocoelous formation, in spite of the coelom being formed in a very peculiar way, all clearly document chaetognaths as deuterostomates. They are not closely related to any other group, and their unique character can be attributed only to a very long and independent evolutionary history.

Phylum Hemichordata

Hemichordates, representing about 75 species, are burrowing or sessile marine worms that are usually found in shallow seas. This phylum is subdivided into two classes, Enteropneusta and Pterobranchia, whose representatives are quite different from each other as indicated in Table 16.2. Both groups are enterocoelous; have a body composed of proboscis, collar, and trunk; and have pharyngeal slits. Since the enteropneusts are more common, are larger, and have a greater species diversity, they will be discussed first and in more detail.

Enteropneusts are rather common in most intertidal and littoral zones, where they usually burrow or live beneath debris, shells, or rocks. Their vermiform elongated body ranges from 2 to 150 cm in length, but most are from 10 to 45 cm long. They are sluggish, slow moving, and typically cream colored. Two of the most common and frequently encountered genera are *Saccoglossus* (Fig. 16.7a) and *Balanoglossus*, both of which construct U-shaped burrows that are lined with mucus (Fig. 16.7b).

All three parts of the enteropneust body are covered by a ciliated epithelium containing numerous mucous glands. The **proboscis** or protosome is usually short and conical and joins the collar by way of a short constriction, the **proboscis stalk;** it is instrumental in burrowing and in suspension feeding. The **collar** or mesosome is short, ringlike, and enlarged and bears the ventral mouth (Fig. 16.7a). The acorn-like shape of the collar and proboscis is the basis for the common name for enteropneusts, acorn worms. The **trunk** or metasome, the major part of the body, is somewhat flaccid and terminates in the anus. Situated along either side of a mid-dorsal ridge on the anterior trunk (branchial) region is a row of **branchial pores.**

Enteropneusts are either deposit-feeders or suspension-feeders. The deposit-feeders consume vast quantities of the sand and mud through which they are burrowing, digest any organic materials, and produce extensive coiled casting at the opening of their burrows (Fig. 16.7b). The suspension-feeders trap minute particles in the mucus of the proboscis; and as the proboscidian cilia carry the particles toward the mouth, most bits are carried ventrally and enter a groove on the proboscis stalk called the **ciliary organ.** The precise function of this organ is not

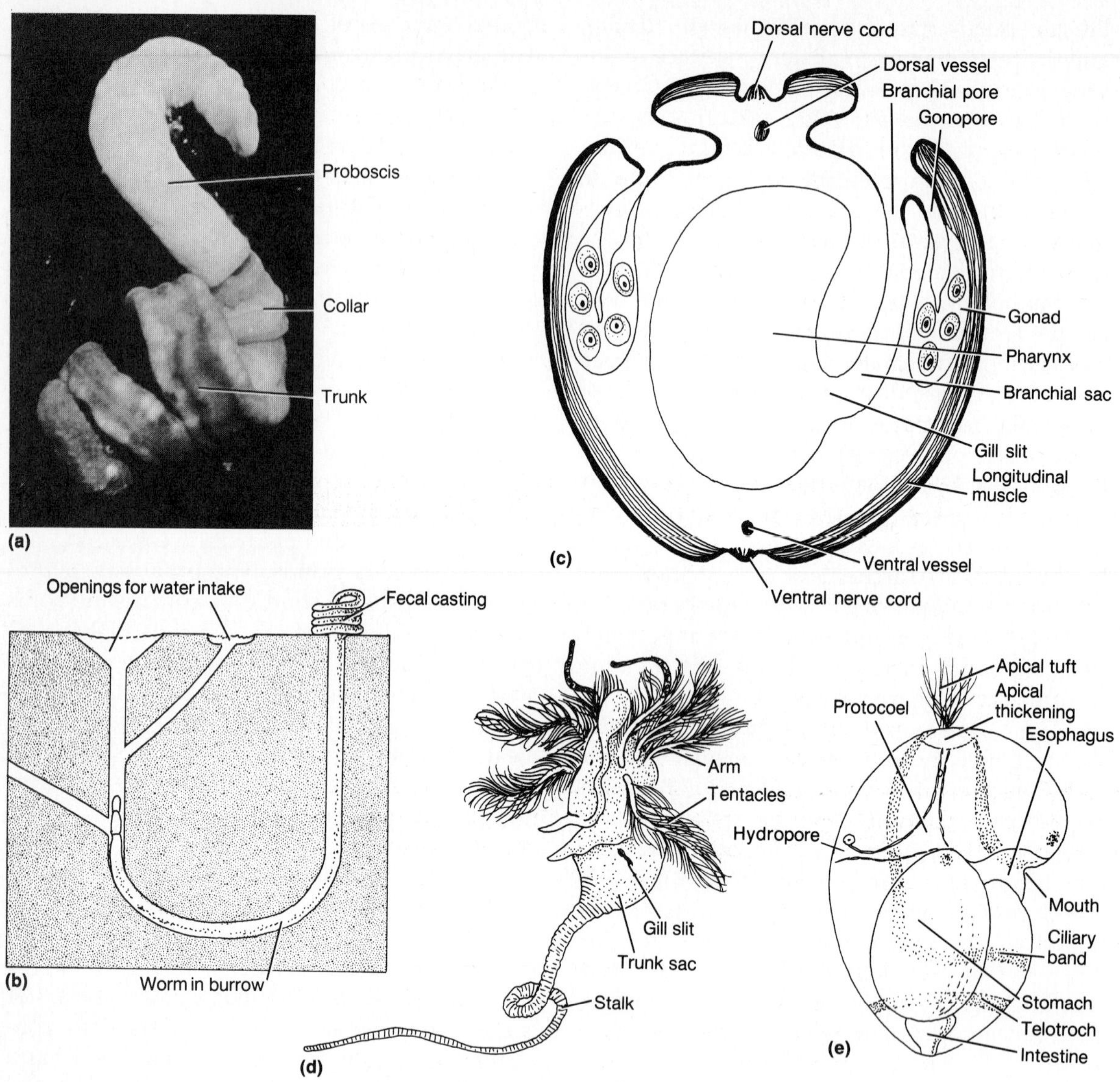

Figure 16.7 *Hemichordates:* ***(a)*** *photograph of* Saccoglossus *(courtesy of Ward's Natural Science, Inc., Rochester, N.Y.);* ***(b)*** *diagram of the U-shaped burrow of* Balanoglossus; ***(c)*** *diagrammatic cross-sectional view of the proboscis of an enteropneust;* ***(d)*** Atubaria, *a pterobranch;* ***(e)*** *a tornaria larva. (parts b, e [from Stiasny], d [from Komai] in L.H. Hyman,* The Invertebrates. *Vol. V:* Smaller Coelomate Groups, *McGraw-Hill Book Co., used with permission.)*

totally clear, but since it contains many sensory cells, it perhaps samples or "tastes" particles and water currents before they enter the mouth.

Internally, there is a tripartite coelom comparted by perforate septa, but it lacks peritoneal linings and is mostly filled with muscle strands and connective tissue. A single **protocoel** is present in the proboscis, and a middorsal pore connects it to the outside. A pair of **mesocoelic spaces** is present in the collar; these chambers likewise open to the exterior by several pores. Paired **metacoelic chambers** without outside openings are found in the trunk. The digestive tract consists of the mouth, the buccal tube in the collar, a perforated pharynx, the esophagus, the long intestine where the majority of digestion and absorption take place, and a posterior anus. Arising from the buccal tube is a dorsal elongated **diverticulum** that extends anteriorly into the proboscis. Since the diverticulum was once thought to be a notochord, hemichordates were considered to be a subordinate group of chordates, all of which have a notochord. But after the alimentary nature of this diverticulum was made clear, the hemichordates were removed from the chordates and given phylum status. The **pharynx** is of particular concern, since its dorsolateral portion bears up to 100 pairs of **gill slits.** Each slit is a U-shaped ciliated cleft or opening in the pharyngeal wall (Fig. 16.7c). On either side of a slit and even between the arms of the cleft, the pharyngeal wall is strengthened by a thickening of the subepidermal basement membrane. Each pharyngeal cleft opens peripherally into a **branchial sac** which in turn opens to the outside by a **branchial pore.** The pharyngeal slits probably evolved as a food-gathering mechanism; water enters the mouth and passes into the pharynx and through the slits, and particulate matter is strained out (Fig. 16.7c). Only secondarily has this mechanism assumed a gas exchange function. The esophagus opens to the outside via several dorsal pores, presumably to provide egress for excess water from the digestive tract.

Internal transport is achieved by an open blood–vascular system that has a dorsal and ventral vessel and a system of interconnecting sinuses. The colorless blood flows anteriorly through the dorsal vessel and enters a central sinus in the proboscis stalk. The muscular ventral wall of the central sinus, the heart, pumps blood anteriorly into a system of blood sinuses known as the **glomerulus,** a complex that probably has an excretory function. Blood then enters a system of channels that convey it to the ventral vessel that extends posteriorly beneath the gut. Sinuses in the digestive tract and body wall convey the blood from the ventral vessel back to the dorsal vessel.

The nervous system is constructed principally as a subepidermal plexus, a middorsal and a midventral longitudinal nerve cord, and nerve rings. One nerve ring is located in the anterior trunk, and it is joined by the two nerve cords into which the ventral cord terminates, but the dorsal cord continues into the collar. This dorsal **collar cord** is hollow, extends anteriorly to join another nerve ring in the proboscis, and may in fact be homologous to the dorsal nerve cord of chordates. Excepting the ciliary organ, no definite sense organs are present, although there are abundant sensory cells on the surface, especially that of the proboscis.

Acorn worms are dioecious, and many paired gonads are found in the midtrunk (genital) region. Some enteropneusts even have lateral trunk wings containing gonads. Each gonad has a separate gonopore through which gametes are spawned, and fertilization is external. Radial indeterminate cleavage, early developmental patterns, and methods of coelom formation are very much like those of echinoderms and chordates. The larva, a **tornaria,** is remarkably similar to the bipinnaria stage of sea stars (see Chapter 17) with a winding ciliary band that later becomes the principal means for locomotion (Fig. 16.7e). After a week or two the larva begins to elongate, develops constrictions that separate the three body regions, sinks to the bottom, and becomes an adult. The common Atlantic coast acorn worm, *Saccoglossus kowalevskii,* has direct development and no larval stage.

The pterobranchs are represented by fewer than ten species of rare, tiny (1–5 mm), sessile, deep-sea hemichordates. A few are solitary, but others form colonies in which each individual (zooid) secretes a tube around itself. The pterobranch proboscis projects anteriorly as a cephalic shield, and the collar bears one to nine pairs of arms. Each arm has a number of hollow, dorsal, ciliated, food-gathering arms or tentacles, the combination of which is similar to a lophophore (Fig. 16.7d). Pterobranchs have a pair of pharyngeal slits (absent in one genus) and a U-shaped digestive tract, are dioecious, and produce a ciliated larva. Colonies are formed by asexual budding.

That hemichordates have a close phylogenetic affiliation with both echinoderms and chordates is very clear, but there is little evidence to enable zoologists to determine precisely where they arose from the deuterostomate line. The pterobranchs are more primitive and may even represent the ancestral form common to both echinoderms and enteropneusts.

Phylum Chordata

The Phylum Chordata is a very large phylum consisting of almost 38,000 species, of which more than 95% are vertebrates and therefore are not considered in this text. What we are concerned with is a heterogeneous group of about 1600 species of nonvertebrate chordates. Representatives of this phylum, including both the vertebrates and the invertebrates, are characterized by three extremely important diagnostic characters that are present at some point in their development: (1) **pharyngeal clefts** similar to those of hemichordates, (2) a **dorsal tubular nerve cord,** a most unusual feature among animals, and (3) a unique cartilaginous supporting rod, the **notochord.** Most have a fourth feature, a **postanal tail** that is employed in locomotion. Great phylogenetic significance is placed on these three (or four) features, since they define a chordate. In addition, the invertebrate chordates are exclusively marine; are mostly benethic, though some are planktonic; and are all filter-feeders. There are two subphyla of nonvertebrate chordates, the Urochordata and Cephalochordata, which will be treated below and the principal taxa of which are characterized in Table 16.2.

Subphylum Urochordata

At least superficially, the urochordates certainly do not appear to fit the ideal or stereotype of what a chordate is, but their aberrance is really a manifestation of a number of primitive features as well as many adaptations to a sedentary life-style. They are commonly known as tunicates, since the body is surrounded by a rather complex, secreted, celluloselike **tunic.** Only in the larval stage do the notochord and dorsal tubular nerve cord develop, and adults possess neither of these features. The three classes Ascidiacea, Larvacea, and Thaliacea are characterized in Table 16.2. Since ascidians represent most of the urochordates, they will be discussed initially, followed by a much briefer treatment of the other two groups.

Class Ascidiacea. The ascidians are rather common sessile urochordates that usually are attached to hard surfaces in littoral zones. They are particularly common on pilings, ship hulls, rocks, shells, and coral reefs, but a minority of tunicates, living in deep-sea habitats, are anchored in mud and sand. Solitary forms are known, but most ascidians are colonial (Fig. 16.8). There are about 1500 species, and they vary in body shape, size, and coloration. Shapes range from spherical and ovate to stalked, and they range from 1 to 12 cm in their largest dimension. Most are gray, brown, or greenish,

(a)

(b)

(c)

Figure 16.8 *Photographic collage of types of ascidians:* ***(a)*** *a colony of* Distaplia occidentalis; ***(b)*** *a colony of* Metandrocarpa; ***(c)*** *a colony of* Ecteinascidia turbinata *(courtesy of C.E. Jenner). (parts a, b courtesy of Ward's Natural Science, Inc., Rochester, N.Y.)*

but some ascidians are beautifully colored with exquisite subtle hues. Some of the more frequently encountered genera include *Molgula, Clavelina, Styela,* and *Ciona.*

The body of an ascidian consists of an anterior **pharyngeal region,** an **abdominal region** housing most of the other viscera, and **postabdominal region** containing the heart and gonads. Most ascidians, however, do not have a postabdomen, and in others, even the abdomen is missing. The body of an ascidian is attached to a surface by its posterior end; at the anterior end is a prominent **buccal siphon,** which is derived from the mouth, and nearby is the equally prominent **atrial siphon** (Fig. 16.9a). Through these two openings in the upper body pass water currents, which are extremely vital to most of the ascidian's physiological functions. When ascidians are collected and squeezed gently between the fingers and thumb, water squirts from one or both siphons, a feature from which ascidians get their common name, sea squirts.

Surrounding all ascidians is a rather thick fibrous matrix, the tunic (Fig. 16.9a). The **tunic,** secreted by the epidermal layer located just beneath it, is composed principally of **tunicin,** a type of cellulose that has also been found recently in the skin of vertebrates. This outer coat also contains some proteins, mineral salts, perhaps some spicules, and even ameboid and blood cells; in a few ascidians the tunic is even

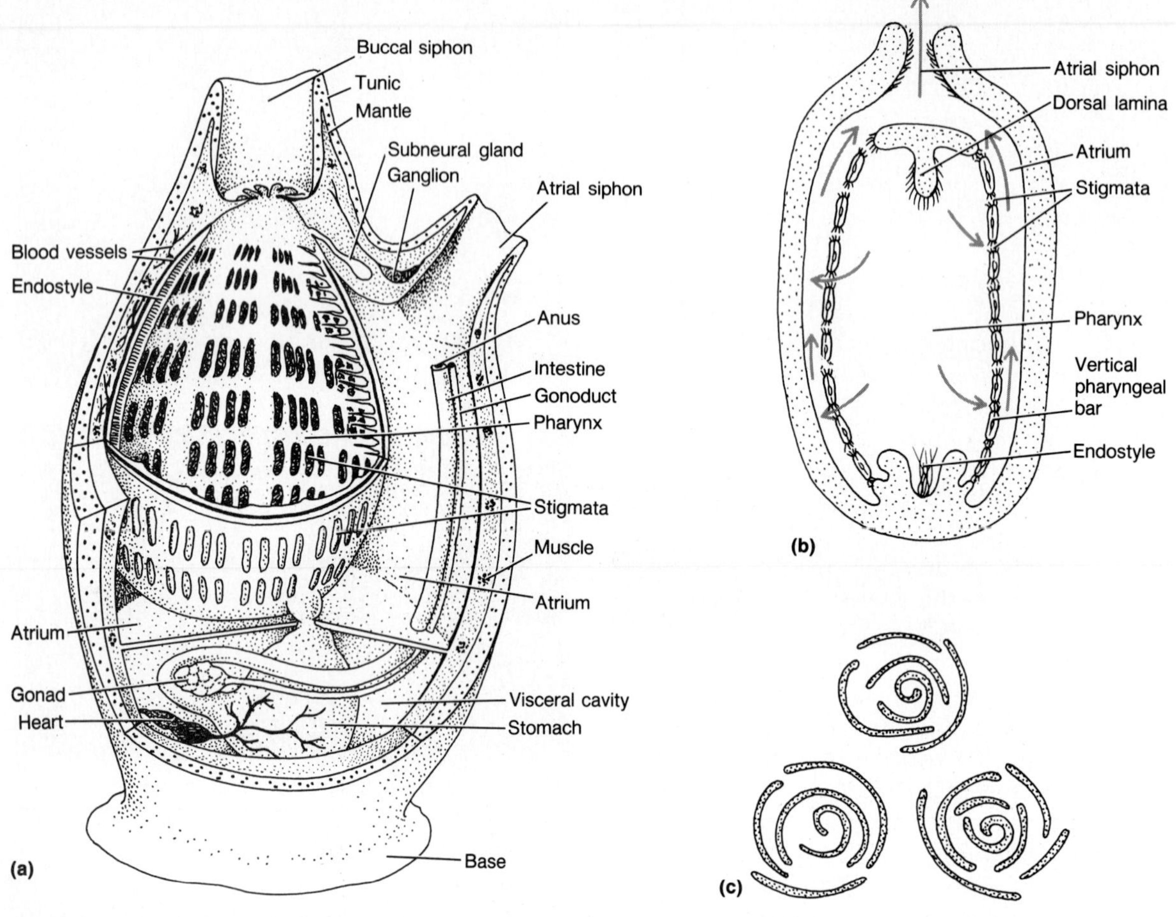

Figure 16.9 *Some ascidian features: **(a)** diagrammatic longitudinal section of a tunicate (from T.I. Storer and R.L. Usinger, 1965,* General Zoology, *4th ed., McGraw-Hill Book Co., used with permission); **(b)** diagrammatic cross-sectional view through the pharyngeal region of a tunicate showing pathways of water currents; **(c)** several spiral stigmata.*

supplied by blood vessels. Therefore the tunic is not entirely inanimate but often takes on the histological appearance and construction of dense connective tissue. Attachment is also a function of the tunic; rootlike **stolons** extend from it to the substratum and provide effective anchorage for the ascidian. The tunic may be transparent but more often is opaque, and it has a quite variable consistency or feel. The role of the tunic in the life of an ascidian is both to protect and to provide some rigidity to the body. Beneath the tunic and epidermis are circular and longitudinal muscle strands.

The pharynx and feeding. Just beneath the body wall is a very large cavity, the **atrium,** which opens to the exterior by way of the atrial siphon. The atrium is lined with epidermis that is continuous with that covering the body. Dominating the atrial space of most ascidians is an unusually large **pharynx,** a basketlike organ surrounded on all sides by the atrium (Fig. 16.9a, b). The buccal siphon is the point of entry into the pharynx and the rest of the digestive tract by external water currents. As the buccal siphon opens into the pharyngeal basket, there is a ring of short inwardly extending **tentacles** that selectively admit or reject incoming particles. The pharynx is typically barrel shaped and bears a very large number of small **pharyngeal clefts** or **stigmata,** which are mostly arranged in both vertical columns and horizontal rows with thickened bars separating individual stigmata from their neighbors (Fig. 16.9). Each stigma is ciliated with frontal and lateral cilia; the latter beat outward so that water passes from the pharynx through the stigmata into the atrium. More advanced ascidians have greater numbers of stigmata, and some, like *Corella,* have groups of stigmata arranged in spiral patterns (Fig. 16.9c). On the ventral inner side and extending the entire length of the pharynx is a rather prominent ciliated groove, the **endostyle,** at whose bottom are very long cilia (flagella) and mucous glands (Fig. 16.9b). Along the entire dorsal aspect of the pharynx and opposite the endostyle is either a **dorsal lamina** or a row of fingerlike projections, the **lanquets,** that extend from the buccal siphon to the esophagus (Fig. 16.9b).

Ascidians are filter-feeders, and they must filter enormous quantities of water to provide themselves with sufficient food; as many as several thousand times the volume of the body pass through an ascidian every day! Water currents, created by the lateral cilia of the stigmata, enter the buccal siphon, pass into the pharynx, exit through the stigmata into the atrium, and flow to the outside through the atrial siphon (Fig. 16.9a, b). Upon this more or less continuous water flow is based the ascidian's nutrition as well as its gaseous exchange, excretion, and even reproduction. The endostyle produces large quantities of mucus, which is then carried dorsally by frontal cilia as a thin sheet over the interior lateral walls of the pharynx and up toward the dorsal lamina. As this continuous sheet moves toward the dorsal lamina, it pushes dorsally the particles trapped by the lateral cilia of the stigmata, and the sheet itself traps additional particles. As a section of the mucous sheet or film arrives at the dorsal lamina, it enters into a gutter formed by lateral rolls of the lamina or the languets and is formed into a roll that is carried posteriorly to the esophageal opening on the dorsal face of the pharynx.

Other viscera. The remainder of the digestive tract is a ciliated, U-shaped tube consisting of the esophagus, stomach, intestine, and anus (Fig. 16.9a). The stomach, situated at the bend of the U, is the locus for most digestion and absorption. The anus is located just below the atrial siphon, and exhalent water currents passing through this aperture help flush fecal materials to the outside. There are no special excretory or gas exchange organs; these functions take place over the general internal body surfaces, aided immeasurably by the water currents passing through the pharynx and atrium. Ascidians lack a true coelom as adults, but certain areas like the pericardium and an unusual epicardium may be coelomic remnants. The **epicardium** is usually a simple tube or a complex of tubes that parallels the digestive tract, and the tubules often surround the gut in a manner like a coelom. The epicardium is also involved in asexual budding.

The open circulatory system is characterized by a rudimentary heart and numerous blood channels. The heart is really a muscular fold of the pericardium lying near the stomach. Its pulsations force blood into a system of ill-defined spaces and channels. The best-defined circulation is that to the pharyngeal walls. A large subendostylar vessel carries blood anteriorly along the ventral pharynx, and numerous small blood channels convey blood to the lateral walls. Pharyngeal blood is collected by a medial dorsal sinus and carried to the viscera before returning to the pericardium. Ascidian circulation is most remarkable for two reasons. First, periodically, normally, and routinely, blood flow is reversed owing to fluctuation in the dominance of two stimulatory regions, one at either end of the heart; other details of the mechanism and significance of this reversal are not known. Second, the blood contains a number of corpuscles including phagocytic amebocytes, morula cells, and storage cells, all of which are derived from lymphocytes. Morula cells may be involved in tunic formation, and in at least two families, green morula cells called **vanadocytes** collect vanadium from seawater, for reasons still unclear, in concentrations half a million times that of seawater! Storage cells or **nephrocytes** collect nitrogenous wastes, mostly uric acid, transport them to the outer walls of the digestive tract or gonads, and deposit them there for the entire life of the ascidian.

The nervous system has a cerebral ganglion lying between the two siphons from which arise several nerves innervating the major body regions such as the siphons, pharynx, and viscera. Beneath the cerebral ganglion lies a glandular body called the **neural** or **subneural gland** (Fig. 16.9a). It contains a duct or tubule that opens as a large ciliated funnel into the pharynx. Its true function is not known, and although some zoologists have speculated that it might be the forerunner of the vertebrate pituitary gland, no endocrine function has been discovered for it. Numerous sensory cells abound on the siphons and buccal tentacles and in the atrium.

Reproduction and development. Ascidians typically live in colonies, and new individuals within the colony are produced by asexual budding, a phenomenon of unusual diversity in ascidians. An asexually produced bud, called a **blastozooid,** often arises from almost any point on a parent but most often from a common stolon or base, and new individuals in some species even share a common tunic. The epicardium often is the point where new buds begin. Many species are so precocious that budding actually begins in the larval stage.

Almost all ascidians are monoecious, and the unpaired gonads lie near the stomach (Fig. 16.9a). The separate gonoducts open as gonopores in the atrium near the atrial siphon, and fertilization is in the open sea. Radial indeterminate cleavage soon produces an embryo in which the blastopore becomes the anus, the archenteron gives rise to the dorsal rodlike **notochord,** the archenteron forms lateral mesodermal masses but no true coelom is ever formed, and the dorsal ectoderm forms the hollow neural tube. As development proceeds, the embryo becomes a "tadpole" larva characterized by a long posterior tail containing both the notochord and the neural tube (Fig. 16.10). This larva also develops the cerebral ganglion and mouth that will become the buccal siphon. A fully formed but nonfunctional digestive tract is present with stigmata and atrium already present. This larva does not feed because the mouth is closed and because the entire larva is en-

Figure 16.10 *Photograph of the tailed larva of* Corella, *an ascidian (courtesy of R.D. Fink).*

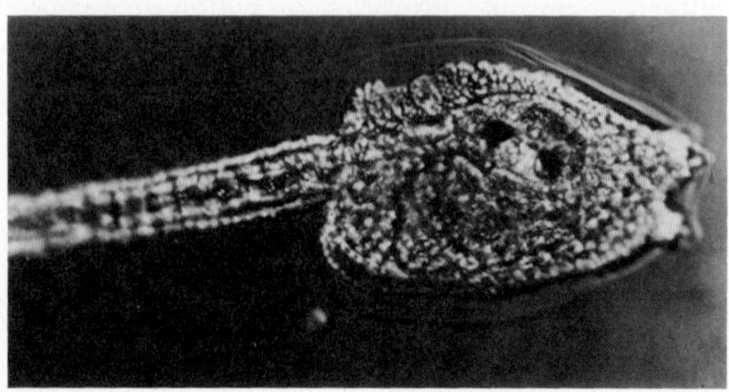

shrouded by the tunic. At the conclusion of the tadpole stage it sinks to the bottom, it attaches by three fixation papillae, and a rather dramatic metamorphosis results. The tail, with its neural tube and notochord are reabsorbed, the entire body is rotated 180°, numbers of stigmata are increased, and the siphons become functional. Somewhat later, the animal matures sexually.

Class Larvacea. The larvaceans are represented by about 70 species of very small (1–5 mm) highly specialized urochordates. They are all planktonic, live near the surface, and are mostly transparent. They often are common in plankton, and population densities of up to 10,000 per cubic meter have been reported. As adults, they have retained many larval features, a phenomenon known as **neotony,** which is the basis for their class name. Perhaps the most outstanding feature of larvaceans is the gelatinous or mucous tunic or **"house"** secreted by the epidermis; it normally surrounds the entire animal and is used as a filter-feeding device (Fig. 16.11a). Water is drawn into the house by the lashing motions of the tail, and minute particles or organisms are trapped on a net located inside the house. In many forms, the house is frequently shed and a new one secreted perhaps in response to the old one becoming clogged with particles. The body has retained the larval tail, which is carried ventrally and contains the notochord and dorsal nerve cord. The pharynx is equipped with only a single pair of pharyngeal slits, and since there is no atrium, each cleft opens directly to the outside. All are hermaphroditic and do not reproduce by budding. Some of the better-known genera are *Oikopleura* (Fig. 16.11a) and *Megalocercus.*

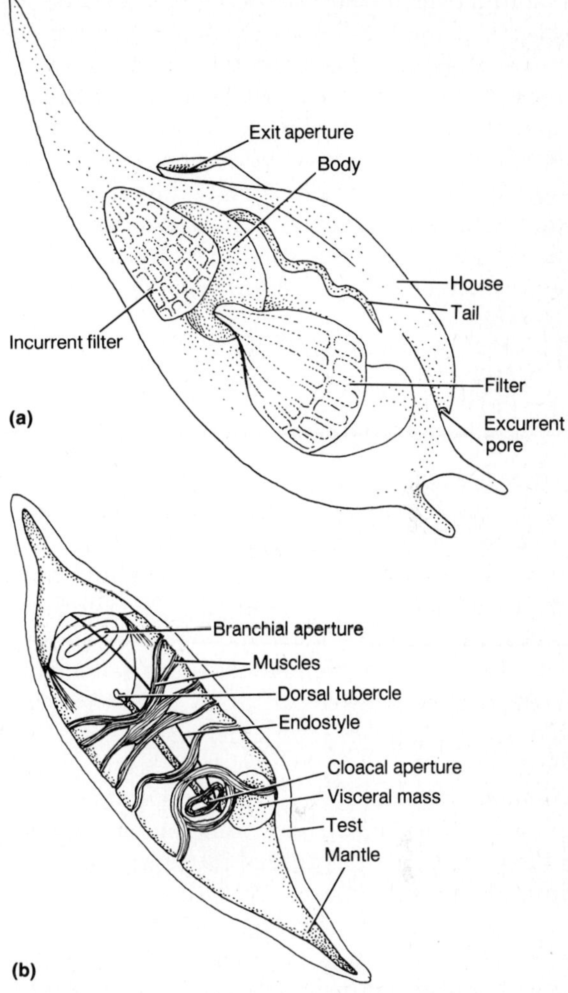

Figure 16.11 *Some other urochordates: **(a)*** Oikopleura, *a larvacean; **(b)*** Salpa, *a thaliacean (from W.A. Herdman, 1910,* Cambridge Natural History, *Macmillan & Co., Ltd., London).*

Class Thaliacea. The thaliaceans, more commonly known as salps, are represented by about 20 species belonging to only six genera. Adults are also planktonic and transparent, and some are luminescent. The adult is barrel shaped and tailless and therefore lacks the notochord and nerve tube present in the larval stage. Salps have an anterior buccal siphon, but the atrial siphon is at the posterior end, and water, forcefully ejected through the atrial siphon, moves the animal forward by jet propulsion (Fig. 16.11b). They have two to many pharyngeal slits. Salps reproduce asexually by forming buds, and thaliaceans have become so specialized that two different types of individuals are formed, those developed from fertilized eggs and those produced by budding, a situation reminiscent of polymorphism in cnidarians (see Chapter 5). Some of the more frequently encountered genera are *Doliolum* and *Salpa* (Fig. 16.11b).

SUBPHYLUM CEPHALOCHORDATA

There are about 35 species of cephalochordates, and representatives all live buried in littoral mud or sand of tropical and warmer temperate seas. They are small fishlike animals that normally are about 5 cm in length. Cephalochordates are unique among the deuterostomates because they are segmented, and their **metamerism** is manifested mostly as paired muscle blocks and nerves that innervate them. The best-known genus is *Branchiostoma,* which is traditionally studied in courses in vertebrate morphology (Fig. 16.12). This same animal is also called amphioxus because of its former but now invalid generic name, *Amphioxus.*

Because their body shape is elongated, flattened laterally, and spindleshaped, cephalochordates are known as lancelets. The mouth and anus are subterminal at the anterior and posterior ends, respectively. The external surface is mostly smooth and bears a low middorsal **fin** that extends around the posterior end (caudal fin) and continues ventrally up to the opening of the atrium (ventral fin). There is a **metapleural fold** along each ventrolateral aspect of the body, and these two folds may aid in gaseous exchange (Fig. 16.12c). Since most lancelets are transparent, many of their internal features are conspicuous and readily seen in the whole animal.

Surrounding the mouth is a **rostrum** bearing a number of small tentacles or cirri that act as feeding strainers (Fig. 16.12b). Lancelets are filter-feeders, and cilia of the tentacles and rostrum draw water through the mouth and into the enlarged pharynx. In the lateral walls of the pharynx are 100 or more **slits,** each of which is separated from its neighbors by skeletal rods. The clefts are ciliated, and the ciliary beat pulls water through the pharynx and its slits and into the atrium. The **atrium** is a large space surrounding the pharynx, and it opens via an **atriopore** located ventrally and anterior to the anus. The pharynx contains a ventral **endostyle** and a dorsal **hyperbranchial groove** (Fig. 16.12c). The endostyle secretes a mucous sheet that is carried laterally and dorsally along the inner surface of the pharynx to the hyperbranchial groove, where the mucus–food complex is concentrated into a roll. Food capture, as in ascidians, is achieved both by the cilia of the pharyngeal slits and by the mucous sheet traversing the walls of the pharynx. The mucus–food complex enters the straight intestine, which, before it terminates as the anal opening, gives rise to a deep narrow pouch, the **cecum,** that extends anteriorly on the right side of the pharynx (Fig. 16.12b). Small bits of food become detached from the mucus and are carried into the cecum, where digestion and absorption take place. The amphioxus cecum may be homologous to the vertebrate liver and/or pancreas.

The circulatory system in cephalochordates is similar in its basic plan to that of primitive chordates but lacks a heart and capillaries. Blood flows anteriorly in an endostylar artery, then dorsally in branchial vessels in the pharyngeal wall to the dorsal aorta, which carries blood posteriorly. Aortic branches carry blood to all regions, and blood is collected by veins that empty into the sinus venosus, which in turn opens into the endostylar artery to complete one circuit. A so-called hepatic portal system, perhaps best described as a cecal portal system, carries blood from the intestine to the cecum, where nutrient levels in the blood may be regulated. The blood is without corpuscles or respiratory pigments.

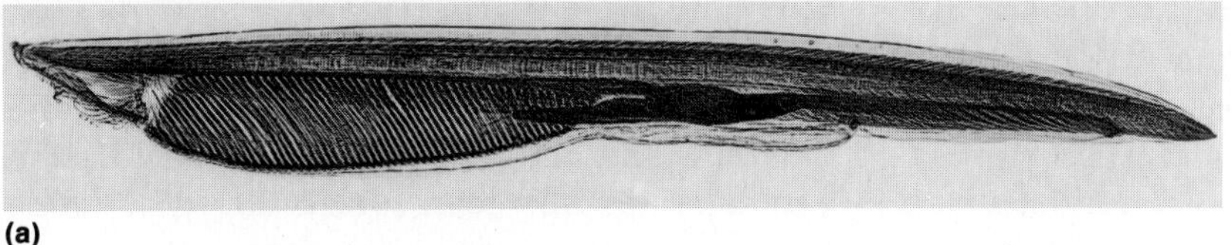

(a)

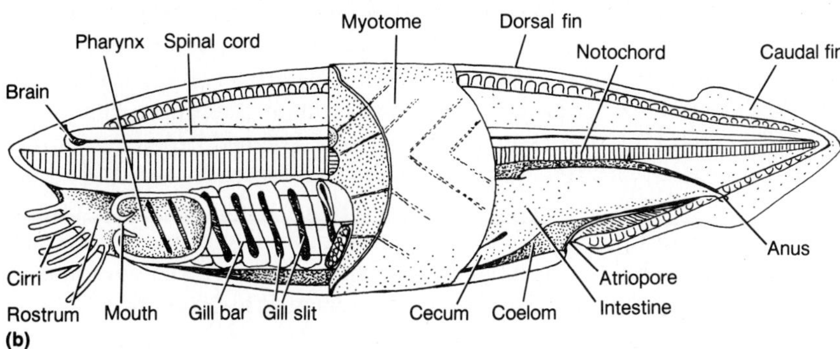

(b)

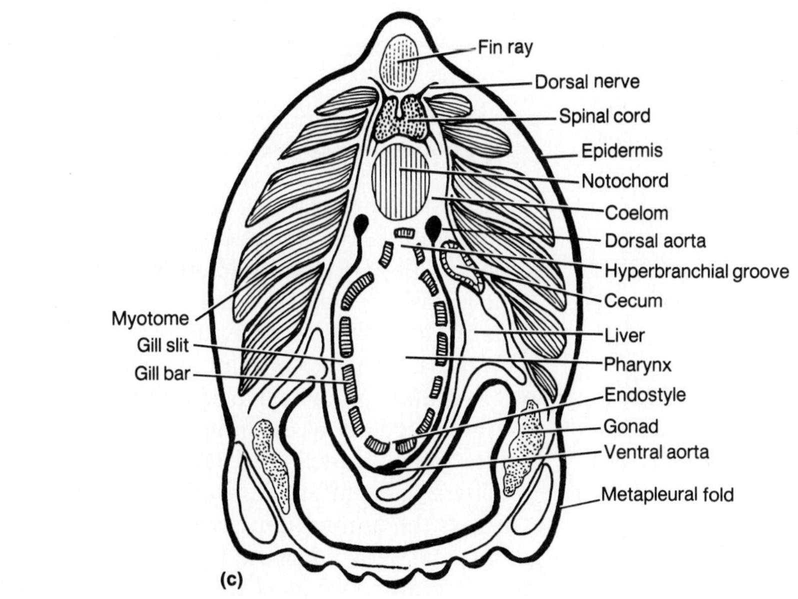

(c)

Figure 16.12 *Cephalochordates:* ***(a)*** *photograph of* Branchiostoma *(photograph by Carolina Biological Supply Company);* ***(b)*** *some of the more prominent features of* Branchiostoma *seen in longitudinal section;* ***(c)*** *cross-sectional view of* Branchiostoma *through the pharyngeal region. (parts b, c [modified] from W.A. Herdman, 1910,* Cambridge Natural History, *Macmillan & Co., Ltd., London.)*

The excretory system consists of a great many protonephridia, each of which bears a number of solenocytes. Nitrogenous wastes are carried by the protonephridial tubule to the atrium and eliminated in the exhalent currents. The coelom is typically tripartite in its ontogeny, but it loses its compartmental nature and is reduced in the adult. The dorsal tubular **nerve cord** lies above the skeletal notochord and extends the full length of the amphioxus (Fig.

16.12b, c). There is no brain as such, and the nerve cord sends both motor and sensory paired metameric nerves to each myotome or muscle unit. A single eyespot may be present, and numerous sensory cells are found in the epidermis.

Prominent in the body wall are metameric longitudinal muscles arranged as chevronlike **myotomes,** which are separated from each other by collagenous connective tissues (Fig. 16.12b). Contraction of these muscles effectively enables the lancelet to swim or burrow by flexing the body from side to side rapidly and forcefully. Propulsion is aided by the myotomic flexion of the dorsal notochord. The rodlike **notochord** consists of a linear series of muscle lamellae or discs much like a stack or roll of coins with fluid-filled spaces between contiguous discs, and the entire complex is surrounded by a collagenous sheath (Fig. 16.12b). As the more or less rigid notochord is bent laterally by the contraction of muscles on that side, it is stiffened, and this rigidity snaps the notochord back to a straight position between the lateral alternating muscle contractions. The notochord and muscles are also instrumental in the burrowing process.

Lancelets are dioecious, but there is no apparent sexual dimorphism. There are 25–33 pairs of minute gonads, and the body wall ruptures to release gametes into the atrium and to the outside, where fertilization takes place. The eggs are isolecithal, and cleavage is total, radial, and indeterminate. A planktonic larval stage ensues; as it develops, it sinks to the bottom and matures into an adult.

Chordate phylogeny

Many chordate characteristics, including most developmental details, point out the similarities of deuterostomates and, more specifically, relate chordates to echinoderms and hemichordates. But there is only speculation as to the nature of the primitive stock from which they all arose. We can say that the ancestral stock was marine, probably benthic and wormlike, coelomate, unsegmented, and above all possessing great adaptive potentiality. Urochordates undoubtedly represent a very early offshoot from the chordate line before metamerism evolved; the ascidians adapted to a sessile life-style, while the larvaceans and thaliaceans developed planktonic adaptations. The line leading to cephalochordates and vertebrates acquired a most important adaptation—metamerism—that was to be immensely important in their locomotion. Cephalochordate evolution, for some reason, carried them into a burrowing, filter-feeding existence that severely limited their phylogenetic plasticity, while their vertebrate cousins evolved an internal skeletal system and a fantastic nervous system whose potential is manifested in higher vertebrates. The skeleton, the nervous system, and an incredible measure of evolutionary plasticity enabled vertebrates to evolve into the dominant animal group today.

SOME PHYLOGENETIC GENERALIZATIONS

Because of the nature of this chapter and its dealing with rather disparate groups, very little can be said apart from the heterogeneity of the eight phyla. This is especially true among the three minor protostomate groups (Priapulida, Tardigrada, Pentastomida); each has had a very long, separate, evolutionary pathway, and each is represented on an evolutionary tree as a very long terminal branch (Fig. 16.13).

There is a little more evidence to group some of the minor deuterostomate phyla together. The Phoronida and Brachiopoda are perhaps distantly related to each other because of the presence of a lophophore and their habitats and habits. Chaetognaths have had a very long and separate evolution apart from any other group. But Hemichordata, Chordata, and Echinodermata (Chapter 17) have a great many similarities that suggest a much closer affiliation for them than that for any other minor phylum (Fig. 16.13).

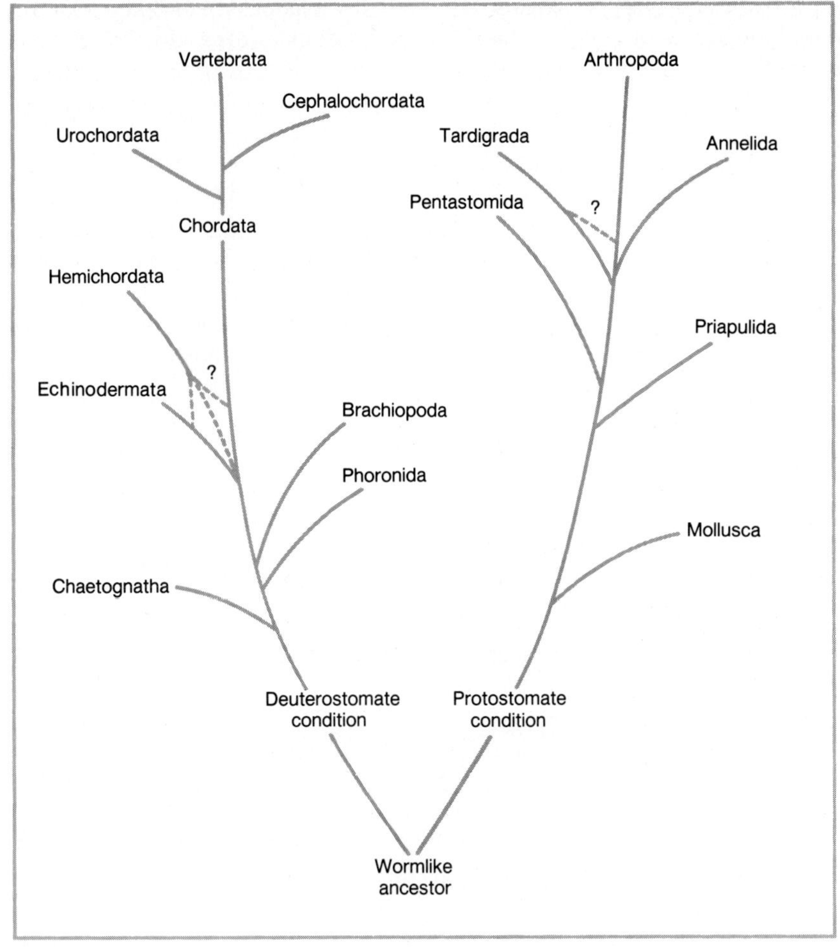

Figure 16.13 *Probable phylogenetic relationships of the eight minor phyla.*

ADDITIONAL READINGS

Alexander, R.M. 1975. *The Chordates,* pp. 25–40. New York: Cambridge University Press.

Alldredge, A. 1976. Appendicularians. *Sci. Amer. 235:*94–102.

Barrington, E.J.W. and R.P.S. Jefferies. 1975. *Protochordates,* 345 pp. Symposium No. 36 of the Zoological Society of London. New York: Academic Press.

Berrill, N.J. 1961. Salpa. *Sci. Amer. 204:*150–160.

Brunton, C.H.C., and G.B. Curry. 1979. *British Brachiopods,* 64 pp. New York: Academic Press.

Cheng, T.C. 1973. *General Parasitology,* pp. 776–785 (pentastomids). New York: Academic Press.

Crowe, J.H., and A.F. Cooper, Jr. 1971. Cryptobiosis. *Sci. Amer. 225:*30–36.

Emig, C.C. 1977. The embryology of Phoronida. *Amer. Zool. 17:*21–37.

Emig, C.C. 1979. *Synopsis of the British Fauna.* New Series 13: *British and Other Phoronids: Keys and Notes for the Identification of the Species,* 57 pp. New York: Academic Press.

Foster, M.W. 1974. *Recent Antarctic and Subantarctic Brachiopods,* 181 pp. Antarctic Research Series, Vol. 21. Washington, D.C.: American Geophysical Union.

Hyman, L.H. *The Invertebrates.* Vol. III: pp. 183–197 (priapulids), 1951; Vol. V: pp. 1–71 (chaetognaths), pp. 72–207 (hemichordates), pp. 228–274 (phoronids), pp. 516–609 (brachiopods), 1959. New York: McGraw-Hill Book Co.

Morris, R.H., D.P. Abbott, and E.C. Haderlie. 1980. *Intertidal Invertebrates of California,* pp. 108–114 (brachiopods, phoronids), pp. 177–226 (urochordates). Stanford: Stanford University Press.

Nielsen, C. 1977. The relationship of Entoprocta, Ectoprocta and Phoronida. *Amer. Zool. 17*:149–150.

Pennak, R.W. 1978. *Fresh-water Invertebrates of the United States,* 2nd ed., pp. 239–253 (tardigrades). New York: Wiley-Interscience Publication.

Rudwick, M.J.S. 1970. *Living and Fossil Brachiopods,* 191 pp. London: Hutchinson University Library.

Symposium on the Biology of the Lophophorates (a number of papers that deal with phoronids and brachiopods). 1977. *Amer. Zool. 17*:1–150.

Symposium on the Developmental Biology of Ascidians. 1982. *Amer. Zool. 22*:751–849.

17

The Echinoderms

OVERVIEW

The echinoderms, a most distinctive and peculiar group, are exclusively marine and constitute the largest number of invertebrate deuterostomates. They possess three unique, interrelated features that are of paramount importance in their functional morphology: symmetry, skeleton, and water vascular system. All adult echinoderms have pentamerous symmetry, which is derived secondarily from a bilateral larva. They develop an internal skeleton of separate or fused dermal ossicles that provides support, protection, and attachment points for muscles. The water vascular system is a canal system equipped with hydraulically operated motor units called tube feet, which are the principal means of locomotion and feeding. Echinoderms also are coelomate, are dioecious, have a subepidermal nervous system but without a brain, and lack excretory and gas exchange systems. Fertilization is typically in the sea, and planktonic larvae are developed. At the conclusion of the larval existence a dramatic metamorphosis ensues so that the bilaterally symmetrical larva becomes a pentamerous benthic adult.

Most crinoids are free swimming and have ten arms; the sea lilies, however, are permanently attached by an aboral stalk. Crinoids live oral side up and capture food particles on an extensive brachial, ambulacral system. They lack a madreporite and distinct gonads and have tube feet in clusters of three.

Asteroids usually have five arms that are not sharply set off from the central disc. Suckers on the tube feet assist in locomotion and food gathering, the latter being assisted by an eversible stomach. Most asteroids prey on large invertebrates. Digestion is achieved by enzymes secreted from paired digestive glands in each arm. Ophiuroids have five solid, jointed arms that are clearly set off from the central disc; these arms are unusually supple and employed in locomotion. The arms and five jaws assist in feeding. The interior of the central disc is mostly filled with stomach pouches and invaginated gas exchange sacs.

Echinoids have a rigid test composed of sutured ossicles and armed with numerous movable spines. Tube feet, stalked pedicellariae, and meridionally arranged ambulacra are all present. Sea urchins have a spherical body and long spines, whereas sand dollars and heart urchins have tertiary bilaterality and short spines. Most bear a distinctive complex masticating lantern with five prominent teeth. Holothuroids are bilateral because they are elongated in their oral–aboral axis, have an anterior mouth and a posterior anus, and lie on one side. Ambulacral tube feet are reduced in number, but buccal podia are greatly modified as food-gathering oral tentacles. A prominent gonad and a complex digestive system are present. A muscular cloaca pumps seawater into a pair of unique dendritic respiratory trees for gaseous exchange.

Our understanding of the phylogeny of echinoderms is far from complete, and their origins are lost in antiquity. It is not at all clear whether ancestral echinoderms were stalked or free moving and whether they were radial or bilateral in symmetry. Both stalked and unstalked fossil echinoderms are known; some were asymmetrical, while others were radial or pentamerous. However, the evidence suggests that the ancestral stem animals were attached aborally and perhaps were pentamerous as exemplified by the primitive crinoids. The shift from being attached with the oral side uppermost to being free moving with the oral side down was a major event that probably happened in several different evolutionary lines. Asteroids and ophiuroids are closely related, and echinoids and holothuroids share a remote common ancestor. All major extant groups had already diverged by the time the first known fossil animals lived.

INTRODUCTION TO THE ECHINODERMS

Among the invertebrates, echinoderms are the most popular symbols of marine life and some of the most celebrated, familiar, and curious creatures. Almost every artistic illustration of the oceanic environment depicts one or more echinoderms such as sea urchins, sand dollars, sea stars, or sea cucumbers. The Phylum Echinodermata is the largest and most important group of the invertebrate deuterostomates; the approximately 6500 species are exclusively marine and inhabit all areas of the oceans from intertidal zones to great abysses. Since almost all are bottom-dwellers, echinoderms often are very important macrofaunal constituents of benthic communities. Most are from 5 to 30 cm in their largest dimension; the largest size attained by modern-day forms is about 1 m, but a fossil sea lily had a stalk 25 m long! Almost every color of the spectrum is present in echinoderms, and their colors and patterns are often strikingly beautiful.

Echinoderms, a group whose origins antedate even the most ancient fossil record, flourished in the Cambrian and Ordovician periods, and their history undoubtedly extends well into the Proterozoic Era (see Table 1.1 in Chapter 1). Even though more than 20,000 fossil species are known, which would suggest a complete and continuous paleontological record, there are many evolutionary gaps, and many phylogenetic details are unclear. Because of the incompleteness of the fossil record as well as various interpretations of that record, a single taxonomic scheme is not accepted by all echinologists. Perhaps the most widely accepted arrangement divides the phylum into four subphyla: Homalozoa (all extinct), Crinozoa, Asterozoa, and Echinozoa. The characteristics of all four appear at the end of this chapter, and the three subphyla with living representatives are characterized in Table 17.1. For our purposes the precise taxonomic framework of all echinoderms, both extinct and extant, is not as important as are the diagnostic features and the functional morphology of living forms. Toward this end, features common to echinoderms will be covered along with those characteristic of the five taxa of extant forms, to which frequent references will be made. The five taxa are the crinoids (feather stars, sea lilies), asteroids (sea

TABLE 17.1. □ **BRIEF CHARACTERIZATIONS OF THE PRINCIPAL TAXA OF LIVING ECHINODERMS**

Taxon	*Common Name(s)*	*Body Shape*	*Location of Ambulacra*	*Other Characteristics*
Subphylum Crinozoa				
Class Crinoidea	Feather stars, sea lilies	Stalked or secondarily stalkless	On branched, food-gathering arms	Oral surface uppermost; most are stalked; arm podia in clusters of 3; lack distinct gonads and madreporite
Subphylum Asterozoa				
Class Stelleroidea				
Subclass Asteroidea	Sea stars (starfishes)	Flattened central disc	Along the 5 arms; open ambulacra	Arms not sharply demarcated from central disc; highly adaptable stomach
Subclass Ophiuroidea	Serpent, brittle, or basket stars	Flattened central disc	Along the 5 arms; closed ambulacra	Arms sharply demarcated from central disc; supple arms used in locomotion; arms jointed because of internal ossicles
Subphylum Echinozoa				
Class Echinoidea	Sea and heart urchins, sand dollars	Globose, cylindrical, or flattened	Equally spaced and meridional	Have sutured ossicles as a rigid test, pedicellariae, Aristotle's lantern
Class Holothuroidea	Sea cucumbers	Cylindrical, elongate	Somewhat equally spaced and meridional	Have oral tentacles, minute nonsutured ossicles, 1 gonad, tertiary bilaterality

stars), ophiuroids (brittle or serpent stars), echinoids (sea urchins, sand dollars), and holothuroids (sea cucumbers). These five are also characterized in Table 17.1, and the most important features of their principal subordinate taxa are highlighted at the end of this chapter.

General Characteristics

Echinoderms form the most readily recognized and distinctive phylum because they have no close relationships with any other phylum and because features characteristic of the group are almost universally possessed by all representatives. All echinoderms are deuterostomate and possess a large coelom, most are dioecious, their nervous system is rather simple and without a brain, and they lack an excretory system or any form of metamerism. Three diagnostic features that are unique to echinoderms sharply differentiate them from all other groups; these are their symmetry, skeleton, and water vascular system. These three distinctive features in

Figure 17.1 *Basic echinoderm symmetry and body orientations: **(a)** photograph of the aboral surface of an echinoderm (brittle star) to illustrate pentamerous symmetry; **(b–f)** stylized drawings of a representative of each of the five principal extant taxa showing body orientation and the locations of mouth (M) and anus (A).*

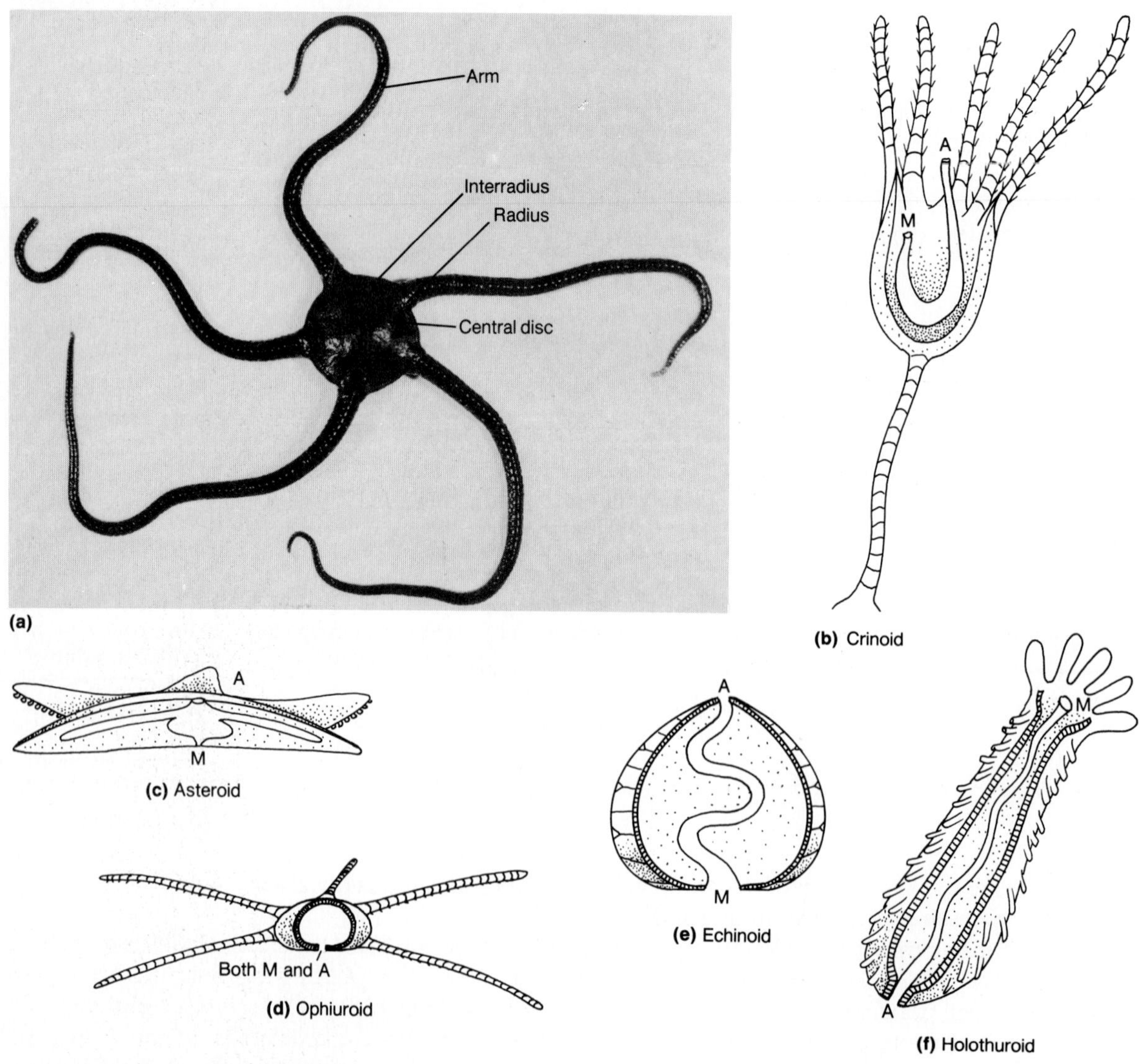

concert with each other are pervasive in most of the structures and functions of these absolutely fascinating creatures.

SYMMETRY

The most obvious feature of adult echinoderms is their symmetry, in which most anatomical parts are present in fives and aligned in five symmetrical axes arranged radially around a central part or pole. Is this not, one might logically ask, a clear example of radial symmetry like that of the radiates (Chapter 5)? No, because an examination of the embryonic development of a typical echinoderm would soon show that echinoderms develop a larval stage that, surprisingly, is bilaterally symmetrical. Further, most biologists believe that echinoderms evolved from very ancient bilateral ancestors. As the larval stage metamorphoses into an adult, a fundamental alteration in symmetry follows so that the adult assumes a radial symmetry that is secondarily derived. This dramatic transformation from bilaterality to radiality is unparalleled in the entire animal world and apparently was an early adaptation to a sessile or almost sessile existence. Since echinoderms have most body parts in fives, their symmetry is also referred to as **pentamerous.** Each of the five basic axes or planes of symmetry is termed a **radius,** and each area between adjacent radii is an **interradius** (Figure 17.1a). Because of their peculiar symmetry, almost every facet of echinoderm morphology has been substantially altered. Since there is neither a head nor a tail, the side of the body that has the **mouth** is called the **oral surface,** and the side opposite the mouth is called the **aboral surface,** which usually carries the **anal opening** (Fig. 17.1b–f). Internally, most echinoderm systems, reflecting pentamerous symmetry, are built on a surprisingly uniform plan of a central ring and radiating canals or parts situated in the five radial axes. This plan is especially evident in the water vascular system, the several systems derived from the coelom, the nervous system, and, to some degree, the digestive and reproductive systems.

Crinoids, the most primitive living group, are oriented with their oral surface uppermost, but other living echinoderms are positioned with the oral surface down next to the substratum (Fig. 17.1b–f). Holothuroids are exceptional in that they lie on one side and the mouth faces forward. Feather stars, sea stars, and brittle stars bear five (or a multiple of five) radial arms. Most echinoids also have very definite pentamerism, but it is of a more subtle nature because they are armless. Still other forms like the holothuroids and some echinoids (sand dollars) exhibit a return to bilateral symmetry. This means that they have a tertiary bilateral symmetry imposed on their secondary radial symmetry, which is in turn imposed on their basic or primary bilateral symmetry. This may be confusing, but this feature and others make echinoderms enormously intriguing.

Why do representatives of this entire phylum have perfect pentamerous symmetry, and why is it such a constant feature within the phylum? There are no concrete answers to these questions, but one plausible hypothesis is elaborated upon in Box 17.1.

SKELETON

The external surface of an echinoderm is a ciliated **epithelium** that covers a connective tissue **dermal layer** of the body wall; it is the dermis that produces many rigid skeletal plates. Since living tissues (epithelium, some dermis) lie on the outside of the skeleton, it is properly called an **endoskeleton.** The echinoderm skeleton, never constructed of bone or cartilage, consists of numerous calcitic plates or **ossicles** composed basically of calcium carbonate but containing considerable amounts of magnesium carbonate. These plates are bound together by connective tissue fibers, and they make up most of the body wall (Fig. 17.3). In asteroids and ophiuroids the ossicles are somewhat loosely articulated, thus permitting parts of the body to be

BOX 17.1 □ A HYPOTHESIS ABOUT THE ORIGINS OF PENTAMEROUS SYMMETRY IN ECHINODERMS

One of the principal distinguishing features of all modern-day echinoderms is their pentamerous symmetry. But why is their symmetry one of fives? In the fossil record, certain ancient echinoderms had no form of radial symmetry, others possessed trimerous or tetramerous symmetry, and still others had pentamerous symmetry. One intriguing hypothesis, set forth by Nichols (1969), warrants some attention.

We can observe the formation of ossicles in modern primitive echinoderms and hypothesize that these events also occurred in ancient ancestral forms. The development of a calcareous theca gave a great deal of protection to the metamorphosing larval echinoderm, which was basically immobile or could move only slowly. The first ossicles to be formed were those at the apex of the aboral or attaching surface. A central plate, usually carrying the anal opening, was formed first, followed by the development of a surrounding ring of five ossicles or plates (Fig. 17.2). But in the development of the larval theca the individual plates were only loosely bound together and were so vulnerable that a nudge from another animal might be of sufficient force to dislodge the plates. In this event the chief lines of weakness would have been the interossicle sutures, which would be very weak structurally. If such an event cracked the ossicles or mashed the larva severely, it would undoubtedly have been lethal. Therefore it was of great survival advantage to minimize the number of sutures and to have the suture lines as short as possible.

If these initial protective plates were to cover the aboral surface of the larval body completely with no holes between them and if the interossicle sutures should be as few and as short as possible, then what would be the optimal number of plates to have around the anal plate? If three ossicles were developed, there would be three main lines of weakness (Fig. 17.2a). But these three would be long and almost straight, and the ossicles would be easily separated. If there were four plates (Fig. 17.2b), there would be two straight, long lines of weakness that would also traverse the anal plate, which is already weakened by the opening for the anus. In a ring of six plates (Fig. 17.2d), there would be three potential fracture lines that would be straight and long and also involve the anal plate. But if the ring of ossicles were five, no suture would be in line with any

flexible. In echinoids, however, the skeletal plates are firmly sutured together to form a rigid, inflexible internal skeleton or **test.** As we shall see, there is a close functional association between the ossicles and the muscles associated with movement and locomotion.

The echinoderms' usage of calcite, however, is not unique among invertebrates. A few sponges form calcite spicules, some protozoans (such as foraminifera) incorporate calcite in their tests, and molluscs certainly use great quantities of calcareous prisms held together by organic fibers to form their shells. What is unique about the echinoderms' calcitic skeleton is the way it is formed and the fact that it is internal in the dermal layer. Each skeletal element or ossicle

other, even on the opposite side of the anal plate (Fig. 17.2c). A suture on one side would in fact be strengthened by a plate opposite it. Therefore with five plates in this initial ring the lines of weakness would be numerically minimized and kept as short as possible. It is quite plausible to conclude that pentamerism was of great adaptive significance to the ancestral stock that gave rise to extant echinoderms.

There are undoubtedly other possible explanations for the origins of pentamerous symmetry in echinoderms. But from the evidence available to us now, these two mechanical demands—lines of weakness being few and spatially short—strongly suggest that they were instrumental in influencing the universal occurrence of pentamerism in modern-day echinoderms.

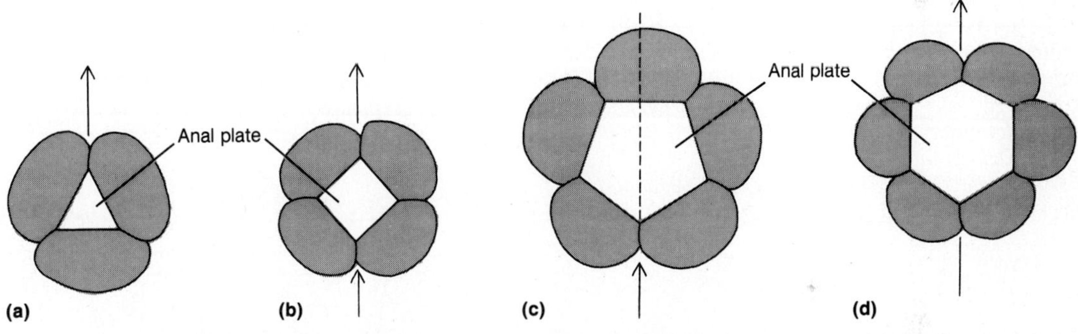

Figure 17.2 *Diagrams of the various possible numbers of plates surrounding the larval anal plate; arrows show lines of weakness:* ***(a)*** *three plates;* ***(b)*** *four plates;* ***(c)*** *five plates;* ***(d)*** *six plates. (parts a–d from D. Nichols, 1969,* Echinoderms, *4th ed., Hutchinson Publishing Group, used with permission.)*

initially is composed of either a single crystal of calcite or an aggregation of many microcrystals on which are deposited additional layers of calcite. By differential accretion these extracellular crystals can grow in any direction as determined by developmental influences to produce ossicles that are variable in shape. Also, some reabsorption or erosion of a developing ossicle can mold and shape the adult plates further. Certain of the plates are perforated to accommodate the tube feet, which are described below. The ossicles are most commonly platelike but may be ovate, rod-shaped, cruciform, or in some other shape (Fig. 17.3).

The skeleton of most echinoderms, especially that of the asteroids, ophiuroids, and echi-

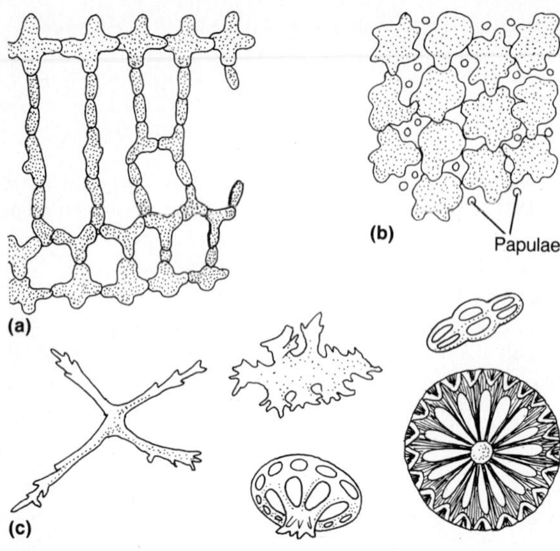

Figure 17.3 *Variations in echinoderm dermal ossicles: **(a)** articulated ossicles (asteroid); **(b)** ossicles only slightly articulated (asteroid); **(c)** a variety of nonarticulated ossicles (holothuroid). (parts a, b [from Fisher], c [from Theel, Danielsson and Koran] in L.H. Hyman,* The Invertebrates. *Vol. IV;* Echinodermata, *McGraw-Hill Book Co., 1955, used with permission.)*

noids, characteristically bears projecting calcareous tubercles, spines, or other protuberances that give the body surface a warty or spiny appearance (Fig. 17.4). It is because of this feature that these creatures are appropriately named (echino = spiny; derm = skin). In the asteroids and ophiuroids the spines are simple, short, and held erect on an underlying ossicle by muscles (Fig. 17.4a,b). But in the echinoids the spines reach their greatest development and are prominent external features that are sometimes longer than the body itself (Fig. 17.4c). In many echinoderms the spines have a certain degree of movement owing to small muscles located at their bases. Coming in a rather wide range of shapes and sizes, spines are used in locomotion, burrowing, protection, breaking up the force of waves, and even the protection of developing embryos.

The skeleton is one of the most important adaptations of echinoderms. As a supporting mechanism, it is relatively light. Its perforations serve as attachment sites for the connective tissue that binds ossicles together, for muscle attachments, and to prevent more serious fractures or cracks that might jeopardize the entire ossicle or test. The numerous spines, prominent externally in most animals, probably evolved initially as devices for defense or protection.

Figure 17.4 *Stylized drawings of echinoderm spines: **(a)** simple spines as parts of ossicles (asteroid); **(b)** larger, separate, immovable spines (ophiuroid); **(c)** movable spines (echinoid).*

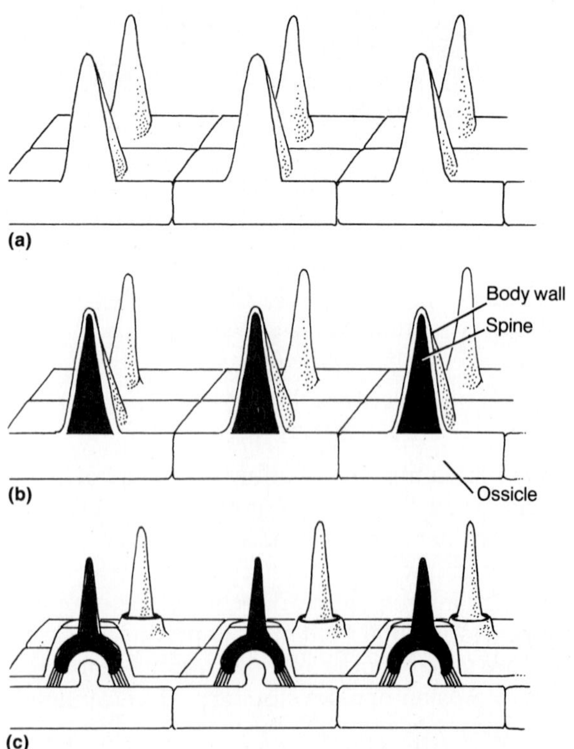

WATER VASCULAR SYSTEM

Perhaps the most remarkable characteristic of echinoderms, certainly the one that has influenced the most aspects of their functional morphology, is their **water vascular** or **ambulacral system.** Early in larval development, a portion of the coelom is partitioned into a series of contiguous canals that form the future water vascular system. These internal canals are connected to a great number of surface appendages, the **tube feet** or **podia,** situated on the oral surface (Fig. 17.5). It is perhaps the tube feet that are the most diagnostic feature for the entire phylum. The entire water vascular system, having a coelomic origin, is lined with a ciliated peritoneum and filled with seawater; by moving this ambulacral fluid through a closed system of canals the animals can hydraulically extend the tube feet and thus facilitate feeding or locomotion. Although a somewhat similar system exists in the tentacular region of sipunculids (see Chapter 9), the water vascular system of echinoderms, a most innovative adaptation, is really a hydrodynamic marvel.

The basic plan for the ambulacral system is a circumoral ring, radial canals, lateral canals, ampullae, and tube feet. The **ring canal** lies just beneath the oral surface and circles the mouth. Associated with the ring canal are one to five large, muscularized sacs called **polian vesicles,** which are reservoirs for ambulacral fluid, and, in some, **Tiedemann's bodies,** which produce coelomic cells (Fig. 17.5a). Joining the ring canal is a single vertical, asymmetrically arranged **stone canal** that typically leads away from the ring canal toward the aboral surface; it is so named because of calcareous deposits in its wall. The stone canal is continuous as a small **aboral ampulla** that connects through many minute pore canals to the external surface. The prominent dermal ossicle bearing these pore canals is the **madreporite,** located aborally in an interradius. The acentric position of the madreporite is a conspicuous indicator of echinoderms' basic bilateral symmetry (Fig. 17.5a).

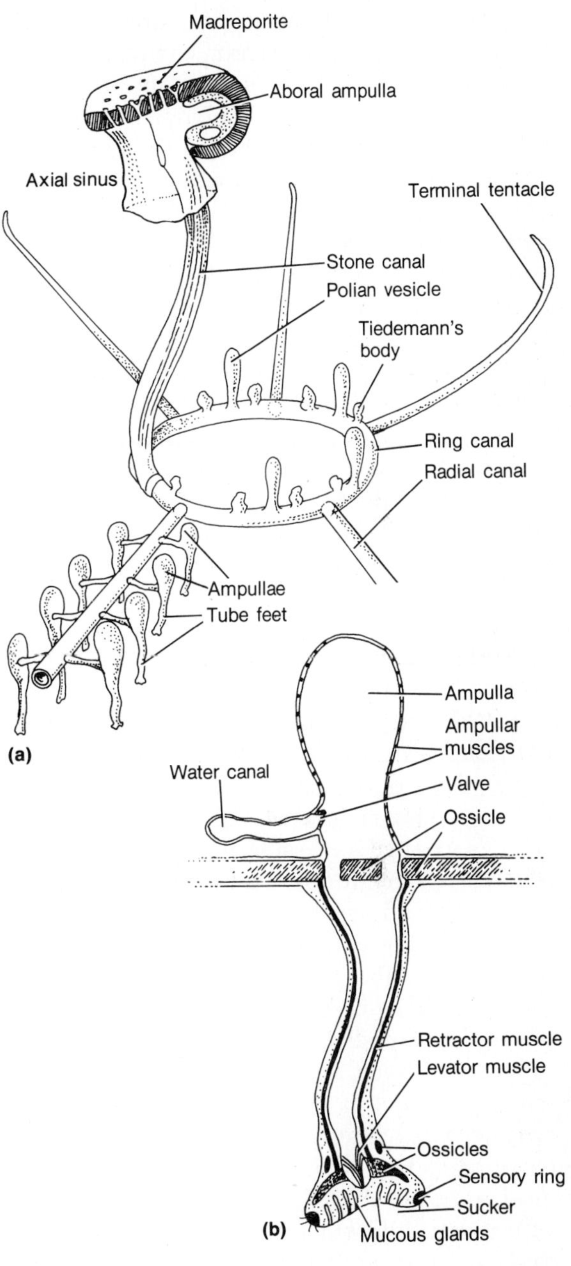

Figure 17.5 *The water-vascular system of echinoderms:* ***(a)*** *a diagrammatic representation of this system (asteroid) (from D. Nichols, in* Physiology of the Echinodermata, *R. A. Boolootian, ed., ©1966, John Wiley & Sons, Inc., used with permission);* ***(b)*** *a longitudinal section through a podial unit (echinoid) (from D. Nichols, 1969,* Echinoderms, *4th ed., Hutchinson Publishing Group, used with permission).*

Arising from the ring canal are five **radial** or **ambulacral canals** that run the length of the arms (crinoids, asteroids, ophiuroids), meridionally inside the test (echinoids), or meridionally just beneath the body wall (holothuroids). Those external areas of an echinoderm's body beneath which run the radial canals are called **ambulacra,** and the area between two adjacent ambulacra is an **interambulacrum.** Along its length, each radial canal has numerous paired **lateral canals,** each of which connects to a blind-end **podial unit** consisting of an ampulla and a tube foot (Fig. 17.5). An **ampulla** is a small sac-like structure whose wall contains a network of muscles and connective tissue; each ampulla is continuous externally as a hollow cylindrical **tube foot** (Fig. 17.5b). The locations of the tube feet mark the external positions of the five ambulacra.

Even though the tube feet are organs of amazing functional variations, they are all built on a rather uniform plan. Each podium is covered externally by an epithelium along whose base runs a nerve plexus, and lining its lumen is the ciliated coelomic peritoneum. Between the two epithelial layers is an outer, supportive, layer of connective-tissue inside of which is a thin layer of **longitudinal retractor muscles** (Fig. 17.5b). These important podial muscles are inserted into the ambulacral ossicles adjacent to where the tube foot passes (stelleroids) or through whose pores the podium passes (echinoids). Each podial unit is separated from the rest of the water vascular system by a **valve;** when the valve is closed, each unit then can operate independently as a single hydraulic effector unit (Fig. 17.5b). With the valve closed, muscles in the ampullar wall contract and force ambulacral fluid out into the tube foot, which easily protracts or extends the podium. But for the tube foot to be withdrawn, the ampullar muscles must relax, and the podial longitudinal muscles then shorten to retract the tube foot. Thus muscles in one part (ampulla) extend the tube foot hydrodynamically, and muscles in another part (tube foot wall) retract the podium by their contraction (Fig. 17.5b). It is important to note that the longitudinal muscles of the tube foot insert into the edge of the pore of the ambulacral ossicle through which the podium itself protrudes. Functionally, this means that the podial muscles would be totally ineffective if they were not anchored into the ambulacral ossicles.

There is not a continuous circulation of seawater from the outside through the madreporite into the water vascular system as one might expect. However, some water may enter to replace water lost as tube feet are injured and thereby leak a small amount of ambulacral fluid. Experimentally blocking the openings in the madreporite apparently does not perceptibly impair the normal functioning of an echinoderm or its ambulacral system. Water is moved throughout the water vascular system by peritoneal cilia lining the entire system. In most forms, cilia in the tube foot lumen are arranged in two longitudinal bands; each band beats in an opposite direction from the other so that efficient circulation is maintained. In sea stars and perhaps other echinoderms as well, water enters each tube foot on the outer side and leaves on the side nearest the ambulacral axis.

Primitively, the tube feet perhaps were simple outpockets from the body wall to facilitate gaseous exchange. It is believed that the podia became specialized somewhat later to collect food and transport it to the mouth, but in most echinoderms the tube feet are most notable for their locomotive functions. Tube feet possess some particular adaptations in keeping with their many possible functions. In asteroids, echinoids, and holothuroids the podia are used in stepping locomotion, which is facilitated by each podium having at its distal end a muscle-operated **sucker** or **disc** (Fig. 17.5b). The podia are used either in levering as the animal moves horizontally or in traction or climbing as it moves vertically.

Tube feet can even be bent along their length; such postural bending is achieved by the regional contraction of the longitudinal muscles on one side. In other echinoderms like the crinoids and ophiuroids the podia are pointed or have sticky secretions and are used in anchoring or feeding. In many holothuroids the tube feet assist the animal in burrowing through the benthic mud or sand. All tube feet have enormous numbers of epithelial sensory cells (up to an estimated 70,000 per mm^2), so the podia are the principal loci for sensory reception. In all living echinoderms the mouth is always surrounded by five (or a multiple of five) **buccal tube feet** that are sensory in function. They apparently "taste" the food before it enters the mouth. The buccal tube feet in holothuroids are highly modified as food-gathering tentacles.

OTHER EXTERNAL FEATURES

As was noted earlier, the integumental surface of an echinoderm is very uneven and equipped with several different kinds of projections; two prominent external structures—spines and tube feet—have already been noted (Figs. 17.4, 17.5b). Two other surface structures—papulae and pedicellariae—are present on many echinoderms. **Papulae** are thin-walled saclike evaginations situated especially on the aboral surface (Fig. 17.6e). These numerous structures, along with the tube feet, are the principal gas exchange structures between the environment and the animal. **Pedicellariae** are small pincerlike organs situated among the spines of echinoids and some asteroids (Fig. 17.6a–d). The pedicellariae are among these animals'

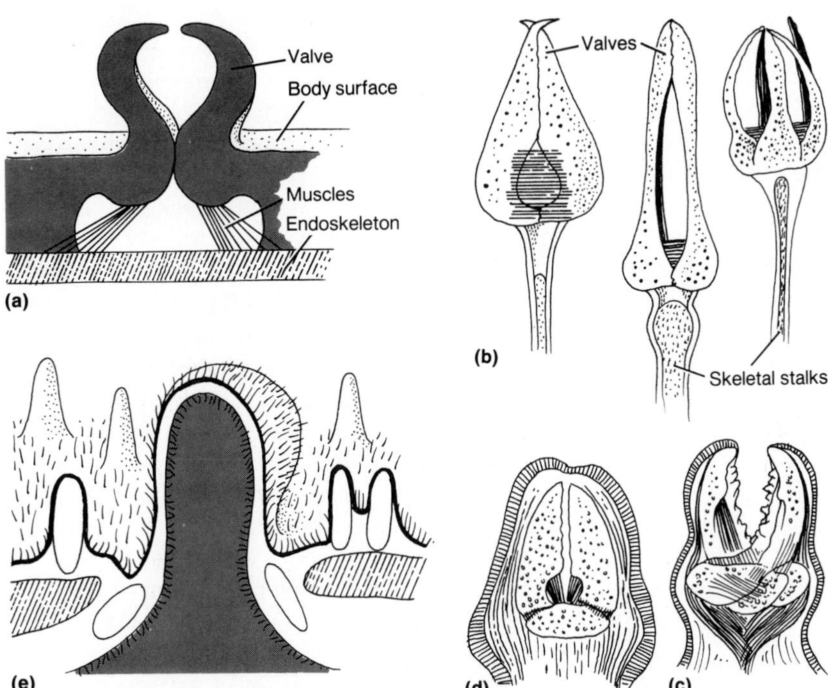

Figure 17.6 *Pedicellariae and papulae:* ***(a)*** *sessile pedicellaria;* ***(b)*** *several different stalked pedicellariae;* ***(c)*** *features of a stalked crossed pedicellaria (from E.W. Macbride, 1909,* Cambridge Natural History, *Macmillan, London);* ***(d)*** *features of a straight pedicellaria;* ***(e)*** *a section through a papula. (parts b [from Mortensen], d in L.H. Hyman,* The Invertebrates. *Vol. IV: 1955,* Echinodermata, *McGraw-Hill Book Co., used with permission.)*

most fascinating features, and they are exceedingly important to many echinoderms.

The external surface of echinoderms is an attractive site on which the larvae of many sessile invertebrates might settle. But these commensals would be most unwelcome visitors to the echinoderm because they would seriously impair the functioning of the papulae, tube feet, madreporite, and general epithelium. As an adaptation that protects them from particles or larvae settling on their surfaces, many echinoderms are provided with pedicellariae, which keep the body surface free from encrusting organisms. Analogous structures also appear in the Bryozoa, in which modified zooids, the avicularia, are remarkably similar to the pedicellariae (see Chapter 15).

In the asteroids, both sessile and stalked pedicellariae are found. In echinoids, where these organs are best developed, they are always stalked (Fig. 17.6b–d). A pedicellaria is composed of two or more short calcareous spines or plates that oppose each other and, by their articulations, act as pincers. The nonsessile pedicellariae have a fleshy movable stalk of variable length with an inner calcite rod and bearing the opposite spines at its distal extremity. The blades or jaws are operated by two sets of antagonistic muscles, the larger adductors and the smaller abductors (Fig. 17.6c, d). When the jaws are touched on the outside, they open; when touched on their inner surfaces, the blades snap shut. The pedicellariae of asteroids are nonpoisonous, but in certain echinoids, poisonous glands are associated with them. The jawlike pedicellariae are most effective in keeping the surface absolutely free of debris and commensal organisms.

Body wall and internal features

The body wall of an echinoderm consists of several layers. These are, in order, beginning with the outside: epidermis, nerve plexus, dermis with ossicles, muscles, and peritoneum. The outer **epidermis** covers the entire surface including that of the spines and consists of ciliated columnar, neurosensory, and mucous cells; the cilia and mucus protect and clean the surface and may be used in food gathering. Beneath the epidermis is a thin subepidermal **nerve plexus,** a part of the ectoneural system that will be discussed more fully below. The **dermis** is a relatively thick layer of connective tissue that contains the ossicles. On the coelom side of the dermis are thin layers of outer **circular** and inner **longitudinal muscles.** The innermost layer of the body wall is the **parietal peritoneum.**

Coelom. In adult echinoderms the coelom is anatomically subdivided into four components of varied functions:

1. water vascular system,
2. perivisceral coelom,
3. hemal system, and
4. perihemal system.

All four coelomic components are filled with fluid that is isosmotic with seawater and contain large numbers of phagocytic cells, the **coelomocytes,** that function in excretion, transport, and immune responses. The coelomic peritoneum is ciliated, and the ciliary beat is responsible for a continuous circulation of coelomic fluid. The principal features of the remarkable water vascular system have already been mentioned. The **perivisceral coelom,** the largest of the four, surrounds the visceral organs as a spacious chamber and extends the length of the arms in crinoids and asteroids. It has been shown recently in some echinoderms that the principal nutrient transport system is the perivisceral coelomic fluid. This same fluid perhaps also has a greater role in internal gaseous transport than once was thought especially in the larger echinoderms. But this rather inefficient means for transporting nutrients and gases is clearly one of the most physiologically limiting features of these animals.

The **hemal system,** which takes several different forms in various taxa, is a system of small, poorly–defined, fluid-filled sinuses basically paralleling the ambulacral system. The hemal system consists of an **oral** and an **aboral ring;** these communicate with each other via an elongated mass of spongy tissue called the **axial gland** which parallels the stone canal (Fig. 17.7). The precise function of the axial gland is debatable, but there is some evidence that a small appendage located in the madreporite, the **dorsal sac,** pulsates. This is suggestive of a circulatory function. Arising from the oral ring are five radial **hemal sinuses** that extend along the ambulacra; from the aboral ring arise five similar radial sinuses or channels that run along the aboral surface. The **perihemal system** is a system of channels, spaces, and sinuses surrounding the various parts of the hemal system like a sleeve. It consists of oral and aboral rings, radial strands arising from both rings, and an **axial sinus** surrounding the axial gland (Fig. 17.7b). The radial aboral strands of the perihemal system are continuous as genital sacs surrounding

Figure 17.7 *Schematic diagrams of the radial or circular arrangement of some coelomic systems and parts of the nervous system: **(a)** the circumoral complex of important coelomic and nervous components; **(b)** the hemal and perihemal systems.*

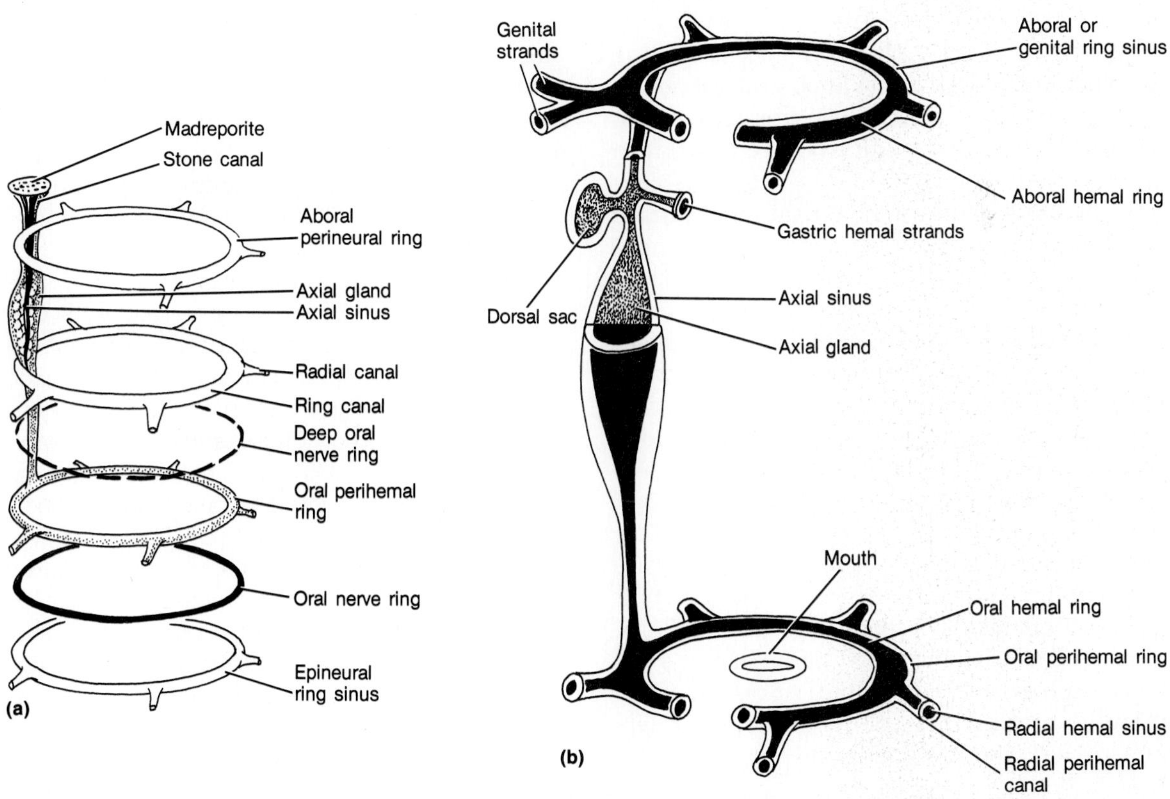

the gonads. The precise functions of the hemal and perihemal systems are poorly known; some circulation within both systems is known to take place, and wandering coelomocytes may aid in the distribution of food materials. Their fluid, sometimes erroneously called blood, is really only coelomic fluid. Since echinoderms have coelomic fluid that is the same osmotically as seawater and lack an effective means of excretion or osmoregulation, echinoderms are forever limited to marine habitats.

Visceral systems. Feeding patterns and the types of food consumed are quite diverse in echinoderms. The primitive method of nutrition was probably by suspension feeding, and tube feet became specialized for directing particles of food along the ambulacra to the mouth. Perhaps a majority of modern-day echinoderms are predaceous, but scavengers, deposit-feeders, and grazers are often found; most echinoderms are generalists in their diets. The alimentary system is composed of a simple short tract extending from the oral mouth to the anus situated on the aboral surface. In the various groups, parts of this tract are often modified to reflect differences in food preferences. The **mouth** is centrally located at the point on the oral surface where ambulacra and interambulacra converge. While masticating structures are absent in many echinoderms, ophiuroids have five small jaws, and echinoids have a complex jaw mechanism surrounding the mouth. A **stomach** is always present but is a highly variable organ depending upon the nutritional habits of the various echinoderms. In those forms that feed on detrital or particulate materials, the stomach is a simple sac, but in the highly predaceous asteroids it is often eversible, complex in construction, and equipped with enzyme-secreting digestive glands. Cilia lining the stomach's lumen aid in circulating food and enzymes. Typically, an **intestine,** either short or coiled, arises from the aboral surface of the stomach and terminates with the aboral **anus** located on the anal plate. In some groups the intestine and anus are either poorly developed or absent altogether, and the stomach is simply a blind-end sac. Echinoderms lack organs of excretion or osmoregulation, but gaseous exchange takes place at the papulae, in the tube feet, or in specialized structures.

The nervous system of echinoderms is rather simple, mostly nonganglionated and lacking a brain, and always subepidermal in location; in fact, it is somewhat reminiscent of that of the radiates but is more organized. The echinoderm nervous system consists of three separate units, the most obvious of which is the oral **ectoneural system** lying just beneath the thin epithelium of the oral surface (Fig. 17.7a). It consists of one to three oral rings, five radial nerves associated with each ring lying adjacent to the ambulacral regions, and a general subepidermal network located just beneath the body wall epithelium. The ectoneural system is mostly a sensory plexus and receives nerve impulses from the tube feet, pedicellariae, and other surface features. A deeper oral **endoneural system** consists of a nerve ring and radial nerves and is associated with the gonads and anus (Fig. 17.7a); it is particularly important in crinoids, in which it innervates essential locomotor muscles. A third system, the aboral or **hyponeural system,** also consists of a nerve ring and radial elements and is associated with the important motor functions of the tube feet. Echinoderm sense cells are mostly simple and exceedingly numerous over the general surface; tactile and chemoreceptors predominate especially on the tube feet, and asteroids have a light-sensitive eye spot at the tip of each arm.

Echinoderm behavior is predicated on the fact that they are plodding, comparatively slow-responding, pentamerous creatures living on the ocean floor. While whole body responses are evident as echinoderms move, feed, and engage in other activities, most of their behavior involves local reflexes in such structures as podia, pedicellariae, and spines or local movement between ossicles. Echinoderm activities will be discussed with each major group.

Almost all echinoderms have separate sexes but lack any form of external sexual dimorphism. Crinoids lack distinct gonads, and gametes are produced by certain areas of the perivisceral coelomic peritoneum. Most all other echinoderms have five (or ten) radially arranged **gonads,** each of which is enclosed in a genital sac. The **genital sacs** are outgrowths of the genital or aboral sinus; thus they are structurally a part of the perihemal system. Each gonad opens to the exterior by a separate aboral **gonopore,** and there is no elaboration in secondary sexual structures or the ductal system. Gametes are spawned into the sea, where fertilization normally takes place. However, the females of some cold-water (antarctic) species in all five extant taxa actually brood their young either internally or, more likely, on their external surfaces. If brooding occurs, the eggs are large and contain rather large amounts of yolk, and the embryos develop directly and do not pass through a separate larval stage. The stimulus for spawning is the presence of gametes in the water from animals of the opposite sex. Also involved is a hormonal influence mediated by the ectoneural system.

Development and Regeneration

Typically, fertilization is external, and development proceeds in the sea as planktonic stages are formed. The ova are homolecithal, and cleavage is both radial and indeterminate, characteristic of deuterostomates (see Chapter 3). The resulting gastrula has a narrow archenteron or primitive gut connected to the exterior by the blastopore, an opening that will develop into the anus. It should be noted here that in echinoderms and all deuterostomates the blastopore gives rise to the future anus, and the mouth develops as a later secondary opening. The developing archenteron grows anteriorly and later connects to the stomodeum to form the mouth. As the archenteron develops, it eventually gives rise to three pairs of lateral pockets whose walls are mesodermal layers (Fig. 17.8). The cavities of these enterocoelous pouches develop into the coelomic spaces in adults, and the mesoderm will eventually produce the skeleton, muscles, and most of the viscera. While specific details about the formation of the coelom and mesoderm vary among the five principal taxa, a general overall developmental picture can be drawn.

Primitively, the two coelomic pouches, formed in the gastrula stage, separate from the archenteron and elongate in a direction parallel to the gut. Each of these lateral coelomic tubes then becomes subdivided into three separate compartments aligned one behind the other; they are the anterior **axocoel,** a middle **hydrocoel,** and a posterior **somatocoel** (Fig. 17.8a). In some echinoderms the three pairs arise individually from the archenteron, and the left axocoel and hydrocoel typically join and eventually form the ambulacral system (Fig. 17.8b, c). The left axocoel forms a dorsal **hydropore** or presumptive madreporite on the surface of the embryo, and the left hydrocoel becomes a scalloped loop around the esophagus that eventually gives rise to the ambulacral ring canal (Fig. 17.8c). The tubular connection between the left axocoel and hydrocoel is the precursor of the stone canal. Usually, the axocoel and hydrocoel on the right side atrophy and never develop into any adult features (Fig. 17.8b, c). The posterior somatocoels fuse above and below the gut and form the perivisceral coelom, and the medial borders of these two cavities unite to form the visceral mesenteries (Fig. 17.8d).

While the coelomic spaces are developing, the gastrula develops into a planktonic, bilaterally symmetrical, ciliated larval stage. The cilia, initially uniformly distributed over the surface of the gastrula, become concentrated into distinct locomotor ciliary bands that wind over

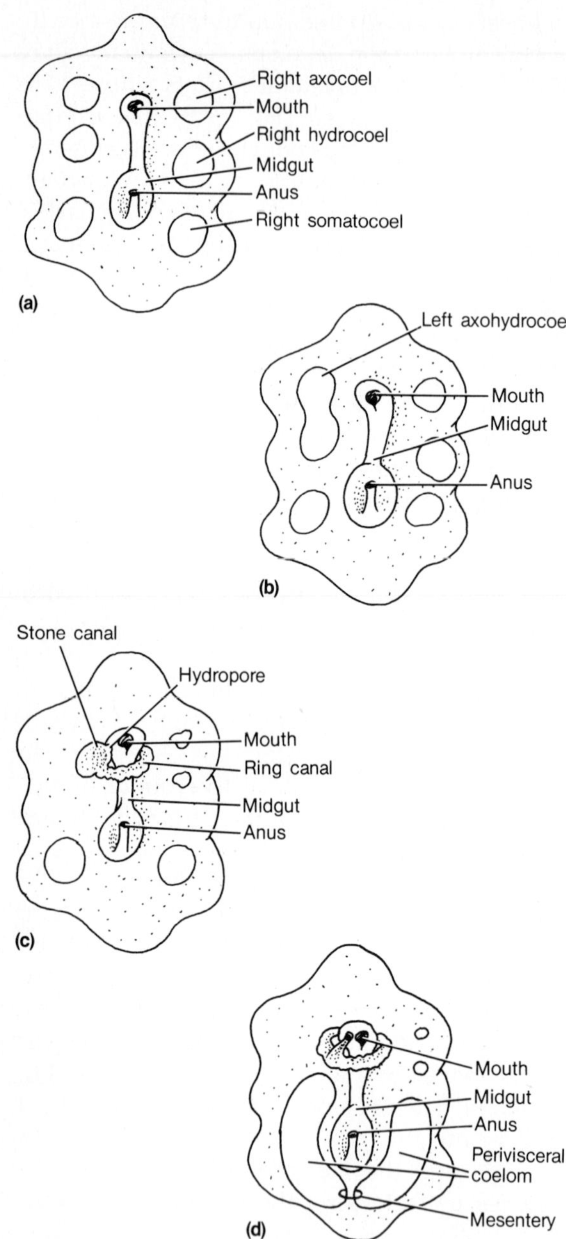

Figure 17.8 *Diagrams of a generalized echinoderm larva showing the development of coelomic pouches and their derivatives:* ***(a)*** *three pairs of coelomic pouches;* ***(b)*** *fusion of the left axocoel and the left hydrocoel;* ***(c)*** *development of the hydropore, stone canal, and ring canal;* ***(d)*** *development of the perivisceral coelom from both somatocoels.*

the larval body. The digestive tract is now functionally developed into a ciliated ventral stomodeum, esophagus, stomach, intestine, and ventral anus. Minute phytoplankton or suspended particles of food are swept into the mouth by the action of the stomodeal cilia. Typically, projections develop from the body wall to form short or long larval **arms** or **lobes,** which are useful in buoyancy and locomotion during the planktonic existence of the larva. These arms, not in any way equivalent to the adult arms of crinoids, asteroids, or ophiuroids, are quite variable in number and shape among the five principal groups; they serve as the principal basis for distinguishing different larval types.

One prevalent theory holds that the basic generalized larval stage, the **dipleurula,** was evolved in echinoderms. From this hypothetical dipleurula stage the various larval forms present in the five taxa of echinoderms may have been derived. Each larval type has been given a particular name. Since larval nomenclature and means for distinguishing between the various larval types are often confusing, Table 17.2 lists the basic larval type for each of the five groups and a succinct account of their most important features. Each larval type is illustrated in Figure 17.9.

Several provocative questions can be raised about echinoderm larvae at this point. What is the underlying reason for such diversity in larval morphology? Since all echinoderm larvae are planktonic in the oceans, what is the functional significance in the variation of ciliary tracts and arms in the various groups? Why do the closely related asteroids and ophiuroids have such different larval stages? Does the presence of similar larvae in both ophiuroids and echinoids mean that they are closely related? Can phylogenetic affinities be established on the basis of similarities or differences in larval types? The definitive answers to these questions are not readily available, but several different hypotheses have been proposed to formulate a generalized or idealized scheme for understand-

TABLE 17.2. □ **CHARACTERISTICS OF THE VARIOUS LARVAE OF ECHINODERMS**

Echinoderm Taxon	*Larval Name*	*Most Important Features*
Crinoidea	Vitellaria	Nonfeeding; barrel-shaped; no arms; 5 ciliary bands run transversely around body
Asteroidea	Bipinnaria	Single ciliated band surrounds both mouth and anus; arms develop late
Ophiuroidea	Ophiopluteus	4 pairs of long ciliated arms strengthened by calcareous rods
Echinoidea	Echinopluteus	4–6 pairs of long ciliated arms strengthened by calcareous rods
Holothuroidea	Auricularia	One ciliated band surrounds both mouth and anus

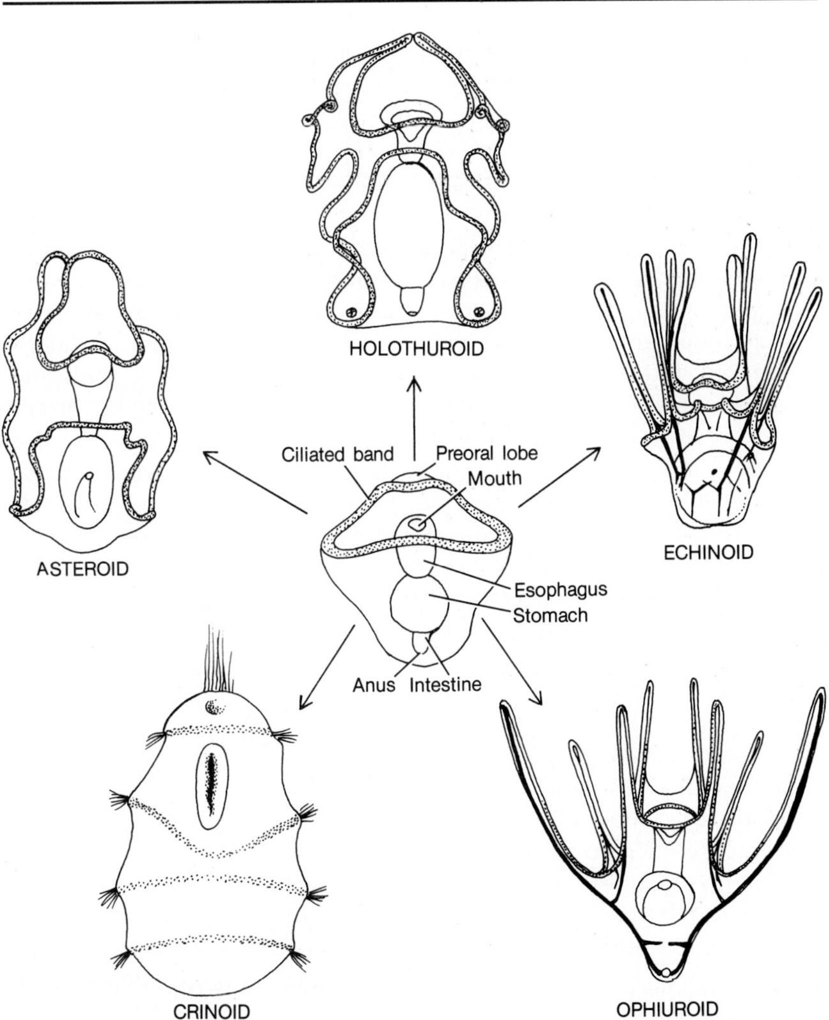

***Figure 17.9** The probable derivation of larval types from the basic generalized dipleurula larva (from G. Ubaghs in R.C. Moore, ed. 1967.* Treatise on Invertebrate Paleontology. *University of Kansas and the Geological Society of America, used with permission of G. Ubaghs).*

ing possible phylogenetic relationships between echinoderm taxa based on the larval types.

One prominent hypothesis has attempted to characterize a generalized echinoderm with the dipleurula larva as the ancestral form, from which were derived the various echinoderms by way of larval modifications (Fig. 17.9). But there are inherent dangers in these approaches, since similar larval stages, by themselves, do not constitute irrefutable proof of close relationships. The entire issue of phylogeny is often compounded by evolutionary convergence between distantly related taxa or divergence within a large taxon. For example, the similarity of larval stages in ophiuroids and echinoids is obviously a case of convergent evolution of the early larval stages of two groups that certainly are not closely related. Larval phylogeny and the possibility of an ancestral dipleurula are provocative issues, but they should be tempered with all other evidence before any attempt to establish a phylogenetic association.

Toward the end of the larval stage, all echinoderms undergo an elaborate metamorphosis from larval bilaterality to adult pentamerism. This transition is a remarkable and fundamental feature of their development. Crinoid and asteroid larvae attach themselves to a suitable substratum during the metamorphic process, but the larvae of other groups remain free-swimming during metamorphosis until the weight of the developing endoskeleton makes them too heavy to remain as plankton. Metamorphosis also involves the degeneration of the larval arms, and in some the mouth and parts of the alimentary system are reorganized and reformed. The left side of the larval body becomes the oral surface, and the right side becomes the aboral surface. Following the developmental lead of the formation of the water vascular system, all other central and radial systems subsequently appear in a precise developmental sequence. The time from fertilization to the development of a benthic adult normally requires from three weeks to six months, and sexual maturity may require several years.

Even though all living things have some capacity for repair, replacement, or regeneration of lost or damaged parts, the capacity for regeneration in echinoderms is especially great and is one of their most notable traits. Damaged arms, spines, podia, and even internal organs are quickly regenerated without any apparent incapacitation to the organism during the regenerative process. Actively involved and essential for regeneration, the coelomocytes migrate to the broken or ruptured area, phagocytize damaged tissues, and carry nutrients to the growing tissues. Tissues adjacent to the wound apparently dedifferentiate, actively grow, and then differentiate into new adult parts, all of which are probably under the control of the nervous system.

Regeneration is perhaps best seen in echinoderms with arms (crinoids, asteroids, ophiuroids). Missing arm parts can be regenerated, and if at least 20% of the central disc remains, the rest of the animal is normally regenerated. Asteroids undergo a type of asexual reproduction called **fissiparity** or self-division in which the central disc normally breaks into two parts, each with two or three arms, and the missing parts are then regenerated. Ophiuroids demonstrate the property of **autotomy** or self-surgery in which an arm can be broken off or self-amputated, and the missing arm portion is regenerated. Some holothuroids can **eviscerate** or void some of their internal organs and regenerate those lost parts. Regeneration is another fascinating property of this remarkable group, the echinoderms.

SUBPHYLUM CRINOZOA, CLASS CRINOIDEA

The crinoids are the most primitive living echinoderms, and modern-day species represent the survivors of very ancient echinoderm stocks. More than 5000 fossil species are known, and their fossil record extends back into the lower Cambrian strata; crinoid origins, however, undoubtedly go well back into the Proterozoic Era. There are two extant groups of crinoids: the more primitive sea lilies, which comprise about 80 species of sessile stalked forms, and the more recent feather stars or comatulids, whose approximately 550 species are mostly stalkless and free-swimming. Crinoids live in most seas but especially in the Indian and Pacific oceans, where they inhabit coastal zones (feather stars) or deeper waters of up to 8200 m (sea lilies). They range in diameter from 3 cm to about 1 m inclusive of the arms, and the average stalk length of sea lilies rarely exceeds 60 cm. Feather stars are variably colored, the arms are frequently banded, and their delicate beauty in both color and graceful movement is outstanding.

External Features and the Body Wall

The crinoid body consists of an attachment stalk and an upper crown with five to many long, pinnate arms (Fig. 17.10). The **stalk,** character-

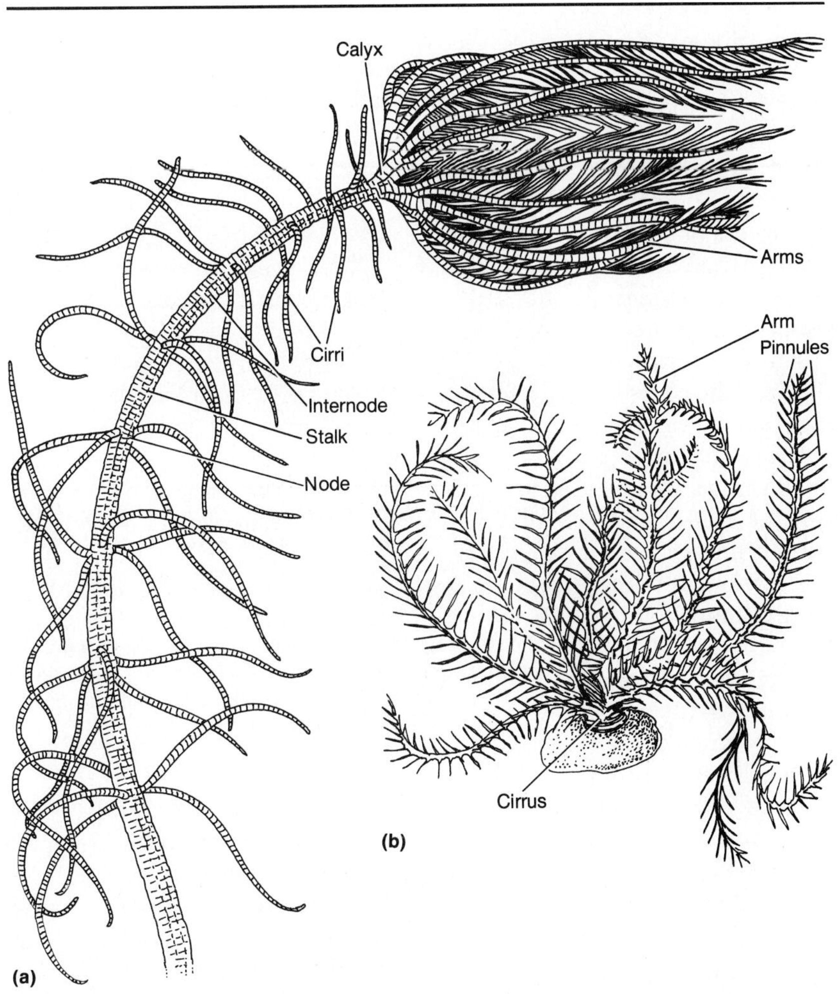

Figure 17.10 *Some examples of crinoids:* ***(a)*** Cenocrinus, *an isocrinid or sea lily;* ***(b)*** Antedon, *a comatulid or feather star, resting on a small stone (parts a, b from Carpenter in L.H. Hyman,* The Invertebrates. *Vol. IV:* Echinodermata, *McGraw-Hill Book Co., 1955, used with permission).*

istic of adult sea lilies and of the early developmental stages of feather stars, is mostly cylindrical in cross section. It consists of a linear series of internal ossicles, which give it a jointed appearance. Typically, the stalk bears whorls of small jointed branches of **cirri** (Figs. 17.10a). Even though feather stars break away from the stalk and are free living, a circlet of cirri is retained at the point where the stem once joined the crown (Fig. 17.10b). The lower terminal end of the stalk bears holdfast cirri, which anchor it into the substratum. The terminal joint of each cirrus is clawlike, a feature that enables crinoids to grasp objects on the bottom temporarily (feather stars) or when the stalk becomes dislodged (sea lilies).

The pentamerous **crown**, the principal portion of the body, is equivalent to the body of other echinoderms and is joined to the stalk at its aboral pole. Thus the body orientation of a crinoid is always oral side uppermost, in contrast to that of all other echinoderms. The parts of the crown are an aboral calyx, an oral tegmen, and lateral arms. The wall of the cuplike **calyx** is constructed mostly of dermis and dermal ossicles. There is a single centrally located, aboral **centrodorsal ossicle** by which the crown is attached to the stalk. Surrounding the centrodorsal ossicle are typically two circles of five ossicles each, the aboral **basals** and an upper circle of **radials** (Fig. 17.11b, c); in some crinoids a third ring is located between the basals and the centrodorsal. The **tegmen** is a dome-shaped membranous layer that may be reinforced with scattered ossicles, and it bears the **mouth** surrounded by five pairs of buccal tube feet. Five **ambulacral grooves** extend from the central mouth across the tegmen and onto the arms, thus dividing the tegmen into five ambulacral and five interambulacral areas (Fig. 17.11a). The tegmen is pierced with numerous (500–1500) small **ciliated funnels** that permit seawater to pass between the outside and the coelom. Also situated on the tegmen is an elevated **anal cone** bearing the **anus** (Fig. 17.11a).

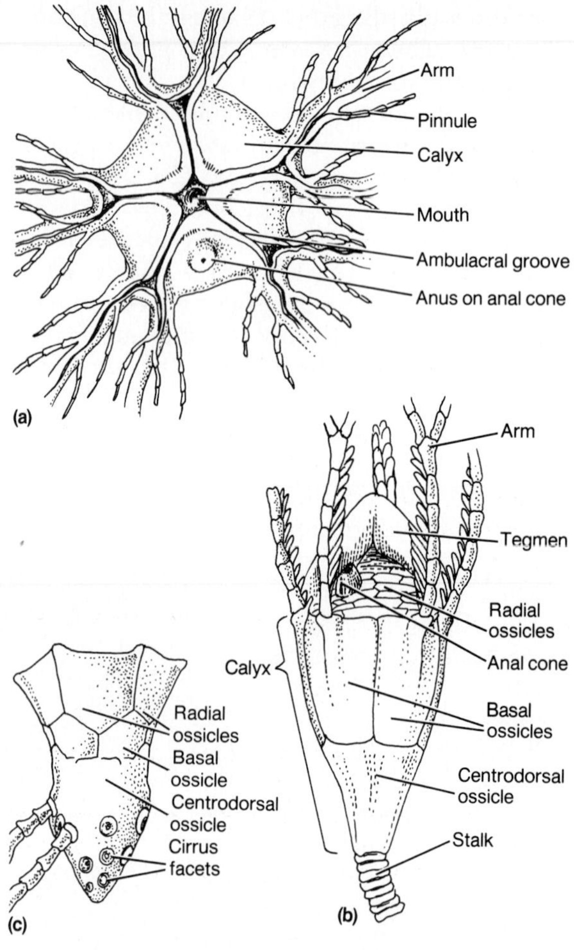

Figure 17.11 *Details of the external features of and ossicles in the crown of crinoids:* ***(a)*** *oral view of* Antedon *(from E.W. MacBride, 1909,* Cambridge Natural History, *Macmillan, London);* ***(b)*** *ossicles of the crown of* Hyocrinus, *a primitive crinoid;* ***(c)*** *ossicles of the crown of* Atelecrinus, *a more advanced crinoid. (parts b [from Carpenter], c [from Clark] in L.H. Hyman,* The Invertebrates. *Vol. IV:* Echinodermata, *McGraw-Hill Book Co., 1955, used with permission.)*

The **arms** originate from the sides of the body at the point where the calyx and tegmen join. Primitively, there are only five arms, but in most crinoids the arms fork near their base to form ten arms; some crinoids have more than ten arms—up to 200 in the comatulid *Comanthina*—formed by repeated forking. Numbers of arms, sometimes varying between individuals of the same species, appear to be a function of water temperature, depth, and even food conditions. Animals living in warm or shallow seas or in areas with limited food supplies have more arms, and the reverse is true for comatulids, which are found in colder or deeper waters or where food organisms are more numerous. The average arm length is 10–12 cm, and arms are usually shorter in animals living in colder water. Owing to the presence of a linear series of mostly solid **brachial ossicles,** the arms are jointed. The arms bear a number of lateral appendages, the **pinnules** or armlets, which give them a feathery appearance and are the basis for the common name, feather stars (Figs. 17.10b, 17.12a). Like the arms and stalk, each pinnule is jointed because it is constructed of a series of smaller ossicles (Fig. 17.12a–c). The ossicles of stalk, arms, and pinnules are so articulated to permit some interossicle movement. Distinct connective tissue ligaments bind arm ossicles together, and the elasticity of these ligaments, especially those on the aboral surfaces, is responsible for extension of arms and pinnules (Fig. 17.12c, e). Paired interossicle muscles, situated on the oral surface, are responsible for the flexion of arms toward the mouth. Thus the aboral brachial ligaments and oral arm muscles are antagonists.

A crinoid has an extensive branching ambulacral system of ciliated grooves extending along the full length of all arms and pinnules on its upper oral surface (Fig. 17.12b, c). *Metacrinus,* with about 55 arms, each of which averages 25 cm in length, has combined ambulacral groove length of 80 m. The margins of the ambulacral grooves are covered by slightly movable plates or **lappets,** which alternately expose and cover the ambulacral grooves. Beneath each lappet is a group of three partially fused tube feet (Fig. 17.12d). The tube feet, bearing numerous mucus-secreting papillae, function primarily in food gathering but also are undoubtedly the loci for significant gaseous exchange; however, crinoid tube feet are not used in locomotion. On the tegmen the lappets are smaller, and each covers a single podium.

With their oral side up and with their extensive system of ambulacral grooves along arms and pinnules, crinoids are well adapted as suspension-feeders. Small planktonic organisms and detritus are entrapped in podial mucus and swept along the ambulacral grooves by the cilia to the mouth. Sea lilies can bend the stalk somewhat so that the crown and arms face into the current to maximize feeding opportunities. Other crinoids have various postures of the arms and crown to facilitate feeding.

Since the mouth and anus are situated so close together on the tegmen, how does a crinoid keep food and fecal materials separate? The slightly acentric anus is located on a flexible cone so that wastes can be directed away from the ambulacral grooves and mouth. Some advanced feather stars send food particles to the mouth on a circuitous route to avoid passing too close to the anus, and other crinoids have nonfunctional food grooves on areas adjacent to the anus.

While movement in the sea lilies is restricted to lateral bending of the stalk and flexion and extension of the arms, the comatulids are much more active and can swim or crawl but always with the oral side uppermost. In the inactive posture either the animal rests on the bottom or the crown is held slightly above the bottom by the arms whose tips are on the substratum. Crawling is achieved by moving about on the tips of the arms; by alternately raising and lowering arms on different sides of the animal, feather stars are able to swim. In most feather stars with multiple arms a set of five arms apparently acts as one unit. The rather common and active *Antedon* has ten arms; every other arm belongs to an opposite locomotor unit, and the arms beat downward sequentially around the animal.

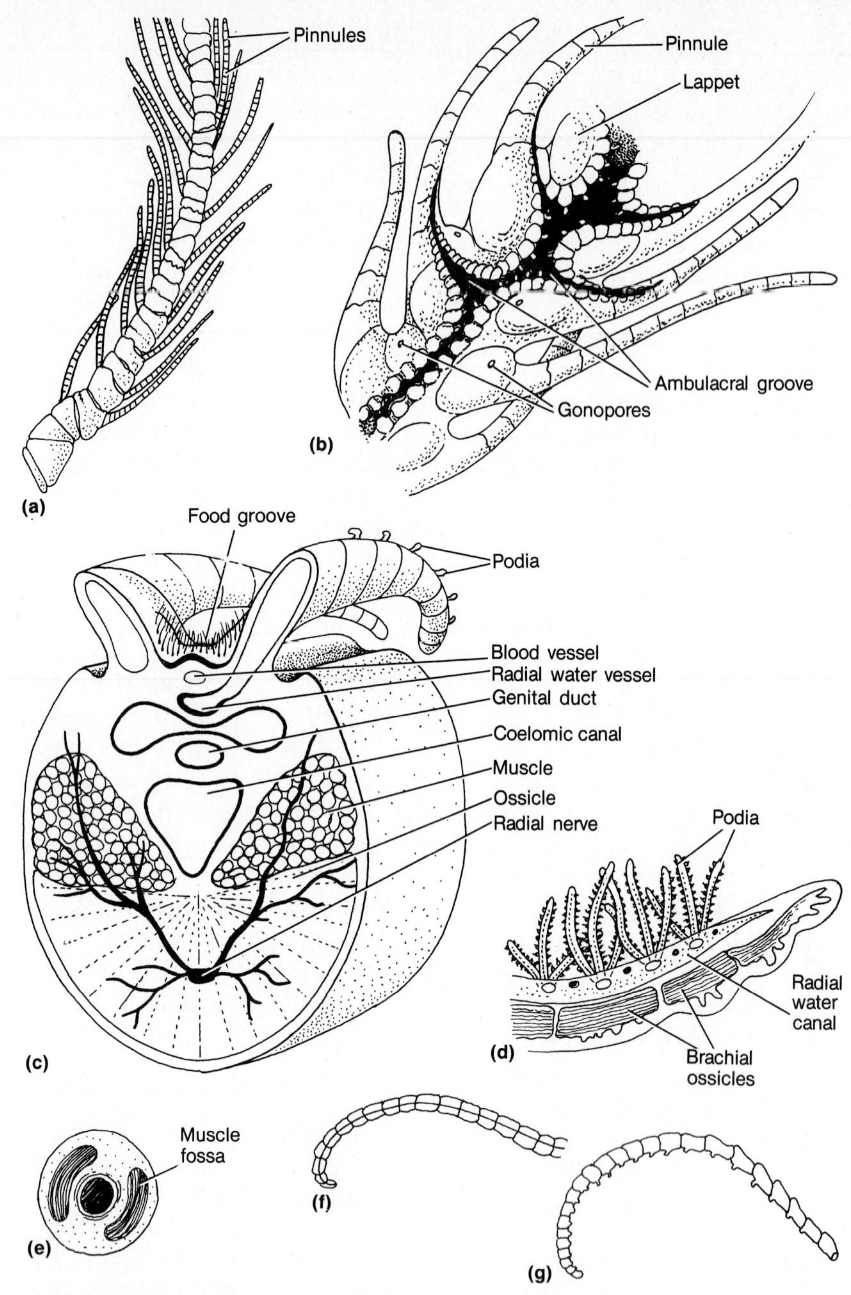

Figure 17.12 The more important features of the arms, pinnules, stem, and cirri of crinoids: ***(a)*** an aboral view of a portion of an arm of a comatulid showing the arrangment of pinnules; ***(b)*** an oral view of a portion of the arm of a comatulid; ***(c)*** a cross-sectional view of a pinnule; ***(d)*** clusters of three podia along an ambulacrum; ***(e)*** a cross-sectional view of a stalk; ***(f)*** a simple cirrus; ***(g)*** a spiny cirrus. (parts a [from Carpenter], b [from Mortensen], d [from Chadwick], e [from Danielsson], f, g [from Clark] in L. H. Hyman, The Invertebrates. Vol. IV, 1955, Echinodermata, McGraw-Hill Book Co., used with permission.)

Internal Features

The digestive tract consists of a short esophagus, a coiled intestine that makes a complete turn around the inner surface of the calyx, a short rectum that extends into the anal cone, and the anus (Fig. 17.13). Numerous small intestinal diverticula are often present, but little is known about their function.

The coelom is somewhat reduced and mostly represented by a series of interconnecting channels and sinuses. The walls of all coelomic channels contain minute ciliated pores to permit both the circulation of coelomic fluid and the passage of coelomocytes into surrounding tissues. The perivisceral coelom extends into each arm and pinnule as a **coelomic canal** (Fig. 17.12c). In crinoids the water vascular system differs from that of other echinoderms in two important ways: there is neither a madreporite nor podial ampullae. Water enters the ambulacral system from the perivisceral coelom through the ciliated pores. The perivisceral coelom, in turn, is connected to the outside by means of the tegmenal ciliated funnels so that the water vascular system is indirectly connected to the outside by way of the perivisceral coelom. Each of the clusters of three podia is

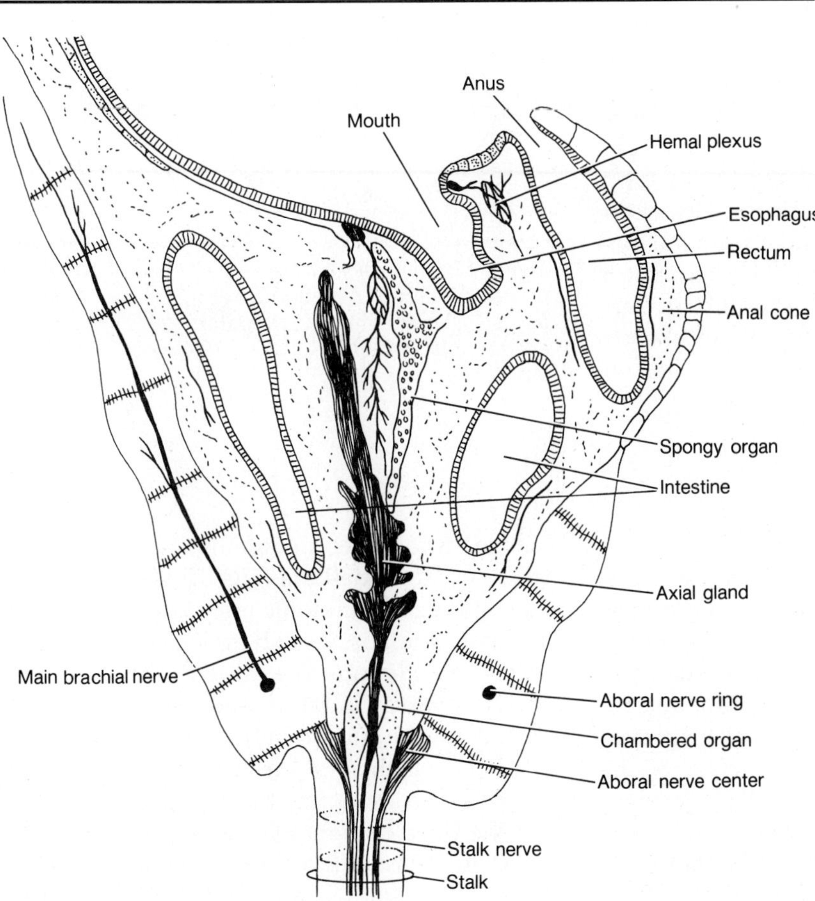

Figure 17.13 *Longitudinal section through the crown of a stalked crinoid to show the major internal parts (from Reichensperger in L. H. Hyman,* The Invertebrates. *Vol. IV:* Echinodermata, *McGraw-Hill Book Co., 1955, used with permission).*

supplied by a small lateral canal of the radial canal; podial extensions are achieved by contractions in the walls of the radial canals. The hemal and perihemal systems are ill-defined. They arise from the circumoral hemal ring and pass aborally in a **spongy body**, which is closely associated with the axial gland and may form coelomocytes. Peculiar to crinoids is another small coelomic structure, the **chambered organ**, located aborally in the calyx, where it sends small branches through the stem to the cirri (Fig. 17.13). Wastes are accumulated by the coelomocytes, deposited in minute saccules along the ambulacral grooves, and subsequently discharged to the exterior.

The crinoid nervous system is notable because it is the endoneural system that predominates. As the principal motor system concerned with movement and posture, the endoneural system is represented by a mass surrounding the chambered organ. From this aboral mass arise stalk and cirral nerves and five brachial nerves that extend through the arms and supply each pinnule and the interossicle muscles of the arm. The sensory plexus of the ectoneural system is not as extensive as in other taxa but is well developed on the tube feet and cirri. Sensory structures are represented only by the podial papillae and epidermal sensory cells of the ambulacral regions.

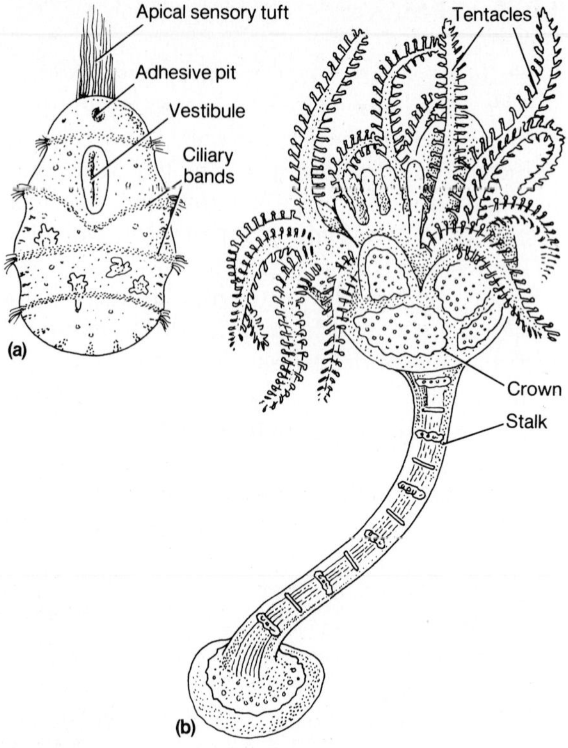

Figure 17.14 *Crinoid larvae:* ***(a)*** *vitellaria of* Antedon; ***(b)*** *pentacrinoid of* Antedon *(parts a, b, [from Thomson] in L.H. Hyman,* The Invertebrates. *Vol. IV:* Echinodermata, *McGraw-Hill Book Co., 1955, used with permission).*

Reproduction and Development

All crinoids are dioecious; in the absence of distinct gonads, gametes develop from peritoneal lining of the coelomic (genital) canals within each arm or proximal pinnule. When gametes are mature, the pinnular walls rupture and release them into the sea. In some, like *Antedon*, the ova adhere to the external surface of the pinnules, and fertilization and early development take place *in situ*.

Following spawning and fertilization, development proceeds rapidly, and the embryo hatches into a free-swimming, nonfeeding larval stage, the **vitellaria** (Fig. 17.14a). After several days as a planktonic organism the vitellaria settles to the bottom and attaches by means of adhesive secretions to an anteroventral gland so that the larval anterior end becomes the attached aboral surface, and the larval posterior end becomes the oral surface. A metamorphosis typical of echinoderms then takes place including the development of a stalk even in feather stars. In *Antedon* the form for which development is best known, this small stalked stage is the **pentacrinoid larva,** which resembles a small sea lily (Fig. 17.14b). After the formation of arms, pinnules, cirri, and internal organs the pentacrinoid breaks loose from its stalk at a point between the upper stalk and the centrodorsal ossicle and begins its free existence.

SUBPHYLUM ASTEROZOA, CLASS STELLEROIDEA

The Asterozoa and representatives of its only extant class, the stelleroids, include the familiar sea stars and brittle stars, all of which are benthic and free moving. Their bodies are star shaped with a central disc and five or more radial arms. Tube feet pass between adjacent endoskeletal ossicles and not through pores within a given ambulacral ossicle.

Subclass Asteroidea

The asteroids or sea stars (starfishes to some) are the most conspicuous of all echinoderms, and details of their biology are much better known than those for any other echinoderm group. Sea stars are all benthic and are mostly coastal; they crawl over rocks, sandy bottoms, or coral reefs. They are predisposed to warmer tropical or subtropical areas and are especially plenteous and diverse on coral reefs; however, in the northern Pacific Ocean from Puget Sound to the Gulf of Alaska, sea stars are surprisingly common and unexpectedly diverse. Almost 1800 species are known; their sizes range from 1 cm to 1 m, and colorations vary from drab yellow to spectacular reds, yellows, blues, and greens. Some examples of asteroids appear in Fig. 17.15. Asteroids are among the most familiar of all invertebrates, and even though most students have had some prior contact with them, the following discussions should heighten the reader's enthusiasm for and interest in these well-known creatures.

EXTERNAL FEATURES AND THE BODY WALL

Asteroids are typically pentamerous with a central disc and five radiating arms, but many species have more than five arms (Fig. 17.15), and the Pacific *Heliaster* has 40 or more arms. Pentamerism, however, is not always manifested with great fidelity in the arms. Numbers tend to vary among sea stars with more arms; often, individuals of the same species will have different numbers of arms mainly because of imperfect regeneration of lost arms. The arms are not usually sharply set off from the central disc and typically grade into it. Some asteroids have arm lengths that are two to three times the diameter of the central disc, but others have such short arms that the entire body assumes a pentagonal shape.

The **central disc** is generally flattened along the oral–aboral axis. On the aboral surface are found both a prominent, often brightly colored **madreporite** located acentrically in an interradius and a central **anus,** which is lacking in some sea stars. Arising from the central **mouth,** a wide furrow or **ambulacral groove,** situated on the oral surface, extends the entire length of each arm and is bounded by rows of tube feet (Fig. 17.16). Characteristic of the asteroids and crinoids but not of any other taxon, the ambulacral grooves are "open," since they are found outside of or oral to the ambulacral ossicles. Along the ambulacral margins are movable spines that close over the groove to protect the vulnerable open ambulacrum. At the tip of each arm is a **tentacle,** an extension of the radial water canal, which bears a terminal red pigment spot or **optic cushion** composed of 75–200 ocelli. The **ocelli** are sensitive to light intensities and thus help the animal respond positively or negatively to light.

The general body surface is rough and knobby owing to the presence of many skeletal spines and tubercles. Most of the aboral **spines** are simple, nonmovable projections of the dermal ossicles, but the oral spines, especially

(a)
(b)
(c)
(d)
(e)
(f)

◀ *Figure 17.15 Photographic collage of types of asteroids:* ***(a)*** Solaster *(spinulosid);* ***(b)*** Pyncopodia helianthoides, *oral view (forcipulatid);* ***(c)*** Pisaster *(forcipulatid);* ***(d)*** Henricia *(spinulosid);* ***(e)*** Asterias *(forcipulatid);* ***(f)*** *another sea star (unidentified). (parts a, b courtesy of Ward's Natural Science, Inc., Rochester, N.Y.; parts c, d, courtesy of S. Armstrong.)*

those bordering the mouth and ambulacral grooves, can articulate with underlying plates by a socket arrangement. Stalked **pedicellariae** are found in members of the Order Forcipulatida including *Heliaster* and the familiar *Asterias,* while sessile pedicellariae are mostly limited to members of the Order Valvatida like *Porania* and *Oreaster.* Incredibly, a few forms like the Pacific *Stylasterias* use their pedicellariae to capture bottom-dwelling fishes that settle on their aboral surface and are later ingested. Numerous **papulae** are scattered over the general body surface between ossicles. Some sea stars have raised aboral ossicles surrounded by a circlet of spines; each ossicle–spine complex is called a **paxilla.** Paxillae are adaptations for burrowing, since the collective paxillae hold sediments away from the surface so that water currents can flow over the papulae even when the animal is buried in soft sediments.

WATER VASCULAR SYSTEM

While this system is basically like the generalized plan in all echinoderms, there are several notable differences or features. The very prominent external madreporite has already been mentioned. On the inner side of the circumoral

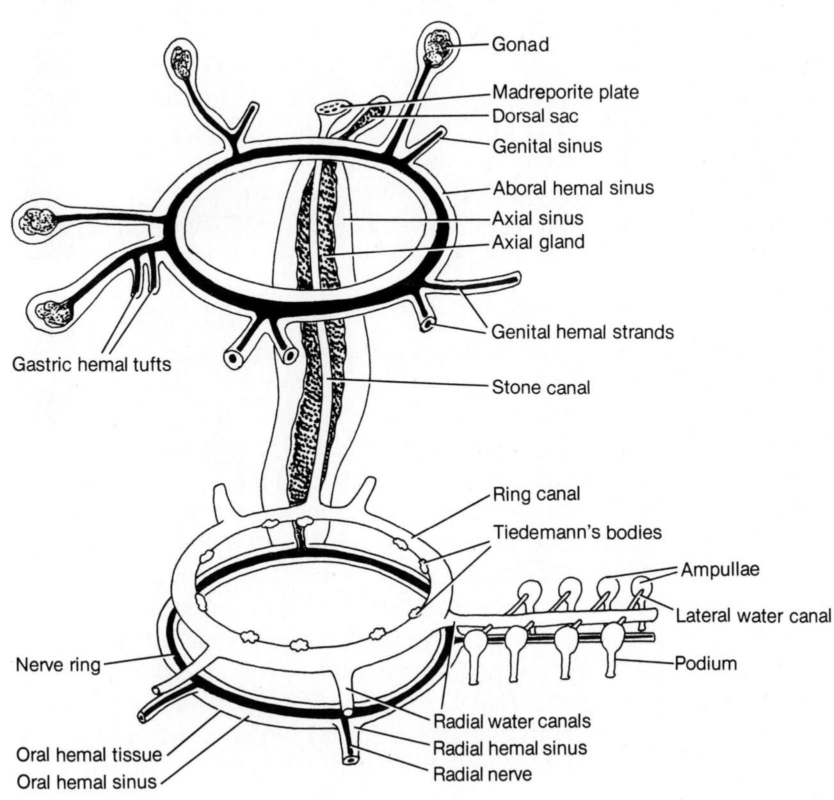

Figure 17.16 A diagrammatic representation of the water vascular and hemal system of an asteroid.

ring canal are typically nine saclike structures called **Tiedemann's bodies** in which coelomocytes are probably produced. The imperfect pentamerous symmetry is due to the absence of one body at the point where the stone canal joins the ring canal (Fig. 17.16). Most asteroids, but not *Asterias,* have polian vessels in addition to the Tiedemann's bodies. The radial water canals extend onto the arms on the outer or oral surface of the ambulacral ossicles in the "open" ambulacral grooves. The tube feet of sea stars are arranged in either one or two rows on either side of the ambulacral groove. If there is one row, then all the lateral canals are of the same length; but if there are two rows, then adjacent lateral canals are alternately long and short.

Locomotion, feeding, and digestion

As in most echinoderms, locomotion is accomplished basically by the tube feet and, to a limited degree, by the bending of the arms. The longitudinal muscle fibers of the body wall are better developed along the aboral surface to facilitate moving the arms aborally. While each podial unit acts independently of other such units, movement is highly coordinated. As a sea star moves from one point to another, one or two arms lead and the others follow, but all of the tube feet move unsynchronously in the same direction. Asteroid podia follow a stepping motion; with the podium extended forward, its longitudinal muscles contract, thereby pulling the animal forward. Then the podium is lifted and set forward, and the cycle is repeated. Studies of asteroid locomotion have indicated that most move at a rate of 30–75 cm per minute. In sea stars with suckered podia the powerful suction created by all the suckers is more than adequate to allow them to climb a vertical surface with ease. Sea stars can right themselves by twisting one or more arms until some of the tube feet make contact with the substratum. These arms are then moved beneath the animal to serve as an anchor, and the entire animal slowly rolls over.

Most asteroids are carnivores; they prey on a wide variety of benthic invertebrates like molluscs, other echinoderms, polychaetes, nemerteans, and crustaceans. Some even capture small fishes using the pedicellariae. There are two general methods of feeding: Sea stars with short arms and suckerless podia ingest small prey whole; asteroids like *Asterias* with long arms and podial suckers prey heavily on large invertebrates, and their overwhelming choice of prey is bivalves.

Those feeding on bivalves such as mussels, clams, and oysters have the perennial problem of how an asteroid can possibly prey upon a comparably sized and shelled bivalve (Fig. 17.17). Perhaps the single most important feeding adaptation in these asteroids is their **stomach.** In these sea stars the stomach can be partially everted through the mouth owing to increased hydraulic pressures within the perivisceral coelom. As the everted stomach is pressed against the prey tissues, powerful digestive enzymes pour over the victim, and digestion proceeds rapidly outside the asteroid. The resulting slurry or broth is transported into the echinoderm's body by ciliated tracts lining the stomach. After feeding is completed, muscles return this organ to its former position.

But how does the sea star overcome the formidable obstacle of the tightly closed valves? Again, the adaptable stomach is the answer. A sea star positions itself astride a bivalve and manipulates the prey so that its ventral gape is near the mouth of the asteroid (Fig. 17.17). The two valves of a bivalve are not perfectly closed but rather have small gaps even when the valves are tightly opposed. Even openings as small as 0.1 mm are large enough to permit entrance of the sea star's stomach! It was once thought that the pull of the asteroid tube feet eventually overcame the action of the bivalve's adductor muscles, but it has now been shown that bivalve fatigue is not a prerequisite for feed-

(a)

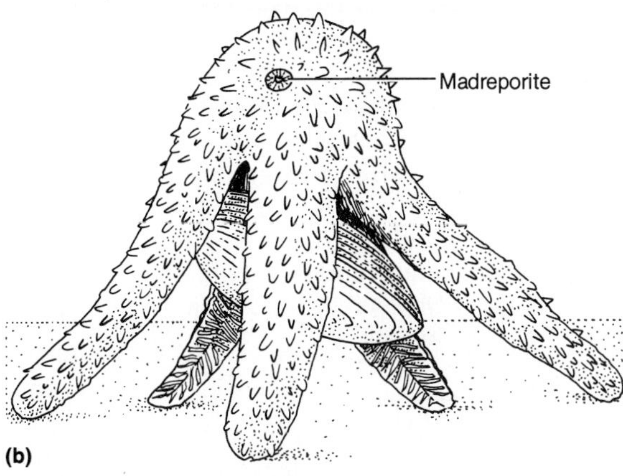

(b)

*Figure 17.17 Feeding in asteroids: **(a)** photograph of an oral view of a sea star feeding on a bivalve (courtesy of Ward's Natural Science, Inc., Rochester, N.Y.); **(b)** sea star astride a mussel before the initiation of feeding (from* E.W MacBride, Cambridge Natural History, *Macmillan, London 1909).*

ing to commence. The pull of the podia against the valves surely aids in the process but is not essential for feeding. In the early stages of feeding, the sea star may relax its pull on the valves from time to time; when this occurs, the valves are closed against the everted stomach, but apparently no damage is done to this organ. Sea stars begin to feed very shortly after capturing a bivalve, and as adductor muscles are digested, the gape of the bivalve obviously widens. Sea stars are particularly an anathema to the shellfish industry, in which considerable economic loss of oyster or mussel beds is due to asteroid predation. Shellfishermen used to hack up sea stars found on such beds to destroy them. Later, to their utter dismay, they learned that every large piece of the mutilated sea stars regenerated all missing parts; their actions did not eliminate the problem—they only made it worse.

Some asteroids prefer sponges and coral polyps as prey. The crown-of-thorns sea star, *Acanthaster planci,* is a rather voracious predator on coral reefs in the Pacific and Indian Oceans. Box 17.2 provides more information about the recent extensive destruction of coral reefs by this asteroid. Other asteroids are scavengers or deposit-feeders. Small organisms, plankton, and detritus are trapped in mucus, swept to the oral surface by epidermal cilia, and carried to the mouth by ambulacral ciliary tracts.

The large stomach is divided by a horizontal constriction into two portions: a larger oral **cardiac stomach,** which is eversible in many asteroids, and a small aboral **pyloric stomach,** which is not usually everted (Fig. 17.18c). Paired gastric ligaments or mesenteries extend from the body wall to the cardiac stomach to control the degree of its eversion. A pair of digestive glands or **pyloric ceca** is present in each arm, and each gland is suspended in place by a dorsal mesentery (Fig. 17.18b). Each cecum consists of a great many lobes whose ducts merge to form a **pyloric duct** that, in turn, opens into the pyloric stomach. Pyloric ducts from both ceca in a given arm often fuse and enter the stomach as a single common duct. The pyloric stomach often is pentagonal in shape because of the five ducts entering it from the digestive glands. The cecal wall consists of secretory cells that produce the digestive enzymes, absorbing cells, and storage cells for food reserves. Digestion, facilitated by cilia lining the entire alimentary canal, is mostly extracellular within the shell of the prey (as in bivalves), both stomachs, and the pyloric ceca. Arising from the aboral surface of the pyloric

BOX 17.2 □ THE CROWN-OF-THORNS SEA STAR: A VILLAIN OR JUST A HUNGRY SEA STAR?

Since 1970, deep concern has been expressed by many tropical marine biologists about both the dramatic increase in numbers of the crown-of-thorns (COT) sea star, *Acanthaster planci,* and the alarming destruction of coral reefs by the grazing of this asteroid on coral polyps. Extensive reef destruction has been reported from many areas of the south Pacific Ocean, especially around Guam and Australia. In attempts to reduce sea star populations, many recent efforts have been made to kill the sea stars wholesale. One team of four divers killed 2500 sea stars in four hours with formaldehyde-loaded syringes; that is an average of more than 2.5 sea stars per diver per minute. Is all this notoriety really deserved by the COT sea star?

The COT sea star is widely distributed in tropical seas throughout the world. Its central disc is about the size of a large dinner plate, but individuals can have an overall diameter of up to 0.5 m. Pacific forms have 16 arms; Red Sea individuals average only about 13 arms. They are also called prickly sea stars because of the poisonous nature of the spines, which can inflict swelling, pain, and nausea. In just 2.5 years, COT sea stars had destroyed more than 90% of the reefs along a 38-km stretch of Guam's coastline and seriously damaged about 500 km of the Great Barrier Reef. In one day an individual *A. planci* usually grazes in an area about twice the size of its central disc or, on an average, about 0.1 m^2. There are many eyewitness accounts of thousands of sea stars moving in close formation over a reef and consuming almost every coral polyp in their path. The denuded reefs are at first almost sterile and completely white, but later they are covered with algae and become a dismal gray color. As the reefs begin to deteriorate, the island coastline becomes exposed to eroding waves and tides. The COT epidemic is of great biological interest; since the South Pacific natives depend heavily on the reefs for food, it is also of great social and economic concern.

Why have the numbers of *A. planci* increased so dramatically, and why have they become such voracious feeders? A number of theories have been advanced to explain the sudden numerical increase: (1) The increased numbers of *A. planci* are due to a decrease in the frequency of its natural gastropod predator, the giant triton. From 1955 to 1975 an estimated 200,000 tritons were taken by collectors in the South Pacific for their exquisite shells. But tritons surely are not the sole predators of this asteroid. (2) Human activities such as blasting for shipping channels and dynamiting for fishes selectively destroyed the gastropod predators. However, this would not completely account for the dramatic increases in sea star populations through the whole area. (3) The testing of nuclear weapons during the 1960s destroyed their predators. (4) Pesticides and other pollutants have decimated the natural predators of *A. planci* larvae. There is, in fact, little substantial evidence to support hypotheses 2, 3, and 4. (5) The dramatic increase in numbers of *A. planci* is just an example of natural fluctuations in population size; numbers of individuals vary considerably from time to time owing to any number of environmental factors. (6) There really has not been a population explosion of the COT sea stars at all. Since more divers are working the area now than ever before and since the development of SCUBA, we are just now learning about a fascinating predator–prey relationship that has gone on for millions of years. In fact, some evidence indicates that the reefs regenerate in time.

None of these six postulates fully satisfies all the questions surrounding this dramatic and serious problem. Do the COT sea stars deserve their bad reputation? Perhaps someday one of you will provide the definitive answer to this perplexing problem.

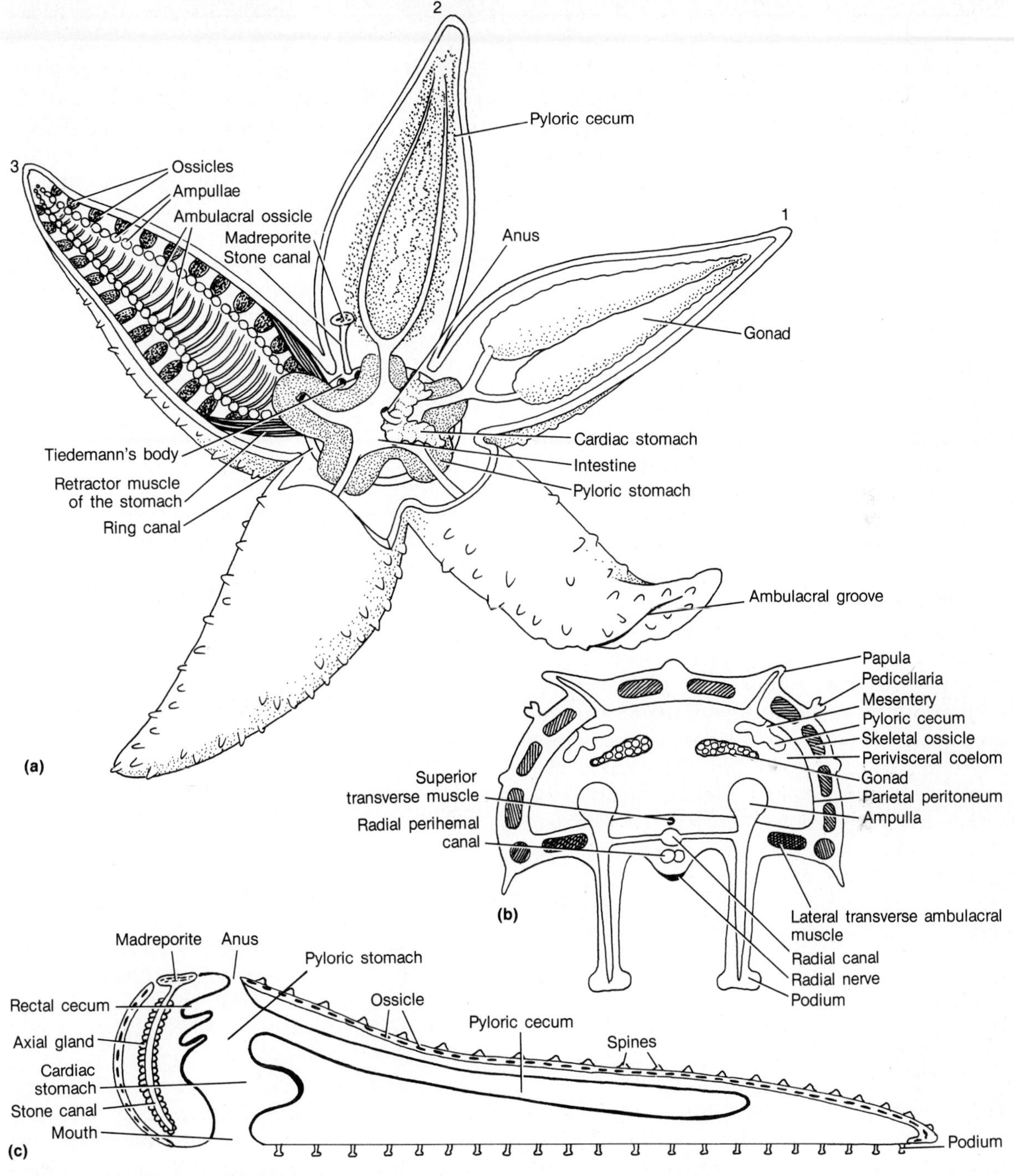

Figure 17.18 *The principal internal features of a sea star:* ***(a)*** *aboral view of the features in the central disc and arms: 1) arm with aboral body wall removed; 2) arm with aboral body wall and gonads removed; 3) arm with aboral body wall and all viscera removed;* ***(b)*** *diagrammatic cross-sectional view of an arm;* ***(c)*** *diagrammatic longitudinal-sectional view of the central disc and along one arm. (parts b, c modified from E.W. MacBride, 1909,* Cambridge Natural History, *Macmillan, London.)*

stomach is a short intestine that leads to the centrally located aboral anus (Fig. 17.18a, c). In many asteroids the intestine and anus are nonfunctional or missing altogether, and undigested materials are voided through the mouth or by everting the stomach.

Other internal features

The perivisceral coelom in sea stars is especially large, surrounds the organs of the central disc and arms, and is the principal transport system (Fig. 17.18b, c); its abundant coelomocytes are involved in phagocytosis and waste removal. Both the hemal and perihemal systems are poorly developed in asteroids. The nervous system is typically like that for echinoderms generally; the principal component is the ectoneural system whose extensive superficial plexus has already been noted. From the oral ring a large radial nerve, V-shaped in cross section and mostly sensory in nature, runs the length of each arm; it is in the bottom of the ambulacral groove just beneath the epidermis and innervates the podia (Fig. 17.18b). The radial elements of both ectoneural and endoneural systems are responsible for the coordinated stepping movements of the podia. Numerous epidermal sensory cells, especially on the tube feet, and the ocelli are the principal sensory receptors.

Reproduction and development

Paired **gonads** lie in each arm, and each is housed in a perihemal **genital sac** (Fig. 17.16). Gonadal size fluctuates widely; at time of spawning, the gonads are enormously enlarged and completely fill the arms, but when the animal is not spawning, they are but small tufts near the base of the arms. There is an individual **gonopore** for each gonad situated either laterally near the base of an arm or on the oral or aboral surface. Gametes are spawned into the seawater, a phenomenon induced by the presence of other gametes and by secretions of neurosecretory cells in the radial nerves. In a few species a form of copulation takes place in which the male lies atop the female with their arms alternately placed and fertilization is internal. Most sea stars have only one reproductive period each year, but numbers of gametes produced by each individual are enormous—sometimes in the millions.

In asteroids with direct development, eggs are brooded by the female beneath her body, in aboral depressions, or at the bases of the arms. No larval stages are produced, and a very small immature sea star hatches after about a month of development. In asteroids that have indirect development, an egg contains little yolk, and a young feeding **bipinnaria** larva is soon produced that is characterized by a ciliated band and later by long arms (Fig. 17.19a). After several weeks of free-swimming existence, three additional anterior arms are developed, and the bipinnaria becomes a **brachiolaria** larva (Fig. 17.19b). The new arms contain branches of the coelom internally and are tipped with adhesive cells. A glandular adhesive area develops at the base of the arms, and these areas are important as the brachiolaria settles to the bottom and attaches to a surface.

Following attachment, metamorphosis takes place with the anterior end degenerating and the posterior end ultimately developing into the adult. The arms develop from five lobes on the larval right side. The various coelomic parts, visceral organs, and dermal ossicles are formed. As tube feet develop, they grip the substratum, and the larval attachment to the substratum is severed. At this point in their ontogeny the asteroids are no more than 1 mm in diameter. Sexual maturity is attained in two to five years. Some forms may live for a third of a century and spawn every year.

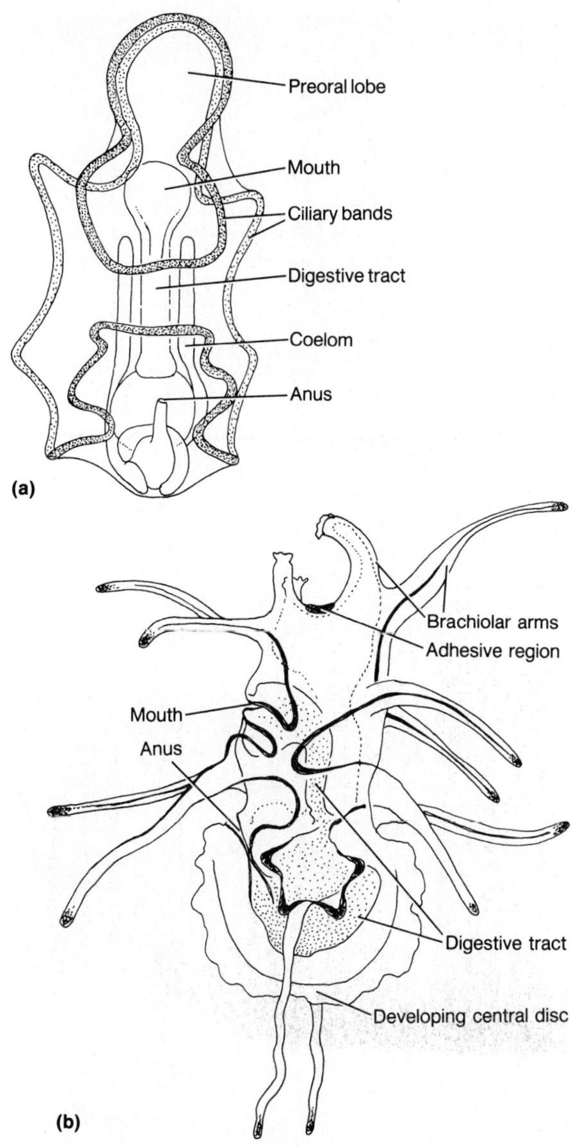

*Figure 17.19 The larval stages of asteroids: **(a)** bipinnaria; **(b)** brachiolaria. (parts a* [from Horstadius], *b [from Mead] in L.H. Hyman,* The Invertebrates. *Vol. IV:* Echinodermata, *McGraw-Hill Book Co., 1955, used with permission.)*

Subclass Ophiuroidea

The ophiuroids, the most successful numerically of all the echinoderm groups with about 2000 modern species, have several common names that reflect some of their important features. They are called serpent stars because of the serpentlike mobility and the banded color patterns of the arms (ophiuroid = serpent-tailed) (Fig. 17.20); they are also called brittle stars because the arms are fragile and easily broken off. A small group of fewer than 100 species are referred to as the basket stars because of the shape of their collective branching arms (Fig. 17.20e). They are found in all types of marine habitats but are surprisingly abundant on the soft deep-sea floor, where they often are dispersed in fairly uniform patterns. But sometimes brittle stars are clumped in unusual aggregations or heaps of up to 2000 individuals per square meter, and densities of up to several hundred per square meter are rather common. Typically, they are found in mucus-lined burrows, buried in the benthic material, or in small crevices, holes, shells, or other retreats. Most ophiuroids are small and are usually less than 10 cm in diameter including the arms, but the basket star *Gorgonocephalus* has an arm spread of 35 cm or more (Fig. 17.20e). Most colors are present, although ophiuroids are not conspicuously colored; the arms typically have banded or mottled patterns. Several species, including *Amphipholis* (Fig. 17.20c), are luminescent.

Even though they are closely related to asteroids, since both groups have a central disc and typically five arms, ophiuroids have a number of basic differences, the most important of which are details of the arms. Therefore we shall consider the functional morphology of the arms first followed by that of the central disc.

Arms

There are five arms, a rather constant number in ophiuroids, although a few species may have six or seven. In the basket stars the arms branch

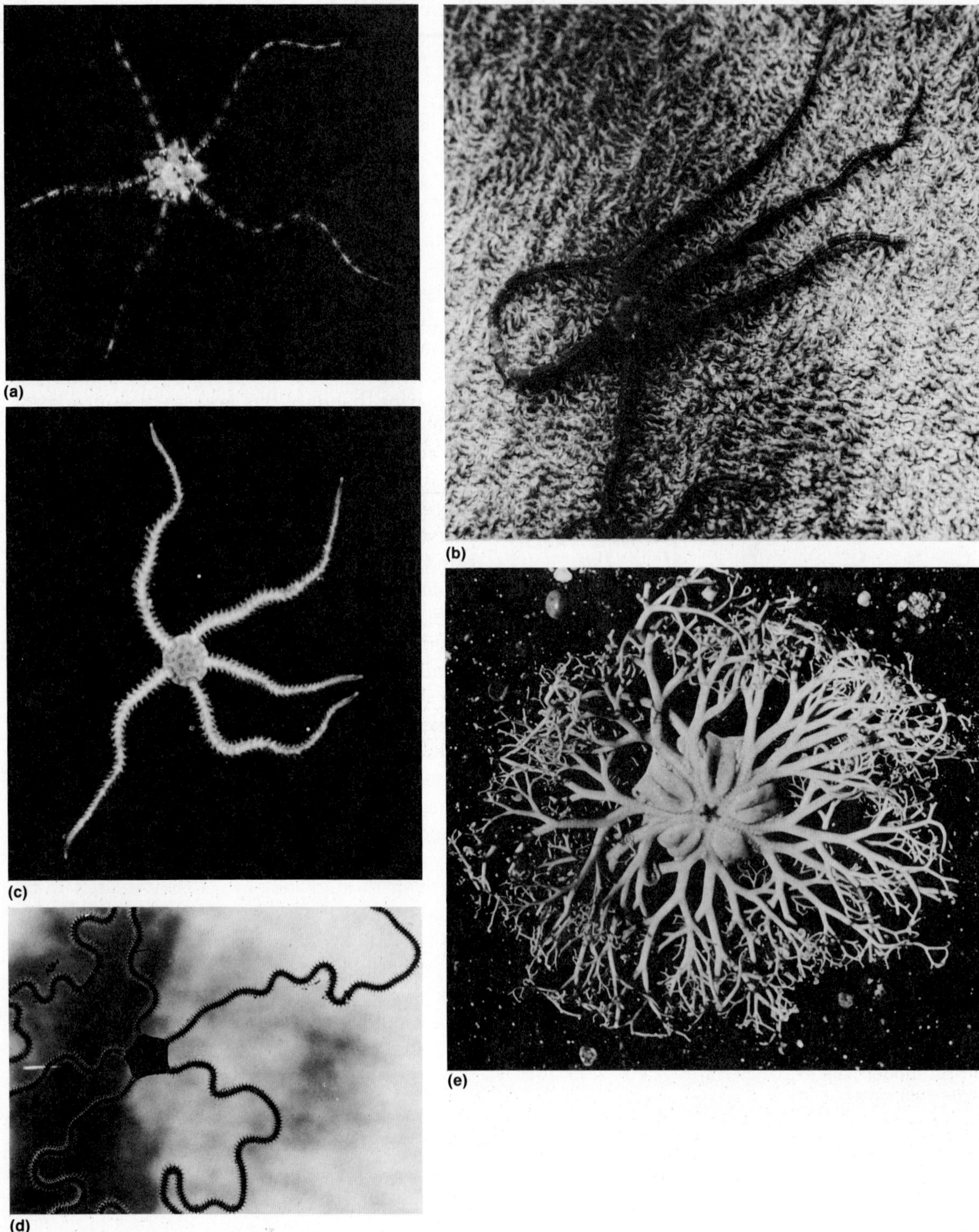
(a)
(b)
(c)
(d)
(e)

◀ ***Figure 17.20*** *Photographic collage of various ophiuroids:* ***(a)*** *a brittle star, probably* Ophiothrix *(ophiurid);* ***(b)*** *a deep-sea brittle star (courtesy of B. Best);* ***(c)*** Amphipholis, *a brittle star (ophiurid);* ***(d)*** *a brittle star, probably* Ophiophragmus *(ophiurid) (courtesy of C.E. Jenner);* ***(e)*** Gorgonocephalus, *a basket star, oral view (phrynophiurid). (parts c, e courtesy of Ward's Natural Science, Inc., Rochester, N.Y.)*

repeatedly so that a complex network or basket results (Fig. 17.20e). The arms of serpent stars are comparatively much longer than those of asteroids and are sharply delimited from the central disc. But perhaps the most striking feature of ophiuroid arms is that they are very flexible and supple and can be moved in sinuous undulations. The arms, not the tube feet, are the primary locomotor organs. Most have sufficient brachial flexibility to grasp or even wrap around an object during locomotion, and many also use the arms as food-gathering organs. The mobility of the arms is a result of their being jointed along their entire length owing to both dermal and internal ossicles articulating against each other.

Subepidermally, each arm bears four longitudinal rows of small dermal ossicles called **shields;** there is a single oral and aboral row and a lateral row on either side. One set of shields (one oral, one aboral, two lateral) completely encircles a small longitudinal segment of the arm; the ossicles in each set articulate with those in the sets immediately proximal and distal to them (Fig. 17.21). The lateral shields are often the largest, and the oral and aboral shields are correspondingly small. Each lateral shield bears a vertical row of 2 to 12 spines, which vary from species to species in number, shape, and arrangement. The spines are effectively used to provide traction in locomotion or in grasping objects. Each oral row of shields, usually covering the oral surface of each arm, continues onto the central disc to the mouth. There is no ambulacral groove as in asteroids or crinoids; rather, the ambulacrum has been displaced internally and is a closed canal. A pair of reduced tube feet is found on each arm segment where each podium emerges between the lateral and oral shields (Fig. 17.21). The tube

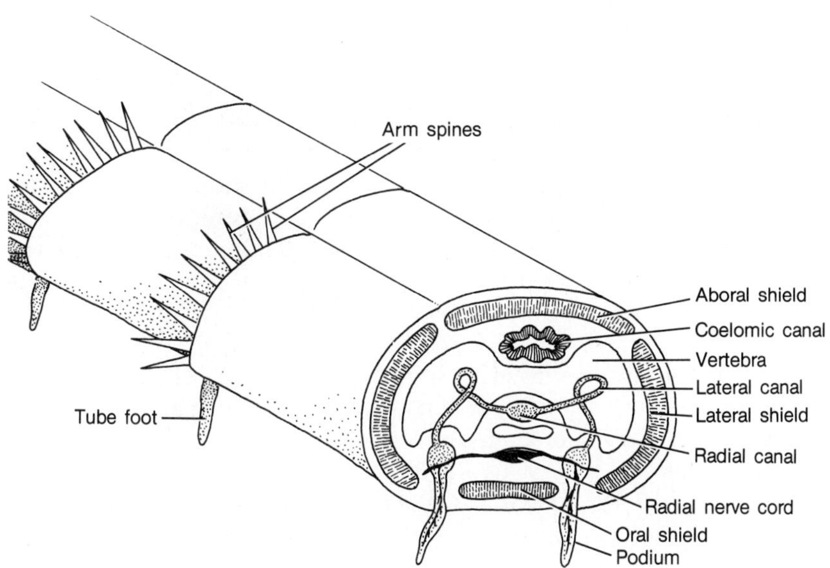

Figure 17.21 *Some details of the ophiuroid arm seen in lateral view and in cross section.*

feet are tentaclelike, lack both suckers and ampullae and are mostly used as sensory structures or in feeding. External cilia, pedicellariae, and papulae are absent in ophiuroids.

Internally, the arms are basically solid because each contains a linear series of internal ossicles. In ophiuroids, ossicles that are homologous to the paired ambulacral ossicles of asteroids have moved internally and become enlarged, and each pair has fused to form a single central ossicle or **"vertebra"** (Fig. 17.21). Since a given vertebra corresponds in position to a set of dermal ossicles, movement takes place at the joints between vertebrae owing to two pairs of intervertebral muscles—one oral and one aboral. Both end surfaces of each vertebra are adapted for articulation. The joint between adjacent vertebrae in most brittle stars is a hinge joint; and because of the restrictive nature of the vertebral articulations, the arms of these ophiuroids can move only in a horizontal (lateral) plane. The arm vertebrae of basket stars articulate by ball-and-socket joints, which permit both vertical and horizontal movements.

Ophiuroids live unobtrusively in crevices; but when they leave these retreats, they are surprisingly motile owing, of course, to the suppleness of the arms. Typically, ophiuroids move using a single arm on either side of the body; one or two arms lead and the others trail, and the central disc is held just above the substratum. Some ophiuroids utilize two pairs of arms in locomotion, which thrust the disc and fifth arm forward. The animal apparently has no preference as to which arms are utilized in locomotion. By using the arms in a rowing motion the animal moves rapidly from place to place, and a few forms can even swim by the rapid lashing of the arms.

The internal ossicles give the arm a basically solid construction, but along the length of each arm are two narrow canals, the radial and coelomic canals (Fig. 17.21). Along the oral face of arm vertebrae is the **radial canal** of the water vascular system. In each ossicle a pair of lateral canals arises from the radial canal, each of which terminates in a tube foot (Fig. 17.21). Ampullae are absent, but a valve between each podium and lateral canal aids in podial extension. Each radial canal terminates distally in a small external **tentacle** that does not contain an optic cushion as in asteroids. Because of the large vertebrae, the coelom is restricted to a small aboral **coelomic canal** (Fig. 17.21); it probably functions in internal transport. Each arm also contains radial nerves from the more superficial, largely sensory ectoneural system and the inner motor hyponeural system.

Often, when an arm is caught or when the serpent star is handled roughly, the animal will simultaneously contract all four longitudinal muscles articulating two of the more proximal vertebrae. The action results in **autotomy** or self-surgery; the arm breaks off, and escape is achieved. Breaks can occur at any arm joint, and the lost portions of the arm are regenerated quickly. Unlike the arms, the central disc normally does not possess regenerative capabilities. However, in some forms with six arms the disc is sometimes cleaved with each half bearing three arms, and the missing disc half and arms are regenerated by both parts.

Central disc

The central disc of ophiuroids is usually flattened, pentagonal, and clearly delineated from the arms. The aboral surface, varying from smooth to rough, contains no openings of any type (Fig. 17.22b). In the center of the oral surface is the **mouth** surrounded by five large, interradial plates or jaws that function as masticating structures. Each **jaw** is triangular in shape with the pointed end projecting inward into the mouth opening (Fig. 17.22a). In fact, each jaw consists of three principal plates: a pair of half-jaws and a large basal oral plate. The first two ambulacral plates, called **half-jaws** or **maxillary plates,** are serrated on their lateral surfaces as cutting edges. The five basal **oral**

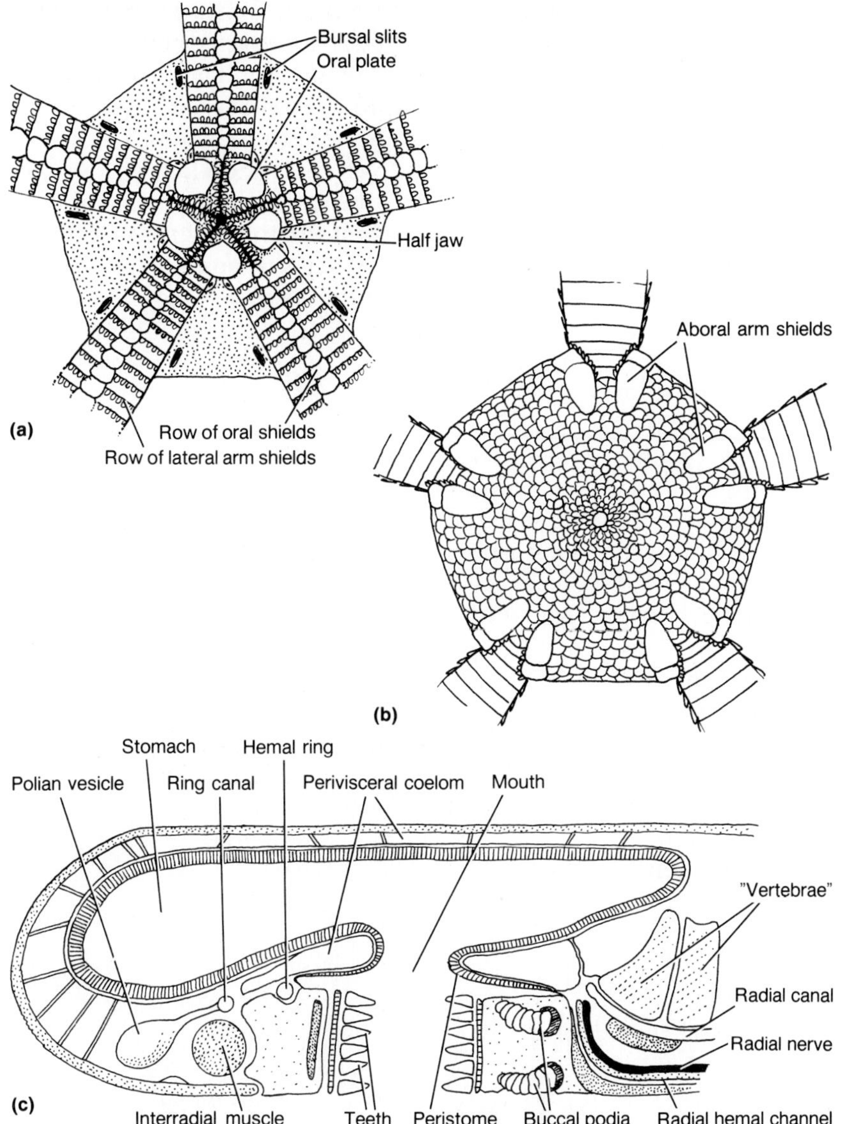

Figure 17.22 *The central disc and arm bases of ophiuroids:* ***(a)*** *oral view;* ***(b)*** *aboral view:* ***(c)*** *schematic longitudinal section through the central disc and basal part of an arm (parts a–c from Ludwig in L.H. Hyman,* The Invertebrates. *Vol. IV:* Echinodermata, *McGraw-Hill Book Co., 1955, used with permission).*

plates (not to be confused with the oral shields) are interradial in position and form the bases for the half-jaws. Two concentric sets of small muscles run from various ossicles around the mouth to the half-jaws to operate them as small cutting plates.

Because of their serrated jaws, flexible arms, and an unusual degree of mobility, one would expect ophiuroids to be strongly predaceous and to feed on a variety of prey. While some are indeed predatory on benthic organisms, many others are scavengers, detritus-feeders, or filter-feeders, and many ophiuroids feed in combinations of these modes. Utilizing the arms and, to some extent, the tube feet, predatory forms capture and ingest coral polyps, tubicolous polychaetes, sponges, and even seaweed like laminarians. Scavengers and detritus-feeders seek out and ingest various forms of organic material including bits of dead and decomposing animals and detritus. Filter-feeding ophiuroids subsist on planktonic organisms such as small crustaceans, polychaetes, and larvae of various animals. Their filtering mechanisms are of two types: In some, particles are trapped in mucous strands running between arm spines; in others, several arms are extended into the water current, and the protracted tube feet form a filter apparatus on either side of each arm. In either method the food particles are collected and compacted by the tube feet and transferred along the midoral surface of the arms to the mouth.

Extending across the oral surface of the central disc toward the mouth are the five rows of oral shields (Fig. 17.22a). The aboral surface has grown orally between the arms and, in so doing, has relocated the **madreporite** on the oral surface, where it is borne by one of the most proximal oral shields. The oral radial location of the madreporite contrasts starkly with its position in asteroids. Situated between these radial oral shields and the interradial oral plates are five (or ten) pairs of **bursal slits,** which open into small invaginated, ciliated gas exchange sacs, the **bursae** (Fig. 17.22a). Water is brought into the bursae either by the ciliary beat, by being pumped into the bursae by raising and lowering the disc wall, or by the action of minute disc muscles.

Internally, the viscera are all housed in the central disc. The alimentary system is unusually simple even for an echinoderm with a short esophagus, a large pouched stomach, and no intestine or anus (Fig. 17.22c). The ten stomach pouches are arranged alternately with ten bursae. Coelomocytes presumably bring excretory materials to the bursae, from which they are eliminated. Ectoneural and hyponeural systems are represented in the central disc by nerve rings, and ophiuroids, lacking definite sensory organs, have a great number of epithelial sensory cells. The perivisceral coelom is not very extensive, since the stomach pouches and bursae take up most of the internal space and the hemal and perihemal systems are mostly restricted to the central disc.

Reproduction and development

Even though most ophiuroids lack sexual dimorphism, several instances are known of dwarf males attached orally and probably permanently to much larger females. The small gonads are borne on the inner surface of the bursae, and when gametes are mature, the bursal

Figure 17.23 *The ophiopluteus larva of ophiuroids (from Narasimhamurti in L.H. Hyman,* The Invertebrates. *Vol. IV:* Echinodermata, *McGraw-Hill Book Co., 1955, used with permission).*

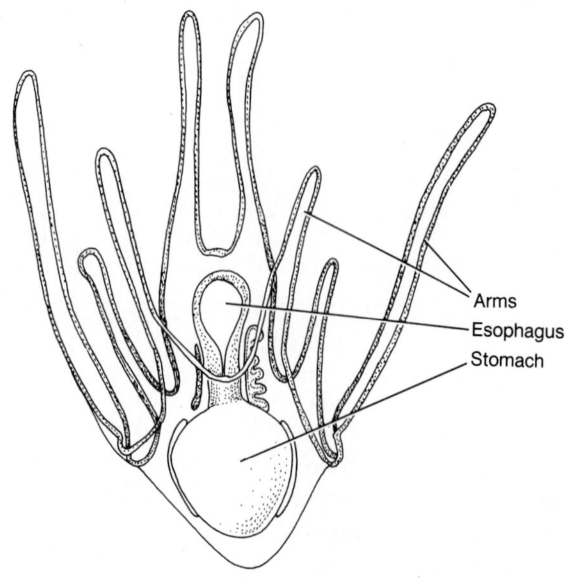

walls rupture to release them. Most ophiuroids spawn gametes into the sea, and the various developmental stages are planktonic. First developed is a free-swimming blastula followed by a gastrula that becomes an **ophiopluteus** larva (Fig. 17.23). This ophiopluteus stage is characterized by having four pairs of very long arms supported internally by calcareous rods and bearing external ciliated bands. Metamorphosis takes place as the larvae remain planktonic, and there is no attachment stage. As adult features are developed, the young brittle stars sink to the bottom and mature, probably within a year.

SUBPHYLUM ECHINOZOA

Representatives of this group of echinoderms, lacking arms, have basically a flattened, rounded, spherical, or ovate body with the oral mouth and aboral anus typically lying at opposite poles; the radial components of the typical echinoderm systems run meridionally. Two distinct extant classes are universally recognized.

Class Echinoidea

Echinoids are represented today by about 900 species of sea urchins, heart urchins, and sand dollars (Fig. 17.24). Echinoids have left a most impressive fossil record with perhaps as many as 2000 fossil species recognized. Varying in size from 0.2 to 30 cm in diameter, they are widely distributed in all seas but are more common in littoral zones. Echinoids possess three important diagnostic features involving their spines, symmetry, and test. The surface is covered with numerous movable spines, a distinctive feature from which is derived the class name (echinoid = hedgehoglike). Second, because of the absence of arms, the echinoid body exhibits basic pentamerous symmetry, but many echinoids tend toward bilaterality. The third important echinoid feature is that the dermal ossicles are sutured together to form a solid inflexible **test.** Each of these features will be elaborated upon more fully in the following discussions.

Basically, echinoids can be subdivided into two groups on the basis of their symmetry, length of spines, and habitats. One group, the regular echinoids or sea urchins, have a rounded body and long spines and are typically found in coastal areas on hard surfaces such as rocks, jetties, and shell beds. The other group, the irregular echinoids, including the heart urchins and sand dollars, have a flattened or ovate body with tertiary bilateral symmetry and shorter spines and are buried in sandy soft bottoms. Because of these significant differences, it will be most instructive to explore the external features of both groups separately followed by a discussion of their common internal features.

REGULAR ECHINOIDS

Some sea urchins are drab colored, but others are bright green, red, purple, or orange. They are characterized by a spherical pentamerous body having an oral–aboral axis with radial parts arranged meridionally in relation to the poles. At the oral pole is the **mouth** surrounded by the **peristomial membrane** (Fig. 17.25c), which is strengthened by small internal ossicles and thickened medially to form a circular lip around the mouth. Associated with the peristomial membrane are several radially arranged structures including five pairs of **buccal podia,** five pairs of small bushy **gills,** and a number of small spines and pedicellariae. Surrounding the mouth are five prominent jaws that are a part of a complex chewing apparatus to be discussed later. At the aboral pole is a membranous **periproct,** which, like the peristomial membrane, contains endoskeletal plates and bears spines, pedicellariae, and the **anus,** usually centric but sometimes eccentric in position (Fig. 17.25a). Extending meridionally from the poles are five ambulacra alternating with five interambulacra.

Figure 17.24 *Collage of photographs (a–e) and a line drawing (f) of some types of echinoids:* ***(a)*** Arbacia, *a sea urchin, aboral view (echinacean);* ***(b)*** Lytechinus, *a sea urchin, oral view (echinacean);* ***(c)*** Strongylocentrotus, *a sea urchin, aboral view (echinacean) (courtesy of S. Armstrong);* ***(d)*** Mellita, *a sand dollar, aboral view (gnathostomate);* ***(e)*** *the test of a sea biscuit, probably* Clypeaster *(gnathostomate);* ***(f)*** Echinocardium, *a heart urchin, aboral view (atelostomate) (from E. W. MacBride,* Cambridge Natural History, *Macmillan, London 1909).*

There are 20 meridional rows of ossicles comprising the rigid test; marking each ambulacrum are two rows of **ambulacral ossicles,** and underlying each interambulacrum are two rows of **interambulacral ossicles** (Fig. 17.25a, b). Adjacent to the anal opening are five interambulacral ossicles called **genital plates,** each bearing a **gonopore,** and one of the five also contains the **madreporite** (Fig. 17.25a).

The surface of a sea urchin is a veritable forest of spines, pedicellariae, tube feet, and other structures. Each of the ossicles of the test bears one or more raised knobs or bosses, the **tubercles,** for the articulation of the movable spines (Fig. 17.25a, b). Each **spine** has a concave base that fits over the tubercle to form a ball-and-socket joint (Fig. 17.25d). Two circular sheaths of muscles run from a marginal depression in the ossicle around the tubercle to the base of the spine; the outer sheath can move the spine in any direction, and the inner muscular sheath is composed of "catch" muscles that lock the spine in an erect position. Spines are straight or curved and are circular, flat, or triangular in cross section. In many sea urchins, spines are of two basic lengths—the longer or primary spines are used in locomotion and defense, and the shorter or secondary spines are used to protect the muscles around the primary spines. Often, spines are longest at the test's equator and shortest at its poles. Sea urchins use the long spines to push off of hard surfaces, to loosen materials to facilitate burrowing, or as wedges in rock crevices to permit them to live in rough wave-swept areas; the shorter spines also provide anchorage against strong waves. Spines are excellent mechanical barriers against many enemies and potential predators. Some sea urchins, especially those living on coral reefs, have hollow spines that contain a poison; the pacific *Asthenosoma* has needle-sharp spines that can easily penetrate the skin, break off, and release a poison that is extremely painful and even dangerous to human beings. The Atlantic

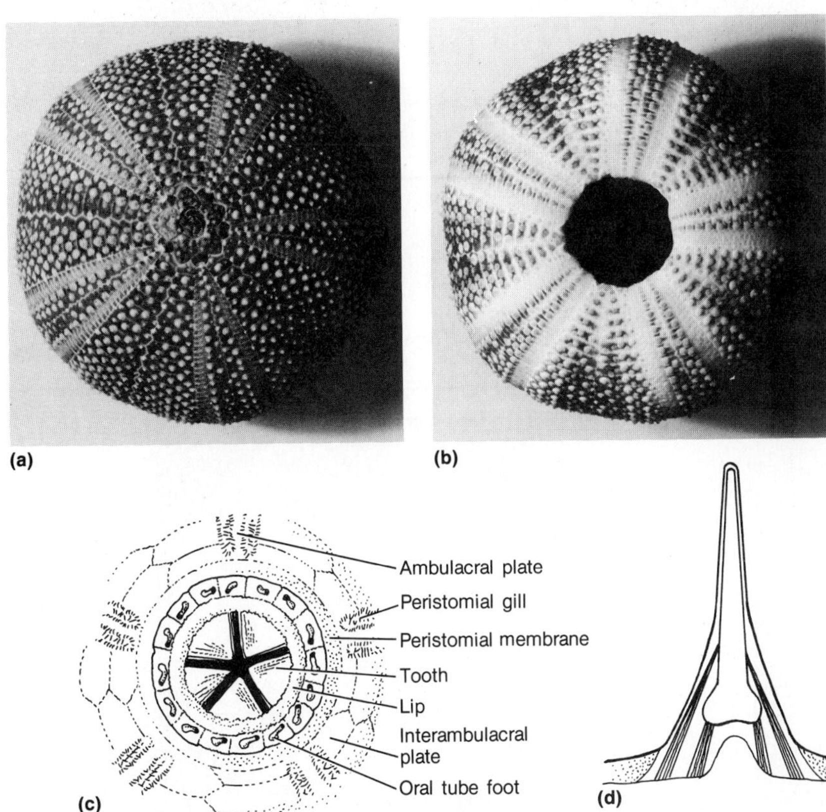

Figure 17.25 *External features of the test of sea urchins (regular echinoids):* ***(a)*** *photograph of a test, aboral view;* ***(b)*** *photograph of a test, oral view (hole is covered in life by the peristome, mouth, and jaws;* ***(c)*** *the peristomial region;* ***(d)*** *the base of a spine.*

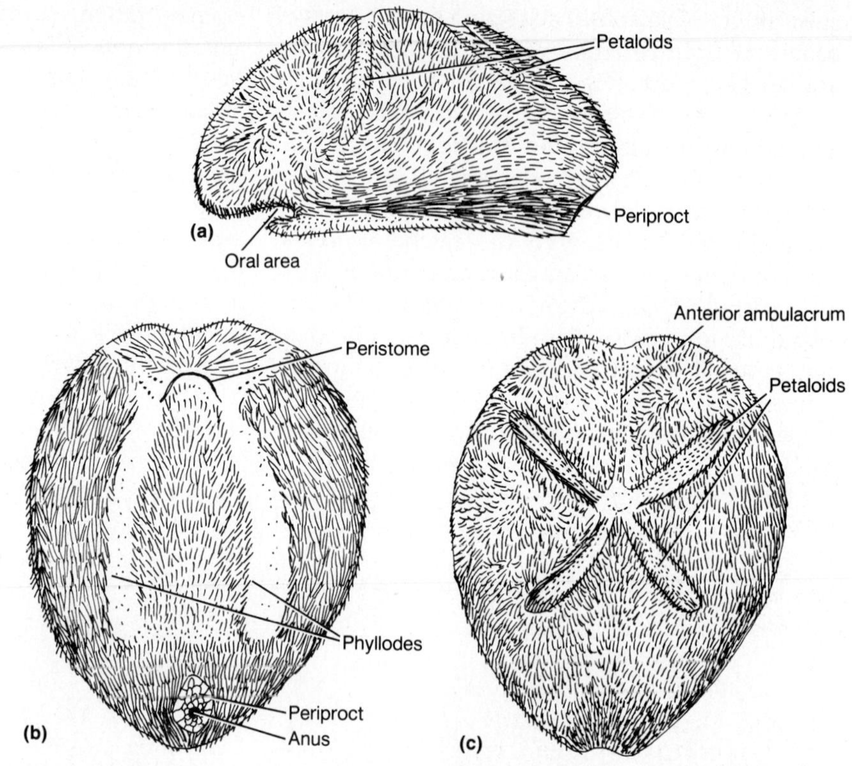

Figure 17.26 External features of the test of heart urchins (irregular echinoids): ***(a)*** lateral view; ***(b)*** oral view; ***(c)*** aboral view (parts a–c from L.H. Hyman, The Invertebrates. Vol. IV: Echinodermata, McGraw-Hill Book Co., 1955, used with permission).

Figure 17.27 Photographs of the external features of the test of Mellita, the keyhole sand dollar (an irregular echinoid): ***(a)*** aboral view; ***(b)*** oral view.

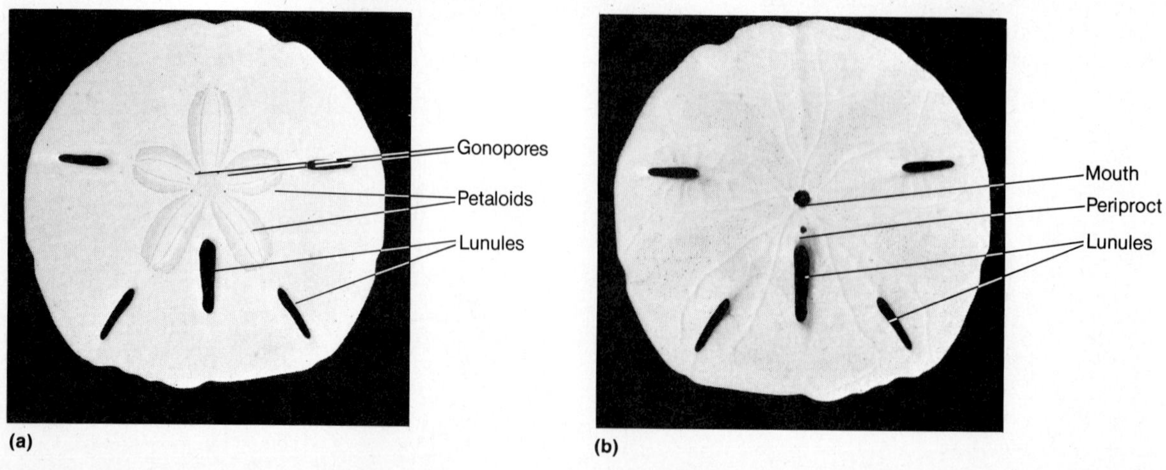

Lytechinus (Fig. 17.24b) uses its spines to place stones and shell fragments on its aboral surface for camouflaging.

Sea urchins also bear a large number of prominent long pedicellariae over the general body surface. Echinoid pedicellariae have three opposing jaws, and the stalk typically contains an internal supporting rod. One type of pedicellaria found in many urchins contains poisonous sacs located at the base of each jaw. This poison has a paralyzing effect on small organisms, drives larger organisms away, and may even, in a few forms, produce pain and aches in people when the echinoid is handled. The ambulacra of most urchins also bear small stalked bodies located near the mouth, the **sphaeridia,** which contain a statocyst and aid the animal in its balance and equilibrium.

Irregular echinoids

All sand dollars and heart urchins are burrowers, and they have several important adaptations for this habit. Being bilaterally symmetrical, they move through the benthic material with the anterior end forward. Since the podia are specialized for gaseous exchange, feeding, and sensory reception, locomotion is achieved by the use of the spines. Water currents are generated through the burrow and over the animal by aboral spines. In keeping with the burrowing habit the irregular echinoids are drab and inconspicuously colored. Since the heart urchins and sand dollars differ substantially in their external features, perhaps the best approach is to discuss their most important external features separately.

Heart urchins, also called spatangoids, have an oval or rounded body. The mouth and peristome have been moved anteriorly, and the anus and periproct, situated outside of the aboral apical system of plates, are located posteriorly along one interradius (Fig. 17.26). On the aboral surface, each conspicuous ambulacral area is petal-shaped and is termed a **petaloid.** The five petaloids are arranged radially around the aboral pole (Fig. 17.26c). Orally, the ambulacra or **phyllodes** are oriented radially around the mouth (Fig. 17.26b). The tube feet of the petaloids are branched, leaflike, and function in gaseous exchange, while the phyllodial podia are strengthened internally and are involved in food gathering and chemoreception. Spines are mostly short and form a dense covering over the body. Certain spines, the **clavules,** are quite small, are flattened at their distal ends, and create a current of water over the body surface. Both pedicellariae, with only two jaws, and sphaeridia, located in the phyllodes, are present.

The sand dollars or the clypeasteroids are usually greatly flattened along the oral–aboral axis and have a rounded, smooth outline (Fig. 17.27). The oral surface is flat, while the aboral surface is somewhat convex. The mouth is located centrally on the oral surface, but phyllodes are missing. Radially arranged ciliated grooves run from the perimeter to the mouth and are used to transport food (Fig. 17.27b). The anus and periproct are situated not on the aboral surface, but on the oral surface in a posterior interambulacrum. The aboral surface contains five prominent petaloids (Fig. 17.27a). Two types of tube feet are present: small podia with suckers that are distributed over both oral and aboral surfaces and aid the spines in locomotion and larger petaloid podia that function in gaseous exchange. Pedicellariae and peristomial sphaeridia are also present. Many sand dollars like *Mellita,* the keyhole sand dollar, bear slitlike openings in the test called **lunules** that perhaps aid in burrowing (Fig. 17.27). During development they begin as marginal indentations, and later these slits are surrounded by growth of the test.

Feeding and the digestive system

Echinoids are mostly scavengers but perhaps can best be termed generalists in food habits and preferences. Most sea urchins feed on algae by scraping away encrusting forms or nibbling on kelp, but some are carnivorous and eat small animals. The irregular echinoids consume detrital particles in the sand through which they burrow.

Characteristic of most echinoids except for the heart urchins is a marvelously adapted chewing mechanism, **Aristotle's lantern,** so named for the Greek philosopher-scientist who first described it as a "lantern with the glass panes left out." The lantern, best developed in sea urchins, is an extremely complex and intricate chewing device composed of 40 ossicles and 60 muscles. The functional units of the lantern are five large, vertically arranged jaws supported by an internal framework of skeletal ossicles and operated by numerous strong muscles (Fig. 17.28). The five **jaws** or **pyramids** are long, slender, calcareous plates situated interradially. Along the inner surface of each pyramid is a long calcareous band that originates in a basal dermal sac and terminates in a very hard **tooth** that projects into the buccal region just beyond the pyramid (Fig. 17.28b). New tooth materials are secreted at the base of the band to replenish the tooth as it is worn away distally by use. Each pyramid is connected to its two adjacent pyramids by muscles so that the five pyramids and their teeth can be adducted forcefully to create an effective cutting,

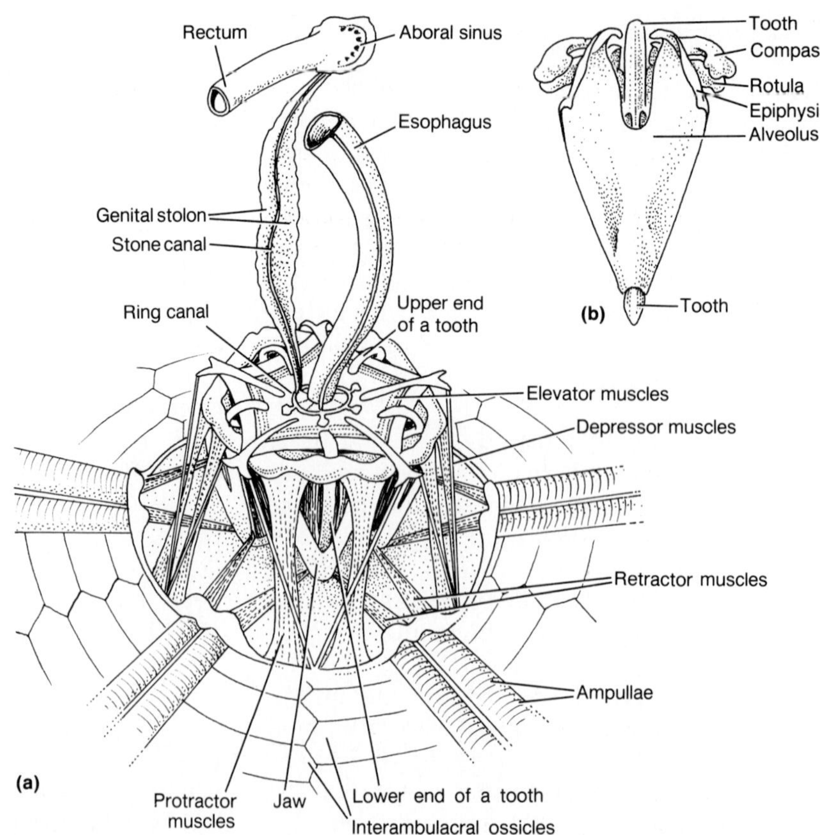

Figure 17.28 *Aristotle's lantern of echinoids:* ***(a)*** *view of the lantern from above and to the side (from E.W. MacBride,* Cambridge Natural History, *Macmillan, London, 1909);* ***(b)*** *a pyramid and its tooth of* Arbacia *(from F.A. Brown, Jr.,* Selected Invertebrate Types, *© 1950, John Wiley & Sons, Inc., used with permission).*

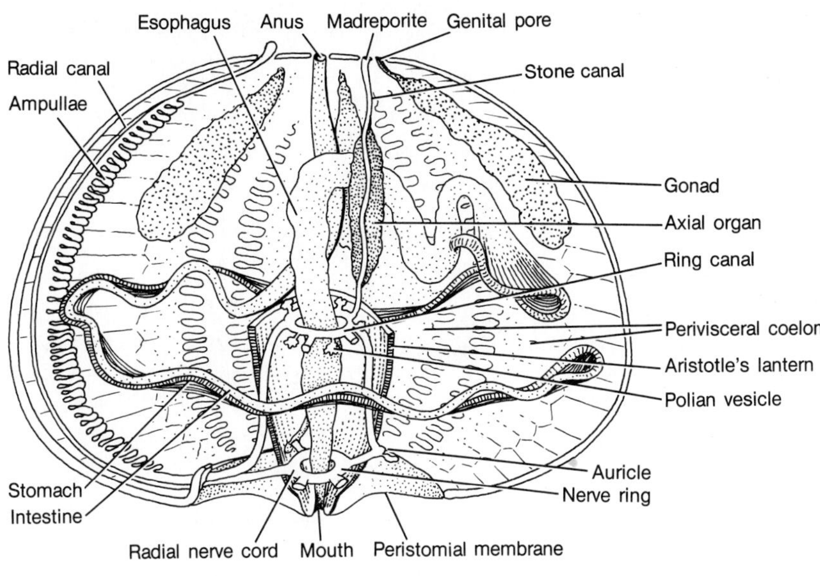

Figure 17.29 *A lateral schematic view of the visceral systems in a sea urchin (from F. A. Brown, Jr.,* Selected Invertebrate Types, *©1950, John Wiley & Sons, Inc., used with permission).*

ripping, tearing, or biting action. At the aboral end of each pyramid there are pentamerously arranged ossicles (rotulas, epiphyses, compasses) and a variety of protractor and retractor muscles with which the entire lantern can be partially protracted through the mouth (Fig. 17.28a). Sand dollars have a modified lantern that is distinguished mostly by its enlarged pyramids and the fact that it cannot be protracted.

The lantern surrounds the oral parts of the alimentary canal including a buccal cavity and pharynx, the latter being continuous as an esophagus. The esophagus first passes aborally and then makes a loop orally to join the intestine; at the esophagus–intestine junction there often is a blind-end cecum. The long intestine makes several loops just inside the test; first, it makes an almost complete counterclockwise circle (viewed aborally) followed by another partial circle in the opposite direction but aborally to the first (Fig. 17.29). In many echinoids a slender tube, the **siphon,** arises from the oral part of the intestine, runs parallel to it for about one-half its length, then rejoins the intestine. The siphon is thought to be a shunt for water in the intestine so that food can be concentrated for digestion. The intestine becomes the rectum, which opens to the outside as the anus located within the periproct. The periproct and anus are located at the aboral pole (sea urchins), at the posterior end (heart urchins), or ventral and just posterior to the mouth (sand dollars).

Other internal features

The perivisceral coelom is spacious in regular echinoids but reduced in others (Fig. 17.29). Certain parts of the water vascular system have already been mentioned including the madreporite and tube feet. From the ring canal arise five radial canals that run meridionally toward the aboral pole and beneath the paired rows of ambulacral ossicles. The connections between an ampulla and tube feet are unlike the arrangement in other echinoderms in two ways: First, the ampulla–podium connection runs through, not between, an ambulacral ossicle; second, these connecting tubes are doubled as they pass through the ossicle (Fig. 17.5b).

Gaseous exchange in sea urchins involves tube feet, gills, Aristotle's lantern, and a special coelomic compartment. Five pairs of evaginations of the body wall, the **peristomial gills,** are the chief centers for gaseous exchange. A unique small coelomic chamber, the **peripharyngeal coelom,** surrounds the lantern and extends into the interior of the gills. Certain lantern muscles contract rhythmically in such a way as to pump coelomic fluid into and out of the gills to facilitate gaseous exchange. In the irregular echinoids, gills are absent, and the petaloid podia become the principal center for respiratory gas exchange.

The nervous system is characterized by the dominant ectoneural system innervating external structures, especially the podia and pedicellariae, and its subepidermal plexus. The hyponeural system basically controls the lantern, and the endoneural system supplies the gonads. Sense organs consist of abundant tactile and chemoreceptor cells on the spines, podia, and pedicellariae; the sphaeridia; and photosensitive cells or ocelli on the aboral surface.

Reproduction and Development

In echinoids the five gonads are suspended by mesenteries in the interambulacral areas, but the posterior gonad is lost in the irregular echinoids. When gametes are mature, muscles in the gonadal walls force them through the gonopore and into the sea, where fertilization takes place. In females that brood, the ova and embryos are retained on the peristome, on the periproct, or among the spines. Fertilized eggs of most echinoids are planktonic and develop quickly into an **echinopluteus** larva (Fig. 17.30). This larval stage, quite similar to the larval stage of ophiuroids, is characterized by six pairs of long arms that apparently increase the buoyancy of the planktonic larva. After as long as several months in this stage, the larva develops ossicles and other adult structures and sinks to the bottom but is not attached to the substratum. Metamorphosis takes place rapidly, sometimes within an hour's time, and the very young miniature juvenile echinoid continues to feed and grow and reaches sexual maturity in about a year.

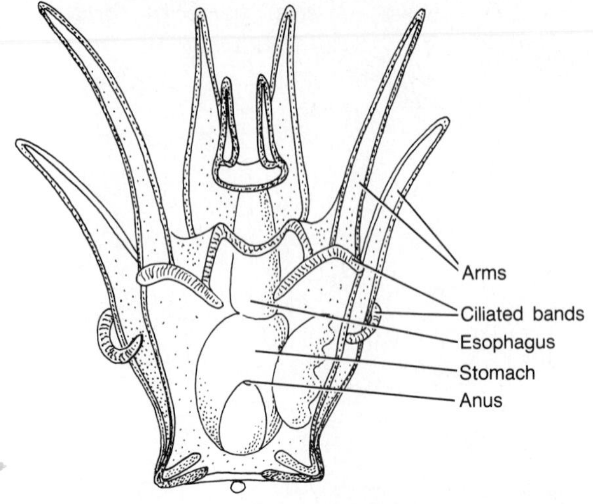

Figure 17.30 The echinopluteus larva of an echinoid (from Mortensen in L.H. Hyman, The Invertebrates. *Vol. IV:* Echinodermata, *McGraw-Hill Book Co., 1955, used with permission).*

Class Holothuroidea

The holothuroids are called sea cucumbers—a term that aptly describes their typical body shape. Holothuroids are the most bizarre and aberrant group of echinoderms, which as a larger group, are themselves peculiar. The most distinctive characteristics of holothuroids are their external shape and symmetry, oral tentacles, microscopic and nonarticulating ossicles, respiratory trees, and ability to eviscerate and still survive; each of these will be elaborated upon in the following discussions. Almost all sea cucumbers are benthic, and most of the approximately 1100 species are found partially buried in mucky bottom materials in shallow or intertidal areas, though some live on hard substrata and others inhabit abyssal regions. Most species are 10–25 cm in length and about 5 cm in diameter. Members of the Philippine *Stichopus* (Fig. 17.31d) may be up to 1 m in length and their diameter one-fourth their length. They have a wide range of colors from dull brown and black to red, yellow, and violet, and many are striped.

EXTERNAL FEATURES AND THE BODY WALL

Most holothuroids are cylindrical and greatly elongated along the oral–aboral axis, but their shapes range from spherical to the wormlike apodids (Fig. 17.31). The **mouth** is located at the oral pole and the **anus** at the aboral pole, although in some forms, either or both openings are somewhat displaced from their terminal positions. Because of their shape, holothuroids have to lie on one side; the underside is the **ventral surface** or **sole,** and the upper side is the **dorsal surface.** The sole, consisting of three ambulacral and two interambulacral areas, has podia arranged either along the three ambulacra or scattered over its entire surface. The dorsal surface contains three interambulacra and two ambulacra, and the podia on this surface are often termed "papillate," since they lack suckers and tend to be generally reduced to warts. Podia on the ventral surface may or may not have suckers, and they are involved mostly with locomotion. Tendencies among the holothuroids

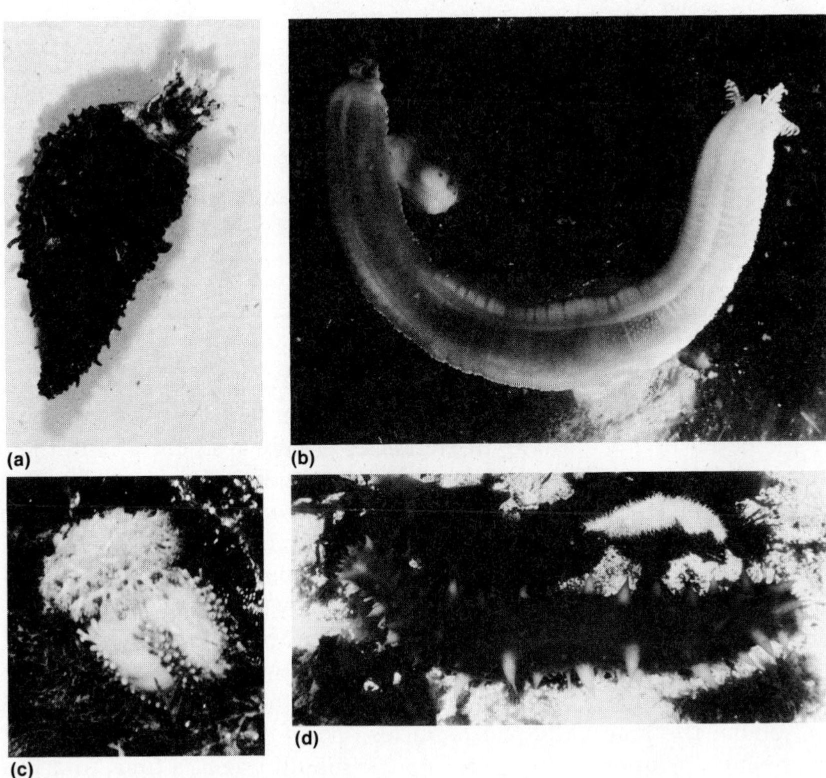

Figure 17.31 *Photographic collage of several different holothuroids:* ***(a)*** Thyone *(dendrochirotacean);* ***(b)*** Leptosynapta *(apodacean) (courtesy of C.E. Jenner);* ***(c)*** Cucumaria *(dendrochirotacean) (courtesy of S. Armstrong);* ***(d)*** Stichopus *(aspidochirotacean) (courtesy of Ward's Natural Science, Inc., Rochester, N.Y.).*

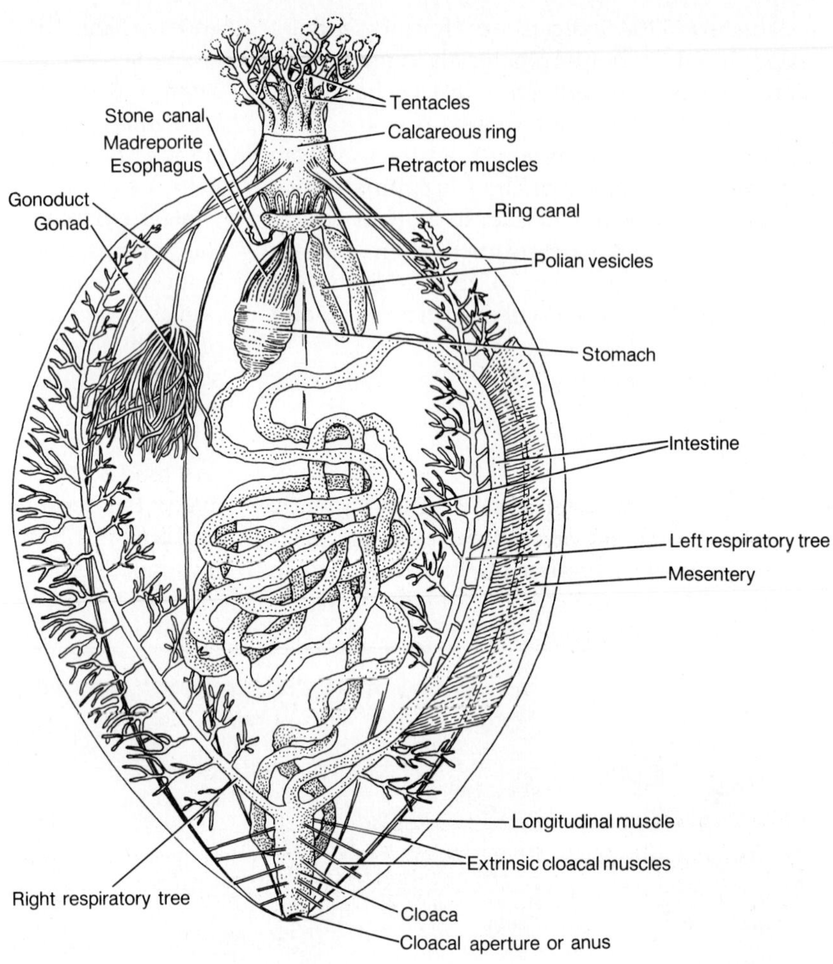

Figure 17.32 *The internal features of* Thyone, *ventral view (from W.R. Coe,* Starfishes, Serpent Stars, Sea Urchins and Sea Cucumbers of the Northeast, *Dover Publications, Inc., New York, N.Y., 1972, used with permission).*

are for podia to be reduced numerically, first on the dorsal surface, and then for podia of the sole to lose their radial distributions and be more randomly distributed; representatives of two orders are missing podia altogether. Because of the orientation of the body, the differentiation of dorsal and ventral surfaces, and an exaggerated oral–aboral axis, a rather well-defined tertiary bilaterality has resulted.

The mouth is always surrounded by a single circle of 10–30 modified buccal podia, the **tentacles** (Figs. 17.31, 17.32). Sometimes simple and fingerlike but otherwise complex and pinnate, the tentacles are always hollow, and their lumens are extensions from the radial water canals. Most sea cucumbers are deposit- or suspension-feeders, and the tentacles are the principal food-gathering organs. Feeding is accomplished either by trapping plankton or particles of food in mucus secreted by special cells in the tentacular walls or by organic material deposited on the surface being swept up by the tentacles. The food-laden tentacles are then put into the mouth one at a time, and the food sucked off. Many deep burrowers consume bottom material, digest its organic contents, and void the undigestible portion as castings.

The body wall is leathery and thick and is characterized by the presence of ossicles and muscle layers. The microscopic ossicles, located in the thick dermis, assume a variety of shapes and forms that are widely used taxonomically (Fig. 17.3c). The ossicles, not articulated with each other except in one small family, are common in the walls of the body, tentacles, many podia, and even internal structures. Circular muscles lying beneath the dermis either make a complete cylinder around the animal or are interrupted by bands of longitudinal muscles. The five single or double bands of meridional **longitudinal muscles** extend orally from the aboral body wall, parallel the ambulacra, and insert into the calcareous ring, a hard supporting feature encircling the anterior pharynx (Fig. 17.32). The contraction of these longitudinal muscles retracts the tentacles completely inside the body and also pulls the anterior body wall in over them.

Sea cucumbers move around but only slowly. Those animals with well-developed ventral podia are most active, each tube foot being involved in a stepping motion much like that in asteroids. Other holothuroids move in a fashion similar to that of an earthworm with alternating contractions of the circular and longitudinal muscles forcing the body forward. Many sea cucumbers use the oral tentacles as a means of burrowing and plowing through the substratum. But incredibly, a few holothuroids are pelagic and either swim by a thrashing of the ends of the body or are suspended in the water using modified podia as flotation devices.

INTERNAL FEATURES

Digestion, gaseous exchange,* and *excretion. The digestive system consists of a mouth, pharynx, esophagus, stomach, intestine, cloaca, and anus, but one or more parts are often missing. On its outer surface the anterior pharynx bears a **calcareous ring** of ossicles bound together by connective tissue (Fig. 17.32). Composed of five radial and five interradial plates, this ring is thought to be homologous to the lantern of echinoids. The calcareous ring has several important functions including supporting the pharynx and ring canal and serving as the insertion point for the longitudinal muscle bands that retract the tentacles and anterior end. The intestine is quite long and loops several times in the perivisceral coelom (Fig. 17.32).

Holothuroids have a posterior cloaca, a structure found in many lower vertebrates and only rarely in invertebrates (Fig. 17.32). The **cloaca** (= sewer) is a chamber that is common to the digestive and some other system, usually the reproductive system, but in holothuroids the cloaca is common to the digestive and gas exchange systems. The cloacal wall contains intrinsic circular muscles and is supplied with extrinsic dilating muscles. Arising from the cloaca is a pair of unique respiratory trees for exchanging gases between the seawater and the animal. Each **respiratory tree** is an extensive invaginated, many-branched system of blind-end tubules, and both respiratory trees may more or less fill the large perivisceral coelom (Fig. 17.32). The pulsating muscular contractions of the cloaca pump seawater into the respiratory trees in the following manner: The cloaca dilates to admit seawater, a sphincter muscle then closes the anus, and contractions of circular cloacal muscles force seawater into the trees. Water is expelled from the respiratory trees by dilating the cloaca, contracting the muscles in the walls of the tree tubules, and opening the anus. Multiple cycles may be required to fill the tree completely, but one expulsion, often vigorous, eliminates all the water. Gaseous exchange via the respiratory trees is not very efficient; the oxygen requirements of sea cucumbers are correspondingly low. Some exchange of respiratory gases also takes place at the general body surface, tentacles, and podia.

Excretion involves the respiratory trees and coelomocytes. Water-soluble nitrogenous wastes like ammonia are voided into the water of the respiratory trees and are subsequently expelled. Other wastes are accumulated by coelomocytes and conveyed to the trees, tentacles, or even gonadal ducts for discharge. Members of the Order Apodida lack respiratory trees but possess a curious system of ciliated funnels or **urns** that are of coelomic origin. Urns, either single or in clusters, are found on various visceral parts but are especially common to mesenteries. The internal cavities of urns contain coelomocytes that have accumulated wastes and have entered by way of the funnels. Later, these coelomocytes transport wastes to the body wall for expulsion.

Two other features associated with the cloaca and respiratory trees should be noted here. Certain holothuroids of the Order Aspidochirotida (*Holothuria* and *Actinopyga*) have small, variously colored blind-end tubules attached to the base of one or both respiratory trees. These **organs of Cuvier** produce defensive threads that dissuade an unwelcome visitor or predator. When agitated or disturbed, these sea cucumbers direct the anus toward the intruder, the body wall forcefully contracts, the cloacal wall is ruptured, and the sticky (and sometimes toxic) threads are forced out onto the threatening animal. The threads wrap around the enemy (mostly decapod crustaceans) and render it totally helpless. New threads and cuvierian organs are quickly regenerated. A more dramatic and drastic phenomenon is **evisceration,** a process by which certain visceral organs are forcefully eliminated through a rupture in the cloaca, oral region, or body wall. Most typically, evisceration occurs through the anus, and the respiratory trees, parts of the alimentary system, and the gonad are voided. Evisceration is caused by a number of environmental conditions (high temperatures, fouled water, crowded conditions), excessive stimulation, or injury, but it may very well be a natural seasonal phenomenon. Regeneration of all lost parts usually follows, and evisceration may in fact have a rejuvenating effect on the organism.

The coelomic systems. The basic plan of the water vascular system is similar to that in other echinoderms, but there are some peculiar features in holothuroids. The prominent **polian vesicles,** numbering from 1 to 50, hang posteriorly from the circumpharyngeal ring canal and serve as expansion chambers to maintain hydraulic pressure in the water vascular system (Fig. 17.32). The **madreporite** is completely internal and has no direct connection whatsoever with the body wall, and some sea cucumbers even have multiple stone canals and madreporites. Each radial canal extends anteriorly from the ring canal, passes inside the pharyngeal calcareous ring, gives off a branch to one or more tentacles, runs posteriorly and meridionally between circular and longitudinal muscles of the body wall, and gives rise to lateral canals each of which supplies a podial unit. Three radial canals are associated with the sole and two with the dorsal surface.

The hemal system, particularly well developed in sea cucumbers, is made up of the typical hemal ring around the pharynx, radially arranged hemal sinuses, and prominent **dorsal** and **ventral sinuses** that are closely associated with the intestine. Containing many hemoglobin-bearing coelomocytes called **hemocytes,** the hemal fluid circulates in many small channels within the intestinal walls and the left respiratory tree. Clearly, the holothuroid hemal system is involved in absorption of foodstuffs from the intestine and in transport of gases to and from the respiratory tree. Future studies on the precise functions of the holothuroid hemal system should be quite fruitful.

Nervous system and sense organs. The nervous system is basically like that for echinoderms generally. A circumoral nerve ring is present at the base of the tentacles, and a radial nerve runs beneath the dermis in each ambulacrum. Strands of the ectoneural system innervate the sensory cells of the body wall and podia, and the hyponeural motor nerves supply the body wall muscles; no endoneural system is present. Sensory structures are represented by eye spots at the bases of the tentacles in some apodids and numerous epidermal sensory cells. Statocysts are present in some of the burrowing apodids in which the oral end is always directed downward.

Reproduction and development. Most holothuroids are dioecious, but some hermaphroditic protandrous apodids are known. The reproductive system is unique among echinoderms in that there is a single **gonad** located beneath a dorsal interambulacrum. Commonly, it is a tuft or cluster of tubules that are united at their bases in a shape that resembles a mop head (Fig. 17.32). A **gonadal duct** runs from the gonad to the **gonopore** situated near the base of the tentacles. In the antarctic forms that brood their embryos the ova are trapped by the tentacles, transferred to the upper or lower surface, and often housed in special brooding chambers: *Thyone rubra* and *Leptosynapta* are most unusual in that ova are brooded within the coelom. All others apparently spawn gametes into the sea, where ova are fertilized. The various embryonic stages are planktonic, and within several days following fertilization an **auricularia** larva is produced (Fig. 17.33a). Similar to the bipinnaria stage of asteroids, it possesses a ciliated locomotor band. Later, the locomotor band of the auricularia is broken up into five girdles or ciliated bands, and the body becomes barrel shaped; this stage is the **doliolaria** larva (Fig. 17.33b). After considerable time as a doliolaria the organism gradually metamorphoses into a small planktonic stage sometimes called the **pentacula** larva and distinguished by the presence of oral tentacles. As growth continues, the juvenile sea cucumber gradually settles to the bottom and becomes a benthic adult.

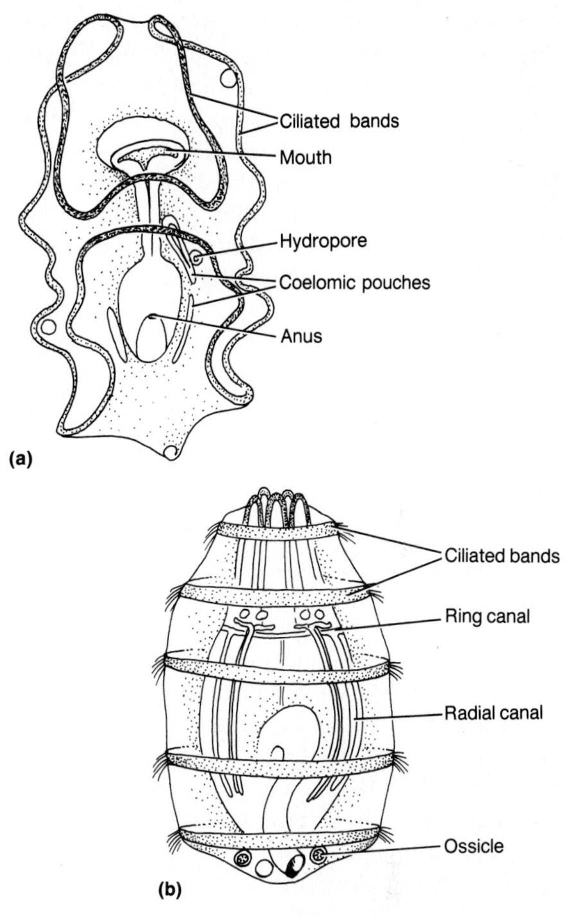

Figure 17.33 *Larvae of holothuroids:* ***(a)*** *auricularia (from Mortensen in L.H. Hyman,* The Invertebrates. *Vol. IV:* Echinodermata, *McGraw-Hill Book Co., 1955, used with permission);* ***(b)*** *doliolaria (from E.W. MacBride,* Cambridge Natural History, *Macmillan, London, 1909).*

PHYLOGENY OF THE ECHINODERMS

Because of the large number of known echinoderm fossils, one might conclude that the evolutionary history of this remarkable group of animals is well known and that the relational lines are quite clear. Unfortunately, that is not the case. True, there is an abundant fossil record that dates back to the early Cambrian period, but by the time the oldest fossil species were being preserved, the phylum had been in existence for 50 or 100 million years or more! Further, the fossil record contains numerous instances of very primitive echinoderms appearing relatively late in geological history—in fact, much later than certain more advanced forms. Therefore while the echinoderm fossil record is extensive and is very helpful in our understanding of their phylogeny, many gaps remain, and many questions are unanswered. Echinoderms are deuterostomates, and most zoologists believe that the fundamental dichotomy between protostomates and deuterostomates took place early in the evolution of animals; there are absolutely no fossil records of early deuterostomates. Logically, the ancestral echinoderm was probably a soft-bodied, bilaterally symmetrical marine worm, and its coelomate condition probably developed early in its history. This conjecture may never be proved concretely because of the scarcity of fossils of soft-bodied animals.

Some zoologists have tried to link the evolution of echinoderms to sipunculids or the lophophorates. Sipunculids (see Chapter 9) are unsegmented and have a coiled intestine and a number of ciliated oral tentacles. The tentacles are hollow, their lumens are continuous with a circumesophageal ring equipped with compensation sacs, and they are extended hydraulically. The bryozoans (see Chapter 15) also have feeding tentacles, the cavities of which are coelomic in origin and connected to a circumesophageal ring. These features in both groups certainly bring to mind the buccal podia and water vascular system of echinoderms. However, both sipunculids and bryozoans are protostomates, while echinoderms are deuterostomates; this constitutes a basic and fundamental difference. In short, if the protostomate–deuterostomate division is as primal a separation as most zoologists hold, then echinoderms could not be related to either sipunculids or bryozoans. Rather, their similarities probably represent convergent evolution of widely disparate groups that have adapted in similar ways in keeping with similar habits. Only a very weak case can even be made for the similarities between the coelomic spaces in the lophophore of phoronids and brachiopods (both deuterostomates) and the water vascular system of echinoderms; this situation may also be a case of convergent evolution.

Evolutionary pathways within the phylum, especially those of ancient extinct groups, are also difficult to evaluate because of the vagaries of the fossil record. Of particular concern to the early phylogenetic history of echinoderms is the ontogeny of their pentamerism and, assuming the ancient forms were stalked, the ways in which the free-living forms were derived with oral surface against the substratum. The most important major, totally extinct echinoderm taxa are characterized in Box 17.3 and illustrated in Fig. 17.34.

There are three extinct groups that appear to be closest to the ancestral echinoderms: eocrinoids and cystoids (both of which are crinozoans) and the carpoids (homalozoans). All three appear to be related, but their lines of relationship are not clear (Figs. 17.34a–c, 17.35). For example, the eocrinoids were pentamerous, but cystoids and carpoids were not. Then the question arises, Were ancient echinoderms asymmetrical (carpoids), trimerous (many cystoids), or pentamerous (eocrinoids)? If we could ascertain which type of symmetry was most primitive, then the early history would be much clearer. Another ancient group, the pentamerous blastoids, were probably derived as evolutionary offshoots from the ancestral line. Certainly a very ancient group, the stalked crinoids arose early in the history of the group, and some representatives persist even today (Fig. 17.35).

All the foregoing groups are distinguished

BOX 17.3 □ A THUMBNAIL SKETCH OF THE PRINCIPAL CHARACTERISTICS OF THE MOST IMPORTANT EXTINCT ECHINODERM GROUPS

SUBPHYLUM HOMALOZOA—the carpoids; all asymmetrical with no evidence of radial symmetry; some stalked, others stalkless; oral surface thought to face substratum; some bore a single arm with an ambulacrum (Fig. 17.34a).

SUBPHYLUM CRINOZOA

Class Eocrinoidea—rather rare, but among oldest fossils; theca was often stemless and bore a crown of grooved brachioles surrounding the apical mouth; no arms; most were pentamerous (Fig. 17.34b).

Class Cystoidea—attached aborally, sometimes by a stalk; oval thecal walls with pores as a part of a gas exchange system; food-gathering brachioles along three or five ambulacra (Fig. 17.34c).

Class Blastoidea—attached aborally either directly or with a short stalk; five ambulacra extended meridionally over the oval theca; bore a peculiar system of folds and pores (hydrospires) laterally along each ambulacrum representing portions of a water route for gaseous exchange (Fig. 17.34d, e).

SUBPHYLUM ASTEROZOA

Class Somasteroidea—earliest asteroids; many skeletal structures were primitive; rod-shaped ossicles; petaloid arms; no anus; madreporite was marginal on small disc; no pedicellariae (Fig. 17.34f).

SUBPHYLUM ECHINOZOA

Class Edrioasteroidea—some were stalked, most were unstalked but attached; upper oral surface with five ambulacra; ossicles probably had spines; no brachioles; ossicles bore pores perhaps for tube feet (Fig. 17.34g, h).

Class Helicoplacoidea—unusual cigar-shaped body that had spiral pleats and a single ambulacrum that made about two circles around body; mouth was at upper end, and aboral end was buried in benthic sand (Fig. 17.34i).

by the fact that they were oriented with the oral side uppermost and their food-gathering activities involved trapping suspended particulate matter from the water above them. But a dramatic change in body orientation and feeding habits occurred one or more times so that the body was turned 180°, they became deposit-feeders, tube feet became adapted for locomotion, and the animal was free to move over the surface. How this new orientation took place, what its intermediate stages were like, and what environmental reasons were behind the development of this adaptation are simply not known. However, each of the four remaining extant groups solved their orientational problems in different ways.

The asterozoans undoubtedly arose from ancient crinoids whose arms bore a double row of ambulacral plates. Ophiuroids evolved from asteroids, and both groups developed certain individual features but retained many common characteristics (Fig. 17.35). The origins of the Echinozoa are more obscure and are open to speculation. Like the asterozoans, the echino-

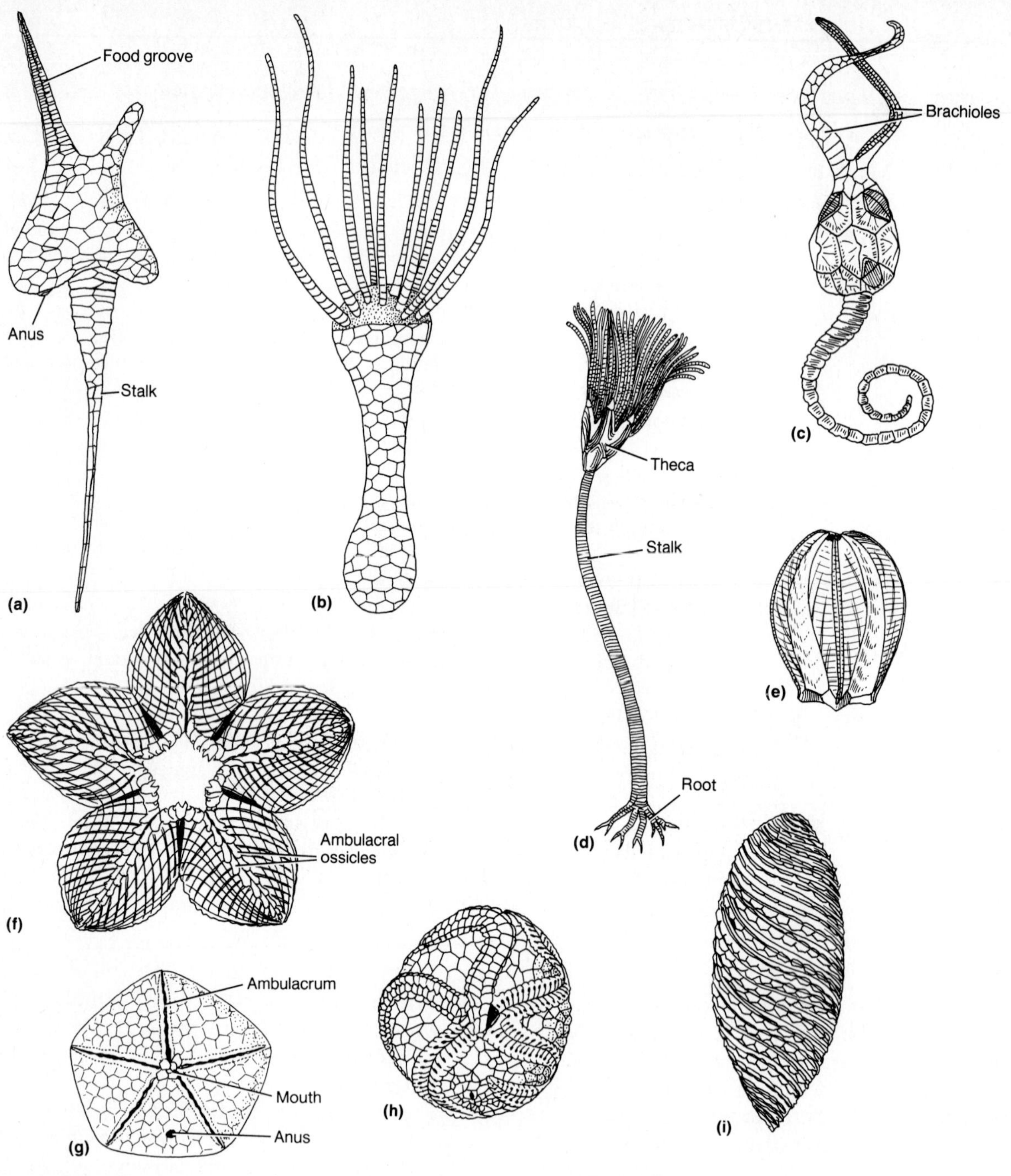

Figure 17.34 *Representatives of the principal extinct echinoderm groups:* ***(a)*** *homalozoan;* ***(b)*** *eocrinoid;* ***(c)*** *cystoid;* ***(d)*** *stalked blastoid;* ***(e)*** *unstalked blastoid;* ***(f)*** *very ancient somasteroid;* ***(g)*** *edrioasteroid;* ***(h)*** *edrioasteroid with curved ambulacra;* ***(i)*** *helicoplacoid (from U. Lehmann and G. Hillmer, 1983,* Fossil Invertebrates, *Cambridge University Press). (parts a, c [from Bather], f [from Spencer] in L.H. Hyman,* The Invertebrates. *Vol. IV,* Echinodermata, *McGraw-Hill Book Co., 1955, used with permission; parts d, e, g, and h from F.A. Bather,* A Treatise on Zoology, *E.R. Lankester, ed., Adam & Charles Black, London, 1900.)*

zoans most likely arose from an ancient stalked crinoid. There are two intriguing fossil groups of echinozoans, the asymmetrical burrowing helicoplacoids and the attached pentamerous edrioasteroids; the latter may be near the ancestral form from which both the echinoids and holothuroids developed (Fig. 17.35).

A second theory holds that primitive echinoids were free living and that the attached habit was evolved secondarily and independently in several different lines. According to this theory, the helicoplacoids would be nearest the ancestral stock, whose representatives were presumably burrowers. One great problem with this theory is the generally held idea that pentamerism was undoubtedly a response to a sessile existence. Since pentamerous symmetry is such a constant character in all living echinoderms, was it not first developed in ancestral forms that were sessile?

Figure 17.35 *A probable phylogenetic tree of echinoderms showing affinities of the five principal extant taxa with the major extinct taxa.*

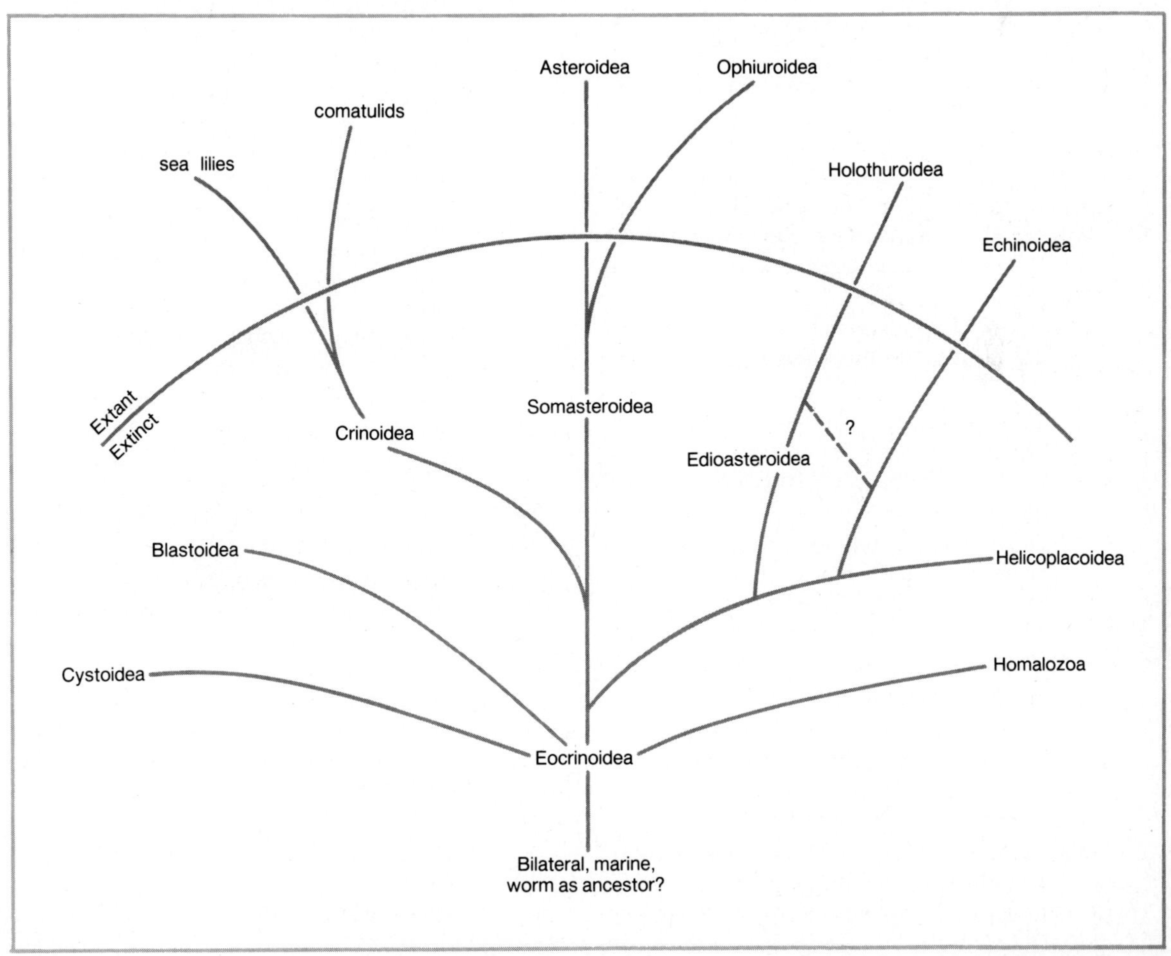

The enigmatic questions of phylogeny are but another intriguing aspect of this absolutely engrossing group of curious animals, the echinoderms.

SYSTEMATIC RESUME

The Major Taxa of Echinoderms

PHYLUM ECHINODERMATA

Subphylum Homalozoa—the carpoids; all extinct; dorsoventrally flattened body with no trace of biradial or pentamerous symmetry; probably the most ancient echinoderms, but their exact phylogenetic affinities are difficult to assess.

Subphylum Crinozoa—globose or cup-shaped body and five or ten long, hollow, food-gathering arms; some are attached by aboral surface or are stalked, but many are free swimming; mostly extinct.

Class Crinoidea—feather stars, sea lilies; only extant class of crinozoans; stalked or secondarily stalkless; oral surface uppermost; crown contains all visceral organs; have branched arms.

Subclass Articulata—all extant crinoids; characteristics are the same as those of the class.

Order Isocrinida—sea lilies; all attached by stalk that bears cirri at regular intervals; *Cenocrinus* (Fig. 17.10a) and *Metacrinus* are representative.

Order Millericrinida—sea lilies; all are attached by a stalk that bears cirri only at lower stalk joints; *Rhizocrinus* is a typical genus.

Order Crytocrinida—sea lilies; lack cirri; *Ptilocrinus* is representative.

Order Comatulida—feather stars; stalkless, unattached, and free swimming; the well-known *Antedon* (Fig. 17.10b) and *Comanthina* are representative.

Subphylum Asterozoa—body flattened, pentagonal, or stellate; central disc plus typically five arms or rays; mouth on undersurface and central in position; tube feet normally restricted to oral surface of arms.

Class Stelleroidea—same as those of subphylum; includes all living asterozoans.

Subclass Asteroidea—sea stars; stellate body composed of flattened central disc and usually five radially arranged arms; open ambulacral grooves on oral side of arms; tube feet in rows paralleling ambulacra; many have pedicellariae.

Order Platyasterida—primitive and mostly extinct; five to many arms; paxillae present; podia without suckers; *Luidia* is rather common.

Order Paxillosida—arms long and pointed; tube feet without suckers; paxillae restricted to aboral surface; many live on soft bottoms, including the well-known *Astropecten*.

Order Valvatida—when present, pedicellariae are valvate and sessile; tube feet with suckers; includes *Oreaster* and *Porania*.

Order Spinulosida—aboral surface covered with short spines; pedicellariae (when present) consisting of grouped spines; tube feet have suckers; more common genera include *Solaster* (Fig. 17.15a), *Asterina*, and *Pteraster*.

Order Euclasterida—mostly deep-sea forms; relatively small central disc; numerous, slender, easily broken arms; have pedicellariae and suctorial podia; well-known members are *Brisinga* and *Odinia*.

Order Zorocallida—domed central disc with regular radially arranged overlapping ossicles; prominent marginal ossicles; *Zoraster* is typical.

Order Forcipulatida—pedicellariae with straight or crossed jaws and stalked; madreporite always on aboral surface; includes most asteroids; *Asterias* (Fig. 17.15e), *Heliaster*, and *Pisaster* (Fig. 17.15c) are frequently encountered.

Subclass Ophiuroidea—serpent, brittle, and basket stars; disc and arms sharply distinct; arms elongate, jointed, and movable; closed ambulacral areas; madreporite on oral surface of disc.

Order Oegophiurida—mostly extinct; many of the typical ophiuroid shields and plates are absent; pyloric ceca enter arms; *Ophiocanops* is the only extant genus.

Order Phrynophiurida—disc and arms covered by skin; aboral arm shields absent; arm joints permit movement in all directions; arms of basket stars are branched; includes some primitive brittle stars like *Ophiomyxa* and *Ophiophrixus*, and all basket stars including *Gorgonocephalus* (Fig. 17.20e), and *Astrodendrum*.

Order Ophiurida—ambulacral grooves closed by growth of lateral shields toward oral surface; usually five arms; arm vertebrae internal and fused; arms capable of only lateral movement; aboral arm shields present; contains about 95% of all ophiuroid species including *Ophiura*, *Ophiocoma*, *Ophionereis*, and *Ophiolepis*.

Subphylum Echinozoa—free living and benthic; basically globose or cylindrical body without arms or rays; most systems arranged into meridional tracts.

Class Echinoidea—sea and heart urchins, sand dollars; most have an elaborate masticatory apparatus (Aristotle's lantern); body mostly globose; have a rigid skeleton or test composed of sutured ossicles; have pedicellariae and movable spines.

Subclass Perischoechinoidea—primitive and mostly extinct; one or two rows of ambulacral and interambulacral plates.

Order Cidaroida—includes all surviving perischoechinoids; considered to be ancestral to all other living echinoids; two or more rows of plates for each ambulacrum and interambulacrum; have widely separated primary spines and smaller secondary spines; includes *Cidaris* and *Psychocidaris*.

Subclass Euechinoidea—contains most living echinoids; test of five bicolumnar, ambulacral rows alternating with five bicolumnar, interambulacral rows of ossicles; lantern and pedicellariae usually present.

Superorder Diadematacea—sea urchins; perforate tubercles; well-developed lantern; gills normally present.

Order Echinothurioida—skeletal plates with interstitial membranous junctions; peristomial ambulacral plates are simple; *Calveriosoma* is typical.

Order Diadematoida—test composed of rigid or flexible plates; five pairs of oral plates on peristomium; gills present; spines cylindrical and hollow; *Diadema* and *Aspidodiadema* are typical.

Order Pedinoida—rigid test with nonhollow primary spines; five pairs of oral plates on peristomium; *Caenopedina* is the only extant genus.

Superorder Echinacea—sea urchins with inflexible test and solid spines; lantern well-developed in adults.

Order Salenioida—periproct displaced eccentrically by suranal plate; *Salenia* is a typical genus.

Order Phymosomatoida—primary tubercles nonperforate; *Glyptocidaris* is representative.

Order Arbacioida—periproct covered by four or five conspicuous triangular anal valves; secondary spines usually absent; includes *Arbacia* (Fig. 17.24a) and *Habrocidaris*.

Order Temnopleuroida—test sculptured usually in adults and always in immature forms; *Lytechinus* (Fig. 17.24b) and *Tripneustes* are common.

Order Echinoida—test not sculptured; tubercles nonperforate and noncrenulate; includes *Echinus*, *Strongylocentrotus* (Fig. 17.24c) and *Echinometra*.

Superorder Gnathostomata—irregular urchins; test is rigid; periproct is eccentric; primary tubercles are usually perforate; spines are hollow; lantern is present.

Order Holectypoida—ambulacra not petaloid; gill slits distinct; *Echinoneus* and *Micropetalon* are only extant genera.

Order Clypeasteroida—sand dollars;

strongly flattened and ovoid body with internal skeletal supports; have petalloid ambulacra; short spines and pedicellariae are numerous; *Clypeaster, Mellita* (Figs. 17.24d, 17.27), and *Fibularia* are commonly encountered.

Superorder Atelostomata—irregular urchins; lantern and branchial slits are absent; primary tubercles are usually perforate; primary spines are hollow.

Order Cassiduloida—mostly extinct; round or oval test; ambulacra located eccentrically; no jaws or gills in adults; *Echinolampas* is cosmopolitan.

Order Holasteroida—oval or elongated; test is thin and delicate; some live in very deep water; *Pourtalesia* is widely distributed.

Order Spatangoida—heart urchins; occur in soft bottoms and are specialized for burrowing; lack jaws and posterior genital plate; possess spines either in uniform or aggregated patterns; some well-known genera are *Hemiaster, Echinocardium, Moira,* and *Spatangus.*

Order Neolampadoida—a poorly known group; nonpetaloid ambulacra; includes *Neolampas.*

Class Holothuroidea—sea cucumbers; body cylindrical and elongated in oral–aboral axis; mouth surrounded by buccal tentacles; armless; madreporite internal; ambulacra represented by closed canals; usually single gonad; skeleton of scattered, microscopic, nonsutured ossicles.

Subclass Dendrochirotacea—tentacles always retractile; usually have tube feet and respiratory trees; gonad in two lateral tufts.

Order Dendrochirotida—tentacles branched and lacking ampullae; tube feet variable in distribution; *Cucumaria* (Fig. 17.31c) and *Thyone* (Figs. 17.31a, 17.32) are well known.

Order Dactylochirotida—tentacles simple and fingerlike; body encased in a flexible test composed of ossicles; *Echinocucumis* is typical.

Subclass Aspidochirotacea—tentacles not retractile; tube feet present; prominent bilaterality.

Order Aspidochirotida—have respiratory trees; intestinal mesentery attached in right ventral interradius; *Holothuria* and *Actinopyga* are unique because of their cuvierian organs.

Order Elasipodida—mostly deep sea; lacking respiratory trees; intestinal mesentery attached to right dorsal interradius; *Pelagothuria* is typical.

Subclass Apodacea—tentacles variable and do not invert into anterior end; tube feet mostly absent.

Order Apodida—wormlike, cylindrical body; lacking respiratory trees and anal papillae; *Leptosynapta* (Fig. 17.31b) and *Synapta* are rather common.

Order Molpadiida—posterior end fusiform and tapered; have respiratory trees and anal papillae; *Molpadia* is typical.

ADDITIONAL READINGS

Binyon, J. 1972. *Physiology of Echinoderms,* 255 pp. New York: Pergamon Press.

Boolootian, R.A., ed. 1966. *Physiology of Echinodermata,* 796 pp. New York: John Wiley & Sons.

Gosner, K.L. 1971. *Guide to Identification of Marine and Estuarine Invertebrates: Cape Hatteras to the Bay of Fundy,* pp. 556–590. New York: Wiley-Interscience.

Highsmith, R.C. 1982. Induced settlement and metamorphosis of sand dollar *(Dendraster excentricus)* larvae in predator-free sites: Adult sand dollar beds. *Ecology 63:*329–337.

Hyman, L.H. 1955. *The Invertebrates.* Vol. IV: *Echinodermata,* 746 pp. New York: McGraw-Hill Book Co.

Millott, N. 1967. *Echinoderm Biology,* 229 pp. Symposium of the Zoological Society of London, No. 20. New York: Academic Press.

Moore, R.C., ed., 1966. *Treatise on Invertebrate Paleontology.* Part S: *Echinodermata 1,* Vols. 1 & 2, 637 pp.; Part T: *Echinodermata 2,* Vols. 1–3, 1002 pp.; Part U: *Echinodermata 3,* Vols. 1 & 2, 672 pp. New York: Geological Society of America and the University of Kansas Press.

Morris, R.H., D.P. Abbott, and E.C. Haderlie. 1980. *Intertidal Invertebrates of California,* pp. 115–176 and plates 40–57. Stanford, Calif.: Stanford University Press.

Mortensen, T. 1977. *Handbook of the Echinoderms of the British Isles,* 448 pp. Reprinted by Clarendon Press, Oxford, England.

Nichols, D. 1969. *Echinoderms,* 184 pp. London: Hutchinson University Library.

Sugar, J.A. 1970. Starfish threaten Pacific reefs. *Nat. Geogr. 137*:340–352.

Glossary

A listing of terms that are most frequently used, are most important, or characterize a particular group.

Aboral opposite the mouth.

Acetabulum a ventral sucker or attachment organ in digenetic trematodes.

Acoelomate a bilaterally symmetrical animal without a body cavity other than the gastrovascular cavity.

Acron the anteriormost non-segmented tip of the body of an annelid or arthropod.

Adaptation any feature that enhances the survival and evolutionary fitness of an organism.

Advanced newer or less like the ancestral condition.

Afferent a vessel or structure leading toward a given point of reference.

Ambulacrum one of five radially arranged series of plates or ossicles in echinoderms that bears rows of podia.

Ameboid forming pseudopodia for locomotion or food capture.

Ampulla a small bladderlike enlargement on the inner end of an echinoderm podium.

Analogous parts that are functionally similar but have different evolutionary ancestries.

Antenna a long, paired, sensory appendage on the anterior end of many annelids and arthropods.

Antennal gland an excretory organ located in the antennal segment of many crustaceans.

Antennule the anteriormost paired appendage on the head of crustaceans.

Anterior toward the head.

Aphotic stratum that deeper part of an ocean or deep lake in which light never penetrates.

Apodeme an inward folding of the cuticle in arthropods, usually for muscle attachments.

Apolysis the separation of the old cuticle from the underlying epidermis in an arthropod in the early stages of molting.

Apterous being wingless, as in some insects.

Archenteron the primitive gut or digestive cavity present in the gastrula stage in metazoan development.

Articular membrane the flexible region between adjacent sclerites or podomeres of an arthropod.

Aschelminth an animal that is generally wormlike and has a pseudocoelom.

Atrium a relatively large chamber opening to the outside; associated with the reproductive, digestive, or gas-exchange systems in many different animals.

Autotomy the self-amputation or voluntary breaking of a body part by reflexive action.

Basal body (kinetosome) organellar base of a flagellum or cilium.

Benthic living on the bottom in aquatic habitats.

Bilateral symmetry a type of symmetry in which a body or body part can be divided into right and left sides, each being a mirror image of the other.

Biradial symmetry a type of symmetry in which parts are arranged in a secondary plan of bilateral symmetry imposed on a primary plan of radial symmetry.

Biramous having two branches, as most crustacean appendages.

Blastocoel the internal cavity of the blastula stage in metazoan development.

Blastomere a cell resulting from cleavage in metazoan development.

Blastopore the opening of the archenteron to the outside in the gastrula stage of metazoan development.

Blastula a hollow sphere of blastomeres (cells) resulting from cleavage in metazoan development.

Botryoidal tissue loosely packed mesenchymal tissue in leeches; resembles a bunch of grapes.

Brachial referring to any appendage or limb.

Branchial referring to the gills.

Brood to care for the developing young by the parents.

Buccal cavity a cavity immediately inside the mouth.

Bud a new individual formed asexually in many different invertebrates.

Bursa any pouchlike invagination; often refers to that part of the female reproductive system specialized for the reception of sperm.

Calcification the deposition of calcium salts in the procuticle thus hardening it; occurs in crustaceans and some millipedes.

Calyx a boatshaped or cuplike central body of an entoproct or crinoid.

Capitulum refers to the enlarged end of a hair or antenna; the anterior end of an acarine; the body of a barnacle surrounded by the mantle.

Carapace the hard dorsal shield arising from a postoral segment of an arthropod and covering part or all of the body.

Cardiac refers to the heart; that portion of the stomach nearest the esophagus or mouth in crustaceans and asteroids.

Caste one of several types of individuals in a colony of termites or hymenopterans; determined by genetic or hormonal influences.

Caudal referring to the tail or posterior part of the body.

Cecum a blind-end pouch or saclike extension of the digestive tract.

Centrolecithal the egg of arthropods in which a large amount of yolk is in the center of peripherally arranged nuclei.

Cephalic pertaining to or toward the head.

Cephalization the tendency to concentrate sense organs and the nervous system in the head.

Cephalothorax a tagma combining the head and thorax as in many arthropods.

Cercus a sensory appendage that projects from the last abdominal segment in many insects and certain crustaceans.

Chaeta (=seta) a chitinous bristle found on each segment of polychaetes and oligochaetes.

Chela (cheliped) a pincer claw on a crustacean.

Chelate clawed or pincerlike.

Chelicera the anteriormost paired appendage in chelicerates, used in food gathering.

Chitin a polysaccharide component of the procuticle of arthropods and present in several other groups.

Choanocyte a flagellated collar cell characteristic of sponges.

Chromatophore a cell containing a pigment responsible for color changes in many different animals.

Cilium a short hairlike process arranged in rows or groups in ciliates and in epidermal cells of many other animals.

Cirrus a slender, flexible, variable appendage present in a wide variety of invertebrates.

Clitellum a midbody glandular swelling in oligochaetes and leeches whose mucous secretions form a cocoon.

Cloaca the terminal portion of the digestive tract into which empties reproductive, excretory, or even gas-exchange systems.

Closed circulatory system a type of vascular circulation in which blood is completely contained within arteries, capillaries, and veins.

Cnidocyte a cnidarian cell that produces an intracellular nematocyst.

Coelenteron the gastrovascular cavity of radiates.

Coelom an internal body cavity lined with peritoneum; found in all higher invertebrates.

Coelomate any animal with a true coelom.

Coelomoduct a mesodermal duct that grows outward from the coelom and carries gametes or excretory products to the outside.

Coiling the process in most gastropods in which the growing visceral mass forms a two-dimensional planospiral coil.

Collocyte an adhesive cell found in ctenophores.

Colony a group of genetically identical, asexually produced individuals living together often with some form of cooperation manifested.

Commensalism a symbiotic relationship in which one species, the commensal, benefits, and the other species, the host, is unaffected by the association.

Commissure a connecting nerve that joins two ganglia.

Compound eye a visual organ of most arthropods; composed of several to many visual units called ommatidia and produces a mosaic image.

Contractile vacuole a cellular organelle concerned with the collection and expulsion of excess water.

Coxa the proximal podomere of an arthropod appendage that attaches the limb to the body.

Coxal gland an excretory gland located at the base (coxa) of a leg in some arthropods.

Crop a thin-walled dilation of the anterior digestive tract in which food is stored temporarily.

Cryptobiosis a state of suspended animation induced by desiccation; occurs in rotifers, nematodes, and tardigrades.

Cryptozoic an animal that hides in or under objects; secretive.

Crystalline style a mucous protein rod in many molluscs that is involved with feeding and enzymatic digestion.

Ctenidium a molluscan gill.

Cuticle a thin, non-living, external covering of many invertebrates.

Cyclomorphosis gradual changes in gross morphology of the population of cladocerans and rotifers during successive seasonal generations.

Cyst a resistant protective covering secreted by many different small invertebrates in response to desiccation or even overcrowding.

Depost-feeder an animal that ingests substrate material including deposited organic materials.

Desiccation the process of drying out.

Determinate cleavage a type of cleavage in which the presumptive fate of all subsequent blastomeres is fixed at the first cell division.

Detritus fine particulate debris or fragments of plant or animal bodies.

Detorsion a process in some gastropods following torsion in which the visceral mass is rotated clockwise.

Deuterostomate any animal in which the blastopore forms the anal opening, and the mouth arises from a new later opening at the anterior end.

Dextral right-handed; the direction of coiling in a gastropod when the aperture is on the right of the shell axis when the shell is held vertically with the apex uppermost.

Diapause a state of developmental arrest induced by photoperiod or temperature in insects and other invertebrates.

Dimorphism existing as two distinct forms as in many cnidarians, acoelomates, and bryozoans.

Dioecious having separate sexes; some individuals are males and others are females.

Direct development having no larval stage; the hatching stage looks like the adult.

Distal farther away from the point of attachment or point of reference.

Dorsal toward or pertaining to the back or upper surface.

Ecdysis the shedding the old cuticle in the process of molting by an arthropod.

Ectoderm that germ layer that gives rise to the outermost portions of an adult animal including the nervous system.

Ectoparasite a parasite which lives on or is attached to its host's external surface.

Efferent a structure leading away from a given point of reference.

Encystment the formation of a protective resistant outer covering at the beginning of a dormant stage in the life cycle of many animals; induced by desiccation or overcrowding.

Endocuticle the innermost untanned layer of the arthropodan procuticle.

Endoderm the innermost germ layer that will give rise to the lining of the digestive tract and associated organs.

Endoparasite a parasite which lives inside of or is attached to an internal site within the host's body.

Endopod the medial branch of a biramous appendage of a crustacean.

Endoskeleton a skeleton that is contained within the body, as in sponges and echinoderms.

Endostyle a ciliated ventral groove in the alimentary system of several different invertebrate groups.

Enterocoelom a coelom formed by the outfoldings or evaginations from the archenteron; found in deuterostomates.

Enteron the digestive tract, alimentary canal, or gut.

Epicuticle the outermost, thin, waxy layer in the cuticle of an arthropod.

Epidermis the outermost cellular layer in the body wall of most metazoans.

Epizoic an animal living on the surface of another animal.

Estuary a partially enclosed, shallow, coastal habitat that is dramatically affected by tidal forces and the inflow of fresh water.

Euphotic stratum the upper lighted stratum of an ocean or lake in which photosynthesis can occur.

Eutely a constant number of cells in a given organ in all adults of the same species; a phenomenon characteristic of a number of small invertebrates.

Evert to protrude by turning inside out.

Exocuticle the outermost sclerotized or calcified layer of the arthropod procuticle.

Exopod the lateral branch of a biramous appendage of a crustacean.

Exoskeleton a skeleton situated on the outside of the body as in some cnidarians, molluscs, and arthropods; functions in movement, support, and protection.

Exuvium the old exoskeleton shed by an arthropod following molting.

Filter or suspension feeder an aquatic animal that feeds on suspended particles (plankton, detritus) filtered from the watery environment.

Flagellum a long, threadlike locomotor organelle occurring singly or in small numbers in mastigophorans, sponges, and the sperm of most animals.

Flame bulb a blind-end, hollow, bulb-shaped structure equipped with one or many flagella and forming a part of the protonephridial system in acoelomates and others.

Frontal gland a mucus-secreting gland at the anterior end of turbellarians; mucus is used in food capture and locomotion.

Gamete a haploid sex cell; a sperm or an egg.

Ganglion an aggregation of cells bodies of neurons.

Gastric pertaining to digestion or the stomach.

Gastrodermis the epidermal layer lining the gastrovascular cavity in radiates and flatworms.

Gastrovascular cavity the central digestive cavity of radiates and flatworms.

Gastrula the developmental stage that is produced by gastrulation; typically consists of a central archenteron surrounded by two layers of cells.

Giant fiber an unusually large neuron that can transmit impulses faster than normal neurons; found in the nerve cord of some molluscs, annelids, and arthropods.

Gill an organ specialized for gaseous exchange in aquatic animals.

Gizzard a region of the foregut that is specialized for grinding food.

Gnathobase the spiny basal process on an appendage of many arachnids and some crustaceans; used for crushing and manipulating food.

Gonoduct a duct that transports sperm or eggs.

Gonopore the external opening of either reproductive system.

Gravid filled with fertilized eggs.

Hemimetabolous development a type of insect development in which incomplete metamorphosis occurs; immature stages are nymphs.

Hemocoel an extensive ill-defined system of extracellular blood spaces present in molluscs and arthropods.

Hemocyanin a copper-containing respiratory blood pigment found in molluscs and many arthropods; is blue when oxygenated.

Herbivorous an animal that feeds chiefly on plant materials.

Hermaphrodite a single individual that has both female and male reproductive systems; both systems can function simultaneously or in sequence; monoecious.

Heterotrophic the process of obtaining organic food by ingesting other organisms or their parts.

Holoblastic cleavage the complete and equal cleavage of blastomeres in early cleavage.

Holometabolous development a type of insect development in which complete metamorphosis occurs; immature stages are larvae and pupae.

Holozoic a type of heterotrophism in which solid particles of food are ingested.

Homologous parts that have a common ancestral origin; they often have different functions.

Host an animal that harbors and sustains another animal, the parasite.

Hydrostatic skeleton a fluid-filled space with a constant volume whose shape can be altered by surrounding muscles.

Hyperosmotic (=hypertonic) a solution containing a relatively larger amount of a dissolved solute compared to that of another solution; gains water by osmosis.

Hyposmotic (=hypotonic) a solution containing a lesser amount of a dissolved solute compared to that of another solution; loses water by osmosis.

Indeterminate cleavage a type of cleavage in which the early blastomeres retain their capability of developing into complete organisms.

Indirect development animals that have an immature or larval stage.

Infraciliature a complex system of basal bodies and fibers lying beneath the pellicle of ciliates.

Integument the outer covering or skin of an animal.

Instar an arthropod during the interval between two consecutive molts.

Interambulacrum (=interradius) that part of an echinoderm body located between adjacent ambulacra.

Introvert an anterior, tentacle-bearing part or proboscis of certain worms that can be retracted into the trunk.

Isosmotic two solutions having the same concentration of solutes.

Kinety a row of closely positioned cilia in ciliates.

Labial pertaining to the lips or other perioral structures.

Labium the postoral lip present in many arthropods; in insects and myriapods it represents the fused second pair of maxillae.

Labrum a fleshy unpaired preoral lip in many arthropods.

Lacuna an extracellular space; often a small part of the hemocoel.

Lamella a thin sheetlike structure.

Larva the early stage in the life cycle of an animal, which is unlike the adult and usually actively feeds.

Lentic a freshwater ecosystem of standing water such as a lake or pond.

Limnetic that open-water zone in a lake in which water depths preclude the occurrence of rooted plants.

Littoral pertains to the shallow parts of lakes and the shallow coastal regions of the oceans.

Lobopodium a blunt, lobelike pseudopodium.

Lophophore an extensible crown of ciliated food-gathering tentacles surrounding the mouth in four different invertebrate phyla.

Lotic a freshwater ecosystem of running water such as a stream or river.

Lumen the internal cavity of a duct, vessel, or organ.

Luminescent the ability of a number of animals to produce light from intracellular chemical reactions.

Macromere one of the group of larger cells resulting from unequal cleavage situated near the vegetal pole.

Macronucleus the larger polyploid nucleus of ciliates that regulates somatic processes.

Macrophagous the process by which large food particles are ingested.

Madreporite the sievelike plate in echinoderms that connects the water vascular system with the exterior.

Malpighian tubule a slender, often branched blind-end tubule that opens into the hindgut of most terrestrial arthropods; functions in excretion of uric acid.

Mandible a paired feeding appendage or jaw found in crustaceans and uniramians.

Mantle a poncholike extension of the body wall in molluscs, barnacles, and brachiopods; secretes the shell or shell plates.

Mantle cavity a space formed between the mantle and body in several different phyla.

Mastax a pharyngeal complex of trophi and muscles forming a crushing organ in rotifers.

Maxilla one of two postoral paired appendages in many arthropods; functions in manipulating food.

Maxillary gland an excretory organ located in the second maxillary segment of many crustaceans.

Maxilliped one of several paired anterior trunk appendages of many crustaceans that is involved with food gathering.

Medusa a free-swimming, bell-shaped, jellyfish stage in the life cycle of many cnidarians.

Meroblastic cleavage a type of cleavage in which the yolk is central and the dividing blastomeres are peripherally located; characteristic of arthropods.

Mesenchyme loosely packed cells of mesodermal origin.

Mesenteron the midgut.

Mesentery a mesodermal fold that suspends the visceral organs; a septum that partitions the gastrovascular cavity in many cnidarians.

Mesoderm a middle embryonic germ layer from which are derived muscles and most visceral organs.

Mesoglea an amorphous material in cnidarians usually containing some cells and lying between the epidermis and gastrodermis.

Mesothorax the second or middle thoracic segment of an insect which bears a pair of legs and usually a pair of wings.

Metamere (=segment) one of a series of body segments in segmented animals.

Metamerism (=segmentation) the segmental repetition of similar metameres; always involves the mesoderm and mesodermally derived organs.

Metamorphosis a substantial change in body form during development or maturation.

Metanephridium a tubular excretory organ that opens into the coelom by a ciliated funnellike nephrostome and to the outside via a nephridiopore.

Metathorax the third or posteriormost thoracic segment of an insect which bears a pair of legs and usually a pair of wings.

Micromere one of the group of smaller cells resulting from unequal cleavage situated near the animal pole.

Micrometer (μm) one millionth of a meter.

Micronucleus the smaller diploid nucleus in ciliates and involved in reproduction.

Microphagous describing the method by which small food particles are ingested.

Molt to shed or cast off the outer cuticle as in arthropods and nematodes.

Monoecious having both male and female gonads in the same individual; hermaphroditic.

Monophyletic a group of organisms that have evolved from a single ancestral source.

Mutualism an obligatory relationship between representatives of two different species in which both organisms benefit by the association.

Naupliar eye a medial unpaired eye present in the nauplius stage and retained by some adult crustaceans.

Nauplius the free-swimming first larval stage of most crustaceans and characterized by three pairs of appendages.

Nektonic a swimming animal.

Nematocyst a threadlike stinging organelle formed within cnidocytes and characteristic of cnidarians.

Neotony a phenomenon in which larval or immature features of ancestors become adult characteristics of descendants.

Nephridiopore the opening of a nephridium to the exterior.

Nephridium a tubular organ that is the excretory and osmoregulatory structure in a wide variety of invertebrates.

Nephrostome the ciliated internal pore by which a metanephridium opens into the coelom.

Neuron a nerve cell consisting of dendrites, cell body, and axons.

Neurosecretory a neuron that produces a hormone.

Nocturnal an animal active at night.

Nymph the immature stage of a hemimetabolous insect.

Oceanic region the open sea in which water depths exceed 200 m.

Ocellus a small cluster of photoreceptors; a simple eye.

Odontophore the cartilaginous material underlying the radular membrane in most molluscs.

Ommatidium an individual photoreceptor unit in the compound eye of an arthropod.

Omnivorous an animal that eats all types of plant and animal food.

Ontogeny the development of the individual.

Open circulatory system a type of vascular circulation found in molluscs and arthropods in which capillaries are absent, and blood flows through hemocoelic spaces.

Operculum any flaplike covering of the gills (many arthropods), entire body (barnacles), or shell aperture (gastropods).

Opisthosoma the abdomen or posterior tagma of a chelicerate.

Oral pertaining to or near the mouth.

Osmoregulation the maintenance of the osmotic concentration of body fluids at a level different from that of the aqueous environment.

Ossicle any calcareous plate or rod making up the echinoderm endoskeleton.

Ostium any incurrent pore as in sponges or in the heart of molluscs, leeches, and arthropods.

Oviduct a tube conveying eggs from the ovary toward the exterior.

Oviparous an animal whose eggs hatch outside the mother's body; egg-laying.

Ovipositor a complex of one or two paired posterior appendages modified for laying eggs.

Ovoviviparous an animal whose eggs develop inside the mother's body but without maternal nutritive support.

Palp any projecting part or appendage, often sensory, on the head or near the mouth in many invertebrates.

Papilla a cone or nipple-shaped projection.

Parapodium a fleshy lobelike appendage occurring in a pair on each segment of a polychaete.

Parasite an organism that lives on or in another (the host) and derives its nutritive support from the host.

Parasitism a type of symbiosis in which one species, the parasite, is benefited and the other, the host, is harmed by the association.

Parietal pertaining to the outer wall of a cavity.

Parthenogenesis the development of an organism from an unfertilized egg.

Pedicellaria a minute pincerlike structure of many echinoderms that helps keep the epidermis free from debris or encrusting organisms.

Pedicle (=pedicel) a stalk as in brachiopods or the slender connection between prosoma and opisthosoma in many arachnids.

Pedipalp the second paired appendage of arachnids.

Peduncle any stalklike structure supporting another part or structure.

Pelagic living in the open sea.

Pentamerous symmetry a type of symmetry present in echinoderms in which parts are arranged in fives or multiples of fives.

Pericardium the membrane lining the cavity in which the heart lies.

Peristalsis the rhythmic involuntary muscular contractions that pass along a hollow organ.

Peritoneum the mesodermal lining of the coelom.

Phylogeny the evolutionary history of a group of organisms.

Plankton small floating plants and animals found in aquatic habitats that cannot swim against a water current.

Pleuron a lateral plate forming the side of a typical metamere in arthropods.

Podium (=tube foot) an organ of locomotion in echinoderms.

Podomere a cylindrical unit of most appendages of an arthropod.

Polymorphism presence of two or more morphological forms in the life cycle of some invertebrates.

Polyp the cylindrical, sessile, asexually-reproducing stage in a cnidarian.

Polyphyletic a group of organisms having evolved from more than one ancestral source.

Posterior toward the rear end.

Predaceous an animal that feeds on other animals.

Primitive unspecialized or more like the ancestral condition.

Proboscis a tubular, anterior, food-gathering process found in many different invertebrates.

Proctodeum the posterior ectodermally lined portion of the digestive tract that becomes the hindgut and anus.

Procuticle the bulk of the arthropod cuticle containing chitin and sclerotized proteins.

Proprioceptor a receptor that provides information to the brain regarding the movement and position of muscles.

Prosoma the anterior tagma of a chelicerate.

Protandry a condition in a hermaphrodite in which sperm are produced earlier than eggs.

Prothorax the anteriormost thoracic segment of an insect that always bears a pair of legs but never wings.

Protonephridium a primitive excretory and osmoregulatory organ featuring tubules that end blindly within the body cavity as flame bulbs.

Protopod the proximal region of a crustacean limb; consisting of the coxa and basis.

Protostomate any animal in which the blastopore forms the mouth opening and the anal opening arises from a later opening at the posterior end.

Proximal closer to the point of attachment or point of reference.

Pseudocoelom a variable body cavity derived from the blastocoel that is not lined with peritoneum; present in the aschelminths.

Pseudocoelomate possessing a pseudocoelom as the aschelminths.

Pseudopodium a locomotory projection of the plasma membrane in sarcodinans.

Pupa the quiescent stage of a holometabolous insect, occurring between the larval and adult stages.

Pygidium the terminal, posterior, non-segmented body part in annelids and arthropods.

Pylorus that part of the stomach of many invertebrates nearer the intestine.

Radial cleavage a cleavage pattern in which the cleavage planes are at right angles or parallel to the polar axis of the egg.

Radial symmetry having similar parts arranged around a common central axis, as in the radiates.

Radula a filelike, protrusible, toothed ribbon located in the buccal cavity of most molluscs.

Ramus a branch or outgrowth of a structure.

Raptorial animals that stalk and capture prey for food.

Regulative development a type of development in deuterostomates in which early blastomeres can regulate their development to replace lost or destroyed blastomeres.

Rhynchocoel the dorsal mesodermal cavity that houses the proboscis in nemerteans.

Saprozoic a type of heterotrophic nutrition in which simple organic materials are absorbed from the surrounding medium.

Schizocoelom a coelom formed from a split in the lateral mesodermal cylinders in protostomates.

Sclerite a hard cuticular plate forming part of the arthropod's exoskeleton.

Sclerotization the process in arthropods in which cuticular proteins are hardened or tanned by the formation of chemical cross linkages.

Sedentary an animal that remains in one place; sessile.

Segment (= metamere) one of a series of body segments in segmented animals.

Segmentation (= metamerism) the segmental repetition of similar metameres; always involves the mesoderm.

Seminal receptacle a sperm-storing chamber in the female reproductive tract of many invertebrates.

Seminal vesicle that part of the male reproductive system that stores sperm.

Septum a membranous partition separating two cavities like that in the gastrovascular cavity of many cnidarians and in the coelom in adjacent segments in annelids.

Sessile permanently attached or fixed; sedentary.

Seta (= chaeta) a chitinous bristle found on each segment of polychaetes and oligochaetes.

Sinistral left-handed; the direction of coiling in a gastropod when the aperture of the shell is on the left of the shell axis when the shell is held vertically with the apex uppermost.

Siphonoglyph a ciliated groove along the pharynx or stomodeum of anthozoans.

Solenocyte a type of flame bulb in a protonephridium with a single long flagellum; found in some annelids and other marine worms.

Specialized adapted in function or morphology for a particular life-style.

Species a fundamental taxonomic unit of similar organisms capable of interbreeding and producing fertile offspring.

Spermatheca a female organ specialized for the reception and storage of sperm.

Spermatophore a packet of sperm that can be conveniently transferred to another animal.

Spiracle the external opening to the tracheae or other gas-exchange organs of arthropods.

Spiral cleavage the type of cleavage in which the cleavage spindles are at an angle to the polar axis, resulting in successive alternating tiers of blastomeres arranged spirally.

Spiraling the three-dimensional coiling of a gastropod into a helical shape.

Spongocoel the internal body cavity of a simple sponge.

Statocyst a sense organ that provides the brain with information about the animal's position or orientation relative to the pull of gravity.

Sternum the ventral sclerite on each segment of an arthropod.

Stomodeum the anterior, ectodermally-lined portion of the digestive tract that becomes the mouth and anterior digestive tract.

Stridulate to produce sounds by rubbing parts of the body together.

Sucker a raised, circular, muscular suction device used as a holdfast or attachment organ in many different invertebrates.

Symbiosis the intimate association of two organisms of different species that live together.

Syncytium a mass of protoplasm or a cell containing several nuclei not separated by cell membranes.

Tactile pertaining to the organs or sense of touch.

Tagma a regional association of several adjacent functionally similar segments of an arthropod's body.

Tagmatization the phenomenon of several adjacent segments being united into a tagma in annelids and arthropods.

Tegument the unciliated living layer on the external surface of trematodes and cestodes.

Telson the anus-bearing segment or a postanal terminal projection in various arthropods.

Tentacle an elongated flexible appendage usually located near the mouth; employed in feeding and sensory reception.

Tergum the dorsal sclerite in each segment of arthropods.

Test an encasement or shell-like skeleton as in some protozoans and echinoderms.

Torsion the 180° counterclockwise rotation of the visceral mass in larval gastropods.

Trachea a cuticular-lined tubule conveying air from the exterior into groups of cells or the hemocoel in many arthropods.

Trichobothrium a long, thin, sensory, cuticular hair typical of many arachnids.

Trochophore larva the characteristic first larval stage of many marine protostomates.

Tubicolous tube building or tube dwelling.

Urosome the abdominal or posterior tagma of some arthropods.

Vas deferens a sperm duct.

Velum a thin circular shelf in hydrozoan medusae; a ciliated swimming organ in the veliger stage of molluscs.

Ventral toward or pertaining to the under surface.

Visceral pertaining to a number of internal organs.

Viviparous an animal whose young develop within the female's reproductive tract and are nourished from maternal sources.

Water vascular system the unique system of fluid-filled canals and tubes in echinoderms concerned with internal transport.

Zooid an individual member of a colony of animals, as in cnidarians and bryozoans.

Zooplankton animals that are planktonic.

Index

Page numbers in **boldface** indicate illustrations.